P9-EMQ-147

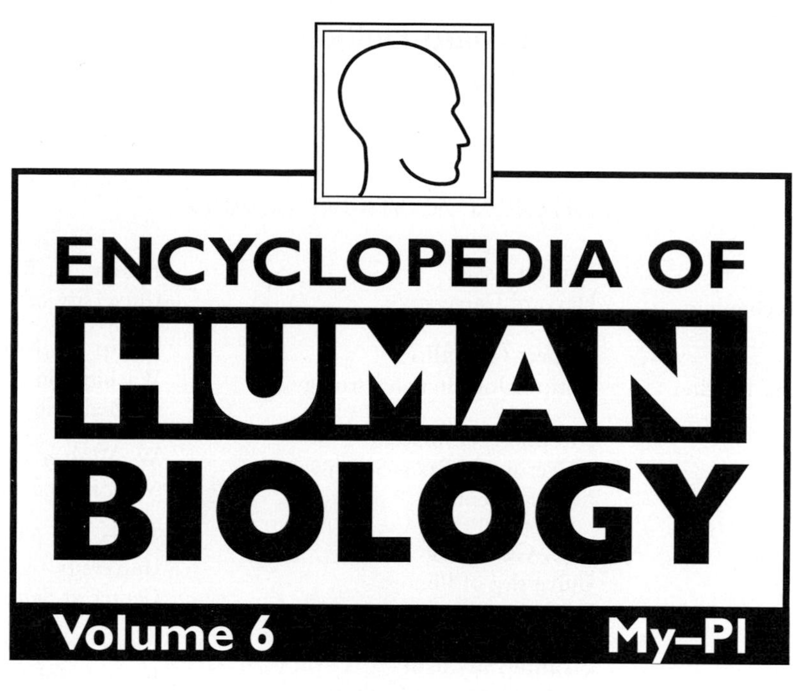

Second Edition

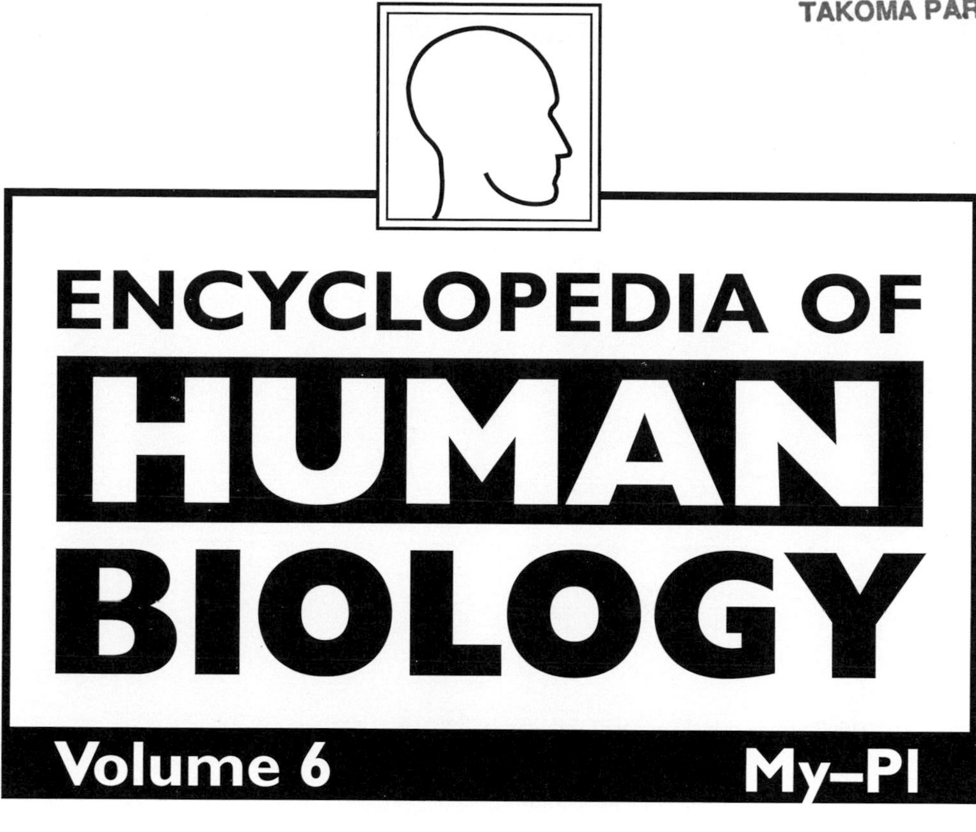

Second Edition

Editor-in-Chief
RENATO DULBECCO
The Salk Institute
La Jolla, California

ACADEMIC PRESS
San Diego London Boston New York Sydney Tokyo Toronto

This book is printed on acid-free paper. ♾

Academic Press
a division of Harcourt Brace & Company
525 B Street, Suite 1900, San Diego, California 92101-4495, USA
http://www.apnet.com

Academic Press Limited
24-28 Oval Road, London NW1 7DX, UK
http://www.hbuk.co.uk/ap/

Library of Congress Cataloging-in-Publication Data

Encyclopedia of human biology / edited by Renato Dulbecco. -- 2nd ed.
p. cm.
Includes bibliographical references and index.
ISBN 0-12-226970-5 (alk. paper: set). -- ISBN 0-12-226971-3 (alk. paper: v. 1). -- ISBN 0-12-226972-1 (alk. paper: v. 2). -- ISBN 0-12-226973-X (alk. paper: v. 3). -- ISBN 0-12-226974-8 (alk. paper: v. 4). -- ISBN 0-12-226975-6 (alk. paper: v. 5). -- ISBN 0-12-226976-4 (alk. paper: v. 6). -- ISBN 0-12-226977-2 (alk. paper : v. 7). -- ISBN 0-12-226978-0 (alk. paper: v. 8). -- ISBN 0-12-226979-9 (alk. paper: v. 9)
1. Human biology--Encyclopedias. I. Dulbecco, Renato, date.
[DNLM: 1. Biology--encyclopedias. 2. Physiology--encyclopedias. QH 302.5 E56 1997]
QP11.E53 1997
612'.003-dc21
DNLM/DLC
for Library of Congress 97-8627
CIP

PRINTED IN THE UNITED STATES OF AMERICA
98 99 00 01 02 EB 9 8 7 6 5 4 3 2

CONTENTS OF VOLUME 6

Contents for each volume of the Encyclopedia appears in Volume 9.

P

PREFACE TO THE FIRST EDITION

We are in the midst of a period of tremendous progress in the field of human biology. New information appears daily at such an astounding rate that it is clearly impossible for any one person to absorb all this material. The *Encyclopedia of Human Biology* was conceived as a solution: an informative yet easy-to-use reference. The Encyclopedia strives to present a complete overview of the current state of knowledge of contemporary human biology, organized to serve as a solid base on which subsequent information can be readily integrated. The Encyclopedia is intended for a wide audience, from the general reader with a background in science to undergraduates, graduate students, practicing researchers, and scientists.

Why human biology? The study of biology began as a correlate of medicine with the human, therefore, as the object. During the Renaissance, the usefulness of studying the properties of simpler organisms began to be recognized and, in time, developed into the biology we know today, which is fundamentally experimental and mainly involves nonhuman subjects. In recent years, however, the identification of the human as an autonomous biological entity has emerged again—stronger than ever. Even in areas where humans and other animals share a certain number of characteristics, a large component is recognized only in humans. Such components include, for example, the complexity of the brain and its role in behavior or its pathology. Of course, even in these studies, humans and other animals share a certain number of characteristics. The biological properties shared with other species are reflected in the Encyclopedia in sections of articles where results obtained in nonhuman species are evaluated. Such experimentation with nonhuman organisms affords evidence that is much more difficult or impossible to obtain in humans but is clearly applicable to us.

Guidance in fields with which the reader has limited familiarity is supplied by the detailed index volume. The articles are written so as to make the material accessible to the uninitiated; special terminology either is avoided or, when used, is clearly explained in a glossary at the beginning of each article. Only a general knowledge of biology is expected of the reader; if specific information is needed, it is reviewed in the same section in simple terms. The amount of detail is kept within limits sufficient to convey background information. In many cases, the more sophisticated reader will want additional information; this will be found in the bibliography at the end of each article. To enhance the long-term validity of the material, untested issues have been avoided or are indicated as controversial.

The material presented in the Encyclopedia was produced by well-recognized specialists of experience and competence and chosen by a roster of outstanding scientists including ten Nobel laureates. The material was then carefully reviewed by outside experts. I have reviewed all the articles and evaluated their contents in my areas of competence, but my major effort has been to ensure uniformity in matters of presentation, organization of material, amount of detail, and degree of documentation, with the goal of presenting in each subject the most advanced information available in easily accessible form.

Renato Dulbecco

PREFACE TO THE SECOND EDITION

The first edition of the *Encyclopedia of Human Biology* has been very successful. It was well received and highly appreciated by those who used it. So one may ask: Why publish a second edition? In fact, the word "encyclopedia" conveys the meaning of an opus that contains immutable information, forever valid. But this depends on the subject. Information about historical subjects and about certain branches of science is essentially immutable. However, in a field such as human biology, great changes occur all the time. This is a field that progresses rapidly; what seemed to be true yesterday may not be true today. The new discoveries constantly being made open new horizons and have practical consequences that were not even considered previously. This change applies to all fields of human biology, from genetics to structural biology and from the intricate mechanisms that control the activation of genes to the biochemical and medical consequences of these processes.

These are the reasons for publishing a second edition. Although much of the first edition is still valid, it lacks the information gained in the six years since its preparation. This new edition updates the information to what we know today, so the reader can be confident of its full validity. All articles have been reread by their authors, who modified them when necessary to bring them up-to-date. Many new articles have also been added to include new information.

The principles followed in preparing the first edition also apply to the second edition. All new articles were contributed by specialists well known in their respective fields. Expositional clarity has been maintained without affecting the completeness of the information. I am convinced that anyone who needs the information presented in this encyclopedia will find it easily, will find it accessible, and, at the same time, will find it complete.

Renato Dulbecco

A GUIDE TO USING THE ENCYCLOPEDIA

The *Encyclopedia of Human Biology, Second Edition* is a complete source of information on the human organism, contained within the covers of a single unified work. It consists of nine volumes and includes 670 separate articles ranging from genetics and cell biology to public health, pediatrics, and gerontology. Each article provides a comprehensive overview of the selected topic to inform a broad spectrum of readers from research professionals to students to the interested general public.

In order that you, the reader, derive maximum benefit from your use of the Encyclopedia, we have provided this Guide. It explains how the Encyclopedia is organized and how the information within it can be located.

ORGANIZATION

The *Encyclopedia of Human Biology, Second Edition* is organized to provide the maximum ease of use for its readers. All of the articles are arranged in a single alphabetical sequence by title. Articles whose titles begin with the letters A to Bi are in volume 1, articles with titles from Bl to Com are in Volume 2, and so on through Volume 8, which contains the articles from Si to Z.

Volume 9 is a separate reference volume providing a Subject Index for the entire work. It also includes a complete Table of Contents for all nine volumes, an alphabetical list of contributors to the Encyclopedia, and an Index of Related Titles. Thus Volume 9 is the best starting point for a search for information on a given topic, via either the Subject Index or Table of Contents.

So that they can be easily located, article titles generally begin with the key word or phrase indicating the topic, with any descriptive terms following. For example, "Calcium, Biochemistry" is the article title rather than "Biochemistry of Calcium" because the specific term *calcium* is the key word rather than the more general term *biochemistry*. Similarly "Protein Targeting, Basic Concepts" is the article title rather than "Basic Concepts of Protein Targeting."

TABLE OF CONTENTS

A complete Table of Contents for the *Encyclopedia of Human Biology, Second Edition* appears in Volume 9. This list of article titles represents topics that have been carefully selected by the Editor-in-Chief, Dr. Renato Dulbecco, and the members of the Editorial Advisory Board (see p. ii for a list of the Board members). The Encyclopedia provides coverage of 35 specific subject areas within the overall field of human biology, ranging alphabetically from Behavior to Virology.

In addition to the complete Table of Contents found in Volume 9, the Encyclopedia also provides an individual table of contents at the front of each volume. This lists the articles included within that particular volume.

INDEX

The Subject Index in Volume 9 contains more than 4200 entries. The subjects are listed alphabetically and indicate the volume and page number where information on this topic can be found.

ARTICLE FORMAT

Articles in the *Encyclopedia of Human Biology, Second Edition* are arranged in a single alphabetical list by title. Each new article begins at the top of a right-hand page, so that it may be quickly located. The author's name and affiliation are displayed at the beginning of the article. The article is organized according to a standard format, as follows:

- Title and author
- Outline
- Glossary
- Defining statement
- Body of the article
- Bibliography

OUTLINE

Each article in the Encyclopedia begins with an outline that indicates the general content of the article. This outline serves two functions. First, it provides a brief preview of the article, so that the reader can get a sense of what is contained there without having to leaf through the pages. Second, it serves to highlight important subtopics that will be discussed within the article. For example, the article "Gene Mapping" includes the subtopic "DNA Sequence and the Human Genome Project."

The outline is intended as an overview and thus it lists only the major headings of the article. In addition, extensive second-level and third-level headings will be found within the article.

GLOSSARY

The Glossary contains terms that are important to an understanding of the article and that may be unfamiliar to the reader. Each term is defined in the context of the particular article in which it is used. Thus the same term may appear as a Glossary entry in two or more articles, with the details of the definition varying slightly from one article to another. The Encyclopedia includes approximately 5000 glossary entries.

DEFINING STATEMENT

The text of each article in the Encyclopedia begins with a single introductory paragraph that defines the topic under discussion and summarizes the content of the article. For example, the article "Free Radicals and Disease" begins with the following statement:

A FREE RADICAL is any species that has one or more unpaired electrons. The most important free radicals in a biological system are oxygen- and nitrogen-derived radicals. Free radicals are generally produced in cells by electron transfer reactions. The major sources of free radical production are inflammation, ischemia/reperfusion, and mitochondrial injury. These three sources constitute the basic components of a wide variety of diseases. . . .

CROSS-REFERENCES

Many of the articles in the Encyclopedia have cross-references to other articles. These cross-references appear within the text of the article, at the end of a paragraph containing relevant material. The cross-references indicate related articles that can be consulted for further information on the same topic, or for other information on a related topic. For example, the article "Brain Evolution" contains a cross reference to the article "Cerebral Specialization."

BIBLIOGRAPHY

The Bibliography appears as the last element in an article. It lists recent secondary sources to aid the reader in locating more detailed or technical information. Review articles and research papers that are important to an understanding of the topic are also listed.

The bibliographies in this Encyclopedia are for the benefit of the reader, to provide references for further reading or research on the given topic. Thus they typically consist of no more than ten to twelve entries. They are not intended to represent a complete listing of all materials consulted by the author or authors in preparing the article.

COMPANION WORKS

The *Encyclopedia of Human Biology, Second Edition* is one of an extensive series of multivolume reference works in the life sciences published by Academic Press. Other such works include the *Encyclopedia of Cancer, Encyclopedia of Virology, Encyclopedia of Immunology,* and *Encyclopedia of Microbiology,* as well as the forthcoming *Encyclopedia of Reproduction.*

Myasthenia Gravis

EDITH G. McGEER
University of British Columbia

GLOSSARY

Anticholinesterase Drug that inhibits acetylcholinesterase and thus prevents the breakdown (metabolism) of acetylcholine

Autoantibodies Antibodies produced to a body's own tissue

α-Bungarotoxin Neurotoxin in the venom of the krait, a snake found in India

Cholinergic Describing neurons that use acetylcholine as a neurotransmitter

Glycoprotein Compound containing a protein and a carbohydrate (sugar) group

Hyperplasia Abnormal increase in the number of cells in a tissue without abnormality in the cells themselves

Ligand Compound active at a receptor

Plasmapheresis Removal of blood, separation of plasma or plasma components, and reinjection of red blood cells in a suitable suspending medium

Synapse Complex consisting of the nerve ending, synaptic cleft, and postsynaptic receptors

Synaptic cleft Gap between a nerve ending and the tissue it innervates; the neuronal message is carried across this gap by a chemical transmitter released by the nerve ending

Thymus Gland lying just below the neck that plays a major role in the production of antibodies

Transmitter One of a group of specific chemicals used by the nervous system to carry messages from one neuron to another, to muscle, or to other innervated tissue

MYASTHENIA GRAVIS IS A RELATIVELY RARE, chronic disorder in which voluntary muscle activity is weak and easily tired. It appears to be an autoimmune disease, that is, one where the body makes antibodies that attack its own tissue. In this case, the tissue attacked is generally the nicotinic acetylcholine receptor, which receives the nerve input to muscles. If these receptors are blocked or missing, the nerve impulse cannot force the muscle to respond. Antibodies to other components of the neuromuscular system may also be involved in some cases. Treatment strategies aimed at suppressing this disadvantageous immune response are now generally so successful that the victims can lead full, normal lives.

I. GENERAL DISCUSSION

Myasthenia gravis is a chronic neuromuscular disorder characterized by weakness and easy fatigability of voluntary muscles. The pattern of weakness often involves the extraocular muscles but may also involve other facial, bulbar, or limb muscles. The prevalence rate is estimated to be about 3 per 100,000 population. The age of onset is variable but is often in the third decade in women and somewhat later in men. Many cases show periods of spontaneous remission as well as periods of exacerbation. The evidence is strong that the disease is commonly an autoimmune disorder (see Section II,B). Epidemiological studies have not yielded many instances of the recurrence of myasthenia gravis within families but have provided evidence that relatives often suffer from some other autoimmune condition. This fact, and some molecular genetic linkage studies, has indicated that an inherited autoimmune susceptibility probably plays a role in many cases. [*See* Autoimmune Disease.]

ENCYCLOPEDIA OF HUMAN BIOLOGY, Second Edition, VOLUME 6.

II. PATHOLOGY

A. Histological Pathology

The most common findings are isolated patches of necrosis of muscle fibers with infiltration of lymphocytes. Electron microscopic studies have indicated abnormalities of the nerve terminals and synapses (Figs. 1 and 2B). None of these changes, however, correlates clearly with the severity or duration of the disease. There is sometimes a tumor of the thymus and, in more than 50% of the patients, hyperplasia of that gland. In many, however, it appear microscopically normal. [*See* Thymus.]

B. Chemical Pathology

The similarity of the myasthenic symptoms to those seen in curare poisoning first suggested that the prob-

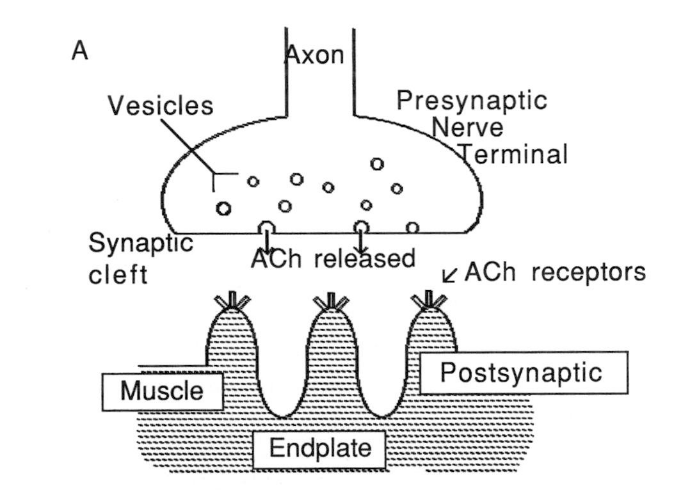

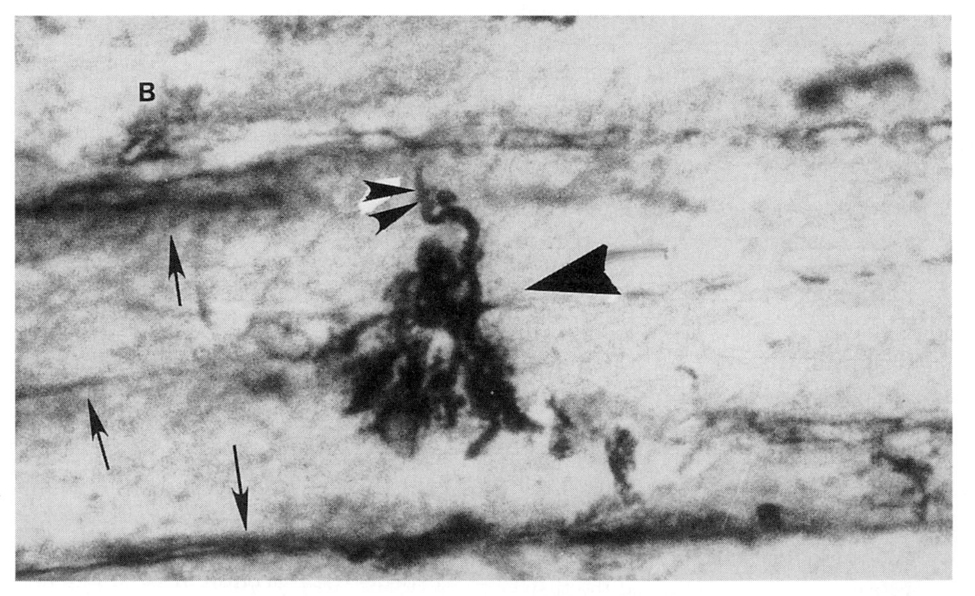

FIGURE 1 The neuromuscular junction. (A) Diagram of a synapse on a muscle. Acetylcholine (ACh) is released by nerve stimulation from vesicles of the nerve ending into the synaptic cleft to act on the ACh receptors on the postsynaptic side. The ACh is destroyed by acetylcholinesterase (not shown), which occurs in both presynaptic and postsynaptic tissue. (B) Photomicrograph of a guinea pig muscle stained for the specific synthetic enzyme that makes ACh and is unique to cholinergic structures. The axon is stained (double arrowhead), as is the multipronged neuromuscular junction (large arrowhead). The black lines (arrows) are muscle striations. (Courtesy of H. Kimura, Kinsmen Laboratory, University of British Columbia.)

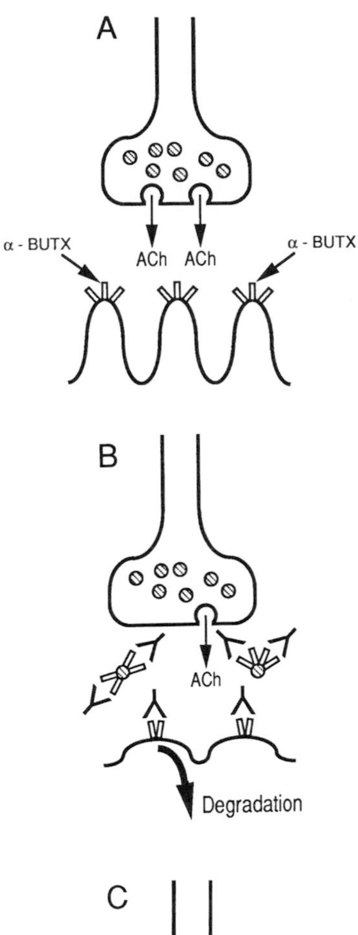

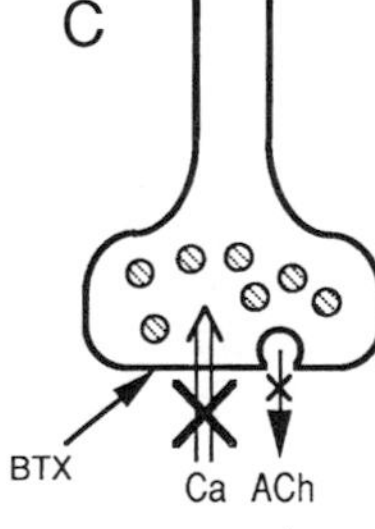

FIGURE 2 Diagrams of the neuromuscular junction showing presumed sites of action of various toxins and antibodies. (A) α-Bungarotoxin (α-BUTX) binds to the acetylcholine (ACh) receptors on muscles, thereby blocking interaction of ACh with the receptors. (B) In myasthenia gravis, antibodies bind to the ACh receptors on muscles to block the action of ACh. Moreover, such binding of antibodies promotes degradation of the receptors and membrane lysis, with extrusion of fragments into a widened synaptic cleft, resulting in a decreased number of receptors. (C) Botulinum toxin (BTX) binds to the nerve ending membrane and blocks the calcium-stimulated release of acetylcholine.

lem was in the neuromuscular junction. The transmitter of motoneurons is acetylcholine, an identification made by Loewi shortly after he first offered proof of the chemical, rather than electrical, nature of synaptic transmission. Thus, the remarkable improvement seen in many myasthenic patients on treatment with anticholinesterases also indicated that the problem probably lay in the neuromuscular junction. It was not known, however, whether the problem was in the presynaptic cholinergic neuron or in the postsynaptic receptor mechanism. Convincing evidence that the fault involved the postsynaptic receptor was provided by measurements, using α-[^{125}I]bungarotoxin (α-BUTX), of the receptor density in muscle biopsies from myasthenic patients and normal controls. Such measurements are possible because α-BUTX, like many other snake neurotoxins, binds tightly to the neuromuscular junction and kills by blocking cholinergic transmission at that site (Fig. 2A). The myasthenic muscles showed a 70–89% reduction in the number of nicotinic acetylcholine receptors (AChR) per neuromuscular junction as compared to muscles from normal controls.

III. CAUSAL THEORIES

Yet the problem remained as to what caused the loss in AChRs. Again, neurotoxins from snake venoms helped to provide an answer. The purification of the AChR from electric eels was made possible using α-cobra toxin, which acts like α-BUTX, attached to a resin to bind the solubilized AChR, which could be subsequently eluted by means of a specific competing ligand. Injection of the purified receptor glycoprotein into rabbits, in an attempt to raise antibodies for further scientific exploration of the AChR, led to the development in the animals of marked muscular weakness closely resembling that seen in the human disease. This serendipitous discovery led to a search for an autoimmune mechanism directed specifically against the cholinergic receptors in the human disease. In the vast majority (80–90%) of cases, circulating antibodies to the AChR itself can be demonstrated and it is believed that the binding of these antibodies may both block the receptors and lead to accelerated degradation (Fig. 2B). Identification of myasthenia gravis as an autoimmune disease in which antibodies to AChRs play a pathologic role was generally accepted and led to treatment by immunosuppression (Section V). However, there are some myasthenic patients in which circulating anti-AChR antibodies cannot be found and, even in those cases where they are found, there is poor correlation between the circulating levels and the severity of the disease. This has led to the hypothesis that myasthenia gravis is a heterogeneous disorder in which autoantibodies to other components of the neuromuscular junction may be involved in addition to, or instead of, the anti-AChR antibodies. A number of such antibodies have been reported, including some to various cytoskeletal proteins found in muscle or to the transmitter acetylcholine itself. The exact roles of the various antibodies in the pathogenesis remain to be determined by future work. [*See* Autoantibodies.]

Mention should be made of the Lambert-Eaton syndrome in which myasthenic symptoms are often seen. This is a rare disorder that is associated, in at least 70% of the cases, with bronchial or oat-cell carcinoma and it has been suggested that such tumors may cause the production of antibodies that act, like botulinum toxin (Fig. 2C), to inhibit acetylcholine release. Repetitive nerve stimulation or treatment with guanidine hydrochloride has a therapeutic effect on both the Lambert-Eaton syndrome and botulinum poisoning by increasing the quantal release of acetylcholine.

IV. ANIMAL MODELS

Injection of purified AChR glycoprotein, or some portions thereof, into various animal species produces symptoms highly similar to those seen in myasthenia gravis. This experimental allergic myasthenia gravis (EAMS) is widely used in studies aimed at a further understanding of the pathologic mechanisms or the development of better therapeutic strategies.

V. TREATMENT

The treatment of myasthenia gravis has progressed to a point where most patients are able to lead a full, productive life. Most treatment strategies presently employed are aimed at suppression of the autoimmune response.

Until this decade, however, the treatment of choice in myasthenia gravis was generally anticholinesterases. These give only symptomatic relief and are short-acting. Moreover, it is difficult to titrate the dose for individual patients, the desirable dose changes with stress, exacerbations, or other phenomena, and the possibility of adverse reactions is high. For these reasons, this therapeutic approach is now rarely used except in some mild cases.

Thymectomies were done more than 50 years ago in a few cases of myasthenia gravis where there was a tumor of the thymus. The beneficial effects reported, and the increasing evidence that this was an autoimmune disease in which the antibody-producing capacity of the thymus might be critical, led to increased numbers of such operations. With the reduction in operative risk to near 0% by improved surgical, anesthetic, and postoperative techniques, thymectomy is now generally recommended in adult cases with progressive or generalized myasthenia. The advantage lies in the possibility of permanent remission (in 35–45% of cases) or substantial improvement (in another 35–45%). However, such improvement is often slow in developing and may take just a few months to more than 10 years. For this reason, many patients continue to require immunosuppressive medication for some time after thymectomy.

The most generally used immunosuppressive drugs are the corticosteroids, which generally give rapid improvement. Other immunosuppressive agents used include azathioprine, cyclophosphamide, methotrexate, and cyclosporin. Plasma exchange or plasmapheresis is sometimes used alone, or with an immunosuppressive drug, to cause a rapid decrease in the level of circulating antibodies.

More selective ways of modulating the immune system are under study in animals with EAMS. One promising approach is to use anti-idiotope antibodies, produced against anti-AChR antibodies, to suppress the immune response to purified AChR protein. When, or whether, such approaches will prove practical in humans remains to be determined.

BIBLIOGRAPHY

Drachman, D. B. (ed.) (1987). Myasthenia gravis: Biology and treatment. *Ann. N.Y. Acad. Sci.* **505**, 1–914.

Drachman, D. B. (1994). Myasthenia gravis. *New Engl. J. Med.* **330**, 1797–1810.

Maselli, R. A. (1994). Pathophysiology of myasthenia gravis and Lambert-Eaton syndrome. *Neurol. Clinics* **12**, 285–303.

Mycotoxins

JOHN L. RICHARD
National Center for Agricultural Utilization Research, USDA/ARS

A. WALLACE HAYES
Duke University Medical Center

GLOSSARY

Agranulocytosis Acute condition characterized by pronounced leukopenia with a reduction in the number of polymorphonuclear leukocytes (white blood cells)

Ataxia Incoordination; an ability to coordinate the muscles in the execution of voluntary movement

Atrophy Wasting of tissues, organs, or the entire body, as from death and reabsorption of cells, diminished cellular growth, pressure, ischemia, malnutrition, decreased function, or hormonal changes

Centromere Nonstaining primary constriction of a chromosome that is the point of attachment of the spindle fiber and is concerned with chromosome movement during cell division

Chromosome One of the bodies in the cell nucleus that is the bearer of genes

Dyspnea Shortness of breath, a subjective difficulty or distress in breathing, usually associated with serious disease of the heart or lungs, and occurring in health during intense physical exertion or at high altitude

Emesis Vomiting

Leukopenia Any situation in which the total number of leukocytes in the circulating blood is less than normal

Mendelian disorder Disorder based on heredity as related to or in accord with Mendel's laws

Mosaic Inlaid; resembling inlaid work; the juxtaposition in an organism or genetically different tissues resulting from somatic mutation (gene mosaicism) and anomaly of chromosome division resulting in two or more types of cells containing different number of chromosomes

Necrotic Pertaining to or affected by necrosis. Necrosis is the pathologic death of one or more cells, or of a portion of tissue or organ

Perinatal period Occurring during, or pertaining to, the periods before, during, or after the time of birth (for a few days, 7 days after delivery in humans)

Polysomes Polyribosomes; conceptually, two or more ribosomes connected by a molecule of messenger RNA; these subcellular organelles are active in protein synthesis

Radiomimetic damage Tissue or cellular damage by a chemical emulating the action of radiation (e.g., nitrogen mustards, which affect cells as high-energy radiation does)

Ring chromosome Chromosome with ends joined to form a circular structure; the normal form of the chromosome in certain bacteria

Sickle-cell anemia Syndrome characterized by the presence of crescent- or sickle-shaped erythrocytes (red blood cells) in peripheral blood

Single-cell protein Protein produced in the form of single cellular organisms (i.e., yeast or algae) from generally nonusable waste material

Thalassemia Any of a group of inherited disorders of hemoglobin metabolism in which there is a decrease in net synthesis of a particular globin chain without change in the structure of that chain; several genetic types exist, and the corresponding clinical picture may vary from barely discernible hematological abnormality to severe and fatal anemia

MYCOTOXIN IS A GENERIC TERM USED TO DESCRIBE a metabolite produced by a mold(s) in foodstuffs or feeds that can cause illness or death by ingestion, skin contact, or inhalation. The term is derived from the Greek words *mykes,* meaning fungus, and *toxicum,* meaning poison or toxin. Several compounds currently classified as mycotoxins were initially studied as potential antibiotics in the 1930s and 1940s, only to be discarded as being too toxic for higher life-forms to be of value in treating bacterial diseases in human populations. Nonetheless, the definition of a mycotoxin encompasses some antibiotics; the difference is one of degree rather than kind. Mushroom poisoning is not included in the definition of mycotoxins because the toxin is not elaborated into a food substrate but is contained within the fungal thallus; the latter is intentionally consumed.

Although ergotism in humans (St. Anthony's fire) was recognized in medieval times as being associated with the ingestion of the sclerotia (hardened masses of fungal tissue) containing toxins of the plant parasitic fungus *Claviceps purpurea,* little concern for toxic compounds produced by fungi was evident outside the veterinary and agricultural communities. Even today there is not a major concern within the medical community for the involvement of mycotoxins in human disease, although experimental evidence from animal studies and epidemiological evidence of certain human diseases would indicate a definite involvement of fungal metabolites as etiological agents of disease in humans. Furthermore, little attention is paid to the potential for agents of mycoses to produce toxic metabolites that could be virulence factors involved in the pathogenesis of the mycotic disease.

I. INTRODUCTION

Although toxic to higher animals and sometimes to plants and microorganisms, mycotoxins are not infectious agents. The diseases caused by these natural chemicals are called *mycotoxicoses.* Mycotoxicoses differ from mycoses or fungal infections such as ringworm because mold growth in the host is not necessary. With the knowledge stimulated from antibiotic screening programs in the 1940s and 1950s, it is surprising that the potential public health hazard of these mold metabolites was not recognized sooner. In fact, until as late as 1962, mycotoxicoses were described as "the neglected disease."

Before the discovery of the aflatoxins, the role these toxins played in diseases of livestock and other animals was uncertain. But with the discovery of the aflatoxins as the cause of acute liver disease among poultry in the United Kingdom and the subsequent demonstration of their carcinogenicity in laboratory animals, the next question asked was, "Are mycotoxins involved in human health?"

Some tribal or ethnic groups occasionally choose moldy substrates for preparing foods or drinks because of the flavor imparted by the fungus. In modern societies, obviously moldy food, or food that has a bad smell or taste, generally is rejected. However, mycotoxins may be present in foods or beverages that do not present such warning signs. As with animals, the element of hunger can be expected to favor ingestion of contaminated food by humans where no alternative is available. Therefore, in developing countries, food shortages, often resulting from weather conditions conducive to mycotoxin formation in crops, can be a cause for consumption of foods normally rejected because of the presence of molds or altered flavor.

Times of war with resulting food shortages and improper food handling also can lead to conditions optimal for mycotoxin production, as exemplified during World War II in the Orenburg Province of Russia. A human disease, alimentary toxic aleukia (ATA), was caused by consumption of toxic, overwintered moldy grain. The mortality rate was as high as 60% in some instances, with 10% of the population being affected. The use of primitive fermentation procedures also favors formation of microbial toxins. Typically, mycotoxins are produced at the time of maximum fungal growth; however, the apparent absence of fungal growth bears no relation to the presence or absence of the toxin. The toxin may remain in the product for years, long after the mold has died. In one instance, corn stored for 12 years, and not obviously moldy, still contained a mycotoxin, probably formed before or shortly after harvest. [*See* Food Microbiology and Hygiene.]

Modern mycotoxicology actually began with the discovery of the aflatoxins, and mycotoxins were considered to be primarily the result of inadequate storage of commodities and other foods. However, it was soon discovered that the aflatoxins were present in preharvest corn and since that time much research effort has been expended in determining the conditions under which the aflatoxins are formed in preharvest corn and other commodities in an effort to eliminate or control their production. Similar experience has occurred with other mycotoxins and their toxin-producing fungi in that they are often agents of plant disease and elaborate their toxins, when there are appropriate climatic conditions, in susceptible host plants during growth and maturation. Thus the mycotoxin problem is not only a storage problem but a problem associated with crops in the field. Also, there are fungi that are closely associated with some plants (e.g., fescue) and exist without killing the host plant. These fungi are called endophytes, and during this type of existence they may produce toxins that, when the plant is consumed by animals, can cause an intoxication. With endophytes there is little or no evidence of the organism's presence without a microscopic search.

When the mycotoxin problem first occurred is difficult to estimate; however, there are reported human mycotoxicoses dating back at least to A.D. 1100. Obviously, a risk assessment of human susceptibility to mycotoxins is difficult if not impossible because we cannot purposely experiment on humans. Although animal data can be extrapolated to humans, it must be appreciated that species variation often renders such data of limited value. Perhaps the most useful estimate of human's susceptibility to a mycotoxin is to be gained by determining the contamination level of human foods by such compounds and by observing the type and severity of diseases associated with the consumption of such foodstuffs.

Although mycotoxicoses of domestic animals have been reported for several decades, few deaths are attributed to these diseases. The mycotoxin problem is quite obscure because they are often consumed at levels that cause only poor weight gains, unthriftiness, weakened resistance to disease, or minor illness that affects the productivity of the animal but are of economic significance to the farmer. Often the mycotoxicosis is difficult to diagnose because there is insufficient material for a chemical analysis of the mycotoxin occurrence in the feed or animal tissues or fluids. Aflatoxins were involved in episodes of toxic hepatitis among domestic animals, particularly swine, in the southeastern United States. This observation was extended by reports that gave brief but incomplete descriptions of the disease in turkeys, swine, cattle, chickens, and ducklings. However, until 1978, the causal chemical had not been reported in a tissue or fluid from a single case of aflatoxicosis in a domestic animal. Necropsies of young pigs from a herd in which 30 of 250 died indicated that death was due to aflatoxins. Histopathologic findings were characteristic of experimentally induced aflatoxicoses, and aflatoxin B_1, the responsible mycotoxin, was found in the serum and liver. The mycotoxin was also in the corn and in the feed made from the locally grown corn, which was fed to these animals.

In addition to the aflatoxins, other mycotoxins have been detected as contaminants of foods and animal feeds. It is also possible to demonstrate trace amounts of various mycotoxins and their metabolites in edible tissues, milk, and eggs of farm animals fed rations containing small amounts of mycotoxins, particularly aflatoxin. The exposure to humans by an indirect route, as compared with direct dietary intake of mycotoxins, does occur. Data relating to the presence of several mycotoxins (aflatoxin, ochratoxin A, T-2 toxin, and zearalenone) in milk and tissues of cattle fed naturally occurring contaminated rations, as well as mycotoxins in tissues after oral dosing with whole cultures of various fungi, have been reported. Farm animals can convert a mycotoxin into a toxic metabolite by as much as 10–15%. An alternate route leading to human consumption is from contaminated meat and meat products. In addition, some of the mycotoxin may pass into the animal waste, which then could be used for single-cell protein production, thus allowing the mycotoxin to reenter the human food supply.

The risk to public health associated with continual, low-level exposure to dietary mycotoxins has not been satisfactorily assessed. However, most countries now control the use of contaminated peanut meal and other contaminated grains for the manufacture of compounded feeds, particularly those destined for dairy cattle. This practice is important because, despite its almost invariable contamination with one of the mycotoxins (aflatoxin B_1), peanut meal is a valuable source of protein and approximately 1% of the ingested mycotoxin appears in milk as a toxic metabolite.

A significant feature of many of the important mycotoxins is that they are immunosuppressive. The majority of the information on the effects of mycotoxins on immunity comes from experimental work conducted in laboratory and domesticated animals, as well as from *in vitro* studies. The natural occurrence of immunosuppression by mycotoxins is difficult to assess because the underlying event, that is, immunosuppression, is likely to be overshadowed by the signs and symptoms of an infectious disease. Medically, the mycotoxin functioning as an immunosuppressant could be important to human health and economically they are of significance to livestock from losses due to the mycotoxicosis, as well as loss in production. The mycotoxins of major significance regarding immunotoxicity are the aflatoxins, certain trichothecenes, and ochratoxin A. Additional mycotoxins are immunosuppressive and could prove to be important agents of disease from this standpoint.

Although a variety of mycotoxins occur naturally in agricultural materials (Table I), most mycotoxins

TABLE I
Natural Occurrence of Mycotoxins in Agricultural Products

Mycotoxin	Producing fungi	Occurrence
Aflatoxin	*Aspergillus flavus, A. parasiticus*	Corn, peanuts, cotton seed, tree nuts
Alternariol, alternariol monomethyl	*Alternaria tenuissima, A. dauci*	Weathered grain, sorghum, pecan pickouts
Citrinin	*Penicillium citrinum, Aspergillus tenuis*	Wheat, barley, peanuts
Cyclopiazonic acid	*Aspergillus flavus*	Corn, kodo millet, sunflower seed, peanuts
Ergot alkaloids (ergotamine, etc.)	*Claviceps* spp.	Ergots, ergot-infected pasture grass
Fumonisins	*Fusarium moniliforme, F. proliferatum*	Corn
Kojic acid	*Aspergillus flavus, A. oryzae*	Moldy corn
Moniliformin	*Fusarium proliferatum*	Corn
Ochratoxin A	*Aspergillus ochraceus, Penicillium viridicatum, P. cyclopium*	Corn, barley, wheat, oats, rice
Patulin	*Aspergillus clavatus, Penicillium patulum, P. urticae*	Silage, apples
Penicillic acid	*Penicillium martensii, P. puberulum, P. palitans*	Corn, beans
Phomopsins	*Phomopsis leptostromoformis*	Lupine
PR toxin	*Penicillium roqueforti*	Silage
Slaframine	*Rhizoctonia leguminicola*	Clover
Sporidesmin	*Pithomyces chartarum*	Dead pasture grass
Sterigmatocystin	*Aspergillus versicolor, A. nidulans, A. ustus*	Wheat, barley, rice, peanuts
Tenuazonic acid	*Alternaria tenuissima, A. alternata, Pyricularia oryzae, Phoma sorghina*	Diseased rice, plants
Trichothecenes	*Fusarium graminearum, F. sporotrichioides, F. nivale, F. sambucinum, F. culmorum, F. equiseti*	Corn, barley, wheat, oats
Zearalenone	*Fusarium graminearum, F. moniliforme, F. nivale, F. oxysporum*	Corn, sorghum, wheat

are not regulated by governments. Control of mycotoxins in food is a complex and difficult task. Sufficient information regarding toxicity, mutagenicity, carcinogenicity, teratogenicity, and the extent of contamination and stability of mycotoxins in foods is lacking for most mycotoxins. Such information is necessary to establish regulatory guidelines, tolerances, and seizure policies. In the absence of tolerances, the U.S. Food and Drug Administration (FDA) has established guidelines for acceptable levels of aflatoxin in food and feed and are based on levels of aflatoxins that are considered safe to feed to different species of animals or for human consumption. In September of 1993 the Associate Commissioner for Regulatory Affairs of the FDA also established advisory levels for deoxynivalenol, one of the trichothecenes.

II. HUMAN DISEASES

The involvement of mycotoxins as etiologic agents of human diseases is difficult to determine. Certainly ergotism and ATA can be attributed to a fungal toxin(s). Evidence suggests that acute cardiac beriberi, common throughout Asia, may be linked to the so-called yellow rice toxins. Examples of human diseases for which evidence suggests involvement of a mycotoxin are summarized in Table II.

The toxicity of mycotoxins that are potentially harmful to humans has also been demonstrated in human tissue culture. The mycotoxins from *Penicillium islandicum* inhibited growth of Chang's liver and HeLa cells and, at sufficient concentrations, led to morphological changes and cell death. Similarly, the endotoxin extracted from the mycelium of *Aspergillus fumigatus* affected respiration and produced morphological changes in kidney cells grown in culture. Mycotoxins such as aflatoxin B_1 and rubratoxin B inhibited a number of human and animal cell culture systems.

In general, mycotoxins are secondary metabolites that perform minor or no obvious function in the metabolic scheme of the organism and are products

TABLE II
Mycotoxicoses in Which Analytic and/or Epidemiologic Data Suggest Human Involvement

Disease	Species	Substrate	Etiologic agent
Aka Kabio-byo	Human	Wheat, barley, oats, rice	*Fusarium* spp.
Alimentary toxic aleukia (ATA or septic angina)	Human	Cereal grains (toxic bread)	*Fusarium* spp.
Balkan nephropathy	Human	Cereal grains	*Penicillium*
Cardiac beriberi	Human	Rice	*Aspergillus* spp., *Penicillium* spp.
Celery harvester's disease	Human	Celery (pink rot)	*Sclerotinia*
Dendrodochiotoxicosis	Horse, human	Fodder (skin contact, inhaled fodder particles)	*Dendrodochium toxicum*
Ergot	Human	Rye, cereal grains	*Claviceps purpurea*
Esophageal tumors	Human	Corn	*Fusarium moniliforme*
Hepatocarcinoma (acute aflatoxicosis)	Human	Cereal grains, peanuts	*Aspergillus* (*A. flavus*)
Kashin Beck disease, "Urov disease"	Human	Cereal grains	*Fusarium*
Onyalai	Human	Millet	*Phoma sorghina*
Reye's syndrome	Human	Cereal grains	*Aspergillus*
Stachybotryotoxicosis	Human, horse, other livestock	Hay, cereal grains, fodder (skin contact, inhaled haydust)	*Stachybotrys atra*

of reactions that branch off at a limited number of biosynthetic pathways, such as those involving acetate, pyruvate, melonate, mevalonate, shikimate, and amino acids. From the standpoint of human and animal health, molds belonging to the genera *Aspergillus, Fusarium,* and *Penicillium* have received the most attention, owing to their frequent occurrence in food and feed commodities. Although *Aspergillus* spp. and *Penicillium* spp. typically are considered storage fungi that do not invade intact grain before harvest, whereas *Fusarium* spp. and *Alternaria* spp. are predominantly field (preharvest) fungi, we now know that this is not a useful criterion for classification. Unfavorable conditions (e.g., drought and damage of seeds by insects or during mechanical harvesting) can enhance mycotoxin production during both growth and storage. Toxin production can take place over a wide range of moisture (10–33%), relative humidity (>70%), and temperature (4–35°C), depending on the fungal organism involved.

After the discovery of aflatoxins and their potent carcinogenicity, the search for mycotoxins in the past three decades has led to the identification of more than a hundred toxigenic fungal organisms and considerably more mycotoxins throughout the world. However, the public health significance of most of these remains unknown. A summary of experimental and domestic animal studies, as well as some current epidemiologic associations of mycotoxins with human disease, is pertinent.

A. Ergot

The history of human mycotoxicoses dates back to the Middle Ages, when ergotism (St. Anthony's fire) was the scourge of Central Europe. Ergotism, which is now rare, was first associated with the consumption of scabrous (ergotized) grain in the mid-sixteenth century. Subsequent studies led to the identification of *Claviceps purpurea* as the fungal agent invading rye, oats, wheat, and Kentucky bluegrass and *C. paspali* invading Dallis grass. Lysergic acid derivatives, the amine and amino acid alkaloids of ergot, were identified as the causative agents of the gangrenous and nervous forms of the disease. Gangrenous ergotism typically is manifested as prickly and intense hot and cold sensations in the limbs, and swollen, inflammed, necrotic, and gangrenous extremities, which eventually sloughed. Convulsive ergotism was characterized by central nervous system (CNS) signs, numbness, cramps, severe convulsions, and death. Both syndromes have been documented in the recent literature in domestic animals and humans (in India) consuming ergotized grains and in humans treated with ergotamine for migraine headaches.

Ergot alkaloids are smooth muscle stimulants, promoting narrowing of blood vessels (leading to gangrenous ergotism) and inducing uterine contractions (oxytocic effect). Ergot alkaloids antagonize serotonin and block both the stimulatory and inhibitory CNS responses of epinephrine. The U.S. Department of Agriculture (USDA) Grains Division has set a tolerance limit of 0.3% (by weight) of contaminated grain in commercial trade. [*See* Alkaloids in Medicine.]

B. Trichothecenes

In the first half of the twentieth century, a large human mycotoxicosis was reported in Russia. The disease, termed *alimentary toxic aleukia,* was characterized by total atrophy of the bone marrow, agranulocytosis, necrotic angina, sepsis, hemorrhagic diathesis, and mortality ranging from 2 to 80%. Later it was linked to the consumption of overwintered cereal grains and wheat or bread made from them. *Fusarium poae* and *F. sporotrichioides,* which grow on these grains, have been shown to produce several trichothecene toxins, including T-2 toxin, neosolaniol, HT-2 toxin, and T-2 tetraol. Signs of ATA were developed in cats given pure T-2 toxin orally. However, because there was lack of proof of the involvement of these toxins in the original outbreak, they remain conjecturally associated with the disease.

Several outbreaks of a seasonal intoxication in horses and cattle caused by consumption of hay contaminated with *Stachybotys atra* (*S. alternans*) were reported from Russia between 1930 and 1960. Two forms of intoxication, the atypical and typical, reflecting acute or chronic exposure are characterized by sudden onset of neurological signs (loss of vision, poor control of movements, and tremors) or signs of dermonecrosis, leukopenia, and gastrointestinal ulceration and hemorrhages, respectively. In these regions, humans exhibited severe dermatitis after handling of or sleeping on contaminated hay. Inhalation of dust from the infected hay can result in inflammation of the nose, fever, chest pain, and leukopenia. Five trichothecene compounds were isolated, of which three belonged to the group of macrocyclics (roridin–verrucarin) containing a conjugated butadiene system attached to the trichothecene structure.

Deoxynivalenol, likely the most frequently occurring trichothecene in grains and foods, has been shown to cause an increase in IgA in the serum of mice and results in kidney damage described as a glomerulonephritis. The latter is a significant disease in the human population and the glomerulonephritis in the mouse model produced by deoxynivalenol is pathologically quite similar to that which occurs in humans. The question of the involvement of deoxynivalenol in human glomerulonephritis is unanswered.

A group of trichothecene toxins, including nivalenol and fusarenon-X, is produced by *F. nivale* in the flowering grainhead of wheat, barley, rice, corn, other cereals, and certain forage grasses. The disease in cereal grains, called *red-mold disease* (akakabi-byo) or *black spot disease* (kokuten-byo), has been associated with intoxications in humans, horses, and sheep in Japan. Symptoms in humans include headaches, vomiting, and diarrhea, and no fatalities. *Fusarium graminearum,* capable of producing deoxynivalenol and its acetylated derivatives on wheat, barley, and oats, was also isolated, suggesting multiple causation.

Fusarium sporotrichioides, known to produce T-2 toxin and neosolaniol, has been isolated from bean hulls incriminated in a disease (bean hull poisoning) in horses characterized by retarded reflexes, decreased heart rate, disturbed respiration, cyclic movements, and convulsions, with a death rate of 10–15%. However, because of some similarities of this disease with equine leukoencephalomalacia, bean hull poisoning has potential to be caused by *F. moniliforme* or perhaps related species.

Other diseases attributable to trichothecenes included dendrodochiotoxicosis in horses, sheep, and pigs in Russia that ingested feedstuffs contaminated with *Myrothecium roridum* (produces roridins and verrucarins); various syndromes reported in the United States and Canada involving corn (moldy corn toxicosis) and cereal grains consumed by farm animals (T-2 toxin and others); and finally the alleged use of trichothecene toxins as a chemical warfare agent (yellow rain) in Southeast Asia. The true involvement of trichothecene toxins in the latter episode remains problematic.

Although many fungal genera such as *Fusarium, Myrothecium,* and *Stachybotrys* can produce these toxins, most trichothecenes of health significance are produced by *Fusarium* spp. Despite the diversity of human and animal diseases associated with this group of toxins, characteristic signs and symptoms of radiomimetic damage (e.g., emesis, feed refusal, irritation and necrosis of skin and mucus membranes, hemorrhage, destruction of thymus and bone marrow, hematologic changes, nervous disturbances, and necrotic angina) are common to all toxic syndromes.

Feed refusal and vomiting are common problems in farm animals, especially swine, in the midwestern United States and are predominately associated with the presence of the trichothecene deoxynivalenol (vomitoxin) in wheat or corn. Although T-2 toxin and diacetoxyscirpenol (DAS) can also cause emesis and feed refusal, their role in the swine feed refusal syndrome appears negligible because of their rare presence in food and feed commodities in the United States. Trichothecenes (T-2 toxin) can cause fetal death and abortions, along with tail and limb abnormalities, in rodent offspring. Although fusarenon-X and T-2 toxin are mutagenic at high doses in bacterial and yeast systems, trichothecenes exhibit no mutagenic effect in most other systems. Carcinogenic effects of trichothecenes are not known.

Metabolism of trichothecenes occurs rapidly through deacetylation and hydroxylation and subsequent glucuronidation in the liver and kidneys. Trichothecenes generally are recognized as potent inhibitors of protein synthesis in eukaryotic systems (animals and plants), inhibiting initiation, elongation, and termination of protein synthesis by way of their inhibition of peptidyl transferase activity, and also their ability to cause disaggregation of polysomes; T-2 toxin, DAS, nivalenol, and fusarenon-X inhibit initiation, whereas trichodermin, crotocin, and verrucarol inhibit elongation or termination. Many of the toxic effects of trichothecenes can be explained by this mechanism. Despite the severe toxic effects of trichothecenes, their low-frequency occurrence in nature (perhaps deoxynivalenol is an exception here), their rapid metabolism to apparently nontoxic metabolites in animals, and their low potential for residue transfer to humans reduce the risk of human disease to extremely low levels. Farm animals, however, are at a higher risk of intoxication than humans because of the greater likelihood of consumption of trichothecenes in moldy feeds.

C. Aflatoxins

It was not until the early 1960s that aflatoxins were discovered as the causative agents of *turkey X disease* in England, which resulted in the death of thousands of turkey poults, ducklings, and chicks that were fed diets containing *Aspergillus flavus* or *A. parasiticus*-contaminated peanut meal. This outbreak, coupled with the reported carcinogenicity of the aflatoxins in experimental animals, helped fuel the scientific curiosity surrounding this group of food contaminants.

The aflatoxins are a group of highly substituted coumarins containing a fused dihydrofuran moiety. Four major aflatoxins, designated B_1, B_2, G_1, and G_2 (based on blue or green fluorescence under ultraviolet light), are produced in varying quantities in a variety of grains and nuts. Commodities most often shown to contain aflatoxins are peanuts, various other tree nuts, cottonseed, corn, and figs. In addition to *A. flavus,* strains of *A. parasiticus* may be capable of producing aflatoxins. Human exposure can occur from consumption of aflatoxins from these sources and the products derived from them, as well as from tissues and milk (aflatoxin M_1, a metabolite of aflatoxin B_1) of animals consuming contaminated feeds.

Aflatoxin B_1 (AFB_1), the most potent and most commonly occurring aflatoxin, has been shown to be acutely toxic (LD_{50} 0.3–9.0 mg/kg) to all species of animals, birds, and fishes tested. Sheep and mice are the most resistant, whereas swine, poultry, cats, dogs, and rabbits are quite sensitive species. Acute effects in animals include death without signs, or signs of anorexia, depression, ataxia, dyspnea, anemia, and hemorrhages from body orifices. In subchronic cases, icterus, hypoprothrombinemia, hematomas, and gastroenteritis are common. Chronic aflatoxicosis is characterized by bile duct proliferation, periportal fibrosis, icterus, and cirrhosis of the liver and is associated with loss of weight and reduced resistance to disease. Dietary levels as low as 0.3 ppm can cause such effects.

Prolonged exposure to low concentrations lead to liver tumors (e.g., hepatoma, cholangiocarcinoma, or hepatocellular carcinoma). In poultry there is a lipid malabsorption syndrome characterized by decreased pancreatic lipase and fats in the feces. Reductions in serum triglycerides, cholesterol, phospholipids, and carotenoids also were seen after AFB_1 exposure. Mutagenicity of AFB_1 has been demonstrated in human cells in culture as well as bacteria. Metabolic activation of AFB_1 was required for the mutagenic effect. AFB_1 is primarily metabolized by enzymes present in the liver and other organs, forming detoxified substances but also generating mutagenic products.

The combination of AFB_1 with DNA or RNA bases leads to formation of repair-resistant adducts, loss of purines, and other changes leading to single-strand breaks, base pair substitution, or frame-shift mutations. Involvement of oncogenes in such interactions may result in oncogene activation. In addition, AFB_1 inhibits the synthesis of DNA, RNA, and protein. Inhibition of protein synthesis may be related to several lesions and signs of aflatoxicoses, including fatty liver (failure to mobilize fats from the liver), alteration of blood coagulation (inhibition of prothrombin synthesis), and reduced immune function. [*See* DNA Synthesis.]

In addition to aflatoxin contamination of foods such as peanuts and corn, aflatoxins and their metabolites can also occur in animal tissues. Especially important is the metabolite aflatoxin M_1 (AFM_1), a product of AFB_1, present mainly in milk of AFB_1-exposed dairy animals. The average daily per capita consumption of AFB_1 and AFM_1 in human populations in the United States has been calculated as 25 and 0.3 ng/kg body weight (BW), respectively. By using the epidemiologic data generated from Asia, Africa, and the United States, males in the United States are twice as resistant to induction of liver cancer by AFB_1 as the males in Asia and Africa. The carcinogenicity of AFM_1 seems to be two orders of magnitude lower than that of AFB_1, and therefore a negligible risk. The effect of AFM_1 on human infants needs further evaluation.

Other less widespread human clinical syndromes in which aflatoxins have been implicated include acute hepatitis (aflatoxicoses) in India, Taiwan, and certain countries in Africa; childhood cirrhosis in India; and possibly Reye's syndrome in many parts of the world. Reye's syndrome is a childhood neurologic disease that resembles viral encephalitis and actually involves a viral syndrome followed by rapid progresses into coma and convulsions leading to death. Characteristic lesions include enlarged, pale, fatty liver and kidneys and severe cerebral edema. Evidence from Thailand and other countries, including the United States, associates aflatoxin consumption or high levels of AFB_1 in tissues from patients with Reye's syndrome. A syndrome strikingly similar to Reye's is produced in monkeys given AFB_1 orally. Epidemiologic evidence, however, also suggested a link between the use of aspirin and Reye's syndrome, prompting the U.S. Surgeon General to advise against the use of salicylates in children with chickenpox or influenza.

Widespread concern regarding the toxic effects of aflatoxins in humans and animals and the possible transfer of residues from animal tissues and milk to humans has led to regulatory actions governing the interstate as well as global transport and consumption of aflatoxin-contaminated food and feed commodities. Action levels of aflatoxins in corn and other feed commodities used to feed mature, nonlactating animals are 100 ppb, although temporary increases in limits are allowed on a case-by-case basis by the FDA in situations such as drought, in which availability of uncontaminated corn is extremely limited. For commodities destined for human consumption and interstate commerce, the action limit is 20 ppb. For milk, the action level of AFM_1 is 0.5 ppb. Among the many approaches tried to limit the aflatoxin contamination of grain, prevention of stress during growth and harvest, use of fungus-resistant varieties of grains, preventing insect damage to grain, avoiding mechanical injury to grain during harvesting, drying of grains to contain less than 15% moisture, and strict control of humidity during storage are important to prevent aflatoxin production.

D. Ochratoxins

In 1957 and 1958, up to 75% of the households in several villages located in the valley floor in contiguous areas of Yugoslavia, Romania, and Bulgaria were found to be affected by chronic nephropathy (kidney disease, Balkan or endemic nephropathy). Although genetic factors appear to be partially involved, evidence was presented that a mycotoxin, ochratoxin A, produced in foodstuffs by *Aspergillus ochraceus* and a number of other aspergilli and penicilli (*P. viridicatum*), was consumed at higher levels more frequently by people in these endemic areas compared with areas free from nephropathy. In addition, a remarkably similar nephropathy was identified in swine (porcine nephropathy) and in bovines fed ochratoxin. Signs include lassitude, fatigue, anorexia, abdominal (epigastric or diffuse) pain, and severe anemia accompanying renal damage. Reduced concentrating ability, reduced renal plasma flow, and decreased glomerular filtration occur subsequently, accompanied by gross and microscopic renal changes. Deaths result from intoxication caused by kidney malfunction.

The ochratoxins are found in barley, corn, wheat, oats, rye, green coffee beans, and peanuts. In experimental animals, ochratoxin A produces renal lesions and hepatic degeneration. In poultry, ochratoxin A causes reduced weight gains and decreased egg shell quality and egg production in addition to renal effects. Teratogenic effects of ochratoxin A in rodents include malformations of the head, jaws, tail, limbs, and heart. Ochratoxin A is not mutagenic in several assay systems tested, but is carcinogenic as it can induce hepatomas and renal adenomas in mice exposed to 40 ppm in the diet.

Ochratoxin A is broken down by carboxypeptidase A and α-chymotrypsin to the less toxic ochratoxin α. Absorbed ochratoxin A distributes mainly to the kidneys and liver and is excreted rapidly in the urine and feces. Ochratoxin A inhibits mitochondrial respiration and reduces ATP levels. These effects are also produced by ochratoxin α, a cleavage product of ochratoxin A. Depletion of glycogen, inhibition of glucose production, a pathway that accounts for 50–60% of the blood glucose in the starved or diabetic stage, and several key cyclic adenosine monophosphate (cAMP)-activated enzymes in this pathway, including phosphoenolpyruvate carboxykinase by ochratoxin A, have received attention in recent years. Whether any of those enzymatic steps is the critical target in the pathogenesis of ochratoxicoses is unknown.

E. Psoralens

Psoralens are furocoumarin compounds that have been used in repigmenting white skin lesions in an acquired disease called *vitiligo*. Psoralens are in some suntan lotions and in drugs used to treat psoriasis. Abuse of such compounds can result in dermatitis after exposure to sun, as well as nausea, vomiting, vertigo, and mental excitation. A phototoxic dermatitis in celery pickers has also been linked to the presence of psoralens in stalks of celery infected with *Sclerotinia sclerotiorum* (pink rot), *S. rolfsii, Rhizoctonia solani,* or *Erwinia aroideae.* In addition to celery, fig, parsley, parsnip, lime, and clove also contain psoralens. Unlike other photosensitizing agents, psoralens seem to act by reacting with DNA in the presence of light and to a lesser extent with RNA. Treatment with 8-methoxypsoralen and ultraviolet light induced squamous cell carcinomas of the ear in mice. Psoralens are rapidly excreted in the urine as fluorescent nontoxic metabolites. One of them, however, 8-methoxypsoralen, appears to undergo change similar to that of aflatoxins and may thus react with DNA in a similar fashion.

The mechanism of psoralen photosensitivity appears to involve its intercalation between the DNA bases followed by cross-linking. Intercalation occurs between two pyrimidines on opposing sides of the helix; then the psoralen forms a chemical complex with one of the pyrimidines after absorption of ultraviolet light. Cross-links are formed by further absorption of ultraviolet light, which links the psoralen–pyrimidine complex with the second pyrimidine.

F. Citreoviridin (Yellow Rice Toxin)

Acute cardiac beriberi (shoshin-kakke), characterized by palpitation, nausea, vomiting, rapid and difficult breathing, cold and bluish extremities, rapid pulse, abnormal heart sounds, low blood pressure, restlessness, and violent mania leading to respiratory failure and death, was observed in Japan in the late 1800s and early 1900s. Strong evidence placed citreoviridin as the causal agent of cardiac beriberi rather than avitaminosis. In fact, the dark yellow toxic metabolite citreoviridin was isolated from *P. citreoviride,* and production of the neurologic syndrome and respiratory failure can be reproduced in rats given extract of *P. citreoviride*-contaminated rice. Other toxins identified failed to produce signs resembling cardiac beriberi. In 1921, the Japanese government passed the Rice Act to reduce the availability of moldy rice in markets. This act resulted in a sharp decrease in the disease during the same year, while maintaining rice as a prominent dietary ingredient. The vitamin-enriched diet of the consumer, plus the improved inspection, has made cardiac beriberi of little importance in modern times. It should be recognized that citreoviridin can occur in corn and other foods.

G. Bishydroxycoumarin (Dicoumarol)

Dicoumarol is a vitamin K antagonist that inhibits the availability of vitamin K. Resulting effects lead to bleeding disorders and death from blood loss from undetectable bleeding sites. The molecules of naturally occurring coumarins in sweet clover (*Melilotus* sp.) hay join together to form dicoumarol as a result of fungal spoilage during curing. Cattle consuming such hay were poisoned in the 1920s in North Dakota and Alberta. Most human poisonings result from therapeutic accidents involving coumarin therapy of clotting disorders.

H. Zearalenone

In addition to zearalenone and zearalenol being contaminants in grains such as corn, wheat, sorghum, barley, and oats, zearalanol, a synthetic analogue of zearalenol (Ralgro), is used as an anabolic agent (promoting body growth) in cattle. Under natural conditions, zearalenone and its derivatives are produced by *Fusarium graminearum,* mostly in ear corn stored in cribs. Zearalenone, despite its structural dissimilarity with estrogens, induces effects similar to those produced by excessive steroidal as well as synthetic estrogens. Among the domestic animals, swine appear to be the most sensitive, exhibiting signs of hyperestrogenic syndrome (i.e., swollen and edematous vulva, hypertrophic myometrium, vaginal cornification, and prolapse in extreme cases).

Zearalenone acts by interacting with estrogen

receptors in the cell cytoplasm. The receptor–zearalenone complex is then translocated to the nucleus, where it combines with receptors in the DNA, causing selective gene transcription, increased water and lowered lipid content in muscle, and increased permeability of the uterus to glucose and RNA and protein precursors. Zearalenone induces biphasic changes in the concentration of luteinizing hormone but not follicle-stimulating hormone in serum. The rapid metabolism of zearalenone and zearalenol to conjugated metabolites, which are then excreted in urine and feces, makes consumption of meat and milk from animals receiving Ralgro an insignificant risk to humans.

Recent evidence from Italy and Puerto Rico, however, suggests that estrogenic substances, especially residues of zearalenone and zearalenol in red meats and poultry, may have caused premature thelarche (development of breast before age 8) and precocious pseudopuberty.

I. Fumonisins

The fumonisins are a group of at least six structurally related mycotoxins that were fairly recently discovered as the result of South African investigators trying to find a cause for the occurrence of esophageal cancer in certain peoples of the Transkei region in South Africa. Their discovery of the fumonisins did not prove the etiology of the esophageal cancer but subsequently they found that the fumonisins, subsequently found to be produced primarily by the fungi *Fusarium moniliforme* and *F. proliferatum* in corn, were the cause of a severe brain disease of horses, which had been known for many years. Prior to this time it was known that *F. moniliforme*-contaminated corn was a cause of this disease but the toxin has escaped detection. Subsequently, they found that the fumonisins could cause liver cancer in the rat and others found that the fumonisins could produce a lung disease in swine. Presently, the fumonisins have been only epidemiologically associated with human esophageal cancer in South Africa and certain areas of China. Although the fumonisins can produce several different diseases from a pathological point of view and the diseases appear to be somewhat different depending on the species ingesting the mycotoxins, the mode of action of the fumonisins may be similar among these species. A unique feature of the fumonisins is that they interfere with sphingolipid biosynthesis by inhibition of certain enzymes in the biosynthetic pathway. The sphingolipids are important components of cells, especially their membranes, and are regulators of cell growth, differentiation, and function. The function of sphingolipids in the body are quite diverse and closely associated with many other important functions such as immune mechanisms; therefore small amounts of fumonisins may be quite detrimental to the recipient.

As a result of the worldwide occurrence of fumonisins in corn, the presence of them in foods is not unexpected as corn is used in a variety of foods from native dishes to snack foods. Presently, there is insufficient information to make an assessment of risk to the human population relative to the levels in foods that may be of significance in causing detrimental effects.

J. Gliotoxin

Gliotoxin is a mycotoxin produced by a wide variety of fungi, including some fungi that are agents of mycoses in humans and other animals. This compound was originally found over a half-century ago by investigators looking for antibiotics produced by microorganisms. Interest in this compound as an antibiotic waned when it was found that the level of toxicity of gliotoxin precluded its use clinically. More recently, investigators found that this compound was produced by the human pathogenic fungus *Aspergillus fumigatus* and that it was a highly immunosuppressive compound. Of likely significance is that this compound is produced during the pathogenic state of the organism in tissues of the infected host. It has been found in experimentally infected mice, in the naturally infected bovine udder, and in both experimentally and naturally infected turkeys. Gliotoxin does not appear to be a significant toxin ingested with food, although a recent report stated that camels became intoxicated when they consumed moldy hay containing this compound. The significance of this compound being produced during pathogenesis could have a number of ramifications, as its immunosuppressive activity could be functional as a virulence factor in the disease. Some current investigations into the involvement of this mycotoxin in avian aspergillosis, an economically important respiratory disease of poultry caused by *A. fumigatus,* are relevant to this concept. The significance of the presence of this compound in edible tissue is not known.

K. Other Mycotoxins

A number of other mycotoxins (Table III) have been identified either as contaminants in foods destined

TABLE III
Miscellaneous Mycotoxins[a]

Mycotoxin	Major producing organisms	Source of fungi	Principal toxic effects
Alternariol and alternariol methyl ether	*Alternaria* sp.	Sorghum, peanuts, wheat	Highly teratogenic to mice; cytotoxic to HeLa cells; lethal to mice
Altenuene, altenuisol	*Alternaria* sp.	Peanuts	Cytotoxic to HeLa cells
Altertoxin I	*Alternaria* sp.	Sorghum, peanuts, wheat	Cytotoxic to HeLa cells; lethal to mice
Ascladiol	*Aspergillus clavatus*	Wheat flour	Lethal to mice
Austamide and congeners	*Aspergillus ustus*	Stored foodstuffs	Toxic to ducklings
Austdiol	*Aspergillus ustus*	Stored foodstuffs	Toxic to ducklings
Austin	*Aspergillus ustus*	Peas	Lethal to chicks
Austocystins	*Aspergillus ustus*	Stored foodstuffs	Toxic to ducklings; cytotoxic to monkey kidney epithelial cells
Beauvericin	*Fusarium moniliforme*	Corn	Ionophore
Chaetoglobosins	*Penicillium aurantiovirens*, *Chaetomium globosum*	Pecans	Toxic to chicks; cytotoxic to HeLa cells
Citreoviridin	*Penicillium citreoviride*	Rice	Neurotoxic, producing convulsions in mice
Citrinin	*Penicillium viridicatum*, *P. citrinum*	Corn, barley	Nephrotoxic to swine
Cyclopiazonic acid	*Penicillium cyclopium*	Ground nuts, meat products, Kodo millet, sunflower seed	Nephrotoxic, enterotoxic, muscle necrosis
Cytochalasins	*Aspergillus clavatus*, *Phoma* sp., *Phomopsis* sp., *Hormiscium* sp., *Helminthosporium dematioideum*, *Metarrhizium anisopliae*	Rice, potatoes, Kodo millet, pecans, tomatoes	Cytotoxic to HeLa cells; teratogenic to mice and chickens
Diplodiatoxin	*Diplodia maydis*	Corn	Nephrotoxic and enterotoxic to cattle and sheep
Emodin	*Aspergillus wentii*	Chestnuts	Lethal to chicks
Fumigaclavines	*Aspergillus fumigatus*	Silage	Enterotoxic to chicks
Gliotoxin	*Penicillium bilaii*, *Aspergillus fumigatus*	Hay	Toxic to mice, rats, and turkeys; immunosuppressive
Kojic acid	*Aspergillus flavus*	Squash, spices	Lethal to mice
Malformins	*Aspergillus niger*	Onions, rice	Lethal to rats
Maltoryzine	*Aspergillus oryzae*	Malted barley	Nepatotoxic and causes paralysis
Oosporein (Chaetomidin)	*Chaetomium trilaterale*	Peanuts	Lethal to chicks
Paspalamines	*Claviceps paspali*	Dallisgrass	Neurotoxic to cattle and horses, causes paspalum staggers
Patulin	*Penicillium urticae*	Apple juice	Lethal to mice; mutagenic; teratogenic to chicks; pulmonary effects in dogs; carcinogenic to rats
Penicillic acid	*Penicillium* spp.	Corn, dried beans	Lethal to mice; mutagenic; carcinogenic to rats
PR toxin	*Penicillium roqueforti*	Mixed grains	Hepatotoxic and nephrotoxic to rats; causes abortion in cattle
Roseotoxin B	*Trichothecium roseum*	Corn	Toxic to mice and ducklings
Rubratoxins	*Penicillium rubrum*	Corn	Causes hemorrhage in animals; hepatotoxic to cattle
Secalonic acids	*Aspergillus aculeatus*, *Penicillium oxalicum*	Rice, corn	Lethal, cardiotoxic, teratogenic, and causes lung irritation in mice
Slaframine	*Rhizoctonia leguminicola*	Red clover	Causes salivation and lacrimation in horses and cattle

(*continues*)

TABLE III (*Continued*)

Mycotoxin	Major producing organisms	Source of fungi	Principal toxic effects
Sporidesmins	*Pithomyces chartarum*	Pasture grasses	Hepatotoxic, causes photosensitization in ruminants
Sterigmatocystin	*Aspergillus flavus*	Mammals	Mutagenic, carcinogenic, and hepatotoxic to mammals
Tenuazonic acid	*Alternaria* sp.	Grains, nuts	Lethal to mice
Terphenyllins	*Aspergillus candidus*	Wheat flour	Hepatotoxic to mice; cytotoxic to HeLa cells
Tremorgenic mycotoxins			
Fumitremorgens A and B	*Aspergillus fumigatus*	Rice	Neurotoxic (prolonged tremors and convulsions)
Paxilline	*Penicillium paxilli*	Pecans	Neurotoxic (prolonged tremors and convulsions)
Penitrems A, B, and C	*Penicillium cyclopium* *P. crustosum*	Cheese, walnuts	Penitrem A: neurotoxic (prolonged tremors and convulsions) to mice, dogs, and horses
Tryptoquivalines	*Aspergillus clavatus*	Rice	Neurotoxic (prolonged tremors and convulsions)
Verruculogen (TR-1)	*Penicillium verruculosum*	Peanuts	Neurotoxic (prolonged tremors and convulsions)
Unidentified toxin(s)	*Aspergillus terrus, Balansia epichloe, Epichloe typhina, Fusarium tricinctum,* and others	Fescue grass	Gangrene (fescue foot); summer slump syndrome, causes fat necrosis and agalactia in cattle
Xanthoascin	*Aspergillus candidus*	Wheat flour	Hepatotoxic and cardiotoxic to mice

[a]For a more complete listing, see Cole and Cox (1981).

for human consumption or as metabolites of fungi isolated from human foods. Although some of these have been associated with outbreaks of domestic animal diseases, no current link between human consumption and disease has been established. Others have been shown to induce toxic and lethal effects in laboratory anaimals with no association between consumption of these toxins by animals or humans and a disease syndrome. Several of these (e.g., cytochalasins and secalonic acid D) have been used to expand our understanding of normal as well as abnormal cellular responses to xenobiotics.

Although it is difficult to assess the significance of consumption of mycotoxins in human foods, it is clear that to perform human risk assessments there is continued need for extensive research into the toxicological aspects of numerous known mycotoxins and the discovery of yet unknown mycotoxins. And though there are a vast number of toxic metabolites of fungi, there appear to be few problems where the food supply and intake is quite varied; problems seem to be more significant where a single staple is relied upon for daily sustenance and where lack of abundant food causes the consumption of moldy food. Much current investigation is concerned with the control of the occurrence of mycotoxins in food crops, and some control has been achieved by control of insects, by avoiding stress to the plants during seed development and harvest, and finally by adequate storage and shipment of commodities entering commercial food marketing channels.

BIBLIOGRAPHY

Beasley, V. R. (1989). "Trichothecene Mycotoxins: Pathophysiologic Effects," Vol. 1 (175), Vol. 2 (198). CRC Press, Boca Raton, Florida.

Bechtel, D. H. (1989). Molecular dosimetry of hepatic aflatoxin B_1–DNA adducts: Linear correlation with hepatic cancer risk. *Regul. Toxicol. Pharmacol.* **10**, 74–81.

Betina, V. (1984). "Mycotoxins: Production, Isolation, Separation, and Purification." Elsevier, Amsterdam.

Bray, G. A., and Ryan, D. H. (eds.) (1991). "Mycotoxins, Cancer, and Health," Vol. 1. Louisiana State Univ. Press, Baton Rouge.

Castegnaro, M., Pleština, R., Dirheimer, G., Chernozemsky, I. N., and Bartsch, H. (eds.) (1991). "Mycotoxins, Endemic Nephropathy and Urinary Tract Tumours." Oxford Univ. Press, New York.

Cole, R. J., and Cox, R. H. (1981). "Handbook of Toxic Fungal Metabolites." Academic Press, New York.

Dvorackova, I. (1990). "Aflatoxins and Human Health." CRC Press, Boca Raton, Florida.

Eaton, D. L., Groopman, J. D. (eds.) (1994). "The Toxicology of Aflatoxins: Human Health, Veterinary, and Agricultural Significance." Academic Press, San Diego.

Eklund, M. E., Richard, J. L., and Mise, K. (eds.) (1995). "Molecular Approaches to Food Safety: Issues Involving Toxic Microorganisms." Alaken, Inc., Fort Collins, Colorado.

Hayes, A. W. (1981). "Mycotoxin, Teratogenicity and Mutagenicity." CRC Press, Boca Raton, Florida.

Kurata, H., and Ueno, Y. (1984). "Toxigenic Fungi—Their Toxins and Health Hazards." Elsevier, Amsterdam.

Marasas, W. F. O., and Nelson, P. E. (1987). "Mycotoxicology." Pennsylvania State Univ. Press, University Park.

Natori, S., Hashimoto, K., and Ueno, Y. (eds.) (1989). "Mycotoxins and Phycotoxins '88: Bioactive Molecules," Vol. 10. Elsevier, Amsterdam.

Pohland, A. E., Dowell, V. R., Jr., and Richard, J. L. (eds.) (1990). "Microbial Toxins in Foods and Feeds: Cellular and Molecular Modes of Action." Plenum, New York.

Richard, J. L., and Cole, R. J. (eds.) (1989). "Economic and Health Risks Associated with Mycotoxins." Council for Agricultural Sciences and Technology, Ames, Iowa.

Richard, J. L., and DeBey, M. C. (1995). Production of gliotoxin during the pathogenic state in turkey poults by *Aspergillus fumigatus* Fresenius. *Mycopathologia* **129**, 111–115.

Richard, J. L., and Thurston, J. R. (eds.) (1986). "Diagnosis of Mycotoxicoses." Martinus Nijhoff, Dordrecht.

Richard, J. L., DeBey, M. C., Chermette, R., Pier, A. C., Hasegawa, A., Lund, A., Bratberg, A. M., Padhye, A. A., and Connole, M. D. (1994). Advances in veterinary mycology: A case for the involvement of gliotoxin in avian aspergillosis. *J. Med. Vet. Mycol.* **32**(1), 169–187.

Seagrave, S. (1981). "Yellow Rain, a Journey through the Terror of Clinical Warfare." M. Evans and Company, New York.

Schiefer, H. B. (1988). Yellow rain. *In* "Comments on Toxicology," pp. 1–62. Gordon & Breach, New York.

Sharma, R. P., and Salunkhe, D. K. (eds.) (1991). "Mycotoxins and Phytoalexins." CRC Press, Boca Raton, Florida.

Smith, J. E., and Henderson, R. S. (eds.) (1991). "Mycotoxins and Animal Foods." CRC Press, Boca Raton, Florida.

Myelin and Demyelinating Disorders

MARJORIE B. LEES
E. K. Shriver Center and Harvard Medical School

OSCAR A. BIZZOZERO
University of New Mexico Medical Center

RICHARD H. QUARLES
National Institute of Neurological Disorders and Stroke

GLOSSARY

Central nervous system Brain and spinal cord

Glia Nonneuronal cells of the nervous system, including cell types that produce myelin; in the central nervous system, the myelin-producing cells are called oligodendrocytes; in the peripheral nervous system, they are referred to as Schwann cells

Leukodystrophies Group of hereditary metabolic disorders characterized by a widespread, severe deficiency of myelin

Multiple sclerosis Common neurological disease characterized by loss of myelin in the central nervous system and leading to motor and sensory deficits

Myelin-associated enzymes Enzymes present in oligodendrocyte and Schwann cell processes, and associated with limited regions of the myelin sheath

Neuropathy Disease of the peripheral nervous system

Node of Ranvier Interruption in the myelin sheath; space between two myelinated segments

Peripheral nervous system Nerves outside the brain and spinal cord, especially those of the limbs

Saltatory conduction Discontinuous conduction in which the nerve impulse jumps from one node to the next

MYELIN IS A MULTILAMELLAR, MEMBRANOUS sheath that surrounds the nerve axon. In the central nervous system (CNS), it is formed by oligodendrocytes, and in the peripheral nervous system (PNS), by comparable cells called Schwann cells. In contrast to any other cell membrane, 70–80% of the dry weight of myelin is lipid and only 20–30% is protein. However, the major proteins are characteristic of myelin and are not found in other membranes. Myelin is characteristic of vertebrates and appears to be a structural adaptation that facilitates rapid conduction of nerve impulses. It serves as an insulator around the axon and speeds conduction of the nerve impulse with conservation of both space and energy. However, its function is more than that of a passive insulator, and the occurrence of active metabolic processes within the membrane is now recognized. The maintenance of myelin structure and activity is required for normal nervous system function, and myelin pathology leads to serious human disorders such as multiple sclerosis and the leukodystrophies.

I. OVERVIEW

A. Evolutionary Appearance

Myelin is a specialization of the vertebrate nervous system, which provides a mechanism whereby the requirement of higher animals for rapid transmission of large numbers of impulses is achieved with an economy of both space and energy. The increase in the speed of conduction along an unmyelinated axon is proportional to the square root of its diameter, whereas that of a myelinated axon is directly proportional to its diameter. Consequently, to conduct at a defined speed, the size of an unmyelinated fiber would have to be dramatically greater than that of a myelinated fiber (Fig. 1). Thus, myelin allows the large number of nerve axons needed to carry out complex activities to be accommodated within a defined space. Furthermore, the periodic interruptions or gaps along the length of the myelin sheath, referred to as nodes of Ranvier, result in saltatory conduction, that is, the nerve current jumps from one node to the next, and the expenditure of energy is much less than in unmyelinated axons, as it is restricted to the nodes rather than being dissipated along the length of the axon (Fig. 2). The area adjacent to the node is referred to as the paranodal region; the region between two consecutive nodes is the internode.

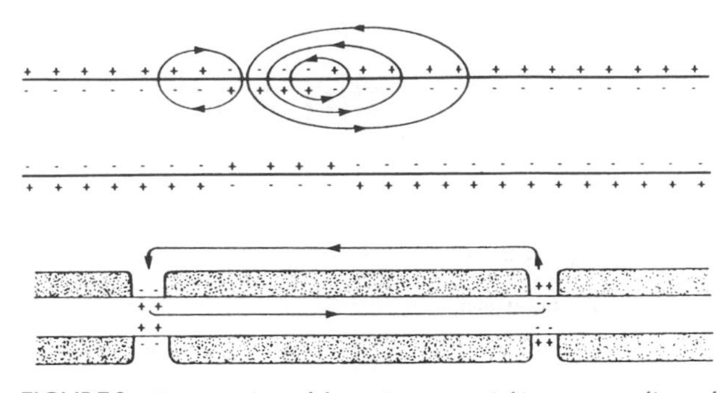

FIGURE 2 Propagation of the action potential in an unmyelinated (top) and a myelinated (bottom) nerve. The arrows indicate the flow of current in local circuits within the active region of the membrane. (+, −) represent the charge stored on the capacitance of the membrane. For unmyelinated fibers, current flows continuously along the membrane, whereas in myelinated fibers, current jumps from one node to the next. [Reprinted from P. Morell *et al.* (1994). Myelin formation, structure, and biochemistry. *In* "Basic Neurochemistry: Molecular, Cellular and Medical Aspects" (G. Siegel *et al.*, eds.), 5th Ed., Chap. 6. Raven, New York.]

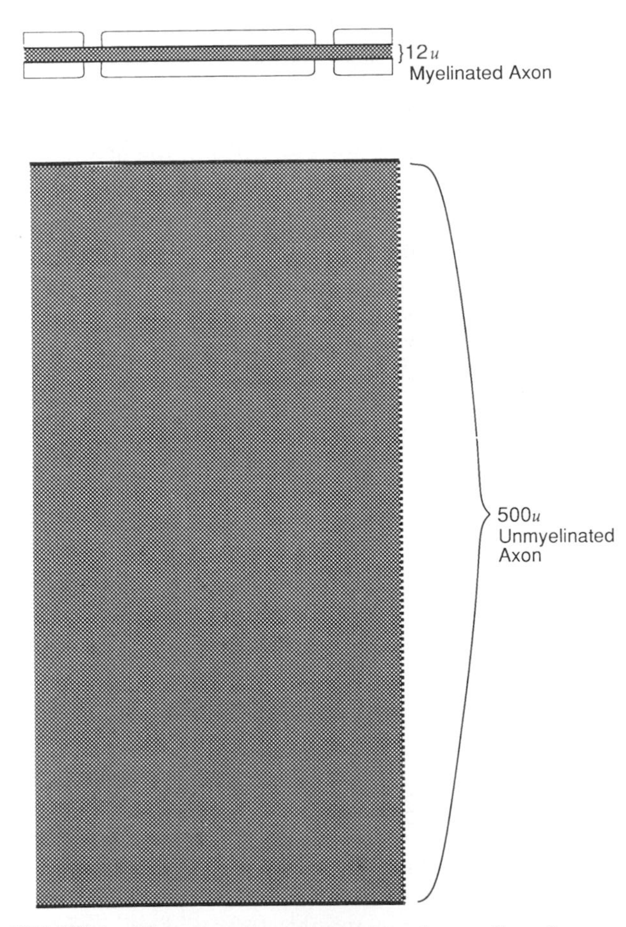

FIGURE 1 Diagrammatic representation of a myelinated nerve fiber (top) and an unmyelinated fiber (bottom) that conduct at the same speed. Many more myelinated fibers can be accommodated in a limited space.

B. Myelin Function

Myelin is commonly referred to as an insulator around the nerve axon and, indeed, the conservation of energy described here derives from the high electrical resistance due to the large amount of lipid present in the myelin membrane. Stimulation of the nerve axon at the nodal region leads to an inflow of sodium ions and an outflow of potassium ions via specific ion channels in the axolemma (axonal membrane). The insulating properties of the myelin prevent depolarization of the adjacent axolemma and a wave of current jumps from one nodal segment to the next (see Fig. 2). The nodal region thus acts as a potassium sink to maintain ion homeostasis. The compact structure of the lamellar membrane and its low water content are additional factors that give rise to the insulating properties of myelin. The stability of the lamellar structure is enhanced by the presence of specialized proteins, which include specific adhesion molecules and proteins capable of binding electrostatically and hydrophobically with lipids and other proteins.

In recent years, it has been recognized that myelin has functions over and above its insulating properties. Because myelin is a part of a living cell, it should come as no surprise that active metabolic processes are occurring. Thus, although myelin is a relatively stable membrane, it undergoes continuing synthesis and degradation, as discussed in Section V. Further, the presence of specific myelin-associated enzymes, the characteristic lipid and protein composition, the restricted localization of certain myelin components, and the evolutionary conservation of protein sequences all point to an interrelated system with important dynamic properties.

C. Methods of Study of Myelin

The gross anatomical distribution of myelin is evident from its glistening white appearance, and large unmyelinated areas can be identified visually on the basis of their grayish color. Also, light and electron microscopy, combined with appropriate stains, clearly identify the multilamellar structure of myelin and its characteristic staining properties (see Section II). The specificity of these microscopy procedures is enhanced by immunocytochemistry in which antibodies to specific myelin components can be used to follow their developmental appearance. [*See* Electron Microscopy.]

For structural and metabolic studies, myelin can be isolated on sucrose gradients as a purified membrane on the basis of its low density. Purified myelin collects as a band at a density of 1.08 g/ml (0.65 *M* sucrose). Myelin can be further separated into subfractions of different densities, presumably corresponding to different regions of the multilamellar structure. Upon isolation, the normal interrelationships between myelin and the cell body and axonal membranes are disrupted, and these may be important for myelin function *in situ*. Brain or nerve tissue can be explanted or dispersed, and oliogodendroglial and Schwann cells grown in culture to follow the initial steps of myelination *in vitro*. These cultures express myelin components and significant amounts of membrane sheets are formed, but only minimal amounts of compact myelin. Nevertheless, the timetable of cellular differentiation is comparable to that *in vivo*.

Molecular biological techniques have provided new approaches to understanding the regulation of CNS and PNS myelin proteins. These have been useful not only for basic studies but also for the diagnosis of genetic disorders involving myelin.

II. MYELIN FORMATION AND STRUCTURE

Myelination is initiated by a signal between the axon and the myelin-forming Schwann cells or oligodendrocytes, which appear to be activated when the axon reaches a critical size (>1 μm). Myelination in the CNS and PNS is similar in principle but has significant differences in detail. Each Schwann cell directly wraps around and ensheaths a single segment of one axon, whereas a single oligodendrocyte sends out multiple processes that expand into broad, flat sheets as they begin to spiral around different axons and different regions of the same axon. After several turns of the spiraled membrane, the extracellular faces of the sheath become apposed to form the intraperiod line. Concomitantly, cytoplasm is extruded and the intracellular faces of the membranes become closely apposed and compacted. However, a rim of cytoplasm remains at the edges of the compacted membranes, and these appear in longitudinal sections as a series of lateral loops (Figs. 3 and 5). Other areas with remaining cytoplasm are the inner and outer mesaxons and incisures known as Schmidt–Lanterman clefts, which are particularly prevalent in PNS myelin sheaths. Some of the metabolic activity associated with myelin undoubtedly occurs in these cytoplasm-rich regions.

The ultrastructural organization of myelin has been elucidated by X-ray diffraction studies and electron microscopy. The compacted membranes appear in the electron microscope as a series of dark and less dark lines between unstained areas (Fig. 4). The darker line is referred to as the major dense line and corresponds to the apposed cytoplasmic faces of the cell, whereas the less dense line is known as the intraperiod or intermediate dense line and represents the apposition of the extracellular faces of the cell membrane (see Fig. 3). The variations in the intensity of staining reflect the differing chemical compositions of the cytoplasmic and extracellular faces of the membrane, particularly the orientation of its proteins.

The ultrastructural organizations has also been studied by X-ray diffraction. In CNS myelin, the unit membrane, with a repeat distance of 80 Å, consists of a single lipid bilayer of approximately 50 Å and protein extending approximately 15 Å on either side of the bilayer. Therefore, two fused bilayers have a repeat distance of 160 Å (see Fig. 3). These values are slightly larger than the electron microscopic dimensions because electron micrograph specimens are prepared differently from X-ray diffraction samples. In the PNS, the myelin periodicity is 10% higher owing to its different protein composition.

Additional differences between the CNS and PNS are seen in the structure of the node of Ranvier (see Fig. 5). In the CNS, the node is bare and the axon is exposed directly to the extracellular compartment. By contrast, the node in the PNS is covered with

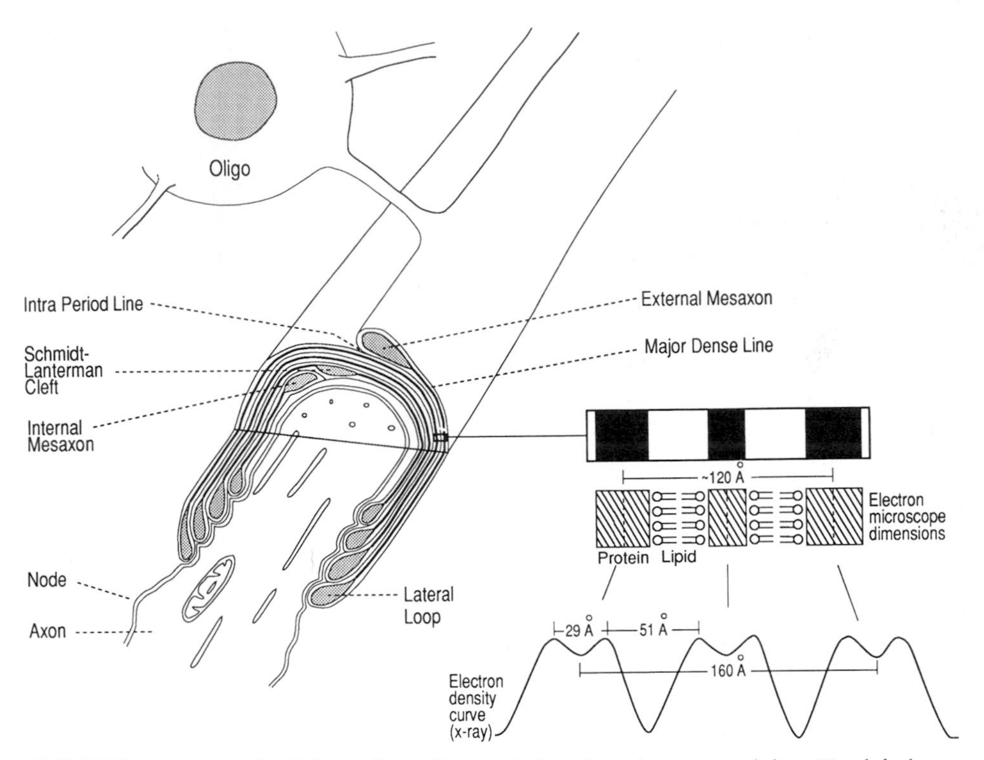

FIGURE 3 Diagram of a CNS myelinated internode based on ultrastructural data. Top left shows an oligodendrocyte with several membranous processes, one of which is connected to the myelin sheath. The cut view of the myelin and axon illustrates the 3-dimensional interrelationships between the two structures. The right-hand side of the figure shows the approximate dimensions of one myelin-repeating unit of protein–lipid–protein as seen by electron microscopy of fixed sections (upper) and by X-ray diffraction of fresh nerve (lower). [Modified and redrawn from P. Morell *et al.* (1994). Myelin formation, structure, and biochemistry. *In* "Basic Neurochemistry: Molecular, Cellular and Medical Aspects" (G. Siegel *et al.*, eds.), 5th Ed., Chap. 6. Raven, New York.]

interdigitated fingers of cytoplasm, and the entire region, including the node, paranode, and internode, is covered with a collagenous basal lamina.

III. MYELIN DEVELOPMENT

Myelination does not occur simultaneously in all regions but begins in PNS, and then proceeds along the neuraxis. In the human, little or no myelination occurs during the first half of gestation. The dorsal and ventral roots begin to myelinate during the third trimester and are, for the most part, completely myelinated at the time of birth (Fig. 6). Some tracts myelinate rapidly and over a short time, whereas others myelinate more slowly and, consequently, myelinogenesis in the latter extends over a prolonged time. Thus, at any single time point, different stages of myelination are occurring simultaneously. During the first postnatal years, active myelination occurs in most tracts. However, myelination continues in many regions during the first decade of life, and some regions may continue to myelinate up to 30 years of age. The last areas to be myelinated are the corpus callosum and the intracerebral commisures, regions involved in higher functions such as pattern recognition and reading skills.

IV. CHEMICAL COMPOSITION

In both the CNS and PNS, the myelin sheath has a low water content and consequently contains relatively more solids than do other tissue membranes. The solids are mainly lipids, with lesser amounts of protein and only small amounts of carbohydrate.

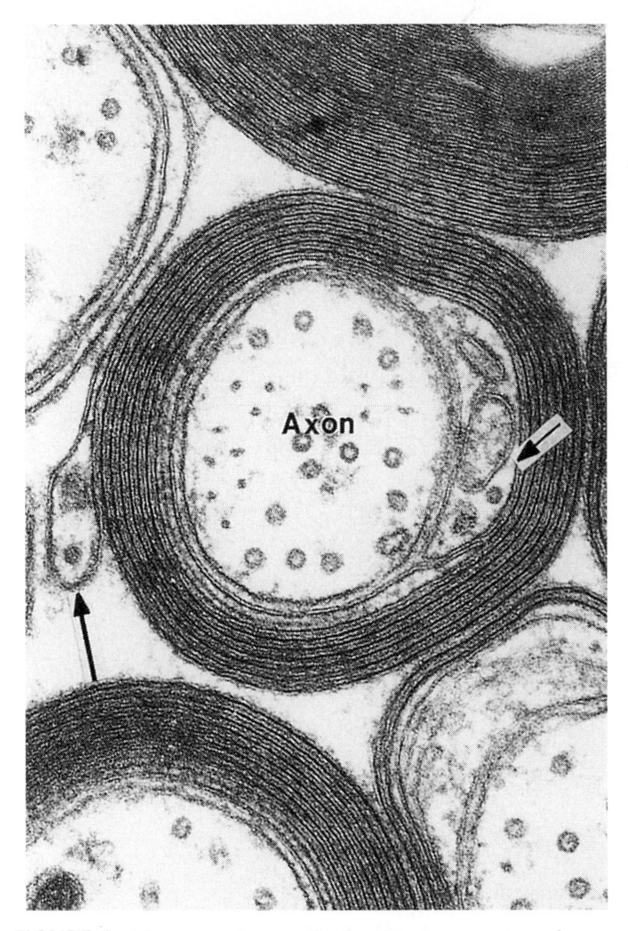

FIGURE 4 Electron micrograph of a transverse section of a myelinated nerve axon in the CNS. The outer (large arrow) and inner (small arrow) mesaxon, containing cellular organelles, the spiral nature of the sheath, and the origins of the intraperiod and major dense lines, are clearly visbile (magnification 150,000×). (Courtesy of Dr. Cedric Raine.)

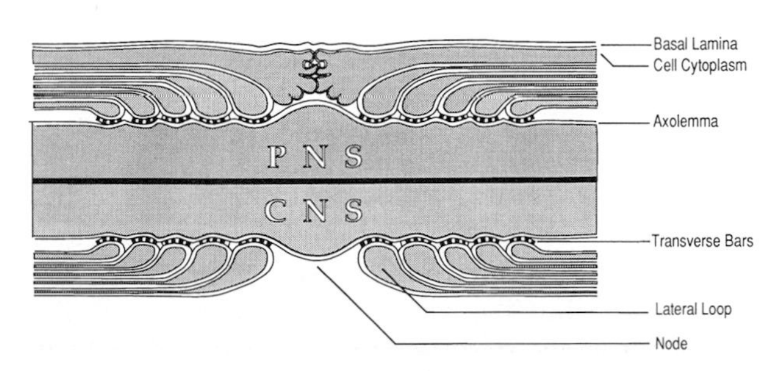

FIGURE 5 Diagrammatic representation comparing the structure of a PNS node with that of a CNS node. Note the interdigitations of Schwann cell processes and basal lamina covering the PNS node. These are absent from the CNS node. Transverse bars connect the lateral loops and the axolemma. [Redrawn from C. S. Raine (1984). Morphology of myelin and myelination. *In* "Myelin" (P. Morrell, ed.), 2nd Ed., Chap. 1. Plenum, New York.]

A. Myelin Lipids

Approximately 70% of the dry weight of myelin is accounted for by lipids, an amount higher than that of other biological membranes. Myelin lipids are not unique to this membrane but, nevertheless, certain characteristic features can be described. Cholesterol, phospholipids, and galactolipids are the major myelin lipid classes (Table I) and, on a molar basis, they occur in an approximate ratio of 2 : 1.8 : 1. Thus, cholesterol is the most abundant lipid. Essentially all the cholesterol is free (nonesterified), and the presence of significant amounts of esterified cholesterol in adults is indicative of myelin pathology and demyelinating disease. This contrasts with other membranes, in which a specific ratio of free to esterified cholesterol is required to maintain membrane structure. The most typical lipid is galactocerebroside, and the increase in its amount parallels that of myelin during development. A part (15–20%) of the galactocerebroside is sulfated, forming an acidic lipid referred to as cerebroside sulfate, or sulfatide. The fatty acid composition of galactocerebrosides includes a relatively high proportion of long-chain fatty acids (C20–C24) and of hydroxy fatty acids. The major phospholipids are phosphatidylcholine, phosphatidylserine, sphingomyelin, and ethanolamine phospholipid, with the latter the most abundant. A unique feature of myelin is that 80% of the ethanolamine phospholipid is present in the plasmalogen form (α, β unsaturated ether). In contrast to other myelin lipids, plasmalogens are synthesized in peroxisomes. The high proportion of plasmalogens results in an increase in the overall unsaturation of the membrane and may be important in maintaining its stability. Other lipids (e.g., polyphosphoinositides) are present in only minor quantities, but they are, nevertheless, enriched in myelin and may be further localized to certain regions of the sheath. In other tissues, polyphosphoinositides are involved in signal transduction processes and their presence in myelin may therefore have functional implications. [*See* Cholesterol; Lipids.]

B. Myelin Proteins

1. Central Nervous System

In contrast to the lipids, which are ubiquitous components, most of the proteins are unique to myelin. Over 80% of the CNS myelin proteins are accounted for by two protein families, namely, proteolipids and myelin basic proteins. These, along with an enzyme

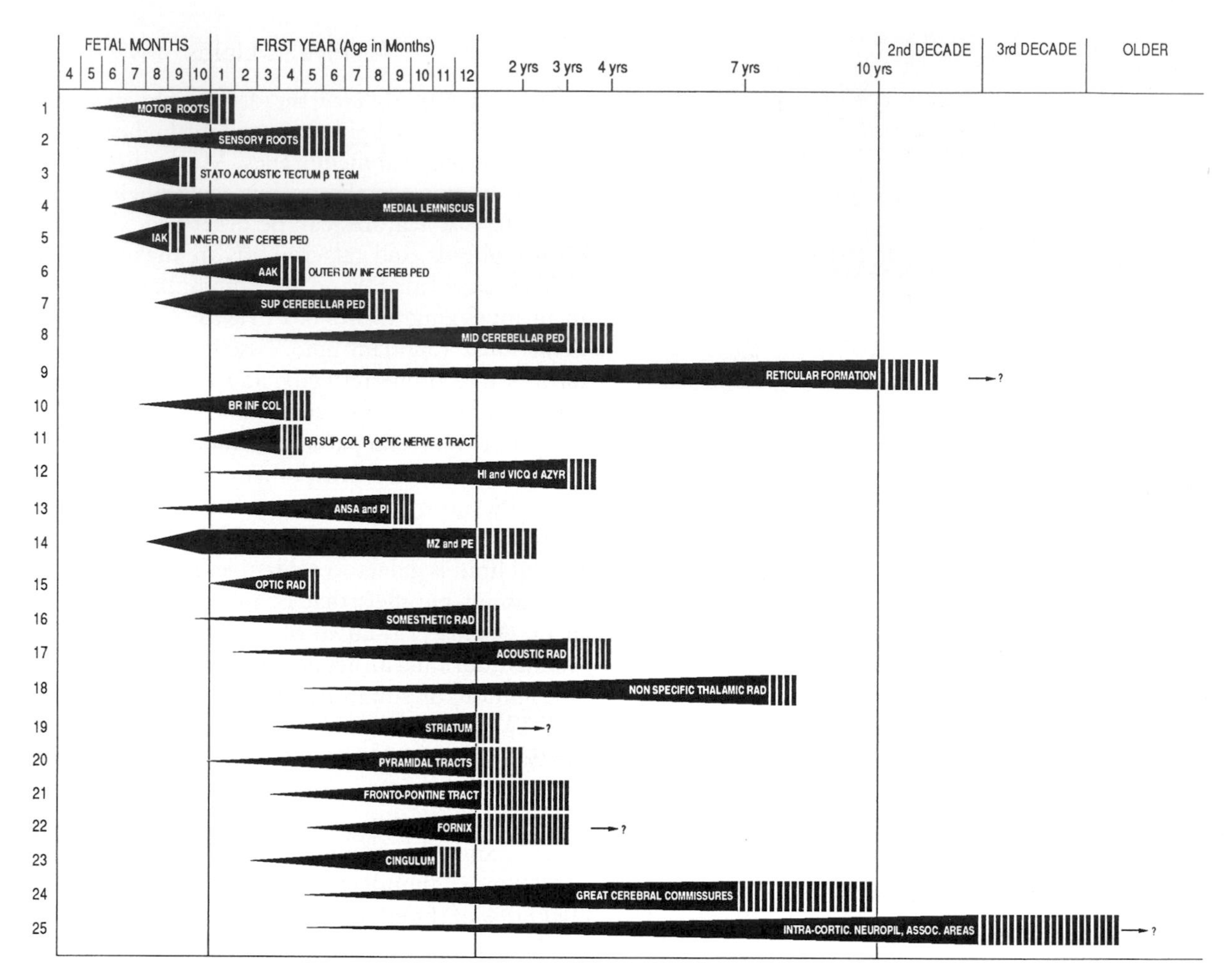

FIGURE 6 Time period of myelinogenesis in different areas of the human brain. The width of the segments denotes the intensity of staining and density of myelinated fibers. Vertical stripes at the end of the bars approximate the age range of the termination of myelination. [Reprinted from P. Yakovlev and A. R. LeCours (1967). The myelinogenetic cycles of regional maturation of the brain. *In* "Regional Development of the Brain in Early Life" (A. Minkowski, ed.). Blackwell Scientific Publications, Oxford, England.]

called 2′3′-cyclic nucleotide-3′-phosphohydrolase (CNPase), which accounts for an additional 2–3% of the myelin proteins, each have basic isoelectric points and can potentially interact electrostatically with acidic lipids. The myelin proteins can be separated on the basis of their molecular weights by polyacrylamide gel electrophoresis in a denaturing detergent (Fig. 7) and further identified by staining with specified antibodies. [*See* Proteins.]

a. Proteolipid Proteins

Myelin proteolipid proteins (PLPs) are a family of hydrophobic integral membrane proteins that account for more than half of the proteins of CNS myelin. They are restricted to the CNS and first appear during evolution in amphibia. However, recent studies indicate small amounts of PLP in PNS myelin and in nonneural tissues. Further, a family of PLP-like molecules with sequence homologies to specific channel proteins has been described in primitive fish, suggesting that PLP could be involved in transport processes in myelin. The amino acid sequence of PLP is highly conserved with identical sequences found for humans and rodents. The major form of PLP has a molecular mass of 30 kilodaltons (kDa). A second, minor protein, designated DM-20, is formed by alternate splicing of the PLP mRNA. The PLP gene, which is located on the X chromosome, has been isolated and shown to contain seven exons and six introns. The normal protein shows a marked domain struc-

TABLE I
Composition of Human CNS Myelin and Brain[a]

Component[b]	Myelin	White matter	Gray matter
Protein	30.0	39.0	55.3
Lipid	70.0	54.9	32.7
Cholesterol	27.7	27.5	22.0
Total galactolipid	27.5	26.4	7.3
Cerebroside	22.7	19.8	5.4
Sulfatide	3.8	5.4	1.7
Total phospholipid	43.1	45.9	69.5
Ethanolamine phosphatides	15.6	14.9	22.7
Lecithin	11.2	12.8	26.7
Sphinomyelin	7.9	7.7	6.9
Phosphatidylerine	4.8	7.9	8.7
Phosphatidylinositol	0.6	0.9	2.7
Plasmalogens	12.3	11.2	8.8

[a]Values from G. J. Siegel *et al.* (eds.) (1994). "Basic Neurochemistry: Molecular, Cellular and Medical Aspects," 5th Ed., Chap. 6, p. 127. Raven, New York.
[b]Protein and lipid figures are expressed as percent of dry weight; all others are percent of total lipid weight.

ture, that is, regions of hydrophobic amino acids alternate with positively charged, hydrophilic regions, and the resulting integral membrane protein traverses the membrane several times (Fig. 8). The precise localization of each of the charged regions in the myelin structure is shown in Fig. 8.

The high proportion of hydrophobic amino acids and the strong domain structure combine to give PLP unusual solubility properties for a protein. PLP is extracted from brain by organic solvents (e.g., chloroform–methanol mixtures), along with the tissue lipids, as a lipid–protein complex. However, after removal of all of the complex lipids, the apoprotein (delipidated protein) retains its solubility in organic solvents). The apoprotein contains 2 moles of covalently bound fatty acid (mainly palmitic, stearic, and oleic acids), and these are attached to cysteine residues. Proteins containing covalently bound fatty acids are referred to as acylated proteins, and acylation is the only known posttranslational modification of PLP. A fraction of PLP molecules undergo a rapid acylation–deacylation cycle. The fatty acid is added, after PLP is inserted into the myelin membrane, by an autoacylation process, that is, a separate enzyme does not appear to be involved. However, deacylation of PLP is catalyzed by a specific, myelin-associated protein : fatty acyl esterase.

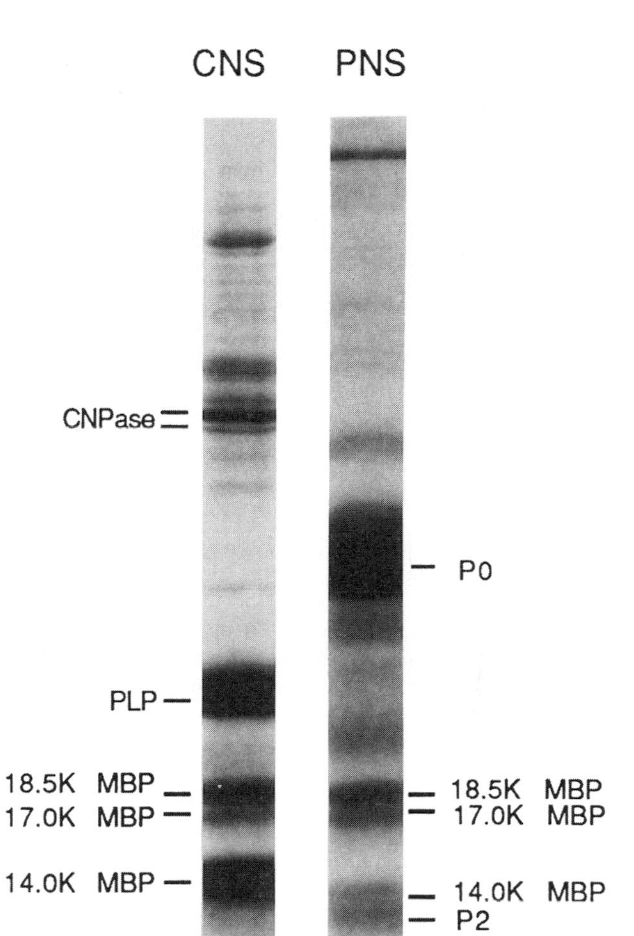

FIGURE 7 Pattern of rat CNS and PNS myelin proteins after electrophoresis in polyacrylamide gels in the presence of the denaturing detergent sodium dodecyl sulfate. Proteins are visualized by staining with Coomassie brilliant blue. In this photograph, the minor PLP, DM-20, is barely evident below the PLP band. The molecular species of the MBPs in human myelin differ from those in the rat.

b. Myelin Basic Protein

Myelin basic proteins (MBPs) are a family of highly positively charged proteins that have been studied extensively in relation to their role in experimental allergic encephalomyelitis (EAE). At least four isoforms of MBP (21.5, 20, 18.5, and 17 kDa) have been identified in human myelin, and these multiple forms arise from alternate splicing of a single gene, located on the distal end of chromosome 18. Mouse mutants with defective MBP have been identified, but no human counterparts have been found thus far. The major isoform in the human adult is the 18.5-kDa species, consisting of 168 amino acids, 24% of which are basic amino acids (lysine, arginine, histidine). This is

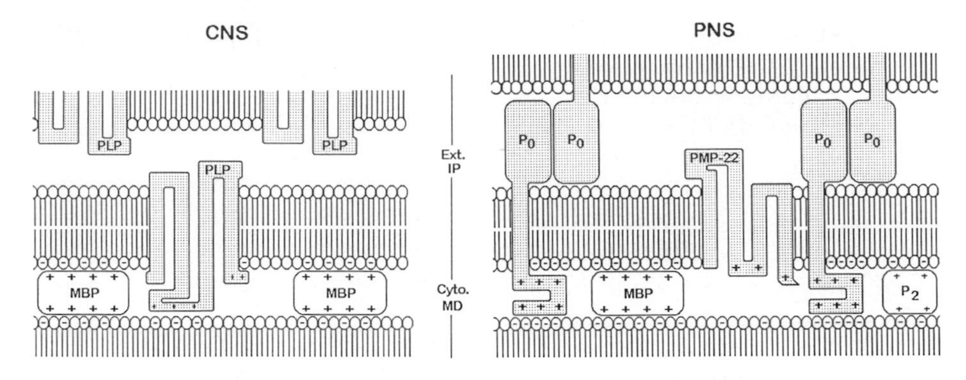

FIGURE 8 Diagrammatic representation of the current model of the molecular organization of compact CNS and PNS myelin depicting the structural roles of the major proteins. The apposition of the extracellular (Ext.) and cytoplasmic (Cyto.) surfaces of the oligodendrocyte or Schwann cell membranes form the intraperiod (IP) and the major dense (MD) lines, respectively. ᑍ represents lipid molecules. The lipid bilayer in the diagram is intended only as a framework for the proteins, and no attempt was made to show the multiplicity of lipid and unsaturated fatty acids or the asymmetric distribution of some lipids. Posttranslational modifications of the major proteins are not shown. Positively charged (+) myelin basic protein (MBP), P2 protein, and intracellular domains of P0 and PLP are located at the cytoplasmic surfaces in the MD line where they interact with the negative (−) charges on the head groups of acidic lipids. In the PNS, the IP line is formed by homophilic interaction of the extracellular domain of P0 molecules from the opposite membrane. (The authors are grateful to Paul Sanchez and Sabine Tetzloff for drawing these models.)

in marked contrast to PLP, which contains only 9% basic amino acids. The abundance of lysine and arginine residues provides many sites for cleavage by the proteolytic enzyme trypsin. The MBPs are hydrophilic, extrinsic membrane proteins and have been localized to the major dense line observed in electron micrographs of myelin (see Figs. 3 and 4). The MBPs undergo several posttranslational modifications, including N-acetylation, phosphorylation, and methylation. These modifications are assumed to contribute to myelin compaction and also to myelin function.

c. Myelin-Associated Glycoprotein

Myelin-associated glycoprotein (MAG) is a 100-kDa protein that appears to play a role in the formation and maintenance of the junction between the myelin-forming glial cell and the axonal surface membrane. It differs from MBP and PLP in that it is absent from multilamellar, compact myelin and is restricted to the inner turn of the oligodendrocyte membrane adjacent to the axon. About one-third of the molecular mass of MAG is accounted for by carbohydrate moieties that are attached by N-glycosyl bonds to asparagine residues in the polypeptide. MAG and other adhesion glycoproteins in the nervous system that function in cell–cell interactions have amino acid sequence similarities with immunoglobulins and other proteins of the immune system. These neural adhesion proteins and related proteins of the immune system make up the so-called immunoglobulin gene superfamily and are believed to have arisen during evolution from a common ancestral protein involved in cell–cell interactions.

d. Myelin-Associated Enzymes

Recognition of the existence of a significant number of myelin-associated enzymes has lent support to the concept of myelin as a dynamic membrane. Isolated myelin contains enzymes involved in ion-transport processes, lipid metabolism, and protein modification. Some of these enzymes are ubiquitous (e.g., carbonic anhydrase), whereas others are highly enriched in myelin (e.g., cholesterol esterifying enzymes) or act on components specific to myelin (e.g., certain protein kinases). Each of these enzymes has a characteristic distribution pattern along the neuraxis, probably reflecting different functional needs.

Of particular interest is the enzyme CNPase, which, although present in low concentrations in other tissues, is characteristic of both oligodendrocytes and Schwann cells and their membranous extrusions. CNPase appears early during development and is often used as a myelin marker, although the relationship is not strictly quantitative. The enzyme has been localized to the cytoplasmic face of the myelin membrane and is restricted mainly to the region of the nodal loops and other cytoplasm-rich regions. CNPase exists as two similar monomers with molecular masses of 46 and 48 kDa, and each of the monomers shows enzyme activity. The primary sequence is relatively conserved among different species. The human enzyme has been cloned and shown to consist of 400 amino acids, with no membrane-spanning domain. The enzyme has a net positive charge and is posttranslationally modified by phosphorylation of serine and threonine residues. The enzyme contains not only phosphoryl binding sites but also nucleotide binding sites, and these may be important for its function.

2. Peripheral Nervous System

PNS myelin does not contain the hydrophobic PLP found in the CNS but contains a 30-kDa integral membrane protein called the P-zero (P0) glycoprotein (see Fig. 7). This protein accounts for over half of the total PNS myelin protein and contains about 5% carbohydrate by weight. Other posttranslational modifications of P0 include acylation, phosphorylation, and sulfation. Like MAG, P0 is a member of the immunoglobulin superfamily, and its extracellular immunoglobulin-like domain is believed to play a role in stabilizing the intraperiod line of compact myelin (see Fig. 8). The P0 protein spans the membrane once, and its positively charged C-terminal, cytoplasmic domain is thought to contribute to stabilization of the major dense line of compact myelin. Another PNS myelin-specific protein is a positively charged fatty acid binding protein called P2, which also contributes to myelin stability. PNS myelin also contains the same myelin basic proteins as found in the CNS, but in lesser amounts. Recently, molecular cloning experiments identified a 22-kDa peripheral myelin protein (PMP-22) that appears to be very important for the formation of normal myelin during development. It is a hydrophobic protein with four transmembrane domains and one oligosaccharide moiety. Although PMP-22 accounts for only about 5% of the total myelin protein, genetic mutations affecting this protein lead to a myelin deficiency of the PNS in experimental animals and humans (see Section VI,C,1). As in the CNS, MAG is in the inner Schwann cell membrane of the PNS myelin, where it forms a junction between glia and axons. In addition, MAG appears to mediate interactions between adjacent "semi-compacted" Schwann cell membranes in Schmidt–Lanterman incisures, paranodal loops, and the inner and outer mesaxons. CNPase is associated with uncompacted regions of PNS myelin sheaths but at a much lower concentration than in CNS sheaths.

C. Molecular Organization of Compact CNS and PNS Myelin

A hypothetical model of the molecular organization of myelin proteins is shown in Fig. 8. Integral membrane proteins are defined as having one or more hydrophobic domains which are embedded in the lipid bilayer, whereas extrinsic proteins are attached to only one side of the bilayer by electrostatic interactions. PLP is an integral membrane protein that passes through the membrane more than once and may stabilize both the intraperiod and major dense lines of central myelin. P0 glycoprotein is the major integral membrane protein in PNS myelin and also appears to stabilize the membrane contacts at both surfaces. The large extracellular domains of P0 are believed to hold membrane surfaces together by homophilic interaction as shown in Fig. 8 and probably account for the well-known greater separation at the intraperiod line in PNS myelin compared to CNS myelin. Both types of myelin contain the extrinsic, positively charged MBPs, which probably help to stabilize the major dense lines of myelin by electrostatic interactions with lipids. P2 protein may function similarly to the MBPs in PNS myelin. MAG and CNPase are not shown in these models of compact myelin, because their localization is restricted to noncompacted membranes of myelin sheaths, as described earlier.

V. MYELIN METABOLISM

Contrary to the earlier concept of myelin as an inert insulator, a variety of active metabolic processes have been shown to occur in this membranous structure. An understanding of how myelin is formed, maintained, and degraded is still far from complete, but most of the current information on assembly and degradation comes from animal experiments using biochemical methods with radioactive precursors, as well as immunocytochemistry.

A. Synthesis

During myelinogenesis, both oligodendrocytes and Schwann cells synthesize large amounts of proteins and lipids and assemble them into extensive sheets of membrane. The synthesis of most myelin proteins and lipids increases rapidly during development, and then declines after the peak period of myelin formation. Increased synthesis, rather than a decrease in catabolism, is mainly responsible for myelin deposition. Intrinsic membrane proteins, such as PLP, P0, and MAG, are synthesized on membrane-bound ribosomes located in the perinuclear region of the glial cells. These proteins are subsequently transported to the Golgi and targeted to the plasma membrane. In contrast, extrinsic proteins (MBP, CNPase) are synthesized on free polysomes and are immediately incorporated into the plasma membrane. From the different time courses of appearance of newly synthesized proteins and lipids, the membrane evidently is not formed as a unit. This concept also apparently applies to the degradation of myelin as discussed in the next section. Furthermore, because of the differences in relative synthetic rates of individual components, the composition of myelin changes during development.

The activity of the enzymes responsible for the synthesis of fatty acids and complex lipids increases severalfold during the period of active myelinogenesis and declines thereafter. On the other hand, those enzymes involved in myelin maintenance and turnover also show increased activity during myelin synthesis but remain elevated in the adult. Enzymes responsible for the formation of cholesterol, phospholipids, and galactolipids are located in microsomes, where the bulk of lipid synthesis occurs. However, certain enzymes involved in the recycling of lipids, such as fatty acyl-CoA synthetase and lysophosphatidylcholine–acyltransferase, are present not only in microsomes but also in myelin, where they may participate in remodeling and maintaining the membrane. Intact phospholipids and their precursors may move from the axon across the axonal membrane to be incorporated into the myelin. The extent to which the latter two mechanisms contribute to myelin deposition and maintenance is unknown.

B. Degradation

Myelin is not degraded as a whole, and even the individual structural components turn over independently. Nevertheless, the turnover of myelin components is relatively slow compared with that of other cell membranes, and half-lives on the order of months have been reported for the major proteins. High-molecular-weight proteins, some probably representing myelin-associated enzymes, turn over more rapidly. How myelin proteins are physiologically degraded is still unclear, but the presence of neutral proteases in myelin has been documented. Degradation can also be carried out by acid proteases located in lysosomes.

Most of the myelin proteins have functional groups that are added during or after synthesis. Some of these groups are metabolically stable with half-lives comparable to those of the polypeptide backbone (e.g., methyl groups in MBP and sulfate and carbohydrate residues in P0 and MAG) and may have a role in maintaining protein configuration. Other side-chain groups turn over more rapidly and at a rate independent of the protein backbone (e.g., phosphate groups in MBP and fatty acyl groups in PLP) and could, therefore, be actively involved in membrane maintenance and function. For example, the turnover of phosphate groups on MBP is accelerated upon repeated nerve stimulation.

As with the proteins, specific lipids turn over at different rates. Cholesterol, sphingolipids, and cerebrosides are the most stable of the myelin lipids with half-lives on the order of months. Glycerophospholipids, particularly phosphatidylcholine and phosphatidylinositol, seem to turn over more rapidly. Different moieties of the same lipid have also been found to turn over independently. Of particular interest is the rapid turnover of phosphate groups in polyphosphoinositides, because these lipids are involved as second messengers in signal transduction mechanisms.

VI. DISEASES INVOLVING MYELIN

The formation and maintenance of myelin are dependent on the normal functioning of oligodendrocytes or Schwann cells, as well as the viability of the ensheathed axons and their neuronal cell bodies. A deficiency of myelin can result either from an inability to produce the normal complement of myelin during development (hypomyelination) or from its breakdown once formed (demyelination). Demyelinating diseases can be divided into two categories: primary demyelination involving early loss of myelin with relative sparing of axons, and secondary demyelination in which myelin loss occurs as a consequence of damage to neurons or axons. Thus, a myelin deficit can result from a multitude of physiological insults that affect either myelin-forming glial cells, neurons, or the myelin sheaths themselves. Hypomyelination or demyelination can be caused by acquired allergic or infectious factors, genetic disorders, toxic agents, or malnutrition. Many of the biochemical changes in neural tissue are similar regardless of the etiology of the demyelinating disease and include a marked increase in water content and a decrease in myelin lipids and proteins.

A. Multiple Sclerosis

Multiple sclerosis (MS) is the most common demyelinating disease of the CNS in humans, causing severe motor and sensory neurological deficits in about 1 out of 1000 persons in the United States. It is generally a chronic disease, with typical onset occurring in the second or third decade of life, and is characterized by exacerbations and remissions, which occur over many years. However, it is sometimes slowly progressive from the beginning and, in a small number of patients, occurs in an acute form that progresses to death within a year. Clinically, slowing of nerve conduction can be demonstrated by measurement of auditory or visually evoked potentials. The pathology is characterized by an inflammatory reaction around small veins with lymphocyte infiltration and demyelinated areas called plaques, which can be observed grossly at autopsy (Fig. 9). The appearance and disappearance of lesions can be visualized in living patients by magnetic resonance imaging. The plaques are characterized by an absence or severe reduction of myelin and oligodendrocytes, with preservation of relatively normal-appearing axons. Electron microscopy indicates that a major mechanism of myelin destruction is the direct removal of myelin lamellae from the surface of intact sheaths by macrophages. This, along with the presence of apparently healthy oligodendrocytes in areas of active demyelination, suggests that the myelin sheath itself is the primary target of the disease rather than the myelin-forming oligodendrocyte. In early lesions, substantial remyelination is observed, but the newly formed myelin is destroyed and oligodendrocytes are eventually lost from the plaques. Older, chronic MS lesions are sharply defined and contain bare, nonmyelinated axons and many fibrous astrocytes. [*See* Multiple Sclerosis.]

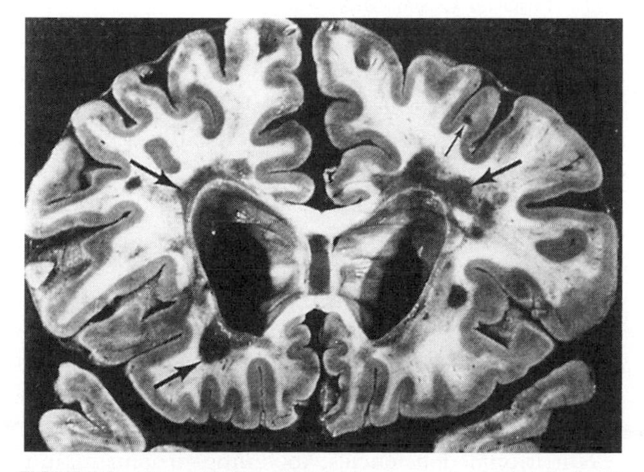

FIGURE 9 Coronal slice of brain from a patient who died with MS. Demyelinated plaques are clearly visible in white matter (large arrows). Small plaques are also observed at the boundaries between gray and white matter (small arrows). [Reproduced from C. S. Raine (1984). The neuropathology of myelin diseases. *In* "Myelin," 2nd Ed., Chap. 8. Plenum, New York.]

Although the cause of MS has not been identified, genetic, immunological, and viral factors are believed to contribute to the pathophysiology. A genetic component is indicated by the increased risk of disease for first-degree relatives of affected individuals and an especially high concordance for the disease in monozygotic twins. Genes coding for histocompatibility molecules are linked to disease susceptibility with a strong association of certain human leukocyte antigen (HLA) types with the disease. However, these determinants in themselves are neither necessary nor sufficient to confer susceptibility.

An environmental factor, possibly an infectious agent, has been suggested by epidemiological studies. Thus, MS has a high prevalence above latitude 40° north in the United States and Europe and a low prevalence near the equator. In the Faroe Islands, no MS cases existed prior to World War II, but after that time a high incidence was reported, suggesting that an infectious agent related to the arrival of British troops might be important. Further evidence for viral involvement comes from experimental animal models of demyelinating diseases that are caused by viruses. Intensive efforts to identify a causative virus in MS patients have led to reports of at least 12 different infectious agents to date, but none has stood the test of time. A human retrovirus has been the most recent addition to the list.

A number of observations suggest an autoimmune basis for the disease with immunological mechanisms of demyelination. Abnormalities of immunoglobulin synthesis, association with HLA types that are in turn linked to immune response genes, and abnormalities in cellular immune function support this concept. In pathological specimens, perivenular infiltration of inflammatory cells is also observed. Further, administration of pharmacological agents that modify immune function, such as corticosteroids and immunosuppressants, often results in favorable clinical responses. Another factor is the similarity of MS to certain forms of an extensively studied autoimmune disease in animals, called experimental allergic encephalomyelitis. In EAE, multiple myelin components can elicit disease, but in MS numerous attempts to identify either a brain or viral antigen that is the target of the pathogenic immune response have been unsuccessful. [*See* Autoimmune Disease.]

It should be emphasized that the genetic, viral, and immunological theories for the pathogenesis of MS are not mutually exclusive. Individuals of certain genetic backgrounds may show a characteristic immune response to a specific virus, and this in turn may be responsible for the demyelination. Effective prevention or treatment will probably not be forthcoming until more is known about the cause of the disease. On the other hand, a recent multicenter clinical trial demonstrated that treatment with β-interferon decreased the frequency and severity of exacerbations in relapsing-remitting MS. Furthermore, the occurrence of some active remyelination in acute MS lesions suggests that it may eventually be possible to stimulate patients' oligodendrocytes to effectively remyelinate demyelinated axons.

B. Acquired Demyelinating Neuropathies

Guillain–Barré syndrome is an acute, monophasic, inflammatory, demyelinating neuropathy often preceded by a viral infection. As in MS, cumulative evidence suggests that the disease is mediated by immunological mechanisms. Humoral immunity is suggested by observations that sera from Guillain–Barré syndrome patients cause demyelination in appropriate test systems and that plasmapheresis (serum exchange) is often an effective therapy. However, the relationship of the viral infection to initiation of the disease and the identity of the principal antigen have not been established with certainty.

Neuropathies also occur in association with benign IgM gammopathies, and these diseases are believed to be caused by the reactivity of monoclonal IgM antibodies against neural antigens, including MAG and various complex glycolipids.

C. Other Disorders of Myelin

1. Genetically Determined Disorders

The leukodystrophies are a group of genetically determined disorders of CNS white matter characterized by a diffuse lack of myelin. Most of the leukodystrophies are familial lipid storage disorders involving failure of lipid degradation in cellular organelles called lysosomes. The genetic defect is in specific lysosomal enzymes and, therefore, specific lipids fail to be degraded. Rather, they accumulate and result in a marked but characteristic perturbation of normal cell metabolism. In the lysosomal disease metachromatic leukodystrophy, the enzyme aryl sulfatase A is missing and, consequently, sufatides are not degraded. Because this lipid is present in both CNS and PNS myelin, its accumulation leads to generalized pathology. The most common leukodystrophy is X-linked adrenoleukodystrophy, in which the degradation of very long chain fatty acids is impaired. The disease is referred to as a peroxisomal disorder, as this metabolic step occurs in cell organelles called peroxisomes.

Other inborn errors of metabolism may also influence developmental processes, including myelination. Among them, genetic defects of amino acid metabolism, such as phenylketonuria, are accompanied by a deficiency of CNS myelin.

In recent years, it has been demonstrated that mutations in some of the major structural proteins of human myelin can also cause a myelin deficiency in humans. X-linked Pelizaeus–Merzbacher disease is a hereditary deficiency of CNS myelin that was first identified at the beginning of this century and is now known to be caused by a variety of point mutations in the PLP gene. Alterations of the PMP-22 gene of PNS myelin, such as duplication, deletion, and point mutations, have been implicated in a number of hereditary neuropathies with myelin abnormalities, including some forms of Charcot–Marie–Tooth disease.

2. Toxic, Nutritional, and Hormonal Disorders

A variety of biological and chemical toxins can interfere with myelination or result in demyelination. Diphtheria toxin causes vacuolation and fragmentation of myelin sheaths in the PNS. Lead is a common environmental toxin that produces a multitude of effects on the nervous system, including retardation of CNS myelination as well as demyelination of the PNS. Experimentally, organotins and hexachlorophene cause reversible, edematous changes in myelin characterized by splitting of the intraperiod line. This type of damage is of clinical significance because of the earlier use of hexachlorophene as an antiseptic agent for human infants.

Nutritional deficiencies, including protein malnutrition and vitamin deficiencies, have widespread effects on the nervous system, especially in the early developmental period. During the vulnerable period of myelin formation, undernourishment can lead to impairment of glial cells with consequent depression of the synthesis of myelin-specific components. Hypomyelination can also be observed in essential fatty acid and certain vitamin deficiencies (thiamine, B_6, or B_{12}).

During development, certain hormonal deficiencies can also lead to hypomyelination. Both hypo- and hyperthyroidism can result in a generalized decrease in myelin synthesis. In diabetics, segmental demyelination of peripheral nerves is observed frequently. However, there is no agreement as to whether the latter effects are directly on myelin or secondary to axonal damage. The archetypical example of secondary demyelination is Wallerian degeneration, which occurs following crush or sectioning of a nerve. The proximal nerve segment survives and regenerates, but both the axon and myelin in the distal segment are lost. Myelin and cellular debris are phagocytosed, and eventually the regenerating axons are remyelinated.

VII. CONCLUDING COMMENTS

Over the last 30 years, the view of myelin as only an inert insulator around the nerve axom has changed drastically. Myelin is now considered a metabolically active plasma membrane, which is part of an interrelated functional system. Yet, thus far, we have only a glimmer of understanding of the role of myelin within the overall activity of the nervous system, and much remains to be clarified. The enormous advances in molecular biology and gene cloning in this decade have just begun to unlock the secrets of the control of myelinogenesis. The new technologies of molecular biology are being effectively utilized to increase our understanding of mechanisms whereby the immune system and viruses may cause myelin loss. They have further been used to develop methods for prenatal diagnosis of leukodystrophies. These procedures, along with the possibility of gene transfer, provide potential approaches to prevent and/or reverse devastating demyelinating diseases.

BIBLIOGRAPHY

Dyck, P. J., Thomas, P. K., Griffin, J. W., and Low, P. A. (eds.) (1993). "Peripheral Neuropathy." Saunders, Philadelphia.

Koetsier, J. C. (ed.) (1985). "Handbook of Clinical Neurology," Vol. 3, Elsevier Science Publishers, Amsterdam.

Lees, M. B., and Sapirstein, V. S. (1983). Myelin-associated enzymes. *In* "Handbook of Neurochemistry" (A. Lajtha, ed.), Vol. 4, 2nd Ed. Plenum, New York.

Martenson, R. E. (ed.) (1992). "Myelin: Biology, and Chemistry." CRC Press, Boca Raton, Florida.

Morell, P. (ed.) (1984). "Myelin," 2nd Ed. Plenum, New York.

Morell, P., Quarles, R. H., and Norton, W. T. (1994). Myelin formation, structure and biochemistry. *In* "Basic Neurochemistry: Molecular, Cellular and Medical Aspects" (G. Siegel, B. Agranoff, W. Albers, and P. Molinoff, eds.), 5th Ed. Raven, New York.

Quarles, R. H., Morell, P., and McFarlin, D. E. (1994). Diseases involving myelin. *In* "Basic Neurochemistry: Molecular, Cellular and Medical Aspects" (G. Siegel, B. Agranoff, W. Albers, and P. Molinoff, eds.), 5th Ed. Raven, New York.

Scriver, C. R., Beaudet, A. L., Sly, W. S., and Valle D. (1989). "The Metabolic Basis of Inherited Disease," Vol. II, Parts 10 Peroxisomes) and 11 (Lysosomal Enzymes), 6th Ed. McGraw–Hill, New York.

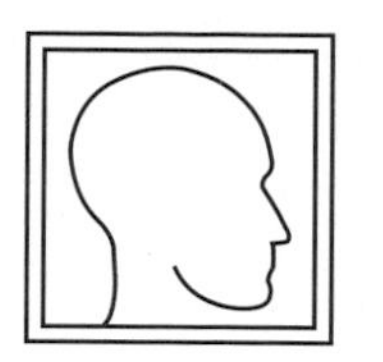

Nasal Tract Biochemistry

JANICE R. THORNTON-MANNING
ALAN R. DAHL
Inhalation Toxicology Research Institute, Lovelace Biomedical and Environmental Research Institute

GLOSSARY

Dehydrogenation Removal of a hydrogen molecule, usually from a carbon

Electrophilic compound Compound that is deficient in electrons; these compounds will react with chemicals that are rich in electrons, such as proteins and nucleic acids

Epoxidation Reaction in which an oxygen atom binds to each of two neighboring carbon atoms, resulting in a three-membered ring

Hydroxylation Formation of a hydroxyl group (-OH), usually on a carbon or nitrogen atom

Isoform Member of a related family of proteins

Xenobiotics Compounds foreign to the body

THE HUMAN NOSE HAS THREE MAJOR FUNCTIONS: (1) to warm and humidify inhaled air, (2) to protect the lung from toxicants, and (3) to receive olfactory stimuli. This combination of unique functions makes the biochemistry of the nasal tract very complex. In addition to proteins required for all living tissues, the nasal tract contains substantial levels of proteins involved in the metabolism of toxicants and for olfaction. Although there are few data to support the contention, it is thought that some individual proteins may be important for both functions.

During the past decade, much progress has been made in learning about enzymes of the nasal tract that metabolize xenobiotics. It has become apparent that, relative to its tissue mass, the nose has a large capacity to metabolize xenobiotics. Less progress has been made in understanding the process of olfaction. However, some proteins that appear to have unique roles in olfaction have been identified.

The anatomy of the human nose is complex. The nasal septum bisects the nose to form right and left cavities. Three cartilaginous structures, called turbinates, project into the nasal cavity on each side. These turbinates are lined with various types of mucosa that contain the proteins required for the diverse functions of the nose. The mucosa is composed of a layer of epithelium, its basement membrane, and the lamina propria, made up of connective tissue, mucosal glands, and blood vessels. It is the epithelium that contacts air directly and, hence, is the initial tissue of contact for inhaled toxicants or odorants. There are four types of epithelia in the human nasal cavity. The most anterior region of the nose is lined with stratified squamous epithelium. Next, moving posteriorally, a small region of transitional epithelium separates the squamous epithelium from respiratory epithelium that covers the greatest surface area and lines the inferior and middle turbinates and septum. In the most posterior portion of the nose, the olfactory neuroepithelium covers much of the superior turbinate. Squamous epithelium appears to have little xenobiotic-metabolizing capacity and does not play a significant role in olfaction. For these reasons, most of the information in this article focuses on olfactory and respiratory epithelia, which for our purposes include the transitional epithelium. This article provides an overview of general nasal biochemistry, emphasizing enzymes that metabolize xenobiotics and proteins that are involved in olfaction.

I. XENOBIOTIC-METABOLIZING ENZYMES OF THE NASAL CAVITY

A. Cytochrome P450 Enzymes of the Nasal Cavity

Because of their capacity to metabolize numerous substrates, the cytochrome P450's may be the most important enzymes of the nasal cavity involved in the metabolism of xenobiotics. Cytochrome P450's are a superfamily of heme-containing proteins that are involved in the oxidative metabolism of numerous endogenous compounds, as well as xenobiotics. Members of this superfamily oxidatively metabolize a variety of compounds, including highly lipophilic polycyclic aromatic hydrocarbons such as benzo(*a*)pyrene and water-soluble compounds such as ethanol and dimethylnitrosamine. Cytochrome P450-mediated metabolism usually involves hydroxylation, epoxidation, and, rarely, dehydrogenation. This metabolism often leads to a more water-soluble chemical that is either further metabolized or excreted. However, in certain cases, metabolism by cytochrome P450's leads to a chemical that can react with cellular macromolecules, such as DNA and/or proteins, resulting in toxicity or carcinogenicity. This type of metabolism, in which a metabolite is more reactive than its parent, is termed bioactivation. Examples of chemicals that are bioactivated by cytochrome P450's in nasal tissue are hexamethylphosphoramide, phenacetin, and benzo(*a*)pyrene—all of which are nasal carcinogens in animals. [*See* Cytochrome P-450.]

To date, over 200 genes that encode different isoforms of cytochrome P450 have been identified. A number of cytochrome P450 isoforms have been identified in the nasal cavity (Table I); like most nasal xenobiotic-metabolizing enzymes, the highest concentrations are found in the olfactory mucosa. Though nasal P450's have been characterized in only a few species, there appear to be species differences in the expression of certain isoforms. For example, cytochrome P450 1A1, the isoform that is commonly associated with the bioactivation of polycyclic aromatic hydrocarbons (PAHs), is present in the rat nose, but not in the rabbit nose.

The rabbit nose has been most thoroughly studied with respect to cytochrome P450's and contains a number of isoforms (see Table I). However, two P450 subfamilies are represented in substantially greater concentrations than the others. CYP2A and CYP2G1 comprise over 90% of the P450 in rabbit olfactory

TABLE I
Cytochrome P450 Isoforms Present in the Nasal Cavity

P450 isoform[a]	Typical substrate	Species	Location in the nasal cavity
1A1	Benzo(*a*)pyrene	Rat	Bowman's gland
1A2	Isosafrole	Rat	Olfactory
		Rabbit	Olfactory
2A	Coumarin	Rat	Olfactory, respiratory, Bowman's gland
		Rabbit	Olfactory, respiratory
		Human	Olfactory, respiratory, Bowman's gland
2B1	Pentoxyresorufin	Rat	Olfactory, respiratory
2B4	Benzphetamine	Rabbit	Olfactory, respiratory
2E1	Dimethylnitrosamine	Rat	Olfactory, respiratory
		Rabbit	Olfactory
2G1	Testosterone	Rat	Olfactory
		Rabbit	Olfactory
3A	Various steroids	Rat	Olfactory, respiratory
4B1	2-Aminofluorene	Rabbit	Olfactory, respiratory

[a]Cytochrome P450 nomenclature is used as described by Nelson *et al.* (1993).

mucosa. Both rat and rabbit CYP2G1, which are active in the metabolism of testosterone, have some activity toward xenobiotics such as hexamethylphosphoramide, phenacetin, and nitrosodiethylamine. Conversely, members of the CYP2A subfamily are more active in the metabolism of these xenobiotics and less active in the metabolism of testosterone. CYP2A is present in the human nasal cavity in both olfactory and respiratory epithelia. In rats and humans, this enzyme is also present in the Bowman's glands, which are located in the subepithelial layer of the olfactory mucosa.

B. Other Enzymes That Catalyze Oxidation, Reduction, or Hydrolysis

Flavin-containing monooxygenases (FMOs) catalyze the oxidation of nitrogen-, phosphorus-, and sulfur-containing compounds. FMOs are present in olfactory mucosa of the rat and in olfactory and respiratory mucosae of the rabbit. The highest concentrations of these enzymes are localized in the olfactory mucosa.

Carboxylesterase activities are among the highest of any enzyme activities found in the nasal cavity. These enzymes catalyze the hydrolysis of carboxylesters, carboxylamides, and carboxylthioesters. Carboxylesterases are a multifamily of enzymes, but very little is known about specific forms in the nasal cavity. Carboxylesterase activity is generally higher in olfactory mucosa than in nasal respiratory mucosa; in some cases; activity is higher in the olfactory mucosa than in the liver.

Aldehyde dehydrogenases catalyze the oxidation of toxic aldehydes to less toxic carboxylic acids. The rat nasal cavity contains formaldehyde dehydrogenase and two isoforms of acetaldehyde dehydrogenase. Formaldehyde dehydrogenase is present in higher concentrations in the olfactory mucosa, whereas the acetaldehyde dehydrogenase are more abundant in the respiratory mucosa. This is particularly evident when the toxicities of acetaldehyde and formaldehyde are examined in the rat nasal cavity. Acetaldehyde induces lesions in the olfactory epithelium, which has little acetaldehyde dehydrogenase activity. Conversely, formaldehyde induces lesions in the respiratory epithelium, which has little formaldehyde dehydrogenase activity.

Experiments using the alcohol dehydrogenase inhibitor 4-methylpyrazole suggest that this enzyme is present in the nasal mucosa of rats and hamsters. Alcohol dehydrogenase catalyzes the oxidation of ethanol and other alcohols and metabolizes inhaled isoamyl alcohol and propranol in the nasal tissue of rodents.

The hydrolysis of epoxides to glycols is catalyzed by a family of proteins known as epoxide hydrolases. These are generally detoxication reactions because glycols are usually less reactive than epoxides. Epoxide hydrolase activity has been detected in the nasal tissue of rats, mice, dogs, and humans. Unlike most other xenobiotic-metabolizing enzyme activities, levels of epoxide hydrolase activity in olfactory and respiratory epithelia of the dog are not markedly different.

Rhodanese is a mitochondrial enzyme that metabolizes cyanide to the less toxic metabolite thiocyanate. This enzyme has been identified in nasal tissue of rats, cows, and humans. Rhodanese may protect against the toxic effects of inhaled cyanide and from cyanide produced from the cytochrome P450 metabolism of inhaled organonitrile compounds.

C. Enzymes That Catalyze Conjugation

Conjugative, or phase II, enzymes are present in the nasal cavities of several species. These enzymes generally transfer a water-soluble moiety onto a hydrophobic substrate. One family of phase II enzymes, glucuronosyltransferases, catalyzes the covalent binding of a glucuronic acid moiety to form glycosides on hydroxylated compounds such as alcohols and phenols. These glycosides are normally water-soluble and, hence, are generally less toxic than the parent compound. Glucuronosyltransferase activity is present in nasal epithelia from rats and dogs, but absent in human respiratory epithelia.

Another phase II reaction, glutathione conjugation, also occurs in nasal tissue. Glutathione *S*-transferases (GSTs) are a superfamily of enzymes that catalyze the nucleophilic addition of the tripeptide glutathione to electrophilic substrates. This enzyme superfamily is composed of four classes: alpha, mu, pi, and theta. These are some of the more important xenobiotic detoxication enzymes, and their substrates are varied. The substrates are usually reactive compounds capable of binding to cellular macromolecules such as DNA and protein, producing toxicity and/or mutagenicity. Thus, in most instances, the actions of GSTs are to protect against cell death or mutation. In some cases, however, GSTs can produce metabolites that are toxic and/or carcinogenic. An example is the formation of an episulfonium ion (a three-atom, positively charged, cyclic sulfur ether) from the actions of the enzyme cysteine β-lyase on the glutathione conju-

TABLE II
Glutathione S-Transferase (GST) Enzymes and Activities in Nasal Mucosa

GST isoform	Substrate	Species
Alpha	Cumene hydroperoxide	Rat
Alpha	1,2-Dichloro-3-nitrobenzene	Cow
Mu	1,2-Dichloro-3-nitrobenzene	Rat
Pi	1,2-Dichloro-3-nitrobenzene	Cow
Pi	1,2-Dichloro-3-nitrobenzene	Human

gate of dibromoethane. It is not known whether cysteine β-lyase is present in the nasal cavity.

Like most other nasal xenobiotic-metabolizing enzymes, GSTs are present in greater concentrations in the olfactory epithelium. These enzyme activities have been identified in the nasal tissues of rats, cows, and humans (Table II).

D. Alterations in Activities of Nasal Xenobiotic-Metabolizing Enzymes

Modifications of hepatic xenobiotic-metabolizing enzyme activities readily occur after starvation, tissue inflammation, and exposure to a variety of chemicals. Although these modifications often involve enzyme induction, inhibition may also result. The xenobiotic-metabolizing enzymes of the nasal cavity are generally resistant to induction by chemicals that readily induce extranasal enzymes. However, these enzymes appear to be inhibited by common inhibitors. For example, the nasal P450's of rats are not affected by common inducers such as phenobarbital benzo(*a*)pyrene, β-naphthoflavone, or 3-methylcholanthrene. Conversely, nasal P450's are inhibited by a number of classic P450 inhibitors such as SKF-525A, metyrapone, α-naphthoflavone, and piperonyl butoxide. 2,3,7,8-Tetrachlorodibenzo-*p*-dioxin and Arochlor 1254, both potent inducers of hepatic P450's, appear to only slightly induce rat nasal P450's. In the rabbit nose, ethanol and acetone have been shown to induce CYP2E1. Though induction of phase II enzymes is not as well characterized, these enzymes are induced by Arochlor 1254, 3-methylcholanthrene, and phenobarbital in the rat nose. Rat olfactory microsomal GST is increased by treatment with *N*-ethylmaleimide similar to that of hepatic microsomal GST. Other enzymes of the nasal cavity known to be inhibited are carboxylesterases, which are inhibited by organophosphates in rats, and rhodanese, which is inhibited by cigarette smoke in human smokers.

Damage caused by environmental toxicants also may alter xenobiotic-metabolizing enzyme activities. In rats exposed chronically to cigarette smoke, a swelling of smooth endoplasmic reticulum of the olfactory epithelium occurs that appears to correlate with an increase in carboxylesterase. In human nasal tissue, carboxylesterase is markedly reduced in the presence of hyperplastic lesions and squamous metaplasia. Inhibition of olfactory GST activity is an early event in rats after exposure to methyl bromide and may contribute to the severe degenerative changes in olfactory epithelium caused by this compound.

Alterations in nasal enzymes may be important for several reasons. First, olfaction may be affected. Methylenedioxyphenyl compounds, potent odorants commonly used in the fragrance industry, readily inhibit nasal P450's *in vitro*. This suggests that interactions between fragrances and nasal xenobiotic-metabolizing enzymes may be important in olfaction. Second, alterations in the levels of toxic chemicals in the nasal mucosa may increase. For example, cyanide may reach higher levels in the olfactory epithelium of smokers because the enzyme that detoxicates this compound, rhodanese, is inhibited in the nasal cavities of these individuals. Third, alterations of nasal enzymes may affect the metabolism of nasally administered pharmaceuticals. Certain decongestants are known to be metabolized by nasal enzymes. Clearly, alterations in nasal enzyme activities may have important toxicological implications and also affect olfaction.

II. EFFECTS OF NASAL XENOBIOTIC-METABOLIZING ENZYMES ON INHALED TOXICANTS AND PHARMACEUTICALS

A. Bioactivation/Detoxication of Toxicants and Carcinogens

Several chemicals are known to be bioactivated or detoxicated by enzymes of the nasal mucosa (Tables III and IV). This may be very important in predicting the toxicological outcome resulting from exposure to these chemicals. Nasal cancer has been reported in experimental animals after exposure to a number of chemicals known to be metabolized by nasal cyto-

TABLE III
Nasal Xenobiotic-Metabolizing Enzymes That Bioactivate Toxicants

Xenobiotic-metabolizing enzyme	Chemical	Species
Cytochrome P450	Dimethylnitrosamine	Human
Cytochrome P450	Acetaminophen	Mouse
Carboxylesterases	Acetate and acrylate esters	Mouse, rat
Cytochrome P450	Ferrocene	Rat
Cytochrome P450	Hexamethylphosphoramide	Rat, human
Flavin-containing monooxygenase	*N,N*-Dimethylaniline	Rat

chrome P450 enzymes. Inhalation exposures to formaldehyde and hexamethylphosphoramide have been reported to cause nasal cancer in laboratory animals. Nasal papillomas occur in rats after oral administration of phenacetin. Also, nasal metabolism has been suggested to be involved in the production of gastrointestinal tumors in laboratory animals caused by inhaled benzo(*a*)pyrene as a result of swallowing mucus laden with nasal P450-generated reactive metabolites.

The effects of bioactivation and detoxication of chemicals in the human nose are not well understood. However, the association of nasal cancer in humans with occupational exposures to selected chemicals suggests that these processes may be involved in the ultimate carcinogenic outcome. Individuals occupationally exposed to isopropyl alcohol, mustard gas, wood, and leather have increased incidences of malignant nasal tumors. Furthermore, a high incidence of nasopharyngeal carcinoma in southeastern China has been correlated with consumption of salted fish, which contains significant amounts of dimethylnitrosamine. Whether the carcinogenicity results from bioactivation of chemicals to reactive carcinogenic metabolites has not been confirmed. However, the induction of nasal tumors in laboratory animals by chemicals known to be bioactivated by nasal enzymes suggests that this may also occur in humans.

TABLE IV
Nasal Xenobiotic-Metabolizing Enzymes That Detoxicate Toxicants

Xenobiotic-metabolizing enzyme	Chemical	Species
Aldehyde dehydrogenase	Formaldehyde	Rat
Epoxide hydrolase	Safrole oxide	Human, rat
Epoxide hydrolase	Styrene oxide	Dog, rat
Rhodanese	Cyanide	Human
Cytochrome P450	Benzo(*a*)pyrene[a]	Rat

[a]The formation of 9-hydroxybenzo(*a*)pyrene, for example, is a detoxication reaction; some epoxide metabolites, however, are carcinogenic.

B. Metabolism of Pharmaceutical Agents in the Nasal Cavity

The intranasal route of drug delivery has become increasingly popular. The ease of this route of delivery offers advantages over intramuscular or intravenous routes and avoids hepatic "first pass" metabolism associated with oral delivery. The presence of xenobiotic-metabolizing enzymes in the nasal mucosa may be problematic for certain drugs and results in substantial metabolism of the drugs before they reach systemic circulation. However, compared to the liver, the number of specific P450 isoforms present in the nose is very low. Hence, certain drugs will not be metabolized in the nose. For example, the β-adrenergic antagonist propranolol has been shown to be cleared from the human nose to the blood without undergoing metabolism; whereas, after oral administration, this drug undergoes extensive hepatic metabolism.

Another potential use for intranasal drug delivery is in the treatment of central nervous system disorders. Certain antibiotics reach higher cerebrospinal fluid levels after intranasal administration as compared to intravenous administration. It has been suggested that pharmaceutical treatment of other disorders, such as Alzheimer's, may best be done intranasally because potential olfactory transport of drugs may deliver them directly to affected tissues.

III. BIOCHEMISTRY UNIQUE TO THE OLFACTORY TISSUE

A. Xenobiotic-Metabolizing Enzymes Unique to the Olfactory Tissue

Cytochrome P450 2G1 is found only in olfactory mucosa in rats and rabbits. Although the significance of this unique localization is not well understood, it may be related to the metabolism of steroids. Like several other P450 isoforms, this enzyme is known to metabolize certain steroids. However, the metabolite profiles produced by 2G1 are unique compared to those of other P450's. Because inhaled steroids play a role in the reproductive process in a number of species, it is possible that CYP2G1 is also important in this process.

Less is known about an olfactory-specific glucuronosyltransferase isozyme. This enzyme has been shown to catalyze the glucuronidation of a number of odorants, suggesting that it may play a direct role in olfaction. It has been suggested that this isozyme plays a role in the clearance of odorant molecules from olfactory receptors to allow for reactivation of the receptor.

B. Other Olfactory-Specific Proteins

The first olfactory-specific protein, olfactory marker protein (OMP), was discovered over 20 years ago. OMP is a soluble, 19-kDa protein that is expressed in olfactory neurons of all species examined. Though this protein has been well studied, its function is still unknown.

Another set of unique olfactory proteins are odorant receptors, proteins that are located on the cilia of olfactory sensory neurons. These receptors are thought to bind to a variety of odorous ligands and to initiate olfactory signal transduction, ultimately leading to the arrival of odorant signals in the brain. These are a family of proteins that traverse the cell membrane seven times and whose function is coupled to the actions of a G protein (guanine nucleotide-binding protein) called G_{OLF}. The transcription factor Olf-1 activates the expression of several olfactory proteins, including G_{OLF} and OMP.

The odorant-binding proteins are another group believed to be important in olfaction. These are a class of soluble proteins that bind to a variety of odorants. They are secreted by various glands of the nasal cavity and released in mucus. These proteins have been identified in humans and other mammalian species. Although the exact function of this group of proteins is unknown, it has been proposed that they act as carriers to move odorant molecules from the nasal mucous to the olfactory receptor cells. [*See* Olfactory Information Processing.]

IV. MUCUS GENERATION AND SECRETION

The mucus covering the olfactory and respiratory epithelium serves an important protective function by the clearance of toxicants. This results primarily through the actions of cilia on respiratory epithelial cells, which move the mucus and associated contaminants over the surface of the epithelium. However, the presence of lysozyme, immunoglobulins, and xenobiotic-metabolizing enzymes in mucus suggests other protective mechanisms of mucus as well.

As discussed earlier, the two primary types of nasal mucosa, respiratory and olfactory, differ with respect to cell composition, biochemistry, and function. Similarly, the mucus covering these types of mucosa also differs. The mucus of the respiratory mucosa is produced by epithelial goblet cells and subepithelial glandular acinar cells. The mucus covering the olfactory mucosa is generated primarily by the acinar cells of the subepithelial Bowman's glands. Although both types of mucus in the human nose are composed primarily of acidic mucopolysaccharides, there are some minor chemical differences in the two types. There is a higher content of sialosulfomucins in respiratory mucus than in olfactory mucus. A major difference is the presence of L-fucose residues attached via a 1,2 linkage to galactose β-1,3-*N*-acetylglucosamine residues in glycoconjugates present in mucus produced by respiratory mucosa but not in that produced by olfactory mucosa. These differences in mucus composition from the two types of nasal mucosa probably reflect their different functions. The primary function of respiratory mucosa is to warm and humidify the air, whereas olfactory mucus provides an environment for sensory transduction. Olfactory mucus also contains proteins involved in immune barrier functions, such as immunoglobulins A and M and lysozyme. This is probably important because of the proximity of the olfactory mucosa to the central nervous system and the potentially perilous consequences of infection.

BIBLIOGRAPHY

Buck, L. (1993). Identification and analysis of a multigene family encoding odorant receptors: Implications for mechanisms underlying olfactory information processing. *Chem. Senses* **18**, 203–208.

Dahl, A. R., and Hadley, W. M. (1991). Nasal cavity enzymes involved in xenobiotic metabolism. Effects on the toxicity of inhalants. *Crit. Rev. Toxicol.* **21**, 345–372.

Getchell, M. L., and Mellert, T. K. (1991). Olfactory mucus secretion. *In* "Smell and Taste in Health and Disease" (T. V. Getchell, R. L. Doty, and J. B. Snow, Jr., eds.). Raven, New York.

Getchell, M. L., Chen, Y., Sparks, D. L., Ding, X., and Getchell, T. V. (1993). Immunochemical localization of a cytochrome P450 isozyme in human nasal mucosa: Age-related trends. *Ann. Otol. Rhinol. Laryngol.* **102**, 368–374.

Lewis, J. L., and Dahl, A. R. (1995). Olfactory mucosa: Composition, enzymatic localization and metabolism. *In* "Handbook of Olfaction and Gustation" (R. L. Doty, ed.). Marcel Dekker, New York.

Lewis, J. L., Nikula, K. J., Novak, R., and Dahl, A. R. (1994). Comparative localization of carboxylesterase in F344 rat, beagle dog, and human nasal tissue. *Anat. Rec.* **239**, 55–64.

Morris, J. B. (1993). Upper respiratory tract metabolism of inspired alcohol dehydrogenase and mixed function oxidase substrate vapors under defined airflow conditions. *Inhal. Toxicol.* **5**, 203–222.

Nelson, D. R., Kamataki, T., Waxman, D. J., Guengerich, F. P., Estabrook, R. W., Feyereisen, R., Gonzalez, F. J., Coon, M. J., Gunsalus, I. C., Gotoh, O., Okuda, K., and Nebert, D. W. (1993). The P450 superfamily: Update on new sequences, gene mapping, accession numbers, early trivial names of enzymes, and nomenclature. *DNA Cell Biol.* **12**, 1–51.

Pelosi, P. (1994). Odorant-binding proteins. *Crit. Rev. Biochem. Mol. Biol.* **29**, 199–228.

Reed, C. J. (1993). Drug metabolism in the nasal cavity: Relevance to toxicology. *Drug Metab. Rev.* **25**, 173–205.

Sarkar, M. A. (1992). Drug metabolism in the nasal mucosa. *Pharm. Res.* **9**, 1–9.

Natural Killer and Other Effector Cells

OSIAS STUTMAN
Memorial Sloan-Kettering Cancer Center

GLOSSARY

Interleukin Cytokine (usually a lymphokine or monokine) that mediates various immunological effects on lymphocytes and other cells, although it may have nonimmunological effects

Ligand Membrane structure(s) recognized by a receptor

Lymphokine Biologically active product released by a lymphocyte; cytokine is the generic term for products released by cells, and monokine for products released by monocytes or macrophages

NATURAL KILLER (NK) CELLS BELONG TO A FAMILY of effector cells in the lymphoid and hemopoietic tissues of most vertebrate species, including humans, which can directly kill certain tumor target cells *in vitro*, without prior immunization. NK cells use killing strategies that are different from those of antigen-specific cytotoxic T cells and from activated nonspecific killer macrophages and similar cells, although they may share the same killing mechanisms.

I. KILLER CELLS

A. Cell-Mediated Cytotoxicity

Cell-mediated cytotoxicity (CMC) is the lysis of target cells by effector killer cells of lymphoid and myeloid origin. [*See* Lymphocyte-Mediated Cytotoxicity.]

1. Basic Model of CMC

Lysis requires intimate contact between killers and targets in three steps: (1) recognition (conjugate formation between killers and targets); (2) programming for lysis (includes triggering of synthesis and/or release of lytic molecules by the effector and the unidirectional production of an irreversible lesion in the membrane of the target); and (3) killer-independent lysis (the killers can be dissociated, and the target with its lesion will eventually die; the released effector can recycle and kill another target cell).

2. Targets

Target cells are usually established cell lines. In most cases, they are derived from human or murine leukemias or other tumors, although a number of nucleated cell types have been used as targets. They range from mitogen-induced blasts to fibroblasts infected by viruses, or even protozoan, plasmodial, or metazoan parasites at various stages of their life cycle. The prototypic targets with which most NK work has been done are the K562 erythroleukemia for the human studies and the murine leukemia virus-induced mouse lymphoma YAC-1 for the rodent studies.

3. Assays

The most common method used to measure cell lysis is the release of chromium-51 (^{51}Cr) from prelabeled target cells in suspension cultures. The ^{51}Cr binds randomly to internal cellular proteins and is released as a consequence of membrane damage by the lytic event. The maximal lysis produced by the effector cells is compared to the maximal release obtained after treating the labeled target cells with a chaotropic detergent; maximal release is about 80% of the intracellular ^{51}Cr. The results are usually expressed as "percent lysis," derived from equations comparing maximal release and release resulting from the effector–target

ENCYCLOPEDIA OF HUMAN BIOLOGY, Second Edition, VOLUME 6.

interaction, corrected for spontaneous release by the target cells alone. Another way of expressing the results is as "lytic units," which are calculated from the linear part of the lysis curve and are expressed as the number of effectors required to produce an arbitrary "percent lysis," such as 20 or 25%. The maximal lysis induced by the killer cells at various effector–target ratios is reached at about 4 hr, although it may require longer assays (i.e., 18 hr), depending on the capacity of the target to repair the lytic damage, the type of effector cell used, and the type of lytic molecules produced by the effectors.

4. Mechanism of Lysis

Depending on the type of effector and target cells, target-killing can be mediated by three nonexclusive mechanisms. These mechanisms include granule exocytosis (the granules contain a variety of lytic molecules), pore formation on the cell membrane (actual holes punched by the lytic molecules in a manner similar to the lysis of cells mediated by complement), and the contact-induced, stimulated nuclear disintegration also known as apoptosis or programmed cell death (a complex event, which has nuclear fragmentation at its end point and is probably mediated by activation of endonucleases). The problem with nuclear disintegration as such is in determining whether it represents the cause or the consequence of the lytic event that programs the cell for death (or suicide). (See Section II,D for more on lytic molecules.)

B. Major Histocompatibility Complex

Major histocompatibility complex (MHC) restriction is required to recognize gene products of the MHC of a given species by some effector cells. Cytotoxic cells can be divided into two categories depending on MHC restriction requirements: (1) T cells that kill in a specific MHC-restricted fashion (i.e., T cells from MHC type A donors immunized to a virus such as influenza will kill only MHC type A-infected targets and not MHC type B-infected targets), and (2) killer cells that do not require MHC recognition for target lysis. T cells have clonally distributed specific receptors for antigen (TcR), which allow recognition of viral and other antigens in the context of self-MHC. The diversity within the repertoire is generated by combinatorial rearrangements, and the interaction with antigen produces a progeny with immunological memory. The TcR structure, with its associated CD3 complex of proteins, binds its specific ligands and transduces signals that trigger cell activation. Activation includes synthesis of a variety of lymphokines, some of them with lytic properties, and cell division. [*See* T-Cell Receptors.]

C. Non-MHC-Restricted Killers

Cytotoxic effector cells, which are independent of recognition of MHC gene products on the targets, include almost any lymphoid or myeloid cell types, such as monocytes, macrophages, B cells, some types of T cells, granulocytes, and even platelets, depending on the type of target cells and/or the stimuli for lytic activation used. Macrophages can kill a wide variety of tumor targets after activation. Many of these effectors will kill preferentially malignant or infected, but not "normal," targets. The nature of such selectivity is still undefined. Resting cells of different types, including T cells, can be made cytotoxic by simply "gluing" them to the target with adhesion lectins or antibodies, thus bypassing the need for the recognition step. Antibody-dependent, cell-mediated cytotoxicity (ADCC) is mediated by cells with receptors for the Fc portion (FcR) of immunoglobulin G (IgG), initially considered a special subset called K cells; however, it was later shown that most of the ADCC effectors belonged to the NK category. NK cells express the surface protein CD16, which is the low-affinity Fc receptor for IgG (FcRIII). In ADCC, antibodies, which are specific for a surface determinant on the target, will bind to the target, and the ADCC effector cells will bind to the antibody via its FcR. Alternatively, FcR+ cells could be "armed" by the specific antibody bound through its Fc portion, leaving the specific binding site available. Thus, the specific antibody will provide the "glue" and the FcR+ cell will provide the killer of the target. [*See* Macrophages.]

NK and other types of effector cells do not require MHC recognition for target lysis, but neither do monocytes and macrophages. As the main property of the NK category, cytotoxic cells are found in fresh blood samples obtained directly from normal donors, and the killing is spontaneous, without any known triggering for activation (thus the "natural" appellative).

D. NK and Other Effector Cells

Terminology in NK studies is not simple. Initially, natural cell-mediated cytotoxicity (NCMC) included a variety of non-MHC-restricted cytotoxic cells obtained directly from blood (in most human studies) and from spleen (in most rodent studies) of normal healthy individuals who were able to kill *in vitro* some selected tumor targets. These effectors included NK, natural cytotoxic (NC) cells, NK-like cells capable of killing virus-infected cells, and a variety of killer cells, which were generated after *in vitro* culture or activation. Almost any cell that could kill the prototypic K562 and YAC-1 target cells was defined as being NK (including macrophages, promonocytes, basophilic granulocyte cell lines, etc.). During 1983–1988, attempts to establish a minimal terminology were made at workshops by researchers in the NK field. The 1983–1986 terminology included two main categories. (1) *NK*—freshly obtained cells with spontaneous cytotoxicity against some prototypic targets, which are lymphoid in appearance, are different from granulocyte–monocyte–macrophage–T cytotoxic cells, and show no MHC restriction. In addition to the canonical NK (which kill YAC-1 or K562 leukemia targets in mice and humans), this group included some NK subsets with broader target specificity and the NC cells (see Section II,E). (2) *Activated killer* (*AK*)—includes all the non-MHC-restricted killers, which are produced after short- or long-term *in vitro* culture with or without added lymphokines such as interleukin-2 (IL-2). Although not stated, the AK category should also include the various types of established cell lines with NK, NK-like, or NC activities. The cloned lines with NK activity in rodents and humans initially created problems when compared to freshly obtained NK cells, however, they have been useful for certain studies, such as to define the NK receptor structures. Because this classification could not exclude nonconventional T cells with NK-like activity from the first category, it generated strong argumentation about NK lineage and confusion regarding the surface-marker phenotype of the effector cells. The 1988 terminology is more restrictive and perhaps more precise. It includes three major types of non-MHC-restricted effectors. (1) *NK*—large granular lymphocytes (LGL) that do not express the TcR-related CD3 surface antigen and do not productively rearrange their alpha, beta, gamma, or delta TcR genes. Based on these criteria, these cells do not belong to the T lineage. (2) *NK-like T or non-MHC-restricted cytotoxic T cells*—a fraction of CD3+ T lymphocytes, which express either the alpha-beta or the gamma-delta TcR, have LGL morphology, and usually kill the same prototypic targets as the NK LGL in the first group. The NK and NK-like T terminology will be used in this text. (3) *Lymphokine-activated killer* (*LAK*) *cells*—activated *in vitro* with IL-2 for a few days, which can be considered as either the NK or NK-like T non-MHC-restricted killers. The relative contribution of the T and NK components to LAK depends on the source of cells and the conditions for *in vitro* activation. [*See* Lymphocytes.]

Stricto sensu, only the first category of cells would qualify for the NK appellative. The second category of cells belongs to a defined lineage (T) and simply expresses the NK type of reactivity. LAK cells would fall within the AK category of the 1983 classification. LAK cells became popular in the nonscientific press around 1986 as a potential new treatment modality for advanced human cancer.

This article focuses on a functional aspect of cellular activity (i.e., the capacity to produce lytic molecules that can kill other cells in close proximity) that is also a property of the cells discussed earlier. The noncytotoxic functions of these cells will be discussed in Section I,G.

E. A Comment on Lymphokines

Since NK cells are lymphokine producers as well as targets for lymphokines produced by other cells, it is worth including a brief comment on lymphokine behavior. The cloning and production of recombinant lymphokines demonstrated without a doubt that a single nonglycosilated recombinant homogeneous protein could have multiple effects on multiple target cells. The basic rules that emerged from studies on the defined recombinant lymphokines and cytokines are: (1) all lymphokines have pleiotropic activities (i.e., a variety of activities that are apparently unrelated to each other); (2) responses usually trigger a cascade of several mediators with synergy (and/or interference) between mediators; (3) targets for each lymphokine are widely distributed among lymphoid, hemopoietic, and other sites (including the central nervous system); (4) lymphokines have a high degree of functional redundancy in spite of no structural homologies [e.g., IL-1 and tumor necrosis factor (TNF) share most functional activities; IL-2, IL-4, IL-6, and IL-7 are growth factors for T cells; IL-1, TNF, IL-2, the interferons, IL-4, IL-5, IL-6, and IL-7 can affect B-cell function; IL-1, IL-3, IL-4, and IL-7 affect hematopoietic progenitors], and some of this redundancy can now be explained by the fact that some of the lymphokine receptors share a chain that transduces the signal [e.g., the gamma chain that is used by the IL-2, IL-4, and IL-7 receptors or the alpha chain shared by the IL-3, IL-5, and granulocyte–macrophage colony-stimulating factor (GM-CSF) receptors, to cite but two examples]; (5) in spite of extensive homologies there are some species restrictions for lymphokine activity (i.e., human IL-1 but not human TNF can act as costimulators for mouse thymocytes); (6) T cells produce several lymphokines after stimulation. Two views prevail: (i) the murine TH1/TH2 concept based on the combination of lymphokines produced (TH1: IL-2, interferon gamma, and lymphotoxin; TH2: IL-4, IL-5, and IL-6; with TNF, GM-CSF and IL-3 produced by both types) and (ii) production of several lymphokines but no clear clustering seen especially with cloned human T cells. Furthermore, the lymphokine production profile of a given T line (especially with human T cells) may not be stable but is convertible by different stimuli. And (7) correlation of *in vitro* optimized models may not correlate *in vivo* function; attempts at defining *in vivo* lymphokine biology using transgenic mice (with the lymphokine gene controlled by constitutive mouse promoters) have given unexpected results. For example, the double transgenic for human IL-2/p55 IL-2R showed reduced T and expanded NK functions, rather than the predicted expanded T. [*See* Interferons; Interleukin-2 and the IL-2 Receptor.]

The redundancy discussed here is also exemplified by the observation that after the knockout of the IL-2 gene by homologous recombination gene disruption in mice, no important abnormalities of the T-cell system were observed, suggesting that other cytokines replaced the missing function. However, unexpectedly these mice develop ulcerative-colitis-like lesions (as also do IL-10 or alpha/beta T-cell receptor knockouts). On the other hand, the knockout of the perforin gene, the main lytic molecule used by NK and T cells, greatly impaired cytotoxicity mediated by those cells.

F. History

The NK acronym was coined in 1975 for the spontaneous killer cell activity found in normal mouse spleen. Prior to the 1970s, NCMC was viewed as an unexpected nuisance during attempts to detect tumor-specific, cell-mediated immunity *in vitro* using lymphocytes from cancer patients, because cells from the normal controls could also kill the tumor targets. Although initially considered an artifact, the unexpected reactivity could not be dismissed by technicalities and became a subject of study. As a measure of acceptance of the NK phenomenon by the immunological community, NK appeared in the subject index of the *Journal of Immunology* in 1981 (lymphocyte appeared for the first time in 1948). Because the expected cytotoxicity in cancer patients had to be directed against tumor-associated antigens, as part of the prevalent views on tumor immunology, the first methodical NK studies in normal mice suggested that killing was directed against antigens coded by endogenous leukemia viruses. However, in 1976 the response clearly was not directed against murine leukemia virus, and the killing showed no MHC restriction. Human NK cells became better characterized, and in 1978 NK activity could be augmented by *in vitro* treatment with interferon in both humans and mice. NC cells were described in 1978, the augmenting effect of IL-2 on NK cells was described in 1981, and LAK cells were described around 1982. Based on the complexities of defining surface phenotype and lineage, as well as on variations on target preference for killing, the notion of activities mediated by a heterogeneous group of effector cells was proposed. Agreement was reached mainly on two NK properties: (1) NK activity was detected in fresh cell samples obtained from normal individuals in the absence of any prior sensitization, and (2) in addition to the independence from MHC restriction, NK cells did not show immunological memory nor clear evidence of clonality in response to a given target, all of which are properties of cytotoxic T cells. Around 1983, the ideas developed that NK and similar cells belonged to the innate or natural type of immunity and that its functions went beyond the potential antitumor activities. In spite of the large amount of information generated (the 1989 review article by G. Trinchieri in the bibliography has 1182 references), questions concerning lineage, receptors, antigens, and functions are still open. All the books cited in the bibliography are multiauthored, containing from 12 to 219 individual chapters or papers and are good examples of the different opinions that still predominate in the field.

G. A Problem

Before describing the properties of classic NK cells, a conceptual problem must be discussed. The potential problem is to consider a given *activity* (such as lysis of certain tumor cells) as the *function* of a specialized cellular system of which the NK cell is the prototype, especially if we keep in mind that the lytic capacity can be triggered under appropriate conditions in most nucleated cells and even nonnucleated platelets. For example, if protein synthesis measured by the classic Lowry method could be used as a functional end point, the differentiation between B cells (which produce immunoglobulin), beta cells of the pancreas (which produce insulin), and hepatocytes (which produce an array of proteins) would be impossible. Based on that activity, one may conclude that the natural protein synthesis cell is heterogeneous and difficult to place within a given lineage. A somewhat similar situation has evolved regarding NK cells. Two views are prevalent. (1) *A unique cell subset*—NK represent a defined category of cells in the myeloid lineage (i.e., of bone marrow origin) with still undefined overall functions, which depend on their killer capacity (this being the more popular view). And (2) *a functional system*—the NK capacity to kill certain tumor targets is an activity expressed by many cells as part of their developmental or functional programs, rather than the function of a particular cell category. Two variations of both views are as follows. (1) *More than just killers*—the functions of the NK subset or system are related to other activities, such as production of γ-interferon (γ-IFN) or other cytokines. And (2) *complex functions*—the functions of the NK subset or system are complex and include, in addition to the original antitumor control of primary and metastatic tumor growth, other functions such as control of some infections (microbial, viral, parasitic, etc., which also include direct killing of certain bacteria), immunoregulatory functions (usually acting as down-regulators or natural suppressors), control of hematopoietic stem-cell growth (and differentiation), and some involvement in rejection of allografts, especially bone marrow grafts.

H. Defense Mechanisms

One of the unique properties of NK and similar cells is that they seem to be naturally activated and, thus, can act instantaneously in response to a foreign stimulus. All of the specific components of the immune system, such as antibody production by B cells or the cell-mediated responses of T cells (including MHC-restricted killing), require time-consuming priming, usually a few days. Priming is followed by clonal expansion and the specific responses against the invading agent as well as important amplification of the specific responses by recruitment of a variety of nonspecific effector cells, such as macrophages or monocytes. However, even the nonspecific effectors, such as macrophages, require activation to become lytic, which is usually provided in a functional cascade by the specific priming of the T-cell system. Thus, NK and related cells may act as early defense mechanisms, which may allow the development of the specific defense response, perhaps as part of a more primitive immune system with broad multispecific receptors for recognition. A variety of vertebrate species tested, from catfish to birds, have NK-like cells.

II. GENERAL PROPERTIES OF NATURAL KILLER CELLS AND OTHER EFFECTOR CELLS

The properties of the classic NK cells will be discussed in Sections II,A–II,D, and those of the other effectors in Section II,E. Unless otherwise stated, the descriptions apply to human cells. The phenomenological and mechanistic studies on NK and other NCMC effector cells in experimental animals and humans are derived, in most cases, from the analysis of at least three variables: (1) the donor providing the effector cells, (2) the type of target utilized, and (3) the type of *in vitro* assay employed. A fourth and often forgotten variable is the type of procedure used for the separation of the effector cells, which in some cases can activate them. The first variable is probably the most difficult to control, because at any given time a variety of stimuli may augment (or depress) NCMC effector activity. This intrinsic variation is well exemplified in a human study where the same group of volunteers were tested periodically for blood NK during a lapse of several years, showing almost all the possible permutations in NK activity levels.

A. Phenotypic Properties of NK

1. LGL Morphology

The identification of large granular lymphocytes as a population enriched for NK activity in humans and rats resulted in the tendency to equate NK to LGL. LGL are detected in peripheral blood using discontinuous density-gradient centrifugation on Percoll and are characterized by size and presence of azurophilic cytoplasmic granules. LGL are nonphagocytic, are usually nonadherent, and represent about 5% of the human peripheral blood lymphocytes. Two caveats: (1) although most of the NK activity is detected in the LGL fraction, not all LGL have killer activity; and (2) cytotoxic T cells obtained after *in vivo* stimulation or after *in vitro* culture, including specific MHC-restricted T-cell lines, may also have LGL morphology. Thus, the LGL morphology may represent a state of functional activation, poised toward the lytic process or secretion rather than the morphological counterpart of a defined cell subset. Using single-cell cytotoxicity assays, with resting or activated blood LGL and a panel of NK susceptible targets, 50–85% of the LGL are capable of killing at least one type of target. The single-cell cytotoxicity assay uses effector–target conjugates transferred to semisolid media and the killing of individual target cells monitored by supravital staining (the dye will penetrate into the damaged cell, whereas healthy cells exclude it). One interpretation of the nonkilling LGL binders is that such cells are at a stage in which they cannot fully express their lytic capacity. Binding is necessary but not sufficient for target lysis, and some targets may lack the capacity to trigger the delivery of the lytic signal by the effector. Another view assumes that the limited number of targets used does not cover the whole repertoire of the NK–LGL cell, which explains why some LGL may not kill any targets.

2. Surface Antigens

The association of a surface marker and a cellular function, such as NK activity, is shown by either negative or positive selection. Negative selection is the elimination of NK activity after treatment of the cells with a given monoclonal antibody in the presence of complement; the positive cells that bind antibody will be killed by the complement. Another type of negative selection is the depletion by immunoaffinity procedures, such as panning, or by fluorescence-activated cell-sorting (FACS) of the fraction, which is positive for a given marker. Positive selection is the enrichment of NK activity in the cell fraction, which is positive for the marker using FACS, panning to antibody-coated plastic dishes, or similar immunoaffinity procedures. Table I summarizes the available information on the expression of surface antigens in human blood lymphocytes with NK activity. There is no single surface marker that is unique for NK cells, and the phenotype is defined by a combination of some markers, such as CD16, CD56, and CD57, in cells that do not express CD3. Blood LGL can be divided into CD3+ and CD3− (i.e., T and non-T cells); most of the CD3− LGL express CD16 (the low-affinity receptor for the IgG Fc or FcRIII). Minor subsets of CD3+ T cells can also express CD16. The CD56 reagent, which detects 8–20% of positive cells in normal blood lymphocytes, defines three subsets that kill NK-susceptible targets: CD16− CD56+ CD3− (NK non-T cell, perhaps a precursor of mature NK), the major NK population of CD16+ CD56+ CD3− cells, and a CD16− CD56+ CD3+ NK-like T cell (about 25% of CD56+ are CD3+ and can also express CD8). With CD16 and CD57, four subsets can be defined in human blood: CD16+ CD57− with high NK, CD16− CD57+ with detectable but low NK, CD16+ CD57+ with wide individual variation of NK, and CD16− CD57− with no NK activity. A fraction of all of these subsets also can express CD3 and/or CD8 (i.e., T cells). CD2 was one of the first markers used to define NK subsets using the formation of rosettes with sheep erythrocytes (it was termed the E receptor or ER); about 50% of human LGL were ER+. The CD2 cluster is the receptor for the LFA-3-antigen (CD58), which is expressed on many cell types and is involved in cell adhesion. CD2 is expressed on most NK and also on thymocytes, T lymphocytes, and activated T cells. Most of the CD56+ cells, including those expressing CD3, are CD2+. Negative selection with antibodies recognizing CD11b (such as OKM1), CD16, and CD56 usually results in the depletion of most of the NK activity. CD69 is an early activation antigen present on activated T, B, and NK cells and other cells including platelets. CD69 is detected on a small fraction of fresh NK cells in blood and appears on all NK cells after activation by IL-2, the interferons, or CD16 cross-linking. On the other hand, most gut-associated T cells express CD69, suggesting *in vivo* activation. One interesting aspect of CD69 is that it belongs to the family surface receptor molecules expressing the calcium-dependent C-type lectin domain, which is also expressed by the most likely candidates to be the NK receptors (discussed in Section II,D,1,b). However, as Table I shows, all of those markers are also expressed on non-NK cell types, including T and non-T lineages. The phenotype of mouse NK is discussed in Section II,E.

In summary, NK cells are a unique population, representing 10–20% of the mononuclear cells in peripheral blood, which are defined by the combined expression of certain surface antigens (CD2, CD7, CD16, CD56) and the absence of specific markers of the T (CD3, TcR) and the B lineages (surface Ig, etc.). Furthermore, the different permutations and surface densities of CD2, CD8, CD16, CD56, and CD57 seem to define subsets or different stages of maturation or activation.

One important point to accent here is that many (probably all) of the surface structures listed in Table I, regardless of the cell type on which they are expressed, are not simple "markers" but molecules with transmembrane and cytoplasmic domains and other attributes that can participate in signal transduction leading to activation, including induction or augmen-

TABLE I
Surface-Marker Phenotype of Human NK Cells[a]

Marker	NK	T	B	MO/M	G
CD2 (LAF-3/CD58 R)[b]	+	+	−	−	−
CD3	−	+	−	−	−
CD4	−	+S	−	+S	−
CD7 (Leu 9)	+	+	−	−	−
CD8	+S, L	+S	−	−	−
CD11a (LFA-1 alpha chain)[c]	+	+	+	+	+
CD11b (C3R alpha chain)[c]	+S, L	+S, L	−	+	+
CD11c (P150/95 alpha chain)[c]	+	−	−	+	−
CD16 (FcRIII; Leu 11)[d]	+	+S?	−	+A	+
CD18 (beta chain of CD11)	+	+	+	+	+
CD25 (IL-2R alpha gp55)	+A	+A	+A	+A	−
CD32 (FcRII)[d]	−	−	+	+	+
CD38 (Leu 17)	+A	+A	−	−	−
CD56 (Leu 19, NKH-1)[e]	+	+A	−	+L	−
CD57 (Leu 7, HNK-1)[f]	+S	+S	+S, L	+L	−
CD64 (FcRI)[d]	−	−	−	+	−
CD69	+A	+A	+A	+A	+A
CD71 (transferrin R)	+A	+A	+A	+	+
4F2[g]	+A	+A	+A	+A	+

[a]MO/M, macrophages–monocytes; G, granulocytes; + or − indicates positive or negative, respectively, for a given marker; R, receptor; S (subset) indicates that only a subset is +; L (low) indicates that + cells express low amounts of antigen and are dull by immunofluorescence; A (activated) indicates markers present only on activated cells; CD stands for "cluster of differentiation," as agreed upon by the four International Workshops on Human Leukocyte Differentiation Antigens (Paris, Boston, Oxford, and Vienna, 1985–1989). At present, 79 CDs are categorized by preferential although not usually exclusive expression into T, activation antigens, B, myeloid, NK and nonlineage specific, and platelet. Leu, NKH, and HNK name various individual monoclonals, which in most cases defined the prototypic antigen in the CD and are still used in the older as well as some of the present literature. A CD is defined by more than one antibody reacting to the antigen. CD3, CD4, and CD8 are included as part of the T and non-T controversy discussed in the text. All are related to T-cell function. CD4 is also the receptor for HIV and is expressed on macrophages in humans and rats, but not mice.

[b]CD2 defines the receptor for the LFA-3 (CD58) antigen expressed on many cell types and involved in cell adhesion. Originally, it was characterized as the receptor involved in rosette formation with sheep erythrocytes (E receptor).

[c]CD11a, CD11b, and CD11c define the alpha chain of molecules, which are always expressed on the cell surface in association with a common beta chain (CD18).

[d]CD16, CD32, and CD64 define the three types of receptors for the Fc portion of IgG, which have different molecular weights and tissue distributions. Only CD16 (the low-affinity FcR, FcRIII) is detected on NK cells. In the 1989 Vienna Workshop, CD16 showed polymorphism with two possible alleles (NA1 and NA2), which until now have been studied only in granulocytes.

[e]CD56 is similar to the 140-kDa isoform of the N-CAM (neural cellular adhesion molecule) and is also expressed on neuroectodermal cells.

[f]CD57 defines a 110-kDa glycoprotein associated to myelin, and the monoclonal antibodies that define it are against the carbohydrate moiety of the molecule.

[g]4F2 is a well-characterized activation marker present in many cell types and has not been clustered because no new specific monoclonal antibodies have been submitted to the workshops.

tation of cytotoxicity and/or lymphokine production, when properly engaged by ligands or by specific antibodies directed against such structures, especially if cross-linked (i.e., CD2 with its internal ligand CD58/LFA-3 or anti-CD2 antibodies; CD11a/CD18, which form the LFA-1 heterodimer with its internal ligands related to ICAM-1; CD4 with its internal ligands related to Class II of MHC and CD8 with Class I of MHC, etc.). The general term of "adhesion" or "accessory" molecules to these types of surface structures denotes their functional importance in signal transduction, cell adhesion, cell locomotion, and recirculation.

B. NK Lineage

The issue of NK lineage can be viewed from two different angles. (1) If NK are actually a defined cell subset, the issue of lineage may have importance (see Section I,D); and (2) if NK are defined as an activity (killing of prototype targets) mediated by a variety of T- and non-T-effector cells, the lineage argumentation becomes less interesting. For example, in both patients and mice with primary immunodeficiencies, such as severe combined immunodeficiency (scid), NK activity is intact, arguing against a lymphoid lineage because scid produces a developmental arrest of both T and B lineages. However, this may also imply that in the absence of the NK-like T activity, only the NK non-T activity is detected. The same interpretation could be applied to the intact NK activity in nude mice and rats (the nude genetic defect produces abnormal differentiation of the thymus and a profound deficiency of classic T cells). Conversely, most of the mouse strains with low NK activity (see Section II,E) have normal T-cell function. In both mice and humans, NK cells are dependent on an intact bone marrow function and can be considered myeloid by definition. Both experimentally induced and genetically determined bone marrow dysfunctions are usually accompanied by low NK function. In humans and rodents, bone marrow contains inactive precursors, which can be induced *in vitro* to express NK activity after cultures with IL-2 (but not with IFN). Such marrow precursors are still not well defined. Another example of the lineage argumentation relates to lytic mechanisms. Around 1982, some studies described that populations enriched for NK cells showed production of superoxide radicals, which are characteristic lytic mechanisms of the professional macrophages (granulocytes and mononuclear phagocytes), prompting editorials with titles suggesting that NK cells were macrophages "in lymphocyte's clothing." However, it was later shown that the detected oxidative bursts were not due to the NK cells but to minor populations of contaminating macrophages.

C. Regulation of NK Activity

1. Augmentation

Short pulses (24 hr or less) of cell populations containing NK cells with the various types of interferons (α-, β-, and γ-IFN) augment lytic activity. Augmentation results from the combination of increased killing efficiency per cell, increased number of cells that bind and kill, and an increased efficiency in the recycling of the killers. NK activity can also be augmented after *in vivo* treatment with either IFN or IFN inducers such as poly I:C. The capacity of IFN to augment NK activity is variable depending on IFN type, because some forms of α-IFN are inactive. Short pulses with IL-2 also augment NK activity, and synergy between IL-2 and IFN can be shown. IL-2 is a growth factor for both NK and NK-like T cells, allowing the long-term growth of cell lines with NK activity. Although recombinant IL-2 has activity, suggesting a direct effect, an indirect role via induction of γ-IFN by the NK cells themselves may also be of importance in the augmentation. From these studies, a theoretical four-stage linear model for development of NK activity is accepted: (1) an undefined progenitor (in marrow), which does not bind nor kill and may or may not have a distinctive surface phenotype (there is controversy about CD57 being such a putative marker); (2) a recognizable precursor, or pre-NK, which binds but does not kill and may bear some distinctive surface markers; (3) the NK effector cell, which binds and kills, may have LGL morphology and the distinctive combination of surface markers; and (4) the activated NK (also known as augmented or boosted) discussed earlier.

2. Inhibition

Inhibition is a less-defined area, and the mechanism of inhibition remains to be defined. Prostaglandins (PG) are potent down-regulators of both spontaneous and IFN-augmented NK activity (especially PGE_2, PGA_1, and PGA_2, with no effect by the PGB and PGF series). A variety of *in vivo* procedures in mice seem to generate cells that can suppress NK activity. The procedures include treatments with various immunoadjuvants [*Corynebacterium parvum*, bacillus Calmette-Gerin (BCG), etc.] and with some anticancer treatment modalities (whole-body irradiation, corticosteroids, adryamycin). The NK suppressor cells are not well characterized but include macrophage-like (perhaps producing PG, as seen in *C. parvum* or steroid models) and nonadherent "null" cells (as seen in the irradiation and BCG models and, by some laboratories, in the *C. parvum* model). The human studies are more limited, but evidence indicates that pulmonary alveolar macrophages can produce a dose-dependent inhibition of blood NK activity in *in vitro* mixing experiments.

3. Interleukin-12 (IL-12)

Initially described as CLMF (cytotoxic lymphocyte maturation factor) or NKSF (NK stimulatory factor), the gene for IL-12 was cloned and expressed in rodents and humans. IL-12 is a unique cytokine heterodimer formed by two chains (p40 and p35 subunits). As with other cytokines, IL-12 has pleiotropic activities that include: (1) increase proliferation of activated NK and T cells, for T cells it promotes especially the development of CD4+ TH1 helper cells; (2) augmentation of cytotoxic activity of NK, T, LAK, and macrophages; (3) increase production of γ-IFN by NK and T cells; (4) up-regulate the expression of many surface molecules and receptors; (5) inhibit IgE secretion; and (6) act as a synergistic factor for hematopoietic stem cells. IL-12 is produced by macrophage/monocytes and B cells in response to bacteria and intracellular parasites. A variety of *in vivo* studies in mice have shown that IL-12 may be an important factor in defense against parasites (Leishmania, toxoplasma, etc.) and some experimental tumors.

D. Mechanisms of NK Activity

1. Recognition, Receptors, and Target Antigens

a. Conjugates

LGL-NK cells can form stable conjugates after short-term incubation with prototypic targets such as K562 in humans and YAC-1 in mice, and a variable fraction of the targets will actually be killed as a consequence of such interaction. A conjugate is one target cell with at least one bound effector cell; usually conjugates include two to five effectors.

b. Receptors

The receptors utilized by NK and NK-like T cells, as well as the ligands recognized in the targets, are beginning to be defined, mostly in rodents. A structure termed NKR-P1 was first isolated from IL-2-activated rat NK cells and its gene cloned. NKR-P1 is a disulfide-bonded homodimer of two ~60-kDa chains and is a type II integral membrane protein with a single transmembrane domain and a relatively short cytoplasmic domain. Interestingly, the extracellular domain shows the features of a calcium-dependent C-type lectin domain, shared by the mannose receptor, the asialoglycoprotein receptor, CD23 CD72 (Lyb-2), and the selectins (CD62-E or endothelial selectin, CD62-L or leukocyte selectin, and CD62-P or platelet selectin; other acronyms such as ELAM-1, LECAM-1, and PADGEM or names such as "homing receptor" or MEL-14 have been applied to these molecules). The functional studies showing that antibodies against NKR-P1 could stimulate NK cells and induce degranulation and transduce phosphorylation signals strongly suggested that NKR-P1 was a receptor-like structure. Three murine NKR-P1 homologues have also been cloned and show 62–75% homology to the rat. One of the murine NKR codes for the NK 1.1 surface antigen, which was used extensively as a marker of NK cells in some strains of mice (see Section II,E,1,f for discussion of murine NK and NC cell phenotype). A human homologue, termed NKG2, has been cloned from activated NK cells and shows less homology to rodent NKR but otherwise similar structure, including the extracellular C-type lectin domain. In mice, an additional molecule termed Ly-49 with several isoforms and with a C-type lectin domain was described and will be further discussed in the following. Both the murine NKR and Ly-49 as well as CD69 (see Section II,A,2 on surface antigens of NK cells) have been mapped to the long arm of chromosome 6, and that region is termed the NK gene complex (NKC). A region that appears syntenic to the mouse NKC could be in the short arm of human chromsome 12, which encodes for NKG2 and CD69. Both the mouse and rat studies have shown that NKR-P1 can make deletional variants by alternate splicing of exons that maintain the same reading frame, and therefore could produce functional proteins. This mechanism could generate receptor diversity without recombination. Accordingly, NK cells could express more than one member of the NKR family with the possibility of heterodimer formation, expanding the receptor repertoire. This is further discussed in the next section on the ligands recognized by the NKR family of receptors.

The finding of the zeta chain of the CD3 complex in freshly isolated NK cells associated with CD16 suggested that it may participate in NK signal transduction. However, even the CD3+ NK-like T cells do not seem to use its TcR for mediating its NK activity. Using cloned NK (and NK-like T) cell lines, which in essence represent expanded unique effectors, a functional heterogeneity has been found when tested against panels of mitogen blasts, tumors, and/or virus-infected targets, with no clear pattern of reactivity being discerned. Using panels of T-cell mitogen-induced blasts as targets and NK clones from single individuals, a pattern of some selectivity for certain HLA-A and HLA-C alleles was detected that regulates recognition and susceptibility/resistance to killing. These and other studies have also suggested that the alloreactive NK activities of these clones are genetically regulated and may depend on the HLA regions that are involved in peptide binding. The main "protective" region of the HLA molecules (i.e., that confers resistance to NK lysis) appears to be located in the alpha 1/alpha 2 domains involved in the groove formation used to present peptide, yet it is still not clear whether the NK clones detect empty or occupied groove, although some experiments suggest that the resistance inhibitory signal is provided by peptide-occupied groove. This type of alloreactive NK specificity appears to be mediated by a different type of still undefined receptors, although some monoclonal antibodies detect receptor-like structures with molecular weights of 58–54 kDa for these allo-MHC-reactive NK clones. The 70-kDa dimer, termed Kp43 and detected in NK cells, a subset of gamma/delta T cells, and a subset of alpha/beta T cells that express CD8 and CD56 could be related to the NKR-P1 family.

In summary, NK cells may interact with targets not through a single type of receptor but by using several surface molecules, including adhesion molecules such as LFA-1, that interact with several corresponding ligands on the targets.

c. Antigens

There is agreement that NK cells can kill targets that lack MHC class I antigens and that NK lysis is often inhibited by the induced expression of MHC class I on the targets. This inverse relationship between expression of MHC class I antigens on target cells and sensitivity to lysis has been extensively described and the transfection of class I antigens seemed to support the hypothesis that NK cells could detect the "absence" of MHC (the so-called "missing self" theory) as an alternative recognition from the MHC-based recognition of the T system. Another theoretical possibility was that MHC could mask or produce interference with the recognition of the specific NK ligands on the targets. However, the inhibitory effect of MHC class I on NK lysis was not seen with all MHC types in mice and humans and, owing to the intimate contact model of lysis (see Section I,A,1), these findings suggested either that the expression of MHC disrupted the target recognition of the NKR-P1 ligands or that, conversely, MHC recognition delivered a negative signal to the NK cell. In mice this seems to be the case and the role of the receptor that recognizes MHC on the target and delivers a negative "no kill" signal to the NK cell is played by Ly-49, structurally related to NKR-P1 and clustered in the NK gene complex. Therefore, target susceptibility to NK lysis will depend on the specific ligand–receptor interaction of the activating receptors such as NKR-P1 and the inhibitory receptors such as Ly-49. The detection of various isoforms of Ly-49 suggests that it could have a broad repertoire of MHC ligands.

Although some target membrane products or antibodies against targets that can partially block killing have been described in humans and rodents, no clear candidates for target antigens have been characterized until recently. The definition of the NKR-P1 type of receptors as belonging to the family of C-type lectin domains suggested that the ligands could be carbohydrates associated to glycoproteins or glycolipids of the target cell membrane. This hypothesis had been put forward by the author in 1980, when it was shown that NK and NC killing could be inhibited with simple sugars. At that time it was proposed that the effector NK and NC cells "could have lectin-like receptors that recognize neutral sugars . . . which are part of complex glycoproteins or glycolipids of the membrane." This hypothesis was confirmed in 1994, when it was shown that the NKR-P1 ligands are oligosaccharides present in a variety of glycolipids and glycoproteins such as the GM2 ganglioside and Heparin I-S neoglycolipid. Interestingly, NK-resistant targets could be rendered susceptible to NK lysis by liposomes containing NKR-P1 ligands such as GM2 or Hep I-S.

2. Lytic Molecules

The basic mechanisms for cell lysis were mentioned in Section I,A and include granule exocytosis, pore formation, and contact-mediated nuclear fragmentation. A soluble mediator from rodent and human NK cells, termed natural killer cytotoxic factor (NKCF), was described in 1981 and studied in several laboratories but is still not characterized biochemically. On the other hand, perforin (or cytolysin) is a protein of 55–75 kDa contained in granules of MHC-restricted cytotoxic T cells and NK and LAK cells that can form cytolytic pore-like lesions on the target cells. Perforin has been cloned in mice and humans and shows some regions of homology with the pore-forming C9 protein of complement. Although perforin seems to be the main lytic molecules of NK cells, as demonstrated by the NK deficiency in the perforin knockout mice, the participation of other still undefined molecules cannot be ruled out. A variety of serine proteases have been described in the granules of cytolytic T cells (known collectively as granzymes), but their actual role in cell killing is still undefined.

E. NC and Other Effectors

1. NC Cells

During studies of specific CMC to anchorage-dependent nonlymphoid fibrosarcomas in mice, lymphoid cells from normal mice were found to kill syngeneic tumors, reminiscent of NK killing. With inbred mouse strains, syngeneic means genetic identity in members of the same strain; allogeneic means genetic disparity between members of two different inbred strains. The natural effector cells against the fibrosarcomas had some properties different fom those of NK (see the following) but shared the same major properties of the NK categories discussed earlier (no MHC restriction, etc.) and were designated natural cytotoxic, or NC cells, in 1978. The prototypic target line in the early NC studies was Meth-A derived from a chemically induced sarcoma, which grew as an adherent monolayer, and was replaced in 1980 by the WEHI-164 cell line, also derived from a chemically induced sarcoma but that grows in suspension and allows the use of standard ^{51}Cr-release assays. The breast cancer cell line BT-20 is the prototype for testing human blood cells with NC activity in ^{51}Cr-release assays.

The properties of murine NK and NC show a number of differences, which will be further discussed: (1) phenotype: NK appear different from NC; (2) regulation: NK are regulated mainly by IFN and IL-2; NC are not affected by IFN and regulated by IL-2 and IL-3; and (3) lytic mechanism: NK cells kill targets by producing a pore-forming perforin; NC cells kill targets by production of tumor necrosis factor. Only points (2) and (3) have been analyzed in human NK–NC comparisons.

a. Kinetics for Lysis

With Meth-A or WEHI-164 targets, it was apparent that the kinetics for lysis by NC was different from that by NK (and from MHC-restricted or unrestricted T cells): whereas NK and T produce a linear increase beginning at time 0, reaching maximal lysis at 4 hr, NC lysis showed a 4- to 8-hr lag followed by a linear increase, reaching maximal lysis at 18 hr. One explanation for the time gap is the capacity of the targets to repair the early lytic lesion. Such counterlysis is dependent on protein synthesis, can be reversed by protein-synthesis inhibitors, and is the usual basis of all the long-term killing assays, regardless of the actual effector cell (some NK-susceptible targets need long-term assays for killing by NK). Because NC cells kill targets by production of TNF in mice and humans, any TNF-producing effector cell, such as monocytes from mice and humans, will kill WEHI-164 in 12–18 hr, as will recombinant human or murine TNF. However, the lytic assays can be reduced to 4 hr by using WEHI-164 targets pretreated with protein-synthesis inhibitors. On the other hand, TNF kills targets mainly by inducing apoptosis. In other models of apoptosis (e.g., steroid-induced apoptosis of thymocytes), protein-synthesis inhibitors will inhibit apoptosis.

b. Target Type

The initial studies with murine NK were done using lymphoma or leukemia targets, whereas the identification of NC used nonlymphoid tumor targets, mostly fibrosarcomas. Although never intended as such, the resultant NK–lymphoid and NC–nonlymphoid association was propagated in the literature. It became so established that for a while investigators defined the type of natural effector cells as NK or NC based on the morphology or anchorage dependency of the target cells, or conversely they described "NK cells that also kill nonlymphoid targets" as if they represented a special category of effectors. However, neither morphology nor anchorage dependency determines the type of effector cell that may kill it. Concerning morphology and tumor type, when a number of murine lymphoid and nonlymphoid tumor targets were screened, they usually fell within three main categories: (1) resistant to killing by fresh cells, (2) sensitive to killing by AK cells of either NK or NC type (a small minority are resistant to AK cells), and (3) sensitive to killing by fresh NK or NC (this being a minority and representing mainly the prototypic tumors). Most of the nonlymphoid tumors tested fall within the first two categories (resistant to fresh NK or NC and killed by AK). This statement applies to established cell lines; targets prepared from fresh tumors are usually resistant to fresh cells and only a fraction of them can be killed by AK cells. A similar picture has been seen with human tumor targets. In mice, the *in vivo* passage of NK- or NC-susceptible tumors through a syngeneic host produces variants that become NK- or NC-resistant when tested again *in vitro*. Resistance can be reversible (target cells recover susceptibility after a few weeks in culture) or permanent (the resistant phenotype remains regardless of time in culture). These types of fluctuations suggest *in vivo* selection toward NK or NC resistance that may include modulation of the susceptible phenotype (the cells become resistant but recover susceptibility *in vitro*) or deletion of the susceptible cells (producing the irreversible resistant variants). Furthermore, resistant variants of TNF-sensitive tumors can be produced *in vitro* and after *in vivo* passage. These are stable variants, suggesting deletion of the susceptible cells.

c. Recognition, Receptors, and Conjugates

The target ligands being recognized by NC are still undefined. Mouse NC form conjugates with their targets and usually kill about 50% of the targets in the conjugates, and TNF mRNA can be detected by *in situ* hybridization on the effector cells. Killing by NC cells can be selectively inhibited by D-mannose, suggesting that these cells also utilize a lectin-like receptor that recognizes mono- or oligosaccharides associated to glycoproteins or glycolipids of the target cell membrane.

d. Ontogeny and Maintenance

Murine NK and NC cells differ in their ontogeny and in their persistence in adult and aged animals. Whereas NK activity in spleen develops at approximately 3–4 weeks of age and declines to low levels in adult mice, splenic NC is present since birth and remains constant for the whole life span of the animal. However, no decline of NK activity is observed in murine peripheral blood lymphocytes. NK in humans is detected in blood since birth and is maintained throughout life, although blood NK levels vary widely, even in serial testings of the same individual within a given time. There are no similar studies done in humans for NC activity.

e. Genetics

In mice, marked differences in strain distribution of activity have been described for NK and NC cells, although in both cases strong influences from a given region (the D-end) of the murine MHC have been detected. A variety of non-MHC-linked genes also seem to affect NK and NC cells. A number of strains show low NK with normal NC activity and can be divided into two groups: (1) strains with low NK activities, which are augmentable (although not totally normalized) by IFN and IFN inducers, and (2) mouse strains with low or undetectable NK activities, which are not augmented by IFN and IFN inducers. Several strains are the prototype of the augmentable NK deficiency, including mice homozygous for the beige mutation, which produces among other effects a generalized lysosomal defect and low NK activity (this is the homologue to the human Chediak-Higashi syndrome, which is also accompanied of low NK activity). Some human examples of nonaugmentable NK deficiencies are beginning to be described.

f. Surface Phenotype

One difference between murine NK and NC cells is that while a variety of cell-surface markers are detectable on NK cells, NC cells were consistently negative for those markers as defined by negative selection with antibodies and complement. Thus, NC cells were "null" cells until 1986, when it was shown that NC activity in murine spleen could be enriched in factions that were positive for a variety of surface markers using positive selection with the FACS. The most used surface markers in murine NK studies are: Thy 1 (expressed on most T cells; the human analog is not expressed on T cells; about 50% of murine resting NK cells are Thy 1+; most of the murine AK cells are Thy 1+), Qa 5 (not characterized biochemically, no human equivalent defined; all resting and activated NK are Qa 5+, a fraction of T cells are also Qa 5+), NK 1.1 (expressed on NK cells; claimed to be a specific marker for NK, although recent studies have shown the existence of CD3+ NK 1.1+ cells, not defined biochemically and no human equivalent), and the ganglioside asialo GM-1 (aGM1; present on NK and T cells). NK 1.1 detects a small fraction of splenic cells (1–5%), where most of the NK activity is found. Comparing cloned NK with cloned MHC-restricted T cells, it is apparent that, in addition to having a similar LGL morphology, both types of clones expressed Thy 1, Qa 5, and NK 1.1, the main difference being that the T clones expressed CD8, whereas the NK clones did not. By FACS analysis, it was shown that NC cells in murine spleen were enriched in the fractions that were positive for Thy 1 (some activity in the Thy 1−), Qa 5, and CD4 and the fractions that were negative for CD8 and surface Ig (sIg), and that both the NK 1.1+ and NK 1.1− fractions expressed NC cells. Thus, by single-parameter FACS analysis, the murine splenic NK phenotype is Thy 1+ or 1−, Qa 5+, NK 1.1+, CD4−, CD8−, sIg−, and aGM1+, whereas the NC phenotype is Thy 1+ (or 1−), Qa

5+, NK 1.1+ or 1.1−, CD4+, CD8−, and sIg− (aGM1 not tested). Both NK and NC cells were enriched in the larger-size fractions as defined by forward light scatter. The conclusion of the NC studies was that NC activity was mediated by cells from various lineages including T and non-T. Two main differences regarding surface phenotype between NK and NC cells are as follows. (1) Although expressing surface antigens, the cells with NC activity appear refractory to negative selection with antibody and complement; and (2) NC activity is detected in a CD4+ T-cell subset not observed in NK (in mice, CD4 is expressed only on T cells; in humans and rats it is also expressed on monocytes and macrophages).

g. Regulation

Cells with NC activity are augmented by short pulses with IL-2 and interleukin-3 (IL-3) and not by any type of IFN in both mice and humans, whereas NK cells are augmented by short pulses with IFN and IL-2 (see Section II,C).

h. Lytic Mechanisms

NC cells in mice and humans kill targets by production of TNF, a well-defined cloned cytokine that is different from the perforin utilized by NK and T cells (see Section II,D). In addition to the cytotoxic–cytostatic activities, TNF has a wide range of other activities in inflammation and immunological regulation (in this latter category, TNF functions overlap with those of interleukin-1).

In summary, NC cells differ from NK in phenotype, regulation, targets for recognition, and lytic mechanisms, although they share with NK the properties of being independent of MHC restriction and detected in fresh samples of blood or spleen.

2. NK Cells That Kill Virus-Infected Targets

Human herpes simplex virus (HSV)-infected fibroblasts become sensitive to lysis by a CD16+ NK-like effector, which shows some minor differences from the K562 killers although it shares most other properties with classic NK. The differences include separate variation of activity (low NK-K562 and normal NK-HSV, or the reverse situation in patients with the Wiscott–Aldrich syndrome) and the requirement for accessory cells for the lysis of the HSV-infected targets, not required by the NK-K562 (the requirement of accessory cells expressing HLA-DR of the MHC has also been described for the lysis of cytomegalovirus-infected fibroblasts by CD16+ NK cells). A somewhat similar system has been described in the mouse against HSV-infected fibroblasts. With other virus-infected targets, such as murine hepatitis virus, natural killing can be mediated by non-NK cells, including B cells, which kill by direct contact. [*See* Herpes Viruses.]

3. AK Cells, Including LAK

As mentioned in Section I,D, a variety of NK-like effector cells have been produced after *in vitro* culture or activation and have been grouped under the AK category. The common AK characteristic, regardless of the procedure used for its generation, is that is usually kills a broader spectrum of targets than the resting NK cells, in some cases including tumor cells obtained from fresh explants in addition to tumor lines. AK cells have been produced by a variety of procedures, including short-term cultures in the presence of heterologous serum (such as fetal calf serum), allogeneic irradiated cells (as in the mixed lymphocyte response), or addition of crude or purified lymphokines such as IL-2. The earlier studies generated a complex terminology with words such as "anomalous" and "promiscuous" for the killer cells detected in culture. Lymphokine-activated killers are among the better-studied AK cells because of their potential use in cancer therapy. LAK are generated from blood lymphocytes after 3–7 days in culture with high dosages of recombinant IL-2 and the phenotype of the progenitors (i.e., the cells at the time of the initiation of the culture) and the effector LAK generated after the culture is identical: mainly associated with LGL, mainly CD16- and CD56-positive, and mostly CD3-negative. It is agreed that LAK is a phenomenon rather than a distinct effector lineage and that in both humans and rodents most of the LAK activity in blood (or spleen in mice) is attributable to IL-2 activated and -expanded NK cells. The contribution of NK-like T cells to the LAK compartment is variable and depends on the type of cells used to initiate the LAK cultures. For example, culture of human thymocytes for 5–8 days in recombinant IL-2 would generate LAK cells that will express CD56 but lack CD16 and CD57 and are mostly CD3−, suggesting derivation from the more immature CD3− thymocyte populations. In cultures of peripheral blood cells with IL-2 and anti-CD3 for up to 20 days, the cells with LAK activity were shown as either CD3+ CD16− NK-like T cells or CD3− CD16+ NK; these studies also showed that the conventional CD3+ CD4 or CD8+ T cells do not seem to contribute to LAK formation, suggesting the T-like LAK progenitor in the CD3+ CD4− CD8− minor subset of blood T cells. The murine studies also support the view that the T-like LAK activity is derived mainly from the CD4− CD8− subset. IL-4, a different lymphokine with various effects on B and T cells, can also induce LAK cells from mouse spleen and synergize with IL-2, and TNF can synergize with IL-2 in LAK generation. Such interactions could be predicted from lymphokine studies in nonlytic systems.

4. What Determines Target Killing?

A puzzling observation is that in spite of the fact that coculture of NK cells with an NK target such as K562 induces the production of both perforin and TNF by the effectors, it is clear that K562 under those conditions is exclusively killed by perforin. Similar examples with T-like NK clones that produce various lytic molecules but use only perforin to kill the NK-susceptible target have been described. Therefore, it is tempting to speculate that to some extent it is the target itself that decides, via a still undefined mechanism, its preferred way of dying. Another interpretation would be that killing mechanisms are exclusive and once the pore-forming killing is initiated it prevents the delivery of other more slowly acting apoptotic killers like TNF. Therefore, it is worth stressing that all of these *in vitro* studies always imply the interaction between the killer and the victim cell.

III. FUNCTION

Functions of NK and other effector cells have been deduced based on the effects of induced or genetic deficiencies of NK activity *in vivo* and in some *in vitro* tests where NK cells have been shown to modify some type of response by another cell type. Most of the studies have been done in experimental animals and at present writing the functions can only be inferred for NK.

The following functional properties of NK can be considered: (1) *Antitumor effects*—good experimental evidence for a role of classic NK in the control of bloodborne metastases, suggestive evidence that NK may influence local growth of primary tumors, and no clear evidence that NK may act as an immunological surveillance mechanism preventing tumor development (i.e., NK deficiencies do not show increased risk for tumor development). (2) *Control of infections*—in addition to the direct lysis of some types of bacteria, NK cells have been implied in resistance to some viral infections (cytomegalovirus, herpes simplex, murine hepatitis, etc.), some parasitic infections (by *Plasmodia, Babesia, Toxoplasma, Giardia,* etc.), some fungal infections (*Candida*), and some bacteria (such as *Brucella* and *Listeria*). It is worth noting that resistance to infections is multifactorial, rarely attributable to a single defense mechanism, and probably is but a component in the response. (3) *Immunoregulatory functions*—of two types: (i) Lymphokine production: The production of IL-1, γ-IFN, and IL-4 by NK cells is well documented; and there are reports of production of IL-2, IL-3, α-IFN, CM-CSF, and TNF by CD3-LGL, which depend on the triggering stimulus used. The production of TNF by NC cells, combined with the pleiotropic nonlytic immunological effects of TNF, could be another example of possible immunoregulation via lymphokine production. And (ii) "direct" effects by cells: The regulation of immune responses by NK cells is supported by some examples where NK-like cells affect antigen presentation or simply down-regulate a response; in the latter case the regulatory cells have been termed "natural suppressors." (4) *Control of hematopoietic stem-cell growth and differentiation*—NK cells have been shown to directly affect the differentiation of lineage-committed stem cells *in vitro,* such as colony-forming units of the granulocyte–macrophage lineage. NK cells can also influence hematopoiesis by production of cytokines. (5) *Transplant rejection*—NK-like cells are involved in the rejection of bone marrow transplants in mice and probably in humans. In mice, NK cells appear to mediate the phenomenon of hybrid resistance. NK also seem to be involved in the development of graft-versus-host disease in mismatched marrow transplantation. The detection of cells with NK markers in organ allograft rejections has also suggested a possible role in organ transplant rejection. (6) *Disease*—the evidence of a possible role in hematological and other diseases is only circumstantial and is mentioned for the sake of completeness. [*See* Immune Surveillance.]

In summary, the participation of NK and related cells in complex multifactorial responses such as the response to an infection or other parasites cannot be denied. However, we must keep in mind that NK cells are just one component, perhaps the first line of defense, but still only one component in a complex cascade of specific and nonspecific immunological and inflammatory events. Therefore, to focus on a given activity, such as γ-IFN production or selected target killing, is a useful approach to define individual mechanisms, but only a stepping-stone to formulate general biological theories. It is probable that, as most of the "immune surveillance" theories suggest, the immune system including the NK branch evolved as a defense system against infections by parasites, bacteria, viruses, and so on, rather than as a defense against tumors or allogeneic bone marrow transplants. The NK responses to tumors and allogeneic bone marrow transplants are certainly worth studying, but probably are by-products of the "natural" functions of the NK and related cells to attack infectious enemies.

IV. RECAPITULATION

These final comments apply to the NK and other effector cells described in Section II,E.

What are NK cells? For the class NK, the definition is reduced to a subset that shares phenotypic properties with both T and myeloid (monocyte–macrophage and granulocyte) lineages and is defined by a combination of markers. The questions about NK as a unique cell subset versus NK as an activity shared by different cell types, including the NK-like non-MHC-restricted T cells and the activated killers generated in culture, are still open. However, NK and AK cells are bone marrow dependent for their ontogeny and maintenance *in vivo* and do not behave as conventional T, B, or macrophage cells.

What do NK cells recognize? The characterization of strong candidates for receptors on NK cells suggests that two main surface structures on the targets are the ligands for such receptors: first, mono- and oligosaccharide components of glycoproteins and glycolipids of the membrane, whose recognition is mediated by lectin-like domains on the receptor molecules; and second, the complex recognition of MHC class I regions, which can give both an activation or a negative signal to the effector cells that favors or prevents killing. This latter area is beginning to be better defined.

How do NK cells work? Concerning the lytic activity, NK, NK-like T, and probably LAK cells kill the targets by production of perforin, a pore-forming lytic molecule. NC lytic activity is mediated by TNF, a lytic molecule that produces cell death mainly by apoptosis. The other potential functions, such as γ-IFN or other cytokine production, are beginning to be defined.

What do NK cells really do? Section III gives a listing of normal and pathological situations where NK and NK-like mechanisms may have an *in vivo* role. Of all the potential functions described, the most "physiological" seems to be those related to cytokine production that may be involved in regulation of hematopoiesis and immune reactions. On the other hand, as a defense system, NK has unique qualities (discussed in Section I,E) to act as the first line of defense against infectious invaders and circulating tumor cells. The understanding of the biological *in vivo* significance of NK and related cells will open possible therapeutic options applicable to infections and cancer.

BIBLIOGRAPHY

Bezouska, K., Yuen, C. T., O'Brien, J., Childs, R. A., Chai, W., Lawson, A. M., Drbal, K., Fiserova, A., Pospisil, M., and Fiezi, T. (1994). Oligosaccharide ligands for NKR-P1 protein activate NK cells and cytotoxicity. *Nature* **372,** 150.

Herberman, R. B. (ed.) (1980). "Natural Cell-Mediated Immunity against Tumors." Academic Press, New York.

Herberman, R. B. (ed.) (1982). "NK Cells and Other Natural Effector Cells." Academic Press, New York.

Herberman, R. B., and Callewaert, D. M. (eds.) (1985). "Mechanisms of Cytotoxicity by NK Cells." Academic Press, Orlando, Florida.

Kägi, D., Ledermann, B., Bürki, K., Seiler, P., Odermatt, B., Olsen, K. J., Podack, E. R., Zinkernagel, R. M., and Hengartner, H. (1994). Cytotoxicity mediated by T cells and natural killer cells is greatly impaired in perforin-deficient mice. *Nature* **369,** 31.

Lattime, E. C., Stoppacciaro, A., Khan, A., and Stutman, O. (1988). Human natural cytotoxic activity mediated by tumor necrosis factor: Regulation by interleukin-2. *J. Natl. Cancer Inst.* **80,** 1035.

Moretta, L., Ciccone, E., Poggi, A., Mingari, M. C., and Moretta, A. (1994). Ontogeny, specific functions and receptors of human natural killer cells. *Immunol. Lett.* **40,** 83.

Nelson, D. S. (ed.) (1989). "Natural Immunity." Academic Press, Sydney.

Stutman, O., and Lattime, E. C. (1986). Natural killer and other effector cells. *In* "Immunology and Cancer, M. D. Anderson Symposium on Fundamental Cancer Research" (M. L. Kripke and P. Frost, eds.), Vol. 38. Univ. of Texas Press, Austin.

Stutman, O., Dien, P., Wisun, R. E., and Lattime, E. C. (1980). Natural cytotoxic cells against solid tumors in mice: Blocking of cytotoxicity by D-mannose. *Proc. Natl. Acad. Sci. USA* **77,** 2895.

Trinchieri, G. (1989). Biology of natural killer cells. *Adv. Immunol.* **47,** 187.

Yokoyama, W. M., and Seaman, W. E. (1993). The Ly-49 and NKR-P1 gene families encoding lectin-like receptors on natural killer cells: The NK gene complex. *Annu. Rev. Immunol.* **11,** 613.

Naturally Occurring Toxicants in Food

STEVE L. TAYLOR
University of Nebraska

GLOSSARY

Hazard Capacity of a chemical substance to produce injury under the circumstances of exposure; since all chemicals are toxic, the dose and duration of exposure and other factors are critical determinants of a chemical's injurious potential

Infection Illness caused by exposure to a viable, pathogenic microorganism

Intoxication Illness caused by exposure to hazardous or injurious quantities of a chemical toxicant

Mycotoxin Toxin produced by molds that naturally contaminate foods and feeds

Naturally occurring toxicant Toxicant that is produced or exists in nature; not artificial or synthetic

Toxicant Chemical that has the potential to produce injury under probable circumstances of exposure

Toxicity Inherent capacity of a chemical substance to produce injury; the central axiom of toxicology is that all chemicals are toxic; the dose and duration of exposure determine the likelihood of injury or harm

FOODS CONTAIN AN ARRAY OF NATURALLY occurring toxicants, chemicals of natural origin that may present a hazard to consumers when exposed under reasonable circumstances of exposure. These naturally occurring toxicants can be typical constituents of the food or natural contaminants of the food. The naturally occurring contaminants include seafood toxicants derived from algae, mycotoxins derived from molds growing on foods, bacterial toxins produced by certain foodborne bacteria, and insect-produced toxicants. Fungi (mushrooms) and plants have the greatest numbers of toxic, naturally occurring constituents. Many naturally occurring toxicants, especially of plant origin, are removed or destroyed by processing, but inadequate processing can create hazardous situations. Other naturally occurring toxicants are insufficiently toxic to affect consumers eating typical diets, but these toxicants may pose risks if the foods that contain them are eaten in abnormally large amounts.

I. INTRODUCTION

The central axiom of toxicology is that all chemicals are toxic at some dose. Thus, all chemicals in foods, both synthetic and naturally occurring, are toxic. Though theoretically correct, it is impractical to view all food constituents as toxicants. For the purposes of this review, naturally occurring toxicants are those chemicals of natural origin in foods that may present a hazard to consumers when eaten under some reasonable circumstances of exposure. Although toxicity is an intrinsic property of all chemicals, hazard is the capacity of a substance to produce injury under the circumstances of exposure, taking into account the dose and frequency of exposure as well as the toxicity of the particular chemical. Illnesses caused by exposure to hazardous levels of chemical toxicants are called intoxications.

A. Definition of Natural

No legal definition exists for the word "natural" in United States food laws or the food laws of most other countries. Thus, there are no legal limitations on the use of the term. Webster's dictionary defines natural as "planted or growing by itself: not cultivated or introduced artificially (e.g., grass); existing in or produced by nature: consisting of objects so existing or produced: not artificial." Most consumers would likely define natural as being anything that is not artificial. Of course, artificial is another term that defies definition. Foods derived from animal or plant materials, whether processed or raw, are considered to be natural. The addition of additives to foods or the synthesis of a food from chemical sources would create a food that is perceived as not natural. To the food industry, additive-free or preservative-free foods are often described as natural. Naturally occurring chemicals from foods would be those chemicals that exist in foods in the raw, unprocessed state, although many of those chemicals would not be altered by processing or preparation practices.

B. Sources of Naturally Occurring Toxicants in Foods

Naturally occurring toxicants in foods can be either naturally occurring constituents of foods or contaminants that are produced in foods by various natural processes. Whereas naturally occurring constituents can be considered to be normal and unavoidable, naturally occurring contaminants are not always present and can be avoided, if contamination is prevented.

Naturally occurring toxicants are present in foods of animal, plant, or fungal origin. The majority of foods of animal origin, including meat, milk, cheese, eggs, fish, molluscs, and crustacea, are not hazardous. The most hazardous chemicals occurring in this category are found in poisonous seafoods. Foods of plant origin include vegetables, fruits, grains, seeds, nuts, and spices. Most plant-derived chemicals are also not especially hazardous, although some have the potential to be hazardous under some conditions of exposure. The major food of fungal origin is mushrooms, many of which are hazardous.

Naturally occurring contaminants can be formed in foods as the result of contamination by bacteria, molds, algae, and insects. The chemicals produced from these biological sources can remain in the food even after the living organism has been removed or destroyed. Such contaminants represent the most important and potentially hazardous chemicals of natural origin existing in foods.

C. Additives versus Natural

Consumers are often concerned about the chemicals added to their foods. These concerns likely arise from government regulatory actions against certain food additives or incidental contaminants such as pesticides. In contrast, natural products are widely viewed as beneficial to health.

U.S. food laws are predicated on the assumption that natural foods are safe, whereas most additives, especially all newly developed ones, must be proven to be safe. Although foods naturally contain many toxicants, including many chemicals that have been shown to be carcinogens in animal studies, government regulatory agencies have rarely taken any action to limit the availability of natural foods as a result of the presence of such toxicants. Meanwhile, the government has banned several food additives (cyclamate, FD&C Red No. 2) and pesticides (DDT, ethylene dibromide) because of concerns about their safety, especially with respect to their possible carcinogenicity. Given this contrast, it is not surprising that consumers have developed a suspicious attitude regarding chemicals added to foods while remaining naive about the potential hazards presented by naturally occurring chemicals in foods.

Although unrecognized by many consumers, naturally occurring chemicals in foods present the greatest risk of any of the chemicals in foods. Thankfully, most of these potentially hazardous naturally occurring toxicants are ingested at doses that are unlikely to cause widespread illness, although the margin of safety is often lower for naturally occurring toxicants than it is for food additives. For example, the majority of the carcinogens in our diets are of natural origin. However, our diets also contain many naturally occurring compounds that have been shown to have anticarcinogenic activity in experimental animals and may also demonstrate similar activity in humans.

Agricultural biotechnology involves the genetic transfer of desirable traits, such as pest resistance, improved quality, and so on, from one species to another. Using biotechnology, specific traits can be targeted and transferred with great precision. Biotechnology can also be used to eliminate undesirable traits such as the production of naturally occurring toxicants. Obviously, biotechnologists must guard against any transfer of genes controlling the development of naturally occurring toxicants along with the genes coding for the desired trait.

II. HAZARDS FROM NATURALLY OCCURRING TOXICANTS

Many naturally occurring, potentially hazardous chemicals exist in foods. These naturally occurring toxicants include acute toxicants (those capable of eliciting symptoms within a few minutes to hours after ingestion) and chronic toxicants (those that may cause symptoms after life-long exposure or many years after exposure). Their ingestion does not invariably cause illness because the dose and other circumstances of exposure are key factors. Chronic illnesses, such as cancer, are particularly difficult to correlate with foods or specific foodborne toxicants since many confounding variables exist that affect the onset and course of such chronic diseases. Many naturally occurring carcinogens are known to exist in foods, but their relationship to cancer in humans remains unclear. A few of these carcinogens will be discussed. With acute toxicants, the relationship between ingestion of the toxicant and the development of symptoms is much stronger. Thus, many of the best examples of naturally occurring toxicants in foods involve acute illnesses.

The Centers for Disease Control keep records on the incidence of foodborne disease. These statistics focus on acute illnesses, because the role of foods in chronic illnesses is uncertain. The most common foodborne illnesses are bacterial infections such as salmonellosis. Toxins produced by bacteria such as the staphylococcal enterotoxins and the botulinal toxins are important in the pathogenesis of some foodborne bacteriological illnesses. Illnesses of chemical etiology, other than bacterial toxins, account for 20–30% of all foodborne disease outbreaks in the United States. Fish and shellfish toxins account for the majority of these outbreaks, although other natural toxicants such as poisonous plants and mushrooms are also involved. The total number of foodborne disease outbreaks, including those of chemical etiology, is unknown because the reporting of such illnesses to the Centers for Disease Control is very incomplete.

As noted earlier, the circumstances of exposure affect the likelihood of intoxication resulting from naturally occurring chemicals in foods. Table I describes some of the circumstances in which naturally occurring chemicals in foods can cause foodborne intoxications.

TABLE I
Intoxications from Consumption of Naturally Occurring Foodborne Toxicants

1. Abnormal though natural contaminants that adversely affect normal consumers eating normal amounts of the food
    - Algal toxins in seafood
    - Staphylococcal enterotoxins in various foods
    - Mycotoxins in various foods
2. Unusual "foods" that adversely affect normal consumers eating normal amounts of this "food"
    - Poisonous mushrooms
    - Poisonous plants
    - Poisonous fish such as pufferfish
3. Normal constituents of food that can cause illness if ingested by normal consumers in abnormally large amounts
    - Cyanogenic glycosides in lima beans, cassava, and fruit pits
    - Glycoalkaloids in potatoes and tomatoes
    - Estrogens in soybeans
    - Goitrogens in cruciferious vegetables
    - Nitrate in spinach, celery, and lettuce
    - Oxalates in rhubarb and spinach
    - Saponins in soybeans
    - Tannins in many plant foods
    - Phytoalexins in many plant foods
4. Normal components of foods that, consumed in normal amounts, are harmful to individuals with genetic abnormalities
    - Food allergies
    - Lactose intolerance
    - Celiac disease
5. Normal foods processed or prepared in an unusual manner and consumed in normal amounts by normal consumers
    - Lectins in underprocessed kidney beans
    - Trypsin inhibitors in underprocessed legumes

A. Naturally Occurring Contaminants

As noted in Table I, the sources of naturally occurring contaminants in foods can include algae, molds, bacteria, and insects.

1. Algal Toxins in Seafoods

Seafoods, both fish and shellfish, can occasionally be contaminated with toxic substances that cause acute illnesses and even death within minutes after ingestion. These toxins are acquired by fish and shellfish through the food chain after being produced by certain species of microscopic, dinoflagellate algae. The fish and shellfish are hazardous only when feeding on toxin-producing algae. Seafood poisonings were once considered a public health problem only in coastal areas, but modern practices involving the shipment of fresh and frozen seafoods over long distances have made these a cosmopolitan concern.

Ciguatera poisoning is the most common cause of foodborne disease of chemical etiology reported to the Centers for Disease Control. Ciguatera poisoning is common throughout the Caribbean and much of the Pacific. Fish that habitate reef and shore areas in temperate regions, such as barracudas, groupers, sea basses, red snappers, and eels, are most commonly implicated in ciguatera poisoning. The identity of all the species of dinoflagellate algae associated with ciguatera poisoning remains unknown, although *Gambierdiscus toxicus* is one of them. Small reef fishes feed on the toxic dinoflagellate algae and are, in turn, consumed by the larger reef fishes. The toxins accumulate in the liver and viscera, but enough can enter the muscle tissues to result in ciguatera poisoning among humans ingesting these fish. Ciguatera poisoning is a neurologic illness initially characterized by tingling of the lips, tongue, and throat followed by numbness. Nausea, vomiting, a metallic taste, dryness of the mouth, abdominal pain, diarrhea, headache, and general muscular pain will ensue in many cases. Weakness may progress until the person is unable to walk. No antidotes are known, but many affected individuals will recover within a few weeks, although deaths from cardiovascular collapse have occurred in a few instances.

Paralytic shellfish poisoning occurs in many coastal areas worldwide, including the Pacific and North Atlantic coasts of North America, Japan, and southern Chile. Many species of shellfish, such as clams and mussels, become poisonous through the consumption of toxic dinoflagellate algae, two of the most common incriminated in paralytic shellfish poisoning being *Gonyaulax catenella* and *G. tamarensis*. The blooms of the toxic dinoflagellates are quite sporadic so shellfish will be hazardous only at certain times. Most shellfish clear the toxins from their system within a few weeks after the end of a dinoflagellate bloom, but some shellfish species, especially the Alaskan butter clam, appear to retain the toxins for long periods. The toxins involved in paralytic shellfish poisoning are known as saxitoxins. These are neurotoxins that act by blocking the passage of sodium ions into the nerve cells, an essential process in nerve transmission. The symptoms of paralytic shellfish poisoning include a tingling sensation and numbness of the lips, tongue, and fingertips, followed by numbness in the legs, arms, and neck with general muscular incoordination. Respiratory distress and muscular paralysis often occur. Death from respiratory failure can occur within 2–12 hours depending on the dose. If the victim survives 24 hours, the prognosis for a full recovery is good. No antidote exists for paralytic shellfish poisoning.

Pufferfish poisoning occurs more rarely because pufferfish are not frequently consumed except in Japan and China. About 30 species are found worldwide, but most are not poisonous. The most poisonous species occur along the coasts of Japan and China. The most choice, edible species belonging to the genus *Fugu* are the most poisonous and the most commonly consumed. The toxin in pufferfish accumulates in the ovaries, liver, intestine, skin, and roe (egg sac), so a carefully cleaned and eviscerated pufferfish may be safe to eat. Pufferfish likely become toxic through the ingestion of toxic dinoflagellate algae, although the food chain relationships have not been completely elucidated. The toxin involved in pufferfish poisoning is a potent neurotoxin called tetrodotoxin, which also acts by blocking the sodium channel in nerve cells. The symptoms resemble those of paralytic shellfish poisoning. Death can result if a sufficient dose of tetrodotoxin is ingested.

2. Mycotoxins

Mycotoxins are produced by a wide variety of molds that can grow on a wide variety of foods. The toxicity of many of the mycotoxins has been recognized by the observation of domestic animals fed moldy animal feeds. Their effects on humans are not so clearly established, although they are potentially hazardous to humans as well. [*See* Mycotoxins.]

Historically, ergotism was the first mycotoxin-associated illness recognized in humans. *Claviceps purpurea* is the responsible mold, which infects the heads of rye and sometimes wheat, barley, and oats. The shriveled, purplish grain kernel contains the mycotoxin. Ergotism is caused by a group of toxins known collectively as the ergot alkaloids and can be manifested in two forms: gangrenous ergotism and convulsive ergotism. In the former, also known as St. Anthony's fire, a burning sensation in the feet and hands is followed by a progressive restriction of blood flow to the hands and feet. This results in gangrene, which can lead to the loss of limbs. Convulsive ergotism involves hallucinations leading to convulsive seizures and sometimes death. No outbreaks of ergotism have been recorded since the early 1950s.

Aspergillus molds are known to produce several types of mycotoxins, most notably the aflatoxins and ochratoxin. The aflatoxins are produced primarily by *A. flavus* and *A. parasiticus* and often contaminate moldy peanuts and corn. Feeding of aflatoxin-contaminated grains or oilseeds to dairy cows can result in aflatoxin contamination of milk. The aflatoxins are potent carcinogens, especially affecting the liver. The role of aflatoxins in human carcinogenesis remains uncertain but they are among the most potent animal carcinogens known. Ochratoxin is produced primarily by *A. ochraceus,* which can contaminate cereals, peanuts, and tree nuts. Ochratoxin affects both the livers and kidneys of domestic animals, with the effects on the kidney being particularly damaging. The toxicity of ochratoxin to humans is unknown.

Fusarium molds produce a number of different mycotoxins, including the trichothecenes, fumonisins, and zearalenone. The trichothecenes are known to cause human illness and can be found as contaminants of grains primarily. Alimentary toxic aleukia (ALA) was observed in the former Soviet Union owing to consumption of grains containing trichothecenes. ALA has four stages. In the first stage, a person experiences a burning sensation in the mouth and throat, which proceeds down the esophagus to the stomach. This is followed by diarrhea, nausea, and vomiting, occurring 1 to 3 days later and ceasing after about 9 days. The second stage occurs from about 2 weeks to 2 months and involves bone marrow destruction, leukemia, agranulocytosis, anemia, and loss of platelets. At the end of this stage, small hemorrhages may occur on the skin. The third stage of ALA lasts from 5 to 20 days and involves total loss of bone marrow with necrotic angina, sepsis, total agranulocytosis, and moderate fever. The hemorrhages become larger and necrotic lesions appear on the skin. Bronchial pneumonia appears along with abscesses and hemorrhages in the lungs. The fourth stage is death, with the mortality rate approaching 80% and dependent on the dose and frequency of exposure to the trichothecenes.

Fumonisins are produced primarily by *F. moniliforme,* which contaminates various grains and soybeans. The fumonisins have been implicated in equine leukoencephalomalacia, a fatal neurotoxic syndrome in horses characterized by extensive necrosis of the white matter in the brain. The fumonisins may also be carcinogenic, although this remains an active area of scientific scrutiny. Their effects on humans remain unknown, although low-level contamination of grains with fumonisins seems to be fairly common. Zearalenone is an estrogenic substance produced in grains by *F. graminearum.* It causes spontaneous abortions in pigs, but its effects on humans are unknown.

Penicillium molds produce a wide variety of mycotoxins, including rubratoxin, patulin, and citrinin. These mycotoxins cause various effects in domestic animals but their effects, if any, on humans are quite uncertain. Many other mycotoxins have been identified, though most of these others have received little scientific study.

3. Bacterial Toxins

Most pathogenic bacteria exert their effects by the infectious route—they invade cells and tissues, multiply, and cause a variety of symptoms. A few bacteria produce exogenous toxins in foods before they are eaten. The ingestion of the toxins initiates the disease process even if the bacteria are destroyed in processing or preparation. The best examples of bacterial intoxications are the staphylococcal enterotoxins and botulinal toxins.

The staphylococcal enterotoxins can be produced in foods by certain strains of *Staphylococcus aureus.* When ingested, these protein enterotoxins cause nausea and vomiting within 1–6 hours. Staphylococcal food poisoning is one of the most common forms of foodborne disease.

The botulinal toxins can be produced in foods under anaerobic conditions by *Clostridium botulinum.* Toxin formation often occurs in canned foods subjected to improper thermal processing. The commercial canning process is predicated on the destruction of this organism and its spores so that the spores will not germinate, grow, and produce toxin on storage of the canned product. Botulinal toxins are among the most potent toxins known to humans and frequently result in respiratory paralysis and death.

4. Insect-Produced Toxins

Insects can also produce and secrete toxic chemicals into foods, yet insect infestation of foods is often considered only in aesthetic terms. Natural toxicants produced by insects in foods have received very little study. Flour beetles are known to secrete benzoquinones into flour, which are mutagenic and carcinogenic. The hazards posed by insect-produced toxicants in foods to humans remain unknown.

B. Naturally Occurring Constituents Eaten in Normal Amounts

Some plants and animals that should not be eaten are consumed intentionally or accidentally on occasion, resulting in foodborne chemical intoxications. Many plants and some animals contain levels of naturally occurring toxicants that are probably not hazardous to humans ingesting typical amounts of these foods.

1. Animals

Very few animal species are poisonous. Several species of poisonous fish exist, the best example being the pufferfish, although recent evidence suggests that the toxin in pufferfish arises from an algal contaminant.

2. Plants

Many plant species are decidedly poisonous. Plants such as hemlock and nightshade were used in ancient times to poison enemies. Consumers can easily avoid poisonous plants by purchasing foods from commercial sources. Intoxication from the ingestion of poisonous plants occurs primarily from misidentification of plants by individuals harvesting their own foods in the wild. The most frequent examples occur when harvesting herbs for herbal tea. For example, an elderly couple succumbed after mistaking foxglove for comfrey while harvesting herbs for tea.

Very rarely, intoxications from poisonous plants occur with products purchased from retail outlets. An example was the contamination of an herbal tea by *Senecio,* a well-known poisonous plant. The tea was fed to infants as an herbal remedy for viral infections. The infants suffered acute symptoms involving liver function from the presence of pyrrolizidine alkaloids in the *Senecio* leaves. Several infants died in this unfortunate incident.

Occasionally, intoxications from poisonous plants occur from the intentional addition of such materials to foods. The most frequently encountered example would be the preparation of brownies laced with marijuana, a potent hallucinogenic plant.

In times of severe food shortages, humans may resort to eating large quantities of plants that contain hazardous components. Lathyrism is an unusual disease that has occurred on such occasions in India and Italy when seeds of *Lathyrus sativus* or chickling vetch are ingested as a principal part of the diet. These seeds contain several lathyrogens, which cause both neurological and skeletal abnormalities when they are eaten in large amounts for an extended period.

3. Fungi

Many species of poisonous mushrooms are known. Harvesting mushrooms in the wild can be a hazardous practice even for those skilled in mushroom identification. Several different types of naturally occurring toxicants exist in poisonous mushrooms.

The most hazardous of the mushroom toxins are the Group I toxins. Amatoxin is the best example and is produced by *Amanita phalloides,* the so-called death cap mushroom. The symptoms of amatoxin poisoning begin 6–24 hours after ingestion of the mushrooms. The first stage of the intoxication involves the gastrointestinal tract with abdominal pain, nausea, vomiting, diarrhea, and hyperglycemia. A short period of remission follows. The third and often fatal stage of the illness involves severe liver and kidney dysfunction with symptoms including abdominal pain, jaundice, renal failure, hypoglycemia, convulsions, coma, and death. Death from hypoglycemic shock occurs 4–7 days after the onset of symptoms. Recovery is possible but requires at least 2 weeks of intensive therapy.

The Group II toxins are hydrazines, with gyromitrin being the premier example, which is produced by *Gyromitra esculenta* mushrooms. Usually, the symptoms include a bloated feeling, nausea, vomiting, watery or bloody diarrhea, abdominal pain, muscle cramps, faintness, and a loss of motor coordination occurring 6–12 hours after consumption.

The Group III toxins, characterized by muscarine, affect the autonomic nervous system. Symptoms include perspiration, salivation, and lacrimation with blurred vision, abdominal cramps, watery diarrhea, constriction of the pupils, hypotension, and a slowed pulse. Death does not usually occur unless the Group III toxins are present in a mushroom such as fly agaric (*Amanita muscarina*), which also contains Group I toxins. With fly agaric, a fatal combination of symptoms can occur.

The classic example of a Group IV toxin is coprine, which causes symptoms only when ingested with alcoholic beverages. Symptoms begin about 30 minutes after consumption of alcohol along with mushrooms containing Group IV toxins such as *Corpinus atramentarinus.* Symptoms include flushing of the face and neck, distension of the veins in the neck, swelling and tingling of the hands, metallic taste, tachycardia, and hypotension, progressing to nausea and vomiting. Symptoms can last for up to 5 days.

The Group V and VI toxins act primarily on the central nervous system to cause hallucinations. The Group V toxins include ibotenic acid and muscimol, which cause dizziness, incoordination, staggering, muscular jerking and spasms, hyperkinetic activity, a coma-like sleep, and hallucinations beginning 30 minutes to 2 hours after ingestion. Fly agaric, in addition to its content of Group I and Group III toxins, is also a good source of Group V toxins. The Group VI toxins include psilocybin and psilocin, whose symptoms include pleasant or aggressive mood, unmotivated laughter and hilarity, compulsive movements, muscle weakness, drowsiness, hallucinations, and sleep. Symptoms usually begin 30–60 minutes after ingestion of the mushrooms and recovery is often spontaneous. *Psilocybe mexicana,* the so-called Mexican mushroom, is a well-known source of the Group VI toxins. Although Mexican mushrooms (also known as magic mushrooms or simply "shrooms") have been used as recreational drugs for their hallucinogenic effects, it must be recognized that the dose of the Group VI toxins in these mushrooms varies widely (more than 100-fold). Exposure to high amounts of the Group VI toxins has led to prolonged and severe side effects, even death. Often, patients experience persistent sequelae and are admitted to mental institutions.

C. Naturally Occurring Constituents Eaten in Abnormally Large Amounts

The dose of a toxic chemical is the major determinant of the degree of hazard that it poses in the diet. Many naturally occurring foodborne toxicants are present at levels that do not constitute a hazard to normal consumers eating a normal diet. However, if unusually large amounts of foods containing these toxicants are consumed then an intoxication may occur. One classic example is the presence of cyanogenic glycosides in lima beans and cassava.

Many plants contain cyanogenic glycosides, which are sugars that can release cyanide on exposure to certain enzymes present in the plant tissues or by acid in the stomach of the consumer. Linamarin in lima beans is one example of a cyanogenic glycoside. Commercial varieties of lima beans contain very little linamarin, but wild lima bean varieties can have substantial amounts. Cyanide is a classic toxicant, which binds to heme proteins in the mitochondria and hemoglobin in the blood. Cyanide prevents oxygen binding to hemoglobin, thus causing cyanosis, noticeable as a bluish discoloration of the skin and mucous membranes. It also inhibits cellular respiration by binding to the mitochondrial heme proteins. The symptoms of cyanide poisoning include the rapid onset of peripheral numbness and dizziness, mental confusion, stupor, cyanosis, twitching, convulsions, coma, and death. Cyanide is rapidly absorbed, and the lethal dose is 0.5–3.5 mg of cyanide per kilogram of body weight. Commercial lima bean varieties release about 10 mg of cyanide per 100 g of beans. Assuming that the lethal dose is 0.5 mg/kg, a 70-kg adult would need to ingest 35 mg of cyanide or 350 g of lima beans. Though this is not impossible, it is certainly very unlikely.

Cassava can release as much as 50 mg of cyanide per 100 g of food. Cassava intake is substantial in Africa and South America in certain locales when other foods are scarce. The major source of cyanide intoxication remains the intentional ingestion of fruit pits that contain amygdalin, also known as laetrile, under the mistaken belief that laetrile counteracts cancer. The level of cyanogenic glycosides in fruit pits is considerable, and several deaths have occurred from their ingestion.

In addition to the cyanogenic glycosides, plants contain an enormous variety of other potentially hazardous chemicals that are not typically evident unless large quantities of these foods are eaten. Examples include glycoalkaloids in potatoes and tomatoes, goitrogens in cruciferous vegetables, nitrate in spinach, celery, and lettuce, oxalates in rhubarb and spinach, estrogens and saponins in soybeans, and tannins and phytoalexins in many plant foods. Symptoms do not occur after ingestion of these foods in typical amounts.

D. Naturally Occurring Constituents Eaten by Consumers with Genetic Abnormalities

Some naturally occurring constituents of foods are hazardous only for consumers with allergies or intolerance to that food. Food allergies and intolerances are the subject of a separate review in this volume. [*See* Food Allergies and Intolerances.]

E. Naturally Occurring Constituents from Foods Processed or Prepared in an Unusual Manner

Many naturally occurring foodborne toxicants are removed or inactivated during processing or preparation. For example, raw soybeans contain trypsin inhibitors, lectins, allergens, amylase inhibitors, saponins, and antivitamins along with other potentially hazardous factors. However, humans never ingest raw soybeans. The trypsin inhibitors and lectins are inactivated by heating of the soybeans. Fermentation in the preparation of tofu destroys certain toxic factors in soybeans.

Intoxications have occurred from the ingestion of undercooked or raw kidney beans. Kidney beans, like soybeans, are legumes. They contan lectins, which can bind to sugar residues on the surfaces of cell membranes, causing hemolysis of red blood cells and intestinal damage. The lectins from kidney beans can be inactivated by thorough cooking. If not inactivated, the lectins will cause nausea, abdominal pain, vomiting, and bloody diarrhea. In England, intoxications have occurred among immigrants unfamiliar with proper cooking practices for this common item in the British diet. These consumers simply soaked the raw beans and ate them with little or no cooking, resulting in the prompt onset of gastrointestinal symptoms.

BIBLIOGRAPHY

Ahmed, F. E. (ed.) (1991). "Seafood Safety." National Academy Press, Washington, D.C.

Beier, R. C., and Nigg, H. N. (1994). Toxicology of naturally occurring chemicals in foods. *In* "Foodborne Disease Handbook, Volume 3: Diseases Caused by Hazardous Substances" (Y. H. Hui, J. R. Gorham, K. D. Murrell, and D. O. Cliver, eds.). Dekker, New York.

Marth, E. H. (1990). Mycotoxins. *In* "Foodborne Diseases" (D. O. Cliver, ed.). Academic Press, San Diego.

Shibamoto, T., and Bjeldanes, L. F. (1993). "Introduction to Food Toxicology." Academic Press, San Diego.

Spoerke, D. G., Jr. (1994). Mushrooms: Epidemiology and medical management. *In* "Foodborne Disease Handbook, Volume 3: Diseases Caused by Hazardous Substances" (Y. H. Hui, J. R. Gorham, K. D. Murrell, and D. O. Cliver, eds.). Dekker, New York.

Taylor, S. L., and Schantz, E. J. (1990). Naturally occurring toxicants in foods. *In* "Foodborne Diseases" (D. O. Cliver, ed.). Academic Press, San Diego.

Natural Toxins

JOHN B. HARRIS
University of Newcastle upon Tyne

GLOSSARY

Natural toxin Poisonous substance produced by a living organism. The term "toxin" should be applied only to a single substance. A toxin is typically a component of a natural poison or venom

Poison Toxin or mixture of toxins produced by an organism. Typically the term would be applied to the secretions of certain fishes or the toxic constituents of poisonous plants

Venom Poisonous mixture of toxic and nontoxic substances inoculated into a victim or prey by a venomous animal such as a snake, scorpion, or bee. Venom is elaborated in a specialized venom gland and although it is typically delivered via a hollow sting or fang, other highly specialized stinging structures may be used, for example, by sea anemones, corals, and cone shells

THE NATURAL ENVIRONMENT INTO WHICH HUmans are born, and in which we live and die, is a delicate and intricate ecosystem wherein animals, plants, and microorganisms struggle to live together. The organisms have evolved a variety of mechanisms to aid their survival and, in many cases, this has resulted in the elaboration of poisonous substances (toxins) that may be used purely passively to inhibit attack by a predator or offensively to help overcome a potential source of food.

These natural toxins have been of great interest to humans for centuries, and the use of such compounds features strongly in mythology and history. Few will be ignorant of Shakespeare's version of the death of Cleopatra following a bite by an "asp" or of the death of Socrates, precipitated by the consumption of a draught containing hemlock. It is only recently, however, that sufficiently reliable statistics have been collected to allow an accurate assessment to be made of the influence of poisonous organisms on humans. It has been claimed that as many as 10% of all hospital beds in parts of West Africa are occupied by victims of snake bite, and in Thailand, snake bite has been estimated to be the fifth most common cause of death. Scorpion stings are responsible for as many as 2000 deaths/year in Mexico, and among the Indians of Ecuador, 70% of the population have a history of being bitten by a venomous snake.

To the cost of human suffering implicit in such figures must be added the enormous cost of producing, storing, and distributing antivenoms, and the economic cost involved in the loss of productive workers. The typical victim of snake bite, for example, is a male small-holder or rural farmer aged 15–35 years. These are the fittest and strongest people in their local community.

It may be claimed that many incidents of poisoning are avoidable. The massive loss of life among infants and very young children as a result of cholera and other diarrheal diseases caused by the ingestion of bacterial toxins could easily be controlled by the provision of clean water and adequate sanitation or, in some cases, by appropriate immunization programs. Such solutions, though technically simple, are in many parts of the world quite beyond reasonable expectation because of their cost.

The influence of natural toxins on humans cannot be defined in terms of human health alone. The losses among livestock are also of considerable economic significance. Diarrheal diseases are common in intensively reared pigs and other animals; salmonella is endemic in the majority of commercially maintained flocks of poultry; and the loss of grazing animals as a result of the consumption of poisonous plants is huge. It has been suggested, for example, that more than 50,000 head of cattle/year die in Brazil alone following the consumption of *Baccharis coridifolia*.

Although it might be expected that a general increase in living standards will result in a major fall in the level of poisoning in both humans and (domestic) livestock, there are other pressures that will possibly become more significant. The widespread move, in the Western world, toward vegetarianism and "natural" as opposed to "scientific and drug-led" medicine has resulted in the more extensive uses of plants for both food and therapies; the unavoidable consequence of this is that poisonous plants may be unwisely used.

Poisoning as a result of the consumption of contaminated food is relatively common. Among the most important contaminants are the fungal toxins produced by fungi contaminating poorly stored seed and grain. Similarly, the consumption of poisonous fishes and shellfish causes outbreaks of poisoning of almost epidemic proportions and yet few fish or shellfish are naturally toxic—most are toxic because they have accumulated toxins from algae and other marine microorganisms.

The recognition of the intense toxicity of many natural toxins and poisons has resulted in their occasional use in incidents of self-poisoning. Examples include the death of a young man who swallowed a poisonous salamander (*Taricha torosa*) as a dare, and the poisoning of a young child who developed the habit of sucking toads. More constructive has been the emerging use of natural toxins in medicine and science.

In summary, the natural toxins elaborated by living organisms have a profound effect on humans. They can determine our standard of health, dictate the way in which we farm and prepare food, and influence the way we design our public buildings and homes.

In the following sections, a number of natural toxins will be described and their known sources identified. However, it is impossible in a short general article such as this to be comprehensive. The mode of action of the toxins will be briefly described, but the descriptions will be limited in breadth and depth. It is also to be noted that this is a rapidly developing field. Detailed references to original work are not provided, but the bibliography lists a series of reviews and major texts that will provide the interested reader with more detailed and critical information.

I. CLASSIFICATION OF NATURAL TOXINS

A. By Chemical Structure

Classification by chemical structure is useful for those interested in the chemical properties of natural toxins, in the mechanisms by which they are synthesized, and in their role as metabolites in the regular metabolic activity of the organism concerned. The plant toxins (such as alkaloids and glycosides) have been subject to this form of classification for many years. Though there are many attractions to this form of classification, it creates problems because there is a considerable variety of structures represented by the natural toxins, and because a particular chemical class of toxins may express a great diversity of pharmacological activity.

B. By Pharmacological Activity

Classification by pharmacological activity relates to observable patterns of activity. It allows the comparison of activity across and within different families, genera, and species. Its weakness is that at the crudest level, such diverse chemical structures can be grouped together that the system has no logical basis. For example, grouping those toxins that cause the postsynaptic blockade of neuromuscular transmission in

FIGURE 1 Natural toxins: these structures are used to illustrate the chemical diversity of some of the more interesting toxic natural compounds produced by living organisms. (A) Palytoxin, (B) tetrodotoxin, (C) aflatoxin B, (D) colchicine, (E) batrachotoxin, (F) a "long" neurotoxin, α-cobratoxin. Palytoxin is isolated from zoanthids of the genus *Palythoa* but is probably elaborated by unicellular microorganisms. Tetrodotoxin is elaborated by marine bacteria and concentrated by a variety of animals, including the puffer fishes, the blue-ringed octopus of Australia, and some frogs, newts, and salamanders. Aflatoxins are fungal toxins. Colchicine is an alkaloid produced by *Colchicum* species. Batrachotoxin is secreted from the skin of poisonous frogs of South and Central America. α-Cobratoxin is representative of the long-chain postsynaptically active snake toxins elaborated by a very large number of elapid snakes. Each compound has had a major impact on human biology.

A

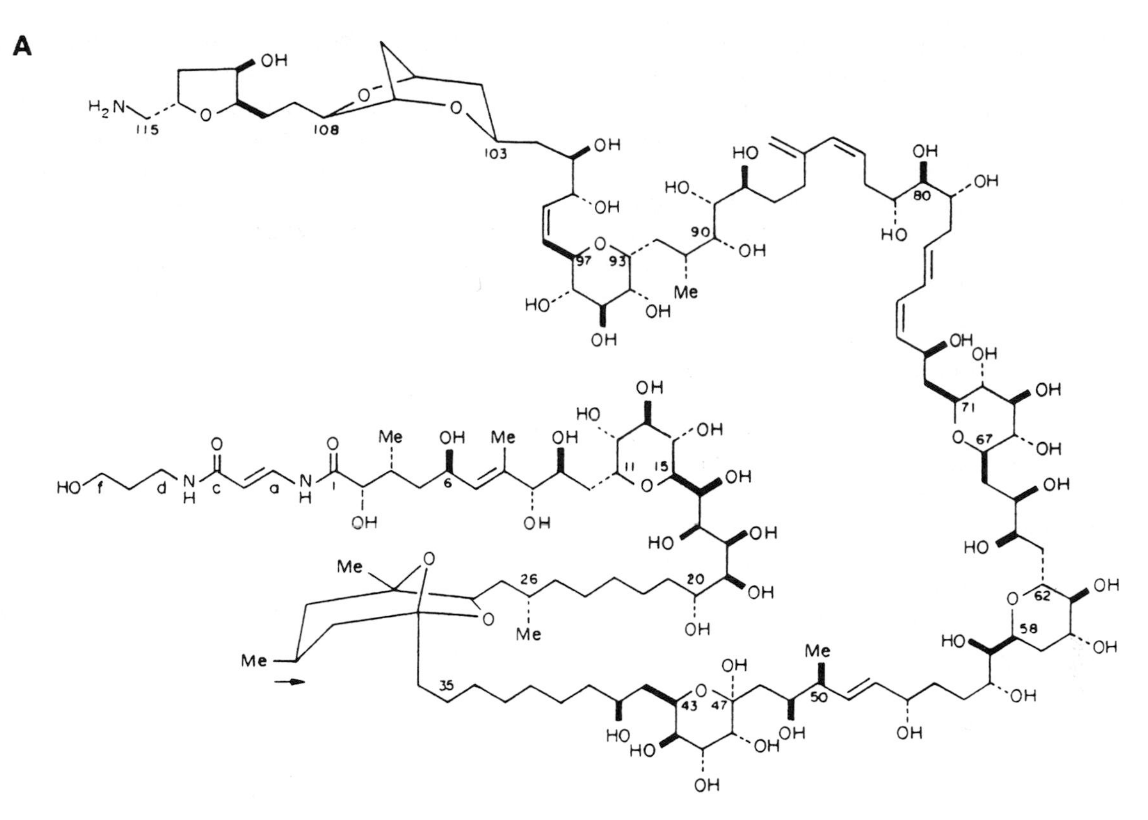

B

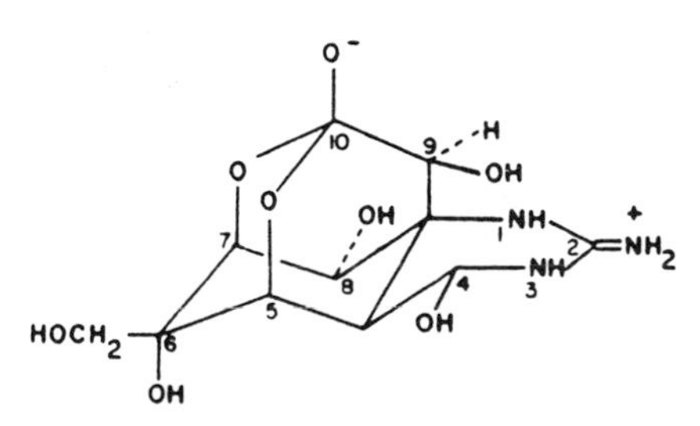
O⁻
HOCH₂
OH
NH₂

C

D

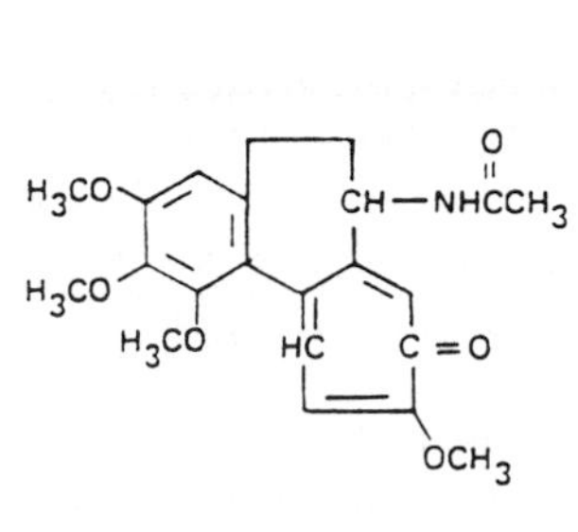
H₃CO
H₃CO
H₃CO
CH—NHCCH₃
C=O
OCH₃

E

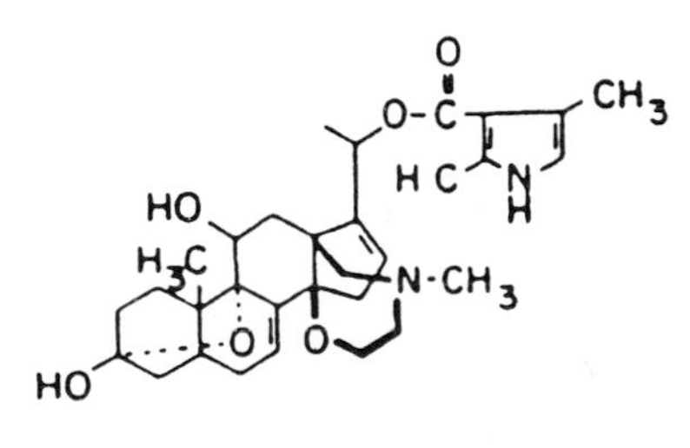
CH₃
HO
N-CH₃
HO

F

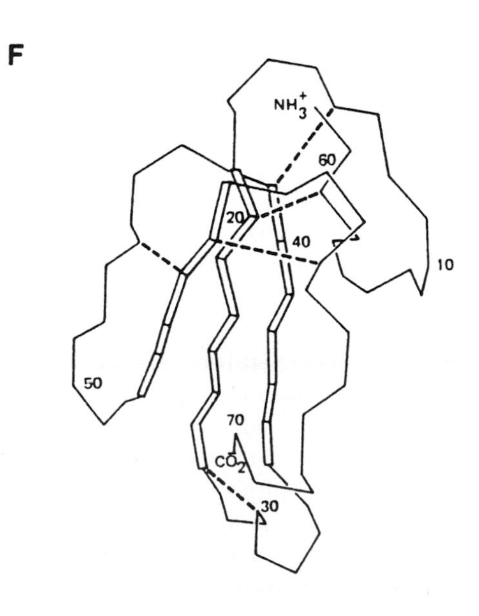
NH₃⁺
60
20
40
10
50
70
CO₂⁻
30

mammals would bring into the same group tubocurarine, a complex alkaloid isolated from the bark or roots of several species of *Strychnos* and *Chondrodendron,* the neurotoxic polypeptides common in the venoms of snakes of the families Elapidae and Hydrophiidae, and the very small polypeptides in the venom of carnivorous cones.

C. By Source

Classification by source is probably the most straightforward way of classifying toxins. The advantages are that the "source" can be used at any one of a number of levels: kingdom, phylum, class, order, family, genus, or species.

In most compilations, a combination of classification systems is used, so that the initial differentiation is based on source, and a second differentiation is based on pharmacological activity or chemical structure. For the purposes of this article, a primary classification based on source is used.

II. SOURCES OF NATURAL TOXINS

A. Bacterial Toxins

Bacterial toxins can be subdivided into two major classes, the endotoxins and the enterotoxins.

The endotoxins are lipopolysaccharides. They constitute a major component of the cell wall of gram-negative bacteria and are often associated with a lipid-associated protein. The endotoxins are implicated in some forms of membrane damage, and in the profound hypotension and renal damage that is associated with gram-negative bacteremia. It is not clear, however, that they are direct causative factors. Rather, it seems probable that most damage is caused by either the release of vasoactive components from mast cells and/or endotoxin-mediated, complement-induced lysis of platelets.

The bacterial enterotoxins are rather large (50–150 kDa) polypeptides of diverse structure and mode of action.

The majority of the highly toxic bacterial enterotoxins comprise two distinct fractions, A and B. The toxins may be elaborated as a single polypeptide that is cleaved (nicked) into its constituent components by proteolytic enzymes, although in many instances the two subunits thus formed are still covalently linked via a disulfide bridge. The toxins in this group are activated by the reduction of this bridge. A few toxins are elaborated as separate A and B subunits that are then combined at the cell surface of the bacterium into a holotoxin consisting of one A unit and a number of B units. Shigella toxin, from *Shigella dysenteriae,* for example, consists of one A subunit and five B subunits.

In all complex bacterial toxins, fraction A is the active subunit and fraction B is a binding subunit. The mechanism of action involves the binding of the binding subunit (B) to the relevant cell membrane, and the internalization of the active (A) subunit. The precise mechanism of transfer across the cell membrane is not known, but almost certainly involves receptor-mediated endocytosis. How the toxins are liberated from endocytotic vesicles into the cytosol is not known. Neither is it clear how the different bacterial toxins are transferred to or recognize their site of action (e.g., prejunctional nerve terminals in the case of botulinum toxin, central nervous system in the case of tetanus toxin, and various areas of the gut in the case of cholera toxin).

The mechanism of action of the bacterial toxins is varied and cannot be adequately covered in this article. The mechanisms may involve, for example, functional damage to specific cell types in which morphological damage to the cell membranes is minimal. Toxins in this category include the botulinum toxins, which inhibit transmitter release from nerve terminals, and cholera toxin, which stimulates a net secretion of electrolytes and water. Other toxins (such as the δ toxins of *Staphylococcus aureus*) act as surfactants and cause the lysis of cells.

The importance of bacterial toxins, and particularly enterotoxins, cannot be overstated. Diarrhea caused by bacterial pathogens affects millions of young children every year. The mortality rate of affected children can be as high as 20% in the poorest parts of the world, resulting in the deaths of as many as 10 million children/year. Infectious diarrhea is also a serious problem in young animals, particularly in those raised under intensive conditions, such as piglets and calves. [*See* Bacterial Infections, Detection.]

Bacterial toxins, especially those elaborated by *Staphylococcus aureaus, Salmonella* species, and *Botulinium* species, are frequently implicated in food poisoning, the incidence of which is certainly underestimated. It has been a salutary lesson to learn that a very large proportion of flocks of egg-laying chickens in Western Europe are infected with *Salmonella* species.

Efforts to control bacterial infection, and to produce effective vaccines against enterotoxic bacteria and their toxins, involve massive resources worldwide.

B. Algal Toxins

Algae are difficult to classify. They have often been considered as primitive or simple plants because they are able to produce their own food by photosynthesis, but some both photosynthesize and consume food particles and others contain no cellulose cell wall (which is characteristic of plant cells). The so-called blue-green algae are closely related to bacteria. The more recent view is that algae are part of the kingdom Protista, which includes bacteria, fungi, and protozoa.

Marine algae often undergo spectacular outbursts of growth (algal blooms), after which red, blue, or green tides are reported. It is not clear what triggers such blooms, but changes in the salinity and temperature of water, and the level of nutrients, have all been considered contributory factors.

The most significant algae involved in poisoning are dinoflagellates of the genera *Gonyaulax, Gymnodinium* (*Ptychodiscus*), and *Gambierdiscus.*

Gonyaulax dinoflagellates are responsible for red tides in temperate regions. These dinoflagellates elaborate a number of toxins, the gonyautoxins, which are accumulated by filter-feeding shellfish. Many of these shellfish are capable of metabolizing the ingested toxins into the more highly toxic saxitoxin. Similar toxins are produced by the tropical dinoflagellate *Pyrodinium bahamense* and by the blue-green alga *Aphanizomenon flos-aquae* (observations that may help to explain reports of toxicity of filter-feeding crabs and other marine animals in tropical waters where blooms of *Gonyaulax* do not occur). Toxins have also been isolated from the large calcareous red alga *Jania,* but it is probable that *Jania* does not produce the toxins: it is more likely that it acts as a host for large populations of toxic dinoflagellates and other algae and a number of toxic marine bacteria. The gonyautoxins are paralytic, blocking fast, voltage-sensitive Na^+ channels in excitable cell membranes. They are among the most toxic compounds known. The toxins are guanidinium derivatives that probably act by "plugging" the sodium channel.

Human poisoning results from the consumption of intoxicated shellfish and other filter-feeding marine invertebrates, but paralytic shellfish poisoning, though not particularly uncommon, is rarely lethal. Restrictions on the collection of shellfish in areas known to be affected by dinoflagellate blooms are widespread and generally respected by commercial fishing organizations.

The various species of *Gymnodinium* are responsible for the red tides of Florida and the Gulf of Mexico; similar blooms have been reported on the coast of South America. The blooms are typically, but not invariably, associated with massive fish and seabird kills, but with relatively low toxicity to humans. The toxins elaborated by the algae are known as brevetoxins. The most significant of the toxins is a large polyether, brevetoxin A, but nothing is known of its synthesis in the dinoflagellate or its metabolism in the shellfish that ingest it.

Gambierdiscus toxicus is possibly the most important of the toxic algae. It is a benthic organism, settling on coral detritus and larger algae. As a result there is no bloom to herald its presence. The toxins elaborated by this dinoflagellate are accumulated first by grazing fish, then by large carnivorous fishes, and are finally ingested by humans when these large fishes are consumed. The resulting syndrome, ciguatera, has been known for several hundred years.

Outbreaks of poisoning can occur without warning anywhere in the Indo-Pacific and Caribbean regions. There are no indications such as appearance or taste of the fish to advise that the fish is poisonous, and no method of preparation will render the animal safe.

Although it was originally thought that a single toxin, ciguatoxin, was responsible for poisoning, this is now known to be wrong. The most positive evidence in favor of multiple toxins is the wide variety of symptoms described by poisoned patients. Several toxins have been isolated from ciguatoxic fishes, including maitotoxin from the stomach of grazing fish, scaritoxin isolated from the flesh of the parrot fish *Scarus gibbus,* and ciguatoxin isolated from the latter source, from *Scarus sordidus,* and from the flesh and viscera of the red snapper, moray eel, and sharks. It is presumed that the toxins are closely related, but although the toxins probably bear a close structural similarity to okadaic acid, their primary structures have not yet been fully elucidated.

The toxigenesis of the algal toxins is still a mystery, and so too is the mechanism of transfer between species. It would appear that metabolic conversion is important and that many algal toxins are secondary metabolites produced as the source material is transmitted through the food chain.

Blue-green algae are the principal cause of freshwater blooms and have been implicated in the poisoning of humans and domestic livestock all over the world. The major toxins are the microcystins (a series of cyclic heptapeptides) and the closely related nodularins. The algae most commonly involved are members of the genera *Anabaena, Aphanizomenon, Microcystis, Nodularia,* and *Oscillatoria.* The microcystins are taken up into susceptible cells (including hepatocytes and enterocytes). Once internalized, the toxins initiate the phosphorylation, aggregation, and breakdown of major cytoskeletal proteins and the disruption and focal deformation of the cell membrane. The damage gives rise to liver damage and gastric disturbances. Some freshwater blue-green algae also elaborate anatoxins, a group of toxins that cause neuromuscular paralysis.

C. Fungal Toxins

Fungi lack the ability to photosynthesize, and so are unable to manufacture their own food. They either live on dead, dying, or decaying organic matter or exist in a symbiotic relationship with plants. Fungi have a major influence on many aspects of human life. Many produce toxins that are implicated in the poisoning of humans and our livestock, some fungi damage our plant foodstuffs, and others produce antibiotics. [*See* Mycotoxins.]

1. Fungal Toxins Involved in Poisoning

The number of fungal toxins so far indentified is enormous, but some are of little real significance outside the laboratory. The more significant are discussed in turn, primarily to indicate the range of toxic effects caused by fungal toxins.

a. The Toxic Peptides of Amanita

Poisoning by mushrooms and toadstools is common in all societies where fungi are used as food items. It has been estimated that the vast majority of fatal poisonings are caused by the consumption of a single species—*Amanita phalloides* (the death cap or destroying angel). The genus *Amanita* includes a number of other poisonous species, including *A. muscarina* (the fly agaric) and *A. pantheria* (the panther cap), although some species of *Amanita* are edible and prized by epicures (e.g., *A. rubescens*, the blusher mushroom). Toxic species of *Amanita* contain two major classes of toxins, the amatoxins and the phallotoxins, and one minor class, the virotoxins (which may be derivatives of the phallotoxins or of a common precursor).

The amatoxins are highly toxic bicyclic octapeptides with an LD_{50} typically around 0.2–0.5 mg/kg (mouse, iv). It has been suggested that all the symptoms of severe poisoning by *Amanita* species can be reproduced by amatoxins, and it is generally considered that these toxins are, therefore, responsible for most, if not all, human fatalities.

Although the amatoxins are toxic to most species following injection, not all species are poisoned by the oral ingestion of the toxins (or indeed the intact fungi). Humans and guinea pigs are particularly sensitive, the dog is less sensitive, rats and mice may be completely insensitive, and rabbits and squirrels have been said to be able to eat the fungi with impunity.

Poisoning by toxic *Amanita* species and by amatoxins is characterized by a delay in onset of 4–12 hr. At this point, nausea, vomiting, colic-like pain, and diarrhea occur. The diarrhea may be sufficiently severe to give rise to dehydration. There then follows a period of respite, which can last for 2–4 days. This phase does not signify recovery: damage to the liver and kidneys continues to develop and the respite gives way to hepatic and renal failure. Death usually occurs a week or so after poisoning. The basis of tissue damage caused by the amatoxins is probably the high affinity of the toxins for RNA polymerase II, the enzyme which catalyzes the synthesis mRNA in transcription.

The phallotoxins are cyclic heptapeptides. They are not as toxic as the amatoxins (typical LD_{50} 1.5–5.0 mg/kg mouse, iv). The phallotoxins, like the toxic cyclic heptapeptides produced by the blue-green algae, are accumulated by susceptible cells and then disrupt the cytoskelton (see Section II,B). Hepatocytes and enterocytes are damaged, giving rise to gastric disturbances and hepatic failure.

The virotoxins are possibly derivatives of the phallotoxins. They have not been studied in such detail as the phallotoxins or amatoxins, but appear to possess similar properties to the phallotoxins.

Other toxic fungi elaborate peptides of the amatoxin and phallotoxin classes. Most are included in the genera *Galerina* and *Lepiota. Amanita* species (especially *A. muscaria*) also elaborate the psychoactive agents ibotenic acid, muscimol, and pantherin, and the fungi have been used to induce an emotionally frenzied state in religious festivals (particularly by the Siberian Koryaks) and warfare (by, it is believed, the Vikings for the induction of *berserksgangr*, the battle fever of the Viking fighters inspired by the rage of Odin).

b. The Ergot Alkaloids

An Assyrian tablet, carved around 600 BC, referred to a "noxious pustule in the ear or grain," a clear reference to the contamination of grain by ergot, *Claviceps purpurea*. The fungus was well known to other early civilizations. A Parsee book refers to "noxious grasses that cause women to drop the womb and die in childbed" and the ancient Greeks referred to the "black malodorous product of Thrace and Macedonia," a reference to contaminated rye.

It was later related to the "holy fire" or "St. Anthony's fire," a disease in which the extremities burn and tingle and blackened gangrenous limbs are shed from the body. These effects are the result of the numerous alkaloids produced by the fungus. Some of the alkaloids, especially ergotamine and ergonovine, are powerful stimulants of uterine smooth muscle, and most of them simultaneously cause peripheral vasoconstriction, depression of vasomotor centers, and blockade of peripheral andrenergic receptors. There is, however, a wide spectrum of pharmacological activity, probably the result of the large number of alkaloids produced by the fungus.

Ergot has had a profound influence on many areas of human activity. It was used in folk medicine nearly 500 years ago, and was introduced into obstetric practice in regular, orthodox medicine in the early nineteenth century. The effects of ergot poisoning have figured in paintings and carvings dating from the Middle Ages, and in much of Europe it resulted in the development of legal requirements for the formal inspection of grain prior to marketing. [*See* Alkaloids in Medicine.]

c. The Aflatoxins

The aflatoxins are secondary metabolites of *Aspergillus flavus* and related fungi. They are chemically defined as bis-dihydrofurans. They react with DNA and ribosomal RNA, and this reactivity is probably causally related to the known carcinogenicity, hepatotoxicity, teratogenicity, and immunosuppressive activity of the toxins.

Since the 1960s the aflatoxins have been implicated in very large losses of poultry, cattle, and pigs as a result of the use of contaminated peanuts, corn, and cottonseed. In all of these cases, hepatotoxicity was the major pathological reaction. The toxins have been shown to be carcinogenic in a wide range of laboratory animals. The target organ is most commonly the liver.

Poisonings in humans have been reported in India following the consumption of moldy corn and it is suspected that aflatoxins are responsible for the high levels of liver damage and hepatoma seen in Uganda, Nigeria, and Indonesia. Poisoning by aflatoxins is not, however, a problem only of poorer countries; there is some evidence, based on the identification of aflatoxins in the circulation, that the toxins have entered the diet of many clinically healthy North Americans, although the primary source of the toxins is unknown.

There also appears to be a possible link between aflatoxin ingestion and encephalopathy and fatty degeneration of the viscera (Reye's syndrome), although a prior or concurrent viral infection may be necessary.

d. The Sporidesmins

The sporidesmins, nine of which have been identified, are complex toxins elaborated by the fungus *Pithomyces chartarum*. The fungus grows on dead vegetation in pastures and growth is favored when periods of warm humid conditions follow long, hot, dry spells. The toxins cause severe hepatic damage and photosensitization in grazing animals. Sheep and fallow deer are particularly severely affected. They exhibit facial eczema, with exudation and swelling. Mortality rates are high, but might be related not only to the liver damage but also to the infestation of open sores by flies and other pests and general stress. Outbreaks are common in New Zealand and have been reported in Australia and South Africa. Calves are susceptible to the disease, but adult cattle and red deer are relatively resistant.

e. The Ochratoxins

The ochratoxins are secondary metabolites of a large number of *Aspergillus* and *Penicillium* species. The toxins have been isolated as contaminants from a variety of seeds and grains used as food for both humans and livestock. The toxins are essentially dihydroisocoumarins linked to phenylalanine. They have been implicated in nephrotoxicity in humans and animals.

In experimental animals, ochratoxins are teratogenic and can cause fetal death. Whether they are teratogenic at the levels consumed when the toxins are contaminants in food is unknown.

2. Fungal Attacks on Plants

It is well known that fungal attack on plants can cause severe damage to human foodstuffs in particular, and the local environment in general. The infestation of potato by the fungus *Phytophthora infestans* was the primary cause of the Irish famine of the mid-nineteenth century, when a quarter of the population of Ireland either died or emigrated. This fungus also attacks soft fruit, eucalyptus, and avocado. The fungus responsible for Dutch elm disease, *Ceratocystis ulmi,* also produces toxic polysaccharides and phenols that are possibly implicated in the disease.

Other classes of toxins elaborated by pathogenic fungi include quinonoids, such as cercosporin, produced by *Cercospora* species, which are involved in leaf spore disease in sugar beet and purple speck in soya bean plants. Pyridines are elaborated by many fungi, especially those of the genus *Fusarium,* and are implicated in wilting diseases of tomato, cotton flax, and banana, for example. Terpenoids are major toxins of *Fusicoccum amygdali,* a fungus that causes severe damage in peach and almond trees.

A great deal of research is needed before a real understanding of fungus/plant interactions emerges. It is probable, for example, that *Bacharis coridifolia,* a shrub responsible for large losses of livestock in South America, is not naturally toxic. It seems more likely that a fungus, *Myrothecium verrucaria,* elaborates a range of toxins. This fungus is associated with the roots of the plant. The toxins are thought to be taken up by the roots and stored in the leaves, although an alternative explanation may be that there is gene transfer between fungus and plant.

3. Antibiotics Produced by Fungi

Although antibiosis is a well-known natural phenomenon, antibiotic is a term more usually used (and used here) in the restricted sense as a definition of a natural toxin used by humans to control bacteria and fungi. The antibiotics can be subdivided according to their broad chemical structure: the β-lactams (penicillins, cephalosporins, and cephamycins), the aminoglycosides (gentamycin, tobramycin, anikacin, kanamycin, streptomycin, and neomycin), the tetracyclines, the macrocyclic antibiotics (rifamycins), and the miscellaneous compounds such as cycloserine, chloramphenicol, and griseofulvin. [*See* Antibiotics.]

The penicillins remain among the most important of the antibiotics, although only two of them, penicillins G and V, are still used in clinical practice. A number of species of *Penicillium* elaborate the antibiotics, but the most useful organism is *P. chrysogenum.* Closely related to the penicillins are the cephalosporins. The first of these, cephalosporins P and N, were isolated from *Cephalosporium acremonium.* Related antibiotics, the cephamycins, were isolated from *Streptomyces lactamdurans.* The significance of the β-lactam antibiotics is that they are amenable to a great deal of manipulation, with the result that a large range of synthetic and semisynthetic β-lactams are now available. The β-lactams inhibit the synthesis of peptidoglycans—essential components of bacterial cell walls.

The aminoglycosides are particularly important because they are used to treat the gram-negative bacteria that are largely resistant to the natural penicillins. Most of the aminoglycoside antibiotics are produced by strains of *Streptomyces,* although gentamycin is obtained from *Micromonospora purpurea.* The aminoglycosidases are bactericidal because, after they are actively transported across the bacterial cell wall, they inhibit protein synthesis.

The tetracyclines and chloramphenicol are also produced by species of *Streptomyces.* These antibiotics, like the aminoglycosides, inhibit bacterial protein synthesis.

The rifamycins, produced by *Streptomyces mediterranei,* are of interest because they are internalized by susceptible bacteria and inhibit RNA synthesis. Natural rifamycins are not used in clinical medicine.

Griseofulvin is an important antifungal agent first isolated from cultures of *Penicillium nigricans.* The antibiotic inhibits hyphal growth.

D. Plant Toxins

The toxins responsible for poisoning by plants may cause acute damage to humans or livestock, but many act slowly and indirectly by inhibiting the adequate utilization of other foodstuffs. So numerous are the examples of toxic plants that only a brief survey of the major classes of plant toxins is possible. The toxins have been classified according to their chemical structure.

1. Polypeptides

The major polypeptide toxins isolated from plants are the protease inhibitors and the lectins. These toxins are principal constituents of beans and cereal grains. The protease inhibitors are particularly important. They are constituents of soybeans, widely used as animal feeds and, after reprocessing, as a meat substitute. The inhibitor(s) can be inactivated by heat treatment. Inadequately treated bean products cause an inhibition of trypsin. This, in turn, leads to an increase in pancreatic activity, hypertrophy and hyperplasia of the pancreas, and, eventually, to the development of pancreatic tumors.

The lectins are very common constituents in seeds and cereals, and most appear to be relatively nontoxic. A few, however, are extremely toxic and ricin (from *Ricinus communis*) and abrin (from *Abrus precatorius*) have been implicated in many fatal poisonings. Many lectins are phytohemagglutinins. They cause the agglutination of red blood cells by interacting with glycoproteins in the red cell membranes. Lectins may also act as antinutritional agents by binding to the lumenal aspect of microvilli in the small intestine. As a result the normal ingestion of food is disrupted, growth of the animal is impaired, and acute gastroenteritis and diarrhea may follow. Finally, some lectins (most importantly those in *Phytolacca americana,* or pokeweed) are mitogens.

2. Amino Acids

Two toxic amino acids are known to be of significant interest. β-*N*-Oxalyl-α,β-diaminopropionic acid is a constituent of the chick pea (or vetch) *Lathyrus sativa.* Consumption of the peas is not uncommon in the Indian subcontinent and gives rise to the neurological condition of lathyrism. The mechanism of action probably involves inhibition of the excitatory transmitter glutamate.

Mimosine is a constituent of *Leucaena leucocephala,* a legume that is used as a foodstuff for many animals. Ruminants convert mimosine into 3,4-dihydroxypyridine, a compound known to cause enlargement and malfunction of the thyroid gland. The effect on the thyroid is not so marked in nonruminants, but these animals lose hair, possibly because mimosine inhibits the conversion of methionine to cysteine.

High levels of consumption of *Leucaena* by pregnant animals can cause abortion, fetal resorption, and fetal abnormalities.

3. Glycosides

Glycosides are common constituents of plants and although many are implicated in some human fatalities and nonfatal poisonings (e.g., the digitalis glycosides), the toxic glycosides commonly implicated in human poisonings are the cyanogenic glycosides of which approximately 30 are known. They have been implicated in major outbreaks of poisoning by lima beans (*Phaseolus lunatus*) but they are also constituents of sorghum, cassawa, and the seeds of bitter almond, peach, and apricot. Poisoning by lima bean is not uncommon in developing countries, and poisoning by the cyanogenic glycoside amygdalin (the major cyanogenic constituent of peach, almond, and apricot seeds) has occurred as the result of using the glycoside as an unorthodox anticancer agent.

4. Alkaloids

The most toxic alkaloids are the pyrrolizidine alkaloids. Approximately 250 are known, of which about 150 are hepatotoxic. They occur in a wide range of small herbs and shrubs in grazing grasslands. Tansy ragwort (*Senecio jacoboea*) and *Crotalaria* species are particularly commonly implicated. The toxic pyrrolizidine alkaloids are typically esters of unsaturated amino alcohols—esters of saturated amino alcohols tend not to be toxic. The toxins cause acute liver necrosis, hyperplasia, and hepatic tumors. These alkaloids have been involved in massive losses of grazing livestock. Since the alkaloids are excreted in the milk, inadvertent exposure of humans to the alkaloids is a considerable risk.

Other alkaloids known to cause significant poisoning of grazing animals are swainosine, an indolizidine alkaloid isolated from locoweed (*Astragalus* species), which gives rise to locoism. This syndrome includes gait disturbances, knuckling over the fetlocks, respiratory distress, and sudden death. Pregnant animals frequently abort. Consumption of the plants have caused massive losses of livestock throughout the world.

Many other poisonous alkaloids are involved in animal losses, including the carcinogenic alkaloids of bracken (*Pteridium aquilinum*), the indole alkaloids reserpine and others from *Rauwolfia* species, vincristine from *Vinca rosa* and physostigmine from *Physostigma venenosum,* and the teratogenic alkaloids anabasine and coniine.

E. Animal Toxins

1. The Invertebrates

Of the million or so known species of animals, approximately 95% are invertebrate—that is, without a backbone. The remaining 5% are vertebrate. This division is clearly unequal and arbitrary. It has resulted in a field of study (invertebrate biology) so diverse and representing such large numbers of animals that only a very broad general description of the toxins elaborated by the invertebrates can be entertained. Two phyla are particularly important: phylum Coelenterata (or Cnidaria) and phylum Arthropoda. The cone shells (family Conidae, phylum Mollusca) are also of special interest. Many of these animals are carnivorous, feeding on small fish and nematodes. The prey animals are captured and paralyzed prior to being ingested.

Phylum Coelenterata includes the jellyfish, sea anemones, and corals. The jellyfish have a life cycle in which the static polypoid form is larval, and the free-swimming medusa form is dominant. They are marine organisms widely distributed throughout the world's oceans.

The jellyfishes are stinging animals that inoculate venoms through a specialized stinging cell, the nematocyst. The nematocyst consists of a coiled or pleated rod or tube held within a capsule, which is closed by a lid or operculum. When appropriately stimulated the operculum opens and hydrostatic pressure discharges the tube to the exterior. The ejected rod or tube penetrates the skin, thus inoculating the paralyzing venom. Other nematocysts may discharge sticky threads solely to capture prey. Most jellyfish are innocuous to humans, but some, especially the sea nettles (genus *Chrysaora*) and the cuboid medusae (box jellyfish) of the family Chironex, are dangerous.

A highly specialized and very dangerous jellyfish, the Portuguese man-of-war, *Physalia physalis,* is a colony of polyps and medusae. These coelenterates are known as the siphonophora. One modified medusa acts as a float. Prey is captured and paralyzed in trailing tentacles armed with batteries of nematocysts.

The sea anemones are classified as anthozoans (i.e., polypoid cnidarians) and also use stinging cells to capture prey. Most sea anemones are innocuous to humans, causing at most mild stinging or tingling and a minor rash, but the very large anemone *Dofleina armata* of Western Australia has been said to be dangerous. Opinion, however, is divided, and there is little documentary evidence either way. The corals are anthozoans that produce a skeleton of calcium carbonate. Corals may exist as solitary, often large polyps, others as colonies of large numbers of very small polyps.

The toxins of the Coelenterata have not been systematically studied. The best characterized are the paralytic toxins of sea anemones. These are typically small polypeptides of 45–50 amino acid residues that are active on Na^+-specific channels in excitable tissues. The toxins are not involved in poisoning humans following contact with anemones—they are important because they are used in studies on the biophysics of excitable tissues.

The toxins of the jellyfish are almost certainly small polypeptides, of varying size, but they have not been properly characterized. They can cause severe necrotic lesions to the skin, and pain (often unbearable and leading to shock) is a common finding with those who come into contact with the jellyfish. Treatment of jellyfish stings is difficult and controversial, and a better understanding of the chemistry and pharmacology of the toxins is needed.

The most significant of the anthozoan toxins is palytoxin, a unique and very complex nonpeptide compound with a molecular mass of 2700 Da. The toxin has been isolated from numerous species of anthozoa of the genus *Palythoa*. It is probable that the toxin is elaborated by a symbiotic microorganism. The toxin is accumulated by coral-grazing fishes and crabs and has been responsible for several incidents of severe poisoning. Its precise mode of action is not well understood, but it inhibits Na^+/K^+-ATPase, causes cation-selective channels to open in excitable membranes, and induces widespread transmitter release from nerve terminals.

Phylum Arthropoda is vast, representing as many as 1 million species. The phylum includes the classes Arachnida (scorpions, spiders, ticks, and mites), Chilopoda and Diplopoda (centipedes and millipedes), and Insecta (the insects). These classes contain the species of particular interest to the person interested in natural toxins.

All 800 species of scorpion are venomous insectivores and 25–30 are medically significant. Prey (typically another small arthropod) is held in the pedipalps and the sting (telson) at the end of the tail is arched over the scorpion's back. Venom, elaborated in venom glands in the telson, is inoculated into the captive prey. The prey is paralyzed by the venom and then consumed. The principal toxins in the venom of scorpions are small polypeptides, typically 60–65 amino acid residues in length. The toxins block cation-specific channels in excitable cells such as neurons.

In humans, the inoculation of venom is intensely painful but the biochemical cause of the pain seems not to be known.

Like the scorpions, spiders are entirely insectivorous, and almost all are venomous. Captured prey is bitten and venom is inoculated from venom glands in the cephalothorax via the fangs that are mounted on the chelicerae. The prey is typically paralyzed by the bite. The toxins responsible for paralysis are polypeptides of variable size. α-Latroxin, from the black widow spider, a species of *Latrodectus,* for example, has a molecular mass of about 125,000 Da; the neurotoxin from the Australian funnel web spider has a molecular mass of about 3800 Da. The majority of neurotoxic toxins block transmission from nerve to muscle by depleting the nerve terminals of chemical transmitter. Hence there may be an early phase of excessive activity, as transmitter is discharged, prior to the onset of paralysis when transmitter stores are emptied. Some of the toxins, however, block transmission by binding with receptors on the muscle cell at the neuromuscular junction, and others by blocking ion-selective channels activated by the excitation of the junctional receptors.

Some spiders cause quite serious local necrotic lesions when they bite nonprey species (humans and larger mammals). Necrotizing toxins have been isolated from a number of such spiders, including the recluse spiders (*Loxosceles* species) and some of the tarantulas. These toxins are also polypeptides, varying in molecular mass from about 19,000 to about 32,000 Da.

The ticks and mites have been said to be of greater economic significance to humans than any other group of invertebrates. Although this is a rather subjective view (*Anopheles* must have a claim for this particular prize), numerous species of ticks and mites are parasitic on our plant foods and livestock, and they are parasitic on humans. Most ticks and mites are not specifically poisonous or venomous. They are important because of tick-borne diseases such as encephalitis and Rocky Mountain fever, or because reactions to their presence may include mange or asthma.

The paralysis ticks secrete a toxin in the saliva that causes neuromuscular paralysis. The most toxic tick is probably *Ixodes holocyclus,* the Australian paralysis tick. A single tick can kill a large dog and three or four ticks can kill a calf. As many as 10,000 calves/year are lost in Australia in this way. The toxin responsible, holocyclotoxin, is a polypeptide with a molecular mass of about 50,000 Da. The toxin has not been fully characterized, but it is probably presynaptically active, blocking transmitter release at neuromuscular junctions. Approximately 40 species of ticks worldwide are thought to cause paralysis and although presynaptic failure seems to be a feature common at least to those studied, the specific toxins responsible have not been isolated.

The members of the class Chilopoda are centipedes. They are predatory animals, feeding mainly on smaller arthropods and, occasionally, earthworms. Prey is captured and held by the poison claws that are carried on the first trunk segment. The venom used by the centipedes is manufactured in a gland situated in the appendage itself. The venom appears to paralyze the prey, but little is known of its composition. Reports that the centipedes can kill humans are exaggerated, though a sting can be painful.

The Diplopoda are the millipedes, which are largely vegetarian and are not directly venomous. Both centipedes and millipedes produce a wide range of noxious defensive secretions, some of which are said to be neurotoxic to predators such as spiders. These secretions are complex mixtures of low-molecular-weight benzene derivatives and alkaloids. Hydrogen cyanide is produced by some species.

The class Insecta (i.e., insects) is generally said to be larger than all other classes of animals combined. At least 750,000 species have been described, and this is almost certainly a substantial underestimate. As a class, insects are of major importance to humans. Flies, bugs, fleas, and mosquitoes bite and sting, transmitting diseases such as sleeping sickness, malaria, plague, and yellow fever, but at the same time many plants are dependent on insect (as opposed to wind) pollination. Many insects produce noxious defensive agents but only a few are directly venomous—bees and wasps, ants, and certain caterpillars.

Bees (superfamily Apoidea) and hornets and wasps (family Vespidae) are characterized by their well-developed stinging apparatus. The sting is used for various purposes. The solitary wasps, for example, use the sting to immobilize and aid the capture of prey. The paralyzed prey—typically spiders or other species of bees or wasps—are then seeded with the eggs of the captor. The captive acts as a food supply for the hatched larvae. The stings of all species are also used defensively to drive away predators. The principal toxic constituents are cytolytic phospholipases, the hemolytic and cytolytic cationic polypeptide mellitin, a number of kinins, and neurotoxins such as the philanthotoxins and apamin (the latter a toxin that blocks Ca^{2+}-activated K^{+} channels), some of which are not well characterized. These insects are responsible for many deaths in humans, but the deaths are not caused directly by the toxins; death is an indirect result of anaphylactic shock. The bees are more important to humans because of their role in pollinating plants and as a source of honey.

The ants, though often delivering painful stings, are not dangerous. They are, however, a significant nuisance in many parts of the world. The venoms delivered by the sting include a range of small alkaloids and a large number of aromatic hydrocarbons.

A few caterpillars carry irritating hairs. In some cases the hairs are hollow and attached to a venom gland. The venom is inoculated when the distal tip of the hair is broken off. Little is known of the precise nature of the toxins.

The cone shells are marine gastropods of the tropical and subtropical Indo-Pacific and Western Atlantic. Many of the cone shells are predatory carnivores, feeding on polychaetes, fishes, and other gastropods. In the predatory cone shells the teeth have evolved into single harpoons, which are stabbed into the prey via the proboscis. The harpoon is connected via a poison duct to a poison gland in the pharynx. The contraction of the gland results in the ejection of the venom into the prey. The prey, paralyzed by the injection of venom, is then drawn into the buccal cavity. The venoms contain small polypeptide toxins, typically comprising 13–30 amino acids. Three classes can be identified: the ω-conotoxins, which block voltage-activated calcium channels, the α-conotoxins, which block nicotinic cholinergic receptors, and the μ-conotoxins, which inactivate voltage-sensitive Na^+ channels of excitable cells.

Although many toxins have been isolated from the crude venoms of the cone shells, the venoms have not been properly classified. It is clear that they produce a number of very important toxins in addition to those that block ion-specific channels in excitable cells. Since cones have been known to kill several adult humans, the toxins they contain are of clinical, as well as scientific, interest.

2. The Vertebrates

The vertebrates that are of particular significance are represented among the fishes, the amphibians, and the reptiles.

a. The Poisonous Fishes

Although some elasmobranchs are poisonous, if consumed in large quantity, the most important poisonous fishes are teleosts. The poisonous fishes can be divided into two groups: the ichthyosarcotoxic fishes (those that are poisonous if consumed, but possess no envenoming apparatus) and the venomous fishes (those that possess an envenoming apparatus). The really important ichthyosarcotoxic fishes are those responsible for ciguatera and puffer fish poisoning.

Ciguatera is now known to be caused by the accumulation and possible metabolic conversion of a toxin produced by the alga *Gambierdiscus toxicus* (see Section II,B).

Puffer fish (*Tetraodontidae* species), urchin fish (*Diodontidae*), and sun fish (*Molidae*) have been known to be poisonous for several hundreds of years. The consumption of the flesh of these fish is popular throughout the Indo-Pacific, but the liver, gonads, and intestines of the fish are dangerous. The toxin responsible for poisoning is tetrodotoxin, a small alkaloid with limited solubility in water. The toxin was long a puzzle, because it is also found in a variety of quite unrelated animals—the blue-ringed octopus of Australia, the goby *Gobius criniger,* the North American salamander *Taricha torosa,* some frogs (especially the genera *Atelopus* and *Brachycephalus*), various filter-feeding crabs, and some other marine invertebrates. The origin of the toxin is unknown, but a number of factors suggest that it is accumulated by the fishes and other animals, perhaps from bacteria. For example, fish become nontoxic if kept in clear water; within a given species of fish, the geographical origin of the fish can determine whether or not it is toxic; toxicity varies with the seasons. This suggestion has since been strengthened by the isolation of tetrodotoxin from bacteria (*Pseudomonas* species) collected from the red alga *Jania*. If confirmed, it may be necessary to reclassify tetrodotoxin as a bacterial toxin, and to reevaluate the mechanism by which such diverse animals become toxic (see Section II,B). The toxin is extremely dangerous. Deaths from the consumption of puffer fishes are well known, and documented cases exist of death following a bite by the blue-ringed octopus and the consumption of one specimen of *T. torosa.* The toxin blocks Na^+ channels in excitable membrane. This leads to rapid and complete paralysis. Tetrodotoxic fishes are largely resistant to tetrodotoxin.

More than 50 species of venomous fishes are known. All the fish are spiny. The venom glands are usually located on the spine and are covered by an integumentary sheath. The sheath is ruptured when the spine enters the flesh, and the venom enters the wound. The fishes most commonly implicated may be broadly classified as the stonefishes, the scorpion fishes, the zebra fishes, the weaver fishes, and the sting rays. None of the fishes is particularly aggressive, although many will defend their territory. Stings tend to be inflicted when victims step on, fall on, or handle the fish. Stonefish, common in many parts of the tropics and subtropics, possess true venom glands. These are discharged when the spines (typically the dorsal spines) are compressed on entry into the flesh of the victim. The entry of the spine forces back the integumentary sheath, which increases pressure on the venom glands situated at the base of the spine. Eventually, when the pressure reaches a threshold, the glands discharge venom into the wound.

The toxins in the venoms of the venomous fishes have not been well characterized beyond the observation that they are polypeptides. Stings are intensely painful and secondary infections are common.

b. The Poisonous Amphibians

The amphibia (frogs, toads, newts and salamanders) have been known for centuries to produce poisonous secretions from cutaneous glands.

By far the most toxic—and interesting—are those secreted by the highly colored neotropical frogs of the genera *Dendrobates* and *Phyllobates.* These are frogs of Central and South America, where native Indians use the skin secretions as arrowhead poisons. The poisons of these frogs are alkaloids, over 200 of which have now been isolated.

The majority of the toxic alkaloids fall into one of five major classes: the batrachotoxins, the histrionicotoxins, the indolizidines, the pumiliotoxins, and the decahydroquinolines.

The batrachotoxins are highly toxic steroidal alkaloids. Only six such toxins have been unequivocally identified. All were isolated from the skin of frogs of the genus *Phyllobates*. The toxins cause a dramatic increase in the permeability of excitable membranes, leading to a Na^+-dependent depolarization and an eventual loss of excitability.

The histrionicotoxins are nonsteroidal alkaloids rather widely distributed in a number of frogs of the genera *Dendrobates* and *Phyllobates*. They have also been isolated from the Madagascan frog *Mantella madagascariensis*. Between 15 and 20 histrionicotoxins have been identified. They bind to the acetylcholine receptor channel and block excitation, but they are relatively nontoxic when compared with batrachotoxins.

The indolizidines (a class that includes toxins originally known as gephyrotoxins) are simple bicyclic alkaloids. Many of these alkaloids have been isolated from *Dendrobates* species but little is known of their toxicity.

The pumiliotoxins were originally isolated from *Dendrobates pumilio*. Related toxins have now been isolated from a number of frogs of the genera *Dendrobates* and *Phyllobates,* and in all more than 30 such toxins are now known. They are of relatively low potency, but little is known of them. The toxins appear to enhance calcium-dependent excitation–contraction coupling.

The decahydroquinolines constitute another class of alkaloids isolated from the skin of certain species of *Dendrobates*. Many of these compounds are now isolated, but little is known of their pharmacology or toxicology.

As well as the alkaloid toxins, frogs and toads secrete a large number of toxic and/or noxious substances including biogenic amines, the hallucinogenic bufotoxins, and small peptides such as bombesin and bradykinin. Though these compounds, and others, cause irritation to many species, they are not life-threatening. Tetrodotoxin has been isolated from some frogs and salamanders.

The origin of tetrodotoxin and other toxic alkaloids in amphibians is difficult to understand. Wild-caught animals seem to remain toxic, however long they are held in captivity, but captive-born amphibia seem to be nontoxic. It must be a possibility that symbiotic microorganisms are the primary source of the toxins.

c. The Venomous Reptiles

Apart from lizards of the genus *Heloderma,* the only poisonous reptiles are snakes. Of the 11 (or 13) recognized families of snakes, 5 contain venomous members. All snakes in four of the families are poisonous (families Elapidae, Hydrophiidae, Viperidae, and Crotalidae). Only a few members of the Colubridae are dangerous to humans.

The venomous snakes have evolved specialized venom glands located in the head of the animal. These glands are attached to grooved or hollow front fangs along which toxic secretions from the muscular venom glands are injected into the bite wounds. The dangerous Colubridae do not possess either a front fang for inoculation of venom or a typical venom gland. In these snakes, the rear fangs convey venom, which is elaborated in Duvenoy's gland. The glands are not muscular, and the toxic secretion oozes along the fang when the snake bites. Clearly, the evolution of a front fang confers a significant advantage to the major groups of venomous snakes.

Snake venom is a complex mixture of toxic and nontoxic polypeptides. The nontoxic polypeptides often possess hydrolytic activity (typically phospholipases, esterases, and proteases) that initiates digestion in prey. Many of the phospholipases are, in addition to their hydrolytic activity, intensely toxic.

The toxins in snake venoms are all polypeptides, ranging in size from 30 to 130 amino acids. They may be broadly classified according to their primary modes of action: α-neurotoxins, cytotoxins, β-neurotoxins, myotoxins, hemorrhagins, and toxins affecting coagulation of the blood. The mambas (snakes of the genus *Dendroaspis*) elaborate neurotoxins that appear to be pharmacologically unique.

The α-neurotoxins block neuromuscular transmission in vertebrate skeletal muscle by binding with the nicotinic acetylcholine (ACh) receptors of the junction, thereby preventing the action of ACh released from the motor nerve terminals. These toxins are typically found in the venoms of snakes of the families Elapidae and Hydrophiidae. The toxins may be classified as short or long neurotoxins. The short toxins are 60–62 amino acids long and are cross-linked by 4 disulfide bonds. The long neurotoxins are 67–74 residues long, cross-linked by 5 disulfide bridges. The molecular organizations of the two classes of toxin are essentially similar and their pharmacological properties are also virtually identical.

The cardiotoxins (also known as cytolysins, direct lytic factors, or cytotoxins) are short, single-chain toxins, 60–62 amino acids in length. They are homologous with the α-neurotoxins, but they are not neurotoxic. The term cardiotoxins is slightly misleading, because few of these toxins show any selectivity for cardiac muscle. The toxins are found almost exclusively in the venoms of the cobras (genus *Naja*). Their mechanism of action is not fully understood. The activity of the cardiotoxins is greatly potentiated by the presence of phospholipases. This is an important observation, because "pure" samples of cardiotoxins are frequently contaminated with traces of phospholipases.

The β-neurotoxins are found in the venoms of snakes of the families Elapidae, Viperidae, and some Crotalidae. They are typically 120–130 residues long and are cross-linked by 6 or 7 disulfide bridges. Though they share many features, they are rather varied in organization. Some are single-chain toxins (e.g., notexin, from *Notechis scutatus*), some consist of two to five subunits, noncovalently linked (e.g., crotoxin from *Crotalus durissus*), and some consist of two covalently linked subunits (e.g., β-bungarotoxin from *Bungarus multicinctus*). All the toxins are phospholipases; in the case of multisubunit toxins, one subunit typically possesses hydrolytic activity. The toxins are homologous with mammalian pancreatic phospholiphase A_2 enzymes. Phospholipase activity is essential for neurotoxicity, but it cannot be the only significant factor because most venom phospholipases are nontoxic. The β-neurotoxins prevent transmitter release from vertebrate motor nerve terminals. Some also prevent transmitter release from effector nerves in the autonomic nervous system. The mechanism of action is not known but is the subject of intense investigation.

Damage to skeletal muscle is not an uncommon consequence of snake bite. Highly active myotoxic toxins have been isolated from many snake venoms. The best characterized are the myotoxic phospholipases. These toxins characteristically are also β-neurotoxins (β-bungarotoxin is an exception, as a β-neurotoxin that is nonmyotoxic). They probably act by hydrolyzing the plasma membrane of skeletal muscle fibers. A second class of myotoxins has been isolated from the venoms of viperid and crotalid snakes. These toxins (the best known of which is myotoxin a from *Crotalus viridis*) cause vacuolation in skeletal muscle, followed by degeneration of the fibers. The mechanism of action is unknown. Some "cardiotoxins" are particularly potent myotoxins. The most potent is probably that from the cobra *Naja kaouthia*. This toxin causes a nonselective increase in ion flux across skeletal muscle fiber membranes.

The hemorrhagins are very important venom constituents, causing the loss of red blood cells from small blood vessels, leading to massive bruising in poisoned victims. The toxins are single-chain polypeptides. Some are without hydrolytic activity and some are proteases. They are major constituents of the venoms of snakes of the families Viperidae and Crotalidae.

Many bites, especially by viperid and crotalid snakes, can give rise to massive and often fatal bleeding from the gums, eyes, and nose. These venoms contain proteolytic enzymes that inhibit blood coagulation. The precise mechanism of action varies. Some of the toxins activate factor X or V, which leads to thrombin formation and the formation of fibrin clots; others cause the formation of soft fibrin clots by hydrolyzing fibrinogen directly. These soft clots are susceptible to fibrinolytic degeneration. The rapid loss of fibrin/fibrinogen leads to a complete loss of coagulability of the blood, which may last for several weeks. Some phospholipases also cause a loss of coagulability by hydrolyzing procoagulant lipids released from platelets during platelet aggregation.

The mambas produce three groups of toxins that do not, at present, appear to be produced by any other snake. They are the (1) dendrotoxins, which enhance transmitter release from vertebrate motor nerve terminals, probably by blocking K^+ channels in the nerve terminal membranes, (2) fasciculins, which inhibit the enzyme acetylcholinesterase, and (3) muscarinic toxins, which block muscarinic ACh receptors.

III. USES OF NATURAL TOXINS

Just as plants and animals use natural toxins as a protection against predators, and to aid the capture of prey, so humans have adopted their use for the same purpose. Early inhabitants of North America used rattlesnake venoms and extracts of black widow spiders as arrowhead poisons, and similar groups of people in other parts of the world used, for example, skin secretions of frogs, *Palythoa* corals, tubocurarine, and other plant products for the same purpose. Tetrodotoxin, consumed with the liver of puffer fish, has been used as an adjunct to suicide, as have snake venoms and numerous draughts prepared from poisonous plants. The best-known examples of such practices are, of course, Cleopatra and Socrates.

Many toxins (and crude venoms and poisons) are used in carefully controlled circumstances in therapy. The ergot alkaloids, the potentially fatal glycosides of *Digitalis,* and tubocurarine are examples of agents in current use, and coagulant and anticoagulant toxins from snake venoms are used in some circumstances for the control of specific problems affecting blood coagulation.

The long α-neurotoxins are used in the diagnosis and study of the disease myasthenia gravis, and numerous toxins are used in laboratory science because their high specificity and potency make them ideal tools and probes for studying the biophysics of excitable membranes.

Bacterial toxins have a special significance. Not only are they used to study basic biological problems, but they are also seen as potentially useful chemical weapons.

IV. FUTURE DEVELOPMENTS

Future work will concentrate on the isolation and characterization of toxins—especially those of invertebrates such as spiders and scorpions because of their potential value to the scientist. The marine toxins will continue to be studied at great length because of the dangers of consuming edible fish and invertebrates that have accumulated algal and other toxins. The molecular biology of snake and scorpion toxins will be increasingly studied in the hope of finding ways of developing a new series of cheap, effective, and safe antivenoms, and the development of vaccines and immunizing programs for protection against the bacterial diseases will receive increasing interest.

BIBLIOGRAPHY

Adams, M. E., and Olivera, B. M. (1994). Neurotoxins: Overview of an emerging research technology. *TINS* **17**(4), 151–155.

Daly, J. W., Myers, C. W., and Whittaker, N. (1987). Further classification of skin alkaloids from neotropical poison frogs (Dendrobatidae), with a general survey of toxic/noxious substances in the Amphibia. *Toxicon* **25**, 1023–1085.

Dolly, J. O. (ed.) (1988). "Neurotoxins in Neurochemistry." Ellis Horwood, Chichester, England.

Hardegree, C., and Tu, A. T. (eds.) (1988). "Handbook of Natural Toxins. 4. Bacterial Toxins." Marcel Dekker, New York.

Harris, J. B. (ed.) (1986). "Natural Toxins, Animal, Plant and Microbial." Clarendon, Oxford, England.

Keeler, R. F., and Tu, A. T. (ed.) (1983). "Handbook of Natural Toxins. 1. Plant and Fungal Toxins." Marcel Dekker, New York.

"Neurotoxins: A Supplement to TINS" (April 1994). Elsevier Trends Journals, Cambridge, England.

Ownby, C. L., and Odell, G. V. (1989). "Natural Toxins, Characterization, Pharmacology and Therapeutics." Pergamon, Oxford, England.

"Receptor and Ion-Channel Supplement: A Supplement to TIPS" (March 1994). Elsevier Trends Journals, Cambridge, England.

Tu, A. T. (ed.) (1984). "Handbook of Natural Toxins. 2. Insect Poisons, Allergens and Other Invertebrate Venoms." Marcel Dekker, New York.

Tu, A. T. (ed.) (1988). "Handbook of Natural Toxins. 3. Marine Toxins and Venoms." Marcel Dekker, New York.

Tu, A. T. (1991). "Handbook of Natural Toxins. 5. Reptile Venoms and Toxins." Marcel Dekker, New York.

Neocortex

BARBARA L. FINLAY
Cornell University

GLOSSARY

Cortical column Fundamental cellular grouping of the neocortex, a collection of cells extending perpendicular to the cortical surface, spanning all cortical layers, and receiving and operating upon a common input

Cytoarchitecture Description of the organization of a structure according to the types, sizes, and arrangements of its cells

Lateralization Property that certain sensory, cognitive, and motor functions are represented preferentially on one side of the neocortex

Modularity Property of the cortex that the circuitry for particular sensory, motor, or cognitive functions is kept physically and computationally separate from other sensory, motor, and cognitive functions

THE NEOCORTEX IS A LAYERED SHEET OF CELLS covering the surface of the forebrain, the largest and most prominent feature of the human brain. Homologues of the cells that make up the neocortex can be found in the forebrain of all extant vertebrates, including fish, amphibians, reptiles, and birds, but only in mammals are these cells found arranged in the six-layered structure of repeating subunits that is termed the neocortex. The cortex's six layers consist of specialized zones for input, for communication with other areas of the cortex both locally and at a distance, and for output to noncortical structures. A column of cells arranged perpendicular to the cortical surface tends to perform a standardized computation on its input; the input can vary greatly. Through this relatively uniform structure passes information for functions as diverse as recognizing faces, conversing, playing the piano, planning for the future, and adjusting emotional displays to the social context. Components of these capacities are represented in paricular parts of the cortical surface, a property termed "modularity," and thus local damage to the cortex will cause disruptions of particular skills and spare others entirely. The human neocortex is lateralized for some computations: in the great majority of individuals, the circuitry for language is located on the left side of the brain. [*See* Brain; Cortex.]

I. EVOLUTION OF THE NEOCORTEX

A. Basic Questions

The neocortex is the single largest structure in the human brain, and because of its prominence and its importance for many capacities viewed as distinctly human, such as language and pronounced manual dexterity, there has been much inquiry into its origin. Three questions have been asked about the evolution of the neocortex. First, when does the neocortex first appear in vertebrate evolution, considering both the neurons that compose it and their organization? Second, what might explain the pronounced enlargement in volume of the cortex seen most notably in primates, but that has occurred independently in a number of mammalian radiations? Finally, in behavioral terms, is the enlargement of the neocortex best understood as the growth of a general learning device, or the accretion of a number of special structures for specialized skills like language and social cognition?

B. Origin and Organization of Cell Types

In many traditional textbooks, the neocortex is presented as structure that is found *de novo* in the mammalian brain. More modern anatomical explorations of vertebrate brains have demonstrated that this is not the case; in fact, the preferred terminology for the neocortex is now "isocortex" since this area of the brain is not new in mammals. All vertebrates possess a number of organized cell masses, termed nuclei, in their forebrains. Even in those extant vertebrates believed to have a brain organization most resembling the most primitive vertebrate condition, the lampreys and hagfishes (class Agnatha), cell groups can be found that have, in part, the approximate location, types of input, patterns of connectivity, neurotransmitters, and neuromodulators that are characteristic of the mammalian forebrain and, specifically, the neocortex. The characteristic mammalian pattern of layering is absent, and many aspects of the connections and organization vary. The neocortex of mammals, and the homologous forebrain structures in vertebrates, should be viewed in much the same way as the relationship between a human arm and bat wing might be viewed. The major bone and muscle masses have similar embryonic origins, similar topologic relationships and attachments, and similar gross functions. However, the absolute and relative size of the components vary considerably; particular bone and muscle groups may be added, deleted, or combined, and details of the attachments of bone and muscle may change.

In reptiles, the structure thought to be homologous to the neocortex is called the dorsal ventricular ridge, and it receives and integrates sensory input through the thalamus and distributes this information to the midbrain and hindbrain, as does the neocortex. In birds, the laminarly arranged cells of the dorsal telencephalon, and also various divisions of cell masses of the fore-

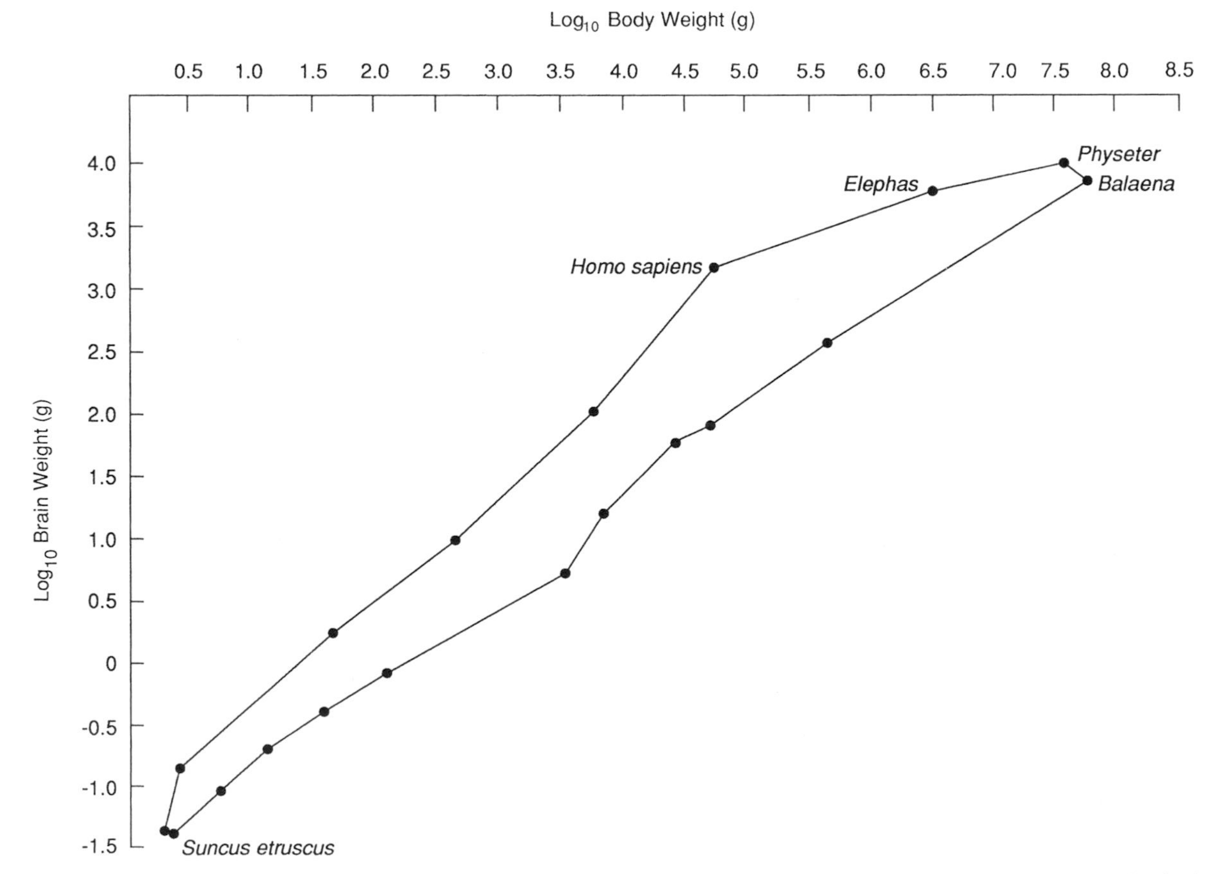

FIGURE 1 Convex polygon bounding the set of mammalian brain/body weight values on a logarithmic scale. Individual species identified represent a shrew, humans, the elephant, and two whales. [Reprinted with permission from J. F. Eisenberg (1981)] "The Mammalian Radiations," p. 227. Univ. of Chicago Press, Chicago.

brain area called the corpus striatum, are thought to be homologous to the neocortex. In that these areas receive and integrate the same sort of information that the mammalian neocortex does, functional homology is found as well as anatomic homology. For example, birdsong, a complex communicative system that involves auditory learning, elaborate sensory integration, and skilled motor performance, depends in part on the areas in its brain that are homologous to the neocortex in mammals and humans.

The difference in the developmental programs that produce a layered arrangement of cells rather than a collection of nuclear masses is not yet known, but a difference in the pattern of migration of cells during development is likely to be involved, which will be discussed next.

C. Change in Volume of the Neocortex across Mammalian Radiations

The entire brain shows striking differences in volume across the mammalian radiations, most pronounced in primates (Fig. 1). Some of this change in volume can be attributed to change in body size: unsurprisingly, bigger bodies are associated with bigger brains. However, at any particular body size, there is a residual variation of at least 10-fold in the relation of whole brain size to body size. The size of the neocortex, in turn, bears an exceedingly regular relationship to whole brain size (Fig. 2): the volume of the neocortex increases exponentially with whole brain size, so that larger brains are composed of an increasingly greater percentage of neocortex. The human brain does not differ from all mammalian vertebrates in this respect. Humans have the largest brain-to-body size ratio of any vertebrate, and a great deal of this hypertrophy is accounted for by the neocortex. However, the amount of cortex in humans is the amount lawfully predicted from whole brain size.

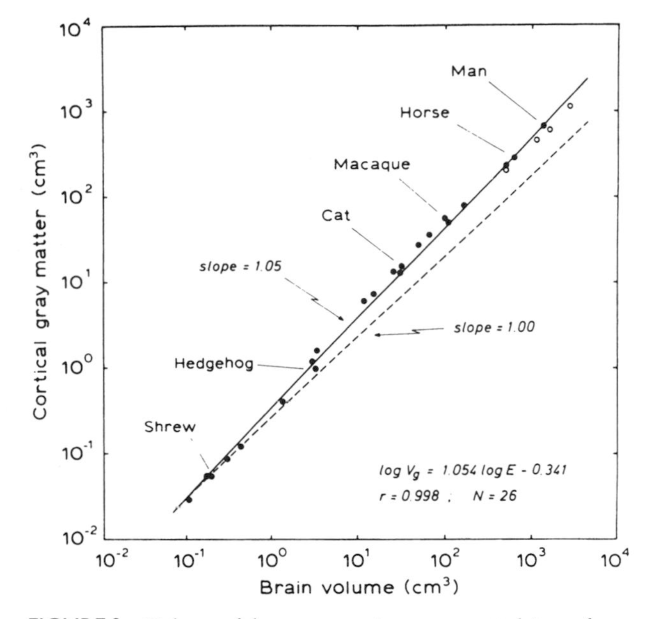

FIGURE 2 Volume of the neocortex (gray matter only) as a function of whole brain volume for a number of mammals on a logarithmic scale. The dashed line represents scaling of cortical volume by the first power, geometric similarity, and shows that the amount of cortex increases at a greater rate than that expected by geometric scaling alone. Dolphins and whales are indicated by open circles. [Reprinted with permission from M. A. Hofman (1989). *Prog. Neurobiol.* **32**, 137–158.]

II. FUNDAMENTAL STRUCTURE AND PHYSIOLOGY OF THE NEOCORTEX

A. Cortical Layers and Columns

Though many different schemes for the naming and ordering of the layers of the neocortex were proposed in the latter half of the nineteenth century when the anatomical organization of the cortex was first described, the one that has persisted is the six-layered scheme laid out by K. Brodmann in his publications on cortical cytoarchitectonics in the period 1903–1920. The cortex consists of a number of distinct cell types and fiber bands, arranged in strata; variations in local areas of cortex as described as having condensations, omissions, or subdivisions of Brodmann's fundamental six strata (Fig. 3). These layers will be discussed not in their numerical order, but in the order that information passes through them (Fig. 4).

The principle input to the cortex comes from the thalamus, a collection of nuclei in the diencephalon that receives information in turn from various sensory domains, from various areas from the midbrain to forebrain, and from the cortex itself. The input from the thalamus distributes to Layer IV and to a lesser extent, to the upper part of Layer VI. Layer IV is termed a "granular" layer in that it is composed of small, relatively symmetric cells with radial dendrites, called stellate cells (Fig. 5). Several types of stellate cells can be further distinguished, as well as multipolar neurons associated with a variety of neuromodulators in this layer. The processes of cells from Layers V and VI also extend through this layer. [*See* Thalamus.]

This information is then relayed up and down, to Layers II and III, and to Layers V and VI. The bulk

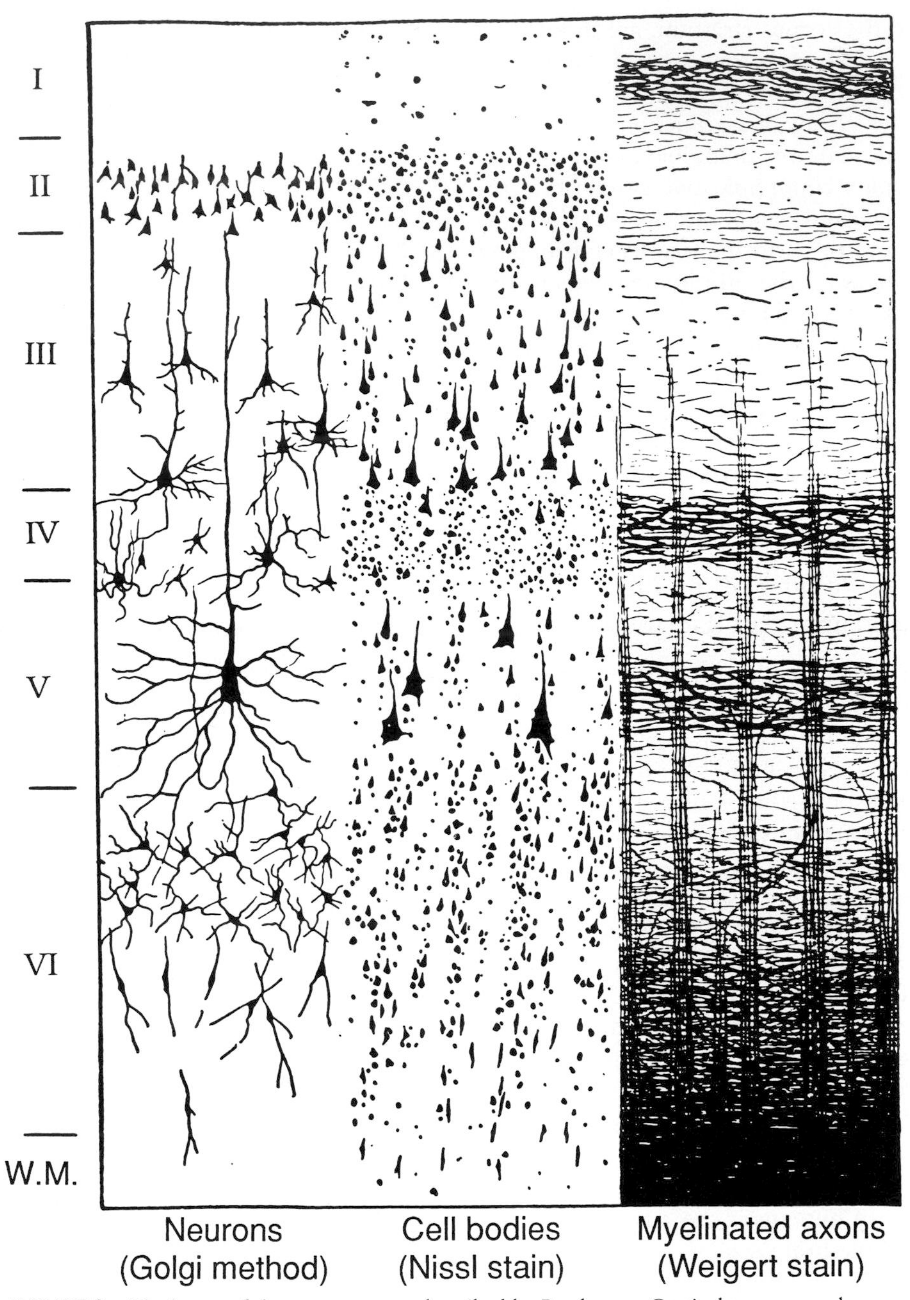

FIGURE 3 The layers of the neocortex, as described by Brodmann. Cortical neurons and axons are arranged in six principal layers designated by Roman numerals. Three types of stains are represented: the Golgi methods, which stains whole cells and all their processes; the Nissl stain, which shows only cell bodies; and the Weigert stain, which stains axons. [Reprinted with permission from J. B. Angevine and C. W. Cotman (1981). "Principles of Neuroanatomy." Oxford Univ. Press, New York.]

of local interactions in the cortex are restricted to a column several cell diameters wide that extends perpendicularly from Layer IV to the cortical surface, and down to Layer VI. It is this local interaction and distribution of information that give rise to the anatomical and physiological unit of the "cortical column." Layer I, the cell-free outermost fiber layer, is composed of the axons and dendrites of these cells

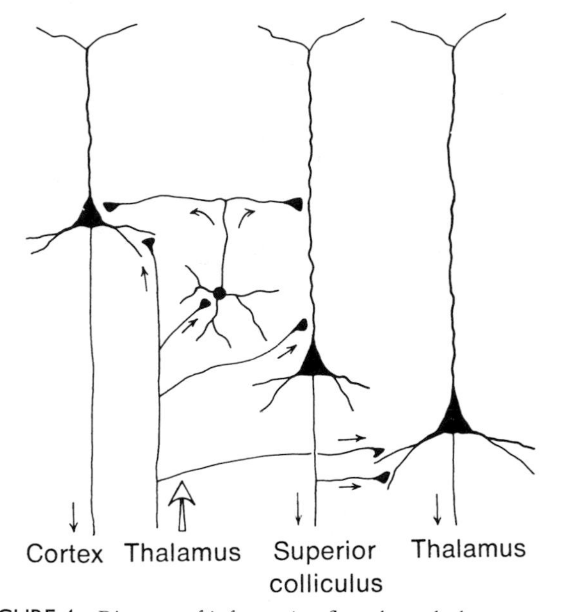

FIGURE 4 Diagram of information flow through the neocortex. Principal input is from the thalamus to the stellate cells of Layer IV (see also Fig. 3). Cortical output distributes to the rest of the cortex, to the thalamus, and to many subcortical structures of which one, the superior colliculus, is shown here. [Reprinted with permission from G. M. Shepherd (1988). "Neurobiology." Oxford Univ. Press, New York.]

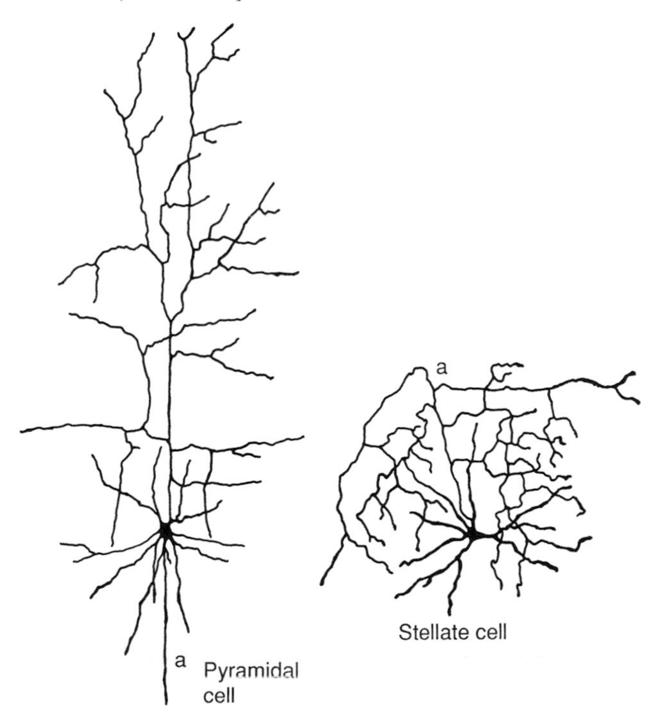

FIGURE 5 Drawings of a typical pyramidal cell and stellate cell. Pyramidal cells are especially abundant in Layers II, III, and V, and stellate cells are found in Layer IV. The letter "a" designates the axon of each cell. [Reprinted with permission from J. B. Angevine and C. W. Cotman (1981). "Principles of Neuroanatomy." Oxford Univ. Press, New York.]

engaging in local interactions. It should be emphasized, however, that longer-range cellular interactions are also important in producing many of the characteristic features of cortical information processing.

Layers II and III are composed of pyramidal cells (see Fig. 5), asymmetric cells with a profusion of dendrites at their base, and a long apical dendrite, all of which receive synaptic input. Multipolar neuromodulatory cells are also found in Layers II and III. Output from Layers II and III is long range and principally intracortical. Axons from these areas distribute to local cortical areas (e.g., from primary to secondary visual cortex), to distant cortical areas (e.g., from secondary visual cortex to visuomotor fields in frontal cortex), and across the corpus callosum to the cortex on the other side of the brain. The connections across the corpus callosum can link both corresponding and noncorresponding cortical areas. These intracortical connections distribute principally to Layers II and III and to Layer V. [*See* Hemispheric Interactions.]

Layer V also consists principally of pyramidal cells. These receive input from Layers II, III, and IV and distribute their axons subcortically. These subcortical areas are quite diverse and include the basal ganglia and amygdala, the prominent nuclear masses of the forebrain; the superior colliculus, a midbrain structure concerned with attention and eye movements; and a variety of brain stem sensory and motor nuclei. Some giant neurons from the motor cortex, named Betz cells, send their axons as far as the spinal cord to terminate directly on motor neurons and associated interneurons.

Layer VI contains cells of a variety of morphologies. A principal output connection of Layer VI is a reciprocal connection to the area of the thalamus that innervates the same cortex.

This same organization of layers and repeating columns is found throughout the neocortex. Local subareas of cortex are modified in such a way that reflects their specialization. Primary visual cortex, which receives a massive thalamic input of visual information, has a large number of cells in Layer IV. In motor cortex, Layer IV is almost absent, and the cells of Layer V, the subcortical output layer, are unusually large and prominent. These local differences can be employed to divide the cortex into "cytoarchitectonic areas," as Brodmann did with his original maps (Fig. 6), the nomenclature of which is still in use today. These cytoarchitectonic divisions, based solely on the visualizable detail of the cellular organization of the cortex, have proved to typically correspond to functional divisions in the cortex as well.

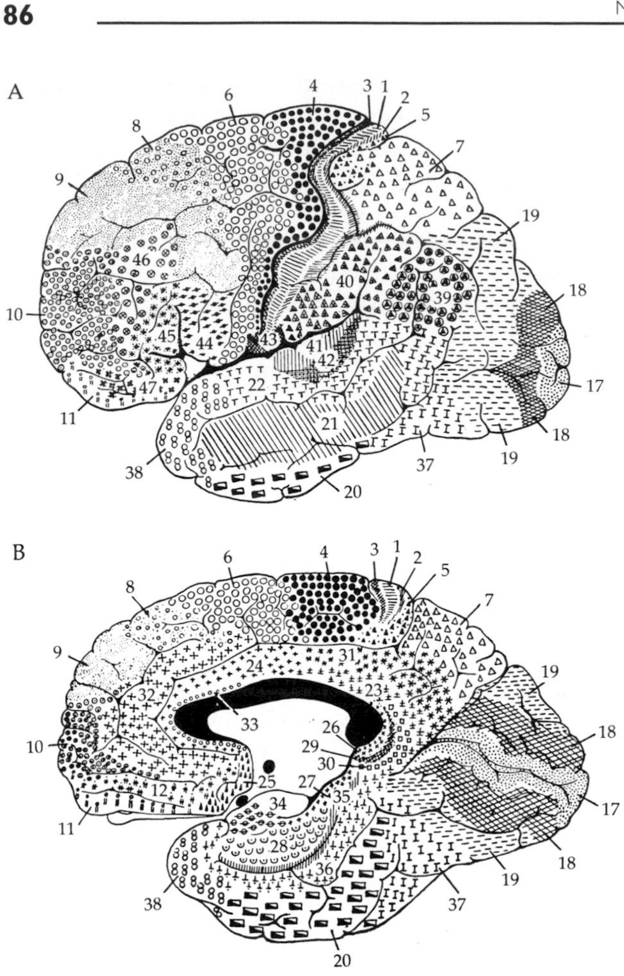

FIGURE 6 Cytoarchitectonic areas of the cortex according to Brodmann (1990). Lateral (A) and medial (B) views of the neocortex are shown. The numbered divisions, based on the thickness, density, and cell size of the cortical layers, also correspond to functional specializations within the cortex. [Reprinted with permission from J. B. Angevine and C. W. Cotman (1981). "Principles of Neuroanatomy." Oxford Univ. Press, New York.]

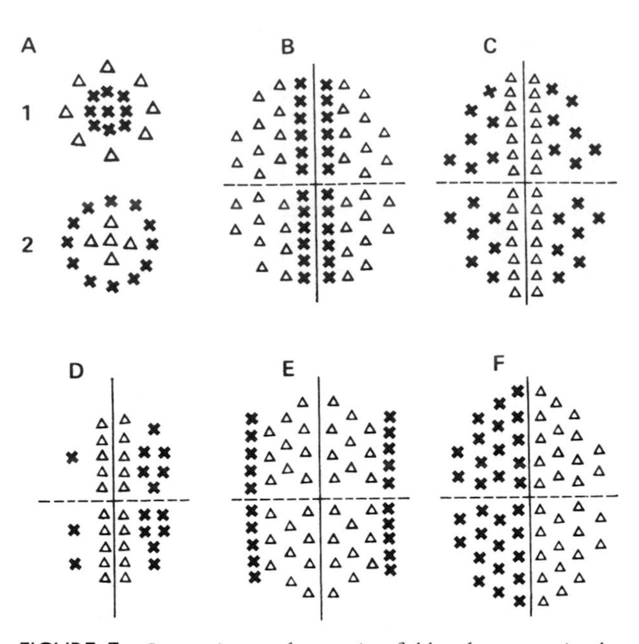

FIGURE 7 Comparisons of receptive fields of neurons in the retina (A1) and lateral geniculate nucleus (A2) with those of neocortical cells in the primary visual cortex (B–F). Whereas the retinal and geniculate cells are symmetric, the cortical cells prefer elongate bars and edges and stimuli for maximal response. Both classes of neurons will respond to stimuli of light or dark contrast. Crosses indicate excitatory response in the receptive field; triangles represent inhibitory areas. [Reprinted with permission from E. R. Kandel and J. H. Schwartz (1985). "Principles of Neural Science," 2nd Ed. Elsevier, New York.]

B. Cortical Physiology

Electrophysiological recording from single neurons in the cortex, first undertaken by V. Mountcastle in the somatosensory cortex and by D. Hubel and T. Wiesel in the primary visual cortex, amplified and extended the neuroanatomical picture of columns and layers in the cortex. First, the best stimulus to excite single cortical cells was typically a complex transformation of the sorts of stimuli that best excite thalamic cells. In the case of the primary visual cortex, the best stimulus for a typical cortical cell would typically be a bar or edge in a particular orientation, in a specific location in the visual field (Fig. 7). Often, cells would respond only to spatially congruent information with the right and the left eye stimulated together. The thalamic input to this area, however, is not binocular and the visual fields of thalamic neurons are spatially symmetric, not elongate. In the somatosensory cortex, a preferred stimulus would typically be a submodality of touch, like light touch or hot or cold, and cells would often be selective for a particular direction of stimulus movement. All of the neurons in a column perpendicular to the cortical surface had similar selectivity for the appropriate stimulus properties, like orientation, binocular integration, touch submodality, or location of the receptive field on the skin surface or visual field. Neurons in the different cortical layers vary systematically on some aspects of receptive field structure, for example, the degree of specificity for location in the visual field, but overall, all neurons in a column process the same type of input. [*See* Somatosensory System; Visual System.]

In areas of cortex that do not receive direct sensory input, the properties of single neurons are often complex combinatorial properties of neurons. These combinations can occur both within and between sensory modalities and can involve aspects of both motor behavior and prior learning and cognition. For example, in the cortex of monkeys and sheep, neurons have been described whose optimal stimulus is the face (or aspects of the face) of the animal's own species. In the area of bat cortex that processes information relating to echolocation, neurons will respond optimally to an auditory stimulus that is the bat's own call, and with a particular delay the echo of that call, which thus specifies a target range. In the parietal cortex of monkeys, which is located between the visual and somatosensory/motor areas of the cortex and which receives input from both, neurons can be found that will fire only when a monkey is looking at and reaching for an object of interest. In the motor cortex, the response patterns of neurons can best be related to movements or limb positions that involve a number of muscle groups, and not a single muscle's contraction. In the frontal cortex, neurons have been described that fire only when the animal is attending to a stimulus that has previously been associated with reward, but that will not fire if the same stimulus is motivationally neutral.

C. Cortical Maps and Functional Modularity

1. Mapping

The dimensions of single-neuron response properties described here are not found randomly distributed around the cortical surface, but are typically found as orderly dimensional maps laid out across the cortical surface. Neighboring cortical columns typically represent progressive changes in the mapped dimensions. The primary visual cortex is the best-described example. In this cortex, the dimensions of location in visual space, preferred eye of activation, and preferred stimulus orientation are all laid out in an orderly way and these maps are superimposed. Location in visual space is represented once over the full extent in primary visual cortex, with the center of gaze represented at the occipital pole and the visual periphery buried in the medial cortex. For each mapped location in visual space, a full range of preferred eye activation (ocular dominance) is represented from left eye dominating to right eye dominating, changing in orderly sequence. For the same location in visual space, all possible stimulus orientations are also represented, also changing in an orderly way from a preferred angle of 0° to 180° across the cortical surface. These last two dimensional maps are arranged roughly perpendicularly to each other, such that for every location in visual space, every possible value of ocular dominance and preferred stimulus orientation is represented. A block of cortex that contains one full cycle of all the mapped dimensions has been termed a "hypercolumn" (Fig. 8). Finally, interposed in this regular map are "color blobs," islands interrupting the regular progression of other dimensions, where calculations involved in color perception are carried out.

Similar maps can be found in other sensory and motor dimensions, such as the map of the body surface in somatosensory cortex, or changing values of best echo delay seen in the part of the bat auditory cortex devoted to echolocation. In many areas of cortex, the dimension that is mapped has remained elusive, but is suggested by the presence of orderly mapping in the neuroanatomical connections between related cortical areas.

Although only the topographic maps of sensory and motor surfaces have been demonstrated directly in the human neocortex, the principle that the cortex

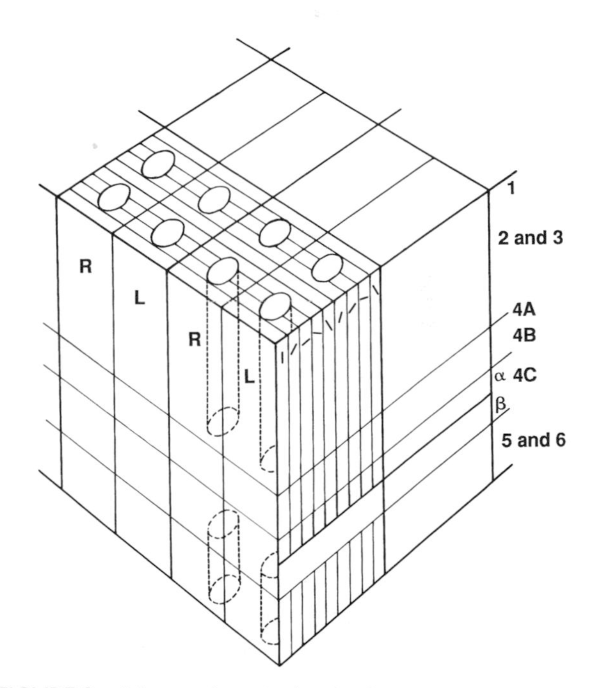

FIGURE 8 A hypercolumn in the visual cortex, representing one location in the visual field. On the left-hand base axis, orderly alternation of preferred activation by the right or left eye is shown. On the right-hand base axis, orderly rotation in the preferred stimulus orientation is shown. On the vertical axis, the Brodmann layers are numbered, with Layer IV broken into three subdivisions. Interposed in this regular array are cylinders called "blobs" that are devoted to color processing. This processing unit is repeated over and over throughout the entire neocortex. [Reprinted with permission from M. S. Livingstone and D. H. Hubel (1984). *J. Neurosci.* 4, 309–339.

represents complex properties of information as changing dimensions that can be mapped incrementally across the cortical surface appears to be a general one. As such, this insight has guided modeling of the cortical mechanism of such particularly human functions as speech perception, which cannot be directly investigated.

2. Relationships between Maps

The mapped stimulus dimensions described earlier for various stimulus dimensions are often represented multiple times in the cortex. For example, in both the monkey and human, the body surface appears to be mapped at least three times over, each representation emphasizing different features of this sensory array. In the monkey, the part of the cortex that analyzes visual information consists of not just one but many separate representations of the visual field (Fig. 9). Connections between these representations are both serial and parallel. For example, the representation of stimulus movement is kept separate from the representation of color through several remappings, and a common representation of the output of these two parallel systems is finally found only in the inferotemporal cortex, an area thought to be important for the identification and naming of objects.

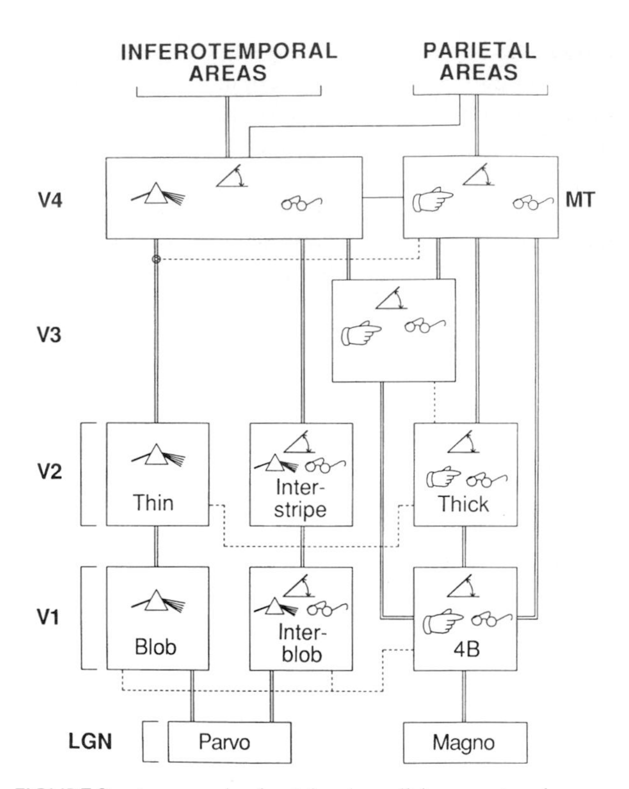

FIGURE 9 An example of serial and parallel connections between the multiple representations of the visual field (V1, V2, V3, V4, and MT) found in the visual cortex. Icons are placed in each compartment to symbolize a high density of neurons showing selectivity for color (prisms), orientation (angle symbols), direction of movement (pointing hands), and binocular depth (spectacles). The input from the thalamus, the lateral geniculate nucleus in this case, is shown with its principal divisions (parvo and magno). The inferotemporal cortex is thought to be involved in recognition and naming; the parietal cortex with pointing, reach, and other operations in space. [Reprinted with permission from G. M. Shepherd (1988). "Neurobiology." Oxford Univ. Press, New York; original figure from E. A. DeYoe and D. C. Van Essen (1987). DeYoe, E. A. and Van Essen, D. C. *Trends in Neurosci.* **11**, 219–226.]

3. Modularity

The property by which the cortex keeps stimulus modalities and particular classes of computation physically separate over at least the initial stages of processing is termed modularity. It is this property that accounts for the often peculiar effects of localized brain damage, in which, for example, an individual may have lost the capacity for language but retains the ability to make the complex inferences required to play chess. These dissociations of function can also occur within sensory modalities; for example, the ability to recognize faces can be lost with damage to a particular area of the parietal cortex while sparing the ability to read and navigate visually in the world. Or an individual with damage to a subarea of frontal language cortex might lose the capacity to appreciate grammatical relationships in sentences but could fully understand and give the meanings of individual words.

The diversity of the functions that depend on the uniform structure of the neocortex should be underscored. We have already discussed a number of the separable functions dependent on the visual areas of the occipital lobe: the sensory categorization of movement, color, and contour; the dissociability of locating objects for purposes of navigation versus naming those objects; and the specialized social skill of facial recognition and analysis of expression. Also represented in the cortex are the complex motor patterns for speech, the syntactic and semantic components of language itself, and aspects of the conceptualization, initiation, and execution of complicated learned motor acts like piano playing. In the frontal lobe are represented a number of dissociable complex capacities that are often difficult to describe. These include, but are not limited to, interrupting and taking account of one's past performance in planning new action; remembering in what context a particular fact was learned; adjusting emotional expression to the social context; linking emotion to judgments of what actions to take; and the execution of social scripts such as proper eye contact and the maintenance of appropriate interpersonal space. Overall, though excellent progress is being made in understanding the nature of cortical analysis in some sensory domains, the nature of the cortical analysis in cognitive and social domains has hardly begun to be understood.

III. LATERALIZATION OF THE HUMAN NEOCORTEX

The human neocortex has the relatively unusual property of showing lateralization for functions that bear no intrinsic relationship to the right or left side of the body. For most sensory and motor functions, the brain is lateralized: for example, the left side of the brain receives sensation from the right side of the body and the right half of the visual field, and controls the motor functions of the right side of the body. However, in most right-handed people, language is represented on the left side of the brain. Damage to the left temporal cortex can cause complete loss of language ("aphasia"), whereas damage to the corresponding area on the right will cause little or no change in language function. "Dexterity" itself is also lateralized; approximately 90% of individuals prefer the use of their right hand and are more skilled in using it for fine motor functions. The clear lateralization of these two prominent functions led to the characterization of the relationship between the hemispheres as "cerebral dominance," meaning the dominance of the left hemisphere for cognitive functions. [*See* Cerebral Specialization.]

Later, however, Roger Sperry and his colleagues showed that the right hemisphere was dominant for some other functions, notably aspects of visuospatial understanding, particularly abilities that have a manipulation component such as drawing, arranging blocks, or recognizing something presented visually by feel. These functions are considerably less lateralized than language is, in that considerable visuospatial function remains after damage to the right side of the brain. This finding that some functions are preferentially lateralized to the right side of the brain has led to a revision of the understanding of the relationship of the cerebral hemispheres, now termed "complementary specialization."

The reason for and genesis of lateralization is not well understood. In experimental situations, some primates will show evidence of lateralization for particular tasks, but not in the systematic way seen in the human brain. The most intriguing case of lateralization in animals is birdsong, which is also controlled by the left half of the brain. The common feature of complex articulation in the case of birdsong and language has led to the hypothesis that the genesis of lateralization might be found in the need for unitary, lateralized motor control of a midline organ, either the various organs of articulation in speech or the syrinx in song.

The pattern of gyral folding in the left hemisphere is different from that on the right. This anatomical asymmetry and behavioral asymmetry are seen in human infants at birth, in the same proportion as there are right-handed individuals in the population at maturity. Ambidextrous or left-handed individuals show a greater likelihood of their language being bilaterally represented, or represented on the right, and their likelihood of an anatomical asymmetry is also less.

IV. DEVELOPMENT OF THE NEOCORTEX

The neocortex is one of the last structures to be formed in the human brain, with the final neurons composing it undergoing their last divisions and migrating into their mature position at about the sixth month postconception. Cells destined for the cortex are generated at a distant location from their adult position from precursor cells surrounding the lateral ventricles in the forebrain and migrate on supporting glial guides to their position on the brain surface. The first cells generated form the innermost layer, Layer VI, and later generated cells bypass these cells and form Layers V, IV, III, and II in turn. Cortical development is further marked by a large transitory population of neurons that are found on the white matter underlying the cortex, and in Layer I early in development, and that appear to die after the migration of neurons to the cortex is completed. The cortex will continue to increase in size well into early childhood by the growth of neurons and elaboration of their processes, but the full complement of neurons and their fundamental layering is established before birth.

Disorders of early cortical development are thus most likely to be caused by disorders of cell generation and migration, which is often visible as unusual cortical thinness, disorders of lamination, or mislocated clusters of cells that have failed to migrate ("ectopias").

V. CORTICAL PATHOLOGIES

Focal cortical damage, of the type caused by strokes, space-occupying tumors, and head trauma, was the first window into the nature of functional cortical organization. Although there are a few deficits that appear to be common to all types of focal brain damage, notably an increased distractibility and inability to screen out irrelevant noise and activity while performing tasks, almost all focal brain damage is followed by inability to perform those particular tasks that correspond to the areas of damage. Particular deficits in sensory systems, language, visuospatial computation, and the like will result, and not a generalized decline in all cognitive functions. Therapy for such dysfunctions typically involves retraining the individual to substitute intact cognitive functions for the lost ones, for example, defining maps by verbal landmarks rather than by spatial relations for an individual with a spatial deficit.

Seizure disorders involve abnormalities in the ongoing electrical activity of the brain and often involve the neocortex, particularly the temporal lobe. Normal electrical activity in the cortex is complex and asynchronous. The cortex has a number of recurrent and repeating circuits, however, that can produce rhythmic and synchronous activation when desynchronizing activity is removed, as normally occurs in sleep. In seizure disorders, cortical activity becomes massively synchronized and rhythmic. Drugs, or in some cases removal of the area generating the abnormal synchrony, are employed as treatments for seizure disorders. [*See* Seizure Generation, Subcortical Mechanisms.]

BIBLIOGRAPHY

Corballis, M. C. (1983). "Human Laterality." Academic Press, New York.

Damasio, A. (1994). "Descartes' Error: Emotion, Reason and the Human Brain." MIT Press, Cambridge, Massachusetts.

Felleman, D. J., and Van Essen, D. C. (1991). Distributed hierarchical processing in the primate cerebral cortex. *Cerebral Cortex* **1**, 1–47.

Finlay, B. L., and Darlington, R. B. (1995). Linked regularities in the development and evolution of mammalian brains. *Science* **1268**, 1578–1584.

Jones, E. G., and Peters, A. (1984). "Cerebral Cortex. Volume 2. Functional Properties of Cortical Cells" Plenum, New York.

Peters, A., and Jones, E. G. (1984). "Cerebral Cortex. Volume 1. Cellular Components of the Cerebral Cortex." Plenum, New York.

Neoplasms, Biochemistry

THOMAS G. PRETLOW
THERESA P. PRETLOW
Case Western Reserve University

GLOSSARY

Carcinoma Malignant neoplasm derived from epithelial cells

Karyotype Chromosomal constitution of the nucleus

Monoclonal Progeny of a single ancestral cell

Mosaic Organism whose tissues are composed of two or more kinds of cells with respect to chromosomes or gene expression

Transformation Conversion of cells in culture to a state in which their growth and morphology are similar to those of some kinds of malignant cells in culture.

Translocation Shifting of part of a chromosome to become attached to and a part of another

Tumor Literally, a swelling; in common parlance "tumor" is used synonymously with "neoplasm"

A NEOPLASM, OR NEW GROWTH, RESULTS FROM the process of neoplasia, which was defined by Perez-Tamayo in 1961 as "a form of abnormal cellular behavior, the result of many properties manifested by anatomic, functional and biochemical changes in tissues." Others might define neoplasia as a pathological alteration of cells that results in abnormal gene expression and abnormal responses to the factors that normally control growth. Subsumed under the general category of neoplasia, there is a diverse array of thousands of benign and malignant neoplasms in a wide variety of species. In that minority of neoplasms for which the etiology is precisely known, the causes are even more numerous than the varieties of neoplasms.

Biochemically, malignant neoplasms or cancers have been studied much more extensively than have benign neoplasms. Based on the small body of data available, the properties, biochemical and otherwise, of most benign neoplasms are intermediate between those of malignant neoplasms and those of their ancestral precursor types of cells. In a few instances, benign neoplasms are thought to be precursors of malignant neoplasms. In many cases the appearance of a malignant neoplasm is associated with a higher than expected frequency of benign neoplasms and "putative preneoplastic" lesions in the same organ. The incomplete state of our understanding of even the definition of neoplasia is exemplified by the fact that our operational definitions of specific neoplasms in animals are based on their gross and microscopic morphological recognition.

I. INTRODUCTION

Often there is marked heterogeneity in the behavior and the biochemistry of neoplasms both within and among even the neoplasms that are viewed as being derived from the same type of cell in the same species. This is true even among neoplasms produced experimentally by single chemically pure carcinogens in littermates of the same sex in highly inbred species fed synthetic diets and housed in carefully controlled environments. This heterogeneity markedly limits the breadth of applicability of generalizations that can be made about neoplasms in a description such as that presented here. Each individual neoplasm must be viewed in the context of the category to which it belongs; that is, its properties show enormous variability, even within the context of the type of cell from which it is derived, its etiology, and the species within which it arises.

As an example of this heterogeneity in a single tissue in humans, there are several dozen kinds of benign and malignant neoplasms of the epidermis (i.e., skin) and its appendages. Among only the neoplasms of the skin that are classified as malignant (i.e., cancers), basal cell carcinomas almost never spread (i.e., metastasize) to grow as tumors (i.e., metastases) at locations distant from their origins. In contrast, malignant melanomas usually metastasize to sites distant from their origins if they are not diagnosed and treated early. The biochemistries of the different cancers of a particular organ are as varied as their behaviors. [*See* Metastasis.]

In the space available here we should emphasize that we can present only an overview. The biochemical alterations that have been found in neoplastic cells include important changes in the cell membrane, the nucleus, the nucleolus, the mitochondria, and others. It would be difficult to find an enzymatic pathway or organelle that has not been shown to be altered in one or many neoplasms. We refer those who want a more detailed treatment of this subject and an exposition of the history of the biochemistry of neoplasia to the bibliography, particularly to the chapters by Bresnick and by Busch *et al.*

II. APPROACHES AND LIMITATIONS THAT AFFECT OUR KNOWLEDGE

A. Selection of Neoplasms for Study

Our knowledge of neoplasia reflects the variety of neoplasms that have been studied extensively to date. Only a small proportion of our knowledge comes from the investigation of primary or metastatic tumors obtained directly from the hosts in which they originated. A large proportion of our knowledge has been derived from the study of neoplastic cells in culture (*in vitro* models) or animal models of neoplasia. Some of the limitations of culture are discussed here. The two most commonly used animal models are neoplasms produced in animals by carcinogens and transplantable tumors.

Tumors produced by carcinogens in the laboratory are different in many ways from most tumors that occur naturally and, in particular, from tumors that occur naturally in humans. Most tumors known to result from the exposure of humans to carcinogens occur only after years or, more often, decades of exposure to carcinogen (e.g., lung cancer in cigarette smokers). In laboratory animals, tumors are generated in months by massive exposure to carcinogens. Though the biochemical characterization of neoplasms remains a complex task, the task is simplified somewhat by virtue of the fact that most neoplasms are derived from a single cell (i.e., they are clonal) (see Section VI). [*See* Tumor Clonality.]

In some cases, transplantable tumors (e.g., the "minimal deviation" hepatomas investigated by Van Potter, Harold Morris, and their collaborators) are relatively stable, have been studied after specified numbers of passages from one host to another, and have been diligently selected for their suitability for tests of specific theories. In contrast, much of our knowledge of the biochemistry of neoplasia has come from study of tumors (e.g., the Walker mammary carcinoma or the Dunning prostatic tumor) that have been transplanted hundreds of times under poorly defined conditions into hosts for which there are frequently no detailed health records for many decades. These tumors have undoubtedly undergone enormous selection such that any cells that elicited a strong immune response protective for the host have been eliminated and cells that grow slowly have been replaced (i.e., outgrown) by more rapidly growing cells. In the process of transplantation through many different hosts, often in many different institutions, it is likely that these tumors have acquired viruses from successive hosts. Although these tumors may have been appropriate for the state of our knowledge in the eras in which they were developed, some of the most commonly studied such tumors even originated in strains of animals that were different from those in which they are now passaged. One must wonder how applicable is the knowledge gained from these systems to neoplasms as they occur in nature.

B. Direct Biochemical Analysis

Our knowledge of the biochemistry of neoplasms is limited by the methods with which we approach the investigation of this subject. The biochemistry of neoplasia can be explored by the direct chemical analysis of neoplasms. When this approach is taken, the resultant data are limited by virtue of the fact that, on the average, at least half of the cells of most neoplasms are nonneoplastic (i.e., blood vessels, inflammatory and/or immunologically active cells, and fibroblasts). In addition, all neoplasms contain varying proportions of extracellular materials and the normal blood cells that are in the blood vessels that provide the circulation of the tumor. The biochemical data obtained from the direct analysis of neoplasms reflect the properties of both the neoplastic and stromal components of the neoplasms.

C. Analysis of Purified Cells

To circumvent the problem that neoplastic cells are diluted by stromal cells in nature, it is often possible to disaggregate neoplasms into suspensions of single cells that can be separated physically by sedimentation, by reactions with specific (often monoclonal) antibodies attached to insoluble supports, or by many other methods. The biochemical data obtained from the analysis of "purified" neoplastic cells are limited by (1) the degree to which the cells in the initial suspensions of cells from the neoplasm represent the cells that were present *in situ* in the neoplasm before the cells were obtained in suspension, (2) the extent to which they were purified, and (3) the proportion of cells lost during the process of purification. The process of obtaining cells in suspension never yields quantitatively all of the cells available *in situ*, and it is likely that there is usually some selection of subpopulations that might or might not be representative of the neoplasm during the process of obtaining cells in suspension and the purification of cells.

D. Analysis of Cultured Cells

The culture of cloned or purified neoplastic cells can be used to circumvent the problems discussed in the preceding sections; however, this is perhaps the most limited approach to the biochemical characterization of neoplasms as they occur in nature. Only a minority of the cells in a neoplasm have the capacity to grow under the conditions for culture that have been developed to date. In most cultures of neoplastic cells from a tumor, there is significant selection of subpopulations of cells that have the capacity to grow in culture and can grow more rapidly than the other neoplastic cells from that tumor that are capable of growth in culture. In addition, neoplastic cells often exhibit marked changes (i.e., modulation) in their biochemical properties as a consequence of the foreign environment that they encounter in culture.

III. GENERALIZATIONS ABOUT THE BIOCHEMISTRY OF NEOPLASMS

George Weber has organized much of our thought about the general biochemistry of neoplasia in recent years. After analyses of a large number of hepatocellular carcinomas (i.e., malignant tumors derived from the hepatocyte, the cell that accounts for the largest proportion of the liver) and a small number of other kinds of tumors studied in his and many other laboratories, he has articulated the concepts that the activities and/or concentrations of key enzymes, isoenzymes, and metabolites are "stringently linked" to the process of the transformation of a normal cell to a malignant cell. The concentrations of the same and/or other enzymes, isoenzymes, and metabolites are similarly linked to the progression of less aggressive malignant cells to more aggressive, invasive, and metastatic states. [*See* Isoenzymes.]

In his masterful reviews of this subject, Weber has pointed out that many dozens of enzymes and metabolites show consistent trends in changes in their concentrations and activities in neoplasms. When one examines successively more aggressive neoplasms of a particular type, it becomes apparent that there are relationships between the increased rates of growth of those particular kinds of neoplasms and the activities of particular enzymes.

In general, the most dangerous malignant neoplasms contain neoplastic cells that divide more rapidly than do normal cells. A few of the enzymes thought to be rate limiting, termed "key" enzymes by Weber, have been studied in detail in many kinds of tumors. For example, glucose-6-phosphate dehydrogenase is the rate-limiting enzyme in one of the two biochemical pathways available to the cell for the production of ribose 5-phosphate, a precursor required for the production of both DNA and RNA. Rapidly growing neoplasms must be able to synthesize DNA and RNA rapidly. This enzyme was shown to be greatly elevated in rapidly growing rat hepatomas as compared to normal liver by Weber, in human breast cancers as compared to normal breast by Russell Hilf, and in human prostatic carcinomas as compared to prostates without carcinoma in our laboratory.

Enzymatic activities that appear to be markedly altered in many different kinds of neoplasms include thymidine kinase, DNA polymerase, lactic dehydrogenase, pyruvate kinase, hexokinase, *N*-acetyl-β-D-glucosaminidase, phosphoserine phosphatase, ornithine decarboxylase, pyrroline-5-carboxylate reductase, and many others. The selection of enzymes mentioned here is almost arbitrary, since many dozens of enzymes have been shown to exhibit marked changes in activity in neoplasms. These enzymes are found in diverse areas of cellular biochemistry, including nucleic acid, amino acid, and carbohydrate metabolism. In many instances, certain isoenzymes (i.e., variants of the same enzyme) are much more affected than others by the changes associated with neoplasia and the progression of neoplasms. Often, the pattern of isoenzymes in neoplasms represents a recapitulation of the patterns observed during the embryonic development of the cell that is the ancestral precursor of the neoplasm. In addition to enzymes, alterations in cell adhesion molecules, growth factors, and the receptors for growth factors are thought to be intimately involved in the transformation of normal cells to malignant ones. Most of the biochemical alterations that affect many kinds of tumors can be postulated to involve what Weber has called key metabolic pathways, the alterations of which are tightly linked to the processes of neoplasia and/or progression without reference to any particular kind of cell or organ.

IV. BIOCHEMICAL CHANGES IN SPECIFIC TUMORS

In addition to the biochemical alterations common to many tumors, there are numerous alterations known to occur in most neoplasms of a particular kind, but not in most of the different kinds of tumors. An example of this type of change is arginase, which is elevated in most human prostatic carcinomas. Similarly, the expression of leucine aminopeptidase is markedly decreased or absent in most human prostatic carcinomas, but it is present in the epithelial elements of prostates without cancer. Colonic and pancreatic cancers frequently demonstrate both qualitative and quantitative changes in their glycoproteins. Although one could speculate about mechanisms that might potentially explain these changes, speculation is useful only when it leads to testable hypotheses; the mechanisms for many of these changes have still not been investigated in detail.

V. PRENEOPLASTIC BIOCHEMICAL ALTERATIONS

There are both systemic and focal abnormalities of cells that are associated with a greater propensity for the development of neoplasia. Systemic conditions (i.e., conditions that affect the entire body) include unrepaired chromosomal breaks (e.g., Fanconi's anemia or ataxia–telangiectasia).

Focal or organ-associated conditions associated with the development of neoplasia are numerous, and we shall give only a few examples. As mentioned, in many experimental systems for carcinogenesis *in vivo* in laboratory animals, the production of malignant neoplasms is associated with the production of benign neoplasms in the same organ system. This strongly suggests that similar etiological processes are important in the pathogenesis of both kinds of neoplasms.

In the same vein, both humans and experimental animals often present focal changes without the gross appearance of tumors in the same organ system that harbors overt neoplasms. Many of these changes are identified morphologically. Such changes include carcinoma *in situ*, or preinvasive cancer, a change that usually can be identified only microscopically; inborn, often genetic, errors (e.g., familial polyposis); abnormal cell kinetics (e.g., the failure of cells in the colon to mature and exhibit terminal differentiation normally); and many others. We are just beginning to be aware of other focal, probably clonal, changes in certain organ systems that might not always be amenable to detection by the usual histological methods. The most studied of these lesions are "enzyme-altered" foci of the liver and the colon.

Enzyme-altered foci, sometimes termed "putative preneoplastic lesions," are particularly important because they represent one of the few extensively studied biochemical alterations associated with the process of carcinogenesis. These lesions were first observed in the liver in the 1940s and have been observed more recently in the colon. The hepatic lesions have been characterized in many laboratories and have been shown to exhibit the increased or decreased expression of many enzymatic activities and/or concentrations demonstrated histochemically or immunohistochemically. These lesions might be associated with subtle alterations of the normal histology; however, they could appear histologically normal and could be apparent only by their histochemically demonstrable biochemical changes.

When sectioned serially and investigated for their expression of different phenotypic markers in successive serial sections, hepatic enzyme-altered foci could show abnormal expression of one or many phenotypic markers. Such markers include γ-glutamyl transpeptidase, benzaldehyde dehydrogenase, glucose-6-phosphate dehydrogenase, and many more.

Similar to enzyme-altered foci in the liver, enzyme-altered foci in the colon could appear histologically normal or could show characteristic morphological abnormalities. The most commonly observed markers that are abnormally expressed are *N*-acetyl-β-D-glucosaminidase (hexosaminidase), glucose-6-phosphate dehydrogenase, α-naphthyl butyrate esterase, diaminopeptidase IV, succinate dehydrogenase, and β-galactosidase. Despite the small numbers of phenotypic markers that have been examined to date in enzyme-altered foci in the colon, it is interesting that some of them are also abnormally regulated in several other kinds of malignant neoplasms. It appears that the abnormal expression of one or more of these markers precedes the overt dysplastic changes commonly viewed by pathologists as preneoplastic and/or neoplastic.

VI. OTHER BIOCHEMICAL FEATURES OF NEOPLASIA

In addition to the biochemical changes already discussed, others have attracted much attention and deserve mention.

A. Antigens in Neoplasia

The genetic alterations present in cancer cells might be expected to be reflected, at least in some cases, in differences in the antigens present in carcinomas compared to normal tissues. In several experimental systems, particularly tumors induced by oncogenic viruses, tumor-associated antigens have been demonstrated, but they appear to be more closely related to the infection by the virus than to the process of neoplasia. With a variety of tests, antigenic differences can be recognized in some human neoplasms. These findings, however, are rare.

Extensive study has concentrated on oncofetal antigens. These antigens are present at relatively high concentrations in some tumors as well as in embryonic tissues during development, but only at low or undetectable levels in adult tissues. These antigens might not cause an immune reaction in the host that carries the neoplasm, but they can be recognized by antibodies raised in other species. Because certain of these antigens are shed from the tumor into the circulation of the patient, their assay in peripheral blood is sometimes useful as a means of monitoring the progress of the tumor.

The antigens are too numerous to list in detail; however, they are well exemplified by carcinoembryonic antigen (CEA). CEA is expressed by the fetal endodermally derived tissues during the first and second trimesters of pregnancy. It is expressed by the colon only at low levels during adult life; however, it can be elevated in the blood of smokers who do not have identified neoplasms. Though CEA is not sufficiently specific to be useful for the screening of populations, it is expressed by several epithelially derived tumors (i.e., carcinomas), particularly colon cancers, and has been useful in monitoring their progress and response to therapy. CEA is produced at high levels in a variable proportion of carcinomas of the mammary gland, in neuroblastomas, and in other neoplasms, as well as in some nonneoplastic diseases.

B. Biochemical Demonstration of Clonality

With few exceptions (e.g., neurofibromas found in familial neurofibromatosis), the overwhelming majority of human neoplasms are clonal in origin; that is, they lack the capability of recruiting contiguous cells and grow by the steady proliferation of cells from an original cell that became neoplastic. The fact that most human neoplasms are monoclonal has facilitated their analysis by electrophoretic techniques that allow one to visualize genetic mutations, since a mixture of genetically different cells would introduce much more heterogeneity into the electrophoretic bands obtained.

Two lines of evidence lead one to the conclusion that most human tumors are monoclonal. The first of these is the fact that karyotypic analysis shows distinctively altered marker chromosomes in all cells of many tumors. The second line of evidence is derived from the study of isoenzymes encoded by genes located on the X chromosome. The Lyon hypothesis states that all X chromosomes, in excess of one, in cells are inactivated randomly such that normal tissues from heterozygous females, in which the two X chromosomes differ at recognizable genes, are mosaics of cells, some of which express the paternal X genes, whereas others express the maternal X genes. In contrast, most human tumors express only the maternal or the paternal gene in the malignant cells, whereas stromal cells from the same tumors, which are non-neoplastic, exhibit the same mosaicism as normal tissues.

The early work in this area was done by observing the patterns of glucose-6-phosphate dehydrogenase by electrophoresis of extracted tissues; however, this approach was restricted to the study of a relatively small subpopulation of people who were heterozygous for the enzyme. With the subsequent discovery of new markers on the X chromosome, one can now test for heterozygosity in tissues from most females. These studies have generally confirmed the earlier observations.

VII. SUMMARY

The biochemistry of most neoplasms is characterized by abnormal gene expression and the defective control of growth. In some cases the biochemical alterations are correlated to the rapid growth rate of the cells. Many biochemical changes are shared by several or most neoplasms; others are specific for only one or a small group of them. In some cases they result from specific chromosome alterations.

It is increasingly apparent that most well-studied neoplasms have changes (e.g., translocations, deletions, or duplications) that are not random. Their investigation has become an increasingly important approach to the systematic study of associated biochemical events. Particular changes appear to be characteristic of specific kinds of neoplasm. Marked differences among different kinds of neoplasms give emphasis to the need for detailed studies of the biochemistry of individual tumors.

BIBLIOGRAPHY

Bresnick, E. (1993). Biochemistry of cancer. *In* "Cancer Medicine" (J. F. Holland, E. Frei, III, R. C. Bast, Jr., D. W. Kufe, D. L. Morton, and R. R. Weichselbaum, eds.), Vol. 1, pp. 121–137. Lea & Febiger, Philadelphia.

Buck, C. A. (1995). Adhesion mechanisms controlling cell–cell and cell–matrix interactions during the metastatic process. *In* "The Molecular Basis of Cancer" (J. Mendelsohn, P. M. Howley, M. A. Israel, and L. A. Liotta, eds.), pp. 172–205. Saunders, Philadelphia.

Busch, H., Tew, K. D., and Schein, P. S. (1985). Molecular and cell biology of cancer. *In* "Medical Oncology. Basic Principles and Clinical Management of Cancer" (P. Calabresi, P. S. Schein, and S. A. Rosenberg, eds.), pp. 3–40. Macmillan, New York.

Kim, Y. S., and Byrd, J. C. (1991). Colonic and pancreatic mucin glycoproteins expressed in neoplasia. *In* "Biochemical and Molecular Aspects of Selected Cancers" (T. G. Pretlow and T. P. Pretlow, eds.), Vol. 1, pp. 277–311. Academic Press, San Diego, California.

Peraino, C., Richards, W. L., and Stevens, F. J. (1983). Multistage hepatocarcinogenesis. *In* "Mechanisms of Tumor Promotion" (T. J. Slaga, ed.), Vol. 1, pp. 1–53. CRC Press, Boca Raton, Florida.

Perez-Tamayo, R. (1961). "Mechanisms of Disease." Saunders, Philadelphia.

Pretlow, T. P. (1994). Alterations associated with early neoplasia in the colon. *In* "Biochemical and Molecular Aspects of Selected Cancers" (T. G. Pretlow and T. P. Pretlow, eds.), Vol. 2, pp. 93–141. Academic Press, San Diego, California.

Pretlow, T. G., Pelley, R. J., and Pretlow, T. P. (1994). Biochemistry of prostatic carcinoma. *In* "Biochemical and Molecular Aspects of Selected Cancers" (T. G. Pretlow and T. P. Pretlow, eds.), Vol. 2, pp. 169–237. Academic Press, San Diego, California.

Tronick, S. R., and Aaronson, S. A. (1995). Growth factors and signal transduction. *In* "The Molecular Basis of Cancer" (J. Mendelsohn, P. M. Howley, M. A. Israel, and L. A. Liotta, eds.), pp. 117–140. Saunders, Philadelphia.

Weber, G. (1983). Biochemical strategy of cancer cells and the design of chemotherapy: G. H. A. Clowes memorial lecture. *Cancer Res.* **43**, 3466–3492.

Neoplasms, Etiology

SUSAN J. FRIEDMAN
PHILIP SKEHAN
National Cancer Institute

IVAN L. CAMERON
University of Texas Health Science Center at San Antonio

GLOSSARY

Autosomal dominant pattern of inheritance In Mendelian genetics, the inheritance of a trait carried on an autosome (i.e., non-sex chromosome) by 75% of the offspring of a mating of heterozygotes [i.e., individuals who have alternative forms of the gene (nonidentical alleles) at that locus]. At the cellular level, a genetic mutation is dominant if it is expressed in the presence of the normal (wild-type) gene

Autosomal recessive pattern of inheritance Inheritance of an autosomal trait by 25% of the offspring of a mating of heterozygotes. At the cellular level, a recessive genetic mutation is expressed in the homozygous state, in which both alleles are identical

Cellular oncogene (protooncogene) Gene that normally functions in growth and differentiation, but can contribute to tumorigenesis when mutated or activated by gene rearrangement

Enhancer DNA sequence that increases the transcriptional activity of a promoter

Genotoxic Directly or indirectly damaging to genetic material

Mitotic chromosome nondisjunction Failure of the replicated chromosomes to separate during mitosis

Neoplastic transformation or conversion Accumulation within a cell of multiple individual transformations, which collectiely give rise to an expanding population of abnormal or inappropriate cells that progressively destablize normal tissue structure and function

Programmed cell death or apoptosis Apoptosis is derived from Greek, meaning "falling off" like petals from a flower. It is a morphologically distinct and genetically regulated form of cell death that is critical for normal development and for normal cell turnover in the adult body and can, under certain conditions, lead to neoplasia. It is characterized by internucleosomal fragmentation of DNA, chromatin compacted into membrane-delineated masses, and shrinkage of the cytoplasm associated with protuberances or "blebs" that pinch off to be phagocytized by adjacent cells. Little or no inflammatory reaction is stimulated by the apoptotic process as opposed to the necrotic form of cell swelling, rupture, and death caused by exposure to a poison

Promoter Region of a gene that binds RNA polymerase and initiates transcription

Retroviral oncogene Transforming gene of an RNA tumor virus that is derived from a cellular gene sequence by transduction

Southern blot hybridization Technique that detects specific DNA sequences in the genome by electrophoresis of restriction endonuclease-digested DNA on agarose gels, denaturation, and transfer of the separated fragments by capillary blotting onto nitrocellulose filters, and hybridization to a specific labeled DNA or RNA probe

Stem cells Early embryonic cells with an extensive capacity to reproduce themselves and to give rise to more differentiated daughter cells

Tissue homeostatic mechanisms Regulatory mechanisms that operate collectively to restrict the degree of variation in the size, function, and cytoarchitecture of normal tissues so as to maintain tissue stability

Transduction Process by which a virus transfers genes from one host cell to another

Transformation Long-term deviation of a cellular trait(s) from the normal range of values for that trait(s); changes in culture in the appearance and the behavior of eukaryotic cells that are produced by oncogenic agents

Transgenic mouse Mouse whose somatic cells contain a foreign DNA sequence that was introduced into the germ line

Transposable element (transposon) Mobile genetic element that can transfer genes from one chromosomal region to another; its structure consists of a defined gene sequence flanked by a long terminal repeat of several hundred bases and a short base repeat of genomic DNA at the insertion site

Tumor suppressor genes Genes that suppress malignant phenotype by negative regulation of growth-promoting genes. Inactivation or loss of function of a tumor suppressor gene usually requires a deletion or point mutation of one allele followed by loss or inactivation of the other allele. These genes are recessive in nature and resulting mutations may be inherited through the germ line

"CANCER" IS A COLLECTIVE TERM FOR SEVERAL hundred diseases characterized by the formation of a tissue mass that is inappropriate for its location, is uncoordinated in its growth and function with surrounding tissues, and persists indefinitely when an evoking stimulus is removed. A unique characteristic of malignant tumors is their ability to invade normal tissues and to metastasize to other sites in the host. The causes of human neoplasms are, for the most part, unknown. It is unlikely that all tumors have a common origin, although, by the process of natural selection during tumor evolution, they might come to acquire common phenotypic features. The most common types of tumors are thought to be nonhereditary, but a subset of these could, nevertheless, share a common etiology with rare inherited predisposition to a particular tumor type.

The mechanisms of tumor development are also uncertain. Cancer is a long-latency multistage process initiated by certain types of viruses, chemicals, radiation, and other tissue-damaging agents. Abnormal cells that arise during tissue injury and adaptation are triggered to begin a series of successive phenotypic changes. They undergo limited self-terminating growth in response to an appropriate signal, accumulate genetic damage by rare mutation-like events, form colonies, and give rise to subpopulations at an increased risk for genetic error.

Neoplastic transformation is a rare event, with an estimated frequency of one in 2×10^{17} mitoses, that occurs when cumulative changes are sufficient to permit growth to a physiologically significant size with disruption of local tissue stability. Changes in the structure and activity of genes involved in the regulation of normal tissue growth and differentiation have been detected in developing tumors and could be necessary for neoplastic conversion. Ultimately, the expression of malignancy is controlled by the specific tissue environment in which these cellular changes have occurred.

I. BIOLOGY OF NEOPLASTIC TRANSFORMATION

A. Classic Pattern

The classic pattern of tumor development is observed in chemically induced tumors of skin, liver, bladder, and several other tissues and can be divided into initiation, promotion, and progression stages. The process begins when normal tissue is exposed to and changed by an initiating stimulus. This event occurs in only a small fraction of the population, on the order of 0.01% or less, which varies directly with the strength of the stimulus.

There is evidence from liver carcinogenesis studies that stem cells, rather than mature hepatocytes, could be the relevant targets for initiation. The initiated state is transient and decays, unless it is stabilized within a critical period by cell proliferation. Initiated cells deviate minimally from normalcy, are often undetectable, and can remain latent for years, even decades.

The distinguishing characteristic of initiated cells is their ability to grow in response to a tumor-promoting stimulus that is ineffective for, or actively inhibits, the growth of the surrounding normal tissue. The selective growth advantage of initiated cells partly depends on their acquisition of resistance to the growth-inhibitory and cytotoxic actions of promoting agents. The recently discovered mammalian genetic stress response could allow initiated cells to respond rapidly to the presence of noxious growth-inhibitory agents by synthesizing a common set of proteins to counteract the proliferation block and minimize genotoxic damage. Included in this group of stress-induced proteins are the heavy metal-binding protein metallothionein, DNA repair and recombination enzymes, a gene amplification-producing factor, proteins with cell type-specific functions, and proteins with suspected growth-associated functions (e.g., ornithine decarboxylase, c-fos, and p53).

If initiated cells are repeatedly exposed to promoting agents, they form multicellular colonies called nodules, foci, polyps, or papillomas. Most colonies regress when the growth-promoting stimulus disappears. The cells assimilate into the normal tissue structure, where they remain dormant but viable for extended periods of time. A small proportion (i.e., 1–3%) of the colonies persist and can remain stable in size for decades without net growth, while the cells undergo progressive and potentially destabilizing changes. With time the number of altered traits per cell and the risk of neoplastic conversion increase. The changes occur randomly and independently of one another with no apparent order or pattern. Recent studies on genetic alterations in human colorectal cancer development suggest that the number of accumulated changes, not the order in which they occur, is associated with malignancy.

B. Nonclassic Pattern

A form of stomach cancer known as intestinal-type gastric carcinoma deviates in several important respects from the classic initiation–promotion model. Certain human populations are at high risk for this disease because of a maternally inherited susceptibility to atrophic gastritis, combined with a high level of exposure to dietary carcinogens. Dietary substances and certain medications chronically irritate the gastric mucosa, which results in an increased turnover of gastric epithelium. With time and under the influence of additional irritating substances, the gastric glands atrophy and the epithelium is replaced, first by small intestine epithelium and later by large intestine epithelium. This atypical epithelium becomes increasingly disorganized and populated by cells that express traits characteristic of less mature gastrointestinal precursor cells. Eventually, a neoplastic growth will arise. [*See* Gastrointestinal Cancer.]

It is not known whether the targets for neoplastic conversion are a small residual population of embryonic cells in the adult gastric epithelium with multiple developmental fates or, alternatively, whether the loss of gastric phenotypic markers and the expression of intestinal properties represent changes in gene expression in the mature gastric epithelium. Throughout much of the preneoplastic period, the development of gastric cancer remains highly susceptible to modulation by dietary influences, both positive and negative. The major apparent differences between this process and classic tumor development are the requirement for specific etiological agents at various stages of preneoplastic and neoplastic progression, and the emergence of neoplastic cells from tissue that no longer expresses the morphological and functional properties of the tissue originally exposed to the "initiating" stimulus.

II. THEORIES OF ONCOGENESIS

Various theories have been proposed to explain the origin of human cancer. A unifying theme that incorporates aspects of these theories is the idea that cancer originates from genetic alterations in somatic cells that act in combination to override the regulation of cell proliferation, differentiation, and death by normal tissue homeostatic mechanisms.

A. Somatic Mutation Theory

According to the somatic mutation theory, cancer is a genetic disease that originates from rare mutational events in somatic cells. Consistent with this idea are the observations that (1) many human tumors appear to be monoclonal (i.e., derived from a single cell), (2) tumor development occurs over many decades and in multiple stages, (3) many environmental carcinogens are mutagenic, (4) inherited DNA repair and chromosome instability disorders greatly increase the risk for certain cancers, and (5) dominantly inherited cancer susceptibility is associated with nonrandom recessive mutations.

1. Monoclonality of Human Tumors

Most human tumors of mesenchymal and hematopoietic tissues, and the few of epithelial origin that have been analyzed are monoclonal in composition. This is usually considered as evidence that the tumor developed from a single neoplastically transformed precursor cell, although the same result would be obtained by the selective outgrowth of a single clone from an initially polyclonal tumor. The monoclonality of many human tumors is consistent with the idea that the cellular changes involved in neoplastic conversion are rare somatically inherited events.

The clonal composition of tumors can be analyzed by several methods. Tumors of B lymphocytes (i.e., antibody-producing cells) are usually examined for the expression of a common surface immunoglobulin molecule or the corresponding DNA sequence; T-cell tumors are analyzed for a common arrangement of antigen receptor DNA; and other tumors for the presence of unique chromosomal markers or for the expression of one of two allelic forms of an X chromosome-encoded gene product in heterozygous females.

The enzyme glucose-6-phosphate dehydrogenase has traditionally been used as a marker for clonal analysis.

Another approach is based on restriction fragment-length polymorphisms. This method relies on the fact that, early in embryogenesis, one of the two X chromosomes of a somatic cell is functionally inactivated in a random and permanent manner. The same inactive X chromosome is inherited by all descendants of that cell. Tumors arising monoclonally in women who are heterozygous for an X-linked gene with allelic forms *a* and *b* express one allele, whereas normal tissue or tumors that originate from multiple cells have active forms of both alleles.

Newly developed methods that allow one to follow the monoclonal origin of tumor cells include cell-specific antigenic markers on B lymphocytes, the ability to follow repeated DNA sequences (trinucleotide expansions) in cells, and reverse transcriptase–polymerase chain reaction to amplify unique gene expression in a specific cell type. Results from all of these newer methods support the monoclonality of tumors.

2. Multistep Nature of Human Tumor Development

The observations that human cancers exhibit a long latency period and an age-dependent increase in incidence and arise from foci of disorganized hyperplastic tissues, rather than from normal tissue, suggest that tumor development is a mutation-like process.

3. Most Environmental Carcinogens Are Mutagens

Most environmental carcinogens can mutate genes and structurally damage chromosomes. However, there are several significant differences between mutagens and carcinogens that suggest that mutagenesis alone might not be sufficient for tumor induction. The most common mutagenic agents—ultraviolet light and ionizing radiation—account for only a small proportion of total cancers; some potent mutagens lack carcinogenic activity (e.g., nitrogen mustards and some nononcogenic viruses): Mutagens also act directly and efficiently to damage their cellular targets, whereas carcinogens damage many more cells than they neoplastically transform and act indirectly to trigger cellular changes that become self-sustaining in the absence of the carcinogen.

4. Disorders of DNA Repair and Chromosome Instability Associated with an Increased Cancer Risk

Rare recessively inherited syndromes that alter chromosome stability and the repair or removal of damaged DNA in somatic cells strongly predispose afflicted individuals to certain types of tumors. These syndromes include xeroderma pigmentosum, ataxia telangiectasia, hereditary nonpolyposis colorectal cancer, Bloom's syndrome, Cockayne's syndrome, and Fanconi's anemia. Bloom's syndrome and ataxia telangiectasia are also associated with impaired immune function. Xeroderma pigmentosum, the best studied of these disorders, is a family of diseases that impairs an important cellular mechanism that repairs DNA damage by excising pyrimidine dimers caused by ultraviolet light. Patients with xeroderma pigmentosum are unusually sensitive to sunlight and develop skin tumors at 10,000 times the normal rate.

Diseases that affect the stability of chromosomes and the repair of DNA damage do not increase the risk for all cancers. Rather, for unknown reasons, each disorder is associated with only one or a few types of cancer. Tissue-specific differences in exposure to particular types of DNA-damaging agents, in the kinetics of cell renewal, in the efficiency of repair of DNA damage, in the accessibility of relevant genomic targets to damage, and in the functions of the cellular processes rendered defective by the mutations, could contribute to the observed patterns of tumor development.

5. Dominantly Inherited Cancer Susceptibility Diseases

Epidemiological and cytogenetic studies of familial patterns of cancer have uncovered as many as 50 forms of cancer with an autosomal dominant pattern of inheritance. That is, family pedigrees reveal that cancer susceptibility is transmitted to 75% of the progeny of a mating between two heterozygotes. These diseases include tumors of gastrointestinal tract (e.g., familial polyposis and colon cancer), nervous system (e.g., retinoblastoma and neurofibromatosis), endocrine system (e.g., multiple endocrine neoplasia syndromes), and kidney (e.g., Wilms' tumor). In retinoblastoma, a rare childhood eye tumor that occurs in hereditary and nonhereditary forms, the genetic locus for susceptibility has been mapped to chromosome 13, band q14, and the gene has been cloned.

In the hereditary form of this disease, one of the wild-type alleles is inactivated in the germ cells of the parent and the mutation is therefore present in all somatic cells of the embryo. Inheritance of the disease follows a dominant pattern, although at the cellular level the retinoblastoma susceptibility gene acts recessively. Only after the second allele has become inactivated or lost in a developing retinal cell (i.e., retinoblast) does cancer develop. The reasons that tumors develop exclusively in retinal tissue are unknown. Nonrandom duplications and amplifications of other chromosomes occur frequently in retinoblastoma cells and may be involved in tumor formation. In nonhereditary retinoblastoma the inactivation of both alleles occurs after birth in a single retinoblast by two independent events; thus, the tumor is even more rare.

Fifty percent of the patients who receive radiation treatment for hereditary retinoblastoma develop the rare bone tumor osteogenic sarcoma 6–8 years later. In these tumor cells both copies of the retinoblastoma gene are functionally inactivated. This suggests either that the wild-type retinoblastoma gene serves an important regulatory function in both tissues or, alternatively, that the mechanisms of inactivation of the wild-type retinoblastoma genes can affect nearby genes that predispose to osteosarcoma. In both the hereditary and nonhereditary forms of retinoblastoma, these mechanisms most frequently involve the loss of the wild-type allele in one of the daughter cells produced at mitosis, because of the failure of replicated chromosomes to separate (i.e., chromosome nondisjunction) or because of an exchange of DNA segments between strands of a chromosome pair (i.e., mitotic recombination). Other rare childhood tumors, including neuroblastoma and Wilms' tumor, are thought to originate by similar mechanisms involving different chromosomal loci.

In contrast to the two-step model proposed for retinoblastoma, the deletion of multiple alleles is necessary for the induction of cancer of the colon and the rectum, which likewise occurs in hereditary and nonhereditary forms. In familial polyposis, a dominantly inherited disorder characterized by the development of multiple colon polyps and a greatly increased risk of colorectal cancer in young adulthood, deletions of the presumptive susceptibility locus (the APC gene locus on chromosome 5) are not found in early adenomatous lesions. Multiple nonrandom deletions of chromosomal segments occur in both hereditary and nonhereditary tumors, most commonly involving putative tumor-suppressor gene loci on chromosomes 17 and 18, but no single deletion is uniquely associated with tumorigenesis.

B. Recombination Theory

In experimental carcinogenesis systems the frequency of cells with altered characteristics is often in the range of 10^{-2} or higher, whereas the spontaneous mutation frequency is estimated to be 10^{-6} per locus per generation. This discrepancy poses difficulties for the somatic mutation hypothesis. It suggests that tumorigenesis might not be initiated by localized mutations in DNA, but by chromosomal rearrangements that cause large-scale changes in gene expression and destabilize the genome, making possible the appearance of cells with altered characteristics. Consistent with this hypothesis is the observed karyotypic (i.e., chromosomal) instability of most solid tumors and the presence of specific chromosomal translocations in hematopoietic tumors that retain a normal chromosome composition.

Two known mechanisms could contribute to the large-scale genetic rearrangements and changes in gene expression that occur during tumorigenesis: chromosomal translocation and genetic transposition.

1. Chromosomal Translocation

Nonrandom chromosomal translocations, in which parts of two different chromosomes are exchanged, are consistently present in certain tumors of B and T lymphocytes and could contribute to the etiology of these diseases. Chromosomal translocations to 14q32, a region on the long arm of chromosome 14 that contains the immunoglobulin heavy-chain locus, are common in a variety of leukemias and lymphomas. One of the best studied of these translocations has an exchange between this locus and a locus on chromosome 8 that contains the cellular oncogene c-*myc*. This exchange occurs in 80% of a rapidly growing tumor, Burkitt's lymphoma, and deregulates the expression of c-*myc*, presumably by bringing it under the control of the regulatory elements of the immunoglobin heavy-chain locus. The translocation event occurs in proliferating B lymphocytes that have undergone the final maturation event (i.e., heavy-chain switching) and would normally enter a resting state. [*See* Lymphoma.]

In the endemic African form of Burkitt's lymphoma, the combined effects of chronic infection of cells with Epstein–Barr virus and malaria could predispose cells to the translocation event. The virus stimulates B lymphocyte proliferation and expands the population at risk for the translocation, while the malaria parasites "swamp" the immune system, preventing the virally infected cells from being attacked by the immune defense mechanisms, as they normally would be.

The precise contribution of the translocation to the tumorigenic process is currently unknown. The translocated c-*myc* is no longer responsive to endogenous signals that suppress transcription of the nontranslocated gene. Experiments in transgenic mice have shown, however, that the deregulated expression of c-*myc* is not sufficient to cause tumors. Under the control of the enhancer of the gene for the immunoglobulin heavy chain, c-*myc* is constitutively expressed (i.e., not regulated) by the B-cell lineage lymphocytes. This results in a fourfold increase in the total pre-B lymphocytes (an early stage of development in this lineage) and an increase in the level of proliferating cells.

One consequence of the expansion of the pre-B-cell population is an increased risk of recombinational errors during the assembly of immunoglobulin genes from their precursor segments. In pre-B cells of non-Burkitt's lymphomas, translocations of chromosomes 11 and 18 to the immunoglobulin heavy-chain locus occur during the process of assembly, at the stage of V–D–J joining (gene domains that are recombined during immunoglobulin gene formation). The analysis of the points on these chromosomes at which the interchromosomal exchange takes place suggests that the recombinase enzyme that mediates this exchange may erroneously join a heavy-chain J segment to a cellular oncogene sequence.

In addition to altering the activity of cellular genes, translocations that interrupt genes can cause the production of abnormal gene products. In the early stages of chronic myelogenous leukemia, the distal end of the long arm of chromosome 9 translocates to chromosome 22 to form an abnormal chromosome (the Philadelphia chromosome) and a fusion gene (c-*abl*/*bcr*) whose product is a mutant protein kinase, an enzyme that is thought to play a significant role in the regulation of gene expression.

2. Genetic Transposition

Transposition is a genetically controlled process in normal tissues that can create new gene combinations by moving a segment of DNA from its normal location on a chromosome to another location on the same or a different chromosome. The segment is inserted at the new location by a recombinational process that does not require extensive homology between donor and recipient DNA. This phenomenon has been studied extensively in bacteria, yeast, and the fruit fly (*Drosophila melanogaster*), which contain special sequences (i.e., transposons) capable of performing these exchanges. By associating structural genes with new regulatory elements, transposition might produce changes in the expression of multiple genetic loci in human cells as well.

If this process were to occur in an uncontrolled random manner, the recombination of unrelated genetic sequences would destabilize the genome and conceivably initiate tumorigenesis. Retroviral sequences with a transposon-like structure are present in many places in the human genome and might serve as transpositional mediators. These sequences insert randomly into DNA, causing gene inactivation, deletion, inversion, and duplications.

A summary of the known types of genetic alterations that predispose to cancer include chromosomal translocation, inversions, and deletions; gene amplifications; aneuploidy and disomy; genetic imprinting; trinucleotide expansion; point mutations; microsatellite instability; and mismatch DNA repair defects.

C. Developmental Theory

The developmental theory of oncogenesis proposes that neoplastic transformation is an adaptive response of cells to an abnormal tissue environment that involves nonmutational changes in gene expression. The tumor cell retains the normal differentiation program of its tissue of origin and can be restored to normalcy in an appropriate tissue environment.

Certain highly malignant tumors of embryonal origin spontaneously differentiate into benign growths, and *in vitro* tumor cell lines can be induced to differentiate with appropriate stimuli. However, in none of these instances has it been demonstrated that normalcy has been restored—only that malignant behavior has been suppressed. So far, teratocarcinoma provides the most convincing evidence that certain cancers might be induced epigenetically and are potentially reversible.

Teratocarcinoma is a germ cell tumor that occurs in mice and can be experimentally induced by transplanting normal early embryos to an inappropriate location (e.g., the peritoneal cavity or a subcutaneous site). The resulting tumor arises from the neoplastic transformation of multipotential embryonic cells. The tumor contains stem cells, which form highly malignant rapidly growing tumors, as well as nonmalignant tissues derived from the differentiation of the stem cells. If a stem cell is introduced into a normal early mouse embryo, which is then allowed to develop *in utero*, its malignancy is suppressed and it can contribute to normal embryonic development.

The offspring of mice whose germ cells carry genetic markers derived from the teratocarcinoma cells develop normally and do not develop teratocarcinomas or other tumors at higher than normal rates. Teratocarcinoma is thought to be nonmutational in origin, because it shows a much higher frequency of induction than can be accounted for by spontaneous somatic mutation, and the loss of malignancy that accompanies cell differentiation is permanent.

As tumors develop they become increasingly disorganized and begin to express phenotypes that are inappropriate for their tissue of origin or stage of development. These changes have been attributed to a disturbance of normal differentiation, in which cells become arrested in their development, revert to an immature highly proliferative developmental stage (retrodifferentiate), convert to a different type of differentiated cell not normally found in the tissue (redifferentiate), lose their differentiated characteristics (dedifferentiate), or express genes that are either foreign to that tissue or are not expressed in adult tissue (abnormally differentiate).

The use of these terms to describe the phenotypic characteristics of a tumor has the unfortunate consequence of implying mechanisms that either have not been or cannot be demonstrated. Normal development is a highly ordered process involving a fixed sequence of progressive changes in gene expression that culminates in the creation of a specific set of precise characteristics for each cell lineage. Tumor progression, by contrast, is random and unpredictable. No two cells undergo the same set of changes nor is there a characteristic end state.

The characteristics of the tumor are determined by several factors: the developmental stage of the target cell(s) at the time of neoplastic conversion, the effect of neoplastic transformation on the process of differentiation, and the rate of intraclonal diversification during tumor progression. This is illustrated by the natural history of chronic granulocytic leukemia, a proliferative disorder of blood cells, which is characterized in its early stages by the overproduction of immature granulocytes. The initial neoplastic transforming event most likely occurs in a pluripotential precursor of the blood cell lineages. The clonal descendants of this cell can be identified by the presence of a unique cytogenetic marker, the Philadelphia chromosome, which, as mentioned in Section II,B,1, results from a translocation involving chromosomes 9 and 22.

In the early chronic stages of this disease, neoplastic transformation does not arrest differentiation nor does it alter the balance between proliferation and differentiation in cell lineages other than granulocytes. The reasons for the selective change in the population dynamics of the granulocyte lineage are not understood. It is only in the later stage of the disease, the blast crisis, that the balance between proliferating and differentiating cells is grossly altered in other lineages and the tumor is overpopulated by immature cells (so-called "blast" cells, from their appearance).

D. Oncogenes and Tumor Suppressor Genes

The contemporary views are that cancer is caused by mutation-like alterations of critical DNA sequences that regulate normal cell growth and differentiation and that tumorigenesis is controlled by both positive and negative mechanisms.

According to the cellular oncogene hypothesis, the positive control over tumorigenesis is exerted by transforming genes (i.e., oncogenes), first found in retroviruses, which derive from protooncogenes after activation by mutation or recombination. [*See* Oncogene Amplification in Human Cancer.]

Approximately 50 cellular oncogenes have been identified, in part by their presence in retroviruses, in part by their ability to cause a morphological transformation of the NIH/3T3 cell culture line, and in part by their increased expression in tumor cells. The gene products of protooncogenes and their derived oncogenes include components of cell signaling pathways (e.g., growth factors and their receptors, protein kinases, and G proteins) and nuclear DNA-binding proteins.

The original version of the cellular oncogene hypothesis viewed cancer as a one-step process caused by the activation of a single protooncogene. Although this was inconsistent with evidence from epidemiological and genetic studies that cancer is a multistep process with a long latency period, the hypothesis was widely accepted after it was known that the transforming sequences of retroviruses able to induce cancer in animals were derived from cellular genes. However, in many cases single activated oncogenes cannot render primary cultures of human or rodent cells *in vitro* tumorigenic or produce tumors *in vivo*. In tumors that consistently expressed a particular oncogene (e.g., colorectal cancer and the mutated *ras* gene), the evidence suggested that oncogene activation was not an obligatory early step in tumorigenesis.

The current hypothesis that cancer results from the synergistic actions of multiple oncogenes conforms to the traditional view of tumor development as a polygenic process. The tumorigenic potential of single oncogenes and oncogene combinations has been tested in transgenic mice and in organ reconstitution systems. Both techniques utilize hybrid genes composed of a truncated viral or cellular oncogene(s) linked to a strong promoter or enhancer. The transgene experiments use regulatory elements (e.g., the mouse mammary tumor virus promoter or the immunoglobulin heavy-chain enhancer) that can be controlled in a tissue-specific manner.

Transgenic mice are produced by microinjecting the gene construct into fertilized eggs. The eggs are then reimplanted into the oviduct of a pseudopregnant mouse and allowed to develop to term. The offspring are then tested for the presence of the transgene by Southern blot hybridization analysis of tail DNA. Transgenic mice carrying each of the two putative oncogenes that are to be tested in combination are mated to produce offspring whose cells carry both genes. At appropriate times, mouse tissues are examined histologically for preneoplastic changes, neoplasms, and oncogene expression.

The organ reconstitution approach uses the fetal mouse urogenital sinus, which, in males, develops into the prostate gland. The sinus can be dissociated, reconstituted with a small number of cells that have been retrovirally transfected with one or more activated oncogenes, and transplanted into the renal capsules of adult isogenic hosts, where it will continue to develop.

An important difference between the two techniques is that, in transgenic mice, all cells contain the transgene, but only certain tissues express it at high levels, whereas in the reconstituted urogenital sinus the proportion of cells in the tissue that constitutively express the gene can be experimentally varied to study the influence of cell–cell interactions on the process of tumorigenesis.

Several conclusions about oncogene cooperativity have emerged from studies on *myc–ras* gene interactions in these experimental models. v-*myc* is the transforming gene of several avian sarcoma viruses and codes for a DNA-binding protein. *ras* is the transforming gene of Harvey (H) and Kirsten murine sarcoma viruses. Its gene product is a GTP-binding protein with GTPase activity.

In transgenic mouse experiments, activation of the cellular *myc* gene (c-*myc*) alone is not sufficient to produce tumors, but it accelerates tumor formation with v-Ha-*ras*. Transgenic animals show an age-dependent incidence for cancer, as do humans. Furthermore, the tumors develop from single foci, even though most of the cells in a tumor-forming tissue express the transgene. In lymphoid tissue, which is amenable to clonal analysis, the tumors that form are monoclonal.

Thus, in addition to oncogene cooperativity, other events are apparently required for tumor induction. The only known instance in which a single activated cellular oncogene is able to induce tumors autonomously and synchronously in all cells of a tissue is the induction of mammary tumors by c-*neu* linked to the mouse mammary tumor virus promoter. c-*neu*, a cellular oncogene derived from a chemically induced rat neuroblastoma, codes for a protein related to the epidermal growth factor receptor and other tyrosine kinases. A homologous cellular oncogene present in human cells, c-*erb*-B2, is amplified and expressed in human breast cancers.

Oncogene experiments with reconstituted urogenital sinus have shown that the prostate gland develops normally when only a small percentage (i.e., 0.01%) of cells express v-Ha-*ras*. However, as the percentage of cells expressing the oncogene is experimentally increased, tissue organization is progressively disturbed. Mesenchymal cells over-proliferate, there is an increased formation of new blood vessels by endothelial cells (angiogenesis), and epithelial cell proliferation is inhibited. The expression of v-*myc* causes a hyperproliferation of epithelial cells in the periphery of the gland, but does not otherwise interfere with tissue organization or cell differentiation. When the sinus is reconstituted with small numbers of cells carrying both oncogenes, epithelial tumors form. The tumors appear to be monoclonal, chromosomally abnormal, undifferentiated, and invasive. These observations suggest that, in addition to oncogene cooperation, other genetic changes occur during tumor outgrowth.

The basis for the synergistic interaction of oncogenes is not yet understood. A possible explanation is that the selective effects of different oncogenes on multiple cell types within a tissue can severely disturb the cell interactions required for tissue homeostasis. Alternatively, the combined expression of oncogenes with different molecular mechanisms of action within a single cell type could produce local changes in cell interactions that would affect the behavior of neighboring normal cells.

Several mechanisms are thought to contribute to the preferential development of tumors in specific tissues (i.e., tissue selectivity) in oncogene experiments.

In transgenic mice, tissue selectivity is partially determined by the type of promoter–enhancer element that controls the transforming gene. The simian virus 40 T antigen, for example, is able to induce tumors in many types of cells in mice. However, when its expression is controlled by the murine promoter–enhancer of a gene expressed normally in the lens of the eye (α-crystallin), it induces lens tumors. Under the control of the insulin-regulatory region, it induces β pancreatic islet cell tumors (the cells that normally produce insulin). With the elastase promoter it induces tumors of the exocrine pancreas, in which the promoter is functional.

In addition, the transforming activity of an oncogene could be restricted to a particular portion of the differentiation program of a cell. Experiments with temperature-sensitive viruses, which express the potential to transform cells at a low, but not high, temperature, show that, as cells progressively become more differentiated at the nonpermissive (i.e., higher) temperature, they could lose their ability to express the characteristics of transformed cells when shifted to permissive (i.e., lower) temperatures.

Certain types of cells must enter into a predifferentiation growth arrest state to begin the differentiation process. The overexpression of certain oncogenes might prevent growth arrest and thereby cause cells to become refractory to normal differentiation signals. For example, it is speculated that the failure of maturing Burkitt's lymphoma B cells carrying the *myc*/Ig translocation to become proliferatively quiescent could be caused by the failure of the translocated *myc* gene to respond to an endogenous signal that turns off the normal allele. In some experimental systems, cells can be induced to differentiate by treatment with growth-inhibiting chemicals or by altering culture conditions to limit their proliferative activity.

Intrinsic differences in the proliferative capacity of specific cell types within a tissue might determine the cell type of origin of the tumor. In studies of v-Ha-*ras*/v-*myc* cooperativity in the urogenital sinus, it was found that prostate tumors arose exclusively in the epithelium, even though the mesenchymal cells appeared to be preferentially transformed by the retroviral oncogene carrier. A possible explanation for the formation of epithelial tumors is that, at the time of viral infection, the tissue was at a developmental stage in which epithelial cells had a significant growth advantage over other cell types.

Tumor suppressor activity could also play a decisive role in tissue-selective patterns of tumor formation. The existence of tumor-suppressing genes was originally deduced from somatic cell hybridization experiments, which showed that when normal and tumor cells were fused, the resulting hybrid cells were no longer tumorigenic. The loss of particular chromosomes occurred with prolonged passage of the hybrids and was associated with the reexpression of malignancy. It was hypothesized that these chromosomes contained dominantly acting tumor suppressor genes.

The inactivation or deletion of both copies of specific chromosomal loci is associated with the development of naturally occurring tumors as well. Studies in *Drosophila melanogaster* flies and *Xiphophorine helleri* and *X. maculatus* fishes suggest that the involved loci encode gene products that regulate developmental processes and act to suppress tumor formation. *Drosophila* flies with homozygous recessive mutations (i.e., mutations expressed when both gene copies are inactive or missing) at the lethal *giant larvae* locus become arrested at the larval stage of development and develop neuroblastoma-like tumors. They can be rescued from both defects by introducing into them the normal gene.

Crosses between two species of *Xiphophorine* fish, *X. helleri* (swordtail) and *X. maculatus* (platyfish), give rise to hybrid fish with an increased incidence of melanomas. The melanomas originate from platyfish-derived pigment cells. The ability of the pigment cells to become neoplastically transformed in these crosses maps genetically to a single gene, *Tu*. *Tu* is regulated by two other classes of genes, *R* and *Co*. *R* restricts the expression of *Tu* to specific tissues, whereas *Co* genes suppress *Tu* in specific regions of the body. Mutation of *R* or its separation from *Tu* by chromosome rearrangement leads to neoplasia. Mutation of a *Co* gene predisposes a fish to malignant transformation of a pigment cell in the part of the body which that *Co* gene regulates.

The loss or inactivation of tumor suppressor genes in human tissues is thought to be a significant factor in human tumor development. A tumor of developing retinal tissue, retinoblastoma, originates from cells in which both copies of the retinoblastoma susceptibility gene have been inactivated or lost (see Section II,A,5). The gene has been cloned and its product identified as a 105-kDa nuclear protein. The activity of the protein might be regulated during the cell cycle by changes in phosphorylation. When transfected into retinoblastoma cells, the wild-type gene inhibits cell growth and tumorigenicity. The gene is expressed by most normal tissues and is dysfunctional or lost in various tumors that are clinically unrelated to retinoblastomas, including breast and small cell lung carcinomas. [*See* Tumor Suppressor Genes.]

A second tumor suppressor gene, *p53*, is located on human chromosome 17p13. It encodes a 53-kDa nuclear phosphoprotein that is proposed to act as a negative growth regulator. The *p53* gene was initially misidentified as an oncogene based on its transforming activity in primary rat fibroblasts. Subsequently it was found that only the mutated gene possessed transforming activity, which was attributed to the ability of the mutant *p53* protein to complex with and presumably inactivate the wild-type protein. *p53* is considered to be a tumor suppressor gene based on its ability to inhibit transformation of primary rat fibroblasts by mutant *p53* and activated *ras* genes, the expression of *p53* by normal cells, and the inactivation/deletion of both *p53* alleles in colorectal and lung cancers. [*See* Tumor Suppressor Genes.]

Progress is being made in the identification and cloning of human tumor suppressor genes. Potentially important mechanisms of tumor suppression include (1) interference with neoplastic transformation by directly antagonizing the expression of a transforming gene, or by inducing cell maturation to a stage at which the cells are no longer at risk for transformation; (2) the inhibition of neoplastic cell growth by normal cells, mediated by the cell surface, extracellular matrix, or diffusible molecules; and (3) the prevention of tumor angiogenesis (i.e., the formation of new blood vessels and their penetration into the tumor), which is required for the growth of neoplastically transformed cells to a physiologically significant size.

Thus both protooncogenes and tumor suppressor genes are causal for cancer. Mutations in tumor suppressor genes cause loss of normal function that holds cell proliferation in check. Mutation in protooncogenes stimulates cell proliferation when the cells should not proliferate. No single mutation of a protooncogene or a tumor suppressor gene is likely to be the cause of any cancer, as each case of cancer can be characterized by multiple mutations. This fact is well illustrated for colorectal cancers (Fig. 1). However, many of the same genetic mutations are repeatedly encountered in different cancers, implying that

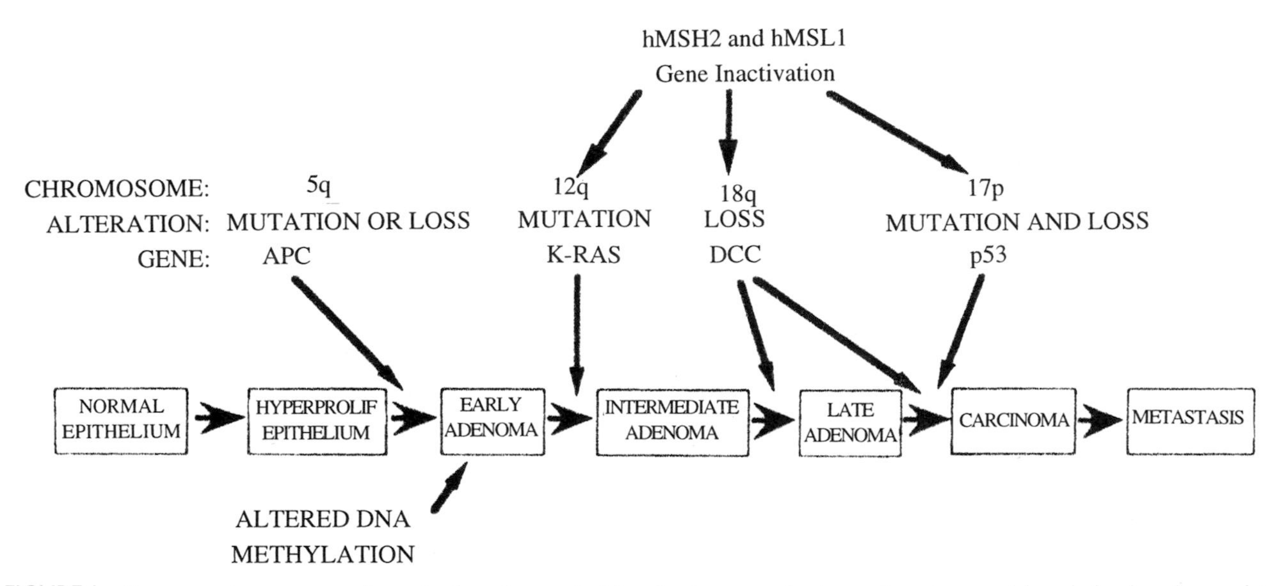

FIGURE 1 Representative sequence of genetic alterations underlying development of colorectal carcinoma; although developed to explain colorectal cancer, it has been generalized to other forms of cancer. This model of the multistep process of cancer, first offered by B. Vogelstein and colleagues, is based on analysis of tumors at progressive structural stages of tumorigenesis. Analyses revealed an accumulation of mutations during stages of tumorigenesis. The loss of tumor suppressor genes, that is, APC, DCC, or *p53*, usually requires two mutations to put out of action both good copies of each suppressor gene. Allelic losses—sometimes called loss of heterozygosity (LOH) because they are detected by restriction fragment-length polymorphisms—appear to occur at later stages of (colorectal) tumorigenesis. The lost genetic loci may contain tumor suppressor genes. An example is DCC (deleted in colorectal carcinoma). The onset of colorectal cancer typically occurs in people older than 55 years of age and involves five or more genetic alterations. The sequence, number, and types of genetic alterations differ between individual colorectal cancers. The hMSH2 and hMSL2 genes are hereditary nonpolyposis colorectal cancer genes and code for repair of regions of DNA where a mismatch in the nucleotide sequence of the strands of DNA has occurred. A loss of *p53* is associated with chromosomal instability and loss of *p53* at an early step in tumorigenesis is expected to hasten occurrence of an overt cancer. Early adenomas rarely have oncogenic mutations of *ras,* but *ras* mutations are commonly found in the later stages of tumorigenesis, perhaps because the later stage loss of *p53* no longer allows repair of a mutated *ras* protooncogene. The rare loss of function of a tumor suppressor gene can be inherited, as is the case with the APC gene. This hereditary loss of APC predisposes to development of multiple adenomas and high risk of cancer at an early age.

there are but a limited number of ways that a normal cell can give rise to a cancer cell through a multimutational process. Recently, attention has diverged from a focus on genes that control cell proliferation in cancer cells to include genes that control cell death.

E. Programmed Cell Death or Apoptosis

Not only is cancer caused by alterations of critical DNA sequences that regulate normal cell growth and differentiation, that is, oncogenes and tumor suppressor genes, but cancer can also be caused by alterations or translocations of critical DNA sequences that act either to induce or to repress a specific type of cell death referred to as programmed cell death or "apoptosis" (Greek for "falling off," like petals from a flower). Apoptosis normally occurs during the course of development in a predictable pattern, for example, during the tidy, timely, and rapid death of cells in the embryonic web tissue between forming fingers of the human hand. Cells that die by apoptosis display characteristic features (Fig. 2), namely chromatin distributes to the inner surface of the nuclear envelope accompanied by nuclear fragmentation, the entire cell shrinks, and the cytoplasm forms "blebs" that bud off as pieces of membrane-bounded cytoplasm containing organelles and nuclear fragments. Neighboring cells, which are not dying, quickly engulf these membrane-enclosed fragments and the dead cell fragments are disposed without induction of inflammation. This tidy process of cell death is distinctly different from necrotic cell death caused by poisoning or trauma, which typically results in cell swelling, rupture, and release of cytoplasmic contents into the extracellular environment, causing inflammation.

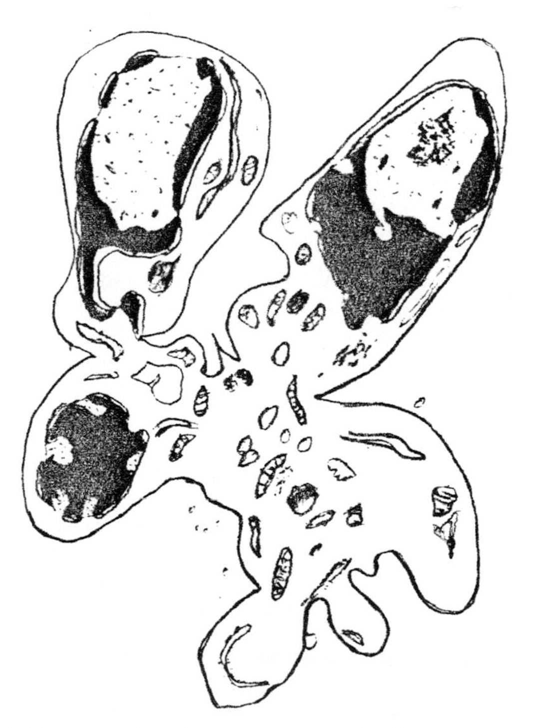

FIGURE 2 Drawing of a photomicrograph of a mammalian cell undergoing apoptosis. Discrete nuclear fragments with condensed chromatin adjacent to the nuclear envelope are apparent. Notice the cytoplasmic "blebs" or protuberances, which eventually pinch off and get phagocytized by neighboring cells.

Apoptosis is a normal and an essential process that occurs in a precise spatial and temporal pattern during development, but is no less important in the regulation of turnover and renewal of cell populations in the adult body. Most cell types in the adult body are capable of cell proliferation and therefore are potentially capable of tumorigenesis. Over time, proliferating cells would result in growth by increase in cell number and mass unless the increase is balanced by cell loss due either to cell death *in situ* or to exfoliation. Several genes that regulate apoptosis have now been identified. Obviously too little cell death could be as effective as too much cell proliferation in the generation of a tumor or neoplasm. Thus understanding the molecular genetic basis of apoptosis is desirable for assessment of risk of neoplasia and in the design of new cancer therapies.

The roundworm *Caenorhabditis elegans* has proved to be an excellent organism for the study of the molecular genetic basis of apoptosis to provide valuable clues regarding its genetic control in humans. The somatic lineage of *C. elegans* is completely known and invariant. Of the 1090 somatic nuclei generated in hermaphrodites, 131 undergo apoptosis in a predictable time and place. Fourteen genes involved in various steps of apoptosis have been identified. Three of the these genes, Ced-3, Ced-4, and Ced-9, control regulation and execution for all 131 cell deaths by apoptosis. Ced-3 and Ced-4 induce death and Ced-9 negatively regulates the pathway for apoptosis. Human genes with homology to apoptosis genes in *C. elegans* also act as inducers or as repressors of apoptosis. These genes include the prototype repressor of apoptosis gene Bc1-2 (B-cell lymphoma-2). This Bcl-2 apoptosis repressor gene was first revealed by its involvement in the human 14:18 chromosomal translocation found in most non-Hodgkin's follicular lymphomas. Overexpression of Bcl-2 caused by this chromosomal translocation results in resistances to apoptosis by a variety of insults and stimuli. The normal Bcl-2 gene product seems to regulate a final common pathway for apoptosis and allows resistance to apoptosis.

Recently, cDNAs have been cloned for several novel human genes, revealing a family of Bcl-2-related proteins. One member of this family, Bax, counteracts the action of Bcl-2 when the two gene products dimerize. The ratio of Bax to Bcl-2 determines cell life or death upon loss of a growth fator or exposure to a toxic stimulus.

Several viral proteins share the Bcl-2-related motif. Among these is a cell-death inhibitor, namely, the cowpox virus protein crmA. CrmA inhibits interleukin-1β-converting enzyme (ICE) (a cysteine protease with an aspartate-X moiety) and inhibits the induction of apoptosis by different stimuli, presumably by blocking the activities of ICE or ICE-like protease. Epstein–Barr virus and African swine fever virus produce proteins (BHRF-1 and LMWF-HL, respectively) that have Bcl-2 family homology and antiapoptotic activity. The mechanism by which these viral proteins prevent or induce apoptosis is not yet known.

Other genes involved in apoptosis but not related to Bcl-2 include c-*myc* and *p53*, but neither of these genes is indispensable for apoptosis of all types. Various oncogene products can suppress apoptosis, including adenovirus E1b, Ras, and v-Ab1. It seems likely that these other genes play a role in the etiology of at least some neoplasms. A number of anticancer therapies induce apoptosis in cancer cells but usually also induce apoptosis in normal cells. Much current research is therefore aimed at the selective manipulation of the genes involved in apoptosis in cancer cells so as to reduce unwanted apoptosis in normal host cells.

The apoptotic pathway can be divided into an induction phase, an effector phase, and a degradation phase. The induction phase is linked to death signals and to the lack of survival signals (removal of a required growth factor). The effector phase precedes the cell's irreversible commitment to death. The degradation phase commits the cell to the apoptosis. The mechanisms causing irreversible death or how those cells that receive death stimuli are protected from death are largely unknown. However, once a cell is committed to apoptotic degradation, the degradation process is the same in all cells. There is growing awareness that different Bcl-2 family members may operate in specific tissues or on different substrates within an individual cell to induce apoptosis.

F. Summary of Genetic Theories of Neoplasm Etiology

Cancer is the accumulation of mutations that make cell proliferation less dependent on the extracellular controls that regulate normal cells. Likewise it appears that cancer cells can also accumulate mutations that enhance pathways involved in cell survival independent of extracellular controls. This lack of dependence on extracellular controls on proliferation and on survival give a survival advantage to cancer cells over normal cells.

BIBLIOGRAPHY

Ames, B. N., Gold, L. S., and Willett, W. C. (1995). The causes and prevention of cancer. *Proc. Natl. Acad. Sci. USA* **92,** 5258.

Correa, P. (1988). A human model of gastric carcinogenesis. *Cancer Res.* **48,** 3554.

Duesberg, P. H. (1987). Cancer genes: Rare recombinants instead of activated oncogenes. *Proc. Natl. Acad. Sci. USA* **84,** 2117.

Fearon, E. R., and Jones, P. A. (1992). Progressing toward a molecular description of colorectal cancer development. *FASEB J.* **6,** 2783.

Greenblatt, M. S., Bennett, W. P., Hallstein, M., and Harris, C. C. (1994). Mutations in the p53 tumor suppressor gene: Clues to cancer etiology and molecular pathogenesis. *Cancer Res.* **54,** 4855.

Herrlich, P., Angel, P., Rahmsdorf, H. J., Mallick, U., Poting, A., Hieber, L., Lucke-Huhle, C., and Schorpp, M. (1986). The mammalian genetic stress response. *Adv. Enzyme Regul.* **25,** 485.

Ishizaki, Y., Cheng, L., Mudge, A. W., and Raff, M. C. (1995) Programmed cell death by default in embryonic cells, fibroblasts, and cancer cells. *Molec. Biol. Cell* **6,** 1443.

Klein, G., and Klein, E. (1986). Conditioned tumorigenicity of activated oncogenes. *Cancer Res.* **46,** 3211.

Levine, A. J. (1993). The tumor suppressor genes. *Annu. Rev. Biochem.* **62,** 1443.

Rubin, H. (1985). Cancer as a dynamic developmental disorder. *Cancer Res.* **45,** 2935.

Sinn, E., Muller, W., Pattengale, P., Tepler, I., Wallace, R., and Leder, P. (1987). Coexpression of MMTV/v-Ha-ras and MMTV/c-myc genes in transgenic mice: Synergistic action of oncogenes *in vivo*. *Cell (Cambridge, Mass.)* **49,** 465.

Sweezy, M. A., and Fishel, R. (1994). Multiple pathways leading to genomic instability and tumorigenesis. *Ann. N.Y. Acad. Sci.* **726,** 165.

Thompsom, T. C., Southgate, J., Kitchener, G., and Land, H. (1989). Multistage carcinogenesis induced by ras and myc oncogenes in a reconstituted organ. *Cell (Cambridge, Mass.)* **56,** 917.

Venitt, S. (1994). Mechanisms of carcinogenesis and individual susceptibility to cancer. *Clin. Chem.* **40,** 1421.

Nerve Regeneration

BERNARD W. AGRANOFF
ANNE M. HEACOCK
University of Michigan

GLOSSARY

Axon Cylindrical extension of the neuron, which carries electrical impulses (calld *action potentials*) away from the cell body of the neuron and toward the synapse, where the axon tip interacts with another neuron, a muscle cell, or a gland cell

Dendrite Extension of nerve cells, often forming tree-like arborizations that carry electrical impulses toward the cell body of the neuron

Extracellular matrix Web-like network of fibrous proteins and mucopolysaccharides in the extracellular space

Glia Nonneuronal cells of the brain (astrocytes and oligodendroglia) or peripheral nervous system (Schwann cells)

Growth cone Motile bulbous structure located at the tip of an elongating extension of the neuron (*axon* or dendrite)

Nerve Bundle of nerve fibers

Nerve fiber Axon or dendrite lined by folds of a glial cell, with oligodendrocytes lining axons in the brain and Schwann cells lining axons in the peripheral nervous system

Neuron Nerve cell consisting of cell bodies, dendrites (afferent extensions), and axons (efferent extensions)

Neuronotrophic factors Substances that neurons recognize in their environment and that are required for their growth and maintenance

INJURY TO A NERVE IS FOLLOWED BY DEGENERAtion of the nerve fibers distal to the injury. Nerve regeneration describes the process by which this injury is repaired and functional reconnection is attained. Successful nerve regeneration requires both a reprogramming of the biosynthetic capacity of the neuronal cell body and the presence of favorable environmental factors. If these requirements are met, new axons can grow out from the cut ends of the injured nerve, elongating sometimes over long distances, to establish the appropriate specific interaction with the original target tissue.

I. INTRODUCTION

The term "regeneration" generally connotes the ability of an organism to replace lost tissue. A starfish or a salamander can regenerate a severed limb. Regeneration in more advanced life-forms is more limited but can occur. For example, some mammals have the ability to regenerate lost liver tissue. Thus, after surgical removal of a hepatic lobe, functioning new liver tissue is produced. In contrast, the nervous system usually does not form new nerve cells (neurons) after injury and therefore does not replace lost tissue. Recovery of lost function in the nervous system, when it occurs, is mediated by a limited regenerative process. After section of a bundle of nerve fibers, axons may regrow from surviving neuronal cell bodies, and eventually appropriate connections with other neurons are reformed (Fig. 1). Given the limitation that we live our lives out with those neurons in our brains and spinal cords that we are born with, remarkable recoveries of function can be seen in favorable instances. Successful regeneration is generally encountered if a nerve bundle is severed in the peripheral nervous system (PNS). The PNS refers to all nervous tissue lying outside the central nervous system (CNS; the brain and spinal

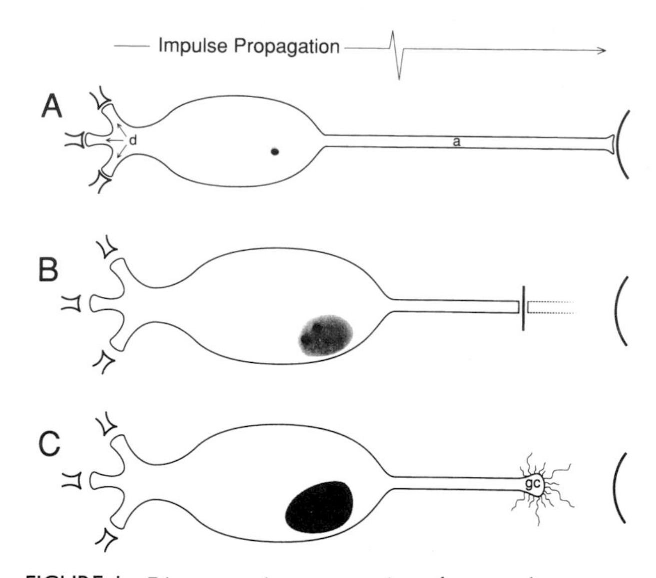

FIGURE 1 Diagrammatic representation of neuronal regeneration. (A) Intact nerve. Nerve impulses from dendritic processes (d) are conducted as a depolarization wave toward the cell body, which contains the nucleus, and continue away from the cell body, along the axon (a). This electrically mediated impulse is transmitted to a neighboring neuron or peripheral target tissue, depicted on the extreme right, by a chemical rather than electrical signaling mechanism, termed *neurotransmission.* (B) Transected nerve. After trauma to a nerve, there is degeneration distal to the lesion, the result of isolation of the axon from the cell body. Characteristic histological (microanatomical) changes in the cell body in response to the injury reflect an increase in RNA and in protein synthesis. (C) The regenerative response. If successful regeneration is to take place, the proximal stump (the end still connected to the cell body) of the severed axon develops a growth cone (gc) from which new growth occurs. When the growth cone retraces its way to its original site of connection, it is transformed into a presynaptic surface. The growth cone structure is lost, having served its purpose. If the nerve injury is in the CNS, the initial histological changes depicted in (B) may be seen, but there is no significant growth at the site of injury. Death of the damaged neuron often ensues.

cord). Thus if a major nerve bundle of the leg—the sciatic nerve—is damaged, there is a good possibility of recovery of function, because the sciatic nerve lies within the PNS, even though some of the cell bodies that give rise to its axons are in the ventral spinal cord, a part of the CNS. The damaged fibers will eventually regrow, to restore both sensation (sensory) and muscular (motor) function. This is in sharp contrast with the course of events after lesions within the CNS. On damage to the brain or spinal cord of warm-blooded vertebrates (birds and mammals, including humans), recovery of function is extremely limited. At the anatomical level, there is no evidence of axonal regrowth, such as seen in the lesioned sciatic nerve in the PNS. Apparent functional recovery after CNS injury seen in young individuals is attributed to the "plasticity" of the young CNS (i.e., its ability to remodel its structure is response to environmental stimuli). The concept of plasticity has also been invoked to provide a theoretical framework to explain adaptive changes in the CNS related to the storage of experiential memory. [*See* Nervous System, Plasticity.]

Why lesions in the PNS can induce successful regeneration whereas those in the CNS cannot, has profound theoretical as well as practical ramifications. An important difference between the local cellular environments of nerve fibers in the PNS and those in the CNS may lie in the nonneuronal supporting cells (glial cells) that are in contact with the nerve cell body and its processes in each milieu. In the CNS, the glial cells (astrocytes and oligodendroglia) provide a structural and metabolic scaffolding for the neurons, whereas in the PNS, this role is fulfilled by distinctly different cells, the Schwann cells. It could thus be that Schwann cells provide a nurturing neuronotrophic factor that CNS glia do not or, conversely, that CNS glial cells in some way inhibit the regenerative response, whereas Schwann cells permit it. Still other hypotheses suggest that the ability of a nerve to regenerate lies within a genetic programming potential within each neuronal cell nucleus. Many experimental approaches are directed at answering this question. Convenient model systems for studies of CNS regeneration are to be found in cold-blooded vertebrates, such as fishes and amphibians. For example, surgical crush of the optic nerve of the goldfish produces blindness, but this is followed by complete recovery of vision within a few weeks. This means not only that the 500,000 or so neuronal axons that constitute the optic nerve of a fish eye regenerate and reconnect, but also that the reconnection is very precise, because the retinal map projected on the brain is restored to its original distribution, each point on the optic tectum of the fish brain representing a specific locus on the retina of the eye. The optic nerve lies within the CNS, so that in humans, as in other warm-blooded vertebrates, blindness produced as a result of such damage to the optic nerve is permanent. [*See* Astrocytes.]

II. PROPERTIES OF REGENERATING NERVES

A. Growth Cones

Growth cones are characteristic organelles that are invariably present on the growing tips of nerve processes, whether such processes conduct impulses into the neuron (dendrites) or out of the neuron (axons). The two kinds of processes are collectively termed "neurites." Seen in still or kinetic images of cell cultures, the growth cones are sites of projections possessing vigorous activity: these are microspikes, also termed *filopodia*, that appear to be randomly searching, or even flailing, like so many fingers, while the growing neurite advances in a relatively straight path, elongating by adding new membrane at the "wrist" (Fig. 2). In the nervous system, neurites seldom travel alone. Instead, multiple neurites fasciculate (i.e., form bundles and travel together) sometimes along a path established by "pioneer" fibers.

B. Pathfinding and Specificity

How growth cones find their way to their specific targets is an important, if poorly understood, process. Although the degenerating stump can appear to redirect growing fibers, it can be demonstrated experimentally that the presence of the distal nerve stump is not necessary. In experimental studies, if a nerve segment is removed and the proximal stump misdirected, the nerve will nevertheless regrow appropriately. Eventual regrowth and reconnection with appropriate neuronal sites appear to involve both fiber–fiber and growth cone–target cell recognition. Speculation as to the nature of such interactions has centered on the chemoattraction hypothesis, first put forth by S. Ramòn y Cajal over a century ago, in which target tissue releases substances that are attractive to the advancing growth cone. Experimental support for such a mechanism has come from studies of cocultures of axon-generating cells and their target tissues, in which axon-attracting effects of target-released factors (some of which have been cloned) could be dem-

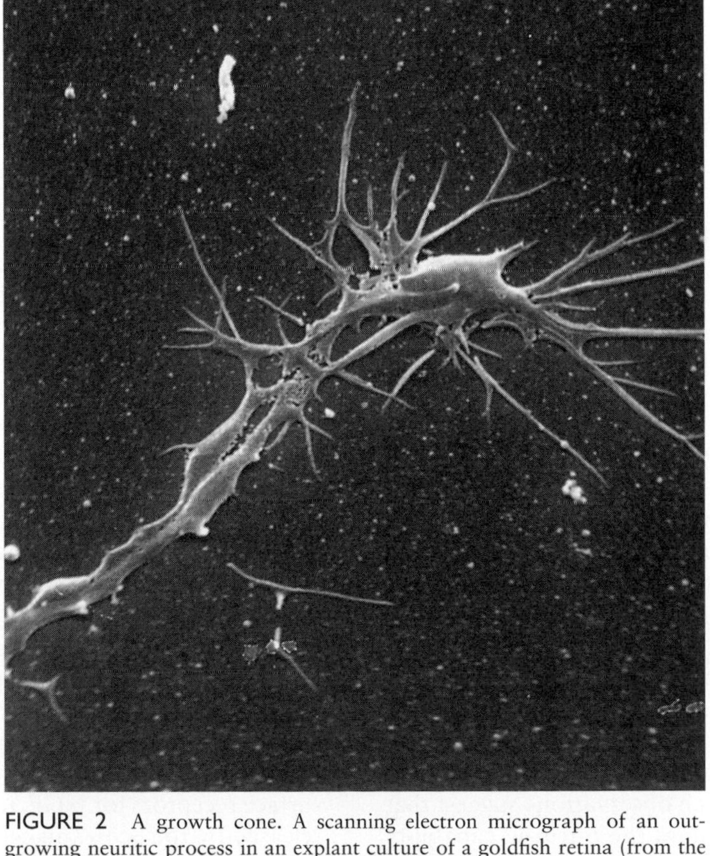

FIGURE 2 A growth cone. A scanning electron micrograph of an outgrowing neuritic process in an explant culture of a goldfish retina (from the authors' laboratory). The nerve fiber represents several neurites that have fasciculated into a single bundle. Magnification: ×4000.

onstrated. Directed collapse and retraction of growth cones may also be important mechanisms in the establishment of target specificity.

C. Axonal Transport

Neurons have the distinctive property (not shared with any other cell types) that they may possess extremely long processes, often thousands of times as long as the cell body. Because the cell bodies and dendrites are the only parts of the neuron that can synthesize proteins, a means must be present to move newly synthesized proteins to the axonal extremities. This is accomplished by the process of axonal flow (also referred to as *axoplasmic transport*). Not only are new neuronal membranes transported to the growth cone, but cytoskeletal elements, such as tubulin and neurofilament proteins, are supplied via anterograde transport (i.e., from the cell body outward). The neuronal nucleus also must receive information from its extremities (e.g., after injury), and this is accomplished by retrograde transport (i.e., movement toward the cell body).

D. Neuronal Cell Body Response: Growth-Associated Proteins

Axonal damage is often followed by characteristic morphological changes in the neuronal cell body. These include cellular hypertrophy, nuclear eccentricity, and increased numbers of nucleoli. A successful regenerative response is accompanied by a redirection of the biosynthetic apparatus of the neuron. Synthesis of proteins involved in neurotransmitter function is shut down, whereas that of cytoskeletal elements, structural components of the axon, may be greatly intensified. A number of proteins have been identified that are specifically expressed during CNS or PNS regeneration, as well as during development of the nervous system. These have been termed *growth-associated proteins* (GAPs). Of these, the most well-characterized is GAP 43, a membrane protein that is localized to the growth cone. Induction of GAP 43 is characteristic of neurons that undergo axonal regeneration. Under certain conditions, however, injured adult CNS neurons express GAP 43 transiently, apparently as part of an abortive regenerative response. Such observations suggest that adult CNS neurons maintain the intrinsic capability to regenerate.

III. EXTRANEURAL ASPECTS

Successful nerve regeneration requires not only that the neuronal cell body be capable of mounting the appropriate biosynthetic response but also that the environment surrounding the site of the nerve injury be capable of supporting the elongation and eventual functional reconnection of the axon. Thus, it has long been recognized that differences in extraneuronal environment between injured PNS and CNS axons contribute to the success or failure of nerve regeneration. The validity of this concept receives direct support from experiments in which mammalian CNS axons were transected and a segment of peripheral nerve (removed from the leg of a donor animal) was apposed, within the CNS, to the cut ends. Under these conditions, extensive regrowth of a small percentage of axons occurred throughout the length of the PNS graft. Molecular studies of PNS and CNS nerves are elucidating the nature of the environmental components that facilitate axonal regrowth. Available evidence suggests that two components are of primary importance: the presence of neuronotrophic factors and of a supportive substratum. Application to transected rat spinal cord of a peripheral nerve bridge with a trophic factor-impregnated glue, resulted in a convincing demonstration of functional regeneration, albeit with limited behavioral recovery.

A. Trophic Factors

Neuronotrophic factors (NTFs) are substances that are required for survival and maintenance of neurons and their processes. Loss of a supply of NTF may result in the death of the neuron. The most well-known example is nerve growth factor (NGF), a well-characterized protein that exerts trophic effects on some populations of PNS and CNS neurons. Isolation and cloning of other related CNS-active trophic factors have led to the discovery that NGF is a member of a larger gene family, the neurotrophins, which are active in a wide variety of CNS neurons. Trophic factors may be produced by the target tissue that is innervated by the nerve fibers. Delivery of the NTF to the neuronal cell body is accomplished after uptake at the nerve ending and subsequent retrograde axonal transport. Alternatively, the supporting nonneuronal cells may be a source of NTFs, as has been demonstrated for Schwann cells, which, in response to injury, produce not only NGF, but also cell-surface receptors for NGF. It is postulated that NGF is released by the Schwann cells, whereupon it binds to NGF receptor sites on the Schwann cell surface, thereby surving to attract growing axons to elongate along this surface. Such a mechanism illustrates that trophic factors may also have tropic effects (i.e., attracting and guiding growing axons in a specific direction). Thus, in some circumstances, NTFs may at least partially function as supportive substrata. However, the latter most commonly refers to web-like components of the extracellular matrix (e.g., collagen, fibronectin, or laminin).

B. Extracellular Matrix

Injury to PNS and CNS nerves of lower vertebrates is often accompanied by a increased amount of laminin in the extracellular environment surrounding the nerve. In the PNS, Schwann cells are the likely source, whereas in the CNs, astrocytes appear to be capable of synthesizing laminin in response to injury. Collagen and fibronectin, but especially laminin, have been found to promote the growth of neurites from cultured neurons or neuronal explants *in vitro*. These extracellular matrix proteins seem to provide an adhesive substratum, which facilitates the forward movement of the growth cone and elongation of the axon. The importance of a supportive extracellular matrix is reinforced by experiments in which a severed nerve tract in the rat brain was induced to grow across an implant of human placenta membrane, a rich source of laminin. [*See* Extracellular Matrix; Laminin in Neuronal Development.]

Other components of the neuronal environment that may contribute to a successful regenerative response are a family of surface membrane neural cell-adhesion molecules (NCAMs), which are found on neurons. NCAMs are thought to play a role in nerve fasciculation. There is also evidence that extracellular degradative enzymes such as proteases, elaborated by the growth cone or by supporting cells, may facilitate the passage of the growing axon through the surrounding tissue. Somewhat paradoxically, endogenous protease inhibitors (nexins) have been found to exert neuronotrophic effects *in vitro*.

C. Transplantation

Transplantation of neural tissue into injured brain has also been attempted as a means to facilitate directed axonal regrowth. Experiments with fetal neural transplants indicate some success in preventing the neuronal death that often results when a CNS neuron is disconnected from its target tissue. This may reflect the elaboration of trophic factors by the transplant. Although axonal sprouting through the transplant is seen to occur, it should be stressed that recovery of function requires regrowth to appropriate specific neuronal sites. Further efforts are aimed at exploiting advances in genetic engineering to construct implants of cultures that have been previously transfected with genes for trophic factors or other molecules, which may enhance the regenerative response.

D. Inhibitory Factors

In addition to the facilitory effects of trophic factors and extracellular matrix components on nerve regeneration, there are also indications that factors elaborated by CNS white matter are inhibitory to axonal outgrowth. Two proteins present in oligodendrocyte membranes and CNS (but not PNS) myelin have been identified as mediators of this inhibitory effect. In addition, myelin-associated glycoprotein (present in CNS and at low levels in PNS) also displays such inhibiting properties. Application of specific antibodies to these myelin-associated proteins neutralizes their axonal outgrowth-inhibiting activity and permits extensive neuritic elongation across a CNS myelin substratum *in vitro*. Treatment with such antibodies *in vivo*, in combination with the appropriate neurotrophic factor, permits substantial axonal regrowth from neurons after spinal cord lesion, although the number of neurons responding is very limited.

Injury to the brain and spinal cord is often accompanied by a glial response to the concomitant axonal degeneration, which results in the formation of scar tissue. Whether or not the glial scar constitutes a physical impediment to regrowth, one must also consider chemical signaling from glial cells as possible positive or negative factors in nerve regeneration.

IV. LIMITED REGENERATION IN THE CNS

A. Collateral Sprouting

Whereas nerve regeneration, as defined by outgrowth from the transected end of an axon leading to functional reconnection with the appropriate target tissue, does not occur in the CNS of higher vertebrates, there is nevertheless a response to injury. Axons from non-injured neurons, either adjacent to the injury or even at some distance from it, may sprout new axons (termed *collateral sprouting*) that will grow into and innervate the regions that had been denervated after the initial trauma. There are abundant examples of such apparently futile CNS responses that are thought to be an injury-induced exaggeration of the normal process of synaptic remodeling, which occurs during learning and memory formation. Whereas lesion-induced collateral sprouting and synaptogenesis have been amply demonstrated, the functional relevance of these processes remains an open question. It is possible that the phenomenon of collateral sprouting may actually compete with the process of terminal sprouting and thus contributes to an abortive regenerative response and lack of functional recovery. These observations of lesion-induced sprouting dramatize the challenge to neuroscientists to discover conditions under which these abortive responses can be converted to purposeful regrowth and recovery of function.

BIBLIOGRAPHY

Bähr, M., and Bonhoeffer, F. (1994). Perspectives on axonal regeneration in the mammalian CNS. *Trends Neurosci.* **17**, 473–479.

Cotman, C. W., Gomez-Pinilla, F., and Kahle, J. S. (1994). Neural plasticity and regeneration. *In* "Basic Neurochemistry" (G. J. Siegel, B. Agranoff, R. W. Albers, and P. B. Molinoff, eds.), 5th Ed. Raven, New York.

Hammerschlag, R., Cyr, J. L., and Brady, S. T. (1994). Axonal transport and the neuronal cytoskeleton. *In* "Basic Neurochemistry" (G. J. Siegel, B. Agranoff, R. W. Albers, and P. B. Molinoff, eds.), 5th Ed. Raven, New York,

Lindsay, R. M., Wiegand, S. J., Altar, C. A., and DiStefano, P. S. (1994). Neurotrophic factors: From molecule to man. *Trends Neurosci* **17**, 182–190.

Seil, F. J. (ed.) (1989). "Neural Regeneration and Transplantation." Alan R. Liss, New York.

Nervous System, Anatomy

ANTHONY J. GAUDIN
California State University, Northridge

GLOSSARY

Autonomic nervous system Portion of the nervous system that controls smooth muscle, cardiac muscle, and glands

Axon Extension of a neuron that carries an impulse away from the cell body

Cerebellum Portion of the hindbrain responsible for coordinating movement

Cerebrum Largest part of the brain, consisting of left and right hemispheres; it receives conscious sensation and controls voluntary motor activity

Dendrites Processes that emanate from the body of a neuron that carry impulses in the direction of the body

Diencephalon Portion of the brain lying between the mesencephalon and the telencephalon, connecting the cerebral hemispheres and the midbrain

Distal Farther, or farthest, from the origin of a structure or the midline of the body

Ganglion Mass of nerve cell bodies localized in structures that lie outside the central nervous system

Motor neuron Nerve cell that carries signals away from the central nervous system to a muscle or a gland

Myelin Fatty material produced in sheaths that wrap around the axons of neurons

Nucleus Group of neuronal cell bodies in the central nervous system, with processes that extend into neighboring nervous tissue

Reflex Involuntary response to a stimulus involving a neural pathway in a muscle or a gland

Sensory neuron Nerve cell that carries signals from sensory cells and organs to the central nervous system

Somatic nervous system Portion of the nervous system other than the brain and the spinal cord

Tract Bundle of nerve fibers that form a pathway in the central nervous system

THE NERVOUS SYSTEM IS AN ELABORATE COMMUNICATION system, composed of a network of cells that extends throughout the body, receiving information about the internal and external environments, assessing that information, and then sending signals to organs that cause an appropriate response. As such, it is one of two major systems used to regulate body processes. In this system, stimuli received by specialized sensory cells (e.g., heat-sensitive cells in the skin or light-sensitive cells in the eye) are transmitted as electrochemical impulses to the spinal column and the brain. When a response is called for, signals are sent to effector organs, glands, and muscles that respond by secreting a hormone or by contracting.

I. ORGANIZATION OF THE NERVOUS SYSTEM

The nervous system consists of two major portions: the central nervous system (CNS) and the peripheral nervous system (PNS). The CNS includes the brain and the spinal cord, and the PNS consists of all of the nervous tissue outside the CNS (Fig. 1).

The brain consists of the nervous tissue contained within the skull. It is in the brain where the evaluation of impulses from sensory organs is performed; where consciousness, personality, and emotion originate; and where the overall general control of other systems occurs. [*See* Brain.]

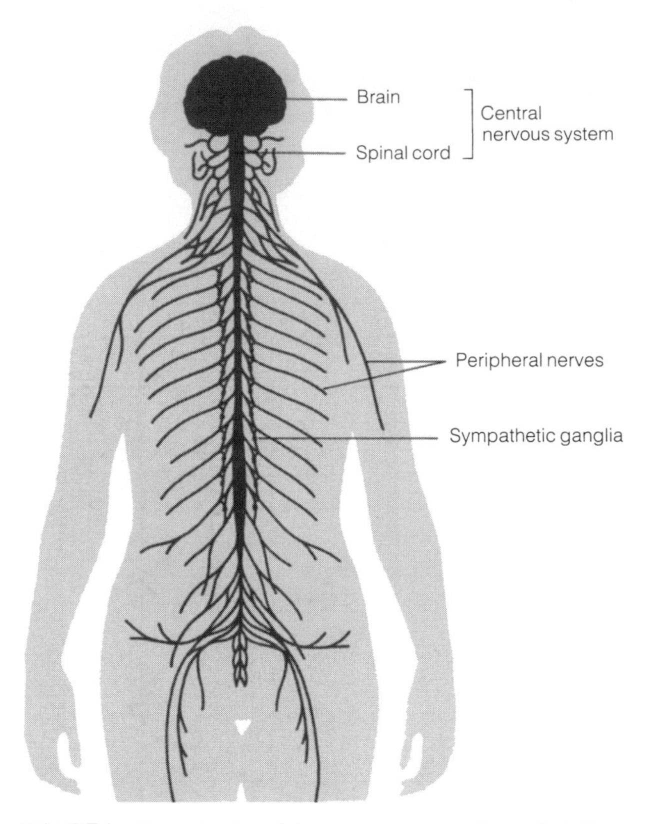

FIGURE 1 Organization of the nervous system. [From A. J. Gaudin and K. C. Jones (1989). "Human Anatomy and Physiology." Harcourt Brace Jovanovich, San Diego, California.]

The spinal cord extends from the brain through the column of vertebrae that compose the backbone, reaching from the lower portion of the brain to the lower spine. Within the spinal cord, nervous tissue is arranged in bundles of cells that provide circuits for the transmission of signals. Such bundles are referred to as tracts. [*See* Spinal Cord.]

The PNS is organized into nerves, bundles of nervous tissue that emanate, cable-like, from the CNS and extend throughout the body, where they provide pathways for signals traveling to and from the CNS. Functionally, the PNS consists of two portions, called the afferent and efferent divisions. The afferent division carries signals to the CNS from sensory organs, whereas the efferent division carries signals from the CNS to effector organs. The efferent division is further subdivided into the somatic nervous system, which carries signals specifically to skeletal muscles, and the autonomic nervous system, which carries signals to glands, smooth muscle, and the heart. Signals carried by the somatic nervous system are often produced by voluntary conscious activity of the brain, whereas signals produced by involuntary subconscious activity of the brain are usually carried by the autonomic nervous system. [*See* Autonomic Nervous System.]

II. ANATOMY OF THE ADULT BRAIN

The brain is organized into four major areas (Color Plate 1). The most anterior portion is the cerebrum, subdivided into two cerebral hemispheres, which fill most of the skull's cranial cavity. Posterior to these are the cerebellar hemispheres tucked under the rear of the cerebrum. Between the cerebellum and the cerebrum lies the diencephalon, composed of the thalamus and the hypothalamus. The remainder of the brain is organized into a rather compact brain stem, which sits at the top of the spinal cord. [*See* Hypothalamus; Thalamus.]

A. Cerebrum

The cerebrum consists primarily of two large hemispheres, each composed of several subdivisions (Color Plate 2). The cerebral cortex forms the outer cerebral layer and is composed of neurons that lack the myelin sheath produced by Schwann's cells. These nonmyelinated cells are dark grayish, so the cortex is frequently referred to as the gray matter of the brain. The neurons in the cortex make numerous interconnections with each other and with fibers in the underlying regions. [*See* Brains, Central Gray Area; Cortex.]

The interior of the cerebral hemispheres is composed mostly of myelinated nerve fibers (i.e., neurons whose fibers are wrapped with a myelin sheath), organized into discrete bundles, or tracts. The corpus callosum in Color Plate 2 is an example of a large tract of the white matter. These tracts connect different areas of the cortex on the same side of the brain and complementary areas on opposite sides of the brain. These myelinated fibers form the so-called white matter of the brain, because they appear white in brain sections. Embedded in the white matter are additional isolated masses of gray matter, the basal ganglia, or basal nuclei (Color Plate 3). The basal ganglia are important in the control and coordination of voluntary muscular movements. A horn-shaped cavity occupies a small space within each cerebral hemisphere, forming two lateral ventricles. These cavities, along with other cavities of the brain and the central canal of the spinal cord, contain cerebrospinal fluid, the special extracellular fluid of the CNS.

The cerebral surface is highly convoluted, consisting of numerous depressions (i.e., sulci) separated by equally numerous ridges (i.e., gyril). Several sulci, noticeably deeper than others, are referred to as fissures. Color Plate 1 illustrates the fissures of the cerebrum and the major areas (i.e., lobes) they define. In addition to these externally visible lobes, named for the cranial bones they underlie, a large internal fold or cortex, the insula, lies deep to the lateral fissure.

The anatomical significance of the gyri and the sulci is that they increase the volume of the cerebral cortex. Functionally, this increased volume increases the number of cells, cellular interconnections (called synapses), and pathways available in the cerebral cortex, thereby increasing the efficiency of the brain in analyzing incoming information and in generating a complex variety of motor impulses responding to sensory input. [*See* Synaptic Physiology of the Brain.]

B. Diencephalon

The diencephalon is almost totally surrounded by the enlarged cerebral hemispheres and is composed of three major regions: the thalamus, epithalamus, and hypothalamus (see Color Plate 1).

The thalamus is shaped roughly like a dumbbell in cross section (Color Plate 4). Two oval masses of nonmyelinated cell bodies form the major portions of the thalamus. These two masses of tissue consist of several nuclei and are joined medially by a short rod-shaped massa intermedia (Fig. 2). The thalamus functions as a relay center for both sensory and motor impulses traveling between the cerebral cortex and other neural areas.

The hypothalamus, a small mass of nerve tissue, is continuous with the inferior end of the thalamus (see Color Plate 1 and Fig. 2). It consists of numerous hypothalamic nuclei, which regulate a variety of homeostatic activities, including the control of the autonomic nervous system, the endocrine system through the pituitary gland, hunger and thirst, body temperature, and wakefulness or sleepiness, as well as emotional feelings (e.g., anger, rage, aggression, and stress). [*See* Pituitary.]

Prominent external landmarks associated with the hypothalamus are the optic chiasma, infundibulum, tuber cinereum, and the mammillary bodies (see Fig. 2). The optic chaisma is an X-shaped area formed by the crossed-over fibers of the optic nerves coming from the eyes. Posterior to this chaisma is the infundibulum, a stalk of neurons that connects several hypothalamic nuclei with the posterior lobe of the pituitary gland (i.e., hypophysis). The tuber cinereum and

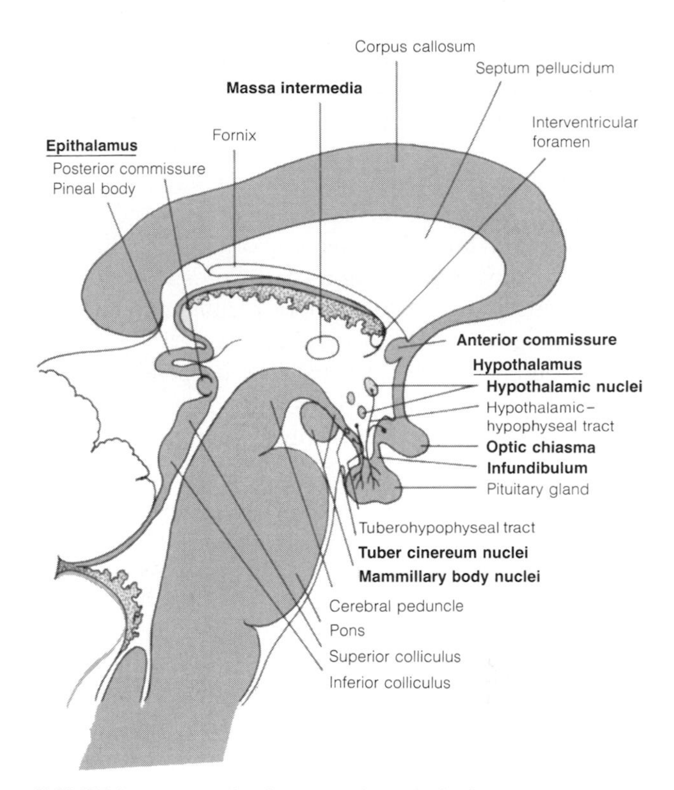

FIGURE 2 Longitudinal section through the brain stem. [From A. J. Gaudin and K. C. Jones (1989). "Human Anatomy and Physiology." Harcourt Brace Jovanovich, San Diego, California.]

mamillary bodies project from the posterior hypothalamic surface.

The most posterior portion of the diencephalon consists of two discrete structures, the pineal body (or gland) and the posterior commissure, collectively referred to as the epithalamus (see Fig. 2). The pineal body is an endocrine gland, whereas the posterior commissure contains fibers that connect the cerebrum and the midbrain.

C. Brain Stem

The brain stem is composed of four anatomical subdivisions: the mesencephalon, cerebellum, pons, and medulla oblongata.

The mesencephalon, or midbrain, is a relatively small structure lying approximately in the middle of the brain (see Fig. 2 and Color Plate 5). A narrow canal called the cerebral aqueduct passes longitudinally through the midbrain. This aqueduct contains cerebrospinal fluid and connects two other fluid-filled cavities: the third ventricle above it and the fourth ventricle below.

A longitudinal section through the midbrain reveals the cerebral aqueduct, dividing the midbrain into posterior and anterior regions. The posterior region (i.e., tectum) contains four rounded masses of nerve tissue (the corpora quadrigemina), and the anterior region (cerebral peduncles) includes many fibers and nuclei. The corpora quadrigemina are organized into two pairs of rounded protrusions (see Color Plate 5). The upper one is called the superior colliculus and the lower bulge is the inferior colliculus.

The cerebral peduncles are paired bulges of nerve fibers that project anteriorly from the brain stem. They are composed primarily of fiber tracts that connect the cerebrum with other parts of the nervous system.

The cerebellum bears a superficial resemblance to the cerebrum, being divided into two cerebellar hemispheres. They are shaped like the shells of an open clam and are joined in the middle by the narrow vermis. The nerve tissue of the cerebellum is organized much like that of the cerebrum. A thin cortex composed of gray matter surrounds an interior composed mostly of white matter. The cerebellar cortex is folded into numerous horizontally oriented ridges, called folia cerebelli. Several prominent fissures subdivide the folia into a number of lobules. A longitudinal section through the vermis reveals a unique branching of the white matter, the arbor vitae, or "tree of life" (see Color Plate 5).

The cerebellum is connected to other parts of the brain through three paired bundles, or peduncles, of fibers. The superior cerebellar peduncles connect with the mesencephalon; the middle cerebellar peduncles, with the pons; and the inferior cerebellar peduncles, with the medulla oblongata.

The pons (pons varolii, meaning "bridge") lies just anterior to the cerebellum. It is approximately the same length as the midbrain (i.e., 2.5 cm) and is separated from the cerebellum by a triangular fluid-filled cavity, the fourth ventricle (see Color Plate 5). The pons contains several reflex centers and serves as a link between parts of the brain and between the brain and the spinal cord.

The medulla oblongata (or just medulla) lies at the base of the brain stem (see Color Plate 5). It is about 3.5 cm in length and forms the connection between the spinal cord and the brain; thus, all sensory and motor tracts connecting the brain and the spinal cord must cross it. Each lateral surface of the medulla contains a prominent swelling, called the olive, which connects the medulla to the cerebellum. The posterior surface of the medulla forms the floor of the fourth ventricle, which continues into the spinal cord as the central canal. The white matter in the interior of the medulla contains fiber tracts, which transmit motor impulses to voluntary muscles, and several reflex centers.

III. VENTRICLES AND MENINGES OF THE CENTRAL NERVOUS SYSTEM

In the embryo the development of the brain involves the differentiation of a hollow neural tube that enlarges anteriorly into the adult brain structures. The internal cavity of the tube enlarges simultaneously with the surrounding structures and persists in the adult brain as a series of cavities filled with cerebrospinal fluid. This fluid is formed through a filtration process from three clumps of highly branched blood vessels, each called a choroid plexus, and the specialized cells that cover these vessels. Each choroid plexus is associated with a specific part of the ventricular system (see Color Plate 5). The constant addition of cerebrospinal fluid from the choroid plexuses to the ventricular system produces a regular circulation of fluid through the ventricles and the spinal cord. Eventually, the fluid makes its way to the exterior surface of the brain, where it is reabsorbed by the tiny fingerlike projections, called arachnoid villi, that cover the brain. Once resorbed, the cerebrospinal fluid returns to the blood from which it was formed. Cerebrospinal fluid functions as a physical "shock absorber," diminishing the effects of any sudden blow or movement of delicate tissues of the brain. In spite of the presence of this fluid, the brain is still susceptible to concussion, especially by violent blows.

A. Ventricles of the Brain

The series of cavities in the brain form four distinct regions, called ventricles (Color Plate 6). The lateral ventricles are U-shaped cavities that occupy the medial portions of the two cerebral hemispheres. The septum pellucidum (see Color Plate 5), a vertical membrane, separates the lateral ventricles and the cerebral hemispheres. A small opening, the interventricular foramen, or foramen of Monro, connects each lateral ventricle and the adjoining third ventricle, allowing cerebrospinal fluid to circulate among these ventricles.

The third ventricle, a thin cavity, separates the thalamic halves. It connects the lateral ventricles via the interventricular foramen and surrounds the massa intermedia of the thalamus. Cerebrospinal fluid enters the third ventricle from above and leaves through the cerebral aqueduct, a narrow canal that connects to the cavities within the pons, cerebellum, medulla oblongata, and spinal cord.

The fourth ventricle is triangular and is located between the cerebellum and the pons, and the medulla oblongata (see Color Plate 6). Located in its roof, inferior to the cerebellum, is one choroid plexus responsible for the production of cerebrospinal fluid. The fourth ventricle has three prominent foramina (i.e., holes) through which cerebrospinal fluid leaves the fourth ventricle and enters the subarachnoid space. Two of these (called the foramina of Luschka) lie in the superior lateral walls of the fourth ventricle. The third foramen (called the foramen of Magendie) lies in its posterior inferior wall. The fourth ventricle is continuous inferiorly with the central canal of the spinal cord.

B. Meninges of the Brain and the Spinal Cord

The brain and the spinal cord are covered by three membranous tissue layers, called meninges, that channel the flow of cerebrospinal fluid around the braiin and the spinal cord and provide additional physical protection to soft nerve tissues. The three meninges are the dura mater, pia mater, and arachnoid (Color Plate 7). [*See* Meninges.]

The dura mater is the outermost of the three meninges. It consists of a double layer of tough ("dura") fibrous connective tissue that covers the brain and attaches to the cranial bones. The external sublayer terminates at the foramen magnum, the large opening at the base of the skull, but the internal sublayer extends through this opening and continues onto the spinal cord. The two dural layers are fused over most of the brain, forming a single tough membrane whose outer surface is the inner lining of the skull bones. In several places these two layers separate, forming dural sinuses that collect venous blood and cerebrospinal fluid from the brain and return them to large veins that drain blood from this area.

The internal layer of dura mater extends into the fissures of the brain, forming prominent partitions that separate the cerebral hemispheres from one another and from the cerebellum.

The pia mater, the innermost meninx (singular of "meninges"), is a thin vascularized connective tissue membrane that adheres closely to the outer surface of the brain, dipping into sulci and fissures (see Color Plate 7). The pia mater extends through the foramen magnum and covers the entire spinal cord medial to the dura mater. At the inferior end of the spinal cord (about the level of the second lumbar vertebra), the pia mater continues as a narrow thread-like extension, the filum terminale, that extends caudally and attaches to the second coccygeal vertebra.

The arachnoid occupies the space between the other two meninges, although it does not attach equally to both. Externally, the arachnoid adheres rather closely to the dura mater, except for the presence of a narrow subdural space between them containing a thin film of cerebrospinal fluid. Internally, the arachnoid is loosely atached to the pia mater by delicate strands of fibrous connective tissue (i.e., trabeculae) that span a larger subarachnoid space filled with cerebrospinal fluid (see Color Plate 7). The arachnoid layer extends into the large longitudinal and transverse cerebral fissures and bridges the sulci and other brain depressions.

IV. ANATOMY OF THE SPINAL CORD

At the lower end of the medulla oblongata, where the brain stem terminates, columns of gray and white matter in the medulla become organized into the spinal cord, which serves two major functions. First, it contains neurons that connect sensory and motor areas of the brain with other parts of the body. These neurons provide pathways for conducting impulses in either direction—from sensory receptors to the brain then back along motor neurons to the muscles and the glands. Second, the spinal cord directly connects sensory neurons with appropriate motor neurons that produce responses independent of brain influences (i.e., spinal reflexes).

The spinal cord is a cylinder of nerve tissue somewhat flattened anteriorly and posteriorly. It is contained within the vertebral canal of the vertebrae and begins at the foramen magnum, passing through the vertebral foramina. In an embryo the spinal cord and the vertebral column are approximately the same length. However, during late embryonic and postnatal growth, the spinal cord elongates at a slower rate than the vertebral column, so that the adult spinal cord does not extend the full length of the vertebral canal. Instead, it terminates in a cone-shaped conus medullaris between the first and second lumbar vertebrae (Fig. 3). The spinal nerves that exit the lower lumbar and sacral regions are long and they angle inferiorly, until they reach their point of exit from the vertebral canal. Early anatomists thought that this

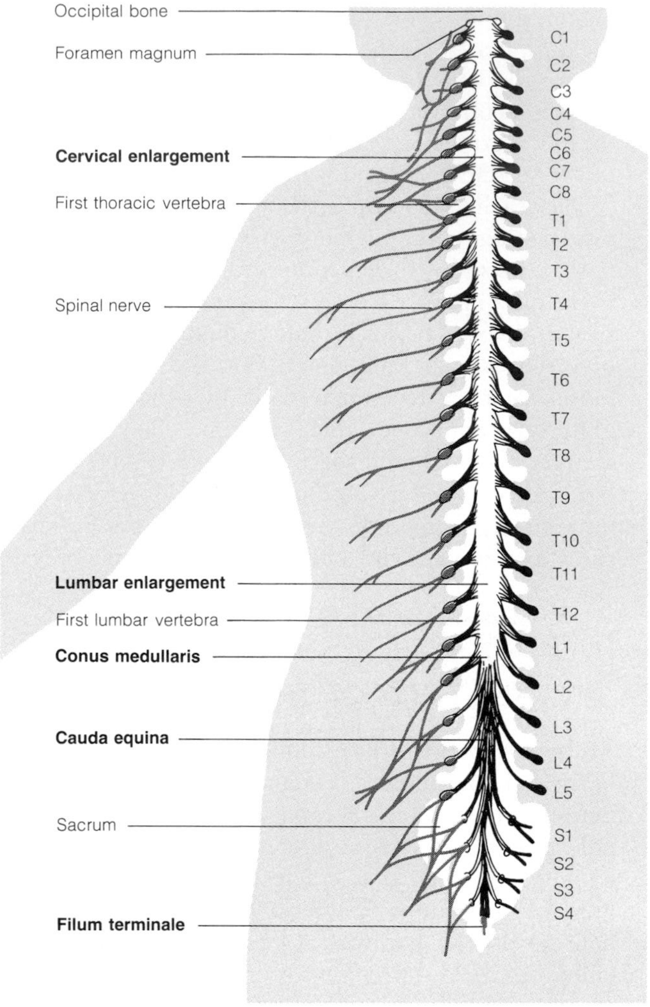

FIGURE 3 Posterior view of the spinal cord and the vertebral column. [From A. J. Gaudin and K. C. Jones (1989). "Human Anatomy and Physiology." Harcourt Brace Jovanovich, San Diego, California.]

group of spinal nerves resembled a horse's tail, so they gave it the name cauda equina.

The anterior surface of the spinal cord has a deep vertical groove, called the anterior medial fissure (see Fig. 3), and the posterior side bears a narrower, yet deeper, slit-like groove, the posterior median sulcus. The lateral sides of the cord show two shallow sulci: the anterior and posterior lateral sulci. An additional groove, the posterior intermediate sulcus, lies between the posterior median sulcus and posterior lateral sulcus in the cervical and upper thoracic regions.

As in the brain, nerve tissue in the spinal cord is divided into gray and white matter. In the spinal cord, however, the locations of the two are reversed, the white matter surrounding the gray matter.

Gray matter of the spinal cord consists mainly of neurons, unmyelinated axons, dendrites, and neuroglia (i.e., specialized supportive cells of the nervous system). In a transverse section of the spinal cord, gray matter forms an "H" in the center of the cord (see Fig. 3). The central canal is a remnant of the channel formed during the embryonic development of the neural tube and is continuous with the fourth ventricle of the brain.

The horizontal "bar" of the spinal gray matter is called the gray commissure, further subdivided by the central canal into anterior and posterior portions. Two anterior columns of gray matter (composed of the cell bodies of motor neurons) extend anteriorly from the horizontal bar. Two narrower posterior columns of gray matter (composed of sensory neurons) extend from the horizontal bar.

The spinal white matter is mainly axons of myelinated nerve fibers embedded in neuroglia. Practically all the fibers are organized into tracts that run lengthwise, except for the white commissure, anterior to the gray commissure. It contains horizontal tracts that cross the spinal cord. The white matter of the cord conducts sensory impulses up the cord and motor impulses down the cord.

V. ANATOMY OF THE PERIPHERAL NERVOUS SYSTEM

The PNS consists of the cranial nerves, spinal nerves, and autonomic nervous system.

A. Cranial Nerves

A total of 12 pairs of cranial nerves extend from the base of the brain. All but the first pair arise from the brain stem (Color Plate 8). They are numbered as they emerge from the anterior to the posterior brain: I, olfactory; II, optic; III, oculomotor; IV, trochlear; V, trigeminal; VI, abducens; VII, facial; VIII, vestibulocochlear; IX, glossopharyngeal; X, vagus; XI, accessory; and XII, hypoglossal.

Some cranial nerves include primarily sensory or motor neuron fibers and are called sensory, or motor, nerves. Others are known as mixed nerves, because they include both sensory and motor fibers. Only the axons and the dendrites form the nerves, whereas the cell bodies of the motor neurons form nuclei within

TABLE I
The Cranial Nerves

Nerve	Origin	Location of skull exit	Functions
Olfactory (I)	Cerebral hemispheres	Cribriform plate of the ethmoid bone	Sensory nerve; smell
Optic (II)	Diencephalon	Optic canal	Sensory nerve; vision
Oculomotor (III)	Midbrain	Superior orbital fissure	Mixed nerve; to the eye muscles
Trochlear (IV)	Midbrain	Superior orbital fissure	Mixed nerve; to the superior oblique eye muscle
Trigeminal (V)			
Ophthalmic	Pons	Superior orbital fissure	Sensory nerve; from the head
Maxillary	Pons	Foramen rotundum	Sensory nerve; from the face
Mandibular	Pons	Foramen ovale	Mixed nerve; facial sensations and chewing motions
Abducens (VI)	Pons	Superior orbital fissure	Mixed nerve; to the lateral rectus eye muscle
Facial (VII)	Pons	Stylomastoid foramen	Mixed nerve; taste sensations and facial muscles
Vestibulocochlear (VIII)	Medulla	Internal acoustic meatus	Sensory nerve; hearing and equilibrium
Glossopharyngeal (IX)	Medulla	Jugular foramen	Mixed nerve; taste and pharyngeal muscles
Vagus (X)	Medulla	Jugular foramen	Mixed nerve; to the head, thorax, and abdomen
Accessory (XI)	Medulla	Jugular foramen	Motor nerve; to the neck muscles
Hypoglossal (XII)	Medulla	Hypoglossal canal	Motor nerve; to the tongue

the brain tissue, and those of sensory neurons are grouped in ganglia outside, but adjacent to, the CNS.

Generally, cranial nerves innervate structures in the head and neck region. However, one exception is the vagus (meaning "wandering") nerve, a mixed nerve, which innervates the palate, neck, thorax, and abdominal cavity.

The 12 cranial nerves and their sites of origin and modes of action are summarized in Table I.

B. Spinal Nerves

Spinal nerves originate in the spinal cord. These nerves leave through foramina of the vertebral column going to the skin, muscles, bones, and joints of the posterior head region, trunk, and appendages (see Fig. 3). In all, 31 pairs of spinal nerves are normal: 8 cervical, 12 thoracic, 5 lumbar, 5 sacral, and 1 (occasionally more) coccygeal.

C. Anatomy of Adult Spinal Nerves

Each spinal nerve originates as two extensions from the spinal cord: an anterior root and a posterior root (Fig. 4). The former includes both voluntary and involuntary motor neurons; the latter includes only sensory neurons.

Voluntary motor neurons in the anterior root are called somatic motor (or somatomotor) neurons. Each

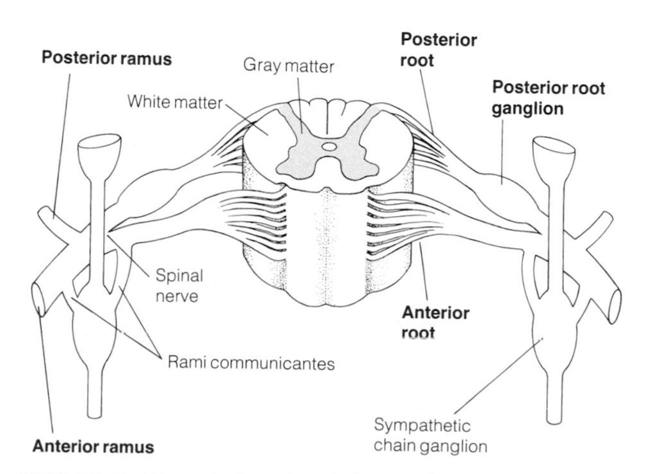

FIGURE 4 The spinal cord and the spinal nerves. [From A. J. Gaudin and K. C. Jones (1989). "Human Anatomy and Physiology." Harcourt Brace Jovanovich, San Diego, California.]

extends its axon from the spinal cord gray matter to a skeletal muscle. Some neurons, such as those innervating the foot, having long axons. These neurons are responsible for voluntary movements produced by contractions of the skeletal muscles attached to bones and certain superficial skin structures in the head and the neck (e.g., the lips and the eyelids).

In addition to these voluntary nerve cells, the anterior roots contain fibers of the autonomic nervous system that control unconscious involuntary activities. The exact positions of these autonomic neurons are discussed in Section V,E.

Posterior roots of spinal nerves have only sensory neurons. Dendrites of these neurons are situated in tissues and organs, and their cell bodies are grouped in a posterior root ganglion (see Fig. 4).

Just distal to the posterior root ganglion, the posterior and anterior roots fuse and form a spinal nerve. Because it includes sensory and motor fibers, each spinal nerve is mixed. Cervical, thoracic, and lumbar spinal nerves extend laterally and exit the neural canal through intervertebral foramina. Immediately distal to these foramina, each spinal nerve divides into an anterior ramus and a posterior ramus (see Fig. 4). Sacral nerves are slightly different from other spinal nerves in that sacral nerves split within the sacral canal, and anterior and posterior rami exit separately through corresponding sacral foramina.

Compared to its anterior counterpart, the posterior ramus of each spinal nerve is relatively short. Posterior rami innervate only restricted areas of the skin and muscles along the back of the head and the sides of the vertebral column. Anterior rami of the spinal nerves innervate the remainder of the trunk and the limbs.

D. Peripheral Regions Innervated by Spinal Nerves

The pattern of spinal nerve innervation of the skin can be mapped as seen in Color Plate 8. A regular sequential pattern is most prominent on the trunk and slightly modified in the limbs. The nerves that innervate the skin, called cutaneous nerves, contain only sensory and autonomic motor neurons and lack somatic motor neurons.

Spinal nerve innervation of skeletal muscles follows a similar sequential pattern; that is, spinal nerves in the cervical and thoracic region innervate skeletal muscles in the neck, arm, and chest, whereas nerves from the lumbar and sacral regions innervate abdominal and leg muscles.

Many areas of both the skin and the muscles are innervated by nerves of mixed spinal origin. Such nerves are produced by a fusion and exchange of neurons from the anterior rami of certain neighboring spinal nerves. The resulting mixed nerves form a plexus (discussed in Section V,D,1).

Nerves supplying the skeletal muscles follow one of two pathways. The first pathway involves the thoracic nerves. These 12 nerves leave the intervertebral foramina and pass between the ribs as intercostal nerves. In general, intercostal nerves T1–T6 innervate the skin and the skeletal muscles in the thoracic region. Fibers from their posterior rami are confined to a narrow area bordering the neural spines of the vertebrae, whereas fibers of the anterior ramus run to the remaining anterior and lateral regions. Intercostal nerves T7–T12 stimulate corresponding intercostal muscles, as well as skin and voluntary abdominal muscles. Nerves T2 also extend into the arms and supply the skin of the axillary (i.e., underarm) region and the surface of the back of the arm.

The second pathway involves anterior rami of the remaining spinal nerves in collections of mixed nerves, or plexuses.

1. Cervical Plexus

The cervical plexus is formed by a mixing of cervical nerves 1–4, with contributions from C5. The pattern of fusion and branching is illustrated in Fig. 5, and the area innervated by each peripheral nerve is summarized in Table II.

2. Brachial Plexus

The brachial plexus is the peripheral nerve supply for the upper extremities. It is formed by the exchange of fibers between the anterior rami of cervical nerves 5–8 and the first thoracic nerve (Fig. 6). Table III summarizes the destinations of the major nerves of the brachial plexus.

TABLE II
The Cervical Plexus

Nerve	Spinal nerve origin	Distribution
Lesser occipital	C2–C3	Scalp behind the ear
Greater auricular	C2–C3	Skin surrounding the ear
Transverse cervical	C2–C3	Anterior skin of the neck
Supraclaviculars	C3–C4	Skin of the upper thorax
Ansa cervicalis	C1–C2	Pharyngeal muscles
Phrenic	C3–C5	Diaphragm

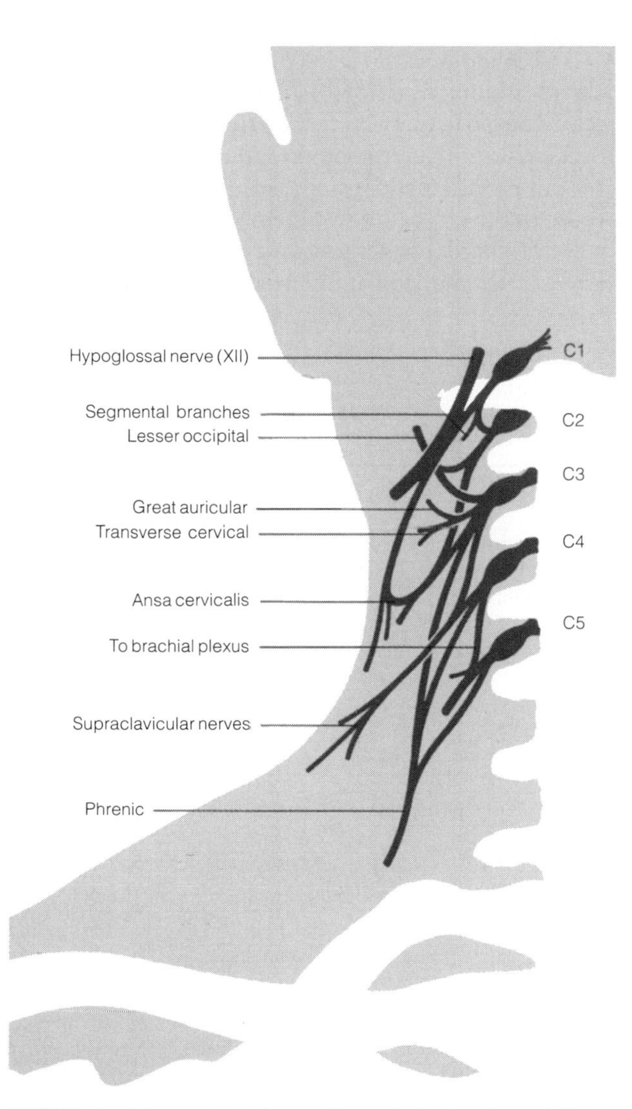

FIGURE 5 The cervical plexus. [From A. J. Gaudin and K. C. Jones (1989). "Human Anatomy and Physiology." Harcourt Brace Jovanovich, San Diego, California.]

3. Lumbosacral Plexus

The lumbosacral plexus combines three groups of spinal nerves: the lumbar plexus, nerves L1–L4; the sacral plexus, nerves L4–L5 and S1–S3; and the pudendal plexus, nerves S2–S4. In a few cases the most inferior thoracic nerve (T12) contributes to the lumbar plexus through nerve L1. The pattern of combinations is shown in Fig. 7; Table IV summarizes the destinations of the major nerves of the lumbosacral plexus.

E. Autonomic Nervous System

The autonomic nervous system is that part of the peripheral nervous system that regulates unconscious

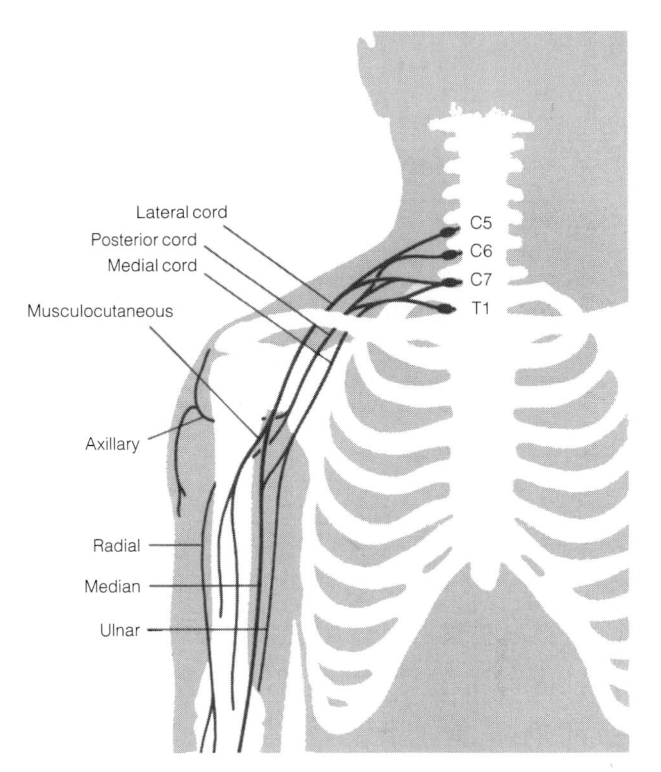

FIGURE 6 The brachial plexus. [From A. J. Gaudin and K. C. Jones (1989). "Human Anatomy and Physiology." Harcourt Brace Jovanovich, San Diego, California.]

TABLE III
The Brachial Plexus

Nerve	Spinal nerve origin	Distribution
Posterior cord		
Axillary	C5–C6	Skin and muscles of the shoulder region
Radial	C5–C8, T1	Skin and muscles of the posterolateral region of the arm, forearm, and hand
Lateral cord, musculocutaneous	C5–C7	Skin and muscles of the lateral region of the forearm
Medial cord		
Ulnar	C8–T1	Skin of the medial portion of the hand, muscles of the front of the forearm and the hand
Median	C5–T1	Skin of the lateral portion of the hand, muscles of the lateral region of the forearm and the hand

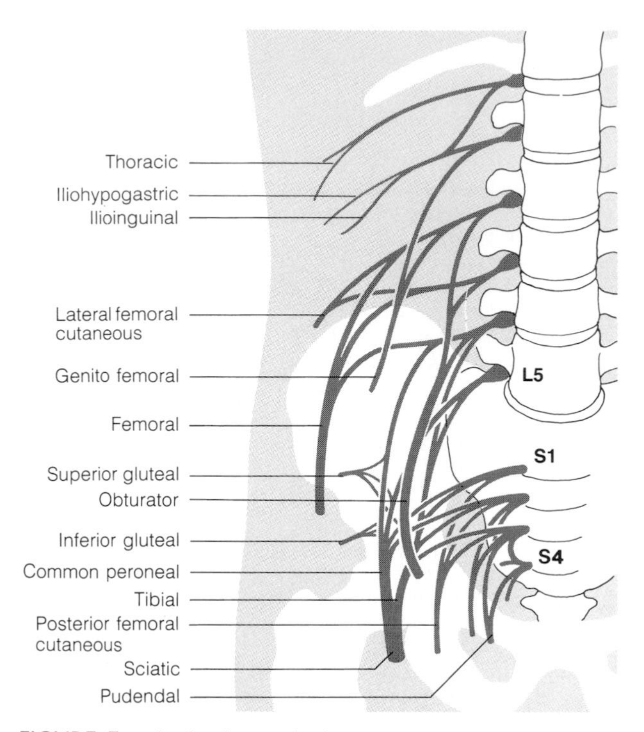

FIGURE 7 The lumbosacral plexus. [From A. J. Gaudin and K. C. Jones (1989). "Human Anatomy and Physiology." Harcourt Brace Jovanovich, San Diego, California.]

involuntary activities, such as the control of the heart beat, movements of the digestive system, and glandular activities. It consists primarily of visceral efferent neurons that carry motor impulses to cardiac muscle, certain glands, and smooth muscles in blood vessels and organs of the thoracic and abdominal cavities (Fig. 8). In the past, physiologists thought that activities that were clearly controlled at a subconscious level were regulated by a functionally separate, hence "autonomous" part of the nervous system. We now know that the autonomic nervous system is not truly autonomous. Indeed, it is regulated by other parts of the CNS, specifically, centers in the cerebral cortex, hypothalamus, and medulla oblongata.

In general, the autonomic nervous system has two distinct anatomic and functional subdivisions: sympathetic and parasympathetic. The sympathetic system, also called the thoracolumbar system, because its neurons emerge from thoracic and lumbar regions of the spine, innervates smooth muscles of the arteries. Just as arteries penetrate all parts of the body, so do sympathetic fibers. Sympathetic fibers also innervate several abdominal organs (Fig. 9). The general effect of the sympathetic nervous system is to prepare the body for action in stressful situations.

The other division, the parasympathetic, or craniosacral, system, functions as the principal nerve supply to certain structures in the head, digestive organs, and other viscera (e.g., the lungs and the heart). Effects of the parasympathetic system are in some ways opposite those of the sympathetic system. The parasympathetic system stimulates activities of the digestsive organs and glands and slows the heart beat and the respira-

TABLE IV
The Lumbosacral Plexus

Nerve	Spinal nerve origin	Distribution
Lumbar plexus		
Iliohypogastric	T12–L1	Skin and muscles of the lower trunk
Ilioinguinal	L1	Skin of the upper thigh and the external genitals in males, muscles of the lower trunk
Genitofemoral	L1–L2	Skin of the external genitals in both sexes and the internal genitals in females, the front of the thigh
Lateral femoral cutaneous	L2–L3	Skin covering the lateral cutaneous region of the thigh
Femoral	L2–L4	Skin of the medial thigh, leg, and foot; certain muscles of the thigh
Obturator	L2–L4	Skin of the medial thigh, certain muscles of the thigh
Sacral plexus		
Superior gluteal	L4–S1	Certain muscles of the thigh
Inferior gluteal	L5–S2	Gluteus maximus muscle
Posterior femoral cutaneous	S1–S3	Skin of the external genitals in both sexes and the internal genitals in females, the posterior surface of the thigh
Sciatic	L4–S3	The tibial and common peroneal nerves
Tibial	L4–S3	Skin of the posterior leg and foot, several muscles of the thigh, leg, and foot
Common peroneal	L4–S2	Skin of the anterior leg and foot, several muscles of the leg
Pudendal plexus		
Pudendal	S2–S4	Skin of the perineum and the genitals

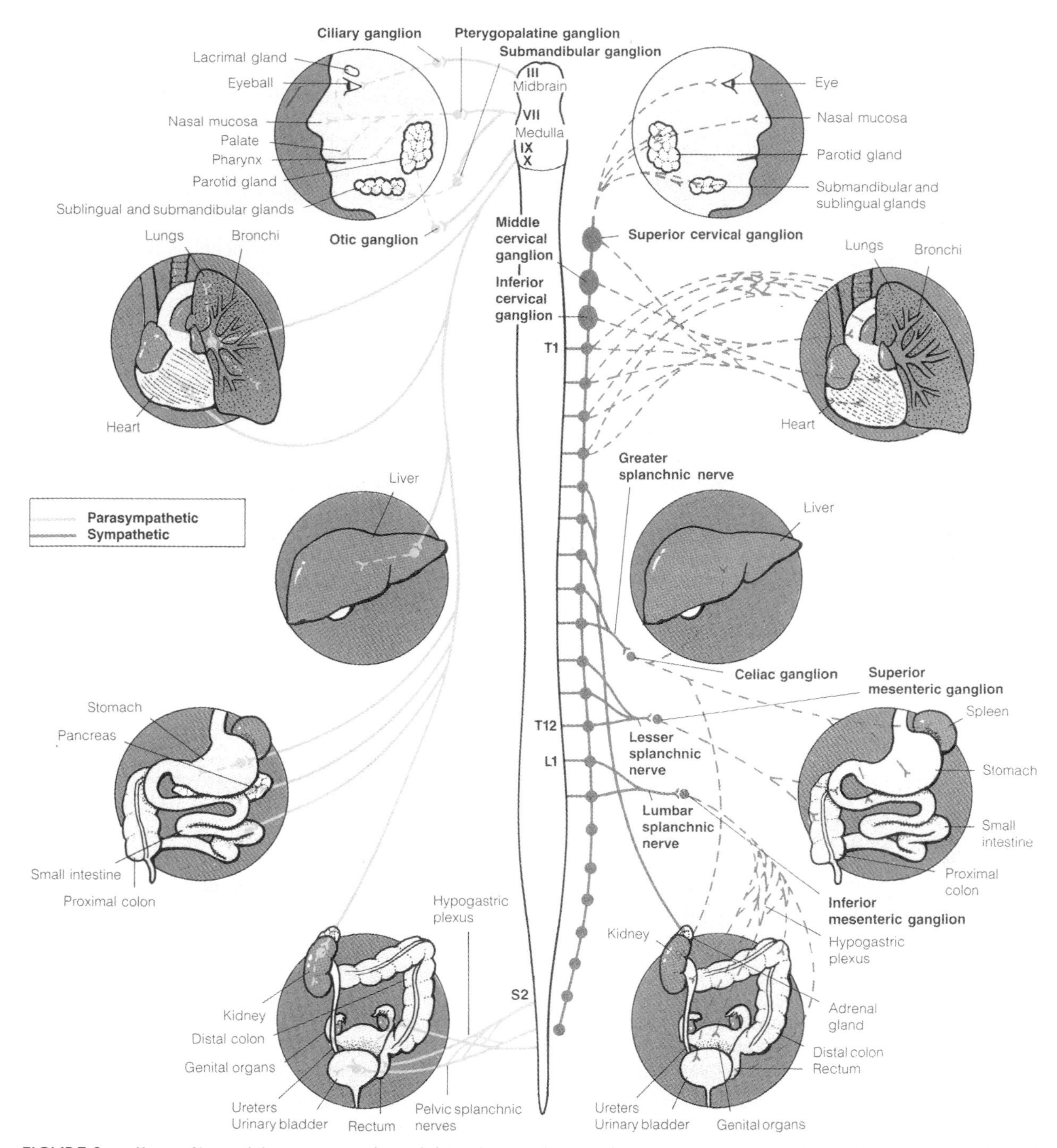

FIGURE 8 Efferent fibers of the parasympathetic (left) and sympathetic (right) autonomic nervous systems. [From A. J. Gaudin and K. C. Jones (1989). "Human Anatomy and Physiology." Harcourt Brace Jovanovich, San Diego, California.]

tory rate. It tends to calm the body after a stress-producing experience, and it promotes activities that maintain life-support systems.

In general, motor neurons of the autonomic nervous system are organized into functional units of two neurons each. Each unit includes a preganglionic, or presynaptic, neuron with a cell body that lies within the CNS (Fig. 10). The axon of this cell synapses with a second, postganglionic (or postsynaptic), neuron. These nerves synapse either within a ganglion that

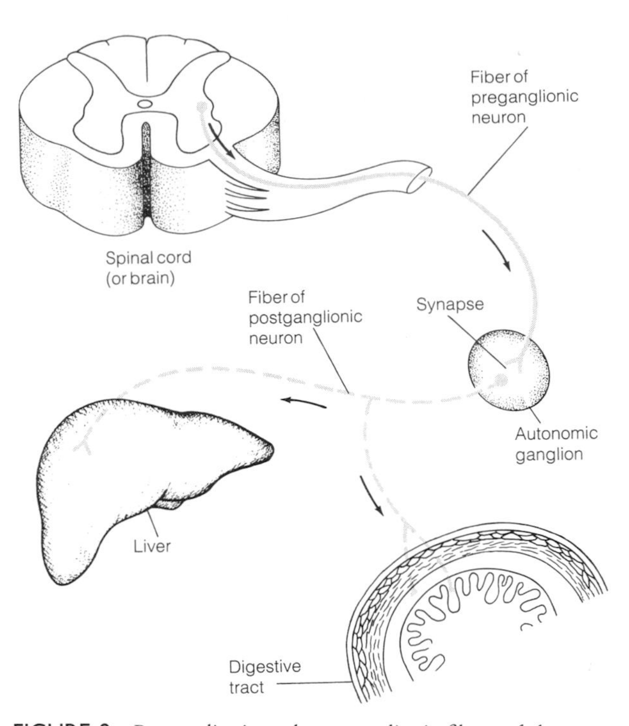

FIGURE 9 Preganglionic and postganglionic fibers of the autonomic nervous system. [From A. J. Gaudin and K. C. Jones (1989). "Human Anatomy and Physiology." Harcourt Brace Jovanovich, San Diego, California.]

lies close to the spinal cord or within one of several ganglia located in the thoracic and abdominal cavities. Axons from postganglionic neurons terminate in a specific predetermined organ or tissue. Impulses carried along these neurons either stimulate or inhibit the metabolic activities of these organs.

1. Sympathetic System

Preganglionic neurons of the sympathetic division are anchored in the lateral columns of the spinal cord and their axons emerge from the cord through anterior roots of all thoracic spinal nerves and through the first two lumbar spinal nerves (see Fig. 8). These axons remain within a spinal nerve for only a short distance, exiting through one of two communicating rami that connect with a chain of ganglia paralleling the vertebral column. These ganglia are the paravertebral ganglia, which compose the sympathetic trunk, or chain. Since preganglionic sympathetic fibers are myelinated, the rami they run through to enter the sympathetic trunk are called white communicating rami. Figure 10 shows pathways available to these motor neurons entering the sympathetic chain.

An axon may synapse within the ganglion it enters, or pass through and synapse in a ganglion at a different level in the chain (see Fig. 10). Other axons run through the sympathetic ganglia and do not synapse until they join neurons in ganglia in or near visceral organs. The nerves formed by these latter preganglionic neurons are called splanchnic nerves.

Impulses in sympathetic neurons travel two different paths to target organs, depending on which synaptic pattern occurs. In the first pattern, impulses are generated in postganglionic fibers that leave the sympathetic trunk ganglia and pass out through spinal nerves. Postganglionic sympathetic fibers are nonmyelinated and pass through gray communicating rami (see Fig. 10). Because this happens along the entire length of the cord, all spinal nerves contain at least some postsynaptic sympathetic fibers. Some of these postganglionic fibers leave the sympathetic trunk ganglia in the cervical region and accompany major arteries in the head until they reach their target organs. Within the head they follow the external carotid, internal carotid, and vertebral arteries. The postganglionic fibers are called plexuses and are referred to in terms of the artery involved. Thus, these fibers are distributed into three plexuses: the external carotid plexus, internal carotid plexus, and vertebral plexus.

A second pattern is formed by preganglionic neurons that run through splanchnic nerves and synapse with their postganglionic partners within abdominal, or collateral, ganglia. Collateral ganglia form masses of cell bodies of postganglionic neurons. Here, synapses occur with preganglionic axons from the spinal cord. Three collateral ganglia (see Fig. 10)—celiac, superior mesenteric, and inferior mesenteric—lie near the large arteries for which they are named. Smaller collateral ganglia lie in the pelvic cavity.

Postganglionic axons exit the collateral ganglia to form a series of extrinsic autonomic plexuses. After reaching target organs, these axons penetrate the organ and are distributed as intrinsic autonomic plexuses within the organ.

One exception to the two-neuron plan in the sympathetic system is known. It involves the adrenal gland, attached to the top of the kidney. Preganglionic sympathetic nerves that innervate cells of the inner portion (i.e., medulla) of the gland do not synapse with postganglionic neurons. Instead, preganglionic cells stimulate secretory cells in the adrenal gland. [*See* Adrenal Gland.]

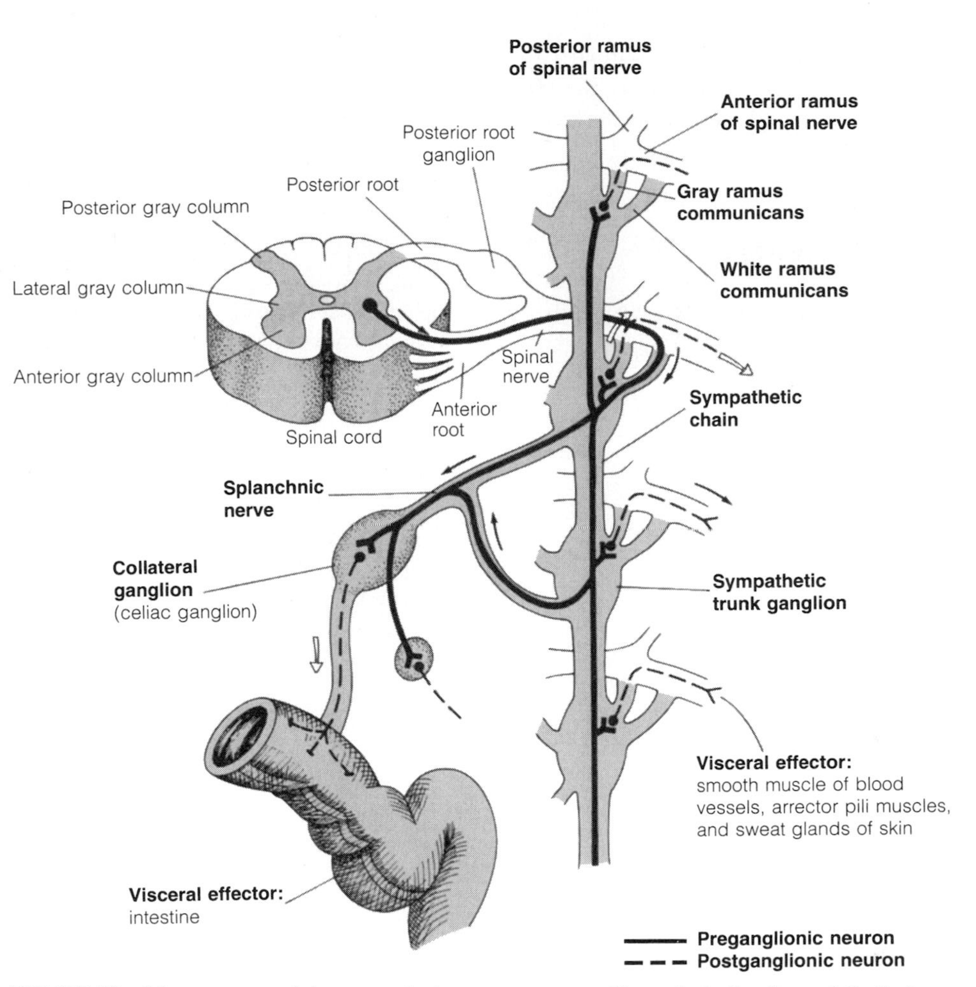

FIGURE 10 Motor routes of the sympathetic nervous system. [From A. J. Gaudin and K. C. Jones (1989). "Human Anatomy and Physiology." Harcourt Brace Jovanovich, San Diego, California.]

2. Parasympathetic System

The second part of the autonomic nervous system is the parasympathetic system, or craniosacral system. Its neurons arise from opposite ends (i.e., the cranial and sacral ends) of the spinal cord (see Fig. 9). It is the principal nerve supply for certain structures in the eye, glands in the head, heart, reproductive organs, smooth muscles, and glands of the digestive system. The fibers of this system accompany four of the cranial nerves and the anterior roots of spinal nerves S2–S4. Spinal nerves proper do not include parasympathetic motor fibers.

The preganglionic parasympathetic fibers that emerge from the head have their cell bodies in nuclei of the brain stem. They extend to ganglia located in the head, where they synapse with postganglionic neurons that pass to several target organs and tissues (see Fig. 8). Preganglionic fibers from the vagus nerve, cranial nerve X, do not pass to a single ganglion. Instead, they innervate the heart, respiratory system, intestinal tract, and their associated glands.

Sacral parasympathetic neurons have preganglionic fibers that arise in the lateral columns of the gray matter in the sacral spinal cord. These axons leave with the anterior roots of sacral nerves 2–4. They form the pelvic splanchnic nerves and travel to several plexuses in the lower abdominal and pelvic regions. From these plexuses, preganglionic axons extend to intrinsic plexuses within the target organs. The primary organs innervated by the sacral parasympathetic fibers are the descending and sigmoid colons, the rectum and the anus, the urinary bladder, and the external genital organs.

BIBLIOGRAPHY

Angevine, J. G., Jr., and Cotman, C. W. (1981). "Principles of Neuroanatomy." Oxford Univ. Press, New York.

Barr, M. L., and Kernan, J. A. (1983). "The Human Nervous System," 4th Ed. Harper & Row, New York.

Gaudin, A. J., and Jones, K. C. (1989). "Human Anatomy and Physiology." Harcourt Brace Jovanovich, San Diego, California.

McMinn, R. M. H., and Hutchings, R. T. (1988). "Color Atlas of Human Anatomy," 2nd Ed. New York Med. Publ., Chicago.

Nauta, W. J. H., and Feirtag, M. (1985). "Fundamental Neuroanatomy." Freeman, New York.

Thompson, R. F. (1985). "The Brain: An Introduction to Neuroscience." Freeman, New York.

Nervous System, Plasticity

FREDRICK J. SEIL
VA Medical Center, Portland and Oregon Health Sciences University

GLOSSARY

Axon Nerve cell process (fiber) that conducts electrical impulses to other cells. Most neurons have a single axon, but the axons may branch many times

Dendrite Neuron process that originates from the soma and that, along with the soma, receives the projections of multiple axon endings (terminals). Dendrites are generally multiple or complexly branched and function as integrators of incoming impulses

Denervate To deprive a target of its incoming axonal projections, such as by severing or otherwise injuring the projecting axons

Innervate To project axons to a target

Neuron Nerve cell, including the cell body (soma) and all its processes

Neurotransmitter Chemical substance, often an amino acid, that is released into the synaptic cleft from the presynaptic terminal in response to an electrical impulse. After crossing the synaptic cleft, the neurotransmitter interacts with a receptor on the postsynaptic membrane to produce a change in either an excitatory or an inhibitory direction

Nucleus Anatomically defined, generally small collection of neurons whose projecting axons tend to be functionally similar

Synapse Site where two neurons communicate. A synapse is composed of a presynaptic element, usually an axon terminal, a postsynaptic element consisting of a specialized membrane on the surface of a nerve cell body, dendrite, or occasionally another axon, and a synaptic cleft, or space between the pre- and postsynaptic elements

BY CURRENT USAGE, NEURAL PLASTICITY IS A TERM that refers to any change in the structure or function of the nervous system. The change may occur spontaneously [e.g., the normal turnover of nerve terminals connecting to muscle fibers (neuromuscular junctions)] or may be induced by some alteration of internal or external conditions (e.g., an injury or disease). Changes may occur at the level of a single cell or may involve the complex neural organization of an organism. Plasticity is a characteristic of the nervous system at any stage of life, including development, maturity, and aging. Plastic changes that occur during development and result in a departure from the normal developmental pattern are usually more dramatic than later occurring changes. The developing nervous system, with its interconnections more "diffusely" organized, has a greater capacity for modification of its connection patterns. Structural changes involve the growth of new nerve fibers or processes or formation of new synapses, resulting in a reorganization of the neural circuitry. Functional changes can vary from expression of a different neurotransmitter by a cell to a local reflex change to a change in behavior. In essence, neural plasticity is the modifiability of the nervous system at a cellular, multicellular, or organismal level to permit an adaptive response to an ongoing or changing set of internal or external circumstances.

I. STRUCTURAL PLASTICITY

A. Axon Collateral Sprouting

A common structural plastic change in response to an injury to the nervous system is collateral sprouting of axons, or the growth of collateral branches from intact, uninjured axons. Sprouted axon collaterals innervate target neurons, muscle fibers, or sensory receptor organs denervated because of damage to or loss of their primary axons. For example, lizards losing tails as a defense or escape reaction may regenerate new tails, but they do not regenerate the sensory ganglia that innervate the tails. Rather, the regenerated tails receive sensory innervation from collateral axonal branches that sprout from the most caudal of the surviving intact ganglia. The surviving ganglia thus extend the territory to which they project by growing additional axonal branches that substitute for the lost axons, thus allowing the development of sensation in the regenerated tails.

Another example of axon collateral sprouting is found in the septal nucleus, one of the deep nuclei of the forebrain. The neurons in this nucleus receive innervation from other parts of the brain via two major incoming nerve fiber bundles [i.e., the fimbria and the medial forebrain bundle (MFB)]. The fimbrial axons synapse only on dendrites, whereas a substantial portion of the MFB axons synapse on neuron cell bodies (somata). If the MFB is experimentally cut, the fimbrial axons undergo collateral sprouting and synapse on the cell somata as well as on the dendrites of septal nucleus neurons, thus maintaining innervation of the cell bodies (Fig. 1).

B. Synaptic Plasticity

Synaptic plasticity usually refers to a modification of synapses (e.g., a change in the number or size of the terminals converging on a target) or a change in the source of the projecting axons, with or without a change in the neurotransmitter of the projecting terminals, the end result of which is an alteration of the original neural circuitry. The structural alteration may be correlated with a functional change (see Section II,B). In parts of the nervous system where synapses turn over normally, the result of the turnover is a maintenance of the existing circuitry. When the synapse turnover is in response to a lesion, then the usual consequence is a change in the synaptic organization of the involved part of the nervous system.

One of the best studied examples of a lesion-induced change in synaptic organization is in the hippocampus, a part of the brain associated with memory. The dendrites of the granule cell neurons of the hippocampus are densely innervated by fibers from a neighboring structure, the ipsilateral (same side) entorhinal cortex (Fig. 2A). The dendrites also receive a dense projection from another area of the hippocampus (CA4) bilaterally, and a sparse projection from the contralateral (opposite side) entorhinal cortex and the

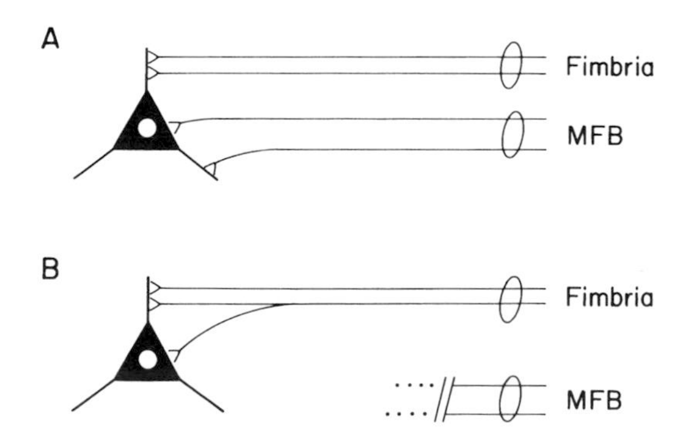

FIGURE 1 Collateral sprouting. (A) Neurons in the septal nucleus receive input from other brain areas by way of two fiber bundles, the fimbria and the medial forebrain bundle (MFB). Fimbrial axons terminate exclusively on dendrites, and MFB axons project to both dendrites and somata of septal nucleus neurons. (B) If the MFB is cut, fimbria axons undergo collateral sprouting and project to somata as well as dendrites of neurons in the septal nucleus. [From Seil (1989), with permission.]

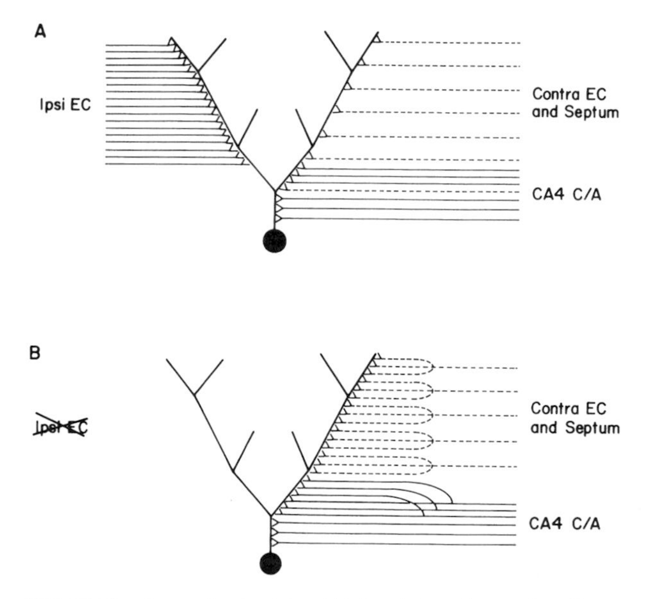

FIGURE 2 Synaptic plasticity. (A) Dendrites of granule cell neurons of the hippocampus form multiple synapses (represented as —<) with axons from the ipsilateral entorhinal cortex (Ipsi EC) of the brain and from the CA4 area of the hippocampus bilaterally (CA4 C/A), and fewer synapses with axons from the contralateral entorhinal cortex (Contra EC) and the septal nuclei. Entorhinal cortex fibers project to the outer 75% of the granule cell dendritic tree, and CA4 fibers project to the inner 25%. (B) If the ipsilateral entorhinal cortex is destroyed (X), CA4 fibers, septal fibers, and fibers from the contralateral entorhinal cortex all undergo collateral sprouting and form new synapses on granule cell dendrites to replace lost synapses. CA4 fibers extend their projection to the inner 35% of the dendritic tree in this scheme of synaptic reorganization. [From Seil (1989), with permission.]

septal nuclei. The entorhinal cortex axons project to the outer three-quarters of the granule cell dendrites, whereas the CA4 fibers project to the inner quarter. [*See* Hippocampal Formation.]

If the ipsilateral entorhinal cortex is destroyed (Fig. 2B), more than 60% of the synapses on the granule cell dendrites are lost. Intact CA4 axons, septal axons, and axons from the contralateral entorhinal cortex all undergo collateral sprouting and form new synapses on the granule cell dendrites that replace the lost synapses. The CA4 fibers extend their territory to innervate the inner 35% of the dendritic tree, whereas septal nucleus and contralateral entorhinal cortex fibers retreat from this area and only reinnervate the denervated territory. The new synapses formed by contralateral entorhinal cortex axons are like those of the ipsilateral entorhinal cortex that they replace (homotypical), whereas the new synapses formed by CA4 and septal nucleus axons are different from the original synapses (heterotypical). Both homotypical and heterotypical synapses are apparently functional, as determined by electrophysiologic studies. The result is a reorganization of the original circuitry in such a way as to compensate for the deficit produced by the lesion.

C. Activity-Dependent Plasticity

Activity-dependent plasticity refers to the role of neural activity in altering patterns of synaptic connectivity. Activity-dependent plasticity has been extensively studied in the developing visual system. In lower vertebrates, a coarse-grained retinotopic map initially develops in the tectum independent of neuronal activity, but the subsequent development of a fine-grained topographic map requires neural activity. Neural activity is similarly required for the formation of ocular dominance columns in mammalian visual cortex. Evidence from other systems, such as mammalian hypothalamus and hippocampus, shows that synaptic rearrangements dependent on different activity states occur throughout adult life.

II. FUNCTIONAL PLASTICITY

A. Neurotransmitter Plasticity

Developing sympathetic neurons are plastic with regard to neurotransmitter expression. Such neurons are normally adrenergic (use norepinephrine as a neurotransmitter) but can be induced to become cholinergic (using acetylcholine as a neurotransmitter) by manipulation of their environment. This can be done *in vivo* by transplanting sympathetic neurons so that they innervate target tissue normally receiving cholinergic innervation or *in vitro* by culturing the sympathetic neurons in the presence of nonneuronal cells or in medium conditioned by nonneuronal cells. The expression of one or the other neurotransmitter, norepinephrine or acetylcholine, profoundly changes the functional properties of the sympathetic neurons. This is a form of neuroplasticity that is the property of a single cell, as opposed to other kinds of changes (e.g., organizational changes) that involve at least a group of neurons.

B. Lesion-Induced Functional Changes

The red nucleus in the brain stem represents an area where inputs from two different parts of the brain converge. Axons from one of the deep nuclei of the contralateral cerebellum (hindbrain) project to the somata of the red nucleus neurons, whereas fibers from the ipsilateral sensorimotor cortex project to the outer portions of the dendritic trees. Red nucleus neurons are large cells that are subject to intracellular electrophysiologic recording. Stimulation of the contralateral deep cerebellar nucleus produces a fast-rising excitatory postsynaptic potential (EPSP) characteristic of a synapse on the cell body and near the recording electrode. Cortical stimulation produces a slow-rising EPSP, consistent with synapses located well out on the dendrites, at a distance from a recording electrode placed in the cell soma. If the deep cerebellar nucleus is lesioned, cortical stimulation 10 or more days later elicits an EPSP with an intermediate rise time, indicative of cortical axons having sprouted and formed new synapses on the inner parts of dendrites or on cell somata. The morphological findings support the electrophysiologic results and provide an elegant example of how structural and functional plastic changes may correlate. The new synapses that are formed in response to lesioning are heterotypical but, as also indicated by the electrophysiologic results, are functional.

C. Latent Synapses

Under some conditions of lesioning or of temporarily blocking the function of somatosensory pathways, rapid changes in the receptive fields mediated by such pathways may occur. The changes may be too rapid or too extensive to be explained by axonal sprouting and formation of new synapses. These kinds of data have raised the possibility of the existence of latent or "silent" synapses, which may be anatomically present but functionally ineffective. A temporary or permanent lesion could induce a change in effectiveness of such structurally static synapses, resulting in a functional change that is not based on a morphological reorganization of the underlying neural circuitry. Thus it is possible that structural changes in synaptic connectivity are not always necessary for the occurrence of functional plasticity.

D. Behavioral Plasticity

Behavioral plasticity is a broad term that encompasses such phenomena as a change of behavior in response to selective experience, conditioning, habituation, sensitization, imprinting, associative learning and memory formation, adaptation to unusual motor or sensory conditions (e.g., the wearing of distorting or inverting lenses), and others. Also generally classified under behavioral plasticity are the morphological changes that are induced in various areas of the brain by experience. Rats raised in an enriched laboratory environment as opposed to standard colony or impoverished conditions have reported increases in brain weight, glial density, cortical thickness, dendritic branching, numbers of dendritic spines and synapses, and increases in areas of individual synapses. Such changes occur at whatever age the enriched environmental condition is imposed. Similar changes occur after application of formal learning procedures.

The substrates for some kinds of behavioral plasticity are known. Clearly, a change such as a structural synaptic reorganization in response to injury along with a circuit change as determined electrophysiologically may have an associated change in behavior. Habituation or a decrement in response to a repeated inconsequential stimulus can, in the gill-withdrawal response of the sea slug *Aplysia,* be associated with a depression of the EPSPs of motor neurons caused by a decrease in the quanta of neurotransmitter released after successive impulses by sensory neuron terminals at their junctions with motor neurons. In the same model, sensitization or intensification of the gill-withdrawal reflex by presentation of a strong stimulus elsewhere on the body surface is associated with an enhancement of neurotransmitter release by sensory neurons. Imprinting, or the behavior of following the first large moving object seen, as occurs in birds (or other animals born in a relatively advanced state of development), is associated with an increase in postsynaptic receptive areas. Memory formation is believed by some to involve a mechanism such as long-term potentiation (LTP). LTP is an increase in the synaptic response of neurons induced by the stimulation of fibers projecting to the neurons with moderate to high-frequency bursts (10 Hz or more). LTP develops after seconds and endures for hours or days. Evidence is available for functional changes at both pre- and postsynaptic sites, as well as for morphological changes at synapses after induction of LTP.

III. PLASTICITY AFTER TRANSPLANTATION

Plasticity associated with transplantation of the nervous system is included under a separate heading not because there is anything basically different about plastic changes associated with transplantation, but because (1) transplantation represents a totally contrived or artifactual phenomenon that does not occur in nature and (2) the changes that occur are generally in the opposite direction to those that occur in response to injury (i.e., the purpose of transplantation is to restore the normal structural/functional state, as opposed to the reorganization of normal patterns that occurs in response to injury or disease).

An initial question that can be raised is whether a nervous system that has undergone reorganizational changes as a consequence of injury, with the formation of new synapses and the establishment of alternate neurotransmitters, is capable of undergoing a second round of reorganization if the missing elements are restored. If so, can the original circuitry be faithfully reconstructed? The answers to these questions are in the affirmative, as animal studies with lesioned adult hippocampus and visual systems, as well as other systems, indicate that damaged and reorganized neural circuitry can be reconstructed by transplantation with fetal tissue. The axons of transplanted neurons find and synapse with appropriate targets in host tissue, and the grafts also receive host fibers. Electrophysiologic and behavioral studies attest to the appropriate functionality of at least some of the connections between graft and host nervous system.

Some insight into how circuit reconstruction might occur is gained from studies with neural tissue cultures. If cerebellar explants derived from neonatal mice are treated with a drug that eliminates one group of cortical neurons, the cerebellar granule cells, one of the surviving neuronal groups, the Purkinje cells, undergo a remarkable sprouting of axon collaterals that form synapses on the dendritic sites normally occupied by the axon terminals of granule cells.The Purkinje cell neurotransmitter is not only different from the granule cell neurotransmitter, but is inhibitory rather than excitatory, which represents a radical change from the normal condition. These abnormal synapses are functional and inhibitory, as indicated by electrophysiologic studies.

If cerebellar granule cells from another source are now introduced into this reorganized system, the sprouted Purkinje cell axon collaterals degenerate or are withdrawn, and the heterotypical synapses that they had formed are replaced with normal synapses formed by the terminals of granule cell axons; normal function is restored. It appears that primary presynaptic elements have priority for the synaptic sites for which they were originally programmed and that substitute terminals give way to primary presynaptic elements when the latter appear, even if the heterotypical synapses have become functional. This kind of hierarchical order in the nervous system facilitates the reconstruction of normal circuitry after transplantation.

Reconstruction of circuitry is not always necessary, however, for the restoration of function after neural transplantation. In some cases, grafting a population of neurons that produce a missing neurotransmitter near a target group of neurons, with resultant diffusion of the target neurons by the neurotransmitter, may be sufficient for functional recovery.

Replacement of damaged parts of the nervous system by transplantation is possible because the nervous system remains plastic throughout life and is amenable to repeated rounds of reorganization or readaptation as new conditions arise. This plastic capability of the nervous system is the basis for hope that further developments in neural regeneration and transplantation are possible and can be extended from the experimental laboratory to humans.

BIBLIOGRAPHY

Cotman, C. W., ed. (1985). "Synaptic Plasticity." The Guilford Press, New York.

Goodman, C. S., and Shatz, C. J. (1993). Developmental mechanisms that generate precise patterns of neuronal connectivity. *Neuron* **10**(Suppl.), 77–98.

Hawkins, R. D., Kandel, E. R., and Siegelbaum, S. A. (1993). Learning to modulate transmitter release: Themes and variations in synaptic plasticity. *In* "Annual Review of Neuroscience" (W. M. Cowan, E. M. Shooter, C. F. Stevens, and R. F. Thompson, eds.), Vol. 16. Annual Reviews, Palo Alto, CA.

Kaas, J. H. (1991). Plasticity of sensory and motor maps in adult mammals. *In* "Annual Review of Neuroscience" (W. M. Cowan, E. M. Shooter, C. F. Stevens, and R. F. Thompson, eds.), Vol. 14. Annual Reviews, Palo Alto, CA.

Madison, D. V. (1991). Mechanisms underlying long-term potentiation of synaptic transmission. *In* "Annual Review of Neuroscience" (W. M. Cowan, E. M. Shooter, C. F. Stevens, and R. F. Thompson, eds.), Vol. 14. Annual Reviews, Palo Alto, CA.

Rosenzweig, M. R., and Leiman, A. L. (1989). "Physiological Psychology," 2nd Ed. Random House, New York.

Seil, F. J., ed. (1989). Neural regeneration and transplantation. *In* "Frontiers of Clinical Neuroscience," Vol. 6. Alan R. Liss, New York.

Theodosis, D. T., and Poulain, D. A. (1993). Activity-dependent neuronal-glial and synaptic plasticity in the adult mammalian hypothalamus. *Neuroscience* 57, 501–535.

Neural Basis of Oral and Facial Function

BARRY J. SESSLE
University of Toronto

GLOSSARY

Periodontal tissues (or periodontium) Supporting tissues of the teeth, which surround the root of each tooth and serve to attach the root of the tooth to its socket in the enveloping alveolar bone; they also contain nerve fibers that control the periodontal blood vessels and, thus, the blood supply of the periodontal tissues, as well as sense organs (receptors) and their associated nerve fibers, which provide the peripheral basis for pain and touch from these tissues

Temporomandibular joint The "jaw joint" immediately in front of the ear; it also has a nerve supply, particularly on the posterolateral aspect of its capsule, that supplies receptors involved in pain, jaw position sense, etc.

Tooth pulp Soft tissues inside each tooth; sometimes referred to as the "nerve" of the tooth because it contains a profuse nerve supply; some of these nerve fibers provide the peripheral basis for pain from the tooth, and some control the blood vessels and blood supply of the pulp

Trigeminal nerve Fifth cranial nerve that has three major branches: the ophthalmic, maxillary, and mandibular; these three branches provide most of the sensory innervation of the face and oral cavity, and the mandibular branch also contains motor axons that supply several muscles of the orofacial region, principally those moving the jaw (mandible)

THIS ARTICLE PROVIDES AN OUTLINE OF THE SENsory and motor neural mechanisms of the face and mouth and, in a more limited sense, of the pharynx and larynx. Few details are provided of some important functions of the face and mouth (e.g., smell, taste, speech) because these topics are covered elsewhere. This article also emphasizes the profuse nerve supplies of the oral–facial tissues and their representation in the brain and the exquisite sensory capabilities provided by this rich innervation and extensive central neural representation. It considers the neural basis of oral–facial touch, temperature, and pain and gives particular emphasis to the latter, because pain commonly occurs in the skin, teeth, muscles, joint, and other tissues of the oral–facial region and humans can have long-term suffering from several pain states or syndromes in the face and mouth. Particular attention is also given to the neural processes underlying the many reflex and other motor functions manifested in the oral–facial region, especially those related to mastication (chewing), swallowing, and associated neuromuscular functions. In addition, this article briefly considers the role of the autonomic nervous system in regulating oral–facial blood flow as well as the limited understanding we presently have of its role in regulating the sensory and neuromuscular functions of the face and mouth.

I. INTRODUCTION

Before reviewing the neural mechanisms underlying the functions of the face and mouth, it first should be noted that the oral–facial region is remarkable in its high level of sensory discriminability and sensitivity. This is probably a reflection of its great innervation density and the large amount of brain tissue devoted to the representation of the oral cavity and surrounding areas. In addition, specialized receptor systems are associated with the periodontal supporting tissues of the teeth and, in lower animals, with the facial whiskers (vibrissae). These receptors provide an added dimension of sensory experience and, together with the tongue and lips, are most important for exploration of the environment and controlling movement and behavior. In addition, some of the most common pains occur in this region (e.g., toothache, headache), and some sensory functions are unique to the region (e.g., taste). Likewise, the oral–facial region is remarkable in the vast array of simple and complex motor activities that are manifested within it. These activities range from relatively simple reflexes, such as the jaw-opening reflex, to the very complex motor activities associated with speech, mastication, and swallowing, which involve the coordinated neuronal activity of many parts of the brain and which provide for social communication and the intake of food and fluid vital for life. It is also important to note that even though this article focuses on human oral–facial functions, studies in animals have been indispensable and crucial to most of the current knowledge of their mechanisms.

II. SENSORY FUNCTIONS

A. Touch

Our ability to sense touch (i.e., tactile sensibility) is extremely well developed in the orofacial region. Tests to measure touch include tactile threshold, stereognosis (a term referring to the ability to recognize the form of objects), and two-point discrimination (Fig. 1), and they have shown that some oral–facial tissues such as the tongue tip and lips have a greater tactile sensitivity than any other part of the body. These peripheral tissues are densely innervated by primary afferent nerve fibers, each of which terminates peripherally at a sensory organ called the receptor. These receptors "sense" stimuli and changes in the environment and transduce this information into electrochemical energy, which is then carried along the afferent fibers, into the brain, as action potentials. [*See* Somatosensory System.]

We also have analogous receptor mechanisms for our ability to detect and discriminate the size of small objects placed between the teeth, their hardness and texture, and bite force; these functions have largely been attributed to receptors that are located in the periodontal tissues around the root of each tooth. It is also becoming increasingly apparent that other receptors, such as those in the jaw joint and even in jaw muscles, make an important contribution (Fig.

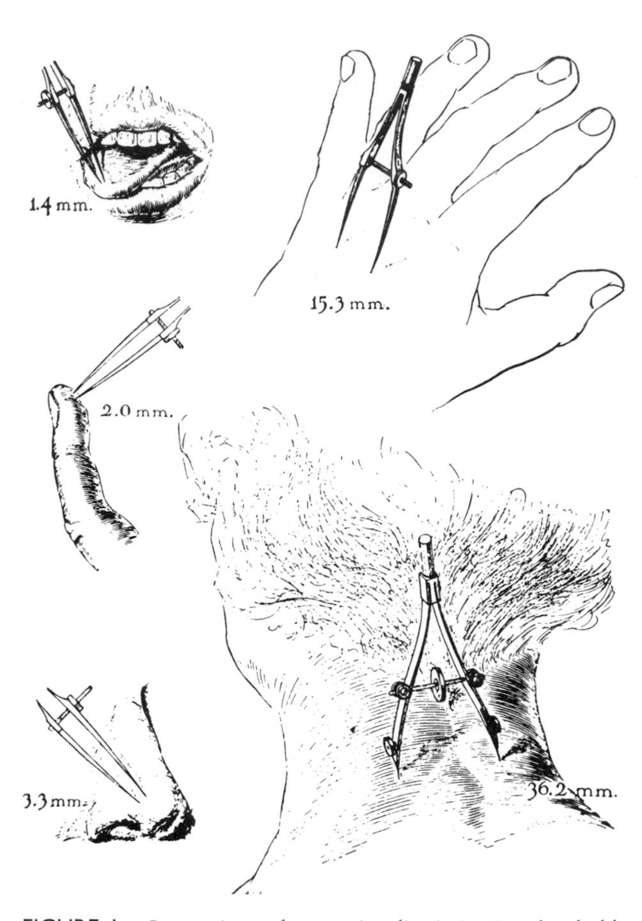

FIGURE 1 Comparison of two-point discrimination thresholds of various parts of the body. In this sensory test, two points are applied simultaneously and with equal force to skin or mucosa; the threshold value is the minimal distance between the points for which they are felt by the subject as two distinct points. [Reproduced, with permission, from L. Langley and E. Cheraskin (1956). "The Physiological Foundation of Dental Practice," 2nd Ed. Mosby, St. Louis.]

2). Receptors in this joint (the temporomandibular joint, or TMJ), as well as those in the jaw muscles, also largely account for our conscious perception of jaw position.

The receptors in the facial skin, oral mucosa, periodontal tissues, and TMJ that are responsible for the sensibility to mechanical stimuli such as touch or pressure stimuli can be broadly categorized into two types: free nerve endings and corpuscular receptors, of which several anatomically distinct examples exist. Functionally, these so-called mechanoreceptors are primarily associated with large-diameter, fast-conducting afferent nerve fibers, and they can respond either transiently (so-called velocity detectors) or throughout (static-position detectors) an innocuous mechani-

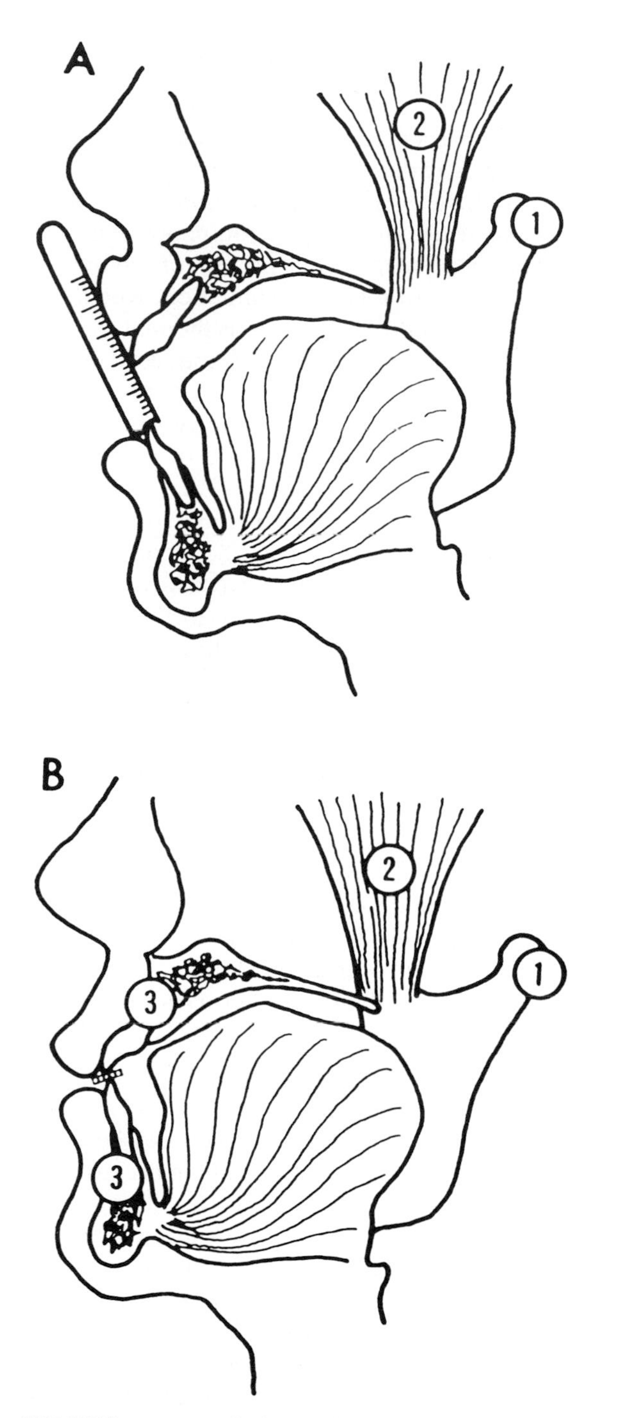

FIGURE 2 Some methods used for determination of mandibular position (A) and interdental size of objects (B), and possible receptor sites involved in these sensory functions. (A) Subjects are asked to reproduce certain magnitudes of jaw opening. (B) Subjects are asked to discriminate between the size of different objects (e.g., foil, wire, disc) placed between the teeth, or the threshold for detection is obtained. Note that the receptor sites include the temporomandibular joints (1), jaw muscles (2), and periodontium (3). [Reproduced, with permission, from A. Storey (1968) *J. Can. Dent. Assoc.* 34, 294–300.]

cal stimulus applied to an oral–facial site. This is reflected, respectively, in a brief or sustained burst of action potentials in their associated nerve fiber, which conducts these action potentials into the brain stem. By these neural signals, the mechanoreceptors collectively can provide the brain with detailed information of, for example, the location, quality, intensity, duration, and rate of movement of an oral–facial tactile stimulus. Many types of mechanoreceptors are exclusively activated by tactile stimuli. This physiological specificity, coupled with the existence of an anatomically recognized receptor structure for some of these mechanoreceptors, and of many neurons in the central relay stations (see the following) that respond exclusively to tactile stimulation, strongly supports the concept of a specificity theory. According to this theory, a specific set of receptors and afferent nerve fibers, and nerve cells and relay stations in the brain, respond exclusively to tactile stimuli and provide the cerebral cortex with neural information related only to touch and not, for example, to pain or temperature.

The major primary afferent pathway carrying the neural signals from the oral–facial mechanoreceptors is the trigeminal nerve (the fifth cranial nerve). The afferent nerve fibers in this nerve pass via the trigeminal ganglion, where their primary afferent cell bodies are located, and the trigeminal sensory nerve root to the trigeminal brain stem sensory nuclear complex (Fig. 3). The neural signals are transferred to nerve cells (i.e., neurons) at all levels of the brain stem complex. This complex can be subdivided into a main sensory nucleus and a spinal tract nucleus; the latter is subdivided further into the subnuclei oralis, interpolaris, and caudalis. These second-order neurons then project to higher sensorimotor centers as well as to local brain stem regions, including those responsible for activating muscles. Thereby, they serve as so-called interneurons involved in reflexes or more complex sensorimotor behaviors (see Section II,B). A major projection from these trigeminal brain stem neurons, however, is concerned with touch perception and passes primarily to the ventroposterior thalamus of the opposite (i.e., contralateral) side (see Fig. 3). After synaptic transmission through third-order neurons, the signals are relayed from here to a particular part of the overlying somatosensory cerebral cortex where the face and mouth are represented and where the cortical neural processing begins that eventually leads to the perception of a touch stimulus.

The second-, third-, and fourth-order neurons in this pathway to the cortex show many functional properties comparable with those of mechanorecep-

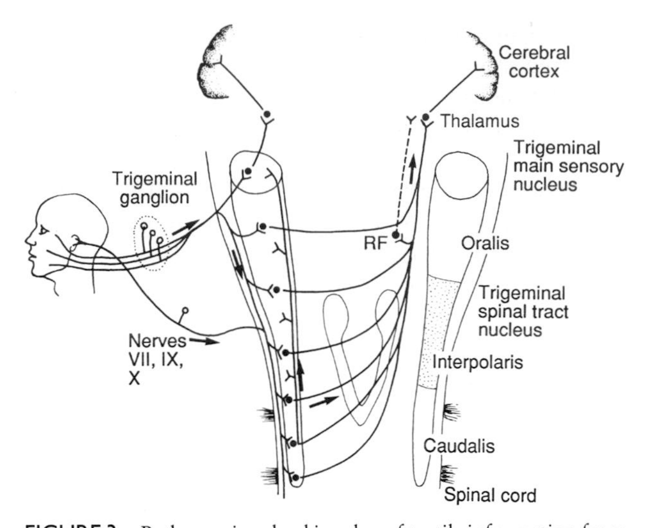

FIGURE 3 Pathways involved in relay of tactile information from the oral–facial region. Mechanoreceptive primary afferent fibers in the three divisions of the trigeminal (V) nerve pass into the brain stem via the trigeminal ganglion (where their cell bodies are located). Some afferent fibers from the ear pass via cranial nerves VII, IX, or X; some afferent fibers also pass to the brain stem via cervical nerves (not shown). The information is relayed on neurons located throughout the trigeminal main sensory and spinal tract nuclei; only a few of the many thousands of these neurons are depicted (solid circles). These neurons then send information mainly to the contralateral thalamus, where it is relayed on neurons located particularly in the medial part of the ventroposterior thalamus. Some of the signals reach the thalamus, however, by more circuitous routes, such as via brain stem reticular formation (RF), and some neurons in the trigeminal main sensory nucleus (and possibly in the spinal tract nucleus) project to the ipsilateral thalamus instead of the contralateral thalamus. From the thalamus, information is relayed directly to the somatosensory cerebral cortex. Note that some neurons in the subnucleus caudalis also send an ascending projection to rostral subnuclei (oralis and interpolaris) of the spinal tract nucleus and to the main sensory nucleus; this projection can modulate relay of tactile information through these regions. The horseshoe-shaped structure near the middle of the caudal brain stem represents the solitary tract nucleus. Arrows indicate direction of signals. [Reproduced, with permission, from G. Roth and R. Calmes (1981). "Oral Biology." Mosby, St. Louis.]

tive primary afferent fibers. They also retain much of the "specificity" of the tactile primary afferents, thus providing further support for the specificity theory of touch. However, by means of the complex ultrastructure and regulatory mechanisms existing at each of the relay sites, considerable modification of tactile transmission can occur, as a result of other incoming sensory signals and descending influences from higher brain centers. This may explain, for example, how distraction or focusing one's attention on a particular task at hand can depress our awareness of a touch stimulus.

B. Thermal Sensation

Our ability to sense the temperature of an object or substance is particularly well developed in the face and mouth, where thermal changes significantly $<1°C$ can be readily detected and discriminated; however, temperature detection and discrimination can vary depending on the magnitude and rate of the thermal changes, the area of thermal stimulation, whether or not the area has undergone previous thermal changes, and the adapting temperature of the skin or oral mucosa.

The receptors for temperature change (i.e., thermoreceptors) are associated with some of the smaller-diameter, slow-conducting afferent fibers. They are specifically activated by a small thermal change in either a cooling (cold afferent fibers) or warming (warm afferent fibers) direction, and they provide the brain with precise information on the location, magnitude, and rate of the temperature shift.

The predominant relay site in the brain stem of the afferent signals carried in thermoreceptive primary afferent fibers appears to be the trigeminal subnucleus caudalis. Neurons in this part of the trigeminal brain stem complex are exclusively activated by thermal stimulation of localized parts of the face and mouth, and this thermal information is relayed to the contralateral thalamus and then to the somatosensory cerebral cortex. The specificity theory thereby appears to be consistent with the properties of peripheral and central neural elements underlying our thermal sensibility.

C. Taste

The special sense of taste is covered elsewhere, but three aspects are briefly mentioned here because they relate to some of the other functions discussed in this article. First, as with pain, taste has affective, cognitive, and motivational dimensions as well as a sensory-discriminative dimension. For example, we find some tastes pleasurable and are motivated to seek them out, whereas other tastes have the opposite effect. Indeed, humans may have innate as well as acquired taste preferences, and the food industry is well aware of our "sweet tooth," an inborn preference for sweetness. Also, taste sensibility is now known not to be confined to specific areas of the tongue; extralingual (e.g., palatal) taste buds may also make an important contribution to taste. Finally, a number of factors have been reported to modify taste; these may include other sensory experiences (e.g., smell), decreased saliva, wearing of dentures, poor oral hygiene, local anesthetics, plant extracts, and perhaps genetic, metabolic, and endocrine factors and the age of the individual. [*See* Tongue and Taste.]

D. Pain

Pain deserves special emphasis because it is the first orofacial sensory experience that comes to mind for most of us. It also causes great human suffering and represents a major economic burden on society through health care costs and time lost from work. Moreover, orofacial pains are particularly noteworthy because they are very common (e.g., toothache) and are often chronic and disabling. Pain is now conceptualized as a multifactorial experience. It includes a sensory-discriminative component, an aspect that allows us to discriminate the quality, location, duration, and intensity of a noxious stimulus (i.e., a tissue-damaging stimulus), but it also encompasses cognitive, motivational, and affective variables, which can modify a person's response to the stimulus. Thus, the reactions to a noxious stimulus can vary from individual to individual. It can depend not only on the magnitude of the noxious stimulus but also on factors such as the meaning of the situation in which the pain occurs, the person's emotional state and motivation to get rid of the pain, and even racial and cultural background.

Of course, this multifactorial nature of pain can complicate diagnosis and treatment of pain for the clinician and also makes the experimental study of pain exceedingly difficult. Nonetheless, pain studies in humans and experimental animals have used a variety of approaches (e.g., behavioral, electrophysiological, anatomical) to give us some insights into the neural mechanisms underlying pain. [*See* Pain.]

1. Pain Transmission

The neural basis for orofacial pain and, indeed, pain from anywhere in the body is still only partly understood, but considerable advances have been made in the last few years. These insights into pain and the mechanisms underlying its control have largely come from studies in animals. The classic concept for explaining pain and the other somatic sensations is the specificity theory; though apparently applicable in other sensations such as touch (see earlier), this theory has been shown to have a number of limitations in trying to explain pain on the basis of a specific peripheral and central system. As a consequence, other theories have been proposed to account for the complexity and multidimensionality of pain. The gate control theory of pain has attracted the most recent interest and research, and although it has limitations, it does provide a good conceptual framework for considering the multifactorial nature of pain. First, this theory emphasizes the sensory interaction that occurs within the brain between the touch-related neural signals carried into the brain by the low-threshold, large-diameter primary afferent fibers and those signals conveyed by the small-diameter fibers; the peripheral terminals of many of the latter fibers respond to noxious stimuli. If, as a result of this interaction, the activity in the small so-called nociceptive fibers prevails, central transmission cells are excited (the "gate" opens) and bring into action the central processes related to the perception of and reactions to noxious stimulation. Second, the theory also emphasizes descending central neural controls (i.e., coming from higher brain centers) related to cognitive, affective, and motivational processes that can modulate the gate.

Recent experimental studies have revealed that some nociceptive primary afferent fibers are specifically sensitive to noxious stimuli, whereas others respond to innocuous stimuli as well. These nociceptive afferents are small diameter and slow conducting, and they terminate in the peripheral tissues as free nerve endings. Those nociceptive primary afferent fibers supplying the face and mouth project via the trigeminal ganglion to part of the trigeminal brain stem sensory nuclear complex. The subnucleus caudalis (see Fig. 3) is especially involved in pain (and temperature) transmission. Many subnucleus caudalis neurons receive the signals from orofacial nociceptive primary afferents and, thus, can respond to noxious stimulation of the face and mouth, TMJ, or jaw and tongue muscles. These neurons are called either wide dynamic range (i.e., they respond to innocuous as well as noxious stimuli) or nociceptive-specific (some responses of a wide dynamic range neuron are shown in Fig. 4). Also noteworthy, these neurons relay nociceptive information to the contralateral thalamus, from where information is relayed to the overlying cerebral cortex or other thalamic regions. Although parts of the thalamus or cortex are involved in the various components of pain behavior (perception, motivation, and so on), the precise function of each region is still uncertain. [*See* Cortex; Thalamus.]

The "rostral" parts of the trigeminal brain stem complex are especially concerned with the relay of tactile information (see earlier), but recent studies also indicate a role for them in orofacial pain mechanisms. For example, afferent fibers from the "nerve" of the

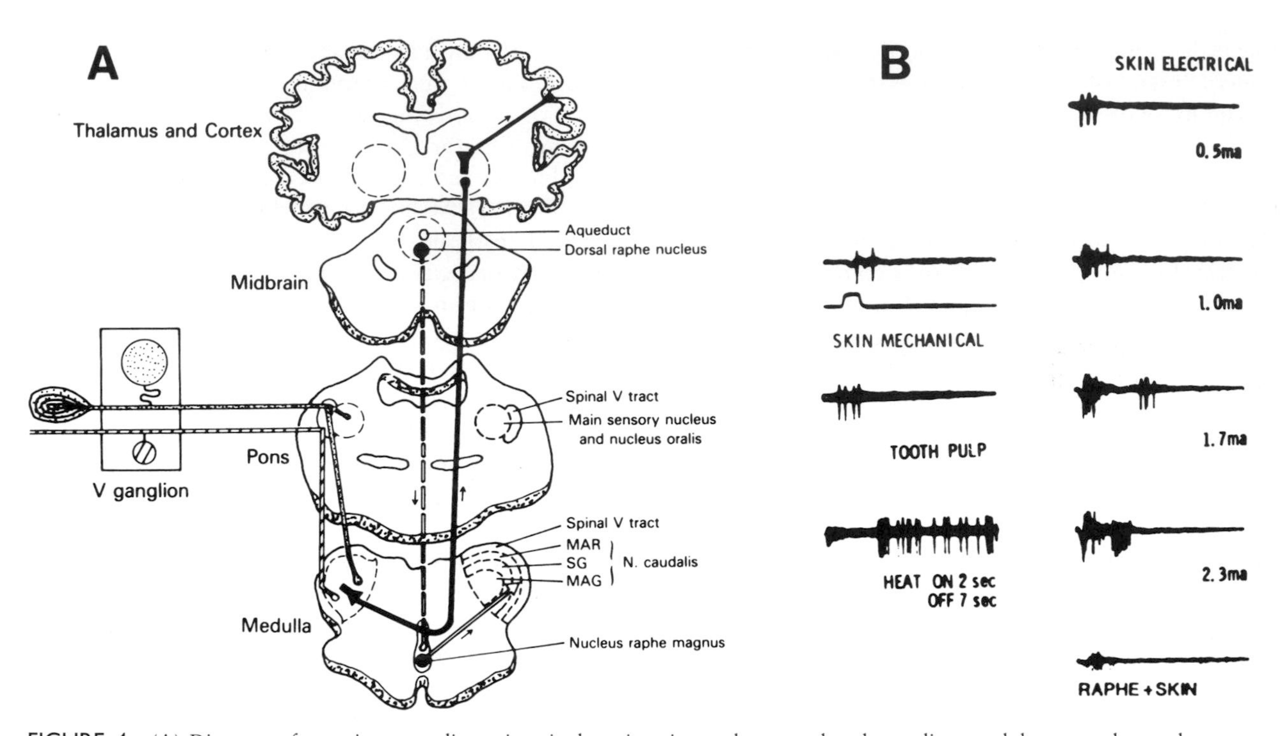

FIGURE 4 (A) Diagram of a major ascending trigeminal nociceptive pathway and a descending modulatory pathway that may suppress activity of neurons in the nociceptive pathway. Nociceptive neurons in the subnucleus caudalis receive and relay information from small-diameter primary afferent fibers (cross-hatched pathway) only, namely, nociceptive-specific neurons, or from small-diameter afferent fibers and from large-diameter primary afferent fibers (stippled pathway) as well, namely, wide dynamic range neurons; they predominate in layers I (marginalis, MAR) and V of the subnucleus caudalis. Substantia gelatinosa (SG) and magnocellularis (MAG) are other layers of the caudalis. Responses of both types of neurons to noxious oral–facial stimuli can be suppressed by descending influences from the dorsal raphe nucleus in the periaqueductal gray and nucleus raphe magnus, as shown (bottom right) for the wide dynamic range neuron illustrated in (B), when the raphe and skin stimuli were interacted. [Reproduced, with permission, from R. Dubner, B. J. Sessle, and A. T. Storey (1978), "The Neural Basis of Oral and Facial Function." Plenum, New York.] (B) This neuron was recorded in the subnucleus caudalis of the anesthetized cat. Its classification as a nicoceptive neuron was based on its responses to various types of oral–facial stimuli. Some of these responses are illustrated in the traces. Note, for example, that while the neuron fired with a brief burst of two action potentials when a light mechanical (tactile) stimulus was applied to a localized skin region of the cat's face, the neuron could also be activated when an electrical stimulus was applied to the cat's canine tooth pulp and when noxious radiant heat (and pinch, not shown) was applied to the same region on the cat's face. Because it responded to innocuous (i.e., tactile) as well as noxious stimuli, this neuron was classified as a wide dynamic range nociceptive neuron. Note on the right that with increasing intensity levels of electrical stimulation of the skin, late as well as early bursts of impulses could be evoked; late discharge probably reflects inputs from nociceptive afferent fibers, and early burst reflects inputs from faster-conducting "tactile" afferent fibers. Time duration of records: 100 msec (except heat record: 10 sec).

tooth (the tooth pulp), generally assumed to represent a nociceptive input, synapse with neurons present not only in subnucleus caudalis but also at the more rostral levels of the complex.

2. Pain Control

A variety of procedures are available for the control of pain, ranging from pharmacologic measures such as local and general anesthetics and analgesic drugs (e.g., aspirin, morphine) to therapeutic procedures such as acupuncture, transcutaneous electric stimulation, hypnosis, and psychiatric counseling. In extreme cases, neurosurgical methods may be employed. All of these procedures are aimed at blocking pain transmission either at the periphery (e.g., aspirin), before nerve impulses enter the brain (e.g., local anesthesia), or within the brain (e.g., general anesthetics and analgesics). As described earlier, the trigeminal brain stem sensory nuclei have a complex structural organization by which they can interact in complex ways with many other parts of the nervous system. Interactions involving inhibitory processes that modulate pain transmission have been extensively documented; indeed, inhibitory modulatory processes are widespread in the brain and are involved, as was noted, in touch as well as in reflex activity and more complex behavioral functions (see Section III). Moreover, facilitatory interactions between these various convergent afferent inputs can also occur. These types of interactions are thought, for example, to contribute to the so-called referral of pain, where pain may be felt not at the site of injury or pathology but also, or instead, at other distant sites.

With respect to inhibitory modulation of pain transmission, evidence indicates (1) sensory interaction and (2) descending central control mechanisms, as the gate theory postulates. The output of trigeminal nociceptive neurons, for example, can be markedly suppressed by large-fiber afferent inputs (i.e., sensory interaction); however, in some situations, small-fiber nociceptive afferent stimulation may also suppress their activity. They can also be inhibited by stimulation of brain sites such as the midbrain periaqueductal gray matter and the nucleus raphe magnus in the lower brain stem (see Fig. 4); stimulation of such descending central controls produces marked analgesic effects in humans and experimental animals. Stimulation of other regions, such as the cerebral cortex, is less effective in suppressing the trigeminal nociceptive neurons, although cortical stimulation does have a profound influence on neurons excited by nonnoxious stimuli.

These suppressive effects on nociceptive transmission are most exciting in terms of enhancing our understanding of pain mechanisms and developing better pain control procedures. The effects have been linked in part to the release of endogenous (i.e., naturally occurring) chemicals that may activate the descending control systems or act relatively directly on the pain-transmission neurons. One of these endogenous substances is enkephalin, a peptide that is pharmacologically similar to the opiate drugs such as morphine; other neurochemicals such as 5-hydroxytryptamine (serotonin) also appear to be involved. When injected into certain parts of the brain, enkephalin produces analgesic effects; if applied locally in the vicinity of pain-transmission neurons in the trigeminal subnucleus caudalis or analogous neurons in the spinal cord, the responses of these neurons to noxious stimuli can be suppressed. Thus, pain-suppressing systems appear to occur naturally within the brain, and a number of important therapeutic procedures have been proposed that may exert their analgesic effects by utilizing such systems. The action of narcotic analgesics such as morphine has been linked to such systems, and therapeutic procedures involving skin, muscle, or nerve stimulation (transcutaneous electric stimulation and acupuncture) may exert their analgesic effect in part by exciting pathways to the brain that ultimately lead to activation of endogenous analgesic systems.

3. Oral–Facial Pain States

The oral–facial region is a particular focus of pain, and a brief description and possible mechanisms are given for several of the most common or interesting pain conditions. The first, formerly known as temporomandibular (or myofascial) pain dysfunction, now appears actually to be a family of disorders and is termed temporomandibular disorders. They present a variety of signs and symptoms, the most common of which are pain in the region of the TMJ or jaw muscles or both, limitation of jaw movement, and crackling (crepitus) or clicking in the joint. The pain can sometimes be referred to other structures such as the teeth and muscles of the jaw or neck. Salivation and lacrimation, possibly reflecting involvement of the autonomic nervous system, are also frequently associated with the disorders. [*See* Autonomic Nervous System.]

The etiology, diagnosis, and treatment of these disorders represent some of the most controversial aspects of dentistry. Until recently, occlusal factors related to the faulty interdigitation of upper and lower teeth were considered to be the most important etiologic factors, but recent studies point to the importance of psychosocial and other central factors (e.g., stress-related) in its etiology in many patients. Treatment procedures are as varied and numerous as the various theories of its etiology, ranging from balancing the bite (i.e., occlusal equilibration) to the administration of muscle relaxants or anxiety-reducing drugs, to even psychiatric counseling.

The etiology of trigeminal neuralgia (or tic douloureux) is also controversial. This disorder, which occurs rarely in persons <45 years of age, manifests as paroxysms of excruciating pain that usually last for a few seconds or minutes, with long periods of remission between attacks. It is said to be the most excruciating pain a human can suffer, yet a most interesting and puzzling feature is that the neuralgic attack is usually triggered not by a noxious stimulus but by a light, nonnoxious stimulus (e.g., puff of air, wisp of cotton) to certain trigger sites in the perioral region. Theories of its etiology relate to either peripheral or central factors. Some researchers believe that peripheral changes, such as compression of the trigeminal sensory nerve root in the vicinity of the trigeminal ganglion by aberrant vessels or bony outgrowths, are the primary etiologic factor. Despite evidence in favor of a peripheral etiology, the signs and symptoms of the disorder are suggestive of central neural changes, particularly in the functional organization of the trigeminal brain stem sensory complex. This view is compatible with the effectiveness of certain anticonvulsant or antiepileptic drugs in depressing the activity of trigeminal brain stem neurons. These drugs are now widely used for the clinical control of the disorder.

Postherpetic neuralgia (shingles) differs from trigeminal neuralgia in that a definitive etiologic factor is known, namely, the herpes zoster virus that has gained access to the trigeminal ganglion. The pain usually involves the skin of the ophthalmic division of the trigeminal nerve (i.e., above and around the eye), and a selective loss of large myelinated fibers is apparent. Although this loss could, within the context of the gate control theory, lead to an imbalance of sensory input and an "opening of the gate" (see earlier), the actual mechanisms responsible for the pain are still unclear.

Another poorly understood condition is so-called atypical facial pain. This pain is diffuse, deep, and dull or throbbing in nature, and can be constant for many days. No trigger zones are associated with the disorder. Its etiology is also uncertain, although a psychogenic mechanism has been suggested for a certain portion of such cases.

Finally, the most common of oral–facial pains is the toothache. This pain is usually associated with trauma or dental decay affecting the "nerve" of the tooth (the pulp) or the overlying hard tissue, the dentine. Much research has centered on the possible peripheral mechanisms of pulp and dentinal sensitivity. Figure 5 illustrates the innervation of the pulp; most of the nerves in the pulp are small-diameter *afferent* fibers associated with sensation, but some are autonomic *efferent* fibers thought to primarily control the blood supply of the pulp. The pulp afferent fibers can be excited by a variety of different types of stimuli (e.g., sugar, hot or cold drinks), as most people can attest to. Though it is generally assumed that their excitation is exclusively related to pain, recent studies suggest that some pulp afferent fibers and their central connections in the brain may be involved in sensory functions other than pain.

Many of the pulp afferent fibers terminate in close proximity to odontoblasts, but some enter the dentinal tubules, occasionally in close contact with the odontoblastic process (see Fig. 5). Although these findings are considered by some researchers as evidence supporting a neural theory of dentinal sensitivity, the role of these intradentinal fibers in sensitivity is still unclear. Indeed, it is still conceivable that intradentinal neural processes contribute to the excitation of pulpal nerve fibers, or that intradentinal elements are involved by means of the odontoblast acting as a transducer, or by hydrodynamic processes, or both. More general acceptance exists for the hydrodynamic theory, which suggests that enamel or dentinal stimuli can cause a displacement of dentinal tubule contents; this, in turn, is thought to bring about a mechanically induced excitation of the nerves.

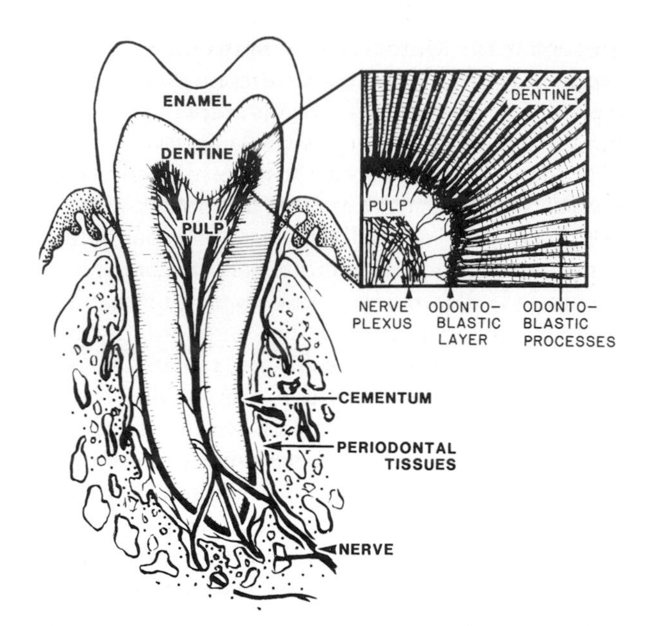

FIGURE 5 The nerves supplying the tooth innervate the periodontal tissues, which are the supporting tissues of the tooth, as well as the tooth pulp. After entering the pulp at the root apex of the tooth, the pulpal nerves arborize extensively, especially in the crown of the tooth. They form a nerve plexus in the periphery of the pulp beneath the layer of odontoblast cells, which have processes extending into the tubules in the overlying dentin (and which are involved in dentin formation). Some individual nerve fibers leave the subodontoblastic plexus; a number enter the dentinal tubules, although many terminate in the odontoblastic layer. [From B. J. Sessle (1987); modified, with permission, from M. R. Byers (1984). *Int. Rev. Neurobiol* **25**, 39–94.]

III. NEUROMUSCULAR FUNCTIONS

A. Muscle

The oral–facial region manifests a vast array of both simple and highly complex motor activities, which have important biological significance to the point of being among the most fundamental behaviors required for survival. Movements of the jaw and the surrounding musculature are integrally involved in human behaviors as diverse as mastication (chewing), drinking and suckling, manipulation of objects with the tongue, cheeks, and lips, communication through facial expressions, and speech production.

The peripheral motor components of these activities are the muscles of the jaw, face, tongue, pharynx, larynx, and palate. Like muscles elsewhere in the body, they consist basically of a passive elastic component (e.g., tendon, ligaments) and an active contractile component; the latter is composed of numerous individual striated muscle fibers. These so-called extrafusal muscle fibers are connected to the axons of alpha motoneurons present in the brain stem: a single alpha motoneuron plus the muscle fibers that it supplies are known collectively as the motor unit. Impulses from the motoneuron are conducted along its axon (the alpha efferent or motor axon) to the muscle fibers and bring about muscle contraction through the process of neuromuscular transmission. [*See* Muscle Dynamics.]

The peripheral *sensory* components of muscle are receptors. For example, muscle contains free nerve endings and these receptors are associated with muscle pain and possibly responses to stretch. In addition, there are also specialized receptors (e.g., the Golgi tendon organ, which is particularly sensitive to muscle tension and the stretch-sensitive muscle spindle). In addition to a dual afferent supply, the muscle spindle receives a motor innervation from the gamma (fusimotor) efferent fibers of small gamma motoneurons, which modify the sensitivity of the afferent fibers to stretch, thereby indirectly assisting or maintaining muscle contraction. In contrast to muscles in most other places of the body, these specialized receptors have a limited distribution in the craniofacial region.

B. Central Mechanisms

The primary muscle afferent pathways and central connections are poorly documented for most of the oral–facial musculature, except for the trigeminal mesencephalic nucleus, which contains the cell bodies of jaw-closing muscle spindle primary afferent fibers (and some other oral–facial primary afferent fibers). This location of primary afferent cell bodies is the only place in the body where primary sensory cell bodies are located within the central nervous system. Other major distinguishing features of the oral–facial motor systems include the following. (1) Many oral–facial muscles lack muscle spindles and Golgi tendon organs, as noted earlier. (2) A fusimotor (gamma-efferent) control system is absent owing to the lack of muscle spindles in many muscles. (3) Reciprocal innervation, where muscle afferent fibers have reciprocally opposite effects on antagonistic spinal motoneurons, is limited owing to the sparsity of muscle spindle and tendon organ afferent fibers. This lack may be compensated for by the powerful regulatory influences afforded by afferent inputs from facial skin, mucosa, TMJ, and teeth (see the following). (4) Coordinating pathways and mechanisms exist to allow for the *bilateral* activity of muscles; although activity of a particular limb or trunk muscle can occur on both the left and right sides of the body in some movements, this bilateral activation (or depression) is particularly prominent in orofacial movements (e.g., chewing, swallowing, speech, coughing).

The muscle afferent fibers, along with cutaneous, joint, and intraoral afferent fibers, make excitatory reflex connections with brain stem motoneurons located within one or more cranial nerve motor nuclei (Fig. 6); these connections are usually indirect and involve interneurons in the trigeminal spinal tract nucleus, the solitary tract nucleus, and the reticular formation. Examples of such "simple" reflexes are the jaw-closing, jaw-opening, and horizontal jaw reflexes; facial muscle reflexes; tongue reflexes; and laryngeal,

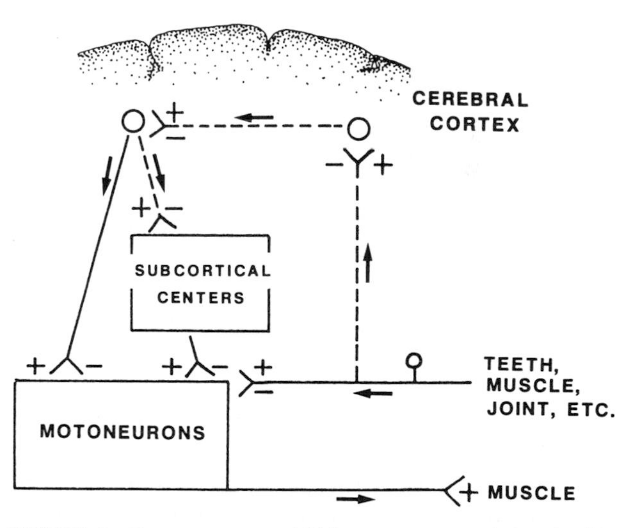

FIGURE 6 Some sensory and higher center controls involved in regulation of brain stem motoneurons supplying muscles of the jaw, face, and tongue. Both alpha and gamma motoneurons can be reflexly influenced (facilitated, +; inhibited, −) by sensory inputs from various receptors. They can also be influenced by descending controls originating from a number of higher brain centers, such as the sensorimotor cerebral cortex. The cortex can exert relatively direct effects on motoneurons or can modulate motoneuron activity indirectly by means of its connections with various subcortical centers (e.g., basal ganglia, cerebellum, and brain stem reticular formation, part of which comprises the "chewing center").

pharyngeal, and palatal reflexes. In addition to these excitatory reflex effects of various oral–facial stimuli, inhibitory effects are also expressed on the reflexes by sensory stimuli and by descending regulatory influences arising from higher brain centers such as the cerebral cortex (see Fig. 6). The excitatory and inhibitory influences are especially involved in protection of the masticatory apparatus (e.g., from biting the tongue during chewing) and in providing much of the neural organization upon which are based more complex motor activities such as the protective reflex synergies of coughing and gagging. An even higher level of complexity of organization is seen with the rhythmic, automated activities of mastication, suckling, and swallowing; speech is another complex sensorimotor behavior utilizing this neural organization.

C. Mastication, Swallowing, and Related Neuromuscular Functions

1. Mastication

Mastication serves to break down foodstuffs for subsequent digestion by means of the masticatory forces generated between the teeth. It is characterized by cyclic jaw movements in three dimensions (vertical, lateral, and anteroposterior) and less rigid facial and tongue movement patterns. These various movements are produced by the coordinated contraction of the jaw, face, and tongue muscles (Fig. 7). Masticatory forces on the teeth are usually in the 5- to 10-kg range, but can vary depending on factors such as the teeth concerned (molars exert the greatest force, incisors the lowest), practice, toughness of the diet, the wearing of dentures, presence of periodontal disease, tooth–cusp configuration, and the distance that the jaws are separated when the forces are applied. Even greater forces are developed during biting: maximal biting forces on a tooth usually range from approximately 20 to 200 kg (the Guinness world record is over 400 kg!).

Although mastication was originally thought to be based on alternating simple jaw reflexes of jaw opening and closing, the current concept is that the cyclic, patterned nature of chewing depends on a subcortical center (see Fig. 6) comprising a central neural pattern generator, the brain stem "chewing center." This generator is sensitive to descending regulatory influences from higher brain centers and to sensory inputs from oral–facial receptors; the sensory inputs may be particularly critical in the learning of mastication and acquisition of masticatory skills, in actively (i.e., reflexly) guiding masticatory movements, and in guiding the position of the jaw when the teeth come into

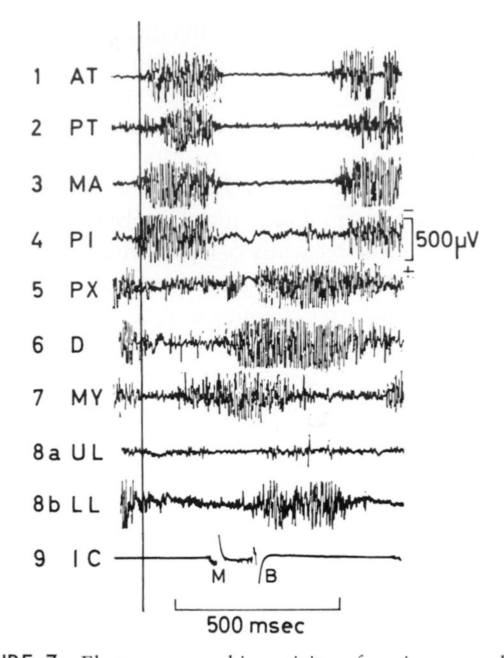

FIGURE 7 Electromyographic activity of various muscles recorded during a natural chewing cycle in a human subject: jaw (1–6); tongue (7); and facial (8). The vertical line indicates onset of activity in right anterior temporalis muscle. M and B indicate, respectively, onset and offset of contact of the opposing incisor teeth. AT, anterior temporalis; PT, posterior temporalis; MA, masseter; PI, medial (internal) pterygoid; PX, lateral (external) pterygoid; D, digastric; MY, mylohyoid; UL, upper lip; LL, lower lip; IC, incisor contact. [Reproduced, with permission, from E. Moller (1966). *Acta Physiol. Scand.* **69**, 1 (Suppl. 280).]

occlusion and when the jaw is at "rest." The clinical significance of the sensory influences is further exemplified by their probable involvement, along with central influences, in the etiology of pathophysiologic conditions such as temporomandibular disorders and bruxism (tooth grinding).

2. Swallowing

In contrast to mastication, swallowing is an innate, reflexly triggered, all-or-none motor activity that is relatively insensitive to sensory or central control. In addition to its obvious alimentary function, swallowing also serves as a protective reflex of the upper airway because it reflexly interrupts respiration and prevents the intake of food or fluid into the airway. Swallowing consists of a rigid, temporal pattern of muscle activities that appears to depend on a brain stem pattern generator, the "swallow center," for its expression. The coordinated muscle activities provide the means for the propulsion of a food or liquid bolus from the oral cavity to the stomach (Fig. 8) and also afford mechanisms for protection of the airway and prevention of reflux (regurgitation). Some of the muscles are "obligate" swallow muscles (i.e., they always are active in swallowing), whereas others show a variable participation in swallowing (the facultative muscles). The latter muscles, which include the jaw and facial muscles, especially may be sensitive to alterations in the oral environment and to maturational changes, and thus their participation can vary depending, for example, on the volume or consistency of a foodstuff, or whether the subject is an infant or adult.

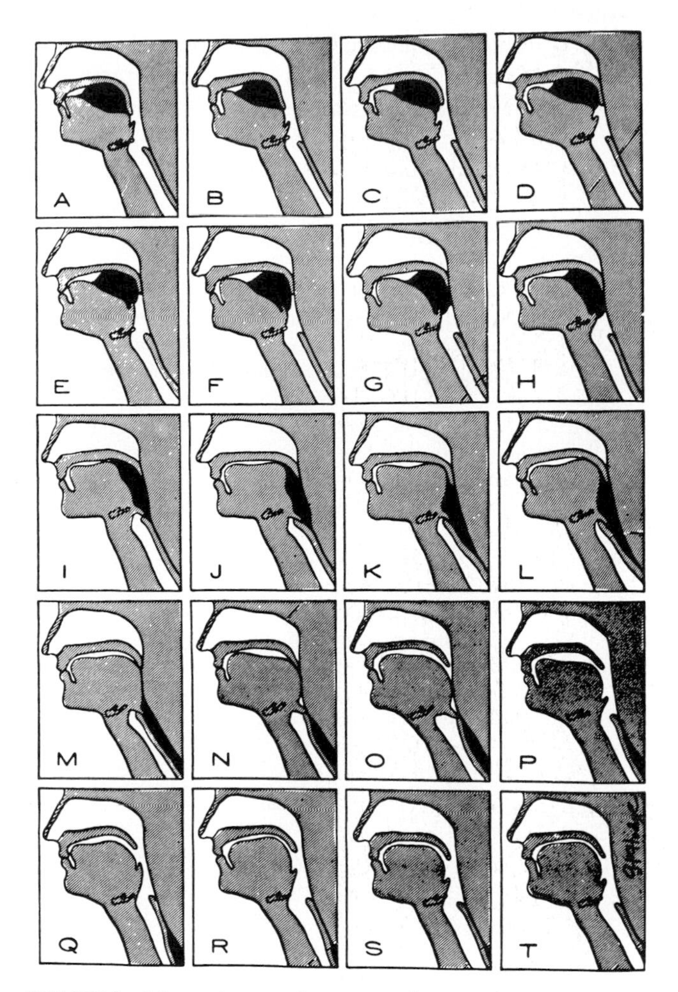

FIGURE 8 The orchestrated sequence of events characterizing a single swallow. At the start of the swallow, the soft palate acts as a partition up to the base of the tongue (A, B) and elevates to engage the posterior pharyngeal wall and close off the nasopharynx as the food bolus (black) moves backward over the tongue surface (C–E). The tongue acts as a piston to squeeze the bolus into the pharynx, and the bolus is conveyed down the pharynx by pharyngeal muscle contractions (F–J); note that the epiglottis is tilted backward [e.g., to protect the entrance into the laryngeal airway (white)]; the glottis (not shown) also serves to close off the entrance into the airway. Then, the bolus is slightly delayed at the upper esophageal sphincter (K), which then relaxes to allow the bolus into the upper esophagus (L, M) before closing again to prevent reflux (regurgitation). As the bolus moves down the esophagus (N–T), the soft palate relaxes and the epiglottis resumes its position and the airway is reopened. [Reproduced, with permission, from R. M. Bradley (1984). "Basic Oral Physiology." Year Book Medical, Chicago; adapted from R. F. Rushner and J. A. Hendron (1951). *J. Appl. Physiol.* 3, 622–630.]

3. Other Functions

Mastication, suckling, and swallowing are themselves components of even more complex behaviors. They are associated with feeding and drinking, which are particularly dependent on the function of higher centers of the brain such as the hypothalamus. Sensory feedback, however, is also utilized for the initiation, maintenance, and cessation of these ingestive behaviors (e.g., a "full stomach" stretches gastrointestinal receptors, which, through their central connections, can inhibit feeding). Many of these higher brain centers are also concerned with other complex functions involving the oral cavity, including oral aggression (e.g., biting), facial expression, and speech.

IV. AUTONOMIC REGULATION

The autonomic nervous system also has regulatory effects on a variety of orofacial functions, but with few exceptions, our knowledge of its role in the orofacial region is quite scant; however, it is well known that smooth muscle (e.g., in the gastrointestinal tract, abdominal and pelvic viscera, walls of blood vessels), cardiac muscle, and glandular structures such as the lacrimal and salivary glands receive autonomic efferent innervation. The two major divisions of the autonomic nervous system, the sympathetic and parasympathetic, have important roles in regulating the contractibility, secretion, or other functions of these various tissues.

As far as autonomic regulation in the orofacial region is concerned, most is known about autonomic control of blood flow and salivary secretion. The following will briefly outline autonomic regulation of orofacial blood flow. The oral mucosa indeed has served as a model system for studying autonomic control of the vasculature. Sympathetic vasoconstrictor regulation of the blood supplies of the mucosa, periodontal, and tooth pulp tissues has been well documented, but whether or not a parasympathetic control also occurs in each of these tissues is still unclear.

Sympathetic vasoconstrictor tone also takes place in the facial skin and, as elsewhere in the body, in involved in thermal regulatory function (e.g., to prevent heat loss from the body surface). Surprisingly however, in the facial skin, vasoconstrictor tone is primarily restricted to the nose, lips, and ears. Thus, heat loss from the head can represent a considerable proportion of total body heat loss, and these three parts of the orofacial region are also very susceptible to cold, a susceptibility that is well reflected in frostbite. Vasodilation can occur through a release of this vasoconstrictor tone, or through "active" vasodilation that is especially expressed in the cheek, forehead, chin, and neck. These four areas correspond to those showing increased blood flow in emotional states, such as the flushing of the face that may occur in an embarrassing situation. The extent to which the parasympathetic nerves are involved in this active orofacial vasodilator tone is still unclear. The parasympathetic nerves also do not appear to have a major function, if any, in regulation of the facial sweat glands. The sympathetic efferents again appear to have a major role here and are involved in the common response of these glands to thermal or emotional stimuli.

In addition, it should be mentioned that the autonomic nervous system may also regulate skeletal muscle function as well as the sensitivity of several receptor systems such as the muscle spindle and some cutaneous receptors. Also, certain pain syndromes may involve dysfunction of the autonomic nervous system, but the mechanisms by which the autonomic efferents modulate pain sensibility are still uncertain. Finally, evidence is accumulating to indicate that the autonomic nervous system can exert so-called neurotrophic influences on several sensory-motor functions. In the orofacial region, these include the taste bud and salivary gland functions.

BIBLIOGRAPHY

Dubner, R., Sessle, B. J., and Storey, A. T. (1978). "The Neural Basis of Oral and Facial Function." Plenum, New York.

Fromm, G. H., and Sessle, B. J. (eds.) (1991). "Trigeminal Neuralgia: Current Concepts Regarding Pathogenesis and Treatment." Butterworths, Stoneham, Massachusetts.

Luschei, E. S., and Goldberg, L. J. (1981). Neural mechanisms of mandibular control: Mastication and voluntary biting. *In* "Handbook of Physiology. The Nervous System. Motor Control," Vol. II, Pt. 2, pp. 1237–1274. American Physiological Society, Bethesda, Maryland.

Sessle, B. J. (1987). The neurobiology of facial and dental pain: Present knowledge, future directions. *J. Dent. Res.* **66**, 962–981.

Wall, P. D., and Melzack, R. (eds.) (1994). "Textbook of Pain," 3rd Ed. Churchill Livingstone, Edinburgh.

Zarb, G. A., Carlsson, G. E., Sessle, B. J., and Mohl, N. D. (eds.) (1994). "Temporomandibular Joint and Masticatory Muscle Disorders." Munksgaard, Copenhagen.

Neural–Immune Interactions

DAVID L. FELTEN
SUZANNE Y. FELTEN
University of Rochester School of Medicine

GLOSSARY

Cytokine Chemical mediator released from a monocyte–macrophage (monokine) or a lymphocyte (lymphokine) that acts on cells of the immune system, providing intracellular signaling to the target cells

Hormone Chemical mediator released from cells in an endocrine gland or from a neuron (neuroendocrine transducer cell) that exerts its effect on target cells via circulation; response of the target depends on the presence of receptors specific for that hormone in or on the target, not on the proximity of the cell releasing the hormone to the target

Mitogen response Proliferation in a lymphocyte resulting from stimulation by a nonspecific agent, usually a plant lectin; this response generally is viewed as indicative of lymphocyte potential to proliferate but should not be equated with the overall responsiveness of the immune system *in vivo,* particularly in ability to respond to a specific antigen

Neuromodulation Process by which a neurally released mediator produces little or no direct functional effect on a target cell (neural, immune, or autonomic effector cell) but significantly alters the responsiveness of that cell to another signal

Neurotransmitter Chemical mediator released from a neuron that acts on receptors on an adjacent target neuron or effector cell, resulting in an intracellular effect in that cell

Norepinephrine Catecholamine present in the periphery and in the central nervous system, used as a neurotransmitter (central neurons, postganglionic sympathetic neurons) or as a hormone (adrenal medullary chromaffin cells); an important neural mediator for communication with autonomic target tissues

Sympathetic nervous system Subdivision of the autonomic nervous system, supplying innervation to smooth muscle, cardiac muscle, secretory glands, some cells related to metabolism, and cells of the immune system, generally involved in an arousal (fight or flight) response

NEURAL–IMMUNE INTERACTIONS CONSIST OF communication channels and meditors that permit bidirectional signaling between the nervous system and the immune system. The nervous system, in response to external and internal cues and inputs, can communicate with cells of the immune system via neuroendocrine secretion and release of neurotransmitters from nerve fibers that innervate lymphoid organs, or tissues where an inflammatory response or immune response is ongoing. Cells of the immune system can secrete cytokines that signal the brain and change its electrical and chemical activity. These bidirectional channels of communication appear to be linked circuitry that permits feedback and feedforward regulation. In addition to the systemic communication channels permitting brain and immune system to signal each other, local signaling within the microenvironment of lymphoid organs permits short-loop communication between cells secreting neurotransmitters and cytokines. These two major classes of mediators likely interact with each other to synergize or counter each other's actions; their actions in turn are regulated by the other mediators present in the microenvironment such as hormones. The consequences of bidirectional neural–immune interactions are a superimposed neural modulation of immune responses and the possibility that psychosocial factors, environmental inputs, and internal cognitive or affective factors may influence the functioning of the immune system. In turn, the brain and its behavior may be responsive to immunologically derived signals that act as "molecular sensory signals." Clearly, the immune system can no longer be considered an autonomous, self-regulating system that operates outside of the context of brain and behavior.

I. EVIDENCE FOR NEURAL INFLUENCES ON IMMUNE RESPONSES

Evidence from many fields points to the ability of the nervous system to modulate immune responses. R. Ader, N. Cohen and colleagues have demonstrated that immune responses can be classically conditioned and that such conditioning can alter the course of a genetic autoimmune disease in rodents. Other investigators have show that numerous cellular activities and responses can be conditioned in the direction of either enhancement or suppression. Numerous stressors, including psychological stressors, can induce alterations in immune responsiveness, although the direction and magnitude of response are highly dependent on the nature, timing, and duration of the stressor. Studies in humans suggest that the stress of examinations in medical students, marital separation, or other aversive circumstances may be accompanied by diminished immune responsiveness. Some investigations of depressed or bereaving individuals suggest that a subset of these people may demonstrate diminished immune responsiveness and increasing incidence of illness, although these two phenomena have not yet been causally linked. A recent study of women with disseminated breast cancer suggests that psychotherapeutic support as an adjunct to standard medical treatment increased the survival of these patients and may permit a better adjustment to their disease. Experimental brain lesions in animals have been shown to cause altered immune responses, with the directionality and duration of the effect dependent on the site and extent of the lesion.

These many lines of investigation provide evidence that external factors that impinge on the brain, or direct manipulation of neural circuitry of the brain, can result in altered immune responses, sometimes of sufficient magnitude to alter the outcome of a genetic or acquired disorder. However, in no instance has the entire circuitry from the stimulus to specific brain sites, to specific central pathways using known neurotransmitters, to specific outflow channels from the brain to organs of the immune system, to specific cellular mechanisms that induce specific intracellular changes that can explain the altered immune response been worked out. The task of unraveling the myriads signal interactions among the nervous system, endocrine system, and immune system, all capable of changing with the animal's behavioral state, brings together the interpretive complexities of systemic biology magnified at least threefold because of the interactions.

However, we do know that the brain has two major routes by which it can send mediators to alter or modulate activity in the periphery: neuroendocrine secretion and direct neural connections from the autonomic nervous system to the viscera. For many years, glucocorticoids have been known to influence immune responses and have been used clinically as immunosuppressive drugs. Since the 1970s, lymphocytes and monocytes–macrophages have been known to express receptors for neurotransmitters on their surface, but the physiological role for these receptors was unknown. The past 10 years has seen an explosion of information about such hormonal and neurotransmitter signaling of cells of the immune system. The present article outlines basic evidence for these mediators playing a modulatory role in immunoregulation. Obviously, we are just beginning to identify the cast of participants in neural–immune signaling; the emerging picture indicates that these mediators can exert a complex array of influences on different cell types of the immune system at different stages of activation and also can interact with cytokine signals from the immune system itself. These neurally derived mediators will likely exert just as complex a regulatory influence over reactivity of lymphoid cells as the cytokines themselves do. [*See* Lymphocytes; Macrophages.]

II. ENDOCRINE EFFECTS ON THE IMMUNE SYSTEM

A. Pituitary

Early studies of hypophysectomized mice and pituitary-deficient dwarf mice reported a resultant diminution in both cellular and humoral immune responses, and involution or hypocellularity of primary and secondary lymphoid organs. The administration of anterior pituitary hormones, particularly growth hormone (GH) and prolactin, led to some degree of restoration of immune function in these animals, suggesting that pituitary hormones could exert a modulatory role over immune responses. [*See* Pituitary.]

B. Growth Hormone

GH deficiencies have been associated with diminished T-cell functions, antibody responses, natural killer (NK) cell activity, and decreased cellularity in bone marrow and thymus; these abnormalities are restored to some extent by administration of GH. GH administration enhances mitogen responsitivity of transformed and normal lymphocytes, restores mitogen responsivity in aged animals with depressed responsivity, and increases the activity of alloantigen-specific cytotoxic T lymphocytes. GH also activates macrophages to produce superoxide anions that nonspecifically kill ingested bacteria. Thymocytes, lymphocytes, and monocytes possess high-affinity receptors for GH.

C. Prolactin

Prolactin inhibition or deficiency results in suppressed antibody responses, suppressed delayed hypersensitivity responses, depressed mitogen responses in B and T lymphocytes that are independent of interleukin-2 (IL-2) receptor expression or IL-2 secretion, suppressed T lymphocyte-dependent activation of macrophages, and suppressed response to bacterial challenge and to its induction of interferon-gamma production. These responses are reversed or restored, for the most part, by administration of exogenous prolactin. The pituitary secretion of both GH and prolactin is increased by the action of thymosin fraction V, a thymic hormone, suggesting a functional thymic–pituitary axis. Interleukin-1 (IL-1) can inhibit release of prolactin from the pituitary, as can glucocorticoids, whereas estrogen can enhance release. Prolactin may act to counter some of the inhibitory effects of the glucocorticoids and may be active in the early phases of a stress response.

D. Adrenal Corticotropin Hormone

Adrenal corticotropin hormone (ACTH), an anterior pituitary hormone, can exert a direct suppressive effect on antibody production, interferon-gamma secretion, and macrophage tumoricidal activity. ACTH is stimulated by corticotropin-releasing factor (CRF) from the hypothalamus and, in turn, stimulates glucocorticoid secretion from the adrenal cortex, thus establishing the classic CRF–ACTH–glucocorticoid axis. This axis is immunosuppressive to many cellular and humoral immune responses.

E. Opioid Peptides—The Endorphins

The endorphins, also produced in the pituitary, have been reported to modulate immune responses, lymphocyte proliferation, NK cell activity, production of interferon-gamma, and phagocyte chemotaxis. The magnitude and directionality of change of these responses vary considerably, depending on the binding capabilities of the opioid peptide under investigation. In addition, studies of opioid receptors indicate that only some of the effects of endorphins can be attributed to direct receptor-mediated interactions, and nonopioid receptor interactions probably mediate some of the endorphin effects on immune responsivity. Beta-endorphin may modulate the expression of the T-cell receptor, the IL-2 receptor, or other receptors on lymphocytes. Alpha-endorphin can bind preferentially to some MHC class I antigens. Thus, endorphin responses may act through numerous mechanisms and may change the responsiveness of the target cells to other signals.

F. Opioid Peptides—The Enkephalins

Met- and leu-enkephalins have also been reported to modulate antibody responses, antibody-dependent cell-mediated cytotoxicity, lymphocyte numbers and proliferation, resistance to viral and tumor challenges, lymphocyte migration, NK cell activity, expression of IL-2 receptors, and production of IL-2. Again, the magnitude and directionality of the responses are highly variable and depend on the dose, route of administration, timing of the mediator with respect to the antigen, and extent of catabolism following administration. Some of these opioid peptides may be present during some forms of intermittent stress, suggesting that the nature and extent of a stressor may determine which hormonal mediators are secreted. In addition, some nerve fibers in the spleen demonstrate met-enkephalin immunoreactivity, suggesting that the enkephalins also may be utilized as neurotransmitters in lymphoid organs, as they are in the brain.

G. Glucocorticoids

Glucocorticoids were shown in early studies to exert anti-inflammatory effects and immunosuppressive effects, resulting in their clinical use. Subsequent studies showed suppressive effects on lymphocyte proliferation, antibody responses, generation of cytotoxic effector cells, mixed leukocyte responses, generation of T helper cells, NK cell activity, and prolongation of allograft rejection and of tolerance. Monocyte and macrophage functions are suppressed, including numbers of circulating monocytes, expression of MHC class II antigen expression, phagocytosis, cytokine production and secretion, chemotaxis, and synthesis of complement components. However, not all effects of glucocorticoids are immunosuppressive; physiological levels of glucocorticoids can suppress nascent lymphocyte proliferation while enhancing proliferation of activated lymphocytes, suggesting that glucocorticoids may enhance signal-to-noise activity and may regulate overproduction of antibody. Also, conditioned immunosuppression, stress-induced lymphocyte hyporesponsivity, and central effects of CRF administration and IL-1 administration can apparently influence immune responsiveness independent of the glucocorticoids. Therefore, not all neural influences on the immune system are mediated by glucocorticoids. Unfortunately, many *in vitro* studies of glucocorticoid effects and many *in vivo* studies using exogenous administration of glucocorticoids have utilized pharmacologic rather than physiologic doses.

The immune system is capable of regulating glucocorticoid secretion; IL-1 can elevate CRF secretion from the paraventricular nucleus of the hypothalamus into the hypophyseal–portal circulation to the anterior pituitary, and also possibly ACTH secretion from the anterior pituitary directly. This apparently occurs at the peak of an antibody response, suggesting that the immune system may be capable of turning on and off the secretion of glucocorticoids as key regulators of the immune response. Some stressors may interact at the level of secretion of CRF in the hypothalamus and dysregulate this feedback circuitry. An additional role played by the glucocorticoids is the modulation of receptors for neurotransmitters on lymphoid cells. Glucocorticoids can up-regulate the expression of beta-adrenoceptors on many cell types, including lymphocytes, thereby enhancing the responsiveness of those cells to a subsequent interaction with catecholamines. These influences may modulate both the magnitude and timing of neurotransmitter signaling.

H. Hormonal Signaling of Lymphoid Cells

The brief discussion of hormonal effects on immune responsiveness has clearly shown that hormones can interact with each other, can exert principal effects on immunologic receptors or mediators in a true modulatory fashion, can influence the secretion of neurotransmitters or the responsiveness of lymphoid cells to them, and can exert direct or indirect effects through hormone receptors expressed by cells of the immune system. The timing and extent of hormonal secretion, the availability and catabolism of the hormone in different compartments of the immune system, and the presence and interactions of many cell types of the immune system may determine the final effect of a given hormone on a specific response from a specific subset of cells at a specific site at a specific time. However, from this complexity the beginning framework has emerged for linked circuitry by which feedback loops can regulate hormonal secretion, similar to how neuroendocrine feedback regulation occurs. To understand how these hormonal mediators actually are utilized by behaving organisms to modulate responsivity of the immune system, we must add to this complexity the superimposed regulation of hormonal secretion by cortical, limbic, and other central neural circuitry that exerts its effect on the releasing-factor neurons, inhibitory-factor neurons, and other neuroendocrine transducer cells of the brain. There is little doubt that such influences exist and are utilized for modulation of immune responses, as demonstrated by many of the behavioral studies; however, our knowledge of the actual circuitry and mechanisms of influence, particularly with the added factor of cytokine feedback, is at a very elementary stage.

III. DIRECT NEURAL CONNECTIONS WITH THE IMMUNE SYSTEM

A. Chemically Specific Innervation of Lymphoid Organs

Both primary lymphoid organs (bone marrow) and secondary lymphoid organs [spleen, lymph nodes, gut-associated lymphoid tissue (GALT)] are innervated directly by noradrenergic (NA) postganglionic sympathetic nerves (Figs. 1–4) and by peptidergic nerves whose origin has not been fully elucidated. Early literature showed that nerve fibers follow the vasculature into a variety of organs, including

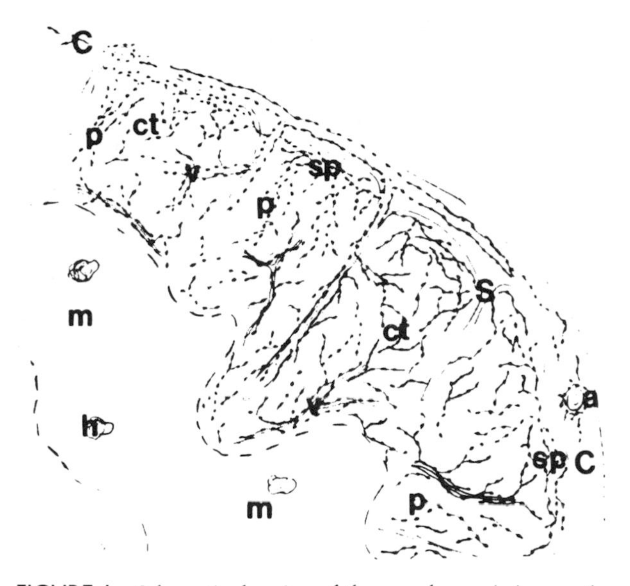

FIGURE 1 Schematic drawing of the noradrenergic innervation of the thymus. Noradrenergic fibers enter the capsule (C) around arterial plexuses (a), travel with vascular plexuses (v) or subcapsular plexuses (sp), and branch into the parenchyma (p) of the thymic cortex. ct, cortex; h, Hassall's corpuscles; m, medulla; S, septa. [From D. L. Felten *et al.* (1985). *J. Immunol.* **135**, 755s–765s, with permission.]

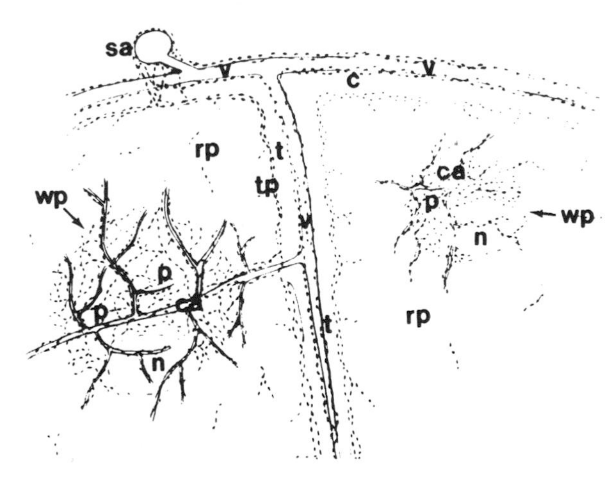

FIGURE 2 Schematic drawing of the noradrenergic innervation of the spleen. Noradrenergic fibers enter the spleen around the splenic artery (sa) as part of the splenic nerve, travel with the vasculature in plexuses (v), and continue into the spleen along the trabeculae (t) in trabecular plexuses (tp). Fibers from both the vascular and trabecular plexuses enter the white pulp (wp), continuing mainly along the central artery (ca) and its branches. Noradrenergic nerve terminals radiate from these plexuses into the periarteriolar lymphatic sheath (p) but mainly avoid the follicular or nodular (n) areas. These parenchymal nerve fibers end among fields of lymphocytes and other cell types. C, capsule; rp, red pulp. [From D. L. Felten *et al.* (1985). *J. Immunol.* **135**, 755s–765s, with permission)

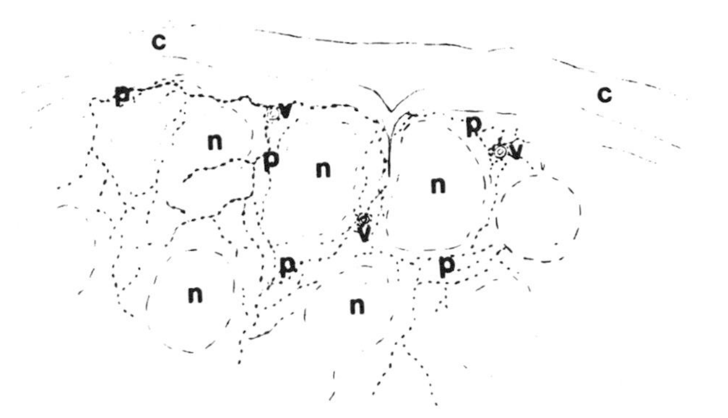

FIGURE 3 Schematic drawing of the noradrenergic innervation of a lymph node. Nerve fibers enter the lymph node in the hilar region (not shown) and travel through the medullary cords with the vasculature (v) and along the sinuses. The fibers terminate mainly in the paracortical regions (p) around the follicles or nodules (n). Some fibers from the subcapsular region distribute into the adjacent cortical zone as well. Noradrenergic fibers distribute with the vessels and also travel into the surrounding parenchyma, where they end among lymphocytes and other cells. c, capsule. [From D. L. Felten *et al.* (1985). *J. Immunol.* **135**, 755s–765s, with permission.]

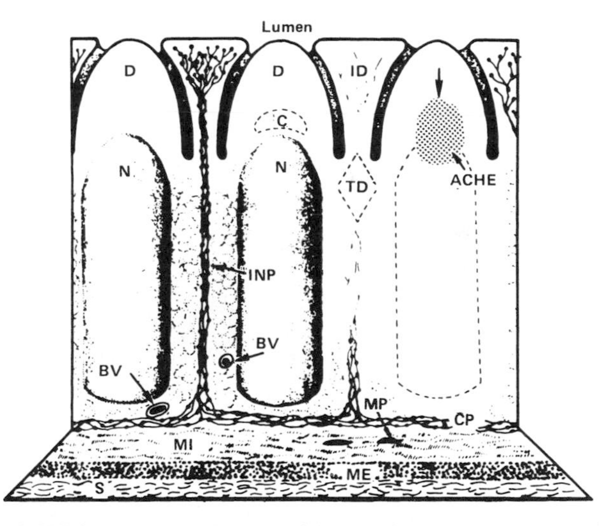

FIGURE 4 Schematic drawing of the noradrenergic innervation of the rabbit appendix as a representative of gut-associated lymphoid tissue. Noradrenergic fibers enter along the outer or serosal (S) surface, form a catecholamine nerve fiber plexus (CP) along the muscular layers (MI and ME) adjacent to Meisner's plexus (MP), and then turn radially to run between the lymph nodules (N) in internodular plexuses (INP). Fibers in this plexus travel directly through the T-dependent zone (TD) and enter the lamina propria, or interdomal region (ID), and then branch profusely in this zone among lymphocytes, enterochromaffin cells, and subepithelial plasma cells. Acetylcholinesterase (ACHE) staining is most abundant (arrows) in the region of the crown (C) in the domes (D) and generally is not associated with nerve fibers here. BV, blood vessels. [From D. L. Felten *et al.* (1985). *J. Immunol.* **135**, 755s–765s, with permission.]

lymphoid organs, and provide regulatory control over blood flow. The first fluorescence histochemical studies of NA innervation of lymphoid organs reported primarily vascular innervation. However, use of more sensitive fluorescence methods and immunocytochemical methods demonstrated extensive parenchymal innervation, with nerve terminals extending among lymphocytes and macrophages (Figs. 5–7).

In addition, numerous neuropeptide-containing nerve fibers have been found in thymus, spleen, and lymph nodes (Figs. 8–10). Some neuropeptides, such as neuropeptide Y (NPY), may be colocalized with norepinephrine (NE) in NA postganglionic sympathetic nerve fibers. Other peptides, such as substance P and calcitonin gene-related peptide (CGRP), may be found in primary sensory fibers, although experimental confirmation of this origin is not yet available. Other peptidergic nerves, such as vasoactive intestinal peptide (VIP) or met-enkephalin-containing fibers, derive from cell bodies whose origin is unknown. Although acetylcholinesterase-positive nerve fibers exist in lymphoid organs, enzymatic studies suggest that they may not be cholinergic nerves; experimental evidence generally is not supportive for cholinergic innervation of lymphoid organs. Tracing studies utilizing retrograde transport of horseradish-peroxidase or other markers have been very difficult to interpret and have produced conflicting results, partly because of the great tendency for these tracers to gradually diffuse from lymphoid organs and give false-positive labeling. Such tracing studies from peripheral organs should also be accompanied by anterograde studies

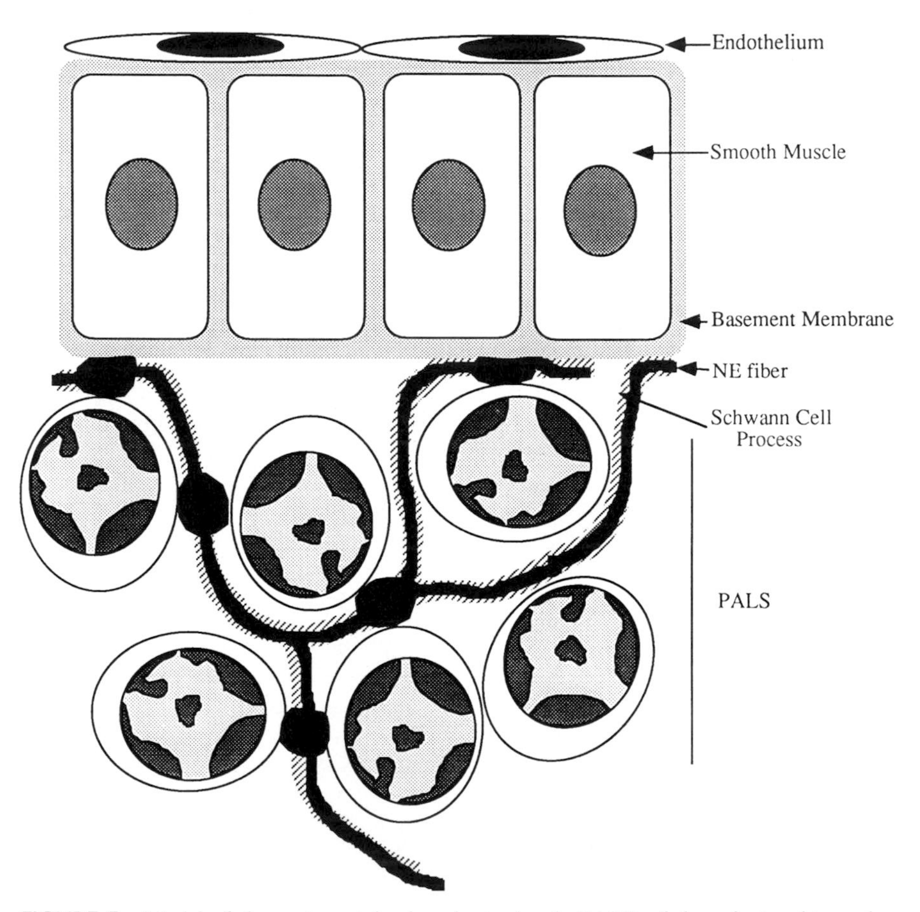

FIGURE 5 Model of the periarteriolar lymphatic sheath (PALS) of the splenic white pulp, demonstrating the presence of noradrenergic nerve terminals along the basement membrane of the smooth muscle of a central arteriole, also in direct contact with lymphocytes, both adjacent and deep to this vessel. These nerve terminals give rise to paracrine-like secretion, with diffusion of the neurotransmitter from the terminal, permitting interaction with receptors on cells that possess them. NE, norepinephrine. [From S. Y. and D. L. Felten (1989). *In* "Frontiers of Stress Research" (H. Weiner *et al.*, eds.), pp. 57–71. Hans Huber Publishers, Toronto.]

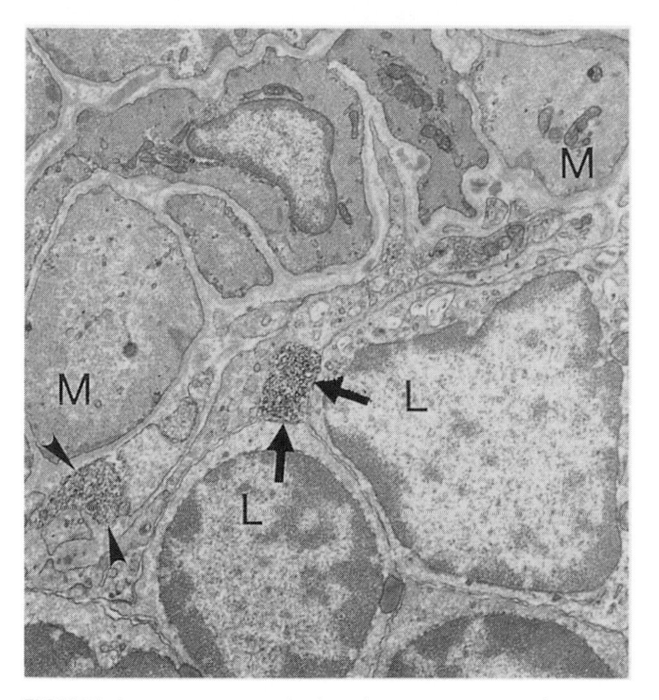

FIGURE 6 Two tyrosine hydroxylase-positive (noradrenergic) nerve terminals at the junction of arteriolar smooth muscle and lymphocytes in the periarteriolar lymphatic sheath of a rat. One terminal (arrowhead) ends adjacent to a smooth muscle cell (M), separated from it by a basement membrane. The second nerve terminal (arrow) ends in direct contact with two lymphocytes (L) in the periarteriolar lymphatic sheath. The contacts are smooth appositions of approximately 6-nm distance. Electron microscopic immunocytochemistry for tyrosine hydroxylase. ×14,000.

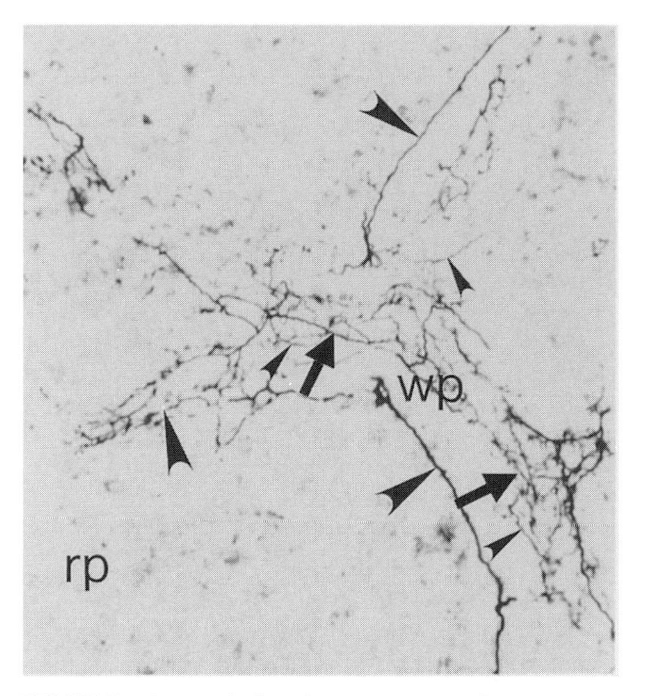

FIGURE 7 Tyrosine hydroxylase-positive (noradrenergic) nerve fibers in the white pulp of the spleen of a rat. These fibers are present along the central artery (arrows), along the marginal sinus (large arrowheads), and within the parenchyma (small arrowheads), where T lymphocytes reside. Immunocytochemistry for tyrosine hydroxylase. rp, red pulp; wp, white pulp. ×160.

and by experimental manipulations such as ablation studies with neurochemical analysis. These have been done extensively only for the NA nerves; the origin of peptidergic innervation of lymphoid organs awaits more detailed and careful study.

In order for a chemical mediator found in nerves to be considered as a true neurotransmitter, it must fulfill several minimal criteria, including: (1) presence and synthesis in neurons whose fibers end in apposition with, or adjacent to, the target cells; (2) release from the nerves and availability for interaction with target cells; (3) presence of receptors for the mediator on target cells, and second-messenger responses or intracellular processes that are affected by the receptor–ligand interaction; and (4) pharmacologically predictable functional consequences of interaction of the mediator with its responsive target cell. These criteria have been met for NE as a neurotransmitter in several organs, such as spleen and thymus, and appear to be mainly fulfilled for some neuropeptides, particularly substance P and VIP. Intensive efforts

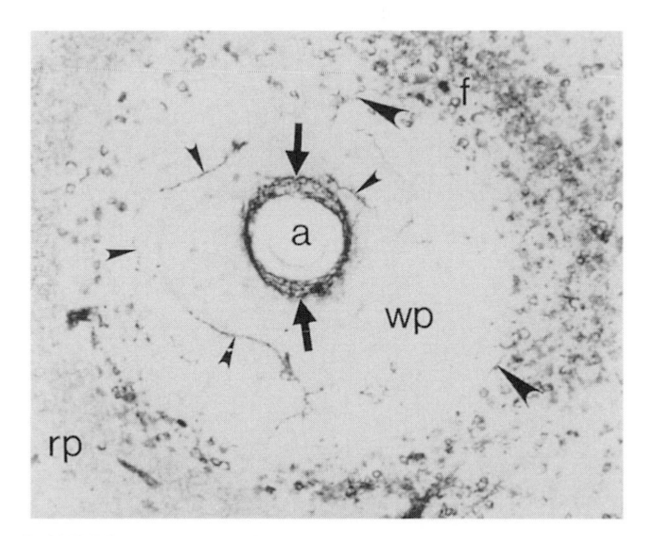

FIGURE 8 Neuropeptide Y-positive nerve fibers in the white pulp of the spleen in a rat. These peptidergic fibers are present around the central artery (arrows), along the marginal sinus (large arrowheads), and within the parenchyma (small arrowheads). This section is counterstained for IgM-positive cells (B lymphocytes), which are present in the follicles and in the marginal zone, a central artery; f, follicle; rp, red pulp; wp, white pulp. Immunocytochemistry for neuropeptide Y. ×250.

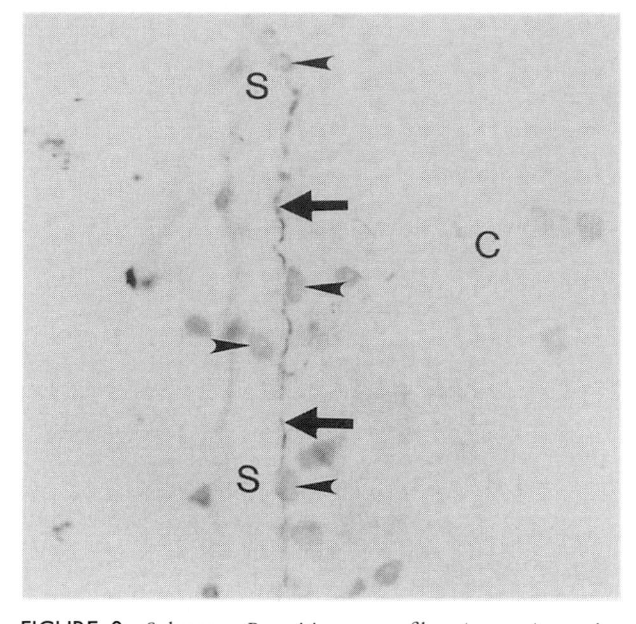

FIGURE 9 Substance P-positive nerve fiber (arrows) running along the margin of a septum in the rat thymus. This fiber is adjacent to mast cells (arrowheads) that stain in this section. C, thymic cortex; S, septum. Immunocytochemistry for substance P. ×250.

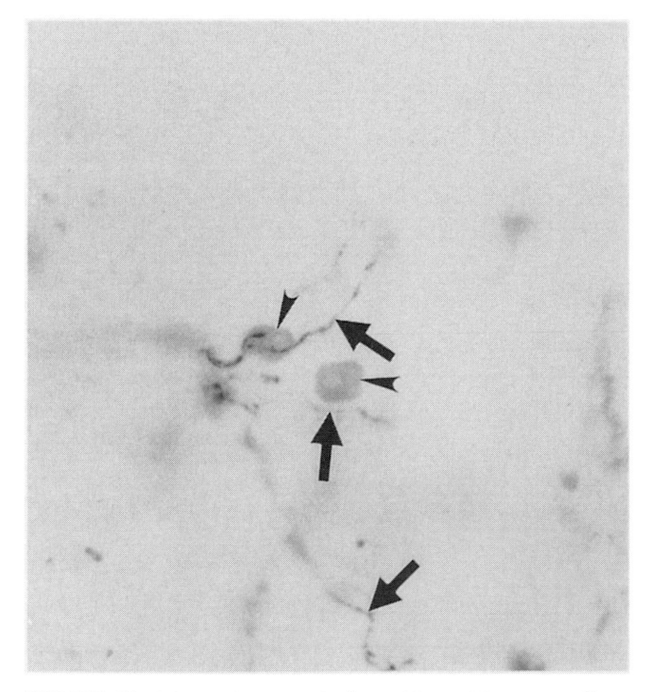

FIGURE 10 Vasoactive intestinal peptide-positive nerve fibers (arrows) adjacent to mast cells (arrowheads) in the outer thymic cortex of a rat. Immunocytochemistry for vasoactive intestinal peptide. ×250.

are now under way in many laboratories working on various aspects of these criteria. The first step in establishing neurotransmission for a substance with cells of the immune system as targets is to demonstrate the presence of nerves in the appropriate compartments of lymphoid organs, thereby providing a possible source of ligand for interaction with receptors on target cells. The following descriptions are drawn mainly from reports from the laboratories of D. Felten and S. Felten.

B. Primary Lymphoid Organs—Bone Marrow and Thymus

Primary lymphoid organs are innervated by NA nerve fibers that travel along the vasculature and also branch into the parenchyma. In the bone marrow, such fibers end among hemopoietic elements. Early studies suggested that sympathetic stimulation released cells into the circulation from the marrow, and beta-adrenoceptor stimulation of these cells triggered stem cells into cycle or shortened the cell cycle. [*See* Hemopoietic System.]

The thymus receives NA innervation from the superior cervical ganglion and from upper-chain ganglia. NA fibers (see Fig. 1) distribute with the vasculature into the cortex, forming the densest plexuses along the vasculature at the cortico–medullary junction, and also branch into the cortex among thymocytes. The medulla and associated epithelial zones are innervated only sparsely, mainly along the vasculature. Thymocytes also possess beta-adrenoceptors and respond to catecholamine stimulation by inhibiting proliferation and enhancing the expression of differentiation antigens. Denervation of the NA nerves from the thymus in newborn rats by chemical sympathectomy results in enhanced proliferative responses of thymocytes at 10 days of age. Thus, key functions in primary lymphoid organs, such as proliferation, differentiation, and migration, may be modulated by NE released from sympathetic NA nerve fibers.

Several groups of peptidergic nerve fibers have been found in the thymus, including NPY (appears to be colocalized with NE), substance P (see Fig. 9) and CGRP (perhaps localized in primary sensory fibers), and VIP (see Fig. 10) (of unknown origin). Thymocytes and lymphocytes possess receptors for numerous neuropeptides, linked with classic second-messenger systems, suggesting that neurotransmitters released from these peptidergic nerves into the microenvironment of lymphoid cells may be available for interaction with these cells.

C. The Spleen

NA innervation of the spleen arises from cells in the superior mesenteric–coeliac ganglion, which in the rodent is supplied by the T6–T12 spinal cord intermediolateral cell column of preganglionic neurons. These fibers travel with the splenic artery and distribute with two major systems. The first is the smooth muscle of the capsule and the trabecular–venous sinus system, probably instrumental in the contractile capabilities of this organ. The second system supplied by NA fibers is the white pulp (see Fig. 2). NA fibers travel with the central artery of the white pulp and its branches and distribute among lymphocytes in the periarteriolar lymphatic sheath (PALS), where T helper and T cytotoxic/suppressor cells are found, along the marginal sinus and in the marginal zone, where macrophages reside, and in the parafollicular zones along the outside of these B lymphocyte-containing follicles (see Fig. 7). NA nerve terminals directly contact T lymphocytes via 6-nm junctional appositions in the PALS (see Figs. 5 and 6) and along the marginal sinus and also contact macrophages and other accessory cells, demonstrated with electron microscopic immunocytochemistry. This is a closer apposition than found between NA sympathetic nerve endings and other target cells in the periphery.

Substance P fibers are also present in the spleen, around the large venous sinuses and some of the arterioles, and along the trabeculae, with scattered fiber profiles extending into the white pulp and red pulp; CGRP-containing fibers are present in the same compartments and appear to be colocalized with substance P. This substance P and CGRP fibers appear to be compartmentalized in zones distinct from the NA fibers. In addition, NPY (see Fig. 8) appears to be colocalized in NA fibers; both types of staining disappear following ganglionectomy or chemical sympathectomy. Other putative neurotransmitters reported to be present in nerve profiles in the spleen include somatostatin, cholecystokinin, neurotensin, met-enkephalin, VIP, and, surprisingly, IL-1. The presence of adrenoceptors and receptors for most of these peptides on lymphocytes and monocytes/macrophages raises the specter of highly complex interactions of many mediators derived from nerves and present in specific compartments of the spleen in varying concentrations.

D. Lymph Nodes

Lymph nodes are innervated by sympathetic ganglion cells of the chain or collateral ganglia, whose fibers enter the nodes with the hilar vasculature, distribute through the medullary cords among lymphocytes and macrophages, extend past the cortico–medullary junctions into the paracortical zones, and branch among T lymphocytes (see Fig. 3). Additional fibers travel in subcapsular plexuses and arborize among lymphocytes in the cortical zones. NA fibers appear not to innervate the follicles, which contain B lymphocytes. Thus, both the spleen and lymph nodes share common patterns of NA innervation, including innervation of sites of lymphocyte entry, sites of antigen capture, sites of antigen presentation and lymphocyte activation, and sites of lymphocyte egress.

E. Gut-Associated Lymphoid Tissue

GALT is innervated by NA and peptidergic nerves. The NA fibers derive from collateral ganglia, enter the gut with the vasculature, traverse the muscular layers, travel radially toward the lumen, distribute through the T lymphocyte-dependent zones, and arborize in the lamina propria among lymphocytes, mast cells, enterochromaffin cells, plasma cells, and other types (see Fig. 4). The NA fibers end along the subepithelial zone alongside the plasma cells. Some peptidergic nerve fibers are localized near immunologically important sites. A plexus of VIP fibers was found by C. Ottaway adjacent to the postcapillary venules, the site of entry of T lymphocytes bearing VIP receptors into the gut; these fibers probably play a roll in the ingress and retention of T lymphocytes in GALT. R. Stead and colleagues found substance P and CGRP-containing nerve fibers adjacent to mast cells in the gut, with preferential terminations on these cells. Physiological evidence suggests that these neuropeptides can modulate the release of mediators from mast cells, and mast cell-derived mediators can stimulate the primary afferents electrically.

IV. FUNCTIONAL ROLES OF NEUROTRANSMITTERS IN THE IMMUNE SYSTEM

A. *In Vitro* Studies of Norepinephrine and Catecholamines

NE and various NA agonists and antagonists have been incorporated into a number of immunological assay systems in an attempt to determine the role of NE in the immune system. The results from these studies are often difficult to interpret due to the wide variability in experimental design, the frequent use of pharmacological rather than physiological doses of drugs, and the unclear relevance of such assays to the living organism in which the lymphoid cell is exposed to a complex local microenvironment. However, a few patterns of action have emerged from these studies.

In general, studies of mitogen-induced proliferation of both T and B lymphocytes have demonstrated that NE, epinephrine, and isoproterenol are catecholamine agonists that act on beta-adrenoceptors and reduce proliferation, an effect that can be blocked by beta-adrenergic antagonists. However, low concentrations of NE in the 10^{-9}–10^{-7} *M* range can act on alpha-adrenoceptors (blockable by phentolamine) and increase proliferation, particularly if the beta-adrenoceptor effect is blocked. The beta-adrenoceptor decrease in proliferation is mediated through stimulatory G proteins in lymphocytes that increase intracellular cyclic adenosine monophosphate (cAMP). Increases in cAMP in lymphocytes can decrease the synthesis and secretion of IL-2, suggesting a possible mechanism for diminished proliferation. Other possible mechanisms include decreased expression of IL-2 receptors and decreased ability of the effector cell to respond to IL-2.

Effects of NE on the ability of cytotoxic cells to lyse target cells have been reported to increase or decrease following incubation with NE. Incubation of NK cells with NE prior to the addition of target cells results in increased killing activity, whereas the presence of NE throughout the entire incubation period results in a dose-dependent decrease in NK activity.

Epinephrine or isoproterenol can decrease phagocytosis by macrophages and can decrease killing of viral-infected cells and tumor cells by interferon-gamma-activated macrophages. These effects probably are mediated through cAMP. Secretion of IL-1 and tumor necrosis factor (TNF) by macrophages has been reported to either increase or decrease after the addition the adrenergic agents, depending on the source of the macrophages.

The conflicting results from these *in vitro* studies make it difficult to predict what effect adrenergic agents should have on more complex responses, such as T-dependent antibody responses. Based on these *in vitro* data, one might be inclined to predict that NE would decrease primary antibody responses, because it appears to decrease many of the processes involved in the response, such as phagocytosis, IL-2 secretion, and proliferation. However, in a series of *in vitro* studies by V. M. Sanders and A. E. Munson, NE was reported to enhance the plaque-forming cell response to sheep red blood cells (RBC) if NE was present during the first 6 hr of culture. This increase was blocked by the beta-antagonist propranolol, but not the alpha-antagonist phentolamine. The effects of NE on this response were additionally complicated by an alpha-2-receptor-mediated (clonidine) decrease in the response seen on Day 5 in culture, and an alpha-1-receptor-mediated (methoxamine) increase in the response unmasked on Day 4 if propanolol blockade of the beta-adrenoceptor effect occurred. Adoptive transfer studies in mice showed the same directional results, with increases in the plaque-forming cell response to sheep RBC if cells were removed from a mouse treated with epinephrine or incubated with epinephrine for 1 hr before transfer into syngeneic irradiated recipients that were then immunized with sheep RBC. These more complex *in vitro* studies are in general agreement with our results obtained from denervation of the spleen and lymph nodes by chemical sympathectomy (reported in Section IV,B). However, what clearly emerges from these studies is the notion that adrenergic effects on immune responses are not simple unidirectional changes that are inhibitory in nature. Even though mitogen-induced proliferation is diminished by NE, primary antibody responses are generally enhanced.

B. *In Vivo* Studies of NA Neural Influences on Immune Responses

The most careful and detailed *in vivo* studies have utilized chemical sympathectomy with 6-hydroxy-dopamine (6-OHDA), an agent that is taken up into NA nerve terminals by the high-affinity uptake carrier and destroys the terminals. Not only is NE removed via this sympathectomy, but so are colocalized neuropeptides. The alternate approach, surgical ganglionectomy, removes NA innervation and colocalized neuropeptides, other peptidergic neurons, and nerve fibers, perhaps of sensory origin, that are merely passing through the ganglion on the way to their target.

Denervation studies of secondary lymphoid organs can be classified into two general categories: (1) long-term denervation of rodents at birth with 6-OHDA or denervation with surgical ganglionectomy in adults, and (2) chemical sympathectomy with 6-OHDA in adult rodents, which leads to a short-term depletion of NE. Long-term sympathectomy generally results in augmented antibody responses in adult mice. The effects of short-term sympathectomy are more complex. A diminution of primary antibody responses in lymph nodes and spleen has been observed in several mouse strains. However, chemical sympathectomy in two mouse strains with a predominant T helper subset response did not reduce antibody responses. Sympathectomy enhanced antibody responses in a mouse strain with a predominantly T-helper-1 response to antigen (interferon-gamma producer), but not in the T-helper-2 predominant strain (IL-4 producer). This suggests that T helper subtype predominance may influence the functional role of the sympathetic nervous system in an intact animal. The effect of sympathectomy can be prevented by treating animals with desipramine prior to 6-OHDA. Desipramine inhibits uptake of 6-OHDA and protects the NA terminals, suggesting that the effect of chemical sympathectomy requires uptake of 6-OHDA into nerve terminals and is not due to toxic actions of 6-OHDA.

Further studies from our laboratories have shown that chemical sympathectomy in adult rodents results in suppression of delayed-type hypersensitivity responses to contact sensitizing agents (by approximately 50%), diminished cytotoxic T-lymphocyte responses to alloantigen accompanied by decreased IL-2 production (by approximately 50%), and enhanced NK cell activity *in vivo* and *in vitro*. B-lymphocyte proliferative responses were augmented from denervated inguinal lymph nodes, consistent with an inhibitory role for NE on proliferation, but T-lymphocyte responses in these nodes and other sites, and B-lymphocyte proliferative responses in spleen and other lymph nodes, were highly complex and sometimes were diminished or showed no response. Removal of some of the other colocalized neuropeptides possibly complicated the response compared with blockade of beta-adrenoceptors in culture.

These findings suggest that acute removal of sympathetic nerves causes dysregulation of immune function, but not in a simple fashion. For example, chemical sympathectomy leads to an apparent disinhibition of B-lymphocyte proliferation from lymph nodes that one might hypothesize would lead to increased antibody production following immunization. Instead, the primary immune response in some strains is virtually abrogated by chemical sympathectomy, probably as a result of effects at an early step in the immune response, such as antigen recognition, processing, or presentation. These observations, and the sometimes conflicting results in the literature, lead to the suspicion that many processes and cell types are affected by NE in different ways, depending on the concentration of NE available to those cells, the other mediators present in the specific microenvironment at that time, and the state of activation of the cells at the time of investigation. In addition to these complications, timing of the exposure to signal molecules may be important. For example, agents that increase intracellular cAMP in B lymphocytes *in vitro* generally diminish antibody production by decreasing the synthesis and secretion of IL-2; however, a transient increase in cAMP is important in the induction of antibody production.

Finally, an interesting connection exists between the integrity of NA nerves in lymphoid organs and the onset of autoimmune disease and immune deficits associated with aging. 6-OHDA-induced sympathectomy in neonatal rats has been demonstrated to increase the severity of T-lymphocyte-mediated experimental allergic encephalomyelitis. Administration of catecholamine agonists can reduce the severity of symptoms in this same disease. We found decreased innervation in lymph nodes prior to the expression of autoimmune disease in NZB mice and (NZB × NZW)F1 mice compared with NZW control mice. We also found that MRL mice homozygous for the *lpr* gene, resulting in fatal lymphoproliferation, have substantially less splenic NE than age-matched heterozygous littermates. Finally, in numerous species, aging is associated with significant deficiencies in cell-mediated immune responses, presumably contributing to the increased incidence of certain tumors, infectious disease, and autoimmune diseases in an aged population. In rats, sympathetic NA innervation of secondary lymphoid organs gradually diminishes with age. These changes in NA innervation may play a role in age-associated deficits in immune function. Because adrenergic agents are readily available and can be administered with minimal adverse response, perhaps a neurotransmitter-based approach to autoimmunity and age-associated immune dysfunction in humans may be possible in the future [*See* Autoimmune Disease.]

C. Effects of Neuropeptides on Immune Responses

As noted earlier, peptidergic nerves (NPY, VIP, substance P, CGRP, somatostatin, etc.) have been identified in the thymus, spleen, lymph nodes, and areas of GALT. Lymphocytes, or subsets of lymphocytes, possess receptors for most of these neuropeptides and can generate intracellular second-messenger responses following stimulation; these intracellular second messengers frequently are utilized by cytokines to achieve an effect on lymphocytes, suggesting the possibility of dual signaling into the same intracellular systems. In keeping with neuropeptides acting as neurotransmitters with lymphocytes as targets, several of the more thoroughly studied neuropeptides have been shown to have functional effects on the immune system.

Substance P has actions in several target cells that directly or indirectly stimulate inflammation. Substance P enhances vascular permeability and increases local vasodilation, both of which contribute to the ability of lymphocytes to migrate into the area. Substance P receptors have been identified on both T (helper and cytotoxic/suppressor subsets) and B lymphocytes, and on macrophages. Substance P is a T-lymphocyte mitogen and can enhance T-cell proliferation to lectins. It also enhances concanavalin A-induced IgA production by lymphocytes from mesenteric lymph nodes, spleen, and Peyer's patches. Substance P also enhances macrophage phagocytosis and chemotaxis of segmented neutrophils. Substance P has been proposed as an important mediator in the expression of rheumatoid arthritis. Adjuvant-induced arthritis can be reduced in severity by removal of substance P with capsaicin, whereas stimulation of these nerves or injection of substance P can exacerbate inflammation and the severity of arthritis. Stress has been observed to exacerbate rheumatoid arthritis, and NE also may play a role. β-2-Adrenergic blockade or NE depletion delays the onset of experimental arthritis and reduce joint injury. Therefore, both substance P and NE may act in concert in this condition.

Somatostatin appears to be inhibitory to many of the actions of substance P. It inhibits the release of substance P from the peripheral terminals of primary afferent nerves. Somatostatin also has a direct receptor-mediated effect on lymphocytes and monocytes; it inhibits PHA-induced human T-lymphocyte mitogenesis, suppresses endotoxin-induced leukocytosis, and suppresses the release of colony-stimulating factor activity by splenic lymphocytes.

VIP also appears to inhibit a variety of immune functions. It is found in peripheral nerves in the thymus, spleen, lymph nodes, and GALT. VIP decreases mitogen-induced proliferation of T lymphocytes but has no effect on B lymphocytes. T lymphocytes possess high-affinity receptors for VIP, related to intracellular cAMP stimulation, generally inhibitory to proliferation in T cells. These VIP receptors on T lymphocytes can be down-regulated by occupancy and internalize considerably more rapidly than they dissociate or are reinserted on the membrane. Ottaway incubated T lymphocytes in VIP to down-regulate the receptors and then injected these cells back into the host; these T lymphocytes failed to home to Peyer's patches and mesenteric lymph nodes but migrated properly to spleen, intestine, liver, and lungs. The internalization of the VIP receptor altered the interaction of these lymphocytes with the specialized high endothelium on the postcapillary venules where lymphocyte ingress occurs. Thus, T-lymphocyte trafficking into Peyer's patches and mesenteric lymph nodes requires the expression of high-affinity receptors of VIP.

The availability of NE and numerous neuropeptides in specific compartments of lymphoid organs suggests that a lymphoid cell in such a compartment may be exposed to a unique combination of neurally derived mediators. An *in vitro* investigation of lymphokine-activated killer (LAK) cell activity and proliferation suggests that such presence of multiple mediators may be extremely important to optimal functioning of the immune system. LAK cell activity is stimulated by IL-2. The maximal effect of a specific concentration of this cytokine can be augmented considerably by the presence of somatostatin and beta-endorphin, both of which are synergistic with each other as well as with IL-2. This synergistic effect is countered or blocked by prostaglandin E2 and by agents that stimulate intracellular cAMP directly or indirectly. Thus, neuropeptides may augment, and NE may inhibit, maximal IL-2-stimulated LAK cell activity directed toward specific tumor or viral-infected targets. As these cells and IL-2 have been injected into humans experimentally as cancer chemotherapy, such synergistic effects are not merely of academic interest, particularly in view of the severe toxicity of IL-2. We suggest that it might be possible to reduce the amount of IL-2 used to stimulate LAK cell activity if somatostatin and an opioid peptide are used to synergize the response, endomethacin and propranolol are used to block elevations in intracellular cAMP, and glucocorticoids are withheld to prevent up-regulation of receptors on the LAK cells, which could be detrimental to the synergistic effects of the neuropeptides and IL-2. It may be possible to exploit the interactions of cytokines and neurotransmitters for therapeutic benefits in the future, thereby reducing the considerable toxicity, including neurotoxicity, that many cytokines exert. [*See* Interleukin-2 and the IL-2 Receptor.]

V. CYTOKINE INTERACTIONS WITH NEURONS

Studies of the effects of immunization have demonstrated that during an immune response some soluble mediators can alter electric activity in neurons in key hypothalamic sites, such as the dorsomedial and paraventricular nuclei, and can alter NE and serotonin levels and turnover in the hypothalamus and limbic sites. These affected regions are the same sites that regulate neuroendocrine and central autonomic outflow to the immune system, thereby establishing complete loops of communication between the nervous and immune system. [*See* Hypothalamus].

Recent investigations with IL-1 have demonstrated that both peripherally injected and intracerebroventricularly injected IL-1 can produce electrical or neurochemical effects similar to those seen with immunization, suggesting that IL-1 may be one mediator of such effects. In addition to its classic effect on thermogenesis and induction of slow-wave sleep, IL-1 enhances the turnover of NE in the hypothalamus and enhances the secretion of CRF, thereby elevating ACTH and glucocorticoids. This action may explain the elevation of glucocorticoids that was observed by H. Besedovsky and colleagues at the peak of an immune response. It is not yet clear whether IL-1 also acts at the pituitary to release anterior pituitary hormones in addition to its effect on releasing factor cells in the hypothalamus. It also is not clear whether IL-1 can cross the blood–brain barrier in sufficient concentrations to produce these effects directly, can cross at circumventricular organs and along pial sleeves around the vasculature, or can initiate a secondary response by other mediators that in turn achieve these effects in hypothalamus and other central nervous system (CNS) sites. Extremely low concentrations of IL-1 in the lateral ventricles, in the 10-femtomolar range, have been reported to alter peripheral NK cell activity and immune responses, apparently via autonomic outflow to the spleen and perhaps other organs.

IL-1 has been reported to exist in hypothalamic neurons in humans and may act within the CNS as a neurally derived mediator. IL-1 immunoreactivity also has been reported in sympathetic (presumably NA) nerves in the spleen. Microglia are capable of synthesizing IL-1 when stimulated by agents such as interferon-gamma. Recently, activated T lymphocytes have been shown to cross through the blood–brain barrier and traverse the brain, providing a possible source for such stimulatory or activating cytokines. A breakdown in the blood–brain barrier also would permit access to the brain by such cytokines, thereby permitting the microglia to synthesize IL-1, up-regulate MHC antigens, and present antigen to initiate an immune response in the CNS. [*See* Blood–Brain Barrier.]

Other cytokines, such as IL-2, IL-4, IL-6, and tumor necrosis factor, have been reported to influence neural or glial responses. Thus apparently numerous immune-derived mediators may provide molecular sensory signaling to the CNS. Recent therapeutic administration of interferons and interleukins for cancer has resulted in severe central side effects, including depression, confusion, and other cognitive or affective side effects, suggesting that some cytokine signaling might profoundly alter behavior and higher functions of the nervous system. [*See* Cytokines and Immune the Response.]

We have only begun to scratch the surface of interactions of mediators of the immune system with the nervous system and interactions of neural mediators with the immune system. However, an important conceptual understanding has emerged: these two systems are in intimate contact with each other, share mediators with each other, and produce signal molecules that may interact with each other. It is no longer possible to view the immune system as an autonomous self-regulated system, and it is no longer possible to ignore immunological mediators when considering neural responses and behavior. Perhaps this field of neural–immune interactions has opened the door for a better understanding of common principles of signal molecules and will lead to a unified conceptual understanding of their actions on target cells, including immunocytes and neurons.

BIBLIOGRAPHY

Ader, R., Felten, D. L., and Cohen, N. (1989). Interactions between the brain and the immune system. *Annu. Rev. Pharm. Tox.* **30,** 561–602.

Ader, R., Felten, D. L., and Cohen N. (1991). "Psychoneuroimmunology," 2nd Ed. Academic Press, San Diego.

Ader, R., Cohen, N., and Felten, D. (1995). Psychoneuroimmunology: Interactions between the nervous system and immune system. *Lancet* **345,** 99–103.

Berczi, I. (1986). "Pituitary Function and Immunity." CRC Press, Boca Raton, Florida.

Blalock, J. E. (1989). A molecular basis for bidirectional communication between the immune and neuroendocrine system. *Physiol. Rev.* **69,** 1–32.

Bost, L. K. (1988). Hormone and neuropeptide receptor on mononuclear leukocytes. *Prog. Allergy* **43,** 68–83.

Carlson, S. L., and Felten, D. L. (1989). Involvement of hypotha lamic and limbic structures in neutral–immune communication. *In* "Neuroimmune Networks: Physiology and Diseases" (E. J. Goetzl and N. H. Spector, eds.), pp. 219–226. Alan R. Liss, New York.

Dinarello, C. A. (1989). Interleukin-1 and its biologically related cytokines. *Adv. Immunol.* **44,** 153–205.

Dunn, A. J. (1989). Psychoneuroimmunology for the psychoneuroendocrinologist: A review of animal studies of nervous system–immune system interactions. *Psychoneuroendocrinology* **14,** 251–274.

Felten, D. L., and Felten, S. Y. (1989). Innervation of the thymus.

In "Thymus Update" (M. D. Kendall and M. A. Ritter, eds.), pp. 73–88. Harwood Academic Publishers, London.

Felten, S. Y., and Felten D. L. (1991). The innervation of lymphoid tissue. *In* "Psychoneuroimmunology" (R. Ader, D. L. Felten, and N. Cohen, 2nd Ed. eds.), Academic Press, San Diego.

Felten, D. L., Felten, S. Y., Bellinger, D. L., Carlson, S. L., Ackerman, K. D., Madden, K. S., Olschowka, J. A., and Livnat, S. (1987). Noradrenergic sympathetic neural interactions with the immune system: Structure and function. *Imm. Rev.* **100,** 225–260.

Felten, S. Y., Felten, D. L., Bellinger, D. L., Carlson, S. L., Ackerman, K. D., Madden, K. S., Olschowka, J. A., and Livnat, S. (1988). Noradrenergic sympathetic innervation of lymphoid organs. *Prog. Allergy* **43,** 14–36.

Fredrickson, R. C. A., Hendric, H. D., Hingtgen, J. N., and Aprison, M. H. (1986). "Neuroregulation of Autonomic, Endocrine, and Immune Systems." Martinus-Nijhof, The Hague.

Goetzl, E. J. (ed.) (1985). Supplement on neuromodulation of immunity and hypersensitivity. *J. Immunol.* **135.**

Goetzl, E. J., Sreedharan, S. P., and Harkonen, W. S. (1988). Pathogenetic roles of neuroimmunologic mediators. *Immunol. Allergy Clin. of North Am.* **8,** 183–200.

Madden, K. S., and Felten, D. L. (1995). Experimental basis for neural–immune interactions. *Physiol. Rev.* **75,** 77–106.

Madden, K. S., Sanders, V. M., and Felten, D. L. (1995). Catecholamine influences and sympathetic neural modulation of responsiveness. *Annu. Rev. Pharm. Tox.* **35,** 417–448.

O'Dorisio, M. S., Wood, C. L., and O'Dorisio, T. M. (1985). Vasoactive intestinal peptide and neuropeptide modulation of the immune response. *J. Immunol.* **135,** 792s–796s.

Payan, D. G., McGillis, J. P., Renold, F. K., Mitsuhashi, M., and Goetzl, E. J. (1987). The immunomodulating properties of neuropeptides. *In* "Hormones and Immunity" (I. Berczi and K. Kovacs, eds.), pp. 203–214. M. T. P. Press, Norwell, Massachusetts.

Smith, E. M. (1988). Hormonal activities of lymphokines, monokines, and other cytokines. *Prog. Allergy* **42,** 121–139.

Weigent, D. A., and Blalock, J. E. (1987). Interactions between the neuroendocrine and immune systems: Common hormones and receptors. *Imm. Rev.* **100,** 79–108.

Wybran, J. (1986). Enkephalins as molecules of lymphocyte activation and modifiers of the biological response. *In* "Enkephalins and Endorphins" (N. P. Plotnikoff, R. E. Faith, A. J. Murgo, and R. A. Good, eds.), pp. 253–282. Plenum, New York.

Neural Networks

LEIF H. FINKEL
University of Pennsylvania

GLOSSARY

Backpropagation Efficient algorithm for training a network to learn by presenting examples and adjusting synaptic efficacies based on their contribution to the output

Credit-assignment problem Problem of how to determine the relative contribution of each node in a network to its overall output

Network System of processing nodes and connections between the nodes in which the state of a node depends on signals it receives along its connections

Neural unit Individual processing unit in a neural network, modeled with greater or lesser accuracy after physiological properties of real neurons

Synaptic rule Formal rule for altering the efficacy of network connections, thereby changing the probability of one unit activating another unit

A NEURAL NETWORK IS A SYSTEM OF INTERconnected, excitable, neuron-like units. Units can be activated by inputs from other units in the network, by external inputs, or by intrinsic processes. Based on these inputs, each unit generates an output that is transmitted to all units to which it projects. The efficacy of connections in transmitting inputs and outputs can be modified, usually in an activity-dependent manner, and various rules have been proposed to govern these synaptic modifications. The most widely used rule, the Hebb rule, modifies connection efficacies according to the correlation of the excitation in the two connected units. The pattern of activity across a network can represent or store information. Ongoing changes in this activity pattern due to the dynamics of excitatory and inhibitory inputs correspond to information processing. Networks can "learn" to generate a response to a particular input stimulus. Memory is encoded by means of changes in synaptic efficacies. Networks can be simulated on computers or directly implemented in hardware such as passive electrical circuits, VLSI chips, or optical components.

In biologically based networks, the anatomical and physiological properties of individual units are closely modeled after real neurons. Model neurons typically have many compartments that can differ in their electrical properties (resistance and capacitance) and that may contain different species of simulated ion channels and neurotransmitter receptors. In applied neural networks, the units are greatly simplified and abstracted. The first such artificial neuron was formulated in 1943 by Warren McCulloch and Walter Pitts. This unit computes a weighted sum of its inputs, subtracts a threshold value from the result, and generates one of two binary output states, 1 or 0, depending on whether the result is positive or negative. Most recent networks employ units with a continuous sigmoidal output function in place of the McCulloch–Pitts step function. In both biologically based and applied networks, the temporal evolution of network activity occurs in discrete time elements, with units in the network evaluated either serially or asynchronously.

The major characteristics that differentiate neural network models are the anatomical architecture of the network, the physiological properties of the individual units, and the rules used for modifying the efficacy of the interconnections between units. Biologically based networks have as their primary motivation the understanding of basic principles of brain function. Applied neural networks make use of some of these principles to solve useful problems involving input/output matching, categorization, memory storage, or optimization. We will review key examples of both types of network and briefly discuss some of the more intriguing applications.

I. BIOLOGICALLY BASED NETWORKS

The concept of a neural network can be traced to the British associationist philosophers and to early work by Nicholas Rashevsky (1938) and Norbert Wiener (1948). In the 1950s, R. L. Beurle and others began to study the dynamical properties of networks of interconnected excitable elements. However, it was only in the 1980s that computational neuroscience developed into an established field, due in large degree to advances in neuroscience and the advent of high-speed computers. Many neural systems have been modeled, including those responsible for vision and flight control in the housefly, swimming in the lamprey, chewing in the lobster, auditory localization in the barn owl, and eye movements in the monkey. Networks have been used to study the problem of how ordered neural maps are created during fetal development (e.g., the map of the retina onto the optic tectum), and how the internal structural organization of these maps arises (e.g., the generation of ocular dominance and orientation columns in the visual cortex). Perhaps the greatest concentration of modeling effort has been applied to studying the operations of the cerebral cortex. [*See* Cortex.]

At present, it is just becoming possible to construct detailed models of small regions of cortex. In one of the most realistic simulations constructed to date, D. C. Somers, S. C. Nelson, and M. Sur modeled a small region of cortex measuring 1.7 × 0.2 mm. In the monkey, such a region would contain roughly 7000 cells through the six-layer depth of cortex. Their model contained 3000 cells with 180,000 synaptic connections. They used only two principal types of cells (a generic excitatory cell and a generic inhibitory cell), whereas cortex contains dozens if not hundreds of distinct cell types. This type of detailed model can be used to study the role of different classes of connections in cortex. For example, the vast majority of connections in cortex are local, recurrent connections between excitatory cells, but the function of these connections is still unknown.

One approach to cortical modeling has been to study the simplest type of cortex, the archicortex, which is composed of only three layers (instead of the six layers of the neocortex). Both the olfactory cortex (responsible for smell) and the hippocampus (implicated in long-term memory) have been modeled. One difficulty in these studies is that the nature of the sensory inputs that activate this ancient cortex is not well defined. Nonetheless, significant progress has been made in understanding the dynamics of the intrinsic cortical circuitry.

Recently, Roger Traub and colleagues have developed a detailed model of the mammalian hippocampus. The network consists of approximately 10,000 neural units, each of which incorporates realistic models of sodium, potassium, and calcium ion channels. In accord with recent anatomical information, the units are sparsely interconnected—each unit contacts less than 3% of the other units in the network. Despite this sparse connectivity, rhythmical firing patterns emerge (the hippocampus displays a prominent firing rhythm known as the theta rhythm at 4–8 firings per second). An interesting aspect of this emergent activity pattern is that no unit fires regularly at this frequency—individual units typically fire much more sporadically. Rather, the rhythm is displayed only by the population of units as a whole; thus, it is a true network phenomenon. This model can reproduce the activity patterns observed in normal hippocampus under a variety of conditions (e.g., application of electric current or various exogenous agents). More interestingly, under conditions of reduced inhibition, the model generates epileptic seizure activity, which very closely mimics that found in humans and monkeys.

The hippocampus has also been shown to be involved in spatial navigation tasks, such as maze learning in rodents. A number of models have proposed network mechanisms for the ability to develop cognitive maps of the spatial environment. One recent approach is to view the CA3 region of the hippocampus as a Hopfield network (see Section II) in which all activity converges toward the spatial goal. One of the most active areas for future modeling will be network models of neurological disease. The hippocampus is involved in a number of neurological diseases; in addition to epilepsy, it is severely affected in Alzheimer's disease and Down's syndrome. With increasing levels of stimulation, the same synaptic processes used in normal learning and memory may be responsible for cell death (excitotoxicity). For this reason, the hippocampus is the brain structure most sensitive to traumatic injury. [*See* Hippocampal Formation.]

Efforts have also been directed at modeling the piriform cortex of the rat, which is the first cortical processing station for olfactory information. In computer simulations, the waves of rostral-to-caudal activity that are observed *in vivo* have been reproduced. Simulations show that these oscillatory waves arise from two types of inhibition (feedforward and feedback) and from a temporal anisotropy in rostral versus caudal connection velocities. [*See* Olfactory Information Processing.]

Another study has modeled how the piriform cortex can categorize olfactory signals into recognized smells, even when multiple odorants are presented together (e.g., one can smell the brewed coffee in a restaurant kitchen). In this simulation, the network performs a temporally distributed analysis of the "odor-scene," with early responses signifying the general class of the odorant (fruity, salty, aromatic, etc.) and later responses of the same network signaling the individual identity of the odorant (orange, grapefruit, etc.). The theta rhythm, which is also displayed by olfactory cortex (and which is an optimum frequency for inducing long-term potentiation in the hippocampus), is in fact the frequency at which rodents sniff the environment. Thus, these models serve as conceptual links between sensory categorization, behavior, and memory.

In constructing realistic simulations of the nervous system, there are various levels of detail at which individual neurons can be modeled. As the number of cells in the simulation grows, computational limitations force the individual units to become simpler and more abstracted—sometimes to the limit of generalized McCulloch–Pitts neurons. A common formulation is to use "integrate and fire" cells, which sum up their inputs and generate a spike of activity whenever the sum exceeds a threshold. Alternatively, some models concentrate at the cellular level. A few individual neurons from the cortex, hippocampus, and cerebellum have been painstakingly reconstructed and modeled with thousands of individual compartments. This type of simulation allows the cellular and subcellular events in neuronal processing to be studied. For example, one can monitor the calcium concentration in various subcellular compartments, or model the action of various biochemical or genetic pathways. Cellular simulations have led to a number of interesting recent findings. One study showed that neurons can be extremely accurate in reproducing their response to repeated instances of a fluctuating noisy input. This suggests that there may be important information in the fine temporal pattern of ongoing activity.

One of the most active areas of current research is in the investigation of these temporal patterns. Since the early days of electroencephalogram (EEG) recordings, brain activity has been associated with temporal oscillations in cell firing. The so-called "alpha rhythm" associated with relaxed behavioral states and used in biofeedback experiments is one of several oscillatory patterns commonly observed. Charles Gray and Wolf Singer found evidence that these temporal patterns may be used by the brain as a means of solving the "binding problem," namely, how to link together activity in different brain regions that relates to the same object. For instance, when we view a bird in flight, separate areas of cortex are engaged in the analysis of its shape, motion, color, position, sound, name, and other properties. There is no "place" in the brain where all this information converges, thus the representation of a bird remains distributed and must be linked through some dynamical mechanism. It was proposed that all neurons concerned with the bird may fire together in a synchronous manner. Thus, there would be a temporal code identifying common information. Many network models have been developed to study this problem—how oscillations can be generated, how synchronization (and, more importantly, desynchronization) can occur, and other aspects of temporal coding. The work of Moshe Abeles, for example, provides a network model that accounts for the observation of repeating "melodies" or patterns of rhythms seen in recordings from frontal cortex in monkeys. Cellular level simulations are particularly important in this regard, as they help to reveal the underlying mechanisms of synchronization.

Using data from the hippocampus, where rhythms are rampant, several recent models have proposed a set of novel mechanisms. John Lisman and colleagues have put forward a network model that makes use of two rhythms: the theta rhythm and the higher-frequency (40–60 Hz) gamma rhythm. In their model, individual short-term memories are encoded in gamma frequency cycles, each cycle representing a different memory. Roughly seven of these memories can be packaged together in one theta cycle, corresponding to the well-known capacity of short-term memory for roughly seven items (hence the number of digits used for phone numbers). A particular memory retains its firing position on successive theta cycles due to the detailed properties of after-depolarization of the cells—and thus the model critically depends on the details of cell function. Several other recent models have used alternative mechanisms to achieve temporal synchronization. Both Roger Traub and Gyorgy Buszaki and colleagues suggest that inhibitory interneurons may be the key to temporal synchrony. A number of modeling studies have had difficulty in getting networks of excitatory units to temporally synchronize. However, inhibitory cells appear to have a natural tendency to phase-lock, provided they carry the appropriate types of active channels, receive the appropriate type of excitatory and inhibitory inputs [fast-acting glutamate and gamma-aminobutyric acid-A (GABA-A) transmitters], and have a threshold level of synaptic density. This area of research has profound implications for understanding brain function and will be an active area in the coming years.

Work on archicortex is closely related to network studies of associative memory (see the following) because the neural architectures are organized in distributed fashion. The neocortex, in contrast, is predominantly organized into well-defined topographic maps in which adjacent regions of cortex respond to adjacent regions of the environment (or to closely related movements in the motor cortex). The somatosensory cortex, for example, which is responsible for the perception of touch, contains a map of the entire body skin surface. Recent physiological recordings have shown that this map is *dynamic* and undergoes continual change and reorganization due to the ongoing tactile inputs received by the skin.

Network simulations of monkey somatosensory cortex can account for the major features of this map plasticity. One recent network models 1500 units that receive topographically organized projections from receptor sheets representing the front and the back of the hand. The network contains excitatory and inhibitory units that are interconnected so that focal cortical stimulation yields a pattern of short-distance excitation and long-distance inhibition (this dichotomy is critical to maintain network stability). Network stability (freedom from oscillation, explosion, or dampening of activity) is also fostered by the use of shunting inhibition (a type of inhibition in which currents are effectively short-circuited at the cell body); similar effects of shunting inhibition have been found in other simulation studies. Excitatory connections in the network are modifiable according to a voltage-dependent Hebb-type synaptic rule. The result of these synaptic modifications is that for a wide range of stimulation protocols, the network organizes itself into neuronal groups—local collections of cells with strong mutual connections that all share similar functional properties. The network also develops a topographic map of the receptor surfaces. Perturbations of stimulation to these receptors cause the maps to reorganize in patterns that closely correspond to *in vivo* results. For example, repeated stimulation of the tip of one finger leads to an expansion, in the cortical network, of the representation of that finger. Conversely, decreased stimulation, as would result from amputation or nerve transection, allows a takeover by the cortical representations of nearby skin surfaces. The simulations show how representational plasticity can emerge from a biologically based synaptic modification rule embedded in a network with simplified anatomical and physiological properties. [*See* Somatosensory System.]

No area of the nervous system has received more attention from neural network modelers than the visual cortex. Physiological studies have shown that in higher animals the visual cortex is composed of multiple, distinct areas (up to two dozen areas in the rhesus monkey). Each of these visual areas is functionally specialized for particular visual tasks, although there is substantial overlap in the properties of different areas. Thus, for example, area V1 (the first visual area) performs a high-resolution analysis of the visual scene, area V4 (the fourth area) is specialized for color and texture vision, and area V5—also called MT, as it is located in the medial temporal cortex—is specialized for discriminating motion. Models have been developed for each of these visual processes. Stereopsis is one such process in which the slight horizontal disparity in the views seen by the two eyes is transformed into a perception of three-dimensional depth. Early network models focused on the problem of determining the exact correspondence between individual points seen by the two eyes. Physiological recordings reveal that disparity-sensitive cells in the visual cortex are tuned for three broad classes of disparity, corresponding to objects located nearer, farther, and in the plane of visual fixation. A recent model of the representation of disparity in the cortex is physiologically based and investigates the basis of distributed representations. In this model, each cell only crudely codes for the actual disparity of an object; however, the population of cells, taken together, is able to discriminate extremely fine depth discontinuities, such as arise in hyperacuity experiments. (The visual system can resolve minute discontinuities that are far smaller than the spacing between cells in the retina.) Such population codes are probably used throughout the nervous system. For example, in the monkey, there appears to be a distributed representation of arm movements in the motor cortex and of eye movements in the superior colliculus. The disparity model exemplifies how relatively simple computational models can elucidate complex principles of brain operation. [*See* Vision, Physiology; Visual System.]

The problem of motion detection has also motivated a number of network models. Most such models use a mechanism of temporally delayed inhibition in which excitation at a spatial location (location 1) inhibits activity from an adjacent location (location 2) a shorter time later. If an object travels from location 1 to location 2 within that time span, it will not result in visual activation due to the inhibition. However, if the object travels in the reverse direction, in some oblique direction, or at a significantly different velocity, it will produce activation. In this manner, a population of cells can represent the direction of motion of an object.

This first stage of motion detection is generally referred to as direction selectivity. In higher animals, it is carried out in the retina. Higher stages of the visual system carry out more difficult computational tasks related to motion processing. A recent network model by N. Grzywacz, M. Pettet, and S. McKee analyzes the contributions of striate cortex to motion processing. They point out that psychophysical estimates of motion improve with longer exposure times, and optimum performance requires upwards of 400 msec. Their model involves a feedforward estimation of motion carried out by long-distance horizontal connections in striate cortex. Such a model has the ability to explain more complicated motion phenomena, such as the ability to pick out objects undergoing smooth motion trajectories (lines, circles) admid a noisy background of randomly moving objects. Striate cortex projects, directly and indirectly, to area MT (also known as V5), which is a major processing center for motion. Recent network models have proposed mechanisms by which MT may solve the so-called "aperture problem," namely, the integration of local motion information into the determination of the actual direction of motion for a complex object.

The solution to the aperture problem requires the realization that certain parts of objects convey more information than others. Likewise, certain elements of a scene may be more important in a given behavioral context. It is well known that attention is directed to various image locations depending on the task at hand, as manifested by the trail of eye movements. Recent network studies have begun to identify the cortical mechanisms underlying the perceptual "salience" of various features in a scene. It has been found that the pattern of long-distance connections between cells in striate cortex can account for many of our innate preferences for certain types of features. For example, the Gestalt laws of perceptual organization describe, phenomenologically, a preference for smooth, continuous contours, for closure, for symmetry, and for other object properties. Models show a tendency for these properties to arise in networks that mimic the anatomical connectivity of cortex. Extensions of these networks are beginning to explain some of the mechanisms of visual attention and visual search.

All of these biologically based simulations share several characteristics. They are primarily motivated by a goal of understanding something about how the nervous system works; they are based, with greater or lesser verisimilitude, on anatomical and physiological data; and they do not include any mechanisms that are biologically unfeasible, such as requiring long times to converge to a result or only working in rigidly defined anatomical networks. Their results are framed in biological terms, and they generate experimental predictions. However, none of these models is able to handle complex scenes or extended time periods, or is capable of being dramatically scaled-up in numbers of neurons. In addition, many of the models are so complicated that it is sometimes not clear how they do what they do. For these reasons, a significant effort has recently been expended in developing simplified network models that are not biologically based, although they do incorporate certain neural-like features. These networks, to which we now turn, are much easier to analyze and have already accomplished some rather remarkable feats.

II. APPLIED NEURAL NETWORKS

Over the last 20 years, a number of different applied neural networks have been used to carry out difficult computational tasks in pattern recognition, associative memory, and learning. Artificial or applied neural networks essentially act as transformation operators that map sets of input vectors into sets of output vectors. Networks can be trained, for example, to "remember" pictures of several hundred human faces. When presented with part of one individual face (e.g., the top half), the network can generate the missing portion of the correct face.

The first attempts to develop applied neural networks used techniques derived from linear algebra. Frank Rosenblatt, Bernard Widrow, and others demonstrated that networks could be developed to handle very sophisticated recognition problems. Initial hopes were crushed, however, with the publication in 1968 of Marvin Minsky and Seymour Papert's book on perceptrons, which pointed out the limitations in these simple networks and argued that artificial intelligence was a more productive route. Network research was only reborn in the 1980s when it was finally shown how the initial problems could be surmounted by new learning algorithms. Several new classes of associative networks were developed, one based on the formalism used in statistical physics to describe spin glasses. The so-called Hopfield networks are rather severely constrained—they must be symmetric (i.e., if neuron A is connected to neuron B, then B must also send a connection to A with the identical synaptic strength). But this unbiological assumption can be relaxed to a certain degree, and it guarantees a very strong learning result. Namely, if the synaptic strengths are modified according to a correlation rule, such as the Hebb rule, then the network will monotonically converge to a learned state. In the Hopfield formalism, the network is viewed as a dynamical system in a high-dimensional space, and learning consists of creating "attractors" that are activity states of the network that attract other activity states. Thus, if the network receives some arbitrary activation from a set of inputs, the activity will rumble around for a bit but will eventually converge on one of these attractors. One can imagine memory space as a kind of evolving golf course, with little hills and valleys. When you create a memory, you make a new hole, and if the "golf ball" is placed anywhere near this hole and is jostled around a bit, it will eventually fall into the hole.

One problem with this scheme is that the system can get caught in local minima (e.g., the golf ball can get stuck in little gullies, preventing it from reaching the hole). A technique for overcoming this problem, called simulated annealing, has been borrowed from metallurgy, in which an analogous problem occurs in the making of metal alloys. Alloys are heated to high temperatures and then very gradually cooled. This corresponds (somewhat forcing the analogy) to giving the golf ball a lot of kinetic energy and then gradually slowing the ball down. The extra energy allows the system to escape from local minima and to find the global minimum. A number of networks use simulated annealing or other gradient descent procedures. One such network, the so-called Boltzmann machine, has been used to perform figure–ground separation problems, such as distinguishing the inside of a maze from the outside.

A label that has been applied to much of current neural network research is "connectionism." Connectionism means that the advanced functions of the nervous system, such as pattern recognition, memory, and speech, can only be carried out in distributed networks. In other words, mental representations are stored in patterns of activity and any memory or action is distributed over populations of cells. This view is opposed to that of artificial intelligence in which both the information and the operators or symbols are assumed to be *directly* represented. One very active area of connectionist research is parallel-distributed processing (PDP), in which simple networks are trained to perform complex tasks. A typical PDP network is shown in Fig. 1 and consists of three layers of cells: an input layer, an output layer, and an intermediate (or "hidden") layer. The network operates in a feedforward manner; each cell is connected to all cells in the next higher layer and in general there can be more than one hidden layer. A special training procedure, known as back-propagation, is used to adjust the connection weights. In back-propagation, a number of different stimuli are presented to the input units, which activate the hidden units, which in turn activate the output units. For each input stimulus, the activity of the output units is compared to the desired output activity pattern and an error signal is computed (error = desired activity − actual activity). This error signal is used to adjust the connection weights of the network such that if a unit is active and the error signal is positive, the weight is increased; otherwise, if the unit is active and the error signal is negative, the weight is decreased. The amount of the change depends on the size of the error (given by an equation known as the delta rule). This rule is a close relative of the original perceptron learning rule intro-

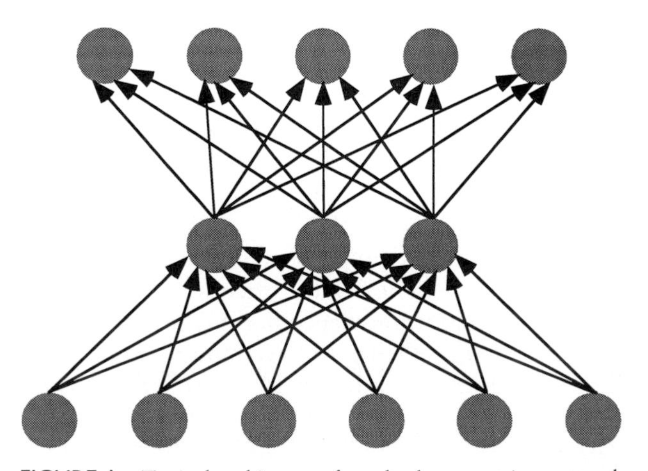

FIGURE 1 Typical architecture for a backpropagation network with three layers. The input layer projects to a "hidden layer" of units, which in turn projects to the output layer. Backpropagation consists of presenting a series of inputs to the first layer, and adjusting the connection strengths between the layers until every input generates the desired output.

duced by Rosenblatt. The major difference is that back-propagation can be applied to multilevel networks in a step-by-step process. The errors are "back-propagated" from output layer(s) through the hidden layers to the input layer. Thus, back-propagation solves the so-called "credit-assignment" problem, which is the determination of the role played by each unit in a multilevel network in arriving at an output.

There is a major difference in how connection strengths are modified in the back-propagation algorithm and in the biologically based networks considered earlier. Back-propagation and related methods have been called "supervised" learning and require a "teacher" that tells each cell whether its response to each stimulus was correct or incorrect. In "unsupervised" learning, no such information is available. For example, in the Hebb rule, synaptic changes occur as a result of correlations in activity patterns *regardless* of the behavioral consequences of these correlations. The Hebb rule and other unsupervised schemes work because the correlations in the system ultimately reflect correlations in the stimulus world. In general, unsupervised schemes deal with lower-level processes (such as early sensory categorization) and implicitly defer any external teaching to higher cognitive tasks. For example, complex behaviors such as talking, reading, or baking a cake require an external teacher, whereas discriminating the orientation of lines or visually tracking a moving object requires exposure to these stimuli but does not explicitly require an external reward system.

One spectacular example of the power of back-propagation is given by the network NETtalk, which learns to read. The network, which was built by T. J. Sejnowski and colleagues, is presented with written text, such as a book or newspaper, and after training on a number of examples it learns to translate the written letters into phonemes, which a voice synthesizer pronounces out loud. The network consists of 309 units [203 input units (29 groups of 7 units), 26 output units, and 80 hidden units, each of which is connected to all the input and output units] and 18.629 connections. At the start of the training procedure, all connection strengths among units in different layers are given random values. As sequences of letters and spaces are presented to the input array, the output units produce patterns of activity that correspond to 1 of 79 possible phonemes. The difference between the correct output pattern and that actually obtained is then fed back to the earlier layers and used to modify the connection strengths of the network via the back-propagation algorithm. After tens of thousands of training cycles, the network achieves excellent accuracy and is able to generalize to read words it has never before seen. Most interestingly, the hidden units develop an organization in their pattern of response. For example, certain hidden units respond only to vowels, others only to consonants, and the response to related vowels (e.g., the "a" in bake and the "e" in beg) is more similar than to unrelated vowels. NETtalk thus illustrates how a network can learn a system of complex and detailed rules, and how these rules can be represented in a distributed fashion.

Applied neural networks have been used to solve several notoriously difficult computational problems. The dynamical properties of some networks can be used to solve optimization problems such as the traveling salesman problem (in which a salesman must visit N cities and wants to find the shortest circuit that allows him to visit each city once and only once). The network finds a good solution quickly; however, it usually does not find the absolute optimal solution and linear programming is still the method of choice for such problems. One other example of current interest is the "cocktail party" problem. Imagine you are at a cocktail party and admid the general noisy hubbub, someone across the room mentions your name in conversation—you immediately tune into that conversation. In general, given N sound sources in a room that are recorded from microphones placed at M different locations, how can one separate and identify the different sources. Network algorithms have been developed for this problem of "blind separation."

Networks have also recently been used to solve such diverse problems as how to balance a pencil on its point, how to read handwriting on bank checks, how to predict the three-dimensional structure of a protein from its amino acid sequence, and how to teach other neural networks. Neural networks offer the hope of being able to rapidly perform difficult computations in fundamentally new ways. The brain is the most complex and powerful computational device known; neural networks, by capturing some of its basic operating principles, provide insight into both brain mechanisms and the power of natural computation.

III. CONCLUSION

Neural networks offer powerful insights into how the nervous system may represent and manipulate information. Applied neural networks have made significant contributions to solving difficult computational problems and promise to gain wider application. The great challenge for the next decade is to integrate neural network studies with basic neuroscientific research, at the molecular, cellular, and systems levels, to help uncover the basic principles of brain function.

BIBLIOGRAPHY

Anderson, J. A., and Rosenfeld, E. (1988). "Neurocomputing: Foundations of Research." MIT Press, Cambridge, Massachusetts.

Bower, J. M., and Beeman, D. (1995). "The Book of GENESIS: Exploring Realistic Neural Models with the General Neural Simulation System." Springer-Verlag, New York.

Churchland, P. S., and Sejnowski, T. J. (1992). "The Computational Brain." MIT Press, Cambridge, Massachusetts.

Hertz, J., Krogh, A., and Palmer, R. G. (1991). "Introduction to the Theory of Neural Computation." Addison–Wesley, New York.

Rumelhart, D. E., and McClelland, J. L. (1986). "Parallel Distributed Processing: Exploration in the Microstructure of Cognition," Vol. 1, "Foundations." MIT Press, Cambridge, Massachusetts.

Traub, R. D., and Miles, R. (1991). "Neuronal Networks of the Hippocampus." Cambridge, Univ. Press, New York.

Neuroendocrinology

JOSEPH B. MARTIN
Harvard University

WILLIAM F. GANONG
University of California, San Francisco

GLOSSARY

Adenohypophysis Anterior portion of the pituitary that secretes six principal hormones

Circumventricular organs Small, highly permeable areas around the third and fourth brain ventricles, where circulating chemicals act to trigger homeostatic responses

Hypothalamic hormones Chemical messengers that are produced by neurons in the hypothalamus and secreted into the portal hypophysial vessels or the general circulation

Hypothalamus Region of brain below the thalamus that is concerned with regulation of food intake and secretion of pituitary hormones

Median eminence Vascular area in the floor of the third ventricle, where hypothalamic hormones enter the portal hypophysial blood vessels

Neurohypophysis Posterior portion of the pituitary, where vasopressin and oxytocin are released into the general circulation

Portal hypophyseal vessels Special vascular system that conducts hypothalamic hormones directly from the median eminence to the adenohypophysis

Trophic hormones Hormones that stimulate the secretion and growth of other endocrine glands; five of the six adenohypophysial hormones are trophic hormones

NEUROENDOCRINOLOGY ENCOMPASSES THE study of the relation between the nervous and endocrine systems, which act in concert to regulate many of the metabolic and homeostatic activities of the organism. Most endocrine glands are innervated and, in many instances, the nerves play a role in regulating the secretion of their hormones. Examples include the secretion of melatonin by the pineal gland, the secretion of insulin and glucagon by the endocrine portion of the pancreas, the secretion of gastrointestinal hormones by cells in the wall of the gastrointestinal tract, the secretion of epinephrine, norepinephrine, and dopamine by the adrenal medulla, and the secretion of renin by the juxtaglomerular cells of the kidneys. However, most attention has been focused on the relation of the pituitary gland to the hypothalamus. Neurons in the hypothalamus secrete hormones that enter the circulation. Two of these hormones, oxytocin and vasopressin, have systemic effects and are transported along nerve fibers to the posterior pituitary, where they are secreted into the general circulation. Six additional hypothalamic hormones regulate the secretion of the anterior pituitary gland, and they enter the special vascular connection between the hypothalamus and the anterior pituitary, the portal hypophysial vessels. The hypothalamic hormones are listed in Table I. Conversely, hormones have important effects on the brain. Some act on the brain and anterior pituitary in a negative feedback fashion, maintaining the constancy of anterior pituitary secretion. Others act on the brain to change patterns of sexual behavior, regulate levels of activity, adjust cardiovascular responses, and regulate food intake.

I. INTRODUCTION

Neuroendocrine mechanisms are involved in the regulation of reproduction, growth and differentiation of

TABLE I
Hypothalamic Hormones

Hypothalamic hormone	Structure	Action
Thyrotropin-releasing hormone	Tripeptide	Stimulates TSH and PRL secretion
Corticotropin-releasing hormone	41-Amino-acid peptide	Stimulates ACTH secretion
Gonadotropin-releasing hormone	Decapeptide	Stimulates LH and FSH secretion
Growth hormone-releasing hormone	44-Amino-acid peptide	Stimulates GH secretion
Somatostatin	14-Amino-acid peptide	Inhibits GH and TSH secretion
Prolactin-inhibiting hormone	Dopamine	Inhibits PRL secretion
Vasopressin	Nonapeptide	Antidiuretic and pressor effects; stimulates ACTH secretion
Oxytocin	Nonapeptide	Milk ejection; uterine contractions

tissues, water and salt balance, the response to stress, regulation of food intake, and a number of other behaviors. Disorders in neuroendocrine regulation can result in deficits in reproduction, alterations in intellectual function, and exacerbation of metabolic diseases such as diabetes mellitus. Overproduction of hormones by the pituitary can lead to serious clinical disturbances, such as Cushing's disease caused by oversecretion of adrenocorticotropic hormone and, consequently, of cortisol by the adrenal, excessive growth caused by hypersecretion of growth hormone, and infertility caused by increased secretion of prolactin. Clinical disturbances may also arise from underproduction of hormones caused by destructive lesions (e.g., tumors arising in either the pituitary or the hypothalamus).

II. EFFECTS OF HORMONES ON THE BRAIN

Steroid hormones from the adrenal cortex and gonads are lipid soluble, and they penetrate the brain with ease. Other substances, including in particular peptides and proteins, penetrate the brain very slowly because of the so-called blood–brain barrier that protects the brain from erratic and potentially dangerous changes in the chemistry of circulating body fluids. However, a number of hormonal peptides act on the brain to produce changes that are of homeostatic importance. [*See* Peptide Hormones and their Convertases.] They are able to do this because they act on the *circumventricular organs,* four small, specialized structures that unlike the rest of the brain have highly permeable fenestrated capillaries. Substances enter these organs and trigger responses via neural pathways to other parts of the brain. The four circumventricular organs are the median eminence, the organum vasculosum of the lamina terminalis, the subfornical organ, and the area postrema (Fig. 1).

Steroids produced by the ovaries and testes act on the brain to stimulate sexual behavior. These effects are most easily demonstrated in rodents, but they also occur in primates and, although modified extensively

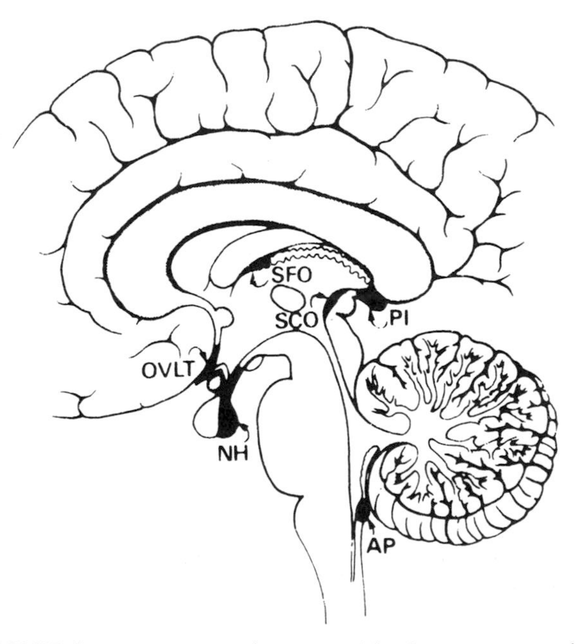

FIGURE 1 Circumventricular organs. The four organs are the neurohypophysis (NH), organum vasculosum of the lamina terminalis (OVLT, supraoptic crest), subfornical organ (SFO), and area postrema (AP). They are shown projected on a sagittal section of the human brain. SCO, subcommissural organ; PI, pineal. [Reproduced, with permission, from W. F. Ganong (1997). "Review of Medical Physiology," 18th Ed. Appleton & Lange, Stamford, Connecticut.]

by experience and social pressures, in humans. They are exerted directly on the hypothalamus and possibly on the adjacent part of the brain known as the limbic system.

Effects of the peptide hormone angiotensin II are good examples of actions on the brain mediated via the circumventricular organs. Angiotensin II acts on the subfornical organ to trigger thirst, and thus to produce a prompt increase in water intake. Circulating angiotensin II is increased by dehydration, and the increase in water intake helps restore plasma osmolality to normal. Circulating angiotensin II probably acts on one or more of the circumventricular organs to stimulate salt appetite, although this point is a matter of some debate. In addition, angiotensin II acts on the area postrema to elevate the set point for the baroreflexes regulating blood pressure. This decreases baroreflex sensitivity and augments the pressor effect of the peptide.

Peptide hormones also appear to act on the brain to regulate food intake and energy expenditure, thus controlling body weight. Food is digested in the stomach and intestine, and the products of digestion cause release from the intestine of gastrointestinal hormones, particularly cholecystokinin, which acts on the hypothalamus to produce a feeling of satiety. Thus, meal size is in part under neuroendocrine control. More long-term control of appetite is exerted in part by leptin, a 16-kDa protein produced by fat cells when they are distended. This hormone probably acts on the hypothalamus, possibly via a circumventricular organ, to produce satiety and increase activity, thus providing feedback control of body weight. Leptin receptors have been cloned and are found in the hypothalamus and the choroid plexus. It is interesting that, in humans, plasma leptin levels are positively correlated with body mass, that is, with the amount of fat in fat depots. Therefore, most cases of human obesity are presumably due to defective leptin receptors (or postreceptor events) rather than decreased leptin production.

III. HYPOTHALAMIC–PITUITARY SYSTEM

The key functional unit of the hypothalamic–pituitary system is the hormone-secreting neuron. Cells secreting substances that act on other cells some distance away can be classified into three types: *exocrine,* which secrete into a duct or lumen that connects to the exterior of the body (e.g., sweat glands, gastrointestinal tract); *endocrine,* which secrete directly into the blood (e.g., insulin from the pancreas); and *neurosecretory,* which refers to release of a hormone into the blood from a nerve terminal, as described earlier.

A. Anatomy of the Hypothalamic–Pituitary System

Visualized from below, the midline ventral surface of the forebrain forms a convex bulge termed the *tuber cinereum.* Arising in the midline from the tuber cinereum is the *median eminence,* recognized by its intense vascularity. The median eminence forms the floor of the third ventricle. This region of the brain is attached by a stalk to the pituitary gland (Fig. 2).

The boundaries of the *hypothalamus* are somewhat arbitrary. Its anterior limits are defined as the optic chiasm and lamina terminalis. It is continuous here with the preoptic area, the substantia innominata, and the septal region. Posteriorly, it is bounded by an imaginary plane from the posterior border of the mammillary bodies ventrally to the posterior commissure dorsally. Caudally, the hypothalamus merges with the midbrain periaqueductal gray and the reticular formation. The dorsal limit of the hypothalamus is defined by the hypothalamic sulcus on the medial wall of the third ventricle. At this junction, the hypothalamus is continuous with the subthalamus and, above it, the thalamus. Laterally, the hypothalamus is bounded by the internal capsule and the basis pedunculi. [*See* Hypothalamus.]

B. Pituitary Gland

The pituitary gland, or *hypophysis,* lies in close proximity to the medial basal hypothalamus and is partially enclosed in humans by a bony cavity called the *sella turcica.* The pituitary is divided into two lobes, the anterior, or *adenohypophysis,* and the posterior, or *neurohypophysis.* The adenohypophysis develops from Rathke's pouch, an evagination extending upward toward the brain from the primitive buccal ectoderm, and consists of three parts. The *pars distalis* is the primary source of the anterior pituitary hormones: adrenocorticotropic hormone (ACTH), thyroid-stimulating hormone (TSH), growth hormone (GH), prolactin (PRL), and the gonadotropins luteinizing hormone (LH) and follicle-stimulating hormone (FSH). Other peptide hormones and growth factors have been purified from anterior pituitary tissues, but their precise physiological functions remain

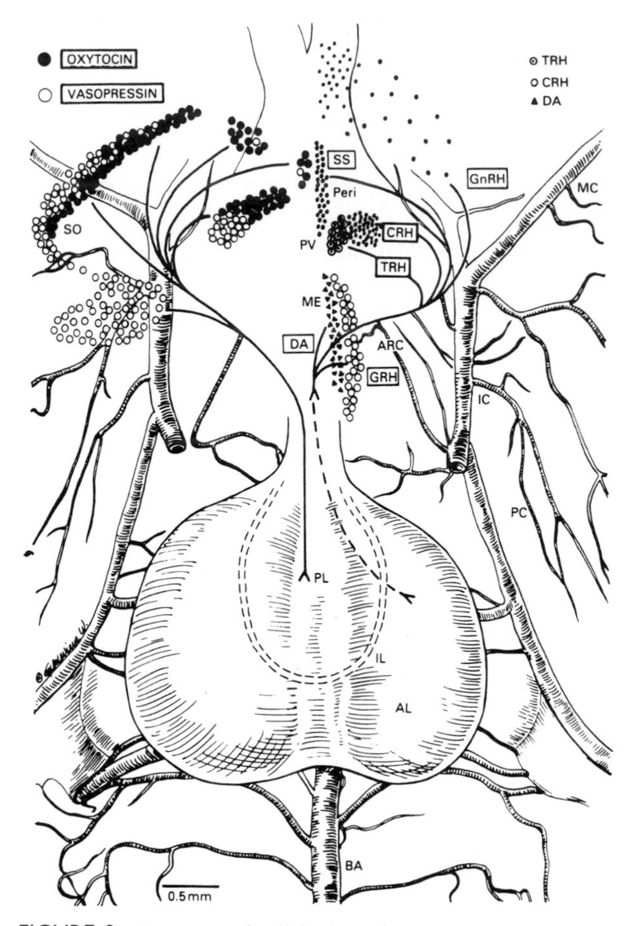

FIGURE 2 Location of cell bodies of hypothalamic hormone-secreting neurons projected on a ventral view of the hypothalamus and pituitary of the rat. AL, anterior lobe; ARC, arcuate nucleus; BA, basilar artery; IC, internal carotid artery; IL, intermediate lobe; MC, middle cerebral artery; ME, median eminence; PC, posterior cerebral artery; Peri, periventricular nucleus; PL, posterior lobe; PV, paraventricular nucleus; SO, supraoptic nucleus. The names of the hormones are enclosed in the boxes. [Courtesy of L. W. Swanson and E. T. Cunningham, Jr.; reproduced, with permission, from W. F. Ganong (1997). "Review of Medical Physiology," 18th Ed. Appleton & Lange, Stamford, Connecticut.]

uncertain. The second part of the anterior pituitary is the *pars intermedia,* which is vestigial in adult humans but is well-defined in the fetus and in rodents and other mammals. It secretes hormones such as melanocyte-stimulating hormone, which has an important function in the regulation of skin pigmentation. The third component of the adenohypophysis is the *pars tuberalis,* which consists of secretory cells similar to those of the pars distalis that envelop the upper pituitary stalk extending over the surface of the median eminence. The physiological function of the pars tuberalis remains uncertain. [*See* Pituitary.]

The neurohypophysis develops as an inferior extension of the hypothalamus, carrying with it, as it grows into the sella turcica, the nerve terminals of hypothalamic neurons that synthesize and secrete vasopressin and oxytocin. The neuronal cell bodies within the hypothalamus that project to the neural lobe are readily visible in microscopic sections by their large size (*magnocellular neurons*) and by their staining for neurosecretory material. They are clustered in two pairs, one located adjacent to the third ventricle, named the *paraventricular nuclei,* and the second straddling the optic tract, called the *supraoptic nuclei* (see Fig. 2). The supraoptic– and paraventricular–neurohypophysial nerve fibers are unmyelinated; they course through the median eminence, enter the pituitary stalk, and terminate on blood vessels in the posterior lobe.

C. Median Eminence

It should be apparent from these descriptions that the median eminence is an important and complex organ that serves several functions. In the first place, it is the principal site of termination of the hypothalamic neurons that regulate the anterior pituitary. Second, the median eminence contains in its deeper layers the axons of nerve cells passing to the posterior pituitary, which produce the peptides vasopressin and oxytocin. In addition, it is covered on its ventricular margin by ependymal cells that separate it from the cerebrospinal fluid. Finally, as noted earlier, the median eminence is a circumventricular organ with fenestrated capillaries that permit access of substances in peripheral blood to the nerve terminals of the hypothalamic–adenohypophysial system.

Some of the vasopressin-secreting cells of the paraventricular nuclei also end in the median eminence, so vasopressin reaches the anterior pituitary, where it stimulates ACTH secretion.

D. Pituitary Portal Blood Supply

The arterial blood supply of the hypothalamus and pituitary are intimately linked. Arterial branches from the internal carotid penetrate the layers of the median eminence to form a capillary bed that lies in direct contact with neurosecretory nerve terminals. The blood is then collected into pituitary portal vessels that traverse the pituitary stalk and form a secondary capillary network bathing the secretory cells of the anterior pituitary. This so-called hypothalamic–pituitary portal circulation is the mechanism by which hypothalamic regulatory peptides reach the anterior pituitary to either stimulate or inhibit secretion of pituitary hormones. The posterior pituitary receives a second blood supply from other branches of the internal carotid that directly penetrate the posterior lobe. Into these capillaries are secreted the hormones of the posterior lobe, vasopressin and oxytocin. Although the arterial blood supply varies somewhat in its detail in different species, it is similar in all mammals in being derived from branches of the internal carotid.

E. Hypothalamic Releasing Hormones

As noted earlier, functional control of hormonal secretion by the anterior pituitary is exerted by six hypothalamic hormones released from nerve endings in the median eminence (see Table I). The release of some of the anterior pituitary hormones appears to be regulated by a single hypothalamic factor: thus the gonadotropins depend on gonadotropin-releasing hormone (GnRH) for stimulation of their release. If GnRH is absent during development, sexual functions fail altogether and puberty fails to occur. In the adult, the development of GnRH deficiency can lead to secondary failure of sexual function with infertility and deficient secretion of sexual steroid hormones (testosterone in the male and estrogen and progesterone in the female). In the case of the other pituitary hormones, a more complex set of regulatory factors is involved. ACTH secretion is stimulated by corticotropin-releasing hormone (CRH) but also by vasopressin. GH secretion is stimulated by growth hormone-releasing hormone (GRH) and is inhibited by somatostatin. Thyrotropin (TSH) is stimulated by thyrotropin-releasing hormone (TRH) and inhibited by somatostatin. PRL secretion is tonically inhibited by the hypothalamus via the action of dopamine. It appears that there is no single prolactin-releasing hormone, but TRH and other peptides such as vasoactive intestinal peptide (VIP) have prolactin-releasing activity.

With the availability of pure synthetic hypothalamic releasing hormones, it has been possible to develop methods to study their precise localization within neuronal subdivisions of the hypothalamus. In rats, TRH and CRH are principally found within neurons of the medial paraventricular nucleus. Somatostatin is found in the periventricular nucleus. GRH and dopamine are found mainly within neurons situated immediately above the median eminence in the arcuate nucleus. GnRH-secreting neurons are located throughout the anterior hypothalamus and preoptic area, with their axons converging on the median eminence (see Fig. 2). In humans, there is an additional group of GnRH neurons in the arcuate nucleus.

The overall regulatory mechanisms required to achieve homeostasis depend on the ability of the brain to sense the circulating levels of the peripheral hormones. This is achieved primarily by negative feedback on the brain and also on the pituitary via hormones released by target organs. Thus the thyroid hormones (thyroxinc and triiodothyronine), the adrenal hormone cortisol, and the gonadal sex steroids each act to suppress further secretion of the trophic hormone that stimulates its production. As levels of cortisol rise, for example, the hypothalamus secretes less CRH, and the pituitary becomes more resistant to the effects of CRH. As cortisol levels fall, these inhibitory effects lessen and ACTH secretion increases.

A few physiological observations further complicate our understanding of the system regulation of the brain–endocrine–target gland axis. One factor is the finding that all hormonal secretions are pulsatile and episodic rather than steady state. In the case of GnRH, this makes a major difference because exposure to GnRH that is at all prolonged down-regulates GnRH receptors and inhibits gonadotropin secretion. Exogenous GnRH stimulates gonadotropin secretion, but only if administered in pulsatile fashion. Treatment with long-acting GnRH preparations is used when the goal is to shut off gonadotropin secretion, for example, in patients with androgen-dependent prostate cancer.

Another important aspect of neuroendocrine regulation is the influence of circadian rhythms on endocrine function. The most apparent of these daily rhythms is that described for ACTH and cortisol. In humans, levels of ACTH rise during the late night hours to peak at about the time of awakening. They decline throughout the day to their lowest point in the early evening and then rise again. The secretion of hormones in a circadian fashion is regulated by paired nuclei located immediately above the optic chiasm in the anterior hypothalamus called the *suprachiasmatic nuclei.* These nuclei have been shown to direct a host of circadian rhythms in addition to their influence on hormones, including motor activity cycles, feeding and drinking behavior, and body temperature regulation. Disruption of these structures experimentally or after brain lesions such as tumors eliminates the coordination of the rhythms of the day–night cycle, which become free-running, that is, they no longer take exactly 24 hr.

IV. NEUROPHARMACOLOGY OF PITUITARY HORMONE SECRETION

In a conceptual sense, the neurosecretory neurons of the hypothalamus can be likened to the motor neurons of the brain stem and spinal cord, which serve as the final common pathway for initiating motor functions by their connections to the muscles. In the case of the neuroendocrine system, the neurosecretory system is the recipient of incoming information from many brain areas, which integrate responses appropriate to the organism. In terms of the paradigms of "fight" or "flight," for example, it is possible to delineate specifically the hormonal responses accompanying these behavioral reactions. The output system comprising the hypothalamic–pituitary axis transduces these neural responses into hormonal outputs. Other examples of a more chronic nature also serve to illustrate this point. The young woman who first discovers the vulnerability of her menstrual cycle to dysregulation after embarking on the new experience of leaving home for the first time to attend college has experienced the temporary shutdown of the hypothalamic drive essential for reproductive capacity. In this instance, the psyche can indicate its perturbation by a dramatic interruption of a previously well-coordinated function. The interruption is almost always temporary, but occasionally may be followed by prolonged absence of menstrual periods (amenorrhea).

These neural influences are mediated over a number of different pathways, many of which are relayed via systems that use neurotransmitters such as glutamate, dopamine, norepinephrine, serotonin, acetylcholine, and γ-aminobutyric acid. These effects are mediated over long pathways in the brain because, at least in the case of the biogenic amines serotonin, epinephrine, and norepinephrine, the cell bodies containing them are located in the brain stem. The effects of the biogenic amines can be demonstrated by the disruptions of pituitary hormone secretion occurring with pharmacological manipulations. For example, administration of drugs that interfere with dopamine synthesis, release, or receptor binding is followed by striking increases in PRL secretion, which in turn may be manifest by the inappropriate secretion of milk from the breast (galactorrhea) and by suppression of gonadotropin secretion, with amenorrhea and infertility. Fortunately, these effects are almost always reversible, resulting in restoration of the normal reproductive cycle with discontinuation of drug administration. Drugs that can induce this kind of effect include antipsychotic drugs such as chlorpromazine and haloperidol, which block D_2 dopamine receptors.

V. REGULATION OF INDIVIDUAL PITUITARY HORMONES

A. Vasopressin and Oxytocin

Vasopressin and oxytocin are small peptides containing nine amino acid residues. They differ from each other by two residues. Each is synthesized within the cell bodies of different magnocellular neurons as preprohormones. Each is associated with a distinct neurophysin during transport of the peptide within neurosecretory granules to nerve terminals in the posterior pituitary. Nerve terminal depolarization results in release of both the peptide and the associated neurophysin.

The peripheral actions of vasopressin are exerted on the collecting tubules of the kidney, where they induce reabsorption of water. In the absence of vasopressin, dilute urine is excreted, leading to stimulation of thirst as plasma osmolality rises. The syndrome of polyuria and polydipsia resulting from a deficiency in vasopressin is called diabetes insipidus, to be distinguished from the much more common association of polyuria with increased glucose in the urine in diabetes mellitus.

Oxytocin acts on smooth muscle cells of the uterus during parturition and on the mammary ducts to bring milk from the breast glands to the nipple. What function oxytocin serves in men is unclear, although it can be detected in both the pituitary and blood.

Increases in plasma osmolality are believed to stimulate the organum vasculosum of the lamina terminalis, and activation of this circumventricular organ (see Fig. 1) stimulates vasopressin secretion. Plasma osmolality is zealously guarded at a set point corresponding to 280 mOsm/kg, and very small increases trigger increased vasopressin secretion. The resulting retention of water dilutes the blood back to its normal value.

Vasopressin secretion is also increased by decreases in blood volume. Receptors mediating this response are located in the left atrium, carotid sinuses, and aortic arch. It is an interesting fact that vasopressin is a potent vasoconstrictor *in vitro,* but it takes relatively large doses of exogenous vasopressin to raise blood pressure. The reason is that *in vivo,* in addition to causing vasoconstriction, vasopressin acts on the area postrema, another of the circumventricular organs (see Fig. 1), to lower the set point of the baroreceptor reflexes. This increases baroreceptor sensitivity and counteracts the pressor effect of the hormone.

Stress and nausea are also able to stimulate the release of vasopressin. Inputs from the gastrointestinal tract and other viscera relayed via the glossopharyngeal and vagal nerves activate neurons in the nucleus of the tractus solitarius of the medulla oblongata. The signals are then relayed to the hypothalamus to trigger vasopressin release.

A common clinical problem in very ill, bedridden patients is the syndrome of inappropriate vasopressin secretion in which small increases in plasma vasopressin levels result in inappropriate water retention and a fall in blood osmolality and serum sodium concentrations. The result can be seizures, coma, and permanent neurological injury.

B. Adrenocorticotropic Hormone

Pituitary secretion of ACTH is both episodic, with alterations in hormone secretion occurring over minutes, and circadian, with levels changing throughout a 24-hr period, resulting in increased levels during the night and failing levels during the daytime. This rhythm is among the most robust of the various circadian rhythms and persists even with change in the daily sleep–wake cycle (as occurs, for instance, in jet travel) for 7–10 days. The disruptions in mental and physical performance that accompany jet lag are believed to be the result primarily of the temporal disruption of the hypothalamic–pituitary ACTH adrenal rhythm (i.e., the rhythm persists in an inappropriate relation to the activity cycle for about 7–10 days).

The hypothalamic regulation of CRH secretion is influenced by a number of central nervous system neurotransmitters, including norepinephrine, serotonin, and acetylcholine. These transmitters are speculated to have a role in the "stress-associated" responsiveness of CRH secretion. Indeed it is the activation of this system by both psychological and physical stresses that is most frequently used to define the very essence of "stress" itself. Disorders of the regulatory set point of the system are known to occur in about one-half of all patients with severe depression, and abnormalities in suppression of ACTH secretion are tested by the use of dexamethasone, a synthetic, highly potent cortisol-like steroid. The dexamethasone suppression test is widely applied in the investigation of depressed patients as a biological marker of the depressed state. Abnormalities characterized by failure of ACTH or cortisol suppression usually revert to normal after recovery of the patient (i.e., the abnormality is state-dependent rather than trait-dependent).

Abnormalities due to hypersecretion of ACTH also occur in Cushing's disease, which is caused by a small benign tumor of ACTH-secretory cells located in the pituitary. The symptoms include central weight gain, caused by excessive deposition of fat, thinning and easy bruising of the skin, and psychological changes including depression. The diagnosis is confirmed by demonstrating increased secretion of cortisol in the urine and by failure of ACTH to suppress with small doses of dexamethasone. The treatment in those cases due to pituitary tumors is surgical, with removal of the tumor resulting in resolution of the symptoms in more than 60% of patients.

C. Gonadotropins

Fundamental to the survival of any species is the ability to propagate its own kind. This capacity is determined by the coordination of sexual functions defined both by successful and appropriate copulatory activities and by hormonal readiness of the partners. This complex set of behaviors and hormonal regulation is accomplished by the neuroendocrine axis. The hypothalamus develops the capacity at the time of puberty to release GnRH episodically, first at night and then throughout the 24-hr circadian cycle. This awakening of the capacity for reproduction can be detected by sleep-time monitoring of LH and FSH secretion. Gonadal steroids rise in response to the gonadotropins, and secondary sex characteristics begin to develop. Full reproductive capacity appears at about 12–13 years in the female with the appearance of menarche (first menstrual period). In the male, puberty occurs slightly later. The same hormones are made in both sexes at both the hypothalamic (GnRH) and the pituitary levels (LH and FSH). The ovaries in the female produce estrogen and progesterone, which act in a negative feedback loop to reduce further secretion of the gonadotropins. In the male, testosterone production by the testis elicits a similar negative effect on the pituitary. The ovaries and the testes also produce peptides called inhibins, which inhibit FSH secretion, but the complexities involved in the hormonal changes of the menstrual cycle are beyond the subject of this short review. [*See* Puberty; Steroids.]

Disorders of reproductive function can occur at several levels. Precocious development of sexual characteristics (precocious puberty) can result from hypothalamic tumors that alter the normal constraint imposed by the brain on endocrine maturation releasing pathways that trigger development. However, destructive lesions of the hypothalamus or tumors that destroy functions of the pituitary can prevent the normal appearance of puberty. These conditions occurring before puberty result in failure of reproductive function to develop in women (primary amenorrhea). They are also sometimes referred to as secondary hypogonadism as opposed to a primary defect located in the gonads themselves. In the adult, pathological processes at either the hypothalamic or pituitary level can result in secondary failure of sexual function (secondary amenorrhea in the female) or secondary hypogonadism in the male. Hormonal replacements with GnRH, which must be administered by a systemic route in a pulsatile fashion (see earlier), have been successful in restoring sexual competence in both primary and secondary forms of amenorrhea and in male infertility caused by hypothalamic or pituitary failure.

The syndrome of hypogonadism caused by excessive PRL secretion was discussed in Section IV.

D. Prolactin

The hypothalamic regulation of PRL secretion differs from that of all the other pituitary hormones by its inhibitory nature. Dopamine secreted into the portal hypophysial system by cells of the arcuate nucleus in the base of the hypothalamus acts on D_2 dopamine receptors on the pituitary lactotrope cells to suppress both the release and the biosynthesis of PRL. This tonic inhibition persists throughout most of the day but is interrupted during the night when pulsatile bursts of PRL occur, resulting in elevation in PRL before morning awakening. PRL rises during the third trimester of pregnancy, preparing the mammary gland for lactation after parturition. PRL continues to be secreted after delivery with stimulation evoked by mechanical stimulation of the nipple of the breast as the infant suckles.

Abnormal secretion of PRL can occur from pituitary adenomas that synthesize abnormal amounts of PRL. These are the most common of all hormone-secreting pituitary tumors. They can arise in either men or women at all ages. In women they are often recognized early because of the associated amenorrhea and galactorrhea. In the case of men they usually grow to a size sufficient to cause visual disturbances or severe headaches before coming to diagnosis. Impotence and infertility are common manifestations of the hyperprolactinemic state in men.

E. Growth Hormone

Normal somatic growth, which requires GH, is mediated to a large degree by growth factors (somatomedins) produced in peripheral tissues such as the liver under the stimulation of the hormone. The principal somatomedin is insulin-like growth factor I (IGF-I). The secretion of GH requires stimulatory actions by hypothalamic GRH. The output of GH is pulsatile, with four to seven pulses of secretion occurring during a 24-hr period, the largest of which is sleep-entrained and evident about 2 hr after onset of sleep. GH secretion can be completely inhibited by somatostatin. Extensive experimental analysis of pulsatile GH secretion indicates that the coordinate hypothalamic secretion of GRH and somatostatin is required for normal secretion. Somatostatin arising from cell bodies in the anterior hypothalamus seems to determine the rhythm of the pulses, whereas GRH determines their timing and magnitude.

Abnormalities of GH secretion occurring as a result of defective production of GRH during childhood lead to growth failure (dwarfism). This can be the result of an inherited defect in the gene that regulates GH biosynthesis in the anterior pituitary or to destructive lesions of either the hypothalamus or pituitary that disrupt the effects of either GRH or of GH secretion. Treatment of growth failure with human GH obtained from cadaver pituitary glands has been suspended after recognition that the agent causing a fatal neurological disease (Creutzfeldt–Jakob disease) could be transmitted with the GH. Fortunately, recombinant DNA technology has led to the *in vitro* production of GH that is free of this agent. Many children are receiving recombinant GH for treatment of GH deficiency.

Excessive pituitary GH secretion in children, usually resulting from the formation of a pituitary adenoma that secretes GH, causes gigantism. In the adults after the epiphyses have closed, hypersecretion leads to a clinical syndrome of broadened hands and fingers (acromegaly) and to altered metabolism, including diabetes mellitus, hypertension, and, in long-standing cases, heart disease. The treatment of pituitary tumors is surgical, accomplished by a procedure called transsphenoidal partial hypophysectomy, in which the pituitary tumor is approached from the nasal cavity through the sphenoid sinus. Successful treatment of early cases can be achieved in more than 75% of cases.

F. Thyroid-Stimulating Hormone

Thyroid-stimulating hormone regulation is achieved primarily via the effects of TRH. In its absence, TSH secretion falls to less than 10% of normal. TSH acts on the thyroid to stimulate synthesis and secretion of the two thyroid hormones thyroxine (T_4) and triiodothyronine (T_3). These hormones act on most tissues in the body to maintain appropriate levels of general metabolic activity. Both hormones can act on the hypothalamus and pituitary to inhibit secretion of TRH and TSH, respectively. However, it appears that the principal site of feedback regulatory control is at the level of the pituitary.

Diminished output of TSH can occasionally follow large destructive lesions of the hypothalamus, particularly those affecting the paraventricular nuclei. More commonly, destructive lesions of the pituitary such as tumors result in deficiency of TSH. It should be recalled that hypothyroidism is most often the result of primary thyroid disease. Excessive thyroid function (hyperthyroidism) is almost always due to Grave's disease, a defect caused by hypersecretion of thyroid hormone induced by an immunological disorder in which antibodies mimic the effects of TSH on the thyroid. In this condition, circulating TSH levels are very low. Pituitary tumors rarely secrete TSH or the gonadotropins, possibly because, unlike ACTH, GH, and PRL, they are glycoproteins, each composed of two separate subunits: an alpha subunit common to all three and a beta subunit that is unique for each and that imparts the specific biological effects of TSH, LH, and FSH. [*See* Thyroid Gland and Its Hormones.]

BIBLIOGRAPHY

Ganong, W. F., and Martini, L. (1990). Neuroendocrinology, 1990. *Frontiers in Neuroendocrinol.* **11**, 1–5.

Greenspan, F. S., and Baxter, J. D. (eds.) (1994). "Basic and Clinical Endocrinology," 4th Ed., Chap. 2. Appleton & Lange, East Norwalk, Connecticut.

Hadley, M. E. (1996). "Endocrinology," 4th Ed., Chaps. 5–8, 20, and 21. Prentice–Hall, Upper Saddle River, New Jersey.

Klavdieva, M. M. (1995–1996). The history of neuropeptides. *Frontiers in Neuroendocrinol.* **16**, 293–321; **17**, 126–153; **17**, 155–179; and **17**, 247–280.

Martin, J. B., and Reichlin, S. (1987). "Clinical Neuroendocrinology," Contemporary Neurology Series, 2nd Ed. F. A. Davis Company, Philadelphia.

Neurological Control of the Circulatory System

CHRISTOPHER J. MATHIAS
University of London at Imperial College School of Medicine at St. Mary's and National Hospital & Institute of Neurology

GLOSSARY

Autonomic nervous system The part of the nervous system that controls involuntary actions.

Baroreceptor A pressure-sensitive receptor organ of the nervous system, found in the blood vessels.

Bradycardia An abnormally slow contraction of the heart muscle, usually defined as a pulse rate below 60 beats per minute; the converse of tachycardia.

Hypertension Abnormally high arterial blood pressure, especially on a persistent basis; the converse of hypotension.

Hypotension Abnormally low arterial blood pressure; the converse of hypertension.

Neurotransmitter Any of a group of chemical substances that transmit nerve impulses across a synapse.

Tachycardia An abnormally rapid contraction of the heart muscle, usually defined as a pulse rate above 100 beats per minute; the converse of bradycardia.

Vagus The tenth and longest pair of cranial nerves, extending from the brain to the heart, lungs, stomach, and various other organs and playing an essential role in many body functions.

Vasoactive Affecting the diameter of the blood vessels; i.e., producing either dilation or constriction.

Vasodilation A widening of the blood vessels.

I. BASIC PRINCIPLES

The prime purpose of the circulatory system is to ensure tissue perfusion. This is needed for the delivery of oxygen and nutrients, for the transport of substances (such as those absorbed from the gut), and for the removal of unwanted metabolites. Vital organs, such as the brain and heart, need to be perfused adequately regardless of the stimulated state; others have specific demands, such as the skeletal musculature during physical exertion and the gastrointestinal tract during food ingestion.

A key factor in perfusion is the level of systemic blood pressure, which is dependent on various factors, structural and others, including the autonomic nervous system (Table I). This system, particularly through the baroreceptor reflex pathways (Fig. 1), is essential for the beat-by-beat regulation of blood pressure and thus the capacity to respond rapidly to ensure satisfactory organ perfusion in different functional states. The baroreflex has three components: afferent pathways conveying information to the brain, central connections within the brain and spinal cord, and efferent pathways in the periphery that supply the blood vessels and heart. The afferents are predominantly in the carotid sinus and aortic arch, and convey information triggered by changes in pulse pressure and distension of vessels. These signals are transmitted through the sino-aortic nerve (in the glossopharyngeal and vagus nerves) into the brain stem. There are major vasomotor centers in the medulla (the nucleus tractus solitarius and the vagal nuclei). These interact with various centers within the brain. From the brain stem, information is conveyed to the heart (via the vagus nerves) and down the spinal cord emerging from the thoracic and upper lumbar seg-

TABLE I
Factors Contributing to Maintenance of Systemic Blood Pressure

Structural
Heart
Resistance vessels (arteries)
Capacitance vessels (veins)
Intravascular volume
Chemicals
Systemically
Renin–angiotensin system
Aldosterone
Locally
Nitric oxide
Endothelin
Autonomic nervous system

ments, via the paravertebral ganglia, to supply the heart and blood vessels (sympathetic efferents). Activation of the cardiac vagi therefore will slow the heart whereas a decrease in activity will raise the heart rate. An increase in sympathetic activity will constrict blood vessels and activate the heart, thus raising blood pressure whereas the reverse will lower blood pressure. It should be noted that afferent stimuli from many parts of the body can influence the autonomic outflow, as can activity generated within the brain itself; cold stimuli in the periphery (such as by immersing the hand in iced water) will rapidly increase sympathetic activity and raise blood pressure and heart rate; similar changes occur through mental stimulation by the activation of cerebral centers. [*See* Cardiovascular System, Physiology and Biochemistry.]

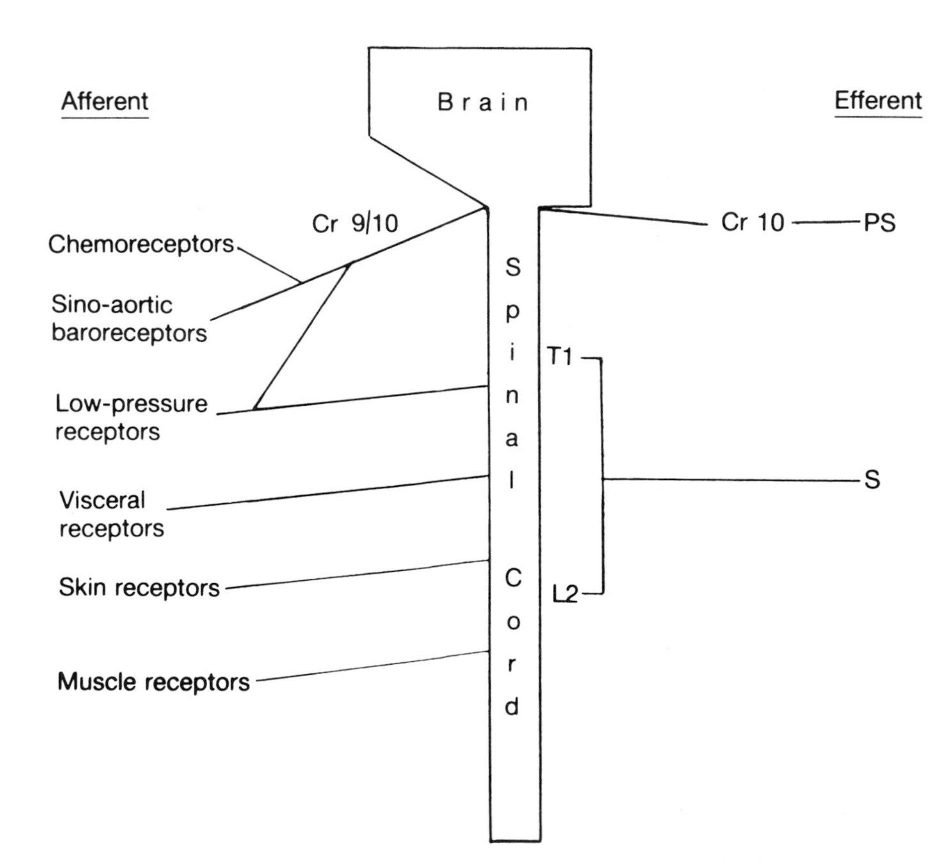

FIGURE 1 Schematic outline of the major autonomic pathways controlling the circulation. The major afferent input into the central nervous system is through the glossopharyngeal (CR 9) and vagus (CR 10) nerves by activation of baroreceptors in the carotid sinus and in the aortic arch. Chemoreceptors and low-pressure receptors also influence the autonomic efferent outflow. The latter consists of the cranial parasympathetic (PS) outflow to the heart through the vagus nerves and the sympathetic outflow from the thoracic and upper lumbar segments of the spinal cord to the heart and blood vessels. Activation of visceral, skin, and muscle receptors, in addition to cerebral stimulation, can influence the efferent outflow. From Mathias and Frankel (1988).

The major transmitters involved in peripheral autonomic activation are denoted in Fig. 2. In the vagus, the predominant postganglionic transmitter, acting on receptors in the heart, is acetylcholine. Its actions can be blocked by drugs such as atropine, a muscarinic cholinergic receptor blocker, as compared to ganglionic blockers that will act on both parasympathetic and sympathetic ganglia via nicotinic receptors. In sympathetic nerves, noradrenaline is the predominant neurotransmitter and is synthesized by a series of steps (Fig. 2b) in nerve endings, from where it is released. Its overflow into the circulation (Fig. 2c) is dependent on a number of factors, including neural activation. The measurement of plasma noradrenaline provides a useful biochemical measure of sympathetic activation in certain circumstances; refinements to biochemical techniques have been devised, and noradrenaline spillover from a particular organ can be derived, providing a measure of regional sympathetic activation. This is of importance when there is differential activation of autonomic neural pathways in varying circumstances; food ingestion is an example where dilatation is needed in the gut vessels, whereas constriction of other vascular beds, by increasing sympathetic activity, is needed to maintain blood pressure. The technique of sympathetic microneurography is a physiological approach utilizing a fine tungsten microelectrode in a peripheral nerve, positioned to record either sympathetic muscle or skin activity (Fig. 3). This provides a beat-by-beat measure of sympathetic neural activation in a particular region. Nonadrenergic, noncholinergic substances also may play key roles as neurotransmitters; vasoactive intestinal polypeptide (VIP) and neuropeptide (NPY) are examples of such peptides coexisting in cholinergic nerves and sympathetic nerves, respectively. [*See* Autonomic Nervous System; Neurotransmitters and Neuropeptide Receptors in the Brain.]

II. STUDYING NEUROCIRCULATORY CONTROL: LESSONS FROM HUMAN AUTONOMIC DISORDERS

In normal subjects the autonomic nervous system responds rapidly, thus ensuring a satisfactory level of blood pressure to maintain the appropriate perfusion of organs in different circumstances. This may mean increasing activity to certain organs and decreasing activity to others, along with regional vascular redistribution, that is done in conjunction with various other factors, including substances that provide local vascular control in different organs. In normal man, the rapidity of action of the autonomic nervous system can cause difficulties in determining the degree of neurological control exerted accurately. One approach is to use drugs with specific effects, which have contributed substantially, except that many have multiple actions, with differential effects on various vascular regions. Of value, therefore, has been the information from studying human subjects with well-defined lesions of the autonomic nervous system; these patients usually have cardiovascular problems which warrant study, and in addition to understanding more about the pathophysiological basis of their clinical state, they provide valuable information on neurological control in normal man. Table II provides a list of some of these disorders. The cause may not be known (primary autonomic failure); one of the two major groups has central autonomic failure (multiple system atrophy, MSA; Shy-Drager syndrome) and the other peripheral failure (pure autonomic failure). There are a number of diseases where autonomic failure is a complication or major problem, including those with a specific neurological lesion (such as high spinal cord transection following injury), a specific deficit (such as deficiency of the enzyme dopamine β-hydroxylase which converts dopamine into noradrenaline), and complex metabolic disorders (such as diabetes mellitus). A number of drugs may cause autonomic failure, either through single or multiple effects. A separate category are those with neurally mediated syncope, in whom there is usually no autonomic abnormality except during attacks that result in collapse and fainting due to abnormal overactivity of the vagus (thus slowing or stopping the heart) and/or the withdrawal of sympathetic activity (thus lowering blood pressure substantially). Some of these disorders will be mentioned later, and examples will be provided of how they have aided the understanding of neurological control of the circulation.

III. BLOOD PRESSURE ABNORMALITIES

A. Low Blood Pressure

Activation of the autonomic nervous system is important for the rapid adjustment of blood pressure;

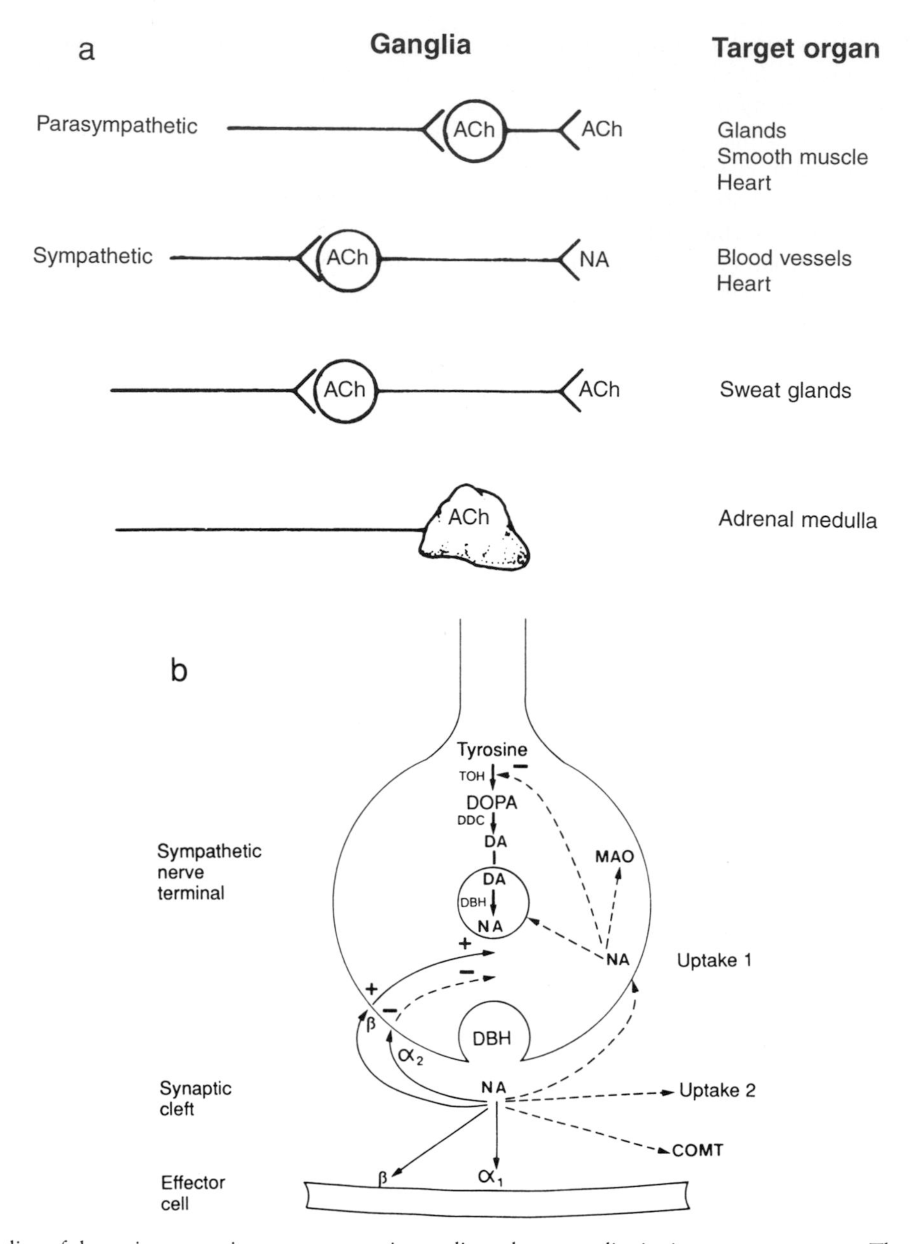

FIGURE 2 (a) Outline of the major transmitters at autonomic ganglia and postganglionic sites on target organs. The ganglionic blockade of nicotinic cholinergic receptors with hexamethonium prevents both parasympathetic and sympathetic activation. Atropine, however, only acts on muscarinic cholinergic receptors at postganglionic parasympathetic and sympathetic cholinergic sites. ACh, acetylcholine; NA, noradrenalin. (b) Schema of some pathways in the formation (expanded in c), release, and metabolism of noradrenaline from sympathetic nerve terminals. Tyrosine is converted into dihydroxyphenylalanine (DOPA) by tyrosine hydroxylase (TOH). DOPA is converted into dopamine (DA) by dopa-decarboxylase. In the vesicles, DA is converted into noradrenaline (NA) by dopamine β-hydroxylase (DBH). Nerve impulses release both DBH and NA into the synaptic cleft by exocytosis. NA acts predominantly on α_1-adrenoreceptors but has actions on β-adrenoceptors on the effector cell of target organs. It also has presynaptic adrenoceptor effects; those on α_2-adrenoceptors inhibit NA release whereas those on β-adrenoceptors stimulate NA release. NA may be taken up by neuronal (uptake 1) processes into the cytosol, which may inhibit further formation of DOPA through the rate-limiting enzyme TOH; it may be taken into vesicles and metabolized by monoamine oxidase (MAO) in the mitochondria; and it may be taken up by a higher-capacity but lower-affinity extraneuronal process (uptake 2) into peripheral tissues such as vascular and cardiac muscle and certain glands. It is also metabolized by catechol-*o*-methyltransferase (COMT). NA measured in plasma is thus the overspill not affected by these numerous processes. From Mathias (1996).

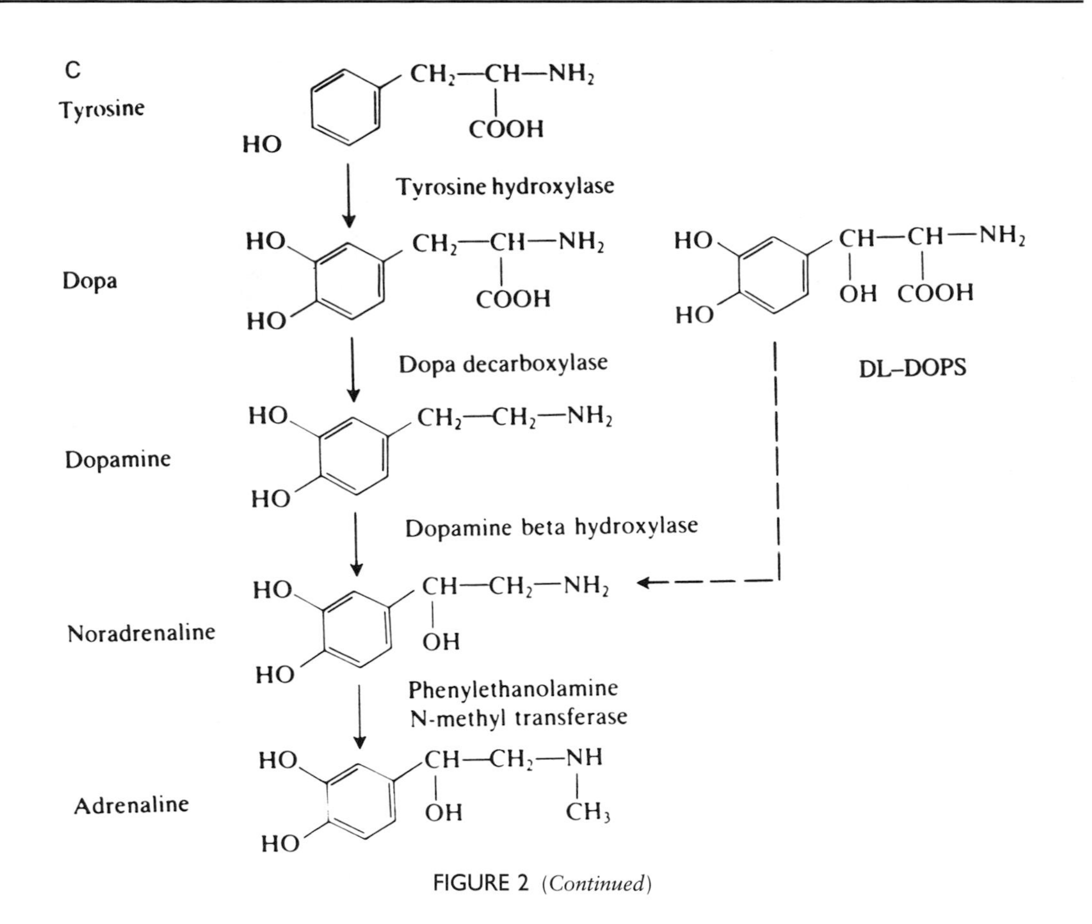

FIGURE 2 (*Continued*)

impairment of such activity, therefore, can result in low blood pressure (hypotension), which is best demonstrated in relation to occurrences in daily life, such as while standing, exercising, and eating.

I. Neurocirculatory Control and Postural Change

A change in posture, from lying to standing, imposes gravitational stress and results in a shift of the intra-

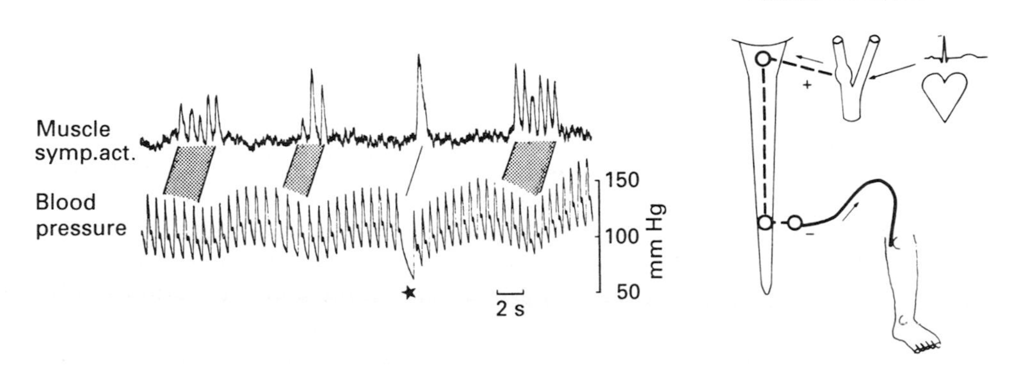

FIGURE 3 Relationship between spontaneous fluctuations of blood pressure and muscle nerve sympathetic activity (left) recorded in the right peroneal nerve. The baroreflex accounts for the inverse relationship between nerve activity and blood pressure fluctuations, with a rise in sympathetic discharge as blood pressure falls. The asterisk indicates the fall in blood pressure due to sudden atrioventricular block, accompanied by a brisk sympathetic response. The stippling indicates corresponding sequences of sympathetic nerve bursts and the heart beat. From Wallin *et al.* (1980).

TABLE II
Outline Classification of Autonomic Disorders Contributing to Our Knowledge of Neurocirculatory Control[a]

Primary autonomic failure
Multiple system atrophy (Shy–Drager syndrome)
Pure autonomic failure
Secondary autonomic failure
Lesions: high spinal cord transection
Specific deficits: dopamine β-hydroxylase deficiency
Disease: diabetes mellitus
Drugs
Neurally mediated syncope

[a]Adapted from Mathias (1996).

vascular volume, with pooling of blood in the periphery. In normal subjects this activates the baroreflex, with a rise in heart rate (because of vagal withdrawal and sympathetic stimulation) and a constriction of blood vessels (because of sympathetic activation); these actions prevent blood pressure from falling. In human sympathetic denervation (primary autonomic failure), this cannot occur and blood pressure falls on head-up postural change; when over 20 mm Hg, the term orthostatic hypotension is used, which is often associated with symptoms arising because of a lack of blood supply to the brain. Cerebral ischemia may result in dizziness, giddiness, visual disturbances (such as blurring of vision), and loss of consciousness, with fainting. Orthostatic hypotension can be tested in the laboratory on a tilt table; head-up tilt to 45° or 60° can result in marked hypotension in autonomic failure (Fig. 5). This can be reversed quickly by returning the subject to the horizontal. In such patients, plasma noradrenaline levels do not rise, confirming the lack of activation of the sympathetic (Fig. 4). The role of other vasoactive neurotransmitters in this response is unclear; substances such as adenosine triphosphate (ATP) and NPY also may contribute, although studies in the rare group with dopamine β-hydroxylase deficiency (with undetectable levels of noradrenaline and adrenaline but otherwise intact sympathetic nerve endings) suggest otherwise. The absence of noradrenaline causes marked orthostatic hypotension; this can be reversed by the drug L-dihydroxyphenylserine (L-DOPS), which bypasses the enzymatic deficit, and is converted to noradrenaline (Fig. 2c). It thus may be that with immediate postural readjustments, noradrenaline is the predominant neurotransmitter, whereas other substances play neuromodulatory or otherwise supportive roles. These roles may change with other stimuli.

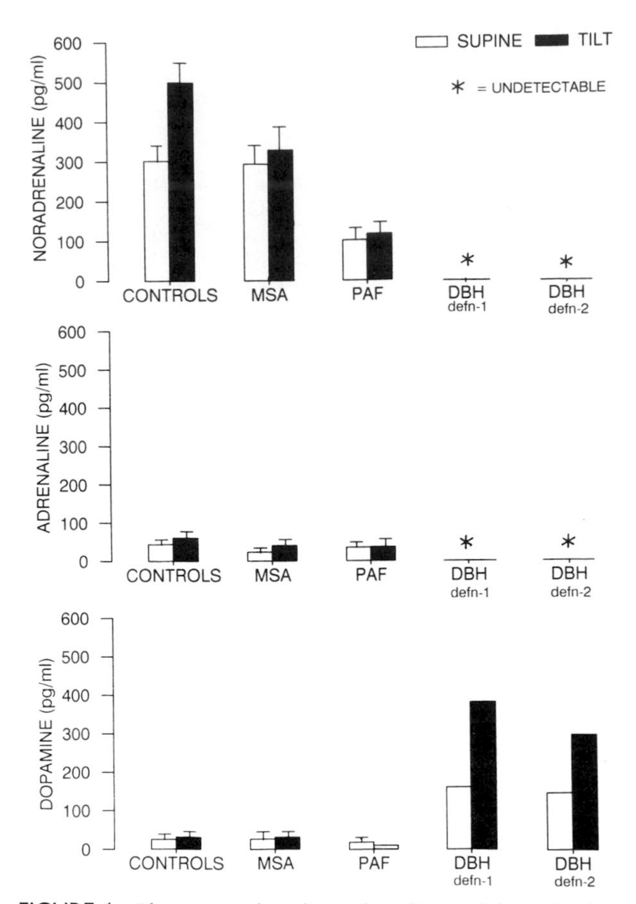

FIGURE 4 Plasma noradrenaline, adrenaline, and dopamine levels (measured by high-pressure liquid chromatography) in normal subjects (controls), in patients with multiple system atrophy (MSA) and pure autonomic failure (PAF), and in two siblings with dopamine β-hydroxylase deficiency (DBH defn). Measurements were taken while the subjects were supine and after head-up tilt to 45° for 10 min. The asterisk indicates levels below the detection limits, which are less than 5 pg/ml for noradrenaline and adrenaline, and less than 20 pg/ml for dopamine. Bars indicate ± SEM. From Bannister and Mathias (1992).

2. Neurocirculatory Control and Exercise

In normal subjects, a modest degree of exercise raises blood pressure through activation of the sympathetic nerves and the adrenal medullae. There is a rise in cardiac output and a readjustment in certain vascular beds to compensate for the vasodilatation in exercising skeletal muscle. Lack of these compensatory mechanisms presumably accounts for the marked fall in blood pressure that occurs during exercise in human sympathetic denervation (Fig. 6). These effects may occur independently of those arising from head-up postural change (as demonstrated in such subjects while exercising horizontally, using a bicycle ergome-

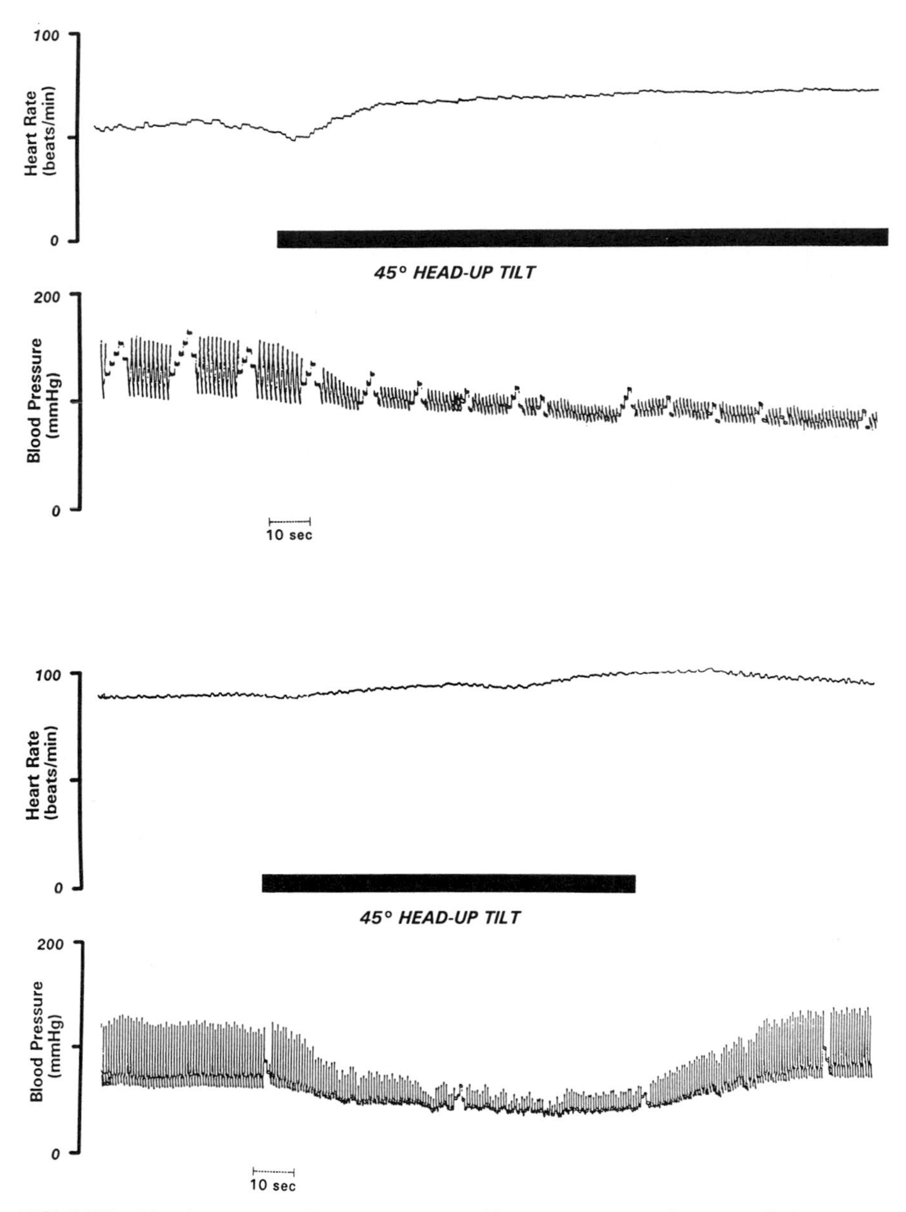

FIGURE 5 Blood pressure and heart rate measured by a noninvasive technique with the Finapres in two patients with autonomic failure. (Top) Blood pressure falls to extremely low levels but the patient could remain tilted for over 20 min with few symptoms; this patient had autonomic failure for many years and could tolerate lower levels of blood pressure. This is in contrast to the patient who had to be put back to the horizontal fairly quickly (bottom). She had developed severe postural hypotension soon after surgery. From Mathias (1996b).

ter). The combination of exercise and standing may be additive and often worsens the symptoms of hypotension in such patients.

3. Neurocirculatory Control and Food Ingestion

The ingestion of food results in a number of processes, involving digestion, absorption, and also associated metabolic demands. In normal man, it causes a marked increase in gut blood flow. Measurements in a major artery supplying the gut (the superior mesenteric artery) indicate blood flow doubling, from 0.5 to 1 liter per minute within 15 min of food ingestion (Fig. 7a). The rise in flow is dependent on the nutrient load of the meal and its composition. Systolic blood pressure, however, is maintained because of various

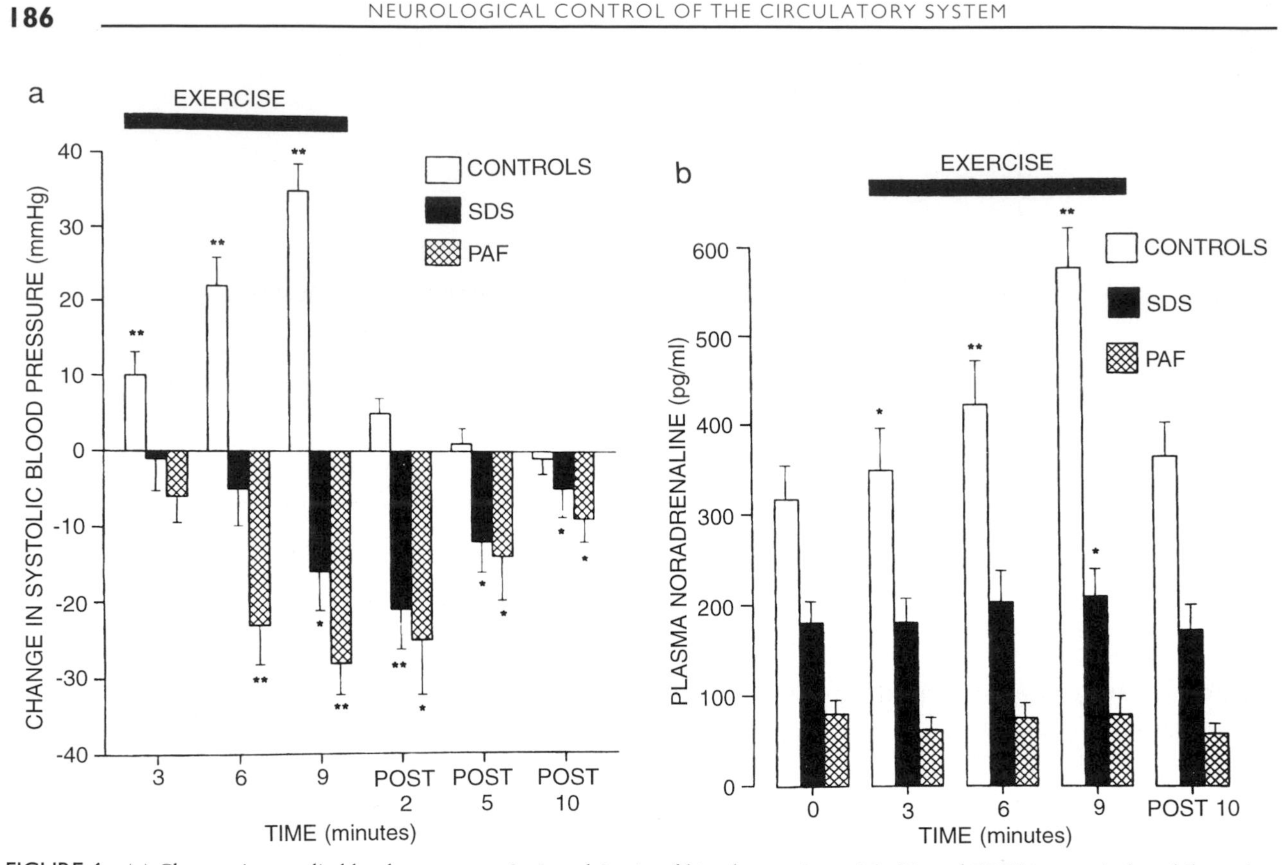

FIGURE 6 (a) Changes in systolic blood pressure at 3, 6, and 9 min of bicycle exercise at 25, 50, and 75 W, respectively, while supine; at 2, 5, and 10 min postexercise in normal subjects (controls); and subjects with sympathetic denervation due to central failure (multiple system atrophy, MSA; Shy–Drager syndrome, SDS) and peripheral autonomic failure (pure autonomic failure, PAF). Asterisks denote significant changes from baseline ($*P < 0.05$, $**P < 0.001$). (b) Venous plasma noradrenaline levels at rest (0) and after 3, 6, and 9 min of exercise and 10 min postexercise in normal subjects (controls), SDS, and PAF. Significant levels are as previously described. From Smith *et al.* (1996).

compensatory adjustments, including an increase in cardiac output and a constriction of blood vessels to skeletal muscle (Fig. 7b). In human sympathetic denervation, the inability to compensate adequately through autonomic nerves results in marked postprandial hypotension (Fig. 8), despite a similar degree of splanchnic vasodilatation to normal subjects. The degree of hypotension is dependent on the composition and caloric value of ingested food. Carbohydrates (in part through their osmotic effects of increasing gut fluid influx and through the release of potentially hypotensive peptides such as insulin) lower blood pressure to a greater extent than a high fat or protein meal. The role of gut peptides in causing splanchnic vasodilatation has been demonstrated using drugs such as the somatostatin analog octreotide, which prevent their release after food ingestion; in normal man (Fig. 7a) the postprandial rise in splanchnic blood flow can be prevented by pretreatment with octreotide. Similar changes occur in subjects with autonomic failure, in whom postprandial hypotension can be effectively prevented by this drug. These responses to food ingestion in normal subjects and subjects with autonomic emphasize the variety of changes involving both systemic and local neural vascular control with stimuli that we are exposed to regularly in daily life.

B. High Blood Pressure

Between 10 and 20% of adults have a mild to moderate elevation of blood pressure (hypertension) that contributes to increased morbidity and mortality through strokes and heart attacks. Increased activity of the autonomic nervous system has been implicated in hypertension, and there are a number of drugs, acting either centrally or peripherally, that reduce sympathetic activity or prevent its effects, thus low-

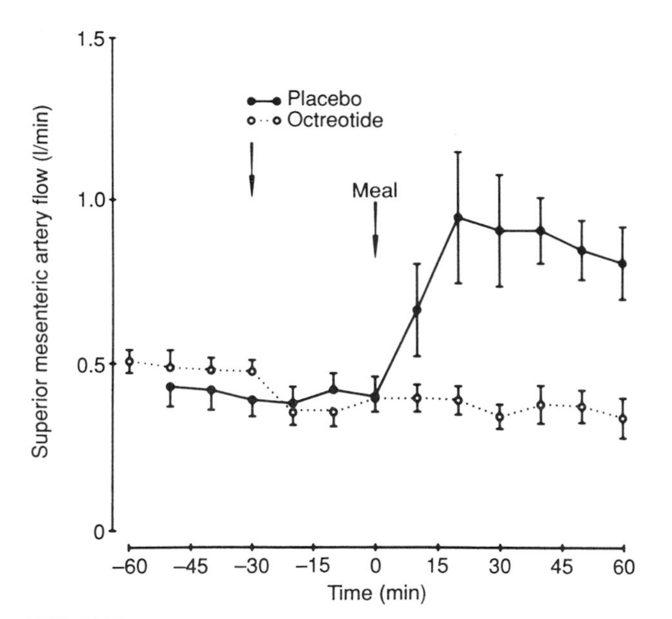

FIGURE 7 Superior mesenteric artery blood flow measured by a noninvasive doppler technique in normal subjects before and after a balanced liquid meal, given either before saline placebo (●) or after 50 mcg of the somatostatin analog octreotide (□), both subcutaneously. The rise in flow postmeal is prevented by the peptide release inhibitor octreotide. From Kooner *et al.* (1989).

ering blood pressure. When antihypertensive drugs were not available in the 1930s and when the prognosis in severe hypotension was extremely poor, some of the initial surgical approaches involved drastic surgical operations involving sections of the splanchnic (autonomic) nerves to the gut; this was, at times, successful. A number of elegant approaches (including sympathetic microneurography) have studied the sympathetic nervous system in hypertensives. Some of these have provided novel information on abnormalities of neurogenic control, even in unusual forms of hypertension such as those complicating pregnancy (preeclamptic toxemia) and when hypertension is induced by the immunosuppressant drug cyclosporin, which is used in cardiac and other organ transplantees. [*See* Hypertension.]

Neurogenic hypertension has been well documented in tetraplegics with cervical or high thoracic spinal cord transection. In such patients the baroreceptor pathways to the brain are intact, as are the vagal efferent pathways; however, the lesion results in disruption of the descending sympathetic pathways. The spinal cord beneath the transection is functionally viable and may operate in an isolated manner, independent of the brain. Stimuli that would normally be uncomfortable or painful (such as an overfull urinary bladder or a decubitus ulcer) can trigger abnormal reflexes at an isolated spinal level, raising blood pressure substantially, often with abnormal autonomic activation of other visceral organs. This syndrome of autonomic dysreflexia (Fig. 9) can be a serious clinical problem, as intracranial hemorrhage may occur as a complication of the severe hypertension. The study of such patients has also emphasized other aspects concerning the plasticity of these neurocirculatory systems. The elevation in blood pressure with autonomic dysreflexia is associated with a significant but small rise in plasma noradrenaline that indicates sympathetic neural activation; this has been confirmed using sympathetic microneurography. The markedly hypertensive response, however, raises the question of supersensitivity. This may in part be due to the inability of brain-directed reflexes to control the distal autonomic outflow, thus lowering blood pressure; also, adrenoceptors on target organs may have either increased in number or may be functionally more effective, thus producing a greater response. These adaptive aspects linked with abnormalities of neurocirculatory control are of clinical importance; with sympathetic denervation, substantially smaller doses of drugs to raise (or lower) blood pressure are needed to avoid excessive hypertension (or hypotension) (Fig. 10).

IV. HEART RATE CONTROL ABNORMALITIES

A. Low Heart Rate

The neural control of heart rate is predominantly through the vagus; increased parasympathetic activity can therefore result in bradycardia or even cardiac arrest, as may occur in certain situations. An example of the former is vasovagal syncope, also referred to as simple faints, which often affect the young, causing intermittent fainting. This is due to a fall in heart rate (resulting from increased cardiac vagal activity) that is often accompanied by a fall in blood pressure (Fig. 11). Bradycardia can be prevented by the use of atropine, which blocks the effects of acetylcholine on cardiac muscarinic receptors, or by a cardiac pacemaker that directly maintains rate. The hypotension may result from the withdrawal of sympathetic tone, as indicated by sympathetic microneurography. In the elderly, an allied condition, although more serious, is carotid sinus hypersensitivity; it also can result in

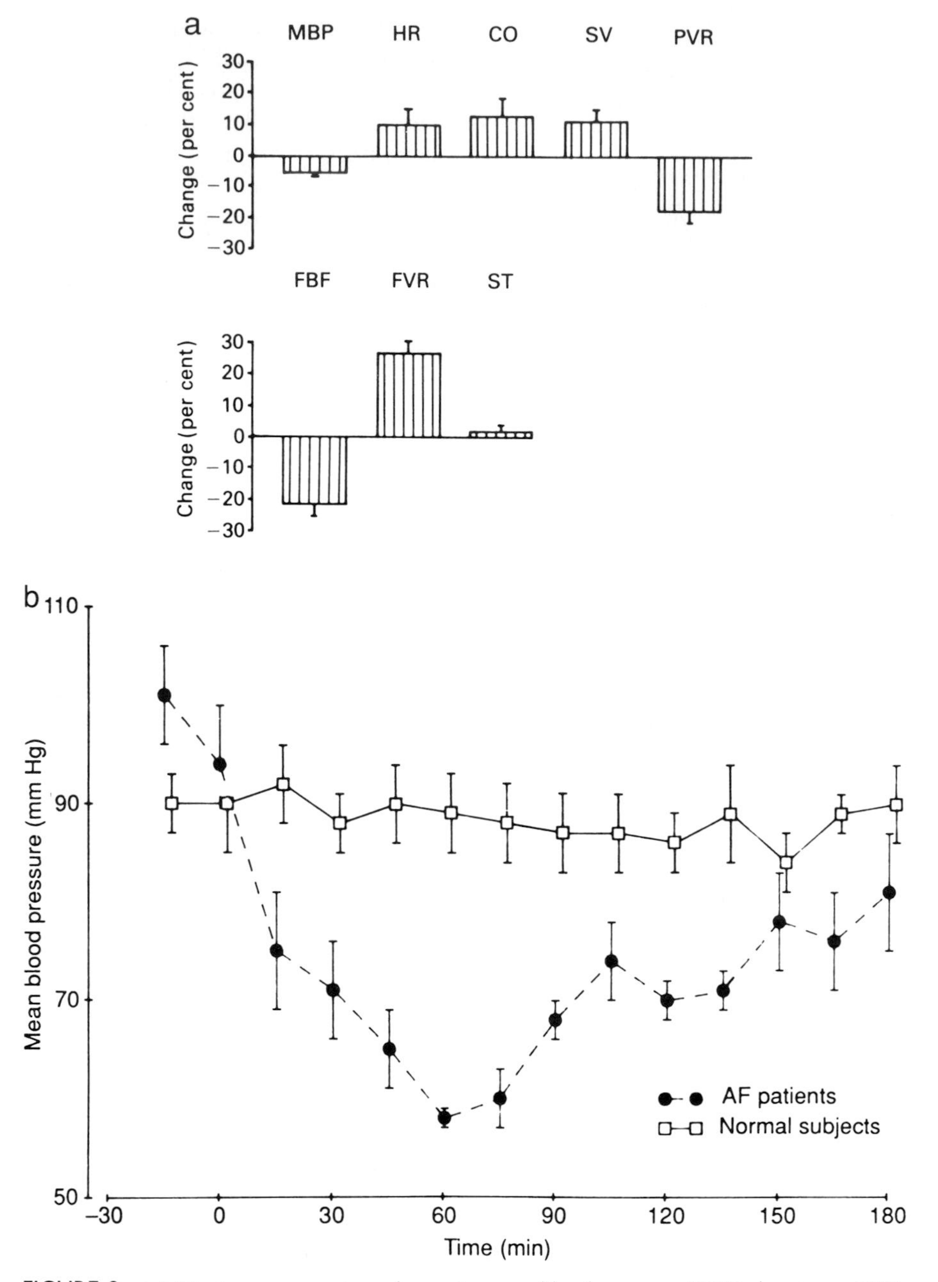

FIGURE 8 (a) Maximum percentage change in mean blood pressure (MBP), heart rate (HR), cardiac outflow (CO), stroke volume (SV), calculated peripheral vascular resistance (PVR), forearm muscle blood flow (FBF), calculated forearm vascular resistance (FVR), and skin temperature to the index finger (ST) in normal subjects in the first hour after food ingestion while lying supine and horizontal. Vertical bars indicate ± SEM. From Bannister and Mathias (1992). (b) Percentage change in mean blood pressure in a group of patients with sympathetic denervation (AF) (●) and in normal subjects (□) before and after food ingestion at time "0". All the recordings were made with the subjects supine, thus excluding gravitational effects. The bars indicate means ± SEM. From Mathias *et al.* (1989).

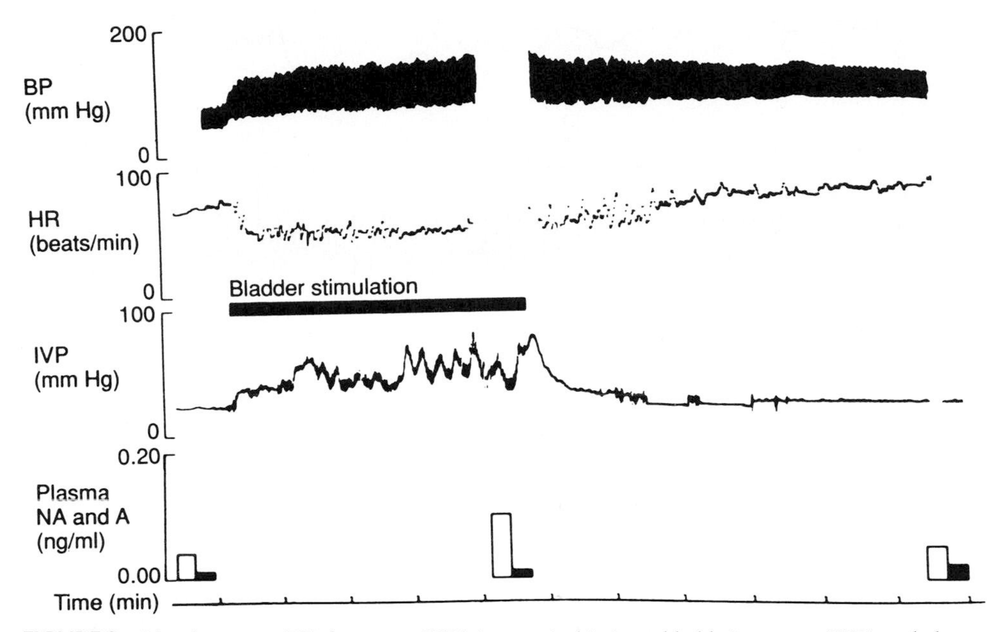

FIGURE 9 Blood pressure (BP), heart rate (HR), intravesical (urinary bladder) pressure (IVP), and plasma noradrenaline (NA, open histograms) and adrenaline (A, filled histograms) in a tetraplegic patient before, during, and after bladder stimulation induced by suprapubic percussion of the anterior abdominal wall. This causes bladder contraction, as shown by the increase in IVP. The rise in BP is accompanied by a fall in HR as a result of increased vagal activity in response to the BP rise. Plasma NA but not A levels rise, suggesting an increase in sympathetic neural activity independently of adrenal medullary activation. From Mathias and Frankel (1986).

severe bradycardia and even cardiac arrest. This can occur with relatively minimal stimulation of the carotid sinus nerves, such as by moving the neck or buttoning one's collar.

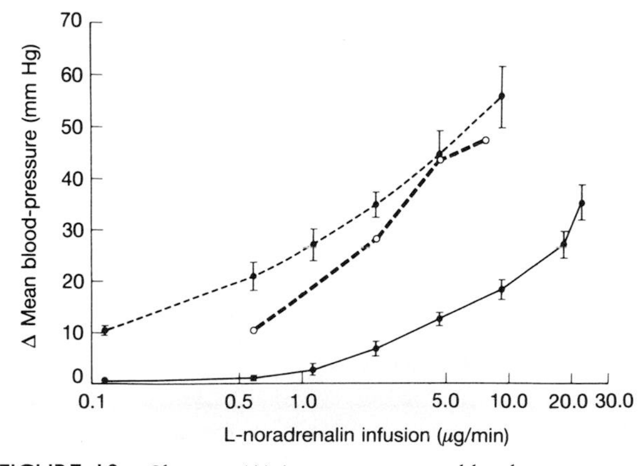

FIGURE 10 Changes (Δ) in average mean blood pressure and heart rate during different dose infusion rates of noradrenaline in 5 chronic tetraplegics (●---●), 3 recently injured tetraplegics (○---○), and 10 normal subjects (●—●). Bars indicate ± SEM. There is an enhanced pressor response to noradrenaline in both groups of tetraplegics over the entire dose range studied. From Mathias *et al.* (1976, 1979).

B. Raised Heart Rate

An elevated heart rate (tachycardia) may result either from withdrawal of vagal tone or by activation of the cardiac sympathetic. Vagal impairment is often seen in autonomic neuropathy complicating diabetes mellitus; the heart rate may be in the region of 100 or 110 bpm and is often "fixed" with little change following maneuvers such as deep breathing, which normally increase heart rate variability (Fig. 12). With increased sympathetic activation (flight or fright), tachycardia may occur; this may also be due to the excessive release of adrenaline as in tumors of the adrenal gland and allied tissue (pheochromocytoma).

V. REGIONAL VASCULAR CONTROL

Each organ has specific perfusion demands that are related to its functions. Regional vascular control is dependent not only on its nerve supply, but usually also on various locally generated substances; these may include the classical neurotransmitters (acetylcholine and noradrenaline) and also a wide range of nonadrenergic, noncholinergic substances, from pu-

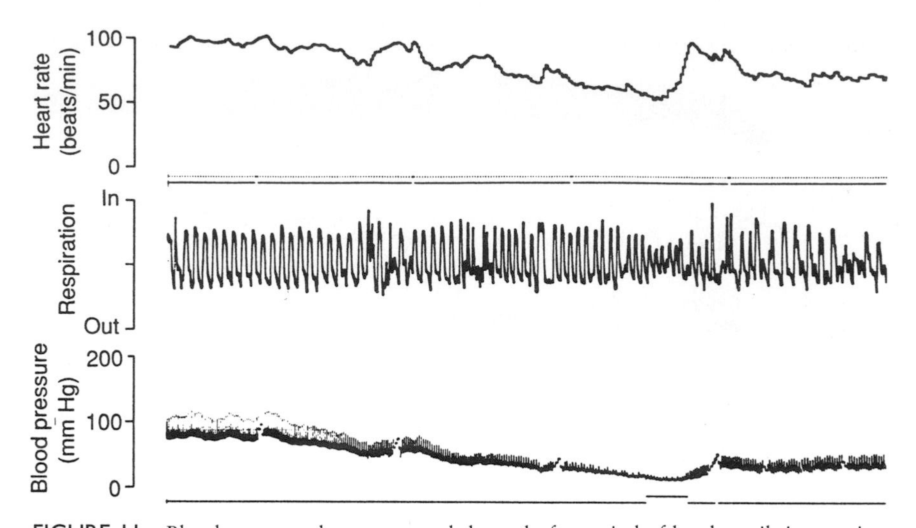

FIGURE 11 Blood pressure changes toward the end of a period of head-up tilt in a patient with recurrent episodes of fainting (syncope). There was no evidence of autonomic failure on testing. Blood pressure, which was previously maintained during head-up tilt, begins to fall. There is also a fall in heart rate. Initially there are relatively minor changes in respiratory rate. Each minor dot, indicated above the respiratory trace, indicates a second whereas bolder marks indicate a minute. When the blood pressure fell to extremely low levels and she was about to faint, the patient was put back to the horizontal (as indicated by the time signal below). Blood pressure and heart rate slowly recovered but still remained lower than previously. Measurements of BP and heart rate were assessed noninvasively by finger plethysmography (Finapres). From Bannister and Mathias (1992).

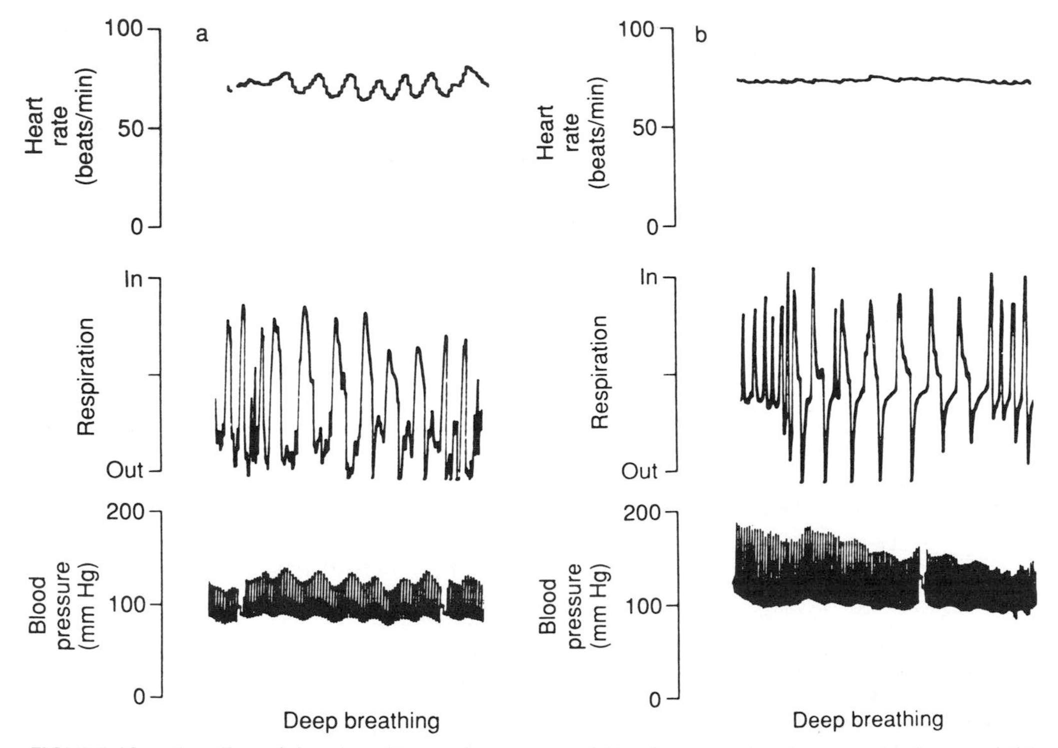

FIGURE 12 The effect of deep breathing on heart rate and blood pressure in (a) a normal subject and (b) a patient with autonomic failure resulting in cardiac parasympathetic impairment. There is no sinus arrhythmia in the patient, despite a fall in blood pressure. Respiratory changes are indicated in the middle. From Bannister and Mathias (1992).

rines such as adenosine triphosphate (ATP), amines such as 5-hydroxytryptamine (5-HT), peptides such as NPY and endothelin, and nitric oxide. In the cerebral circulation, cholinergic vasodilator mechanisms may predominate in conjunction with others such as the 5-HT system; abnormalities may result in conditions such as migraine. In the penile circulation, locally produced nitric oxide may play a major role; there is a reduction in nitric oxide in diabetics with erectile failure. In the gastrointestinal tract, a variety of pancreatic and gut peptides have vascular and also neuromodulatory effects. The release of some of these can be prevented by the somatostatin analog octreotide and explains its ability in preventing the postprandial rise in splanchnic blood flow (Fig. 8a). In skin, the peptides, calcitonin gene-related peptide, and substance P are involved in vasodilatation; this may be mediated by local (axon) reflexes. Regional circulatory control, therefore, is exerted in a multiplicity of ways, with a close interaction between the nervous system and other locally produced vasoactive or neuromodulatory substances.

VI. CONCLUSION

The autonomic nervous system thus plays a key role in the beat-by-beat regulation of systemic blood pressure; it also is involved in the control of heart rate and cardiac function and in the appropriate redistribution of blood to different vascular regions, thus influencing the perfusion of various organs, depending on their functional needs. The nervous system, therefore, maximizes the efficiency of the circulation in a variety of situations and, importantly, enables rapid adaptation when needed.

BIBLIOGRAPHY

Bannister, R., and Mathias, C. J. (eds.) (1992). "Autonomic Failure: A Textbook of Clinical Disorders of the Autonomic Nervous System," 3rd Ed. Oxford University Press, Oxford.

Bennett, T., and Gardiner, S. M. (eds.) (1996). "Nervous Control of Blood Vessels." Harwood Academic Publishers, The Netherlands.

Kooner, J., Peart, W. S., and Mathias, C. J. (1989). The peptide release inhibitor octreotide (SMS201-995) prevents the haemodynamic changes following food ingestion in normal human subjects. *Q.J. Exp. Physiol.* **74**, 569–572.

Loewy, A. D., and Spyer, K. M. (eds.) (1990). "Central Regulation and Autonomic Function." Oxford University Press, New York/Oxford.

Mathias, C. J. (1996a). Disorders of the autonomic nervous system. *In* "Neurology in Clinical Practice" (W. G. Bradley, R. B. Daroff, G. M. Fenichel, and C. D. Marsden, eds.), 2nd Ed., Vol. 82, pp. 1953–1981. Butterworth-Heinemann, Boston.

Mathias, C. J. (1996b). Disorders affecting autonomic function in parkinsonian patients. *In* "Advances in Neurology" (L. Battistin, G. Scarlato, T. Caraceni, and O. Ruggieri, eds.), Vol. 69, pp. 383–391. Lippincott & Raven, Philadelphia.

Mathias, C. J., Christensen, N. J., Frankel, H. L., and Spalding, J. M. K. (1979). Cardiovascular control in recently injured tetraplegics in spinal shock. *Q.J. Med.* **48**, 273–287.

Mathias, C. J., da Costa, D. F., Fosbraey, P., Bannister, R., Wood, S. M., Bloom, S. R., and Christensen, N. J. (1989). Cardiovascular, biochemical and hormonal changes during food-induced hypotension in chronic autonomic failure. *J. Neurol. Sci.* **94**, 255–269.

Mathias, C. J., and Frankel, H. L. (1986). The neurological and hormonal control of blood vessels in heart and spinal man. *J. Autonom. Nerv. Syst.* **Suppl.**, 457–464.

Mathias, C. J., and Frankel, H. L. (1988). Cardiovascular control in spinal man. *Annu. Rev. Physiol.* **50**, 577–592.

Mathias, C. J., Frankel, H. L., Christensen, N. J., and Spalding, J. M. K. (1976). Enhanced pressor response to noradrenaline in patients with cervical spinal cord transection. *Brain* **99**, 755–770.

Mathias, C. J., and Kooner, J. S. (1989). Neurogenic aspects of blood pressure control in man: With an emphasis on renovascular hypertension. *In* "Concepts in Hypertension—Festschrift for Sir Stanley Peart" (C. J. Mathias and P. S. Sever, eds.), pp. 67–92. Steinkopff Verlag, Darmstadt.

Rowell, L. (1992). "Human Cardiovascular Control." Oxford University Press, New York/Oxford.

Smith, G. D. P., Watson, L. P., and Mathias, C. J. (1996). Neurohumoral, peptidergic and biochemical responses to supine exercise in two groups with primary autonomic failure: Shy-Daager syndrome/multiple system atrophy and pure autonomic failure. *Clin. Autonom. Res.* **6**, 255–262.

Wallin, G., *et al.* (1980). Baroflex mechanisms controlling sympathetic outflow to the muscles in man. *In* "Arterial Baroreceptors and Hypertension" (P. Sleight, ed.). Oxford University Press, Oxford.

Neurology of Memory

LARRY R. SQUIRE
University of California, San Diego and Veterans Affairs Medical Center

GLOSSARY

Anterograde amnesia Loss of the ability to learn

Diencephalic midline Area of the diencephalon of the brain surrounding the ventricles, including the medial thalamus and hypothalamus

Magnetic resonance imaging Noninvasive technique for visualizing the structure of living brain

Medial temporal lobe Inner area of the temporal lobe away from the lateral surface, including the hippocampal formation, amygdala, parahippocampal gyrus, entorhinal cortex, and perirhinal cortex

Neocortex More recently evolved outer region of the brain, which occurs as a large thin sheet of highly infolded tissue

LEARNING IS THE PROCESS OF ACQUIRING NEW information, and memory refers to the persistence of learning in a state that can be revealed at some time after learning is completed. Memory is localized in the brain as physical changes produced by experience. Damage to particular brain regions causes amnesia (i.e., loss of the ability to acquire new information). A major principle of the organization of memory in the brain is the distinction between conscious and nonconscious memory systems. The ability to store and use conscious memories depends on the integrity of the hippocampus and adjacent, anatomically related structures.

I. MEMORY AS A TOPIC FOR NEUROSCIENCE AND PSYCHOLOGY

Prior to the technological developments of neuroscience, memory was a problem studied primarily by psychologists. In the past few decades, however, the successes of basic neuroscience have made it possible to study the nervous system in increasing detail and to obtain information about cellular and molecular events within single neurons. It has thus become possible to investigate memory at many levels of analysis, from the analysis of synaptic plasticity to the analysis of brain systems and cognition.

The neurology of memory can be advantageously investigated by combining the approaches of neuroscience and psychology. This is because many of the important questions about the neurology of memory are directed at a relatively broad, global level of analysis, in which the subject matter of neuroscience and psychology significantly overlap. What are the learning processes and memory systems whose neurobiological mechanisms we want to understand? Is there just one kind of memory or are there many? How and where are memories represented? What are the brain systems and connections involved in memory, and what jobs do they do?

II. MEMORY STORAGE

The brain is highly specialized and differentiated, organized so that separate regions of neocortex simultaneously carry out computations on separate features or dimensions of the external world (e.g., the analysis of visual patterns, location, and movement). Although the evidence is not yet definitive, memory for an event, or even for something as apparently simple as a single visual object, is considered to be stored in component parts, within the same specific processing areas that are ordinarily engaged during perception. That is, memory is stored in the same neural systems that ordinarily participate in perception, analysis, and processing. Stated differently, memories are stored as outcomes of processing operations and in the same cortical regions that are involved in the analysis and perception of the items and events to be remembered. This leads to the idea that memory is *localized* in the sense that particular brain systems represent specific aspects of each event, and it is *distributed* in the sense that many neural systems must participate in representing a whole event. [*See* Learning and Memory.]

III. MEMORY DYSFUNCTION

A. Anterograde Amnesia

Having emphasized this close association between information processing and information storage, it is important to appreciate that there are brain structures and connections that are not themselves permanent repositories of information but that nevertheless play an essential role in the formation and establishment of enduring memory. These structures lie within the medial temporal lobe and the diencephalic midline. Bilateral damage in either of these regions causes an amnesic syndrome (i.e., a global impairment in the ability to acquire new memories regardless of modality, and a loss of some memories, especially recent ones, from the period before amnesia began). The memory deficit occurs in the absence of deficits in perception or other intellectual functions.

The fact that circumscribed amnesia can occur at all indicates directly that memory storage is not inextricably linked to the cognitive processing that ultimately leads to memory. If it were, amnesia could not occur as it does against a background of intact intellectual functions, intact immediate memory, intact personality and social skills, and intact ability to recall memories acquired long ago., The point is that there are brain structures that when damaged cause amnesia and that are especially important for memory functions.

The deficit in amnesia is easily detected with tests of paired-associate learning or delayed recall. For paired-associate learning, one presents pairs of unrelated words (e.g., army—table) and later asks for recall of the second word of each pair when the first word of the pair is presented. For delayed recall, one presents information to be remembered (e.g., a passage of text, a list of words, or geometric designs) and, after a delay of several minutes, asks the subject to reproduce the originally presented material. The deficit is present regardless of the sensory modality in which information is presented and regardless of whether memory of a prior occurrence is tested by unaided recall, recognition (e.g., presenting alternatives and asking the subject to choose the previously encountered one), or cued recall (e.g., asking the subject to recall an item when a hint is provided). Moreover, the memory impairment involves not just difficulty in learning about specific episodes and events that occur in a certain time and place, but also difficulty in learning factual information.

B. Preserved Learning

It is now appreciated that memory is not a unitary mental faculty but depends on the operation of many separate systems. Some of the most compelling evidence for this idea is the finding that amnesia spares certain important kinds of learning and memory. Among the kinds of learning that can be accomplished *normally* by amnesic patients are perceptuomotor skill learning, word priming, adaptation-level effects, simple classical conditioning, habit learning, and prototype learning. Perceptuomotor skills refer to hand–eye coordination skills (e.g., learning to trace a figure that has been reversed in a mirror). Amnesic patients can learn to read mirror-reversed text at a normal rate, although they do not later recollect the words that they read. In other words, they learn the skill but not the material that was used in the learning of the skill.

Word priming refers to a change in the facility for identifying specific words, which is caused by recent exposure to the words. For example, if the word *baby* is presented, the probability is more than doubled for both normal subjects and amnesic patients that this word (*baby*) will later be elicited by instructions to free-associate a single response to the word *child*. Similarly, presentation of the word *income* produces a strong tendency (about 50%) to produce the word again if the word stem *inc* is given along with instructions to complete the stem with the first word that comes to mind. (The probability that *income* will be produced when it was not presented in the first place is about 10%.) These priming effects occur normally in amnesic patients despite the fact that the patients fail conventional memory tasks that ask them to recall recently presented words or to recognize them as familiar. One view of priming is that these effects operate at a relatively early stage in the analysis of information.

Adaptation-level effects refer to changes in judgments about stimuli (e.g., their heaviness or size) that are caused by recent exposure to stimuli of particular qualities. For example, experience with light-weighted stimuli subsequently causes new stimuli to be judged as heavier than if the lighter stimuli had not just been encountered.

Simple classical conditioning refers to the development of a conditioned response to a previously neutral stimulus as a result of its having been temporally paired with an unconditioned stimulus. For example, after temporal pairing of a tone and an air puff, which elicits an eyeblink, the tone itself will come to elicit the eyeblink. Habit learning refers to the gradual learning of reinforcement contingencies. Often, such tasks can be solved by memorizing the structure of the task, and amnesic patients are disadvantaged in comparison to normal subjects. However, tasks can be constructed that resist memorization, for example, when the outcomes are probabilistic. In such a case, amnesic patients and normal subjects learn at the same gradual rate. Prototype learning refers to the ability to acquire information above averageness or typicality by repeated experience with examples. For example, through experience with individual instances, normal subjects acquire information that defines category membership (e.g., categories like birds, chairs, and trees), such that they can accurately classify novel instances. Amnesic patients appear able to acquire category-level knowledge as well as normal subjects although they are severely impaired at recollecting the specific examples they have encountered.

The kind of memory that is affected in amnesia is explicit and accessible to conscious memory. It has sometimes been termed *declarative memory* (also termed *explicit*) because it can be declared (i.e., brought to mind consciously as a proposition or an image). Declarative memory thus includes the facts, events, faces, and routes of everyday life that comprise conventional memory experiments. By contrast, *nondeclarative memory* (also termed *implicit memory*) comprises a heterogeneous collection of learning and memory abilities, all of which afford the capacity to acquire information implicitly and nonconsicously. For example, in the case of motor skill tasks and perceptual skill tasks (e.g., mirror reading), the knowledge that is acquired cannot be declared. Thus, we do not know what we have learned when we demonstrate a new tennis backhand, at least not in the same sense that we remember the practice sessions themselves. Knowledge of the skill is embedded in procedures and is expressed in performance engaged by the procedures. Skill learning does not require the integrity of the brain structures damaged in amnesia. In the case of priming, already existing cognitive operations are tuned or biased and for a time thereafter can facilitate behavior in a specific way. Thus, the brain has organized its memory functions around fundamentally different information storage systems, some of which are impaired in amnesia and some of which are not (Fig. 1).

C. Retrograde Amnesia

Additional information about the organization of declarative memory comes from the phenomenon of *retrograde amnesia*, which refers to the inability to remember information that occurred before the onset of amnesia. Retrograde amnesia affects both facts and episodes, particularly those that were encountered close to the time when amnesia began. It can be relatively brief or quite extensive, but it is usually temporally graded. That is, very old (remote) memory is affected less than recent memory. For example, in a middle-aged adult, retrograde amnesia can affect memories from the two or three decades preceding the onset of amnesia and leave relatively intact the memories of childhood and adolescence. Because very old memories are intact, the brain regions damaged in amnesia cannot be the permanent repository of declarative memory. The critical structures damaged in amnesia perform a time-limited function. For a period of time after learning, the storage of declarative memory and its retrieval depend on an interaction

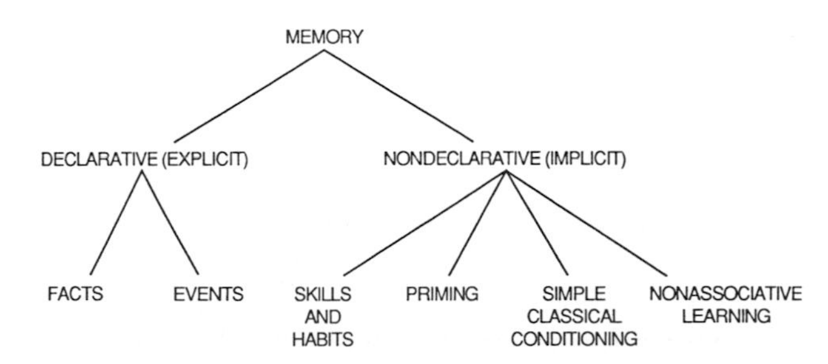

FIGURE 1 Classification of memory. Declarative (explicit) memory refers to conscious recollections of facts and events and depends on the integrity of the medial temporal lobe (see text). Nondeclarative (implicit) memory refers to a collection of abilities and is independent of the medial temporal lobe. Nonassociative learning includes habituation and sensitization. In the case of nondeclarative memory, experience alters behavior nonconsciously without providing access to any memory content.

between the neural systems damaged in amnesia and memory storage sites located elsewhere, presumably in neocortex. After sufficient time has passed, memories no longer require the participation of these structures. Memories appear to undergo some lengthy process of reorganization and consolidation within neocortex whereby they become independent of the structures damaged in amnesia.

IV. ANATOMY OF MEMORY

Information about which neural connections and structures belong to the functional system damaged in amnesia comes from two sources: well-studied cases of human amnesia and recent successes at developing an animal model of amnesia in the monkey. New techniques for imaging living brain have also made it possible to detect some kinds of pathological change directly. For example, with magnetic resonance imaging it is possible to detect abnormalities in the hippocampal formation of some amnesic patients. In one carefully studied case of amnesia (R.B.), in which both behavioral and neuropathological data were available, the only significant damage was a bilateral lesion confined to the CA1 field of the hippocampus. Thus damage limited to the hippocampus itself is sufficient to cause amnesia. These findings show that the hippocampus is an essential component of the neural system necessary for establishing declarative memory. Cumulative work with animal models suggests that the full medial temporal lobe memory system consists of the hippocampus and adjacent, anatomically related structures, including entorhinal cortex, parahippocampal gyrus, and perirhinal cortex (Fig. 2). When several of these structures are damaged together, the severity of amnesia is greater than when only the hippocampus itself is damaged [*See* Hippocampal Formation.]

The critical regions in the diencephalon that when damaged produce amnesia have not yet been identified with certainty. Most likely, the important structures include the mediodorsal thalamic neucleus, the anterior nucleus, the internal medullary lamina, the mammillothalamic tract, and the mammillary nuclei. Damage to the mammillary nuclei alone seems not to be sufficient to produce severe and long-lasting amnesia. Because diencephalic amnesia resembles medial lobe temporal amnesia in many ways, these two regions together probably form an anatomically linked, functional system. It is also likely that the two regions contribute in different ways to the functional system.

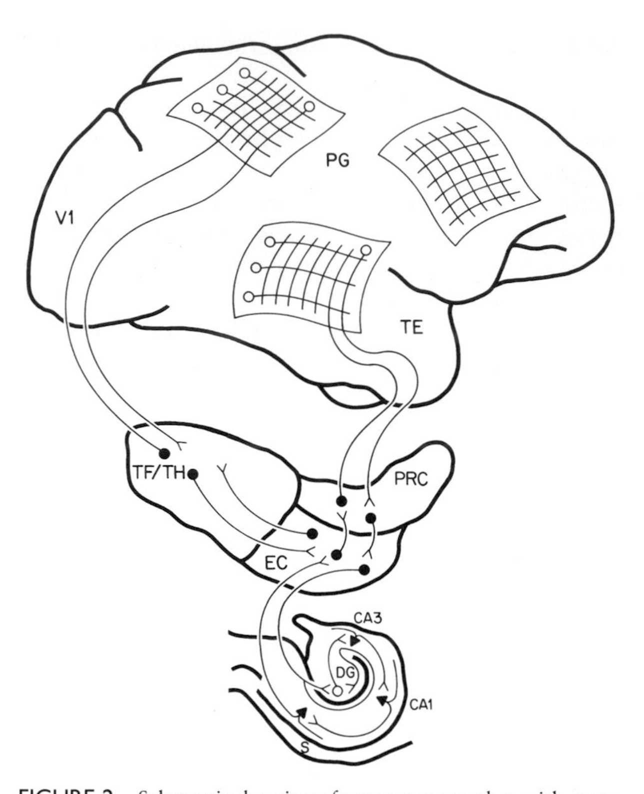

FIGURE 2 Schematic drawing of neocortex together with structures and connections in the medial temporal region believed to be important in the transformation of perceptions into memories. The networks in cortex show putative representations concerning visual object quality (in area TE) and object location (in area PG). If this disparate neural activity is to cohere into a stable long-term memory, convergent activity must occur along projections from these regions to the medial temporal lobe. Projections from neocortex arrive initially at the parahippocampal gyrus (TF/TH) and perirhinal cortex (PRC) and then at entorhinal cortex (EC), the gateway to the hippocampus. Further processing of information occurs in the several stages of the hippocampus, first in the dentate gyrus (DG) and then in the CA3 and CA1 regions. The fully processed input eventually exits this circuit by way of the subiculum (S) and the entorhinal cortex, where widespread efferent projections return to neocortex. The hippocampus and adjacent structures appear to bind together or otherwise support the development of representations in neocortex, so that subsequently memory for a whole event (e.g., representations in both TE and PG) can be revivified even from a partial cue. Damage to this medial temporal system causes anterograde and retrograde amnesia. The severity of the deficit increases as damage involves more components of the system. (This diagram is a simplification and does not show, e.g., the diencephalic structures involved in memory functions.)

Information is still accumulating about how memory is organized and what structures and connections are involved. The disciplines of both psychology and neuroscience continue to contribute to this enterprise. Better understanding of the neurology of memory can be expected to result eventually in quantitative approaches that can provide computer-based models and tests of specific mechanisms. In addition, information obtained at this level of analysis should be directly relevant to neurobiological studies of learning and memory that seek cellular and molecular information. For example, information about the brain systems involved in memory can suggest where to look for cellular and molecular changes and can indicate the functional importance of these changes. Finally, a fundamental understanding of the neurological foundations of memory should lead to better diagnosis, treatments, and prevention of neurological diseases that affect memory.

BIBLIOGRAPHY

Squire, L. R. (1987). "Memory and Brain." Oxford Univ. Press, New York.

Squire, L. R., and Zola-Morgan, S. (1991). The medial temporal lobe memory system. *Science,* **253,** 1380–1386.

Squire, L. R., Knowlton, B., and Musen, G. (1993). The structure and organization of memory. *Ann. Rev. Psychol.* **44,** 453–495.

Squire, L. R. (1995). Biological foundations of accuracy and inaccuracy in memory. *In* "Memory Distortion: How Minds, Brains, and Societies Reconstruct the Past," (D. L. Schacter, J. T. Coyle, G. D. Fischbach, M.-M. Mesulam, and L. E. Sullivan, eds.). Harvard University Press, Cambridge.

Neuropharmacology

ANTHONY DICKENSON
University College London

GLOSSARY

Agonist Chemical that binds to a receptor, initiating a biological response

Antagonist Chemical that binds to a receptor and does not elict a response, but blocks the effects of an agonist

Neurotransmitter Naturally occurring chemical that is released from the terminal into the synaptic gap and acts as a receptor agonist

Receptor Specific binding site on a cell onto which drugs attach to evoke a response (i.e., agonists) or to prevent access of the agonist (i.e., antagonists)

Synapse Gap between a terminal of a neuron and the postsynaptic cell

NEUROPHARMACOLOGY IS THE STUDY OF THE ACtions of drugs on nervous tissue, whether the nervous tissue is the brain or the nerves within the body. The drugs can be the synthetic compounds used for medical or recreational purposes or the natural chemicals that influence nervous tissue as part of the normal functioning of the body. Many of the synthetic drugs given to patients are part of therapies attempting to counter disorders of the natural chemicals. It is obvious, then, that two major challenges of neuropharmacology arise from these dual aspects: (1) increasing our knowledge of neuropharmacology will inevitably lead to a better understanding of how the nervous system functions in health and disease and (2) therapies or improvements on current therapies will surely follow. Perhaps it is best to put aside the paradox that we are using the neuropharmacological systems within the brain in studying these very systems and to consider instead that chemicals are released from cells in the nervous system and, by diffusing to other cells, they change activity in parts of the brain; the integrated whole of these events somehow results in human consciousness.

I. CHEMICAL TRANSMISSION

It is worth considering first the ways in which nerve cells communicate with each other. The electrical signals, or action potentials, are generated in nerve cells (i.e., neurons) by changes in ionic balance across the membrane of the cell. These action potentials travel from the cell body along the extended process of the cell (i.e., the axon) until they arrive at the specialized end of the axon: the terminal (Fig. 1). The terminal contacts another cell by coming into close proximity to it. It is estimated that a cell can be contacted by several thousand others, as the brain contains several billion neurons. There is, however, no direct contact, and a gap (the synaptic cleft, which is about 20 nm across) lies between the presynaptic terminal and the postsynaptic cell. Synapses can also be made on other parts of the neurons besides the cell body (i.e., the dendrites, terminals, and axons of neurons). In certain lowly invertebrates a cell communicates with the next neuron in the line by direct electrical events. In higher animals the action potential is transferred via release across the synaptic cleft of a chemical, a process known as chemical transmission. The arrival of the action potential in the terminal depolarizes the membrane, calcium ions flow into the cell, and then a chemical, the neurotransmitter, is released from the

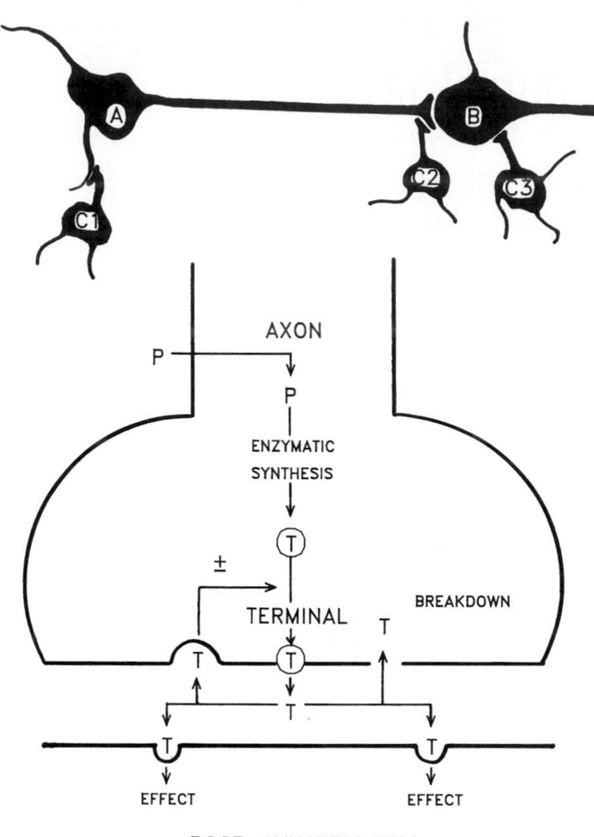

FIGURE 1 The processes of synaptic transmission. The upper portion shows a neuron A contacting the cell body of neuron B via a synapse made by the terminal of the axon of neuron A. C1 is another neuron making a synapse on the dendrites of cell A, C2 synapses onto the terminal of neuron A, and C3 is yet another cell contacting neuron B. Thus, chemical transmission can occur at all of these sites. The lower portion shows an expanded scheme of what happens at a synapse. The precursor (P) is taken up by the cell and converted enzymatically to the transmitter (T), which is then released into the synapse. Effects are produced by binding to receptors in the postsynaptic cells. Autoreceptors on the presynaptic terminal can increase or, more commonly, decrease (±) the transmitter release mechanism. Finally, the transmitter is often taken back up by the terminal prior to breakdown or recycling.

presynaptic terminals into the synapse. Neurotransmitters are enclosed in vesicles, and fusion of the vesicle with the membrane of the terminal releases the transmitter. The transmitter then diffuses across the synapse and binds to a specialized part of the postsynaptic membrane: the receptor. This binding process initiates a chain of events in the postsynaptic cell that has been likened to a lock-and-key concept, and the message passes on in the form of another action potential.

The neurotransmitter is synthesized in the neurons from either a simple dietary precursor or a large precursor produced in the cell body in the case of the peptide transmitters. The synthesis is enzymatic and usually consists of a few steps. The active transmitter so produced is then available for release (see Fig. 1).

There are many different chemicals acting as transmitters and a greater number of receptors. A transmitter can produce different effects, depending on the receptor type it diffuses to. There is, however, a high degree of specificity, so a transmitter chemical can activate only a limited number of receptor types, in many cases only one or two. The transmitter is then rapidly removed from the synapse or broken down, so turning off the message. The various transmitters used by the nervous system are synthesized in the neuron, from which they are released. Although it was once thought that one neuron contained one transmitter, more than one transmitter per cell now appears to be the rule, a sort of dual-key approach. It is this chemical transmission that allows us to experience the outside world and to respond to our environment in terms of movement, memory, or introspective thought and that forms the basis for nervous system functions ranging from the involuntary control of heart rate to abstract thought. However, it is obvious that higher mental functions have a complexity far beyond our present knowledge, a case in which the relationship between the sum and the parts is hazy. Nevertheless, our knowledge of neuropharmacology provides a good basis for the understanding of neuronal function and a better rationale for therapy.

The receptor for the neurotransmitter is a key target site for synthetic drugs used in clinical medicine. The neurotransmitters bind to their receptor and act as agonists (e.g., they produce an effect by activation of the receptor). This effect, however, can be excitatory (the production of action potentials in the postsynaptic cell), inhibitory (the reduced firing of action potentials), or something between these two extremes (a gentle shift in the excitability of the neuron, making it more or less susceptible to other inputs). These subtle effects clearly allow flexibility of function within the brain. [*See* Neurotransmitter and Neuropeptide Receptors in the Brain.]

In addition to these postsynaptic receptors, there are receptors in the presynaptic terminals, so that one transmitter can regulate the amount of another transmitter released, as well as autoreceptors on either the terminal or cell body that allow a transmitter to control its own release. Certain clinically used drugs (e.g., morphine) are agonists; that is, they bind and produce a response at receptors, so enhancing the function of the natural neurotransmitter. Other drugs, however (e.g., those used for schizophrenia), are antagonists. Antagonists bind to the receptor, but produce no activation and also prevent an agonist from binding to the receptor. They therefore block the action of a neurotransmitter. Another target for drugs is the process of removal of the transmitter from the synapse. This removal can occur by reuptake by the terminal or by the breakdown of the transmitter by specific enzymes present in the synaptic cleft. Drugs used to treat depression interfere with transmitter removal processes. Reuptake is most important in the central nervous system, and the transmitter once taken up by the terminal is broken down and becomes available for new synthesis (see Fig. 1).

Recently, it has become clear that neurotransmitters, via receptor-mediated events, can elicit profound changes in target neurons, including influencing genes. Switching genes on and off can cause chemical and even structural changes in neurons that can persist for long time spans. In addition, novel transmitters, including the gas nitric oxide (NO), can be produced following receptor activation and internal neuronal events, and by diffusion, NO can alter adjacent neuronal function. New therapies may arise.

What, then, are the transmitters in the nervous system? It is best to consider the nervous system in two parts, although both influence each other: the central nervous system (i.e., the brain and the spinal cord) and the autonomic nervous system, the controlling system for the unconscious workings of the organs of the body. With the exception of the control of muscles for movement, the autonomic nervous system is the major output path of the central nervous system. [*See* Autonomic Nervous System.]

II. AUTONOMIC NERVOUS SYSTEM

As discussed earlier, the central nervous system is balanced by the activity of inhibitory and excitatory neuronal systems and so, too, is the autonomic nervous system. Two branches—the sympathetic and the parasympathetic—exist, which act in concert to control the body. In both cases the cell body of a first neuron is contained within the central nervous system, and this neuron synapses onto a second cell outside the central nervous system: the ganglionic cell. This second cell then sends an axon to the target organ. Hence, the two neurons involved are termed the pre- and postganglionic cells. The ganglionic cells for both systems use acetylcholine as the transmitter, and the released acetylcholine acts on an acetylcholinergic receptor, termed the nicotinic receptor. However, the postganglionic cells of the parasympathetic nervous system use acetylcholine again, but noradrenaline is the transmitter in the sympathetic nervous system. The noradrenaline acts on a variety of noradrenergic receptors, and the acetylcholine released from the postganglionic cell binds to a different cholinergic receptor: the muscarinic receptor. [*See* Adrenergic and Related G Protein-Coupled Receptors.]

The parasympathetic and sympathetic nervous systems innervate the organs of the body (e.g., the heart, lungs, gastrointestinal tract, blood vessels, and glands). The aim of this article is to concentrate on the central nervous system, yet from this account of the autonomic system it is easy to see how drugs acting as agonists or antagonists at cholinergic or adrenergic receptors can be used as therapy for cardiovascular, respiratory, gastrointestinal, and other disorders.

III. NEUROMUSCULAR FUNCTION

Many analogies have been drawn between the neuropharmacologies of the autonomic and central nervous systems. Further parallels exist between the brain and the neuromuscular junction, the synapse between the motor nerves that pass out from the spinal cord and the muscles. Here, the transmitter is acetylcholine, and the receptor is the nicotinic cholinergic receptor. Thus, drugs that act as antagonists at this site are useful agents for producing muscle relaxation during surgery, since the release of acetylcholine from the terminals causes the muscle contraction for movement. This neuromuscular system, even more than the autonomic nervous system, is too simple to be an adequate model for the brain. In the brain, many terminals, transmitters, and receptors converge onto single cells, which are themselves part of the massive networks of neurons.

IV. TRANSMITTERS IN THE CENTRAL NERVOUS SYSTEM

When one considers the enormous complexity of the central nervous system, it is not surprising that there are many transmitters. The criteria for a chemical's being considered as a transmitter are (1) location in appropriate areas; (2) the ability of the candidate transmitter to be released from the terminal by some form of stimulation; (3) application of the candidate transmitter should mimic the physiological events at that synapse; (4) antagonism; and (5) processes for breakdown or removal, so that synaptic events can be terminated.

Applying these criteria, the following can be considered as major transmitters in the central nervous system: for the excitatory amino acids, glutamate and aspartate; for the inhibitory amino acids, γ-aminobutyric acid (GABA) and glycine; for the monoamines, noradrenaline, serotonin (i.e., 5-hydroxytryptamine), and dopamine; for acetycholine and for the neuropeptides, enkephalins, substance P, somatostatin, cholecystokinin, and others.

Thus, not only is this list much longer than the two major transmitters in the autonomic nervous system—namely, noradrenaline and acetylcholine—but the two also have major functions in the brain; conversely, others on the list have important roles elsewhere in the body, implying a conservation in these biological systems, so that transmitters act in both the central nervous system and the periphery, but play different roles. This illustrates a major problem in drug therapy, in that at the present time it is next to impossible to target a drug onto a specific site. Many side effects arise from drugs acting on transmitter receptors other than those at the primary therapeutic site.

The central nervous system transmitters fall into two major classes, depending on the speed of onset and the duration of their effects. Thus, the excitatory and inhibitory amino acids have fast actions, turning neurons on or off, and the rest of the transmitters have slower, longer, or more subtle effects. They then change the level of excitability of the neurons, so that when a fast excitation or inhibition is evoked, the background state of the recipient cell is altered. Clearly, this type of organization imparts flexibility and adaptability to the functioning of the central nervous system.

Added to this system of fast and slow transmitters is the organization of the central nervous system networks of neurons. Certain neurons send long axons from one area of the brain to another and are involved therefore in the transmission of information from one region to another or the modulation of one zone by another. Here, the transmitters are generally excitatory amino acids or monoamines, but some peptides in sensory nerves are also in long axon neurons. Within restricted zones some neurons send only a short axon. These so-called interneurons are involved in the local control of transmission, and the inhibitory amino acids and, again, neuropeptides tend to be found in these types of neurons.

The following sections discuss the roles of these transmitters in a variety of disorders, since it is really only when a transmitter system is thrown into disarray that we can then assign a role to a particular transmitter in the workings of the human brain, based on the actions of clinically effective drugs.

A. Depression

The first hints as to the causal factors in depression came from the use of two drugs for entirely different medical reasons about 40 years ago. One was used to treat elevated blood pressure; although effective, the patients showed depressed mood and suicidal thoughts.

The other, an antituberculotic drug, caused a marked elevation of mood in the patients. These odd findings were reconciled by the finding that both agents could alter the levels of noradrenaline in the brain. From these chance observations a theory was created from the results of studies using behavioral reward in animals, with manipulations of brain levels of noradrenaline. The central tenet was that elevations of noradrenaline are pleasurable, rewarding, or produced enhanced mood, whereas the converse leads to depression. Support for this idea comes from the clinically used present generation of antidepressant drugs and the effects of cocaine. [*See* Antidepresssants.]

The antidepressants increase levels of noradrenaline in the synapse either by preventing its breakdown or by blocking its reuptake back into the terminal. Cocaine also shares this latter mode of action. The problems with the use of cocaine (and the antidepressants) come from the enhanced levels of noradrenaline in the sympathetic nerve terminal synapses on the heart, lungs, and blood vessels. It is impossible to administer a drug and to target it on the brain, excluding the rest of the body and so circumventing these side effects. However, there are other antidepressants that seem to act on 5-hydroxytryptamine, and this transmitter appears to balance the rewarding effects of noradrenaline and to act within a punishment system. Drugs such as fluoxetine have demonstrated that 5-hydroxytryptamine is important since they have effectiveness against depression via their ability to block uptake of this transmitter. Agents such as these can also be used to break panic and anxiety. Long-term therapy with antidepressants may alter the receptor characteristics by altering the synaptic levels of the transmitter, and those receptor changes may underlie the need for the therapy to continue before improvement is produced, the so-called therapeutic lag.

B. Anxiety

A sense of anxiety improves human performance when the anxiety is mild, but seriously impairs daily life when the level of anxiety is too great. Anxiety is treated by a class of drugs known as the benzodiazepines (e.g., librium and valium). These drugs, however, can also be used as sleeping pills, as muscle relaxants, and to treat epilepsy. Since the benzodiazepines can elicit a continuum of central nervous system depression—depending on their potency, dose, and duration of effect—particular drugs can be selected for particular problems. Yet these agents all share the same basic neuropharmacological effect, which is unique among all of the drugs used to treat disorders of the brain. GABA is the main inhibitory transmitter in the brain and is found mostly in interneurons controlling the excitability of other cells. GABA can activate two types of receptor: the $GABA_A$ and $GABA_B$ types. The benzodiazepines enhance the action of GABA on the $GABA_A$ site, thus increasing inhibitory influences. The means by which they do this is somewhat mysterious; by binding to a site, the benzodiazepine receptor, these drugs make $GABA_A$ receptor activation by GABA itself more effective.

The increased GABA activity causes a greater degree of inhibition, which is important in reducing 5-hydroxytryptamine function in anxiety, relating to the idea that 5-hydroxytryptamine neurons act as a punishment system. The relative safety of the benzodiazepines seems to be due to their enhancement of a preexisting inhibition by GABA. This acts as a ceiling on their effects; that is, they do not produce any effects in the absence of GABA. Does this receptor or site for the benzodiazepines mean that a natural benzodiazepine uses this site, mimicked by the synthetic drug? Several groups are investigating this possibility, but the evidence is equivocal as yet.

C. Schizophrenia

Schizophrenia, a profound disorder of thought processes, mood, and social functioning, has been cruelly described as the epitome of madness. A triumph of modern neuropharmacology has been the ability to control the symptoms of this disorder by the use of drugs, to the extent that straightjackets, padded cells, and ice baths are now no longer used in psychiatric hospitals. The variety of effective drugs all share a common mechanism; namely, they are antagonists of dopamine.This block of dopamine is able to reduce the schizophrenic symptoms, but does not appear to influence the cause of the disorder, since medication must be continued if relapse is to be avoided. These drugs have some other pharmacological actions that produce side effects (e.g., antihistaminic and anticholinergic effects), which probably give rise to their sedative effects. It should be pointed out that these effects are not necessary for antischizophrenic action, but that it has not yet been possible to produce drugs that are completely selective for dopamine receptors. Recent demonstrations of more dopamine receptors than was first realized may allow better antischizophrenic therapy and provide an explanation for why so-called atypical drugs are effective. [*See* Schizophrenic Disorders.]

With high doses or long-term treatment, some motor problems can arise that relate to the roles of dopamine in movement, which are discussed in the next section. The role of dopamine in affective disorders is likely to reside in the dopamine pathways arising from midbrain areas and innervating the cortex, the motor function from another dopamine system running from the substantia nigra to the motor-integrating areas of the striatum. This is a good example of the different functions of a transmitter in different areas of the brain. Since there is evidence for atrophy of some brain areas in schizophrenics, it could be that the primary lesion removes a control on dopamine, giving rise to the schizophrenic symptoms.

D. Parkinson's Disease

Parkinson's disease is, in most cases, a disease of old age and, with increasing life expectancy of the world population, it will become more and more prevalent. The deficit is, again, amenable to treatment, and the primary cause is a selective loss of cells containing dopamine in the substantia nigra–striatum pathway. The cause for this cell loss is unknown, but the compound 1-methyl-4-phenyl-1,2,3,6-tetrahydropyridine (MPTP), a by-product of a flawed designer drug, has been shown to exactly mimic the lesion. It is highly unlikely that an environmental toxin such as this produces Parkinson's disease, since rural populations and preindustrial societies still suffer from the disease. Since the loss is of dopamine, it appears that this transmitter exerts a controlling influence on the motor command networks in the striatum; a loss of control causes the rigidity, tremor, and akinesia typical of Parkinsonism. [*See* Parkinson's Disease.]

Therapy is based on attempting to replenish this transmitter deficit, either by direct dopamine agonists (e.g., bromocriptine), which are moderately successful, or by building up the levels of dopamine in the remaining cells. Dopamine itself cannot be given, because it does not gain access to the brain, so a related compound, L-dopa, is used that does get into the brain. L-Dopa is a precursor for dopamine, is taken up into the dopamine neurons, and is converted to dopamine, which is then available for release. This effectively replenishes the deficit, and normal movement is restored. Unfortunately, with too much dopamine, schizophrenic-like symptoms occur, and as the dopamine levels fluctuate near the end of the dose, so-called "on–off" episodes occur in which movement is uncontrolled, the rigidity sets in, and a cycle ensues; this is obviously upsetting for the patient.

In animals the Parkinson's disease-like syndrome can be alleviated by transplanting peripheral or neural tissue containing dopamine, and there have been a few patients in whom this has been attempted. The results remain controversial, and the moral and ethical aspects require debate, but if cells containing dopamine, which are easily grown in culture in the test tube, were shown to be effective, this could be a useful therapy. At present this looks unlikely.

E. Alzheimer's Disease

Alzheimer's disease, or senile dementia, is another disease in which a transmitter deficit seems to cause the symptoms, but, unlike Parkinson's disease, there is no effective replacement therapy. There are several symptoms, but losses of memory, insight, and severe cognitive disorders typify the syndrome. The major lesion seems to be a loss of cholinergic input into the cortex from the basal nucleus, situated just below the cortex itself. It is known that anticholinergic agents can impair memory, so this lesion is consistent with the amnesia symptomatic of Alzheimer's disease.

There is also loss of another transmitter, this time a peptide, somatostatin, which could result from a loss of control of excitatory amino acid mechanisms; this overexcitation might be the cause of death of the somatostatin-containing cells. There is no evidence as to what symptoms result from the loss of these peptide cells. A small number of Alzheimer's disease patients have the familial disease and there is great interest in locating the gene locus, together with attempts to understand the plaques, tangles, and other aberrant neural elements that occur in the disease. Advances in therapy are delayed by our inability to replenish the loss of acetylcholine by drugs acting effectively as cholinergic muscarinic agonists. As in Parkinson's disease, transplants in animals seem to have potential for developing therapy. [*See* Alzheimer's Disease.]

F. Epilepsy

The overexcitability of certain areas of the brain leads to the uncontrolled movements, convulsions, and other disturbances underlying epilepsy, which takes certain forms depending on the degree or the severity of the attack. From the points already made in this article, it is obvious that the strategy for therapy will be either decreasing excitation or increasing inhibitory influences. The former approach could become possible with the development of antagonists for the excitatory amino acid receptors, thus reducing the effects of glutamate and/or aspartate. This reduction could be a heavy price to pay, as these chemicals have ubiquitous actions throughout the brain and antagonism might bring the central nervous system to a halt. Most treatments rely on increasing inhibitions, by using either benzodiazepines (see Section IV,B) or other drugs (e.g., barbiturates) that open the channel associated with the GABA receptor. The other approach is to prevent the spread of abnormal activity; some drugs do this, presumably because they can reduce abnormal patterns of activity, leaving normal functioning of the cells elsewhere in the brain. [*See* Epilepsies.]

G. Pain

Pain is not a disorder of the central nervous system, but is an important signal for protection of the organism from impending or actual tissue damage. However, an excess of pain (e.g., from chronic diseases or injuries) produces suffering and distress, which overwhelm the need for a warning signal. [*See* Pain.]

The dramatic examples of pain seemingly arising from an area of the body amputated decades beforehand, the transformation of touch stimuli into pain after nerve damage, and the difficulties of treating some types of pain compared to others underlie the complexities of pain transmission and indicate the lack of a specific pain pathway, immutable and hard-wired. Nevertheless, we know a reasonable amount regarding both pain transmission and its pharmacological control.

Pain is generated by nociceptors (receptors for chemicals and heat embedded in the skin, viscerae, and muscles), which in turn activate peripheral nerve fibers, known as C fibers. Waves of action potentials pass up these nerves and release a variety of transmitters into the spinal cord, the site of the first synapse in the system. These transmitters include glutamate and a number of peptides, including substance P, which excite dorsal horn neurons in the upper half of the spinal cord. These cells, in turn, are modulated, and their responses are integrated and modified prior to transmitting messages to both higher centers of the brain (producing the sensation of pain) and the lower (ventral) part of the spinal cord, where motoneuron activation causes withdrawal from the stimulus.

The activation of the spinal neurons responding to noxious stimulation can be reduced or abolished by the action of drugs such as morphine, which act to inhibit the release of the transmitters from the C fibers. This action is via receptors for opioid drugs on the terminals of the sensory fibers and so is presynaptic inhibition. Just over a decade ago a paradox regarding morphine and related drugs was solved: Why should extracts from a plant (i.e., the opium poppy) produce such profound effects on the brain as the abolition of pain? The reason is now clear; natural opioids exist in the central nervous system and the periphery as transmitters. Like any other transmitter system, there are receptors for these chemicals, and morphine, by a chemical similarity, can activate these receptors. There are at least three types of receptor—the μ, δ, and κ subtypes; morphine acts on the μ receptor. Since along with the morphine analgesia are severe unwanted side effects (e.g., respiratory depression, constipation, cough suppression, and dependence), there is great interest in drugs that act on the δ and κ receptors as potential alternatives to morphine with the desirable analgesia, but not the side effects. The so-called endogenous opioids are peptides, and three families exist that correspond to three types of opioid receptors: β endorphins correspond to μ receptors, enkephalins to δ receptors, and dynorphins to κ receptors. [*See* Opioids.]

There is great interest in the roles of these peptides, since their presence in the brain might relate to a natural painkilling function or even a built-in dependence system. The enkephalins, in particular, are quickly broken down, so drugs that prolong their actions have been produced, which reveal that enkephalins are released during the production of a painful stimulus and so could reduce pain but not abolish it. The other roles of opioids remain elusive but are being pursued, and, in general terms, the study of opioid peptides and nonopioid peptides could provide an important addition to neuropharmacology, since present therapies are all aimed at the nonpeptide transmitters. Peptides such as cholecystokinin might be involved in Parkinson's disease and schizophrenia; angiotensin in hypertension; and somatostatin in Alzheimer's disease. There is much interest in devising drugs that manipulate peptide transmission.

H. Stroke

A major consequence of disruption of the blood supply to parts of the brain is death of the neurons in the area deprived of their normal blood supply. Stroke can result in marked neurological deficits, including paralysis, amnesia, and sensory loss, and although some recovery can occur (presumably, some neurons are injured, but they do not die), there are long-term effects that cannot be alleviated. There is great interest in the therapeutic use of some antagonists of the excitatory amino acid receptors for glutamate. It might be that the vascular insult and resulting ischemia cause a loss of control of glutamate release. It appears that the massive release of glutamate can provoke overexcitation of neurons, which leads to death of the neurons, because of the resultant huge imbalances in ions within the cells. There is accumulating evidence that blocking one of the glutamate receptors or, indeed, the ion channels associated with the receptor can protect some of the neurons at risk. Remarkably, treatment after the event could be effective. If the potential of these drugs is verified in humans, a treatment that could partially alleviate the consequences of stroke could emerge. This glutamate receptor might also become overactive in epilepsy and, under normal conditions, seems to be involved in memory, learning, and other long-term events. A consequence of therapy via this route could be amnesia, but this might be a worthwhile price to pay for reducing the problems of cerebral ischemia. [*See* Stroke.]

V. THE FUTURE

Given that we now have a reasonable idea of how some of the transmitters in the central nervous system function in health and disease, the future must be determined. It seems to be easier, at present, to block the effects of transmitters by using direct antagonists or by giving inhibitory drugs such as the opiates. The opposite approach, such as the treatment of Alzheimer's disease and Parkinson's disease (in this case after the initial few years of therapy), in which one is trying to increase transmitter levels, is more difficult. The idea of using transplants to augment the transmitter is feasible from animal studies, but the ethical aspects need to be considered apart from the longer-term effects of the transplant within the delicate balanced structures of the brain. Another area of great interest is definition of the roles of the multitude of peptide transmitters found in the brain. These compounds have been found to coexist with other transmitters, yet the present therapies, because of the lack of drugs for the manipulation of peptides, concentrate only on the well-known or classic transmitters, such as those described earlier. More complete and therefore better clinical effects might be achieved by manipulating both peptide and nonpeptide transmitters. Nerve growth factors might provide protection or a survival aid to damaged or lost neurons and could be used in syndromes or diseases in which cell death and damage occur. An area of great interest is the realization that neurotransmitters can influence long-term changes in neuronal function.

Another large step forward will be simply improving our knowledge of pharmacology, since at present we have a reasonably broad idea of neuropharmacology, but many areas remain hazy and, in most cases, details are lacking. Thus, we know certain roles of transmitters, such as dopamine in movement and psychoses, opioids in pain states, and noradrenaline in mood, but exactly how, where, and when do these systems function? Answers will hopefully be forthcoming, and such basic knowledge will surely aid the clinical treatment of central nervous system disorders.

BIBLIOGRAPHY

Cooper, J. R., Bloom, F. E., and Roth, R. H. (1991). "The Biochemical Basis of Neuropharmacology," 6th Ed. Oxford. Univ. Press, New York.

Dickenson, A. H. (1995). Novel pharmacological targets in the treatment of pain. *Pain Rev.* **2**, 1–12.

Scientific American (1992). "Mind and Brain." Special issue. **267**(3).

Thompson, R. F. (1993). "The Brain," 2nd Ed. Freeman, New York.

Neurotransmitter and Neuropeptide Receptors in the Brain

SIEW YEEN CHAI
G. PETER ALDRED
FREDERICK A. O. MENDELSOHN
University of Melbourne

GLOSSARY

Adenylyl cyclase Enzyme associated with the inner surface of the cell membrane, which converts ATP into cyclic AMP. The enzyme is regulated by stimulatory (G_s) and inhibitory (G_i) G-proteins coupled to activated receptors

G-protein GTP-binding regulatory protein composed of three subunits, α, β, and γ. The α subunit binds and hydrolyzes GTP and the $\beta\gamma$ complex anchors the G-protein to the cytoplasmic face of the membrane. In the resting state, the protein exists as a trimer to which GDP is bound. When activated, the G-protein binds to the hormone–receptor complex, allowing GTP to bind in place of GDP. The GTP-coupled α subunit of the G-protein then dissociates from the hormone–receptor complex and triggers second-messenger systems

***In vitro* autoradiography** Technique that involves the binding *in vitro* of a specific radioligand to receptors on tissue sections. These are then exposed to photographic emulsion of X-ray film to obtain an image of the binding site distribution and density

Ion channels Channels within the cell membrane that regulate the passage of selected ions into and out of the cell. These channels are regulated by membrane potential, ligand occupancy, or second messengers

Oncogene Genes encoding normal proteins that have become incorporated, often in modified forms, into retroviruses. These genes may induce uncontrolled cell proliferation in the infected cell and thus produce tumors. Protooncogenes are the normal cellular genes from which the oncogene is derived

Phosphoinositide cycle Cascade that is initiated by the binding of a hormone/neurotransmitter to a cell-surface receptor, which then activates phospholipase C to hydrolyze phosphoinositide 4,5-diphosphate to inositol 1,4,5-triphosphate and diacyl glycerol, two separate second-messenger systems, which induce increased intracellular Ca^{2+} and phosphorylation of specific proteins via protein kinase C

RECEPTORS ARE CELL-SURFACE, CYTOPLASMIC, OR nuclear molecules that bind specific neurotransmitters, hormones, growth factors, or other biochemicals, and then transmit signals that cause the cell to respond in an appropriate manner. All classic neurotransmitter and neuropeptide receptors are located on the cell surface. Many of the genes for these proteins have now been cloned, numerous families of receptors have been recognized, and insights into signal transduction mechanisms and second messengers are now becoming available.

I. INTRODUCTION

Many receptors were recognized to contain distinct subtypes on the basis of ligand specificity, distribution, and mode of signal transduction. With the remarkable progress in analyzing the genes encoding receptors, two major facts have emerged. First, on the basis of their primary structures and predicted transmembrane topologies, many receptors and ion channels belong to superfamilies. Second, the degree of receptors heterogeneity within a given neurotransmitter receptor type is much greater than previously suspected. This diversity of receptor types for one neurotransmitter may endow neural cells with greater information-handling ability.

The term receptor is also used in neuroscience to denote specialized neural structures involved in transduction of sensory information into neural impulses. This is not the sense in which receptors will be discussed here. However, it is intriguing to note that at least two forms of special sense receptors—those involved in light transduction and in olfaction—use G-proteins in their transduction process, as do many of the cell-surface receptors involved in chemical neurotransmission, which are discussed here.

II. CLASSES OF NEUROTRANSMITTERS

The classic neurotransmitters include biogenic amines, such as the catecholamines (adrenaline, noradrenaline, and dopamine), indoleamines (serotonin and histamine), and esters (acetylcholine). Following these, the excitatory (glutamate and aspartate) and inhibitory (γ-aminobutyric acid, or GABA, and glycine) amino acids were recognized. The largest family is the neuropeptides (examples are enkephalins, substance P, cholecystokinin (CCK), neuropeptide Y, neurotensin, and many others). A particular group of polypeptides are the neurotrophins: nerve growth factor (NGF) and brain-derived neurotrophic factor (BDNF), neurotrophins-3 and -4/5, and ciliary neurotrophic factor (CNTF), which are involved in the control of growth in the nervous system. Finally, some steroids, previously thought to be synthesized only in peripheral endocrine glands, are now known to be produced in the central nervous system (CNS) and are named neurosteroids. All of these compounds interact with one, or usually several, specific receptors. Recently, simple molecules such as nitric oxide have been recognized to carry out important roles in the CNS.

TABLE I
Major Classes of Receptor Families

1. Ligand-gated ion channels
2. G-protein-coupled receptors
3. Enzyme-linked receptors
4. Nuclear receptors

III. FAMILIES OF RECEPTORS

A summary of the major families of receptors is given in Table I.

A. Ligand-Gated Ion Channel Receptors

Members of the superfamily of ligand-gated ion channel receptors are shown in Table II. The first member to be characterized was the nicotinic acetylcholine receptor (nAChR). Later the $GABA_A$, inhibitory glycine, 5-hydroxytryptamine ($5\text{-}HT_3$), and several glutamate receptors were found to belong to this group. They share the feature of four membrane-spanning domains in their molecular structure and show sequence homology.

The nAChR is one of the best-characterized receptors because it was isolated early. This was facilitated by an abundant source in the electric organ of the fish *Torpedo californica* and the availability of a snake venom toxin (α-bungarotoxin), which binds tightly and specifically to the receptor. The receptor consists of various combinations of five homologous but distinct subunits (α, β, γ, δ, and ε) to form a pentameric rosette that spans the membrane to form a central ion channel. In the absence of acetylcholine, the channel is closed, but after binding acetylcholine to the α subunit, the channel opens for a few milliseconds to permit passage of sodium and potassium ions. Studies of single-channel currents of chimeric receptors and of receptors, expressed after site-directed mutagenesis, reveal the importance of the various subunits and of charged and uncharged residues in the function of the nAChR (Fig. 1). At least 10 neuronal nicotinic receptor subunit genes have been identified and many different combinations are possible to form receptors with a diversity of properties.

Glycine is the major inhibitory neutotransmitter in the spinal cord and brain stem, whereas GABA is the main inhibitory transmitter in the rest of the central nervous system. Binding of both amino acids to their cognate receptor causes hyperpolarization of the cell

TABLE II
Ligand-Gated Ion Channels

Receptor	Subtype	Channel
1. Nicotinic acetylcholine	Muscle	Intrinsic $Na^+/K^+/Ca^{2+}$
	Ganglionic	Intrinsic $Na^+/K^+/Ca^{2+}$
	Neuronal	Intrinsic $Na^+/K^+/Ca^{2+}$
	α7 neuronal	Intrinsic $Na^+/K^+/Ca^{2+}$
2. Glutamate	NMDA	Intrinsic $Na^+/K^+/Ca^{2+}$
	AMPA	Intrinsic Na^+/K^+
	Kainate	Intrinsic Na^+/K^+
3. $GABA_A$	—	Intrinsic Cl^-
4. Strychnine-sensitive glycine	—	Intrinsic Cl^-
5. 5-Hydroxytryptamine	$5\text{-}HT_3$	Intrinsic cation
6. P_2 purinoceptor	P_{2T}	Intrinsic cation
	P_{2X}	Intrinsic cation
	P_{2Z}	Intrinsic cation

by increasing Cl^- conductance. The inhibitory glycine receptor resembles the nAChR in both amino acid sequence and structural organization. It is pentamer of α subunits, which are ligand binding, and β subunits. Four different subtypes of α subunit and one β subunit have been identified. The drug strychnine acts by binding to the inhibitory glycine receptor and inhibiting Cl^- channel opening. A number of different amino acids, glycine, β-alanine, taurine, serine, and proline, can all activate the inhibitory glycine receptor.

The GABA/benzodiazepine ($GABA_A$) receptor is of particular interest because it contains binding sites for a number of important drugs that allosterically modify GABA binding or the Cl^- channel. These in-

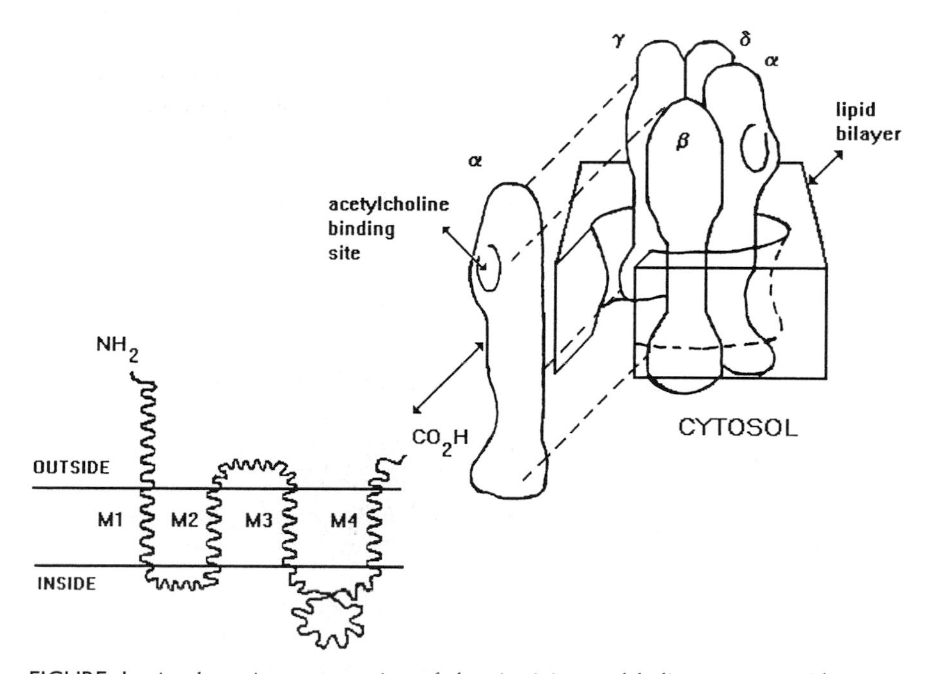

FIGURE 1 A schematic representation of the nicotinic acetylcholine receptor with two α subunits and one each of β, γ, and δ that form a pentameric rosette. The five subunits surround the ion pore. Each of the subunits comprises four membrane-spanning regions.

clude anxiolytics (benzodiazepines), anticonvulsants, convulsants, and steroids. In addition, ethanol, like barbiturates and benzodiazepines, interacts with the $GABA_A$ receptors to potentiate GABA actions and thereby exert its depressant actions.

Like other members of this superfamily, $GABA_A$ receptors are heteropentameric glycoproteins. The genes encoding 16 isoforms of the five types of subunits (α, β, γ, δ, and ρ) have been characterized. Differential expression of these isoforms probably explains regional variations of physiology and pharmacology of $GABA_A$ receptors in different populations of neurons. The amino acid sequences also show homology with the nAChR and identify the $GABA_A$ receptor as part of this superfamily.

The amino acids glutamate and aspartate are the major excitatory neurotransmitters in the brain and exert their actions via two broad types of receptors, the ligand-gated ion channels (ionotropic receptors) and G-protein-coupled receptors (metabotropic receptors), which are discussed in Section III,B. Three types of ionotropic glutamate receptors are known: the *N*-methyl-D-aspartate (NMDA), α-amino-3-hydroxy-5-methyl-4-isoxazole propionic acid (AMPA), and kainate (KA) receptors. There are currently 16 molecularly characterized subunits of the inotropic glutamate receptor channels, GluRA–GluRD for AMPA, KA1–KA2 and GluR5–GluR7 for kainate, NR1 and NR2A–NR2D for NMDA, and 2 as yet unassigned. These subunits can be reconstituted in different combinations *in vitro* to generate functional receptor channels.

The NMDA receptor is regulated by numerous ligands, which alter the probability of ion channel opening, including recognition sites for two agonists, glutamate and glycine, a polyamine recognition site, and sites for the divalent metal cations Zn^{2+} and Mg^{2+}, which inhibit ion flux. Binding of both glutamate and glycine is needed for activation of the NMDA receptor, and this glycine recognition site is distinct from the inhibitory strychnine-sensitive glycine receptor discussed earlier. In addition to major roles in normal neurotransmission, NMDA receptors may mediate neurotoxic effects of glutamate in many forms of brain injury and some neurodegenerative disorders. Numerous agonists and antagonists of NMDA receptors are now available.

The AMPA receptor, previously called "quisqualate receptor," is widespread throughout the CNS in a majority of excitatory synapses. The AMPA receptor differs from the NMDA receptor in that it is not activated by aspartate and has fast channel kinetics and low Ca^{2+} permeability.

The kainate receptor is pharmacologically distinct from the AMPA and NMDA receptors and is characterized by fast desensitization in peripheral sensory ganglia.

B. G-Protein-Coupled Receptors

The G-protein-coupled receptors are very large and diverse family characterized by amino acid sequences that contain seven hydrophobic domains that represent transmembrane-spanning regions (Fig. 2). These receptors interact with heterotrimeric guanyl nucleotide binding proteins (G-proteins), which transmit signals in the interior of the cell by interaction with a number of signaling pathways. [*See* G Proteins.]

These receptors share many features in parallel. They are composed of single polypeptide chains, 400–500 amino acids long, which bear seven hydrophobic regions separated by alternating extracellular and cytoplasmic loops. The hydrophobic regions of 20–25 amino acids form transmembrane-spanning α helices. The N-terminal extracellular domains show one to three sites that may undergo N-linked glycosylation. This glycosylation is not essential for ligand binding or for coupling to G-proteins, but is involved in expression of the receptor on the cell surface and its trafficking through the cell.

Cysteine disulfide bridges within and between the extracellular loops are important for both agonist and antagonist binding. Many of these receptors are palmitoylated via a cysteine residue in the cytoplasmic tail.

The genes for the G-protein-coupled receptors fall into two groups, those with introns in the coding region (e.g., the dopamine D_2–D_4 receptor genes) and those without, such as the β-adrenergic receptors.

The first member of this family identified was the visual pigment rhodopsin, which is embedded in the membrane of discs in rods of the retina. On exposure to light, rhodopsin undergoes a change in its conformation, which enables it to interact with the G-protein transducin, thereby triggering cyclic GMP phosphodiesterase, which hydrolyzes cGMP, thereby terminating its activation of Na^+ channels to mediate photoreception.

All of the classic neurotransmitters interact with G-protein-coupled receptors in addition to the ligand-gated ion channels discussed earlier. G-protein-coupled receptors for the classic neurotransmitters

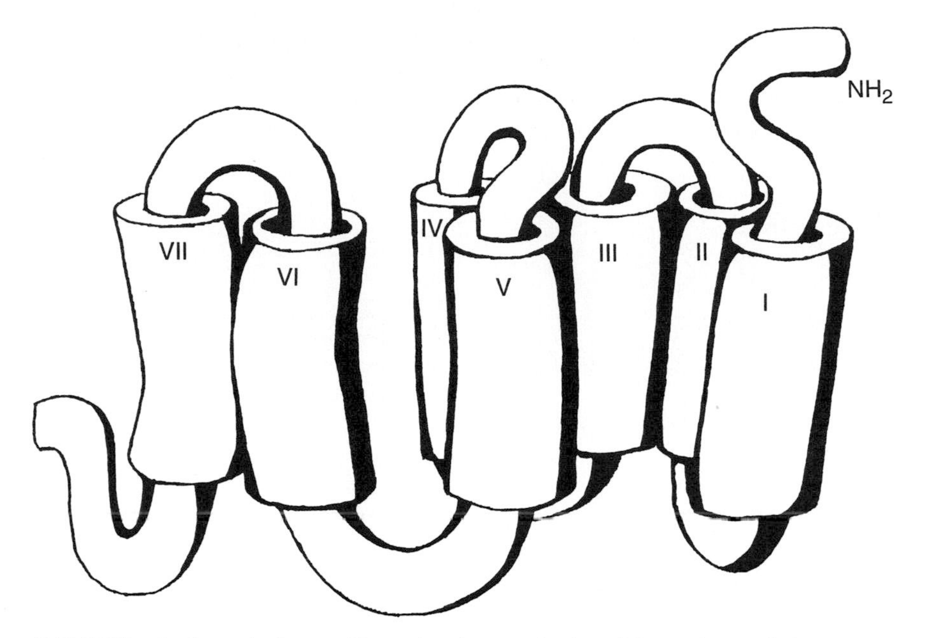

FIGURE 2 A schematic diagram illustrating the organization of G-protein-coupled receptors. The receptors have seven putative membrane-spanning regions with alternate extracellular and intracellular domains. The extracellular and membrane regions form the ligand binding pocket, and the intracellular loops are involved in coupling to the G-proteins.

include the α- and β-adrenoceptors, muscarinic acetylcholine, dopamine, metabotropic glutamate receptors, and the $GABA_B$ receptor (Table III).

The α_1-adrenoceptors are coupled via $G_{q/11}$ to phospholipase C (PLC), whereas the α_2-adrenoceptors employ $G_{i/o}$ to inhibit adenylyl cyclase (AC) and the β-adrenoceptors employ G_s to stimulate AC.

Dopamine receptors were initially classified as D_1 (which stimulate AC) and D_2 (which inhibit AC) but are now known to include five gene products: two D_1-like (D_1, D_5) and three D_2-like (D_2, D_3, D_4), with variation in mRNA splicing providing further subtypes. The D_2 receptor gene has a very large first intron and alternate splicing of exon 6 gives the forms D_{2L} and D_{2S}, both of which inhibit AC. The D_3 receptor also is subject to alternate splicing of mRNA to form shorter variants, some of which may not be functional. The D_3 receptor has higher affinity for dopamine than the D_2 receptor and displays a relative preference for atypical neuroleptics. The D_4 receptor has a relatively small gene, does not undergo alternative slicing, and is highly polymorphic in populations as determined by the number of 16 amino acid repeats located in the third cytoplasmic loop of the receptor protein. The D_4 receptor appears to be the target for the antipsychotic drug clozapine.

Serotonin interacts with a very large and complicated group of receptors. Apart from the 5-HT_3 receptor, which is a ligand-gated ion channel, the remaining subtypes all belong to the G-protein-coupled receptor superfamily. The 5-HT_1 family has five subclasses, all coupled via $G_{i/o}$ to inhibition of AC. Humans do not have 5-HT_{1B} receptors, which are the rodent analogue of human 5-HT_{1D} receptors, although these two receptors are pharmacologically distinct. The 5-HT_2 family of receptors are all coupled via $G_{q/11}$ to PLC, whereas the 5-HT_4 receptors are coupled via G_s. Two subtypes of 5-HT_5 receptor have been proposed on the basis of molecular cloning. The 5-HT_6 and 5-HT_7 subtypes have recently been identified and they are both coupled via G_s to stimulate AC.

In addition to its interaction with three types of ionotrophic receptors, the excitatory transmitter glutamate interacts with seven different subtypes of metabotropic receptors: $mGlu_1$ to $mGlu_7$, all of which are seven-transmembrane proteins. $mGlu_1$ and $mGlu_5$ activate the phosphoinositide pathway via PLC through $G_{q/11}$, whereas $mGlu_2$, $mGlu_3$, $mGlu_4$, $mGlu_6$, and $mGlu_7$ inhibit AC through $G_{i/o}$.

Receptors for peptides are usually G-protein-coupled (Table IV) and include those for: angiotensin (AT_1, AT_2), bombesin (BB_1, BB_2), bradykinin (B_1, B_2),

TABLE III
G-Protein-Coupled Receptors for the Classic Neurotransmitters[a]

Receptor	Subtype	G-protein	Specific agonist	Specific antagonist
Muscarinic acetylcholine	M_1	$G_{q/11}$	—	Pirenzepine/telenzepine
	M_2	$G_{i/o}$	—	Methoctramine/AFDX116
	M_3	$G_{q/11}$	—	Hexahydrosiladifenidol
	M_4	$G_{i/o}$	—	Tropicamide
	M_5	$G_{q/11}$	—	—
α_1-adrenoceptor	α_{1A}	$G_{q/11}$	Noradrenaline	WB4101
	α_{1B}	$G_{q/11}$	Adrenaline, noradrenaline	Spiperone
	α_{1D}	$G_{q/11}$	Adrenaline, noradrenaline	BMY7378
α_2-adrenoceptor	α_{2A}	$G_{i/o}$	Oxymetazoline	—
	α_{2B}	$G_{i/o}$	—	Prazosin/ARC 239
	α_{2C}	$G_{i/o}$	—	Prazosin/ARC 239
β-adrenoceptor	β_1	G_s	Noradrenaline/xamoterol	CGP20712A/betaxolol
	β_2	G_s	Procaterol	ICI118551/butaxamine
	β_3	G_s	BRL37344/CL316243	—
Dopamine	D_1	G_s	SKF 38393	SKF 83566/SCH 23390
	D_2	$G_{i/o}$	N0437/fenoldopam	(−)-Sulpiride/YM091512
	D_3	$G_{i/o}$	Bromocriptine	—
	D_4	$G_{i/o}$	—	—
	D_5	G_s	—	—
Excitatory amino acid (metabotropic)	$mGlu_1$	$G_{q/11}$	DHPG	4CPG/MCPG
	$mGlu_2$	$G_{i/o}$	DCG-IV	MCCG
	$mGlu_3$	$G_{i/o}$	DCG-IV	MCCG
	$mGlu_4$	$G_{i/o}$	L-AP4	MAP4
	$mGlu_5$	$G_{1/11}$	DHPG	4CPG
	$mGlu_6$	$G_{i/o}$	L-AP4	MAP4
	$mGlu_7$	$G_{i/o}$	L-AP4	MAP4
Serotonin	$5\text{-}HT_{1A}$	$G_{i/o}$	8-OH-DPAT	WAY100635
	$5\text{-}HT_{1B}$	$G_{i/o}$	CP93129	SDZ21009
	$5\text{-}HT_{1D}$	$G_{i/o}$	L694247	GR127935
	$5\text{-}HT_{1E}$	$G_{i/o}$	—	—
	$5\text{-}HT_{1F}$	$G_{i/o}$	—	—
	$5\text{-}HT_{2A}$	$G_{q/11}$	α-Methyl-5-HT	Ketanserin/MDL100907
	$5\text{-}HT_{2B}$	$G_{q/11}$	α-Methyl-5-HT	SB200646
	$5\text{-}HT_{2C}$	$G_{q/11}$	α-Methyl-5-HT	Mesulergine
	$5\text{-}HT_4$	G_s	BIMU8/renzapride	GR113808/SB204070
	$5\text{-}HT_{5A}$		—	—
	$5\text{-}HT_{5B}$		—	—
	$5\text{-}HT_6$	G_s	—	—
	$5\text{-}HT_7$	G_s	—	—

[a]AFDX116: 11-([2-{(diethylamino)methyl}-1-piperidinyl]acetyl)-5,11-dihydro-6*H*-pyrido[2,3-*b*][1,4]benzodiazepine-6-one; ARC239: 2-(2,4-[O-methoxyphenyl]-piperazin)-1-yl; BMY7378: 8-{2-[-(2-methoxyphenyl)-1-piperazinyl]ethyl}-8-azaspiro[4,5]decae-7,9-dione dihydrochloride; BRL37344: sodium-4-(2-[2-hydroxy-{3-chlorophenyl}ethylamino]propyl)phenoxyacetate; CGP20712A: 2-hydroxy-5-(2-[{2-hydroxy-3-(4-[1-methyl-4-trifluoromethyl-2-imidazolyl]phenoxy)propyl}amino]ethoxy)benzamide; CL316243: disodium (*R*,*R*)-5-(2-{[2-(3-chlorophenyl)-2-hydroxyethyl]-amino}propyl)-1,3-benzodioxole-2,2-dicarboxylate; ICI118551: erythro-DL-1-(7-methylindan-4-yloxy)-3-isopropyl-aminobutane-2-ol; WB4101: *N*-[2-(2,6-dimethoxyphenoxy)ethyl]-2,3-dihydro-1,4-benzodioxan-2-methanamine; N0437: 2(*N*-*n*-propyl-*N*-2-[2-phenyl]ethylamino)-5-hydroxytetralin; SCH23390: 7-chloro-2,3,4,5-tetrahydro-3-methyl-5-phenyl-1*H*-3-benzazepine-7-ol; SKF38393: 2,3,4,5-tetrahydro-7,8-dihydroxy-1-phenyl-1*H*-3-benzazepine HCL; SKF83566: (+)7-bromo-8-hydroxy-3-methyl-1-phenyl-2,3,4,5 tetrahydro-1*H*-3-benzazepine HCl; YM091512: *cis*-*N*-(1-benzyl-2-methylpyrrolidin-3-yl)-5-chloro-2-methoxy-4-methylamino benzamide; L-AP4: L-amino-4-phosphonobutanoate; 4CPG: *s*-4-carboxyphenylglycine; DCG-IV: (2*s*,1′*R*,2′*R*,3′*R*)-2-(2′,3′-dicarboxycyclopropyl)glycine; DHPG: 3,5,dihydrophenylglycine; MAP4: α-methyl-L-AP4; MCPG: α-methyl-4-carboxyphenylglycine; CP93129: 5-hydroxy-3-(4-1,2,5,6-tetrahydropyridyl)-4-azaindole; 8-OH-DPAT: 8-hydroxy-2-(di-*n*-propylamino)tetralin; GR113808: [1-[2-[(methylsulphonyl)amino]ethyl]-4-piperidinyl]methyl-1-methyl-1*H*-indole-3-carboxylate; L694247: 2-(5-[3-{4-methylsulphonylaminobenzyl}-1,2,4-oxadiazol-5-yl]-1*H*-indole-3-yl)ethylamine; SB200646: *N*-(1-methyl-5-indolyl)-*N*′-(3-pyridyl)urea HCl; SB204070: 1-butyl-4-piperidinylmethyl)-8-amino-7-chloro-1,4-benzodioxan-5-carboxylate; WAY100135: *N*-*tert*-butyl-3-4-(2-methoxyphenyl)piperazin-1-yl-2-phenylpropanamide dihydrochloride.

TABLE IV
G-Protein-Coupled Receptors for Neuropeptides[a]

Receptor	Subtype	G-protein	Specific agonist	Specific antagonist
Angiotensin	AT_1	$G_{q/11}$	—	Losartan/EXP3174
	AT_2		CGP42112A	PD123177
Bradykinin	B_1	$G_{q/11}$	[des-Arg^{10}]kallidin	[Leu^9][des-Arg^{10}]kallidin
			Sar[DPhe^8][des-Arg^9]BK	[des-Arg^{10}]HOE140
	B_2	$G_{q/11}$	[Phe^8,ψ(CH$_2$-NH)Arg^9]BK	HOE140
			[Hyp^3, Tyr(Me)^8]BK	WIN64338
Cholecystokinin and gastrin	CCK_A	$G_{q/11}$	A71623	Devazepide/lorglumide
	CCK_B	$G_{q/11}$	[*N*-Methyl-Nle^{28,21}]desulfated CCK_8	L365260
			Desulfated CCK_8	YM022
Neuropeptide Y	Y_1	$G_{i/o}$	[Pro^{34}]NPY	BIBP3226
			[Leu^{31},Pro^{34}]NPY	SR120107A
	Y_2	$G_{i/o}$	NPY_{13-36}	
			PYY_{13-36}	
Opioid	μ	$G_{i/o}$	DAMGO/sufentanil	CTOP
	δ	$G_{i/o}$	DPDPE	ICI174864
			DSBULET	
	κ	$G_{i/o}$	U69593	Nor-binaltorphimine
			Cl977	
Vasopressin and oxytocin	V_{1A}	$G_{q/11}$	—	SR49059
	V_{1B}	$G_{q/11}$	d[D3Pal^2]VP	—
	V_2	G_s	d[DArg^8]VP	OPC31260
	OT	$G_{q/11}$	[Thr^4,Gly^7]OT	*cyc*(D1-Nal,Ile,DPip,Pip,DHis,Pro)
Tachykinin	NK_1	$G_{q/11}$	SP methyl ester	CP99994
			[Sar^9,Met(O$_2$)^{11}]SP	SR140333
	NK_2	$G_{q/11}$	[β-Ala^8]NKA_{4-10}	SR48968
			GR64349	GR94800
	NK_3	$G_{q/11}$	Senktide	SR142802
			[MePhe^7]NKB	PD157672

[a]CGP42112A: nicotinic acid-Tyr-(*N*-benzoylcarbonyl-Arg)Lys-His-Pro-Ile-OH; EXP3174: *n*-butyl-4-chloro-1-([2′-{1*H*-tetrazol-5-yl}-biphenyl-4-yl]methyl)imidazole-5-carboxylic acid; PD123177: 1-(4-amino-3-methylphenyl)methyl-3-diphenyl-acetyl-4,5,6,7-tetrahydro-III-imidazo(4,5-O)pyridine-6-carboxylic acid; HOE140: DArg[Hyp^3,Thi^5,DTic^7,Oic^8]BK; WIN64338: ([4-{(2-[{bis(cyclohexylamino)memthylene}amino]-3-[2-napthyl]-1-oxopropyl)amino}phenyl]methyl)tributylphosphonium chloride monohydrochloride; A71623: Boc-Trp-Lys(O-Me-Phe-NH)-Asp-(NMe)Phe-NH$_2$; L365260: 3*R*(+)-*N*-(2,3-dihydro-1-methyl-2-oxo-5-phenyl-1*H*-1,4-benzodiazepin-3-yl)-*N*′-3-methylphenyl urea; YM022: (*R*)-1-(2,3-dihydro-1-[2′-methylphenacyl]-2-oxo-5-phenyl-1*H*-1,4-benzodiazepin-3-yl)-3-(3-methylphenyl)urea; BIBP3226: *R*-*N*^{2}-(diphenylacetyl)-*N*-(4-hydroxyphenyl)methyl-argininamide; Cl977: (5*R*)-(5α,7α,8β)-*N*-methyl-*N*-(7-[1-pyrrolidinyl]-1-oxaspiro[4,5]dec-8-yl)-4-benzofuranacetamide monohydrochloride; CTOP: DPhe-Cys-Tyr-DTrp-Lys-Thr-Pen-Thr-NH$_2$; DAMGO: Tyr-DAla-Gly-[*N*MePhe]-NH(CH$_2$)$_2$-OH; DPDPE: [DPen^2,DPen^5]enkephalin; DSBULET: Tyr-DSer(OtBu)-Gly-Phe-Leu-Thr; ICI174864: *N*-*N*-diallyl-Tyr-Aib-Aib-Phe-Thr; U69593: 5α,7α,β-(−)-*N*-methyl-*N*-[7-(1-pyrrolidinyl)-1-oxaspiro(4,5)dec-8-yl]benzene acetamide; OPC31260: 5-dimethylamino-1-(4-[2-methylbenzoylamino]benzoyl)-2,3,4,5-tetrahydro-1*H*-benzazepine; SR49059: (2*S*)1-([2*R*,3*S*]-[5-chloro-3-{chlorophenyl}-1-[3,4-dimethoxysulphonyl]-3-hydroxy-2,3-dihydro-1H-indole-2-carbonyl]-pyrrolidine-2-carboxamide); CP99994: (+)-(2*s*,3*s*)-3-(2-methoxybenzylamino)-2-phenylpiperidine; GR94800: PhCO-Ala-Ala-DTrp-Phe-DPro-Pro-Nle-NH$_2$; PD157672: Boc(*S*)Phe(*R*)aMePhe-NH(CH$_2$)$_7$NHCONH$_2$; SR48968: (*S*)-*N*-methyl-*N*-(4-acetylamino-4-phenylpiperidino)-2-(3,4-dichlorophenyl)butylbenzamide; SR140333: (*S*)-1-(2-{3-{3,4-dichlorophenyl}-1-{3-isopropoxyphenylacetyl}piperidin-3yl}ethyl)-4-phenyl-1-azoniabicyclo(2.2.2); SR142802: (*S*)-(*N*)-(1-(3-(1-benzoyl-3-(3,4-dichlorophenyl)piperidin-3-yl)propyl)-4-phenylpiperidin-4-yl)-*N*-methylacetamide.

calcitonin, calcitonin gene-related peptide (CGRP), CCK/gastrin (CCK_A, CCK_B), endothelin (ET_A, ET_B), glucagon, gonadotropin-releasing hormone (GnRH), melanocortins [adrenocorticotropic hormone (ACTH) receptor, melanocyte-stimulating hormone (MSH) receptor], neuropeptide Y (NPY, types Y_1 and Y_2), neurotensin, opioid peptides (μ, δ, κ), parathyroid hormone (PTH), somatostatin (sst_1, sst_2, sst_3, sst_4, sst_5), tachykinins (NK_1, NK_2, NK_3), vasoactive intestinal peptide (VIP) family (VIP_1, VIP_2, PACAP), vasopressin (V_{1A}, V_{1B}, V_2), and oxytocin. The glycoprotein pituitary hormones, follicle-stimulating

TABLE V
Other G-Protein-Coupled Receptors[a]

Receptor	Subtype	G-protein	Specific agonist	Specific antagonist
Adenosine	A_1	$G_{i/o}$	N^6-Cyclopentyladenosine 2-Cl-N^6-cyclopentyladenosine	DPCPX 8-Cyclopentyltheophylline
	A_{2A}	G_s	CGS21680/PAPA-APEC	ZM241385/CP66713
	A_{2B}	G_s	—	—
	A_3	$G_{i/o}$	APNEA/IB-MECA	I-ABOPX
Histamine	H_1	$G_{q/11}$	2-(*m*-Fluorophenyl)histamine	Mepyramine
	H_2	G_s	Dimaprit/impromidine	Ranitidine
	H_3		*R*-α-Methylhistamine/imetit	Thioperamide
P_2 purinoceptor	P_{2U}	$G_{q/11}$	UTP-γ-S	
	P_{2Y}	$G_{q/11}$	2-Methylthio ATP ADPβS	Suramin

[a]APNEA: N^6-2-(4-aminophenyl)ethyladenosine; CGS21680: (2-*p*-carboxyethyl)phenylamino-5′-*N*-carboxamidoadenosine; CP66713: 4-amino-8-chloro-1-phenyl-[1,2,4]triazolo[4,3-α]-quinoxaline; DPCPX: 1,3-dipropyl-8-cyclopentylxanthine; I-ABOPX: 3-(3-iodo-4-aminobenzyl)-8-(4-oxyacetate)phenyl-1-propylxanthine; IB-MECA: N^6-iodobenzyl-*N*′-methylcarboxamidoadenosine; PAPA-APEC: 2-(4-[2-{(4-aminophenyl)methylcarbonyl}ethyl]phenyl)ethylamino-5′-*N*-ethylcarboxamidoadenosine; ZM241385: 4-(2-[7-amino-2-{2-furyl}-[1,2,4]triazolo{2,3-a}{1,3,5}triazin-5-yl-amino]ethyl)phenol.

hormone (FSH), luteinizing hormone/human chorionic gonadotropin (LH/CG), and thyroid-stimulating hormone (TSH), also interact with specific G-protein-coupled receptors, as do the hypothalamic-releasing peptides corticotropin releasing factor (CRF) and GnRH.

There are also a number of other G-protein-coupled receptors summarized in Table V. Receptors that bind purines were previously divided into two groups, the P_1 purinoceptors, which have high affinity for adenosine, and P_2, which preferentially bind ATP. Four subtypes of the adenosine receptor have been characterized, all of which are G-protein linked, the A_1 and A_3 coupled to AC via $G_{i/o}$ and the A_{2A} and A_{2B} via G_s. The P2 purinoceptors are divided into two categories: the P_{2T}, P_{2X}, and P_{2Z}, which belong to the ion channel superfamily of receptors, and the P_{2U} and P_{2Y}, which are members of the G-protein-coupled seven-transmembrane group.

In addition, G-protein-coupled receptors are employed by a wide range of stimuli and ligands: vision (rhodopsin) has been mentioned, and olfactory receptors are also G-protein-coupled. Receptors for leukotrienes, prostanoids, platelet activating factor (PAF), cannabinoids, and some viral proteins for example from cytomegalovirus, belong to this family.

Two remarkable members of the family are the thrombin receptor, which is activated by proteolytic cleavage by thrombin of the receptor's N terminus to liberate a peptide that interacts with the receptor to induce activation, and the calcium receptor, which has a very long N terminus that interacts with extracellular calcium to regulate PTH secretion from parathyroid cells. Finally, there is a group of "orphan receptors" identified as G-protein-coupled receptors on the basis of their structures, but for which the endogenous ligands are unknown.

For many small ligands, the binding pocket is within the transmembrane domains, whereas for peptides and glycoproteins the extracellular regions are also involved. Within various receptor families, it has been shown that certain key residues have a role in agonist binding; for example, in the adrenoceptors, a conserved aspartate in the third transmembrane domain is involved in binding catecholamines. Binding of endogenous peptides and nonpeptide antagonists appears to involve distinct sites, for example, in the NK_1 and AT_1 receptors.

Rapid desensitization of G-protein-coupled receptors during continued exposure to agonist often involves the phosphorylation of key serine and threonine residues on intracellular regions by protein kinase A or protein kinase C.

G-protein-coupled receptors interact with G-proteins via regions of the second and third intracellular loops and the proximal part of the C-terminal intracytoplasmic tail. G-proteins are heterotrimers composed of α, β, and γ subunits encoded by distinct genes. Each subunit type represents a family of genes with varying degrees of complexity. The most varied are those for α subunits, for which at least 17 genes have been identified. However, these α subunits form four main classes: G_s, G_i, G_q, and G_{12}.

G-proteins are linked to various effector and second-messenger systems, specific for their α subunit. These include activation (G_s) or inhibition (G_i) of adenylyl cyclase to modulate intracellular cyclic AMP. Subunits of the G_i type also open K^+ channels. Activation of phospholipase Cβ occurs by coupling with $G_{q/11}$ to liberate inositol 1,4,5-trisphosphate and diacyl glycerol, which elevate cytosolic Ca^{2+} and activate protein kinase C.

A cyclic GMP phosphodiesterase in retinal photoreceptors is activated by photoactivated rhodopsin via the G-protein transducin and thereby leads to a fall in cyclic GMP levels from closure of the cGMP-gated ion channel and membrane hyperpolarization.

In addition to these transduction mechanisms, there is also evidence that some ion channels may be directly coupled to G-proteins, such as the muscarinic-gated K^+ channel from cardiac atria. Neurotransmitters activate a variety of K^+ and Ca^{2+} channels in neurons by mechanisms involving G-proteins, and the abundant G-protein of brain, G_o, can reconstitute this activity. However, in these latter cases it is not known if this is a direct action or whether second messengers are involved.

C. Enzyme-Linked Receptors

There are a group of receptors that bear intrinsic enzyme catalytic sites on their cytoplasmic domains that mediate their action. The enzymes involved include tyrosine kinases, tyrosine phosphatases, serine/threonine kinases, and guanylyl cyclases.

Receptors for the natriuretic peptides, ANP, BNP, and CNP, interact with two receptors containing intrinsic guanylate cyclase activity, NPR_A and NPR_B, and one that does not, the NPR_C. The transforming growth factor β (TGF-β) receptor is a receptor with serine/threonine kinase activity, and the CD45 protein has intrinsic tyrosine phosphatase activity.

The largest group of receptors that bear intrinsic enzyme activity are the tyrosine kinase receptors, which include a group of growth factor-related receptors, those for insulin, insulin-like growth factors I and II (IGF-I/IGF-II), platelet-derived growth factor (PDGF-α and -β), epidermal growth factor (EGF), fibroblast growth factor (FGF R1–4), and nerve growth factor (NGF low-affinity receptor, Trk), and other neurotrophins. Most of these receptors consist of a single polypeptide chain that traverses the membrane once and bear tyrosine-specific protein kinase activity on their cytoplasmic domains, with the exception of insulin and IGF-I and IGF-II receptors. Activation of many of these receptors probably involves formation of a dimer generated by two chains of the molecule.

The first identified neurotrophic factor was NGF and now a family of related factors are known, including brain-derived neurotrophic factor, neurotrophin-3 (NT-3), neurotrophin-4/5 (NT-4/5), neurotrophin-6 (NT-6), and ciliary neurotrophic factor.

The neurotrophins bind to two different types of receptors, the p75 (low-affinity neurotrophin receptor) and the Trk tyrosine kinases; TrkA, TrkB, and TrkC were initially described as receptors for NGF, BDNF, and NT-3, respectively, although both TrkA and TrkB also bind NT-3 and NT-4/5. The CNTF receptor is not related to the tyrosine kinase Trk receptors but rather to hemopoietic cytokines. The CNTF receptor complex comprises the CNTF receptor, which binds CNTF, and gp130 and LIFR-β [which are shared with leukaemia inhibiting factor (LIF) and interleukin-6 (IL-6)] which are involved in signal transduction. Neurotrophins are involved in growth and survival of specific populations of neurons and are key molecules in neuronal development, maintenance, and regeneration.

Some oncogenes encode abnormal receptors that have catalytic domains that are constitutively active in the absence of receptor occupation. These include the *v-erbB* oncogene, which encodes a truncated version of the EGF receptor, the *sis* oncogene, which encodes an abnormal form of the PDGF receptor, and the *neu* oncogene, detected in some chemically induced nervous system tumors, which encodes a tyrosine-specific kinase, but the ligand for the normal receptor is not known.

D. Neurotransmitter Transporters

In the case of monoamine transmitters (e.g., dopamine, noradrenaline, and serotonin), amino acids (e.g., glutamate, aspartate, and glycine), and certain other amino acids and precursors, there are transporters that are responsible for high-affinity reuptake of the transmitters back into the nerve terminal coupled to sodium cotransport. This process is responsible for terminating synaptic action of transmitters, as well as conserving transmitter stores in the nerve terminal.

Neurotransmitter transporters belong to a family of Na^+- and Cl^--dependent symporters that have 11–13 transmembrane domains and are targets for antidepressant drugs, which often block the 5-HT reuptake process (e.g., fluoxetine, sertraline, paroxetine), and drugs of abuse, such as cocaine and amphetamines, which block the dopamine transporter. By blocking the reuptake transporter, these drugs lead to elevated synaptic concentrations of serotonin or dopamine, respectively, and this effect is believed to underlie their actions.

E. Voltage-Sensitive Ion Channels

These membrane proteins mediate rapid, voltage-gated changes in ion permeability in excitable cells, and also modulate membrane potential and thereby excitability. They are not receptors in the usual sense but are included here because they provide important components of the cellular response to receptor-mediated events, and they act as receptors for some drugs and toxins. Three main classes of voltage-sensitive channels have been studied: those for Na^+, Ca^{2+}, and K^+ (Fig. 3).

Voltage-sensitive Na^+ channels mediate rapid depolarization during the initial phase of the action potential. The channel proteins have been isolated with the use of five classes of neurotoxins, which bind tightly to separate sites on the channel and modify their functions. The Na^+ channel is a heterotrimeric complex that spans the membrane and shows voltage-dependent changes in ion conductance. The α subunit determines the major functional characteristics of the channel. It is a single polypeptide chain containing four homologous repeats, each with six transmembrane-spanning segments. Although it is associated with other regulatory subunits, the α subunit performs the basic functions of the voltage-gated ion channel. Multiple types of Na^+ channel have been distinguished by the structure of their α subunits and on the basis of their sensitivity to the toxins tetrodotoxin, saxitoxin, and μ-conotoxin.

Voltage-sensitive Ca^{2+} channels belong to multiple classes. One class, the L-type channel, is blocked by dihydropyridine drugs and has been isolated from skeletal muscle, although channels in the brain are probably structurally similar. This channel is a heteropentamer of which the α subunits of the Na^+ and Ca^{2+} channels show substantial homology, having four homologous repeats, each containing six transmembrane-spanning segments.

Ca^{2+} channels are activated by phosphorylation of their α_1 and β subunits by cAMP-dependent protein kinase. This phosphorylation probably underlies Ca^{2+}-channel regulation by β-adrenergic agonists.

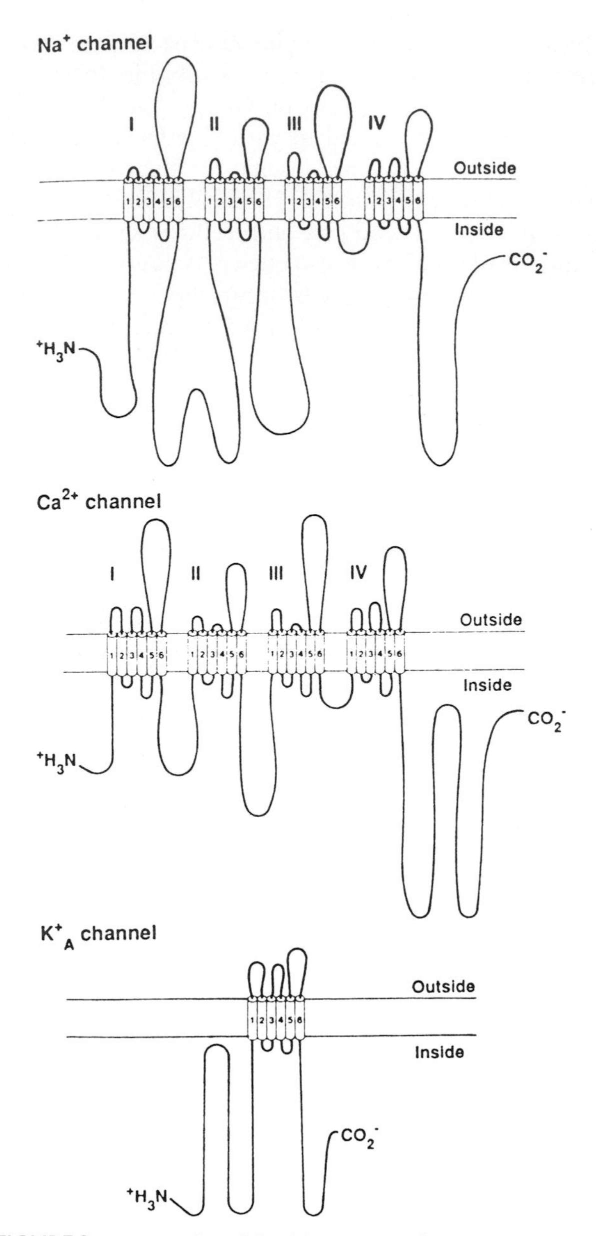

FIGURE 3 Proposed models of the transmembrane arrangements of the principal subunits of the Na^+, Ca^{2+}, and A-current K^+ channels. Both the Na^+ and Ca^{2+} channels have four domains with highly conserved amino acid sequences, and each domain contains multiple hydrophobic segments that form the transmembrane α helices. [Reproduced, with permission, form W. A. Catterall (1988). Structure and function of voltage-sensitive ion channels. *Science* 242, 50–61.]

The α_1 subunits provide the site for binding blockers of the channel, vary in type in skeletal, cardiac muscle, and brain, and are derived from separate genes.

N-type calcium channels occur only in neurons, are sensitive to ω-conotoxin, and appear to be involved in transmitter release. T-type calcium channels show slow deactivation and are involved in spike activity in neurons and endocrine cells. P-type channels occur in some CNS neurons, are non-inactivating, and are sensitive to ω-agatoxin.

Voltage-sensitive K^+ channels terminate the action potential by mediating the outward movement of K^+, which repolarizes the cell. K^+ channels also set the resting membrane potential and thereby modify the excitability of the cell. Its protein consists of a single polypeptide that has homology with the principal subunits of the Na^+ and Ca^{2+} channels. Six types of voltage-dependent K^+ channel have been identified and are distinct from Ca^{2+}-activated K^+ channels, receptor-operated channels (e.g., muscarinic acetylcholine) and ATP-regulated K^+ channels.

F. Nuclear Receptor Superfamily

Receptors for steroids, thyroid hormones, vitamin D_3, and retinoids are present in the cell cytosol. When occupied by the cognate ligand, they translocate to the nucleus, where they act as transacting DNA regulatory factors and regulate transcription of specific genes.

The steroid receptor family includes receptors for estrogens (ER), androgens (AR), progesterone (PR), glucocorticoids (GR), and mineralocorticoids (MR). Retinoid receptors include the retinoic acid receptor (RAR α, β, γ, and subtypes) and the RXR receptors (RXR α, β, γ_1, and γ_2). Although they are not neurotransmitter receptors as such, they play important roles in longer-term effects on cells and gene transcription.

IV. FUNCTIONAL IMPORTANCE OF DIFFERENT RECEPTOR CLASSES

The various classes of receptor discussed earlier have different kinetic properties and signaling mechanisms: the ligand-gated ion channels have relatively low affinity for their ligands and this enables rapid on and off rates, which are ideally suited to fast neurotransmission.

The G-protein-coupled seven-transmembrane receptors have functionally separated receptor binding and subsequent effector activation by G-proteins. This enables relatively high affinity (thereby enabling responses to low levels of transmitter), with a relatively rapid response due to amplification of the signal by second-messenger pathways. This system has also evolved great flexibility by virtue of coupling to a range of G-proteins to activate multiple different signaling pathways.

The receptor tyrosine kinases generally display high affinity and consequently a slow off rate, which is appropriate for roles in cell proliferation, differentiation, and as neurotrophic maintenance and cell survival factors. The cytosolic receptors of the steroids/thyroid/retinoic acid/vitamin D family are able to regulate the transcription of multiple specific genes in a coordinated fashion to regulate the activity of many neuronal functions and transmitter systems.

The ability of a single neurotransmitter to activate multiple structurally distinct receptor families, and the colocalization of neurotransmitters and neuropeptides, enables highly complex signals to be transmitted. When this signal is combined with a number of divergent and convergent second-messenger systems, a vast array of actions, all with different time scales, can be generated.

V. INVESTIGATION OF NEURORECEPTORS IN THE LIVING HUMAN BRAIN

Using single-photon emission tomography (SPECT) or positron emission tomography (PET) with suitably radiolabeled ligands, it is possible to image and quantify a range of neuroreceptors in the living human brain. These include SPECT and PET ligands for the dopamine D_1 and D_2 families, serotonin 5-HT_2 receptors, $GABA_A$/benzodiazepine receptor complex, presynaptic dopamine reuptake transporters, and even neurotransmitter-metabolizing enzymes such as monoamine oxidase A and B (Table VI).

VI. LOCALIZATION AND PROPERTIES OF RECEPTORS IN THE HUMAN BRAIN

Receptors were originally characterized pharmacologically using biological responses to specific agonists and antagonists. On this basis, many subtypes of receptors were defined. The distribution of receptor subtypes can also be investigated using radiolabeled agonists or antagonists. Early work on receptor mapping involved determination of receptor sites by

TABLE VI
Neuroreceptor Ligand for Single-Photon Emission Tomography and Positron Emission Tomography

Receptor	Subtype	Ligand
Acetylcholine	Muscarinic	$[^{123}I]$IQNB
		$[^{123}I]$Iododexetimide
		$[^{11}C]$- and $[^{18}F]$Dexetimide
		$[^{11}C]$TBR
		$[^{11}C]$Scopolamine
		$[^{11}C]$Benzotropane
	Nicotinic	$[^{11}C]$Nicotine
Dopamine	D_1	$[^{11}C]$SCH23390
		$[^{76}Br]$SKF83566
		$[^{123}I]$SCH38840
	D_2	$[^{11}C]$Raclopride
		$[^{11}C]$-*N*-Methyl-spiperone
		$[^{76}Br]$-4-Bromospiperone
		$[^{18}F]$-(3-*N*-Methyl)benperidol
		$[^{76}Br]$Lisuride
		$[^{123}I]$Iodobenzamide
		$[^{123}I]$Iodospiperone
		$[^{123}I]$Iodolisuride
5-Hydroxytryptamine		$[^{123}I]$- and $[^{76}Br]$LSD
		$[^{76}Br]$- and $[^{11}C]$-*N*-Methyl-Br-LSD
		$[^{11}C]$- and $[^{123}I]$Ketanserin
		$[^{18}F]$Setoperone
		$[^{18}F]$Altanserin
Benzodiazepine		$[^{11}C]$Flumazenil
		$[^{123}I]$Iomazenil

the binding of radioligand to membrane fractions. A refinement of radioligand binding, *in vitro* autoradiography, involves labeling receptor sites on slide-mounted tissue sections. This provides receptor localization in much greater detail and cellular resolution. More recently, the availability of nucleic acid sequences of many receptors enables the use of cDNA and oligonucleotide probes to determine the sites of synthesis of receptors by *in situ* hybridization histochemistry.

Studies of the localization of receptors in the human brain are not as advanced as those in lower mammals such as the rat. The receptors discussed in the following sections have all been studied in the human brain. In general, each receptor subtype shows a highly characteristic distribution. These patterns give clues to the roles in the overall brain, of neurotransmitter systems, as well as the effects of drugs that interact with these receptors.

A. Acetylcholine Receptors

Nicotinic cholinergic receptors are highest in the striatum, nucleus basalis of Meynert, and substantia nigra, with lower levels in the hippocampus, thalamus, and neocortex.

The cholinergic muscarinic M_1 receptor is concentrated in the external layers of the cerebral cortex (the highest density in layers I and II and the lowest levels in layer V), in the innermost layers of the primary visual cortex, in the CA fields and dentate gyrus of the hippocampus, and in the basal ganglia. In contrast, the M_2 receptor subtype is more densely distributed in the caudate nucleus, thalamus (particularly the anterior group of nuclei), amygdala, substantia innominata, hypothalamus, and brain stem, with much lower levels in the cortex and hippocampus. [*See* Brain.]

Deficits in the cortical and hippocampal cholinergic nicotinic and muscarinic M_2 receptors occur in Alzheimer's dementia, and cholinergic agonists may improve memory in Alzheimer's patients. These observations suggest a role for the central cholinergic system in cognitive and behavioral functions. High concentrations of muscarinic and nicotinic receptors in the human cortex, hippocampus, and substantia innominata support this hypothesis. Moreover studies of nicotinic neurotransmitter activity using PET and SPECT show a positive correlation between cognitive function and uptake of S-^{11}C-nicotine in the temporal cortex of Alzheimer's patients. [*See* Alzheimer's Disease.]

B. Adrenoceptors

The α_2-adrenergic receptor subtype is present in high concentrations in the frontal cortex and in the hypothalamus, with lower levels in the hippocampus and cerebellum. The locus coeruleus and the nucleus of the solitary tract also contain significant densities of α_2 receptors. These brain nuclei, together with certain hypothalamic areas, are important sites for the central regulation of blood pressure and may be important targets for the centrally acting antihypertensive agents. Most of these α_2-adrenergic receptors are predominantly α_{2A}, since α_{2B} receptors are detected only in the striatum and globus pallidus.

The β-adrenergic receptor occurs in high densities in the external cortical layers, hippocampus, striatum, lateral posterior thalamus, and the granular layer of the cerebellum. Lower levels are found in the pallidum, medial thalamic nucleus, substantia nigra, and the red nucleus. The receptors in the hippocampus and cerebellum are predominantly β_2, whereas those in the frontal cortex and the basal ganglia contain mostly the β_1 receptor subtype. The β_1 receptor in the striatum occurs in the matrix compartment and is increased in schizophrenic brains.

Alterations in the central adrenergic system have been implicated in the pathogenesis of certain psychiatric disorders, including schizophrenia and manic-depressive disorder. The high levels of adrenergic receptors in the cortex and hippocampus may mediate these effects. [*See* Mood Disorders; Schizophrenic Disorders.]

C. Dopamine Receptors

The D_1 and D_2 receptors, and their mRNAs, are concentrated in the basal ganglia, where the D_1 receptor occurs predominantly in the striatum, internal globus pallidus, and substantia nigra, whereas the D_2 occurs in the striatum, nucleus accumbens, and compact part of the substantia nigra. In addition, high densities of D_2 receptors are also present in the olfactory tubercle. Lower levels of D_1 receptors occur in the amygdala, mammillary bodies, and cerebral cortex, and both D_1 and D_2 receptors are present in the hippocampus. D_3 receptors are found in the ventral striatum and nucleus accumbens, with lower levels in the striatum and cerebral and cerebellar cortices.

D_2 receptors are primary targets for drugs used in the treatment of psychosis (the neuroleptics). The neuroleptic drugs are thought to elicit their inhibitory actions on the dopaminergic mesolimbic system, which includes the nucleus accumbens and olfactory tubercle, brain regions that are rich in D_2 receptors.

D. 5-Hydroxytryptamine Receptors

The 5-HT receptors are widely distributed in the human brain and have been characterized into many different subtypes depending on their pharmacology, molecular structure, and signaling pathways, as discussed earlier.

High concentrations of 5-HT_{1A} receptors occur in the external layers of the cerebral cortex, layers V of the anterior entorhinal cortex, pyramidal and lacunosum molecular layers of the hippocampus, cortical and granular nuclei of the amygdala, and dorsal, central, and linear raphe nuclei. The high concentrations of 5-HT_{1A} receptors in the limbic system suggest involvement of these receptors in emotion and visceral functions. This is supported by the observation of anxiolytic properties of certain 5-HT_{1A} agonists.

The 5-HT_{1B} receptor is found only in rat and mouse and the human equivalent, with 97% homology, is 5-$HT_{1D\beta}$. High densities of 5-HT_{1D} receptors occur in the basal ganglia, with a lower, homogeneous distribution throughout all the layers of the cortex and hippocampus. Although this receptor subtype is the predominant 5-HT_1 class of receptor in the human brain, it occurs only in higher mammals and is absent in rodents.

The highest levels of 5-HT_2 receptors are found in layers III and VA of the isocortex and layer III of the entorhinal cortex. A lower density of receptors occurs in the lateral nucleus of the amygdala, claustrum, striatum, mammillary bodies, substantia innominata, and CA_1 of the hippocampus. The presence of 5-HT_2 receptors in the cortex and in components of the limbic system supports a role of these receptors in anxiety, depression, and dementia. These receptors are decreased after chronic antidepressant treatment and in patients with Alzheimer's dementia.

5-HT_{2C} receptors are present in high concentrations in the choroid plexus and globus pallidus, with lower levels in the striatum, nucleus accumbens, pyramidal layer of the hippocampus, ventromedial nucleus and mammillary bodies of the hypothalamus, and layers II and IV of the cortex.

5-HT_3 receptors are distributed in a distinct pattern with high concentrations in the dorsal vagal complex, area postrema, and substantia gelatinosa of the medulla oblongata. Lower densities of receptors are present in components of the limbic system. 5-HT_3 antagonists are effective antiemetics.

E. GABA Receptors

GABA acts on two receptor classes, the bicuculline-sensitive $GABA_A$ receptor (a ligand-gated Cl^- channel) and the G-protein-linked, bicuculline-insensitive $GABA_B$ receptor. Benzodiazepines bind to a modulatory site on the $GABA_A$ receptor and there are other sites that bind barbiturates, alcohol, and neurosteroids.

The $GABA_A$ receptor occurs in high densities in layer IV of the primary visual cortex and layers III and IV of the cerebral cortex and cerebellum, with lower levels in the basal ganglia and the pyramidal layer and dentate gyrus of the hippocampus. Similarly, benzodiazepine binding sites are found in high concentrations in the cerebral cortex, hippocampus, amygdala, nucleus accumbens, and mammillary bodies, with lower levels in the basal ganglia and thalamic and hypothalamic nuclei. Because GABA is a major inhibitory neurotransmitter in the human brain, and the $GABA_A$ receptor complex includes a binding site for benzodiazepines, a range of clinically important drugs has been found to act via this receptor complex. Alterations in central $GABA_A$ receptor function have been implicated in disorders such as epilepsy, anxiety, and insomnia.

$GABA_B$ receptors were initially characterized in peripheral sites such as autonomic nerve terminals but have since been identified in mammalian brains, including the human brain, where it occurs in high concentrations in frontal cortex and hippocampus.

F. Glycine Receptors

Receptors for glycine occur in high concentrations in the periaqueductal gray and the oculomotor nuclei of the midbrain, and in the motor and sensory trigeminal, facial, hypoglossal, cuneate, and gracile nuclei of the medulla oblongata. The concentration of glycine receptors in the brain stem is in agreement with its role as the major inhibitory neurotransmitter in these brain regions. These are possible sites where drugs such as strychnine, which binds to the glycine receptor, may act. Toxic doses of strychnine cause death by convulsions.

G. Excitatory Amino Acid Receptors

There are at least four different groups of excitatory amino acid receptors, the ionotropic receptors—NMDA, AMPA, and kainate—and the metabotropic receptors discussed earlier. High densities of the NMDA and AMPA receptors are present in the external layers of the cerebral cortex and in the CA_1 and dentate gyrus of the hippocampus. High concentrations of NMDA receptors also occur in the primary visual cortex, the amygdala, the mediodorsal nucleus of the thalamus, and the granular layer of the cerebellum, with lower levels in the striatum. AMPA receptors are present in CA_3 and CA_4 of the hippocampus and the molecular layer of the cerebellum.

In contrast, kainate receptors are concentrated in the deep and superficial layers of the cortex. They are also present in the CA_3 and dentate gyrus of the hippocampus and in the granular layer of the cerebellum. Moderate levels of the kainate receptors are found in the striatum.

H. Opioid Receptors

Drugs that act on the opioid receptors elicit effects on mood, motor and autonomic control, and pain perception. There are three well-characterized opioid receptors—the μ, δ, and κ receptors—and within these three groups there appear to be pharmacologically distinct subtypes.

All opioid receptor subtypes occur in high densities in the cerebral cortex, where the μ and δ receptors are concentrated in superficial and the κ in the deep layers of the cortex. The μ and δ receptors are also densely distributed in the visual and auditory cortex.

The μ receptor subtype is the most widespread opioid receptor in the human central nervous system and displays high affinities for morphine and methadone. The μ receptor is also present in high levels in the amygdala, the molecular layer of the cerebellum, and striatum, with lower levels in the pallidum, bed nucleus, nucleus basalis of Meynert, and laterodorsal, medial, and pulvinar nuclei of the thalamus. Messenger RNA for the μ receptor is highly expressed in the hypothalamus, thalamus, and subthalamic nucleus, with lower levels in the amygdala and caudate nucleus.

High densities of δ receptors occur in the dentate gyrus of the hippocampus, with moderate levels in the basal ganglia, amygdala, claustrum, and thalamus. The mRNA for the δ receptor is expressed in most of the regions where the protein is found except for the thalamus.

The κ receptors are found in high concentrations in the amygdala, claustrum, medial and ventrolateral nucleus of the thalamus, ventromedial and posterior nuclei of the hypothalamus, and the cerebellum. [*See* Opioids.]

Sigma receptors, which were initially classified as opioid receptors, include binding sites for phencyclidine and are found in high concentrations in the substantia nigra pars compacta and cerebellum, with lower levels in the neocortex (laminae II–IVA), granular layer of the hippocampus, and striatum.

I. Other Peptide Receptors

Many peptide receptors have been identified in the human brain. One of the best-studied peptide receptors, the CCK_B receptor, occurs in high densities in the glomerular and external plexiform layer of the olfactory bulb, layer IV of the cortex, and pontine nuclei, with lower levels in the nucleus accumbens and striatum. CCK_8, the primary form of CCK in the brain, is thought to regulate satiety but exerts many influences on brain peptide and monoamine systems and modulates dopamine release and turnover in the striatum, which has high concentrations of the CCK_B receptor.

Somatostatin receptors are highly concentrated in the deep layers of the cortex, in the basal ganglia, and in the hippocampus, amygdala, and habenula nucleus. Lower levels of this receptor occur in the substantia gelatinosa. The discrete distribution of somatostatin suggests that the peptide may be involved in cognitive and sensory processing. Multiple somatostatin receptor subtypes have been characterized and most of them are expressed in the cortex and hippocampus. However, in some brain regions, these receptor subtypes are differentially expressed—the spinal cord has abundant sst_1 mRNA, whereas sst_3 mRNA is predominantly found in the olfactory bulb and cerebellum.

Angiotensin II AT_1 receptors are found in high densities in the forebrain circumventricular organs, in the medium preoptic, paraventricular, and arcuate nuclei of the hypothalamus, in the striatum and compact part of the substantia nigra, and in nuclei associated with the autonomic nervous system in the medulla oblongata. The presence of high concentrations of angiotensin II receptors in many regions of the hypothalamus and in the medulla oblongata is consistent with the proposed role of central angiotensin II. The cerebellar cortex contains both AT_1 and AT_2 receptors.

Calcitonin receptors are concentrated in the anterior and posterior hypothalamus, median eminence, inferior colliculus, and substantia nigra, with lower levels in the basal ganglia and hippocampus. Calcitonin is thought to inhibit appetite and to influence gastric acid and pituitary hormone secretion. The dense distribution of its receptors in the hypothalamus may represent sites where calcitonin can elicit these actions.

Calcitonin gene-related peptide occurs from the alternate splicing of the calcitonin gene. Receptors for CGRP are found in high concentrations in the cerebellum, inferior olivary complex, pontine nuclei, the basal ganglia, dorsal vagal complex, and locus coeruleus. CGRP is thought to regulate sympathetic noradrenergic outflow and may influence somatosensory function, actions that may be mediated via these sites.

Neurotensin receptors occur in high concentrations in the substantia nigra and hippocampus, where they are concentrated in the presubiculum and entorhinal area. Complementary to the receptor distribution is demonstration of expression of neurotensin receptor mRNA in the substantia nigra pars compacta, inferior colliculus, red nucleus, and striatum.

The predominant form of neuropeptide Y receptor in the human brain appears to be the Y_2 subtype and is localized in the pyramidal layer of the hippocampus, caudate nucleus, locus coeruleus, and substantia nigra. The Y_1 receptor occurs in conjunction with the Y_2 in the molecular layer of the dentate gyrus, in the oriens and radiatum layers of the hippocampus, and in the frontal cortex.

The distribution of vasopressin receptors differs from that of oxytocin receptors in the human forebrain. The vasopressin V_1 receptors are concentrated in the lateral septum, midline thalamic nuclei, dentate gyrus, and basal amygdaloid nucleus, whereas the oxytocin receptors are found in the basal nucleus of Meynert, vertical limb of the diagonal band, preoptic and anterior hypothalamus, globus pallidus, and ventral pallidum. However, the receptor distributions overlap in the brain stem.

J. Adenosine Receptors

Receptor sites for adenosine A_1 occur in high concentrations in the hippocampus, cerebral cortex, striatum, and anterior and medial nuclei of the thalamus. Moderate levels occur in the amygdala, olfactory tubercle, and nucleus accumbens.

Adenosine is a potent sedative and must act as an endogenous acticonvulsant. The high concentration of A_1 in the cortex and hippocampus may represent sites where adenosine can mediate these actions.

K. Histamine Receptors

The histamine H_1 receptor occurs in high concentrations in the cerebral cortex and in parts of the limbic system such as the cingulate and orbital cortex, amygdala, and uncus. Moderate levels of this receptor are also present in the striatum, hippocampus, septum, and the hypothalamus. Some of the most potent inhibitors of the H_1 receptors are tricyclic antidepressants. The brain H_1 receptors are thought to be involved in arousal and regulation of appetite, and one of the side effects of H_1 antagonist antidepressants is drowsiness.

Histamine H_2 receptor-mediated release of histamine has been reported from human cortical brain slices, suggesting the presence of H_3 receptors in the human central nervous system.

VII. CONCLUSION

The human central nervous system employs multiple classes of chemical transmitters, and in general each interacts with more than one type of cell-surface receptor.

Neuroreceptors have evolved from a discrete number of molecules to generate recognizable families. Initial classification of these receptors was based on their physiological ligands, later on pharmacological differences in agonist and antagonist actions, and most recently on their structures as revealed by molecular cloning and their signalling mechanisms and pathways. The resulting reclassification of receptors has provided a more rational and robust understanding of these important molecules.

Quite often various receptors for a given transmitter belong to different families and differ markedly in their transmembrane signaling, second-messenger systems, and final intracellular response. However, the converse is also true—different classes of neurotransmitter may have similar intracellular effects via different receptors that converge in their second-messenger pathways or gene regulation.

Many neurological and psychiatric disorders may result from abnormalities of these chemical communication systems. Of the large number of drugs that affect brain function, most act by binding to receptors and either stimulating or blocking their function. Although a bewildering number of receptors are already known, the list is almost certainly incomplete—new members of known receptor families are being rapidly discovered by the techniques of molecular biology. Furthermore, there is no reason to believe that the range of known neurotransmitters, particularly neuropeptides, represents the complete list. Extrapolation from studies on the molluscan nervous system suggests that many more neuropeptides may be discovered in mammals. Therefore the field of brain receptors is likely to continue its rapid development.

The ability to image and quantify neuroreceptors and enzymes in the living human brain using SPECT and PET technology, plus the rapid advances in basic neurosciences at the molecular level, presages exciting advances in understanding the functions of the normal brain and its neurological and psychiatric disorders.

BIBLIOGRAPHY

Berridge, M. J., and Irvine, R. F. (1989). Inositol phosphates and cell signalling. *Nature* **341**, 197–205.

Catterall, W. A. (1988). Structure and functions of voltage-sensitive ion channels. *Science* **242**, 50–61.

Caulfield, M. P. (1993). Muscarinic receptors: Characterization, coupling and function. *Pharmacol. Ther.* **58**, 319–379.

Gingrich, J. A., and Caron, M. G. (1993). Recent advances in the molecular biology of dopamine receptors. *Annu. Rev. Neurosci.* **16**, 299–321.

Humphreys, P. P. A., Hartig, P., and Hoyer, D. (1993). A proposed new nomenclature for 5-HT receptors. *Trends Pharmacol. Sci.* **14**, 233–236.

Kobilka, B. K. (1992). Adrenergic receptors as models for G protein-coupled receptors. *Annu. Rev. Neurosci.* **15**, 87–114.

Neer, E. J., and Clapham, D. E. (1988). Roles of G-protein subunits in transmembrane signalling. *Nature* **333**, 129–134.

Nicoll, R. A. (1988). The coupling of neurotransmitter receptors to ion channels in the brain. *Science* **241**, 545–551.

Reisine, T., and Bell, G. I. (1993). Molecular biology of opioid receptors. *Trends Neurosci.* **16**, 506–510.

Sargent, P. (1993). The diversity of neuronal nicotinic acetylcholine receptors. *Annu. Rev. Neurosci.* **16**, 403–443.

Watson, S., and Arkinstall, S. (1994). "The G-Protein Linked Receptor—Facts Book." Academic Press, London.

Watson, S., and Girdlestone, D. (1995). 1995 receptor and ion channel nomenclature supplement. *Trends Pharmacol. Sci.* Seventh edition.

Neurotrophic Factors

THEO HAGG
Dalhousie University

GLOSSARY

Axonal regeneration New outgrowth of a severed nerve fiber (axon); compare to collateral sprouting, which is the formation of a new axon from an existing intact axon

Factor Soluble protein for signaling between cells (intercellular), similar to a cytokine, which is a term from the immunology field; they are larger than neuropeptides, which are involved in the modulation of synaptic transmission

Neural Pertaining to nervous system tissue or cells (neurons, oligodendrocytes, astrocytes, microglia, Schwann cells)

Neurite Long, thin extension (process) emanating from a nerve cell (neuron) body in tissue culture; the *in vivo* equivalents are axons and dendrites

Neuronal Pertaining to nerve cells (neurons)

Neurotrophins Family of the related neurotrophic factors: nerve growth factor, brain-derived neurotrophic factor, and neurotrophin-3, -4/5, and -6

Trophic Maintaining or promoting growth in cell body size (as in trophic factor), generally also indicating support of neural cell survival; compare to "growth" (as in growth factor), which indicates growth of cell population, that is, proliferation

NEUROTROPHIC FACTORS ARE SPECIALIZED PROteins that act as intercellular signaling molecules and, by activating cell-surface receptors, regulate neuronal cell behaviors, including survival, size, process outgrowth, and function. They are generally found in the nervous system or target areas to which the neurons project, but their activity is not limited to neurons or to the nervous system. The discovery of nerve growth factor over 40 years ago was a major evolution in neurobiology. It has had an enormous impact on the understanding of biological processes as diverse as normal nervous system development and the potential to use these factors in various neurological disorders. The neurotrophic factor field is rapidly expanding and newly formulated biological concepts are likely to continue to evolve and change over the coming years.

I. HISTORY OF NEUROTROPHIC FACTORS

As early as the beginning of this century, the now famous Spanish neuroscientist Santiago Ramon Y Cajal had formulated the concept that "trophic" substances in the nervous system could influence the behavior of neurons (nerve cells). To better understand the mechanisms that underlie poor regrowth of injured nerve fibers in the central nervous system (CNS: brain, spinal cord, and optic nerve), Cajal and his colleague F. Tello implanted peripheral nerve grafts into the brain of adult rabbits. They observed that over time the endings of severed nerve fibers in the brain would sprout, change direction, and grow toward these grafts. Cajal interpreted this to mean that "trophic" agents had diffused from the peripheral nerve environment into the brain.

In 1947, Rita Levi-Montalcini from Italy joined the laboratory of Viktor Hamburger in St. Louis, Missouri, to resolve and investigate the developmental

ENCYCLOPEDIA OF HUMAN BIOLOGY, Second Edition, VOLUME 6.

mechanisms that cause a correct number of peripheral nervous system neurons to innervate the peripheral targets. She repeated one of her previous studies of development of spinal ganglia neurons of a chick embryo. She showed that the number of neurons increased during early developmental stages, but during the time their neurites were reaching their target, the number of neurons began to decrease again to a final number. Moreover, several days after removal of a wing bud, she observed many degenerating neurons in the contributing ganglia. This led Hamburger and Levi-Montalcini to conclude that "the periphery provides for conditions necessary for continued growth and maintenance of neurons in stages following the first outgrowth of neurites." This was the basis for what has become a general neurotrophic hypothesis, that is, during development, when more than the required number of neurons innervate their target tissue, the latter produces limiting amounts of trophic factors that are essential for neuronal survival. Those neurons that do not make an optimum connection with the target will die by a process called programmed or developmental cell death. It is believed that this mechanism guarantees the formation of sufficient and optimum connection patterns.

In the following years, Hamburger and Levi-Montalcini set out to isolate what they assumed to be a chemical agent provided to the neurites by the tissues of the limb bud. A former student of Hamburger, Elmer Bueker, had identified two sarcoma tumor cell lines that caused ingrowth of sensory fibers and enlargement of ganglia when implanted into embryos. He concluded that these tumors were producing and releasing a chemical substance. In 1952, Levi-Montalcini found that cultured sensory ganglia would grow a dense halo of neurites when placed near a piece of this tumor. In 1953, the biochemist Stanley Cohen joined Hamburger and Levi-Montalcini and, using this bioassay, he isolated an active nucleoprotein fraction from mouse sarcoma tumor. Cohen used snake venom as a source of phosphodiesterase to remove nucleic acids from the fractions and showed that the activity was in the protein fraction. However, to their surprise, the venom itself was 1000 times more active than the sarcoma fractions. By 1956 they had purified a protein from snake venom that they named "nerve growth promoting factor" for its obvious nerve fiber outgrowth-promoting activities. It was subsequently abbreviated to what we now know as nerve growth factor or NGF. The further isolation and biological characterization of NGF was greatly facilitated by the insight of Cohen that the organ that produces venom in the snake is similar to the mammalian salivary gland. They found that male (not female) mouse (not rat) submaxillary glands were an even richer source of NGF than snake venom. This abundant and readily available source of NGF also determined the further history of the neurotrophic factor field. Until the recent emergence of modern molecular biological tools, the purification and identification of other factors was hampered by the very low amounts of such substances in the brain or periphery. For instance, the amount of NGF in brain and target tissues is 100 million times lower than in the salivary glands.

Levi and Cohen went on to establish that the NGF protein was responsible for the development of the spinal sensory and sympathetic neurons. NGF injected into newborn mice caused hypertrophy and injection of antibodies resulted in atrophy and degeneration of the neurons. During his purification steps, Cohen had injected a submaxillary gland fraction into newborn mice and observed that their eyelids opened and their teeth erupted earlier than normal. After further purification, the effect disappeared and he concluded that there were two activities in the earlier fraction. Cohen later followed up this observation, which led to his discovery of epidermal growth factor, which stimulates the proliferation of epidermal cells and therefore had caused the accelerated eye development and tooth growth in the mice. The paramount importance of Levi-Montalcini and Cohen's discoveries, and the implication that specific proteins can act as intercellular signals to regulate cell behavior (survival, differentiation, proliferation, etc.), was appropriately recognized by awarding them the 1986 Nobel Prize in Medicine.

It took about 15 years after its purification and the emergence of molecular biology to determine the amino acid sequence of mouse NGF and another 12 years to determine the human sequence. Similarly delayed was the cloning and sequencing of the "second" factors, CNTF (ciliary neurotrophic factor) and BDNF (brain-derived neurotrophic factor) in 1989. Only then did these "new" factors, whose activity had been discovered many years before, become available in sufficient amounts for many more investigators to become involved and for *in vivo* experimentation in adult animals. Also, since then, several family members, their respective receptors, and other new factors have been cloned and the neurotrophic factor field has gained tremendous momentum (Fig. 1). [*See* Nerve Regeneration.]

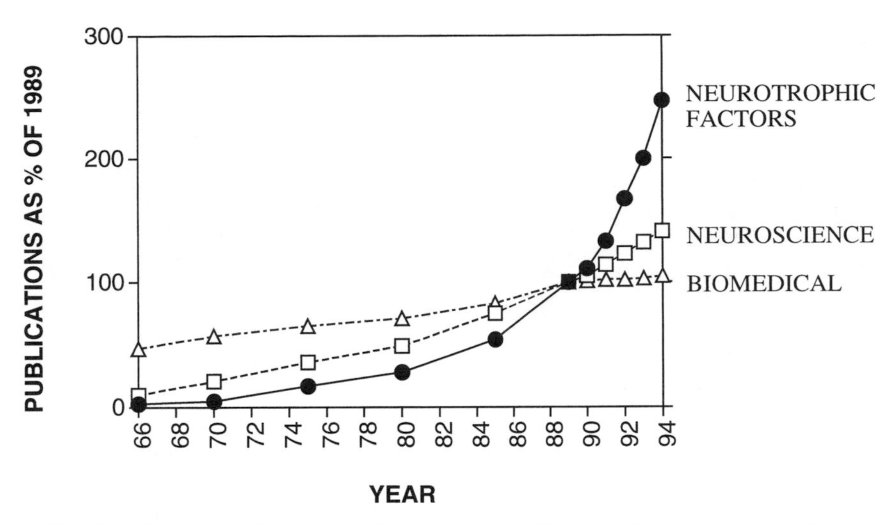

FIGURE 1 Recent rapid increase in the number of publications about neurotrophic factors. Numbers represent those publications listed on MEDLINE that contain terms related to neurotrophic factors and are expressed as a percentage of those in 1989. For comparison, the relative percentages of publications with terms pertaining to neurons since 1989 are given as an indication of the steady growth of the neuroscience field as a whole. Also illustrated is the relative percentage of all publications on MEDLINE since 1989 as an indication of activity in the biomedical sciences.

II. NEUROTROPHIC FACTOR FAMILIES

A. Neurotrophins

The determination of the amino acid sequence of BDNF revealed its homology to NGF and led to the subsequent rapid discovery of other members of the "neurotrophin" family that have similarly conserved sequences. These are neurotrophin-3 (NT-3), neurotrophin-4/5 (NT-4/5; initially distinguished as *Xenopus* neurotrophin-4 and mammalian neurotrophin-5), and the most recent addition, neurotrophin-6 from fish.

These members have similar protein structures and biological activities for neural cells. They consist of two identical protein strands that are attached (bonded) to each other and form a so-called homodimer. Each strand is around 120 amino acids long (any two neurotrophins are ~50–60% identical) and migrates on Western blots at around 13 kDa; the molecules are highly basic (pI 9–10.5). Six regions in the monomers have cystines, which form three disulfide bonds to provide a three-dimensional structure and several loops (folds) to the molecules. Different regions are important for specific binding of the neurotrophins to their preferred receptor or to a common receptor (see the following). The neurotrophin protein strands are processed from their individual precursor peptide, which derives from individual mRNA and genes. All neurotrophins, with the exception of NT-6, have signal sequences that are needed for classic secretion from cells through the endoplasmic reticulum and Golgi apparatus. The human genes for NGF, BDNF, NT-3, and NT-4/5 are located on, respectively, chromosomes 1, 11, 12, and 19.

Each member of the receptor family, which is activated by the neurotrophins and transduces the signals into the cell, is a transmembrane tyrosine kinase or Trk (pronounced Track) (Fig. 2), a product of the trk protooncogenes. There is some overlap between the binding abilities of the neurotrophins to their receptors. NGF binds specifically to and activates a receptor named TrkA (often referred to as Trk), BDNF and NT-4/5 preferentially activate TrkB, and NT-3 primarily activates TrkC and in some cell types TrkB. TrkB and TrkC also exist as truncated receptors without an intracellular domain, that is, they can bind the particular neurotrophins but do not transduce their intracellular signals. These truncated "binding proteins" may function as buffers to create stable neurotrophin concentrations and thus keep them in a physiological range. The homodimers of the neurotrophins induce dimerization of

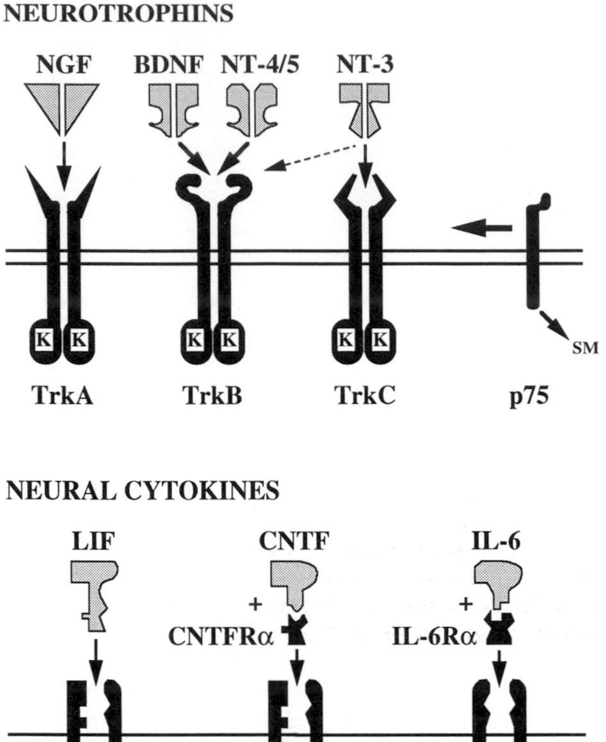

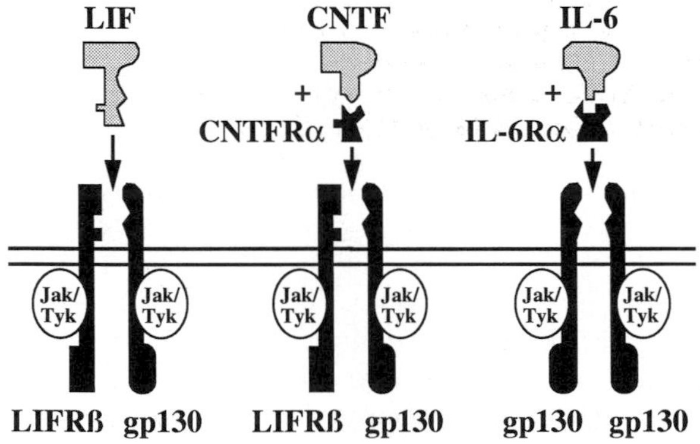

FIGURE 2 Illustration of receptors for neurotrophins and neural cytokines. Neurotrophins bind with high affinity to their specific extracellular Trk receptor domains and cause receptor dimerization that results in receptor autophosphorylation and activation. NT-3 has some affinity for TrkB. K indicates intracellular tyrosine kinase domains. Neural cytokines induce autophosphorylation of receptor complexes consisting of shared components that interact with Jak/Tyk-type intracellular tyrosine kinases. CNTF uses a specifying soluble CNTF α receptor to bind to a heterodimer of LIF β receptor and gp130. IL-6 uses IL-6α receptor to bind to a homodimer of gp130.

the two identical Trk receptors and cause increased autophosphorylation of their intracellular tyrosine kinase portions. This causes subsequent activation of a number of second-messenger systems. Through targeted mutations of the amino acid composition of the neurotrophins and the determination of the crystal structure of NGF, the various binding domains of the neurotrophins are being discovered. This new information is helping to elucidate various aspects of neurotrophin receptor interaction and is a target to develop therapeutic agents for human (neurological) diseases. The human genes for TrkA, TrkB, and TrkC are located on, respectively, chromosomes 1, 9, and 15.

All the neurotrophins can also bind to another single transmembrane receptor that was initially named low-affinity NGF receptor (LNGFR) and is the product of a different gene than the trk family (see Fig. 2). The predicted structure of this receptor is similar to that of a class of receptors that induces one of the tumor necrosis factor receptors, the Fas receptor, and CD40. This receptor has subsequently been named p75 for its 75 kDa "weight" (not to be confused with the p75 receptor for tumor necrosis factor), $p75^{LNGFR}$, and LANR (low-affinity neurotrophin receptor) to indicate its binding ability for all neurotrophins. The physiological relationship of p75 to the Trk receptors has not been fully resolved. The consensus at present is that although p75 is not needed for Trk transduction, it can enhance severalfold the autophosphorylation of Trk in the presence of ligand (e.g., NGF). In the absence of NGF ligand, p75 may function to decrease the tendency of TrkA to autophosphorylate, thus preventing ongoing signaling. The p75 receptor also appears to have Trk-independent signaling abilities, that is, binding by neurotrophin activates the sphingomyelin cycle and subsequent ceramide (a second messenger) formation. Through this pathway, p75 can mediate death (apoptosis) of certain cells. The human gene for p75 is located on chromosome 17.

B. Neural Cytokines

The "second" neurotrophic factor was discovered during the late 1970s and was named ciliary neurotrophic factor for its survival-promoting activity for ciliary ganglion neurons, which had been used as a bioassay to identify the activity. However, the amino acid sequence of CNTF was not determined until 1989 and showed CNTF to be completely unrelated to the neurotrophins. CNTF has very little amino acid homology with other factors and was initially thought to constitute a whole new class or type. However, subsequent prediction of its tertiary structure and the elucidation of its receptor complex revealed that CNTF is a distant relative of a cytokine family that includes LIF (leukemia inhibitory factor), IL-6 (interleukin-6), and OSM (oncostatin-M). Most of these cytokines were discovered for their activities in the immune system, where most of their normal functions occur, although some have been found to also have activities on neurons. CNTF, on the other hand, is found almost exclusively in the nervous system, to which under physiological conditions its activity is limited.

CNTF is an ~22-kDa slightly acidic protein with ~200 amino acids. In contrast to most neurotrophins, CNTF has no hydrophobic leader sequence, which is needed for the classic pathway of secretion, and in tissue culture experiments CNTF is not released from several types of cells. CNTF could be secreted under certain conditions through other pathways, as is the case for other proteins without a signal sequence. In contrast to the neurotrophins, CNTF is a single-stranded molecule but can dimerize at high concentrations. The gene for human CNTF is located on chromosome 11.

The specific receptor for CNTF has been identified and shown to lack a transmembrane domain, that is, it cannot transduce the signals of CNTF to the intracellular compartment of the cell (see Fig. 2). This CNTF α receptor associates in a tripartite receptor complex with the functional transmembrane LIF β and gp130 receptors. In the presence of the CNTF-related cytokine LIF, the LIF β receptor and gp130 dimerize, autophosphorylate, and activate a set of tyrosine kinases named Jak/Tyk. CNTF binds to the specific CNTF α receptor and together they associate with and activate the LIF β–gp130 dimer. This explains why CNTF and LIF have almost identical biological activities. Of great interest is the fact that CNTF α receptor is soluble and under certain circumstances can diffuse from one cell to the other, thereby inducing a responsiveness to CNTF (provided those cells contain LIF β and gp130 receptors). The gene for human CNTF α receptor is located on chromosome 9.

C. Other Protein Factors with Activities for Neurons

Very recently, another potent neurotrophic factor has been discovered and, because it was isolated from a glial cell line, was named glial cell line-derived neurotrophic factor or GDNF. GDNF has attracted a great deal of interest because of its very high potency and effectiveness to promote neuron survival compared to other neurotrophic factors. GDNF is a member of the TGF-β (transforming growth factor) superfamily whose members are known for their multifunctional nature and as regulators of cell proliferation and differentiation. GDNF is an ~134-amino-acid protein (~15 kDa; ~20 kDa when glycosylated) and, similar to the neurotrophins, normally exists as a homodimer. Some forms appear to be synthesized by alternative splicing of a single GDNF gene. GDNF synthesis is not unique to the nervous system. The gene for human GDNF is located on chromosome 5. A specific non-transducing GDNF α receptor interacts with the transmembrane-transducing receptor Ret, possibly in a tripartite complex.

Following is a brief list of some of the proteins that have demonstrated neurotrophic activity, including their commonly used abbreviations. This list is not intended to be complete, but presents important examples and gives a sense of the variety. The number of newly identified factors with neurotrophic activity is steadily increasing.

Neurotrophins	
nerve growth factor	NGF
brain-derived neurotrophic factor	BDNF
neurotrophin-3, -4/5, and -6	NT-3, NT-4/5, NT-6
Neural cytokines	
ciliary neurotrophic factor	CNTF
leukemia inhibitory factor	LIF
Glial cell line-derived neurotrophic factor (member of TGF family)	GDNF
Cytokines	
interleukin-1, -2, -3, and -6	IL-1, IL-2, IL-3, IL-6
erythropoietin	EPO
granulocyte–macrophage colony stimulating factor	GM-CSF
Growth factors	
transforming growth factor-α	TGF-α
transforming growth fator-β1–3	TGF-β1–TGF-β3
acidic fibroblast growth factor	aFGF or FGF-1
basic fibroblast growth factor	bFGF or FGF-2
fibroblast growth factor-5	FGF-5
epidermal growth factor	EGF
insulin-like growth factor-1 and -2	IGF-1, IGF-2

It should be noted that biological activities of *neurotrophic* factors are not necessarily specific for neurons; for example, NT-3 has a clear role for the development of oligodendrocytes (the myelin-producing glial cells of the brain and spinal cord). Neurotrophic factor activities are also not limited to the nervous system; for example, NGF regulates B lymphocytes and is critically involved in gamete formation in the reproductive system. On the other hand, some factors such as CNTF and its receptor are essentially found and active only in the nervous system. As will be detailed in the next section, neuro*trophic* factors can also have activities that are not trophic in nature; for example, NGF can stimulate proliferation of certain cells. Thus neurotrophic factors (and cytokines in general) are or can be multifunctional and multitargeting, that is, they can modify behaviors of many types of cells (Table I).

III. ROLES IN NEURONAL DEVELOPMENT

Most of the knowledge about the types of biological activities of neurotrophic factors on neurons has been obtained from tissue culture (*in vitro*) experiments using developmental tissue from various animal species. The peripheral nervous system has been studied most extensively because of the relative ease of dissecting and obtaining such neurons during early embryonic stages. The various activities have been confirmed in developing whole animals. In particular, confirmation for some of the ideas has come from the recent generation and characterization of transgenic mice with a targeted null (knock-out) mutation of individual neurotrophic factors or their receptors. These mice have opened up a whole new era of investigation of the endogenous roles of neurotrophic factors. It should be noted that individual neurotrophic factors do not necessarily have all of the following activities.

TABLE I
Neurotrophic Factors Can Affect Different Cells and Different Events[a]

Neurotrophic factor	Cell type	Biological event
NGF	Neuron	Survival
		Differentiation
		Process formation
		Transmitter synthesis
		Membrane channel functions
	B lymphocyte (immune)	Proliferation
		Immunoglobulin secretion
	Mast cell (immune)	Histamine release
	Melanoma cell	Degradative enzyme synthesis
	Pancreatic islet cell	Development
	Prostatic epithelia	Growth
	Sertoli cell (testis)	Spermatogenesis
CNTF	Neuron	Similar to NGF
	Liver cell	Acute-phase protein synthesis
	Microglia	Activation
	Oligodendrocyte	Survival
	Oligodendrocyte precursor	Differentiation to astrocyte-2
	Photoreceptor	Protection against UV light
	Skeletal muscle	Size maintenance
bFGF	Neuron	Similar to NGF
	Adrenal chromaffin cell	Transdifferentiation to neurons
	Endothelia	Proliferation
	Fibroblast	Proliferation
	Melanoma cell	Proliferation

[a]Only some factors are used to illustrate the variety of cells and activities.

A. Proliferation

Neurons derive from neuronal precursor cells, which are the product of proliferation of progenitor and stem cells. Some evidence has been found that neurotrophic factors such as NT-3, NGF, and bFGF stimulate the proliferation of early-developing neuronal progenitors. For instance, NT-3 can induce proliferation of neural crest cells that migrate from the trunk and differentiate into satellite cells and neurons of the dorsal root ganglion. Of interest are the findings that such factors can also stimulate the proliferation of stem cells that are present in the adult brain. On the other hand, factors such as NGF are well known for their induction of neuronal differentiation, which includes the inhibition of further mitosis. It is possible that during early development, stem cells and progenitor cells respond to a particular factor with mitosis and at later stages with differentiation. Such dual functions of factors is an example of the economizing nature of biological agents or events in general, that is, more effects with fewer genes.

B. Differentiation

Differentiation means the process or event that change a precursor cell into a neuron or into a neuron with a particular phenotype. This phenotype generally includes the ability to synthesize the transmitter of the neuron, which is an indication that the neuron has obtained the ability to become functional. After differentiation the neuron matures, that is, it changes morphology and initiates process formation. Many neurotrophic factors have been found to have differentiating effects for many types of neurons, perhaps because this activity is readily recognized in tissue culture. It should be noted that differentiation during development could also be a result of novel expression of the receptors for the neurotrophic factors that were already present during earlier times.

C. Survival

During very early development, neurons appear to be independent of neurotrophic factors for their survival, at least from evidence that dissociated neurons can survive without factors *in vitro*. It has not been resolved whether these neurons provide survival-promoting factors to themselves through an autocrine mechanism. According to the neurotrophic factor hypothesis and several lines of evidence, it appears that neurons become dependent on neurotrophic support around the time when their neuritic processes (axon) reach the target. The timing of the onset of this dependence may be programmed and may be dependent on the type of neuron and the expected length that the axon has to grow. For example, in culture, cranial sensory neurons, which in animals have to grow over short distances, become dependent on BDNF sooner than those that have to grow over longer distances. Interestingly, neurons may be dependent on different neurotrophic factors during different stages of their development or may use more than one. For instance, spinal and facial motor neurons in culture and *in vivo* respond to and can be rescued by BDNF, CNTF, or GDNF. On the other hand, it is clear that individual types of neurons do not respond to all neurotrophic factors. Limited amounts of such critical factors in the target tissue may support only those neurons that have made proper connections. This mechanism appears to be critically involved in regulating the total number of neurons that persist into adulthood.

D. Process Outgrowth and Synapse Formation

The earliest observation about NGF was the effect on neurite (or nerve) outgrowth from neurons, therefore the name nerve growth (promoting) factor (see Section I). In addition, those early experiments revealed the potent effects on axonal sprouting and growth in embryos. Not all factors appear to have neurite-promoting effects and neurite outgrowth is also dependent on other molecules such as the substrate formed by extracellular matrix and adhesion molecules. One characteristic action of NGF is its chemoattractive properties, that is, NGF can cause a turning response of the neurite front, the so-called growth cone, and can cause a change in course of the axon toward the NGF source. NT-4/5 and BDNF can modify innervation patterns in the visual system of young cats, suggesting that they are involved in activity-dependent control of axonal branching and synapse formation during development of the central nervous system. The role of NGF in developmental neurite outgrowth of certain central neurons is evident in transgenic mice that lack the gene for NGF. For instance, the NGF-responsive cholinergic basal forebrain neurons of these animals survive but have severely underdeveloped axonal projections to their targets, the hippocampus, and cerebral cortex.

IV. ROLES IN NORMAL ADULT NERVOUS SYSTEM

The presence and cellular localization of the proteins and mRNA have been established to various degrees for different neurotrophic factors and their receptors. Because of the normal presence of their receptors, such factors are expected to have a physiological role in the nervous system, although some (e.g., CNTF) appear to become more involved during pathological situations. One example for a direct role of NGF has been established for the cholinergic neurons of the basal forebrain neurons in laboratory animals. NGF and its mRNA are present in postsynaptic neurons of the target area (hippocampal formation and cerebral cortex), as well as in neurons closer to the cholinergic cell bodies. These cholinergic cell bodies express the specific TrkA receptor for NGF. When NGF is injected into the target, it binds to the TrkA and p75 receptors and is retrogradely transported to the cholinergic cell bodies. Moreover, when NGF is administered to the normal cholinergic neurons, they show increased production of choline acetyltransferase (ChAT), the rate-limiting enzyme in the synthesis of their transmitter acetylcholine, as well as a slight increase in size. More convincingly, if the availability of NGF is reduced by neutralizing antibodies or if the hippocampal formation is removed, these cells undergo decreases in ChAT enzyme and in size, but do not die. Thus, when analyzed on an individual level, neurotrophic factors appear to be important only for the regulation of cell performance and not essential for cell survival. However, since most neurons use more than one factor, it is possible that total factor deprivation would lead to neuronal cell death. Another indication of the functional effects of neurotrophic factors are several findings that neurotrophins can cause a rapid and persistent modulation of the strength and activity of synaptic transmission. Such factors may also be involved in more extensive structural rearrangements of synapses; for example, CNTF can induce sprouting (outgrowth) of the endings of adult rat motor axons that are located on muscles.

The information about the localization and transport of neurotrophic factors in living animals and data from tissue culture experiments have revealed different modes of delivery of these endogenous factors to the cell (Fig. 3). Many of these factors are typically produced by glial cells such as astrocytes (e.g., bFGF, GDNF) and Schwann cells (CNTF), others by neurons (e.g., NGF), and others by both (e.g., BDNF). Such cells can release these factors for local action on the neuronal processes or for retrograde or anterograde transport to the neuronal cell body. Other factors may be provided by neurons to neighboring neurons (paracrine) or may be used by the neuron itself (autocrine). For example, BDNF is produced by dopaminergic neurons of the substantia nigra, which also have functional TrkB receptors that respond to BDNF. The release mechanism of neurotrophic factors is not completely understood but may occur by the postsynaptic neuron or neighboring glial cell upon the release of transmitter in the synapse or after an action potential. This is particularly suggested by the fact that epileptic activity increases the synthesis of many neurotrophic factors in the hippocampal formation. On the other hand, factors such as CNTF seem to be stored in large quantities and may be released only under special circumstances.

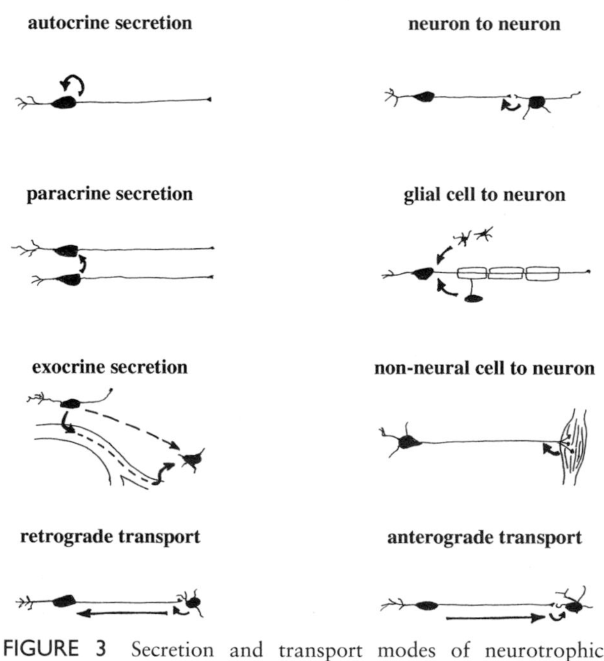

FIGURE 3 Secretion and transport modes of neurotrophic factors.

V. THERAPEUTIC POTENTIAL FOR NEUROLOGICAL DISORDERS

A. Historical Perspective: NGF, Cholinergic Basal Forebrain Neurons, and Alzheimer's Disease

Many neurological disorders are characterized by neuronal cell death and loss of connectivity with the target. The recognition of the potent survival- and outgrowth-promoting activities of neurotrophic factors in tissue culture spawned the search during the 1970s and early 1980s to prove that these factors can be beneficial in adult animal models of neuronal damage. Of special interest was the central nervous system, since in general the extent of neuronal dysfunction and death is more severe there than in the periphery. As is clear from the previous sections, much of the knowledge about neurotrophic factors has been gained by investigation of NGF, especially since, until recently, it was the only clearly identified neurotrophic factor. This is even more evident for *in vivo* studies in which large quantities of neurotrophic proteins were available only for NGF. The likely involvement of NGF in various cell behaviors had been established for cholinergic neurons of the basal forebrain, that is, they had been shown to transport NGF and respond to NGF in culture. These neurons degenerate after transection of the fiber bundle or tract through which they project to the hippocampus. In 1986 and 1987, the first reports that administration of exogenous NGF into the brain close to the cholinergic cell bodies could almost completely prevent the axotomy-induced degeneration were published. Since then many investigators have used this model to investigate particular questions about the biology of NGF. Soon afterward, NGF was shown to reverse cholinergic-related memory deficits in aged rats and to reverse the atrophy (size decrease) of these cholinergic neurons.

NGF can induce axonal sprouting of these injured neurons and promote their regrowth or regeneration back into the original target areas. This helped formulate the hypothesis that the poor recovery in the central nervous system (as compared to the peripheral nervous system) is in large part due to the lack of sufficient stimulatory molecules such as neurotrophic factors. At the same time, specific outgrowth inhibitory molecules were identified in the central nervous system, suggesting that outgrowth (and potentially synaptic plasticity) is regulated by the balance of stimulatory and inhibitory molecules. In many ways, the mechanisms that regulate regeneration appear similar to those that play a role during development. Similar studies have confirmed that NGF has neurotropic or chemoattractant properties for the regenerating cholinergic axons, a property that had been observed for peripheral neurons in culture and in embryos. NGF (endogenous and exogenous) can also stimulate (so-called collateral) sprouting and synapse formation of normal uninjured cholinergic axons in hippocampal regions when other projections are removed. This again suggests that neurotrophic factors are likely to be involved in synaptic rearrangement and therefore in plasticity of the brain.

Because of the exciting animal results with NGF and the cholinergic systems, it is presently considered a candidate drug to treat Alzheimer's disease. In this chronic progressive degenerative disease, these neurons become dysfunctional and die, a process that is thought to contribute to the characteristic memory loss (dementia). Of interest is the recent finding that NGF levels in Alzheimer's disease are not decreased in the projection regions (cortex regions) but are decreased in the nucleus basalis, a major site of cholinergic neuron loss. This does not mean that such degenerative disorders are caused by a decrease in neurotrophin levels. However, providing extra neurotrophic factor support for these neurons may enable them to better survive the underlying disease process. See Table II for an overview of neurological disorders and neurotrophic factors that could potentially be used as therapeutic agents. [*See* Alzheimer's Disease.]

B. Other Neurological Indications

Amyotrophic lateral sclerosis (ALS) or Lou Gehrig's disease (named after the famous baseball player) is a severe and progressive neurological disease. It is characterized by the loss of motor function and death of spinal and brain stem motor neurons that innervate the skeletal muscles. NGF has no effect on these neurons, but CNTF, BDNF, and IGFs, and recently GDNF, were tested and found to have very potent survival-promoting effects in culture. Subsequently, these factors were shown to rescue motor neurons and promote regeneration and functional recovery in various acute and more chronic animal models for motor neuron degeneration. The initial promising results with CNTF led to its clinical testing in ALS patients, but one trial was stopped because of side effects related to systemic effects on other organs. Still under consideration is IGF-1, BDNF, and a combination treatment of BDNF and a lower dose of CNTF, since these factors may enhance each other's potency. The finding that GDNF is even more potent than the other two has also made it a prime candidate drug for ALS. Disorders such as ALS, which affect neurons that project through peripheral nerves, are considered

TABLE II
Neurological Disorders and Potential Neurotrophic Factor Applications[a]

Disease	Affected neuron	Neurotrophic factor
ALS	Motor	CNTF, BDNF, NT-4/5, IGF-1, GDNF
Alzheimer's	Cholinergic basal forebrain	Neurotrophins
Epilepsy	Hippocampal	Methods that neutralize factors
Malignancies	Neuroblastoma	Neurotrophins
Parkinson's	Dopaminergic substantia nigra	GDNF, BDNF, NT-3, NT-4/5
Peripheral neuropathies	Primary sensory	Neurotrophins
Stroke	Various	bFGF, neurotrophins
Spinal cord injury	Primary sensory	Neurotrophins
	Corticospinal motor	CNTF, NT-3

[a]Some are in clinical trials (see text).

favorable target diseases. This is predominantly because the large protein factors can be given systemically from where they have access to the nerve, perhaps also through the muscle compartment. As will be discussed in the next section, access of these factors to the brain and spinal cord is much more problematic because of their inability to cross the blood–brain barrier. [*See* Amyotrophic Lateral Sclerosis; Blood–Brain Barrier.]

For the same reasons of access, neurotrophic factors are being considered or are in clinical trials for peripheral neuropathies, many of which are characterized by degeneration of sensory nerve fibers. Examples are neuropathy caused by diabetes and those induced by cancer chemotherapy. Because of its clear protective and regeneration-promoting effects on sensory neurons both *in vitro* and in animal models, NGF is presently in clinical testing.

The symptoms of Parkinson's disease are caused in large part by the severely reduced release of dopamine in the extrapyramidal system (primarily the caudate putamen region), which is a result of the progressive loss of dopaminergic neurons in the substantia nigra. In animal models, these neurons can be rescued by various neurotrophic factors (CNTF, BDNF, NT-3, and NT-4/5), including GDNF, which appears to be the most potent. As of 1995, GDNF is considered the prime candidate for clinical testing and is likely to be tested in humans after more preclinical information has been gathered, including results from Parkinson models in primates. [*See* Parkinson's Disease, Biological Basis and Therapy.]

Recent studies have provided evidence that the effects of neurotrophic factors are not always desired. In an animal model of epilepsy, blocking the activity of neurotrophins delayed the onset of electrically induced seizures in the hippocampus and at the same time blocked sprouting of mossy fibers in the hippocampal formation. Thus, future clinical indications might also include the neutralization or blocking of endogenous neurotrophic factors or their receptors.

Neuronal cells can become deregulated, start proliferating, and thereby become malignant. As described in previous sections, neurotrophic factors can induce differentiation of such tumor cell lines in culture and potentially also in a clinical setting.

Not covered here are psychiatric disorders such as affective disorders and schizophrenia, which may be caused by over- or underactivity or connectivity of particular sets of neurons, and therefore could become targets for neurotrophic factor treatments. [*See* Schizophrenic Disorders.]

C. Emerging Strategies to Deliver Protein Factors to the Central Nervous System

Neurotrophic factors are moderately large proteins, probably because their specialized intercellular signaling function requires a large number of different ligands with highly specialized receptor interactions. In general, such large proteins cannot cross the blood–brain barrier. Therefore, straightforward systemic administration fails to provide significant levels of neurotrophic factor in the central nervous system, where the pathological process of most neurological disorders is located. A second problem of treatments of central nervous system disorders is constituted by the size of the affected area. For instance, multiple direct

TABLE III
Emerging Strategies to Deliver Protein Factors to the Central Nervous System

Strategy	Advantages	Potential Problems
Direct administration	Local, modifiable, and terminable	Chronic surgical intervention, local
Blood–brain barrier breakage or crossing carriers	Systemic, global, modifiable, terminable	Altered protective blood–brain barrier, changing nondiseased areas
Viral transfection of resident cells	Local or global, chronic	Changed normal cell functions
Injectable cell factories	Local, chronic	Other cell products, potential for tumors, local
Stimulation of endogenous sources	Systemic, global, modifiable, terminable	Changing nondiseased areas
Small agonist or antagonist molecules	Systemic, global, modifiable, terminable	Changing nondiseased areas

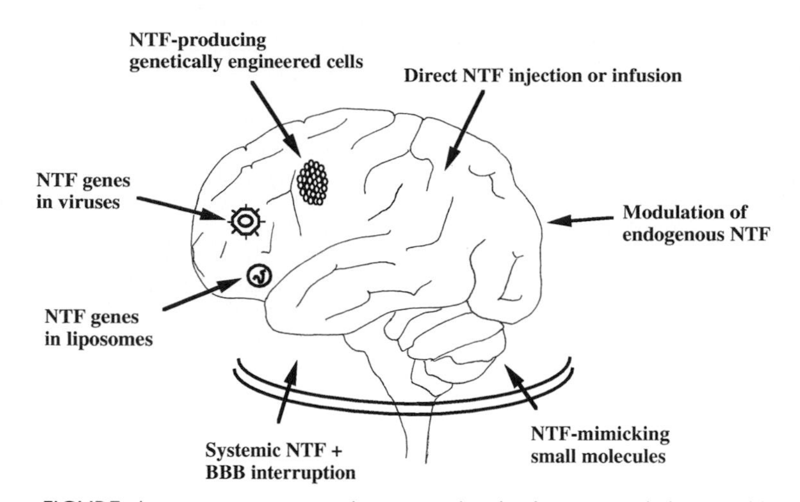

FIGURE 4 Various strategies that are under development and that would enable access of neurotrophic factors to the central nervous system, despite the impermeable blood–brain barrier (BBB).

injections would be needed to support an adequate area in cases of disorders that encompass large brain regions (global). On the other hand, treatments that result in global diffusion of factors beyond smaller affected areas (local) might have unwanted side effects. Another issue of treatment in general is the need for control of delivered amounts (dose) and timing (short, intermittent, chronic, or terminable, i.e., the possibility to interrupt treatment). Table III and Fig. 4 present the treatment strategies that are under development for neurotrophic factors at present.

VI. OUTLOOK

Over the past several years, an increasing awareness has developed about the similar nature of cell biological events and mechanisms in the nervous and immune systems. The knowledge base in immunology is farther advanced. The flow of information will increase and continue to facilitate the identification of biological principles in the nervous system. Also, the emerging recognition that many protein factors and cytokines are multifunctional and have many target cells will expand.

In the next decade we should see an evolution in the understanding of the biology of neurotrophin synthesis and release, and a better understanding of the spatial and temporal cues that underlie the normal and abnormal function in the nervous system. Similarly, the complex nature of factor signaling and the multitude of secondary message systems will be unraveled and will help identify those signaling pathways that are common as well as unique to classes of neurotrophic factors. This should also help in understanding the role of specific sets of genes that are involved in specific cell behaviors.

Finally, the frontier of neurotrophic factors is also found in the exciting potential for their use in (therapeutic) modulation of neuronal behavior, that is, the enhancement of specific brain functions and a more direct or orchestrated regulation of genes. The expectation is that the need for neurotrophic factor administration will be replaced by small molecules that mimic the receptor binding domains of neurotrophic factors or their secondary messengers and/or by small molecules that regulate the synthesis of endogenous neurotrophic factors.

BIBLIOGRAPHY

Barbacid, M. (1995). Neurotrophic factors and their receptors. *Curr. Opin. Cell Biol.* **7**, 148–155.

Bothwell, M. (1995). Functional interactions of neurotrophins and neurotrophin receptors. *Ann. Rev. Neurosci.* **18**, 223–253.

Davies, A. M. (1994). The role of neurotrophins in the developing nervous system. *J. Neurobiol.* **25**, 1334–1348.

Hagg, T., Louis, J. C., and Varon, S. (1993). Neurotrophic factors and CNS regeneration. *In* "Neuroregeneration" (A. Gorio, ed.), pp. 265–287. Raven, New York.

Hamburger, V. (1993). The history of the discovery of the nerve growth factor. *J. Neurobiol.* **24**, 893–897.

Hefti, F. (1994). Neurotrophic factor therapy for nervous system degenerative diseases. *J. Neurobiol.* **25**, 1418–1435.

Henderson, C. E. (1996). Role of neurotrophic factors in neuronal development. *Curr. Opin. Neurobiol.* **6**, 64–70.

Ip, N. Y., and Yancopoulos, G. D. (1996). The neurotrophins and CNTF: Two families of collaborative neurotrophic factors. *Ann. Rev. Neurosci.* **19**, 491–515.

Levi-Montalcini, R. (1987). The nerve growth factor 35 years later. *Science* **237**, 1154–1162.

Lewin, G. R., and Barde, Y. A. (1996). Physiology of the neurotrophins. *Annu. Rev. Neurosci.* **19**, 289–317.

Lindsay, R. M., Wiegand, S. J., Altar, C. A., and DiStefano, P. S. (1994). Neurotrophic factors: From molecule to man. *Trends Neurosci.* **17**, 182–190.

Loughlin, S., and Fallon, J. (eds.) (1993). "Neurotrophic Factors." Academic Press, San Diego.

Segal, R. A., and Greenberg, M. E. (1996). Intracellular signaling pathways activated by neurotrophic factors. *Ann. Rev. Neurosci.* **19**, 463–489.

Sendtner, M., Carroll, P., Holtmann, B., Hughes, R. A., and Thoenen, H. (1994). Ciliary neurotrophic factor. *J. Neurobiol.* **25**, 1436–1453.

Neutrophils

THOMAS P. STOSSEL
Harvard Medical School and Brigham & Women's Hospital

NEUTROPHILIC POLYMORPHONUCLEAR LEUKOcytes (neutrophils; PMNs) are bone marrow-derived cells that defend human tissues against bacterial and fungal infections. PMNs circulate briefly in the blood, leave the circulation by active penetration of vascular walls, and seek out invading microorganisms or damaged host cells. The accumulation and degeneration of PMNs at such sites accounts for the collection of viscous greenish material known as pus. PMNs crawl through tissues by a kind of amoeboid locomotion in response to gradients of molecules called chemoattractants. They ingest objects that they recognize by utilizing specific membrane receptors. PMNs kill and digest prey and surrounding cells using a variety of hydrolases, reactive oxygen and nitrogen species, and membrane-permeabilizing proteins. Persons deficient in PMNs or with PMNs defective in certain of their antimicrobial functions are susceptible to bacterial and fungal infections.

I. LIFE CYCLE OF PMNs

A normal adult produces over 100 billion PMNs per day. PMNs mature from a sequence of precursors beginning with the pluripotent hematopoietic stem cell, some progeny of which commit to PMN development (Fig. 1). The earliest recognizable PMN precursor, the myeloblast, derives from the granulocyte-macrophage stem cell. Further maturation of the myeloblast to a mature PMN involves, on the one hand, amplification in numbers and acquisition of motile and antimicrobial functions and, on the other, loss of the capacity to self-replicate. Hematologists arbitrarily classify PMN progenitors according to their morphology on specially stained films (Wright-Giemsa). Myeloblasts are oval cells with large nuclei containing lacy chromatin and well-demarcated nucleoli. Similar cells containing blue-purple (azurophilic) granules are promyelocytes. These granules become less visible in more numerous myelocytes, which are smaller and have an eccentric nucleus and a prominent Golgi apparatus. These cells are the most mature PMN precursors capable of dividing and represent the main amplification step in PMN production. Indentation of the nucleus defines a metamyelocyte; separation of the nucleus into nearly two lobes, a band or stab form, and segmentation of the nucleus in three or more distinct lobes define the mature PMN. The segmentation of the PMN nucleus is what gives the cells their name and may serve to permit the highly deformable PMN to negotiate narrow apertures as it crawls through tissues. [*See* Hemopoietic System; Macrophages.]

To some extent the distribution of hematopoietic cells among lineages is stochastic, but glycoprotein hormones determine their growth and maturation programs. In addition to promoting mitosis of early progenitors and differentiation of both early and late myeloid precursors, these factors prevent the phenomenon of programmed cell death, also known as apoptosis. Early myeloid maturation is driven by interleukin-3 (IL-3), by granulocyte colony-stimulating factor (G-CSF), and by granulocyte-macrophage colony-stimulating factor (GM-CSF) and is modulated by many other cytokines. These factors stimulate cytokine receptors that activate intracellular signal transduction mechanisms involving, among other reactions, metabolic pathways mediated by cellular oncogenes. PMN maturation from myeloblast to PMN takes approximately 1 week. Following release from the bone marrow, PMNs circulate in the peripheral blood for 6–12 hr and then enter the tissues. In infec-

ENCYCLOPEDIA OF HUMAN BIOLOGY, Second Edition, VOLUME 6.

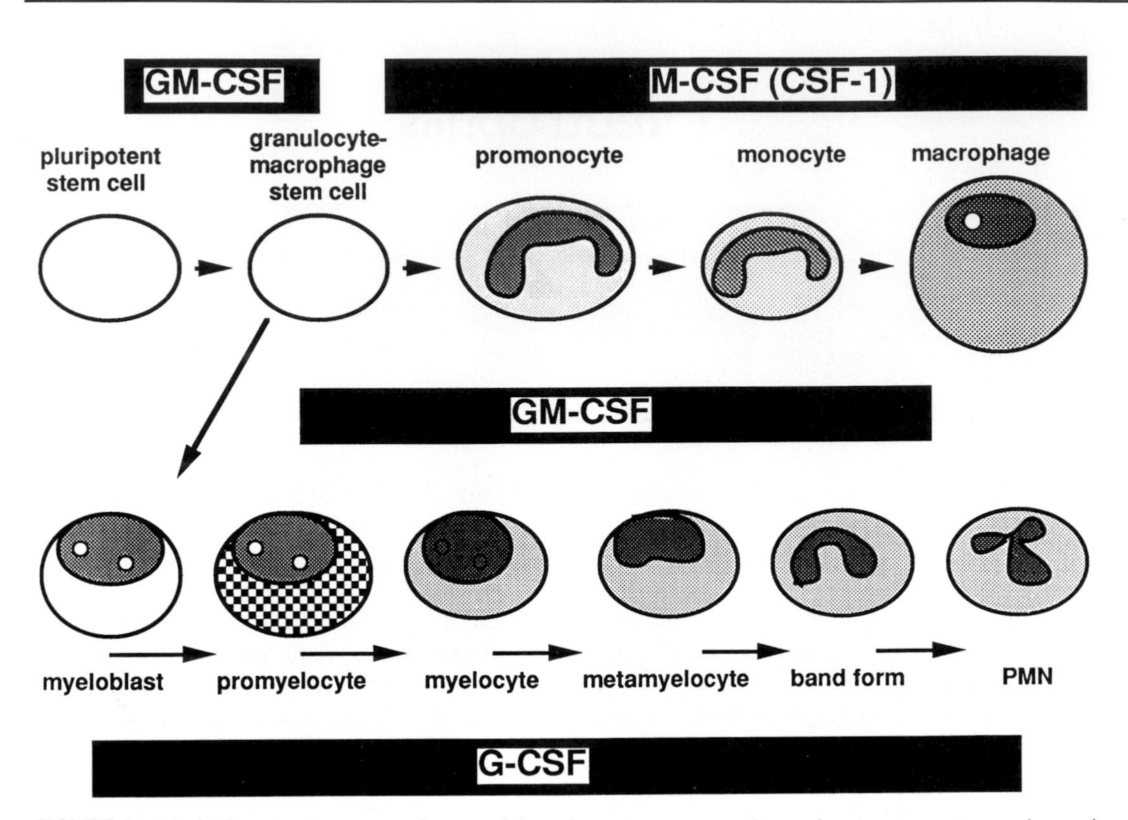

FIGURE 1 Myeloid maturation stages of neutrophils and monocyte–macrophages showing approximate places of regulation by cytokines.

tious or noninfectious inflammatory states, PMN production rates may increase up to 10-fold. The metabolic demands of myeloid cell production require considerable DNA synthesis. Therefore deficiency of folic acid, needed for DNA metabolism, or interference with cell replication by high-energy radiation or by chemicals (such as agents used against cancer cells) result in impaired PMN production. [*See* Cytokines and the Immune Response.]

II. FUNCTIONS OF PMNs

A. Adhesion, Emigration (Diapedesis), Chemotaxis, Phagocytosis, and Degranulation

The activities of PMNs—adhesion, emigration (diapedesis), chemotaxis, phagocytosis, and degranulation—are responsible for getting PMNs out of the bone marrow and into the tissues where they can surround microorganisms. When a PMN leaves the bone marrow it acquires a morphology characteristic of ameboid motility. The cell becomes elongated and extends a flat sheet of membrane and peripheral cytoplasm, a lamellipod, in the direction of movement. This lamellipod attaches transiently to the surface, and the cell body is contracted forward toward it. Repetition of these steps, mediated by the remodeling of intracellular networks composed of actin filaments, results in translational locomotion. Once in the peripheral blood, PMNs become round with multiple projecting surface pleats containing ligands (selectins). These ligands recognize partners that arise on endothelial cells of lung capillaries or elsewhere on postcapillary venules during inflammatory states associated with liberation of proinflammatory cytokines by tissue macrophages (Fig. 2). The binding of these complementary receptors causes the PMN to roll along the vessel wall.

As inflammation intensifies, the rolling cells stop, flatten, and become highly adherent to the endothelial surface. The PMN then sheds its selectins, up-regulates expression and adhesive function of another class of adhesion receptors, the β_2 integrins, and then crawls between endothelial cells, across the vessel basement membrane, and into the connective tissue toward the inflammatory source. The transendothelial

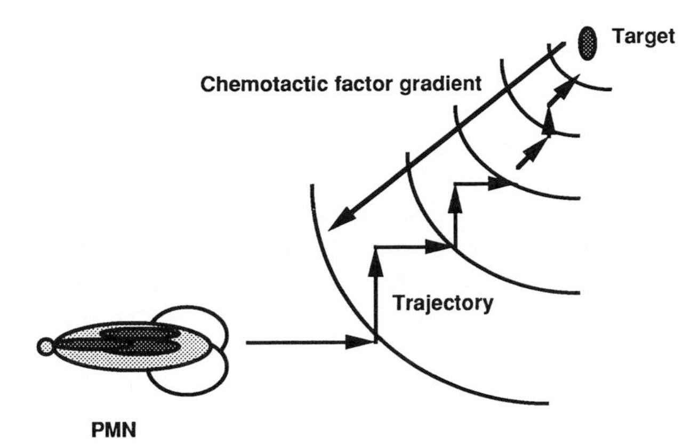

FIGURE 2 Migration of a neutrophil of a chemotactic gradient.

migration step requires accommodation of the endothelial cells, which utilize a contractile mechanism mediated by as yet uncharacterized factors released by the PMNs to raise intracellular endothelial cell calcium levels.

When PMNs arrive at their targets, they recognize them with receptors for immunoglobulin G (IgG) (Fc receptors) and for the C3b and C3bi fragments of the third component of plasma complement (CR1 and CR3 receptors, respectively). Ligation of these receptors causes pseudopodia that spread around the targets. As this phagocytic event takes plae, the PMN granules fuse with the plasma membrane and secrete their contents into the phagosome, which now becomes a phagolysosome, and the PMN is said to have "degranulated."

B. Microbicidal, Digestive, and Inflammatory Functions of PMNs

The effector mechanisms are usually divided into oxidative and nonoxidative. In the former, perturbation of chemoattractant and phagocytic receptors results in the assembly of an electron transport apparatus known as the respiratory burst oxidase. The oxidase contains multiple components, some of which (p47 phox, p67 phox, and the GTPases *rac* 1 and *rap*) reside in the cytoplasm of the resting cell but associate on activation with a membrane-bound two-chain *b*-type cytochrome, which is the ultimate electron transport species that reduces molecular oxygen (O_2) to the superoxide anion (O_2^-). O_2^- spontaneously dismutes to hydrogen peroxide (H_2O_2), whence it becomes a substrate for granule-derived myeloperoxidase. Myeloperoxidase is a heme-containing protein, the abundance of which in PMN accounts for the greenish hue of pus. Myeloperoxidase-mediated degradation of H_2O_2 in the presence of chloride leads to formation of long-lived reactive species such as hypochlorous acid (HOCl), which is capable of oxidizing amines to chloramines. H_2O_2 can also react with O_2^- to generate the highly reactive hydroxyl radical ($OH\cdot$); this interaction, known as the Fenton reaction, is catalyzed by iron provided by granule-associated lactoferrin.

An independent oxygen-related system of PMNs depends on an arginine-oxidizing flavoprotein, nitric oxide synthetase (NOS), which generates NO. Yet another is the lipoxygenase cascade that oxidizes arachidonic acid to a class of bioactive mediators, the leukotrienes. The oxygen-related products of activated phagocytes collectively exert considerable antimicrobial and inflammatory potential mediated by the oxidation of protein side chains, scission of DNA, and lipids. The lipid mediators and NO have potent vasoactive effects.

The nonoxidative products of PMNs consist of hydrolytic enzymes, including elastase, gelatinase, phosphatases, lipases, glucosidases, and lysozyme, as well as myeloperoxidase mentioned earlier. Cationic proteins with antimicrobial activities are the defensins, granzymes, and a bacterial permeability-inducing factor.

C. Clinical Situations That Emphasize Antimicrobial Effects of PMNs

In cancer that spreads to the bone marrow, leukemia, cancer therapy with cytotoxic drugs or radiation, or aplastic anemia, impairment of the PMN-producing capacity of the bone marrow can be severe, and when the PMN peripheral blood count falls below 500 per milliliter, the affected patients suffer from infections caused by bacteria and fungi; the lower the PMN count the greater the risk of infection. The infections arise mainly on the skin, in the lungs, and in the mouth and lower gastrointestinal tract, places where the host is exposed to most environmental microorganisms. Treatment of this serious situation requires removal of the underlying cause (abolition of cancer, leukemia, etc.) and supportive care using cytokines to stimulate PMN production, and in some cases bone marrow transplantation to restore PMN-manufacturing capacity.

In genetic PMN adhesion deficiency (LAD), inherited as an autosomal dominant disorder, PMNs fail to express the β_2 integrins required for optimal PMN adhesion to vessel walls and for migration. Chronic granulomatous disease features PMNs that are unable to mount a respiratory burst. Genetic defects of the *b* cytochrome (transmitted as an X-linked trait) or of the activating phox complex (autosomal recessive) lead to this condition. Patients with these disorders and a variety of others suffer from recurrent tissue infections and illustrate the need for optimal PMN function to fight off microbial invasion.

D. PMN in Inflammation

The contribution of PMNs to inflammation is exemplified by the well-known association of pus with redness, pain, heat, and tissue destruction. The inflammatory role of PMN has also been demonstrated by selective PMN depletion using antibodies in a variety of animal models of disease. Inflammation is the price to be paid for effective antimicrobial defense. Vasodilatation, which accounts for the redness and some of the swelling observed, brings PMN, antibodies, and complement proteins to the invasion sites. Tissue destruction by PMN may serve a function resembling surgical debridement of dead tissues that provide food for microorganisms. In chronic inflammation, the cost exceeds the benefits of PMN activity. Foreign bodies in tissues permit bacteria to hide from PMN and persist despite constant bombardment with microbicidal agents. In certain disorders, such as ulcerative colitis, the basic cause of the disease is unknown, but PMNs mercilessly attack the large intestine, causing bloody diarrhea, scarring, and eventually even cancer. It has been proposed that PMN-derived oxidants cause DNA mutations that account for the high incidence of neoplasia in chronic inflammatory disorders. Unfortunately, there are presently no PMN-specific retardants to apply against these cells in chronic inflammation. Corticosteroids reduce PMN migration but have many other side effects. [*See* Inflammation.]

BIBLIOGRAPHY

Babior, B. M., and Stossel, T. P. (1995). Structure, function and functional disorders of the phagocytic system. *In* "Blood, Principles and Practice of Hematology" (R. I. Handin, S. E. Lux, and T. P. Stossel, eds.), pp. 575–620. Lippincott, Philadelphia.

Newborn Screening

LINDA McCABE
EDWARD R. B. McCABE
University of California, Los Angeles

GLOSSARY

Autosomal Refers to 22 of the 23 chromosomal pairs that are not the sex chromosomes

Carriers Individuals who are phenotypically normal and have only one mutant gene of a pair for a recessive trait; also known as heterozygotes; expression of the recessive trait requires two mutant genes of the same pair for that trait, one inherited from each carrier parent

Diagnosis In the context of a newborn screening program, this refers to the definitive assignment of a disease classification to an individual after pursuing a positive screening test by taking a history, performing a physical examination, and carrying out the appropriate definitive tests; screening is not the equivalent of diagnosis

Dried blood specimens Samples taken by pricking the skin and allowing blood to soak into a filter paper, where it dries; the blood-impregnated filter paper is then used for newborn screening

False-negative result Newborn screening test result indicating a low risk for the disease, although the individual does have the disease; also referred to as a "missed" patient

False-positive result Newborn screening test result indicating a high risk of the disease, although the individual does not have the disease; also known as a "false alarm"

Negative result Newborn screening test result suggesting that the individual has a low risk for disease

Phenylketonuria Recessive genetic disease due to decreased phenylalanine hydroxylase; treatment consists of a diet restricted in phenylalanine; untreated, this disease can lead to mental retardation

Positive result Newborn screening test result suggesting that the individual has a high risk for disease

Recessive Refers to a genetic trait that requires both genes of a pair to be mutant before expression of the trait is observed; that is, both parents must be carriers and must pass their mutant gene to the child in order for the child to have the disorder

NEWBORN SCREENING INVOLVES THE APPLICAtion of tests or examinations to newborn infants in order to separate those who are likely to have a disease from those who are not. Every newborn should undergo this testing, as the tests are rapid and inexpensive. Positive results on these tests are not a diagnosis of disease, but rather are an indication that the infant needs further testing and evaluation before treatment is initiated. The purpose of newborn testing includes the detection of disease in order to prevent irreparable damage by the early initiation of therapy, the identification of disease-free individuals that carry the gene(s) (i.e., heterozygotes or carriers) for a disease that could appear in their children, and the accumulation of information regarding the genetic composition of the population tested. [*See* Genetic Testing.]

I. INTRODUCTION

A. Purpose

The goals of newborn screening are to minimize or prevent the effects of untreated or late-treated genetic diseases and other birth defects. The use of automated laboratory techniques facilitates rapid cost-effective

ENCYCLOPEDIA OF HUMAN BIOLOGY, Second Edition, VOLUME 6.

identification of populations at significant risk for disorders that can have devastating outcomes, including fatality. These disorders can be treated more effectively when affected newborns are diagnosed early and receive appropriate therapy. The diseases are not readily recognized during the newborn period with normal well-child care. Follow-up includes further medical evaluation and treatment when a more extensive examination and testing are diagnostic of the disorder.

B. Principles

To provide screening services efficiently to large populations, the tests must be rapid and reliable, amenable to automation, and inexpensive.

1. The Entire Target Population Should Be Screened

Unless 100% of the newborns are tested, cases will be missed because affected infants are not tested. A child might not be tested in a newborn screening program for various reasons, including lack of information or incorrect information regarding the timing of specimen collection, specific policies of a hospital or practitioner that discourage comprehensive screening, and an individual's disinterest in screening. Populations more likely to escape testing include the home-born and infants born to transient or homeless families or to parents working or traveling abroad. Other newborns might not be tested because of events during the perinatal period, such as prematurity, early hospital discharge (especially before 24 hours of age), illness, parenteral feeding, and transfer from one hospital to another or from one unit to another within the same hospital.

2. False-Negative Results Should Be Minimized

A child who has the disease, but is classified as being at low risk for the disease by the newborn screening test, will not receive follow-up evaluation and will not be diagnosed during the newborn period. This result is termed "false negative." Certain characteristics influence the number of false-negative results, including the reliability and validity of the test and the quality control of the laboratory running the test. There are also biological explanations for false-negative results, such as an inadequate specimen (e.g., insufficient blood or a mishandled specimen), an inappropriate specimen (cord blood, because the metabolite, such as phenylalanine, has not had time to become elevated; blood collected after a plasma or red blood cell transfusion, because normal protein has been added by the transfusion; exchange transfusion; or dialysis, because the metabolite has been removed), and a specimen obtained from a newborn before physiological changes have occurred to allow the screening test to detect the high risk.

To decrease the possibility of false-negative results, some testing programs use a second newborn screening test, administered when the child is 2 weeks of age. To minimize the frequency of false-negative results, other programs adjust the level of the test result used to distinguish infants at high risk from those at low risk for the disease. This might increase the number of false-positive results and the time and effort in follow-up testing and/or treatment. These decisions are made by the public health authorities responsible for supervising the newborn screening program.

3. False-Positive Results Should Be Minimized

A child who does not have a disease, but is labeled by the screening test as having a high risk for the disease, has received a "false-positive" result, also known as a "false alarm." This child would receive follow-up evaluation to determine the presence or absence of the disease. A high rate of false-positive results produces unnecessary increases in the cost of follow-up and in the anxiety of the parents of infants receiving such results and diminishes the responsiveness of the health professionals contacted for follow-up evaluations.

4. Adequate Expeditious Follow-up Should Be Available to All Infants with Positive Test Results

Each newborn with a positive newborn screen must receive confirmatory diagnostic follow-up testing without delay. Time could be critical in the prevention of mental retardation, neurological damage, or death. In addition, treatment is necessary for children whose diagnosis is confirmed by the follow-up tests. Unfortunately, many newborn screening programs provide only for the newborn testing, but do not fund personnel, follow-up testing or required treatment.

5. Benefits of Newborn Screening Should Exceed Costs of the Program

The U.S. General Accounting Office estimated that for each $1 spent on newborn screening, $24 was saved in lifetime care costs for untreated patients. There are additional benefits to newborn screening that cannot be readily assigned a dollar value. These include saving a life, preventing mental retardation, counseling a couple regarding the risk for having other affected children, and improving understanding of the frequency and treatment of these diseases.

C. Outcomes

Newborn screening test results could indicate either a high (positive) or low (negative) risk for having a disease. Infants with a high risk are given follow-up testing. Depending on the disease, the follow-up is a repeat of the newborn screening test or a test that is diagnostic for the disease. Usually, treatment is delayed until these test results are available. However, if the disease is life-threatening and treatment for a brief period of time is safe for unaffected infants, therapy might be instituted immediately upon the initial positive screening result. Once the diagnosis is confirmed and treatment is initiated, most patients remain on treatment for life. This continued therapy is required to prevent the life- or intellect-threatening consequences of the disease.

D. Medical Legal Issues

When a newborn screening test does not identify an infant with the disease, the child has a false-negative result and will not receive the appropriate follow-up to newborn screening. In the absence of follow-up testing and treatment, the infant will develop the undesirable consequences of the disease. If the child is diagnosed as having the disease after developing mental retardation or after death from the disease, the family might seek legal recourse.

II. PHENYLKETONURIA AS A MODEL

A. Development

Phenylketonuria (PKU) is an autosomal-recessive biochemical genetic disease that leads to mental retardation if untreated. The incidence of PKU is one in 10,000 to one in 25,000 in the United States. Individuals with this disease do not metabolize phenylalanine, an amino acid found in protein foods. Treatment involves restricting the amount of phenylalanine consumed. The newborn screening test involves measurement of the level of phenylalanine in the blood. [*See* Phenylketonuria, Molecular Genetics.]

The initial newborn screening test for PKU, which has been abandoned, involved the measurement of phenylpyruvic acid in the urine. Phenylpyruvic acid is produced in the body from phenylalanine, and is therefore an indirect indicator of the level of phenylalanine. This test had a high false-negative rate, classifying one-third to one-half of the infants with the disease as low risk.

The currently used Guthrie test is a bacterial inhibition assay that measures the amount of phenylalanine in the blood; this test is more reliable and valid. The Guthrie test has the additional advantage of using dried blood on a filter paper, which is obtained by pricking the heel of the newborn; this simplifies obtaining the specimen and facilitates specimen transport to regionalized screening laboratories. It is a semiquantitative screening test and is not diagnostic. This test uses a mutant bacteria that requires phenylalanine for growth; the higher the phenylalanine in the dried blood spot, the larger the bacterial growth zone on the test plate.

Another semiquantitative method is paper chromatography, which separates the amino acids as spots on a sheet of paper, one end of which is immersed in a solvent. In addition, there are two tests that provide quantitative diagnostic measures of phneylalanine in the blood: the fluorometric method (based on the fluorescence of phenylalanine with other chemicals) and the column chromatographic method (based on separation and quantitation of each amino acid). Newborn screening is typically administered by each state's Department of Health. Currently, all states in the United States screen for PKU.

B. Treatment Programs

Dietary restriction of phenylalanine with regular monitoring of phenylalanine blood levels is the treatment for PKU. Treatment should begin as early as possible, definitely before 4 weeks of age, to prevent the mental retardation associated with PKU. The initial strategy involved the use of special formulas containing little or no phenylalanine. It was then determined that a small amount of phenylalanine was required for normal growth and development, so a small amount of phenylalanine was added to the diet; for infants this phenylalanine is added in the form of regular infant formula. In the past these practices meant that a breast-fed infant with a positive newborn screening for PKU and positive follow-up diagnostic testing was abruptly weaned from breast milk and switched to the formula diet low in phenylalanine. Subsequent research has demonstrated that phenylalanine supplementation can be provided by human breast milk; hence, breast milk and phenylalanine-free formula are now used to maintain blood levels of phenylalanine at therapeutic levels.

When PKU was initially treated with diet, the expectation was that treatment was necessary only during the first 4–6 years of life, while the nerve cells were undergoing full development (i.e., myelination). After this it was expected that the child could eat a normal diet, without phenylalanine restriction. However, this approach led to poorer school performance and a deterioration in behavioral control. The current recommendation is for a person with PKU to continue on a phenylalanine-restricted diet for his or her lifetime. This is particularly true for females with PKU.

C. Areas of Concern

Females of child-bearing age who have PKU need to be on a phenylalanine-restricted diet prior to conception and continuously throughout their pregnancy to protect the fetus. The fetus of a woman on an unrestricted diet might have microcephaly, mental retardation, or congenital heart disease. Currently, there is a widespread search for women with PKU who might not be aware of this potential problem and who are no longer on phenylalanine-restricted diets.

III. OTHER DISEASES FOR WHICH NEWBORN SCREENING IS AVAILABLE

A. Congenital Hypothyroidism

Congenital hypothyroidism is caused by inadequate production of thyroid hormone (i.e., thryoxine, or T_4). Patients who are not identified and treated promptly with supplementary thyroid hormone will be mentally retarded, fail to grow, suffer constipation, and have a slow metabolic rate. The incidence of congenital hypothyroidism is one in 3600 to one in 5000 in the United States. In this country, routine newborn screening uses dried blood spots, obtained at the same time as those for the PKU test, for immunoassay measurement of T_4 and/or thyroid-stimulating hormone. All states currently screen for congenital hypothyroidism as well as PKU, although the programs vary in the other diseases screened.

B. Sickle-Cell Disease

Sickle-cell disease, an autosomal-recessive disease, is caused by an abnormality in the β chains of hemoglobin, which transports oxygen in the blood. Untreated patients could have overwhelming infection, chronic anemia, "crises" involving the blockage of blood vessels, and pooling (or accumulation) of blood in the spleen. The incidence of sickle-cell disease is one in 400 among African-Americans. Treatment for these patients includes prophylactic antibiotics, immunization against specific bacterial infections, and rapid access to appropriate medical care. Newborn screening is performed on cord blood or dried blood spots and involves separation of the hemoglobin proteins by electrophoresis or some other method. [*See* Sickle-Cell Hemoglobin.]

C. Galactosemia

Galactosemia, an inherited disorder of galactose metabolism, results in failure to thrive and in vomiting, liver disease, cataracts, and mental retardation. The disease could even be lethal, if untreated, owing to overwhelming infection or liver disease in the newborn period. The frequency of galactosemia is one in 60,000 to one in 80,000 in this country. Treatment involves a galactose-free diet, which should be started as early as possible and continued throughout life. Newborn screening involves a test for either elevated blood galactose levels or deficiency of the enzyme that causes galactosemia.

D. Cystic Fibrosis

Cystic fibrosis is an autosomal-recessive disease that results in abnormalities of the lungs, pancreas, and sweat glands. Some affected infants have small-bowel obstruction due to meconium ileus or impaction of the fetal stool with intestinal blockage. Others die from malabsorption and malnutrition. Young adults die from pulmonary disease and infection. The incidence of cystic fibrosis is one in 2000 among whites in the United States. Treatment involves special dietary supplementation, pulmonary management, and antibiotics. The primary benefit of newborn screening appears to be the prevention of early malnutrition; the possibility of benefits to pulmonary function and long-term survival is the subject of current research. The newborn screening test for cystic fibrosis involves an immunoassay measurement of trypsin in dried blood specimens. [See Cystic Fibrosis, Molecular Genetics.]

E. Maple Syrup Urine Disease

Maple syrup urine disease is an autosomal-recessive disorder that results in lethargy, irritability, and vomiting and progresses to coma and death in the first 1–2 months of life if untreated. The incidence of this disease is one in 250,000 to one in 300,000 in this country. Treatment involves the dietary restriction of branched-chain amino acids. Newborn screening for this disorder is similar to that for PKU, but the amino acid leucine is measured by the bacterial inhibition assay.

F. Homocystinuria

Homocystinuria is an autosomal-recessive disorder involving a block in the pathway for breakdown of methionine, an amino acid. It has an incidence of one in 50,000 to one in 150,000 in the United States. Untreated patients develop a marfanoid physique (tall and thin, with long limbs, toes, and fingers), thromboembolism, dislocated lenses, and osteoporosis. Between 65 and 80% of untreated individuals are mentally retarded, and there is also an increased frequency of seizures, psychiatric disturbances, and myopathy. The treatment for homocystinuria depends on the underlying cause of the disease and involves vitamin treatment and/or dietary management. Newborn screening is similar to that for PKU, but methionine is measured by the bacterial inhibition assay.

G. Biotinidase Deficiency

Biotinidase deficiency is an autosomal-recessive disorder due to a block in the recycling of biotin, which in turn results in deficiencies of the carboxylase enzymes. Untreated patients can have convulsions, skin rashes, baldness, ataxia, and coma or might even die. Survivors are generally developmentally delayed. The estimate of the incidence of this disorder from limited patient series is one in 70,000. Treatment for this disorder consists of oral biotin. Newborn screening for biotinidase uses a colorimetric enzymatic assay on the dried blood specimen.

H. Congenital Adrenal Hyperplasia

Untreated patients affected by congenital adrenal hyperplasia, an autosomal-recessive metabolic abnormality in steroid hormone synthesis, might have life-threatening episodes with dehydration, salt imbalance, and shock in the neonatal period. Females have ambiguous genitalia with masculinization, frequently leading to the incorrect assignment of sex at birth. The incidence is approximately one in 12,000 in the United States, but this number varies substantially among racial and ethnic groups, being up to one in 680 in Yupik Eskimos. Medical management involves steroid replacement, but surgical treatment is frequently required for female infants. Approximately 90–95% of the patients with the disease can be detected by newborn screening, which is performed by immunoassay measurement of an abnormally elevated steroid metabolite in the dried blood specimens.

I. Duchenne Muscular Dystrophy

Duchenne muscular dystrophy is an X chromosome-linked recessive disease that primarily affects males and leads to progressive deterioration of muscles, beginning in infancy, with death in the second or third decade. No effective medical therapy is available. Physical therapy and bracing might prolong function and ambulation. Screening involves the measurement of creatinine kinase in dried blood specimens and currently requires a specimen drawn after 4 weeks of age. Screening is limited to pilot programs. [*See* Muscular Dystrophy, Molecular Genetics.]

J. Toxoplasmosis

Toxoplasmosis is not a genetic disease, but a congenital parasitic infection contracted by the mother primarily from raw meat and cat litter. Severely affected neonates suffer intrauterine growth retardation, brain calcifications, convulsions, retinal abnormalities, and severe developmental delay. More mildly affected children might have a later onset of developmental abnormalities. This milder form could be amenable to treatment. Newborn screening for this disorder remains investigational, while questions of incidence and benefit from early intervention are being addressed.

K. Acquired Immunodeficiency Syndrome

Acquired immunodeficiency syndrome (AIDS) is a disorder caused by the human immunodeficiency virus, which can be acquired congenitally by transmission from the mother. To date, newborn screening is completely anonymous and used solely for epidemiological purposes. [*See* Acquired Immunodeficiency Syndrome, Virology.]

IV. FUTURE DIRECTIONS

A. New Diseases

Currently, all states and the District of Columbia have newborn screening programs for PKU and congenital hypothyroidism. A 1987 National Institutes of Health consensus conference recommended universal screening for sickle-cell disease. Newborn screening programs continue to evolve, with different states and regions carrying out groups of tests generally from the foregoing list.

B. New Technology

Molecular genetic techniques enable researchers to obtain DNA from dried blood specimens. Using a technique called the polymerase chain reaction, small amounts of DNA are taken from the dried blood spots and amplified to amounts that can be analyzed more easily. The amplified DNA can be used to establish the genotype for many disorders, including sickle-cell disease, cystic fibrosis, and Duchenne muscular dystrophy. A number of states use DNA testing as a diagnostic test for all of their positive sickle-cell screening tests. This has significantly improved the time to antibiotic prophylaxis and reduced morbidity and death from infection. Molecular genetic testing is part of the pilot cystic fibrosis screening programs for infants with positive immunoassays. For both sickle-cell disease and cystic fibrosis, DNA diagnosis is performed using the original newborn screening sample, eliminating the need for a second sample, thus speeding diagnosis and reducing follow-up costs. DNA testing for Duchenne muscular dystrophy has been incorporated into pilot newborn screening programs for that disorder to provide an early diagnosis so that the family can plan for the care of their affected child and for additional children.

BIBLIOGRAPHY

American Academy of Pediatrics, Committee on Genetics (1982). New issues in newborn screening for phenylketonuria and congenital hypothyroidism. *Pediatrics* **69,** 104.

American Academy of Pediatrics, Committee on Genetics (1987). Newborn screening for congenital hypothyroidism: Recommended guidelines. *Pediatrics* **80,** 745.

American Academy of Pediatrics, Committee on Genetics (1989). Newborn screening fact sheets. *Pediatrics* **83,** 449.

American Academy of Pediatrics, Committee on Genetics (1992). Issues in newborn screening. *Pediatrics* **89,** 345.

Andrews, L. B. (ed.) (1985). "Legal Liability and Quality Assurance in Newborn Screening." American Bar Foundation, Chicago.

Andrews, L. B. (1985). "State Laws and Regulations Governing Newborn Screening." American Bar Foundation, Chicago.

Descartes, M., Huang, Y., Zhang, Y.-H., McCabe, L., Gibbs, R., Therrell, B. L., and McCabe, E. R. B. (1992). Genotypic confirmation from original dried blood specimens in a neonatal hemoglobinopathy screening program. *Pediatr. Res.* **31,** 217.

Jinks, D. C., Minter, M., Tarver, D. A., Vanderford, M., Hejtmancik, J. F., and McCabe, E. R. B. (1989). Molecular genetic diagnosis of sickle cell disease using dried blood specimens from newborn screening blotters. *Hum. Genet.* **81,** 363.

McCabe, E. R. B., McCabe, L., Mosher, G. A., Allen, R. J., and Berman, J. L. (1983). Newborn screening for phenylketonuria: Predictive validity as a function of age. *Pediatrics* **72,** 390.

Newborn screening for sickle cell disease and other hemoglobinopathies. (1987). *JAMA* **258,** 1205.

Newborn screening for sickle cell disease and other hemoglobinopathies. (1989). *Pediatrics. Suppl.* **83,** 813.

Seltzer, W. K., Accurso, F., Fall, M. Z., Van Riper, A. J., Descartes, M., Huang, Y., and McCabe, E. R. B. (1991). Screening for cystic fibrosis: Feasibility of molecular genetic analysis of dried blood specimens. *Biochem. Med. Metab. Biol.* **46,** 105.

Therrell, B. L., Jr. (ed.) (1987). Advances in neonatal screening. *Int. Congr. Ser.—Excerpta Med.* **741.**

Zhang, Y.-H., McCabe, L., Wilborn, M., Therrell, B. L., and McCabe, E. R. B. (1994). Application of molecular genetics in public health: Improve follow-up in a neonatal hemoglobinopathy screening program. *Biochem. Med. Metab. Biol.* **52,** 27.

Nitric Oxide

LOUIS J. IGNARRO
University of California, Los Angeles, School of Medicine

GLOSSARY

Cyclic GMP An intracellular chemical that acts as a secondary messenger in mediating the biological effects of hormones

Free radicals Chemically unstable and highly reactive molecules that can cause destruction of other chemicals and cells

Guanylate cyclase An enzyme that catalyzes the conversion of GTP to cyclic GMP in the intracellular compartment

Prostaglandins Polyunsaturated fatty acids formed in the body from arachidonic acid which function as autacoids

Receptors Molecules located outside or inside cells that interact selectively with chemicals or first messengers to trigger a cellular response

ENDOTHELIUM-DERIVED RELAXING FACTOR (EDRF) is an unstable chemical substance synthesized within and released from vascular endothelial cells, which line the inner, or intimal, surface of blood vessels. The two main biological actions of EDRF are the relaxation of vascular smooth muscle (i.e., vasodilation) and the inhibition of platelet clumping (i.e., aggregation) and adherence to the vascular endothelial surface. Various chemically diverse tissue hormones are capable of interacting with the endothelium and provoking the release of EDRF, which then diffuses into the underlying smooth muscle and nearby blood platelets. EDRF from artery, vein, and cultured vascular endothelial cells has been identified chemically and pharmacologically as nitric oxide (NO). NO is lipophilic (i.e., lipid soluble), rapidly diffuses through cell membranes, has a biological half-life of only several seconds, and undergoes spontaneous destruction in the presence of oxygen or certain oxygen-derived free radicals.

EDRF is a potent relaxant of vascular smooth muscle and an inhibitor of platelet function. NO is formed from the basic amino acid L-arginine and elicits its biological actions by activating the cytoplasmic enzyme guanylate cyclase, thereby stimulating the intracellular formation of cyclic GMP. In turn, cyclic GMP triggers subsequent steps down the cascade of biochemical pathways, leading to a decrease in the intracellular level of free calcium. Low intracellular free calcium levels are generally associated with smooth muscle relaxation and reduced platelet function, whereas high free calcium provokes smooth muscle contraction and platelet clumping. NO plays a physiological role in the local regulation of blood flow, blood pressure, and thrombus formation. Hypertension, myocardial infarction, and cerebral stroke are pathological states that may be attributed, at least in part, to a deficiency of NO formation resulting from vascular endothelial damage.

I. DISCOVERY AND UNDERLYING THEORY

A. Background

Although it had been appreciated for decades that the neurotransmitter acetylcholine causes vascular smooth muscle relaxation and a decrease in systemic blood pressure upon intravenous injection into mam-

ENCYCLOPEDIA OF HUMAN BIOLOGY, Second Edition, VOLUME 6.

mals, isolated strips of arterial smooth muscle mounted in special bath chambers usually contracted upon the addition of acetylcholine. The reason for this difference in responsiveness was provided by R. F. Furchgott in 1980, who discovered that the relaxation of arterial smooth muscle by acetylcholine required the presence of an intact functioning endothelium on the arterial strips. Normal healthy blood vessels have a single layer of squamous epithelial cells (i.e., endothelium) lining their intimal surface (i.e., the inner surface in contact with blood), which functions to transport substances between smooth muscle and blood and to provide a vascular surface that retards platelet adherence and blood clotting. Acetylcholine interacts with endothelial cells and triggers the formation and/or release of EDRF. Vascular tissue with damaged or nonfunctional endothelium displays contractile instead of relaxant responses to acetylcholine. Thus, acetylcholine is an endothelium-dependent vasodilator, and one of the chemical substances released from the endothelium was termed endothelium-derived relaxing factor. [*See* Smooth Muscle.]

As investigators learned more about the chemical and pharmacological properties of EDRF, it became increasingly clear that there were many similarities between EDRF and NO. EDRF was proposed to be NO in 1986 and was conclusively demonstrated in 1987. Due to the lack of absolute specificity of current chemical procedures for NO, it has not yet been possible to distinguish unequivocally between authentic NO and unstable nitroso compounds that decompose spontaneously and liberate NO. Although it is possible that EDRF is a labile nitroso compound, it is clear that the principal actions of EDRF are attributed to NO itself.

TABLE I
List of Endothelium-Dependent Vasodilators

Amines and related agents
Acetylcholine
Histamine
Norepinephrine
Serotonin
Peptides
Angiotensin
Bradykinin
Platelet-activating factor
Substance P
Thrombin
Vasoactive intestinal polypeptide
Vasopressin
Polyunsaturated fatty acids
Arachidonic acid
Leukotrienes
Basic polyamino acids
Polyarginine
Polylysine
Polyornithine
Other chemicals
ATP
Calcium ionophore A23817
Melittin
Thimerosal

B. Endothelium-Dependent Vasodilation

Numerous substances possessing diverse chemical properties cause endothelium-dependent vasodilation of arteries through the actions of NO. Using a similar mechanism, some of these substances relax veins as well. A partial list of endothelium-dependent vasodilators is presented in Table I. Although their chemistries differ, each chemical agent works in a similar fashion by interacting with selective receptors on the endothelial cell surface, thereby triggering a cascade of biochemical reactions, culminating in the formation and release of NO. Many endothelium-independent vasodilators exist, some of which are categorized as nitrovasodilators and include several clinically important drugs, such as nitroglycerin and sodium nitroprusside. These drugs, used clinically to reduce high blood pressure, elicit their effects in smooth muscle cells after biotransformation to NO, a potent endothelium-independent vasodilator.

C. Inhibition of Platelet Function

Another important biological action of NO is inhibition of platelet clumping and adherence to the vascular endothelium. NO diffuses from the endothelial cells to the intimal surface in contact with the circulating blood and interacts with nearby or adherent platelets to prevent further platelet adherence and clumping. This action serves to retard local thrombus, or blood clot, formation. Endothelium-dependent vasodilators thus not only relax vascular smooth muscle, but also inhibit platelet function via the actions of NO.

D. Localized Actions

Many endothelium-dependent vasodilators are naturally occurring tissue hormones that elicit localized actions in the body. Since NO is chemically unstable, possessing a biological half-life of only several seconds, the actions of NO are confined to the immediate vicinity of its formation and release. Because of its local and fleeting action, NO may also be categorized as a tissue hormone. Thus, NO is important in the local regulation of blood flow, perhaps in small resistance vessels (i.e., arterioles) and capillary beds, and thrombus formation. Such localized actions of NO serve to maintain adequate tissue perfusion with blood and complement the body's more diffuse mechanisms of maintaining blood flow, especially in vital organs such as the heart and the brain.

II. BIOLOGICAL ACTIONS

A. Demonstration of Release

The labile nature of NO initially made it difficult to demonstrate unequivocally its existence and release from vascular endothelial cells. Using bioassay techniques, in which an intact arterial segment is perfused through the lumen and the perfusate is allowed to bathe a nearby isolated and mounted strip or ring of endothelium-denuded artery, NO release from the perfused artery in the presence of added endothelium-dependent vasodilators can be easily demonstrated. Modified bioassay procedures, in which the perfusate is directed over a series of several arterial or venous strips arranged in a cascade, such that each strip is separated by about 3 sec in flow time, reveal the short biological half-life of NO (Fig. 1). This technique has been used to demonstrate NO release from arteries, veins, and cultured arterial endothelial cells and has wide applications in biological research.

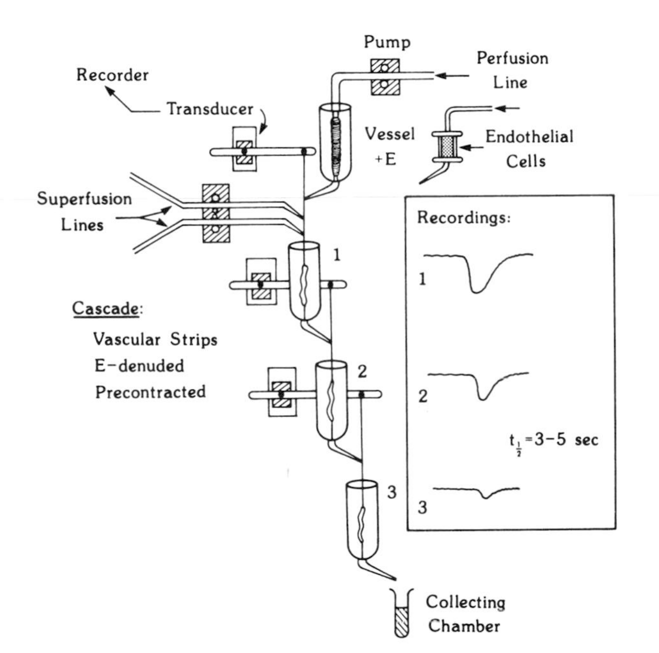

FIGURE 1 Bioassay cascade procedure used to study the release and properties of EDRF. Physiological salt solution is perfused by a roller pump through a segment of endothelium (E)-intact blood vessel or a column of isolated endothelial cells. The perfusate is superfused over three strips of E-denuded artery or vein that are precontracted by agents added to a separate superfusion line. Strips are mounted in a series (cascade), separated from one another by 3 sec of flow time. Changes in smooth muscle tension are measured by transducers and recorded on chart paper. EDRF release from blood vessels or endothelial cells results in decremental relaxation responses down the cascade of strips due to the short half-life ($t_{1/2}$) of EDRF. (Inset) Recordings of typical relaxant responses, downward deflections signifying a loss of muscle tone or relaxation.

B. Chemical Properties

Even before the identification of the relaxing factor as NO, it was clear that it was a labile lipophilic chemical of low molecular weight capable of diffusing rapidly through several cell membrane barriers in order to reach its site of action in smooth muscle and platelets. In an environment rich in oxygen, NO is labile and undergoes spontaneous chemical inactivation. Inactivation is caused not only by oxygen, but also by certain oxygen-derived free radicals, such as superoxide anion, that are spontaneously generated in oxygen-containing solutions. In experimental situations, the addition of superoxide dismutase, an enzyme that catalyzes the destruction of superoxide anion, protects NO against rapid inactivation and thereby prolongs its duration of action. Hemoglobin and oxyhemoglobin have a high binding affinity for NO and attract and trap it, thereby preventing its entry into vascular smooth muscle or platelets. [*See* Hemoglobin; Smooth Muscle.]

C. Pharmacological Properties

NO released from perfused blood vessels (i.e., arteries or veins) or from cultured vascular endothelial cells causes direct endothelium-independent relaxation of superfused isolated strips or rings of artery and vein in the same or different species. The relaxing factors from numerous sources possess the same chemical properties and also inhibit platelet clumping and adherence to vascular endothelial surface: they are all NO.

D. Mechanism of Action

NO elicits its two principal actions by the same mechanism. After diffusion into nearby vascular smooth muscle cells and locally circulating or adherent platelets, it activates the cytoplasmic enzyme guanylate cyclase, which catalyzes the conversion of GTP to cyclic GMP. In turn, cyclic GMP causes vascular smooth muscle relaxation and the inhibition of platelet function.

Like hemoglobin, cytoplasmic guanylate cyclase is a heme-containing protein from which the heme groups can be detached. In experiments designed to compare the properties of the heme-containing and heme-deficient forms of guanylate cyclase, it can be observed that NO activates guanylate cyclase by heme-dependent mechanisms. The activation of heme-containing guanylate cyclase by NO can be inhibited by the addition of excess heme or hemoglobin to enzyme reaction mixtures.

III. PHARMACOLOGY OF NITRIC OXIDE

A. Activation of Guanylate Cyclase

NO is a colorless, odorless gas that is soluble in aqueous and organic solvents and is a decomposition product of numerous unstable nitroso and nitro compounds. NO gas is rapidly oxidized to the brown pungent gas nitrogen dioxide (NO_2). In solution, NO is rapidly oxidized to NO_2^- (nitrite) and NO_3^- (nitrate). In the mid-1970s, NO gas was first shown to activate guanylate cyclase present in the soluble or cytoplasmic fraction of cells and to stimulate the intracellular accumulation of cyclic GMP in various tissues and species. Chemical agents that decompose to liberate NO also activate guanylate cyclase, as would be expected. Such chemical agents include the nitrovasodilators nitroglycerin, sodium nitroprusside, and amyl nitrite. The knowledge that nitrovasodilators liberate NO and stimulate cyclic GMP formation led to the discovery in 1979 that NO is a potent vasodilator that works through the actions of cyclic GMP. NO activates guanylate cyclase by heme-dependent mechanisms involving the binding of NO to the heme group of the enzyme, with consequent formation of the NO–heme–enzyme complex. This enzyme form represents the activated state of guanylate cyclase and catalyzes the conversion of GTP to cyclic GMP at a rate that is 50- to 200-fold greater than that of the unactivated enzyme. Guanylate cyclase activation by NO can be inhibited by methylene blue and related oxidizing agents.

B. Vascular Smooth Muscle Relaxation

The vascular smooth muscle relaxation induced by NO in the whole animal causes vasodilation, which can lead to a decrease in blood pressure. A more localized action of NO may improve regional blood flow without necessarily decreasing systemic blood pressure. Solutions of sodium nitroprusside spontaneously liberate NO, whereas nitroglycerin and amyl nitrate permeate smooth muscle cells and engage in biotransformation reactions, resulting in the formation of NO.

C. Inhibition of Platelet Function

The mechanism of action of NO on platelets is identical to that by which NO relaxes vascular smooth muscle: through cyclic GMP accumulation. In turn, cyclic GMP triggers other intracellular events that result in the inhibition of platelet function. Those nitrovasodilators that spontaneously liberate NO in solution (e.g., sodium nitroprusside) or that can be biotransformed to NO within platelets (e.g., isosorbide dinitrate and amyl nitrite) also inhibit platelet function.

D. Other Actions

NO is produced not only by the vascular endothelium but also by other cell types. EDRF signifies the NO that is produced exclusively by vascular endothelial cells by an enzyme termed endothelial NO synthase. NO is produced also by peripheral and central neurons belonging to the class termed "nonadrenergic-noncholinergic neurons" (NANC neurons), where NO serves the function of a neurotransmitter to cause target vascular and nonvascular smooth muscle relaxation. One well known neurotransmitter function for NO is as a physiologic mediator of penile erection in the mammalian species. The NO present in neurons is synthesized by neuronal NO synthase and may also serve certain central functions in facilitating memory and learning. NO is produced also by various other cell types that function in host defense and inflammation including macrophages, neutrophils and hepatocytes. In the latter cell types the quantity of NO produced is generally much greater than that produced by the vascular endothelium or NANC neurons. Indeed, the relatively large quantities of NO produced by macrophages and related cells appear to play a pathophysiological role in causing target cell injury, such as cytostatic or cytocidal actions on invading tumor cells, bacteria, fungi, and viruses. Whereas the mechanism of physiological action of small quantities of NO on smooth muscle and platelets involves the secondary production of cyclic GMP, large amounts of NO cause cytotoxic effects by direct and indirect chemical interaction with certain components of key proteins involved in cell proliferation and function.

IV. SECONDARY MESSENGER ROLE OF CYCLIC GMP

A. Secondary Messenger Concept

The secondary messenger concept pertains to the manner in which a signal is communicated from the extracellular to the intracellular compartment for the purpose of initiating a cellular response to an external stimulus. Often, the first messenger, such as a hormone, comes into contact with target sites or receptors on the cell surface. The hormone–receptor interaction triggers a response within the cell to generate a second signal, usually by increasing the concentration of another chemical, the secondary messenger, which then triggers the cellular response, thereby mediating the action of the first messenger. The secondary messenger may be cyclic GMP, cyclic AMP, calcium, or another chemical. NO acts in an even simpler way because it is lipophilic and diffuses into the cell to interact with an intracellular, rather than a membrane, receptor. The intracellular receptor for NO is the heme group bound to guanylate cyclase, and the secondary messenger is cyclic GMP.

B. Cyclic GMP

The information on cyclic GMP as a secondary messenger is relatively limited, essentially causing vascular smooth muscle relaxation and the inhibition of platelet function in response to NO. Cyclic GMP may also, however, play a role in regulating leukocyte function and cellular proliferation in general, but definitive conclusions cannot be drawn.

V. FORMATION, RELEASE, AND METABOLISM

A unique enzymatic pathway is involved in the biosynthesis of endothelium-derived NO. The single enzyme protein, NO synthase, catalyzes the oxidation of L-arginine to NO plus L-citrulline. Catalysis involves the oxidative cleavage of one of the two basic guanidino amino nitrogen atoms to NO and oxidation of the remaining molecule to L-citrulline. The enzymatic reaction is complex and requires molecular oxygen as well as at least five different cofactors, including nicotine adenine dinucleotide phosphate, flavin adenine dinucleotide, flavin mononucleotide, calmodulin, and tetrahydrobiopterin. These cofactors function to generate and shuttle electrons through NO synthase so that the five-electron oxidation of L-arginine to NO can proceed. There are at least three different isoforms of NO synthase. One is in vascular endothelial cells, another is in neurons, and the third is in activated macrophages and related cells. The endothelial and neuronal NO synthases are constitutive and are regulated by calcium, which activates the enzymes. The macrophage NO synthase is not regulated by calcium and must be newly synthesized whenever it is needed, and this is often stimulated by endotoxin and cytokines. This form of NO synthase, which is often referred to as inducible NO synthase, generates much larger quantities of NO than the endothelial or neuronal forms of NO synthase.

NO, being lipophilic and of low molecular weight, exits the endothelial cell by simple diffusion through the membrane. This mechanism requires no expenditure of energy and allows the NO to exit the cell in all directions, thereby reaching the underlying smooth muscle and nearby platelets on the opposite side as well. NO produced by macrophages and other cells in which NO synthase can be induced also diffuses to nearby blood vessels and platelets.

NO is rapidly metabolized. In the presence of its natural environment of oxygen and certain oxygen-containing compounds, any NO formed will be oxidized rapidly to NO_2^- and NO_3^-, which are relatively inactive on blood vessels and platelets.

Figure 2 is a schematic illustration of the formation, release, action, and metabolism of endothelium-derived NO.

VI. PHYSIOLOGICAL SIGNIFICANCE

A. Healthy State

Although the existence, identification, and pharmacological actions of the relaxing factor are clear, less than dogmatic conclusions can presently be drawn concerning its precise role in normal physiology.

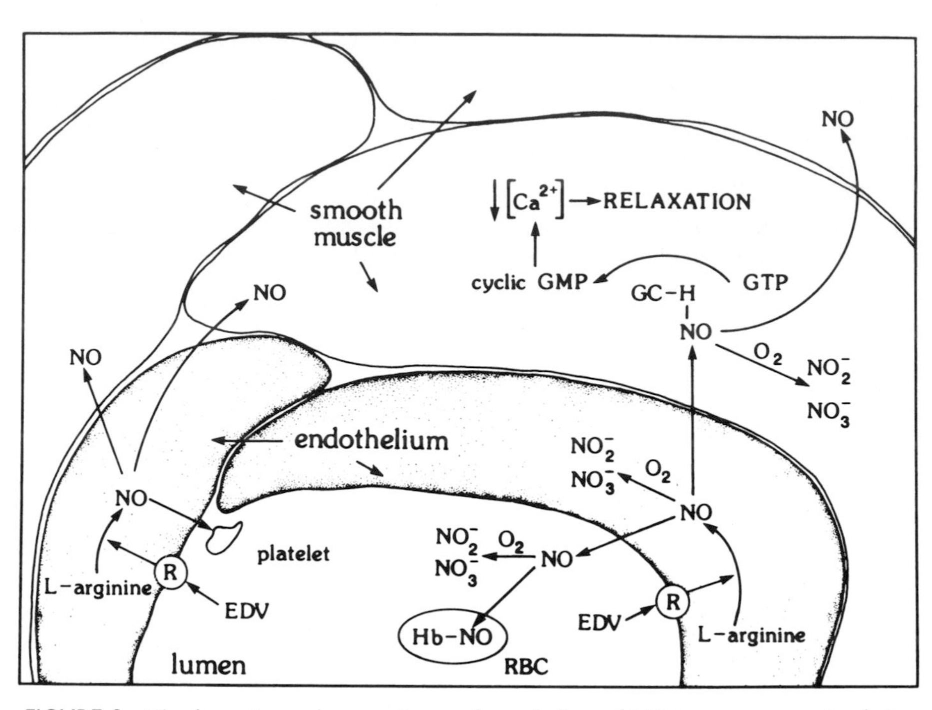

FIGURE 2 The formation, release, action, and metabolism of NO in response to circulating endothelium-dependent vasodilators (EDV). EDV interacts with receptors (R) on the luminal surface of endothelial cells, triggering the formation of NO from L-arginine. NO diffuses locally into the underlying smooth muscle to cause relaxation and into circulating or adherent platelets to cause the inhibition of platelet aggregation and adherence to the luminal surface of the endothelium. NO causes its effects in both smooth muscle and platelets by interacting with the heme (H) group of cytoplasmic guanylate cyclase (GC), thereby activating the enzyme and stimulating the formation of cyclic GMP from GTP. Cyclic GMP causes a lowering of intracellular free calcium levels, thereby resulting in vascular smooth muscle relaxation and the inhibition of platelet function. The actions of NO are rapidly terminated by spontaneous oxidation of NO_2^- and NO_3^- and by binding to hemoglobin (Hb) present in circulating red blood cells (RBC).

However, our knowledge of its chemical and biological properties, together with our current understanding of the actions of other regulatory molecules, allows some logical and informative speculation on its physiological significance.

NO plays a regulatory role in influencing local tissue blood flow and blood clot formation. A small quantity of NO is released constantly from the vascular endothelium, maintaining tissue perfusion by keeping arterioles open and free of blood clots. The release of NO is determined by stimuli originating in the body, such as the molecules listed in Table I, local increases in blood flow, and pressure, or shear, at the endothelium–blood interface. The platelets release ADP during clumping, and ADP can act locally to release NO. Thus, NO is likely to be involved in local, rather than more diffuse, or systemic, regulation of blood flow and platelet function. NO works in concert with a second endothelium-derived chemical, prostaglandin I_2 (i.e., prostacyclin), that, though the secondary messenger action of cyclic AMP, also causes vasodilation and inhibits platelet aggregation. When the two substances are present together, there is great enhancement of the inhibition of platelet aggregation, thereby ensuring minimal local blood clotting, if any.

B. Pathological or Disease States

In light of the biological actions of endothelium-derived NO, it is clear that cardiovascular disease states such as hypertension, myocardial infarction, and cerebral stroke may be attributed to its deficiency. Damage to vascular endothelial cells will result in a greatly diminished capacity to generate NO (as well as prostacyclin). Hypertension, atherosclerosis, and diabetes are often characterized morphologically by the abnormal appearance of the vascular endothelium and pathologically by elevated blood pressure and blood clot formation. Local vasoconstriction or vasospasm, with or without local thrombus formation, greatly curtails regional blood flow, with severe consequences when it occurs in the heart or the brain. [*See* Atherosclerosis; Hypertension; Stroke.]

The high-output production of NO by inducible NO synthase present in activated macrophages and related cells in the host defense system can lead to local host tissue inflammation and injury as such cells attempt to protect the host against foreign or invading cells or particles. This is characteristic of acute inflammation, where NO is now considered to be a major player. The high-output production of NO can occur also in cells that are not typically considered to be part of the immune or host defense system such as vascular smooth muscle cells. NO synthase can be induced in vascular smooth muscle by endotoxin and certain cytokines, a condition known as endotoxin or septic shock, which is characterized by marked and life-threatening vasodilation and hypotension. These symptoms are mediated by the large quantities of NO produced directly in the vascular smooth muscle, which becomes relaxed in response to the locally synthesized NO. Efforts are in progress to develop potent inhibitors of inducible NO synthase for the symptomatic treatment of endotoxin shock.

BIBLIOGRAPHY

Bassenge, E., and Busse, R. (1988). Endothelial modulation of coronary tone. *Prog. Cardiovasc. Dis.* **30**, 349.

Bredt, D. S., and Snyder, S. H. (1994). Nitric oxide: A physiologic messenger molecule. *Annu. Rev. Biochem.* **63**, 175–195.

Feldman, P. L., Griffith, O. W., and Stuehr, D. J. (1993). The surprising life of nitric oxide. *Chem. Engineer. News* December 20, 26.

Forstermann, U. (1986). Properties and mechanisms of production and action of endothelium-derived relaxing factor. *J. Cardiovasc. Pharmacol.* **8**, S45.

Furchgott, R. F. (1983). Role of endothelium in responses of vascular smooth muscle. *Circ. Res.* **53**, 557.

Ignarro, L. J. (1989). Endothelium-derived nitric oxide: Actions and properties. *FASEB J.* **3**, 31.

Ignarro, L. J. (1989). Biological actions and properties of endothelium-derived nitric oxide formed and released from artery and vein. *Circ. Res.* **64**, 1.

Moncada, S., Palmer, R. M. J., and Higgs, E. A. (1991). Nitric oxide: Physiology, pathophysiology and pharmacology. *Pharmacol. Rev.* **43**, 109–142.

Nathan, C. (1992). Nitric oxide as a secretory product of mammalian cells. *FASEB J.* **6**, 3051.

Vanhoutte, P. M., Rubanyi, G. M., Miller, V. M., and Houston, D. S. (1986). Modulation of vascular smooth muscle contraction by the endothelium. *Annu. Rev. Physiol.* **48**, 307.

Waldman, S. A., and Murad, F. (1987). Cyclic GMP synthesis and function. *Pharmacol. Rev.* **39**, 163.

Xie, Q.-w., and Nathan, C. (1994). The high-output nitric oxide pathway: Role and regulation. *J. Leucocyte Biol.* **56**, 576–582.

Nonnarcotic Drug Use and Abuse

LESTER GRINSPOON
JAMES B. BAKALAR
Harvard Medical School

GLOSSARY

Acute effect Effect caused by a single dose of a drug

Chronic effect Effect caused by long-term and repeated use of a drug

Cross-tolerance Tolerance to one drug that develops through repeated use of another drug with similar neurophysiological activity

Metabolic tolerance Tolerance that arises because the liver breaks down the drug and eliminates it from the body faster

Pharmacodynamic tolerance Tolerance arising from adaptation of nerve receptors in the brain to a drug

Substance dependence (drug dependence) Psychiatric disorder that involves symptoms such as repeatedly taking a drug while knowing that it does harm but being unable to stop, often taking more of a drug than intended, and developing tolerance or a withdrawal reaction to a drug, or both

Tolerance Loss of sensitivity to a drug that makes it necessary to take more to preserve the original effect

Withdrawal reaction Physical and psychological symptoms that may arise when a dependent user stops taking a drug

SUBSTANCE ABUSE IS DEFINED AS A MALADAPTIVE pattern of substance use indicated by continued use despite the persistent social, occupational, psychological, or physical problems it causes or recurrent use when it is hazardous. Interpretations of drug abuse are vulnerable to cultural prejudice; for the purposes of this article, it means taking drugs at dose levels and in circumstances and settings that significantly augment their potential for harm.

Physical dependence on a drug implies a biochemical or physiological change in the body that makes the continued presence of the drug necessary to avoid a withdrawal reaction. Tolerance refers to a declining effect of the drug on repeated administration of the same dose and the consequent necessity to increase the dose to obtain the original euphoric effect. A widely accepted definition of substance dependence involves at least three of the following: (1) much time spent thinking about the drug or arranging to get it, (2) often taking more than intended, (3) tolerance, (4) withdrawal symptoms, (5) use to avoid or relieve withdrawal symptoms, and (6) persistent or repeated desire or effort to cut down use.

Although neither substance abuse nor substance dependence is now classified as a personality disorder, those who develop a dependency or abuse problem often need a euphoriant to experience pleasure or a respite from pain. People with character disorders, anxiety, depression, feelings of inadequacy, or intolerable life situations are susceptible to abuse of a number of drugs in addition to alcohol and opiates. Just as some adults use alcohol when they feel depressed, anxious, and inadequate, some young people may try to deal with emotional problems through the use of illicit drugs. When drug abuse is symptomatic, the abuser can often make good use of psychotherapy. Once a substance has been abused to the point of compulsive use or physical dependency, treatment becomes more difficult and complicated. In these cases, community resources concerned with drug abuse are an important complement to the traditional psychiatric approach.

I. AMPHETAMINES

Amphetamines and amphetamine congeners are a large group of central nervous system stimulant drugs. Among the best known are dextroamphetamine (Dexedrine), methamphetamine (Methedrine), and methylphenidate (Ritalin). Racemic amphetamine sulfate (Benzedrine) was first introduced as a medicine in 1932 and became available in tablet form in 1937. Soon, amphetamine began to receive much sensational publicity, and many physicians came to regard it as a versatile remedy second only to a few other extraordinary drugs such as aspirin. In 1971, the annual legal United States production of all kinds of amphetamines reached more than 10 billion tablets. From the mid-1960s on, considerable growth occurred in illicit laboratory synthesis and black market diversion of legitimately produced drugs.

Since 1970, amphetamine abuse has declined, partly because of legal restrictions. Amphetamines now have accepted therapeutic applications only in treating narcolepsy (a sleep disorder) and attention deficit disorder, and as an occasional adjunct to tricyclic antidepressants in the treatment of depression. Few physicians now prescribe amphetamines for weight loss, because of well-founded doubts concerning efficacy and safety.

Both acute amphetamine intoxication and chronic use have numerous adverse physical effects. Symptoms of acute poisoning include flushing, pallor, cyanosis (bluish skin), fever, high heart rate, nausea, vomiting, difficulty in breathing, tremor, lack of coordination, loss of sensory capacities, elevated blood pressure, hemorrhages, strokes, convulsions, and loss of consciousness. Death from overdose is usually associated with high fever, convulsions, and cardiovascular shock. Intravenous abuse can produce diseases of the liver, lung, heart, and arteries.

Adverse psychological effects include restlessness, excessive talking, insomnia, irritability, hostility, tension, confusion, anxiety, panic, and, quite commonly, psychosis. Clinical experiments using volunteers show that a psychosis may be induced in essentially normal people by even short-term administration of dextroamphetamine.

The symptoms of amphetamine psychosis strikingly resemble those of paranoid schizophrenia. Restlessness, increased irritability, and heightened perceptual sensitivity often precede delusions of persecution and bizarre visual and auditory hallucinations. Some authorities believe that predominance of visual hallucinations, appropriate mood, clear consciousness, hyperactivity, hypersensitivity, or absence of thought disorder distinguishes amphetamine psychosis from an acute schizophrenic episode. Others consider a strictly clinical differentiation all but impossible. Laboratory tests for amphetamine in urine are technically complex and are negative after 48 hr. Physicians often recognize amphetamine psychosis only in retrospect, by the rapid disappearance of the symptoms—usually within days or, at most, weeks. Suspiciousness and ideas of reference sometimes persist for months after the overt psychosis has ended. Psychosis usually occurs when an abuser who is already taking large doses takes even larger amounts than usual for a period of time.

Chronic amphetamine abusers often find that the drug begins to dominate their lives through a craving severe enough to be called a compulsion. Their irritability and paranoia may cause unprovoked violence and drive their friends away; their preoccupation with the drug has a disastrous effect on their family relationships and work. A high degree of tolerance develops; an abuser may eventually need up to twenty times as much to recover the original euphoric effect.

A letdown or crash often occurs when an amphetamine abuser is forced to stop using the drug for a time because it is producing agitation, paranoia, and malnutrition. A debilitating cycle of runs (heavy use for several days to a week) and crashes is a common pattern of abuse. The physical symptoms of withdrawal include headache, sweating, muscle cramps, stomach cramps, and hunger. The characteristic psychological symptom is a lethargic depression, suicidal at times, which peaks at 48–72 hr after the last dose of amphetamine but may persist at lesser intensity for weeks. Often a vicious cycle develops. Patients who suffer from chronic depression or feelings of inadequacy take amphetamine for relief and become dependent on the drug. When they try to stop using it, they become depressed even further and feel a need to start again.

II. COCAINE

Cocaine is an alkaloid derived from the shrub *Erythroxylum coca,* a plant indigenous to Bolivia and

Peru, where peasants chew its leaves for their stimulating effect. Cocaine was isolated in 1860 and became important after 1884 as the first effective local anesthetic (also the only current medical use). In 1914, cocaine was subjected to the same laws as morphine and heroin and since then has been legally classified with narcotics. Since about 1970, cocaine has steadily been gaining popularity and is now the second most popular illicit drug after marijuana.

Street cocaine varies greatly in purity; it is usually adulterated with sugars, procaine, or other substances. Users rarely take cocaine by mouth, because the effect is regarded as too mild to warrant the expense. Three widespread methods of cocaine ingestion are prevalent: snuffing (snorting), subcutaneous or intravenous injection, and freebasing. Freebasers convert cocaine hydrochloride, cocaine alkaloid (the free base), by treating it with a basic solution such as baking soda or ammonia. They then smoke the alkaloid in a water pipe. Freebase cocaine is now sold on the streets under the name "crack." Freebasing and injection allow the drug to reach the brain faster, providing an ecstatic rush and subsequent letdown that create particularly serious abuse problems.

The adverse effects of cocaine resemble those of amphetamine. One common acute effect is an anxiety reaction with symptoms including high blood pressure, racing heart, anxiety, and paranoia. High doses or repeated use can produce a state resembling mania, with impaired judgment, incessant rambling talk, and hyperactivity. Because cocaine increases energy and confidence and can produce irritability and paranoia, it may also lead to physical aggression and crime. In high doses, especially when smoked or injected, cocaine can produce cardiac symptoms, nausea, headache, cold sweat, tremors, and fast, shallow breathing. People who have high blood pressure or damaged arteries may suffer strokes. High doses can also cause seizures and death from cardiac or respiratory arrest. Deaths from opiates and cocaine injected in combination intravenously (a speedball) occur frequently. The number of deaths from cocaine and emergency room admissions for drug reactions involving cocaine has increased greatly over the past decade.

Cocaine, along with some amphetamines, is the drug most eagerly consumed by experimental animals; they will kill themselves with voluntary injections. Craving can also become a serious problem for human beings who have constant access to cocaine, especially for freebasers and intravenous abusers. The financial as well as physical and psychological costs of compulsive cocaine abuse can be devastating. A damaging habit usually develops over a period of several months to several years. A compulsive user thinks about the drug constantly, cannot turn it down, and may borrow, steal, or deal to pay for it. Cocaine produces substantial acute tolerance and some long-term tolerance. Experimenters report cross-tolerance to amphetamines. On the other hand, sensitization to the seizure-inducing effect may occur.

A common pattern of cocaine abuse consists of binges followed by crashes—the same pattern found in amphetamine abuse and observed in laboratory animals as well as human beings. A weeklong binge produces extreme acute tolerance and exhaustion from lack of food and sleep. The user then falls into several days of severe depression, excessive sleep, overeating, and sometimes chills or muscle pains. Craving is likely to continue after this acute phase is over and may reemerge even many months or even years later, especially when certain moods or situations have come to evoke a conditioned response.

Some effects of chronic cocaine abuse are weight loss, insomnia, perceptual disturbances, paranoid thinking, and psychoses. Users often take alcohol, sedatives, or narcotics to calm cocaine-induced agitation, and this practice may create a dual dependency that is particularly hard to treat. Intensive and compulsive users become jittery, irritable, and self-absorbed. They may also become hypersensitive to sound, light, and touch. At first, cocaine may be sexually stimulating, but chronic abusers tend to lose sexual interest. Like amphetamines, cocaine may also cause absorption in apparently meaningless, repetitive activities—scribbling, counting, pacing, sorting and ordering, and taking apart and reassembling mechanical objects.

Each method of ingestion has its own dangers. Snorting causes noses to become runny or clogged and sometimes inflamed, swollen, or ulcerated; heavy users occasionally develop perforated septa. Freebasing may damage the surface of the lung and reduce its capacity to exchange gases. Injection may lead to infection or embolism. [*See* Cocaine and Stimulant Addiction.]

III. BARBITURATES

The barbiturates most commonly used are the long-acting (12–24 hr) phenobarbital, the intermediate-range (6–12 hr) amobarbital (Amytal), and the short-acting (3–6 hr) pentobarbital (Nembutal) and secobarbital (Seconal). Barbiturates are used mainly as sedatives, hypnotics, tranquilizers, anticonvulsants, and anesthetics. The illicit market meets its needs by diverting shipments from legitimate manufacturers and by robbing drug warehouses. The most common barbiturates on the black market are secobarbital ("reds," "red devils," "downers"), pentobarbital ("yellow jackets," "nembies"), and a combination of secobarbital and amobarbital ("reds and blues").

But illicit barbiturates are only part of the problem. For many years these drugs were prescribed rather indiscriminately, and many persons were unwittingly maintained in the habit by family doctors. This problem is becoming less common as benzodiazepines replace barbiturates in many medical uses. Pentobarbital, secobarbital, and amobarbital are now under the same federal legal controls as morphine. Both legal and illicit barbiturate use seems to be declining.

A. Patterns of Abuse

Chronic intoxication occurs especially in middle-aged, middle-class people who obtain the drug from their doctors as a prescription for insomnia or anxiety. Now that refills are limited by law, these abusers may visit many physicians and get a prescription from each one. Their drug dependence may go unnoticed for months or years, until their work begins to suffer or they show slurred speech and other physical signs. This pattern is becoming less common as medical use of barbiturates declines.

A second pattern is periodic intoxication. Users in this category are generally adolescents or young adults who take barbiturates for the same reason they use alcohol, to produce a high or a sense of well-being. Personality, set (expectations), and setting determine whether the effect is regarded as a sedative or euphoriant. Barbiturates may become a fairly constant part of their lives, as alcohol is for the social drinker.

The most dangerous pattern is intravenous barbiturate use, mainly practiced by young adults intimately involved in the drug culture. Their drug experience has usually been extensive, including intravenous heroin and amphetamines. They often use barbiturates because the habit is less expensive to maintain then a heroin habit, even at a level of 2000–3000 mg/day. The rush is described as a pleasant, warm, drowsy feeling. Like amphetamine abusers (speed freaks), these barbiturate abusers tend to be irresponsible, violent, and disruptive. The physical dangers of injection include cellulitis, vascular complications from accidental injections into an artery, infections, and allergic reactions to contaminants. Users can often be identified by skin abscesses. Barbiturates are also used incidentally to dependence on other drugs—by heroin addicts to boost the effects of weak heroin, by alcoholics to enhance the intoxication or relieve the symptoms of alcohol withdrawal, and by amphetamine and cocaine abusers as a sedative to avoid paranoia and agitation.

B. Effects

Mild barbiturate intoxication (both acute and chronic) resembles alcohol intoxication. Symptoms include sluggishness, difficulty in thinking, poor memory, slowness of speech and comprehension, faulty judgment, narrow range of attention, emotional lability, and exaggeration of basic personality traits. The sluggishness usually wears off after a few hours, but judgment may remain defective, mood distorted, and motor skills impaired for as long as 10–22 hr. Other symptoms are hostility, quarrelsomeness, moroseness, and, occasionally, paranoid ideas and suicidal tendencies. Neurological effects include nystagmus, diplopia, strabismus, ataxic gait, positive Romberg's sign, hypotonia, dysmetria, and decreased superficial reflexes. The diagnosis may be confirmed by a number of laboratory tests.

All patterns of use present dangers to health. Acute intoxication can produce death from suicide, accident, or unintentional overdose. Barbiturates are the drugs most commonly taken with suicidal intent. In many doubtful cases, automatism must be considered: people whose judgment and memory are already impaired by the drug may forget or disregard previous doses and automatically take another dose (overdose) in order to fall asleep, without any conscious intent of committing suicide. The effects of alcohol and barbiturates are additive, and this combination is especially dangerous. Barbiturate-induced death arrives through a sequence of deep coma, respiratory arrest, and cardiovascular failure. The lethal dose varies with the route of administration, excitability of the central nervous system, and acquired tolerance. The ratio of lethal to effective dose can be as low as 3 to 1 or as high as 50 to 1 for the most commonly abused barbiturates.

C. Tolerance and Withdrawal

Like many other drugs, barbiturates produce pharmacodynamic or central nervous system tolerance: nerve cells adapt to the drug and require more of it to respond. By inducing drug-metabolizing microsomal enzymes in the liver, they also produce metabolic tolerance (and incidentally reduce the effectiveness of a number of other drugs, especially anticoagulants and tricyclic antidepressants). Cross-tolerance with alcohol develops.

A withdrawal reaction occurs when the drug has become necessary to maintain the proper functioning of the body. This usually requires 1–2 months or more at doses well above the recommended therapeutic level. The barbiturate withdrawal reaction ranges from mild symptoms such as anxiety, weakness, sweating, and insomnia to seizures, delirium, and cardiovascular collapse leading to death. At its worst, it is the most severe of the drug abstinence syndromes. Pentobarbital or secobarbital users with 400 mg/day habits experience only mild withdrawal symptoms. Those taking 800 mg/day experience weakness, tremor, and anxiety; about 75% have convulsions. Users on even higher doses may suffer from anxiety, anorexia, and psychoses, as well as convulsions resembling grand mal epilepsy. The psychosis is clinically indistinguishable from alcoholic withdrawal delirium (delirium tremens); its main features are confusion, agitation, delusions, and hallucinations that are usually visual but sometimes auditory. Most of the symptoms appear on the first day of abstinence, but seizures generally do not occur until the second or third day, when the symptoms are at their worst. Psychosis, if it does develop, starts in the third to the eighth day. The various symptoms may last as long as 2 weeks.

Dependent users often take an average daily dose of 1.5 g of short-acting barbiturate, and some have been reported to take as much as 2.5 g/day for months. Tolerance to the lethal effect does not develop nearly to the same extent as tolerance to the desired psychoactive effect. Thus, withdrawal in a hospital may eventually become necessary to prevent accidental death from overdose.

IV. BENZODIAZEPINES

The benzodiazepines include diazepam (Valium), flurazepam (Dalmane), oxazepam (Serax), chlordiazepoxide (Librium), alprazolam (Xanax), and about 20 other drugs. They are used mainly in treating anxiety but also serve as sedatives, muscle relaxants, and anesthetics, and in the treatment of alcohol withdrawal. Introduced in the 1960s, benzodiazepines soon became the most popular prescription drug in the United States. A recent substantial decline in medical use suggests that physicians are becoming more cautious about them. Benzodiazepines produce little respiratory depression, and the ratio of lethal to effective dose is very high, estimated at 200 : 1 or more. Very large amounts (>2 g) taken in suicide attempts produce drowsiness, lethargy, confusion, and ataxia but rarely cause permanent damage. However, chlordiazepoxide in high doses has been known to induce coma. The adverse effects of lower doses include drowsiness, unsteadiness, and weakness. Some benzodiazepines have a disinhibiting effect, which may cause hostile or aggressive behavior in susceptible people subject to frustration.

Benzodiazepines produce less euphoria than other tranquilizing drugs, so the risk of dependence and abuse is relatively low. They do not stimulate the liver enzymes that cause metabolic tolerance nearly as much as barbiturates do, but both tolerance and withdrawal symptoms can develop. The withdrawal reaction may occur not only at very high doses but at therapeutic doses, depending on which benzodiazepine is used and for how long. Symptoms include anxiety, numbness in the extremities, dysphoria, intolerance for bright lights and loud noises, nausea, sweating, muscle twitching, and sometimes convulsions. Withdrawal is not usually accompanied by craving for the drug. Because benzodiazepines are eliminated from the body slowly, symptoms may continue to develop for several weeks. To prevent seizures and other problems, withdrawal is accomplished by gradual reduction of the dose.

V. VOLATILE SOLVENTS AND AEROSOLS

Some abused volatile solvents and aerosols are gasoline, varnish remover, lighter fluid, airplane glue, rubber cement, cleaning fluid, spray paints, deodorants, and hair sprays. The active ingredients include toluene, acetone, benzene, and halogenated hydrocarbons. Because these substances are legal, cheap, and accessible, they are used mostly by the young (ages 6–16 years) and poor, who inhale them from a tube, can, plastic bag, or rag held over the nose. The intoxication usually lasts 15–30 min. They have a central depressant effect, characterized by euphoria, excitement, a floating sensation, dizziness, slurred speech, ataxia, and a sense of heightened power. Like alcohol, solvents may cause impaired judgment leading to impulsive and aggressive behavior, and amnesia may occur for the period of intoxication. Other acute effects are nausea, anorexia, and, in high doses, stupor and even unconsciousness. Deaths may be caused by central respiratory depression, asphyxiation, or accident.

Substantial tolerance develops after repeated sniffing. A serious risk is irreversible damage to the liver, kidney, and other organs from benzene or halogenated hydrocarbons. Peripheral neuritis has also been reported. Permanent neuromuscular and brain damage must be considered a possibility. Some clinicians report brain atrophy and chronic motor impairment in toluene users.

VI. MARIJUANA

Marijuana has been known for thousands of years as a medicine and intoxicant, and it was widely used in the nineteenth century as an analgesic, anticonvulsant, and hypnotic. Recently, interest has developed in using it to treat glaucoma and the nausea produced by cancer chemotherapy. It may also have antitumoral and antibiotic properties. One of its nonpsychoactive constituents, cannabidiol, may prove to be useful as an anticonvulsant. But marijuana has been valued throughout history mainly as a euphoriant. Drug preparations from the hemp plant (*Cannabis sativa*) vary widely in quality and potency, depending on the type (possibly three species or, alternatively, various ecotypes), climate, soil, cultivation, and method of preparation. A resin that covers the plant's flowers and leaves contains the active substances, the chief of which is delta-1-tetrahydrocannabinol. The drug can be taken in the form of a drink or in foods, but in the United States it is usually smoked, either in a pipe or in a cigarette called a "joint."

An extensive literature describes the psychological effects of cannabis (a general term for the psychoactive products of the plant). The intoxication varies a great deal, but common symptoms are calmness, euphoria, laughter, rapid flow of ideas, loss of short-term memory, and visual pseudohallucinations, often followed by drowsiness. The intoxication heightens sensitivity to external stimuli, reveals details that would ordinarily be overlooked, makes colors seem brighter and richer, and enhances the appreciation of art and music. Time seems to slow down, and more seems to happen in each moment. Appetite increases. The effects of smoking last for 2–4 hr; the effects of ingestion last 5–12 hr.

Curiously, a splitting of consciousness may occur: the smokers, while experiencing the high, may at the same time objectively observe their own intoxication. They may, for example, have paranoid thoughts yet at the same time laugh at them and, in a sense, enjoy them. This ability to retain objectivity may explain why many experienced users manage to behave soberly in public even when highly intoxicated.

Sometimes referred to as a hallucinogen, marijuana can produce some of the same effects as lysergic acid diethylamide (LSD) and LSD-type substances: distorted perception of body parts, spatial and temporal distortion, depersonalization, increased sensitivity to sound, synesthesia, heightened suggestibility, and a sense of deeper awareness. Marijuana can also cause anxiety and paranoid reactions, but the nightmarish reactions that even the experienced LSD user may endure rarely afflict experienced marijuana smokers, who are using a far less potent drug and have more control over its effects. Marijuana lacks the powerful consciousness-altering qualities of LSD, mescaline, and psilocybin. Some investigators question whether the doses normally used in the United States ever produce true hallucinations. Furthermore, cannabis tends to sedate, whereas LSD and the LSD-type drugs often induce wakefulness and even restlessness. Unlike LSD, marijuana does not dilate the pupils or heighten blood pressure, reflexes, and body temperature. Tolerance develops rapidly with LSD-type drugs but much more slowly with marijuana. For these reasons, some clinicians doubt its credentials for inclusion among the hallucinogens.

The significance of tolerance and withdrawal with regular heavy use of cannabis remains uncertain. Some indications of tolerance and a mild withdrawal reaction after frequent use of high doses exist, but little clinical evidence indicates that a withdrawal syndrome or a need to increase the dose presents serious problems to users or causes them to continue using cannabis. Craving or difficulty in stopping use may rarely occur as part of a pattern of regular heavy use.

Cannabis probably tends to suppress inclinations toward violence by inducing serenity or lethargy and reducing aggressiveness. Release of inhibitions is expressed in word, thought, and fantasy rather than in action.

Although many users report that marijuana enhances the enjoyment of sexual intercourse, as well as food, art, and music, little evidence exists that it weakens or stimulates sexual desire or potency.

Chronic heavy use is sometimes said to cause an amotivational syndrome of sloth and apathy. Reports by many investigators, particularly in Egypt and in parts of Asia, indicate that long-term users of the potent versions of cannabis are passive, unproductive, and lacking in ambition. This suggests that chronic use of the drug in its stronger forms may have debilitating effects, as prolonged heavy drinking does.

There is substantial evidence that moderate use of marijuana does not cause physical or mental deterioration. For example, in the LaGuardia study conducted in New York City and published in 1944, an examination of chronic users who had averaged about seven marijuana cigarettes a day over a mean period of 8 years showed that they had suffered no mental or physical decline as a result. The report of the National Commission on Marijuana and Drug Abuse (1972), although it did much to demythologize cannabis, cautioned that 2% of American marijuana users were at risk as chronic abusers, but it did not make clear exactly what risk was involved. Controlled studies of chronic heavy use have not found personality deterioration, amotivational syndrome, or other psychological harm.

The official diagnostic manual of the American Psychiatric Association refers to a cannabis delusional disorder, but some clinicians question its existence. In the form of "hemp insanity," or cannabis psychosis, this disorder has been reported mainly in India, Egypt, and Morocco, more often in the late nineteenth and early twentieth centuries than today. It is described as a prolonged psychosis caused mainly by chronic heavy use of the drug. This phenomenon has proved to be peculiarly elusive as a clinical entity. The symptoms said to identify it also occur in acute schizophrenia, mania, and toxic states associated with malnutrition and endemic infection.

One explanation for such psychoses is that persons maintaining a delicate balance of ego-functioning may be overwhelmed and precipitated into a psychotic reaction by a drug that alters their consciousness, however mildly. This concatenation of factors—an ego struggling to manage overwhelming anxiety and to prevent distortion of perception and body image, which is subjected to a drug that, in some persons, promotes these effects—may, indeed, be the last straw in precipitating a psychotic break. Several clinical reports also suggest that cannabis can cause a relapse or worsen symptoms in patients with schizophrenia or mood disorders.

In susceptible people, cannabis may precipitate several other types of mental dysfunction. The most serious is the reaction called toxic psychosis or acute brain syndrome. It is caused by the presence in the brain of substances that interfere with cerebral functions. Delirium usually ends when the toxins disappear from the brain. The symptoms include clouding of consciousness, restlessness, confusion, bewilderment, disorientation, dream-like thinking, fear, illusions, and hallucinations. Delirium occurs only with a large, ingested dose of cannabis and rarely when it is smoked.

Cannabis users may also suffer short-lived acute anxiety states, sometimes accompanied by paranoid thoughts. The anxiety may become intense enough to be called panic. Although uncommon, this is probably the most frequent adverse reaction to the moderate use of smoked marijuana. Sufferers may believe that body-image distortions mean that they are ill or dying; they may interpret the psychological changes as an indication that they are losing their sanity. Rarely, the panic becomes incapacitating, usually for a relatively short time. Simple reassurance is the best treatment; it is dangerous to mistake this for toxic psychosis and subject the user to physical intervention that heightens the panic. The anxiety reaction is not psychotic, because the ability to test reality remains intact. Any paranoid ideas can ordinarily be dispelled by reassurance.

Anxiety reactions and paranoid thoughts may occur in those taking the drug for the first time or in unpleasant or unfamiliar settings. Such reactions are less common in experienced users. The likelihood varies directly with the dose and inversely with the user's experience; thus, the most vulnerable person is the inexperienced user who inadvertently (often precisely because of unfamiliarity with the drug) takes a large dose that produces unexpected perceptual and somatic changes.

One rather rare reaction to cannabis is the flashback, or spontaneous recurrence, of drug symptoms while not intoxicated. Although some reports suggest that this may occur in marijuana users even without prior use of any other drug, in general it seems to arise only when people who have used more powerful hallucinogenic or psychedelic drugs smoke marijuana at a later time.

Among new users of marijuana, an acute depressive reaction occasionally occurs. It is generally rather mild and transient but sometimes requires psychiatric intervention. This reaction is most likely in a user with underlying depression.

Much recent research on cannabis has been concerned with possible adverse physical effects of chronic use. Researchers have studied cerebral atrophy, seizure susceptibility, chromosome damage and birth defects, impairment of immune response, and effects on testosterone and the menstrual cycle. The findings are complicated, but the commonly accepted view is that they are inconclusive. Clinical observations of marijuana users, including recent studies of long-term users in the Caribbean and Greece, show no evidence of disease or organic pathology attributable to any of the causes mentioned here.

The only well-documented adverse effects of chronic marijuana use are produced in the lungs. Mild airway constriction is reported in studies of both animals and human beings. Marijuana smoke also contains many of the same carcinogenic hydrocarbons as tobacco smoke. Although not yet confirmed clinically, chronic respiratory disease and lung cancer must be considered dangers for long-term heavy users. [*See* Marijuana and Cannabinoids.]

There is growing interest in the potential medical uses of marijuana and increasing public pressure to legalize its medical use. Tetrahydrocannabinol has been available since 1985 in pill form as a treatment for nausea in cancer chemotherapy and the wasting syndrome of human immunodeficiency virus (HIV) infection, but many patients prefer smoked cannabis. They have found it useful not only as an antinauseant and appetite stimulant but also for glaucoma, muscle spasms, chronic pain, depression, and other symptoms.

VII. PHENCYCLIDINE

Phencyclidine (PCP), 1-(1-phenyicyclohexyl)piperidine, was first investigated as a surgical anesthetic and analgesic. Because of disorientation, agitation, and delirium on emergence from anesthesia, it is now medically available only for veterinary use. Illicit PCP may be taken by snuffing, orally, or intravenously, but it is usually smoked in a cigarette, because that is the best means of titrating the dose. The most popular street names are "angel dust," "crystal," and "hog." PCP is occasionally misrepresented as mescaline, psilocybin, tetrahydrocannabinol (THC), cannabinol, cocaine, or methaqualone, although this is less common now than in the past.

PCP is relatively cheap and easy to synthesize in garage laboratories. About 30 chemical analogues exist, some of which have appeared on the illicit market. Another related drug is ketamine (Ketalar), a short-acting anesthetic with psychoactive properties similar to those of PCP. PCP is one of the more popular drugs for a new generation of drug users, among blacks and the white working class as well as the middle class; some people have used it daily for as long as 6 years.

One gram of PCP may be used to make as few as four or as many as several dozen cigarettes. This variability, together with the extreme uncertainty of PCP content in street samples, makes it difficult to predict the effect, which also depends on the setting and the user's previous experience. PCP doses <5 mg are considered low, whereas doses >10 mg are considered high. The effects of 2–3 mg of smoked PCP begin within 5 min and plateau in 30 min. In the early phases, users are often uncommunicative and lost in fantasy. They experience "speedy" feelings, euphoria, body warmth, tingling, peaceful floating sensations, and occasional feelings of depersonalization or isolation. Striking alterations of body image, distortions of space and time perception, and delusions or hallucinations may occur. Thought may become confused and disorganized. The user may be sympathetic, sociable, and talkative at one moment, hostile and negative at another. Anxiety is often the most prominent symptom in an adverse reaction. Head-rolling movement, stroking, grimacing, and repetitive chanting may occur. The intoxication lasts about 4–6 hr, sometimes giving way to a mild irritable depression. Users may find that it takes from 24 to 48 hr to recover completely from the high. Laboratory tests show that PCP may remain in the blood and urine for more than a week.

Mild cases of adverse PCP reaction or overdose usually do not come to medical attention, and when they do they may often be treated as an emergency in the outpatient department. Symptoms at low doses may range from mild euphoria and restlessness to increasing levels of anxiety, fearfulness, confusion, and agitation. Patients may exhibit difficulty in communication, a blank staring appearance, disordered thinking, depression, and occasionally self-destructive or belligerent and irrationally assaultive behavior. Accidents, suicide, and homicide are the main causes of death in PCP use.

Like the other effects of PCP intoxication, neurological and physiological symptoms are dose-related. Among the common symptoms in cases brought to emergency rooms are hypertension and involuntary movements of the eyeballs. At low doses there may be loss of coordination and muscle rigidity, particularly in the face and neck. Heightened deep tendon reflexes and diminished response to pain are commonly observed. Higher doses may lead to high fever, agitated and repetitive movement, jerking of the extremities, and occasionally involuntary assumption of fixed body postures. Patients may become drowsy, stuporous, or even comatose. Vomiting, hypersalivation, and sweating are common. Seizures and death from respiratory arrest may also occur.

Two to three days may elapse before psychiatric help is sought, because friends are trying to deal with the psychosis by providing resources and support; patients who lose consciousness are seen earlier. Although most patients will recover within a day or two, some will remain psychotic for as long as 2 weeks. Patients who are first seen in a coma often show disorientation, hallucinations, confusion, and difficulty in communication upon regaining consciousness—symptoms also seen in noncomatose patients. Other symptoms of PCP psychosis are staring, posturing, sleep disturbances, paranoid ideation, and depression. The behavioral disturbance may include inappropriate laughing and crying, public masturbation, urinary incontinence, and violence. Often there is amnesia for the entire period of the psychosis.

The proportion of psychotic reactions to PCP occurring in people with a preexisting disposition to psychosis is unclear. Many patients who appear at hospitals with PCP psychoses return later with acute psychiatric reactions not related to the drug. Schizophrenics are very sensitive to PCP, and apparently they cannot easily distinguish between its effects and an intensification of their symptoms. Many PCP psychoses possibly involve mainly an aggravation of underlying psychopathology, whereas the physical symptoms are more specifically drug-related. Some observers believe that PCP psychosis is seriously underdiagnosed, because toxic symptoms indicating the presence of a drug are often not obvious and because the most commonly used tests for PCP in blood and urine are unreliable.

The term "crystallized" is sometimes applied to chronic PCP users who seem to suffer from dulled thinking and reflexes, loss of memory and impulse control, depression, lethargy, and difficulty in concentration. No clinical evidence of permanent brain damage exists, but some clinicians report neurological and cognitive dysfunction in chronic users even after 2–3 weeks of abstinence. Users may also develop tolerance and a withdrawal reaction consisting of lethargy, depression, and craving.

VIII. PSYCHEDELICS

There are many psychedelic or hallucinogenic drugs, some natural and some synthetic. The best known are mescaline, which is derived from the peyote cactus, psilocybin, found in about 100 mushroom species, and the synthetic drug lysergic acid diethylamide or LSD, which is related to psychoactive alkaloids found in morning glory seeds, the lysergic acid amides. Other psychedelic drugs include the natural substances harmine, harmaline, ibogaine, and dimethyltryptamine (DMT), as well as a large number of synthetic drugs with a tryptamine or methoxylated amphetamine structure. A few of these are diethyltryptamine (DET), dipropyltryptamine (DPT), 3,4-methylenedioxyamphetamine (MDA), 3,4-methylenedioxymethamphetamine (MDMA), and 2,5-dimethoxy-4-methylamphetamine (DOM, also known as STP). The average effective dose varies considerably: for example, 75 μg for LSD, about 1 mg for lysergic acid amides, 3 mg for DOM, 6 mg for psilocybin, 30 mg for DMT, 100 mg for MDA, and 200 mg for mescaline. Only LSD, psychedelic mushrooms, and to some extent MDA and MDMA (discussed later) are now available in any quantity on the illicit market.

The subjective effects of these drugs differ somewhat in quality and duration, but LSD produces the widest range of effects and can be taken as a prototype. The reaction varies with personality, expectations, and setting even more than the reaction to other psychoactive drugs, but LSD almost always produces profound alterations in perception, mood, and thinking. Perceptions become unusually brilliant and intense: colors and textures seem richer, contours sharpened, music more emotionally profound, smell and taste heightened. Normally unnoticed details capture the attention, and ordinary things are seen with wonder, as if for the first time. Synesthesia ("hearing" colors and "seeing" sounds) is common. Changes in body image and alterations of time and space perception also occur. Intensely vivid, dream-like, kaleidoscopic imagery appears before closed eyes. True hallucinations are rare, but visual distortions are common. Emotions become unusually intense and may change abruptly and often; two seemingly incompatible feelings may be experienced at once. Suggestibility is greatly heightened, and sensitivity to nonverbal cues is increased. Exaggerated empathy with or detachment from other people may arise. Other features that often appear are seeming awareness of internal body organs, recovery of lost early memories, release of unconscious material in symbolic form, and regression and apparent reliving of past events including birth. A heightened sense of reality and significance suffuses the experience. Introspective reflection and feelings of religious and philosophical insight are common. The sense of self is greatly changed, sometimes to the point of depersonalization, merging with the external world, separation of self from body, or total dissolution of the ego in mystical ecstasy.

People sometimes maintain that a single psychedelic experience or a few such experiences have given them increased creative capacity, new psychological insight, relief from neurotic and psychosomatic symptoms, or a desirable change in personality. For many years, and especially from 1950 to the mid-1960s, psychiatrists showed great interest in LSD and related drugs not only as possible drug models for schizophrenia but also as therapeutic agents in a wide variety of diagnoses. Since 1966, obtaining the drugs legally for therapeutic purposes has become impossible, and professional interest has declined.

The most common adverse effect of LSD and related drugs is the "bad trip," which resembles the acute panic reaction to cannabis but can be more severe and occasionally produces true psychotic symptoms. The bad trip ends when the immediate effect of the drug wears off—in the case of LSD, in 8–12 hr. Psychiatric help is usually unnecessary; the best treatment is protection, companionship, and reassurance, although occasionally tranquilizers may be required.

Another common effect of hallucinogenic drugs is the flashback, a spontaneous transitory recurrence of drug-induced experience in a drug-free state. Most flashbacks are episodes of visual distortion, time expansion, physical symptoms, loss of ego boundaries, or relived intense emotion lasting usually a few seconds to a few minutes but sometimes longer. Probably about a quarter of all psychedelic drug users have experienced some form of flashback. As a rule they are mild, sometimes even pleasant, but occasionally they turn into repeated frightening images or thoughts resembling a traumatic neurosis; in that case they may require psychiatric attention. Flashbacks decrease in number and intensity with time. They are most likely to occur under stress or at a time of diminished ego control; thus, they can be induced by fatigue, drunkenness, marijuana intoxication, or severe illness. Marijuana smoking is possibly the most common single source of LSD flashbacks.

Prolonged adverse reactions to LSD present the same variety of symptoms as bad trips and flashbacks. They have been classified as anxiety reactions, depressive reactions, and psychoses; often they resemble prolonged and more or less attenuated bad trips or flashbacks. Most of these adverse reactions end after 24–48 hr, but sometimes they last weeks or even months. Psychedelic drugs are capable of magnifying and bringing into consciousness almost any internal conflict, so there is no typical prolonged adverse reaction to LSD, as there is a typical amphetamine psychosis. Instead, many different affective, neurotic, and psychotic symptoms may appear, depending on an individual's forms of vulnerability. This lack of specificity makes it hard to distinguish between LSD reactions and unrelated pathological processes, especially when some time passes between the drug trip and the onset of the disturbance.

The most likely candidates for adverse reactions are schizoid and prepsychotic personalities with a barely stable ego balance and a great deal of anxiety, who cannot cope with the perceptual changes, body-image distortions, and symbolic unconscious material produced by the drug trip. People hospitalized for LSD reactions have a high rate of previous mental instability. In the late 1960s, a number of adverse reactions occurred because LSD was being promoted as a self-prescribed psychotherapy for emotional crises in the lives of seriously disturbed people.

Long-term psychedelic drug use is not very common. There is no physical addiction, and psychological dependence is rare because each LSD experience is different and the drug does not produce a reliable euphoria. Tolerance to these drugs develops very quickly but also disappears quickly—in 2 or 3 days. There is no clear evidence of organic brain damage, drastic personality change, or chronic psychosis produced by long-term LSD use, although there is some controversy about this.

The religious and other experiences produced by psychedelic drugs have the capacity, in certain circumstances, to catalyze a transformation in beliefs and ways of life, which may be mistaken for a drug-induced personality or organic change. Drastic changes in cultural and metaphysical identity after taking psychedelic drugs were most common in the 1960s, when the hippie movement promised to build a new society on the use of these drugs.

A persistent issue has been genetic damage and birth defects. The available evidence suggests that LSD produces no serious chromosome damage in reproductive cells; the same is true of other psychedelic drugs to the extent that they have been tested. Nor is there evidence of teratogenicity in human users at normal doses. Nevertheless, all drugs should of course be avoided if possible during pregnancy, especially in the early stages, when the embryo is most vulnerable.

IX. MDMA

A word should be said about MDMA (3,4-methylenedioxymethamphetamine), a drug that became widely used and known in the 1980s. Although structurally related to both amphetamine, a stimulant, and mescaline, a psychedelic drug, MDMA produces neither a stimulant effect nor perceptual distortions, body-image alterations, and changes in the sense of self. It evokes a gentler, subtler, and more controllable experience than LSD—one that invites rather than compels self-exploration and the intensification of feelings. In particular, it reduces defensive anxiety. The effects last 2–4 hr. MDMA has been used in psychotherapy and proposed as a catalyst for insight or an aid in establishing a therapeutic alliance.

Bad trips, psychotic reactions, and flashbacks of the kind produced by LSD and mescaline almost never occur. A few users have reported prolonged adverse reactions such as anxiety or depression for a week or two. There is no craving or withdrawal reaction. With repeated use, tolerance to the psychological effects develops, and physical side effects—dry mouth, jaw clenching, rapid heartbeat, and sometimes a mild hangover—become more uncomfortable. Thus, long-term heavy use is rare. But serious questions have been raised about possible neurotoxic effects of chronic use. The consequences of such toxicity, if it exists, are uncertain.

BIBLIOGRAPHY

American Psychiatric Association (1990). "Benzodiazepine Abuse and Dependence: A Task Force Report of the American Psychiatric Association." American Psychiatric Press, Washington, D.C.

American Psychiatric Association (1994). "Diagnostic and Statistical Manual of Mental Disorders," 4th Ed. American Psychiatric Press, Washington, D.C.

Grinspoon, L. (1977). "Marihuana Reconsidered," 2nd Ed. Harvard Univ. Press, Cambridge, Massachusetts.

Grinspoon, L., and Bakalar, J. B. (1979). "Psychedelic Drugs Reconsidered." Basic Books, New York.

Grinspoon, L., and Bakalar, J. B. (1985). "Cocaine: A Drug and Its Social Evolution," 2nd Ed. Basic Books, New York.

Grinspoon, L., and Bakalar, J. B. (1986). Can drugs be used to enhance the psychotherapeutic process? *Am. J. Psychother.* **40**, 393–404.

Grinspoon, L., and Bakalar, J. B. (1992). Marihuana. *In* "Substance Abuse: A Comprehensive Textbook" (J. H. Lowinson, P. Ruiz, R. B. Millman, and R. B. Langrod, eds.), 2nd Ed. Williams & Wilkins, Baltimore.

Grinspoon, L., and Bakalar, J. B. (1997). "Marihuana, the Forbidden Medicine." 2nd Ed. Yale Univ. Press, New Haven, Connecticut.

Grinspoon, L., and Hedblom, P. (1975). "The Speed Culture: Amphetamine Use and Abuse in America." Harvard Univ. Press, Cambridge, Massachusetts.

Institute of Medicine (1982)."Marijuana and Health." National Academy Press, Washington, D.C.

Jacobs, B. L. (ed.) (1984). "Hallucinogens: Neurochemical, Behavioral, and Clinical Perspectives." Raven, New York.

Peroutka, S. J. (ed.) (1990). "Ecstasy: The Clinical, Pharmacological, and Neurotoxicological Effects of the Drug MDMA." Kluwer Academic Publishers, Boston.

Rubin, V., and Comitas, L. (1975). "Ganja in Jamaica." Mouton, The Hague.

Schultes, R. M., and Hofmann, A. (1980). "The Botany and Chemistry of Hallucinogens," 2nd Ed. Charles C. Thomas, Springfield, Illinois.

Strassman, R. J. (1984). Adverse reactions to psychedelic drugs: A review of the literature. *J. Nervous Mental Dis.* **172**, 577–595.

Szara, S. I., and Ludford, J. P. (eds.) (1980). "Benzodiazepines: A Review of Research Results, 1980," NIDA Research Monograph No. 33. U.S. Government Printing Office, Washington, D.C.

Tims, F. M. (ed.) (1993). "Cocaine Treatment: Research and Clinical Perspective," NIDA Research Monograph No. 135. U.S. Government Printing Office, Washington, D.C.

Wesson, D. R., and Smith, D. E. (1977). "Barbiturates: Their Use, Misuse, and Abuse." Human Sciences Press, New York.

Wesson, D. R., Smith, D. E., and Seymour, R. B. (1992). Sedative-hypnotics and tricyclics. *In* "Substance Abuse: A Comprehensive Textbook" (J. H. Lowinson, P. Ruiz, R. B. Millman, and R. B. Langrod, eds.), 2nd Ed. Williams & Wilkins, Baltimore.

Zukin, S. R., and Zukin, R. S. (1992). Phencyclidine. *In* "Substance Abuse: A Comprehensive Textbook" (J. H. Lowinson, P. Ruiz, R. B. Millman, and R. B. Langrod, eds.), 2nd Ed. Williams & Wilkins, Baltimore.

Nuclear Pore, Structure and Function

JENNY E. HINSHAW
JOHN A. HANOVER
National Institutes of Health

GLOSSARY

Nuclear envelope Double membrane surrounding the nucleus of eukaryotic cells

Nuclear pore complex Organized structure found at the junction of inner and outer nuclear membranes and involved in transport into and out of the nucleus

Nucleoplasm Part of the cytoplasm that is surrounded by the nuclear envelope

O-linked *N*-acetylglucosamine *N*-Acetylglucosamine linked via an O-glycosidic linkage to a Ser or Thr residue in a protein; it is present on glycoproteins of the nuclear pore complex

Ribonucleoproteins Macromolecular complexes composed of both polypeptide backbone and ribonucleic acid

THE NUCLEAR PORE COMPLEX IS PRESENT IN ALL eukaryotic cells and is the portal through which transfer of materials between cytoplasm and nucleus occurs. Nuclear pore complexes appear in electron micrographs as an octagonal collection of ribosome-sized particles. The pore is some 1200 Å in diameter and traverses both bilayers that make up the nuclear envelope, thereby creating a connection between nucleus and cytoplasm. The pore complex allows bidirectional transport: proteins enter the nucleus from the cytoplasm, and RNA is exported from the nucleus. Molecules approximately <40Å can diffuse freely across the pore complex. Larger proteins require specific targeting signals before they cross the pore. The mechanisms involved in protein and RNA transport across the pore complex are poorly understood.

I. INTRODUCTION

The two major compartments of the eukaryotic cell, the nucleus and cytoplasm, are morphologically distinct and separated by a membrane barrier. This structural compartmentalization also reflects a segregation of function. Transcription of RNA from DNA occurs exclusively in the nucleus; translation of RNA into protein occurs on cytoplasmic ribosomes. Exchange of molecules between the nucleus and cytoplasm is indispensable for gene regulation and normal cell growth and development. RNA is exported from the nucleus into the cytoplasm, where the protein-synthesizing machinery translates the genetic information. The proteins needed for the replication of DNA or factors required for transcription are made in the cytoplasm and must enter the nucleus to perform their functions. Moreover, the nuclear envelope breaks down during mitosis, and the nuclear proteins must also have the ability to reenter the nucleus after the nuclear envelope is re-formed. The mature nuclear protein must be capable of nuclear localization. Their transport, therefore, is different from many other intracellular sorting mechanisms, which involve precursor sequences removed during transport. [*See* DNA and Gene Transcription.]

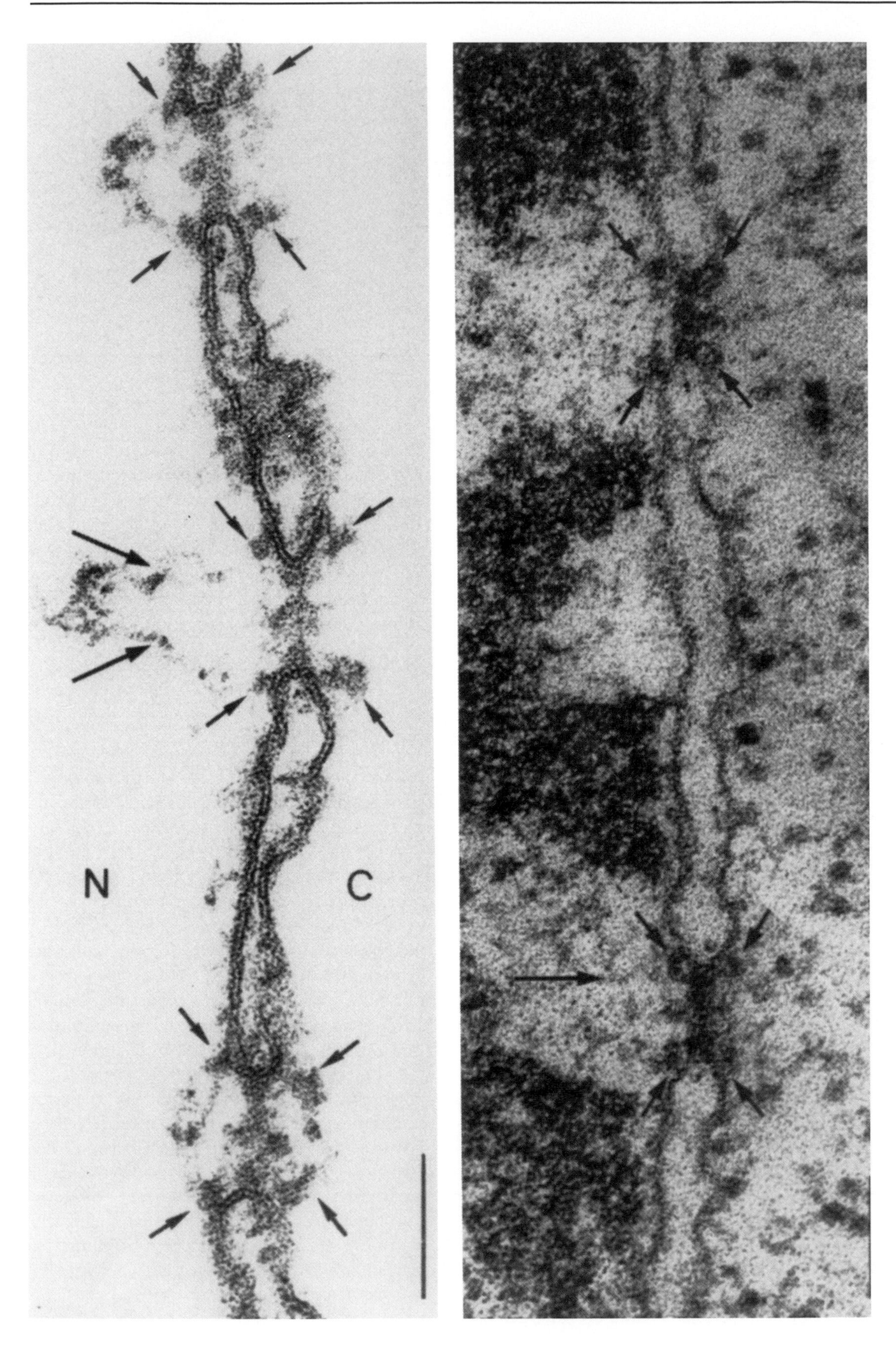
N
C

The nuclear envelope is unlike any other membrane system in the eukaryotic cell. It consists of two lipid bilayers enclosing a cisternal space. The outer nuclear membrane, which is continuous with the endoplasmic reticulum, has attached ribosomes; the inner membrane may form attachment points for chromatin. The nuclear envelope is traversed by nuclear pore complexes, which form a visible link between nucleus and cytoplasm. These pores allow passive diffusion of electrolytes and other small molecules, as well as some small macromolecules. However, dextrans with a molecular weight (MW) >22,000 and globular proteins with a MW >65,000 are excluded from the nucleus. Measurements such as these have established that the functional radius of the pore for simple diffusion is approximately 35 Å. Larger proteins require specific stretches of amino acid (nuclear localization sequences) for their proper transport across the nuclear membrane. A number of such sequences have been identified. As for protein transport across the nuclear pore, simple diffusion does not account for the export of RNA-containing particles from the nucleus. RNA typically leaves the nucleus in a mature, fully processed form. The transport event is energy-dependent and involves release of the RNA from some component of the nuclear substructure followed by movement across the nuclear pore. The mechanism of these steps, still under investigation, is poorly understood. Current evidence suggests that one pore complex may be capable of concurrent RNA transport *out of* and protein uptake *into* the nucleus. [*See* Cell Membrane Transport.]

II. ULTRASTRUCTURE OF THE NUCLEAR PORE COMPLEX

When viewed by electron microscopy, the nuclear pore complex is a prominent feature of the nuclear envelopes of all eukaryotic cells. It forms the junction of the inner and outer nuclear membranes and has an outer diameter of about 1200 Å. It forms a clear morphological connection between the nucleus and the cytoplasm and is thought to be the route of most nucleocytoplasmic traffic. The number of pores per nucleus varies greatly, from 1×10^2 to 5×10^7. The density of pores seems to depend on the metabolic or developmental stage of the cell. Nucleated red blood cells or lymphocytes, for instance, have about 3 pores per μm^2, rat liver nuclei have 14–16 pores per μm^2, and the giant nucleus of *Xenopus* oocytes (the germinal vesicle) contains about 50 pores per μm^2. The degree to which the morphological characteristics of the nuclear pore are conserved throughout eukaryotic evolution is even more striking. Indeed, by electron microscopy, a pore complex from a human liver sample is very hard to distinguish from that of a yeast or fungus. Images of the nuclear envelopes with several pore complexes, as seen by thin-section electron microscopy, are shown in Fig. 1. The inner and outer nuclear membranes are clearly continuous with each other where the nuclear pore complexes form. Also apparent in these images are the filamentous structures or fibrils that emanate from the cytoplasmic and nucleoplasmic surface of the nuclear pore complex. Over the years, various interpretations of the ultrastructure of the pore complex have been made and some of these are shown in Fig. 2. It was clear, even from the initial structural studies of the pore complex, that it exists as a ring of eight globular subunits or spokes.

In more recent years, finer detail of the pore complex has been elucidated by high-resolution electron microscopy. Color Plate 9 represents three-dimensional maps of the pore complex, showing internal organization of the eight globular domains or spokes. Color Plate 9A and 9B are tilt and cut-away views, respectively, of the basic framework of the pore complex that were calculated from negatively stained images (see background images in Color Plate 9A). At the resolution of this map, 50 Å, the framework of the pore complex is eight-fold symmetric about the central channel and two-fold symmetric about the plane of the membrane. The pore complex can be broken into four domains, ring, column, annular, and lumenal. These are two rings, one on the cytoplasmic side and one on the nucleoplasmic side of the pore complex. The column domain represents the bulk of each spoke, whereas the annular domain connects neighboring spokes and the two halves of the complex. The predicted positioning of the nuclear membrane, as shown by the dotted line in Color Plate 9B,

FIGURE 1 Section through the nuclear envelope examined by electron microscopy. The inner and outer nuclear membranes are clearly visible, as are the complexes, indicated by the small arrows. Nucleoplasmic filaments or baskets are indicated by large arrows. [Reproduced from "The Liver: Biology and Pathology" (1988), with permission, from Raven Press, New York.]

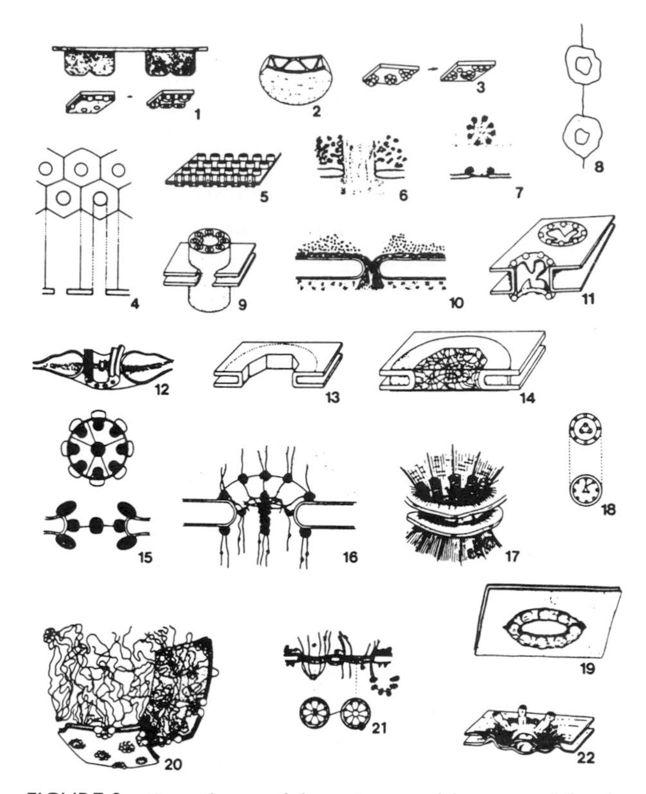

FIGURE 2 Compilation of the various models proposed for the morphological appearance of the nuclear pore. The models (except 8) are arranged in the same orientation, with the nucleoplasmic face up and the cytoplasmic surface down. Models 1, 3, 12, and 21 should be oriented upside down to obtain the perspective intended by the authors. The models are roughly in chronological order and were proposed from 1950 to 1985. (Reproduced with permission from Dr. G. H. Bahr.)

positions the lumenal domain of the complex within the nuclear envelope and suggests that peripheral channels exist between the membrane and the inner annular regions (see double-headed arrows in Color Plate 9B). These peripheral channels may provide an alternative route for passive diffusion, especially during heavy nucleocytoplasmic traffic. The diameter of the peripheral channels is ~100 Å, which correlates well with the size limitations of passive diffusion. Color Plate 9C represents a three-dimensional map of the nuclear pore complex that includes the central plug or transporter. The central plug may represent material in transit or be a permanent component of the pore complex, a transporter, with a channel for nucleocytoplasmic transport.

High-resolution scanning electron microscopy has been used to visualize the surface of the pore complex with its peripheral attachments. Basket structures, seen on the nuclear side of the pore complex, extend into the nucleoplasm 20–40 nm and are composed of eight filaments that are anchored at the periphery by a nuclear coaxial ring (see Fig. 3, Color Plate 10, and Fig. 4). The coaxial rings are joined to each other by a lattice structure. Evidence suggests that the coaxial ring of the basket anchors mRNA particles on the nuclear side and then opens to allow mRNA particles to pass through to the pore complex channel. On the cytoplasmic side there are eight cytoplasmic filaments extending from the pore complex, which have been suggested to be involved in directing nuclear proteins to the pore complex (Color Plate 10). Several proteins have been localized to the cytoplasmic and nucleoplasmic filaments, as well as to regions of the basic framework. These proteins will be discussed in following sections.

III. BIOCHEMISTRY OF THE NUCLEAR PORE COMPLEX

Central to the identification of the proteins making up the pore was the initial observation that many of the nuclear pore polypeptides contained a novel carbohydrate modification consisting of O-linked *N*-acetylglucosamine (GlcNAc) residues covalently attached to Ser and Thr residues. A plant lectin, wheat-germ agglutinin (WGA), binds to these proteins by associating with the O-linked GlcNAc residues. This group of glycoproteins are peripheral membrane proteins, that is, they do not cross the membrane bilayer, because they can be dissociated from the nuclear membrane by high-salt extraction. They are exposed only on the cytoplasmic and nucleoplasmic surface of the pore; they are not components of the space between the outer and inner nuclear membrane (cisternae). The enzyme that adds the O-linked GlcNAc moiety to nuclear pore proteins has been molecularly cloned and characterized. It is present in both the nucleus and cytoplasm and contains numerous 34-amino-acid repeats that have been termed TPR (tetratricopeptide repeats). The precise role this modification plays in nuclear transport is still under investigation.

Monoclonal antibodies, which react with the nuclear pore proteins, recognize a family of proteins of 45–210 kDa in mass, which all seem to bear O-linked GlcNAc. Immunoelectron microscopy has demonstrated that the antigens recognized by the antibodies are located exclusively within the nuclear pore complex at both the nucleoplasmic and cytoplasmic sur-

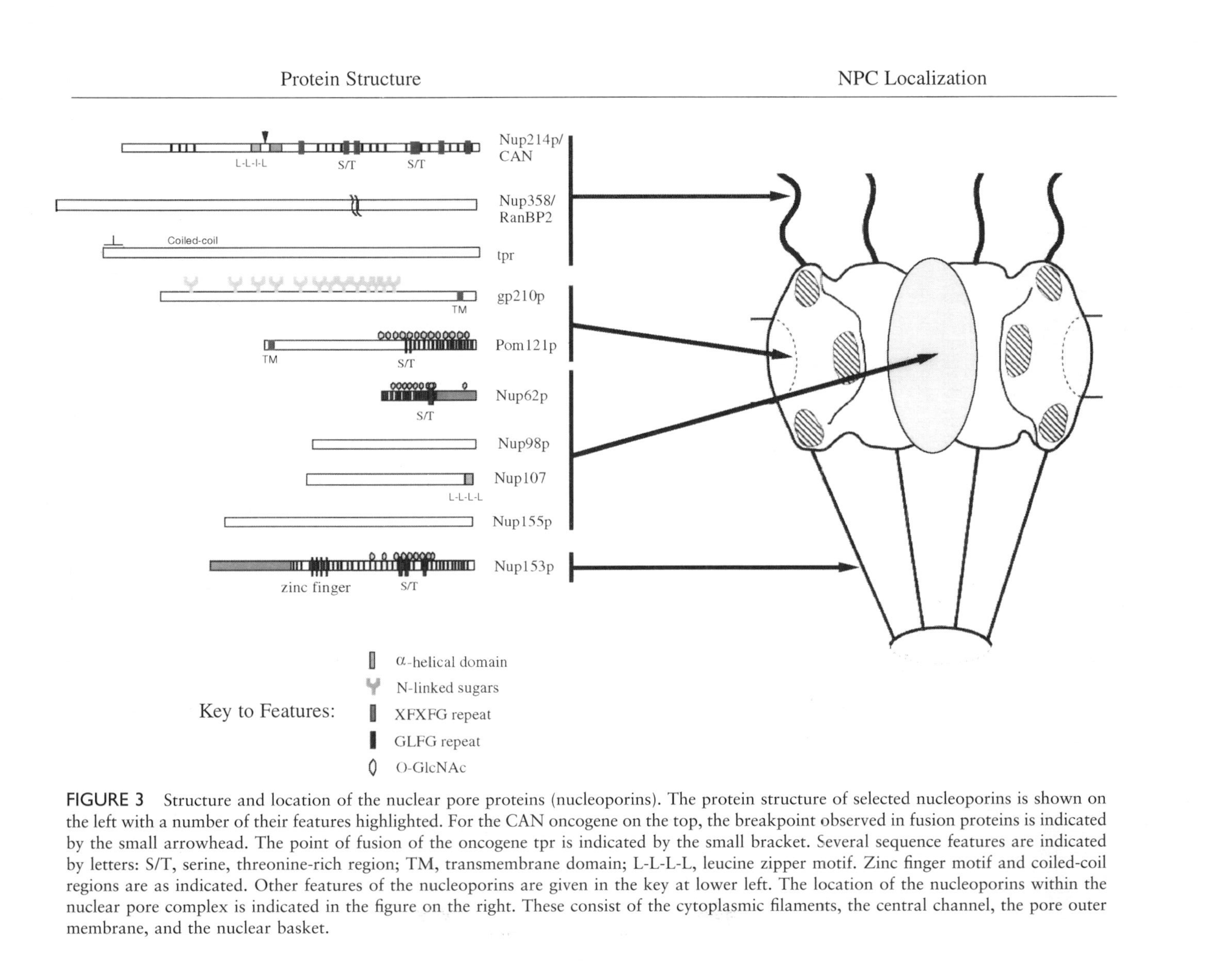

FIGURE 3 Structure and location of the nuclear pore proteins (nucleoporins). The protein structure of selected nucleoporins is shown on the left with a number of their features highlighted. For the CAN oncogene on the top, the breakpoint observed in fusion proteins is indicated by the small arrowhead. The point of fusion of the oncogene tpr is indicated by the small bracket. Several sequence features are indicated by letters: S/T, serine, threonine-rich region; TM, transmembrane domain; L-L-L-L, leucine zipper motif. Zinc finger motif and coiled-coil regions are as indicated. Other features of the nucleoporins are given in the key at lower left. The location of the nucleoporins within the nuclear pore complex is indicated in the figure on the right. These consist of the cytoplasmic filaments, the central channel, the pore outer membrane, and the nuclear basket.

- HIV-1 Rev LQLPPLERLTLD
- TFIIIA LPVLENLTLK
- PKI LALKLAGLDIN

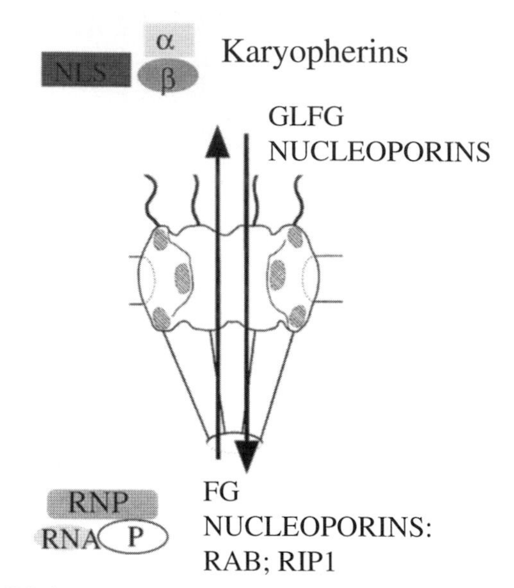

FIGURE 4 Nuclear export signals in shuttling proteins involved in RNA export. The export of molecules such as HIV Rev, TFIIIA, and PKI is mediated by a short leucine-rich motif as indicated. Rev or Rev-like molecules, which themselves bind to RNA, would mediate export of RNA species into the cytoplasm. Little is known about the precise mechanism of this process, but by analogy with protein import it has been suggested that repetitive elements in nucleoporins may be involved in binding these molecules and through a series of association–dissociation reactions mediating their vectorial transport. Abbreviations used are: RNP, ribonucleoprotein; RNA, ribonucleic acid; RAB, Rev activation domain-binding protein; RIP1, Rev interacting protein 1; NLS, nuclear localization sequence.

faces. When examined in light microscopy by immunofluorescence, the antibodies give a highly punctate pattern. This pattern changes dramatically during nuclear envelope breakdown, reflecting the dispersal of pore components. [*See* Antibody–Antigen Complexes: Biological Consequences; Monoclonal Antibody Technology.]

Another component believed to be part of the pore complex is a high-MW glycoprotein tightly associated with the nuclear membrane (gp210p). This antigen was first identified biochemically and antibodies were subsequently raised that recognize the protein. In contrast to the other components of the pore, this glycoprotein bears N-linked oligosaccharides of the high-mannose type and is an integral membrane protein, that is, it crosses the membrane. This glycoprotein has been suggested to anchor the other pore components to the nuclear envelope.

IV. MOLECULAR CHARACTERIZATION OF NUCLEAR PORE PROTEINS

Molecular genetics techniques have recently been applied to identify individual nuclear pore proteins. The first of these to be molecularly characterized was the abundant nucleoporin Nup62p, which is modified by numerous O-linked GlcNAc residues. A large number of other peripheral nucleoporins have now been characterized (rat Nup155p, Nup107p, tpr/p180), some of which are characterized by the presence of FG repeats (vertebrate Nup62p, 98p, 153p, 214p/CAN, 358p/RanBP2, Rip1p). These features are summarized in Fig. 3. A brief examination of these structures reveals a number of likely sites involved in protein–protein interaction or protein–DNA/RNA interaction. These include heptad repeats, leucine zippers, zinc fingers, and other motifs. This is not surprising considering the interaction of individual components within the nuclear pore complex required to form an entity of this size and complexity (see Fig. 3). A second class of nuclear pore components consists of the pore membrane proteins (POMs), which contain a transmembrane domain allowing them to insert specifically into the pore membrane region (vertebrate gp210, Pom121p). Yeast homologues of all of these proteins have been identified and studied genetically. We will focus here on the vertebrate nucleoporins.

A. Proteins Making up the Cytoplasmic Filaments

A number of nucleoporins are found in association with filaments that have been visualized to extend well into the cytoplasm. Nup214p is a WGA-binding protein with numerous XFXFG repeats that is associated with these filaments. Interestingly, Nup214p was found to be identical to a domain of the CAN oncogene product associated with certain forms of leukemia. RanBP2 or Nup358 is a rather complex nucleoprotein consisting of numerous XFXFG repeats, four Ran-GTP binding sites, zinc fingers a cyclophilin A homologous domain, and a leucine-rich region. Surprisingly, four of the zinc fingers found in the cytoplasmically exposed Nup358p are identical to those found in the nucleoplastically localized Nup153p (see Section IV,C). Tpr/p265 is a large coiled-coil protein exclusively localized to the cytoplasmic filaments of the nuclear pore complex. The amino-terminal part of the human gene product Tpr (*t*ranslocated *p*romoter *r*egion) has been identified as part of a fusion protein involving the kinase domains of certain protooncogenes.

B. Integral Proteins of the Pore Membrane Domain

Rat gp210p is an integral glycoprotein containing N-linked high-mannose-type oligosaccharides. Studies using limited proteolysis as well as domain-specific antibodies revealed that gp210p consists of a large cisternal N terminus that may contain an EF-hand Ca^{2+} binding motif, a transmembrane segment of 21 residues, and a short C-terminal tail that may serve as an anchor for other nuclear pore proteins. An area adjacent to the transmembrane region has been found to have some similarity to the equivalent region of rat Pom121p. Since this domain is predicted to be exposed to the pore side, this region could be involved in directing the proteins to the pore. Pom121p contains FG pentapeptide motifs and one or two hydrophobic domain(s), predicted to insert this protein into the pore membrane domain. The bulk of the protein is exposed to the cytoplasm and may anchor other pore components to this specialized area.

C. Repeat-Containing Nucleoporins (GLFG and GFXFG)

In general, nuclear pore proteins that contain repetitive FG motifs are also modified by addition of O-linked *N*-acetylglucosamine and bind WGA. Rat Nup62p, a prototypic member of the repeat family, consists of a repetitive N-terminal domain predicted to form a β-barrel structure and a C-terminal α-helical rod domain. O-linked GlcNAcs are clustered in an S/T-rich linker region responsible for its ability to bind to WGA and are also present on the numerous TTPST repeats present near the amino terminus interspersed with GFXFG repeats. Nup62p is thought to be localized to the central channel of the nuclear pore. Another of this class, Nup153p, contains a zinc finger domain and was localized to the inner ring of the nuclear basket. The zinc finger domain and its ability to bind DNA *in vitro* gave rise to speculation that Nup153p binds DNA and may play a role in the organization of the genome. Nup98p is a WGA-reactive protein with an N-terminal repeat region composed predominantly of GLFG/FG and XFXFG units.

D. Nonrepeat Nuclear Pore Proteins

Rat Nup155p and Nup107p are urea- and salt-extractable proteins of the nuclear envelope originally identified owing to their inability to bind WGA. Cloning and subsequent cDNA analysis revealed that both proteins lack repetitive sequences of the XFXFG type. Antibodies raised against unique sequences were used to locate Nup107p and Nup155p to the nuclear pore complex. Like Nup214p, Pom121p, Nup155p, and other nucleoporins, Nup107p contains numerous kinase consensus sites. Nup214p and Nup107p are both predicted to form a leucine zipper motif that could be crucial for intermolecular interactions.

V. FUNCTIONS OF THE NUCLEAR PORE: NUCLEAR PROTEIN TRANSPORT

The nuclear pore confers upon the nuclear membrane properties of a molecular sieve: small molecules diffuse across while larger molecules are excluded. Large nuclear proteins have sorting signals that lead to their proper nuclear compartmentalization. The existence of these signals was established in dissected germinal vesicles form amphibian oocytes of *Xenopus laevis*. When cytoplasmic and nuclear proteins were injected into the oocyte, cytoplasmic nuclear proteins accumulated in the nucleus; cytoplasmic proteins remained in the cytoplasm. By proteolytic cleavage of nucleoplasmin, the most abundant nuclear protein, it was possible to show that the property of nuclear localization resided in a distinct part of the protein molecule, the tail carboxyl-terminal fraction. The inserted tail fragment itself rapidly accumulated in the nucleus. This approach localized the signal for nuclear entry to a 12,000-Da protein fragment.

To date, the best-studied nuclear protein is the large T-antigen of SV40 virus. The T-antigen subunit has a MW of 94,000, and it is too large to enter the nucleus by passive diffusion. Recombinant DNA techniques have been used to identify the nuclear localization signal. The conversion of a Lys within the sequence Pro-Lys-Lys^{128}-Lys-Arg-Lys-Val into any amino acid other than Arg completely abolishes nuclear localization. If the sequence is attached to other

TABLE I
Nuclear Targeting Sequences

SV40 large T	$_{126}$**PKKKRKV**
Nucleoplasmin	$_{155}$**KR**PAATKKAGQA**KKKKLDKEDES**
N1	$_{534}$**KRK**TEEESPLKD**KDAKKSKQE**
GR (rodent)	$_{479}$**RK**CLQAGMNLEA**RKTKK**
Estrogen R	$_{250}$**RK**CYEVGMMKGGI**RKDR**
Androgen R	$_{432}$**RK**CYEAGMTLGA**RKLKK**
P53	$_{305}$**KR**ALPNNTSSSPQP**KKK**
Abl	$_{595}$**KK**TNLFSALI**KKKKK**TA
c-fos	**RR**ERNKMAAKC**RNRRR**
c-jun	$_{267}$**KR**MRNRIAASKC**RKRK**
DNA polymerase α	$_{25}$**KK**SKKGRQEALE**RLKKA**

proteins, it causes their localization to the nucleus. Similar sequences, found in nuclear SV40, contain two independent sequence elements that contribute to its nuclear proteins specified by other viruses, such as polyoma virus and adenovirus.

The primary sequences of other nuclear localization signals are summarized in detail in Table I. Most consist of several highly basic and hydrophilic amino acids with a structure-perturbing amino acid, such as Gly or Pro on the N- or C-terminal side. Many of the sequences conform to the so-called "bipartite motif" consisting of two basic residues a variable number of other residues (10–15) and a string of 3–5 additional lysines and arginines. Given their diversity, overall shape and charge rather than the exact amino acid sequence could be responsible for nuclear import.

In addition to containing a nuclear localization sequence (NLS) intrinsic to the protein, nuclear proteins may cross the nuclear pore by associating with another nuclear protein bearing a localization signal; for instance, antibodies against SV40 large T-antigen are cotransported with the antigen into the nucleus. Therefore, it is unlikely that protein unfolding occurs during transport of proteins into the nucleus. Nuclear protein transport via localization sequences has a size exclusion limit: uptake occurs up to a MW of 465,000 (ferritin), but not with IgM (MW 900,000). There is also reason to believe that the nuclear targeting sequence has to be exposed on the surface of the molecule to interact with a receptor for nuclear localization sequences. Thus, the SV40 T-antigen sequence can be introduced into several different sites within pyruvate kinase and still function correctly. However, when inserted into a part of the molecule that is buried according to crystallographic studies, the signal is inactive. Moreover, monoclonal antibodies against the localization signal of SV40 bind to *in vitro* translated SV40 T-antigen and to the T-antigen expressed in cultured cells. The exposure of the nuclear localization signal in proteins may be developmentally regulated. Thus, in *Xenopus*, the "late migrating nuclear antigens" are nuclear in the oocyte, cytoplasmic in the early embryo, and nuclear again in later stages of development. Exposure of a localization signal can also occur in response to external stimuli. For instance, the glucocorticoid receptor may enter the nucleus only when it is occupied by the hormone.

The components that recognize the nuclear localization signal and the transport machinery are under active investigation using biochemical and molecular tools (see the following). Morphologically, electron microscopic studies showed that gold particles coated with nuclear proteins readily enter the nucleus regardless of their size (between 50 and 200 Å in diameter). In content, colloidal gold particles coated with cytoplasmic control proteins are excluded from the nucleus and only at very high concentration are associated with the nuclear envelope.

The association of nucleoplasmin–gold particles may occur along fibrillar structures that are associated with the nuclear pore and extend into both the cytoplasm and the nucleoplasm. These filaments are composed of a number of proteins some of whose molecular identity has recently been established (see the foregoing). If gold particles are coated with intact nucleoplasmin, they become associated with the cytoplasmic side of the fibrils and inside the nuclear pores, but not on the nucleoplasmic side of the fibrils.

Nuclear uptake is time-dependent. Microinjection studies with nucleoplasmin or proteins conjugated with the SV40 large T-antigen nuclear localization sequence have shown that they rapidly bind around the nuclear envelope and that accumulation in the nucleus follows in a time-dependent manner. The number of nuclear localization peptides has an influence on the speed of uptake; the more peptide moieties, the faster the uptake. Nuclear protein transport exhibits saturable kinetics, that is, it reaches a maximum rate with increasing protein concentration and is temperature-dependent. The optimal temperature for nuclear transport *in vitro* is 30–33°C, probably due to a loss of stability of the nuclear membrane at 37°C. In microinjected living cells, uptake of conjugates bearing a nuclear localization sequence does not occur at 0°C and can be induced by raising the temperature to 22°C, and even more at 37°C. Nuclear uptake is adenosine triphosphate (ATP)-dependent. If

ATP pools are lowered using poisons such as sodium azide and deoxyglucose, transport into the nucleus is inhibited. An intact nuclear membrane is also required for nuclear transport. Nuclear uptake can be inhibited by molecules that bind to pore proteins, such as WGA and nuclear pore-specific antibodies. However, electron microscopy has shown that binding of nuclear proteins to the pore is not inhibited by WGA, suggesting that some nuclear pores are specifically involved in protein transport into the nucleus. [*See* Adenosine Triphosphate (ATP).]

The available experimental evidence suggests that the nuclear pore is the principal route of nuclear protein import. The pore complex must accommodate large nuclear proteins and selectively exclude cytoplasmic proteins. Uptake requires ATP, yet it is unlikely that this energy is involved in the unfolding of nuclear proteins prior to uptake. A model that summarizes much of the current knowledge of nuclear protein import is shown in Color Plate 10. Several cytosolic factors necessary for nuclear protein import have been purified recently. A docking complex was identified consisting of karyopherin α (the p54/p55 NLS receptors) and karyopherin β (p97, importin β). This complex binds to the NLS-containing protein in the cytosol and then moves to the filamentous components present on the cytosolic surface of the nuclear pore. The translocation of this complex through the nuclear pore is more complex. It is now clear that the small GTP-binding protein Ran/TC4 and a stimulatory dimer of 10 kDa participate in the translocation step. It is likely that the Nup358p nucleoporin, which is a Ran-binding protein, is also involved in this interaction. Furthermore, another cytosolic Ran-binding protein, a guanine nucleotide exchange protein (RCC1), and a guanine nucleotide activator protein are involved.

As Color Plate 10 indicates, there is also a requirement for ATP and Ca^{2+}. Ca^{2+} is required both in the lumen of the endoplasmic reticulum/nuclear envelope and in the cytosol. Recently we have shown that the cytoplasmic Ca^{2+} requirement is mediated by calmodulin. An intriguing possibility is that the EF-hand binding domain of gp210p in the lumen of the nuclear envelope may mediate the regulation by lumenal Ca^{2+}. The EF-hand domain of calmodulin has been shown to mediate the cytosolic Ca^{2+} requirement. We have proposed that the release of intracellular calcium stores upon cell activation inhibits GTP-dependent nuclear transport; the elevated cytosolic calcium then acts through calmodulin to stimulate a novel GTP-independent mode of import. Taken together these findings have led to the conclusion that active nuclear import may be regulated by such common and ubiquitous mediators of cellular signaling as the guanine nucleotides and Ca^{2+}. Although many of the molecules required for nuclear import are now identified, the precise mechanism for how the nuclear pore mediates this process is still in doubt. Some envision the process occurring by a series of association and dissociation reactions with the highly repetitive motifs of the nucleoporins. Both Ran and calmodulin can serve as molecular switches that could mediate this kind of reversible association. Others envision the filaments themselves mediating the transport by flexing or bending to deliver the proteins to a transporter that translocates nuclear proteins to the nuclear interior.

VI. FUNCTIONS OF THE NUCLEAR PORE: RNA TRANSPORT

All known RNA species, mRNA, tRNA, and rRNA, are transcribed from the DNA sequences in the nucleus. The RNA species undergo a series of posttranscriptional modifications within the nucleus, including processing, splicing, and polyadenylation, and must ultimately be transported to the cytoplasm. Although RNA species are single-stranded, they may have substantial secondary structure arising from folding of the chain into double-stranded regions (hairpins). The extent to which the secondary structure contributes to the recognition, processing, and transport of RNA species is, at present, unknown. The tRNAs are the smallest RNA molecules, containing between 73 and 93 nucleotides and having a molecular mass of about 25 kDa. They might be expected to diffuse through the nuclear pore passively; however, the recent identification of transport-defective mutants of tRNA species strongly argues against passive diffusion. A single G-to-U substitution at position 57 in the vertebrate $tRNA_{Met}$ molecule was found to reduce its transport rate 95%. This highly conserved region of the tRNA molecule is therefore thought to be critical for recognition by the transport mechanism. This mechanism has been studied in the oocyte of *X. laevis* using nuclear microinjection and microdissection techniques. Transport behaves as a saturable, carrier-mediated translocation process rather than diffusion through a simple pore or channel. It is possible that ribosome-like components surrounding the nuclear pore function as a tRNA translocation motor. Transport of tRNA species has also been examined morphologically in *Xenopus* oocytes using gold particles of different sizes. Each pore seems to be capable of bidirectional transport of RNA out and protein in. Whether or not the mechanisms involved in these coordinated transport processes are similar is unknown.

The synthesis of other ribonucleoproteins (RNPs) is often much more complex than that of the tRNAs. The small nuclear RNAs (snRNAs) involved in mRNA processing appear to be rapidly exported to the cytoplasm after transcription, where they associate with other proteins. After posttranscriptional modifications such as cap hypermethylation to convert their m^7GpppN 5′ ends to $m^{2,2,7G}$pppN, the RNPs are imported back into the nucleus, where they are completed by addition of other RNA species. A similar biosynthetic pathway is observed for ribosomes. Ribosomal proteins are imported into the nucleus, where they associate with rRNA prior to export to the cytoplasm as intact subunits. Since human cells have about 10 million ribosomes, about 600,000 ribosomal proteins are imported and 15,000 ribosomal subunits must be exported each *minute*. Each nuclear pore must then carry 100 ribosomal proteins and 3 ribosomal subunits per minute.

The signals that mediated RNA export are only beginning to be defined. Most RNA species are exported only when they are posttranscriptionally modified. Until this processing takes place they are retained. At least one step in the export process is the release from retention in the nucleus. It is also clear that nearly all RNA leaves the nucleus in the form of RNP particles. Proteins that bind to RNA may be directly involved in the export. It is also likely that a series of partially redundant signals may be required to export large RNPs into the cytoplasm. The only well-defined RNA export signal is the cap of RNA polymerase II transcripts. Cap-binding proteins are critical for the export of certain U snRNA species.

By far the best understood mediator of RNA export is the Rev protein of human immunodeficiency virus (HIV). HIV-1 transcripts contain an element known as the RRE (Rev response element) to which Rev binds. When bound to the RRE, Rev mediates export of incompletely spliced mRNA. Rev contains not only an RNA binding domain but also an "activation domain" that is an export signal (NES). The Rev NES, a 10-amino-acid hydrophobic sequence (LPPLERLTLD), is characterized by leucine residues, and removal of a single leucine inhibits Rev export (see Fig. 4). A number of other proteins have been found to contain this leucine-rich motif and these may also be involved in nuclear export. What binds to these exported molecules that may mediate RNA export? A number of candidate proteins have been identified, including small nucleoporin-like molecules that have been designated RIP1 or RAB (see Fig. 4). These proteins contain the FG repeats that characterize nucleoporins. Other FG nucleoporins have been shown to interact with the NES, suggesting that the interaction may be with the entire family of GFXFG nucleoporins. In the model depicted in Fig. 4, RNPs may interact with Rev or Rev-like cellular proteins, which may then interact with the GFXFG nucleoporins in a fashion involving successive association–dissociation reactions. As mentioned previously, a similar model has been proposed to account for the directed nuclear import of proteins (compare Color Plate 10 and Fig. 4).

But what are the cellular mediators of RNA export? The Rev-like proteins appear to carry out only the transport of 5*S* rRNA and U snRNA, although similar mechanisms may hold for the export of mRNA and tRNA. Future experiments may shed additional light on these export processes.

VII. SUMMARY

Communication between the nucleus and cytoplasm is mediated by the nuclear pore complex. The nuclear pore is an elaborate structure consisting of integral and peripheral membrane glycoproteins, whose primary structure has recently been elucidated. The nuclear pore must be capable of discriminating between proteins that must remain in the cytoplasm and those that must enter the nucleus. This uptake process might be tightly regulated and involves recognition of signals encoded in the structure of nuclear proteins. In addition, the pore appears to play a critical role in the controlled release of RNA species from the nuclear interior to the cytoplasm. Understanding the structure of the pore complex should lead to a more complete understanding of its role in mediating these important cellular processes.

BIBLIOGRAPHY

Adam, E. J., and Adam, S. A. (1994). Identification of cytosolic factors required for nuclear location sequence-mediated binding to the nuclear envelope. *J. Cell Biol.* **125**, 547–555.

Akey, C. W., and Radermacher, M. (1993). Architecture of the *Xenopus* nuclear pore complex revealed by three-dimensional cryo-electron microscopy. *J. Cell Biol.* **122**, 1–19.

Bogerd, H. P., Fridell, R. A., Madore, S., and Cullen, B. R. (1995). Identification of a novel cellular cofactor for the Rev/Rex class of retroviral regulatory proteins. *Cell* **82**, 485–494.

Davis, L. I. (1995). The nuclear pore complex. *Annu. Rev. Biochem.* **64,** 865–896.

Davis, L. I., and Blobel, G. (1986). Identification and characterization of a nuclear pore complex protein. *Cell* **45,** 699–709.

Dworetzki, S. I., and Feldherr, C. M. (1988). Translocation of RNA-coated gold particles through the nuclear pore of oocytes. *J. Cell Biol.* **106,** 575–584.

Fabre, E., and Hurt, E. C. (1994). Nuclear transport [published erratum appears in *Curr. Opin. Cell Biol.* **6,** 616]. *Curr. Opin. Cell Biol.* **6,** 335–342.

Goldberg, M. W., and Allen, T. D. (1992). High resolution scanning electron microscopy of the nuclear envelope: Demonstration of a new, regular, fibrous lattice attached to the baskets of the nucleoplasmic face of the nuclear pores. *J. Cell Biol.* **119,** 1429–1440.

Gorlich, D., and Mattaj, I. W. (1996). Nucleocytoplasmic transport. *Science* **271,** 1513–1518.

Gorlich, D., Vogel, F., Mills, A. D., Hartmann, E., and Laskey, R. A. (1995). Distinct functions for the two importin subunits in nuclear protein import. *Nature* **377,** 246–248.

Hanover, J. A. (1988). Molecular signals controlling membrane traffic. *In* "The Liver: Biology and Pathobiology," pp. 189–205. Raven, New York.

Hanover, J. A., Cohen, C. K., Willingham, M. C., and Park, M. K. (1987). O-linked *N*-acetylglucosamine is attached to proteins of the nuclear pore. Evidence for cytoplasmic and nucleoplasmic glycoproteins. *J. Biol. Chem.* **262,** 9887–9894.

Hinshaw, J. E., Carragher, B. O., and Milligan, R. A. (1992). Architecture and design of the nuclear pore complex. *Cell* **69,** 1133–1141.

Kiseleva, E., Goldberg, M. W., Daneholt, B., and Allen, T. D. (1996). RNP export is mediated by structural reorganization of the nuclear pore basket. *J. Mol. Biol.* **257,** 848–865.

Kraemer, D., Wozniak, R. W., Blobel, G., and Radu, A. (1994). The human CAN protein, a putative oncogene product associated with myeloid leukemogenesis, is a nuclear pore complex protein that faces the cytoplasm. *Proc. Natl. Acad. Sci. U.S.A.* **91,** 1519–1523.

Melchoir, F., and Gerace, L. (1995). Mechanism of nuclear protein import. *Curr. Opin. Cell Biol.* **7,** 310–318.

Melchoir, F., Paschal, B., Evans, J., and Gerace, L. (1993). Inhibition of nuclear protein import by nonhydrolyzable analogues of GTP and identification of the small GTPase Ran/TC4 as an essential transport factor [published erratum appears in *J. Cell Biol.* **124,** 217]. *J. Cell Biol.* **123,** 1649–1659.

Miller, M. W., and Hanover, J. A. (1994). Functional nuclear pores reconstituted with beta 1–4 galactose-modified O-linked *N*-acetylglucosamine glycoproteins. *J. Biol. Chem.* **269,** 9289–9297.

Newport, J. W., and Forbes, D. J. (1987). The nucleus: Structure, function and dynamics. *Annu. Rev. Biochem.* **56,** 535–565.

Pante, N., and Aebi, U. (1993). The nuclear pore complex. *J. Cell Biol.* **122,** 977–984.

Park, M. K., D'Onofrio, M., Willingham, M. C., and Hanover, J. A. (1987). A monoclonal antibody against a family of nuclear pore proteins (nucleoporins): O-linked *N*-acetylglucosamine is part of the immunodeterminant. *Proc. Natl. Acad. Sci. U.S.A.* **84,** 6462–6466.

Powers, M. A., and Forbes, D. J. (1994). Cytosolic factors in nuclear transport: What's importin? *Cell* **79,** 931–934.

Smith, A. E., Kalderon, D., Roberts, B. L., Colledge, W. H., Edge, M., Gillett, P., Markham, A., Paucha, E., and Richardson, W. D. (1985). The nuclear location signal. *Proc. Roy. Soc. London* **226,** 43–58.

Snow, C. M., Senior, A., and Gerace, L. (1987). Monoclonal antibodies identify a group of nuclear pore complex proteins. *J. Cell Biol.* **104,** 1143–1156.

Wen, W., Meinkoth, J. L., Tsien, R. Y., and Taylor, S. S. (1995). Identification of a signal for rapid export of proteins from the nucleus. *Cell* **82,** 463–473.

Wozniak, R. W., Barnik, E., and Blobel, G. (1989). Primary structure analysis of an integral membrane glycoprotein of the nuclear pore. *J. Cell Biol.* **108,** 2083–2092.

Zapp, M. L. (1995). The ins and outs of RNA nucleocytoplasmic transport. *Curr. Opin. Genet. Dev.* **5,** 229–233.

Nucleotide and Nucleic Acid Synthesis, Cellular Organization

CHRISTOPHER K. MATHEWS
Oregon State University

GLOSSARY

Channeling Kinetic facilitation of a metabolic pathway, usually involving preferential transfer of an intermediate from one enzyme catalytic site to the next, in preference to its release to the surrounding milieu

Compartmentation Maintenance within a cell of two or more distinct pools of a particular metabolite; pools may be separated either physically (present within different organelles or substructures) or kinetically (different rates of utilization in a particular metabolic process)

dNMP, dNDP, dNTP 2′-Deoxyribonucleoside 5′-mono-, di-, and triphosphate, respectively

Metabolon Relatively loosely bound enzyme aggregate that catalyzes a sequential series of reactions

Pool Intracellular or intraorganellar content of a metabolite; pool size refers to the amount of a particular metabolite in a cell or organelle

Protein machine Group of functionally related proteins, usually physically associated, that functions in an organized way to coordinate and catalyze a complex process (usually in macromolecular metabolism), such as DNA replication or repair or cell movement

Reconstitution Duplication of part or all of a metabolic process *in vitro,* achieved by mixing separately purified components, including enzymes and other proteins

Replication factory Discrete area within the nuclei of mammalian cells, within which DNA chains are being extended in about two dozen replication forks

rNMP, rNDP, rNTP Ribonucleoside 5′-mono-, di-, and triphosphate, respectively

EVEN THOUGH MOST OF THE ENZYMATIC REACTIONS of nucleotide and nucleic acid synthesis are catalyzed by enzymes that are readily soluble proteins, the processes and pathways are organized within cells. Organization is brought about by intracellular compartmentation of the enzymes involved, by the frequent use of multifunctional enzymes that catalyze more than one reaction, by multienzyme complexes that involve noncovalent protein–protein interactions, and by dynamic associations (sequential interactions among proteins, nucleic acids, and membranes, which generate structures essential for initiation of replication and transcription). Because deoxyribonucleoside triphosphates (dNTPs) find virtually their sole functions as DNA precursors, a close coordination exists between dNTP biosynthesis and DNA replication. The importance of this coordination can be seen from the severe genetic consequences of loss of control of one or more of the biosynthetic reactions.

I. GENERAL IMPORTANCE OF METABOLIC ORGANIZATION

Until recently, enzymes were characterized as particulate or soluble. Enzymes isolated in particulate form were assumed to represent binding of enzyme protein to a membrane in an organelle, whereas soluble enzymes were thought to exist within cells in a less highly structured form, either as a highly concentrated solution or in a gel-like phase. Processes such as biological electron transport, fatty acid oxidation, and photosynthesis are catalyzed by enzymes firmly bound to membranes in specific organelles, namely, mitochondria or chloroplasts for the examples cited. Thus, the cellular organization of one of these processes could be defined experimentally as the process itself was elucidated. For example, analysis of mitochondrial oxidative phosphorylation, in which energy from oxidation reactions drives the energy-requiring synthesis of ATP, revealed the existence of five specific enzyme complexes within the inner mitochondrial membrane, each of which carries out a specific reaction sequence within the overall process. Learning how to solubilize each enzyme in such a process required the investigator to learn something about the structure within which each enzyme was embedded inside the cell.

By contrast, pathways and processes that are catalyzed largely by soluble proteins have resisted attempts to define intracellular structures with which these processes are associated. The ease of isolation of many such soluble enzymes tends to lead the researcher past the question of cellular organization to focus upon issues such as mechanisms of catalysis, three-dimensional structure of the enzyme, allosteric regulation, and the quantitative role of the enzyme in a pathway. Glycolysis is an excellent example. All of the enzymes that ferment glucose to lactate (or ethanol in many microorganisms) were crystallized decades ago, and their catalytic mechanisms are well understood. Only recently has it become apparent that these soluble proteins interact with one another *in vitro,* albeit weakly and at very high concentrations. At the same time, it has been learned that intracellular concentrations of these enzymes are very high, often exceeding the molar concentrations of low-molecular-weight glycolytic intermediates within the cytosol of eukaryotic cells. Such observations point strongly toward mechanisms to assemble and organize the enzymes *in vivo,* for catalytic facilitation of this central energy-generating pathway. However, the evidence supporting a glycolytic "metabolon," or loosely associated enzyme complex, is controversial, in part because of the difficulty in demonstrating such associations *in vitro* at the vastly lower protein concentrations used in standard enzymological experiments.

A similar situation exists for the processes that we discuss in this article—the biosynthesis of nucleic acids and their precursors, the nucleoside triphosphates. Like glycolytic enzymes, most of the enzymes involved are readily isolated as soluble proteins; however, they also form complexes. With a few exceptions, protein–protein associations in this group of enzymes are relatively weak. Furthermore, some associations may be transient in nature, existing only during the instant that two enzymes simultaneously bind one metabolite molecule during its transit from one enzyme to the next.

Recent studies have yielded great insight into the structures of protein machines that carry out nucleic acid synthesis. The interactions of proteins in these machines with nucleic acids provide stabilization of a type not seen in loosely associated complexes in nucleotide synthesis, where only protein–protein associations are available to provide stabilization. Dramatic advances in our understanding of the protein machines in nucleic acid synthesis have come from reconstitution experiments with purified proteins and defined DNA substrates.

II. OVERVIEW OF NUCLEOTIDE BIOSYNTHETIC PATHWAYS

The pathways of nucleotide biosynthesis are well understood. Figure 1 summarizes the *de novo* synthetic pathways leading to nucleoside triphosphates, which are virtually identical in all organisms. These pathways use amino acids, pentose phosphates, carbamoyl phosphate, folic acid coenzymes, and ATP. Corresponding salvage pathways, which utilize nucleobases or nucleosides containing preformed purine or pyrimidine rings, vary considerably among different organisms, and these variations provide convenient targets for therapy of diseases, such as infections by viruses or parasites. [*See* Purine and Pyrimidine Metabolism.]

In mammalian cells, the branch point between RNA and DNA precursor biosynthesis is at the level of ribonucleoside diphosphates (rNDPs). All four common rNDPs are substrates for a multifunctional enzyme, ribonucleotide reductase, which reduces adenosine diphosphate (ADP), cytidine diphosphate (CDP), guanosine diphosphate (GDP), and uridine diphosphate (UDP) to the corresponding deoxyribonucleoside diphosphates (dNDPs). Beyond this stage, metabolites are destined to be used for DNA synthesis. In

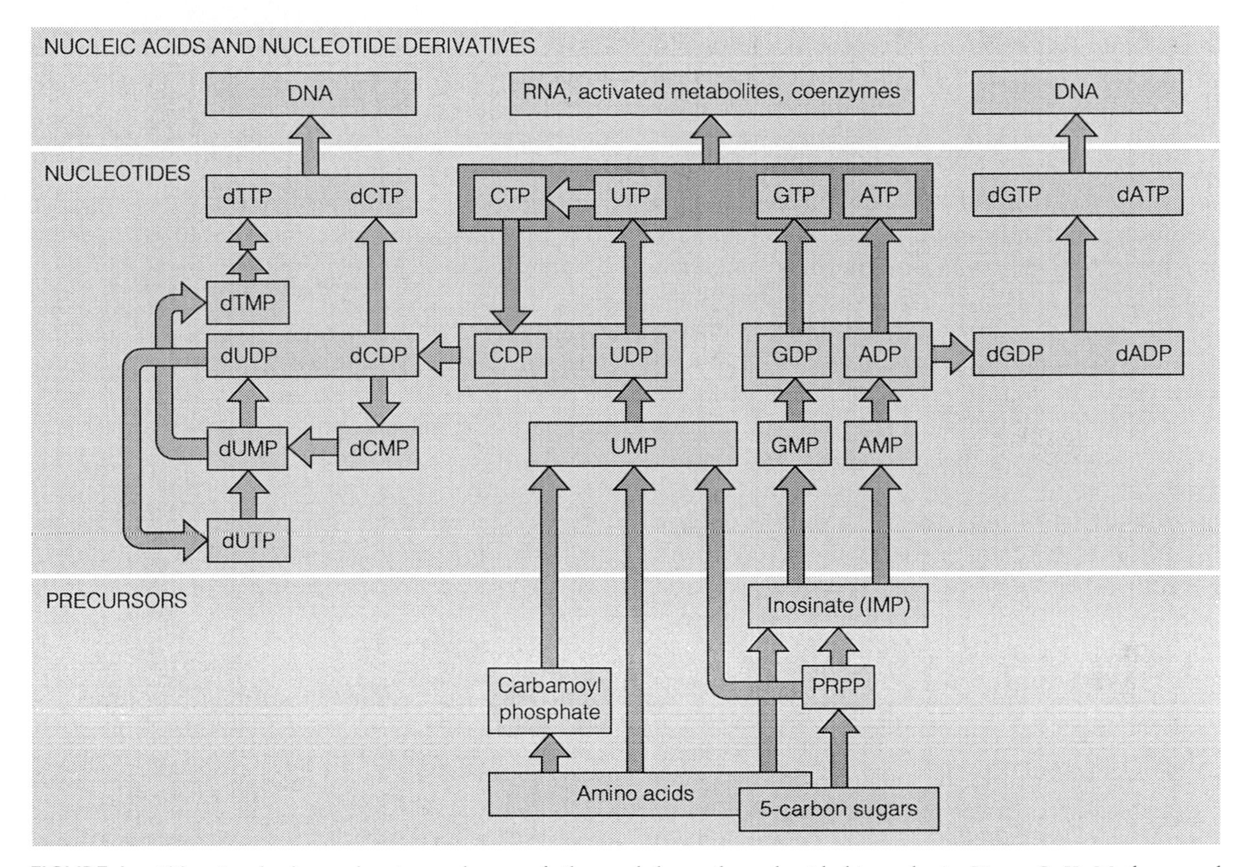

FIGURE 1 Abbreviated scheme showing pathways of ribo- and deoxyribonucleotide biosynthesis. [From C. K. Mathews and K. E. van Holde (1996). "Biochemistry," 2nd Ed. Copyright © 1996 Benjamin/Cummings Publishing Company, Redwood City, California. Reprinted by permission.]

mammalian cells, dNTPs have no known roles other than those of DNA precursors and regulators of their own biosynthesis. dNTPs are substrates for the action of DNA polymerases, which function along with other proteins to catalyze the template-guided insertion of nucleotides into polymeric chains. Similar processes occur for RNA synthesis, although the protein machines involved in transcription are quite different. In addition, the ribonucleoside triphosphate (rNTP) precursors for RNA synthesis have numerous additional metabolic roles—as energy cofactors (e.g., ATP), as precursors of coenzymes and metabolic regulators like cyclic AMP, and in the synthesis of activated metabolic intermediates, such as uridine diphosphate glucose and acetyl-coenzyme A. Because of these additional metabolic roles, and because most cells contain 5 to 10 times as much RNA as DNA, at least 90% of the metabolic flux through the *de novo* nucleotide synthetic pathways is destined for ribonucleotide, not deoxyribonucleotide, synthesis.

III. THE NEED FOR ORGANIZATION OF DNA METABOLISM

Because of their diverse roles in intermediary metabolism and transcription, ribonucleotides are utilized in all compartments of the cell and in all phases of the cell cycle. By contrast, except for a small amount of DNA synthesis in mitochondria, chloroplasts, and other organelles, DNA replication is confined to the nucleus, occurs only in dividing cells, and is limited to the S phase of the cell division cycle. Therefore, it is appropriate that the cell organize and regulate the synthesis of deoxyribonucleotides so that they are delivered to replication sites on demand. Moreover, because deoxyribonucleotides play extremely limited additional roles in mammalian cells, their rates of synthesis must be controlled, so that unwanted pool imbalances do not develop; the four dNTPs must be provided at the same relative rates at which they are used for DNA synthesis. When a regulatory abnormality occurs, leading to unbalanced dNTP pools, the consequences can include enhanced mutagenesis, chromosomal abnormalities, induction of endogenous viruses, enhanced genetic recombination, and cell death. [*See* Apoptosis (Programmed Cell Death).]

Another factor that must be considered is the low intracellular concentrations of deoxyribonucleotides. From data on pool sizes and intracellular volumes, one can estimate that rNTP concentrations are in the millimolar range in mammalian cells. ATP, the most abundant ribonucleotide, is present at concentrations ranging from 3 to 10 m*M* in different cells and organelles. By contrast, dNTP pool sizes are one to two orders of magnitude lower. These pools must be maintained in the face of vastly increased demands as proliferating cells move from the G1 phase of the cell cycle to S phase. Although the DNA chain growth rate is about 10-fold slower in mammalian than in bacterial cells, a typical somatic cell will double its DNA content within 6 hours, using between 1000 and 10,000 replication forks per cell. Since it seems likely that replicative DNA polymerases operate at maximum velocity, it is essential that the cell be organized so that dNTPs can be delivered to these sites fast enough to meet the demand and to maintain saturating concentrations in the vicinity of DNA polymerases. The demand is even higher in early development, where a cleavage-stage embryo may double its DNA content in a half hour or less. Thus, DNA precursor biosynthesis must be organized and coordinated with DNA replication. Much of this coordination involves allosteric regulation of the function of enzymes of deoxyribonucleotide biosynthesis by precursors or intermediates. The relevant pathways, critical enzymes, and control sites are identified in Fig. 2. The key regulated enzyme is ribonucleotide reductase, which catalyzes the first step committed to DNA replication. As noted earlier, a single enzyme reduces ADP, CDP, GDP, and UDP to the respective 2′-deoxy derivatives. The enzyme protein, which is an $\alpha_2\beta_2$ tetramer, is controlled by interaction of nucleoside triphosphates with two classes of regulatory sites. One site controls primarily the catalytic activity of the enzyme, and the other controls its specificity, the efficiency with which it acts upon each of its four substrates. Together, these interactions help to maintain balanced rates of synthesis of the four DNA precursors—dATP, dCTP, dGTP, and dTTP.

As noted earlier, the demand for ribonucleotides is more generalized, and rNTPs must be made available continuously throughout the cell. From the standpoint of intracellular organization, the most remarkable feature of RNA metabolism is the splicing of exons to one another, as introns are excised during posttranscriptional RNA processing. This process involves an intricate structure called the "spliceosome," which forms in the nucleus from small ribonucleoprotein particles. The process is not driven by kinetic factors; it is quite slow, but very precise. [*See* RNA Replication.]

IV. MECHANISMS AND PROCESSES

Most of the mechanisms known to organize nucleotide synthesis and coordinate it with nucleic acid synthesis are comparable to processes that organize other areas of metabolism—subcellular compartmentation and the use of multienzyme complexes and multifunctional proteins to channel scarce or unstable intermediates.

A. Intracellular Compartmentation

Most of the reactions of purine and pyrimidine synthesis take place in the cytosol of mammalian cells. Obviously, both replication and transcription must occur in the nucleus, where the DNA template resides. Because the nuclear membrane contains large pores, transport of nucleotides to their sites of utilization should be straightforward, involving passive diffusion through the pores. However, under some conditions, the intranuclear deoxyribonucleotide pools seem to be metabolically distinct from the much larger cytosolic pools. Some of the compartmentation of deoxycytidine nucleotides may result from the fact that dCTP is a precursor to the deoxy analogs of the well-known CDP-choline and CDP-ethanolamine, intermediates in phospholipid biosynthesis. The metabolic significance of these "lipodeoxynucleotides" is not yet known.

A significant amount of the cytosolic ribonucleotide output is transported into mitochondria to participate in energy-generating pathways. The mitochondrial inner membrane contains an adenine nucleotide "translocase" system, which pumps ADP in, concomitantly with the efflux of ATP. In cultured HeLa cells, about 15–20% of the total content of each rNTP is in mitochondria. Mitochondria also contain several deoxyribonucleotide biosynthetic enzymes, including deoxyguanosine kinase, thymidine kinase, thymidylate synthase, and ribonucleotide reductase. It is not yet clear to what extent dNTPs for mitochondrial DNA replication are generated inside the mitochondrion,

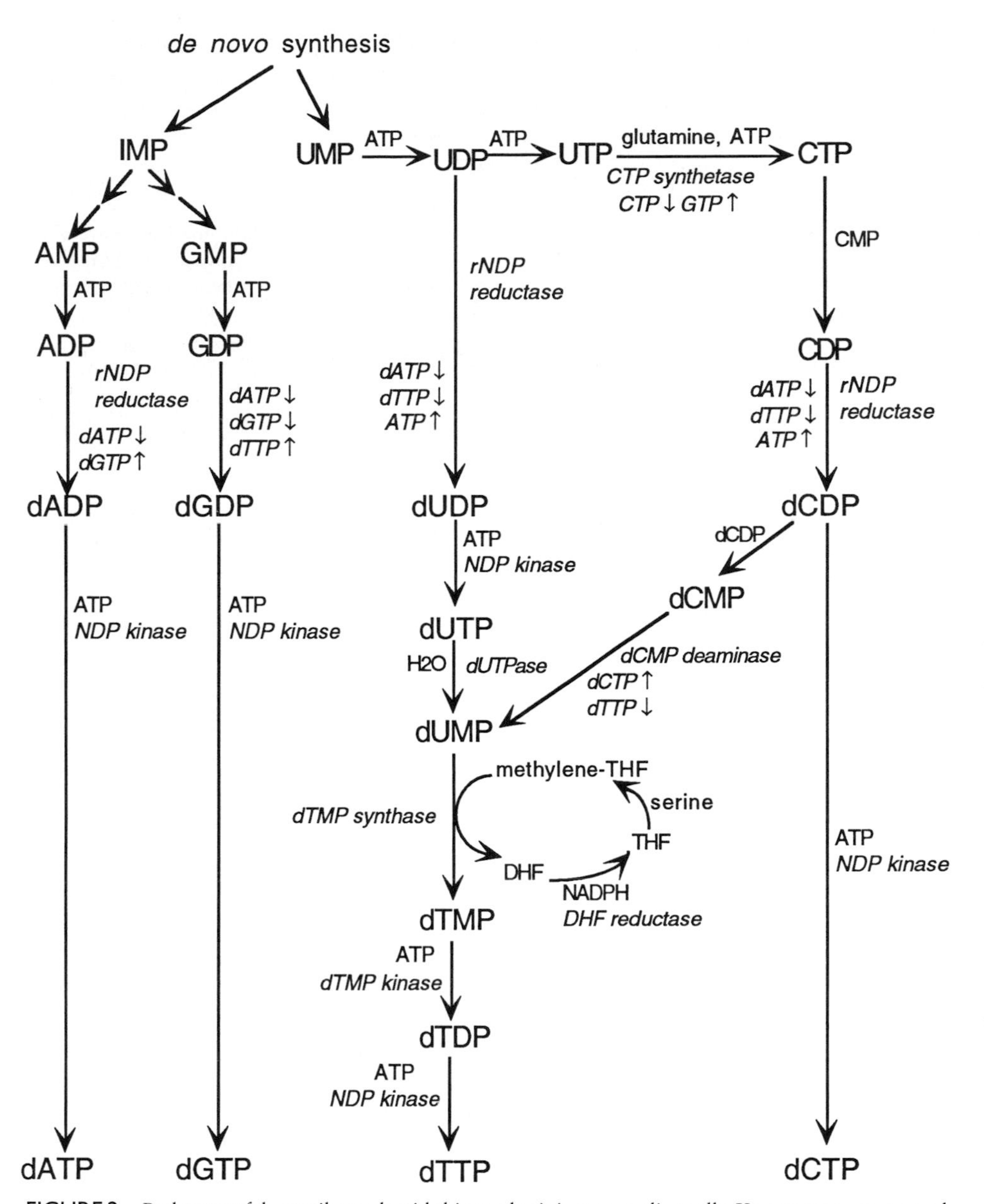

FIGURE 2 Pathways of deoxyribonucleotide biosynthesis in mammalian cells. Key enzymes are named, and allosteric effectors are identified with an upward arrow (for activators) or a downward arrow (for inhibitors).

or else transported inward as completed dNTP molecules. The intramitochondrial pools of DNA precursors behave quite distinctly from those used in nuclear DNA replication, but the mechanisms of the compartmentation involved are not yet known.

From the standpoint of cellular organization, one aspect of nucleotide compartmentation seems paradoxical. The *de novo* biosynthesis of pyrimidines involves six reactions leading to uridylate, the first fully formed pyrimidine nucleotide:

$$CO_2 + H_2O + ATP \rightarrow \text{carbamoyl phosphate} \rightarrow \text{carbamoyl aspartate} \rightarrow \text{dihydroorotate} \rightarrow \text{orotate} \rightarrow \text{orotidylate} \rightarrow \text{UMP}$$

The first three reactions are catalyzed by a trifunctional enzyme in eukaryotic cells. The last two reactions are catalyzed by a bifunctional enzyme (see next section); both of these multifunctional enzymes are cytosolic. However, the fourth enzyme on the pathway, dihydroorotate dehydrogenase (a flavoprotein enzyme), is located on the mitochondrial inner membrane. Therefore, the pathway begins and ends in the cytosol, but intermediates must travel into, and then out of, the mitochondrion. Because the intermediates are not used for other known processes within the mitochondrion, the metabolic logic behind this subcellular organization remains elusive.

B. Multifunctional Proteins

An effective way to organize metabolic pathways is to juxtapose the active sites of enzymes catalyzing sequential reactions, so that intermediates are transferred directly form site to site, equilibrating either little or not at all with bulk pools. This phenomenon, called metabolic channeling, can in principle improve cell functions in various ways. First, by preventing diffusion of intermediates away from their sites of utilization, channeling provides for more efficient utilization of substrates and delivery of end products to sites of ultimate disposition. Second, by keeping average concentrations of most intermediates low, channeling prevents the accumulation of intermediates to levels where they would adversely affect osmotic pressure in the cell. Finally, a multifunctional protein allows for joint regulation of activities that act in one pathway, obviating a requirement for several distinct control mechanisms.

Table I lists several multienzyme complexes and multifunctional proteins involved in nucleotide biosynthesis. Although most of these have been observed in mammalian cells, some have not. In fact, enzymes such as the bifunctional dihydrofolate reductase–thymidylate synthase found in protozoans are so different from their counterpart enzymes in mammalian cells that they present attractive targets for chemotherapy of malaria and other protozoal infections. Similarly, the herpes virus deoxypyrimidine kinase is an actual chemotherapeutic target because, by its broad specificity, it permits the intracellular activation of drugs only in infected cells. For example, acycloguanosine (Acyclovir) is activated to the inhibitory nucleotide acycloguanosine triphosphate, through the action of the virus-specified kinase.

Tetrahydrofolate coenzymes participate in the synthesis of purine nucleotides and thymine nucleotides. These coenzymes are unstable and also present at very low intracellular concentrations. Activities that interconvert these coenzymes are, perhaps not surprisingly, covalently linked, particularly the widely distributed formylmethenylmethylenetetrahydrofolate synthetase, a single protein with three catalytic activities—formyltetrahydrofolate synthetase, methylenetetrahydrofolate dehydrogenase, and methenyltetrahydrofolate cyclohydrolase. Perhaps surprisingly, these juxtaposed active sites do not always create metabolic channeling, as shown by *in vitro* studies of some reaction sequences.

Another class of multifunctional enzymes does not catalyze sequential reactions, but instead acts upon different substrates in the same active site. An excel-

TABLE I
Multifunctional Enzymes and Multienzyme Complexes in Nucleotide Biosynthesis

Enzyme or complex	Reactions catalyzed	Source
	Multifunctional Enzymes	
H_2-folate reductase/dTMP synthase	Synthesis of dTMP[a] from dUMP; reduction of H_2-folate	Protozoa
Deoxypyrimidine/dTMP kinase	Phosphorylation of dTMP and many nucleosides	Herpes virus-infected cells
GART protein	Three reactions of *de novo* purine synthesis	*Drosophila* and other eukaryotes
CAD protein	First three reactions of pyrimidine synthesis	Eukaryotes
UMP synthase	Orotate → orotidylate → uridylate	Eukaryotes
Formylmethenylmethylenetetrahydrofolate synthetase	Three reactions shown in Fig. 3	Eukaryotes
FIGLU transformylase/formimino-H_4-folate cyclodeaminase	Formiminoglutamate → formimino-H_4-folate → 10-formyl-H_4-folate	Bacteria
	Multienzyme Complexes	
Purine transformylase complex	Transformylase reactions in purine synthesis (Fig. 3)	Eukaryotes
Carrot thymidylate synthesis complex	dTMP synthase, H_2-folate reductase, serine transhydroxymethylase	Carrot cells
T4 dNTP synthetase complex	rNDPs →→→ dNTPs; dNMPs →→→ dNTPs	T4 phage-infected *E. coli*

[a]dTMP, thymidine monophosphate; dUMP, deoxyuridine monophosphate.

lent example is ribonucleotide reductase, which, as noted earlier, catalyzes reactions leading to all four DNA precursors. It seems certain that the architecture of this enzyme molecule is configured so as to coordinate the four activities, permitting balanced synthesis of the four dNTPs. Another example, is deoxycytidine kinase, which in vertebrate cells has a low K_M for deoxycytidine, but which acts also upon deoxyadenosine or deoxyguanosine. Affinities for the purine substrates are much lower than for deoxycytidine, meaning that the enzyme will preferentially phosphorylate deoxycytidine under most physiological conditions.

C. Multienzyme Complexes and Metabolons

In many cases, enzyme active sites that catalyze sequential reactions are not covalently joined on single-protein molecules, but reside instead on separate proteins that interact with one another strongly enough to allow the isolation of a multienzyme complex. Three examples from nucleotide metabolism are listed in Table I. The one that has been best characterized in vertebrate cells is a four-protein complex containing the two transformylases of the purine nucleotide biosynthetic pathway and two enzymes of tetrahydrofolate interconversion (Fig. 3).

Multienzyme complexes that carry out the synthesis of deoxyribonucleotides have been described in both prokaryotic and eukaryotic systems. The only one that has been highly purified and characterized kinetically is a complex from T4 bacteriophage-infected *Escherichia coli.* This complex demonstrates substrate channeling *in vitro,* in both crude and highly purified preparations. Moreover, a number of viral gene mutations that affect individual enzymes in the complex also disrupt the integrity of the complex as isolated *in vitro.* This type of genetic evidence, which goes hand in hand with the kinetic studies that show channeling, suggests that the complex is formed by specific protein–protein interactions, and not by artifactual protein aggregation that occurs only after cells are broken open for isolation of enzymes.

In general, the demonstration that a multiprotein aggregate functions as an organized complex requires the gathering of quite diverse lines of evidence—from enzyme purification, enzyme kinetics, intracellular localization studies, analysis of interactions between purified proteins, and reconstitution of active complexes or subassemblies. Multienzyme aggregates, comparable to the T4 "dNTP synthetase complex," have been described in mammalian cells, but to date none has been characterized by the full range of genetic, enzymological, and cell biological approaches mentioned here.

Interest in dNTP-synthesizing complexes is focused on the possibility that they participate as part of a substrate-channeling mechanism for delivery of DNA

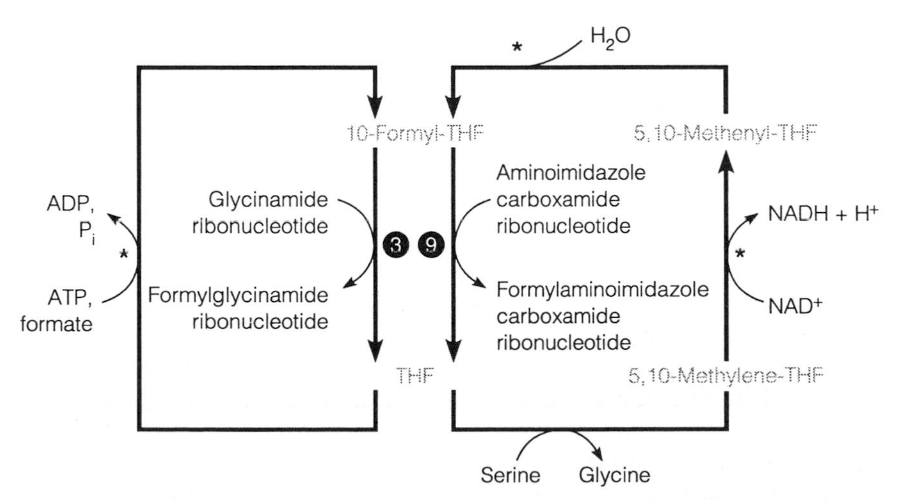

FIGURE 3 Reactions catalyzed by a multienzyme complex for transformylation reactions in purine nucleotide synthesis. The three enzyme activities marked with asterisks are carried on a single multifunctional protein molecule. [From C. K. Mathews and K. E. van Holde (1996). "Biochemistry," 2nd Ed. Copyright © 1996 Benjamin/Cummings Publishing Company, Redwood City, California. Reprinted by permission.]

precursors to replication sites. Evidence for such a model is accumulating in prokaryotic systems, where replicative DNA polymerases are characterized by high K_M values (50–100 μM) and high V_{max} values (500–1000 nucleotides per second). Diffusion of dNTPs from distant sites of synthesis probably does not suffice to maintain dNTP concentrations at saturating values near replication sites. In fact, one can calculate that in T4-infected *E. coli* the pools of dNTPs close to replication sites must completely drain and refill ten times per second in order to sustain observed rates of DNA replication.

In eukaryotic cells, however, direct coupling of dNTP synthesis to DNA replication seems unlikely, if only because the enzymes of dNTP synthesis are found largely in the cytosol (although recent evidence suggests that they may be clustered along the outer edge of the nuclear membrane). In addition, the kinetic properties of eukaryotic replicative DNA polymerases are quite different from those of prokaryotes. V_{max} and K_M values for mammalian DNA polymerases are at least an order of magnitude lower than those of corresponding prokaryotic enzymes. For example, K_M values for dNTPs with calf thymus DNA polymerase α are in the range of 2–5 μM. At the same time, estimations of intranuclear dNTP concentrations in mammalian cells give values for dGTP, the *least* abundant dGTP, of about 10 μM. Thus, one need not invoke a channeling mechanism to understand how replicative DNA polymerases can operate at or near substrate saturation in mammalian cells. However, little is known at present about the transfer of dNTPs from cytoplasmic sites of synthesis to the interior of the nucleus. In studies where radiolabeled deoxyribonucleosides were microinjected into mammalian cell cytoplasm, radioactivity was rapidly and efficiently incorporated into DNA. Whether movement into the nucleus occurs by simple diffusion or by a facilitated process is not yet known.

V. PROTEIN MACHINES IN NUCLEIC ACID SYNTHESIS

It has long been known that DNA replication and transcription require enzymes and proteins other than the polymerases catalyzing formation of the phosphodiester bonds that link nucleotides together. However, isolation of multiprotein complexes capable of carrying out all of the steps in nucleic acid synthesis—initiation, chain elongation, and termination—is usually quite difficult. A complementary approach has been far more successful—purification of individual proteins and reconstitution of systems, which Bruce Alberts has termed protein machines, capable of carrying out one or more steps in the process. Difficulties in isolating such machines intact arise from the dynamic nature of the processes involved.

Transcription, which is discussed at length elsewhere, is a good example. [*See* DNA and Gene.] We know that transcription normally begins with binding of transcription factors (usually proteins) near promoter sites, to facilitate the binding of RNA polymerase. Once RNA polymerase binds, it unwinds template DNA strands locally and then moves away from the complex of transcription factors at the promoter. At this time one of the polypeptide subunits of RNA polymerase (the σ subunit in bacteria) dissociates. When a termination signal is reached, additional proteins may come into play to facilitate the dissociation of template DNA, nascent RNA transcript, and RNA polymerase. Because of the dynamic nature of the process, isolation of transcriptional elongation complexes or termination complexes in their native form is all but impossible. Compare this situation with a complex in intermediary metabolism, such as the pyruvate dehydrogenase complex. Here the proteins can retain fixed positions with respect to each other, whereas the low-molecular-weight intermediates move from site to site within the complex. Hence, forces holding the proteins together can be strong enough to allow isolation of the complex (although, in metabolons such as the T4 dNTP synthetase complex, the interactions are often much weaker).

On the other hand, great insight into transcriptional mechanisms has come from isolation of the proteins involved and their interactions with specific DNA templates. For example, by using a DNA template with one promoter site, and incubating this with RNA polymerase plus a mixture of three rNTPs, the investigator can stall all of the elongation complexes at the same point—where the DNA template dNMP residue is complementary to the missing ribonucleotide. It turns out that such complexes, formed *in vitro,* are extremely stable and have been used to analyze how the enzyme is bound to its template and to the nascent transcript.

In several respects, DNA replication is harder to analyze than transcription, partly because both DNA strands are copied and partly because replication is both faster and more accurate than transcription. In addition, while transcription involves copying single genes or relatively short multicistronic operons, replication involves entire chromosomes. Once replication has commenced, the machine carrying it out must remain bound to the template over many thousands of nucleotide base pairs copied, a property called processivity.

Our earliest insights into the operation of protein machines in replication came from studies of the chain elongation process. For simple microbial systems—bacteria and bacteriophages—where extensive genetic analysis was available, the existence of mutations in many genes that affected replication made it apparent that about a dozen proteins are involved in the replication process. By using simple biochemical assays for DNA synthetic activity and appropriate replication-defective mutants, it became possible to purify all of the proteins involved and to identify their roles, a process that is still continuing. Within the past decade, comparable studies have become possible in mammalian systems, and a remarkably similar pattern is developing.

As shown in Fig. 4, two DNA polymerase molecules must function as parental DNA strands unwind in a replication fork, because each of the parental DNA strands is being copied. In addition, because of the antiparallel nature of the DNA double helix, one polymerase molecule moves in the direction of fork move-

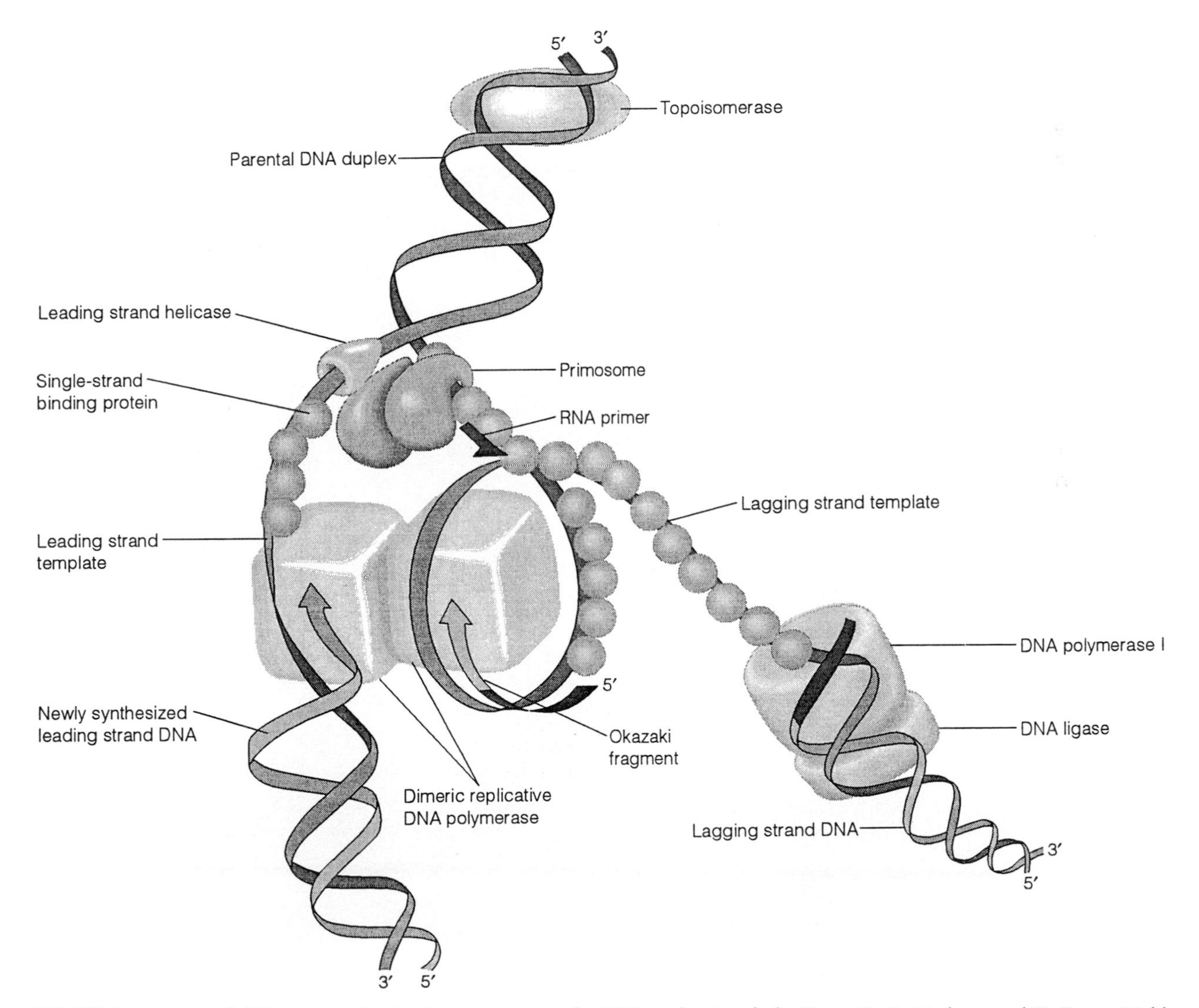

FIGURE 4 Actions of different proteins in the propagation of a DNA replication fork. [From C. K. Mathews and K. E. van Holde (1995). "Biochemistry," 2nd Ed. Copyright © 1995 Benjamin/Cummings Publishing Company, Redwood City, California. Reprinted by permission.]

ment (the "leading strand" polymerase), while the second, the "lagging strand" polymerase, moves in the opposite direction. Other proteins involved include a "primosome," or helicase–primase complex. This complex serves two functions—the energy-dependent unwinding of the parental DNA duplex and the synthesis of short RNA molecules along the lagging strand. These RNA molecules provide primers, or short nucleic acid molecules that can be extended by the polymerase-catalyzed incorporation of deoxyribonucleotides. A different DNA polymerase is involved in removal of the RNA primers and replacement of the ribonucleotide residues by deoxyribonucleotides. Another protein, the single-strand DNA-binding protein, or SSB, stabilizes template strands in the single-stranded form, so that they are free to base-pair with incoming nucleotides. Ahead of the fork, topoisomerases act like a swivel, to relieve the torsional stress created by the strand unwinding process. Finally, a group of "polymerase-accessory proteins" helps the DNA polymerases remain bound to the template.

The means by which these accessory proteins enhance processivity has become clear in quite recent developments. In *E. coli* a protein called β is part of the multisubunit DNA polymerase III holoenzyme, the principal replicative polymerase. This doughnut-shaped, dimeric molecule functions like a clamp, encircling the DNA duplex and making it all but impossible for the enzyme to dissociate, once bound (Fig. 5). In mammalian cells a related protein called PCNA plays the same function.

How does the clamp bind to DNA? Important experiments, reported in 1995, identify a role as "clamp loader" for another protein complex, which in *E. coli* is the five-protein γ complex. As schematized in Fig. 6,

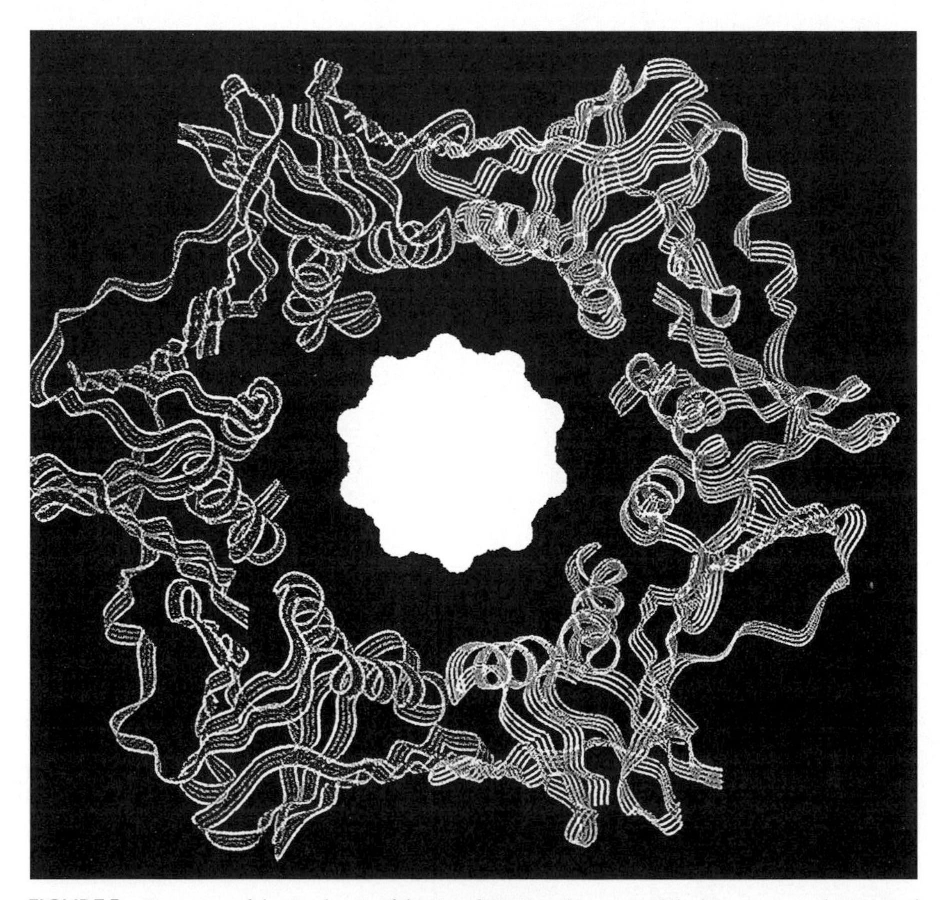

FIGURE 5 Structure of the β subunit of the *E. coli* DNA polymerase III holoenzyme, as determined by X-ray crystallography. The doughnut-shaped dimeric protein molecule wraps about double-stranded DNA, shown in profiles as the cross section of a B-form DNA duplex. [Figure provided by Mike O'Donnell. From J. Kuriyan and M. O'Donnell (1993). *J. Mol. Biol.* **234**, 915. Reprinted by permission of copyright owner.]

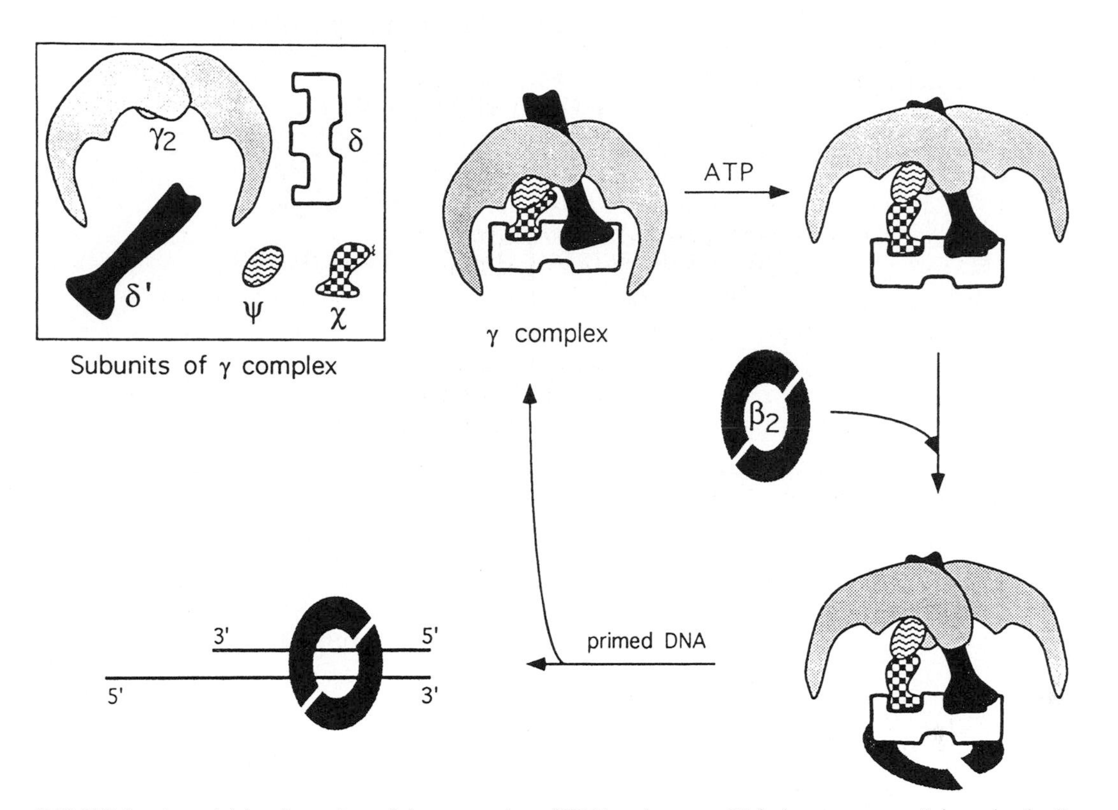

FIGURE 6 A model for the action of the γ complex of DNA polymerase III holoenzyme as a "clamp loader," facilitating the binding of the β dimer to DNA. An ATP-dependent conformational change exposes a surface on the δ subunit of the γ complex, which allows it to bind β and transfer it to a DNA template–primer combination. [Figure provided by Mike O'Donnell. From V. Naktinis, R. Onrust, L. Fang, and M. O'Donnell (1995). *J. Biol. Chem.* 270, 13358. Copyright © 1995 American Society for Biochemistry and Molecular Biology. Reprinted by permission.]

this complex evidently undergoes an ATP-dependent conformational change, which allows it to bind β in a form that partially separates the two polypeptide subunits. An additional interaction, with a DNA template bound to a primer, leads to more conformational changes that wrap β about the DNA and dissociate it from γ, which remains behind as DNA polymerase then advances.

With regard to cellular organization of DNA and RNA synthesis, there is an important traffic control problem. Since replication and transcription share the same DNA template, what happens when a replication fork and a transcription complex arrive at the same region of DNA at the same time? Recently reported experiments indicate that DNA replication has priority, as you might expect if replication is ordered so that each chromosome is replicated once and only once during each cell division cycle, whereas a particular gene might be transcribed hundreds or thousands of times in a generation. However, when a fast-moving replication fork catches up with a slower-moving transcription complex moving in the same direction, the transcription complex is not broken down. Rather, as schematized in Fig. 7, the RNA polymerase and bound transcript are temporarily pushed aside, but they remain bound to the template. Once the replication fork has passed, the original transcription complex can reform and continue, as if nothing had happened.

VI. THE NUCLEAR MATRIX AND IMMOBILIZED REPLICATION AND TRANSCRIPTION COMPLEXES

The classic view of both transcription and replication is that the polymerase, along with its associated proteins, moves along the DNA template, copying one or both template strands as it moves. Starting in the

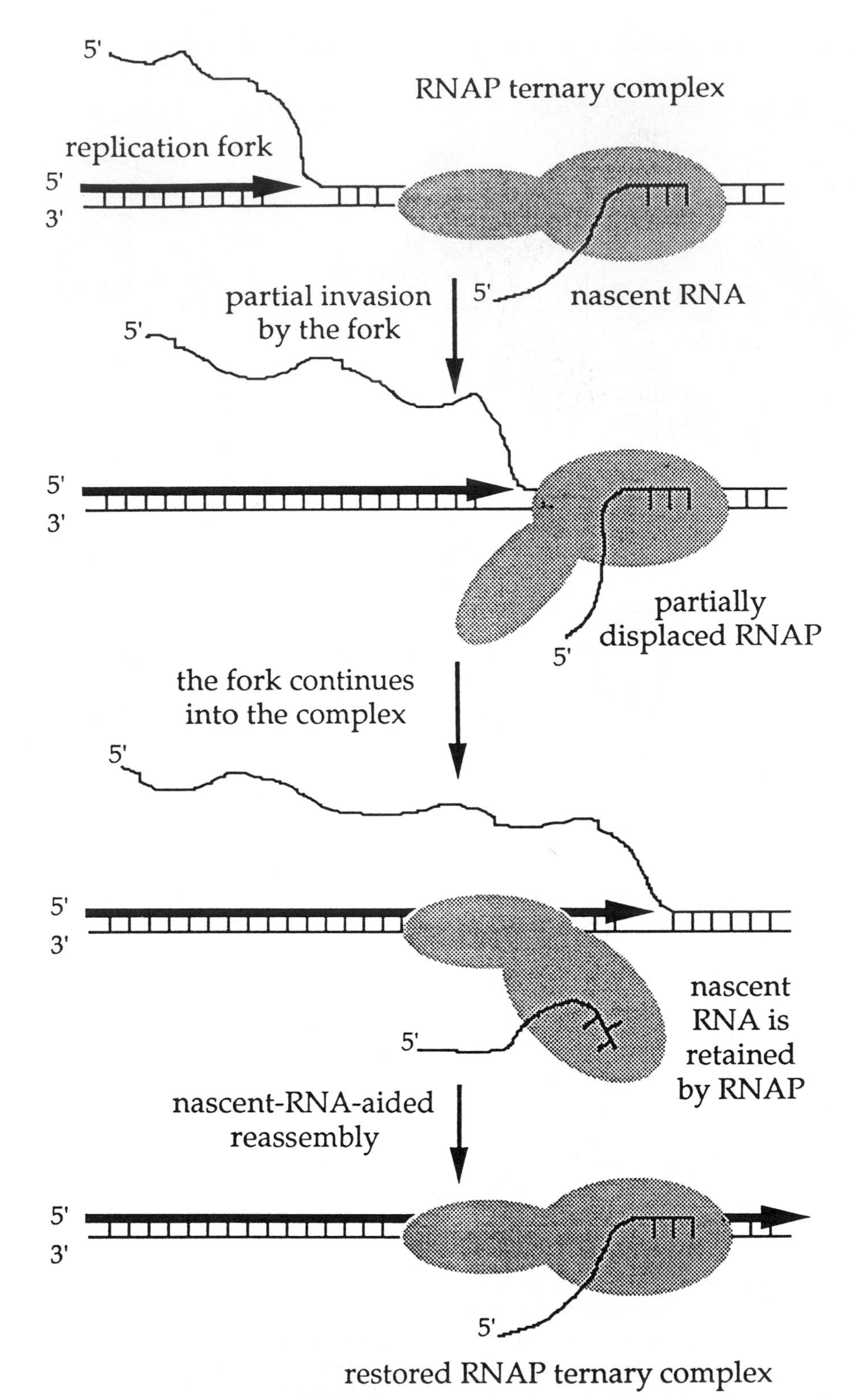

FIGURE 7 A model to explain how a replication fork can pass a transcribing RNA polymerase molecule without displacing the enzyme and its nascent transcript. For simplicity the diagram shows just one of the two DNA strands in the replication fork and does not show the replication complex. Specific base pairing between the 3′ end of the polymerase-bound nascent RNA with its DNA template is thought to be involved in reformation of the transcription complex in precise register. [Figure provided by E. Peter Geiduschek. From B. Liu, M. L. Wong, R. L. Tinker, E. P. Geiduschek, and B. M. Alberts (1993). *Nature* **366**, 33. Reprinted by permission of copyright owner.]

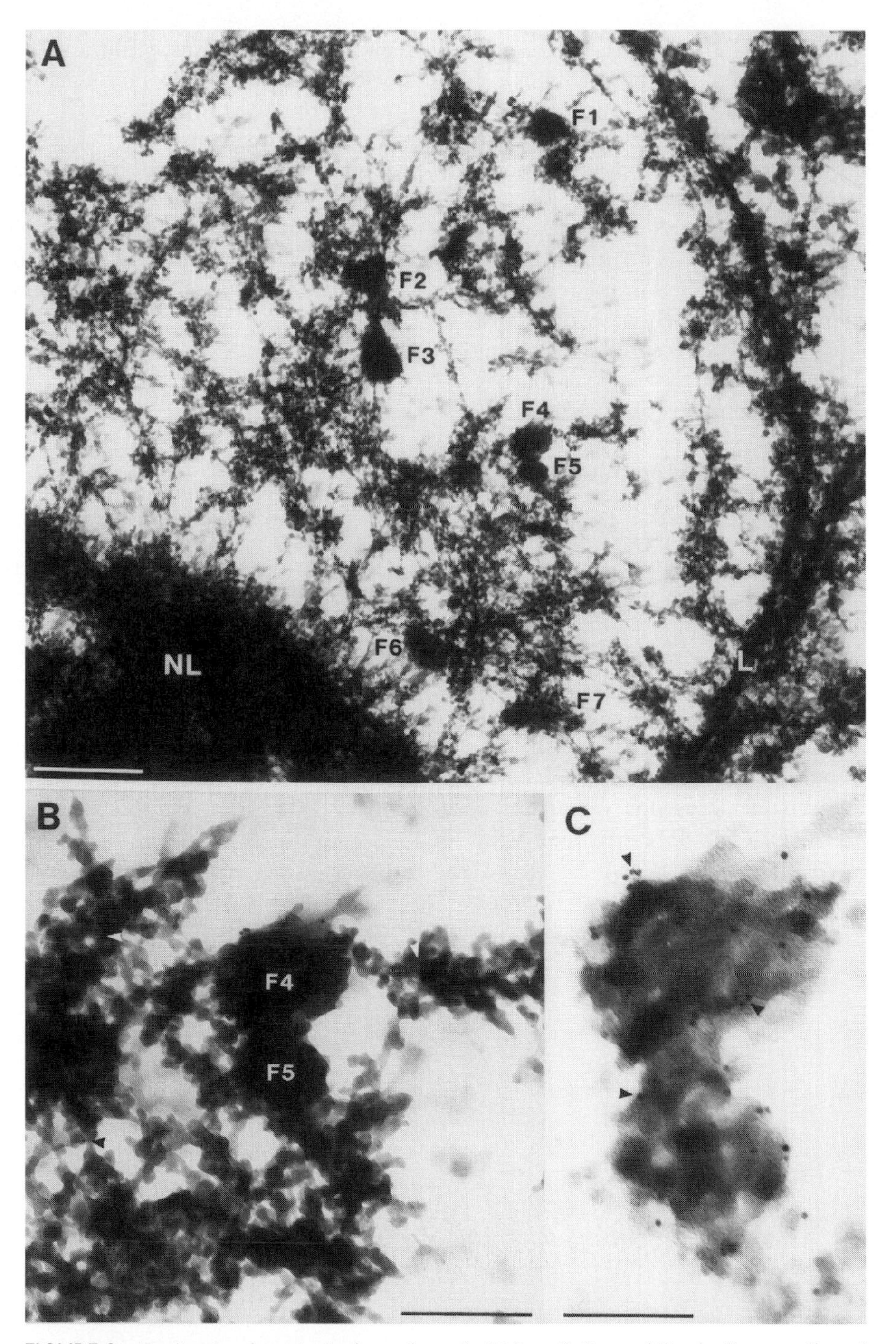

FIGURE 8 Replication factories in the nucleus of a HeLa cell. Permeabilized cells were allowed to incorporate biotinylated dUTP into DNA, as described in the text, and these sites were immunolabeled with gold. (A) An electron micrograph of part of the nucleus, showing five electron-dense areas identified as factories (F1–F5). (B) A higher-power view of F4 and F5. (C) An underexposure of one factory, showing multiple electron-dense gold particles that represent sites of DNA synthesis; arrowheads point to three of them. [Figure provided by Peter Cook. From P. Hozák, A. B. Hassan, D. A. Jackson, and P. Cook (1993). *Cell* 73, 361. Copyright © 1993 Cell Press. Reprinted by permission.]

late 1970s and still gathering momentum is an alternative view, namely, that the enzymatic machinery is immobilized and that nucleic acid is synthesized as the DNA template is threaded through a stationary complex. This picture began to develop with studies on the nuclear matrix, that material left after isolated nuclei are treated sequentially with nucleases, salt, and detergent. Though it might seem that material surviving such rough treatment plays more of a structural than an enzymatic role, labeling studies with radioactive thymidine show preferential incorporation into DNA associated with the nuclear matrix, suggesting that replication complexes are also associated with the matrix. Also this model is consistent with the fact that chromosomes appear to contain multiple loops of chromatin; it has been suggested that each loop contains DNA that is immobilized at each end by attachment to a stationary polymerase.

Recently, new techniques have been devised for visualizing replicating DNA in mammalian cells. After careful permeabilization, most of the chromatin is removed by restriction nuclease treatment and washing. Then the nuclei are incubated with a dNTP mixture that includes biotin-modified deoxyuridine triphosphate, a dTTP analog that can be immunolabeled with gold particles. Electron microscopic visualization of such nuclei in HeLa cells (Fig. 8) reveals that replication occurs in electron-dense areas called "replication factories," each of which contains at least 20 replication forks. Thus, replication occurs at structurally discrete sites. The fact that multiple replicons undergo replication in the same area may be related to the mechanisms used by the cell, still not well understood, which ensure that each replicon is copied once and only once during the cell division cycle. The structures of these factories and the routes by which they are provided with DNA precursors are important questions for future investigations in this area.

BIBLIOGRAPHY

Cook, P. R. (1994). RNA polymerase: Structural determinant of the chromatin loop and the chromosome. *BioEssays* **16**, 425.

French, S. (1993). Replication meets transcription: Who's got the right of way. *ASM News* **59**, 437.

Goodsell, D. S. (1992). A look inside the living cell. *Am. Sci.* **1992**, 457.

Jackson, D. A. (1995). Nuclear organization: Uniting replication foci, chromatin domains, and chromosome structure. *BioEssays* **17**, 587.

Kornberg, A., and Baker, T. A. (1992). "DNA Replication," 2nd Ed. Freeman, San Francisco.

Mathews, C. K. (1993). The cell: Bag of enzymes or network of channels? *J. Bacteriol.* **175**, 6377.

Mathews, C. K. (1993). Enzyme organization in DNA precursor biosynthesis. *Progr. Nucl. Acid Res. Mol. Biol.* **44**, 167.

Mathews, C. K., and van Holde, K. E. (1996). "Biochemistry," 2nd Ed., Chaps. 22, 24, and 26. Benjamin/Cummings, Redwood City, California.

Nutrients, Energy Content

G. B. STOKES
Murdoch University

GLOSSARY

Biosphere/biomass Biosphere refers to the total habitat of living things, from the ocean depths to mountaintops; biomass refers to all of the matter associated with living things, including an elephant's leg and the limb of a tree

Calorie Defined as the amount of heat required to raise 1 g of water from 15 to 16°C; the term used in dietetics is the Calorie, which refers to 1000 cal or 1 kcal and is equivalent to 4.186 kJ (kilojoules)

Carbohydrate A simple sugar or complex molecule derived from the polymerization of sugar molecules and their derivatives (such as malt and starch); the general empirical formula is $(CH_2O)_n$

Digestible energy Difference between the energy of the food consumed and that excreted in the feces; sometimes corrected for metabolic fecal energy, which is the energy excreted during fasting

Heat of combustion Also referred to as caloric value; heat evolved (kJ/mol, kJ/g) when a compound or substance is fully burned

Macronutrients Fat, carbohydrate, and protein, the major dietary compounds; distinct from micronutrients such as vitamins and trace elements

Metabolizable energy Difference between digestible energy and the energy excreted in urine; that portion of dietary energy actually available for supporting metabolism

Metabolism Biochemical reactions of living things, made up of the interaction between synthetic reactions (anabolism) and the reactions of degradation or breakdown (catabolism)

Redox In intermediary metabolism (the set of biochemical reactions concerned with the processing of food molecules for energy), electrons are transferred in certain reactions only (the exchangeable electrons are also called reducing equivalents); the reactions in which the donor molecule contributes electrons to the acceptor molecule are termed redox reactions and are catalyzed by a family of enzymes termed dehydrogenases; in the transfer, the donor is oxidized and the acceptor is reduced; the electrons are passed between donor and acceptor molecules via cofactors, which may be free or, in some cases, are bound to the particular dehydrogenase enzyme

Respiratory quotient Ratio between carbon dioxide breathed out to oxygen breathed in; used to provide an indication of which depot molecules (carbohydrate, fat, or protein) are being catabolized to support the body's energy needs

TABLES THAT LIST THE CALORIC VALUE OF FOODS afford a ready means to estimate the effect on body weight of eating various amounts and combinations of foods. Fat has a higher caloric value (kJ/g) than carbohydrate or protein, so the calorie-conscious consumer pays particular regard to limiting fat consumption. But why are fats so much higher in calories? Where are the calories to be found in foods?

In this article, the characteristics of molecular structure that determine the caloric value of food molecules are discussed, and arithmetic methods are given for estimating the energy content of macronutrients based on their structure or empirical formula.

The major energy transformations in the biosphere are also examined in order to appreciate the flow of energy that passes from incident solar radiation, through biomass, to final heat release during biological work.

Oxygen is released into the atmosphere during the photosynthetic reaction and reabsorbed in biological oxidation reactions when fuels are used up. The amount of energy captured during synthesis and released when food is eaten (measured as the caloric value or heat of combustion of the nutrient) is directly proportional to the amount of oxygen produced or consumed. The approach presented for understanding the energy content of nutrients also allows a simple explanation of the concept of respiratory quotient.

When nutritionists consider the body's capacity to utilize the digestible energy available from macronutrients, a special case must be made for protein, because about 20% of the energy in protein is excreted in urine. The basis for this energy loss is also discussed.

The biological energy depot molecules of most interest to humans—the macronutrients fat, carbohydrate, and protein—are also reviewed.

I. ENERGY FLOW THROUGH THE BIOSPHERE

Energy enters the biosphere principally as reducing equivalents (electrons) derived from the photolysis of water by incident solar radiation during the process of photosynthesis. The immediate products of photolysis are molecular oxygen, protons, and reducing equivalents. The electrons released during photolysis end up mostly in the food chain as reduction products of carbon dioxide (carbohydrates and fats).

A simplified diagram of the major energy transformations in the biosphere is presented in Fig. 1. Electrons released from water are transferred in complex biochemical pathways via cofactors such as nicotinamide adenine dinucleotide phosphate ($NADP^+$) and reduce oxidized acceptor molecules (CO_2, N_2, NO_3^-, and intermediate products). The ultimate products of these reductive pathways are the food chain fuels.

On subsequent reoxidation of fuels, the electrons end up in an electron sink (usually transferred to molecular oxygen that is reduced back to water). The energy released during this reoxidation process is coupled to the phosphorylation of adenosine diphosphate (ADP) to ATP. ATP serves throughout the plant and animal kingdoms as the main energy currency of the cell. When ATP is hydrolyzed back to ADP and inorganic phosphate (P_i) as part of the coupled biochemical sequences that drive the reactions of the cell, the energy captured initially by photosynthetic cells is finally released from the biosphere as heat. [*See* Adenosine Triphosphate (ATP).]

During photosynthesis, plants take up carbon dioxide and release oxygen (described in the left half of Fig. 1). In contrast, during the process of respiration (described in the right half of Fig. 1), animals breathe in oxygen and breathe out carbon dioxide. Similarly, oxygen is consumed and carbon dioxide and water vapor are released when fossil or other fuels are burned in engines or fires. Hence, the total amounts of free and dissolved oxygen and carbon dioxide vary reciprocally and depend on the relative global rates of photosynthesis and combustion, as well as on changes in the amount and composition of organic molecules in the biomass and fossil fuel reserves.

II. MOLECULAR STRUCTURE AND HEATS OF COMBUSTION

Table I shows the heats of combustion for a series of simple compounds from propane to carbon dioxide. The compounds are listed in order of decreasing number of reduced bonds, representing oxidation due to

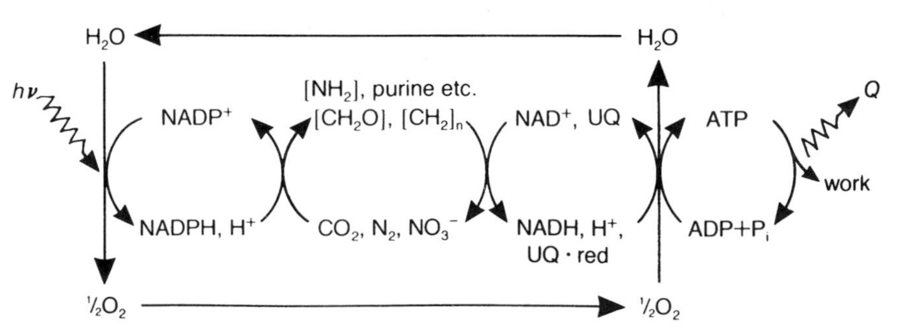

FIGURE 1 Major energy transformations in the biosphere. [Reprinted, with permission, from G. B. Stokes (1988). *Trends Biochem. Sci.* **13**, 422–424, Elsevier Publications, Cambridge.]

TABLE I
Heats of Combustion of Some Simple Compounds

Number of C—C and C—H bonds (k)	Compound	Heats of combustion (kJ/mol) Observed[a]	Predicted ($220k$)
10	$CH_3-CH_2-CH_3$	2,222	2,200
9	$CH_3-CH=CH_2$	2,050	1,980
9	$CH_3-CH_2-CH_2OH$	2,021	1,980
8	CH_3-CH_2-CHO	1,816	1,760
7	CH_3-CH_2-COOH	1,527	1,540
7	CH_3-CH_3	1,540	1,540
6	$CH_2=CH_2$	1,410	1,320
6	CH_3-CH_2OH	1,372	1,320
5	CH_3-CHO	1,167	1,100
4	CH_3-COOH	874	880
4	CH_4	891	880
3	CH_3OH	728	660
2	CH_2O	569	440
1	$H-COOH$	255	220
0	CO_2	0	0

[a]Data obtained from the "CRC Handbook of Chemistry and Physics." CRC Press, Boca Raton, Florida.

the sequential loss of electrons from their bonds. In this article, the bonds that store chemical potential energy (C—C, C—H, N—C and N—H) will be designated as "reduced bonds."

It is apparent from Table I that the energy in a C—C bond is similar to that in a C—H bond (in acetic acid, 874 kJ/mol, and in methane, 891 kJ/mol, respectively). Also, only minor changes occur when alkenes are hydrated (propene 2050, and propanol 2021 kJ/mol), thus a double bond represents about twice the potential energy of a single bond. The oxidized bonds (C—O and O—H) contribute no heat on combustion of the parent compound because they are preserved in CO_2 and H_2O, the fully oxidized products of combustion, which cannot be burned further.

The heats of combustion of many organic compounds containing C, H, and O are described by the relationship

$$Q = 110n + C \tag{1}$$

where Q is the heat of combustion expressed in kJ (110 kJ is the net energy change per mole electron), n is the number of reducing equivalents, and C is a constant characteristic of the functional group in the compound. The effect of the functional group is apparent from Table I if, for example, the oxidation of an alcohol to an aldehyde is compared with the oxidation of an aldehyde to an acid.

One consequence of the oxygen–water cycle illustrated in Fig. 1 is that the amount of oxygen in the atmosphere provides a rough estimate of the energy reserves (reduced bonds) stored in the biosphere. Furthermore, the amount of oxygen taken up in catabolism will be proportional to the number of bonds that are oxidized. Equation (1) also amounts to stating that the heat of combustion of a compound is proportional to the oxygen taken up. This may be seen from the following equations:

$$H_2O + -\overset{|}{\underset{|}{C}}-\overset{|}{\underset{|}{C}}- + \tfrac{1}{2}O_2 \rightarrow -\overset{|}{\underset{|}{C}}-OH + HO-\overset{|}{\underset{|}{C}}- \tag{2}$$

$$-\overset{|}{\underset{|}{C}}-H + \tfrac{1}{2}O_2 \rightarrow -\overset{|}{\underset{|}{C}}-OH \tag{3}$$

In the reaction shown in Eq. (2), one C—C bond is oxidized with the uptake of one atom of oxygen, whereas in Eq. (3), one C—H bond is oxidized, again with the uptake of one oxygen atom.

Equation (1) can be extended to nitrogen-containing compounds. Thus, the change in enthalpy (ΔH) or heat content following combustion, when expressed in kJ/mol, is

$$\Delta H = 220k + 105l \tag{4}$$

where k is the number of C—C and C—H bonds, and l is the number of C—N and N—H bonds. The constant for the C—N and N—H bonds (i.e., 105 kJ/mol) is close to one-half of that for C—C and C—H bonds (220 kJ/mol). This corresponds with the fact that in the oxidation of compounds to CO_2, H_2O, and N_2, each C—N and N—H bond is oxidized with one-half the oxygen uptake of a C—C or C—H bond, as follows:

$$H_2O + 2[-\overset{|}{\underset{|}{C}}-\overset{|}{\underset{|}{N}}] + \tfrac{1}{2}O_2 \rightarrow 2[-\overset{|}{\underset{|}{C}}-OH] + 2\overset{|}{\underset{|}{N}} \tag{5}$$

$$2[\overset{|}{\underset{|}{N}}-H] + \tfrac{1}{2}O_2 \rightarrow 2\overset{|}{\underset{|}{N}} + H_2O \tag{6}$$

Nitrogen in the products is represented as unbonded, since N_2 contains no bond to another element.

Thus the oxidation of any compound can be represented as the oxidation of its constituent C—C, C—H, C—N, and N—H bonds. Hence heats of combustion (kJ/mol) can be calculated either by counting the number of reduced bonds and applying Eq. (4) or by determining the oxygen uptake from the overall equation of oxidation

$$C_aH_bN_cO_d + (2a + b/2 - d)/2\ O_2 = a\ CO_2 + c/2\ N_2 + b/2\ H_2O \quad (7)$$

and multiplying the total oxygen uptake (2a + b/2 − d)/2 by $110n$ [according to Eq. (1)], where $n = 4$ for each mol of oxygen.

During oxidative phosphorylation in the mitochondria of cells, fuels are first oxidized by the enzymatic transfer of electrons to the cofactors nicotinamide adenine dinucleotide (NAD^+) and ubiquinone (UQ). The electrons are then passed into the electron transport chain and ultimately reduce oxygen to water. When electron transport is coupled to the phosphorylation of ADP, two or three ATP molecules are formed per electron pair, depending on the nature of the cofactor carrying those electrons. Reduced NAD^+ introduces electrons into the "top" of the chain and three ATP molecules are formed (P/O ratio = 3), whereas UQ introduces electrons into the chain at a lower potential and only two molecules of ATP are formed per pair of electrons (P/O ratio = 2). Therefore, it is interesting to note from Table I that the change in enthalpy for biological oxidation reactions where UQ accepts the pair of electrons via a flavoprotein (alkane–alkene) is substantially less and approximately two-thirds of the value for reactions where NAD^+ accepts the pair of electrons (alcohol–aldehyde; aldehyde–acid). [*See* ATP Synthesis by Oxidative Phosphorylation in Mitochondria.]

The enthalpy of combustion of nutrients has biological interest merely because it can provide a rough guide to the number of ADP molecules that can be phsophorylated to ATP.

III. COUNTING REDUCED BONDS AND CALCULATING ATP YIELD

A useful approximation for students of bioenergetics is that three ADPs are phosphorylated with the energy released from the oxidation of each reduced C—H and C—C bond in substrate molecules. This estimation, based on counting the number of reduced bonds, does not specify the immediate source of the ATP, which may come both from substrate-level phosphorylation and from oxidative phosphorylation using electrons donated by reduced NAD^+ (P/O = 3) and/or reduced UQ (via a flavoprotein) (P/O = 2).

Overall, 12 ATP molecules can be generated from the oxidation of one molecule of acetyl-CoA ($CH_3CO \sim$ CoA). The sites of energy storage in the molecule responsible for the 12 ATPs are the single C—C and the three C—H bonds.

Glucose ($C_6H_{12}O_6$) contains 5 C—C bonds and 7 C—H bonds, a total of 12 reduced bonds. We can expect that all 12 reduced bonds will be preserved in the products of glucose fermentation (glycolysis), because no net oxidation occurs in this anaerobic process, which occurs in the absence of oxygen. In muscle, glycolysis produces lactate (CH_3—CHOH—COOH), whereas in yeast the products of glycolysis are ethanol (CH_3—CH_2OH) and CO_2. A count of C—C and C—H bonds shows that lactate contains four C—H and two C—C bonds, whereas ethanol contains five C—H bonds and one C—C bond. Because 2 mol of lactate (or ethanol) are produced per mol of glucose, the 12 reduced bonds of glucose have been preserved in the 6 bonds in each pair of product molecules.

This simple calculation provides a useful check. At each redox step in a metabolic pathway, the number of reduced bonds will change by one, whereas no change will occur at other steps, such as those involving hydration (addition of water across a bond), phosphorylation (formation of a phosphate ester bond), or association or dissociation (gain or loss of a proton). Thus, the oxidized product of a dehydrogenase reaction will always contain one less reduced bond than the substrate (six for glyceraldehyde-3-phosphate and five for 1,3-bisphospho-glycerate; six for lactate and five for pyruvate). Furthermore, we would expect to generate 18 ATPs from the oxidative metabolism of lactate or ethanol (6 bonds × 3 ATP per bond).

The bond count approach can also be used to quickly assess a redox sequence of reactions. For example, the parasitic helminth *Ascaris* metabolizes glucose anaerobically to malate (via PEP carboxy kinase) and the malate is dismutated into succinate (via fumarate reductase) and pyruvate (via malic enzyme). Hence, the 12 bonds of glucose are found in the 7 bonds of succinate plus the 5 bonds of pyruvate. In this way, *Ascaris* is able to recycle its redox cofactors under anaerobic conditions.

This viewpoint may also help one understand why more energy can be stored as depot fat than as an equal weight of glycogen: because these fuels contain different densities of reduced bonds. The point is illustrated by considering the oxidation of the repeating unit of carboyhydrate and of fat:

$$\text{Carbohydrate: } -[-\overset{\displaystyle H}{\underset{\displaystyle OH}{C}}-]- + \quad O_2 \rightarrow CO_2 + H_2O \tag{8}$$

$$\text{Fat: } -[-\overset{\displaystyle H}{\underset{\displaystyle H}{C}}-]- + 1\tfrac{1}{2}\, O_2 \rightarrow CO_2 + H_2O \tag{9}$$

The -CHOH- group of carbohydrate has two reduced bonds (one C—H and one C—C, counting half for each of the two C—C bonds shared with its two neighbors) and therefore requires two oxygen atoms for complete oxidation. It has a molar mass of 30 g/mol, so at 220 kJ/mol for each reduced bond, it yields $220 \times 2 \div 30$ kJ/g, or 14.7 kJ/g. The -CH_2- group of fat has three reduced bonds (requiring three oxygen atoms for complete oxidation) and a molar mass of 14 g/mol, yielding $220 \times 3 \div 14$ kJ/g, or 47 kJ/g. Hence, to a first approximation, fat on a dry weight basis contains about three times as much energy (47/14.7) as carbohydrate.

IV. UNDERSTANDING THE RESPIRATORY QUOTIENT

These same Eqs. (8) and (9) also show the chemical basis of the respiratory quotient (RQ). The RQ has been found particularly useful for estimating the nature of the body's fuel reserves (e.g., carbohydrate or fat), being used to support metabolism during exercise and rest. The RQ is calculated as the moles ratio of carbon dioxide produced to oxygen consumed. The ratio for pure carbohydrate metabolism is 1.0 [see Eq. (8)] and for fat metabolism it is 0.7 [see Eq. (9)]. Experimentally, the RQ at the end of a brief sprint is found to be close to 1.0, showing that carbohydrate is the major fuel source. In contrast, the RQ is closer to 0.7 at the end of a marathon run, indicating that metabolism is being supported principally by the oxidation of fat.

The RQ for protein metabolism is 0.8. This ratio is more complicated to explain because the constituent amino acids of protein are more variable in their empirical formula than carbohydrate or fat molecules, and they are incompletely oxidized during catabolism. This is because protein nitrogen is excreted in a reduced form as ammonia or urea. A more detailed explanation is given in the following section, where the difference between digestible and metabolizable energy is discussed.

V. CALCULATING THE CALORIC VALUE OF NUTRIENTS

One test of the validity of this approach for understanding the energy content of nutrients would be its capacity to predict the caloric value of macronutrients. All that should be required is the empirical formula (for estimating the oxygen uptake) or the structure (for counting the number of reduced bonds), and the molecular weight according to Eqs. (4) and (7).

The heats of combustion for the building blocks of fat, carbohydrate, and protein have been calculated using this approach. Values determined experimentally in a bomb calorimeter (and which appeared in the popular press in units of kcal as Calories) and those predicted by the bond-count approach [using Eq. (4)] are listed in Table II. The comparisons are remarkably close and justify confidence in the application of this method.

In studies of nutrition, the total energy of the diet is measured, and the digestibility of that energy is calculated after subtracting the energy excreted in the feces. When protein is fed, it is found that approximately 20% of the digestible energy is subsequently lost to the body through urinary excretion. Thus, the metabolizable energy of protein is about 80% of digestible energy. This difference between digestible energy and metabolizable energy is not observed for dietary carbohydrate or fat.

This effect presents another important concept that can be explained using the approach of counting reduced bonds. Carbon and hydrogen in digested nutrients are excreted from the body in a fully oxidized state (CO_2 and H_2O), whereas nitrogen is excreted from the body in a reduced state (as ammonia, urea, or uric acid). Hence, the 105 kcal/mol contributed to digestible energy by each reduced nitrogen bond in the food is not available to support metabolism.

In the example for alanine calculated in Table II, the metabolizable energy should be approximately 81% of digestible energy [$(220 \times 6) \div (220 \times 6 + 105 \times 3)$]. Thus, for a completely digested protein,

TABLE II
Predicted and Observed Heats of Combustion of Macronutrients

Macronutrient	M_r	Number of reduced bonds	Heats of combustion kJ/mol Predicted	kJ/mol Observed	kJ/g Predicted	kJ/g Observed
Fat component (e.g., palmitic acid) $CH_3(CH_2)_{14}COOH$	256	C−C = 15 C−H = 31 Σ = 46	220 × 46 = 10,120	10,031	10,120 ÷ 256 = 39.5	39.3
Carbohydrate component (e.g., glucose) $CHO(CHOH)_4CH_2OH$	180	C−C = 5 C−H = 7 Σ = 12	220 × 12 = 2,640	2,815	2,640 ÷ 180 = 14.7	15.7
Protein component (e.g., alanine) $NH_2CH(CH_3)COOH$	89	C−C = 2 C−H = 4 Σ = 6 + N−C = 1 N−H = 2 Σ = 3	220 × 6 + 105 × 3 = 1,635	1,623	1,635 ÷ 89 = 18.4	18.2

we would predict a metabolizable energy value of 14.9 kJ/g (18.4 × 0.81).

VI. CONCLUSION

The examination of the chemical basis for the energy content of macronutrients leads to the conclusion that the source of energy in the foods that nourish humankind is the reducing equivalents or valence electrons that bond their carbon and hydrogen atoms together. These electrons are initially derived from water and are sequestered during the reactions of photosynthesis. They move through the food chain until their release to sustain metabolism. Their ultimate fate is to reduce molecular oxygen to water, thereby completing the cycle. Energy release is directly proportional to oxygen consumption during combustion, whether the combustion occurs within a cell or an automobile engine.

With this understanding, many of the difficult concepts of bioenergetics can be unraveled. Moreover, two approaches become available for calculating the energy content of macronutrients, based only on a knowledge of the structure or empirical formula and the molecular weight of the material of interest.

In answering the question "Why are fats so much higher in calories than carbohydrate?" one can now reply with confidence: "because fat contains three times as many reduced bonds per unit weight!"

BIBLIOGRAPHY

Kharasch, M. (1929). Heats of combustion of organic compounds. *Bur. Standards J. Res.* **2**, 359–430.

Saz, H. J. (1981). Energy metabolism of parasitic helminths: Adaptions to parasitism. *Annu. Rev. Physiol.* **43**, 323–341.

Stokes, G. B. (1988). Estimating the energy content of nutrients. *Trends Biochem. Sci.* **13**, 422–424.

Nutritional Modulation of Aging Processes

EDWARD J. MASORO
University of Texas Health Science Center at San Antonio

GLOSSARY

Age-associated diseases Diseases that become more prevalent with advancing age or occur primarily during a specific age range

Gerontology Science of the study of aging and of the special problems of the aged

Life expectancy Mean length of life remaining for a population of a given age

Life span of a species Length of life of the longest-lived members of a species

Primary aging processes Intrinsic basic biological processes underlying aging

IT IS GENERALLY HELD THAT AGING IS AN INTRINSIC property of most, if not all, living organisms. Although the primary processes underlying aging have yet to be identified, an often debated issue is whether and to what extent aging can be influenced by extrinsic or environmental factors. Of these, nutrition is viewed by many as being a particularly important factor influencing the primary aging processes. The currently available information on the interaction of nutrition with the aging processes is the subject of this article.

I. CRITERIA INDICATING MODULATION OF THE AGING PROCESSES

A. Current State of Knowledge of Aging

What is aging? In basic biological terms we do not have an answer. This statement is hard for many to believe because most of us have an intuitive understanding of aging. For example, most observers have no problem in distinguishing between young animals and people and the old members of the species. However, this distinction is based on observable characteristics, of which some do indeed relate to the primary aging processes; others may be due to the prolonged exposure to environmental and/or life-style factors.

In the absence of knowledge of the primary aging processes, gerontologists have resorted to defining aging in terms of the decreasing ability of an organism to survive with advancing age. Although the probability of dying does increase with increasing adult age, factors other than age can also be, and usually are, involved in the death of an organism. Thus, a definition of aging based on mortality suffers from a lack of specificity. Indeed, the almost universal use of mortality by gerontologists when defining aging underscores our lack of knowledge of the nature of the basic aging processes.

Nevertheless, mortality data, in spite of these conceptual problems, are of great use to gerontology in general and to the area of focus of this article in particular. *Life expectancy* at birth (i.e., the mean length of life of a population of a given species born at a particular moment in time) is the most used in this regard. Unfortunately, life expectancy is of limited value for gerontology because it is markedly influenced by many factors in addition to the aging processes. For example, the increase in life expectancy of the United States population during the first half of the twentieth century appears to have resulted primarily from protecting the young from infectious disease and therefore bears little relationship to the aging processes. In contrast, the *maximum life span* of a species (the age at death of the longest-lived members) is felt to strongly relate to the aging processes. For example, the maximum life span of humans is about 110 years, of elephants about 70 years, of dogs about 20 years, and of the house mouse about 3 years. These data are interpreted by gerontologists as indicating that humans age the most slowly and mice the most rapidly; that is, it is felt that the rate of aging of a species inversely relates to the maximum life span of the species. Although the life expectancy of Americans increased markedly during the first half of the twentieth century, the life span of Americans did not. This fact is interpreted as indicating that the medical and technological advances during this time did not influence the aging processes of humans. However, the validity of using the maximum life span as an index of aging has recently been challenged. It is believed that the increase with advancing postmaturation age in the *age-specific mortality rate* (fraction of the population of a given age dying during a given time interval) is a better index of the rate of aging.

Physiological changes with age (most of which are of a deteriorative nature) have long been believed to be an inevitable consequence of the aging processes. However, recent studies suggest that the contribution of intrinsic aging processes to age-associated physiological deterioration may have been exaggerated. Factors such as diseases, life-style, and a variety of environmental influences may play major roles in the frequently observed age changes in the physiological systems. Indeed, the current challenge is to design studies that can sort out the contribution of the primary aging processes to the physiological and psychosocial changes observed with advancing age. Until this is accomplished, physiological activities cannot be viewed as valid markers of biological age. The current lack of valid nonchronologic markers of age is unfortunate, since their availability would greatly assist in the evaluation of the effects of nutrition on human aging.

Many diseases are age-associated in that they are more prevalent at advanced ages or in a specific age range. Examples are coronary heart disease, cerebrovascular disease, many kinds of tumors, Type II diabetes, Alzheimer's disease, and Parkinson's disease. However, the relationship of these diseases to the primary aging processes remains to be defined. Do aging and a particular disease process share the same time frame but relate in no other way? Is the disease process promoted by the aging processes? Is the disease process a part of the aging processes? Currently these questions cannot be answered for any age-associated disease. Research aimed at yielding answers should be of high priority. [*See* Alzheimer's Disease.]

B. Longevity as a Criterion

From the preceding discussion, it is evident that an increase in life expectancy from birth (or in the median length of life) in response to a manipulation does not provide strong evidence that the aging processes have been retarded. Yet the claim is often made that aging has been retarded on the basis of an increased life expectancy. For example, it is claimed that exercise retards the aging processes in humans, apparently because of the emerging evidence that exercise may result in an increased life expectancy. The basis of this increase in life expectancy (if indeed it does occur) appears to be the retarding influence that long-term exercise has on the occurrence of coronary heart disease and other chronic cardiovascular disease processes. Since the relationship of these disease processes to the aging processes has yet to be defined, the claim that exercise or any other manipulation retards the aging processes on the basis of an increase in life expectancy due to retardation of cardiovascular disease seems inappropriate. [*See* Longevity.]

If exercise were also to be shown to increase the maximum life span of humans, the case for its retarding action on aging processes would be much stronger. However, since the information on life expectancy and exercise has only recently emerged, it will be many years before the effect of exercise on maximum life span will be known. Moreover, even in the case of maximum life span of a species, it is possible that there may be factors other than aging that underlie an increase. Thus, although an increase in the maximum life span provides strong evidence that a manipulation has decelerated the aging processes, it alone cannot be considered to provide unequivocal proof. Showing that a manipulation decreases the increase in age-specific mortality rate would provide very strong evidence of retardation of aging processes.

C. Age-Associated Physiological Changes as a Criterion

Although many age-associated physiological changes occur, the extent to which a particular change is secondary to primary aging processes is not known. Thus, the ability of a manipulation to retard a particular age-associated physiological change is not strong evidence that the manipulation has influenced the primary aging processes. However, retarding a broad spectrum of age-associated physiological changes by a manipulation does indicate that the primary aging processes have been slowed.

D. Age-Associated Disease Processes as a Criterion

The relationship of a particular age-associated disease process to the aging processes has not been determined for any of the age-associated diseases. For this reason, retardation by a manipulation of a particular disease process (e.g., atherogenesis) cannot be used as a criterion that the aging processes have been decelerated. However, the retardation of a broad spectrum of age-associated diseases by a manipulation is indicative of the aging processes in general having been affected rather than the pathogenesis of only a particular disease.

E. Difficulties in Determining if Human Aging Has Been Slowed

To determine if a manipulation has slowed the aging processes is particularly difficult in the case of humans. For example, the effect on the maximum life span of humans is almost impossible to learn because of its length, being the longest life span of any mammalian species. Also the effect on the increase with age of the age-specific mortality rate has not been determined in humans for any nutritional manipulation and to do so will be difficult. Although providing less compelling evidence, it has been necessary with humans to rely on the effects on the physiological and disease processes as the major vehicles for evaluating the effects of an intervention on aging. Moreover, even this information is often difficult to obtain reliably because of the lack of control of, indeed the lack of an accurate knowledge of, the lifelong environment and life-style of human subjects.

II. NUTRITION AND THE AGING PROCESSES

A. Nutrition and Human Aging

The statement is often made that nutrition influences the aging processes of humans. Unfortunately, there is little hard evidence in support of or against this claim.

There is evidence that several age-associated diseases can be retarded by nutritional means. The progression of atherosclerosis appears to be influenced by dietary lipids and by caloric intake relative to energy expenditure. Inadequate dietary calcium and a lack of vitamin D have been proposed as contributors to osteoporosis; however, it has been questioned whether nutritional factors are major players in this disease process. The incidences of many types of cancer are age-associated and some (not totally convincing) evidence indicates that nutrition may play some role in the occurrence. Nutrition plays a role in several other age-associated pathologies, for example, dietary protein in renal failure, dietary sodium and calcium in hypertension, and caloric intake in adult-onset diabetes. [*See* Atherosclerosis; Lipids.]

If these disease processes are part of or influenced by the primary aging processes, it can be concluded that nutrition influences aging in humans. However, not enough is known at this time about the relationship of the aging processes to these diseases to permit such a conclusion to be drawn. Indeed, the specificity of the nutritional manipulation required to influence these diseases indicates that it is the pathogenesis of the specific disease process rather than aging in general that is being affected.

B. Food Restriction and the Aging of Laboratory Rodents

That nutrition can influence the primary aging processes of mammals is clearly shown by the action of long-term food restriction on laboratory rodents. Restricting the food intake of rats, mice, and hamsters to levels significantly below (30 to 60% below) that of those fed *ad libitum* (i.e., animals with continuous access to food) markedly decelerates the aging processes. The evidence supporting this claim includes the effects on longevity, on a wide spectrum of age changes in the physiological systems, and on a broad range of age-associated disease processes.

Probably the strongest of this evidence is the effect of food restriction on the maximal life span and on the increase with age in age-specific mortality rate. This is illustrated by the survival curves in Fig. 1 for a group of 115 *ad libitum*-fed male Fischer 344 rats (Group A) and a group (Group R) of 115 food-restricted male Fischer 344 rats (fed 60% of the mean intake of the Group A rats from 6 weeks of age on). Although the median length of life and the life expectancy of the Group R rats were much greater than for the Group A rats, the finding of most importance

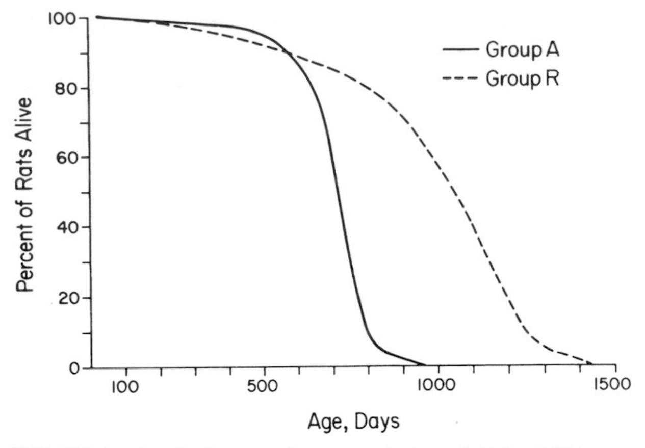

FIGURE 1 Survival curves for a population of 115 *ad libitum*-fed male Fischer 344 rats (Group A) and 115 food-restricted (60% of the *ad libitum* intake) male Fischer 344 rats (Group R). [Reproduced, with permission, from B. P. Yu *et al.* (1982). *J. Gerontol.* 37, 130.]

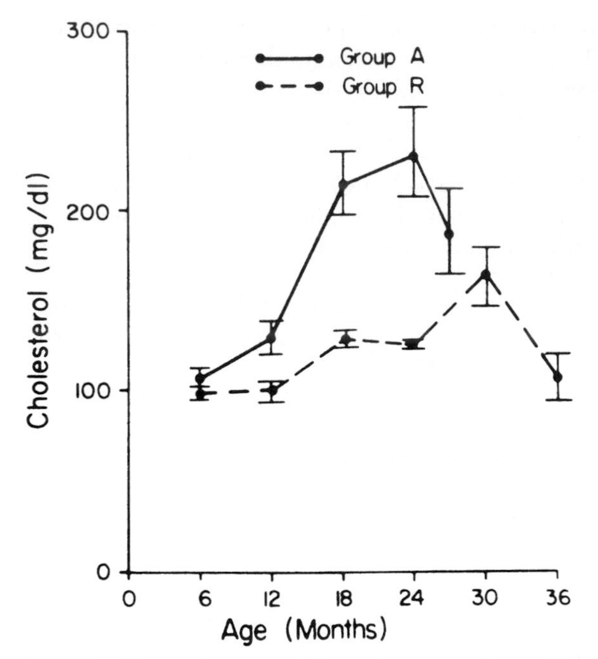

FIGURE 2 Changes in serum cholesterol concentrations with age in *ad libitum*-fed male Fischer 344 rats (Group A) and in food-restricted (60% of *ad libitum* intake) male Fischer 344 rats (Group R). [Reproduced, with permission, from G. U. Liepa *et al.* (1980). *Am. J. Physiol.* **238**, E253.]

in regard to the aging processes is the marked increase in the maximum life span. Indeed, when the last of the 115 Group A rats died, approximately 70% of the Group R rats were still living. Assessment of the increase in age-specific mortality rate with increasing postmaturational age reveals that food restriction markedly slows the rate of increase.

Food restriction also retards most age-associated changes in the physiological systems. Typical of this action are the data in Fig. 2 on the serum cholesterol concentrations of male Fischer 344 rats *ad libitum*-fed or food-restricted from 6 weeks of age on. Although food restriction did not influence the serum cholesterol levels of young rats (6 months of age), it markedly decreased the increase in serum cholesterol that occurs with advancing age. A broad range of physiological processes are similarly protected from age change. Examples are the loss of gamma crystallins from the lens of the eye, the loss of neurotransmitter receptors from the brain, the loss in immune function, the loss in the ability to learn a complex maze, the loss in the ability of fat cells to respond to hormones, and the loss of reproductive function. The breadth of these effects provides strong evidence for the view that food restriction is slowing the primary aging processes. [*See* Cholesterol.]

Food restriction has also been found to delay the occurrence and in some cases prevent the clinical expression of many of the age-associated diseases of rodents. Examples are chronic kidney disease, heart disease, cancer, benign tumors, and autoimmune diseases. The breadth of these effects points to an action on the primary aging processes rather than on the specific pathogenesis of a particular disease process.

Research at several laboratories indicates that these many actions of food restriction are due to the reduction in caloric intake and not to the reduced intake of specific dietary components. It has further been shown that the actions of food restriction are not due to a retardation of growth and development or to a reduction in body fat content or to a reduction in metabolic rate. Current research is being focused on the biochemical and neuroendocrinologic mechanisms by which food restriction retards aging processes in rodents. It is felt that knowledge of these mechanisms will provide insights in regard to the nature of the primary aging processes.

Although other nutritional manipulations in rodents have been found to influence specific age-associated problems (e.g., reducing dietary protein retards the development of kidney failure in rats), none of these manipulations of specific nutrients has the broad influence on longevity and the many other aspects of age changes as food restriction. Thus, of these dietary manipulations, food restriction appears to be unique in its ability to decelerate the primary aging processes.

C. Relevance of Rodent Findings to the Human Situation

If the premise is correct that the primary aging processes are similar in all mammalian species, it is likely that food restriction will decelerate the aging processes in all mammals including humans. However, food restriction has been studied adequately only in rodent species because of the long life span of most other mammalian species. This long life span makes it difficult to execute the relevant research because of the large financial resources that would be required and because of the unwillingness of investigators to invest the large amount of time (much of their scientific life span) to complete such a study. There are, however, two ongoing studies on nonhuman primates that if completed will provide such information in ten years or so for species closely related to humans.

Since on theoretical grounds it seems likely that food restriction would be effective in retarding aging in humans, should this regimen be employed by people? The answer is no, because of the risks of malnutrition. For food restriction regimens in rodents to be successful in retarding the aging processes without any adverse consequences, it is necessary to produce undernutrition without causing malnutrition. Such can be readily accomplished with mice and rats because of our knowledge of their lifelong nutritional needs and because of the extensive experience that has been gained in the experimental use of food restriction in these species. Neither the knowledge nor the experience is available in regard to human use, thus posing the risk inherent in using any untested procedure. [*See* Malnutrition.]

Does this mean that the extensive research that has been done on the effects of food restriction on rodent aging is of no value for humans? No, but its value resides in learning the mechanisms by which food restriction slows the aging processes of the rodent models. An understanding of these mechanisms is likely to provide the knowledge base needed for the development of safe and effective interventions for human aging. Moreover, such interventions are likely to be more palatable for humans than the restriction of food intake for most of a person's life span.

III. CONCLUSIONS

Whether proper nutrition can slow the aging processes of humans is difficult to establish because of our lack of knowledge of the nature of the primary aging processes. On the basis of current knowledge, the following criteria should be met for it to be concluded that a manipulation has decelerated the aging processes: an increase in life expectancy; an increase in maximum life span; slowing the increase with age in age-specific mortality rate; retardation of most age-associated changes in the physiological systems; and retardation of most age-associated diseases. Based on these criteria, there is little hard evidence to support the often made claim that nutrition can decelerate the aging processes of humans. The only manipulation that has been shown to slow the aging processes in a mammalian species is that of food restriction (apparently because of caloric restriction) in rodent species. Indeed, the findings on food restriction clearly show that nutrition can markedly influence the aging processes of mammals. Although food restriction is likely to have similar actions in all mammalian species including humans, it has not been directly tested in long-lived species because of the high cost in resources and time. Fortunately, tests on nonhuman primates are now under way. At this time, the use of food restriction by humans should be discouraged because of the possible adverse effects of malnutrition that might be encountered during the lifelong use of an uncharted procedure. Exploration of the mechanisms by which food restriction retards the aging processes of rodents is encouraged because the knowledge from such studies may well lead to the development of safe and palatable interventions of human aging.

BIBLIOGRAPHY

Ausman, L. M., and Russell, R. M. (1990). Nutrition and aging. *In* "Handbook of the Biology of Aging" (E. L. Schneider and J. W. Rowe, eds.), 3rd Ed. Academic Press, San Diego, California.

Committee on Chemical Toxicity and Aging (1987). "Aging in Today's Environment." National Academy Press, Washington, D.C.

Finch, C. E. (1990). "Longevity, Senescence, and the Genome." Univ. of Chicago Press, Chicago.

Masoro, E. J. (1993). Dietary restriction and aging. *J. Am. Geriatr. Soc.* **41**, 967.

Rowe, J. W., and Kahn, R. L. (1987). Human aging: Usual and successful. *Science* **237**, 143.

Yu, B. P. (ed.) (1994). "Modulation of Aging Processes by Dietary Restriction." CRC Press, Boca Raton, Florida.

Nutritional Quality of Foods and Diets

DENNIS D. MILLER
Cornell University

GLOSSARY

Carbohydrates Compounds containing carbon, hydrogen, and oxygen in the approximate proportion of 1 : 2 : 1 (CH_2O); food carbohydrates include sugars (sucrose, glucose, fructose, lactose), starches, and most forms of dietary fiber

Cholesterol Lipid present in foods of animal origin that can also be synthesized by the body; it is necessary for healthy cell membranes and is a precursor for steroid hormones and bile salts

Essential mineral elements Elements other than carbon, hydrogen, oxygen, and nitrogen that are constituents of plant or animal tissues and are found in the ash remaining after incineration of the tissue; they are also necessary for the normal metabolic functioning of the body

Essential nutrients Chemical substances required for metabolism that cannot be synthesized within the body and must be present in the diet; they include vitamins, minerals, and essential amino and fatty acids

Fats Lipids that contain fatty acids; food fats are composed primarily of triacylglycerols

Fatty acids Lipids composed of molecules containing a hydrocarbon chain and a carboxylic acid group; linoleic and α-linolenic acids are fatty acids that are essential nutrients

Kilocalorie Unit used to quantify the energy value of food (1 kilocalorie = 1000 calories); in common usage, calorie refers to the kilocalorie

Lipids Substances in foods and living organisms that are insoluble in water and soluble in organic solvents such as chloroform and diethyl ether; they include triacylglycerols (esters composed of three fatty acids and one glycerol residue) and cholesterol

Monounsaturated fatty acids Fatty acids with one carbon–carbon double bond per molecule

Nutritional quality Capacity of foods or diets to provide essential nutrients, supply energy, support growth, promote health, and prevent disease

Polyunsaturated fatty acids Fatty acids with two or more carbon–carbon double bonds per molecule

Proteins Large molecules composed of amino acids linked together by peptide bonds; they contain the elements carbon, hydrogen, oxygen, nitrogen, and sulfur

Saturated fatty acids Fatty acids with no double bonds between carbons

Vitamins Organic substances necessary for the normal metabolic functioning of the body; dietary requirements for vitamins are small, falling in the milligram range or less

THE IMPORTANCE OF DIET IN HEALTH PROMOtion and disease prevention is widely recognized by consumers, physicians, nutritionists, and policy makers. In the United States and other highly developed, industrialized countries with abundant food supplies, the nutritional quality of diets and national food supplies has a larger impact on health than quantity of available food. The relationship between diet and health is complex, and this makes it difficult to define nutritional quality. Clearly, however, nutritional quality is determined by the nutrient composition of the diet. In the first half of this century, nutrition

ENCYCLOPEDIA OF HUMAN BIOLOGY, Second Edition, VOLUME 6.

research identified the nutrients essential for the prevention of nutrient deficiency diseases and established recommended daily intakes for many of them. Government policy makers, the food industry, and nutrition educators developed policies and programs based on this research that resulted in the virtual elimination of the deficiency diseases that were so prevalent in the early part of the century. For example, fortification of salt with iodine and milk with vitamin D is credited with preventing goiter and rickets, respectively. Nutrition educators used the concept of the basic four food groups to effectively teach people how to make appropriate food choices. Sadly, in recent years, the alarming increases in poverty, hunger, substance abuse, school dropout rates, and teen-age pregnancy may cause a return of nutrient deficiency diseases in the United States. Nevertheless, with the conquest of nutrient deficiency diseases and infectious diseases, chronic diseases such as heart disease, cancer, and osteoporosis have become the principal causes of morbidity and mortality in the developed countries. Appropriately, the focus of nutrition research has shifted to questions related to the relationships between diet and chronic disease. As these relationships are being clarified, policies and programs are being developed to guide food industry practices and consumer food choices in the direction of improved nutritional quality of foods and diets. Current dietary guidelines issued by the government and health agencies stress the need for reduced intakes of total fat, saturated fat, cholesterol, and salt and increased intakes of calcium, iron, and complex carbohydrates and fiber. Nutritional quality of a food or diet, then, is determined by the concentrations of essential nutrients and other beneficial food constituents and the relative amounts of carbohydrate, fat, cholesterol, and sodium. Chemical form as well as concentration of food components may have significant effects on nutritional quality.

I. PROBLEMS WITH THE AMERICAN DIET

A. Prevalence of Diet-Related Diseases

Cause-of-death statistics provide clues about imbalances in national diets. Numbers and causes of deaths in the United States in 1990 are listed in Table I. Though factors associated with deaths in most of the categories listed are multiple, strong evidence indicates that diet is a factor in heart disease, cancer, cerebrovascular disease (stroke), and diabetes mellitus. J. M. McGinnis and W. H. Foege estimated that about half of the total U.S. deaths have attributable environmental or behavioral causes. Their estimates of the number of deaths caused by various factors are shown in Table II. Tobacco causes more deaths than any other attributable cause, but diet/activity pattern is a close second. Clearly, hundreds of thousands of premature deaths could be prevented or delayed if people would quit smoking, improve their diets, and get more physical exercise.

TABLE I
Causes of Death in the United States in 1990

Cause	No. of deaths
Heart disease	720,000
Cancer	505,000
Cerebrovascular disease	144,000
Accidents	92,000
Chronic obstructive pulmonary disease	87,000
Pneumonia and influenza	80,000
Diabetes mellitus	48,000
Suicide	31,000
Chronic liver disease and cirrhosis	26,000
HIV infection	20,000
Total deaths	2,148,000

Source: J. M. McGinnis and W. H. Foege (1993). Actual causes of death in the United States. *J. Am. Med. Assoc.*, Nov. 10, pp. 2207–2212.

Poor diets may also contribute to conditions that, although not directly life-threatening, result in a lower quality of life. In the United States, obesity, iron deficiency, osteoporosis, and dental disease are other significant public health problems that are diet related. The problem of obesity is especially alarming because of its high and increasing prevalence (approximately one-third of U.S. adults are more than 20% above their ideal weight and the mean body weight of U.S. adults has increased nearly 8 pounds in the last 15 years). Obese people are at greater risk for high blood pressure, high blood cholesterol, and diabetes and therefore are at greater risk for heart disease.

If diet is a factor in so many diseases, what specifically is wrong with the American diet? Hundreds of research studies have been conducted to answer this question and we now have good evidence that our diets are too rich in some components and lacking in others. Using the results of these studies, the Joint Nutrition Monitoring Evaluation Committee

TABLE II
Actual Causes of Death (for Deaths with Attributable Causes) in the United States in 1990

Cause	Deaths	
	Estimated no.[a]	Percentage of total deaths
Tobacco	400,000	19
Diet/activity patterns	300,000	14
Alcohol	100,000	5
Microbial agents	90,000	4
Toxic agents	60,000	3
Firearms	35,000	2
Sexual behavior	30,000	1
Motor vehicles	25,000	1
Illicit use of drugs	20,000	<1
Total	1,060,000	50

Source: J. M. McGinnis and W. H. Foege (1993). Actual causes of death in the United States. *J. Am. Med. Assoc.*, Nov. 10, pp. 2207–2212.

[a]Composite approximation drawn from studies that use different approaches to derive estimates, ranging from actual counts (e.g., firearms) to population attributable risk calculations (e.g., tobacco). Numbers over 100,000 are rounded to the nearest 100,000; over 50,000, rounded to the nearest 10,000; below 50,000, rounded to the nearest 5000.

(JNMEC) identified nutrients that are consumed in either excessive or inadequate amounts by significant numbers of the population. In a 1986 report titled "Nutrition Monitoring in the United States," the JNMEC singled out the following dietary components as being consumed in either excessive or inadequate amounts:

High dietary consumption	Low dietary consumption
Food energy	Vitamin C
Total fat	Calcium
Saturated fatty acids	Iron
Cholesterol	Fluoride
Sodium	
Alcohol	

Although there have undoubtedly been changes in the American diet since 1986, these nutrients remain problems for many Americans.

II. CRITERIA FOR ASSESSING NUTRITIONAL QUALITY

Criteria for assessing nutritional quality may be established for the evaluation of individual foods, diets, or national food supplies. Evaluation of diets is most appropriate, because nutritional status (nutritional well-being) is determined by total nutrient intake. Individual foods contain varying amounts and proportions of nutrients. No single food, with the exception of breast milk for infants <6 months of age and a few specially formulated liquid formulas, contains all of the essential nutrients in the proportions and amounts necessary to meet nutrient needs. Nevertheless, nutritional comparisons among individual foods are useful and scientifically valid, particularly when evaluating the effects of food processing or when comparing foods within a food group. Monitoring of the nutritional quality of the national food supply is useful because it provides information about the contributions of various food groups to nutrient supplies. Moreover, data describing the nutritional composition of national food supplies have been extremely valuable to epidemiologists who study relationships between diet and health. For example, the positive correlation between the concentration of saturated fat in national food supplies and the prevalence of coronary heart disease in different countries is strong evidence in support of the hypothesis that diets high in saturated fat increase risk of heart disease.

Of course, just knowing the concentrations of nutrients in food is of little value if we do not know the nutrient requirements of people consuming the foods. Many nutrients are present in only trace amounts (micrograms or milligrams per serving) and yet these amounts may be perfectly adequate if the requirement is low. Thus, assessments of nutritional quality must be based on nutritional standards or guidelines. The most widely used standards in the United States are the Recommended Dietary Allowances and the "Dietary Guidelines for Americans."

A. Recommended Dietary Allowances

An adequate diet must contain all of the essential nutrients in the proper amounts and proportions. When intakes are below requirements, signs and symptoms of nutrient deficiency disease develop. On the other hand, excessive intakes of nutrients may also result in adverse health consequences. The relationship between the intake of one class of nutrients (trace minerals) and the concentration of these nutrients in the body has been described (Fig. 1). When intakes are deficient, the concentrations in the body are below levels necessary for optimal physiological function. For example, when intakes of dietary iron are below the requirement for an extended period of time, the supply of iron to the bone marrow becomes insufficient to support hemoglobin synthesis at an adequate rate and iron deficiency anemia develops. On the other hand, when iron intakes are excessively high for an extended period of time, a condition of iron overload may develop. Iron overload is the term used to describe the toxic accumulation of iron in tissues. Iron overload can damage the liver, heart, and other vital organs. Fortunately, the range of safe and adequate intakes for most nutrients is relatively wide. This is because physiological mechanisms operate to allow the body to adapt to low or high intakes of nutrients. To use the iron example again, when iron intakes are low, the body adapts by absorbing a higher percentage of dietary iron from the gastrointestinal tract. Conversely, individuals with adequate or high levels of body iron absorb iron less efficiently. The net result is that concentrations of nutrients in body tissues are maintained within a narrow range even when intakes may vary quite widely. The adaptive processes responsible for maintaining steady-state levels of nutrients are collectively referred to as homeostatic mechanisms. Without them, intakes would have to be held to very narrow ranges. It is important to understand, however, that homeostatic regulation has its limits, and it is possible to overwhelm the system with intakes that are either too high or too low (see Fig. 1).

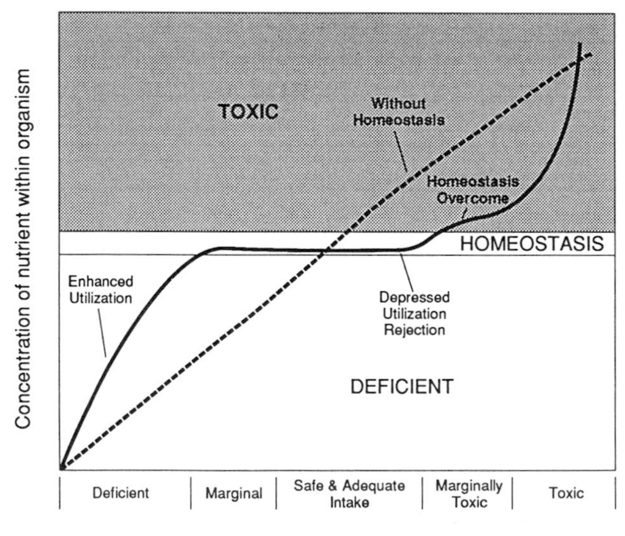

FIGURE 1 Body processes collectively referred to as homeostasis operate to maintain tissue levels of nutrients within an acceptable, often narrow, range, even when dietary intakes vary. High levels of nutrients may be toxic, and low levels result in impaired function. [From W. Mertz (1983). Our most unique nutrients. *Nutrition Today,* **18**(2), 6–10, 27–29, and 32–33. © by Williams & Wilkins, Baltimore.]

Nutrient requirements are determined by measuring the amount of a nutrient that must be ingested to meet the nutrient needs of an *individual.* Nutrient requirements vary among individuals. Therefore, recommendations for nutrient intakes, which are based on research on individual subjects, are set higher than the average requirement to ensure that the needs of most individuals will be met. The Food and Nutrition Board of the National Research Council, National Academy of Sciences, has established Recommended Dietary Allowances (RDAs) for the U.S. population (Table III). The Food and Nutrition Board defines RDAs as "the levels of intake of essential nutrients that, on the basis of scientific knowledge, are judged by the Food and Nutrition Board to be adequate to meet the known nutrient needs of practically all healthy persons." This means that individuals with nutrient intakes below the RDA do not necessarily have an inadequate intake, because individual requirements are usually lower than the RDA. However, when intakes are significantly below the RDA, the probability that intakes by a given individual are inadequate is increased. Because of an inadequate research base, RDAs have been established for only about half of the known essential nutrients. In its two latest revisions, the Food and Nutrition Board published "estimated safe and adequate intake" levels for several additional essential nutrients (Table IV). The Food and Nutrition Board revises the RDAs periodically to reflect new research findings. The most recent edition of the RDAs at the time of this writing was published in 1989.

B. Dietary Guidelines

Criteria based on the RDAs are useful for evaluating the contribution that a food or diet will make toward meeting requirements for protein, vitamins, and minerals. Expanding knowledge in the second half of this century has made it clear that criteria based solely on RDAs is insufficient for evaluating the potential impact that diet may have on health. As a result, emphasis in dietary advice has shifted from protein, vitamins, and minerals to food components that are associated with chronic diseases such as heart disease, cancer, osteoporosis, and hypertension. These include fat, saturated fat, cholesterol, sodium, calcium, complex carbohydrates, dietary fiber, and antioxidants. Since the 1960s, numerous publications by a variety of government and private health organizations have recommended adjustments in the typical American diet to improve public health. Generally, these recommendations have been to reduce fat, saturated fat, cholesterol, sodium, alcohol, and sugar and to increase dietary fiber and complex carbohydrates. Tables V, VI, and VII summarize recommendations published by the Surgeon General of the U.S. Public Health Service; the Committee on Diet and Health of the Food and Nutrition Board, Commission on Life Sciences, National Academy of Sciences; and the U.S. Department of Agriculture (USDA)–U.S. Department of Health and Human Services (DHHS). The recommendations in the first two cases are based on a comprehensive review of the literature by expert committees. Readers interested in more detail on relationships between diet and health should consult the "Surgeon General's Report on Nutrition and Health" and "Diet and Health, Implications for Reducing Chronic Disease Risk."

III. NUTRITIONAL QUALITY OF MAJOR FOOD GROUPS

This section is organized according to the food groups in the Food Guide Pyramid (see Fig. 3). Foods representative of each group are listed in Table VIII.

A. Bread, Cereal, Rice, and Pasta Group

Foods in this group are prepared from cereal grains. Important food cereals include wheat, rice, maize (corn), oats, barley, and rye. Cereals have been major dietary staples in most cultures throughout recorded history. Cereal grains are compact, stable, and relatively easy to ship and store. They may be processed into a wide variety of food products.

Whole-grain cereal products are good sources of energy (mainly as starch); protein (cereals contain 7–15% protein by weight); trace minerals include iron, zinc, and magnesium; vitamins, especially thiamin, riboflavin, niacin, and folic acid; and dietary fiber. Cereals contain very little fat, calcium, or vitamins A, D, C, and B_{12}.

A large proportion of wheat and rice is milled before it is consumed. Milling involves the removal of bran and germ from the intact kernel leaving the white, starchy endosperm. This process results in a marked reduction of B vitamins, trace minerals, and dietary fiber, which are concentrated in the bran and germ portions of the kernel. Iron, riboflavin, niacin, and thiamin are added back in enriched cereal products. The nutrient compositions of selected cereal foods are listed in Table IX.

B. Vegetable Group

The vegetables group encompasses a wide variety of plant species, whose nutrient composition varies considerably. All vegetables are sources of dietary fiber and, as a group, vegetables contain vitamins and minerals that are not found in sufficient concentrations in other food groups.

Dark green vegetables are good sources of vitamin A and C, riboflavin, folic acid, iron, calcium, magnesium, and potassium. Deep yellow vegetables are rich in vitamin A. Dry beans and peas contain thiamin, folic acid, iron, magnesium, phosphorus, zinc, potassium, protein, and starch. Starchy vegetables are high in starch and contain varying levels of niacin, vitamin B_6, zinc, and potassium. Table X lists the nutrient composition of some representative vegetables.

Foods in this group, along with foods from the fruit group, have been the focus of extensive research in recent years because of their health promotion and disease prevention effects. There is now widespread agreement that adequate intakes of fruits and vegetables reduces risks for many cancers and possibly heart disease. Unfortunately, intakes by the majority of Americans of vegetables and fruits are far below recommendations.

C. Fruit Group

Most fruits are low in calories and are good sources of fiber and some vitamins and minerals. Citrus fruits are particularly rich in ascorbic acid (vitamin C) and deep yellow fruits such as apricots, cantaloupes, and mangos are good sources of vitamin A. Table XI lists the nutrient composition of some representative fruits.

D. Milk, Yogurt, and Cheese Group

The most striking nutritional attribute of dairy products is their calcium content. Three-fourths of all

TABLE III

Food and Nutrition Board, National Academy of Sciences–National Research Council Recommended Dietary Allowances,[a] Revised 1989; Designed for the Maintenance of Good Nutrition of Practically All Healthy People in the United States

		Weight[b]		Height[b]			Fat-soluble vitamins			
Category	Age (yr) or condition	kg	lb	cm	in	Protein (g)	Vitamin A (μg re[c])	Vitamin D (μg)[d]	Vitamin E (mg α-te[e])	Vitamin K (μg)
Infants	0.0–0.5	6	13	60	24	13	375	7.5	3	5
	0.5–1.0	9	20	71	28	14	375	10	4	10
Children	1–3	13	29	90	35	16	400	10	6	15
	4–6	20	44	112	44	24	500	10	7	20
	7–10	28	62	132	52	28	700	10	7	30
Males	11–14	45	99	157	62	45	1,000	10	10	45
	15–18	66	145	176	69	59	1,000	10	10	65
	19–24	72	160	177	70	58	1,000	10	10	70
	25–50	79	174	176	70	63	1,000	5	10	80
	51+	77	170	173	68	63	1,000	5	10	80
Females	11–14	46	101	157	62	46	800	10	8	45
	15–18	55	120	163	64	44	800	10	8	55
	19–24	58	128	164	65	46	800	10	8	60
	25–50	63	138	163	64	50	800	5	8	65
	51+	65	143	160	63	50	800	5	8	65
Pregnant						60	800	10	10	65
Lactating	1st 6 months					65	1,300	10	12	65
	2nd 6 months					62	1,200	10	11	65

Source: Subcommittee on the Tenth Edition of the RDAs (1989) by the National Academy of Sciences, National Academy Press, Washington, D.C.

[a]The allowances, expressed as average daily intakes over time, are intended to provide for individual variations among most normal persons as they live in the United States under usual environmental stresses. Diets should be based on a variety of common foods to provide other nutrients for which human requirements have been less well defined.

[b]Weights and heights of reference adults are actual medians for the U.S. population of the designated age, as reported by NHANES II.

the calcium in the American diet comes from dairy foods. Dairy products are also good sources of protein, riboflavin, vitamins A and B_{12}, and thiamin. Virtually all of the fluid milk sold in the United States is fortified with vitamin D. Many dairy products contain significant amounts of fat, saturated fat, cholesterol, and sodium. Table XII lists the nutrient composition of some representative dairy products.

E. Meat, Poultry, Fish, Dry Beans, Eggs, and Nuts Group

Foods in this group are rich in a wide variety of nutrients. They are particularly good sources of protein, iron, zinc, niacin, and vitamin B_6. Foods of animal origin are the only source of B_{12} in the diet. The iron in meat, poultry, and fish has high bioavailability, and these foods enhance the absorption of iron from other foods when consumed together in a meal. The iron in eggs and legumes, on the other hand, is less bioavailable, and eggs and legumes may actually depress iron absorption from other foods consumed along with them. The fat content of foods in this group is highly variable. Lean cuts of meat and poultry with the skin removed are quite low in fat. Hamburger, luncheon meats, and sausages tend to be quite high in fat. In addition, the fat is more saturated than most vegetable fats. Egg yolks are very high in cholesterol and also contain a considerable amount of fat. The cholesterol content of most meat products is similar on a weight basis because the cholesterol content of lean and fat tissues is approximately the same. Dry beans and peas are very low in fat, but nuts, seeds, and peanut butter may be higher in fat than many meat

Water-soluble vitamins							Minerals						
Vitamin C (mg)	Thiamin (mg)	Riboflavin (mg)	Niacin (mg ne[f])	Vitamin B_6 (mg)	Folate (μg)	Vitamin B_{12} (μg)	Calcium (mg)	Phosphorus (mg)	Magnesium (mg)	Iron (mg)	Zinc (mg)	Iodine (μg)	Selenium (μg)
30	0.3	0.4	5	0.3	25	0.3	400	300	40	6	5	40	10
35	0.4	0.5	6	0.6	35	0.5	600	500	60	10	5	50	15
40	0.7	0.8	9	1.0	50	0.7	800	800	80	10	10	70	20
45	0.9	1.1	12	1.1	75	1.0	800	800	120	10	10	90	20
45	1.0	1.2	13	1.4	100	1.4	800	800	170	10	10	120	30
50	1.3	1.5	17	1.7	150	2.0	1,200	1,200	270	12	15	150	40
60	1.5	1.8	20	2.0	200	2.0	1,200	1,200	400	12	15	150	50
60	1.5	1.7	19	2.0	200	2.0	1,200	1,200	350	10	15	150	70
60	1.5	1.7	19	2.0	200	2.0	800	800	350	10	15	150	70
60	1.2	1.4	15	2.0	200	2.0	800	800	350	10	15	150	70
50	1.1	1.3	15	1.4	150	2.0	1,200	1,200	280	15	12	150	45
60	1.1	1.3	15	1.5	180	2.0	1,200	1,200	300	15	12	150	50
60	1.1	1.3	15	1.6	180	2.0	1,200	1,200	280	15	12	150	55
60	1.1	1.3	15	1.6	180	2.0	800	800	280	15	12	150	55
60	1.0	1.2	13	1.6	180	2.0	800	800	280	10	12	150	55
70	1.5	1.6	17	2.2	400	2.2	1,200	1,200	320	30	15	175	65
95	1.6	1.8	20	2.1	280	2.6	1,200	1,200	355	15	19	200	75
90	1.6	1.7	20	2.1	260	2.6	1,200	1,200	340	15	16	200	75

[c]Retinol equivalent: 1 re = 1 μg retinol or 6 μg β-carotene.
[d]As cholecalciferol: 10 μg cholecalciferol = 400 IU of vitamin D.
[e]α-Tocopherol equivalent: 1 mg d-α-tocopherol = 1 α-te.
[f]Niacin equivalent: 1 ne = 1 mg niacin or 60 mg of dietary tryptophan.

TABLE IV

Estimated Safe and Adequate Daily Dietary Intakes of Selected Vitamins and Minerals[a]

		Vitamins		Trace elements[b]				
Category	Age (yr)	Biotin (μg)	Pantothenic acid (mg)	Copper (mg)	Manganese (mg)	Fluoride (mg)	Chromium (μg)	Molybdenum (μg)
Infants	0–0.5	10	2	0.4–0.6	0.3–0.6	0.1–0.5	10–40	15–30
	0.5–1	15	3	0.6–0.7	0.6–1.0	0.2–1.0	20–60	20–40
Children and adolescents	1–3	20	3	0.7–1.0	1.0–1.5	0.5–1.5	20–80	25–50
	4–6	25	3–4	1.0–1.5	1.5–2.0	1.0–2.5	30–120	30–75
	7–10	30	4–5	1.0–2.0	2.0–3.0	1.5–2.5	50–200	50–150
	11+	30–100	4–7	1.5–2.5	2.0–5.0	1.5–2.5	50–200	75–250
Adults		30–100	4–7	1.5–3.0	2.0–5.0	1.5–4.0	50–200	75–250

Source: Subcommittee on the Tenth Edition of the RDAs (1989) by the National Academy of Sciences, National Academy Press, Washington, D.C.

[a]Because there is less information on which to base allowances, these figures are not given in the main table of RDA and are provided here in the form of ranges of recommended intakes.

[b]Because the toxic levels for many trace elements may be only several times the usual intakes, the upper levels for the trace elements given in this table should not be habitually exceeded.

TABLE V

Dietary Recommendations from "The Surgeon General's Report on Nutrition and Health"

Issues for Most People

Fats and cholesterol: Reduce consumption of fat (especially saturated fat) and cholesterol. Choose foods relatively low in these substances, such as vegetables, fruits, whole-grain foods, fish, poultry, lean meats, and low-fat dairy products. Use food preparation methods that add little or no fat.

Energy and weight control: Achieve and maintain a desirable body weight. To do so, choose a dietary pattern in which energy (caloric) intake is consistent with energy expenditure. To reduce energy intake, limit consumption of foods relatively high in calories, fats, and sugars, and minimize alcohol consumption. Increase energy expenditure through regular and sustained physical activity.

Complex carbohydrates and fiber: Increase consumption of whole-grain foods and cereal products, vegetables (including dried beans and peas), and fruits.

Sodium: Reduce intake of sodium by choosing foods relatively low in sodium and limiting the amount of salt added in food preparation and at the table.

Alcohol: To reduce the risk for chronic disease, take alcohol only in moderation (no more than two drinks a day), if at all. Avoid drinking any alcohol before or while driving, operating machinery, taking medications, or engaging in any other activity requiring judgment. Avoid drinking alcohol while pregnant.

Other Issues for Some People

Fluoride: Community water systems should contain fluoride at optimal levels for prevention of tooth decay. If such water is not available, use other appropriate sources of fluoride.

Sugars: Those who are particularly vulnerable to dental caries (cavities), especially children, should limit their consumption and frequency of use of foods high in sugars.

Calcium: Adolescent girls and adult women should increase consumption of foods high in calcium, including low-fat dairy products.

Iron: Children, adolescents, and women of childbearing age should be sure to consume foods that are good sources of iron, such as lean meats, fish, certain beans, and iron-enriched cereals and whole-grain products. This issue is of special concern for low-income families.

Source: DHHS Public Health Service (PHS) Publication No. 88-50210 (1988).

TABLE VI

Dietary Recommendations of the Committee on Diet and Health, Food and Nutrition Board, Commission on Life Sciences, National Research Council (1989)

- Reduce total fat intake to ~30% of calories. Reduce saturated fatty acid intake to <10% of calories and the intake of cholesterol to <300 mg daily. The intake of fat and cholesterol can be reduced by substituting fish, poultry without skin, lean meats, and low- or nonfat dairy products for fatty meats and whole-milk dairy products; by choosing more vegetables, fruits, cereals, and legumes; and by limiting oils, fats, egg yolks, and fried and other fatty foods.
- Every day eat five or more servings of a combination of vegetables and fruits, especially green and yellow vegetables and citrus fruits. Also, increase intake of starches and other complex carbohydrates by eating six or more daily servings of a combination of breads, cereals, and legumes.
- Maintain protein intake at moderate levels.
- Balance food intake and physical activity to maintain appropriate body weight.
- The committee does not recommend alcohol consumption. For those who drink alcoholic beverages, the committee recommends limiting consumption to <1 ounce of pure alcohol in a single day. This is the equivalent of two cans of beer, two small glasses of wine, or two average cocktails. Pregnant women should avoid alcoholic beverages.
- Limit total daily intake of salt (sodium chloride) to ≤6 g. Limit the use of salt in cooking and avoid adding it to food at the table. Salty, highly processed salty, salt-preserved, and salt-pickled foods should be consumed sparingly.
- Maintain adequate calcium intake.
- Avoid taking dietary supplements in excess of the RDA in any one day.
- Maintain an optimal intake of fluoride, particularly during the years of primary and secondary tooth formation and growth.

Source: Committee on Diet and Health, Food and Nutrition Board, Commission on Life Sciences, National Research Council (1989). "Diet and Health—Implications for Reducing Chronic Disease Risk." National Academy Press, Washington, D.C.

TABLE VII
USDA–DHHS Dietary Guidelines for Americans

- Eat a variety of foods
- Balance the food you eat with physical activity—maintain or improve your weight
- Choose a diet with plenty of grain products, vegetables, and fruits
- Choose a diet low in fat, saturated fat, and cholesterol
- Choose a diet moderate in sugars
- Choose a diet moderate in salt and sodium
- If you drink alcoholic beverages, do so in moderation

Source: USDA–DHHS (1995). "Nutrition and Your Health—Dietary Guidelines for Americans." Fourth Edition. Home and Garden Bulletin No. 232. U.S. Department of Agriculture, Human Nutrition Information Service, Hyattsville, Maryland.

products. Table XIII lists the nutrient composition of selected foods from this group.

F. Fats, Oils, and Sweets Group

The foods in this group are notable for their high calorie density and low levels of nutrients. The USDA recommends that foods from this group be used sparingly. The nutrient compositions of selected foods from this group are listed in Table XIV.

IV. NUTRITIONAL QUALITY OF FOOD COMPONENTS

As discussed earlier, different foods, even within a given food group, may have markedly different nutrient compositions and, therefore, selecting the recommended number of servings from each food group does not necessarily assure a balanced diet. Moreover, composition alone does not always give a clear picture of the potential physiological effect that a food or diet may have. For example, it is well known that coconut oil will produce increases in serum cholesterol levels when it is consumed in large amounts. Corn oil, on the other hand, appears to have the opposite effect. Both oils are nearly 100% triacylglycerol (fat), but the fatty acids that make up the triacylglycerol are different. Therefore, it is necessary to characterize foods not only according to the amount of a given nutrient but, often, by the chemical form of that nutrient as well.

A. Protein

The nutritional quality of a protein is determined by its amino acid composition and digestibility. The amino acid pattern in proteins of high nutritional quality matches the pattern of amino acids required for protein synthesis in the cells of the animal or human consuming the protein. If just one essential amino acid in a food protein is present in suboptimal amounts, the quality of the protein will be lower than that of an ideal protein. Therefore, protein quality is determined by the limiting amino acid in the protein. In general, proteins in animal foods such as meat, milk, and eggs have high quality, whereas proteins from plant sources usually contain suboptimal levels of one or more amino acids. Fortunately, the limiting amino acids in plant proteins vary with the source, so it is generally possible to "complement" proteins by mixing foods with different limiting amino acids. For example, cereal proteins are low in lysine but relatively high in sulfur amino acids, whereas legumes are low in sulfur amino acids but high in lysine. Therefore, mixing a cereal with a legume will result in a combination of relatively high protein quality.

B. Lipids

Lipids are defined as the class of naturally occurring compounds that are soluble in nonpolar organic solvents such as chloroform or diethyl ether. Lipids in foods include triacylglycerols, phospholipids, cholesterol, fat-soluble vitamins, and plant sterols. [*See* Lipids.]

TABLE VIII

Food Groups and Some Foods They Contain

Breads, Cereal, Rice, and Pasta Group

Whole Grain			Enriched		
Brown rice	Oatmeal	Whole wheat bread and rolls	Bagels	Farina	Muffins
Buckwheat groats	Popcorn	Whole wheat cereals	Biscuits	French bread	Noodles
Bulgur	Pumpernickel bread	Whole wheat crackers	Corn bread	Grits	Pancakes
Corn tortillas	Ready-to-eat cereals	Whole wheat pasta	Corn muffins	Hamburger rolls	Pasta
Graham crackers	Rye crackers		Cornmeal	Hot dog buns	Ready-to-eat cereals
Granola			Crackers	Italian bread	Rice
			English muffins	Macaroni	White bread and rolls

Fruit Group

Citrus, Melons, Berries			Other Fruits		
Blueberries	Honeydew melon	Raspberries	Apple	Grapes	Pineapple
Cantaloupe	Kiwifruit	Strawberries	Apricot	Guava	Plantain
Citrus juices	Lemon	Tangerine	Banana	Mango	Plum
Cranberries	Orange	Watermelon	Cherries	Nectarine	Pomegranate
Grapefruit			Dates	Papaya	Prune
			Figs	Peach	Raisins
			Fruit juices	Pear	

Vegetable Group

Dark Green			Deep Yellow	Starchy	
Beet greens	Dandelion greens	Romaine lettuce	Carrots	Breadfruit	Lima beans
Broccoli	Endive	Spinach	Pumpkin	Corn	Potatoes
Chard	Escarole	Turnip greens	Sweet potatoes	Green peas	Rutabaga
Chicory	Kale	Watercress	Winter squash	Hominy	Taro
Collard greens	Mustard greens				

Other Vegetables			
Artichokes	Cabbage	Green beans	Radishes
Asparagus	Cauliflower	Green peppers	Summer squash
Beans and alfalfa sprouts	Celery	Lettuce	Tomatoes
Beets	Chinese cabbage	Mushrooms	Turnips
Brussels sprouts	Cucumbers	Okra	Vegetable juices
	Eggplant	Onions (mature and green)	Zucchini

Meat, Poultry, Fish, Dry Beans Eggs, and Nuts Group

Meat, Poultry, and Fish				Dry Beans, Eggs, and Nuts	
Beef	Ham	Pork	Veal	Dry beans and peas (legumes)	Nuts and seeds
Chicken	Lamb	Shellfish	Luncheon meats	Eggs	Peanut butter
Fish	Organ meats	Turkey	Sausage		Tofu

Milk, Yogurt, and Cheese Group

Low-Fat Milk Products		Other Milk Products with More Fat or Sugar			
Buttermilk	Low-fat plain yogurt	American cheese	Chocolate milk	Fruit yogurt	Swiss cheese
Low-fat milk (1%, 2%)	Skim milk	Cheddar cheese	Flavored yogurt	Processed cheeses	Whole milk

Fats, Oils, and Sweets Group

Fats		Sweets		
Bacon, salt pork	Mayonnaise	Candy	Jam	Popsicles and ices
Butter	Mayonnaise-type salad dressing	Corn syrup	Jelly	Sherbets
Cream (dairy, nondairy)	Salad dressing	Frosting	Maple syrup	Soft drinks and colas
Cream cheese	Shortening	Fruit drinks, ades	Marmalade	Sugar (white and brown)
Lard	Sour cream	Gelatin desserts	Molasses	
Margarine	Vegetable oil	Honey		

TABLE IX

Nutrient Composition of Bread, Cereal, Rice, and Pasta Group[a]

Quantity	Item	Grams (g)	Cals	Fat-T (g)	Fat-S (g)	Fat-M (g)	Fat-P (g)	Chol (mg)	Sod (mg)	Potas (mg)	Carb (g)	Fiber (g)	Sugar (g)
1 piece	Whole wheat bread	35	86	1	0	1	0	0	184	88	16	2	1
1 piece	Soft white bread	28	76	1	0	0	0	1	143	30	14	1	1
1 oz-wt	Kelloggs Corn Flakes cereal	28	110	0	0	0	0	0	290	26	24	1	2
.5 cup	Long-grain brown rice-cooked-hot	98	108	1	0	0	0	0	5	42	23	2	0
.5 cup	Long-grain white rice-cooked-hot	103	134	0	0	0	0	0	1	36	29	1	0
4 each	Saltine crackers	12	52	1	0	1	0	0	156	15	9	0	0
2 each	Plain pancakes-mix-prepared	54	105	1	0	0	0	6	340	95	20	1	4
1 each	Butter croissant bread	57	231	12	0	3	1	43	424	67	26	1	2
1 each	Cake doughnut	50	211	12	2	5	4	19	273	64	25	1	8
1 each	Fruit-filled Danish pastry	94	335	16	3	10	2	19	333	110	45		
2 each	Rich chocolate chip cookies	20	96	5	2	2	0	0	63	27	13	1	9
1 piece	Apple cobbler—3 × 3 inch	104	199	6	1	3	2	1	304	87	35	1	24

Quantity	Item	Prot (g)	A-Tot (RE)	Vit C (mg)	Calc (mg)	Iron (mg)	Vit D (μg)	E-αTE (mg)	B1 (mg)	B2 (mg)	B3 (mg)	B6 (mg)	Fola (μg)
1 piece	Whole wheat bread	3	0	0	25	1	0	0	.1	.1	1.3	.1	17.5
1 piece	Soft white bread	2	0	0	24	1	0	0	.1	.1	.9	0	9.8
1 oz-wt	Kelloggs Corn Flakes cereal	2	375	15	1	2	1	0	.4	.4	5.0	.5	100
.5 cup	Long-grain brown rice-cooked-hot	3	0	0	10	0	0	0	.1	0	1.5	.1	3.9
.5 cup	Long-grain white rice-cooked-hot	3	0	0	10	1	0	0	.2	0	1.5	.1	3.1
4 each	Saltine crackers	1	0	0	14	1	0	0	.1	.1	.6	0	3.7
2 each	Plain pancakes-mix-prepared	3	5	0	68	1	1	0	.1	.1	.9	.1	4.9
1 each	Butter croissant bread	5	78	0	21	1	0	0	.2	.1	1.3	0	16
1 each	Cake doughnut	3	9	0	22	1	0	1	.1	.1	.9	0	4
1 each	Fruit-filled Danish pastry	5	24	2	22	1		1	.2	.2	1.8	.1	15
2 each	Rich chocolate chip cookies	1	0	0	5	1		1	.0	.1	.5	0	1.8
1 piece	Apple cobbler—3 × 3 inch	2	76	0	31	1	0	1	.1	.1	.7	0	3.3

Quantity	Item	B12 (μg)	Phos (mg)	Iodin (μg)	Magn (mg)	Zinc (mg)	Copp (mg)
1 piece	Whole wheat bread	0	80		30	1	.1
1 piece	Soft white bread	0	28	2	6	0	0
1 oz-wt	Kelloggs Corn Flakes cereal	0	18		3	0	0
.5 cup	Long-grain brown rice-cooked-hot	0	81		42	1	.1
.5 cup	Long-grain white rice-cooked-hot	0	44	5	12	1	.1
4 each	Saltine crackers	0	13		3	0	0
2 each	Plain pancakes-mix-prepared	.1	180		11	0	.1
1 each	Butter croissant bread	.2	60		9	0	0
1 each	Cake doughnut	.1	135		10	0	.1
1 each	Fruit-filled Danish pastry	.2	69		14	0	.1
2 each	Rich chocolate chip cookies	0	22		6	0	0
1 piece	Apple cobbler—3 × 3 inch	0	45	2	6	0	.1

Source: The Food Processor Plus Nutrition Software (1994). ESHA Research, Salem, Oregon.

[a]Cals, kilocalories; Fat-T, total fat; Fat-S, saturated fat; Fat-M, monounsaturated fat; Fat-P, polyunsaturated fat; Chol, cholesterol; Sod, sodium; Potas, potassium; Carb, carbohydrate, Prot, protein; A-Tot, total vitamin A activity in retinol equivalents; Vit C, vitamin C; Calc, calcium; Vit D, vitamin D; E-αTE, vitamin E in α-tocopherol equivalents; B1, thiamin; B2, riboflavin, B3, niacin; B6, vitamin B6; Fola, folic acid; Phos, phosphorous; Iodin, iodine; Magn, magnesium; Copp, copper.

TABLE X
Nutrient Composition of Vegetable Group[a]

Quantity	Item	Grams (g)	Cals	Fat-T (g)	Fat-S (g)	Fat-M (g)	Fat-P (g)	Chol (mg)	Sod (mg)	Potas (mg)	Carb (g)	Fiber (g)	Sugar (g)
1 each	Peeled potato-boiled	135	116	0	0	0	0	0	7	443	27	2	1
58 g	French fries-frozen-restaurant fried	58	183	10	3	2	4	0	125	425	23	2	0
75 g	Whole carrots-raw	75	32	0	0	0	0	0	26	243	8	2	5
75 g	Carrots-raw slices-boiled	75	34	0	0	0	0	0	50	170	8	2	3
75 g	Carrots-canned-drained	75	17	0	0	0	0	0	181	135	4	1	2
.5 cup	Yellow corn-boiled	82	89	1	0	0	0	0	14	204	21	2	2
.5 cup	Yellow corn-canned-drained	82	66	1	0	0	0	0	265	160	15	2	2
.5 cup	Yellow corn-frozen-boiled	82	66	0	0	0	0	0	4	114	17	2	1
.5 cup	Green peas-raw-boiled	80	67	0	0	0	0	0	2	217	13	4	5
.5 cup	Green peas-canned-drained	85	59	0	0	0	0	0	186	147	11	3	3
.5 cup	Green peas-frozen-boiled	80	62	0	0	0	0	0	70	134	11	4	4
.5 cup	Broccoli pieces-raw	44	12	0	0	0	0	0	12	143	2	1	1
.5 each	Medium whole tomato	62	13	0	0	0	0	0	6	137	3	1	2
.5 cup	Iceberg/crisphead lettuce-chopped	28	4	0	0	0	0	0	3	44	1	0	0
.5 cup	Spinach-raw-chopped	28	6	0	0	0	0	0	22	156	1	1	0
1 each	Small sweet potato-peeled after baking	60	62	0	0	0	0	0	6	209	15	2	7

Quantity	Item	Prot (g)	A-Tot (RE)	Vit C (mg)	Calc (mg)	Iron (mg)	Vit D (μg)	E-αTE (mg)	B1 (mg)	B2 (mg)	B3 (mg)	B6 (mg)	Fola (μg)
1 each	Peeled potato-boiled	2	0	10	11	0	0	0	.1	0	1.8	.4	12
58 g	French fries-frozen-restaurant fried	2	0	6	11	0	0	0	.1	0	1.9	.1	16.8
75 g	Whole carrots-raw	1	2109	7	20	0	0	0	.1	0	.7	.1	10.5
75 g	Carrots-raw slices-boiled	1	1840	2	23	0	0	0	0	0	.4	.2	10.4
75 g	Carrots-canned-drained	0	1033	2	19	0	0	0	0	0	.4	.1	6.9
.5 cup	Yellow corn-boiled	3	18	5	2	1	0	0	.2	.1	1.3	0	38
.5 cup	Yellow corn-canned-drained	2	13	7	4	1	0	0	0	.1	1.0	0	39.9
.5 cup	Yellow corn-frozen-boiled	2	20	2	2	0	0	0	.1	.1	1.1	.1	18.7
.5 cup	Green peas-raw-boiled	4	48	11	22	1	0	0	.2	.1	1.6	.2	50.5
.5 cup	Green peas-canned-drained	4	65	8	17	1	0	0	.1	.1	.6	.1	37.7
.5 cup	Green peas-frozen-boiled	4	53	8	19	1	0	0	.2	.1	1.2	.1	46.9
.5 cup	Broccoli pieces-raw	1	68	41	21	0	0	0	0	.1	.3	.1	31.3
.5 each	Medium whole tomato	1	38	12	3	0	0	0	0	0	.4	0	9.3
.5 cup	Iceberg/crisphead lettuce-chopped	0	9	1	5	0	0	0	0	0	.1	0	15.7
.5 cup	Spinach-raw-chopped	1	188	8	28	1	0	0	0	.1	.2	.1	54.5
1 each	Small sweet potato-peeled after baking	1	1309	15	17	0	0	0	0	.1	.4	.1	13.6

Quantity	Item	B12 (μg)	Phos (mg)	Iodin (μg)	Magn (mg)	Zinc (mg)	Copp (mg)
1 each	Peeled potato-boiled	0	54	4	27	0	.2
58 g	French fries-frozen-restaurant fried	0	54		20	0	.1
75 g	Whole carrots-raw	0	33		11	0	0
75 g	Carrots-raw slices-boiled	0	23		10	0	.1
75 g	Carrots-canned-drained	0	18		6	0	.1
.5 cup	Yellow corn-boiled	0	85		26	0	0
.5 cup	Yellow corn-canned-drained	0	53		16	0	0
.5 cup	Yellow corn-frozen-boiled	0	39		15	0	0
.5 cup	Green peas-raw-boiled	0	94	2	31	1	.1
.5 cup	Green peas-canned-drained	0	57		15	1	.1
.5 cup	Green peas-frozen-boiled	0	72	2	23	1	.1
.5 cup	Broccoli pieces-raw	0	29	1	11	0	0
.5 each	Medium whole tomato	0	15		7	0	0
.5 cup	Iceberg/crisphead lettuce-chopped	0	6	1	3	0	0
.5 cup	Spinach-raw-chopped	0	14	1	22	0	0
1 each	Small sweet potato-peeled after baking	0	33	2	12	0	.1

Source: The Food Processor Plus Nutrition Software (1994). ESHA Research, Salem, Oregon.
[a]See Table IX for abbreviations.

TABLE XI
Nutrient Composition of Fruit Group[a]

Quantity	Item	Grams (g)	Cals	Fat-T (g)	Fat-S (g)	Fat-M (g)	Fat-P (g)	Chol (mg)	Sod (mg)	Potas (mg)	Carb (g)	Fiber (g)	Sugar (g)
1 each	Medium apple w/peel	138	81	0	0	0	0	0	0	159	21	3	17
1 each	Banana	114	105	1	0	0	0	0	1	451	27	2	21
1 each	Medium orange	131	62	0	0	0	0	0	2	237	16	2	12
24 each	Thompson seedless grapes	120	85	1	0	0	0	0	6	222	21	1	21
.75 cup	Grape juice-bottled/canned-unsweetened	190	116	0	0	0	0	0	2	251	29	0	28
.75 cup	Orange juice-unsweet-frozen conc + water	187	84	0	0	0	0	0	6	355	20	0	20
.75 cup	Apple juice-canned/bottled	186	88	0	0	0	0	0	8	221	22	0	21
.5 cup	Peaches in heavy syrup-canned	128	95	0	0	0	0	0	5	118	26	1	24
.5 cup	Peaches in juice-canned	124	55	0	0	0	0	0	7	159	14	2	13
.5 cup	Cantaloupe/muskmelon-cubes	80	28	0	0	0	0	0		247	7	1	7

Quantity	Item	Prot (g)	A-Tot (RE)	Vit C (mg)	Calc (mg)	Iron (mg)	Vit D (μg)	E-αTE (mg)	B1 (mg)	B2 (mg)	B3 (mg)	B6 (mg)	Fola (μg)
1 each	Medium apple w/peel	0	7	8	10	0	0	1	0	0	.1	.1	3.9
1 each	Banana	1	9	10	7	0	0	0	.1	.1	.6	.7	21.8
1 each	Medium orange	1	27	70	52	0	0	0	.1	.1	.4	.1	39.7
24 each	Thompson seedless grapes	1	9	13	13	0	0	1	.1	.1	.4	.1	4.7
.75 cup	Grape juice-bottled/canned-unsweetened	1	2	0	17	0	0	0	0	.1	.5	.1	4.9
.75 cup	Orange juice-unsweet-frozen conc + water	1	15	73	17	0	0	0	.1	0	.4	.1	81.8
.75 cup	Apple juice-canned/bottled	0	0	2	13	1	0	0	0	0	.2	.1	.2
.5 cup	Peaches in heavy syrup-canned	1	43	4	4	0	0	1	0	0	.8	0	4.1
.5 cup	Peaches in juice-canned	1	47	4	7	0	0	0	0	0	.7	0	4.2
.5 cup	Cantaloupe/muskmelon-cubes	1	257	34	9	0	0	0	0	0	.5	.1	13.6

Quantity	Item	B12 (μg)	Phos (mg)	Iodin (μg)	Magn (mg)	Zinc (mg)	Copp (mg)
1 each	Medium apple w/peel	0	10		7	0	.1
1 each	Banana	0	23	9	33	0	.1
1 each	Medium orange	0	18		13	0	.1
24 each	Thompson seedless grapes	0	16	1	7	0	.1
.75 cup	Grape juice-bottled/canned-unsweetened	0	21		19	0	.1
.75 cup	Orange juice-unsweet-frozen conc + water	0	30		19	0	.1
.75 cup	Apple juice-canned/bottled	0	13		6	0	0
.5 cup	Peaches in heavy syrup-canned	0	14		6	0	.1
.5 cup	Peaches in juice-canned	0	21		9	0	.1
.5 cup	Cantaloupe/muskmelon-cubes	0	14	3	9	0	0

Source: The Food Processor Plus Nutrition Software (1994). ESHA Research, Salem, Oregon.
[a]See Table IX for abbreviations.

TABLE XII
Nutrient Composition of Milk, Yogurt, and Cheese Group[a]

Quantity	Item	Grams (g)	Cals	Fat-T (g)	Fat-S (g)	Fat-M (g)	Fat-P (g)	Chol (mg)	Sod (mg)	Potas (mg)	Carb (g)	Fiber (g)	Sugar (g)
1 cup	Whole milk	244	150	8	5	2	0	33	120	371	11	0	11
1 cup	2% lofat milk	244	121	5	3	1	0	18	122	376	12	0	12
1 cup	1% lofat milk	244	102	3	2	1	0	10	123	381	12	0	12
1 cup	Nonfat skim milk	245	86	0	0	0	0	4	126	407	12	0	12
1 cup	Lofat yogurt-plain	245	155	4	2	1	0	15	172	573	17	0	17
1 cup	Lofat yogurt-fruit	245	250	3	2	1	0	10	143	478	47	0	47
1.5 oz-wt	Cheddar cheese-diced	43	171	14	9	4	0	45	264	42	1	0	1
1.5 oz-wt	American processed cheese	43	160	13	8	4	0	40	608	69	1	0	1
.5 cup	Creamed cottage cheese-small curd	105	108	5	3	1	0	16	425	89	3	0	3
.5 cup	Vanilla ice cream	66	133	7	4	2	0	29	53	131	16	0	15
.5 cup	Healthy Choice lofat ice cream-vanilla	71	100	2	1		0	5	50	254	18	0	17

Quantity	Item	Prot (g)	A-Tot (RE)	Vit C (mg)	Calc (mg)	Iron (mg)	Vit D (μg)	E-αTE (mg)	B1 (mg)	B2 (mg)	B3 (mg)	B6 (mg)	Fola (μg)
1 cup	Whole milk	8	76	2	290	0	2	0	.1	.4	.2	.1	12.2
1 cup	2% lofat milk	8	139	2	298	0	2	0	.1	.4	.2	.1	12.4
1 cup	1% lofat milk	8	144	2	300	0	2	0	.1	.4	.2	.1	12.4
1 cup	Nonfat skim milk	8	149	2	301	0	2	0	.1	.3	.2	.1	12.7
1 cup	Lofat yogurt-plain	13	39	2	448	0	0	0	.1	.5	.3	.1	27.4
1 cup	Lofat yogurt-fruit	11	27	2	372	0	0	0	.1	.4	.2	.1	22.8
1.5 oz-wt	Cheddar cheese-diced	11	129	0	307	0	0	0	0	.2	0	0	7.7
1.5 oz-wt	American processed cheese	9	123	0	261	0	0	0	0	.2	0	0	3.3
.5 cup	Creamed cottage cheese-small curd	13	51	0	63	0	0	0	0	.2	.1	.1	12.8
.5 cup	Vanilla ice cream	2	77	0	85	0	0	0	0	.2	.1	0	3.3
.5 cup	Healthy Choice lofat ice cream-vanilla	3	60	2	100				.1	.2			

Quantity	Item	B12 (μg)	Phos (mg)	Iodin (μg)	Magn (mg)	Zinc (mg)	Copp (mg)
1 cup	Whole milk	.9	228	56	33	1	0
1 cup	2% lofat milk	.9	232	59	33	1	0
1 cup	1% lofat milk	.9	235	59	34	1	0
1 cup	Nonfat skim milk	.9	247	56	28	1	0
1 cup	Lofat yogurt-plain	1.4	353	87	43	2	0
1 cup	Lofat yogurt-fruit	1.1	292	45	36	2	.2
1.5 oz-wt	Cheddar cheese-diced	.4	218	16	12	1	0
1.5 oz-wt	American processed cheese	.3	316	13	9	1	0
.5 cup	Creamed cottage cheese-small curd	.7	138	27	6	0	0
.5 cup	Vanilla ice cream	.3	69	27	9	0	0
.5 cup	Healthy Choice lofat ice cream-vanilla		141				

Source: The Food Processor Plus Nutrition Software (1994). ESHA Research, Salem, Oregon.
[a]See Table IX for abbreviations.

TABLE XIII
Nutrient Composition of Meat, Poultry, Fish, Dry Beans, and Nuts Group[a]

Quantity	Item	Grams (g)	Cals	Fat-T (g)	Fat-S (g)	Fat-M (g)	Fat-P (g)	Chol (mg)	Sod (mg)	Potas (mg)	Carb (g)	Fiber (g)	Sugar (g)
3 oz-wt	Beef filet mignon steak-broiled-trimmed	85	179	9	3	3	0	71	54	357	0	0	0
3 oz-wt	Ground beef-lean-fried-medium done	85	234	16	6	7	1	71	66	254	0	0	0
3 oz-wt	Ground beef-regular-fried-medium done	85	260	19	8	8	1	76	71	255	0	0	0
3 oz-wt	Skinless chicken breast-roasted	85	140	3	1	1	1	72	63	218	0	0	0
3 oz-wt	Chicken breast-roasted	85	167	7	2	3	1	71	60	208	0	0	0
3 oz-wt	Whitefish fillet-baked/broiled	85	146	6	1	2	2	66	55	345	0	0	0
3 oz-wt	Channel catfish fillet-breaded-fried	85	195	11	3	5	3	69	239	289	7	0	0
.5 cup	Red kidney beans-dry-boiled	89	113	0	0	0	0	0	1	357	20	7	2
.5 cup	Refried beans/frijoles-canned	127	136	1	1	1	0	0	536	497	23	7	2
1 each	Large whole egg-fried in margarine	46	92	7	2	3	1	211	162	61	1	0	1
2 tbs	Smooth peanut butter	32	188	16	3	8	5	0	153	231	7	2	3

Quantity	Item	Prot (g)	A-Tot (RE)	Vit C (mg)	Calc (mg)	Iron (mg)	Vit D (μg)	E-αTE (mg)	B1 (mg)	B2 (mg)	B3 (mg)	B6 (mg)	Fola (μg)
3 oz-wt	Beef filet mignon steak-broiled-trimmed	24	0	0	6	3		0	.1	.3	3.3	.4	5.9
3 oz-wt	Ground beef-lean-fried-medium done	21	0	0	9	2	0	0	0	.2	4.1	.2	7.7
3 oz-wt	Ground beef-regular-fried-medium done	20	0	0	9	2	0	0	0	.1	5.0	.2	7.7
3 oz-wt	Skinless chicken breast-roasted	26	5	0	13	1	0	0	.1	.1	11.7	.5	3.4
3 oz-wt	Chicken breast-roasted	25	23	0	12	1	0	0	.1	.1	10.8	.5	3.4
3 oz-wt	Whitefish fillet-baked/broiled	21	33	0	28	0		0	.1	.1	3.3	.3	14.5
3 oz-wt	Channel catfish fillet-breaded-fried	15	7	0	37	1	11	2	.1	.1	1.9	.2	14.1
.5 cup	Red kidney beans-dry-boiled	8	0	1	25	3	0	0	.1	.1	.5	.1	115
.5 cup	Refried beans/frijoles-canned	8	0	8	58	2	0	1	.1	.1	.6	.1	106
1 each	Large whole egg-fried in margarine	6	114	0	25	1	1	1	0	.2	0	.1	17.5
2 tbs	Smooth peanut butter	8	0	0	11	1	0	2	0	0	4.2	.1	25

Quantity	Item	B12 (μg)	Phos (mg)	Iodin (μg)	Magn (mg)	Zinc (mg)	Copp (mg)
3 oz-wt	Beef filet mignon steak-broiled-trimmed	2.2	202		26	5	.2
3 oz-wt	Ground beef-lean-fried-medium done	1.9	135		17	4	.1
3 oz-wt	Ground beef-regular-fried-medium done	2.3	145		17	4	.1
3 oz-wt	Skinless chicken breast-roasted	.3	194		25	1	0
3 oz-wt	Chicken breast-roasted	.3	182		23	1	0
3 oz-wt	Whitefish fillet-baked/broiled	.8	294		36	1	.1
3 oz-wt	Channel catfish fillet-breaded-fried	1.6	184		23	1	.1
.5 cup	Red kidney beans-dry-boiled	0	126		40	1	.2
.5 cup	Refried beans/frijoles-canned	0	107		49	2	.5
1 each	Large whole egg-fried in margarine	.4	89	29	5	1	0
2 tbs	Smooth peanut butter	0	103		50	1	.2

Source: The Food Processor Plus Nutrition Software (1994). ESHA Research, Salem, Oregon.
[a]See Table IX for abbreviations.

TABLE XIV
Nutrient Composition of Fats, Oils, and Sweets Group[a]

Quantity	Item	Grams (g)	Cals	Fat-T (g)	Fat-S (g)	Fat-M (g)	Fat-P (g)	Chol (mg)	Sod (mg)	Potas (mg)	Carb (g)	Fiber (g)	Sugar (g)
1 tsp	Butter	5	34	4	2	1	0	10	39	1	0	0	0
1 tsp	Fleischmann's Corn Oil margarine-stick	5	34	4	1	2	1	0	32	2	0	0	0
1 tsp	Fleischmann's Corn Oil margarine-tub	5	34	4	1	2	1	0	32	2	0	0	0
1 tsp	Corn oil	5	40	5	1	1	3	0	0	0	0	0	0
1 tbs	Mayonnaise	14	99	11	1	3	6	8	78	5	0	0	0
1 tbs	Ranch salad dressing	15	55	6	1	2	2	6	65	20	1	0	1
1 oz-wt	Cream cheese	28	99	10	6	3	0	31	84	34	1	0	1
1 tsp	Jelly	6	16	0	0	0	0	0	2	4	4	0	4
12 fl oz	Diet Cola/Coke w/aspartame-bottle/can	355	4	0	0	0	0	0	21	0	0	0	0
12 fl oz	Cola-type soda pop	370	152	0	0	0	0	0	15	4	39	0	39
1 each	Milk chocolate candy bar	44	226	14	8	4	0	10	36	169	26	2	23

Quantity	Item	Prot (g)	A-Tot (RE)	Vit C (mg)	Calc (mg)	Iron (mg)	Vit D (μg)	E-αTE (mg)	B1 (mg)	B2 (mg)	B3 (mg)	B6 (mg)	Fola (μg)
1 tsp	Butter	0	36	0	1	0	0	0	0	0	0	0	.1
1 tsp	Fleischmann's Corn Oil margarine-stick	0	47	0	1	0	1	1	0	0	0	0	.1
1 tsp	Fleischmann's Corn Oil margarine-tub	0	47	0	1	0	1	1	0	0	0	0	.1
1 tsp	Corn oil	0	0	0	0	0	0	1	0	0	0	0	0
1 tbs	Mayonnaise	0	12	0	2	0	0	4	0	0	0	.1	1.1
1 tbs	Ranch salad dressing	0	11	0	15	0	0	1	0	0	0	0	.8
1 oz-wt	Cream cheese	2	124	0	23	0	0	0	0	.1	0	0	3.7
1 tsp	Jelly	0	0	0	0	0	0	0	0	0	0	0	.1
12 fl oz	Diet Cola/Coke w/aspartame-bottle/can	0	0	0	14	0	0	0	0	.1	0	0	0
12 fl oz	Cola-type soda pop	0	0	0	11	0	0	0	0	0	0	0	0
1 each	Milk chocolate candy bar	3	21	0	84	1	2	1	0	.1	.1	0	3.1

Quantity	Item	B12 (μg)	Phos (mg)	Iodin (μg)	Magn (mg)	Zinc (mg)	Copp (mg)
1 tsp	Butter	0	1		0	0	0
1 tsp	Fleischmann's Corn Oil margarine-stick	0	1		0		
1 tsp	Fleischmann's Corn Oil margarine-tub	0	1		0	0	
1 tsp	Corn oil	0	0		0	0	0
1 tbs	Mayonnaise	0	4		0	0	0
1 tbs	Ranch salad dressing	0	13		2	0	0
1 oz-wt	Cream cheese	.1	29		2	0	0
1 tsp	Jelly	0	0		0	0	0
12 fl oz	Diet Cola/Coke w/aspartame-bottle/can	0	32		4	0	0
12 fl oz	Cola-type soda pop	0	44		4	0	0
1 each	Milk chocolate candy bar	.2	95		26	1	.2

Source: The Food Processor Plus Nutrition Software (1994). ESHA Research, Salem, Oregon.
[a]See Table IX for abbreviations.

Triacylglycerols (also known as triglycerides) are esters of fatty acids and glycerol. The fatty acids may vary in length from 4 carbons to more than 24 carbons. Food fatty acids may be saturated (no double bonds), monounsaturated (one double bond per molecule), or polyunsaturated (two or more double bonds). Table XV lists the fatty acids commonly found in foods. Concentrations of free fatty acids (i.e., not esterified to glycerol) in foods are generally very low. Triacylglycerols make up the bulk of the lipid fraction in most foods, approaching 100% of the total weight of vegetable oils. Other lipid components make up a much smaller fraction but may still have a significant nutritional impact.

Phospholipids are made up of fatty acids, glycerol, phosphate, and, with few exceptions, a nitrogen-containing group. Lecithin is an example. Phospholipids play a structural role in cell membranes and lipoproteins. [*See* Phospholipid Metabolism.]

Cholesterol is an alcohol that is structurally unrelated to triacylglycerol. It is found only in foods of animal origin. Egg yolks, liver, and brain are especially high in cholesterol. [*See* Cholesterol.]

There is substantial evidence that fatty acid composition of dietary fat as well as total dietary fat is related to serum cholesterol levels. Saturated fatty acids, particularly C14:0 and C16:0, tend to raise serum cholesterol. Polyunsaturated fatty acids (PUFAs) and monounsaturated fatty acids (MUFAs) have the opposite effect and tend to lower serum cholesterol. In general, animal fats tend to be more saturated and vegetable oils more unsaturated with a few exceptions. Coconut oil, for example, is a highly saturated vegetable oil. Though PUFA is effective in lowering serum cholesterol levels, there is concern that high intakes of PUFA may not be desirable. Some studies with animals have shown that PUFA acts as a promoter for some cancers. Recent studies have shown that MUFAs are as effective as PUFAs in lowering serum cholesterol and may be preferable to PUFAs. The fatty acid compositions of some fats and oils are listed in Table XVI.

C. Carbohydrates

Carbohydrates are composed of the elements carbon, hydrogen, and oxygen in the approximate ratio of 1 carbon:2 hydrogens:1 oxygen (CH_2O). They are frequently classified as simple or complex. Simple carbohydrates are sugars and include primarily mono- and disaccharides. Glucose, fructose, and galactose are examples of monosaccharides. Disaccharides are

TABLE XV
Fatty Acids Commonly Found in Foods

Common name	Shorthand representation	Food source
Saturated fatty acids		
Butyric	$C_{4:0}$[a]	Butterfat
Caproic	$C_{6:0}$	Butterfat
Caprylic	$C_{8:0}$	Coconut oil
Capric	$C_{10:0}$	Coconut oil
Lauric	$C_{12:0}$	Coconut oil
Myristic	$C_{14:0}$	Butterfat/coconut oil
Palmitic	$C_{16:0}$	Most fats and oils
Stearic	$C_{18:0}$	Most fats and oils
Monounsaturated fatty acids		
Palmitoleic	$C_{16:1}$, ω-7	Fish oils
Oleic	$C_{18:1}$, ω-9	Most fats and oils
Polyunsaturated fatty acids		
Linoleic	$C_{18:2}$, ω-6[b]	Most vegetable oils
Linolenic	$C_{18:3}$, ω-3	Soybean oil, canola oil
Arachidonic	$C_{20:4}$, ω-6	Lard
Eicosapentaenoic	$C_{20:5}$, ω-3	Fish oils
Docosahexaenoic	$C_{22:6}$, ω-3	Fish oils

[a]Example: $C_{4:0}$ indicates 4 carbon atoms and no double bonds in the fatty acid molecule.
[b]Example: $C_{18:2}$, ω-6 indicates 18 carbons and 2 double bonds, with the first double bond on the number 6 carbon counting from the methyl end of the molecule.

TABLE XVI
Fatty Acid[a] and Cholesterol Content of Selected Fats and Oils

Fat/oil	Cholesterol (mg/tbsp)	Saturated fat	Polyunsaturated fat	Monounsaturated fat
Coconut oil	0	92	2	6
Butterfat	33	66	4	30
Beef tallow	14	52	4	44
Lard	12	41	12	47
Chicken fat	11	31	22	47
Peanut oil	0	18	33	49
Soybean oil	0	15	61	24
Corn oil	0	13	62	25
Sunflower oil	0	11	69	20
Canola oil	0	6	32	62
Stick margarine	0	18–24	14–35	45–67
Tub margarine	0	17–21	27–44	36–55

[a]Percentage of total fatty acids.

composed of two monosaccharides chemically linked together. They include sucrose (glucose + fructose), maltose (glucose + glucose), and lactose (glucose + galactose). Sugars occur naturally in a wide variety of foods and are also available in highly refined form. Ordinary white table sugar is one of the most chemically pure foods available, consisting of more than 99% sucrose. [*See* Nutrition, Carbohydrates.]

Complex carbohydrates are composed of monosaccharides chemically linked to form polysaccharides. Complex carbohydrates may be further divided into available (digestible) and unavailable (indigestible) carbohydrates.

Starch is the most common available polysaccharide in foods and contains several hundred to several thousand glucose units in each molecule. Molecular weights of starches range from 50,000 to several million. Starches are hydrolyzed to glucose by digestive enzymes in the gastrointestinal tract.

Unavailable carbohydrates are more commonly known as dietary fiber. Dietary fiber consists of a wide variety of polysaccharides and lignin, which, technically, is not a carbohydrate. Because of its complexity and problems with measurement and characterization, fiber is difficult to define precisely. A useful definition of dietary fiber follows: components of plant material that are resistant to digestion by endogenous enzymes of the mammalian gastrointestinal tract. Dietary fiber is present only in foods of plant origin. The composition of dietary fiber varies with the food. Components include cellulose, noncellulose polysaccharides (hemicellulose, pectins, mucilages, gums), and lignin (a nonpolysaccharide). The physiological effects of fiber components appear to be related to water solubility. Therefore, another possible classification of fiber is based on water solubility. The water-insoluble fractions (e.g., cellulose, hemicellulose) aid in laxation and increase stool weight. The water-soluble fractions (e.g., pectin, gums) have little effect on stool weight but appear to be effective in reducing serum cholesterol levels. Total dietary fiber is defined as the sum of the soluble and insoluble fractions. [*See* Nutrition, Dietary Fiber.]

Carbohydrates are the major source of energy in most diets. Sugars and starches provide about 4 kcal/g. Although dietary fiber is not digested in the small intestine and, therefore, cannot be absorbed, it is partially degraded by the microflora of the large intestine. The microflora metabolize the products of fiber degradation to short-chain fatty acids (acetic, propionic, and butyric), which are rapidly absorbed and used for energy.

Health implications related to dietary carbohydrates have been hotly debated by professional nutritionists and the general public for decades. Refined sugars have been maligned as contributors to health problems ranging from hyperactivity in children to heart disease in adults. However, little clear evidence links refined sugars to any health problem other than dental caries. On the other hand, the addition of refined sugar to the diet clearly will cause nutrient dilution, because refined sugars provide calories but no nutrients and may displace more nutritious foods from the diet.

Interest in dietary fiber has increased dramatically in the past two decades. Many organizations are currently recommending that dietary fiber intakes be increased to a total intake of 25–50 g/day. Putative benefits of dietary fiber range from prevention or relief of constipation to prevention of colon cancer. Recently, several studies have been published showing that soluble dietary fiber, when consumed as part of a low-fat, low-cholesterol diet, reduces serum cholesterol levels. Food sources of soluble dietary fiber include oat bran, legumes, and fruits. Most nutritionists agree that fiber should be obtained from food, not in concentrated forms present in fiber-supplement pills.

V. FACTORS AFFECTING NUTRITIONAL QUALITY

A. Food Choices

The most important factor affecting the nutritional quality of individual diets is food choice by individuals, families, or institutional food service managers. Because no single food or food group contains all the essential nutrients necessary for an adequate diet, variety is essential. Nutrition education efforts have used the concept of food groups to assist consumers with making appropriate food choices. The adequacy of diets based on any food guide depends on the nutrient composition of the individual foods that are chosen. If the nutrient composition of foods within a food group is lower than expected, then diets may be nutritionally inadequate even when food guides are followed. As it turns out, many factors can affect the nutrient composition of foods. Therefore, we must understand these factors and strive to minimize nutritional quality losses in all foods.

B. Agricultural Practices

Many agricultural practices can affect the nutrient composition of foods. For plant foods these include geographic location, soil type, fertilization, moisture availability, sunlight exposure, maturity at harvest, method of harvest (mechanical versus hand picking), season of planting and harvest, and genetics. For animal foods, feeding practices, age at time of slaughter, season, sex, hormone use, and genetics may affect nutrient composition. Because the chemical composition of both plant and animal tissues is dictated by the genetic code, genetics is probably the one factor that has the largest influence on nutrient composition at the time of harvest or slaughter.

C. Food Storage, Processing, and Preparation

1. Storage

Storage conditions after harvest can have a marked effect on the nutrient composition of foods, particularly fruits and vegetables. Most plant tissues continue to transpire and respire after harvest up until they are processed or cooked. Control of storage conditions can markedly extend the storage life of fresh fruits and vegetables. Improper storage can result in wilting, decay due to bacterial and fungal growth, weight loss, overmaturity, and chilling injury. The deterioration that results from improper storage adversely affects appearance, flavor, and texture, but it may also cause nutrient loss. For example, studies have shown that the vitamin C content of wilted spinach is much lower than in spinach that has been stored to minimize wilting. Approaches to improve storage conditions include rapid cooling after harvest, control of storage temperature and humidity, and manipulation of the storage atmosphere (this usually involves lowering the O_2 and increasing the CO_2 in the air in the storage chamber). For optimum retention of quality, storage conditions must be tailored for each individual product.

2. Food Processing and Preparation

Modern food-processing methods have increased the variety and year-round availability of foods. In addition, many processed foods offer a high level of convenience. Although it is true that in most cases foods fresh from the garden, farm, or sea have superior nutritional quality, few of us can rely directly on these sources for most of our foods. On the other hand, the greater variety and convenience afforded by processed foods would be expected to improve the overall nutritional quality of the diet by increasing the variety of foods that people actually eat.

Food-processing methods can be divided into nine basic categories:

1. Heat processing: blanching, pasteurizing, sterilizing
2. Low-temperature processing: refrigerating, freezing
3. Moisture removal: drying, concentrating

4. Chemical processing: addition of salt, sugar, antioxidants, emulsifiers, stabilizers, thickeners, nutrients, etc.
5. Mechanical processing: milling of cereals, removal of peels, etc.
6. Food irradiation: exposure of foods to ionizing radiation (usually gamma rays)
7. Microbiological processing: cheese-making, brewing, etc.
8. Packaging: enclosing in a container
9. Final preparation: peeling, chopping, thawing, heating (boiling, broiling, microwaving, baking, frying), holding (steam table, etc.)

Virtually all of the processes listed will alter, to a greater or lesser degree, the nutritional quality of foods. A brief description of some of the more important processing effects follows. The reader is referred to Karmas and Harris (1988) for an in-depth treatment of processing effects on nutritional quality.

VI. MECHANISMS BY WHICH NUTRITIONAL QUALITY MAY BE AFFECTED

A. Changes in Chemical Structure

Changes in the chemical structure of some nutrients may result in loss of nutritional activity. Rates of these chemical reactions are affected by several parameters, including temperature, pH, concentration, oxygen, light, prooxidants (e.g., iron ions), and water activity (the amount of "free water" available in the food). A primary objective of the food-processing engineer is to design food-processing systems that will meet the objectives of microbiological safety and organoleptic quality and at the same time will result in optimal retention of nutrients.

1. Vitamins

The class of nutrients most susceptible to destruction is vitamins. Many vitamins are rather fragile organic molecules, which lose their biological activity when even very small changes in their chemical structures occur. Some of the more significant factors affecting vitamin stabilities in foods are listed in Table XVII. Fortunately, conditions that result in accelerated rates of destruction are similar for most nutrients. Thus, if processing conditions are chosen to protect one nutrient, most other nutrients will be protected as well.

2. Minerals

Minerals, unlike vitamins and proteins, are virtually indestructible. Losses may result from leaching, milling, and trimming but not from actual chemical destruction. The bioavailability of minerals, however, may be altered by processing. Bioavailability may be defined as the percentage of a nutrient in a food that is potentially available for absorption from the gastrointestinal tract in a physiologically active form. Processing may change the oxidation state of a mineral, and this may affect its bioavailability. For example, iron may be oxidized from the ferrous form (Fe^{2+}) to the ferric form (Fe^{3+}) during processing. Ferrous iron

TABLE XVII
Factors That Affect Nutrient Stabilities in Foods

Nutrient	Factors that protect against losses	Factors that accelerate losses
Protein		Heat, especially in the presence of sugars; high pH
Carbohydrates		No nutritionally significant losses occur
Polyunsaturated fatty acids	Antioxidants	Air, oxygen, light, prooxidants such as iron ions; hydrogenation
Ascorbic acid	Low pH, sulfite, low moisture	Heat, oxygen, light, high pH, iron and copper ions, some plant enzymes
Thiamin	Low pH	Heat, neutral and high pH, sulfite, thiaminases (enzymes in raw fish)
Riboflavin	Low pH	Light, especially in liquid foods such as milk
Niacin		Stable under most conditions; leachable
Vitamin B_6		Heat, light (stability varies depending on the form of the vitamin)
Folacin		Heat (stability varies depending on form)
Vitamin A	Antioxidants	Heat, oxygen, iron and copper ions, low pH

is thought to be more bioavailable than ferric iron. Processing may also affect mineral bioavailability indirectly by altering other nutrients in the food. For example, ascorbic acid is a potent enhancer of iron bioavailability. If a food process destroys ascorbic acid, iron bioavailability from that food and other foods consumed with it in the same meal will be diminished. Conversely, fortifying a food with ascorbic acid will improve iron bioavailability. [*See* Minerals in Human Life.]

3. Proteins

Processing may alter protein quality by promoting chemical reactions between essential amino acid residues and other food components. In general, amino acid residues in intact protein molecules are quite stable to most processing and storage conditions. The exception to this is the amino acid lysine, which contains a free amino group on the epsilon carbon of the side chain. Under some conditions, this amino group can react with reducing sugars (e.g., glucose, lactose) to form brown products. This is the familiar browning that occurs when bread is baked or breakfast cereal flakes are toasted. It is known as nonenzymatic browning, or Maillard browning (after the French chemist who first described the reaction). When this browning occurs, the nutritional activity of the lysine is destroyed and protein quality is diminished. The nutritional impact is the greatest in cereals, which are low in lysine to begin with. Fortunately, processing effects on protein quality have little significance in the United States, where protein intakes exceed requirements by substantial margins. Moreover, the class of foods in which protein quality losses are greatest (i.e., ready-to-eat toasted breakfast cereals) are normally consumed with milk, which is rich in lysine and therefore compensates for the lost lysine in the cereal.

In many cases, processing improves protein quality by enhancing protein digestibility. Legumes, for example, contain antinutrients called trypsin inhibitors. These substrates interfere with the activity of trypsin, a digestive enzyme that breaks down proteins in the small intestine. It is well known that raw soybeans, which are a rich source of a relatively high quality protein, will not support growth in rats and other animals. When the soybeans are treated with moist heat, growth improves dramatically. This is because the trypsin inhibitor is heat-sensitive and heat destroys it, thereby preventing it from interfering with the digestion of the protein in the soybeans. [*See* Nutrition, Protein.]

4. Fats

Processing may change the nutritional quality of fats by altering the structure of the fatty acids present in triacylglycerols. These changes may destroy the essential fatty acid activity of linoleic and linolenic acids and/or alter the effects of the fat on blood cholesterol levels.

Over the course of this century, consumption of vegetable fats (cooking and salad oils, margarine, vegetable shortening, deep-frying fats) has increased while consumption of animal fats (butter, lard, and tallow) has decreased. Substitution of vegetable fats for animal fats has been made possible by a technology called hydrogenation. In this process, refined vegetable oils are mixed with hydrogen gas and a catalyst in a pressurized vessel for an appropriate period of time. Under these conditions, some of the polyunsaturated fatty acids (linoleic and linolenic acids) are converted to monounsaturated fatty acids and some are converted to saturated fatty acids. In addition, molecular rearrangements occur during hydrogenation that result in the formation of trans fatty acids, giving a mixture of cis and trans fatty acids in the hydrogenated fat:

$$CH_3(CH_2)_4CH{=}CHCH_2CH{=}CH(CH_2)_7COOH + H_2$$

Linoleic acid (both double bonds are cis)

$$\downarrow$$

$$CH_3(CH_2)_7CH{=}CHCH_2(CH_2)_7COOH$$

Oleic acid (cis)

$$+$$

$$CH_3(CH_2)_7CH{=}CHCH_2(CH_2)_7COOH$$

Eliadic acid (trans)

$$+$$

$$CH_3(CH_2)_{16}COOH$$

Stearic acid

Hydrogenation changes the physical characteristics of the oil. Depending on the extent of the hydrogenation, the liquid oil will be converted to a soft or a hard fat (the most familiar examples are tub and stick margarines, soft and hard fats, respectively).

Even though partially hydrogenated vegetable oils are more saturated than the oils from which they are made, they are less saturated than most animal fats (see Tables XIV and XV). Therefore, substitution of margarine for butter in the diet will usually result in a lowering of serum cholesterol levels. However, the formation of trans fatty acids during hydrogenation is a concern. In their natural form, all fatty acids in plant lipids and most of the fatty acids in animal lipids contain only cis double bonds. Several studies have shown that trans fatty acids behave more like saturated fatty acids than cis-monosaturated fatty acids in the body. Thus the question, "which is better, butter or margarine?", remains controversial. At present, the best advice is to eat a low-fat diet and use oils rather than hydrogenated vegetable oils or animal fats whenever possible.

B. Physical Separation of Nutrients from Foods

In many food processes, nutrients are physically removed from the edible portion of foods. Probably the most dramatic example of this is the nutrient changes that occur when wheat is milled into white flour. The wheat kernel may be divided into three basic parts: (1) the endosperm is the starchy interior of the kernel, making up about 83% of the total weight, (2) the bran includes the outer layers of the kernel, making up about 14.5% of the intact kernel; and (3) the germ is the smallest part, making up only about 2.5% of the total. Whole-wheat flour includes all three parts of the kernel. White flour is mostly endosperm. Many of the nutrients in the intact wheat kernel are concentrated in the bran and/or germ layers, so removal of these in milling causes dramatic losses. Four nutrients—iron, thiamin, riboflavin, and niacin—are routinely restored to white flour through fortification (enrichment), but other nutrients, such as zinc, magnesium, and vitamin B_6, are not. A comparison of the nutrient compositions of whole wheat and white bread (see Table IX) reveals that iron, thiamin, riboflavin, and niacin concentrations are comparable in the two, whereas fiber and B_6 are much lower in the white bread.

Leaching may also separate nutrients from the edible portion of the food. This is mainly a problem with the water-soluble nutrients such as vitamin C and potassium. It occurs when foods are blanched, canned, or cooked in water. Losses are greatest when the food is finely chopped, when a large water-to-food ratio exists, when the food is processed at high temperatures for long periods, and when processing or cooking water is discarded. A comparison of the sodium and potassium contents of fresh, canned, and frozen vegetables provides an example of how leaching can affect the content of two different minerals in opposite directions (see Table X). Notice that potassium is substantially lower in the canned and frozen corn and peas compared to the fresh. Processing causes potassium losses because potassium is a highly soluble mineral nutrient and leaches out when the vegetables are blanched or canned in water. In many cases, processing results in an increase in the sodium content of the edible portion of the food. This is because salt is added during processing. Salt is not usually added to frozen vegetables. Peas are an exception, because a brine solution is used to separate peas by size prior to blanching and freezing.

Unfortunately, because of suspected relationships between sodium and potassium intakes and hypertension, both the decreases in potassium and the increases in sodium are nutritionally undesirable.

C. Nutrient Dilution

Nutrient dilution is a term used to describe the addition of sugars or fats to foods. Food-grade sugars include sucrose, high-fructose corn syrup, corn syrup, lactose, and honey. These are highly purified carbohydrates that contain only insignificant concentrations of nutrients. Sugars do contain approximately 4 kcal/g dry weight. Food-grade fats and oils include butter, margarine, shortening, lard, beef tallow, and vegetable oils. Fats and oils contain varying amounts of the fat-soluble vitamins A, D, and E as well as essential fatty acids; however, they are highly concentrated sources of calories (9 kcal/g).

Both sugars and fats are added to a wide variety of foods either during processing and preparation or at the table. It was estimated that in 1988, fats and oils provided 20% and sugars provided 17% of the available calories in the food supply. The net effect of this nutrient dilution is that more calories have to be consumed to get the quantities of nutrients that would be available in the foods before sugar or fat was added. As an example, compare the nutrient compositions of peaches canned in heavy syrup with those canned in juice (see Table XI). Notice that the nutrient compositions are similar but a serving of the peaches canned in heavy syrup has nearly twice the calories as those canned in juice. For some people, choosing a diet adequate in all essential nutrients may be difficult if the foods contain excessive amounts of added sugar and fat. In addition, evidence indicates that diets high in fat may lead to obesity, heart disease, and/or cancer in some people. [*See* Nutrition, Fats and Oils.]

D. Addition of Nutrients to Foods

Several terms have been used to describe the practice of adding nutrients to foods, including enrichment, fortification, and restoration. Although these terms initially had slightly different meanings, in practice they are frequently used interchangeably. Here, nutrification will be used to indicate any addition of nutrients to foods. Nutrification has a long history in the United States. Iodinization of salt began in 1924, vitamin D was first added to milk in 1933, and a program for enrichment of flour and cereal products began in 1941. Originally, nutrification was instituted as a public health measure for reducing the widespread prevalence of nutrient deficiency diseases such as goiter, rickets, and pellagra. The early successes of nutrification efforts are well documented and have led to the nutrification of a much broader spectrum of food products. Today, foods ranging from breakfast cereals to candy and soda pop are being nutrified. Although it is generally felt that the addition of nutrients to foods enhances nutritional quality, many nutritionists are beginning to view with concern the widespread nutrification of the U.S. food supply without apparent regard to potentially adverse nutrient–nutrient interactions. Federal regulations do not specify which foods may be fortified, but the FDA has recently issued a fortification policy in the form of a series of guidelines. The FDA urges food manufacturers to adhere to its policy when adding nutrients to foods. The following quote from this policy statement reflects the feelings of many nutritionists and food scientists on this important issue:

> The addition of nutrients to specific foods can be an effective way of maintaining and improving the overall nutritional quality of the food supply. However, random fortification of foods could result in over or under fortification in consumer diets and create nutrient imbalances in the food supply. It could also result in deceptive or misleading claims for certain foods. The FDA does not encourage indiscriminate addition of nutrients to foods, nor does it consider it appropriate to fortify fresh produce; meat, poultry, or fish products; sugars; or snack foods such as candies and carbonated beverages.

VII. DIETARY GUIDANCE FOR CONSUMERS

The dietary guidelines outlined in Tables V, VI, and VII are designed primarily for food and health professionals and may not be particularly effective for helping consumers to select foods and diets with high nutritional quality. Thus, the federal government has taken two important steps to assist consumers in choosing healthy diets: the issuance of revised nutrition labeling regulations and the Food Guide Pyramid. The new nutrition label makes it easier for consumers to compare the nutritional quality of individual foods and the Food Guide Pyramid helps consumers understand the importance of variety, moderation, and proportionality in making food choices.

A. The Nutrition Label

The Nutrition Labeling and Education Act (NLEA) was signed into law by President Bush on November 9, 1990. This act mandated, for the first time, nutrition labeling on nearly all packaged foods and established new guidelines to govern the format and content of the nutrition label. As required by this legislation, the FDA wrote detailed new regulations for nutrition labeling. These are published in the Code of Federal Regulations and became effective in May of 1994. The food industry acted quickly to either replace old nutrition labels or include labels on foods that did not previously carry a nutrition label.

A generic example of the nutrition label, recognizable by the title "Nutrition Facts," is shown in Fig. 2. Whereas the old label highlighted the protein, vitamin, and mineral composition of foods, the new label emphasizes fat, cholesterol, sodium, and carbohydrate, including dietary fiber. This new emphasis reflects the current consensus that these dietary components are associated with the chronic diseases that cause most of the morbidity and mortality in the United States today. Americans tend to overconsume most of these (calories, fat, saturated fat, cholesterol, and sodium). The list of required vitamins and minerals has been reduced to include only vitamin A, vitamin C, calcium, and iron, reflecting the continuing problem of low intakes and deficiencies of these nutrients. Of course other vitamins and minerals are still important and may be listed voluntarily by food manufacturers.

Nutrient concentrations in foods are expressed as %Daily Value (%DV) per serving. The weight (g or mg) per serving is also listed for fat, cholesterol, sodium, carbohydrate, and protein. Fat content is expressed three ways: calories from fat, grams total fat, and %DV. Daily values are recommended daily intakes and are based on two sets of dietary standards. Values for fat, saturated fat, cholesterol, sodium, carbohydrate, and dietary fiber are based on dietary guidelines and are calculated assuming a caloric intake of 2000 calories per day. For example, most dietary

The New Food Label at a Glance

The new food label will carry an up-to-date, easier-to-use nutrition information guide, to be required on almost all packaged foods (compared to about 60 percent of products up till now). The guide will serve as a key to help in planning a healthy diet.*

Serving sizes are now more consistent across product lines, stated in both household and metric measures, and reflect the amounts people actually eat.

New title signals that the label contains the newly required information.

Calories from fat are now shown on the label to help consumers meet dietary guidelines that recommend people get no more than 30 percent of their calories from fat.

% Daily Value shows how a food fits into the overall daily diet.

The list of nutrients covers those most important to the health of today's consumers, most of whom need to worry about getting too much of certain items (fat, for example), rather than too few vitamins or minerals, as in the past.

Daily values are also something new. Some are maximums, as with fat (65 grams or less); others are minimums, as with carbohydrates (300 grams or more). The daily values on the label are based on a daily diet of 2,000 and 2,500 calories. Individuals should adjust the values to fit their own calorie intake.

The label will now tell the number of calories per pgram of fat, carbohydrates, and protein.

Nutrition Facts

Serving Size 1/2 cup (114g)
Servings Per Container 4

Amount Per Serving

Calories 90 Calories from Fat 30

	% Daily Value*
Total Fat 3g	**5%**
Saturated Fat 0g	**0%**
Cholesterol 0mg	**0%**
Sodium 300mg	**13%**
Total Carbohydrate 13g	**4%**
Dietary Fiber 3g	**12%**
Sugars 3g	
Protein 3g	

Vitamin A 80%	•	Vitamin C 60%
Calcium 4%	•	Iron 4%

* Percent Daily Values are based on a 2,000 calorie diet. Your daily values may be higher or lower depending on your calorie needs:

	Calories	2,000	2,500
Total Fat	Less than	65g	80g
Sat Fat	Less than	20g	25g
Cholesterol	Less than	300mg	300mg
Sodium	Less than	2,400mg	2,400mg
Total Carbohydrate		300g	375g
Fiber		25g	30g

Calories per gram:
Fat 9 • Carbohydrates 4 • Protein 4

* This label is only a sample. Exact specifications are in the final rules.
Source: Food and Drug Administration 1992

FIGURE 2 Explanation of the nutrition label.

guidelines recommend no more than 30% of calories from fat. Thus the DV for fat is calculated as follows: 2000 calories × 30% = 600 calories from fat; 600 calories divided by 9 calories per gram of fat = 66.7 g fat. This is rounded down to 65 as shown in the lower panel of the label and in Table XVIII. DVs for saturated fat and total carbohydrate are calculated in a similar manner. Since these values are based on calorie intake, individual values will be higher or lower depending on caloric intake. Values for the other nutrients are not based on caloric intake and will be the same for all individuals. Daily values for protein, vitamins, and minerals are derived from the Recommended Dietary Allowances. Table XIX lists the values used.

The Nutrition Facts label shows at a glance whether a food is high or low in a particular nutrient (foods with less than 5% DV are usually considered low). The goal should be to consume about 100% of the DV for each nutrient over the course of a day. For nutrients that may have adverse effects if overconsumed (e.g., fat, cholesterol, sodium), 100% DV should be considered a maximum. For other nutrients (e.g., calcium and vitamin C), the goal should be at least 100% DV.

Nutrition labeling regulations also provide rules to govern content claims. *Content claims* are expressed or implied statements on the label that portray the level of the nutrient in the food (e.g., "high," "low," "free," etc.). To ensure that content claims are meaningful and not misleading, the FDA has established strict definitions for terms that portray content (Table XX). Food manufacturers must follow these regulations carefully if they choose to use these terms on their labels.

Nutrition labeling regulations also permit food manufacturers to include health claims on their labels to emphasize food components that have been shown to benefit health. A *health claim* is defined by the FDA as follows: "*Health claim* means *any claim made* on the label or in labeling of a food, including a dietary supplement, *that expressly or by implication,* including 'third party' references, written statements (e.g., a brand name including a term such as 'heart'), or vignettes, *characterizes the relationship of any substance to a disease or health related condition.*" To prevent misleading or unsupported claims, only claims specifically approved by the FDA are permitted. These claims are listed in Table XXI. The FDA allows some flexibility in the wording of health claims provided they are not false or misleading but they do provide model claims that manufacturers may use.

TABLE XVIII
Reference Values for Calculating %DV of Other Food Components of Nutrition Labels

Food component	Reference value
Fat	65 g
Saturated fatty acids	20 g
Cholesterol	300 mg
Total carbohydrates	300 g
Fiber	25 g
Sodium	2400 mg
Potassium	3500 mg
Protein	50 g

Source: Code of Federal Regulations, 21, parts 100–169, April 1, 1995. United States Government, Washington, D.C.

TABLE XIX
Reference Values for Vitamins and Minerals for Calculating %DV on Nutrition Labels

Nutrient	Amount
Vitamin A	5000 IU[a]
Vitamin C	60 mg
Thiamin	1.5 mg
Riboflavin	1.7 mg
Niacin	20 mg
Calcium	1.0 g
Iron	18 mg
Vitamin D	400 IU
Vitamin E	30 IU
Vitamin B_6	2.0 mg
Folate	0.4 mg
Vitamin B_{12}	6.0 μg
Phosphorus	1.0 g
Iodine	150 μg
Magnesium	400 mg
Zinc	15 mg
Copper	2.0 mg
Biotin	0.3 mg
Pantothenic acid	10 mg

Source: Code of Federal Regulations, 21, parts 100–169, April 1, 1995. United States Government, Washington, D.C.
[a]International Units.

TABLE XX
Definitions of Some Content Descriptors Allowed on Food Labels

Descriptive term	Definition of term as defined by FDA regulations
Sugar free	Less than 0.5 g sugar per serving
Reduced sugar	At least 25% less sugar per serving than reference food
Calorie free	Fewer than 5 calories per serving
Low calorie	40 calories or less per serving, and if the serving is 30 g or less or 2 tablespoons or less, per 50 g of the food
Reduced calories	At least 25% fewer calories per serving than reference food
Fat free	Less than 0.5 g of fat per serving
Low fat	3 g or less fat per serving, and if the serving is 30 g or less or 2 tablespoons or less, per 50 g of the food
Reduced fat	At least 25% less fat per serving than reference food
Cholesterol free	Less than 2 mg cholesterol and 2 g or less of saturated fat per serving
Low cholesterol	20 mg cholesterol or less and 2 g or less saturated fat per serving
Reduced cholesterol	At least 25% less cholesterol and 2 g or less saturated fat per serving than reference food
Sodium free	Less than 5 mg sodium per serving
Low sodium	140 mg or less per serving, and if the serving is 30 g or less or 2 tablespoons or less, per 50 g of the food
Very low sodium	35 mg or less per serving, and if the serving is 30 g or less or 2 tablespoons or less, per 50 g of the food
Reduced sodium	At least 25% less per serving than reference food
High fiber	5 g or more fiber per serving
Good source of fiber	2.5 to 4.9 g fiber per serving

Source: Code of Federal Regulations, 21, parts 100–169, April 1, 1995. United States Government, Washington, D.C.

TABLE XXI
Health Claims Authorized by FDA for Use on Food and Dietary Supplement Labels

Allowed relationship	Model health claim
Calcium and osteoporosis	Regular exercise and a healthy diet with enough calcium helps teen and young adult white and Asian women maintain good bone health and may reduce their high risk of osteoporosis later in life.
Dietary lipids and cancer	Development of cancer depends on many factors. A diet low in total fat may reduce the risk of some cancers.
Sodium and hypertension	Diets low in sodium may reduce the risk of high blood pressure, a disease associated with many factors.
Dietary saturated fat and cholesterol and risk of coronary heart disease	While many factors affect heart disease, diets low in saturated fat and cholesterol may reduce the risk of this disease
Fiber-containing grain products, fruits, and vegetables and cancer	Low-fat diets rich in fiber-containing grain products, fruits, and vegetables may reduce the risk of some types of cancer, a disease associated with many factors.
Fruits, vegetables, and grain products that contain fiber, particularly soluble fiber, and risk of coronary heart disease	Diets low in saturated fat and cholesterol and rich in fruits, vegetables, and grain products that contain some types of dietary fiber, particularly soluble fiber, may reduce the risk of heart disease, a disease associated with many factors.
Fruits, vegetables, and cancer	Low-fat diets rich in fruits and vegetables (foods that are low in fat and may contain dietary fiber, vitamin A, and vitamin C) may reduce the risk of some types of cancer, a disease associated with many factors. Broccoli is high in vitamins A and C, and it is a good source of dietary fiber.
Folate and neural tube defects	Women who consume adequate amounts of folate, a B vitamin, daily throughout their childbearing years may reduce their risk of having a child with a neural tube birth defect. Such birth defects, while not widespread, are very serious. They can have many causes. Adequate amounts of folate can be obtained from diets rich in fruits, dark green leafy vegetables and legumes, enriched grain products, fortified cereals, or a supplement. Folate consumption should be limited to 1000 μg per day from all sources.

Source: Code of Federal Regulations, 21, parts 100–169, April 1, 1995. United States Government, Washington, D.C.

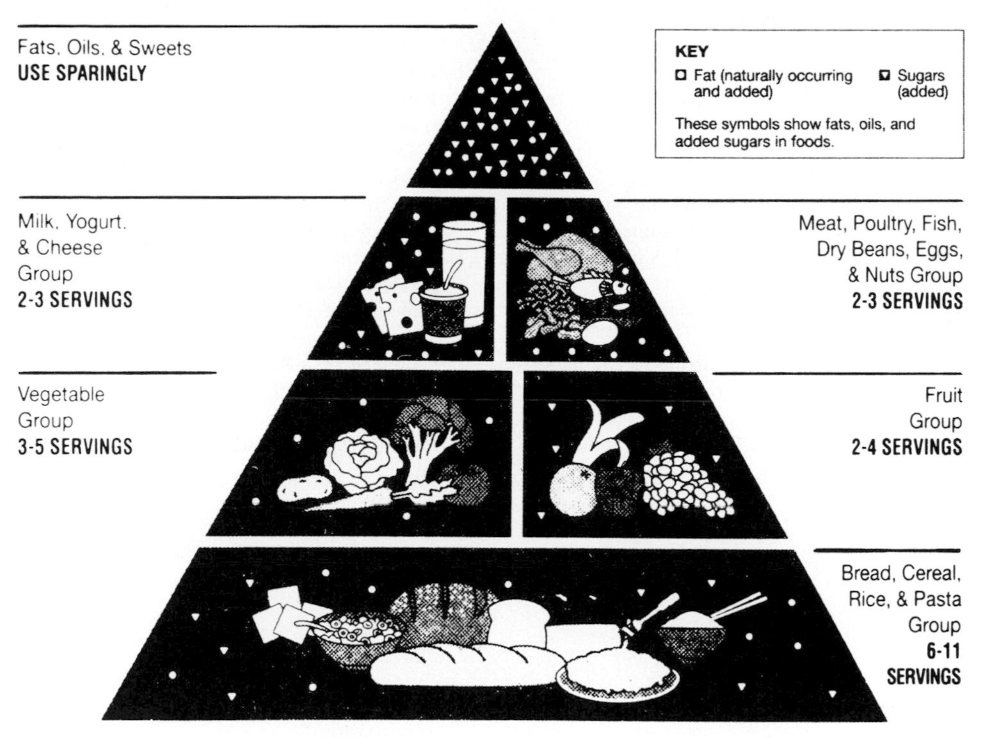

FIGURE 3 The Food Guide Pyramid.

B. The Food Guide Pyramid

The Food Guide Pyramid was developed by the USDA as a guide to help consumers choose diets that conform with the dietary guidelines (Fig. 3). It replaces the old "Basic Four Food Groups" and, along with the Dietary Guidelines for Americans, forms the basis of federal nutrition policy.

The pyramid is divided into five major food groups plus a sixth group called fats, oils, and sweets. It incorporates the key principles of variety, moderation, and proportionality and focuses on fat, because most American's diets are too high in fat. *Variety* is emphasized through recommended numbers of servings from the five food groups as well as by including a selection of foods within each food group. *Moderation* is portrayed by the admonition to "use sparingly" in the fats, oils, and sweets group and by the density of the small circles and triangles that remind people that fat and added sugar are present in many foods. Because the fat content of foods within a food group may vary over a wide range, consumers should rely on food labels for a more accurate indication of fat content. *Proportionality* is conveyed by the relative area designated for each food group in the pyramid and by the recommended number of servings for each food group.

VIII. SUMMARY

Our national food supply is a precious resource that is essential for the health and well-being of every American. As our understanding of relationships between diet and health has evolved, it has become apparent that the nutritional quality of our national diet, though good, needs to be improved if we are to continue to increase life expectancies and improve the quality of life. Reponsibility for maintaining and improving the nutritional quality of foods and diets rests with all of us. Individuals are responsible for making healthy food choices from the broad array of available foods in the marketplace. Governmental, educational, and health organizations are responsible for educating consumers about good nutrition and for sponsoring research to expand our knowledge about diet–health relationships. The agricultural and food industries are responsible for producing, processing, and marketing safe and affordable foods of high nutritional quality.

BIBLIOGRAPHY

Committee on Diet and Health, Food and Nutrition Board, Commission on Life Sciences, National Research Council (1989). "Diet and Health—Implications for Reducing Chronic Disease Risk." National Academy Press, Washington, D.C.

Committee on Technological Options to Improve the Nutritional Attributes of Animal Products (1988). "Designing Foods—Animal Product Options in the Marketplace." National Academy Press, Washington, D.C.

Hotchkiss, J. H., and Potter, N. N. (1995). "Food Science," 5th Ed. Van Nostrand–Reinhold, New York.

Karmas, E., and Harris, R. S. (eds.) (1988). "Nutritional Evaluation of Food Processing," 3rd Ed. Van Nostrand–Reinhold, New York.

Subcommittee on the Tenth Edition of the RDAs, Food and Nutrition Board, Commission on Life Sciences, National Research Council (1989). "Recommended Dietary Allowances," 10th Ed. National Academy Press, Washington, D.C.

U.S. Department of Agriculture (1986–1989). "Composition of Foods: Raw, Processed, Prepared," Agriculture Handbook 8, revised, Vols. 8-1–8-17. USDA, Washington, D.C.

U.S. Department of Agriculture–Human Nutrition Information Service (1990). "Nutrition and Your Health—Dietary Guidelines for Americans," Home and Garden Bulletin No. 232-1. USDA, Hyattsville, Maryland.

U.S. Department of Agriculture–Human Nutrition Information Service (1992). "The Food Guide Pyramid," Home and Garden Bulletin No. 252. USDA, Hyattsville, Maryland.

U.S. Department of Health and Human Services (1988). "The Surgeon General's Report on Nutrition and Health." U.S. Government Printing Office, Washington, D.C.

Wardlaw, G. M., Insel, P. M., and Seyler, M. F. (1994). "Contemporary Nutrition: Issues and Insights," 2nd Ed. Mosby, St. Louis.

Nutrition, Carbohydrates

CAROL N. MEREDITH
University of California, Davis

GLOSSARY

Cariogenicity Capacity of a food to produce dental caries

Complex carbohydrates Polysaccharides such as starch, cellulose, and hemicellulose; only starch can be digested to glucose in the gut, whereas the others make up dietary fiber

Corn sweeteners Sweet mixture of fructose, glucose, and other carbohydrates made by the enzymatic breakdown of corn starch

Glycemic index Ratio between the increase in blood sugar after consuming a test carbohydrate and the increase obtained after consuming an equivalent amount of carbohydrate from a standard food such as white bread

Invert sugar Liquid mixture of glucose and fructose made by chemical hydrolysis of sucrose

Nonsugar sweeteners Natural and manufactured simple sugars such as fructose, sorbitol, and xylitol that are absorbed less efficiently than sugar or glucose

NUTRITIONAL CARBOHYDRATES ARE ORGANIC compounds of carbon, hydrogen, and oxygen, which can be digested, absorbed, and broken down in the tissues to water and carbon dioxide. They make up the bulk of human diets throughout the world; starch-rich foods are the least expensive to produce and store. The main function of carbohydrates is to provide energy. They also contribute to the taste and texture of food, with sucrose and fructose used as sweeteners. Diets rich in carbohydrates, especially complex carbohydrates, are healthier than high-fat diets and are not necessarily linked to obesity, diabetes, atherosclerosis, or hyperactivity.

I. CONSUMPTION PATTERNS IN THE UNITED STATES AND THE WORLD

Primitive nomadic peoples subsisted on meat and milk-based diets providing protein and fat as the main sources of energy. The domestication of plants with starch-rich seeds, such as wheat, rice, and corn, allowed humans to settle communities with an assured supply of food, which was the basis of civilization. A diet where about 80% of the energy came from starch was the norm before the 20th century in the United States and is still typical in poor countries. Today, in the United States and other western nations, only 50% of the daily energy intake comes from carbohydrates.

The types of carbohydrates in the diet have also changed over the years. The intake of starch in the United States from foods such as potatoes, bread, and legumes has declined about 50% over this century, while the intake of sugar and other sweeteners has increased. Since the early 1960s, the consumption of sweeteners has further increased due to a greater use of sugar in manufactured foods (e.g., baked goods, breakfast cereals, ketchup, canned fruit) and the consumption of sweetened beverages (e.g., sodas, fruit juices) (Fig. 1). New sweeteners other than sugar have emerged, such as fructose and corn sweeteners, both manufactured from corn starch, and the nonsugar sweeteners such as xylitol and sorbitol, used in chew-

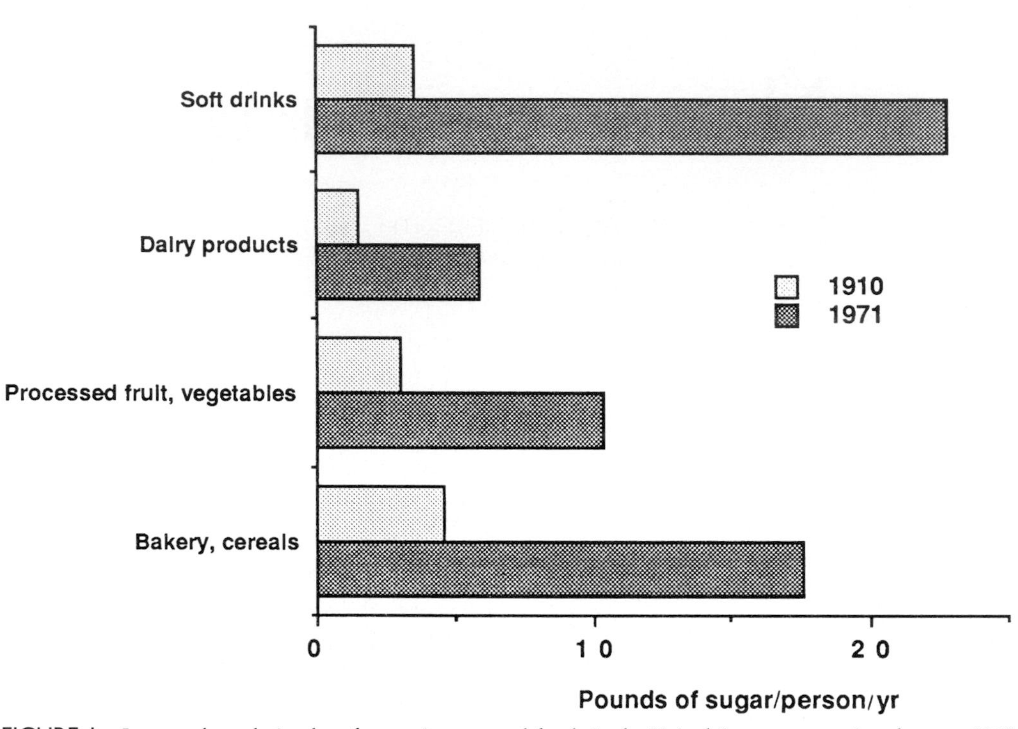

FIGURE 1 Increased yearly intake of sugar in prepared foods in the United States, comparing the year 1910 with 1971. Household use of sugar has declined.

ing gum and candy. These new sweeteners may be digested and absorbed slightly differently from other carbohydrates, but they provide about the same amount of calories per unit of weight.

Carbohydrates are not essential nutrients because they can be made in the body from other compounds. However, a diet with little or no carbohydrates is perceived by most people as extremely unappetizing, and adaptation to such a diet is associated with fatigue, headaches, and lack of well-being. National guidelines for a healthier diet promote an increased intake of carbohydrates, especially as complex carbohydrates, and a lower intake of sugar.

II. CHEMICAL COMPOSITION OF NUTRITIONAL CARBOHYDRATES IN FOODS

Carbohydrates are made up of one or more units of simple sugars. According to the number of sugar units, they are classified as monosaccharides (1 sugar), disaccharides (2 sugars), oligosaccharides (3–10 sugars), and polysaccharides (10 to several thousand sugars). The nutritional carbohydrates are mostly converted to glucose in the body and can be enzymatically broken down to carbon dioxide and water, producing an average 4 kcal/g carbohydrate oxidized.

A. Monosaccharides

The simple sugars, or monosaccharides, that are nutritionally important are six-carbon compounds with alcohol (—OH) groups in different positions and spatial arrangements. They are highly soluble in water and have a sweet taste (Table I). Glucose is the most important because it is the carbohydrate that enters the tissues and is converted to energy or other products. Glucose is also called dextrose because in solution it bends a beam of polarized light to the right. Fructose is also known as levulose because it bends polarized light to the left. Fructose is sweeter than sugar and enhances the taste of fruit products. The simple sugar galactose is found only as a component of the disaccharide lactose and does not exist in free form in foods.

Polysaccharides and disaccharides must be hydrolyzed to simple sugars before being absorbed in the gut. Patients who must be fed with intravenous solutions can only utilize simple sugars (usually only

TABLE I
Sweetness of Different Sugars Found Naturally in Food or Added to Foods

Name	Composition	Relative sweetness[a]	Origin
Starch	[glucose]n	—	Cereals, tubers
Sucrose	Fructose + glucose	100%	Sugar cane, beet
Lactose	Galactose + glucose	30%	Milk
Maltose	Glucose + glucose	40%	Germinating seeds
Glucose	$C_6H_{12}O_6$	70%	Fruit
Fructose	$C_6H_{12}O_6$	120%	Fruit
Galactose	$C_6H_{12}O_6$	—	Not found free
Sorbitol	$C_6H_{14}O_6$	54%	Artificial sweetener
Xylitol	$C_5H_{12}O_5$	134%	Artificial sweetener

[a]Compared to sucrose.

glucose) because these are the only carbohydrates that can cross cell membranes to enter the tissues.

B. Disaccharides

The carbohydrates made up of two similar or different sugar units are the disaccharides. Lactose, the carbohydrate in milk, is made up of glucose + galactose. Maltose, a product of starch breakdown during germination of seeds or during digestion in the gut, is made up of two glucose units. Sucrose, or table sugar, is made up of glucose + fructose. The disaccharides are soluble in water.

C. Polysaccharides

The main polysaccharide in the diet is starch, made up of 300 to many thousand glucose units bound to each other in linkages denoted α-glucosidic bonds. While the polysaccharide cellulose is also a polymer of glucose, they are linked in a way that resists hydrolysis by the enzymes of the digestive system of humans or other mammals. Herbivores can feed off grass and leaves not through the action of their own digestive enzymes, but because bacteria in the gut produce enzymes that can break the glucose–glucose bonds in cellulose.

The glucose chains in starch are about 25% straight chains forming spirals, called amylose, and about 75% branched chains, called amylopectin. Starch is a stable product, almost insoluble in cold water, as suggested by its use in cosmetic dusting powders. It is an important source of stored energy in plants. The starch in seeds, tubers, and other plant parts is found in granules. The starch in unbroken granules is different to digest, but during moist cooking, the granules swell and burst, facilitating digestion.

Starches that are modified chemically or enzymatically are used industrially to improve the texture, appearance, and keeping qualities of processed foods (Table II).

The animal polysaccharide is glycogen, similar in structure to amylopectin. Although the main store of energy in humans is as fat, the body of an adult man contains about 300 g glycogen in liver and muscle. Liver glycogen is a reserve of energy, but more importantly it is a store of easily released glucose units that can nourish tissues that are totally glucose dependent (such as the brain) during times of fasting. Muscle

TABLE II
Examples of the Use of Modified Starches and Starch-Derived Sweeteners in the Food Industry

Processed food	Effect of adding various starch-derived products
Canned food	Improved resistance to heat and acid; lower freezing temperature
Frozen foods	Improved stability
Ice cream	Decreased formation of large ice crystals
Jams and jellies	Decreased formation of sugar crystals; enhanced fruit flavor
Confections	Changes in viscosity, sweetness, shine, and transparency
Fruit juice	Enhanced fruit flavor; lower freezing temperature

glycogen can be broken down to lactate and other products, producing energy that can be used for physical work, but muscle cells lack the enzymes for regenerating glucose from stored glycogen. In meats such as steak, liver, or fish, the amounts of glycogen are insignificant, as these stores disappear before the food is eaten.

The carbohydrates containing 3–10 different simple sugars are known as oligosaccharides and are found in beans and other legumes. They are considered undesirable products because they are not digestible and lead to flatulence.

Carbohydrates in foods are often not measured directly but, rather, as the amount remaining after accounting for water, protein, fat, and minerals. In older food tables, the term "carbohydrate" may include starch as well as the nonabsorbable cellulose, sugar as well as nonabsorbable oligosaccharides, other types of plant fiber, organic acids, and minor compounds. The accurate analysis of each type of carbohydrate in a typical food is a major undertaking given the variety of compounds involved.

III. DIETARY SOURCES OF CARBOHYDRATES

Traditional staple foods (bread, rice, tortillas, corn, porridge) usually provide starch as the main nutrient, but in our diet about 25% of the carbohydrates are sweeteners (sugar, glucose, fructose). Refined sugar foods are the most concentrated form of carbohydrate and are termed "empty calories" because they provide no other useful nutrients. In hard candies, artificially flavored fruit drinks, and sodas, >95% of the energy is provided as sugar or high-fructose syrup, with trivial amounts of essential nutrients or nonnutritive components such as fiber.

Honey is a natural sweetener that is a mixture of free fructose and glucose. It is not superior to sugar, although it provides some minerals and vitamins. Honey contains 70% more caloried per spoonful than sugar, and both sweeteners enter the circulation as fructose and glucose.

Foods with a high content of sugar are jams, jellies, and dried fruits. Fresh fruits and vegetables provide a wide range of sucrose, glucose, starch, and nondigestible carbohydrates, depending on the species, variety, and state of ripeness.

Dried cereal grains such as wheat, rice, and corn contain 33–37% carbohydrates by weight. These foods are generally the cheapest source of calories. Because they make up such a large part of the diet, they provide substantial amounts of protein, fiber, minerals, and vitamins to the daily ration in most of the world.

IV. DIGESTION OF CARBOHYDRATES

All carbohydrates made up of more than one sugar unit must be hydrolyzed before being absorbed or utilized by cells. The breakdown of carbohydrates to smaller units or to simple sugars sometimes begins with food-processing and cooking.

When a food is eaten, digestion of carbohydrates begins in the mouth, where the salivary enzyme amylase splits starch into smaller units. In the stomach, the acid environment inhibits amylase but favors some chemical breakdown of sugar and starch. In the small intestine, the well-mixed liquid products of gastric digestion encounter the enzymes of pancreatic juice and of the intestinal cells. Pancreatic juice contains amylase that breaks down starch to short, straight chains of glucose units (maltose, maltotriose) and to short, branched chains of 5–10 glucose units, known as α-dextrins. The enzymes of the brush border of the cells lining the intestine complete the breakdown of the disaccharides, α-dextrins, and maltotriose. The brush-border enzymes include maltase, sucrase, isomaltase, and lactase, which hydrolyze maltose and maltotriose, sucrose, α-dextrins, and lactose, respectively; the simple sugars released (glucose, fructose, and galactose) are readily absorbed.

Carbohydrate digestion is generally efficient. Impaired digestion and absorption due to low activity of one or more digestive enzymes allows the carbohydrate to pass to the large bowel, where it becomes a nutrient for bacteria. The fermentation of the CHO by the bacteria leads to the production of gas, acid products, and toxins, which irritate the gut and lead to pain and diarrhea. The activity of lactase declines during childhood in about 80% of the population, except for Caucasians of Northern European origin, producing lactose intolerance. Among older children and adults of non-Caucasian ancestry, a dose of >10 g lactose, equivalent to a glass of cow's milk, quickly produces gastric pain, flatulence, and diarrhea. The only dairy products containing <10 g lactose per serving are butter, cream, cheese, cottage cheese, and some ice creams. Even in persons or infants who normally tolerate lactose, an episode of gastrointestinal infection and diarrhea can produce a transient decline in lactase activity and an intolerance for milk. [*See* Lactose Malabsorption and Intolerance.]

Young infants should not be fed starch because amylase activity is low in pancreatic juice until 1 or 2 months of age, leading to starch intolerance.

Some digestive enzymes adapt to the dietary supply of their substrate. Diets rich in sugar increase sucrase and maltase activity, but feeding more lactose does not increase lactase activity or improve lactose digestibility.

The passage of simple sugars from the gut mucosa to the blood is by diffusion and, for some sugars, also by active transport. The actively absorbed sugars are glucose and galactose, while fructose is absorbed more slowly by facilitated diffusion. A dose of >100 g pure fructose is only slowly absorbed, and the excess forms a concentrated solution in the gut that draws water from its surroundings by osmotic pressure and leads to diarrhea. The nonsugar sweeteners such as xylitol and sorbitol can produce a similar effect when consumed in large amounts.

V. GLUCOSE UPTAKE AND UTILIZATION BY TISSUES

After the digestion and absorption of a carbohydrate-containing meal, the liver takes up the glucose and other simple sugars transported from the gut. The main end product of carbohydrate digestion is glucose, as it makes up 70–90% of all the carbohydrates we eat. Of all the sugars, glucose has the tightest metabolic control, by hormones and other regulators. The liver can use fructose and galactose as fuel or can convert them to glucose and other products. Glucose is oxidized for energy. It can be stored as the polysaccharide glycogen, the excess can be converted into fat, and a carefully regulated amount is released into the circulation to be transported to other tissues. Most of the tissues obtain energy from the oxidation of fat and its breakdown products plus glucose, but the brain, red blood cells, and developing fetus are almost entirely dependent on glucose for fuel.

After a meal, in response to blood glucose levels >100 mg/100 ml, insulin is released from the pancreas. This hormone has a key role in regulating blood glucose and the metabolism of many tissues. Insulin increases the uptake of glucose by muscle and adipose tissue, promoting the storage of glycogen, protein, and fat. If no food is consumed for several days, the tissues gradually adapt to utilizing the fuels obtained from the breakdown of fat stores and tissue proteins, but the body still needs about 130 g glucose per day. Initially, glucose is obtained by breaking down glycogen stores in the liver, mediated by the hormone glucagon or the stress hormones. After 1 or 2 days, liver glycogen stores are exhausted and new glucose is made in the liver from the carbon chains produced in the partial breakdown of fats and proteins. Glucose must also be manufactured from other substrates in the liver when the diet provides less than about 130 g glucose per day. The provision of enough carbohydrate in the diet is protein-sparing because it prevents the use of amino acids for the synthesis of glucose. [*See* Insulin and Glucagon.]

Under stress conditions, hormones that increase blood glucose levels are released. Glucose is the most important fuel for muscles during rapid and intense exercise. The effect of a surge of anger and fear is to fuel the muscle for "flight or fight." Unfortunately, in a civilized context where persons do not immediately respond to stress by violent movement or running away, the increased glucose is channeled into the synthesis of triglycerides.

Glucose absorbed into the cells of a tissue can follow several pathways. It can be used to produce energy, by hydrolysis to lactate in the absence of oxygen, as occurs in muscles during intense contractions, or by complete oxidation to carbon dioxide and water. Glucose can also be used for making other substrates or products for energy storage. In muscle and liver, it can be stored as the polysaccharide glycogen. In adipose tissue and the liver, it can be converted into fat and stored. It is directly or indirectly a substrate for making dispensable amino acids, cholesterol, nucleic acids, mucous secretions, and components of cartilage and for detoxifying certain drugs in the liver.

VI. UTILIZATION OF DIETARY CARBOHYDRATES

A. Characteristics of the Consumer

1. Age

The digestion of carbohydrates and the utilization of glucose by the tissues are affected by age. Lactose digestibility is greatest in infants, declines sharply in childhood in most ethnic groups, and declines further in old age. The sensitivity of tissues to insulin declines in old age, especially in persons who are obese and inactive. A drink containing 75 g glucose can produce high levels of blood glucose (120–180 mg/100 ml) for at least 2 hr. Although this is not diabetes, it is defined as glucose intolerance.

2. Physical Activity

The energy for intense, brief exercise is produced by the anaerobic breakdown of glucose to lactate, whereas prolonged low-intensity exercise is fueled by the oxidation of glucose and fat. The muscles of endurance-trained athletes contain more glycogen and have a greater capacity for taking up glucose from the blood, yet during aerobic exercise they produce energy from a fuel mix that is richer in fat compared with the muscles of sedentary persons. The increased ability of trained muscles to take up glucose is a transient effect of individual bouts of exercise, lasting for 2–3 days.

3. Genetic Background

Digestion and absorption of certain carbohydrates can be impaired because of a genetic deficiency in specific digestive enzymes. The problem of impaired utilization of glucose and inadequate regulation of blood glucose has a strong genetic component. In diabetes mellitus, high-circulating glucose coexists with poor nutrition of tissues because the lack of insulin prevents adequate entry of glucose into cells. Juvenile diabetes is a serious disease requiring insulin administration. The more common form of diabetes is a milder disease that emerges during middle age, especially in persons who are excessively fat and whose body fat is accumulated in the trunk rather than the hips and thighs. Obesity is found in 80% of patients with adult diabetes. In the United States, persons of Native American ancestry, including most Mexican Americans, have a three to four times greater risk of developing diabetes than Caucasians or Blacks. Maintaining appropriate body weight by dieting and exercise decreases the risk of diabetes in susceptible persons. Chronic high blood glucose in diabetics produces accelerated atherosclerosis, nerve damage, cataracts, and capillary damage that can lead to kidney disease and retinal damage.

Abnormally low blood glucose, or hypoglycemia, can occur with an overdose of insulin, with very prolonged exercise, or, more rarely, because of an inherent inability to increase blood glucose during fasting. True hypoglycemia is a rare disease. Its symptoms are headache, fatigue, confusion, seizures, and even loss of consciousness brought on by fasting 8–14 hr. These symptoms can be avoided by eating small frequent meals.

B. Characteristics of the Diet

Overnutrition leading to fatness impairs glucose utilization, whereas weight loss in obese subjects improves glucose utilization. The composition of the diet has immediate and long-term effects. The increase in blood glucose and accompanying insulin is different for the various carbohydrates and is affected by the physical composition of food, the presence of other nutrients, and the presence of different sorts of dietary fiber. After consuming glucose or sugar, blood glucose immediately increases. The response to eating complex carbohydrates is slower but varies for different starches. The glucose and insulin response to carbohydrate consumed as potatoes is greater than the same amount of carbohydrate consumed as rice. The glucose response to apple juice is greater than for applesauce, which in turn is greater than for a raw, unpeeled apple, reflecting the difference in rate of digestion and absorption due to fiber.

Changing to a high-carbohydrate diet improves glucose utilization by the tissues. Some scientists attribute this effect to the increased fiber in diets that are rich in complex carbohydrates. Conversely, eating a high-fat diet impairs glucose utilization, probably because it involves lowering carbohydrate intake. Excess sugar is not a cause of diabetes, but it is harmful for persons who have the disease. The easiest way to increase the carbohydrate content of the diet is to choose low-fat foods, as shown in the examples of Table III.

The micronutrients affect the efficiency of glucose utilization in cells because many enzymes required for glucose metabolism derive from the B vitamins and require mineral cofactors such as magnesium, zinc, and chromium.

VII. EFFECTS OF A VERY HIGH-CARBOHYDRATE DIET

A. Athletic Performance

Endurance-trained persons who compete in events lasting several hours have found that an increased store of muscle glycogen can delay fatigue by about 23% or allow a greater speed toward the end of a race. Experimentally, athletes can increase muscle glycogen by eating a diet where about 70% of the calories come from carbohydrates, during about 3 days before

TABLE III
Examples of Daily Meals for a High-Carbohydrate or a Low-Carbohydrate Diet

High-carbohydrate diet	Low-carbohydrate diet
Breakfast	
English muffin (large)	Whole-grain bread (thin slice)
Jelly	Butter
Cornflakes	Fried egg and bacon
Coffee and skim milk	Coffee and cream
Sugar	Artificial sweetener
Lunch	
Grilled chicken (no skin)	Tuna in oil
Steamed rice	Crackers
Steamed zucchini	Avocado and lettuce
Bread roll	Creamy salad dressing
Lemonade	Diet soda
Grapes	Chocolate chip cookies
Dinner	
Pea soup	Hamburger on a roll
Spaghetti with tomato sauce	French fried potatoes
Roasted apple with syrup	Ice cream
Soda	Diet soda
Snacks	
Popcorn	Peanuts
Nonfat yogurt with sugar	Full-fat yogurt, no sugar
Crackers	Doughnut
Skim milk	Whole milk

competition, while reducing training intensity. A typical precompetition diet is shown by the high-carbohydrate meals in Table III. However, a large intake of carbohydrate immediately before exercising is not advisable, especially in the form of rapidly absorbed glucose or sugar. This will produce a surge of insulin at a time when exercise itself is causing the rapid removal of glucose from the blood to the muscles. The combination of an insulin peak together with exercise can drive down blood glucose to levels that cause fatigue, discomfort, and reduced endurance. At the end of an exercise bout, consuming dilute solutions of sugar or other easily assimilated carbohydrates helps rapidly restore body water and muscle glycogen. This is important for sports involving repeated events, such as triathlons or cross-country cycling.

B. Caries

A linear relationship exists between a country's per capita sugar intake and the average number of dental caries in its inhabitants. The bacteria that cause caries reproduce on any food particles adhering to teeth, and starch as well as sugar can support the formation of plaque and bacterial growth. The most cariogenic foods, however, are rich in sugar and sticky, such as chewy candy, cookies, or raisins, while the dissolved sugar in sodas or milk-moistened breakfast cereals is less harmful. [*See* Dental Caries.]

C. Sugar and Hyperactivity in Children

In recent years, sugar has been blamed for the hyperactivity of some children, despite little experimental evidence. Properly designed double-blind studies, in which neither the children nor the investigators could tell if sugar or a fake dose of sugar was given before examining behavior, have not shown that sugar has any effect on hyperactivity.

D. Atherosclerosis and Cardiovascular Disease

High-carbohydrate, low-fat diets reduce plasma cholesterol and blood lipids only if the diet is also rich in fiber. Diets rich in complex carbohydrates are made up of a variety of plant foods, which include high amounts of different types of fiber. [*See* Atherosclerosis.]

BIBLIOGRAPHY

Dobbing, J. (1989). "Dietary Starches and Sugars in Man: A Comparison." Springer-Verlag, London.

Gracey, M., *et al.* (1991). "Sugars in Nutrition." Raven, New York.

MacDonald, I. (1988). Carbohydrates. *In* "Modern Nutrition in Health and Disease" (M. E. Shils and V. R. Young, eds.). Lea & Febiger, New York.

National Research Council. (1989). "Recommended Dietary Allowances," 10th Ed. National Academy of Sciences, Washington, D.C.

Sipple, H. L., and McNutt, K. W. (1974). "Sugars in Nutrition." Academic Press, New York.

Nutrition, Dietary Fiber

ROBERT RASIAH SELVENDRAN
BBSRC Institute of Food Research, Norwich Laboratory

GLOSSARY

α- and β-linkages Aglycone (glycosidically linked group) on carbon-1 is axial when α-linked and equatorial when β-linked; this confers helical structure on starch and linear structure on cellulose; alimentary enzymes are α-endoglucosylases and, therefore, hydrolyse starch but not cellulose.

Cation-exchange capacity Ability of a fiber preparation to exchange some of the cations it contains for others in solution

Cellulose Linear polymer of β-(1→4)-linked D-glucose and the only truly fibrous component of plant cell walls

Dietary fiber Mainly composed of plant structural materials and resistant to digestion by endogenous enzymes

Hemicelluloses Complex cell wall polysaccharides composed mainly of neutral sugars; they are not water soluble but are soluble in cold alkali

Lignin High molecular weight aromatic polymer associated with woody and some specialized tissues

Nonstarch polysaccharides Dietary polysaccharides other than starch, and mainly plant cell wall in origin

Pectin Branched polymer containing a backbone of partially methyl-esterified D-galacturonic acid interspersed with rhamnose residues, some of which carry short side chains of neutral sugars

Resistant starch Starch in cooked and processed foods that is resistant to amylolytic degradation

Starch Plant storage polymer of glucose, which exists in two forms: amylose, an unbranched molecule containing α-(1→4) linkages, and amylopectin, a branched molecule containing α-(1→4) and α-(1→6) linkages

Water-holding capacity Amount of water that can be taken up by 1 g of dry fiber

DIETARY FIBER (DF) MAY BE CONSIDERED A STRUCtural component of plant foods unavailable to digestion by human intestinal enzymes during normal gut transit. The cell walls of vegetables, fruits, and cereals are important as sources of DF. The composition and physicochemical properties of DF polymers influence the rate of digestion and absorption of nutrients, particularly in the small intestine. The DF polymers eventually pass from the small intestine to the colon, which contains vast numbers of anaerobic bacteria that can break down the polymers to release simpler molecules, some of which can be absorbed to provide energy. The fermentable components of DF also contribute to bacterial growth and, therefore, to fecal biomass, whereas some materials resist fermentation and only provide physical bulk to the feces. Both these effects have a laxative effect which influences the long-term health of the large bowel.

I. UNDERLYING HYPOTHESIS AND DEFINITION OF DIETARY FIBER

The role of DF in human nutrition and health has become increasingly topical in recent years. This revival of interest can be traced to the hypothesis that the consumption of low-fiber diets is a common etiological factor in many metabolic and gastrointestinal diseases of the Western world. Although the amount of fiber ingested is small in relation to the total diet, fiber appears to exert a major influence on the metabo-

ENCYCLOPEDIA OF HUMAN BIOLOGY, Second Edition, VOLUME 6.

lism of the gastrointestinal tract, and these effects are dependent on the physicochemical properties of the fiber and its components. The following two statements outline the "DF hypothesis."

1. A diet rich in foods that contain plant cell walls (e.g., high-extraction cereals, fruits, vegetables) is protective against a range of diseases, in particular those prevalent in affluent Western communities (e.g., constipation, diverticular disease, large bowel cancer, coronary heart disease, diabetes, obesity, gallstones).
2. In some instances, a diet providing a low intake of plant cell walls is a causative factor in the etiology of the disease, and in others it provides the conditions under which other etiological factors are more active.

The hypothesis, as stated, implies that the essential difference between protective diets and nonprotective diets is the amount of plant cell wall material they provide, and the protection is, or is derived from, the plant cell walls in the diet.

DF was initially defined as "the skeletal remains of plant cells, in our diet, that are resistant to hydrolysis by the digestive enzymes of man." As this definition did not include the polysaccharides present in some food additives, such as pectins, guar gum, alginates, and modified starches, the definition was later extended to include "all the polysaccharides and lignin in the diet that are not digested by the endogenous secretions of the human digestive tract." Accordingly, for analytical purposes, the term DF refers mainly to the nonstarch polysaccharides (NSP) and lignin in the diet. While the revised definition is generally accepted, it should be kept in mind that the polysaccharides of food additives generally constitute only a very small proportion (<2–3%) of the DF component of most diets. Hence, most of our DF intake comes from the cell walls of vegetables, fruits, cereal products, and other seeds.

The principal components of DF are complex polysaccharides, some of which are associated with polyphenolics (including lignin) and proteins. The other noncarbohydrate components of cell walls, such as cutin, waxes, suberin, phenolic esters, and metal ions (e.g., Ca^{2+}), are quantitatively minor constituents of most plant foods, but some of them have significant effects on the properties and physiological effects of DF. Although most of the DF constituents may survive digestion in the proximal gastrointestinal tract, a significant proportion of them are degraded by the microorganisms of the human colon.

The average intake of DF in the United Kingdom (and in most Western countries) is about 20 g per person per day and, of this, about one-third comes from cereal sources. There is considerable variation between individuals; those with a high consumption tend to obtain more from cereal foods. The British intake of DF is small compared with that of a rural African diet, which might contain 60–150 g DF per day.

II. MAIN COMPONENTS OF DIETARY FIBER

A diversity of plant tissues and organs constitute the vegetables, fruits, and cereals in the diet, but three main types of tissue predominate: (1) ground tissue (parenchyma), (2) supporting and conducting tissue (vascular), and (3) covering or protecting tissue (epidermal). Some supporting and strengthening tissue (sclereids) may sometimes be present. Parenchymatous tissues are the main source of vegetable and fruit fibers because the vascular bundles and sclereids of many vegetables are still relatively immature and are only slightly lignified when the vegetables are harvested and eaten, and the lignin content of the cell wall material is usually <5% (w/w). Lignified tissues are of greater importance in cereal sources—such as wheat bran and bran-based and dehulled oat products. Wheat bran contains the lignified outer layers of the grain and the aleurone layer, and the lignin content of the cell wall material is about 12% (w/w). Wheat bran makes a significant contribution to cereal fiber intake. Because the nature of the carbohydrate polymers associated with various types of tissue from dicotyledonous and monocotyledonous plants are different, the polymers present in these species will be discussed separately. Water is also an important component of the cell wall and is present in various amounts—high (~80%, w/w) in the parenchyma cell walls of most tissues, except mature dry seeds, but low (~10–15%, w/w) in secondary walls.

Although several methods have been used to estimate the DF content of foods, the two methods now most commonly used are the AOAC (Association of Official Analytical Chemists) method and the one based on NSP content. The AOAC method, which is used in the United States, is an enzymatic gravimetric method in which starch and protein are first degraded by enzymes, ethanol is added to give a final concentration of 80% (v/v), and the insoluble material is then dried and weighed. After suitable correction for residual protein and minerals, a value for DF is obtained. The NSP method, which is used in the United Kingdom, is an enzymatic chemical method. This method uses a different combination of starch-degrading enzymes and treatments, and ethanol precipitation, but the insoluble residue is analyzed chemically for nonstarch polysaccharides, which are calculated by summing the individual neutral sugars and uronic acid. Values for DF by the AOAC method are generally greater than those obtained by the NSP method.

Both the AOAC and NSP methods have advantages and disadvantages. The AOAC method has the advantage of causing much less degradation of pectic polysaccharides, and therefore of giving more realistic values for soluble fiber content of vegetables and fruits. It also has the advantage of including lignin, cutin, and suberin in the DF value. However, it has the disadvantage of also including Maillard reaction products in certain processed foods, and some starch resistant to the enzymic treatment. Furthermore, there is no chemical characterization of the DF. The advantages of the NSP method are that the NSPs are at least partially characterized and there is almost a complete removal of starch. The disadvantages are (1) the conditions used for obtaining values for soluble fiber cause much degradation of pectic polysaccharides, giving elevated values for soluble fiber, about three to four times the actual values; (2) about 5–10% of the sugar residues are degraded during acid hydrolysis and are therefore not measured; and (3) only NSPs are determined; lignin, cutin, and suberin are not included in the values. Although the following sections give DF values obtained by the NSP method, the relative merits of the methods should be kept in mind.

Some processed starch-rich products such as potato powder and cornflakes contain a significant but variable level of retrograded or "resistant" starch. Resistant starch is a term used to describe starch that is resistant to hydrolysis by the amylolytic enzymes that are used in the analysis of DF. It is formed when gelatinized starch retrogrades when cooled and dried, and could arise from a strong hydrogen bond formation between OH groups on adjacent starch molecules. This starch becomes susceptible to enzymatic hydrolysis only after vigorous chemical treatment, usually with alkali. Most of the resistant starch of processed products would survive small intestinal transit to be degraded by colonic bacteria.

A. Cell Walls of Dictotyledonous Plants

The nonstarch polysaccharides are the major contributors to the DF content of fruits and vegetables and usually comprise between 2 and 2.5% of the plant organs' fresh weight. These organs consist mainly of parenchymatous and growing tissues. Leguminous seeds have a more variable NSP content (e.g., dried peas, 18.6%; haricot beans, 17.1%; chick peas, 9.9%; butter beans, 15.9%) on a dry weight basis.

1. Parenchymatous and Growing Tissues

The parenchyma cell wall is a specialized form of extracellular matrix surrounding the protoplast and exterior to the plasmalemma of the cell. Each cell interacts with its neighbors through the intercellular layers of the middle lamellae, which are rich in pectic substances and are cross-linked by Ca^{2+} to form various tissue types. Removal of the Ca^{2+} leads to cell separation. The dry matter of the cell walls of soft tissues is composed of pectic substances (40–45%), cellulose (30–35%), hemicellulosic polymers (10–15%), proteins (5–10%), and polyphenolics (5%). The proteins can occur as glycoproteins or as proteoglycans, in which the protein component carries polysaccharide substituents, and these appear to serve as cross-linking agents within the walls. Simple and polyphenolics can also serve to cross-link the wall polymers. By virtue of the various types of cross-links, the primary cell wall components are organized into a complex covalently linked "macromolecular" matrix. Within this matrix are two separate morphological phases; a complexed continuous matrix, containing mainly noncellulosic polysaccharides, which are ester cross-linked and in which distinct microfibrillar structures, composed mainly of cellulose, are dispersed. These appear to be cross-linked to the noncellulosic polysaccharides via glycoproteins and highly branched pectic polysaccharides. The whole structure is bathed in an aqueous medium. It is, therefore, essential to recognize that DF in food should not be regarded as mainly the sum of the component cell wall polymers—the organization of the components into discrete structures confers the properties on DF.

a. The Microfibrillar Component: Cellulose

Cellulose is the major structural polysaccharide of the cell walls of all higher plants. Cellulose is a polymer characterized by long chains of β-(1→4)-linked D-glucopyranosyl residues, whose degree of polymerization is of the order 10,000. Because of the high molecular chain length and considerable capacity of the free hydroxyl groups to hydrogen-bond, both inter- and intramolecularly, cellulose forms fibers of remarkable strength. The fibers are arranged in an ordered manner within the microfibrils, which have crystalline and amorphous regions. The relative proportions of these regions may account for the differences in properties between cellulose in the primary and secondary walls. The amorphous regions may incorporate other polymers, which probably link the microfibrils to the matrix polysaccharides. This inference is based on the fact that the α-cellulose residues isolated from the cell walls of all parenchymatous tissues of vegetables and fruits contain small but significant amounts of associated pectic polysaccharides, wall glycoproteins, and polyphenolic compounds.

Cellulose is insoluble in water but is soluble in 72% (w/w) sulfuric acid because sulfonation takes place at the hydroxyl group on carbon-6 (C-6), breaking the hydrogen bonds. In primary cell walls, the cellulose microfibrils are "coated" with a layer of xyloglucans (bound by hydrogen bonds), enabling the insoluble cellulose to be dispersed within the wall matrix. A proportion of the C-6 hydroxyl groups can be chemically replaced with carboxyl groups to give carboxymethyl-cellulose (CMC), which is water soluble and is used as a food additive. The water-holding capacity (WHC) of cellulose is ~5 g water/g cellulose, whereas that of CMC is 50–60 g water/g CMC. This shows how the properties of DF polymers can be dramatically altered.

b. The Matrix Components: Noncellulosic Polymers

The major *hemicellulosic polymers* of parenchymatous tissues are xyloglucans, glucomannans, and proteoglycan complexes. The xyloglucans and proteoglycans constitute approximately 7–10% and about 5% of the dry weight of the cell walls, respectively. Xyloglucans contain a cellulosic β-D-glucan backbone to which short side chains are attached at C-6 of at least one-half of the glucosyl residues. α-D-Xylopyranosyl residues appear to be directly linked to glucosyl residues, but these side chains may be extended by the apposition of β-D-galactopyranosyl, α-L-arabinofuranosyl, or α-L-fucopyranosyl-(1→2)-β-D-galactopyranosyl residues. As mentioned before, the xyloglucans are hydrogen bonded to cellulose microfibrils, which aids cellulose dispersion through the wall.

Growing evidence indicates the occurrence of small but significant amounts of proteoglycans—complexes containing polysaccharide, protein, and polyphenolic moieties—in cell walls of soft tissues. Glycoproteins are also present in small amounts, and both polymers seem to cross-link the wall polysaccharides. The hydroxyproline contents of both types of protein complex vary; the glycoproteins generally contain higher levels. Parenchymatous tissues of most vegetables contain relatively small amounts of wall glycoproteins, although the cell walls of legume parenchyma, such as runner bean pods, contain fairly high levels of hydroxyproline-rich glycoproteins. [*See* Proteoglycans.]

The *pectic substances* are a complex mixture of colloidal polysaccharides, which can only be partially extracted from the cell walls with hot water, hot dilute mineral acid, or hot aqueous solutions of chelating agents such as ethylenediaminetetraacetate or ammonium oxalate. The solvent action of the chelating agents depends, in part, on their ability to complex with Ca^{2+} held in the walls by pectins. Recent work has shown that the above conditions of extraction result in considerable breakdown of the pectins by β-eliminative degradation, giving rise to two distinct polysaccharide fractions—one enriched in galacturonic acid with relatively small amounts of neutral sugar residues (usually designated rhamnogalacturonans), and the other low in acidic residues but rich in neutral sugar residues (e.g., pectic arabinans, galactans).

Rhamnogalacturonans are the major constituents of pectic substances in which a proportion of the galacturonic acid residues are present as methyl esters. The rhamnogalacturonan backbone consists of chains of α-(1→4)-D-galacturonosyl residues interspersed with (1→2)- and (1→2,4)-linked L-rhamnopyranosyl residues. Attached to the main chains, mostly through C-4 of rhamnosyl residues, are side chains consisting primarily of D-galactose and L-arabinose, and also lesser amounts of D-xylose, D-mannose, L-fucose, D-glucuronic acid, and some rare methylated as well as unusual sugars.

Recent studies have shown that most of the "neutral" pectic polysaccharides such as arabinans and galactans that have previously been isolated are in fact degradation products of more complex pectic polysaccharides. Arabinans contain mainly α-(1→5)-linked L-arabinofuranosyl residues to some of which terminal arabinofuranosyl residues are attached through C-2 and/or C-3, and galactans contain primarily chains of β-(1→4)-linked D-galactopyranosyl residues. Using nondegradative conditions of extraction, only very small amounts of neutral pectic polysaccharides have been solubilized from a variety of vegetables and fruits.

Recent work on pectic polymers has shed new light on the nature and distribution of the pectic polymers within the walls of parenchymatous tissues. The main conclusions are as follows: (1) The pectic polymers of the middle lamellae are rich in galacturonic acid and are highly methyl esterified, and a significant proportion of them can be solubilized by a dilute solution of cyclohexanediamine-tetraacetate (CDTA, Na salt) at 20°C. These pectic polymers are only slightly branched and serve to cross-link adjacent cells mainly through Ca^{2+}-pectate bridges, and probably give rise to rhamnogalacturonans under degradative conditions of extraction. (2) The bulk of the pectic polymers of primary cell walls can only be solubilized (after extraction of cell walls with CDTA) under mild alkaline conditions, which suggests that in addition to removing the bridging Ca^{2+}, the ester cross-links between the galacturonic acid residues and hydroxyl groups of sugar residues, and phenolics elsewhere in the wall matrix must be hydrolyzed. (3) A small but significant proportion of the highly branched pectic polymers is associated with the α-cellulose residue. (4) The ratio of rhamnose to galacturonic acid is ~1:40–50 for the bulk of the pectic polymers from the middle lamellae and ~1:10–20 for the pectic polymers from the primary cell walls; the relative amount of branched rhamnosyl residues is much higher in the latter. These comments serve to emphasize the extent to which the pectic polysaccharides are cross-linked within the parenchyma wall matrix.

2. Lignified Tissues (Secondary Cell Walls)

During maturation, considerable thickening of certain primary cell walls occurs, mainly through the preferential deposition of cellulose, hemicelluloses, and lignins, which constitute about 40, 25–30, and 20–25%, respectively, of the dry weight of the secondary walls. The hemicelluloses deposited are mainly acidic xylans and small amounts of glucomannans. Acidic xylans contain primarily β-(1→4)-linked D-xylopyranosyl residues, ~10–15% of which carry 4-O-CH_3-glucuronic acid or glucuronic acid residues on C-2. Lignins are high molecular weight aromatic polymers and consist, for the most part, of variously linked phenylpropane residues. The relative proportions of these residues vary in the lignins of different plants. Lignin can account for up to 25–30% of the dry weight of the walls of most supporting structures and appears to be covalently linked, mainly through ester links, with the glucuronic acid residues of xylans. This has the effect of cementing together the cell wall polysaccharides to form a rigid matrix and stratifying the wall. In commonly eaten vegetables and fruits, the overall lignin content of the cell walls is usually <5%. When the lignin content is >8–10%, the texture becomes unacceptably tough.

3. Seeds

The seeds of dicotyledonous plants can be classified as those that are free of an endosperm [i.e., nonendospermic (e.g., pea, bean)] and those that have an endosperm [i.e., endospermic, as in certain leguminous species (e.g., guar, locust bean)]. The former type of seed usually has starch as the main storage polysaccharide and their cell walls are derived mainly from the tissues of the cotyledons with some contribution from the testa. The cell wall polysaccharides of the cotyledons are similar to those of parenchymatous tissues and are composed mainly of peptic substances, cellulose, hemicelluloses (e.g., xyloglucans), and glycoproteins. The notable difference is that pectic arabinans (in the case of peas) and arabinogalactans (in the case of soy beans), both probably linked to rhamnogalacturonans, tend to predominate in the seeds.

In contrast to nonendospermic seeds, most endospermic seeds contain galactomannans, which are located in the endosperm cell walls. The galactomannans are solubilized during germination of the seeds. The galactomannans are essentially linear molecules but are highly substituted on C-6 of the β-(1→4)-linked D-mannosyl residues with single galactopyranosyl residues, which renders them hydrophilic and greatly enhances their ability to bind water.

4. Epidermal Protective Coverings of the Wall

The outer walls of epidermal cells of leaves, fruits, and many other aerial organs of plants are covered with a protective layer of waxes and cutin, and these are quantitatively minor constituents of DF. Generally the cutin penetrates and intermingles with the outer polysaccharides of the wall. Underground organs (e.g., tubers) are protected by another type of lipid-derived polymeric material—suberin. Both cutin and suberin are embedded in and overlaid with a complex mixture of relatively nonpolar lipids, which are called waxes. Because cutin and waxes are resistant to bacterial degradation, cutinized tissues may serve an important role in restricting the access of intestinal bacteria (and bacterial enzymes) to the cell wall polysaccharides of some vegetables and fruits, especially when they are not cooked. This phenomenon may be particularly significant in the case of leafy vegetables (e.g., lettuce, cabbage), which have spongy parenchyma cells sandwiched between cutinized layers.

B. Cell Walls of Monocotyledonous Plants (Cereals)

Dietary sources of monocotyledonous plants are mainly cereal grains, and these are an important source of DF. However, some vegetables, notably onions and leeks, are monocotyledons, and these have a NSP content and composition similar to the dicotyledonous fruits and vegetables discussed previously. The NSP content of cereal grains, on a dry weight basis, is variable (e.g., brown rice, 2.1%; pearl barley, 7.8%; whole wheat, 10.4%; porridge oats, 7.1%). The NSP content of cereal products also depends on the extent of refinement; high-extraction wheat products (e.g., whole meal bread, whole meal breakfast cereals) contain more DF than products from low-extraction white flours (e.g., whole wheat flour, 10.4%; whole wheat bread, 10%; wheat bran, 41.7%; 72% extraction white flour, 3.3%). Products derived from grains are also commonly used as breakfast foods in the form of bran-based products, flakes, or porridge. The milling process of wheat is a complex operation, involving separation into three main fractions: bran or pericarp with the attached testa and aleurone layer (in all about 12–17%), the endosperm itself (about 80–85%), and the wheat germ (about 3%). The removal of the bran may result in a loss of up to 75% of the total DF content of the grain. Therefore, a knowledge of the cell walls of the endospermous and lignified tissues of the grain is important, and this will be considered for some important cereals: wheat, barley (which is comparable to oats), and rice.

1. Endospermous Tissues and Aleurone Layer

Unlike parenchymatous tissues of dicots, the endosperm and aleurone cell walls are virtually free of pectic polysaccharides, contain very small amounts of cellulose (~2%), and are rich in hemicelluloses (~80–85%). Protein, containing negligible amounts of hydroxyproline, accounts for ~8–10% of the walls. The endosperm cell walls consist of an amorphous matrix of hemicelluloses in which microfibrillar structures are dispersed. The hemicelluloses of wheat endosperm are mainly highly branched arabinoxylans (~80%), associated with small amounts of phenolics such as ferulic acid, and the mixed linkage β-D-glucan content is only ~1–2%. Arabinoxylans contain primarily β-(1→4)-linked xylopyranosyl residues to some of which terminal L-arabinofuranosyl residues are attached through C-2 and/or C-3. In contrast, barley and oat hemicelluloses are rich in unbranched β-glucans (~75%) and are relatively poor in arabinoxylans (~15–20%). The ratio of (1→3) to (1→4) linkages in the barley β-glucans is ~3 : 7, and most of the β-glucans seem to be proteoglycans (polysaccharide–protein complexes). The hot water-soluble β-glucans of oat endosperm have an inherent viscosity not unlike that of pectins and specialist food gums such as guar gum.

Unlike wheat and barley endosperm cell walls, rice endosperm cell walls contain appreciable amounts of cellulose (~48%) and galacturonic acid-containing pectic substances rich in arabinose and xylose (~10%), acidic and neutral arabinoxylans, and some β-glucans. Therefore, the composition of rice endosperm cell walls appears to be intermediate between that of parenchymatous tissues of dicotyledons and that of wheat and barley.

The walls of wheat and barley aleurone cells are comparable and contain mainly arabinoxylans (~65–70%), with some β-glucans (~30%) and proteins. Small amounts of cellulose (~2%) and glucomannans are also present, and the wall polysaccharides are cross-linked by phenolic esters (e.g., ferulic acid). Lignin is absent. The arabinoxylans are neutral and are relatively slightly branched. The absence of glucuronic acid in these arabinoxylans and the absence of lignin indicate that the cross-linking of the wall polymers is not extensive. This scarcity of cross-linking contributes to their relative ease of degradation by bacteria.

2. Lignified Tissues of Wheat Grain (Beeswing Wheat Bran)

Beeswing wheat bran consists mainly of the outer coating of the grain and is about three cells thick, containing the cuticle, epidermis, and hypodermis. The cell walls contain about 53% hemicelluloses, 30% cellulose, 12% phenolics and lignin, and 5% protein. The hemicelluloses are of three types (in decreasing order of abundance): highly branched acidic arabinoxylans, slightly branched acidic arabinoxylans, and arabinoxylan–xyloglucan complexes. Cross-linkages have been found between the acidic arabinoxylans, xyloglucans, phenolic esters, and lignin; the glucuronic acid component of the arabinoxylans is also involved in cross-link formation. This high degree of cross-linkage limits the solubility of beeswing bran and, hence, accounts for its low degradability by bacteria.

C. Polysaccharide Food Additives: Gums and Mucilages

1. Food Gums

A range of polysaccharides, the majority of which are heteroglycans with branched structures, are used in small amounts in the food industry to give the desired texture to processed products. The food gums are included under the broader definition of the term DF, but their overall contribution to the total DF intake is not significant, except in certain special cases (e.g., guar-based breads for diabetics). Among the leading materials, in decreasing order of use in food, are pectins, gum arabic, alginates, guar gum, CMC, carrageenan, locust-bean gum, and modified starches. With the exception of modified starches and gum arabic, all of these polymers are derived from plant cell walls.

2. Seed Husk (Mucilage)

The husk (epidermal layers) of seeds of a number of species of the plant genus *Plantago*, referred to as ispaghula or pysllium, is obtained from seeds milled and suitably processed to give a fiber-enriched pharmaceutical preparation. The polysaccharides of the husk exhibit heterogeneity, and the main neutral sugars of the husk mucilage are D-xylose, L-arabinose, and L-rhamnose, which are present in the molar ratio 10 : 3.2 : 1, and the major polysaccharide has a highly substituted xylan backbone. A significant proportion of the rhamnosyl residues in the side chains carry terminal galacturonosyl residues, and these structural features of the constituent polysaccharides enable them to bind much water. As a consequence, the finely subdivided husk becomes readily dispersible in water, giving apparently viscous, gel-like mucilage, which retains many times its own weight of water. The mucilage has important physiological effects in both the small and the large intestines and, unlike most soluble fibers, such as pectins, is only partially degraded by colonic bacteria. Experimental evidence shows continued fermentation of the ispaghula husk in the distal colon. These properties of the husk are probably some of the factors responsible for its laxative action, and the husk is widely used for treating large bowel disorders such as constipation and diverticular disease.

III. PROPERTIES OF DIETARY FIBER

A. General Considerations

Diets rich in fiber exhibit characteristic properties depending on (1) structure of the tissue types, (2) the nature of the intracellular compounds, and (3) the form in which the food is taken—fresh, cooked, or processed. Because DF is ingested in the diet mostly as clusters of cells rather than as extracted purified polymers, the analysis of the extracted polymers can go only a part of the way in explaining the properties of DF. Nevertheless, an appreciation of how matrix polysaccharides behave during treatment with aqueous inorganic solvents helps one to understand how the wall matrix might behave during gut transit.

Several levels of organization can be described for cell walls and their constituent polymers within plant organs. The higher levels of organization have important implications for the study of the effects of DF, which are generally not well appreciated. Before being eaten, plant foods contain whole assemblies of cells; some are weakly attached to their neighbors while others are attached more strongly. During food preparation, mastication, and passage through the gastrointestinal tract, various events act to disrupt the tissue into small clumps of cells, which then form the food particles of the bolus. The degree to which cell–cell adhesion is reduced by the preparation of the food has a marked effect on the ease with which cells separate during mastication and gut transit. The cell walls, which act as physical barriers for the diffusion of cell contents and the entry of digestive secretions, therefore have a significant effect on the food in the gastrointestinal tract. The particle size of milled cereal products has an important effect on the properties, such as the WHC and solubility of the constituent DF polymers, by virtue of the increased surface area as the particle size decreases. There is also an increase in the degradability of the starch as particle size decreases. These considerations serve to show the importance of considering the physical state of the food product when assessing the role of DF. Some of the properties of DF that may be of importance in relation to its gastrointestinal function and that have definite or plausible links with the etiology of the diseases associated with low-fiber intake are discussed in the following sections.

B. Physicochemical Properties

1. WHC

The WHC of a fiber preparation is taken as a measure of its ability to immobilize water within its matrix. Water outside the matrix can be removed by filtration or by centrifugation and is referred to as "free" water. The water molecules that absorb to the DF polymers hydrogen bond at positions not otherwise involved in intra- and intermolecular bonding of the polysaccharide molecules. The WHC of pectins and wood cellulose are ~55–80 and 5, respectively; the WHC was determined by a dialysis bag method and is expressed as gram water/gram dry polymer after exposure to water at ambient temperature for 48 hr. The WHC of vegetable and wheat bran cell wall preparations measured by the centrifugation or filtration methods range from 20 to 25 g and 3 to 4 g/g dry cell wall material, respectively. Vegetable fiber contains much pectic material, and the reduced WHC relative to free pectin is due to the association of the pectic polysaccharides with each other through the formation of Ca^{2+} bridges and ester cross-links to form a "compact" structure within the walls. The WHC of wheat bran is low primarily due to (1) the presence of significant amounts of lignified fibers, which bind water poorly, and (2) the fact that the cells of bran are much smaller compared with parenchymatous tissues of vegetables and, therefore, "imbibe" much less water on hydration.

2. Viscosity-Enhancing and Gel-Forming Properties

Viscosity-enhancing and gel-forming properties of certain DF components (e.g., guar gum, oat gum, pectins) are important for two main reasons: (1) viscous components can, under appropriate conditions, delay gastric emptying and (2) viscous components possibly reduce absorption rates in the small intestine. The rate at which food is emptied from the stomach determines the availability of nutrients for absorption in the small intestine. Hence, a delay in gastric emptying contributes to slower transit and absorption. Mixing viscous polymers with food may modify the release of gastrointestinal and pancreatic hormones, gastrointestinal mobility, and morphology. Within the small intestine, viscous polymers may slow absorption by trapping nutrients, digestive enzymes, or steroids within the gel matrix and by slow mixing and diffusion. Recent work has shown that viscous polymers increase the apparent thickness of the "unstirred" layer at the mucosal surface, which slows the rate of transport of water-soluble nutrients and cholesterol across the mucosal wall. These properties are probably responsible for some of the potential therapeutic applications of guar gum, citrus pectin, oat gum, rolled oats, and oat bran in the management of hyperglycemia (in diabetics) and hypercholesterolemia. Fermentation of viscous polysaccharides in the hind gut may also have some role in the hypocholesterolemic effects of viscous polymers through the potential inhibitory effect of propionate on hepatic synthesis.

3. Bile Salt-Binding Characteristics

DF preparations from different sources have been shown to have variable capacity to bind bile salts, and the binding was estimated from the change in bile salt concentration on exposure of the solution (usually 1–5 m*M*) to the adsorbent. These and other related studies suggest that some sources of DF bind bile salts and cause an increase in their excretion in feces, which results in an increased demand for cholesterol conversion to bile salts to maintain bile salt pool sizes. Conversion of cholesterol to bile salts, but without a concomitant increase in the rate of cholesterol synthesis, is a possible contributory factor in the reduction of serum cholesterol levels (hypocholesterolemic effect). Deconjugated bile salts were more strongly bound than conjugated bile salts, and the binding was shown to be dependent on time of exposure and pH of the medium; maximal binding occurred at acidic pH mainly due to the "unionized" pectic polysaccharides, and the binding was considerably reduced at alkaline pH. Also, lignified fiber preparations from spent barley grains, wheat bran, and rice bran bound significantly higher levels of some bile salts compared with the delignified preparations. In the case of fiber from barley grains and alfalfa, the binding of bile salts was significantly reduced (~75%) after delignification; somewhat similar results were obtained with cell walls from heavily lignified tissues of mature runner bean pods. These results suggest that lignin may also play a role in the binding of bile salts.

Vegetable and cereal fibers, and some food gums, also have the capacity to interact with and bind lipid components of micelles and partially disrupt them and may consequently inhibit lipid absorption. The ability of "fiber" preparations to bind lipids from micelles decreases in the order guar gum > lignin > alfalfa > wheat bran > cellulose. For a realistic view of the effects of DF on lipid metabolism, the previously mentioned properties of fiber should be considered in conjunction with those discussed in Section III,B,2.

4. Cation-Exchange Capacity and Mineral Absorption

The main functional groups of cell wall polymers that can bind cations are the carboxyl groups of uronic acids. However, the nature of the uronic acid-containing polymers, their mode of occurrence within the cell wall complex, and the form in which the food is taken (fresh, cooked, or processed) are important considerations when assessing the cation-exchange capacity (CEC) of DF. Phytates, which are present in significant amounts in some plant foods, particularly cereal products (e.g., wheat bran, bran-based products), can also bind a range of divalent ions such as Ca, Mg, Fe, and Zn and, thus, influence their bioavailability. A few examples will be given to illustrate these points.

1. In the case of vegetable and fruit fibers, the nonesterified regions of the pectic polysaccharides of the middle lamellae and primary cell walls, most of which are involved in the formation of "egg box" junction zones with Ca^{2+}, have the potential to bind cations. The CEC of the fibers from immature cabbage leaves and apples is ~1.5 meq/g fiber. This value is less than the total uronic acid content of the fiber preparations, which is ~2.0 meq/g fiber, because some of the galacturonic acid residues of the pectic polysaccharides are esterified. While pectic polysaccharides of vegetable and fruit fibers may bind divalent metallic ions *in vivo*, the effect of such fibers on mineral bioavailability *in vivo* is not significant. In fact, in human studies, isolated pectin was shown to have no effect on Ca, Fe, and Zn balance, and vegetable and fruit fibers had a small variable effect. The latter observation is probably due to degradation of the pectates in the colon by bacteria and subsequent absorption of the associated minerals in the large bowel.

2. Cell walls of endospermous tissues of cereals contain negligible amounts of uronic acids, and the cell walls of lignified tissues (e.g., wheat bran) have only small amounts of uronic acid associated with the hemicelluloses. The associated glucuronic acids are ester linked to lignin and phenolics and are not available for cation exchange. However, phytates are present in the aleurone cells of wheat bran, and the acidic phosphate groups of potassium phytate would be available for cation binding, provided the molecules are exposed. Processing (e.g., milling) and cooking bran-containing products would be expected to disrupt the cellular structures of the aleurone cells and expose the phytates, which can then cause malabsorption of Ca, Fe, and Zn. Recent work has confirmed that phytates have a marked inhibitory effect on the absorption of nonheme iron, which accounts for 85–90% of the total dietary iron intake. It is of interest that only small amounts of phytates (5–10 mg phytate phosphorus) are needed in a meal to reduce iron absorption by half and that the effect of phytates is much less marked in a meal supplemented with ascorbic acid and/or meat. High intake of phytate-rich foods can also reduce the availability of zinc, while low-phytate, fiber-rich foods have no effect on zinc absorption. In a balanced Western diet, where fiber-rich foods account for 20–30% of the daily energy and zinc intake, the effect of phytate is easily counteracted by the animal protein content of the diet. Phytates have also been shown to bind calcium and reduce its absorption. However, for Ca, as for many other minerals, the net retention is determined not only by variations in absorption but also by variations in losses. The effect of high intakes of fiber, from bran-enriched products, on Fe, Zn, and Ca absorption may be undesirable for infants, children, and young adolescents, and recommendations for DF intakes in these groups should be different from those of adults.

5. DF and the Human Colon

a. Site of DF Breakdown and Action of Colonic Bacteria

Substantial evidence now indicates that the polysaccharides of DF are broken down during passage through the human alimentary tract. The site of breakdown of DF is undoubtedly the large intestine where the polysaccharides are degraded by the anaerobic bacterial flora. In recent studies, using ileostomy subjects as a model, it has been shown that the NSP from oats, wheat, and corn escape digestion in the small intestine; other studies using a similar model have shown complete recovery of NSP from pectin and wheat bran. In the first study, it was also found that the breakdown of starch in unprocessed cereal foods (rolled and steamed oats) was complete but that in processed foods (white bread and cornflakes) some "resistant" starch escaped digestion by the enzymes of the small intestine. In another investigation, the same workers studied the digestibility of starch in unripe and ripe bananas by the ileostomy subjects. Among other things, their results showed that with unripe bananas the amount of starch not hydrolyzed and adsorbed from the small intestine and, therefore, passing into the colon, may be up to eight times more than the NSP present. The starch in ripe bananas is degraded to a greater extent in the small intestine.

These studies demonstrate that not only the NSP of DF escape breakdown in the small bowel, but that a proportion of the starch also does. In addition to exogenous polysaccharides, endogenous polysaccharides such as mucins and mucopolysaccharides are also available to bacteria in the colon. The bacteria in the human colon are mostly (99%) anaerobic, saccharolytic organisms, which may be present at levels 10^{10}–10^{11} cells/g in feces, making up 25–55% of fecal solids; saccharolytic organisms derive their energy primarily from carbohydrates and their derivatives. The number of bacteria in the colon (10^9/g) is less than in feces. From fecal studies, it appears that the five major genera of the colon are *Bacteroides, Eubacteria,*

Bifidobacteria, Peptostreptococci, and *Fusobacteria*. Of these, the *Bacteroides* are thought to account for about one-third of the total organisms.

The main products of bacterial degradation are short-chain fatty acids (SCFA) (acetate, propionate, and butyrate), H_2, CO_2, and methane. In all regions of the large intestine, the SCFA are present in relatively high concentrations (90–130 mmol) and can be absorbed from the colonic lumen and metabolized. The SCFA contribute to normal energy metabolism and represent 60–70% of the energy available had the carbohydrate been absorbed in the small intestine. The gases produced by the bacteria diffuse into the bloodstream for expiration in the breath but are mostly expelled through the rectum as flatus.

b. DF and Fecal Weight

The well-established properties of DF are those of increasing fecal bulk and reducing transit time, and many studies have shown that while cereal fiber (enriched with bran) is very effective, vegetable fiber is less so. An increase in digesta bulk may cause colonic propulsion and a consequent reduction in transit time, but there are also other nondietary determinants of transit time. In a mixed diet, an increase in fecal weight is probably the result of a combination of three main factors: (1) increased mass from undegraded fiber, (2) increase in bacterial mass as a result of fermentation of the fiber, and (3) increase in water retained by the residual fiber–bacteria complex. This component will contain water entrapped by the degraded fiber and bacterial cells. An appreciable proportion of the water will be contained within the bacterial cells, which are ~80% water. With a typical British type of diet, bacteria account for about 55% of the total mass of fecal solids; undigested fiber and water-soluble material account for 17% and 24%, respectively. The water-soluble material consists mainly of minerals, fats, and nitrogen-containing compounds. It has been shown that the fermentability of a DF preparation is a major factor in its effect on fecal bulk. The fiber of vegetables and fruits (e.g., cabbage, carrots, apples) is far more rapidly degraded by colonic bacteria than fiber of cereal products containing lignified tissues (e.g., wheat bran); in the case of wheat bran, the polysaccharides of the aleurone layer are preferentially degraded. This is because vegetable fiber contains much higher levels of soluble cell wall polysaccharides, which allow the bacteria to penetrate the fiber matrix, and even the cellulose, which is not impregnated with lignin, can be readily degraded.

It has been shown in clinical trials that cabbage fiber is extensively degraded (~92%), stimulates bacterial growth, and leads to a modest increase in fecal weight. In contrast, fiber from wheat bran is only ~36% degraded due to the low solubility and extensive cross-linking of the component polysaccharides, which significantly reduce access to bacteria and subsequent breakdown. As a consequence, with bran the incompletely degraded fiber (which has appreciable WHC), associated water, and bacteria gave rise to a significant increase in fecal weight, about twice that of the vegetable fiber. However, the importance of the physical characteristics of the DF is illustrated by the fact that grinding wheat bran to a fine particle size can greatly reduce its fecal bulking ability. This is a result of the increased available surface area and also increased packing density of smaller particles. These and related studies show that the mechanisms by which fiber affects colonic physiology are dependent on the fermentability of the fiber, and this is related to the type of fiber ingested.

c. Mechanisms by which DF May Protect against Colorectal Cancer

Experimental studies on animals and humans have shown that dietary fibers may mediate protection against the development of colorectal cancer. The protective mechanisms proposed can be divided into two main groups: those in which the DF may be acting directly, and those which are a consequence of the DF being degraded by colonic bacterial enzymes and the products of fermentation. The first mechanism involves the binding of carcinogens by undegradable fiber components. If carcinogens are bound to undegradable fibers (soluble and insoluble) and associated bacteria, the possibility of effective interaction of the carcinogens with the colonic mucosal cells will be reduced and the carcinogens may pass out in the feces. Also, residual fiber can increase the bulk of the feces and hence reduce the concentration of carcinogen, reducing the possibility of effective interaction with the mucosal cells. Increased fecal bulk also results in shortened transit times through the gut which results in decreased times for interaction of carcinogens and mucosal cells. Shortened transit times generally result in reduced water absorption by the colon and hence wetter stools.

The second mechanism involves increased bacterial numbers and their possible consequences. The fermentation products of DF, such as SCFAs, lower the pH of the colon contents. With less readily degradable fibers, such as wheat bran and ispaghula husk, higher concentration of SCFAs are found in the distal colon, which may serve as nutrients for the epithelial cells, reducing the incidence of cancer known to be more common in the distal colon. The increased production of butyrate in the distal colon may have a particular importance in preventing later stages of cancer development because the butyrate is the preferred energy substrate for normal colonocytes, and *in vitro* butyrate has been shown to slow proliferation of colon cancer cells.

IV. CONCLUDING REMARKS

DF is derived primarily from the cell walls of edible plant organs; therefore, an attempt has been made in this chapter to describe the chemistry of some of the better characterized cell wall polymers of edible plants with particular emphasis on how this knowledge enhances our understanding of (1) the chemical and physical properties of DF, and the DF content of a range of plant foods, and (2) the mode of action of DF in the human alimentary tract, particularly the human colon, which suggests that most of the established physiological effects of DF are probably protective ones.

As it is becoming increasingly clear that the nature of the cell wall polymers associated with various tissue types are different, an effort has been made to describe the chemistry of cell wall polymers from different tissues. In general, the parenchymatous tissues of vegetables and fruits are rich in pectic polysaccharides and cellulose, whereas those of cereals are rich in hemicelluloses (β-glucans and/or arabinoxylans), and the constituent polymers are not highly cross-linked by phenolics. This is in contrast to the hemicelluloses of lignified tissues (of cereals) in which the polymers are highly cross-linked by phenolics and the cellulose is impregnated with lignin. Disruption of the fibrous structure results in increased access of the pancreatic enzymes to cell contents and, hence, increases the digestibility of starch, etc. Cooking encourages cell separation and increases the amount of soluble DF polysaccharides (e.g., pectin). Pectins and soluble food gums (e.g., guar, β-glucans) have a high viscosity in aqueous solution and can alter gastrointestinal function so as to slow the absorption of sugars, which is beneficial for diabetics. The ispaghula husk binds much water and forms a mucilage that alters gastrointestinal function significantly and in ways different from the previously mentioned soluble food gums.

One of the major roles of DF is the provision of NSP as substrates for the colonic bacteria, and these together with unabsorbed sugars and starch are the main sources of energy for the bacteria. The DF not degraded by bacteria, together with associated bacteria and entrapped water, make up the major components of feces, so that enhanced excretion of fecal matter is a characteristic of high-fiber diets. This property of DF may mediate protection against colorectal cancer. Several good reviews and books are devoted to various aspects of cell walls of edible plant organs and DF, and the reader is referred to them for further information.

BIBLIOGRAPHY

Eastwood, M. A. (1992). The physiological effects of dietary fiber: An update. *In* "Annual Review of Nutrition" (R. E. Olson, D. M. Bier, and D. B. McCormick, eds.), Vol. 12, pp. 19–35.

Harris, P. J., and Ferguson, L. R. (1993). Dietary fibre: Its composition and role in protection against colorectal cancer. *In* "Mutation Research," Vol. 290, pp. 97–110.

Kritchevsky, D., Bonfield, C., and Anderson, J. W. (eds.) (1988). "Dietary Fibre: Chemistry, Physiology and Health Effects." Plenum Press, New York.

Selvendran, R. R., and O'Neill, M. A. (1987). Isolation and analysis of cell walls from plant material. *In* "Methods of Biochemical Analysis" (D. Glick, ed.), Vol. 32. Wiley, New York.

Selvendran, R. R., Stevens, B. J. H., and DuPont, M. S. (1987). Dietary fiber: Chemistry, analysis, and properties. *In* "Advances in Food Research" (C. O. Chichester, E. M. Mrak, and B. S. Schweigert, eds.), Vol. 31. Academic Press, New York.

Selvendran, R. R., Verne, A. V. F. V., and Faulks, R. M. (1989). Methods for analysis of dietary fibre. *In* "Modern Methods of Plant Analysis: Plant Fibers" (H. F. Linskens and J. F. Jackson, eds.), Vol. 10. Springer-Verlag, Berlin/New York.

Southgate, D. A. T. (1989). Dietary fibre and the diseases of affluence. *In* "A Balanced Diet?" (J. Deobbing, ed.), Springer-Verlag, Berlin/New York.

Southgate, D. A. T., Waldron, K., Johnson, I. T., and Fenwick, G. R. (eds.) (1990). "Dietary Fibre: Chemical and Biological Aspects," Special Publication No. 83. Royal Society of Chemistry, London.

Taylor, T. G., and Jenkins, N. K. (eds.) (1986). "Proceedings of the XIII International Congress of Nutrition." John Libbey & Co. Ltd., London and Paris.

Trowell, H., Burkitt, D., and Heaton, K. (eds.) (1985). "Dietary Fibre, Fibre-Depleted Foods and Disease." Academic Press, London.

Vahouny, G., and Kritchevsky, D. (eds.) (1986). "Dietary Fibre: Basic and Clinical Aspects." Plenum Press, New York.

Nutrition, Fats and Oils

PATRICIA V. JOHNSTON
University of Illinois at Urbana–Champaign

GLOSSARY

Eicosanoids Biologically active 20-carbon compounds derived from the essential fatty acids

Essential fatty acids Long-chain unsaturated fatty acids that cannot be synthesized in the body and must be provided in the diet for the maintenance of good health

Lipoproteins System of soluble proteins by which lipids are transported in the blood; they are combinations of lipids and proteins that are more readily transported in blood than lipids alone

P/S ratio Ratio of polyunsaturated fatty acids to saturated fatty acids (excluding the monounsaturated fatty acids)

Triglycerides Predominant chemical form of the constituents of fats and oils

FATS AND OILS ARE PART OF A LARGE GROUP OF mainly water-insoluble compounds known as lipids. They are composed of triglycerides in which three fatty acid chains are linked to a glycerol backbone. The difference between fats and oils is their physical state at room temperature: fats are solid and oils are liquid. Dietary fat is mainly in the form of triglycerides from fats and oils, with smaller amounts provided by phospholipids. Food sources of fat can be either animal or vegetable. Animal fats contain more saturated fatty acids than most vegetable sources. Many vegetable oils are rich in polyunsaturated fatty acids and are sources of the essential fatty acids. Fat is digested and absorbed in the small intestine and most of it enters the lymphatic circulation in the form of fatty particles known as chylomicrons. These are transformed in the liver into the water-soluble lipoproteins that transport lipids in the blood to the tissues. Fat serves as a concentrated form of energy, increases the satiety value of foods, and improves palatability. Fat also serves as a carrier for the fat-soluble vitamins. In the body, fat serves as an energy reserve, as an insulating material, and (in the form of fat pads) as a protector of vital organs such as the kidneys. Many vegetable oils are rich in the essential fatty acids, linoleic and α-linolenic acids. These fatty acids are metabolized in the body to the precursors of the eicosanoids, the 20-carbon compounds that are potent regulators of many physiological functions.

I. CHEMICAL COMPOSITION OF FATS AND OILS

A. Triglycerides and Phospholipids

Fats and oils are part of a broad group of compounds called lipids, a word derived from the Greek *lipos* meaning "fat." Lipids comprise a diverse group of compounds that are primarily insoluble in water but soluble in organic solvents such as hexane, diethyl ether, and chloroform. Fats and oils are among the simplest of lipids and are composed mainly of triacylglycerols or triglycerides as they are commonly known. The triglycerides consist of three fatty acid chains joined in an ester bond to a glycerol backbone as shown below:

ENCYCLOPEDIA OF HUMAN BIOLOGY, Second Edition, VOLUME 6.

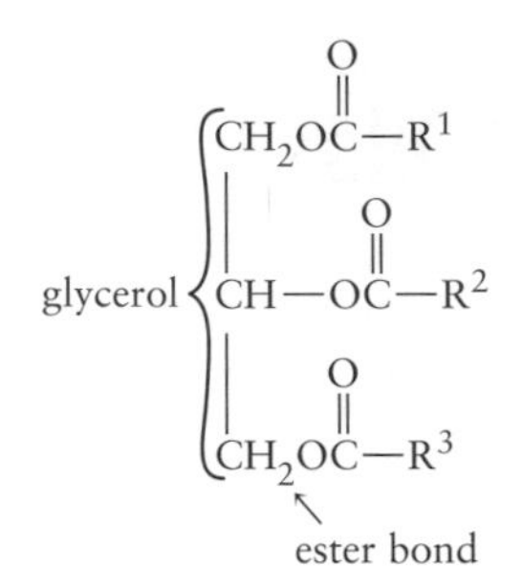

where R^1, R^2, and R^3 are fatty acyl chains that may be the same but are more often different. Fats and oils are mixtures of triglycerides; fats differ from oils in their physical states at room temperature: fats are solid and oils are liquid. It is usual when referring to dietary fats and oils to simply use the term *dietary fat* and it is assumed that this refers to the mixture of fats and oils derived from various food sources. The fat we consume therefore is primarily in the form of triglycerides, although small amounts of mono- and diglycerides having one and two fatty acyl chains only are usually present as additives. Sterols, such as cholesterol from animal fat sources and phytosterols from plants, are also consumed. Cholesterol is important as a constituent of cell membranes and as a precursor of some hormones. It is also synthesized by the liver. Depending on the source, varying amounts of phospholipids will also be consumed. These lipids are a major constituent of all cell membranes. They consist of the glycerol backbone with only two fatty acids linked to it; the third is occupied by a phosphate group joined to an alcohol:

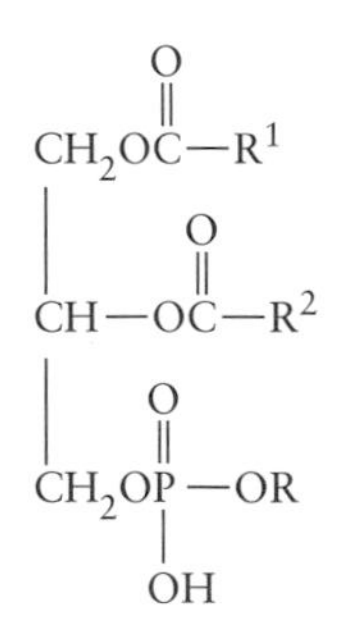

where R^1 and R^2 are fatty acyl chains and R is choline, ethanolamine, serine, or the carbohydrate inositol. Traces of other lipids may be present but they contribute little to the nutritional value of the dietary fat. [*See* Lipids.]

B. Fatty Acid Constituents of Glycerides and Phospholipids

Fatty acids are chains of carbon atoms that have a methyl group (CH_3) at one end and a carboxylic acid group (COOH) at the other. Carbon atoms have a valence of four and therefore are attached to hydrogen atoms, another carbon atom, and a hydroxyl (OH) or oxygen. Thus a typical fatty acid has this structure:

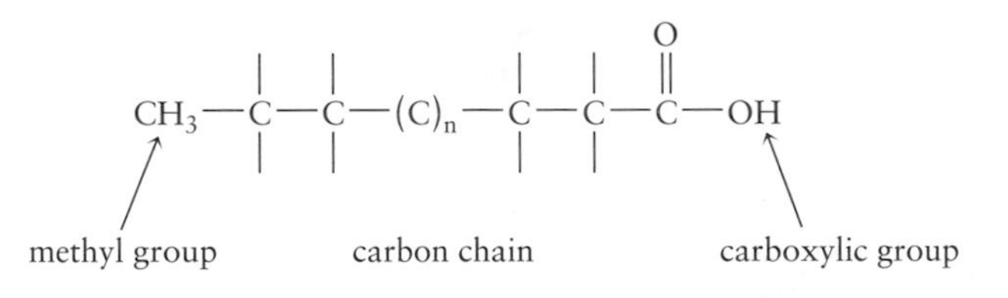

For all practical purposes, fatty acids in foods have an even number of carbon atoms ranging from 4 to 22.

There are two main classes of fatty acids: saturated and unsaturated. In saturated fatty acids all four valence bonds of all carbons are satisfied. In unsaturated fatty acids some hydrogens along the chain are missing and carbons form bonds with each other, giving rise to olefinic or double bonds, such as

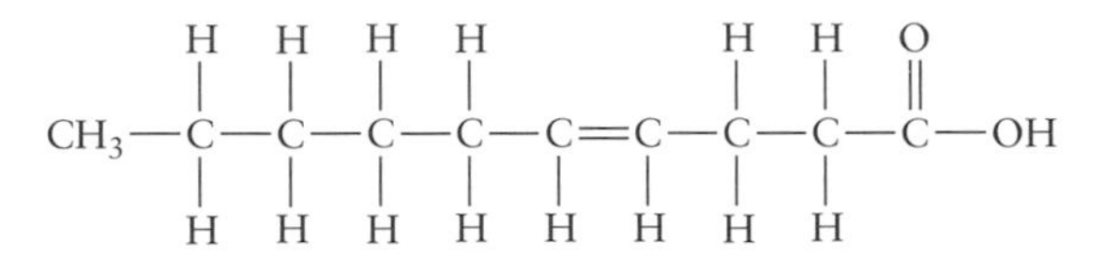

or more simply

$$CH_3(CH_2)_4{-}CH{=}CH{-}(CH_2)_2COOH.$$

1. Fatty Acid Nomenclature

Names of fatty acids are derived from the appropriate parent hydrocarbon. For example, if the fatty acid has eight carbon atoms the parent hydrocarbon is octane. Remove the terminal "e" and replace it with the suffix "oic" to give octanoic acid, $CH_3(CH_2)_6$ COOH. If an acid has one double bond the parent hydrocarbon is octene and the fatty acid is octenoic acid. Fatty acids with one double bond are known as monenoic acids; those with two are dienoic acids. Similarly, we have trienoic, tetraenoic, pentaenoic, and hexaenoic for those acids with three, four, five, and six double bonds, respectively.

There are two ways in which the position of the double bond is indicated. In the first, the terminal carboxyl(COOH) group is termed carbon 1. Counting back toward the methyl (CH_3) end, oleic acid $CH_3(CH_2)_7CH{=}CH(CH_2)_7\overset{1}{C}OOH$ becomes 9,10-octadecenoic acid or simply 9-octadecenoic acid. Linoleic acid with two double bonds, $CH_3(CH_2)_4CH{=}\overset{12}{C}H{-}CH_2CH{=}\overset{9}{C}H{-}(CH_2)_7\overset{1}{C}OOH$, is 9,12-octadecadienoic acid. The most common unsaturated fatty acids are shown in Table I.

Most naturally occurring fatty acids have unconjugated double bonds, that is, the double bonds are separated by one or more single-bonded carbon atoms. In most cases the double bonds are interrupted by a single methylene ($-CH_2-$) group. Conjugated double bonds, those adjacent to each other, do occur but they are rare. Because most naturally occurring fatty acids are unconjugated and their double bonds are interrupted by single methylene groups, it is possible to use a shorthand system for naming fatty acids. In this system the terminal methyl group carbon is termed carbon 1 and given the designation omega (ω) or n-. Counting toward the carboxyl end, the first double bond from the ω is numbered and the series of fatty acids is named by the position of this first double bond. For example, if a fatty acid is described as $18:2\omega6$ (or n-6), we have all the information necessary to draw its structure. The number before the colon gives the number of carbons in the chain (18). The number after the colon gives the number of double bonds (2), and the number after the ω gives the position of the first double bond from the methyl end (6). Since the double bonds are interrupted by a single methylene group, we can write the structure as

$$\overset{\omega}{C}H_3(CH_2)_4\overset{6}{C}H{=}CHCH_2CH{=}CH{-}(CH_2)_7COOH$$

or, named the other way, 9,12-octadecadienoic acid. Another example is $18:3\omega3$, which therefore has the structure

$$\overset{\omega}{C}H_3CH_2\overset{3}{C}H{=}CH{-}CH_2CH{=}CH{-}CH_2CH{=}CH{-}(CH_2)_7COOH$$

or 9,12,15-octadecatrienoic acid.

Most natural fatty acids have double bonds in the *cis* configuration. Compounds containing double bonds with atoms or groups on either side can have different spatial arrangements. If the atoms or groups on each side of the double bond are on the same side, the compound is said to be *cis;* if they are on the opposite side, the compound is *trans*. This is known as *geometric isomerism*. It is assumed unless stated otherwise that all double bonds are cis. *Trans* double bonds arise on biohydrogenation or commercial hydrogenation. If a fatty acid has *trans* double bonds it should be stated. For example $18:2\omega6$ *cis*, *trans*, or $18:3\omega3$ all *trans*.

This shorthand system for naming fatty acids not only is convenient but also has biochemical significance. Animals cannot desaturate toward the methyl end. Elongation and desaturation can only take place toward the carboxyl end of the chain. Plants and microorganisms can, however, place double bonds toward the methyl end. It so happens that most natural fatty acids fall into one of four families when named using the ω system. Natural fatty acids fall into one of the series shown in Table II. Once a fatty acid is in a family, any further elongations and desaturations in animal systems keep it in that family because of the lack of enzymes to desaturate toward the methyl end. Animals can synthesize the $\omega7$ and $\omega9$ series *de novo*:

TABLE I
Structures and Names of Common Unsaturated Fatty Acids

Structure	Systematic name	Trivial name
$CH_3(CH_2)_7CH{=}CH(CH_2)_7COOH$	9-Octadecenoic acid	Oleic
$CH_3(CH_2)_4CH{=}CH{-}CH_2{-}CH{=}CH(CH_2)_7COOH$	9,12 Octadecadienoic acid	Linoleic
$CH_3CH_2CH{=}CH{-}CH_2CH{=}CHCH_2CH{=}CH{-}(CH_2)_7COOH$	9,12,15-Octadecatrienoic acid	α-Linolenic
$CH_3(CH_2)_4CH{=}CH{-}CH_2CH{=}CH{-}CH_2CH{=}CH{-}CH_2CH{=}CH(CH_2)_3COOH$	5,8,11,14-Eicosatetraenoic acid	Arachidonic

TABLE II
Families of Naturally Occurring Fatty Acids

Series	Methyl terminal of fatty acid
ω3	$CH_3CH_2CH{=}CH-$
ω6	$CH_3(CH_2)_4CH{=}CH-$
ω7	$CH_3(CH_2)_5CH{=}CH-$
ω9	$CH_3(CH_2)_7CH{=}CH-$

16 : 2ω7 (from 16 : 1ω7, to 18 : 2ω7)

16 : 0 → 16 : 1ω7 → 18 : 1ω7 → 18 : 2ω7 → 18 : 3ω7 → 20 : 3ω7 → 20 : 4ω7

18 : 0 → 18 : 1ω7 → 18 : 2ω9 → 20 : 2ω9 → 20 : 3ω9

TABLE III
Some Common Fatty Acids

Trivial name	Omega nomenclature
Lauric	12 : 0
Myristic	14 : 0
Palmitic	16 : 0
Stearic	18 : 0
Palmitoleic	16 : 1ω7
Oleic	18 : 1ω9
Linoleic	18 : 2ω6
α-Linolenic	18 : 3ω3
γ-Linolenic	18 : 3ω6
Dihomo-γ-linolenic	20 : 3ω6
Arachidonic	20 : 4ω6
Timnodonic[a]	20 : 5ω3

[a]Also known as EPA (for eicosapentaenoic acid).

Animals lack the enzymes for synthesis of the ω6 and ω3 families and these must be ingested in the diet. Because of this the dietary precursors of these families are termed essential fatty acids (EFA). The EFA are very important nutritionally and they are discussed in detail in Section IV. The precursors of the ω6 and ω3 families are 9,12-octadecadienoic acid (18 : 2ω6 or linoleic acid) and 9,12,15-octadecatrienoic acid (18 : 3ω3) or α-linolenic acid. When ingested by animals, including humans, they undergo elongation and desaturation as follows:

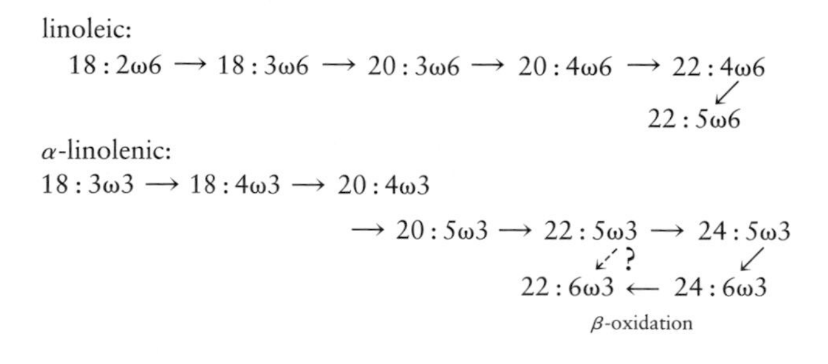

Because of the biochemical and nutritional significance of the omega naming system it will be the predominant system used throughout this text. Most fatty acids have trivial names and the use of these has persisted. Table III lists common fatty acids using their trivial names and omega nomenclature. Many other types of fatty acids occur in nature, including ones having hydroxy, keto, and cyclic groups, and branched chains. These are of little significance from a nutritional viewpoint.

2. Distribution of Fatty Acid among Triglycerides

The simplest system for numbering a triglyceride molecule will be used here. In this system the carbons on the glycerol moiety are numbered 1, 2, and 3, and 1 and 3 are regarded as interchangeable. There is a stereospecific numbering system in which positions 1 and 3 are not regarded as interchangeable. Using the simple system a typical triglyceride will look like

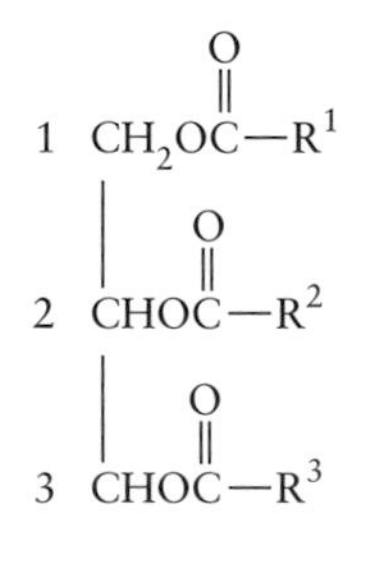

Fats and oils are mixtures of such triglycerides. Early in the development of lipid chemistry, researchers asked questions regarding the distribution of fatty acids among the triglycerides. The average fat or oil contains several different fatty acids. Was there some general rule by which one could predict the positions the various fatty acids would occupy? Several theories were proposed but technical difficulties hampered efforts to prove or disprove them. The most accepted theory, and one closest to determined values, is a modified version of the 1, 3, and 2 random distribution theory. This theory makes the 1 and 3 positions on the triglyceride identical and predicts that these will be occupied by identical kinds and proportions of acyl groups distributed at random. Similarly, in the 2 position acyl groups are distributed at random although a modification of the theory states that there is a preferential occupation of this position by 18-carbon unsaturated fatty acids. This theory is close to determined distribution patterns although some fats and oils are more random than others. The distribution pattern of the fatty acids confers particular properties on fats and oils that make them more or less suitable for particular food uses. Modification of natural fats and oils by hydrogenation, interesterification (or redistribution of fatty acid groups), and winterization to render them more useful for food use is common. For example, partial hydrogenation of vegetable oils for use in margarines and shortenings and winterization of some vegetable oils to remove trisaturated species of triglycerides (by lowering the temperature to 5°C) for use as salad oils are usual in the edible oil industry. Winterization involves holding an oil at 5°C, allowing the trisaturated species of triglycerides to precipitate, and removing them by filtration.

The hydrogenation process leads to the transformation of some *cis* double bonds in fatty acids to the trans form. The amount of *trans* fatty acids formed depends on the conditions of hydrogenation. Under present processing conditions and eating patterns only about 5–8% *trans* fatty acids are present in the diet. Some of this arises from nonprocessed sources such as milk and butter due to the biohydrogenation of fatty acids in the rumen of the cow. In the past there have been concerns regarding the possible health effects of trans fatty acids. Most health professionals now regard *trans* fatty acids as being similar to saturated fatty acids in that they have higher melting points than their cis counterparts. It should be noted, however, that the essential fatty acids 18:2ω6 and 18:3ω3 must have all their bonds in the cis form to exhibit activity as essential fatty acids (see Section IV,B,3). Hydrogenation conditions leading to high trans contents are much less common than several years ago. The winterization process merely removes more saturated triglyceride species from the salad oil and makes it remain liquid at refrigeration temperatures. Tempering and interesterification of fats are used to improve plasticity properties and have no apparent effect on nutritional value.

II. FOOD SOURCES OF FATS AND OILS

A. Trends in Fat Consumption

Throughout this century there has been an increase in the amount of fat consumed. For example, in 1987 the percentage of kilocalories from fat in the U.S. food supply was 43%, whereas the desirable dietary goals suggest 30–35%. The amount of animal fat consumed has, however, decreased as the use of red meat, eggs, and dairy products has declined. Over half the available fat still comes from animal products. In the last decade, both men and women in the United States have reduced their average daily fat intake by several grams. Moreover, there has been a shift to increased use of margarine, cooking oils, nondairy creamers, and skim milk rather than butter, shortening, and whole milk.

Some fats are readily recognized as visible fats; examples are butter, margarine, salad oils, and the fat surrounding meats. Such sources account for less than half of dietary fat, with the rest coming from invisible sources such as fat around meat fibers (marbling), homogenized milk, egg yolks, nuts, and whole-grain cereals. Total fat intake and types of dietary fat are influenced by many factors, including economical, social, and cultural. More affluent societies tend to consume more fat. Many Asian countries have a lower fat intake than the United States. Some Mediterranean countries (such as Greece) consume considerable amounts of olive oil. Though the efforts to reduce total fat and especially saturated fat are commendable, it should be remembered that intake of some fat is essential. Without fat in the diet we may lack essential fatty acids, calories, and sufficient fat-soluble vitamins.

B. Food Sources

The fat content and type of fat for some selected foods are shown in Table IV. Meats and foods of animal origin, such as many cheeses, are higher in fat than most foods of vegetable origin. Cheddar, edam, swiss, and colby cheeses contain as much as 8 g of fat per ounce; feta had 6 g; and cottage and skim milk ricotta have as little as 1 g per ounce.

It was reported recently that, contrary to what was

TABLE IV
Fat Content and Type of Fat in Some Selected Foods

Foods	Total fat %	Fatty acid % Saturated	Polyunsaturated[a]
Butter	81	62	4
Vegetable shortening	100	25–30	12–26
Poultry	5–20	30	20
Beef	10–40	50	4
Pork	20–35	39	11
Salmon	9	17	51
Tuna	5	41	30
Egg yolk	33	30	13
Peanuts	51	18	27
Avocado	18	19	12

[a]Mainly linoleic acid except in fish.

TABLE V
P/S and P–M/S Ratios for Selected Fats, Oils, and Foods

Foods	P/S ratio	P–M/S ratio
Fats and oils		
Corn	4.5	6.4
Soybean	4.1	5.8
Peanut	1.9	4.6
Canola	4.9	12.7
Olive	0.6	5.9
Coconut	0.02	0.16
Butter	0.05	0.51
Margarines		
Tub (safflower)	4.9	7.4
Stick (corn)	1.4	4.9
Animal fats		
Beef	0.06	0.78
Poultry	0.7	1.9
Fish		
Salmon	2.0	3.0
Tuna	0.5	1.0
Nuts		
Walnuts	7.0	9.5
Peanuts	1.9	5.6

previously thought, not all saturated fat has an adverse effect on serum cholesterol and lipoproteins. Stearic acid (18 : 0), found in beef fat and hydrogenated vegetable oils, does not have the undesirable cholesterol-raising effect of other saturated fatty acids. Nevertheless, there are strong recommendations from health professionals' groups that people decrease their overall intake of saturated fat. A useful guide to the saturation level of a fat is the P/S ratio, which is the ratio of polyunsaturated fatty acids to saturated fatty acids after exclusion of the monounsaturated fatty acids. There is now evidence that monounsaturated fatty acids have beneficial effects on serum lipids. It is probable that the P/S ratio will be replaced by the P–M/S ratio, in which the ratio of polyunsaturated plus monounsaturated fatty acids to saturated fatty acids is calculated. The higher the P/S and P–M/S ratios, the more desirable the fat. The ratios for some selected fats and oils and some foods are shown in Table V.

The potential usefulness of the P–M/S ratio as a health guide is particularly apparent in some cases. Note that the P/S ratio of peanut oil is relatively low (1.9) whereas its P–M/S ratio is 4.6. This is a reflection of the high level of the monounsaturated fatty acid oleic acid (18 : 1ω9) in peanut oil. The same situation pertains to canola (low erucic acid rapeseed) and olive oils. Also note that selection of a margarine could be significant in that tub or soft-serve products made with a highly unsaturated oil like safflower have more desirable P/S and P–M/S ratios.

It will be noted that the two fish listed have relatively low P/S and P–M/S ratios. This should not detract from the importance of including and perhaps increasing the use of fish in the diet. With some exceptions, fish can contribute greatly to maintaining a lower total fat intake, particularly when the fish are baked or broiled with little or no added fat. Moreover, fish oils may have added health benefits in that they contain ω3 fatty acids—not in the form of the parent fatty acid of the family, 18 : 3ω3, but as the elongated desaturated products, eicosapentaenoic acid (20 : 5ω3 or EPA) and docosahexaenoic acid (22 : 6ω3 or DHA). More will be discussed about the ω3 family and these fatty acids in particular in Section IV.

III. DIGESTION, ABSORPTION, AND TRANSPORT OF FAT

A. Digestion and Absorption

The absorption process to be described is the one taking place in the nonruminant animal. There are several differences in the process in ruminants.

Lipids are ingested largely in the form of triglycerides. Also ingested are plant and animal sterols, some phospholipids, and monoglycerides in the form of emulsifiers.

The absorption process is given in Table VI. In the stomach the food is churned by the gastric motion, and a coarse oil-in-water emulsion of large (10,000 Å) particles is formed. Here also the lipoproteins undergo proteolysis. The absorption process is initiated in the duodenum, where bile salts and pancreatic lipase enter. The enzymatic action on the triglycerides quickly forms diglycerides, slowly forms monoglycerides, and is very slow to give glycerol and free fatty acid (FA). Using a shorthand version of triglycerides without showing the components of glycerol backbone, this is the process:

$$\begin{array}{l}1\text{—OR}\\2\text{—OR}\\3\text{—OR}\end{array} \xrightarrow[\text{fast}]{\text{lipase}} \begin{array}{l}1\text{—OH}\\2\text{—OR}\\3\text{—OH}\end{array} + \text{FA} \begin{array}{l}\text{—OH}\\\text{—OR}\\\text{—OH}\end{array}$$

$$+\ \text{FA} \xrightarrow[\text{slow}]{\text{very}} \begin{array}{l}\text{—OH}\\\text{—OH}\\\text{—OH}\end{array} +\ \text{FA}$$

Note that the monoglyceride formed is the 2-monoglyceride. This is important because the fatty acid in this position will largely persist throughout metabolism. The phosphoglycerides are also broken down by phospholipase A to give the lysophosphatides. These and the monoglycerides pass on down the intestine. [*See* Phospholipid Metabolism.]

The major area of absorption is the upper jejunum, where mixed micelles of 50–100 Å are formed. This is facilitated by the detergent action of bile salts, lysophosphatides, and free fatty acids. Lysophosphatides are phosphoglycerides from which one fatty acid has been removed. Such compounds have a strong detergent action. The important molecular penetration phase then takes place. At the border of the intestinal epithelial cells the micelles are disrupted and the 2-monoglycerides, free fatty acids, and lysophosphatides enter the cell. This process is probably by diffusion as it is independent of energetic and enzyme effects.

The intracellular phase of absorption then follows. In this process triglycerides are resynthesized. There is then addition of proteins and phospholipids to give lipid–protein particles known as *chylomicrons*. These pass out into the intercellular space into the lymph and via the thoracic duct to the blood. Thus most fatty acids ingested find their way to the blood via the lymph. Only short-chain (less than 10 carbons) fatty acids are directly absorbed into the portal blood.

A number of disorders can lead to malabsorption of fat. These include genetic enzyme defects, pancreatic disease, and inflammatory disorders such as colitis. When fat is not properly absorbed, light-colored fatty stools will be excreted. This condition is known as *steatorrhea*. Such a condition also leads to loss of fat-soluble vitamins and can lead to a deficiency of linoleic acid, an essential fatty acid.

Cholesterol is absorbed free as a micelle with subsequent reesterification before transport into lymph in chylomicrons. [*See* Cholesterol.]

TABLE VI
The Absorption of Lipids

Section of gastrointestinal tract	Events
Stomach[a]	Churning to form oil-in-water emulsion of lipids
Duodenum	Entry of bile salts and enzyme pancreatic lipase
Upper jejunum	Initiation of absorption; lipase starts to break down triglycerides; formation of mixed micelles
Lower jejunum	Penetration phase; micelles are disrupted on intestinal mucosal cell surface
Intestinal mucosal cell interior	Triglycerides are reformed from monoglycerides and chylomicrons leave cell to lymph (short-chain FA to portal blood)
Ileum	Resorption of bile salts
Colon	Any undigested fat to stool

[a]Lingual lipase, an enzyme at the base of the tongue, initiates some hydrolysis of triglycerides to diglycerides.

B. Transport of Lipids in the Blood

Only chylomicron remnants reach the liver because under the influence of the plasma enzyme, lipoprotein lipase (clearing factor), they undergo partial loss of triglycerides. The lipoproteins then undergo progressive changes in the liver and extrahepatic sites.

Plasma lipoproteins are classified according to density and they are separated by differential centrifugation. This classification is purely one of convenience. In reality the plasma lipoproteins are a continuum. They are divided into two main classes, low- and high-density lipoproteins (LDL, HDL), which are further subdivided into very low-density lipoproteins

(VLDL), intermediate low-density and low-density lipoproteins (IDL, LDL), and various HDL fractions such as HDL_1 and HDL_2. This classification is shown in Table VII, together with some physical and chemical characteristics of the fractions.

The VLDL are mainly formed in the liver but some are formed in the intestinal mucosal cell. The apoprotein B (apo B) formed in the liver is essential for the removal of the VLDL from the liver by exocytosis. The VLDL are then changed by the action of lipoprotein lipase in extrahepatic tissues with the release of free fatty acids. This gives IDL and then by an unclear mechanism yields LDL. Part of the IDL are changed to give nascent HDL, which exists in the form of a discoidal bilayer. This is believed to be the preferred shape for the enzyme lecithin-cholesterol acyl transferase (LCAT). Under the influence of LCAT, lysolecithin (phosphatidylcholine minus one fatty acid) is formed and the fatty acid released esterifies to cholesterol. This leads to a change in shape of the HDL to spherical. Note in Table VII that the amount of cholesterol increases as this process proceeds. The HDL is removed to the liver, where the cholesterol carried by it is transformed into bile salts.

HDL is of particular interest in risk of coronary heart disease. As noted, HDL can transport cholesterol to the liver, where it is excreted as bile salts. This is not the case with LDL. The cholesterol in HDL is often referred to as "good cholesterol" because it is so effectively removed from the bloodstream and cannot contribute to atheromatous plaque formation in the intimal (inner) layer of the arteries, which is one of the features of atherosclerosis. LDL contains the highest level of cholesterol and its esters (Table VII). It is transported to many tissues where it is recognized by cells and taken up for degradation. Thus the cholesterol in LDL ends up mainly in the structural units of cell membranes. The infiltration theory of atherosclerosis states that normally the gaps between cells lining the arteries are too small for LDL to pass through. However, if the cells are damaged by trauma, LDL can infiltrate and build up at these sites. The smooth muscle cells of the arteries are stimulated, multiply, and readily take up LDL. Continued proliferation and uptake by the cells lead to the formation of fibrous plaques, which occlude the artery. LDL is, therefore, considered to be particularly atherogenic. HDL, on the other hand, is not recognized by receptors on arterial cells, so little of it is taken up and most of it goes to the liver, where the cholesterol is metabolized to bile salts. High HDL levels are, therefore, considered to be a protective factor against atherosclerosis. [*See* Atherosclerosis.]

Note in Table VII that the HDL level of premenopausal females is higher than that of males. After menopause it tends to decrease. This is consistent with the fact that premenopausal females have less heart disease than males in the same age group. Postmenopausally their risk increases as their HDL decreases.

Obesity, untreated diabetes, cigarette smoking, and hypertriglyceridemia are associated with lower HDL levels and increased incidence of atherosclerosis. Exer-

TABLE VII
Classification and Some Properties of Plasma Lipoproteins

	Low-density class				High-density class	
	Chylomicrons	VLDL	IDL	LDL	HDL_1	HDL_2
Hydrated density	0.93	0.94–1.01	1.01–1.05		1.09–1.14	
Average conc., mg/100 ml						
(male)	12 ± 2	129 ± 59	439 ± 99		300 ± 83	
(female)	13 ± 3	122 ± 63	389 ± 79		457 ± 115	
% protein	2	8	21		50	59
% lipid	98	92	79		50	41
Lipid composition (%)						
Cholesterol esters	3	13	47		40	25
Free cholesterol	1	8	10		4	5
Triglyceride	88	55	14		8	10
Phospholipids	8	20	28		48	56
Free fatty acids	1	2	1		—	3

cise and moderate daily alcohol consumption are associated with higher HDL levels.

As the lipoproteins enter the various cells of the body the fat portion is split off and is hydrolyzed again into glycerol and fatty acids. The fatty acids then undergo one of the following fates:

1. They are used as an immediate source of energy giving rise to carbon dioxide and water.
2. They are stored in cells, especially the specialized white adipocytes, in adipose tissue as a reserve form of energy.
3. They are incorporated into cell structures either in their original form or as elongated and desaturated products.
4. They are used in synthesis of essential metabolic compounds.

Depending on the tissue, released glycerol may be used in resynthesis of triglycerides or to form glucose, thus giving rise to energy.

C. Control of Lipolysis in the White Adipocyte

We are concerned here with the white adipocyte (fat cell) as distinct from the brown fat cell, a specialized form of adipocyte occurring in embryonic life, the neonate, and hibernators. [*See* Adipose Cell.]

Plasma clearing factor or lipoprotein lipase (LPL) is bound to endothelial cells of capillaries supplying blood to tissues. It liberates free fatty acids from triglycerides, chylomicrons, and lipoproteins arriving from the liver and via the lymph from intestinal absorption. Heparin in mast cells in the connective tissue assists in LPL release from capillaries. LPL is high in the fed state and low in fasting or starvation, when the mobilization of fat is required as an energy source. Insulin increases LPL activity and promotes lipogenesis. The hydrolysis of fat prior to its release from the adipocyte is a tightly controlled series of reactions. Because of this, all reactions leading to the release of stored lipid for formation of fatty acids and glycerol are separated from those pathways leading to lipogenesis and increased fat deposition. The reason for this tight control is that the specific function of the fat cell is to store fat in times of plenty and release it during periods of deprivation. It is important that these reactions related to storage and release are integrated with the needs of the whole organism.

The key factor is the hormone-sensitive lipase (HSL), which catalyzes triglyceride hydrolysis in adipose tissue. The HSL is high in fasting and low in fed animals. A brief exposure of the adipocyte to epinephrine and other hormones like glucagon and ACTH increases the activity of HSL. Exposure to insulin and prostaglandin E_1 (see Section VI) decreases the activity. HSL is, therefore, the rate-determining enzyme in lipolysis. Its activity controls the overall rate of lipolysis. [*See* Insulin and Glucagon.]

D. Fatty Acid Oxidation, Ketosis, and Low-Carbohydrate Reducing Diets

Fatty acid oxidation is the major pathway by which fatty acids are degraded to provide energy. The major mechanism is β-oxidation, that is, oxidation at the β carbon (position 2 from the carboxyl end) to yield a β-ketoacid, which is then cleaved. Two carbon units at a time are systematically removed this way until the whole molecule is broken down to yield molecules of acetyl coenzyme A (acetyl CoA) or a 2-carbon unit bound to the coenzyme A. Thus, palmitic acid (16 : 0) would yield eight molecules of acetyl CoA. The units of acetyl CoA then enter the tricarboxylic acid (TCA) cycle to be metabolized to carbon dioxide and water.

The liver has the capacity to divert some acetyl CoA from fatty acid or pyruvate oxidation to acetoacetate and β-hydroxy butyrate. Ultimately one can get the formation of free acetoacetic, β-hydroxybutyric acids, and acetone. These are known collectively as *ketone bodies.* They can diffuse out of the liver into the bloodstream to peripheral tissues. Extrahepatic tissues like muscle can oxidize ketone bodies in the mitochondria via the TCA cycle. Normally the amount of ketones is low but in conditions of increased lipolysis (fasting and diabetes mellitus) massive production can occur.

Most acetoacetyl CoA arising in the liver is from head-to-tail condensation of two molecules of acetyl CoA from fatty acid oxidation:

$$\text{2 acetyl CoA} \longrightarrow \underset{\text{acetoacetyl CoA}}{CH_3\,\underset{\substack{\| \\ O}}{C}{-}CH_2{-}\underset{\substack{\| \\ O}}{C}{-}SCoA}$$

Further reactions can then occur to give β-hydroxybutyric acid and acetone, the other ketone bodies.

Normally there are low blood and urine levels of ketone bodies. When the rate of formation in the liver exceeds the capacity of the peripheral tissues to utilize them a condition of ketosis results. The blood levels will rise (ketonemia) as will urine levels (ketonuria). The brain has the capacity to use ketone bodies for energy, especially in starvation.

High organic acid levels in the blood can be harmful and even fatal. Hydrogen ions from acetoacetate and β-hydroxy butyrate lead to a condition of acidosis in which blood pH decreases (becomes more acidic). Sodium and other electrolytes can be lost by chelation. If this is excessive there is water loss, dehydration, and coma—which if unchecked results in death. Uric acid has to compete with ketone bodies for excretion via the kidney. If this is excessive, a condition of gout will arise.

Conditions favoring ketosis are any in which fat breakdown (catabolism) is increased and in which carbohydrate catabolism is decreased. Juvenile onset diabetes is one such condition. Others include low-carbohydrate/high-protein diets for weight reduction. Such diets rely on putting the dieter into ketosis (which possibly has an anorexic effect). This obviously can be dangerous and such diets should be used only under medical supervision.

IV. FUNCTIONS OF FAT IN THE DIET AND IN THE BODY

A. Role of Fat in the Diet

1. As a Source of Energy

Each gram of fat regardless of its food source provides 9 kcal compared to only 4 kcal for each gram of carbohydrate or protein. Even in lean cuts of meat with less than 30% fat, the fat will provide nearly 80% of the calories.

2. As a Carrier of Fat-Soluble Vitamins

There are four fat-soluble vitamins: A, D, E, and K. Dietary fat serves as a carrier for these vitamins and aids in their absorption. Much of our vitamin A supply comes from provitamins A in vegetable foods (such as carrots) that contain carotenoids. About 10% dietary fat is required for efficient absorption of these vitamins. Elimination of fat in the diet would eventually lead to fat-soluble vitamin deficiencies. Any situation that leads to malabsorption of fat—inflammatory bowel disease, obstruction of the bile duct, or ingestion of peroxidized fat—can cause fat-soluble vitamin deficiency.

3. As a Satiety Value

The time taken for a meal to leave the stomach depends on its fat content. The higher the fat content the longer it takes the stomach to empty. Fat is, therefore, an important contributor to the feeling of satiety after eating. This is of importance in planning weight-reducing diets. It is easier to stick to a diet that is low in calories but allows fat in the form of salad oils or spreads on bread. The dieter feels more satisfied with such meals.

4. In the Palatability of Foods

Fat in food carries much of the flavor and aroma. Anyone who has tried to eat a very low fat diet becomes acutely aware of the role of fat in palatability. Very lean cuts of meat lack the flavor and tenderness of nicely marbled cuts. Much of chicken fat and cholesterol is in the skin or just below it. Removing it assists in lowering total fat and cholesterol but many individuals find that this greatly detracts from the taste.

B. Role of Fat in the Body

1. As an Energy Reserve

Body fat is the main store of energy in the body. As noted previously, much of the fat for energy is stored in specialized cells called adipocytes and the release of this stored energy by lipolysis is a tightly controlled process under the regulation of hormone-sensitive lipase. It is considered normal and desirable for women to have 18–24% body fat and men 15–18%. People with reserves above these levels are considered overweight. Those with reserves 20–30% above these levels are considered obese. Excessive leanness, especially in females, can cause problems. Extremely thin females often have amenorrhea and may also have difficulty in maintaining a pregnancy.

This bulk of storage fat is in white adipocytes or white fat cells. Brown fat cells, which have an extensive blood supply, are also sources of energy. Brown fat is more important in neonates, infants, and hibernating animals. Its purpose is to provide energy from oxidation of fat at a much faster rate than from white fat. An important triggering mechanism for brown fat oxidation is exposure to cold temperatures.

2. As an Insulator and Protector of Vital Organs

Deposits of fat beneath the skin serve to insulate the body and protect against changes in the environmental temperature and prevent excessive heat loss. Deposits around certain organs such as the kidneys, adrenals, and heart protect them from physical shock.

3. As a Source of the Essential Fatty Acids

When G. O. Burr and M. M. Burr first described the essentiality of some fatty acids in 1929 they considered linoleic, α-linolenic, and arachidonic acids to be essential for growth and the prevention of dermatitis in the rat. It was realized later that arachidonic acid (20 : 4ω6) was formed from linoleic acid (18 : 2ω6), so really only linoleic was truly essential. Later the essentiality of α-linolenic acid was questioned and its status wavered between essential and nonessential for many years. Here we shall use a broad concept of essentiality of a nutrient. An essential nutrient is one that the organism cannot synthesize, that appears to have a function, and that causes the organism to function less effectively without it. The absence of such essential nutrients in the diet may or may not manifest itself in overt clinical symptoms.

a. Deficiency and Function of Linoleic Acid, 18 : 2ω6

Linoleic acid deficiency has been demonstrated in all laboratory species examined as well as in humans. In the rat the symptoms are poor growth rate or failure to maintain weight and dermatitis in the later stages. There are also various metabolic effects such as cholesterol accumulation in some tissues. Other tissues (e.g., the thymus) weigh less than normal. The deficiency is also associated with excessive water loss through the skin and excessive water intake (*polydipsia*). Males suffer testicular degeneration and in a severe deficiency reproduction is impossible. If deficient females are mated with normal males they can bear and suckle young if they are not too deficient. Extremely deficient animals will fail to lactate adequately to maintain viable litters. Poor wound healing is another symptom of the deficiency.

Most of the symptoms of linoleic acid deficiency can be reversed by feeding the fatty acid. The dermatitis in particular clears up quite readily. If α-linolenic acid (18 : 3ω3) is also absent from the diet there will be an increase in the so-called deficiency triene, 20 : 3ω9, in serum and in tissues. This is because the Δ^6 desaturase enzyme involved in the further desaturation of 18 : 2ω6, 18 : 3ω3, and 18 : 1ω9 prefers the ω3 as a substrate followed by the ω6 and finally the ω9. When both 18 : 2ω6 and 18 : 3ω3 are absent the enzyme is left to desaturate the ω9 series, leading to a buildup of 20 : 3ω9, which normally is present in very small amounts. The desaturated, elongation product of 18 : 2ω6, arachidonic acid (20 : 4ω6), is normally a major tissue fatty acid. When both EFAs are absent there is a fall in 20 : 4ω6 levels and a rise in 20 : 3ω9, so the ratio 20 : 3ω9/20 : 4ω6 can be used as a measure of the severity of the deficiency. Ratios slightly above 0.4 indicate a marginal deficiency and ratios of 4.0 a severe deficiency.

Linoleic acid deficiency was not seen in humans until the advent of total parenteral (intravenous) feeding (TPN). Patients on TPN are often infants with intestinal problems and/or cancer patients. Both have poor stores of linoleic acid in their tissues. When the infusion has no lipid component to provide either EFA, the deficiency results. The ω6 fatty acid in serum will fall and the deficiency triene will increase, giving an elevated 20 : 3ω9/20 : 4ω6 ratio. There will be dermatitis, failure to gain weight, poor wound healing, and increased risk of infections. The condition is reversed by the addition of an EFA-containing emulsion to the infusion. A common one in the United States is a soybean oil, which therefore contains both linoleic and linolenic acids. Prior to the introduction of this emulsion several cases of TPN-induced EFA deficiency were reported. The FDA did not permit the use of lipids in TPN until several years ago because initial use of lipids had led to body temperature elevation and thrombi problems.

The function of linoleic acid is fairly well understood although a few of its deficiency symptoms have no clear explanation as yet. There appear to be two main functions of linoleic acid, one purely structural and the other as a precursor of a very important group of regulatory compounds, the prostaglandins and their related metabolites.

The structural roles are concerned with the function of long-chain, desaturated metabolites of linoleic acid in the maintenance of the integrity, fluidity, and permeability properties of cellular membranes. Of extreme importance is the role of fatty acids in membrane-based lipids in maintaining the optimum environment for enzymes. In EFA deficiency many enzyme activities are changed, including that of ubiquitous enzymes like Na^+-K^+-ATPase, and the changes are reversed on feeding EFA.

Much interest in linoleic acid now centers around its role as a dietary precursor of some of the most active prostaglandins. The desaturated, elongated products of linoleic acid are the direct precursors of these types of compounds. Dihomo-γ-linolenic acid (20 : 3ω6) gives the prostaglandin of the 1-series, arachidonic acid (20 : 4ω6) the prostaglandins of the 2-

series, and timnodonic acid (20 : 5ω3) the 3-series (see Section IV,B,3,c).

b. α-Linolenic Acid, 18 : 3ω3

As noted earlier, the original discoverers of EFA reported that α-linolenic acid was also essential. It was then noted that α-linolenic acid did not cure the dermatitis of rats on fat-free diets whereas linoleic acid did. Over the years some investigators have continued trying to prove that α-linolenic was or was not essential. Some confusion arose early in that EFA deficiency was usually brought about by feeding a fat-free or hydrogenated fat diet. Thus, such diets were devoid of both linoleic and α-linolenic acids. Later, as diets were refined, diets adequate in linoleic acid and lacking α-linolenic acid were fed. Such diets were fed through several generations of rats but no ill effects were observed on growth or reproductive performance even though tissues showed marked depletion of fatty acids of the ω3 series. There were some reports of neurological changes in primates on α-linolenic-deficient diets and of learning disability in rats. The latter observation is probably explained by effects on the retina and is due therefore to lessened visual acuity. The retina is rich in ω3 fatty acids and the absence of α-linolenic acid in the diet interferes with cell repair in the retina. Nevertheless, no really clear-cut syndrome recognizable as a nutrient deficiency was ever established. [*See* Retina.]

Many nutritionists and food scientists dismissed α-linolenic acid as being nonessential. Others held the view that if a function can be defined for a nutrient that cannot be synthesized by the organism it should be regarded as essential. In 1982 it was reported that a young female placed on TPN using an emulsion of safflower oil developed neurological symptoms that were alleviated by changing to a soybean oil emulsion. Safflower oil is very high in linoleic acid (70+%) and very low in α-linolenic acid (ca. 0.5%). The Δ^6 desaturase thus was flooded with ω6 fatty acids and had very little of the preferred ω3 acids. The consequence was a depletion of ω3 stores and an elevation of the ω6. At about the same time investigators began showing that at least one function of α-linolenic acid may be to modulate synthesis of the 1- and 2-series prostaglandins from linoleic acid. As noted earlier, α-linolenic acid is the precursor of timnodonic acid, 20 : 5ω3, which is the substrate for production of prostaglandins of the 3-series.

Since the publication of the first case of apparent 18 : 3ω3 deficiency several other cases in humans have been described. Studies using monkeys have also shown neurological defects.

α-Linolenic acid, therefore, appears to have two functions, one in the maintenance of membrane properties (especially in cells in which ω3 fatty acids are high) and the other in modulation of prostaglandin synthesis and the synthesis of other eicosanoids.

c. Prostaglandins and Related Metabolites: The Eicosanoids

The prostaglandins (PGs) and their related metabolites are now such an extensive field that they could be the subject of many books. Here we shall consider enough about them to illustrate why and how they are becoming of interest to food scientists and nutritionists.

Prostaglandins are compounds based on the skeleton of prostanoic acid, a 20-carbon cyclic compound. They are essentially water-soluble, cyclic fatty acids. They were discovered over 50 years ago but they are produced in such small amounts that they could not be studied for the first 30 or more years after their discovery. The technology was just not available. The PGs are produced by all cells studied except the erythrocyte. They are produced in response to some stimuli like hormones or trauma and they are usually destroyed at the site of their production. In other words, cells usually have the enzymatic machinery to both produce and catabolize the PGs. It is important to remember that PGs are not stored in tissues. One cannot speak of "PG content" but only of "capacity to synthesize PGs."

As we have seen, the dietary precursors of PGs are the essential fatty acids linoleic and α-linolenic and the direct precursors arise from the elongation and desaturation of these:

linoleic: 18 : 2ω6 ⟶ 18 : 3ω3 ⟶ 20 : 3ω6 ⟶ 20 : 4ω6
20 : 3ω6 ↓ PGs of 1-series
20 : 4ω6 ↓ PGs of 2-series

α-linoleic: 18 : 3ω3 ⟶ 18 : 4ω3 ⟶ 20 : 4ω3 ⟶ 20 : 5ω3
20 : 5ω3 ↓ PGs of 3-series

The 1, 2, and 3 of the series name arise from the number of double bonds in the side chains of the PGs.

When the cyclooxygenase enzyme acts on 20 : 3ω6, 20 : 4ω6, or 20 : 5ω3 to produce PGs, the first products are very short-lived PGs called endoperoxides and given the names PGG and PGH. These endoperoxides of all three series undergo further reactions to produce

PGs designated PGA through PGF and the very potent prostacyclin (PGI). Just as the three series of PGs differ only in the number of double bonds in their side chains, these various PGs differ little in structure (e.g., constituents on five-membered rings) but they may differ markedly in biological action and may even have opposing effects in the same physiological system.

Prostaglandins are regulatory compounds that have many functions. The most important appears to belong to the 2-series and to a lesser extent the 1-series. (Less is known about the 3-series.) It is important to distinguish between physiological and pharmacological effects of prostaglandins. They may have one effect physiologically and another when given at drug levels. Some of the known involvements of prostaglandins are:

- Control of intrarenal blood flow and blood pressure
- Regulation of microcirculation
- Control of vascular tone
- Regulation of neurotransmission
- Regulation of cell differentiation and division
- Regulation of the immune response
- Initiation of parturition
- Acceleration of luteolysis
- Facilitation of calcium binding to membranes
- Bronchoconstriction and bronchodilation
- Neuromuscular activity regulation
- Control of platelet aggregation and blood clotting
- Regulation of natriuresis

Many of the effects of prostaglandins are mediated via the cyclic nucleotides, cAMP and cGMP.

In addition to the classic PGs, other metabolites of arachidonic acid were gradually discovered. These include the thromboxanes A and B (TXA, TXB). In addition to this pathway another catalyzed by lipoxygenase was discovered and this produces a series of hydroperoxy and hydroxy eicosatetraenoic acids (HPETEs and HETEs) and substances known as leukotrienes and hepoxylins. Collectively, they are known as eicosanoids. Probably all the compounds produced from arachidonic acid can be produced for the corresponding 1- and 3-series from $20:3\omega6$ and $20:5\omega3$. It should be noted that different cells produce different products in various amounts and not all corresponding substances from different series are equally biologically active. Most PGs and the other eicosanoids are highly biologically active, hence they do not exist for long or they would be destructive to the cell and the whole organism. They are produced, do their work, and are destroyed at the site or on passage through the lung or kidney. They are "hormone-like" in their action but are not true hormones. Many of the lipoxygenase products are now known to have powerful physiological effects. Some of them inhibit the synthesis of the cyclooxygenase products, others are chemotactic. Some of the leukotrienes are the slow-reacting substances of anaphylaxis (sRSA). Anaphylaxis is an antigen–antibody reaction in which there is an unusual or exaggerated response to a foreign protein or other substance. sRSA is, therefore, of great interest in allergic responses.

The hepoxylins are the latest lipoxygenase products to be discovered. Two have been isolated and characterized, namely, hepoxylin A_3 and hepoxylin B_3. The term hepoxylin was derived from their structural feature, Hydroxy EPOXide, and their first reported biological activity, insuLIN release. Several other biological activities have since been reported. The picture of the role of all of these compounds is far from complete but it is clear that they exert profound regulatory effects on many body functions.

d. Modulation of Eicosanoids Synthesis by Dietary Fats and Oils

Eicosanoid synthesis can be modified by drugs. For example, aspirin and other nonsteroidal anti-inflammatory drugs inhibit the cyclooxygenase and this, at least in part, explains their action against pain. Steroids inhibit the release of free arachidonic acid by the enzyme phospholipase A_2 and lead to less arachidonic acid entering the lipoxygenase pathway, explaining why they have therapeutic effects in diseases like rheumatoid arthritis and psoriasis. Since the precursors of the eicosanoids are the EFA it was obvious to investigate whether or not eicosanoid synthesis can be modulated by changing dietary fats and oils. There are now many examples of the success of this approach in both laboratory animals and humans. If the $18:2\omega6$ intake is lowered there is less $20:4\omega6$ in tissues and the production of PGs of the 2-series is decreased compared to that in subjects receiving higher intake of $18:2\omega6$. This is an extensive subject in itself. Just a few examples will suffice to illustrate the potential of this approach of intervening therapeutically via dietary fat.

A dietary intervention of popular interest in recent years has been the use of increased levels of $\omega3$ fatty acids for the prevention of risk factors for heart disease and stroke.

It has been known for some time that Eskimos living a true Eskimo life-style have fewer heart attacks. They also have prolonged bleeding times and less sticky platelets than other population groups. Prolonged bleeding was noted by early Arctic explorers. An obvious connection is their high consumption of foods of marine origin. Marine oils are rich in $\omega 3$ fatty acids, particularly in $20:5\omega 3$, the precursor of the 3-series PGs.

There have now been many studies in this area. $\omega 3$ fatty acids from both an $18:3\omega 3$ source or a fish oil source ($20:5\omega 3$) have been shown to lower serum cholesterol, although the $20:5\omega 3$ is usually more effective than the $18:3\omega 3$. Significantly it has been shown that $20:5\omega 3$ from marine sources or $18:3\omega 3$ transformed to $20:5\omega 3$ in the body is effective in decreasing the stickiness of platelets, which can form white thrombi in blood vessels leading to heart attack or stroke. The mechanisms appear to be as follows. The endoperoxides from $20:4\omega 6$ in platelets (PGG_2 and PGH_2) are highly aggregatory to the platelets, as is the thromboxane A_2, which also constricts blood vessels. Prostacyclin I_2, on the other hand, disaggregates platelets and dilates blood vessels. There is, therefore, a balance between the actions of the PGG_2, PGH_2, and TXA_2, and PGI_2. If this balance is tipped in favor of the aggregatory eicosanoids, say by a damaged arterial wall where PGI_2 is produced, then thrombi may form, leading to heart attack or stroke. If the diet is high in $\omega 3$ fatty acids this chance appears to be lessened for the following reasons. The $20:5\omega 3$ leads to less $20:4\omega 6$ in tissues. It also competes with $20:4\omega 6$ for the cyclooxygenase enzyme. As a result synthesis of the 2-series PGs is decreased and some 3-series are synthesized. It so happens that the thromboxane from $20:5\omega 3$ is not aggregatory to platelets but the PGI_3 is just as disaggregatory as, and a vasodilator like, PGI_2. Thus, the balance is tipped toward less sticky platelets, a tendency to bleed, but less chance of heart attack.

Another example is the use of $\omega 3$ fatty acids in the treatment of psoriasis. There have been several reports that the ingestion of $\omega 3$ fatty acids, either $18:3\omega 3$ or $20:5\omega 3$ (the latter more often), alleviates the inflammation and skin proliferation in the disease. It has been known for some time that patients with psoriasis have excess free $20:4\omega 6$ in their skin. Skin also has an active lipoxygenase enzyme that transforms $20:4\omega 6$ to leukotriene B_4 (LTB_4). This leukotriene is highly chemotactic for neutrophils (it attracts white cells). These white cells enter the skin from capillaries below the surface and cause an inflammatory response. Additionally, LTB_4 is hyperproliferative to skin cells. Consequently, there is redness of the skin and accumulation of immature skin cells that pile up to form the psoriatic plaques. Leukotriene B_5 formed from $20:5\omega 3$ is less chemotactic than LTB_4, which is produced in lesser amounts on an $\omega 3$ diet; as a result the psoriatic process is modified.

Thus we have seen that dietary fats and oils are not only important as sources of energy, enhancers of food flavor and aroma, and carriers of some vitamins but also have important functions in the body and can be used to modify health. Judicious choice of dietary fat and oils can lower serum cholesterol, a heart disease risk factor, and have potential for intervention in alleviating or preventing other diseases. A word of caution should be added. Dramatic changes in type of fat intake and especially the use of fish oil supplements should be done under medical supervision only. Fish-oil capsule use may lead to excessive bleeding on injury and may, depending on their source, contain oxidative polymers that could be deleterious to health.

BIBLIOGRAPHY

Guthrie, H. A. (1989). "Introductory Nutrition." Times Mirror/Mosby, St. Louis.

Hunt, S. M., and Groff, J. L. (1990). "Advanced Nutrition and Human Metabolism." West Publishing Co., St. Paul/New York.

Kinsella, J. E. (1987). "Seafoods and Fish Oils in Human Health and Disease." Dekker, New York/Basel.

Simopoulos, A. P., Kifer, R. R., and Martin, R. E. (eds.) (1986). "Health Effects of Polyunsaturated Fatty Acids in Seafoods." Academic Press, New York.

Willis, A. L. (1984). Essential fatty acids, prostaglandins, and related eicosanoids. *In* "Present Knowledge in Nutrition." The Nutrition Foundation Inc., Washington, D.C.

Nutrition, Protein

CAROL N. MEREDITH
University of California, Davis

Revised by
PETER PELLETT
University of Massachusetts

GLOSSARY

Digestibility Percentage of total protein intake that is absorbed by the gut

Protein Chains of amino acids found in all cells, containing an average 16% nitrogen. Dietary protein is often expressed as nitrogen intake

Protein quality Capacity of a dietary protein to promote growth in the young or nitrogen retention in the adult. Quality is determined by digestibility and the pattern of indispensable amino acids, which ideally is similar to the pattern of tissue proteins

Protein requirement Minimum protein intake that can sustain normal growth in the young or no net loss of body nitrogen in the adult. In healthy young adults it is 0.60 g/kb when the diet provides adequate levels of all other nutrients

Protein sparing Effect that adequate or excess energy intake has on increasing the efficiency of nitrogen retention in the body, thus decreasing the protein requirement

Protein turnover Rate of new protein synthesis and of breakdown of body proteins, occurring at about three times the rate of protein intake in adults

Recommended dietary allowance Protein intake meeting the nutritional needs of practically all healthy persons, equal to 0.8 g of protein per kilogram body weight per day

Reference protein Protein of known good quality with a pattern of essential amino acids that can be used as a standard of comparison. Examples include whole egg protein for the adult and human milk protein for the infant

IN THE AVERAGE AMERICAN DIET, 12–14% OF ENergy comes from protein, providing adults with more than 75 g of protein per day. Much of the dietary protein in developed countries is from animal products (e.g., meat, milk, eggs, and cheese), supplying the highest-quality protein but also substantial and undesirable amounts of fat. In less-developed countries, most protein is provided by one or more staple foods of plant origin (e.g., rice in Asian countries and maize and cassava in sub-Saharan African countries). The rice-based diet of rural people in Thailand, for example, supplies 8% of energy as protein, which is only marginally enough for an adult. As countries become more prosperous and urbanized, protein intake increases, especially from animal foods. When income does not limit dietary choice, people tend to eat a diet with about 16% of total calories as protein, coinciding with the amount recommended by the U.S. National Research Council.

I. CHEMICAL COMPOSITION OF DIETARY PROTEINS

Proteins in nature are made from different combinations of 20 amino acid building blocks. Each amino acid has at least one nitrogen-containing amino group

ENCYCLOPEDIA OF HUMAN BIOLOGY, Second Edition, VOLUME 6.

attached to a hydrocarbon backbone of 2–11 carbons. The 20 amino acids function like letters in an alphabet, and the genetic code of each living organism determines how they will be put together to form a vast variety of proteins, analogous to the words and sentences in the books of a library. When we consume and digest proteins in food, the free amino acids enter the tissues and are knitted together through peptide bonds to make new proteins that are as different as hair, the transparent lens of the eye, or blood-clotting enzymes. Protein synthesis cannot occur at a normal rate unless each and every amino acid is present in the cell in the amounts demanded by the genetic code.

A. Indispensable or Essential Amino Acids

Eight amino acids cannot be made by animals and are considered essential or indispensable. In approximate order of abundance in animal muscle protein, they are lysine, leucine, valine, isoleucine, threonine, phenylalanine, methionine, and tryptophan. If a young animal is fed a diet in which one indispensable amino acid is absent, normal growth cannot occur. There are almost no natural food proteins that are totally lacking in one or more indispensable amino acids, with the exception of gelatin, a partially processed protein, which lacks tryptophan. However, many natural proteins supply only small amounts of one or more of the indispensable amino acids. When an essential amino acid is present in an amount too low to sustain normal growth, the protein is called *incomplete* and that amino acid is termed the *limiting amino acid*. The poor quality of many plant proteins is due to a pattern of indispensable amino acids that is markedly different from the pattern needed to make body proteins. It should be noted that incomplete proteins do not *lack* an essential amino acid, they simply provide a disproportionately small amount with respect to the body's needs.

B. Dispensable or Nonessential Amino Acids

Twelve amino acids can be made from other amino acids and carbon compounds and are considered dispensable or nonessential. In approximate order of abundance in animal muscle proteins, they are arginine, glutamate, glutamine, aspartate, asparagine, alanine, glycine, serine, tyrosine, histidine, proline, and cystine.

The ability to make all the dispensable amino acids at a rate allowing rapid growth is limited. The first intravenous feeding solutions made from mixtures of crystalline L-amino acids provided all the essential amino acids but only one or two dispensable amino acids, usually glycine.These formulas, leading to toxic levels of ammonia in infants and not efficiently used for growth, were improved by adding other dispensable amino acids such as arginine, alanine, glutamic acid, and proline. The most efficiently used artificial amino acid mixtures provide nearly all the dispensable amino acids, similar to natural food proteins.

II. SOURCES OF PROTEIN

A. Quantity in Foods and Diets

The amount of dietary protein is usually calculated from its nitrogen content multiplied by a conversion factor of 6.25, based on the fact that most proteins contain 16% nitrogen and nearly all dietary nitrogen is in the form of proteins or amino acids.

Protein is present in all unprocessed foods, and only highly purified foods (e.g., cooking oil or white sugar) lack protein. Table I shows the content and quality of protein in some foods. Prepared foods, containing more than 70% water, usually have between 1 and 12% of protein. Vegetables and fruit are almost protein-free, but plant seeds have more protein. Foods of animal origin tend to be richer in proteins. The most concentrated natural sources of dietary protein are lean meats (e.g., fish or skinned poultry).

B. Quality in Foods and Diets

Because the dispensable amino acids can be made in the body, quality is mainly determined by the amount and pattern of indispensable amino acids in the diet. Quality is tested in relation to growth or nitrogen retention. In the United States, the principal method for experimentally measuring the nutritive value of proteins has been the Protein Efficiency Ratio (PER), which relates the weight gain of young male rats to the amount of protein eaten. An international expert consultation (FAO/WHO, 1991) has recommended replacement of the PER by the use of amino acid composition and digestibility data. In humans, quality is measured as the percentage of protein intake that is retained in the body. Nitrogen balance, or nitrogen retention,

TABLE I
Content and Quality of Some Protein Foods[a]

Food	Protein concentration (g/100 g)		Dig. (%)	Nut. Val. (%)	Lim. AA
	Raw	Ready to eat			
Meat	25	25	100	70	Methionine
Egg	12	12	98	98	None
Cow's milk	3	3	98	83	Methionine
Human milk	1	1	100	100	None
Soy milk	3	3	92	60	Methionine
White beans	22	5	77	40	Methionine
Bread	9	9	97	37	Lysine
Rice	7	2	97	55	Lysine, threonine
Corn grits	9	1	86	51	Tryptophan, lysine
Potatoes	2	2	82	60	Methionine

[a]Dig., digestibility in rats = (N absorbed/N intake) × 100; Nut.Val., nutritive value in rats = (N retained/N intake) × 100; Lim. AA, limiting amino acid, or the indispensable amino acid present in lowest amounts with respect to the need for that amino acid.

is determined by subtracting nitrogen excretion from nitrogen intake, after adapting to the test diet for 5–10 days. Most minor nitrogen losses (skin, hair, nails, secretions) are estimated and not usually measured directly. The balance method overestimates nitrogen retention and thus protein quality, as it is hard to measure all nitrogen losses accurately.

1. Amino Acid Patterns

By knowing the amount of each indispensable amino acid in a food protein, its quality can be predicted by calculating its "chemical score." This expresses the concentration of each indispensable amino acid as a percentage of the concentration in an "ideal" or "reference" protein (e.g., egg protein). The lowest percentage obtained identifies the amino acid that is "limiting"' and is a measure of the biological usefulness of the food protein. There is generally good agreement between the chemical score and biological measurements of protein quality. In the case of infants, the reference protein is human milk protein.

2. Complementarity

Most children and adults eat a variety of foods with different amounts and qualities of protein. Eating different proteins is beneficial, as proteins containing low amounts of one indispensable amino acid are eaten together with those proteins having abundant amounts of that amino acid, and the mixture absorbed from the gut has a more balanced composition.

In most of the world, cereals are the source of both protein and calories. Although the protein content of prepared cereals is not much lower than in milk-based formulas, the pattern of indispensable amino acids of cereals makes them a poor protein source during early childhood. Cereals are lmiting in lysine (wheat) and often low in threonine (rice) and in tryptophan (maize). There are two ways of improving the quality of cereal proteins. One is by *supplementing* the cereal food with a small amount of an animal product rich in high-quality protein. For example, a dish prepared from rice, vegetables, and some egg provides good-quality protein. However, in poor rural communities, foods such as eggs, meat, or milk are expensive and scarce. Another method is by *complementing* the protein in a cereal food with other plant proteins that are richer in the cereal's limiting amino acid. In poor communities or for strict vegetarians, this is the only way to improve the protein quality of the diet. Compared with cereals, legumes provide more and better protein and are richer in lysine, thus increasing the lysine content of the mixture to the point that it does not limit protein utilization. As Table I shows, the quality of corn protein is limited by low lysine and tryptophan, whereas beans are limited by methionine and cystine. Experiments in young rats show that corn by itself has an efficiency of utilization of 56%, white beans by themselves have an efficiency of utilization of 76%, but a mixture of half beans and half corn, as in succotash, provides a protein that has an 82% efficiency of utilization (using the milk protein casein as a reference protein). Soybeans are a particularly good protein source and are used to make infant foods. Their high lysine content supplements the quality of cereals.

3. Amino Acid Digestibility

The amino acids in food proteins must enter the circulation before they can be used. Plant proteins are less easily broken down to amino acids because their structure is more resistant to digestive enzymes. Some plant foods (e.g., legumes) contain enzymes that inhibit protein digestion. Cooking improves digestibility by inactivating toxic enzymes and facilitating the attack of enzymes, by loosening protein structure.

4. Amino Acid Availability

Although moderate heating increases the availability of amino acids, the hot temperatures of toasting can destroy cystine and reduce the availability of lysine. In dry, nonacid foods, storage in hot climates or heat treatment favors a reaction between lysine and free sugars, forming yellowish or brown compounds, where lysine cannot be digested or absorbed. The loss of available lysine becomes nutritionally significant when the food is already limiting in that indispensable amino acid (i.e., in cereal products) or it is a major dietary source of lysine (i.e., powdered milk for infants and children).

C. Commercial Supplements of Proteins and Amino Acids

In recent years, a thriving market has been developed for protein and amino acid supplements, directed especially to strength-trained athletes. Weight lifters can spend hundreds of dollars on these products, although studies show that strength gain and muscle enlargement occur with training even if energy and protein intake are curtailed. The argument is that persons who are increasing muscle size and strength through resistance training need more protein to "bulk up" and that commercial protein supplements are more effective than dietary protein. Most supplements are expensive mixtures of milk protein and soy protein, not likely to be harmful in healthy persons consuming a varied diet and adequate amounts of fluids, but neither have they been proven effective. More highly touted and expensive are the pure amino acid supplements, sold as enhancers of protein synthesis and promoters of growth hormone release. Single amino acid supplements are potentially more hazardous and less useful for protein synthesis, as they can disrupt the normal entry of other amino acids into tissues.

III. DIGESTION AND DISTRIBUTION

Adults and children tend to eat during the day and fast during the night, whereas infants are usually constantly in the fed state. After a protein-containing meal, proteins are degraded by proteolytic enzymes secreted by the lining of the stomach and by the pancreas, producing free amino acids and small peptides made up of strands of two to four amino acids. The cells lining the upper small intestine take these up and further break them down to free amino acids before they enter the blood. The concentration of free amino acids in the blood flowing from the gut to the liver increases severalfold. On reaching the liver, some of the amino acids are used for protein synthesis, some are broken down to small carbon compounds and urea, and the remainder are released to the general circulation. The liver contains enzymes for degrading all the amino acids except for leucine, isoleucine, and valine, which are mainly broken down in muscle. Insulin release following meals favors the uptake of amino acids from the blood, especially by muscle, which is the largest tissue in the body. All tissues constantly take up and release amino acids, but after meals a net uptake of amino acids and net increase in protein content is caused by greater synthesis than breakdown of tissue proteins.

In the fasted state, the flow of amino acids from the gut and liver to peripheral tissues diminishes, whereas the flow from the peripheral tissues to the visceral organs increases. The rate of protein synthesis decreases, and there is a net breakdown of proteins and net release of amino acids from muscle. Alanine and glutamine account for more than 50% of the amino acids released by muscle. These two dispensable amino acids are sources of carbon chains for glucose synthesis in the liver. During fasting, the glucose supplied by the liver feeds the tissues that have an obligatory and constant need for glucose as an energy source (e.g., the nervous system and red blood cells).

IV. ELIMINATION OF EXCESS PROTEIN

The main function of amino acids is to provide units for the synthesis of protein, and other pathways are less important. Amino acids react with the enzymes of protein synthesis 20–100 times more avidly than with the enzymes of amino acid breakdown. This allows a highly efficient retention of amino acids present in low concentrations and provides a mechanism for breaking down excessive amounts of each amino acid. The activity of the enzymes that break down amino acids also adjusts to the amount of each amino acid in the diet. For example, an animal fed a low-lysine diet (e.g., wheat flour) reduces the activity of the enzyme that oxidizes lysine, conserving it for protein synthesis. However, the efficiency of indispensable amino acid or protein conservation is not total, and a certain amount must be supplied daily in the diet to prevent a gradual loss of body proteins. Infants born with a deficiency of any of the amino acid breakdown enzymes accumulate toxic amounts of that amino acid and its breakdown products in the blood. The tissue that suffers the greatest damage is the central nervous system. The brain is susceptible because a large excess of a single amino acid can competitively inhibit the uptake of other amino acids, preventing the entry of the full range of amino acids needed for normal synthesis of proteins and neurotransmitters. In phenylketonuria, for example, the liver enzyme that converts phenylalanine to tyrosine is lacking, and unless a diet low in phenylalanine is provided, the child will become mentally retarded.

The enzymes that break down amino acids produce short-chain carbon compounds that can be used for energy metabolism or to make fats, glucose, and glycogen, while the potentially toxic nitrogen molecule is made into urea by the liver. The enzymes that make urea rapidly adjust their activity, increasing within hours of a protein-rich meal, or decreasing in response to a high-carbohydrate, low-protein meal. Urea is excreted in the urine.

In newborn infants the capacity to make urea is limited. The diet should supply enough protein for growth, but not so much that amino acids and ammonia accumulate and produce toxic effects. In adults, a large protein intake with formation and excretion of urea is not harmful unless there is kidney damage, in which case a high-protein diet accelerates the loss of renal function.

V. UTILIZATION OF DIETARY PROTEINS

A. Protein Synthesis and Breakdown

In infants, protein synthesis and breakdown occur at rates more than eight times greater than the rate of protein intake, and rapid growth results from a greater synthesis than breakdown. There are periods of enhanced protein synthesis in the development of various organs. The timing of these growth spurts determines which tissues will be most affected by dietary deficiencies. For example, the high rates of decay in the "baby teeth" of poor 6-year-old children may be a consequence of malnutrition during weaning that coincides with the formation of deciduous teeth.

Meals do not greatly affect the rate of whole-body protein breakdown, but protein synthesis increases after meals to an extent that depends on the adequacy of dietary protein and energy.

B. Dietary Protein Requirements

There is a daily need for total protein and each of the essential amino acids. There are no stores of amino acids. If the diet provides too little protein, protein of poor quality, or insufficient calories so that amino acids are diverted toward energy production, there is growth faltering in young children and gradual loss of lean tissue in adults (negative nitrogen balance).

How much protein is enough? The adult's need for dietary protein is established as the smallest amount that allows all the nitrogen consumed as protein to equal all the nitrogen excreted, thus allowing nitrogen equilibrium. In healthy young American men fed egg or other high-quality protein, nitrogen equilibrium is achieved with a daily intake of 0.60 g of protein per kilogram of body weight. In developing countries, men consuming mixed animal and cereal proteins typical of the local diets achieve nitrogen equilibrium with about 0.75 g of protein per kilogram of body weight. The Recommended Dietary Allowance of protein in the United States is 0.8 g/kg, an amount that considers the lower digestibility and quality of mixed dietary proteins and the varied needs of a healthy population.

For the young child, satisfactory growth and good health are the criteria for dietary adequacy. The safe level of protein intake is based on the amount of cow's milk formula consumed by infants growing at a maximal rate. However, breast-fed infants consume much less protein, tend to grow more slowly, yet are healthier than formula-fed infants. In unsanitary and poor communities where infectious diseases are common, children who are no longer breast-fed may need more protein to allow rapid growth during periods of convalescence.

C. Effects of the Consumer

Age: The rate of growth, protein synthesis, and protein breakdown are greatest in newborns, coinciding with the highest need for dietary protein per unit of body weight. Protein needs decline with age, as reflected by the "safe levels of protein intake" in Fig. 1. Elderly men and women seem to need as much protein as young people, despite their lower lean body mass.

Sex: There is no difference in the dietary protein needs of men and women per unit of body weight, although men have greater body protein stores. During pregnancy, a woman is advised to increase protein intake by 6–30 g/day to sustain the growth of her own tissues and of the fetus. The placenta actively abstracts amino acids from the mother's blood, providing an abundant supply for fetal protein synthesis.

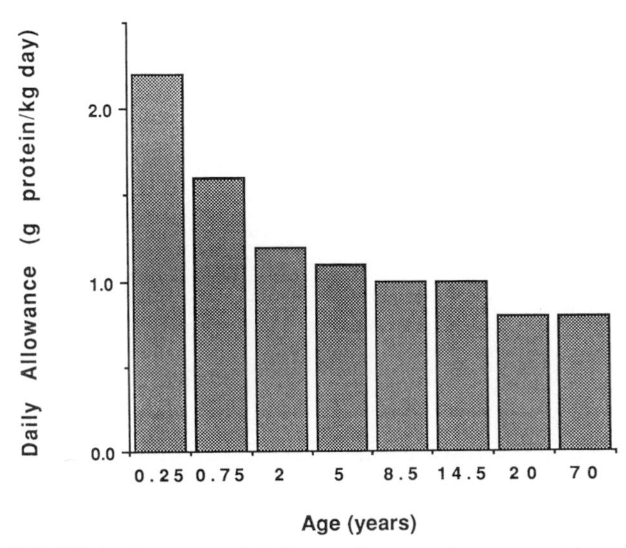

FIGURE 1 Recommended dietary allowance for protein. (Data from the U.S. National Research Council, 1989.)

During lactation, women are advised to eat an additional 12–16 g of protein per day, by consuming a more protein-rich diet, although in most cultures women do not change the foods they eat during pregnancy or lactation. Animal studies show that dietary protein quantity and quality affect the rate of milk protein synthesis and total volume of milk secreted. Women consuming more than 100 g of protein per day produce more milk than those consuming low or marginally adequate amounts of protein.

Nutritional status: Dietary protein is more efficiently retained after a period of low total food intake, low protein intake, or low intake of an indispensable amino acid. During recovery from undernutrition, children gain weight at more than 20 times the normal rate for their age, as long as they get sufficient energy and protein. This is called *catch-up* growth. In adults, nutritional recovery is shown by positive nitrogen balance.

Physical activity: Activity can range from nearly complete immobility to rigorous physical training. Extremes of complete bed rest or exhaustive exercise each lead to a loss of protein stores, whereas moderate activity improves dietary protein use. Endurance athletes have a higher protein requirement (0.95 g/kg/day), but their large food intake generally supplies enough protein. Beginning a program of moderate exercise while consuming an adequate diet increases nitrogen retention in young or old adults and increases linear growth in young children. A resistance exercise program enlarges muscles by markedly increasing protein synthesis and slightly increasing protein breakdown, but strength-trained men do not appear to need more dietary protein to maintain nitrogen equilibrium.

Disease: Bed rest, infection, surgery, trauma, and even emotional distress lead to a loss of protein stores, shown by negative nitrogen balance and muscle wasting. There is increased protein breakdown, whereas protein synthesis can decline or increase, depending on whether protein intake is low or high. Enhanced breakdown of muscle proteins provides an increased supply of amino acids to the liver, white blood cells, and other tissues needing substrates to fight infection.

D. Effects of Diet

Energy intake: Changes in energy intake affect body weight, but effects on nitrogen retention are even more rapid. When energy intake is lower than energy expenditure, nitrogen retention progressively decreases. Obese adults on weight loss diets lose fat but also lose lean body mass (i.e., protein). Conversely, nitrogen retention increases when more calories are consumed, for reasons that are not clear. In infants, nitrogen retention is poor if either protein intake or energy intake are too low; above 70 kcal/kg/day, the use of protein for growth improves as a function of both energy and protein intake. Additional energy as carbohydrate appears to be more protein-sparing than fat in short experiments. [*See* Energy Metabolism.]

Vitamins are cofactors in many enzymatic reactions involving amino acids. The need for thiamin depends on protein intake because thiamin is a cofactor in amino acid breakdown. Pyridoxine is needed for the synthesis of dispensable amino acids. Folate and vitamin B_{12} are involved in the metabolism of methionine and cystine, and B_{12} is also important for the oxidation of valine, methionine, and isoleucine. Minerals such as potassium and zinc are cofactors in the synthesis of nucleic acids and proteins and thus can limit protein utilization for growth.

VI. PROTEIN UNDERNUTRITION

The most likely victims of protein undernutrition are poor children younger than 2 years of age, living in unsanitary urban slums or poor rural areas, and, most especially, the victims of war. In developed countries, protein undernutrition is less common or severe. It accompanies socioeconomic deprivation among families, often where there is alcoholism, drug addiction, or psychiatric disease. [*See* Malnutrition.]

A. Children

Protein and energy undernutrition worldwide is most serious in infants and young children, as they have a large daily requirement for protein and a rapid growth rate. In 1992 a FAO/WHO global assessment reported that there were 192 million children below the age of 5 years defined as severely underweight (i.e., minus 2 S.D. below the international reference). Many more than this number could be defined as malnourished. This is caused by a combination of low supply and increased needs. Poverty and ignorance may lead a mother to feed her child with food that is too dilute, not sanitary, or providing low-quality protein. Food intake may be low because of an impaired appetite or the inappropriate withdrawal of food during infection or diarrhea. Protein needs may be increased because of frequent disease.

Two extremes of protein-energy malnutrition are described: kwashiorkor and marasmus. Kwashiorkor occurs when a young child is weaned from breast milk to a low-protein, high-starch, or high-sugar gruel in an unsanitary environment. The kwashiorkor child is irritable and lethargic, has thin limbs, a pot belly caused by edema, skin lesions reminiscent of pellagra, thin discolored hair, and a fatty liver. Death is common, mainly from severe diarrhea and bacterial infections. Mild to severe marasmus is more common. The child with marasmus is starved of all nutrients, including protein, with infection being a major cause. In response to food deprivation, the child stops growing and uses up its reserves of fat and muscle as fuel to keep alive. The child with marasmus has a "skin and bones" body and the wizened face of an old man.

After treating life-threatening infections and imbalances of fluids and electrolytes, the refed undernourished child can gain weight rapidly, but it is more difficult to attain normal height.

B. Adults

In adults, protein-energy malnutrition is less severe or common. In the United States it can be found in persons with psychiatric problems (anorexia, senile dementia) or in hospitalized patients maintained for long periods on low-protein foods or amino acid-free intravenous feeding solutions. Underlying protein-energy malnutrition is often associated with prolonged convalescence after surgery or illness, hip fracture in the elderly, and the disorders caused by alcoholism and drug addiction.

BIBLIOGRAPHY

FAO/WHO/UNU Expert Consultation (1985). "Energy and Protein Requirements," WHO Technical Report Series 724. WHO, Geneva.

FAO/WHO Expert Consultation (1991). "Protein Quality Evaluation," Food and Nutrition Paper No. 51. FAO, Rome.

FAO/WHO (1992). Nutrition and development—A global assessment. International conference on nutrition ICN/92/INF/5. FAO, Rome.

Munro, H. N., and Crim, M. C. (1988). The proteins and amino acids. *In* "Modern Nutrition in Health and Disease" (M. E. Shils and V. R. Young, eds.). Lea & Febiger, New York.

Shils, M. E., and J. A. Olson (eds.) (1993). "Modern Nutrition in Health and Disease," 8th Ed. Lea & Febiger, New York.

Torun, B., and Viteri, F. E. (1988). Protein-energy malnutrition. *In* "Modern Nutrition in Health and Disease" (M. E. Shils and V. R. Young, eds.). Lea & Febiger, New York.

Waterlow, J. C. (1984). Protein turnover with special reference to man. *Q. J. Exp. Physiol.* **69**, 409–438.

Nutrition, Trace Elements

FORREST H. NIELSEN
U.S. Department of Agriculture, Agricultural Research Service, Grand Forks Human Nutrition Research Center

GLOSSARY

Anemia Any condition in which the number of red blood cells, the amount of hemoglobin, and the volume of packed red blood cells are less than normal

Cardiomyopathy Heart muscle disease; *cardio*, heart; *myo*, muscle; *pathy*, disease

-emia Suffix meaning blood; for example, hypercholesterolemia means high blood cholesterol

Erythropoiesis Formation of red blood cells

Hematopoiesis Formation of blood

Homeostasis Condition in which a substance (e.g., trace element) is maintained at a steady optimal level in the body, that is, an equilibrium in which intake equals utilization and loss by the body so that the concentrations of a specific trace element in various tissues remain rather constant

Hyper- Prefix meaning high, or elevated; for example, hypercholesterolemia means high blood cholesterol

Hypo- Prefix meaning low, small, or depressed; for example, hypouricosuria means depressed urinary excretion of uric acid

Myelinization Development or formation of a myelin (lipid and protein material) sheath around a nerve fiber

Necrosis Pathologic death of one or more cells, or of a portion of tissue or organs, resulting from irreversible damage

Neuropathy Disease of the nervous system

Osteo- Prefix meaning bone; for example, osteoporosis means porous bone, osteosclerosis means bone hardening, and osteomalacia means bone softening or bending

Pharmacological Relatively high oral intake of a substance that alleviates an abnormality caused by something other than a nutritional deficiency of that substance, or that alters some biochemical function or biological structure in a manner that may be construed as beneficial

Total parenteral nutrition Receiving all nutrition by some other means than through the gastrointestinal tract; nutrients introduced into veins or subcutaneous tissues

-uria Suffix meaning urine; for example, hypouricosuria means depressed urinary excretion of uric acid

TRACE ELEMENTS ARE THOSE ELEMENTS OF THE periodic table that occur in the body in μg/g amounts or less. They may be essential (i.e., indispensable for growth and health) or they may be nonessential, fortuitous reminders of our geochemical origins or indicators of environmental exposure. Some of the nonessential trace elements can be beneficial to health through pharmacological action. All the trace elements are potentially toxic when intake is excessive. Trace elements known to be essential to humans are cobalt, copper, iodine, iron, manganese, molybdenum, selenium, zinc, and possibly boron and chromium. Based on findings with animals, several other trace elements most likely are essential to humans; they include aluminum, arsenic, lithium, nickel, silicon, and possibly vanadium. An element with a well-known beneficial function is fluorine. Other elements suggested to be essential, based on limited evidence include bromine, cadmium, lead, rubidium, and tin; acceptance as being essential awaits confirmation of essential properties by additional research groups.

I. BACKGROUND

In the earlier part of this century, scientists were able to qualitatively detect small amounts of several elements in living organisms. In reports, these elements were described as present in traces or trace amounts. Within a short time, these elements became known as trace elements.

The trace elements essential for health and well-being are usually required by animals in parts per million (ppm) amounts, or μg/g diet. For humans, the requirements for these elements are indicated by mg/day. Recently, the term "ultratrace element" has appeared in the nutritional literature. These are elements with estimated dietary requirements of usually <1 μg/g and often <50 ng/g [parts per billion (ppb)] dry diet for animals. For humans, the estimated requirements are indicated by μg/day.

An element is considered nutritionally essential if an organism can neither grow nor complete its life cycle in the absence of the element. A severe deficiency of an element resulting in death is difficult to produce, particularly if the element is required in a very low amount, because of contamination. Thus, a broader definition of essentiality has been widely accepted: An element is nutritionally essential if a dietary deficiency consistently results in a suboptimal biological function that is preventable or reversible by physiological amounts of the element. In this definition, physiological is construed as those quantitites usually found in biological material.

II. BIOLOGICAL ROLES

Trace elements have five known roles in living organisms. (1) In close association with enzymes, some trace elements are an integral part of the catalytic centers at which the reactions of biological chemistry occur. Working in concert with a protein, and frequently with other organic coenzymes, the trace element attracts substrate molecules and facilitates their conversion to a specific end product. (2) Some trace elements donate or accept electrons in reactions of reduction or oxidation. These redox reactions are of primary importance in the generation and utilization of metabolic energy through the "burning" of foods in cells. Chemical transformations of molecules frequently involve redox reactions. (3) Some trace elements (especially iron) bind, transport, and release oxygen in the body. (4) Some trace elements have structural roles, that is, imparting stability and three-dimensional structure to important biological molecules. (5) Some trace elements have a regulatory role. They control important biological processes through such actions as inhibiting enzymatic reactions, facilitating the binding of molecules to receptor sites on cell membranes, and inducing genes to form important molecules.

III. HOMEOSTATIC REGULATION

The responses of a living organism to various intakes of a trace element are shown in Fig. 1. When an essential trace element is absent or too low for adequate activity of an essential function, death occurs. As the intake of the trace element increases, the following occurs. (1) The organism survives but with suboptimal health and well-being. (2) An intake is reached at which optimal health and well-being are maintained. The range of intakes at which this occurs usually is quite large because of powerful homeostatic mechanisms. (3) A decline in health and well-being, and finally death, as regulatory mechanisms are overcome by increasing intakes that become toxic.

Homeostatic regulation involves the processes of absorption, storage, and excretion. The relative importance of these three processes varies among the trace elements. The amount absorbed from the gastrointestinal (GI) tract often is the controlling mechanism for positively charged trace elements such as copper, iron, and zinc. If the body is low in the trace element, the percentage of the element absorbed from the GI tract is increased, and vice versa. Trace elements absorbed as negatively charged anions such as boron and selenium are usually absorbed quite freely and completely from the GI tract. Excretion through the urine, bile, sweat, and breath is, therefore, the major mechanism for controlling the amount of these trace elements in the organism. By being stored at inactive

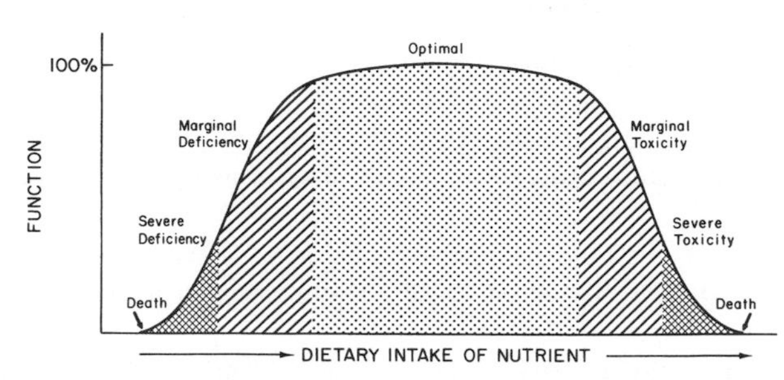

FIGURE 1 Dose–response curve for a typical trace element.

sites, some trace elements are prevented from causing adverse reactions when present in high quantities. Examples of this homeostatic process are the binding of copper by the cystine-rich protein called metallothionein and the storage of iron as ferritin. Release of a trace element from storage forms also can be important in preventing deficiency.

IV. FACTORS AFFECTING NUTRITIONAL IMPORTANCE

Although trace elements play key roles in a variety of processes necessary for life, except for iodine and iron, the occurrence of overt, simple, or uncomplicated deficiencies of any of the trace elements in humans is relatively uncommon. This fortunate circumstance is probably the result of two factors. The first factor is the presence of powerful homeostatic mechanisms described earlier. The other factor is that, especially in industrialized countries like the United States, the diverse sources and types of food in diets are very effective in providing the trace elements in the amounts and forms required. Situations other than a simple acute deficiency, however, may be possible, which would make a trace element of practical nutritional importance. Trace elements show potential clinical importance in at least four areas. These areas include genetically caused errors of metabolism that affect absorption, retention, or excretion; alterations in metabolism and/or biochemistry as a secondary consequence to malnutrition, disease, or injury; inadvertent omission from a total parenteral nutrition solution; and an enhanced requirement caused by some nutritional, metabolic, hormonal, or physiological stressors.

V. ESSENTIAL AND POSSIBLY ESSENTIAL TRACE ELEMENTS

Table I shows the trace elements generally accepted as essential for animals and humans. Table I also indicates the biological functions and rich dietary sources for specific trace elements. The following summaries indicate the nature and likelihood of abnormal nutrition of specific trace elements.

A. Aluminum

Signs of aluminum deficiency have not been described for humans. Reported signs of aluminum deprivation for the goat are increased spontaneous abortions, depressed growth, uncoordination and weakness in the hind legs, and decreased life expectancy. In the growing chick, aluminum deprivation depressed growth.

Aluminum has a low order of toxicity when administered orally. This is exemplified by the fact that numerous aluminum compounds are used as food additives. Baked goods prepared with chemical leavening agents (e.g., baking powder) and processed cheese are among the foods that contain luxuriant amounts of aluminum as a food additive. High amounts of aluminum are found in some pharmaceutical products, including some antacids, buffered analgesics, antidiarrheals, and antiulcer drugs. It has been estimated that daily intakes of aluminum of 840 to 5000 mg in antacids and 126 to 728 mg in buffered analgesics are possible for individuals using large amounts of these pharmaceuticals. Although not inherently toxic, these high dietary intakes may have an adverse effect in other ways. High amounts of orally administered aluminum compounds have been shown to reduce the absorption of other nutrients, including iron, phosphorus, and calcium; this could lead to a deficiency of these elements. For example, large doses of aluminum over an extended period of time can cause severe phosphate depletion resulting in osteomalacia. In addition to this, high dietary aluminum has been associated with Alzheimer's disease; however, this association is controversial. Aluminum toxicity is of most concern when the normal barriers to aluminum absorption are circumvented, that is, when aluminum is administered parenterally through total parenteral nutrition or through dialysis (as performed on kidney failure patients) with an aluminum-contaminated dialysate. This type of aluminum overload leads to neurological and skeletal pathology. The neurological syndrome, called dialysis encephalopathy or dialysis dementia, is characterized by seizures, hallucinations, personality changes, and death within 6 to 8 months after onset of symptoms. Skeletal lesions include bone pain, osteomalacia, and fractures.

Because of the recency of the suggestion that aluminum is an essential nutrient, and because an aluminum deficiency has not been recognized in humans, aluminum nutrition other than toxicological aspects has not received any attention. Whether aluminum needs more attention is questionable. The presence of high amounts of aluminum additives in commonly consumed pharmaceuticals and foods, compared to its apparent small requirement, suggests that pathology caused by aluminum deficiency would be rare.

TABLE I
Essential and Possibly Essential Trace Elements for Human Nutrition

Trace element	Adult human body content	Recommended daily intake	Dietary sources	Essentiality status for humans	Biological function
Aluminum (Al)	45 mg	1–5 mg[a]	Grains, herbs, tea, foods with aluminum additives (e.g., processed cheese, products with baking powder)	Possibly essential; signs of deficiency described for animals but not for humans	Unknown; may be a structural component or cofactor of an enzyme or a regulatory molecule
Arsenic (As)	1–2 mg	12–15 μg[a]	Fish, seafood, meat, poultry, grains, cereal products	Possibly essential; signs of deficiency described for animals but not for humans	Unknown; may be involved in phospholipid, labile methyl group, and/or polyamine metabolism
Boron (B)	10–20 mg	1–3 mg[a]	Fruits, nuts, vegetables, legumes	Possibly essential; apparent signs of deficiency described for animals and humans	Unknown; affects macromineral (calcium, magnesium) metabolism
Chromium (Cr)	2 mg	50–200 μg[b]	Brewer's yeast, beer, cheese, meat products, whole grains, condiments	Possibly essential; apparent signs of deficiency described for animals and humans	Unknown; involved with normal glucose and lipid metabolism
Cobalt (Co)	1.5 mg	3 μg, as[c] vitamin B_{12}	Green leafy vegetables, organ meats	Essential	Component of vitamin B_{12}, a known cofactor for three human enzymes
Copper (Cu)	80 mg	1.5–3 mg[b]	Shellfish, organ meats, nuts, beans	Essential	Cofactor of several oxidative enzymes in iron metabolism, and the synthesis of cartilage, bone, and myelin
Fluorine (F)	3 g	1.5–4 mg[b]	Seafood, fluoridated water	Weak evidence for essentiality; pharmacologically beneficial	None known; beneficial in preventing dental caries
Iodine (I)	11 mg	150 μg[c]	Iodized salt, seafood, food from high-iodine soils	Essential	Component of thyroid hormones, thyroxine (T_4) and triiodothyronine (T_3)
Iron (Fe)	3–5 g	10–15 mg[c]	Red meat, organ meat, whole grains, vegetables	Essential	Oxygen transport as hemoglobin and myoglobin; electron transport in enzyme reactions
Lithium (Li)	1–3 mg	50–200 μg[a]	Leafy and root vegetables, eggs, organ meats, red meat, fish	Possibly essential; signs of deficiency described for animals but not for humans	Unknown; may be involved in endocrine, especially thyroid, metabolism
Manganese (Mn)	1 g	2.0–5.0 mg[b]	Whole grains, nuts, legumes	Essential	Cofactor for enzymes involved in protein and energy metabolism, antioxidant action, and mucopolysaccharide synthesis
Molybdenum (Mo)	10 mg	75–250 μg[b]	Legumes, cereals, milk and milk products, vegetables, organ meats	Essential	Cofactor of enzymes catalyzing the hydroxylation of substrates using the elements of water

TABLE I (*Continued*)

Trace element	Adult human body content	Recommended daily intake	Dietary sources	Essentiality status for humans	Biological function
Nickel (Ni)	1–2 mg	75–150 μg[a]	Nuts, chocolate, dried legumes, grains	Possibly essential; signs of deficiency described for animals but not for humans	Unknown; enzyme cofactor in plants and microorganisms
Selenium (Se)	6–12 mg	55–70 μg[c]	Fish, eggs, meat, grains grown on high-Se soil	Essential	Cofactor for enzymes involved in antioxidant action and thyroid hormone metabolism
Silicon (Si)	2–3 g	5–20 mg[a]	Whole grains, cereal products, root vegetables	Possibly essential, signs of deficiency described for animals but not for humans	Unknown; most likely a cross-linking agent or has a role in bone and collagen biosynthesis
Vanadium (V)	100 μg	6–20 μg[a]	Shellfish, mushrooms, condiments	Possibly essential; signs of deficiency described for animals but not for humans	Unknown for mammals; enzyme cofactor in bacteria, algae, and lichens
Zinc (Zn)	2–3 g	12–15 mg[c]	Meat and meat products, organ meats, shellfish, dairy products	Essential	Cofactor or activator of >100 enzymes

[a]Recommended intake estimated by author through extrapolation of data from animals.
[b]Safe and adequate intake estimated by the National Academy of Sciences.
[c]Recommended dietary allowance established by the National Academy of Sciences.

B. Arsenic

Signs of arsenic deficiency have not been described for humans. In the goat, miniature pig, and rat, the most consistent signs of arsenic deprivation described are depressed growth and abnormal reproduction characterized by impaired fertility and elevated perinatal mortality. Other notable signs of deprivation described for goats are depressed serum triglycerides and death during lactation. Myocardial damage was present in lactating goats. Other signs have been reported. Specific deficiency signs cannot be listed because studies with chicks, rats, and hamsters have revealed that the nature and severity of the signs of arsenic deprivation are affected by several dietary manipulations, including variations in the concentration of zinc, arginine, choline, methionine, taurine, and guanidoacetic acid, all of which can affect methyl metabolism.

Because mechanisms exist for the homeostatic regulation of arsenic, the toxicity of arsenic through oral intake is relatively low; it is actually much less toxic than selenium, a trace element with a well-established nutritional value. Toxic quantities of inorganic arsenic generally are measured in milligrams, and the ratio of the toxic to nutritional dose for rats is apparently near 1250. The organic forms of arsenic that occur naturally in foods are virtually nontoxic. Briefly, the signs of subacute and chronic high exposure of arsenic in humans include the development of dermatoses of various types, hematopoietic depression, liver damage characterized by jaundice, liver cirrhosis, sensory disturbances, peripheral neuritis, anorexia, and weight loss.

Numerous epidemiological studies have suggested an association between chronic overexposure to arsenic and cancer. However, the role of arsenic in carcinogenesis remains controversial. Recent studies indicate that arsenic is not a primary carcinogen and is either an inactive or extremely weak mitogen.

Specific disorders in which arsenic deprivation is a contributing factor cannot be identified until more is known about its biological function. Thus, at present, the human health concerns of arsenic are involved with toxicity. The ingestion of toxic amounts of arsenic usually occurs as the result of homicidal, pharmaceutical, or accidental consumption of arsenic compounds.

C. Boron

Definitively stating the signs of boron deficiency is difficult. The response to boron deprivation in animals is affected by variables that affect macromineral (magnesium, calcium, phosphorus) metabolism. In other words, the reported signs of boron deficiency vary in nature and severity as the diet is varied in content of calcium, phosphorus, magnesium, potassium, cholecalciferol, aluminum, and methionine. However, although the nature and severity of the changes vary with dietary composition, many findings indicate that boron deprivation impairs calcium utilization, energy metabolism, and brain function. In humans, boron supplementation after depletion enhanced the elevation in serum 17β-estradiol and plasma copper caused by estrogen ingestion, altered electroencephalograms such that they suggested improved behavioral activation (e.g., less drowsiness) and mental alertness, and improved psychomotor skills and the cognitive processes of attention and memory. Boron deprivation also affected erythropoiesis and hematopoiesis and variables associated with macromineral, energy, and nitrogen metabolism.

Boron has a low order of toxicity when administered orally. Toxicity signs generally occur in animals only after the dietary boron concentration exceeds 100 μg/g. A study involving the feeding of pigs weighing near 60 kg showed that 8 mg boron/kg/day detrimentally affected calcium metabolism. The approximately 500 mg/day dose apparently induced an osteoporosis associated with a reduction in parathyroid activity. In humans, the signs of acute toxicity include nausea, vomiting, diarrhea, dermatitis, and lethargy.

Accumulating evidence indicates that boron deprivation may be a practical nutritional problem. Because boron is a dynamic trace element that affects macromineral metabolism, it may have a role in some disorders of unknown etiology that exhibit disturbed mineral metabolism (e.g., osteoporosis). Current research will probably soon clarify the importance of boron in human nutrition. Since the use of boron as a food preservative was discontinued, the likelihood of finding boron toxicity in the general population has been low. It occurs mainly when individuals accidentally or intentionally consume borate, boric acid, or substances containing boron as an active ingredient (e.g., insecticides).

D. Chromium

Chromium deficiency in humans was first recognized in Jordanian infants with severe protein-calorie malnutrition. Glucose clearance of the infants was severely impaired, and the impairment was alleviated by chromium supplementation. Chromium was also shown to improve growth of Turkish infants who were recovering from protein-calorie malnutrition.

Chromium deficiency has also been observed in patients given low-chromium total parenteral nutrition for long periods. Signs of deficiency included decreased sensitivity to insulin, glucose intolerance, impaired amino acid utilization, and peripheral neuropathy. However, until a specific biochemical function is found for chromium, skepticism will remain about the conclusion that it is an essential instead of beneficial element.

Trivalent chromium has a low order of toxicity; a wide margin of safety exists between amounts ordinarily ingested and those likely to induce deleterious effects. For example, cats tolerate 1000 mg/day and rats showed no adverse effects from consuming a diet containing 100 mg Cr/kg. Hexavalent chromium is much more toxic than trivalent chromium. Feeding 50 mg chromate/kg diet causes growth depression and liver and kidney damage in experimental animals.

Chromium toxicity is not a major concern in nutrition. However, this judgment may be altered by the recent advent of chromium supplements touted as useful for building muscle and strength, inducing fat loss, and reducing blood cholesterol. Although these claims are based mostly on overinterpreted findings that have not been reproducibly and reliably demonstrated or reported in the peer-reviewed scientific literature, large amounts of bioavailable chromium in the form of supplements are being consumed. The potential toxicity of these supplements has not been adequately assessed. On the other hand, marginal dietary chromium intakes may be more widespread than is currently acknowledged. Most likely, chromium nutrition is of concern in individuals whose normal insulin-dependent metabolism of carbohydrate, protein, and fat is disturbed. Stress, including trauma, infection, surgery, and intense heat or cold, elevates the secretion of hormones, which alters glucose metabolism and apparently the need for chromium. The practical significance of chromium nutrition is an active area of research.

E. Cobalt

The only known function for cobalt is as the metal ion at the center of the corrin ring of vitamin B_{12}. Thus, the discussion of cobalt in human nutrition is essentially one centered around vitamin B_{12}, which is made in nature only by microorganisms.

The major consequence of cobalt deficiency is anemia, because vitamin B_{12} is necessary for erythropoiesis. Other signs of vitamin B_{12} deficiency include pallor, weakness, fatigue, difficult or rapid breathing with exertion, and a sore tongue. Deficiency also causes patchy or diffuse and progressive demyelinization resulting in neuropathy.

Cobalt has a low order of toxicity in all species studied, including humans. Signs of cobalt toxicity include elevated number of red blood cells (polycythemia) accompanied by increased cells in the bone marrow and increased blood volume. Other signs in humans include vasodilation and flushing. Oral intakes of about 250 mg/kg diet are necessary to induce a significant increase in red blood cells. Cobalt toxicity is suspected to be the cause of several occurrences of cardiac failure in heavy beer drinkers. Cobalt had been added to beer to improve its foaming qualities. Apparently, high cobalt (8 mg/day) and high alcohol are both necessary to induce the distinctive cardiomyopathy.

Cobalt deficiency is more of a concern than cobalt toxicity in nutrition. The prevalence of cobalt deficiency is that of vitamin B_{12} deficiency, which can be caused by inadequate ingestion, absorption, or utilization, and is not uncommon. Adequate vitamin B_{12} nutrition is of special concern for strict vegetarians.

F. Copper

Manifestations of copper deficiency in humans are best known in infants. Signs of deficiency include pallor, lethargy, and anemia unresponsive to iron therapy. Bone changes also occur, including osteoporosis, flaring of the anterior ribs, cupping and flaring of the growth zones of the long bones, and spontaneous fractures of the ribs.

Copper deficiency in experimental animals causes many abnormalities, which may have implications for human health; they include impairment of glucose metabolism, abnormal lipid metabolism and hypercholesterolemia, cardiac necrosis, disecting aneurysms, and impaired myelinization.

Chronic copper toxicity is unusual. It occurs mainly in liver diseases in which bile flow is impaired or with genetic disorders including Wilson's disease and Indian childhood cirrhosis. In these diseases, copper accumulates in the liver, renal tubules, cornea, brain, and other organs, resulting in damage to those structures.

Severe copper deficiency in humans also is unusual. It has been observed in premature infants, infants receiving only cow's milk subsequent to recovery from protein-calorie malnutrition, persons receiving copper-deficient total parenteral nutrition, and infants with the genetic disease known as Menke's syndrome. Some recent human studies have produced findings that indicate variables that are predictors of cardiovascular disease are adversely affected by mild copper deficiency. If this is true, copper nutrition may be of more practical concern than currently acknowledged. [*See* Copper, Iron, and Zinc in Human Metabolism.]

G. Fluorine (Fluoride)

At present, no evidence shows unequivocally that fluoride is an essential nutrient for either animals or humans. The known beneficial effects from the oral ingestion of large amounts of fluoride are probably the result of pharmacological action; these effects include prevention of dental caries and osteoporosis. [*See* Dental Caries.]

Toxicity is the major nutritional concern for fluoride. Chronic toxicity via excessive fluoride in water supplies, or from industrial exposure, occurs in many parts of the world. Toxicity is manifested in dental fluorosis (mottled enamel) and skeletal fluorosis resulting in osteosclerosis, osteoporosis, and osteomalacia. Severe skeletal fluorosis can be crippling.

H. Iodine

The most common manifestation of iodine deficiency is goiter, or enlargement of the thyroid gland. With severe and prolonged iodine deficiency, hypothyroidism, or myxedema, occurs. Myxedema is characterized by reduced metabolic rate and the accumulation of a jelly-like substance, mucin, in the body; these, in turn, cause a great number of the characteristic abnormalities. A deficiency of iodine during pregnancy, infancy, or early childhood may result in cretinism. Cretinism is characterized by mental retardation, retarded growth, deafness, deaf-mutism, delayed psychomotor development, and various neurological abnormalities.

Excessive intake of iodine may cause hyperthyroidism (thyrotoxicosis). Symptoms of hyperthyroidism include nervousness, tremor, heat intolerance, goiter, bulging eyes, rapid heart beating, fatigue, weakness, increased appetite, and weight loss. Iodine intakes of above 2000 μg/day have been suggested as being excessive.

Both iodine deficiency and toxicity are practical nutritional concerns. However, iodine deficiency seldom occurs in the United States because of the use of iodine-containing compounds in the dairy and baking industries and the availability of iodized salt. Iodine deficiency continues to be a significant problem in some mountainous, developing countries found in the Andes chain, parts of Africa, and Southeast and Central Asia. Goiter occurs as a consequence of large intakes of dietary iodine in Japan and other Asian countries where seaweed containing high amounts of iodine is commonly eaten.

I. Iron

The most recognizable consequence of iron deficiency is anemia. However, prior and subsequent to the appearance of anemia, iron deprivation may have other functional consequences. These include poor growth and development; impaired neuropsychological function, physical work performance, and thermoregulation; and increased susceptibility to certain infections.

Iron overload results in the accumulation of iron in several organs including the liver, heart, and spleen. The accumulation of chemically active iron results in tissue damage and, if severe enough, death. Recently, high iron stores in the body have been associated with an increased risk to ischemic heart disease.

Toxicity through the dietary intake of iron is uncommon because encountering food with sufficient bioavailable iron to cause toxicity is rare. In some individuals, iron overload damage possibly could result from high ascorbic acid (vitamin C) supplements, especially with iron supplementation. Vitamin C enhances iron absorption and the release of chemically active iron from stores. Iron overload mainly occurs as the result of prolonged treatment of certain anemias by transfusion and with genetic disorders resulting in abnormal hemoglobin formation (thalassemia, hemochromatosis).

From the nutritional point of view, iron is the most important of the trace elements. Iron deficiency is one of the most common deficiency diseases in the world. It is the major cause of anemia. Persons at risk of iron deficiency are rapidly growing infants, women during the childbearing years, and adolescents. Persons consuming low amounts of meat and related products have an increased risk of iron deficiency.

J. Lithium

Signs of lithium deficiency have been described for the rat, goat, and Japanese quail. In rats, lithium deprivation of dams resulted in smaller litter sizes, litter weight at birth and weaning, and percentage of pups weaned. In goats, lithium deprivation resulted in impaired reproduction characterized by decreased fertility and increased abortion rate, less milk production, and decreased life expectancy. Lithium deficiency also depressed the activity of plasma isocitrate dehydrogenase, malate dehydrogenase, glutamate dehydrogenase and aldolase, and liver monoamine oxidase. Feeding lithium-deficient diets to quail resulted in eggs with decreased shell quality and hatchability. Signs of deficiency have not been described for humans; however, significant inverse correlations have been made between the lithium content of drinking water and the admission rates of patients with psychoses, neuroses, and personality disorders into mental hospitals, and the incidence of crimes such as homicide, rape, robbery, burglary, theft, and suicide. Like fluoride, in relatively high amounts lithium has beneficial pharmacological properties. It has been used effectively in the treatment of manic-depressive disorders.

From a nutritional point of view, lithium is a relatively nontoxic element. The dosages required to raise serum lithium to effective therapeutic concentrations of about 3.5–10.4 mg/liter in the treatment of manic-depressive disorders usually are around 500 mg of the element daily as a lithium salt. This dosage is approximately 2500 times the normal intake, or possible recommended intake (see Table I). However, toxic symptoms have been reported in some patients with serum concentrations as low as 11.0 mg/liter. In some patients, particularly women, lithium treatment depresses thyroid function. Diarrhea, vomiting, drowsiness, muscular weakness, and lack of coordination are early signs of lithium intoxication. Serum lithium concentrations above 21 mg/liter can produce a complex clinical picture involving multiple organs and organ systems.

From the preceding discussion, it is obvious that lithium toxicity is a concern in the clinical setting, but not in nutrition. Whether low intakes of lithium are of practical nutritional concern remains to be determined. The association of low intakes with violent behavior and mental illness suggests that further study of this issue is warranted.

K. Manganese

The essentiality of manganese for animals has been known for over 50 years. Deficiency causes testicular degeneration (rats), slipped tendons (chicks), osteodystrophy, severe glucose intolerance (guinea pigs), ataxia (mice, mink), depigmentation of hair, and seizures. However, only one description of an unequivocal case of human manganese deficiency has been reported. A child with a postoperative short bowel receiving over 90% of her nutrition parenterally, which was low in manganese, exhibited short stature and brittle bones. Manganese supplementation resulted in catchup growth and improved bone composition and function.

Manganese is often considered to be among the least toxic trace elements if consumed orally. The most common form of manganese toxicity occurs as the result of the chronic inhalation of large amounts of airborne manganese found in mines, steel mills, and areas of some chemical industries. This toxicity causes symptoms similar to Parkinson's disease.

Because both dietary manganese deficiency and toxicity have been difficult to find in humans, manganese nutrition has not been considered to be of practical nutritional concern.

L. Molybdenum

A patient receiving prolonged parenteral nutrition therapy developed a syndrome described as acquired molybdenum deficiency. This syndrome, exacerbated by methionine administration, was characterized by hypermethioninemia, hypouricemia, hyperoxypurinemia, hypouricosuria, and very low urinary sulfate excretion. In addition, the patient suffered mental disturbances that progressed to coma. Supplementation of the patient with ammonium molybdate improved the clinical condition, reversed the sulfur-handling defect, and normalized uric acid production.

Large oral doses are necessary to overcome the homeostatic control of molybdenum. Thus, molybdenum is a relatively nontoxic element; in nonruminants, an intake of 100–5000 mg/kg food or water is required to produce clinical toxicity symptoms. Many of the signs of toxicity are similar or identical to those of copper deficiency or indicate abnormal sulfur metabolism. Both occupational and high dietary exposure to molybdenum have been linked through epidemiological methods to elevated uric acid in blood and increased incidence of gout.

Because dietary molybdenum deficiency is difficult to induce, and dietary molybdenum toxicity is uncommon, molybdenum has not been considered to be of much practical importance in human nutrition. However, xanthine oxidase and aldehyde oxidase, both molybdoenzymes, may be involved in the *in vivo* detoxification of xenobiotic compounds. The exposure to high amounts of xenobiotics, including carcinogens, may be the stress necessary to make a low intake of molybdenum of consequence. This possibility needs further research.

M. Nickel

Nickel deficiency signs have been described only for animals. The following signs apparently are representative of nickel deficiency. If the nickel deficiency is severe, growth and hematopoiesis are depressed, especially in marginally iron-adequate, or in methyl-depleted, animals. Iron utilization is impaired. The mineral element profiles of various organs change. Moreover, nickel deprivation affects several variables associated with the metabolism of arginine. As with other elements, the extent and severity of nickel deprivation signs are markedly affected by diet composition.

Life-threatening toxicity of nickel through oral intake is unlikely. Generally, 250 μg or more of nickel per gram of diet are required to produce signs of nickel toxicity (such as depressed growth) in rats, mice, chickens, rabbits, and monkeys. If animal data can be extrapolated to humans, a daily dose of 250 mg soluble nickel would produce toxic symptoms in humans. However, some evidence suggests that the ingestion of small amounts of nickel may be of greater importance than external contacts in maintaining eczema caused by nickel allergy. An oral dose as low as 0.6 mg nickel as nickel sulfate given with water (nickel much more absorbable than in food) to fasting subjects produced a positive reaction in some nickel-sensitive individuals. That dose is only a few times higher than the human daily requirement postulated from animal studies (0.15 mg/day).

Because dietary nickel deficiency has not been recognized, and dietary nickel toxicity is uncommon, nickel in human nutrition has not received much attention. However, because recent research has shown nickel to be a dynamic trace element in organisms other than humans, further research is needed to assure that nickel is not of nutritional or toxicological importance for humans.

N. Selenium

Manifestations of selenium deficiency in livestock include muscular dystrophy (white muscle disease), exudative diathesis (edema and hemorrhaging), and liver necrosis. The nutritional importance of selenium is most evident in vitamin E-deficient animals.

Keshan disease, a juvenile cardiomyopathy that is endemic in certain areas of China, has been associated with poor selenium nutrition. Otherwise, reports of selenium deficiency have been limited to patients on long-term total parenteral nutrition. The signs of deficiency in these patients included bilateral muscular discomfort and a dilated cardiomyopathy that histopathologically resembled Keshan disease.

Selenium toxicity in livestock has long been recognized. Chronic intoxication is characterized by emaciation, loss of hair, loss of hoofs, joint erosion, atrophy of the heart, cirrhosis, and anemia. The signs of toxicity to humans have not been as clearly defined. Dermatitis, loose hair, mottled teeth, and pathological nails have been associated with the excessive intake of selenium. Selenium is a very toxic element; chronic toxicity occurs in experimental animals and livestock with concentrations >2 μg/ml drinking water or 3 μg/g diet. This suggests that the consumption of 2–3 mg selenium daily might have toxicological consequences in humans.

The practical importance of selenium in nutrition is currently being established. Because selenium deficiency seems to be most important when humans are exposed to some other stressor, research is being done to identify those stressors that enhance the need for selenium. For example, researchers are currently investigating the possibility that higher intakes of selenium reduce the susceptibility to some forms of cancer. Studies establishing intakes of selenium that are not toxic are also being done. [*See* Selenium in Nutrition.]

O. Silicon

Signs of silicon deficiency have not been described for humans. Most signs of silicon deficiency in chickens and rats indicate aberrant metabolism of connective tissue and bone. Silicon-deficient chicks exhibit skull structure abnormalities associated with a depressed collagen content in bone and long-bone abnormalities characterized by small, poorly formed joints and defective endochondrial bone growth. Tibial bone from silicon-deficient chicks contain depressed amounts of articular cartilage, water, hexosamine, and collagen.

Silicon is essentially nontoxic when taken orally. Evidence for its nontoxicity is the observation that magnesium trisilicate, an over-the-counter antacid, has been used by humans for more than 40 years without obvious deleterious effects. Other silicates are food additives used as anticaking or antifoaming agents.

More work is needed to clarify the consequences of silicon deprivation in humans. This need has not prevented speculation that the lack of silicon is involved in several human disorders, including atherosclerosis, osteoarthritis, and hypertension, as well as in wound healing and the aging process. Those speculations indicate the need for studying the importance of silicon in nutrition, especially aging humans.

P. Vanadium

Many of the deficiency signs reported for vanadium are questionable. Apparently, the reported differences between vanadium-deprived and vanadium-supplemented animals were the consequences of high vanadium supplements (10–100 times the amount normally found in the diet) that resulted in pharmacological changes in suboptimally performing animals fed imbalanced diets. Recently, however, vanadium deprivation studies probably have found some true deficiency signs; these include suboptimal reproductive performance characterized by increased spontaneous abortions, death preceded by convulsions shortly after birth, and skeletal deformations.

Vanadium can be a relatively toxic element; units used for toxic amounts of vanadium are a few tens of mg/day or μg/g of diet. A number of substances ameliorate vanadium toxicity, including ascorbic acid, chromium, protein, ferrous iron, chloride, and perhaps aluminum hydroxide. Age and animal species also influence vanadium toxicity. Thus, not surprisingly, a variety of signs of vanadium toxicity have been reported. Some of the more consistent signs include depressed growth, diarrhea, depressed food intake, and death. Excessive *in vivo* vanadium has been suggested to be a factor in manic-depressive illness.

The recent discovery in lower forms of life of two enzymes that require vanadium to be active lends credence to the possibility that vanadium has a similar role in higher animals. These enzymes are nitrogenase in bacteria, which reduces dinitrogen to ammonia, and bromoperoxidase in algae and lichens, which catalyzes the oxidation of halide ions by hydrogen peroxide. Nonetheless, whether or not vanadium has some practical nutritional importance beyond its toxicological and pharmacological aspects remains to be determined.

Q. Zinc

Major manifestations of zinc deficiency in humans include dermatitis, impaired growth and development, delayed wound healing, depressed resistance to infections, abnormal nervous system function, impaired taste and smell, and loss of dark adaptation.

Little is known about the consequences of excessive intake of zinc. The habitual consumption of 160 mg zinc was found to depress plasma high-density lipoprotein cholesterol. Copper deficiency has been observed in patients treated with 150 mg zinc daily. A high intake of zinc apparently can exert negative effects on the metabolism of other trace elements.

Because zinc has a low order of toxicity and highly regulated mechanisms of absorption and excretion, the toxic potential of zinc is of little concern. Zinc deficiency has been found in humans throughout the world. In developing countries, the occurrence of zinc deficiency is associated with diets high in cereals and low in animal products and with parasitic diseases. In industrialized countries, zinc deficiency is found among persons with inflammatory bowel diseases, intestinal malabsorption, liver disease, alcoholism, renal failure, and protein-calorie malnutrition. Zinc deficiency causes the severe symptoms of the rare genetic disease acrodermatitis enteropathica. Pregnant women, rapidly growing infants, children, the poor, and the elderly have an increased risk of deficiency. [*See* Zinc Metabolism.]

VI. CONCLUDING STATEMENT

Progress in showing the nutritional significance of trace elements other than cobalt as vitamin B_{12}, iodine, iron, and zinc has been slow. There are, however, indications that progress may be more rapid in the future. Findings with boron, chromium, copper, and selenium suggest that the right situations and questions are just beginning to be addressed for many of the trace elements. An increasing number of studies have been described that examined the interactions of specific trace elements with various forms of nutritional, metabolic, hormonal, and physiological stressors in animals and humans. These studies indicate that some of the trace elements are more important in human nutrition than is now generally acknowledged.

BIBLIOGRAPHY

Casey, C. D., and Robinson, M. F. (1983). Some aspects of nutritional trace element research. *In* "Metal Ions in Biological Systems, Vol. 16, Methods Involving Metal Ions and Complexes in Clinical Chemistry" (H. Sigel, ed.) pp. 1–26. Marcel Dekker, New York.

Frieden, E. (ed.) (1984). "Biochemistry of the Essential Ultratrace Elements." Plenum, New York.

Mertz, W. (1981). The essential trace elements. *Science* **213,** 1332–1338.

Mertz, W. (ed.) (1986 and 1987). "Trace Elements in Human and Animal Nutrition," 5th Ed., Vols. 1 and 2. Academic Press, Orlando, Florida.

Mertz, W., Abernathy, C. O., and Olin, S. S. (eds.) (1994). "Risk Assessment of Essential Elements," ILSI Press, Washington, D.C.

Nielsen, F. H. (1984). Ultratrace elements in nutrition. *Annu. Rev. Nutr.* **4,** 21–41.

Nielsen, F. H. (1988). Nutritional significance of the ultratrace elements. *Nutr. Rev.* **46,** 337–341.

Nielsen, F. H. (1988). Possible future implications of ultratrace elements in human health and disease. *In* "Essential and Toxic Trace Elements in Human Health and Disease" (A. S. Prasad, ed.), pp. 277–292. Alan R. Liss, New York.

Nielsen, F. H. (1993). Ultratrace elements of possible importance for human health: An update. *In* "Essential and Toxic Trace Elements in Human Health and Disease: An Update" (A. S. Prasad, ed.), pp. 355–376. Alan R. Liss, New York.

Saltman, P., Hegenauer, J., and Strause, L. (1984). For the want of a nail. . . . Trace elements in health and disease. *In* "Metabolism of Trace Metals in Man, Vol. 1, Developmental Aspects" (O. M. Rennert and W.-Y. Chan, eds.), pp. 1–16. CRC Press, Boca Raton, Florida.

Shils, M. E., Olson, J. A., and Shike, M. (eds.) (1994). "Modern Nutrition in Health and Disease," 8th Ed. Lea & Febiger, Philadelphia.

Smith, K. T. (ed.) (1988). "Trace Minerals in Foods." Marcel Dekker, New York.

Obesity

GEORGE A. BRAY
Pennington Biomedical Research Center, Baton Rouge

GLOSSARY

Computed tomography An X-ray procedure which gives very accurate pictures of the body

Dual energy X-ray absorptiometry Method of determining body fat

Hypothalamus Region of the brain, located just above the pituitary and below the thalamus, that regulates food intake, body temperature, pituitary function, and other vital functions

^{40}K Radioactive isotope of potassium, which occurs naturally and can be measured to determine that amount of lean tissue, as most of the potassium in the body is in muscle

Magnetic resonance imaging A procedure involving magnetism for obtaining accurate pictures of the body

Norepinephrine Chemical neurotransmitter

Nucleus Group of cells within a region of the brain

Overweight An excess of weight relative to some standard for height

Paraventricular nucleus Collection of brain cells located near the ventricle but rostral to the third ventromedial nucleus

Satiety State of feeling satisfied

Serotonin Neurotransmitter also called 5-hydroxytryptamine

Ventromedial nucleus Nucleus in the hypothalamus that is near the ventricle

Visceral fat Fat surrounding internal organs

ALTHOUGH OVERWEIGHT AND OBESITY ARE FREquently used interchangeably, obesity is an excess of body fat, not strictly an excess of body weight. Both excess body fat and increased abdominal or visceral fat pose health hazards. When overweight is sufficiently great, it almost certainly implies obesity. Near the upper end of the normal range of body weights, however, some individuals may be obese but not overweight, whereas others are overweight but not obese. For this reason, measurements of total body fat and its distribution are important.

Underlying the development of obesity is an imbalance between food intake and energy expenditure, which results in expansion of the stores of fat in adipose tissue. After reviewing the relationship between food intake and energy expenditure, the types of obesity and the risks associated with obesity will be reviewed. Finally, treatments for obesity will be briefly discussed. For more details on any aspect of this major public health problem, the reader is referred to several monographs.

I. MEASUREMENT OF BODY FAT

Accurate measurement of body fat requires sophisticated techniques, but simpler measures can be used reliably for large populations. These sophisticated techniques include the following:

1. Measurements of body density determined by weighing an individual out of water and after a submersion with correction for the air in the lungs.
2. Isotopic or chemical measure of total body fat or water content from which fat is calculated.

3. Measurement of the naturally occurring isotope of potassium (^{40}K) which is present in all human beings. Because most potassium is in muscle, the amount of this isotope in the body reflects lean body mass. Body fat can be determined by subtracting lean body mass from total body weight.

4. Because of its ease, dual energy X-ray absorptiometry has largely replaced the preceding methods.

Simpler methods used to determine overweight and estimate fat include

1. The measurement of height and weight
2. Determination of skinfold thickness.

The term overweight can be expressed in several ways, including relative weight, which is the weight of an individual in relation to their height compared to the median weight for individuals of the same height. It can also be expressed in terms of various ratios of body weight and height. Based on the lowest rates of associated health risks, Table I provides good body weights for most adult men and women.

The distribution of fat is an important predictor of risks for heart disease, diabetes, and high blood pressure. Fat distribution can be estimated from thickness of fat skinfolds measured alone or as the ratio between the trunk and the limbs. Intraabdominal fat can be measured using sophisticated scanning techniques known as computed tomography or magnetic resonance imaging. Practically, however, regional fat can be estimated by determining the circumference of the waist. When waist circumference is higher than 40 inches (100 cm) in men and 38 inches (95 cm) in women, the degree of central fat is unhealthy.

TABLE I
Good Weight in Pounds[a]

Height[b] (in feet/inches)	Women (19 to 34 years) Men all ages		Women over 35 years	
	Good weight[c]	Range	Good weight[c]	Range
5 0	112	97–128	123	108–138
5 1	116	101–132	127	111–143
5 2	120	104–137	131	115–148
5 3	124	107–141	135	119–152
5 4	128	111–146	140	122–157
5 5	132	114–150	144	126–162
5 6	136	118–155	148	130–167
5 7	140	121–160	153	134–172
5 8	144	125–164	158	138–178
5 9	149	129–169	162	142–183
5 10	153	132–174	167	146–188
5 11	157	136–179	172	151–194
6 0	162	140–184	177	155–199
6 1	166	144–189	182	159–205
6 2	171	148–195	187	164–210
6 3	176	152–200	192	168–216
6 4	180	156–205	197	173–222
6 5	185	160–211	202	177–228
6 6	190	164–216	208	182–234
Body mass index kg/m^2	22	19–25	24	21–17

[a]Derived from the National Research Council (1989). There is an age adjustment for women but not for men.
[b]Without shoes.
[c]Without clothes.

TABLE II
Prevalence of Obesity in Americans

NCHS survey[a]	Percentage of overweight	
	Men	Women
1960–1962	22.8	25.7
1971–1974	23.7	26
1976–1980	24.1	26.5
1988–1991	31.7	34.9

[a]From Kuczmarski, R. J., Flegal, K. M., Campbell, S. M., and Johnson, C. L. (1994). Increasing prevalence of overweight among U.S. adults. *JAMA* 272(3), 205–211.

Obesity is a common problem in the United States. The National Center for Health Statistics estimates that 35% of adult Americans are overweight. Since the mid-1970s, the percentage of American men and women who are overweight has increased sharply. This increase is in the opposite direction from information published by the U.S. government and action is clearly needed (Table II).

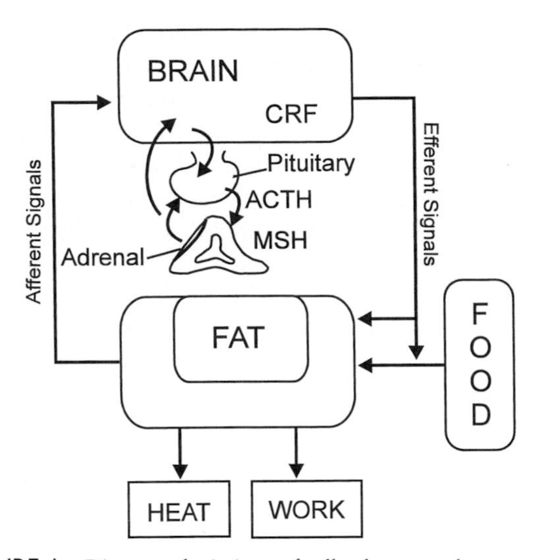

FIGURE 1 Diagram depicting a feedback or regulatory system for the regulation of food intake. Food enters the system on the right. Signals from the ingestion of food can act to reduce future short-term food intake by one of several mechanisms, including hormones, nutrients, or neural signals. Information (efferent signals) about the status of energy in the body and food intake is integrated in the brain, where efferent signals are generated that lead to food seeking and ingestion or termination of meals. Other signals activate various systems for storage or release of nutrients from fat tissue. The pituitary and adrenal glands are depicted with their connections to the brain. This is included because the development of obesity appears to depend on the presence of adrenal steroids to a much greater extent than any other endocrine system. ACTH, adrenocorticotropin; CRF, corticotropin-releasing factor; MSH, melanocyte-stimulating hormone.

II. PATHOGENESIS OF OBESITY

Body fat is stored primarily in fat tissue. The average adult male eats nearly 1 million kcal/year and expends essentially the same amount. An error of 1.0% (10,000 kcal/year more intake than expenditure) would increase the body weight by 1 kg (2.2 lb) every year or 10 kg (22 lb) per decade. Because this obviously does not happen to most people, the regulation of body weight over time for most individuals must be quite good. This regulatory system is depicted schematically in Fig. 1. To maintain a stable body weight, the ratio of fat to carbohydrate consumed in the diet (called food quotient) must equal the ratio of fat to carbohydrate used by the body. If more fat is eaten than is burned in the body, the excess will be stored and body weight will increase. Thus the relative amount of fat and carbohydrate in the diet can influence body weight. A high fat diet leads to obesity more easily than a low fat diet.

Food intake is initiated by a variety of factors, including social environment, availability of food, and internal hunger drives, which lead to food seeking. Once initiated, intake of a single food item occurs at a decelerating rate, i.e., we eat rapidly and then slow down. However, if a second food item is provided, there is usually "room" for it. This phenomenon of sensory-specific satiety related to different foods is an important item in short-term regulation of the amount humans eat. The decelerating rate of food intake implies an initial stimulation to food intake produced by the smell, taste, and physiologic effects of food followed by inhibitory events collectively called satiety, meaning the events associated with termination of a meal.

Inhibition of food intake may be produced by the nutrients themselves as well as hormones released when food enters the gastrointestinal tract. Of these hormones, cholecystokinin has received the most study, but several other important candidates exist. Nutrients themselves may directly initiate feedback signals after transport into the brain. Alternatively, nutrients may trigger signals transmitted over the vagus or sympathetic nerves as the nutrients are absorbed from the gastrointestinal tract or pass through the liver. For example, as glucose is absorbed and passes through the liver, the activity of the vagus nerve slows, and this signal is transmitted to the brain.

Several centers in the brain are involved in the feeding process. Afferent messages from the vagus nerve enter the vagal center (nucleus of the tractus solitarius) and are relayed from there to the hypothalamus. The hypothalamic coordinating centers also receive information from other parts of the brain that monitor sight, smell, and taste of food. Internal metabolic signals such as low glucose or high insulin levels may also act directly on areas in the medial or lateral hypothalamus. Destruction of the medial hypothalamus (specifically the paraventricular and ventromedial nucleus) is associated with an increase in food intake and obesity in a wide variety of animals and birds. Alternatively, damage to the more lateral hypothalamic area is associated with reduced food intake and weight loss in these same species. [*See* Hypothalamus.]

The central integration of messages about feeding involves two monoamine neurotransmitters (norepinephrine and serotonin), whose effects may be modulated by a variety of small peptides released from nerve endings. In the medial hypothalamus, norepinephrine stimulates feeding, whereas in the more lateral hypothalamus it can inhibit feeding. Conversely, serotonin (5-hydroxytryptamine) in the medial hypothalamus lowers food intake. Several peptides, including the opioids, neuropeptide Y, and galanin, are known to stimulate food intake when injected into appropriate brain areas. Several other peptide hormones are known to depress food intake. These inhibitory and stimulatory systems operate on the motor control of food seeking and food intake and indirectly on the metabolic processing of food.

The ingestion of food can activate the sympathetic nervous system as it leaves the brain. This alteration in sympathetic activity associated with food intake may also modulate the release of insulin from the pancreas. Insulin is a key circulating hormone involved with the storage of nutrients from the diet. Insulin receptors on liver, muscle, and adipose tissue are important in facilitating the response to insulin expressed as an increased storage of fat, glycogen, and protein. [*See* Insulin and Glucagon.]

Human beings can store fat by increasing the size of individual cells or by increasing the number of cells. In individuals who are modestly overweight, most of the increase in fat stores occurs by increasing the size of preexisting fat cells. However, individuals who are markedly overweight almost always have an increase in the total number of fat cells as well as enlargement of the individual cells. With weight loss, fat cells shrink in size but, in most instances, the number of fat cells does not decrease. Thus, individuals who become markedly overweight in childhood with an increase in the total number of fat cells find it difficult to lose weight in adult life.

III. TYPES OF OBESITY

The distribution of fat can be used to classify individuals as either upper body, or android, obesity (male-like) versus lower body, or gynoid, obesity (female-like). The number of fat cells is another basis on which types of obesity can be classified. Some individuals are fat and have a normal number of cells, which are enlarged (hypertrophic obesity), whereas others have both an increased size and an increase in the number of fat cells (hyperplastic, hypertrophic obesity). The age at which obesity begins is a third way in which individuals can be classified. For some individuals obesity begins in childhood, whereas for others it begins in adult life.

Several defined causes of obesity have been identified. Both single and polygenic inheritance is involved in the transmission of obesity in human beings. In rare circumstances, specific genetic diseases are associated with obesity. Genes have been cloned for two types of animals that inherit obesity. In most individuals, an underlying genetic predisposition is essential for the development of obesity. Within the same family, some individuals will become fat whereas others will not. Environmental and nonhereditary factors account for about one-half to two-thirds of this difference in individuals within families, whereas genetic factors account for the remaining 30 to 50%. The best estimates suggest that genetic factors may be of equal importance to environmental ones in the overall determination of body fat and that genetic factors may be more important than environmental ones in determining fat distribution.

Obesity can be caused by a variety of neuroendocrine diseases. Damage to the medial hypothalamus, as noted earlier, can cause obesity in humans as well as most other mammals. An ovarian disease known as polycystic ovaries is frequently associated with increased body fat. Likewise, hypersecretion of steroids from the adrenal glands can produce obesity. Increased insulin secretion or administration of excess insulin, which may occur during the treatment of diabetes, may also lead to obesity.

Diet also plays an important role in the development of obesity. A high-fat diet in animals and probably humans enhances the likelihood of becoming obese. Certain drugs taken for other diseases can also increase body fat (e.g., tricyclic antidepressants, glucocorticosteroids).

Finally, social, economic, and psychological factors play a role in the development of obesity. In the United States, obesity is more prevalent in lower social–economic groups than in higher ones.

IV. RISKS OF OBESITY

Regardless of its cause, obesity may be associated with a variety of risks. Life insurance data and studies of large populations have shown that overweight and obesity are associated with increased risks of premature death. In addition to the direct effects on mortality, obesity increases the risk of a variety of diseases. Several studies suggest that obesity may be a determinant for the risk of developing heart attacks and other cardiovascular diseases. Obesity is also a primary risk factor for the development of noninsulin-dependent diabetes (type 2 diabetes). Risks of gallbladder disease and some types of cancer are also increased in the overweight. In addition to these significant sources of increased risks, evidence indicates that obesity influences a variety of bodily functions, including heart, lung, and metabolic systems.

Increased abdominal, visceral, or upper body fat, like total body fat, also increases the risks of mortality, heart disease, stroke, and diabetes. This increased risk from abdominal fat can occur even in individuals who are of normal weight.

V. TREATMENT FOR OBESITY

Many different treatments have been used for obesity. At one extreme are the high-risk treatments such as surgical intervention with gastric or intestinal operations; at the other extreme are diets, behavior modification, and minor changes in exercise patterns. Except for individuals who are massively obese, surgery is inappropriate. For most individuals wishing to lose weight, behavioral modification, including changes in eating behaviors as well as changes in exercise behavior, should be the centerpiece. To the extent that changes in both eating and exercise patterns represent long-term adaptations in life-style, changes in body weight are a realistic expectation. For individuals who want rapid changes in weight loss without changing the behaviors associated with maladaptive eating and exercise patterns, the prognosis for maintenance of weight loss is low.

BIBLIOGRAPHY

Blackburn, G. L., and Kanders, B. S. (eds.) (1994). "Obesity: Pathophysiology, Psychology and Treatment." Chapman and Hall, New York.

Bouchard, C. (eds.). (1993). "The Genetics of Obesity." CRC Press, Boca Raton, FL.

Bray, G. A. (1989). Overweight and fat distribution: Basic consideration and clinical approaches. *Dis. Month.* **35**, 451–537.

Ciliska, D. (1990). "Beyond Dieting. Psychoeducational Interventions for Chronically Obese Women: A Non-dieting Approach." Brunner/Mazel, New York.

Garrow, J. S. (1988). "Obesity and Related Diseases," pp. 1–329. Churchill Livingstone, Edinburgh/London.

Grundy, S. M., and Barnett, J. P. (1990). Metabolic and health complications of obesity. *Dis. Month.* **36**(12), 641–696.

Perri, M. G., Nezu, A. M., and Viegener, B. J. (1992). "Improving Long-Term Management of Obesity: Theory, Research and Clinical Guidelines." Wiley, New York.

Stunkard, A. J., and Wadden, T. A. (eds.) (1992). "Obesity: Theory and Therapy," 2nd Ed. Raven Press, New York.

Obesity, Genetics

ENRIQUE SAEZ
YAACOV BARAK
The Salk Institute

EXCESSIVE ACCUMULATION OF ADIPOSE TISSUE results in obesity. Feeding and exercise patterns are an important component of the pathogenesis of obesity. Nonetheless, individuals respond in different ways to the caloric surplus that is thought to precipitate obesity development. Some people become morbidly obese, whereas others do not. An individual's genetic makeup is a major determinant in the body's reaction to unnecessary energy intake. The study of animals that spontaneously develop obesity has allowed the identification of genes and weight-regulation mechanisms that may constitute the molecular basis of this differential response in humans. Progress has also been made at elucidating the molecular details of how fat cells form. This article reviews our current understanding of the genetic mechanisms that outline the hereditary component of obesity development. An effort is also made to provide a molecular understanding for how obesity-associated insulin resistance and diabetes arise.

I. GENETIC EPIDEMIOLOGY OF OBESITY

The proportion of obese and overweight individuals in the general population has surged in the last decades. The dramatic pace of this process, especially in industrialized societies, suggests that obesity is most likely brought about by rapid environmental changes related to modernization, such as increased nutritional affluence and the prevalence of sedentary activity. The Pima Indians of southern Arizona are a relatively genetically homogeneous population that is afflicted with one of the highest incidences worldwide of morbid obesity and its associated complication, type II diabetes. The emergence of the obese-diabetic phenotype in Pimas within the last few generations is a clear manifestation of drastic shifts in diet and activity patterns, as is suggested by comparison of Arizonian Pimas to their Pima relatives in Mexico. Mexican Pimas diverged from their Arizonian kinsmen only a few centuries ago; consequently, they are endowed with a similar genetic background. However, thanks to the maintenance of a traditional life-style, Mexican Pimas rarely succumb to obesity or diabetes. Although environmental circumstances can undoubtedly explain part of the disease incidence difference between these two Pima groups, a role for genetic factors is suggested by two observations. First, the incidence of obesity-diabetes in Arizonian Pimas is unusually high compared to neighboring populations exposed to analogous environmental influences. Second, a small but persistent number of Arizonian Pimas remain mildly obese and do not develop diabetes.

The emerging preponderance of a genetic predisposition for excess weight gain seeks explanation in evolutionary theory. One evolution-inspired hypothesis suggests that today's human population is the result of millennia of natural selection in favor of individuals who could sustain prolonged periods of food restriction and famine. The winners in this evolutionary struggle were those who, thanks to a higher metabolic efficiency, could rapidly accumulate caloric depots during rare spurts of food abundance. Several "thrift" genes that helped the organism achieve these aims might have evolved independently; their combination could have conferred further advantage. Though the effects of these genes are overshadowed in life-styles involving moderate energy expenditure and restricted nutrition, in the circumstances of an industrial society thrift genes unfold their potential when they are no longer in demand, thereby promoting obesity and diabetes.

The contribution of genetics to obesity is made clear by studies of identical twin cohorts, including pairs that were reared apart. These studies have found that the genetic makeup of individuals is more important for the development of obesity than the environmental setting. A significant correlation between genetics and the propensity to develop obesity was also found among first-degree relatives, but not among nonrelated individuals. Multiple obesity-related parameters could be associated with the genetic background of the subjects, including body-mass index (BMI), skinfold thickness, and percentage body fat. Most importantly, the disposition to gain weight and develop insulin-resistance in response to experimental overfeeding was very similar in identical twin pairs as compared to unrelated subjects who had a much broader range of response.

In addition to population studies, several well-documented single-gene human disorders, such as Prader-Willi, Bardet-Biedl, Borjeson, Ahlstrom, and Cohen syndromes, are also associated with obesity. However, these diseases are characterized by pleiotropic defects that often include mental retardation, deficiencies in the visual system, facial anomalies, and additional syndrome-specific aberrations. Most obese individuals do not share the multiple abnormalities seen in these syndromes. Moreover, linkage studies of obese families have failed to associate the chromosomal loci of these diseases with those responsible for common obesity. In fact, most types of human obesity display a strong genetic component but do not follow single-gene inheritance patterns. Common human obesity is a multifactorial trait, resulting from an elaborate interplay between several gene products. Thus, the discovery of genes responsible for the aforementioned syndromes, though providing essential physiological information, may yield limited knowledge regarding the causes of obesity in the general population. [*See* Human Genetics.]

II. RODENT MODELS OF GENETIC OBESITY

The identification of genes associated with the development of obesity in humans is hampered by the complex nature of the disease and by the difficulties inherent in the dissection of the genetic components from the environmental factors that influence the evolution of the disorder in human populations. The genetic basis of obesity is more amenable to study in animals, for which the environmental conditions (diet, exercise frequency, etc.) can be rigorously controlled and the transmission pattern of genes predisposing to obesity can be discerned through breeding experiments. The study of rodents has been particularly fruitful in this regard, as over the years many mutants have been identified that constitute valuable tools in the exploration of the pathogenesis of obesity. Basic concepts derived from the molecular analysis of these animal models may provide some insight into the physiological pathways that underlie human obesity.

There are eight naturally occurring rodent models of obesity in which a single gene is responsible for the disease phenotype (Table I). In addition, the advent of techniques that allow to functionally inactivate genes at will has resulted in the generation of novel strains of mice that develop obesity as a consequence of intentional genetic lesions. These models suggest that obesity can develop by at least two different genetic mechanisms: (1) a metabolic abnormality can predispose the organism to conserve energy by lowering its rate of energy expenditure, thus excess calories accumulate in the form of white adipose tissue; and (2) mutation of genes that regulate feeding behavior can lead to increased food consumption in the context of a normal metabolic state; without a compensatory rise in the pace of energy depletion, surplus food is stored as fat. The study of genetically obese rodents has led to the discovery of several components of the body's weight-regulation circuitry. Alterations in the expression or the function of the proteins that form part of these weight-control systems can lead to abnormal weight gain by either of the mechanisms mentioned.

A. The *Obese* and *Diabetes* Mice: Unraveling the Leptin Loop

The *obese* (*ob/ob*) mutant arose spontaneously in 1950 in the large mouse stock housed at the Jackson Laboratory in Bar Harbor, Maine (Fig. 1). Some years later, the *diabetes* mutant mouse (*db/db*) was described by several laboratories. Both of these mutants develop early-onset massive obesity, with profound abnormalities in glucose and insulin metabolism. Depending on the genetic background of the mutant animals, the severity of the defect in glucose and insulin homeostasis can range from chronic elevated levels

TABLE I
Single-Gene Obesity Syndromes in Rodents

Mutation	Gene	Mutated gene product	Result of the mutation
Mice			
obese	*ob*	Leptin	Absence of fat-secreted hormone that signals the hypothalamus
diabetes	*db*	Leptin receptor	Absence of the receptor that mediates the leptin signal in the hypothalamus
tubby	*tub*	Tub, a hypothalamic protein	Unknown
fatty	*fat*	Carboxypeptidase E	Incomplete processing of insulin and neuropeptides that affect feedi.g behavior
Agouti yellow	A^y	Agouti	Mimicry of the effects of MCH induces increased food consumption?
Adult	*ad*	Unknown	Unknown
Rats			
fatty	*fa*	Leptin receptor	Defect in leptin signaling (similar to *db*)
corpulent	*cp*	Leptin receptor	Defect in leptin signaling (similar to *db*)

of insulin (hyperinsulinemia) with transient high levels of circulating glucose (hyperglycemia) to overt non-insulin-dependent diabetes (type II diabetes). *Obese* and *diabetes* animals display many of the additional conditions associated with morbid obesity, including aberrant thermoregulation, infertility, and shortened life span. The mutant animals are hyperphagic—they chronically overeat. However, these mice are not

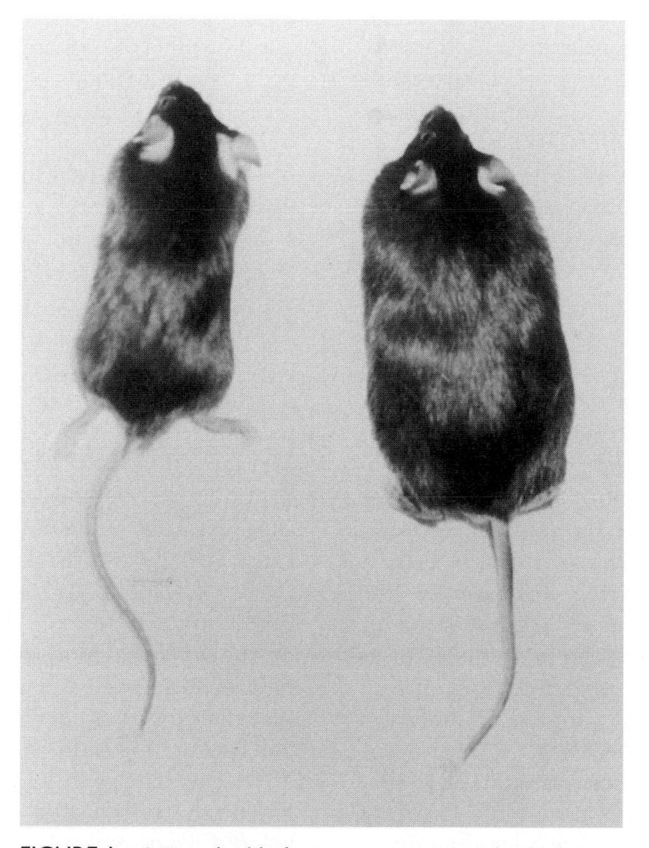

FIGURE 1 A 6-week-old *obese* mouse mutant (right) and a normal littermate (left). Notice the pronounced effect of the mutation at this early age.

obese simply as a consequence of their overeating, for they continue to gain weight and accumulate fat even if their food intake is restricted to half that of a normal mouse.

The molecular basis of the remarkable metabolic efficiency of *ob* and *db* mice and of their obese-diabetic phenotype has been the subject of intense study. Early parabiosis experiments where the circulation of *ob* and *db* mice was linked to that of each other and to that of normal mice suggested that the *ob* phenotype was the result of the loss of a bloodborne satiety factor. On the other hand, *db* mice produced vast quantities of this same putative satiety factor but seemed unable to respond to it.

Advances in molecular genetics allowed the isolation in 1994 of the gene responsible for the *ob* mutation. The *ob* gene encodes a novel protein that has been named leptin, from the Greek root *leptós,* meaning "thin." The *obese* phenotype is due to the lack of a functional form of this protein. Leptin is a protein of 167 amino acids that is secreted into the circulation exclusively by adipose tissue. Treatment of *ob/ob* mice with injections of leptin caused the mutant mice to lose significant amounts of weight, up to 40% of their original mass in just a few weeks. This weight loss was the result not only of decreased food intake but also of increased energy expenditure. Leptin treatment also reversed the insulin resistance and diabetes in *ob/ob* mice, as well as the infertility phenotype. In contrast, leptin injections had no effect on *db/db* mice, substantiating the hypothesis that the gene mutated in *db* encodes an important component of the leptin response cascade.

A year after the isolation of *ob,* the gene accountable for the *db* mutation was identified. This gene, termed *OB-R,* encodes a transmembrane protein that has homology to cytokine receptors. Alternative splicing of the *OB-R* message gives rise to several forms of the protein. The *OB-R* mRNA found in most tissues codes for a membrane-bound receptor with a very short cytoplasmic domain of 34 amino acids (OB-R short). A longer form of OB-R is produced in the hypothalamus, a region of the brain that is critical for weight regulation. This isoform has an extracellular domain identical to that of OB-R short, but a cytoplasmic domain that is 268 residues longer (OB-R long). The mutation in *OB-R* responsible for the *diabetes* phenotype truncates the long form of the protein, such that *db/db* mice express only the naturally occurring short form. The loss of the intracellular domain of OB-R long is thought to render the hypothalamic form of the receptor incapable of signaling in response to leptin. Indeed, *in vitro* studies indicate that OB-R long is part of a signal transduction pathway that resembles those utilized by interleukin 6-type receptors. Signaling through this type of receptors drives members of a family of receptor-associated Janus kinases (JAKs) to activate the transcriptional ability of Signal Transducer and Activator of Transcription (STATs) proteins that control the expression of specific target genes. No signal transduction pathway has been conclusively shown to be activated by the short form of OB-R. In light of its broad pattern of expression, this observation has led to the speculation that OB-R short may function as a leptin transport protein, rather than as a signaling receptor.

The discovery of functional leptin receptors in the hypothalamus (OB-R long), along with the striking adipose-specific pattern of expression of this hormone, suggests that leptin is an important component of a feedback loop that signals the brain to stop eating once enough energy has been accumulated in fat depots. This notion is supported by the finding that the levels of circulating leptin are strongly correlated to the percentage of body fat, the body-mass index, and basal insulin concentrations. Leptin secretion has also been shown to increase after food intake and to decrease during fasting. Moreover, insulin dramatically up-regulates leptin production, mimicking the effects of food intake. Given that leptin levels accurately reflect the size of energy stores, and that leptin secretion indicates the fed state of the organism, it is possible to envision a leptin-based weight-regulation scheme. Under a simple scenario, food intake would lead to a rise in insulin, surging insulin levels would stimulate leptin secretion, and increased circulating leptin would incite the hypothalamus to curb appetite and increase energy expenditure. [*See* Appetite; Energy Metabolism; Hypothalamus.]

How does leptin induce the hypothalamus to suppress food intake and augment the metabolic rate? Though the search for genes that mediate the effects of leptin has only begun, it is conceivable that leptin may function by regulating the expression of the hypothalamic gene neuropeptide Y (NPY). Neuropeptide Y is a small brain-derived peptide that stimulates appetite, decreases thermogenesis, and prompts insulin and glucocorticoid secretion. NPY levels are high in all rodent models of obesity. Furthermore, repeated injections of NPY elicit increased food intake, hyperinsulinemia, and suppression of brown adipose tissue function, leading to the rapid development of morbid obesity. Leptin treatment can suppress hypothalamic NPY production, suggesting that one mechanism whereby leptin may regulate food intake and metabolism is by inhibiting neuropeptide Y synthesis and release. In *ob/ob* mice, no leptin is produced and NPY levels are high. Treatment of *ob/ob* mutants with leptin decreases hypothalamic NPY concentration and normalizes their metabolism and body weight. Although this is an attractive regulatory framework, the precise role of NPY as a mediator of leptin's effects is muddled by the finding that mice that lack NPY as a result of an induced genetic lesion (a gene knockout mutation) respond to leptin and regulate feeding and body weight normally. Hence there must be additional mediators of leptin's effects on energy homeostasis.

Even though mutations in leptin and its receptor can cause massive obesity in rodents, it is not quite clear that the leptin signal evolved as a mechanism to control obesity. Recent work suggests that leptin can also regulate the organism's reaction to starvation. Food deprivation is a threat to energy homeostasis that spurs rodents and humans to activate the stress response, decrease thermogenesis, and limit procreation. If the fall in leptin levels that is seen during weight loss and fasting is neutralized by injection of recombinant leptin, these adaptations to starvation are blunted. The dose of leptin required to achieve this effect is smaller than that needed to induce weight loss. Since an effective retort to nutritional deprivation is likely to have greater survival value than the prevention of obesity, it is possible that leptin's principal function may be to modulate the neuroendocrine response to starvation.

Regardless of what its main physiological role might be, the significance of leptin for the development of human obesity has come under intense evaluation. Mutations in the leptin receptor have been found in genetically obese rats (the Zucker *fatty* and *corpulent* mutants), suggesting that the leptin cascade may be commonly mutated in obese mammals other than mice. To date, no mutations in either leptin or its receptor have been detected in human patients, though it is premature to conclude that none exist. As is the case in mice, the level of circulating leptin in humans is an accurate reflection of the amount of adipose tissue in the body: obese individuals have significantly higher leptin levels than normal individuals. Furthermore, serum leptin levels decrease notably during weight loss. The high levels of leptin associated with human obesity may mean that obese people frequently become resistant to the effects of leptin, or they may be a further indication that leptin cannot function on its own as a signal to prevent obesity.

B. The *Tubby* Mouse

The study of *ob* and *db* has resulted in the discovery of the leptin pathway, perhaps the first weight-regulation feedback mechanism in which adipose tissue plays a clear role. However, the early appearance and the severity of the obese-diabetic phenotype in *ob/ob* and *db/db* mice do not faithfully recapitulate how the disease usually develops in humans. *Tubby* mice (*tub/tub*) manifest a form of obesity that is more similar to human obesity in terms of onset, fat distribution, and disruption of insulin and glucose metabolism. These mice gain weight later in life and develop a form of insulin resistance that usually does not progress to frank diabetes. The gene mutated in *tubby* mice encodes a novel 505-amino-acid protein (Tub) that may function in the cytoplasm of the cell. The *tub* message is aberrantly spliced in *tubby* mice, resulting in the production of a protein with a different carboxy terminus. Consonant with its presumed role in biochemical pathways controlling appetite and energy metabolism, the *tub* gene is abundantly expressed in the hypothalamus. Nonetheless, though Tub has weak homology to phosphodiesterases, the function of this protein and the mechanism whereby the mutant protein musters obesity remain to be characterized. It is also unknown whether *tub* is mutated in obese humans.

C. The *Fatty* Mouse

Fatty mice (*fat/fat*) have a similar phenotype to that of *tubby* animals. They accumulate fat late in life, primarily in the axial/inguinal regions, becoming insulin resistant but not diabetic. Mutations in the gene that encodes the enzyme carboxypeptidase E are responsible for the *fatty* phenotype. This enzyme is involved in processing of precursors to several hormones, including insulin. Because the carboxypeptidase E they produce has little enzymatic activity, *fatty* mice accumulate incompletely processed proinsulin, a weak form of insulin. It is possible that hyperproinsulinemia causes obesity by a yet-to-be-identified mechanism. But the finding that *fat/fat* mice also have high levels of immature hypothalamic peptides (particularly proopiomelanocortin) has led to the suspicion that obesity in these animals may be due to a defect in the processing of neuropeptides with broad effects on appetite and energy expenditure. A similar disorder (obesity associated with high levels of immature insulin and incompletely processed neuropeptides) has been described in humans, though it is not known whether carboxypeptidase E is the gene responsible for the human disease.

D. The *Agouti* Mouse

Whereas obesity in *ob, db, tub,* and *fat* mice is a recessive trait, *Agouti yellow* mice carry a mutation that acts in a dominant fashion to induce adult obesity, insulin resistance, diabetes, and an increased predisposition to tumor development. The molecular lesion that underlies *Agouti* disrupts the pattern of expression of a protein that is normally expressed exclusively in the skin of neonatal mice. Obesity-causing mutations at the *Agouti* locus are the result of expression of the Agouti protein in various tissues of the adult animal. Agouti influences hair pigmentation by antagonizing the binding of melanocyte-stimulating hormone (MSH) to melanocortin receptors in melanocytes. In normal mice, MSH competes with melanin-concentrating hormone (MCH) for binding to this receptor. MCH is a hypothalamic neuropeptide that stimulates feeding in an analogous manner to neuropeptide Y. In *Agouti* mice, ectopic expression of the Agouti protein may mimic the effects of MCH, thus antagonizing MSH and instigating an increase in food consumption. Mutations in the human homologue of *Agouti* have yet to be associated with hyperphagia or obesity.

E. Targeted Mutants of Significance for Human Obesity

Strains of mice have been generated by genome-manipulation techniques in which the disruption of single genes altered the ability of the animals to achieve energy homeostasis. Some of these induced mutants are worthy of mention, for the genes that they lack have been shown to be important for the development of obesity in humans. One of the more interesting mutants is the 5-HT_{2C} serotonin receptor knockout mouse. Serotonin (5-HT) is a neurotransmitter that is believed to modulate multiple sensory and behavioral processes, including appetite and food intake. Mice lacking 5-HT_{2C}, one of the many receptors that mediate serotonin's actions, are obese as the result of their inability to control feeding behavior: 5-HT_{2C}-deficient mice eat significantly more than their normal littermates. These mice constitute a compelling argument that serotonin signaling through the 5-HT_{2C} receptor is important for appetite regulation, corroborating results from human studies in which drugs that potentiate serotonin function were shown to suppress appetite. Indeed, a recently approved drug for the treatment of obesity (dexfenfluramine, known commercially as Redux) quells appetite by stimulating serotonin release. The hyperphagic phenotype of 5-HT_{2C}-deficient mice may also suggest an explanation as to why dieting is such a hard exercise that seldom results in lasting weight loss. Dieting lowers the levels of tryptophan, the amino acid precursor of serotonin synthesis. If this decrease in tryptophan levels translates into reduced serotonin production, less signaling would be expected to occur through the 5-HT_{2C} receptor. Hence dieting may approximate the situation in the mutant mice, where the absence of the receptor prevents signaling. Given that 5-HT_{2C}-deficient mice overeat, if dieting impairs serotonergic signaling, dieters may also feel an impetus to increase their food intake, an urge that if carried out would make weight loss difficult.

The effect of mutations in the β-3 adrenergic receptor is another example of the similarity between the physiological pathways that regulate body weight in rodents and humans. A large system of adrenergic receptors mediates the effects of catecholamines on energy expenditure. These stress-response hormones mobilize energy stores by stimulating lipolysis in white fat cells and thermogenesis in brown adipose tissue. β-3 adrenergic receptors are expressed predominantly in white and brown adipose tissue. The phenotype of mice lacking the β-3 adrenergic receptor demonstrates that this receptor is critical for catecholamine-induced thermogenesis, and for leptin-independent appetite suppression. β-3 mutant mice have modestly increased fat stores, suggesting that selective agonists of this receptor may be useful as antiobesity drugs. Mutations in the β-3 adrenergic receptor have been associated with obesity in Pima Indians, Finns, and morbidly obese French patients, highlighting the universality of the adrenergic system in the regulation of energy balance.

Rodent studies have also suggested that numerous neuropeptides and hormones control satiety and feeding behavior. Whereas neuropeptide Y, melanin-concentrating hormone, and galanin stimulate food intake, molecules such as glucagon-like peptide-1

(GLP-1), cholecystokinin (CCK), corticotropin-releasing factor (CRF), and urocortin act as satiety signals, prodding the animal to stop eating. Dysregulation/mutation of the genes involved in the synthesis or processing of appetite-controlling proteins can lead to the development of obesity in rodents. It is likely that these genes also participate in the complex web of metabolic and behavioral interactions that ultimately lead to obesity in humans.

III. MOLECULAR MECHANISMS OF ADIPOGENESIS

Obesity is defined as a disproportional increase in the mass of adipose tissue. Mild cases usually involve an increase in the size of individual adipocytes (hypertrophy), but not in their number. Morbid forms, including childhood-onset obesity, are frequently associated with both hypertrophy of adipocytes and hyperplasia (increase in cell number) of adipose tissue. Insight into the development of obesity can be gained by understanding the genetic control of adipose tissue formation. [*See* Adipose Cell.]

Adipose tissue can be divided into two functionally and morphologically distinct classes: white adipose tissue (WAT), whose main role is to serve as an accessible energy depot (though it also serves as an endocrine organ as evidenced by the production of leptin), and brown adipose tissue (BAT), which functions as a site for regulated energy dissipation through thermogenesis. This unique activity of BAT is brought about by tissue-specific expression of the uncoupling protein (UCP), which, faithful to its name, uncouples oxygen-derived energy in the mitochondria from the production of chemical energy in the form of ATP. The net result of this process is the loss of energy as heat. Interestingly, mice engineered to have a diminished amount of BAT through toxin-mediated ablation of this tissue develop marked obesity and the associated complications of insulin resistance, dyslipidemia, and hyperphagia.

In the whole organism, adipocytes are derived from fibroblastic precursors through the process of adipogenesis. Adipogenic differentiation has been recapitulated in cultured cell systems, allowing the identification of several transcription factors implicated in the control of this process. These include peroxisome proliferator-activated receptor gamma (PPARγ), members of the CCAAT-box/enhancer-binding protein family (C/EBPs), and the sterol-regulated transcription factor SREBP1/ADD1.

A. PPARγ

PPARγ is a member of the extended superfamily of nuclear hormone receptors, which includes the receptors for steroid hormones, vitamin A and D derivatives, and thyroid hormone, as well as a battery of "orphan" receptors with no known ligand. Nuclear hormone receptor ligands, typically small lipophilic molecules, bind to a cryptic transcriptional activation domain in their receptors and convert it into a potent transactivator. Receptors recognize specific DNA sequences in the promoters of target genes through their DNA-binding domain. In this manner they can induce the expression of specific subsets of genes in response to hormonal stimuli. DNA-binding properties serve to functionally classify nuclear hormone receptors. PPARγ and its homologues PPARα and PPARδ belong to the class of heterodimeric receptors. Members of this branch of the superfamily, which also includes the thyroid hormone, retinoic acid, and vitamin D receptors, bind to their cognate DNA motifs as heterodimers with the retinoid-X receptor (RXR).

Binding sites for PPARγ–RXR heterodimers are found in the promoter regions of several genes that are central to the function of adipocytes. These include the genes for adipocyte fatty-acid-binding protein (aP2), the glycolytic enzyme phosphoenolpyruvate carboxykinase (PEPCK), and the UCP gene. The PPARα receptor binds to regulatory sites of genes encoding the bifunctional enzyme, liver fatty-acid-binding protein, and a host of other genes involved in hepatic fatty acid oxidation. Neither targets nor biological function have been ascribed to PPARδ. Nevertheless, PPARδ–RXR heterodimers bind to sequences resembling those for PPARγ and α, and it is conceivable that they too may be involved in regulating fatty acid metabolism.

PPARs derive their name from the ability of PPARα to mediate the physiological response to a diverse group of compounds collectively termed peroxisome proliferators (PPs). These chemicals include industrial plasticizers, as well as hypolipidemic drugs used to treat coronary heart disease. PPs share an ability to promote the proliferation of peroxisomes and their lipolytic enzyme constituents in the liver. A similar response can be induced by conditions of metabolic stress, such as starvation, high-fat diet, and diabetes. In this respect, the use of the name PPAR for the γ

and δ genes is a misnomer that derives merely from their homology to the PPARα receptor.

PPARγ is expressed very abundantly in adipose tissue. It is also expressed at significant levels in several other tissues, such as hematopoietic cells, the gut epithelium, the placenta, and the adrenal gland. In addition to the PPARγ1 isoform, which is the major form of this receptor in most tissues, adipose tissue expresses, through the use of an alternative promoter, a PPARγ2 transcript, encoding a receptor with a 30-amino-acid N-terminal extension.

A dramatic increase in the expression of PPARγ2 is observed early in the process of *in vitro* adipogenesis. Most importantly, PPARγ is capable of inducing this entire process when expressed ectopically (i.e., expressed from a vector introduced into cells that do not normally express PPARγ). Interestingly, forced expression of PPARα can also induce some adipogenic activity, although its limited extent and the low levels of endogenous PPARα in adipocytes argue against a possible role for this receptor in adipogenesis *in vivo*.

Like other members of the nuclear hormone receptor superfamily, the transcriptional activity of the PPARs is stimulated by specific ligands. The quest for PPAR ligands concentrated initially on fatty-acid-derived metabolites. This was the result of the postulated relevance of these molecules for the activity of these receptors, along with the attractive notion that products or substrates of the metabolic pathways affected by PPARs could have a feedback effect on their regulators. Indeed, polyunsaturated fatty acids and eicosanoids (metabolic derivatives of arachidonic acid) can induce transcriptional activation through various PPARs. However, the concentrations required to trigger a response suggest that they are unlikely to serve as physiological activators. Considerable progress in the hunt for PPARγ ligands comes from the discovery that a prostaglandin J2 (PGJ2) derivative—15-deoxy-$^{\Delta12,14}$-PGJ2 (15d-PGJ2)—is a potent and specific PPARγ ligand at micromolar concentrations. It awaits to be seen whether this eicosanoid or a closely related molecule is present in organisms at concentrations sufficient to activate endogenous PPARγ. A more intriguing finding is that thiazolidinediones (TZDs), an emerging class of drugs developed to treat non-insulin-dependent diabetes, are high-affinity ligands and potent activators of PPARγ. Both 15d-PGJ2 and TZDs dramatically enhance the adipogenic activity of PPARγ in *in vitro* differentiation assays. Moreover, both can promote adipogenesis in preadipocytes expressing low endogenous levels of PPARγ.

B. C/EBPs

Members of the CCAAT/enhancer-binding protein (C/EBP) family of transcription factors have been implicated in the regulation of many key metabolic processes, particularly some that take place in the liver. A marked induction of three C/EBP genes is observed during *in vitro* adipogenesis. C/EBPβ and δ are induced very early in this process, even prior to PPARγ. In fact, ectopic expression of C/EBPβ can activate PPARγ expression, and coexpression of C/EBPδ seems to aid in this process. On the other hand, C/EBPα, which like PPARγ regulates multiple fat-specific genes, is induced only following PPARγ expression. The central role of C/EBPα in adipogenesis is demonstrated by its ability to activate adipocyte differentiation upon ectopic expression. Coexpression of C/EBPα can greatly enhance the adipogenic activity of PPARγ, especially in cells that are not readily responsive to either one alone. In contrast, C/EBPβ can support differentiation only in the presence of a PPARγ ligand. These observations suggest that different C/EBPs may play distinct roles as activators, effectors, and/or partners of PPARγ in fat cell differentiation. Nonetheless, it should be noted that mice lacking the C/EBPα gene, though manifesting a clear compromise in the lipid content of their fat cells, still exhibit fat cell differentiation. These mutant mice die during the first few hours of life owing to severe metabolic deficiencies.

C. ADD1/SREBP1

ADD1/SREBP1 is one of a pair of structurally related sterol response element-binding proteins of redundant function that regulate the cholesterol content of cells. These transcription factors are attached to the endoplasmic reticulum in the basal state, but are cleaved upon sterol deprivation and rendered free to go to the nucleus. There they activate various genes encoding cholesterol biosynthetic enzymes and the low-density lipoprotein (LDL) receptor, as well as enzymes involved in fatty acid synthesis and desaturation. Thus, SREBPs apparently coordinate the sterol and fatty acid biosynthetic pathways.

The expression of the transcription factor ADD1, also termed SREBP1 (sterol response element-binding protein 1), is notably up-regulated during adipogenesis. Ectopic expression of ADD1/SREBP1 can promote adipocyte differentiation under conditions favoring this process, whereas under less favorable conditions, ADD1 overexpression can induce the expression of genes for fatty acid metabolic enzymes without prompting genuine differentiation. The adipogenic effect of ADD1/SREBP1 may be attributed, at least in part, to a stimulation of the transcriptional activity of PPARγ.

In conclusion, a network of transcription factors, part of which has been elucidated, controls the adipogenic program. It is likely that direct and indirect influences on the activity of these proteins may be held accountable for metabolic disturbances culminating in obesity.

IV. MOLECULAR LINKS BETWEEN OBESITY AND INSULIN RESISTANCE

The most prevalent complication associated with obesity is insulin resistance and its development into non-insulin-dependent (type II) diabetes (NIDDM). During the initial stages of NIDDM, a reduction in the patient's response to insulin surfaces, a phenomenon labeled insulin resistance. Glucose and lipid disposal by muscle, fat, and liver are impaired despite supranormal circulating insulin levels; resistance to insulin's actions results in hyperglycemia and hyperlipidemia. Although NIDDM also develops in nonobese individuals, it is 50-fold more frequent in obese people. Furthermore, NIDDM can be alleviated to some degree by weight-reduction measures. The obesity–diabetes connection is recapped in rodent models of obesity. Amelioration of their obese state (e.g., in *ob/ob* mice by leptin administration) is followed by a parallel decrease in insulin resistance.

The molecular basis of the link between obesity and insulin resistance has been the subject of much speculation. The existence of a secreted molecule manufactured by adipose tissue and overproduced in obese subjects that interferes with insulin-mediated signaling has been proposed. Tumor necrosis factor alpha (TNFα) is a protein that fulfills these criteria. It is produced by adipose tissue, its plasma levels are higher in obese patients, and it can inhibit the tyrosine kinase activity of the insulin receptor. Similar obesity-related overproduction of angiotensinogen (the precursor for the blood pressure-regulating hormone angiotensin) and plasminogen activator inhibitor I (PAI-I) has been observed, but no solid molecular link to insulin resistance has been suggested. The involvement of leptin in down-modulation of insulin receptor activity has also been postulated.

The opposite concept, insulin resistance being the result of reduced expression of certain adipose tissue genes in the obese state, has also been considered. The expression of two molecules secreted by fat, adipsin and adipoQ/ACRP30, is markedly down-regulated in obese individuals. The biological function of these molecules is somewhat obscure at the moment, though both are thought to be components of the complement system. Whether these proteins are functionally linked to the systemic response to insulin is still unclear. [*See* Endocrine System.]

Finally, an interesting connection between obesity and type II diabetes is brought to light by the dual action of TZDs as adipogenic agents and as effective insulin-sensitizing therapeutics. TZDs can efficiently promote fat cell formation via PPARγ stimulation. Therefore, the potential of TZDs to ameliorate a condition that is frequently an undesirable correlate of obesity is rather paradoxical. The ability of TZDs to improve NIDDM suggests that insulin resistance and type II diabetes may be a direct consequence of a reversible down-regulation of PPARγ function in obese individuals, or that a compensatory increase in PPARγ activity can overshadow the underlying causes of obesity-associated NIDDM. It remains to be seen where the PPARγ pools relevant for TZD-mediated treatment of NIDDM reside. Because of their significance for the pathogenesis of NIDDM, adipose tissue, muscle, and liver are good candidates for this role. Muscle is the most important of these organs for insulin and glucose disposal, but it expresses only minute amounts of PPARγ. Once the organ that responds to TZDs is identified, the challenge will be to pinpoint the transcriptional targets of PPARγ that are critical for insulin resistance and NIDDM. This molecular information will deepen our insight into the development of obesity-diabetes, stimulating the search for new therapeutics to treat what is one of the most common disorders in developed nations.

BIBLIOGRAPHY

Billingson, C. J., and Levine, A. S. (1996). Appetite regulation: Shedding new light on obesity. *Curr. Biol.* **6**(8), 920.

Bouchard, C., and Bray, G. A. (eds.) (1996). "Regulation of Body Weight: Biological and Behavioral Mechanisms." John Wiley & Sons, New York.

Coleman, D. L. (1982). Diabetes–obesity syndromes in mice. *Diabetes* **31**(Suppl. 1), 1.

Friedman, J. M., Leibel, R. L., and Bahary, N. (1991). Molecular mapping of obesity genes. *Mammalian Genome* **1**, 130.

Spiegelman, B. M., and Flier, J. S. (1996). Adipogenesis and obesity: Rounding out the big picture. *Cell* **87**, 377.

Olfactory Information Processing

DETLEV SCHILD
University of Göttingen

GLOSSARY

Action potential = spike Electrical pulse generated by a nerve cell and traveling on nerve fibers

Ethmoid bone Section of the skull base above the nasal cavity separating it from the forebrain

Stereotype behavior Behavior that is triggered by a particular stimulus and rigidly performed in a schematic way (i.e., always the same sequence of actions)

Sustentacular cell Nonneuronal cell located between neurons and in functional contact with neighboring neurons

Turbinate Folds of mucous membrane in the nasal cavity holding the olfactory mucosa

ALL LIVING CELLS ARE ABLE TO RECOGNIZE chemical cues from their environment. In this general sense all cells are chemosensitive. Protozoa are attracted by certain chemical stimuli, lymphocytes react and respond to an enormous variety of antigens, growth and metabolism of most cells of an organism are influenced or controlled by hormones, nerve cells are stimulated by neurotransmitters, specialized cells that are involved in the control of cardiovascular function can measure the plasma partial pressures of O_2 and CO_2, and finally, the senses of taste and smell endow animals with the capability of sampling and reacting to chemical stimuli they are exposed to.

Chemoreception of protozoa and lymphocytes, the action of hormones and neurotransmitters, the influence of O_2 and CO_2 receptors in cardiovascular control, and the taste system are dealt with in other articles of this encyclopedia. This article describes the processing of olfactory chemosensory information: the action of odors on olfactory receptor cells, the subsequent processing of this information by higher brain regions, and its impact on behavior. The basic principles of olfactory coding as compared with the coding in other sensory systems are analyzed in a separate section.

I. ANATOMY

A. Receptor Cells in the Olfactory Mucosa

In vertebrates and humans, the olfactory receptor cells are situated in the epithelium in the nasal cavity; each receptor cell is tightly surrounded by sustentacular cells. In humans, this epithelium, the olfactory mucosa, is a few square centimeters large. It is located in the upper part of the superior turbinate, the nasal septum, and the roof between. It is composed of three major cell types: olfactory receptor neurons, sustentacular cells, and basal cells. Olfactory receptor cells have two unique properties: (1) they are probably the only neurons that communicate directly with the environment of the organism, and (2) they have the capability of regeneration throughout life. Basal cells are progenitor cells of olfactory neurons; they divide, differentiate, and replace dying receptor neurons. The sustentacular cells separate olfactory receptor neurons; they have supporting, glia-like, and secretory functions. The lamina propria underlying these cells contains the axons of the olfactory neurons and secretory glands. Secretion of these glands together with secretion of sustentacular cells provide the mucus, which forms a 20- to 30-μm-thick layer above the olfactory tissue. Environment and mucus are separated from the extracellular compartment by a net of tight junctions, which surround all mucosal cells at their apical part.

Olfactory receptor cells are bipolar primary neurons with one long dendrite from the top of which, the so-called olfactory knob, several cilia issue. The cilia are relatively long (up to 150 μm) as compared with the thickness of the mucus layer (i.e., they extend mainly horizontally as a dense long-haired carpet in the mucus layer). The cell body (soma) of an olfactory neuron is usually fairly small; it has an oval form with diameters of about 4 and 7 μm. As primary neurons the cells have an axon attached to the soma. The unmyelinated axons form bundles ensheathed by Schwann cells. All bundles together correspond to the first cranial nerve (olfactory nerve), which passes through small openings of the cribriform plate of the ethmoid bone into the brain. Here the axon terminals make synaptic contact in the glomerula of the olfactory bulb. The nerve passage through the cribriform plate is vulnerable: head traumata often lead to injuries of the olfactory nerve with subsequent complete or partial anosmia.

B. Olfactory Bulb

The olfactory bulb is the second stage in the olfactory pathway. In many animal species, it is also the last stage where the entire olfactory information is processed and from where signals diverge to many other parts of the brain. Unfortunately, little is known about the human olfactory bulb; the anatomical and morphological similarities found in other higher vertebrates (e.g., rat, mouse, and rabbit) suggest, however, that the knowledge obtained from these species can be applied to humans.

The olfactory bulb in higher vertebrates is a neatly stratified structure with distinct layers: the olfactory nerve layer (ONL), the glomerular layer (GL), the external plexiform layer (EPL), the mitral cell layer (MCL), the internal plexiform layer (IPL), and the granule cell layer (GCL). The outermost layer is the ONL, where bundles of receptor cell axons distribute on the way to the glomerula. The next layer is the GL. Glomerula are spherically shaped morphological entities in which the primary receptor fibers make synaptic contact with the secondary neurons, the mitral cells (MC). MCs and many tufted cells are the only cells that send their output to higher brain structures. There is as yet no explanation for the fact that in most vertebrates the synaptic contacts between primary fibers and MCs form glomerula. However, in some lower vertebrates (e.g., goldfish) there are no typical glomerula, and it seems that glomerula formation begins in synchrony with the appearance of interneurons, which mediate information between glomerula. These interneurons, the periglomerular cells, are excited by primary fibers and act predominantly on MC dendrites, whereby this action is suppressive (inhibitory) in most of the periglomerular cells and excitatory in only about 20% of the periglomerular cells. The information processing of periglomerular cells thus appears to be an important and intermediate step in the information channel receptor neurons $\rightarrow$ MCs, and it seems to require the transition of a homogeneous, plexus-like GL into separate glomerula.

Mitral cells in higher vertebrates form a narrow and sharp layer, the MCL. These cells have several dendrites, only one of which, the long primary dendrite, reaches one glomerulum. The other secondary dendrites are oriented mainly orthogonally with respect to the primary dendrite. It is important to note the number of receptor neurons (N_r), glomerula (N_g), and MCs (N_m). Although these numbers differ from species to species, there are fairly constant ratios. The relation $N_r \gg N_m < N_g$ seems to hold in all species, whereby the ratio $N_r : N_g : N_m$ is approximately given by 1000 : F : 1, with F being in the range between 10 and 50.

Along the length of the primary MC dendrites, there is a layer between the GL and the MCL. In this layer, the EPL, information is conveyed vertically in the primary MC dendrite, from a glomerulum to a MC soma. The second type of relay neurons of the olfactory bulb, the tufted cells, are distributed in the EPL. Most of the rest of the information transfer in this layer consists in local lateral interactions between secondary dendrites of different MCs, mainly through granule cell dendrites but also through other cell types such as short axon cells.

Following the EPL and the MCL, there is the IPL, consisting mainly of numerous fibers, particularly axon collaterals of mitral cells. The latter can also be observed in the MCL as well as in the following layer, the GCL, where the granule cells are distributed. These GABAergic interneurons make reciprocal synapses on MC dendrites, thereby mediating self-inhibition and lateral inhibition between MCs.

The olfactory bulb also receives centrifugal afferent input from several brain regions. The innervation pattern of these centrifugal fibers within the bulb is correlated within their origin. The most conspicuous sites of origin are the ipsilateral olfactory cortex, ipsilateral horizontal limb of the diagonal band, locus ceruleus, raphe nucleus, and the contralateral anterior olfactory nucleus.

C. Central Structures

The output fibers of the olfactory bulb, together with the centrifugal afferent fibers, which reach the olfactory bulb from many parts of the brain, form the olfactory tract. The centripetal fibers make direct contact in the anterior olfactory nucleus, the piriform cortex, the olfactory tubercle, parts of the amygdala, and the lateral entorhinal cortex. These areas, which are considered as the primary olfactory cortex, are interconnected by numerous associational fibers. Projections from the primary olfactory cortical areas run to the thalamus, the neocortex, the hypothalamus, the hippocampus, and other nuclei of the amygdala. The thalamus probably does not have the relay function it has in other sensory systems because the direct projections from the primary olfactory cortex to the neocortex show a considerably higher density than the transthalamic pathway. The fact that only a few cells from many primary olfactory cortical areas project to the thalamus indicates further that this pathway presumably does not reflect a typical sensory information transfer through the thalamus. The involvement of thalamic cells might therefore rather serve to compare olfactory information with neocortical activity. The projection to the hypothalamus is probably concerned with odor-induced integrative control of autonomic functions; the fibers from the anterior cortical nucleus, which (in the rat) project to the preoptic/anterior hypothalamic area as well as to the medial hypothalamus, might provide a direct influence on reproductive and endocrine functions. [*See* Hippocampal Formation; Hypothalamus; Neocortex; Thalamus.]

II. PHYSIOLOGY

A. Receptor Neurons

Electrophysiological studies on single olfactory receptor cells have proven to be difficult because of their small size. Much of our knowledge about vertebrate olfactory receptor cells is therefore based on recordings from amphibia species, which have relatively large olfactory receptor cells.

Odorants, which are always airborne in higher vertebrates, enter the nose through the nostrils and reach the mucus of the olfactory mucosa. The odorous molecules enter the mucus according to the partition coefficient of the molecule type for the air–mucus interface. Once in the mucus, there is a certain probability for the odorous molecules to hit a molecular receptor molecule on a cilum of an olfactory neuron and to bind to it.

The mucus is an osmotic barrier between air and the nerve cell processes; in addition, there are proteins in the mucus that bind odorant molecules, thereby enhancing the probability of detection. Such odor-binding proteins would also transport the odor molecules in the mucus flow away from their site of action.

The activity characteristics of olfactory neurons can be summarized as follows:

1. The neurons show a low spontaneous discharge rate; in many species less than one action potential (or spike) per second. This feature is advantageous because a relatively small change in firing rate indicates a significant response.
2. When maximally stimulated with natural stimuli, the neurons respond with 20–30 spikes/sec.
3. When the intervals between spikes are analyzed, they show an interval distribution that suggests that the spikes are distributed randomly in time (i.e., only the mean firing rate of a response is characteristic for the response).
4. Responses to odors show a characteristic and well-reproducible pattern: the spike rate increases within a few seconds to a maximum and then decreases to a fairly constant niveau (phasic–tonic response).
5. A typical vertebrate olfactory receptor neuron responds to a large number of stimuli, whereby every stimulus brings about a different response spike rate. Another olfactory receptor neuron would also respond to a large number of stimuli, maybe partly overlapping with those of the first-mentioned cell, but a particular stimulus would usually cause different responses in the two cells. A receptor class is formed by neurons that respond identically to any sequence of stimuli. A particular class of receptor cells is presently thought to have one or only very few receptor proteins.

The events occurring between the binding of stimulus molecules to the cilia and the generation of action potentials (i.e., the details of stimulus transduction) are only partially known. Binding of a stimulus molecule to a ciliar receptor protein seems to trigger one or more intracellular enzyme cascades, which eventually lead to a net inward current, the so-called generator current, into the cilia. Evidence shows that various G proteins and cyclic adenosine monophosphate (cAMP), cycloguanosine monophosphate (cGMP), and inositol trisphosphate (IP_3) are involved in the transduction processes. The precise roles of these second-messenger pathways remain to be established in detail, however. The generator current loads the capacitance of the cell, thereby depolarizing (i.e., making it less negative inside) the membrane potential to the firing threshold of the cell. It has been shown that olfactory receptor cells are extremely sensitive, responding with spikes to current injections of only a few (<10 pA) picoamperes. The generator current is modulated and processed by the voltage-gated channels of the olfactory receptor neurons. There seem to be two voltage-gated inward currents and three different outward currents in all studied species: (1) a fast inward current carried by Na^+ ions that is responsible for the initiation of action potentials; in higher vertebrates, it is blockable by tetrodotoxin (TTX); (2) a small Ca^{2+} current that partially inactivates (i.e., its conductance decreases with time); (3) a Ca^{2+}-dependent outward current carried by K^+ ions; (4) a fast K^+ current that inactivates within a second; and (5) a third K^+ current that inactivates slowly in the range of tens of seconds. The physiological role of the outward currents is partially to return the membrane potential to or below the original level (repolarization and hyperpolarization), but also the adaptive regulation of the sensitivity of the cell as a function of its activity state and recent activity history.

Future understanding of the olfactory system will crucially depend on methodological progress, allowing the study of the stimulus selectivity of many cells in the same mucosa.

B. Olfactory Bulb

Four different approaches have been undertaken to investigate the physiological behavior of olfactory bulb (OB) neurons on natural stimulation of the olfactory system: (1) extracellular single- or double-unit recording from OB cells, (2) the distribution of 2-deoxyglucose (DOG) in OB cells after exposure to stimuli, (3) the activity pattern recorded by many electroencephalogram (EEG) electrodes, and (4) optical recording with video or photodiode techniques.

1. Extracellular Recordings

Extracellular recordings from OB neurons have mostly been obtained from MC/tufted cells and more rarely from granule cells. Although MC/tufted cells fire regular action potentials, granule cells generate potentials of longer duration and varying amplitude. In lower vertebrates, the temporal response pattern of MCs on stimulation is of the phasic–tonic type and seems to reflect the response behavior of the receptor neurons. In higher vertebrates, the temporal response patterns appear to be more complex and additionally correlated with the respiratory rhythm. A typical MC responds to more or less one-third of all stimuli, which means that a large number of MCs are usually involved in the coding of every stimulus. Different MCs respond in a different manner to the same stimulus. Recordings with two electrodes from two MCs have shown that the cell responses are positively correlated if one cell is in the neighborhood of the other. In a certain range beyond this neighborhood, the correlation is negative. Adjacent MCs or MCs within a certain neighborhood seem thus to be responsive in a similar way to a given stimulus, whereas an MC that is excited by a certain stimulus suppresses other MCs that are located in a ring around it (lateral inhibition). These phenomena can be explained (1) by the fact that adjacent MCs get similar inputs from glomerula and (2) by the lateral inhibition mediated mainly by granule cells.

For the investigation of such lateral interactions between OB cells, intracellular recordings are particularly appropriate and have in fact revealed details such as time constants of excitation and inhibition of MC/tufted cells. However, the fact that a large number of MC/tufted cells are involved in the coding of most stimuli also emphasizes the main drawback of single-unit recordings in the OB: as coding is done by sets of MC activities, little can be concluded from single-cell recordings. Obviously, to achieve this goal, a recording technique with a sufficiently high spatial resolution is necessary.

2. 2-Deoxyglucose

DOG is taken up as glucose by active nerve cells and glial cells but is not metabolically degraded. Labeled 2-deoxyglucose is thus stored only in metabolically active cells. Stimulation of the olfactory system with a certain odor results in a DOG uptake in the active cells, which can be determined in slices and histological determination, determining which regions of the OB were involved in the stimulus response. Because a sufficient DOG uptake takes a much longer time than one short stimulus presentation, stimuli must be applied repeatedly. The DOG staining pattern corresponds then to the average activity pattern over all stimulus applications. The stained areas do not exclusively indicate MCs but all kinds of cells including glia. With this method, a number of important findings evolved: Almost every odor is coded at the same time at many areas of the OB. Activation patterns can be altered experimentally by a preceding stimulus deprivation. In any case, the staining pattern for two different odorants usually overlaps considerably but are well distinguishable. Odors similar in concentration are mapped to similar DOG activation patterns. In particular, low odorant concentrations lead to only a few stained foci corresponding to single glomerula or small groups of glomerula. When higher or saturation concentrations were used, the densely stained foci were spread over large areas of the GL. The density of labeling within every stained glomerulum tends to be uniform so that the glomerula are currently thought of as functional units. Because the spatial resolution of the method does not reach the single-cell level, no cellular activation patterns can be assessed.

3. EEG Electrodes

Experiments have been performed in which 64 EEG electrodes have been placed on the surface of the OB of the rabbit to record the change of electrical potential at the surface. Contrary to the DOG method, the time resolution of this method is excellent but the spatial resolution is poorer. Nonetheless, some more interesting features of the signal processing in the OB, which are different from those formerly described, could be found in this way.

Every EEG electrode signal is oscillatory. The oscillation frequency is not characteristic for any odor, whereas the amplitudes are characteristically influenced by stimulus application. These amplitudes are therefore observables of the OB mass activity. They can be plotted in the conventional way as time functions; what is more instructive is to plot them in state space. As a simple example, take the three-dimensional state space: three appropriate EEG signals are chosen as the x, y, and z axis. At a particular time these three signals correspond to a point in this state space. If the points for all recorded time samples are connected, a characteristic curve results. This can be a simple point (fixed point), a closed curve standing for a periodic oscillation, or a chaotic attractor (i.e., a bunch of nonperiodic trajectories in state space that form a certain structure with nonvanishing volume to which all trajectories converge). The EEG approach has shown that all three classes of OB activity states exist and that they can change from one into another according to the consciousness states of the animal. Furthermore, different odors are clearly characterized by different OB activity states. One of the nice points of this approach is that time does not play the role of a variable; it is simply a parameter, and the activity state as a whole is time independent, corresponding to the time-independent characteristics of the applied odor.

4. Optical Recording

It has become feasible to stain nervous tissue with voltage-sensitive dyes which, in a way not completely understood, reflect the electrical potential of the plasma membrane by absorbing transmitted light or emitting fluorescent light in a manner that is proportional to the potential across the membrane. These optical signals can be detected by a video camera or by an array of presently up to 500 photodiodes. Time resolution of a video camera is video frequency, while, with fewer photodiodes, a much higher time resolution can be achieved. These methods have thus an excellent time resolution and at the same time a fairly high spatial resolution, although single-cell activities cannot be resolved. In a sense, the optical methods incorporate the advantages of both the DOG and the EEG approach: They reach the same time resolution of the EEG method and the spatial resolution of the DOG method. The results obtained with these methods show clearly how waves of activity spread over the OB when stimuli are applied. It can also be observed how different odor stimuli lead to different activation patterns, although one picture element still corresponds to thousands of cells.

On the whole, the electrophysiological and optical recording techniques show clearly that large parts of the OB are involved in the processing of most stimuli. The activation of MCs can thus be understood as an image of a chemosensory stimulus. However, many important details of this chemosensory map (e.g., how concentration coding is accomplished or whether the olfactory bulb contains an associative memory) (i.e., a memory not addressed by its address, as in computers, but by parts of its contents) remain to be investigated.

The physiology of higher brain structures of the olfactory pathways leave even more room for speculation because surprisingly little work has been done in this field. There is, however, evidence that the piriform cortex shows the typical features of an associative memory.

III. BEHAVIOR

The sense of smell is considered as one of the so-called "lower senses," and in most situations of human adult life, other senses, above all vision and audition, are of greater importance for (fast) behavioral responses. However, this would not exclude that olfaction might influence behavioral responses on a slower time scale and the general motivational state. Olfactory stimuli do not seem to enter consciousness in the way visual or auditory stimuli do. Many of the olfactory stimulus responses might even bypass consciousness. Olfactory behavioral responses are not well controllable by interference with consciousness or conscious memory. From an evolutionary point of view, this makes olfactory-guided behavior stable and also stereotype.

It is remarkable that olfactory cues are most often described by comparisons, analogies, or metaphors (i.e., they are classified almost exclusively by invoking associations). Our knowledge in this field stems from direct studies of human olfactory-guided behavior and also from olfactory dysfunctions in humans. Comparisons with animal behavior, especially in vertebrates, have proven to be most instructive because they lead to new experiments and have suggested fairly general and widespread mechanisms of olfactory behavior. It must be mentioned in this context that our conceptual framework of behavior and olfaction should not only be based on our present cultural understanding and attitude concerning the existence and effects of olfactory stimuli. During the past centuries (i.e., in the very recent past), olfaction played a remarkably different role in human behavior, so that in comparison, our age seems odorless and largely deodorized.

Olfactory stimuli can lead to a variety of behavioral responses that can be grouped into four categories: genotype recognition, reproduction, homing, and food intake control. The latter is obvious and well known. A host of compounds can be differentiated, and with some training, which enhances the system's resolution considerably, humans are capable of distinguishing between many tens of thousands of odor notes. This ability is particularly developed in wine tasters and perfumers. Olfaction-guided homing is certainly present in some fish such as salmon, which find "their" river arm after a migration cycle by the chemical cues they were exposed to before the first migration. Olfaction-controlled homing might also occur in other less studied species of the animal kingdom (e.g., birds). Amphibia are among the best studied species as far as olfaction is concerned, particularly when morphological or electrophysiological questions were addressed. Amazingly, the mucosa of many frogs undergo characteristic changes when the animals move from one medium to the other (water/air), but nobody really knows what frogs smell in their terrestrial life. An attractive hypothesis is that the sense of smell is important during spawning time; the animals find the pond in which they grew up (a special case of homing). For instance, salamanders displaced from their usual habitat found "their brook" among many others in the immediate neighborhood, although, for humans, there were no essential differences in such things are vegetation, climate, or soil characteristics between the brooks. It seems to be a common feature of olfaction-guided homing that the animals always return to their reproduction grounds; they seem to be imprinted with odors of this place. Imprinting with odors immediately after birth is also well established in mammals [e.g., rodents and rabbits appear to find their mothers' nipples by smelling a pheromone (a hormone-like acting substance) that is secreted at the nipples]. When the nipples are odorized with a different odor (e.g., a perfume) before the first contact, rabbit pups searching for the nipples approach objects with that perfume odor.

A host of examples show clearly that many types of reproductive behavior are induced or modulated by odors. For example, in many mammals, sex attraction, sex status recognition, and the advertisement of sex status are communicated by odors, and odors can even induce estrus and/or ovulation. Further, courtship and mating are often under the control of odor stimuli. Mice, for example, show clear odor-induced mating preferences; the animals prefer some individual members of the species over others, and this preference depends on the individual odors of the different animals. Genotypic variations are reflected in characteristic (body) odors that, in principle, allow the distinction between different members of the same species. This fact was suggested long ago as a hypothesis in the context of territory demarcation. The major histocompatibility complex on chromosome 17, which plays a key role in the immune response, is the genetic basis not only for the individuality of the body's cells but also for the fact that gene differences can be identified by scent. Genotype recognition thus plays an important role in reproduction behavior. Another remarkable example of the interference of olfactory genotype recognition in reproduction behavior is the Bruce effect discovered in 1959: In the mouse and other species, pregnant females abort with a high probability if they are exposed to a genetically strange mate or its urine. Injection of prolactine counteracts this pregnancy block.

Little is known about the connection between reproduction and olfaction in humans. There is, however, no doubt that some olfactory thresholds in women are lower around ovulation and during pregnancy. It is also interesting that the sense of smell in humans changes in a characteristic way with age. Human neonates detect odors within hours of birth as observed by consistent variations of salivation and respiration. A 2-week-old infant orients reliably toward his or her mother's breast pad as opposed to that of another woman. There are probably two fairly different age-related phases of odor perception, the first lasting until the age of 5 or 6 years and the second throughout the rest of the life whereby a decline of thresholds occurs in the last decades. A classical study presented children with the odors of amyl acetate (fruity banana-like smell), feces, and sweat. Reliable differentiation between the pleasant and unpleasant odors did not occur until the age of about 5. Most younger children (more than 95%) showed no displeasure at odors that adults normally find unpleasant. After age 6, adults and children show the same pattern of reactions.

Although a normal sense of smell remains almost unnoticed by humans, disorders pose severe problems to those who have them. They reduce the quality of life and can even cause anorexia, stress, or depression. Dysfunctions are described by the following terms: (partial) anosmia, absence of the capability of smelling (some) odors; hyposmia, decreased sensitivity of the sense of smell; and dysosmia, distortion of normal smell.

The most frequently encountered reasons for a loss of odor perception are viral infections, normal aging, and head traumata, which affect the fibers passing through the cribriform plate. However, many other diseases can cause a diminished or even abolished odor perception as well. These include Alzheimer's disease, multiple sclerosis, Parkinson's disease, cirrhosis of the liver, and vitamin B_{12} deficiency. Some typical changes of nerve cells in Alzheimer's disease have been shown to also appear in olfactory receptor neurons. The easier access to these neurons than to other neurons of the central nervous system might facilitate the diagnosis at the cellular level. The sense of smell is further diminished by many drugs, although the precise actions are only rarely established. Many of the diseases mentioned and some drugs might slow the rate of receptor neurons turnover (i.e., the continuously occurring degeneration and regeneration). Other olfactory symptoms in diseases such as olfactory hallucinations as part of the aura of epileptic seizures are probably not related to morphological alterations. [*See* Alzheimer's Disease.]

IV. OLFACTORY CODING

A. Topological Aspects

Contrary to other sensory systems, the basic coding mechanisms in the olfactory system have withstood all attempts of explanation during the past decades. One reason for this is probably the topology of the map stimulus $\rightarrow$ neuronal response: In the visual black/white system, for example, the stimuli are images (i.e., intensity functions on two spatial dimensions); the same is true for the sense of touch. In audition, the spectra of sound waves are detected by the hair cells of the inner ear; a one-dimensional cell array codes one parameter, the frequency, and spectral intensities are coded by spike rates. In these sensory systems, the stimuli can be described by one or two ordered parameters (frequency in audition and the two spatial coordinates in vision); the stimulus space, including intensities, is thus at most three dimensional. In principle, a one-to-one or point-to-point map of the stimuli to the activities of a one-dimensional chain or a two-dimensional grid of neurons is therefore straightforward.

In contrast, if the stimuli of a sensory modality must be described by more than two ordered parameters, the stimulus space is higher than three dimensional and a simple one-to-one map of the stimulus space to the activities of a two-dimensional grid of receptor cells is not possible. This seems to be the situation for olfaction and taste. Accordingly, receptor cells in the olfactory mucosa cannot be placed in an ordered way with respect to stimulus parameters. In fact, no such order has as yet been found. It has rather to be expected that there are classes of receptor cells, each of them being responsive to one or some molecular stimulus properties, and that the responses of these cells reflect the high dimensionality of the stimulus space. More than 100 receptor proteins have been cloned, possibly corresponding to an equal number of receptor classes. Assuming linear dependencies among these classes, the dimensionality of the olfactory input space would be lower than the number of receptor classes. The principal analysis of receptor responses suggested a dimension higher than 10.

B. Glomerula, the Core of Olfactory Coding

The cell layer that follows the receptor neurons are MC/tufted cells. The axons of these primary cells of the OB project directly to higher brain structures. MCs are neatly placed on a two-dimensional surface so that they can be viewed as a two-dimensional grid. At this level of the system, olfactory stimuli are coded as the activities or amplitudes of all MCs (i.e., as a function on a two-dimensional grid). Such a function can also be interpreted as an image whereby activities are coded on a gray scale. Activity images are nothing unusual in the nervous system and may be even the most common form of nervous information. A good comparison to the MC image is probably the activity image of ganglion cells in the retina. The crucial difference between these systems is that the input to the retinal network is a real image in the everyday sense, whereas the input to the olfactory network is chemical substances with no definite spatial ordering.

The most marked morphological structures between olfactory neurons and MCs are the glomerula. This means that synaptic connections in the glomerula (i.e., the direct ones from olfactory axons to MC dendrites and the indirect ones from primary nerves through periglomerular cells to MC axons) accomplish the transformation that is characteristic for the olfactory system. It has to transform vectors in the high-dimensional receptor activity space into an ordered image of MC/tufted cell activities.

The term *order* needs an explanation. MCs can be imagined in two different ways: The first is simply the geometrical one, where every cell occupies a certain place on a two-dimensional grid. It turns out, however, to be convenient to imagine MCs as a function of their inputs. For the sake of simplification we assume that there are only two classes of olfactory receptor cell neurons and that every MC is connected to both in varying proportions. Then an MC can be described by two numbers: the connection strength to class I and the connection strength to class 2. In this picture, MCs are described in their *input parameter space*. The example can easily be extended to many dimensions. MCs are adjacent in their input parameter space if they are connected in a similar way to the receptor cell classes. MCs that are adjacent in the input parameter space are generally not adjacent in the geometrical space on the grid. However, in the special case in which MCs are neighbors in the input parameter space and in the geometrical space, the map from receptor neurons to MCs and the MC sheet structure itself is called *ordered*. Ordered maps are essential for the nervous systems because the neurons of the second and higher nerve cell layers behind the sensory cells cannot obtain any information about relations between the stimuli if these relations are not mapped in an ordered way. The experimentally established fact that spatially adjacent MCs respond in a similar way to the same stimulus suggests that the projection to MCs is in fact ordered. It follows that glomerula have the function of connecting the primary fibers to MCs in such a way that an ordered projection from a high dimensional space (the receptor activity space in which odors are represented) onto an activity function on a two-dimensional grid (the activities of MCs) results. Odor images of this form can subsequently enter the olfactory cortex, the amygdala, the hypothalamus, and other central structures to lead to an adequate behavioral response.

BIBLIOGRAPHY

Corbin, A. (1982). "Le miasure et la Jonquille. L'odorat et l'imagiuaine social XVIIIe-XIXe sièceles." Aubier Gontainge, Paris.

Hudson, R. (1993). Olfactory imprinting. *Curr. Opin. Neurobiol.* **3**, 548–552.

Restrepo, D., Okada, Y., Teeter, J. H., Lowry, L. D., and Cowart, B. (1993). Human olfactory neurons respond to odor stimuli with an increase in cytoplasmic Ca^{2+}. *Biophys. J.* **64**, 1961–1966.

Restrepo, D., Teeter, J. H., and Schild, D. (1996). Second messenger signalling in olfactory transduction. *J. Neurobiol.* **30**, 37–48.

Schild, D. (1988). Principles of odor coding and a neural network for odor discrimination. *Biophys. J.* **54**, 1001–1011.

Schild, D. (1996). Olfaction and Taste. *In* "Comprehensive Human Physiology" (R. Greger and U. Windhorst, eds.), pp. 847–860. Springer, Berlin.

Stoddard, M. D. (1980). "The Ecology of Vertebrate Olfaction." Chapman and Hall, London/New York.

Omega-3 Fatty Acids in Growth and Development

ARTEMIS P. SIMOPOULOS
The Center for Genetics, Nutrition and Health

GLOSSARY

AA Arachidonic acid (20-carbon)
DHA Docosahexaenoic acid (22-carbon)
EFA Essential fatty acid
EPA Eicosapentaenoic acid (20-carbon)
LA Linoleic acid (18-carbon)
LNA Alpha-linolenic acid (18-carbon)
PUFA Polyunsaturated fatty acid

OMEGA-3 AND OMEGA-6 FATTY ACIDS ARE THE two classes of essential polyunsaturated fatty acids (PUFAs). The omega-3 class is represented by $18:3\omega 3$, alpha-linolenic acid (LNA), and the omega-6 class by $18:2\omega 6$, linoleic acid (LA) (Table I). Animals and humans do not have the capacity to synthesize either $18:3\omega 3$ or $18:2\omega 6$; therefore, both LNA and LA must come from the diet.

In 1918, Aron was the first to suggest that fats have nutritional functions in addition to being a good source of energy from food, providing 9 kcal/g. In growth and development, nutritional thinking has been dominated for a long time by concepts concerning protein and body growth. In 1981, M. A. Crawford suggested that it is quite possible that lipids and essential fatty acids, such as LNA and LA, and their long-chain metabolic products, particularly docosahexaenoic acid (DHA), were of greater significance to early human development than proteins. This paper presents information on omega-3 fatty acids: their sources and metabolism; elongation and desaturation of LNA; and the role of omega-3 fatty acids in pregnancy, fetal growth, human milk, infant feeding, childhood, and aging; as well as the evolutionary aspects and the omega-3–omega-6 balance and dietary recommendations. [*See* Fatty Acids.]

I. OMEGA-3 FATTY ACID SOURCES AND METABOLISM

A. Sources

LNA is the predominant terrestrial omega-3 fatty acid, and eicosapentaenoic acid (EPA) and DHA are the predominant aquatic omega-3 fatty acids. LNA is found in the chloroplast of green leafy plants (Table II) and in a few vegetable oils, specifically linseed,

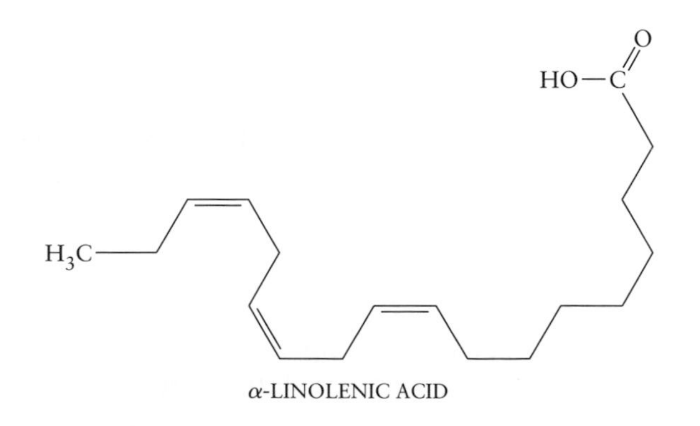

α-LINOLENIC ACID

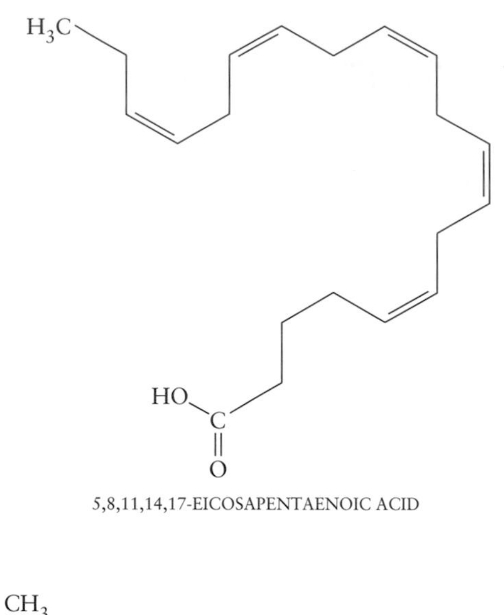

5,8,11,14,17-EICOSAPENTAENOIC ACID

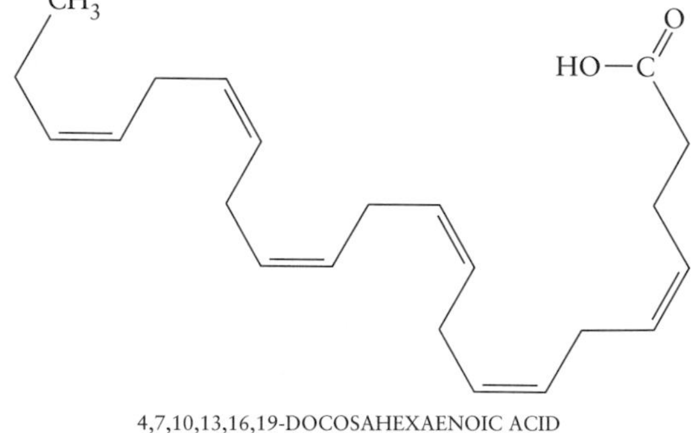

4,7,10,13,16,19-DOCOSAHEXAENOIC ACID

rapeseed, walnut, wheat germ, and soybean (Table III). LA is widely distributed in the vegetable kingdom and occurs in large quantities in most, but not all, vegetable seeds and in the oils produced from the seeds (with coconut oil, cocoa butter, and palm oil being exceptions).

Although food-selection patterns of land mammals vary and the diet of some herbivorous species may include more leaf or more seed material, most mammals will obtain LNA and LA from their food.

Plankton, on which fish feed, is rich in LNA, EPA, and DHA. Thus, fish get EPA and DHA from eating plankton or from metabolizing LNA to EPA and DHA. Although plankton mainly provide dietary omega-3 fatty acids for aquatic animals, it is sometimes eaten by land animals. Both marine and freshwater algae contain LNA, EPA, and DHA. The fatty acids in marine algae vary according to species, season of collection, and nutrient supply. Zooplankton, such as krill (whale food), has very little LNA (1%) but is quite rich in EPA and DHA. Its contents of EPA and DHA are 12–22% and 5–11%, respectively. Zooplankton in the North Atlantic and in the Mediterranean contains 7–16% EPA and 16–24% DHA, respectively.

Oily, or fatty, fish from deep, cold water (ocean) are rich sources of omega-3 fatty acids (LNA, EPA, and DHA) (Table IV). Although all fish and shellfish contain omega-3 fatty acids, the amount that is in a single serving of one species may vary significantly from that contained in a single serving of another, due to differences in the total oil content. In general, a single serving of lean fish, because of its lower oil content, provides a lesser amount of omega-3 fatty acid than does a single serving of oily fish. Seafood and their lipids are important to human nutrition. Unfortunately, most white fish that are popular in the United States are low in fat content and, therefore, contain minimal amounts of fish oil and, thus, minimal amounts of omega-3 fatty acids. [*See* Nutrition, Fats and Oils.]

B. Elongation and Desaturation

With the exception of carnivores such as lions and cats, who obtain EPA and DHA directly from the flesh of other mammals, animals and humans can convert LNA from the diet to EPA and DHA. This process results in the two classes of the omega-3 and omega-6 fatty acids with 20 and 22 carbon atoms, respectively, and four, five, or six double bonds (see Table I). The amounts of the elongated polyunsaturated derivatives are dependent on the dietary source. Because most edible vegetable oils contain smaller amounts of LNA, their higher concentration of LA may depress synthesis of EPA and DHA from LNA, except selectively in tissues such as retina, brain, and testis, which are rich in DHA. Humans who consume large amounts of EPA and DHA, which are present in fatty fish and fish oils, have increased levels of these two fatty acids in their plasma and tissue lipids at the expense of LA and arachidonic acid (AA). Alternatively, vegetarians, whose intake of LA is high, have more elevated levels of LA and AA and lower levels of EPA and DHA in plasma lipids and in cell membranes than do omnivores.

Elongation and desaturation of LNA to EPA and DHA occurs in human leukocytes and in the liver of both humans and rodents (e.g., rats). Omega-3 and omega-6 fatty acids compete for the desaturation enzymes. But both Δ^4 and Δ^6 desaturates (the enzymes involved in desaturation) prefer the omega-3 to the

TABLE I
EFA Metabolism Desaturation and Elongation of Omega-3 and Omega-6[a]

Linolenate series	Linoleate series
C18 : 3ω3 LNA	C18 : 2ω6 LA
↓ Δ^6 desaturase	↓ Δ^6 desaturase
C18 : 4ω3	C18 : 3ω6 gamma-linolenic acid
↓	↓
C20 : 4ω3	C20 : 3ω6 dihomo-gamma linolenic acid
↓ Δ^5 desaturase	↓ Δ^5 desaturase
C20 : 5ω3 EPA	C20 : 4ω6 AA
↓	↓
C22 : 5ω3 docosapentaenoic acid	C22 : 4ω6
↓ Δ^4 desaturase	↓ Δ^4 desaturase
C22 : 6ω3 DHA	C22 : 5ω6 docosapentaenoic acid

[a]The first number (18) refers to the number of carbon atoms in the molecule, the number after the colon (3 for 18 : 3ω3 and 2 for 18 : 2ω6) refers to the number of double bonds in the molecule, and the ω3 and ω6 refer to the position of the double bond closest to the methyl end (CH_3) of the molecule. The use of the omega system of designation is based on the fact that the characteristics of unsaturated fatty acids from a nutritional standpoint depend on what exists near the methyl end, not the carboxyl end, of the fatty acid. Dr. Ralph Holman called such a designation the omega (ω) system. Omega, the last letter of the Greek alphabet, implies that the counting of the carbon atoms begins at the methyl end of the fatty acid molecule. The omega system nomenclature is now more popular than the n-3 or n-w nomenclature.

omega-6 fatty acid. Retroconversion of DHA and of AA to shorter-chain fatty acids has been shown to occur in suspended rat hepatocytes by beta-oxidation (see Table I).

EPA and DHA are found in membrane phospholipids of practically all cells of individuals who consume omega-3 fatty acids. DHA is found mostly in phospholipids, whereas LNA is found mostly in triglycer-

TABLE II
Fatty Acid Content of Plants[a,b]

Fatty acid	Purslane	Spinach	Buttercrunch lettuce	Red leaf lettuce	Mustard
14:0	0.16	0.03	0.01	0.03	0.02
16:0	0.81	0.16	0.07	0.10	0.13
18:0	0.20	0.01	0.02	0.01	0.02
18 : 1ω9	0.43	0.04	0.03	0.01	0.01
18 : 2ω6	0.89	0.14	0.10	0.12	0.12
18 : 3ω3	4.05	0.89	0.26	0.31	0.48
20 : 5ω3	0.01	0.00	0.00	0.00	0.00
22 : 6ω3	0.00	0.00	0.001	0.002	0.001
Other	1.95	0.43	0.11	0.12	0.32
Total fatty acid content	8.50	1.70	0.601	0.702	1.101

[a]Source: A. P. Simopoulos and N. Salem, Jr. (1986). *N. Engl. J. Med.* **315**, 833.
[b]Milligrams per gram of wet weight.

TABLE III
Sources of Omega-3 Fatty Acids[a]

Oils	18:3[b]
Linseed oil	53.3
Rapeseed oil (canola)	11.1
Walnut oil	10.4
Wheat-germ oil	6.9
Soybean oil	6.8
Tomato seed oil	2.3
Rice bran oil	1.6

[a]Source: Provisional Table on the Content of Omega-3 Fatty Acids and Other Fat Components in Selected Foods, U.S. Department of Agriculture, Washington, D.C., February 1986.
[b]Edible portion, raw, 100 g.

ides, cholesteryl esters, and in very small amounts in phospholipids. EPA is found in cholesteryl esters, phospholipids, and triglycerides.

Human and other mammals' cerebral cortex, retina, testis, and sperm are exceptionally rich in DHA. DHA is one of the most abundant components of the brain's structural lipids, namely, phosphatidylethanolamine (PE), phosphatidylcholine (PC), and phosphatidylserine (PS). As indicated earlier, DHA can be obtained directly from the diet by, for example, eating fish or synthesizing it from dietary LNA.

C. Metabolism

EPA and DHA are the precursors of the prostaglandins and thromboxanes of the 3-series and leukotrienes of the 5-series. AA is the precursor of the 2-series of prostaglandins and thromboxanes and of the leukotrienes of the 4-series. Prostaglandins, thromboxanes, and leukotrienes derived from EPA have different biological properties from those derived from AA. Prostaglandins, thromboxanes, and leukotrienes derived from AA are potent platelet aggregators and vasoconstrictors and are powerful inducers of inflammation, whereas those derived from EPA prevent platelet aggregation and vasoconstriction and are weak inducers of inflammation. Competition also exists between EPA and DHA in prostaglandin formation and leukotriene synthesis. Because omega-3 fatty acids are found in all cell membranes, they have a variety of biological functions that influence inflammation, platelet adhesion, plasma lipid levels, blood pressure, and growth and development.

Ingestion of omega-3 fatty acids, particularly of increased amounts of EPA and DHA omega-3 fatty acids, replaces the omega-6 fatty acids in cell-membrane phospholipids. The fatty acid composition, the cholesterol content, and the phospholipid class of biomembranes are critical determinants of membrane physical properties and have been shown to influence a wide variety of membrane-dependent functions, such as integral enzyme activity, membrane transport, and receptor function. This ability to alter both the lipid composition and function *in vivo* by diet, even when essential fatty acids are adequately supplied, demonstrates the importance of diet in growth and development and in health and disease. [*See* Nutrition and Immunity.]

II. ANIMAL STUDIES IN GROWTH AND DEVELOPMENT

A. Animal Studies

The term "essential fatty acids" was first coined by G. O. Burr and M. M. Burr in 1930. Diets deficient in fatty acids are associated with clinical symptoms such as skin rash, retardation of brain growth, incomplete brain-cell division, and a high neonatal mortality in second-generation rat pups. Subsequent work by Burr and Burr showed that the addition of fatty acids to the diet, particularly LA, eliminated the symptoms caused by the fat-free diet in the deprived animals. The EFAs discovered by these researchers include both LNA and LA, but their effects differ. LA restores healthy skin, successful growth, reproduction, and lactation, whereas LNA supports growth but does not prevent the skin lesions of EFA deficiency or support reproduction.

Research on the role of omega-3 fatty acids during pregnancy and lactation has expanded over the past 30 years. As a result the work of B. L. Walker, it became possible to produce DHA deficiency in the rat for the first time. Walker published his studies in 1967 and demonstrated that restriction of dietary LNA in the maternal diet of rats is reflected in a lowering of the $22:6\omega3$ in the brain lipids of the pups at birth. Postnatally, even low levels of linolenate in the diet of the lactating dam result in rapid accumulation of DHA in the brain. This remarkable affinity of brain lipids for DHA raised the possibility of a functional requirement for linolenic acid. Walker's work is indeed a milestone because it made possible an animal model—the rat—that was finally deficient

TABLE IV
Relative Content of Omega-3 Fatty Acids and Cholesterol in Fish[a]

Species	% Oil in flesh[b]	% Omega-3 fatty acids in oil[b,c]	% Omega-3 fatty acids in flesh[d]	% Cholesterol in flesh[e]
Haddock	00.5	39.6	0.198	0.060
Snapper	01.1	23.0	0.253	0.040
Tuna, canned	01.0	30.0	0.300	0.063
Shrimp	01.1	28.5	0.314	0.180
Cod, Atlantic	00.7	45.9	0.321	0.050
Pollock	00.8	48.4	0.387	0.071
Sole, lemon	01.4	31.0	0.434	0.050
Ocean perch	02.0	22.0	0.440	0.050
Flounder	01.3	35.0	0.455	0.050
Squid	01.0	53.3	0.533	0.241
Mullet	03.0	19.1	0.573	0.021
Halibut	02.0	36.0	0.720	0.050
Shad	02.8	26.1	0.731	0.038
Whiting, Pacific	03.0	33.3	0.999	0.066
Swordfish	04.4	25.7	1.131	0.057
Trout, rainbow	07.0	17.6	1.232	0.050
Tuna, raw	05.1	30.0	1.530	0.046
Whitefish, lake	07.0	22.2	1.554	0.060
Sardine, canned	06.3	26.8	1.688	0.140
Salmon	09.3	23.0	2.139	0.053
Sablefish	10.0	22.9	2.290	0.040
Mackerel, Atlantic	13.0	19.0	2.470	0.065
Herring	15.0	18.4	2.760	0.085
Dogfish shark	14.1	24.5	3.455	0.039

[a]Source: Barton, M., and Emerson, J. A. (1986). Seafood in your diet—A choice of recipes. *In* "Health Effects of Polyunsaturated Fatty Acids in Seafoods" (A. P. Simopoulas, R. E. Kifer, and R. R. Martin, eds.). pp. 403–430. Academic Press, Orlando.

[b]Data derived from the National Marine Fisheries Service and the scientific literature.

[c]Total percentages of 13 : 3ω3, 20 : 5ω3, 22 : 5ω3, and 22 : 6ω3.

[d]Percent oil in flesh × percent omega-3 fatty acids in oil.

[e]Cholesterol percentages represent total cholesterol values. Note: the levels presented here are average figures. These levels may vary widely due to seasonality, fish diet, age, size, and the processing methods employed for the various product forms in which these fish species are sold to the public. Therefore, these figures should not be considered as absolute.

in DHA and the deficiency could be restored with even small amounts of dietary omega-3 fatty acids.

C. Galli *et al.* in 1971 and R. Paoletti and C. Galli in 1972 concluded that the dietary EFA deficiency was found to affect the *developing* central nervous system of the developing rat; this nutritional stress could be of comparable severity to that obtained by protein and/or calorie malnutrition. In LNA deficiency, 22 : 5ω6 increases and LNA supplementation raises the proportion of 22 : 6ω3 in brain glycerophosphatides. C. Galli *et al.* proposed to consider the ratio of 22 : 5ω6/22 : 6ω3 in tissue lipids as an index of relative LNA deficiency.

In 1973, R. M. Benolken *et al.* discovered that dietary LNA influences the electroretinogram (ERG) response in the rat. In 1975, T. G. Wheeler and R. M. Benolken measured the ERG response in the rat as a function of dietary supplements of purified ethyl esters of LNA, LA, and oleic acids. Dietary LNA affected the ERG amplitudes to a greater extent than did LA.

The electrical response of photoreceptor cell membranes appears to be a function of the position of the double bonds as well as a function of the total number of double bonds in fatty acid supplements. These findings firmly establish a selective functional role in the visual system for omega-3 fatty acids and suggests that the observed electrical alterations are associated with fatty acid substitutions in the plasma membrane of the photoreceptor cells. Therefore, polyunsaturated fatty acids derived from LNA and LA appear to be important functional components of photoreceptor cell membranes. The photoreceptor cells of vertebrates consist of an inner and an outer segment. DHA is the dominant PUFA of the phospholipids of vertebrate membranes.

The studies of B. L. Walker, R. M. Benolken, T. G. Wheeler, C. Galli, and C. Galli and R. Paoletti were critical for the understanding of the role of omega-3 fatty acids on retina and brain function. By 1975 progress had been made in the development of a model—the DHA-deficient rat—by using omega-3 fatty acid-deficient diets during pregnancy and postnatally, and the use of the ERG response in the rat had established that DHA deficiency in retina could be reflected by measuring ERG amplitude changes.

M. S. Lamptey and B. L. Walker extended this work further. These investigators carried out a study in which female rats were fed a diet high in linoleic/linolenic acid ratio prior to mating and during pregnancy and lactation. The offspring were weaned to the same diet and their physical, neuromotor, and neurophysical development assessed and compared to that of pups from dams fed a diet containing soybean oil rich in omega-3 fatty acids. The brain lipids in the two groups of animals were compared also. The soybean-fed progeny had higher levels of 22 : 6ω3 and lower levels of 22 : 5ω6 in the brain PE, and their performance in the discrimination-learning test was superior to that of progeny fed safflower oil rich in omega-6 fatty acids, providing further support for the essential role of dietary linolenic acid for the young rat.

While these studies were going on, R. N. T. W. Fiennes *et al.* carried out the first experimental studies of LNA requirement in primates (capuchins). The capuchins suffered from symptoms that closely resemble those of EFA deficiency, yet their diet contained adequate amounts of LA. When linseed oil (55.7% LNA) was added to this diet, their symptoms improved. Although the capuchins were on a purified diet (24–28 months) that contained very little LNA, the liver and red blood cell PE still contained about 6% of their fatty acids as LNA metabolites, mostly DHA. Their red blood cell fatty acids had the profile expected from the studies in rats fed a diet high in LA and poor in LNA, specifically, elevated levels of 22 : 4ω6 and 22 : 5ω6 and low levels of 22 : 6ω3.

Beginning in 1984, M. Neuringer *et al.* carried out a series of studies in primates, which, of course, are closer to human beings in retinal structure and visual function. Studies were carried out in the infant rhesus monkey that had been DHA-deficient during pregnancy and lactation, caused by a LNA-deficient diet. The maternal omega-3 fatty acid deprivation led to abnormally low levels of DHA in the tissues of the near-term fetus and newborn infant. Vulnerability to dietary omega-3 fatty acid deprivation was even greater after birth. The monkeys were followed up to 22 months. In the control monkeys the DHA (22 : 6ω3) had increased in the cerebral cortex and retina, twice as high as at birth, whereas the deficient monkeys failed to show an increase in both the cerebral cortex and retina by 22 months. As was shown previously in the rat studies and in the studies with the capuchins, the deficient animal tissues had a compensatory increase of 22 : 5ω6. Subnormal visual acuity and prolonged recovery time of the dark-adapted ERG after a saturating light flash were noted in the deficient monkeys at 4–12 weeks of age. According to M. Neuringer *et al.*:

> The fatty acid compositions of the cerebral cortex described here for control newborn and juvenile rhesus monkeys are very similar to those reported by Svennerholm (1968) for human newborns and adolescents, respectively. However, the brain and retina of human infants are less developed at birth than those of rhesus monkeys (Cheek 1975), so that human infants might be even more vulnerable to postnatal dietary deprivation of ω3 fatty acids.
>
> . . . Our findings provide evidence that dietary ω3 fatty acids are essential for normal prenatal and postnatal development of the retina and brain. Further research will be required to determine the relative contributions of prenatal versus postnatal deprivation to the observed functional deficits and to determine the degree to which the biochemical and functional effects of ω3 fatty acid deficiency are reversible.

B. Aging

N. P. Rotstein *et al.* in 1987 studied the effects of aging on the compositional and metabolic aspects of retinal phospholipids in the rat. The levels of DHA and other omega-3 hexaenoic acids were decreased in retinal glycerophospholipids of aging rats, particularly in those containing choline and serine. *In vitro* labeling of retinal lipids with [2-^{3}H] glycerol and [1-^{14}C]DHA in young and aged animals showed that most retinal lipid classes incorported DHA. The incorporation of DHA was most marked when DHA was decreased and further stimulated by aging. This indicates that the decrease in the DHA content of retinal phospholipids is simply due to the decreased availability of DHA in the retina rather than to an impaired activity of the enzymes involved. The levels of the omega-3 pentaenoic acids in retinal lipids were much less affected by aging than those of the omega-3 hexaenoic acids, which indicates no defect in either the availability of 18 : 3ω3 or its metabolic products up to 22 : 5ω3.

These findings are important because they indicate a marked difference between the effects of aging and EFA deficiency in the retina. In LA (18 : 2ω6) deficiency, the AA (20 : 4ω6) is decreased in tissue lipids and is replaced by 20 : 3ω9 (from oleic acid). In LNA (18 : 3ω3) deficiency, the DHA (22 : 6ω3) decreases in lipids and is replaced by 22 : 5ω6 (from LA). In the aging retina, none of these compensatory mechanisms was shown. This suggests that an impairment of the Δ_4 desaturase enzyme system is most likely responsible for the decreased levels of 22 : 6ω3 (and 22 : 5ω6) observed in retinal lipids as a consequence of aging. Both 22 : 5ω6 and 22 : 6ω3 fatty acids require Δ_4 desaturase for their synthesis.

A decrease in DHA possibly plays a significant role in visual impairments that accompany old age, because DHA is required for normal function of photoreceptors in rats and primates. If one were to speculate as to the appropriate dietary ω3 fatty acids for the aged, it would be dietary DHA rather than LNA that might maintain and possibly improve visual function in the elderly.

III. HUMAN STUDIES IN GROWTH AND DEVELOPMENT

A. Pregnancy and Fetal Growth

Since the Second World War, the role of maternal nutrition on fetal growth and development has been extensively studied in the context of protein-calorie malnutrition. The role of omega-3 fatty acids has only recently come into focus, despite the evidence of its importance having been demonstrated long ago in studies involving rats and primates. Lipid nutrition during pregnancy and lactation is of special relevance to human development, because brain development in the human takes place during fetal life and the first two years after birth. DHA is found in large amounts in the gray matter of the brain and in the retinal membranes, where it accounts for 30% or more of the fatty acids in ethanolamine and serine phospholipid. DHA accumulates in the neurons of the brain between Weeks 26 and 40 of gestation in humans.

The EFA requirements for pregnancy have been estimated by M. Crawford *et al.* (1981) to be between 600 and 650 g in terms of both LA and LNA. This figure is about 1% of the nonpregnant woman's dietary energy; another 0.5% of energy should come from AA and DHA.

Determinations carried out on the fatty acids in PC and PE in maternal plasma, fetal cord blood, fetal liver, and brain from the human fetus at mid-term abortion show a decrease in LA and LNA and a progressive increase of AA and DHA in the PC and PE from maternal liver to cord blood, to fetal liver, and finally to fetal brain. Crawford *et al.* termed this sequence a process of biomagnification, and considered it responsible for the high content of AA and DHA originating from LA and linolenic acid, respectively, in the brain. Thus, there is a preferential transfer of DHA and AA to the fetal side of the placenta. *In vitro* studies with human placenta show simultaneous accumulation of both omega-3 and omega-6 chain elongation products.

During the third trimester of human development, rapid synthesis of brain tissue occurs, in association with increasing neuromotor activity. The increase in cell size, number, and type requires *de novo* synthesis of structural lipids. Both M. Martinez and M. T. Clandinin noted the accumulation of DHA in the brain of the human infant during the last trimester. The levels of LNA and LA were consistently low in the brain, whereas marked accretion of long-chain desaturation products, specifically DHA and AA, occurred. Subsequent studies by M. Martinez and A. Ballabriga (1987) confirmed these findings. More recent data indicate that the main developmental changes in the brain seem to be an increase in DHA at the end of gestation and a decrease in oleic acid (18 : 1ω9) and AA in PE. Similar changes occurred in the liver. Therefore, a prematurely born infant (prior to 37 weeks) has much lower amounts of DHA in the brain and liver and is at risk of becoming deficient unless DHA is supplied in the diet. In the full-term newborn, about half of the DHA accumulates in the brain before birth and the other half after birth.

There is epidemiologic evidence that the birth weights of newborns in the Faroe Islands (where fish intake is high) are higher than those in Denmark, and so is the length of gestation: 40.3 ± 1.7 weeks for the Faroese versus 39.7 ± 1.8 weeks for the Danish. The average birth weight of primiparas was 194 g higher for the Faroe Islands. The higher dietary omega-3 fatty acid intake quite possibly influenced endogenous prostaglandin metabolism. It is hypothesized that the dietary omega-3 fatty acids inhibit the production of the dienoic prostaglandins, especially PGF_{2a} and PGE_2, because they are involved in the mediation of uterine contractions and the ripening of the cervix that lead to labor and delivery. These important observations need to be further investigated, as the prevention of prematurity is one of the most critical issues to be overcome in perinatal medicine.

B. Human Milk and Infant Feeding

A number of studies from around the world indicate that human milk contains both LNA and LA and their long-chain omega-3 and omega-6 fatty acids, whereas cow milk and infant formula do not; therefore, it seems reasonable to recommend their inclusion in infant formulas. Crawford *et al.* (1973) analyzed 32 samples of human milk and reported the presence of LNA and LA and the long-chain omega-3 and omega-6 fatty acids. At that time, they recommended that all of these fatty acids should be considered essential, because they can be classified as structural lipids in the human brain. Infant formulas still do not contain long-chain fatty acids (Table V).

TABLE V
Fatty Acid Composition of Human Milk and Formulas (molar percent)[a]

Fatty acid[b]	Human milk (n = 11)	Portagen®[c]	Enfamil® Premature	Similac® Special Care
8.0	0.35 ± 0.00	60	24.5	24.1
10:0	1.39 ± 0.14	24	14.1	17.7
12:0	6.99 ± 0.70	0.42	12.2	14.9
14:0	7.96 ± 0.88	Trace	4.7	5.8
16:0	19.82 ± 0.37	0.19	7.5	6.8
16:1	3.20 ± 0.21		0.1	0.2
18:0	5.91 ± 0.3	0.47	1.7	2.3
18:1	34.82 ± 1.4	4.1	12.4	10.0
18:2n6	16.00 ± 1.3	8.1	22.4	17.4
18:3n3	0.62 ± 0.04	Trace	0.6	0.9
20:1	1.10 ± 0.2		0.3	0.1
20:2n6	0.61 ± 0.1			
20:3n6	0.42 ± 0.04			
20:4n6	0.59 ± 0.04			
20:5n3	0.03 ± 0.00			
22:1	0.10 ± 0.00			
22:4n6	0.21 ± 0.00			
22:5n6	0.22 ± 0.00			
22:5n3	0.09 ± 0.03			
22:6n3	0.19 ± 0.03			

[a]Source: S. E. Carlson *et al.* (1986). *Am. J. Clin. Nutr.* **44**, 798–804.
[b]Values are expressed as mean ± SEM.
[c]"Pediatric Products Handbook," 1983 Ed., Mead Johnson Nutritional Division, Evansville, Indiana.

The long-chain fatty acid composition of red cell membrane phospholipids may reflect the composition of phospholipids in the brain. Therefore, determination of red cell membrane phospholipids has been carried out by many investigators to determine the long-chain PUFA content in breast-fed and bottle-fed infants. As expected, the fatty acids 22:5ω3 and 22:6ω3 were higher in the erythrocytes from breast-fed infants than in those from bottle-fed babies and the 20:3ω9 was lower in the erythrocytes of the breast-fed infants.

Following birth, the amount of red blood cell DHA in premature infants decreases, and thus the amount of DHA available to the premature infant assumes critical importance. Although premature infants convert LNA to DHA, this amount of DHA is not adequate for normal function (see Table I); therefore, a number of studies have been carried out on the DHA status of the premature infant. Premature babies have decreased amounts of DHA, but human milk contains enough DHA to support normal growth of the premature baby. Clandinin *et al.* were concerned that infant formula that does not contain DHA is fed to premature babies and did studies in which they developed infant formulas supplemented with DHA from two different sources, fish oil and hen egg yolk oil (Table VI).

When fish oil is incorporated into formula, DHA is absorbed by the infants' gastrointestinal tract, as seen by measuring DHA in plasma PE. A single dose of 71 mg/kg/day of fish oil DHA in a bolus is absorbed, as well as 11 mg/kg/day fish oil DHA dispersed in the formula. Dispersed fish oil appears to be absorbed as much or even more despite the much lower dose. DHA at a dose of 11 mg/kg/day results in 0.2% DHA in the total dietary fatty acids in the formula, which is within the range of 0.1–0.3% found in human milk. This dose of DHA does not decrease plasma phospholipid AA and appears to be a physiologic amount that could prevent declines in membrane phospholipid DHA following preterm delivery.

The amount of omega-3 fatty acids in human milk, particularly DHA, is lower in vegetarians than in om-

TABLE VI
Fatty Acid Composition of Human Milk and Formulas Prepared by Blending Egg Yolk Lipid or Fish-Oil Products with Other Oils[a]

		Formula	
Fatty acid (% w/w)	Human milk	Egg yolk oil	Fish oil
Short chain	2.0	1.00	1.95
Medium chain	11.0		
16:0	20.0	22.70	19.21
16:1	03.2	3.94	2.44
18:0	05.8	11.94	8.08
18:1ω9	38.0	38.97	39.30
18:2ω6	12.0	11.49	12.22
18:3ω3	00.9	0.46	1.68
20:4ω6	00.5	0.25	0.13
20 and 22ω6	01.0	0.50	0.21
20 and 22ω3	0.1–0.7	0.11	3.30

[a]Source: M. T. Clandinin and J. E. Chappell (1985). *In* "Composition and Physiological Properties of Human Milk" (J. Schaub, ed.), p. 221. Elsevier Science Publishers, Amsterdam.

nivores. One can increase the amount of DHA in human milk by giving fish oil rich in DHA to the mother. In one study, 5 g/day of fish oil raised the levels from a baseline of 0.1 ± 0.06% to 0.5 ± 0.1% ($P < 0.001$); 10 g/day raised DHA levels to 0.8 ± 0.1% ($P < 0.001$); and 47 g/day led to a DHA level of 4.8%. Thus, relatively low intakes of dietary DHA can significantly elevate milk DHA content.

The need to supplement infant formula with omega-3 fatty acids and, particularly, DHA for the premature is now recognized. Studies are in progress comparing the growth and development of premature infants who are fed mother's milk with those who are receiving formula supplemented with omega-3 fatty acids and those whose formula is not supplemented.

C. Childhood

The first case of human LNA deficiency was reported in 1982. The LNA deficiency was induced by long-term intravenous hyperalimentation. The patient was a 6-year-old white female who had 266 cm of her small intestine, the ileocecal valve, and 34 cm of her large bowel removed as a result of a 22-caliber rifle wound to her abdomen. For 7 months she was on total parenteral nutrition (TPN) solution, which was high in LA and low in LNA. She developed neurological symptoms consisting of episodes of distal numbness, paresthesia, weakness, and blurring of vision. Analysis of fatty acids of serum lipids indicated marginal LA deficiency but marked LNA deficiency. Because neurological abnormalities do not occur with omega-6 fatty acid deficiency, omega-3 fatty acid deficiency was suspected and the patient was given an emulsion that contained LNA; the neurological symptoms disappeared. Based on studies of this patient, the requirement for LNA was estimated to be about 0.54% of calories. Recently, another child who was on TPN without LNA supplementation developed neurological symptoms that subsided with LNA supplementation.

Over the past 20 years, data from other studies involving humans on enteral or TPN, which were low or deficient in omega-3 fatty acids, have focused attention on the importance of omega-3 fatty acids in vision and central nervous system function.

D. Aging

K. S. Bjerve in 1987 described four patients with LNA deficiency as a result of long-term gastric tube-feeding that included large amounts of skim milk without LNA supplementation. These patients, who were in nursing homes, developed skin lesions diagnosed as scaly dermatitis, which disappeared with EFA supplementation. A number of other patients were reported to have omega-3 fatty acid deficiency, again patients on long-term gastric tube-feedings or prolonged TPN because of chronic illnesses. Bjerve estimated that the minimal daily requirements of LNA and of long-chain omega-3 fatty acids in diets are equivalent to 0.2–0.3% and 0.1–0.2%, respectively, of total energy intake. If a deficiency of total omega-3 fatty acid intake is suspected, its concentration in plasma should be measured. A decrease in the concentration of 20:5ω3, 22:5w3, and particularly 22:6ω3 in plasma or erythrocyte lipids indicates that the dietary intake of omega-3 fatty acids has been low. The presence of clinical symptoms, along with the biochemical determinations, provide additional support for the diagnosis. To verify the diagnosis, it is essential that the clinical symptoms disappear upon supplementation of the deficient diet with omega-3 fatty acids.

With the increase in the number of elderly persons in the population, and the proliferation of nursing homes, particular attention must be given to the nutritional requirements of the elderly, especially those who are fed enterally or parenterally. The optimal and minimal dietary requirements of LNA have been estimated to be 860–1220 mg/day and 290–390 mg/day, respectively. Similarly, the optimal and minimal dietary requirements of long-chain omega-3 fatty acids are estimated to be 350–400 mg/day and 100–200 mg/day, respectively. Long-chain fatty acids appear to be two to three times more effective than LNA in curing and probably preventing clinical symptoms of omega-3 fatty acid deficiency.

IV. IMPLICATIONS FOR DIETARY RECOMMENDATIONS

A. Evolutionary Aspects and the Omega-3–Omega-6 Balance

The development of agricultural food production took place 10,000 years ago and was one of the main steps toward human cultural evolution. Although precise information does not exist about the diet of humans, the development of agriculture occurred independently in the Middle East and other parts of the world, such as Central and South America and Southeast Asia. Human beings usually ate wild plants, wild animals, and fish from the rivers, lakes, and oceans. Wild animals and birds who feed on wild plants (rich in omega-3 fatty acids) (see Table II) are very lean, with a carcass fat content of only 3.9%, and contain about five times more PUFAs per gram than is found in domestic livestock. Furthermore, 4% of the fat of wild animals contains EPA, whereas domestic beef contains a very small amount, if any, because cattle are fed grains rich in omega-6 fatty acids and poor in omega-3 fatty acids. Similarly, eggs from free-ranging chickens are rich in omega-3 fatty acids with a ratio of omega-6 to omega-3 of 1.3, whereas the standard USDA egg is poor in omega-3 with a ratio of omega-6 to omega-3 of 19.4 (Table VII).

From all available information, human beings appear to have evolved on a diet rich in omega-3 fatty acids and balanced between omega-3 and omega-6 fatty acids, which is a more physiological state. Over the past 100 years, the Western diet has shifted dramatically toward increased amounts of omega-6 fatty acids at the expense of omega-3 fatty acids. This shift has led to a diet relatively deficient in omega-3 fatty acids, with a ratio of omega-3–omega-6 equal to 1 : 14 instead of 1 : 1 as is the case with wild animals and presumably early humans, although the optimal ratio of omega-3–omega-6 in the diet is not precisely known.

The competition between omega-3 and omega-6 fatty acids in their elongation and desaturation products and in prostaglandin metabolism, the increase of

TABLE VII
Fatty Acid Levels in Chicken Egg Yolks[a,b]

Fatty acid	Greek egg	Supermarket egg
	milligrams of fatty acid[c]	
Saturated fats		
14:0	1.10	0.70
15:0	—	0.07
16:0	77.60	56.66
17:0	0.66	0.34
18:0	21.30	22.88
Total	100.66	80.65
Monounsaturated fats		
16:1n-7	21.70	4.67
18:1	120.50	109.97
20:1n-9	0.58	0.68
22:1n-9	—	—
24:1n-9	—	0.04
Total	142.78	115.36
n-6 Fatty acids		
18:2n-6	16.00	26.14
18:3n-6	—	0.25
20:2n-6	0.17	0.36
20:3n-6	0.46	0.47
20:4n-6	5.40	5.02
22:4n-6	0.70	0.37
22:5n-6	0.29	1.20
Total	23.02	33.81
n-3 Fatty acids		
18:3n-3	6.90	0.52
20:3n-3	0.16	0.03
20:5n-3	1.20	—
22:5n-3	2.80	0.09
22:6n-3	6.60	1.09
Total	17.66	1.73
Ratio of fatty acids to saturated fats	0.4	0.44
Ratio of n-6 to n-3	1.3	19.4

[a]Source: A. P. Simopoulos and N. Salem, Jr. (1989). n-3 fatty acids in eggs from range-fed Greek chickens (letter to the editor). *N. Engl. J. Med.* **321**, 1412.
[b]The eggs were hard-boiled, and their fatty acid composition and lipid content were assessed as described elsewhere.
[c]Per gram of egg yolk.

the AA metabolites in cardiovascular disease, hypertension, and autoimmune disorders, and their decrease with ingestion of LNA or EPA and DHA strongly suggest that a balance is needed between omega-3 and omega-6 fatty acids for platelet function, blood viscosity, monocyte function, membrane function, and central nervous system function.

Over the past 20 years, studies in the rat and the rhesus monkey indicate that dietary restriction of omega-3 fatty acids during pregnancy and infancy interfere with normal visual function and may even impair learning ability in the offspring. These findings and the presence of omega-3 fatty acids in human milk provide evidence for the essentiality of omega-3 fatty acids in growth and development and in health and disease.

B. Dietary Recommendations

In making dietary recommendations about fat intake, particular attention should be paid to the relative amounts of omega-3 and omega-6 fatty acids in the diet. The amounts of omega-3 fatty acids for infant nutrition should be consistent with the amounts found in human milk. After infancy, the ratio of omega-3–omega-6 should be either 1 : 1 or 1 : 5.

The type of omega-3 fatty acid to be considered varies with the age of the individual. Because premature infants, and possibly the elderly, are limited in their ability to convert LNA to EPA and DHA, DHA should be included in their diet. Preliminary evidence indicates that some diabetics and hypertensives lack this ability as well. For others, LNA by itself or LNA, EPA, and DHA from vegetables and fish or fish oils (for those who cannot obtain or will not eat fish) should be included in the diet. Americans today consume 15 lb of fish per person per year. To improve the balance of omega-3–omega-6, fish must be eaten a minimum of two times per week, significantly substituting meat consumption. Vegetarians who do not eat fish should increase the amount of omega-3 fatty acid sources in their diet by eating soybeans, walnuts, and omega-3 fatty acid-containing vegetable oils.

BIBLIOGRAPHY

Bjerve, K. S., Fougner, K. J., Midthjell, K., and Bonaa, K. (1989). N-3 fatty acids in old age. *J. Int. Med.* **225**, 191–196.

Carlson, S. E., Rhodes, P. G., and Ferguson, M. G. (1986). Docosahexanoic acid status of preterm infants at birth and following feeding with human milk or formula. *Am. J. Clin. Nutr.* **44**, 798–804.

Clandinin, M. T., Chappell, J. E., Leong, S., Heim, T., Swyer, P. R., and Chance, G. W. (1980). Intrauterine fatty acid accretion rates in human brain: Implications for fatty acid requirements. *Early Hum. Develop.* **4**, 121–129.

Crawford, M. A., Sinclair, A. J., Msuya, P. M., Munhambo, A. (1973). Structural lipids and their polyenoic constituents in human milk. *In* "Dietary Lipids and Postnatal Development" (C. Galli, G. Jacini, and A. Pecile, eds.). Raven, New York.

Farquharson, J., Cockburn, F., Ainslie, P. W. (1992). Infant cerebral cortex phospholipid fatty-acid composition and diet. *Lancet* **340**, 810–813.

Galli, C., and Simopoulos, A. P. (eds.) (1989). "Dietary ω3 and ω6 Fatty Acids: Biological Effects and Nutritional Essentiality." Plenum, New York.

Hoffman, D., and Uauy, R. (1992). Essentiality of dietary omega-3 fatty acids for premature infants: Plasma and red blood cell fatty acid composition. *Lipids* **27**, 886–895.

Holman, R. T., Johnson, S. B., and Hatch, T. F. (1982). A case of human linolenic acid deficiency involving neurological abnormalities. *Am. J. Clin. Nutr.* **35**, 617–623.

Neuringer, M., Connor, W. E., Lin, D. S., Barstad, L., and Luck, S. (1986). Biochemical and functional effects of prenatal and postnatal ω-3 fatty acid deficiency on retina and brain in rhesus monkeys. *Proc. Natl. Acad. Sci. USA* **83**, 4021–4025.

Rotstein, N. P., Ilincheta de Boschero, M. G., Giusto, N. M., and Aveldano, M. I. (1987). Effects of aging on the composition and metabolism of docosahexaenoate-containing lipids of retina. *Lipids* **22**, 253–260.

Salem, N., Jr. (1989). Omega-3 fatty acids: Molecular and biochemical aspects. *In* "New Protective Role for Selected Nutrients" (G. Spiller and J. Scala, eds.). Alan R. Liss, New York.

Simopoulos, A. P. (1991). Omega-3 fatty acids in health and disease and in growth and development. *Am. J. Clin. Nutr.* **54**, 438–463.

Simopoulos, A. P., Kifer, R. R., and Martin, R. E. (eds.) (1986). "Health Effects of Polyunsaturated Fatty Acids in Seafoods." Academic Press, Orlando, Florida.

Simopoulos, A. P., Kifer, R. R., Martin, R. E., and Barlow, S. M. (eds.) (1991). Health effects of omega 3 polyunsaturated fatty acids in seafoods. *In* "World Review of Nutrition and Dietetics." Karger, Basel, Switzerland.

Uauy, R. D., Birch, D. G., Birch, E. E., and Uauy, R. (1990). Effect of dietary omega-3 fatty acids on retinal function of very-low-birth-weight neonates. *Pediatr. Res.* **28**, 485–492.

Oncogene Amplification in Human Cancer

BARRY M. KACINSKI
Yale University School of Medicine

GLOSSARY

Oncogenes and tumor suppressors Genes which control normal cellular proliferation and differentiation. Disruption in their structure, function, or level of expression can generate neoplastic cells in which normal controls of cellular proliferation and differentiation have been disrupted from totally normal, nonneoplastic precursors

I. INTRODUCTION

Cancer is a complex set of related disorders in which inherited germline or acquired somatic mutations in the function or level of expression of genes which control cellular proliferation and differentiation produce clonal outgrowths of cells whose control of cellular proliferation and differentiation is significantly abnormal. In many cases, those genes whose function or expression has been altered are known as oncogenes or tumor suppressors. [*See* Oncogenes and Proto-oncogenes; Tumor Suppressor Genes.]

Point mutations in oncogenes or tumor suppressors can result in the synthesis of abnormal protein products whose function is altered quantitatively or qualitatively to promote abnormal cellular proliferation and differentiation. A qualitative increase in oncogene function can result from overexpression of a wild-type, functionally normal protein consequent to the translocations of a structurally normal oncogene to another chromosomal site where it can now be actively transcribed. Alternatively, overexpression can be the consequence of other genetic events which increase the copy number of an otherwise wild-type gene; this article focuses primarily on the latter.

A. Chromosomal Abnormalities Associated with Oncogene Amplification

Although functional gene amplification occurs "physiologically" in the polytene chromosomes of *Drosophila* and in the amplified rRNA sequences of amphibian oocytes, gene amplification rarely, if ever, occurs in the somatic cells of higher vertebrates. Gene amplification was first recognized to occur in cancer when it was demonstrated that certain specific chromosomal aberrations—double minutes (DMs) and heterogeneous staining regions (HSRs)—were composed of multiple copies of the same DNA sequence. [*See* Chromosome Anomalies.]

DMs and HSRs are also frequently observed in cells exposed to cytoxic agents. In these cells, the DNA sequences which have been amplified often encode proteins which confer resistance to these agents. Because HSRs are incorporated into chromosomes with centromeric sequences and DMs are not, HSRs are stably maintained in subsequent cell divisions. In contrast, DMs are randomly distributed to daughter cells during successive mitoses and are gradually lost after exposure to the toxic agent has been discontinued.

However, both DMs and HSRs are also commonly observed in tumor cells before any treatment has been administered. In such cases, the DNA segments which have been amplified (i.e., DNA amplicons) often incorporate one or more genes whose presence in multiple copies confers a proliferative or survival advantage on the neoplastic cells which carry them.

One of the first oncogenes found to be amplified in neoplastic cells was c-*myc*. Cells of the HL60 cell line as well as the leukemic cells from which the cell lines were established were shown to carry between 8 and 32 copies of the c-*myc* gene. Similar levels of c-*myc* amplification were found in the COLO30 neuroendocrine colon carcinoma cell line where the multiple c-*myc* copies were present either as DMs or as HSRs depending on the subline studied. Because DMs had also been observed in the cells of the tumor from which the cell line was established, the observed c-*myc* amplification had apparently occurred *in vivo*.

Another set of genes frequently amplified in human cancers map to band 13 of the long arm of chromosome 11. Amplicons derived from this region incorporate several different genes (int-2, hst, cyclin D1) whose overexpression could contribute to the neoplastic phenotypes of cells in which they have been amplified.

Gene amplification can also occur in the absence of gross cytogenetic abnormalities. The recent application of comparative genomic hybridization has demonstrated that small regions of amplification and loss are far more common than had been previously suspected. For example, low level amplification of c-*myb* has been observed in a characteristic marker chromosome of a colon carcinoma without any evidence of HSRs or DMs.

B. Mechanisms of Gene Amplification

To date, the mechanisms by which DNA sequences become amplified remain unclear. Some models of gene amplification propose that multiple rounds of DNA synthesis at a given site generate a multistranded, multicopy structure that is "resolved" by recombinational mechanisms to produce an integrated (HSR) or an extrachromosomal (DM) tract of multiple copies of the same segment of DNA. Other models suggest that recombinogenic sequences (such as microsatellite tracts) recombine with each other and "loop out" the intervening sequence to produce an extrachromosomal circular DNA molecule. Multiple rounds of DNA replication and recombination of these circles produce multicopy structures or DMs which integrate to produce HSRs. [*See* Gene Amplification.]

That gene amplification may depend on specific chromosomal sites or DNA sequences is suggested by the observation that certain chromosomal regions are far more prone to amplification than others. Also, experimental studies demonstrate that the amplification of randomly integrated genes can vary by as much as two orders of magnitude depending on the site of integration.

The frequency of amplification of genes which confer drug resistance can be enhanced as much as 100-fold by pretreatment with certain carcinogens or cytoxic agents before drug selection or by exposure to mitogens and/or tumor promoters during selection. Those cytoxic agents which promote gene amplification are, in general, agents which damage DNA or inhibit DNA synthesis, thereby fostering aberrant replication and/or stimulating recombination. Tumor promoters or mitogens may drive DNA synthesis under adverse conditions to promote aberrant DNA replication or they may merely enhance the survival of resistant clones under the conditions of drug selection.

II. TUMOR SPECIFICITY OF ONCOGENE AMPLIFICATION

A. Neuroblastoma: Amplification of the N-*myc* Oncogene

Neuroblastoma, a pediatric malignancy, arises from primitive neuroectodermal cells in the adrenals and at other sites such as the retroperitoneum and mediastinum. The N-*myc* protooncogene was originally isolated from a neuroblastoma cell line as a *myc* homologous gene present in increased copy number. From 3–300 fold amplification of N-*myc* is observed in about 40% of advanced neuroblastomas and in many neuroblastoma-derived cultured cell lines. For example, the HSR on chromosome 1 of IMR-32 neuroblastoma cells contains about 3 Mb of chromosomal DNA, including the N-*myc* locus and several distinct domains of chromsome 2.

N-*myc* amplification is detected in more than half of disseminated (stage IV) neuroblastomas but rarely in the tumors of patients with localized disease (stage I or II) and excellent prognosis. Amplification of N-*myc* has been clearly shown to be an important negative prognostic variable, independent of clinical stage (i.e., extent of disease at presentation) as nearly two-thirds of stage IV patients without N-*myc* amplification are without evidence of disease progression 9 months after diagnosis. In contrast, disease progresses in more than 90% of neuroblastoma patients whose tumors have more than 10 copies of the N-*myc* gene.

It is not clear how amplification of N-*myc* relates to neuroblastoma progression since even within an individual tumor, different cells have a different N-*myc* copy number, with the highest number of copies found in histologically undifferentiated neuroblasts. It is possible that the level of N-*myc* gene expression may be a better prognostic marker than DNA amplification as the former can be elevated even in the absence of gene amplification and may more accurately reflect the presence of increased amounts of functional N-*myc* transcription factor in neoplastic cells.

B. Small Cell Lung Carcinoma: Amplification of Various *myc* Family Oncogenes

Amplification of all three *myc* family oncogenes, c-*myc*, N-*myc*, and L-*myc*, can be observed in human small cell lung carcinomas (SCLC) which represent approximately one-fourth of human lung neoplasms. Amplification of one of these three oncogenes is observed in ≈40% of cell lines derived from SCLCs, but in no cell line are two different *myc* genes amplified at the same time.

Amplification of *myc* genes is associated with a high level *myc* transcript and protein expression and appears to correlate with certain specific cellular phenotypes such as rapid growth, improved cloning efficiency, loss of antigenic markers characteristic of SCLCs, decreased adhesiveness, and so-called variant morphology (i.e., resembling large cell carcinomas).

In contrast to the high frequency of *myc* amplification observed in SCLC-derived cell lines, c-, N-, and L-*myc* amplifications are observed in 2, 7, and 12% of SCLC tumor specimens, even less than the 15% frequency of c-*myc* amplification reported in human squamous cell carcinomas. The reason for this discrepancy is unclear. However, the so-called "variant" histologic features characteristic of SCLC cell lines with *myc* gene amplification are apparent in only 10% of SCLC at presentation but in 30% at autopsy. Such results suggest that most SCLCs contain small populations of "variant" cells with *myc* gene amplification whose fraction increases with time as a consequence of treatment or other factors which give SCLC cells with *myc* amplification a selective advantage *in vivo*. Alternatively, the increased *in vitro* growth rate and cloning efficiency conferred upon SCLC cells by the amplification and increased expression of one of the three *myc* oncogenes may render those cells best able to proliferate *in vitro* and establish a cell line.

As mentioned earlier, experimental studies of gene amplification have demonstrated that many agents employed to selectively kill neoplastic cells by damaging DNA or disrupting DNA synthesis also stimulate gene amplification manyfold *in vitro* and, perhaps, *in vivo*. Therefore, it is at least plausible that treatment of SCLC patients with agents which promote gene amplification will result in the amplification of many different genes. Selection will then favor those cells which have acquired amplifications which confer more rapid growth or improved cloning efficiency, such as in one of the *myc* oncogenes, and cells with such advantageous amplifications will eventually predominate in the tumor.

C. Glioblastomas and Squamous Cell Carcinomas: Amplification of c-erbB1, the Epidermal Growth Factor Receptor Gene

Epidermal growth factor (EGF) is a polypeptide mitogen that induces a variety of responses in normal and neoplastic cells that express its receptor (EGF-R) encoded by the c-erbB1 (HER1) locus. These responses are triggered by binding of the ligand, EGF, to the receptor EGF-R, receptor dimerization, and receptor autophosphorylation on specific tyrosines. The latter events couple the receptor to a variety of intracellular "second messengers" which transduce the signal from the membrane through the cytoplasm to the nucleus to specifically alter gene expression.

A truncated version of the EGF-R is one of the transforming viral oncogenes transduced by the avian erythroblastosis virus; although similar truncated or mutated versions of the EGF-R are rare in human neoplasms, a 6- to 60-fold amplification of the wild-type erbB1 gene and overexpression of the EGF-R protein are common in human squamous cell carcinomas and glioblastomas.

D. Breast and Other Adenocarcinomas: Amplification of the *neu* Oncogene

The *neu* (erbB2, HER-2) oncogene, first identified in an ethylnitrosourea-induced rat neuroblastoma, encodes a receptor tyrosine kinase related in sequence to EGF-R and to two other receptor tyrosine kinases: HER-3 and HER-4. Although no specific ligand has yet been identified for the *neu*-encoded receptor kinase, *neu* protein monomers readily form heterodimers with other HER family receptors after they have been activated by their specific ligands. Such activated heterodimers then couple with a variety of second messengers and activate intracellular signal transduction pathways in ways similar, but not identical, to those observed after activation of EGF-R, HER-3, or HER-4 by their cognate ligands. Also, the spontaneous formation of *neu* protein homodimers and tyrosine autophosphorylation can be observed in cells which express very high levels of *neu* protein in the absence of any ligand.

Amplification and overexpression of *neu* are observed in a significant fraction (10–30%) of human breast, ovarian, and a variety of other adenocarcinomas; a high level of overexpression of *neu* can even be observed in the absence of gene amplification. Although the original form of *neu* isolated from rat neuroblastomas was mutated (glutamic acid to valine) in its transmembrane domain in a way which facilitates spontaneous homodimer formation and autophosphorylation on tyrosine, mutations of *neu* have yet to be convincingly demonstrated in any spontaneous human neoplasm.

In invasive breast and ovarian carcinomas, *neu* gene amplification or overexpression has been associated with poorer relapse-free and overall survival, but the prognostic importance of *neu* amplification or overexpression relative to other clinicopathologic indicators of outcome in breast and ovarian cancer patients is still the subject of some debate. Perhaps ongoing studies with antibodies able to discriminate the tyrosine phosphorylated, "activated" form of the *neu*-encoded receptor from its inactive precursor in histologic sections will allow investigators to determine whether the presence of an overexpressed and functionally activated *neu* protein does indeed imply a poor prognosis. [*See* Breast Cancer Biology.]

In transgenic mice engineered to overexpress the *neu* protein in the mammary epithelium, mice develop multifocal tumors if the construct expresses a "mutated" *neu* protein.

Amplification of oncogenes other than *neu* is also observed in human breast carcinomas. In particular, a region of 11q13 which incorporates the *int-2*, *hst*, and *prad* (cyclin D1) oncogenes is commonly amplified in breast carcinomas. Such amplifications have been associated with the overexpression of cyclin D1 and with poor prognosis.

E. Other Types of Sequence Amplification: Microsatellite Instability and Dynamic Mutation

In addition to the types of gene amplification described earlier, other sequences appear to be unstable in carcinomas, particularly tracts of mono-, di-, tri-, and tetranucleotide repeats, so-called "microsatellites" sequences.

Low level amplification and contraction of microsatellite sequences occur in gastrointestinal and gynecological malignancies, particularly in the syndrome of hereditary nonpolyposis coli carcinomas which includes patients with hereditary colon, other gastrointestinal, uterine, and ovarian carcinomas. Such patients have been shown to inherit defective genes that encode proteins responsible for the repair of base mismatches in DNA, homologous to the *mutL* and *mutS* genes of enterobacteria. [*See* Colon Cancer Biology.]

Larger scale instability or "dynamic mutation" was first described in a variety of hereditary neurologic syndromes such as Huntington's chorea. In these diseases, a microsatellite tract undergoes progressive enlargement in successive generations, eventually becoming so large as to disrupt the expression or function of genes in which it is located. In some syndromes, such as fragile X disease, the expanded tract becomes the site of frequent chromosomal breakage. [*See* Huntington's Disease.]

Because the repetitive sequence tracts become larger with successive generations, the disruption in gene expression or function becomes progressively greater, leading to the phenomenon of "anticipation," i.e., more severe disease, often with an earlier age of onset of symptoms, in successive generations. Similar anticipation is observed in several different hereditary human neoplasms, and evidence shows that similar large-scale expansions of repetitive sequences also occur in some of these disorders. Whether these expansions lie in or near specific oncogenes or tumor suppressors or occur in many different microsatellite tracts scattered throughout the genome is currently the subject of active investigation.

III. SIGNIFICANCE OF SEQUENCE AMPLIFICATION AND GENETIC INSTABILITY IN CARCINOGENESIS AND TUMOR PROGRESSION

The information summarized in this article demonstrates that DNA sequence amplification is a common event in many neoplasms. In particular, the recent application of comparative genomic hybridization technology has made it clear that sequence amplification (and loss) is far more common than had been previously suspected. When such amplified sequences include protooncogenes that encode such things as transcription factors and receptor tyrosine kinases and lead to unregulated, increased levels of expression of their protein products, they may directly contribute to neoplastic cell phenotypes and tumor progression.

Because specific gene amplification, either transient or stable, can be observed in normal cells at certain phases of development, it is possible that some examples of gene amplification observed in human neoplasms are merely the aberrant expression of a normal cellular phenotype. However, insights into the role of DNA mismatch repair genes in genetic instability suggest that mutations in these or other genes with similar functions might result in other forms of genetic instability, including those that result in the amplification of large pieces of genomic DNA.

If sequence amplification is initially random, it is reasonable to expect that tumor cells that have amplified a segment of DNA encoding genes whose overexpression confers a growth advantage will overgrow those that do not. Therefore, it should not be surprising that genes that are often amplified (*myc* family oncogenes, HER family receptor tyrosine kinases, cyclin D1) are genes that confer rapid growth, better cloning efficiency, and other advantageous phenotypes when overexpressed. Amplifications of other oncogenes, such as those of the *ras* family, where overexpression of a wild-type protein alone does not confer an advantageous phenotype, are much less frequent.

In most tumors, gene amplification alone is insufficient to produce a neoplastic cell from a benign precursor. In fact, elegant studies of colorectal carcinomas demonstrate that carcinogenesis is a multistep process in which mutation, deletion of tumor suppressors, and mutation and/or amplification of oncogenes all play important roles. The challenge will be to define the mechanisms by which these fascinating and very important events occur.

BIBLIOGRAPHY

Alitalo, K., and Schwab, M. (1986). Oncogene amplification in tumor cells. *Adv. Cancer Res.* **47**, 235–281.

Alt, F. W., DePinho, R., Zimmerman, K., *et al.* (1986). The human *myc* gene family. *Cold Spring Harbor Symp. Quant. Biol.* **51**, 931–941.

Biedler, J. L., Meyers, M. B., and Spengler, B. A. (1983). Homogeneously staining regions and double minute chromosomes, prevalent cytogenetic abnormalities of human neuroblastoma cells. *Adv. Cell. Neurobiol.* **4**, 268–301.

Bishop, J. M. (1983). Cellular oncogenes and retroviruses. *Annu. Rev. Biochem.* **52**, 301–354.

Bishop, J. M. (1985). Viral oncogenes. *Cell* **42**, 23–38.

Ionov, Y., Peinado, M. A., Malkhosyan, S., Shibata, D., and Perucho, M. (1993). Ubiquitous somatic mutations in simple repeated sequences reveal a new mechanism for colonic carcinogenesis. *Nature* **363**, 558–561.

King, B. L., Carcangiu, M. L., Carter, D., Kiechle, M., Pfisterer, J., Pfleiderer, A., and Kacinski, B. M. (1995). Microsatellite instability in ovarian neoplasms. *Br. J. Cancer* **72**, 376–382.

Kolodner, R. D., Hall, N. R., Lipford, J., Kane, M. F., Morrison, P. T., Finan, P. J., Burn, J., Chapman, P., Earabino, C., Merchant, E., and Bishop, D. T. (1995). Structure of the human *MLH1* locus and analysis of a large hereditary nonpolyposis colorectal carcinoma kindred for mlh1 mutations. *Cancer Res* **55**, 242–248.

Leach, F. S., Nicolaides, N. C., Papadopoulos, N., Liu, B., Jen, J., Parsons, R., Peltomäki, P., Sistonen, P., Aaltonen, L. A., Nyström-Lahti, M., Guan, X. Y., Zhang, J., Meltzer, P. S., Yu, J. W., Kao, F. T., Chen, D. J., Cerosaletti, K. M., Fournier, R. E. K., Todd, S., Lewis, T., Leach, R. J., Naylor, S. L., Weissenbach, J., Meckin, J. P., Järvinen, H., Petersen, G. M., Hamilton, S. R., Green, J., Jass, J., Watson, P., Lynch, H. T., Trent, J. M., de la Chappelle, A., Kinzler, K. W., and Vogelstein, B. (1993). Mutations of a mutS homolog in hereditary nonpolyposis colorectal cancer. *Cell* **75**, 1215–1225.

Minna, J. D., Battey, J. F., Brooks, B. J., *et al.* (1986). Molecular genetic analysis reveals chromosomal deletion, gene amplification, and autocrine growth factor production in the pathogenesis of human lung cancer. *Cold Spring Harbor Symp. Quant. Biol.* **51**, 843–853.

Schimke, R. T., Hoy, C., Ricc, G., Sherwood, S. W., and Schumacher, R. I. (1988). Enhancement of gene amplification by perturbation of DNA synthesis in cultured mammalian cells. *In* "Cancer Cells" (T. Kelly and B. Stillman, eds.), Vol. 6, pp. 317–322. Cold Spring Harbor Laboratory, Cold Spring Harbor, NY.

Schwab, M. (1985). Amplification of N-*myc* gene in human neuroblastomas. *Trends Genet.* **1**, 271–275.

Sicinski, P., Donaher, J. L., Parker, S. B., Li, T., Fazeli, A., Gardner, H., Haslam, S. Z., Bronson, R. T., Elledge, S. J., and Weinberg, R. A. (1995). Cyclin D1 provides a link between development and oncogenesis in the retina and breast. *Cell* **82**, 621–630.

Slamon, D. J., Godolphin, W., Jones, L. A., *et al.* (1989). Studies of the HER-2/*neu* proto-oncogene in human breast and ovarian cancer. *Science* **244**, 707–711.

Tlsty, T. D., Briot, A., Gualberto, A., Hall, I., Hess, S., Hixon, M., Kuppuswamy, D., Romanov, S., Sage, M., and White, A. (1995). Genomic instability and cancer. *Mutat. Res.* **337**, 1–7.

van de Vijver, M., Perterse, J. L., Moor, W. J., *et al.* (1988). *Neu*-protein and comedo-type ductal carcinoma *in situ* and limited prognostic value in stage II breast cancer. *N. Engl. J. Med.* **319**, 1239.

Venter, D. J., Kumar, S., Tuzi, N. L., and Gullick, W. J. (1987). Overexpression of the c-*erb* B-2 oncoprotein in human breast carcinomas: Immunohistological assessment correlates with gene amplification. *Lancet* **II**, 69.

Oncogenes and Proto-oncogenes

JAMES R. THOMPSON
Cornell University Medical College

GLOSSARY

Immortalization Process by which cells having a defined life span or generation time in tissue culture are altered so that they can be propagated indefinitely

Kinase Enzyme that can catalyze the transfer of a phosphate group to another protein or biomolecule

Nude mice Mutant strain of hairless mice that lack thymus tissue and is used to study tumorigenicity. In the absence of a thymus, these mice do not have the ability to generate an appropriate immune response. When injected with transformed cells from a foreign source, a tumor can be generated without rejection by the host

Oncogene Gene that has been mutated or misexpressed in some way so that it causes cellular transformation. Oncogenes located in viral genomes are designated by the prefix *v*- and cellular oncogenes are designated by the prefix *c*-

Proto-oncogene Normal unmutated cellular version of an oncogene that has the potential to become oncogenic

Retrovirus Class of RNA viruses that can integrate their genomes into that of the host cell. A subset of retroviruses have acquired oncogenes and thus have the ability to cause cellular transformation

Transformation Process by which normal cells are altered to a tumorigenic phenotype. Typical characteristics of transformed cells are loss of contact inhibition, loss of anchorage dependence, immortalization, and loss of growth factor dependence

Tumorigenic Ability of transformed cells, when introduced into an immunocompromised host such as a nude mouse, to generate a tumor

MALIGNANCY OCCURS AS A RESULT OF A CELL'S inability to control its growth and differentiation pathway. Regulation of this process can be mediated at multiple steps, which are summarized in Fig. 1. Cells can be stimulated from a quiescent state to begin proliferating by an external signal or mitogen. Likewise, proliferating cells can be signaled to stop growing and to differentiate. The recognition of an external signal is dependent on a specific receptor, expressed by the cell, which can bind to and interact with the signaling factor. The activated receptor can then in turn activate secondary cellular messenger systems to transmit the growth or differentiation signal to nuclear proteins. Nuclear proteins may then activate or repress numerous target genes that result in phenotypic changes in the cell's proliferative state. Mutations in a gene that encodes any one of the proteins involved in this signaling cascade may result in uncontrolled cellular proliferation. These mutated genes are termed *oncogenes,* whereas the unmutated normal versions are termed *proto-oncogenes.*

I. INTRODUCTION

The discovery of oncogenes originally stemmed from research in the field of virology, where researchers demonstrated that a number of animal viruses had the ability to alter cells grown in tissue culture. This alteration or transformation is marked by several phenotypic changes: (1) loss of dependence on growth

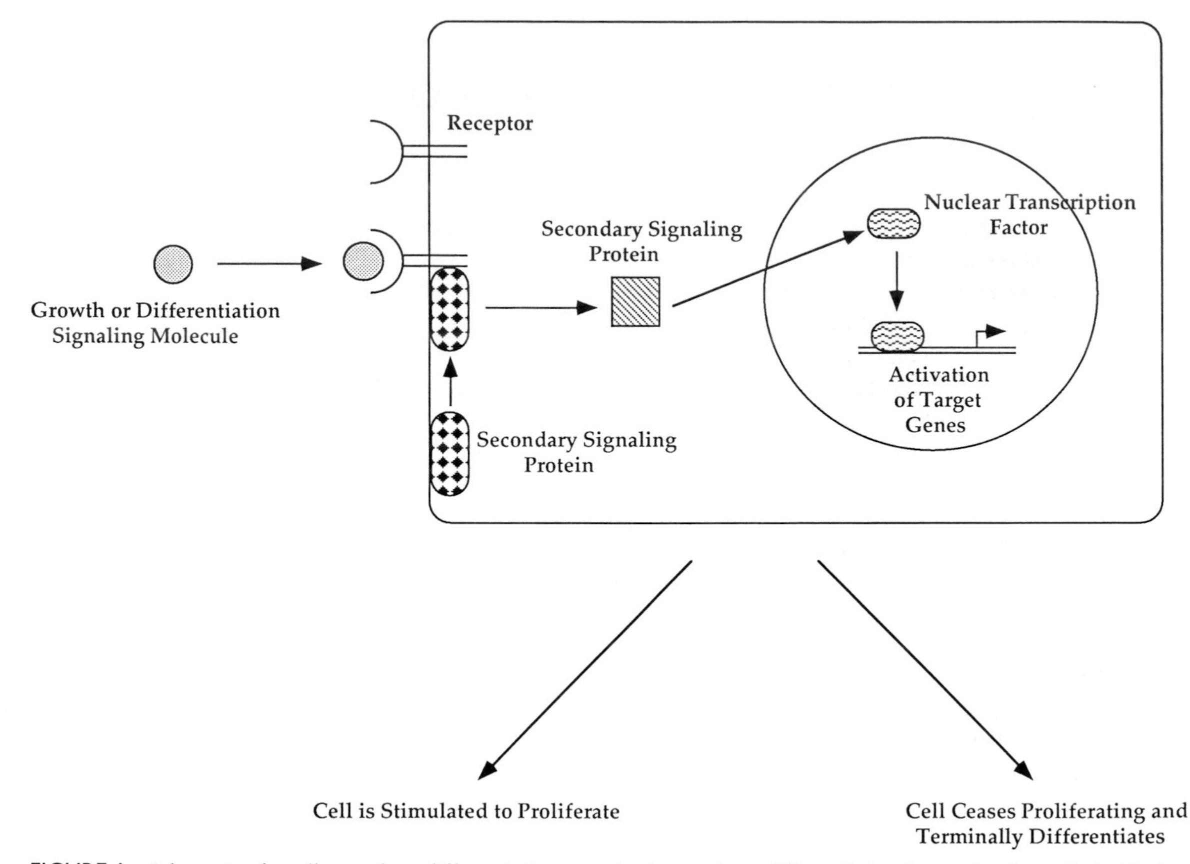

FIGURE 1 Schematic of a cell growth or differentiation cascade. A growth or differentiation factor signals a cell via binding to a cell-surface receptor. The receptor then activates secondary cellular signals such as kinases or guanine nucleotide binding proteins. Nuclear transcription factors are then activated and result in the activation or repression of target genes. This change in gene expression results in a phenotypic change in the cell, that is, proliferation or differentiation depending on the original signaling factor.

factors, (2) loss of anchorage dependence, (3) loss of contact inhibition, and (4) immortalization. Cells grown in tissue culture often rely on the presence of growth factors in their media to proliferate. Tissue culture cells infected with a transforming virus lose their dependence on growth factors and continue to grow in their absence. Normal cells grown in culture rely on a substrate for adherence and growth; when the cells reach a density at which they begin to touch each other, they cease proliferating. In contrast, transformed cells are no longer anchorage dependent and contact inhibited. In tissue culture, this behavior is characterized by the formation of dense colonies called foci. Finally, primary cells isolated from an animal and grown in culture usually have a defined life span. These cells, when virally transformed, become immortalized and will continue to proliferate indefinitely.

Some transformed cells have the ability to form tumors when injected into an immunocompromised host such as a newborn mouse or a nude mouse. Newborn mice lack a developed immune system and nude mice are a strain of mice that are athymic. The absence of a thymus in nude mice causes them to be defective in generating an appropriate immune response. When transformed cells are introduced into an immunologically competent host, these cells are recognized as foreign and eliminated. Introduction of transformed cells into an immunocompromised host does not induce an immune response; thus tumors can arise.

The first transforming virus discovered was Rous sarcoma virus (RSV). RSV was shown to cause tumors in chickens and in newborn animals. The cause of the tumors was initially unclear since the tumor tissue showed no new viral particles being produced. Analysis of the genome of the virus determined that a single gene, called *v-src,* was responsible for the virus's ability to transform cells in culture and to cause tumors *in vivo.* Further analysis showed that this gene was not unique to the virus but was present in the DNA genome of uninfected chickens and was normally expressed by the host without causing tumorigenesis. This was the first indication that the oncogenic effect of a transforming virus might be the result of an alteration in a normal cellular gene. Since the discovery of RSV, numerous transforming viruses have been characterized and the genes that are responsible for the virus's transforming ability have been identified. These genes are called oncogenes, whereas their normal cellular counterparts are called proto-oncogenes. The function of proto-oncogenes in the host appears to be in the regulation of cell growth and differentiation. Evidence indicates that these cellular proto-oncogenes or c-oncogenes were transduced from the host DNA into the viral genome. As part of the viral genome, proto-oncogenes became oncogenes through mutation events that alter protein activity or via misexpression by virtue of coming under transcriptional control of a viral promoter. Oncogenes have been found to encode for proteins that are involved in various steps in cellular growth or the differentiation cascade (summarized in Fig. 1), ranging from soluble growth factors to nuclear transcription factor. Examples of each will be described. [*See* Viruses and Cancer.]

II. GROWTH FACTORS AS ONCOGENES

An example of a secreted cell signaling molecule that can be activated to become an oncogene is platelet-derived growth factor (PDGF). The oncogenic potential of PDGF was first noted when it was found to be encoded in the genome of simian sarcoma virus (SSV). SSV was originally isolated from the fibrosarcomas of wooly monkeys. Purified virus was able to induce fibrosarcomas in uninfected monkeys as well as transform fibroblasts in tissue culture. This transforming activity was mapped to an oncogene, called *v-sis,* which encodes a 28-kDa protein. Sequence comparison showed that *v-sis* was identical to a previously identified cellular gene called PDGF.

The function of normal cellular PDGF is thought to be the promotion of wound healing through stimulation of cell growth. Animal sera derived from unclotted blood does not support the growth of cells when compared to sera derived from clotted blood. PDGF is the mitogen present in clotted blood that is responsible for the stimulation of cell growth. Purified PDGF consists of a dimer of two polypeptide chains, A and B, which are encoded by two different genes. Both A and B forms of the protein possess biological activity. *In vivo,* PDGF is stored in the granules of blood platelets. Following injury, clotting and platelet adhesion occur. PDGF is released by platelets and stimulates the migration of smooth muscle cells to the site of injury where the cells proliferate. In tissue culture, PDGF has been shown to stimulate the growth of cells of mesenchymal origin such as smooth muscle cells, glial cells, and fibroblasts. [*See* Growth Factors and Tissue Repair.]

The virally encoded *v-sis* oncogene is the B form of PDGF and differs in two ways from the cellular version of PDGF B. First, *v-sis* is a fusion protein; the coding region of PDGF has replaced a large portion of the viral envelope gene (*env*) in the SSV genome. The result is a fusion protein that contains the amino terminus of *env* and the carboxy region of PDGF. The *env* amino terminus contains a signal sequence that directs the fusion protein to the cell surface, where it is secreted. The second difference is that the fusion *v-sis* gene is controlled by a strong viral promoter termed a long terminal repeat (LTR). The LTR directs strong constitutive expression of the fusion protein. Evidence indicates that it is this second difference in the viral oncogene, deregulated expression, that is the critical feature by which *v-sis* transforms cells. When the *c-sis* proto-oncogene is placed downstream of a viral LTR and expressed in fibroblasts, this construct is able to transform cells in culture.

Although *v-sis* can transform fibroblasts, it cannot transform epithelial cells. The reason for this is that epithelial cells do not possess receptors for PDGF. Stimulation of cell growth by PDGF requires that PDGF bind to its cellular receptor. A cell that constitutively expresses both PDGF and the PDGF receptor can continuously stimulate its own proliferation, resulting in transformation. In summary, *v-sis* functions as an oncogene by its overexpression of a fusion protein containing the gene for PDGF. This transforming ability, however, requires that the cells that express *v-sis* also express PDGF receptor.

III. GROWTH FACTOR RECEPTORS AS ONCOGENES

An example of a cell-surface growth factor receptor that can be mutated to become an oncogene is the

EGF receptor. Epidermal growth factor (EGF) is a mitogen that has the ability to stimulate the proliferation of cells of ectodermal and mesodermal origin. The cellular receptor that binds EGF (EGFR) and mediates this proliferative signal was purified and found to be the human proto-oncogene form of the viral oncogene *erbB*. *erbB* is the transforming gene of the avian erythroblast virus (AEV) and is responsible for AEV's ability to cause erythroleukemia in chickens and to transform erythrocytes and fibroblasts in tissue culture.

The wild-type cellular EGF receptor is an 1186-amino-acid transmembrane glycoprotein and consists of three domains: (1) a 621-residue amino terminus portion that is glycosylated, located outside the cell, and is responsible for binding EGF; (2) a small 23-amino-acid transmembrane portion; and (3) a 542-residue carboxy terminal portion that is located within the cell. The carboxy-terminal cytoplasmic domain contains tyrosine kinase activity, that is, the EGF receptor can catalyze the transfer of the terminal phosphate group from ATP to a tyrosine residue of a substrate protein. Upon binding of EGF, dimerization of EGF receptors occurs and the cytoplasmic tyrosine kinase activity of EGFR is activated. The result is autophosphorylation, that is, EGFR phosphorylates itself on tyrosine residues, and transphosphorylation of other cellular target proteins. Following EGF binding, the EGF–EGFR complex undergoes receptor-mediated endocytosis and is degraded while the phosphorylation of the other substrate proteins mediates the mitogenic signal.

The oncogene *v-erbB* of AEV differs from the protooncogenic EGFR in that it lacks most of the amino terminus EGF-binding domain; *v-erb* cannot bind EGF. Also, 30 carboxy-terminal residues are deleted and there are 8 point mutations in the *v-erb* oncogene. The carboxy-terminal deletion in *v-erbB* deletes one of the tyrosine autophosphorylation sites of EGFR. Although *v-erbB* has lost its ability to bind EGF, the oncogene is in a constitutively activated state. In the absence of EGF, *v-erbB* continuously signals the cell to proliferate, leading to a cellular transformation.

IV. PROTEIN KINASES AS ONCOGENES

Information from extracellular factors regulating cell growth and differentiation cascades is communicated from the cell surface to nuclear proteins via secondary messengers. One pathway by which signals are transduced is through the phosphorylation of proteins. An example of an oncogene that functions as a protein kinase is *v-src*. *v-src,* the first oncogene discovered, was identified as part of the genome of Rous sarcoma virus. RSV has been shown to transform cells in tissue culture and to induce tumors in animals.

The cellular homologue of *v-src,* called *c-src,* is a 60-kDa phosphoprotein (designated pp60). The *c-src* protein is myristylated at its amino terminus and is located on the inner surface of the plasma membrane. Like the EGF receptor, pp60 is a tyrosine kinase. A number of substrate proteins that can be phosphorylated by pp60 *in vivo* has been identified (e.g., vinculin, p36, enolase, phosphoglycerate mutase, lactate dehydrogenase). However, the significance of the phosphorylation of these substrates and the role they play in normal cell growth and transformation are unclear. The ability of pp60 to act as a tyrosine kinase is regulated by the phosphorylation state of pp60 itself. A tyrosine residue is located at amino acid 527 of pp60 (Tyr 527). When this amino acid is phosphorylated, pp60, adopts a "closed conformation" and is unable to function efficiently as a kinase. Dephosphorylation of this residue results in an "open conformation" whereby pp60 can phosphorylate its substrates. In addition to Tyr 527, phosphorylation of another tyrosine residue at position 416 also appears to play a role in regulating pp60 kinase activity. [*See* Protein Phosphorylation.]

The most notable difference between virally encoded *v-src* and cellular *c-src* is that *v-src* is 526 amino acids in length whereas *c-src* is 533 amino acids in length. Therefore the oncogene *v-src* does not contain Tyr 527, and its kinase activity cannot be regulated via phosphorylation of this site. The result is that *v-src* has a constitutively active kinase activity. Confirmation that this mutation is responsible for the transforming activity of *v-src* was shown by mutagenesis of the proto-oncogene *c-src*. Tyr 527 of *c-src* was changed to a phenylalanine residue that is unable to be phosphorylated. This change resulted in at least a 10-fold increase in the protein's kinase activity and was sufficient to convert the proto-oncogene to an oncogene. In summary, phosphorylation of substrates by *c-src* leads to cellular proliferation. When *src* kinase activity is constitutively activated, as is seen with *v-src,* cellular transformation occurs.

V. GTP BINDING PROTEINS AS ONCOGENES

Another pathway by which cell-surface signals are translated to nuclear proteins is through guanine nucleotide binding proteins. These proteins appear to play a central role in linking cell-surface receptors through a protein kinase cascade to nuclear proteins that alter gene expression and change cell morphology. An example of a GTP binding protein that can be mutated to an oncogene is H-*ras-1*. Although H-*ras-1* is encoded by a transforming virus, the oncogene was also identified independently from a spontaneously occurring human bladder carcinoma. A cell line termed EJ cells was derived from a human bladder carcinoma. To identify the transforming activity in the EJ cell line, DNA prepared from these cells was transfected into mouse cells in culture. Some of the mouse cells became transformed, as demonstrated by the formation of foci. DNA was then isolated from transfected cells that became transformed and the human gene responsible for the transforming activity was characterized. The gene turned out to be H-*ras-1,* the cellular homologue of the oncogene identified in Harvey sarcoma virus.

H-*ras-1* has a normal cellular counterpart, *c-ras*. The proto-oncogene *c-ras* is a 21-kDa protein (p21) that localizes to the plasma membrane. In addition to binding guanine nucleotides, p21 also has been found to be a GTPase, that is, p21 can hydrolyze GTP to GDP. Although regulation of *c-ras* is complex and involves numerous proteins and cofactors, in general, p21 is activated through G-coupled protein receptors. Activated p21 contains bound GTP. Activated p21 in the "on" state can modulate the function of other proteins and kinases to regulate nuclear proteins and gene transcription. In response to stimuli, p21 hydrolyzes GTP to GDP. p21 containing bound GDP is in the "off" or inactive state.

The oncogene H-*ras-1* differs from the proto-oncogene *c-ras* by a single-nucleotide change. This single-nucleotide change results in the substitution of the amino acid glycine at position 12 of *c-ras* to valine in H-*ras-1*. This mutation does not affect the GTP binding ability of H-*ras-1;* however, it drastically inhibits the protein's GTPase activity. The result is that H-*ras-1* protein cannot hydrolyze GTP and thus is constitutively active. Interestingly, a large family of *ras*-related GTPase proteins have been identified. In addition, activated *ras* oncogenes have been identified in 10–20% of all human tumors. Sequence analyses of these genes show that mutation of amino acid 12 is a common mechanism by which a *ras* protein is converted to an oncogene. Mutagenesis of amino acid 12 to any other amino acid except proline is sufficient to convert the proto-oncogene to an oncogene.

VI. NUCLEAR PROTEINS AS ONCOGENES

An example of a nuclear protein that can be altered to become an oncogene is *c-myc*. The *myc* gene was first identified as part of the genome of a number of avian transforming viruses. These viruses are able to cause a variety of cancerous phenotypes, including sarcomas and myelocytomas. In tissue culture, these viruses can readily transform fibroblasts, epithelial cells, and myeloid cells. The transforming ability of these viruses is due to the presence of a gene called *v-myc*. The *v-myc* gene, as encoded in the different avian viral genomes, shows a variety of mutations as compared to its cellular proto-oncogene homologue. However, the critical route by which *v-myc* transforms cells is by overexpression of *myc* protein following viral infection. In transforming viruses, *v-myc* is overexpressed as a result of transcriptional control through a strong viral promoter.

The wild-type version of *myc, c-myc,* is a phosphoprotein and is expressed exclusively in the nucleus. Two forms of the polypeptide, with molecular masses of 64 and 67 kDa, have been identified in cells. Evidence indicates that *c-myc* plays a role in cellular proliferation. Quiescent cells show a low level of *c-myc* mRNA. When these cells are stimulated by mitogens, *c-myc* expression is rapidly induced. Conversely, when cells cease proliferating and terminally differentiate, *c-myc* levels decrease. Increased *c-myc* levels are constant in proliferating cells and do not fluctuate throughout the cell cycle. Experiments have shown that artificially increasing *c-myc* levels in F-MEL erythroid precursor cells can inhibit their differentiation in response to chemical factors (e.g., DMSO, hypoxanthine). Also, artificially decreasing *c-myc* levels in proliferating myeloid HL-60 cells results in a decrease in proliferation and the induction of terminal differentiation. In normal cells, *c-myc* functions as a transcription factor. The *c-myc* protein contains basic helix–loop–helix leucine zipper motifs typical of many transcription factors. Although on its own *c-myc* appears not to bind DNA, *c-myc* can heterodimerize with several different proteins. These heterodimers have the ability to bind DNA and, depending on which protein partner *c-myc* is associated with, the heterodimers can either activate or repress the transcription of target genes.

Besides overexpression of *myc* via its integration into a viral genome, it has been reported that proto-oncogene *c-myc* can be activated in three other ways to become an oncogene: (1) retroviral insertion, (2) translocation, and (3) gene amplification. For retroviral insertion, some retroviruses, such as avian leukosis virus (ALV), have no significant transforming ability but, in rare cases, may cause tumors *in vivo*. Retroviruses insert into the host cell genome and, in the case of the majority of bursal lymphomas caused by ALV, the viral genome has been found integrated near the *c-myc* gene. This proviral insertion of a strong viral promoter near the *c-myc* gene leads to deregulated overexpression of *c-myc* and may result in tumorigenesis.

An example of *c-myc* activation by DNA translocation is seen in Burkitt's lymphomas, which are B-cell tumors. One feature of these tumors is that of reciprocal chromosomal translocations in which DNA from one chromosome is joined to that of another. In the majority of these tumors, this translocation results in the juxtaposition of immunoglobulin genes and the *c-myc* gene such that the expression of the latter is increased. (The mechanism by which *c-myc* is up-regulated appears to be dependent on the type of translocation that has taken place.)

Finally, elevated levels of *c-myc* have been found in numerous tumor types as a result of amplification of the *c-myc* gene. Experiments suggest that this amplification of gene copy number and subsequent overexpression of *c-myc* is not the original cause of these tumors, but rather is a secondary effect associated with tumor progression. In malignancy, the cellular genome becomes destabilized and gene amplification may result. Cells in which the *c-myc* gene is amplified demonstrate enhanced proliferative ability. Interestingly, *c-myc* has been shown to be a member of a gene family consisting of the two related genes L-*myc* and N-*myc*. Although *c-myc* is the only family member shown to be part of a viral genome, both L-*myc* and N-*myc* have also been found to be amplified in certain tumors. [*See* Oncogene Amplification in Human Cancer.]

VII. CHROMOSOMAL TRANSLOCATIONS IN THE GENERATION OF NOVEL ONCOGENES

As in the case of *c-myc,* chromosomal translocations may introduce a strong promoter upstream of a proto-oncogene and result in increased gene expression and cellular transformation. Another method by which chromosomal breakage may cause transformation is by the fusion of two protein-coding regions to generate a novel protein with oncogenic activity. An example of this phenomenon is the PML–RARα fusion protein. In 90% of patients with acute promyelocytic leukemia (APL), a reciprocal chromosomal translocation has taken place between chromosomes 15 and 17. The site of this reciprocal translocation occurs in the retinoic acid α receptor (RARα) gene on chromosome 17 and the PML gene on chromosome 15. The result of this reciprocal translocation is to generate two novel chimeric fusion proteins—RAR–PML, which contains the amino terminus of RARα and the carboxy terminus of PML, and PML–RAR, which contains the amino terminus of PML and the carboxy terminus of RARα.

RARα normally functions as a transcription factor that is localized in the nucleus. RARα binds its ligand, all-trans retinoic acid (RA). When RA is bound to the receptor, RARα can activate target genes. RARα is a DNA binding protein that contains zinc finger motifs that mediate its binding to DNA elements in the promoters of its target genes. RARα also forms heterodimers with the protein RXR. The normal function of PML protein is unclear. However, PML contains a number of motifs that are homologous to proteins that are (1) involved in the repair of ultraviolet-damaged DNA, (2) involved in protein dimerization, (3) involved in binding RNA, and (4) are transcription factors. PML is also expressed in the nuclei of cells but, unlike RARα, its expression pattern appears as discrete punctate dots, typical of proteins known to be associated with the nuclear matrix.

The form of the fusion protein thought to be responsible for acute promyelocytic leukemia is the PML–RAR form. This chimeric protein contains most of the wild-type PML protein fused to the DNA-binding and ligand (RA)-binding portion of RARα. It is unclear how this fusion protein is oncogenic, however, evidence suggests that the PML–RAR fusion protein blocks normal promyelocytic differentiation by interfering with either PML- or RA-mediated transcriptional pathways. One possibility is that the chimeric protein has altered transactivation activity as compared to either RARα or PML. In one experiment, the chimeric PML–RAR fusion was able to activate a promoter containing an RAR response element (RARE) to a much higher level than RARα alone. Another possibility is that the fusion protein retains its ability to dimerize with other proteins, thereby acting in a transdominant fashion to alter the transcriptional activity of these other proteins. Retinoic acid can effectively treat APL by presumably inducing differentiation and inhibiting the growth of the pro-

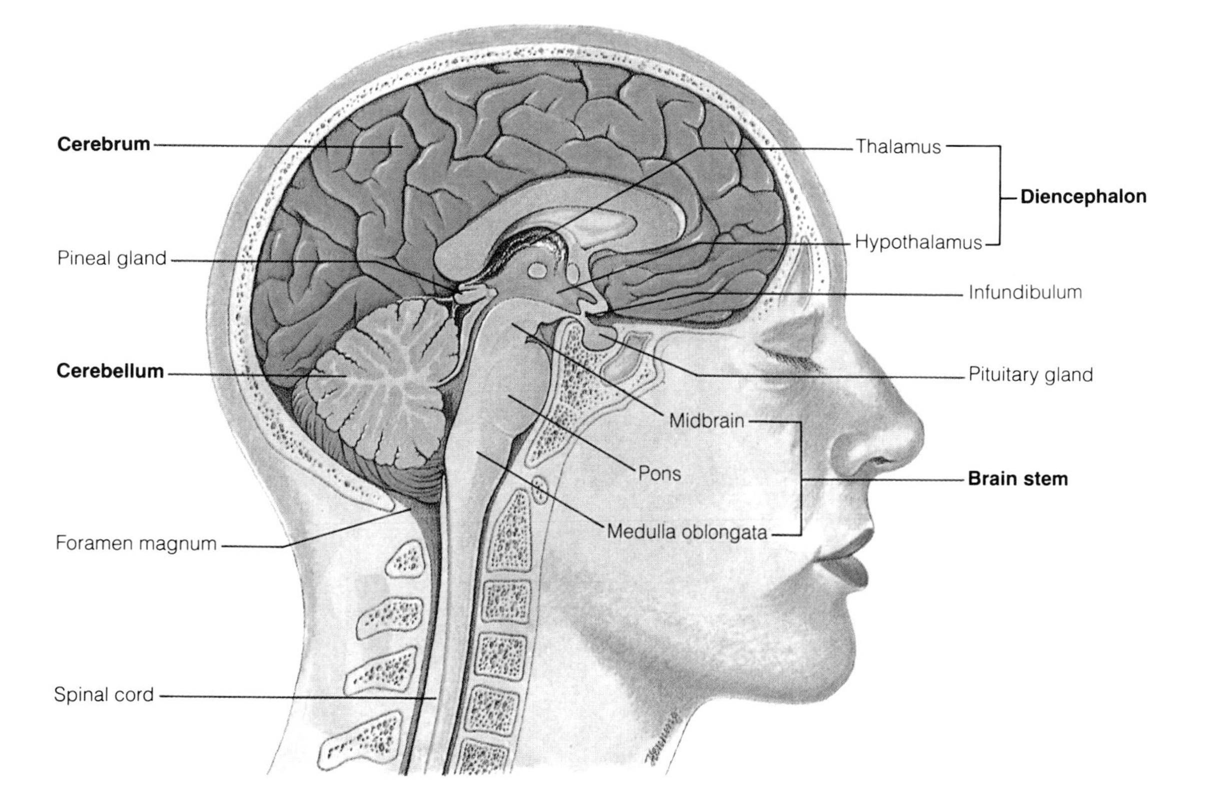

COLOR PLATE 1 Longitudinal section of the brain and the spinal cord. [Source: Gaudin, A. J., and Jones, K. C. (1989). "Human Anatomy and Physiology." Harcourt Brace Jovanovich, San Diego, p. 291. Reproduced with permission.] [*See* Nervous System, Anatomy.]

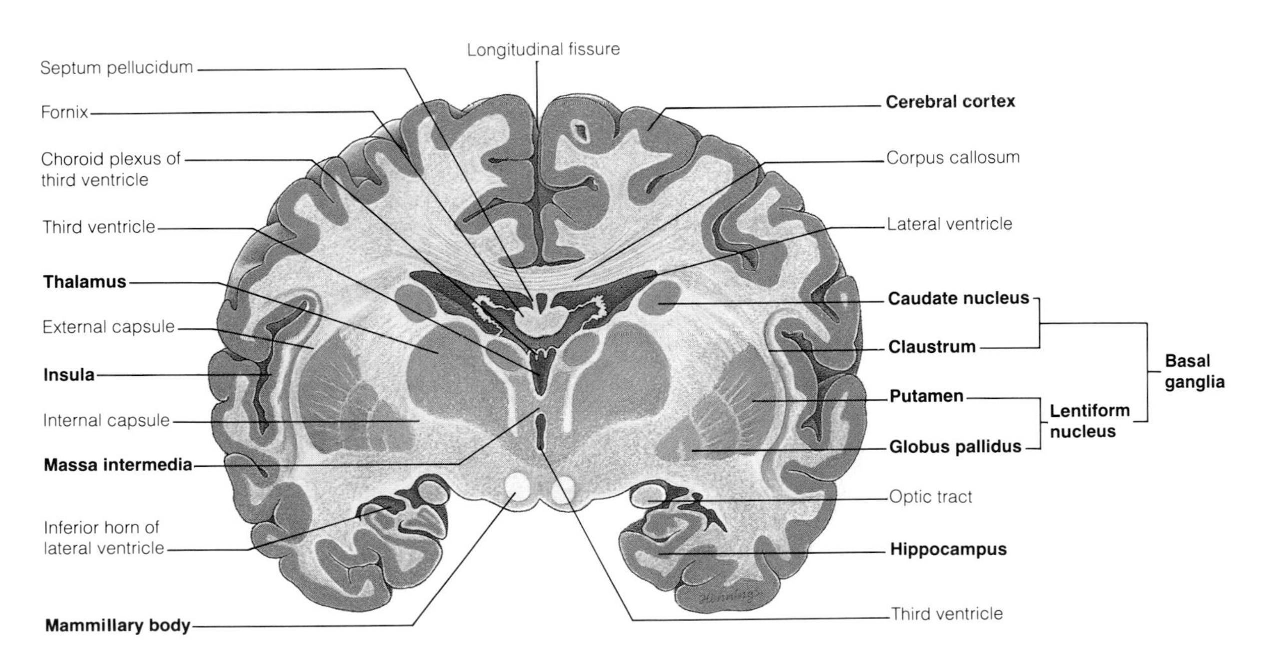

COLOR PLATE 2 Frontal section through the cerebral hemispheres of the brain. [Source: Gaudin, A. J., and Jones, K. C. (1989). "Human Anatomy and Physiology." Harcourt Brace Jovanovich, San Diego, p. 297. Reproduced with permission.] [*See* Nervous System, Anatomy.]

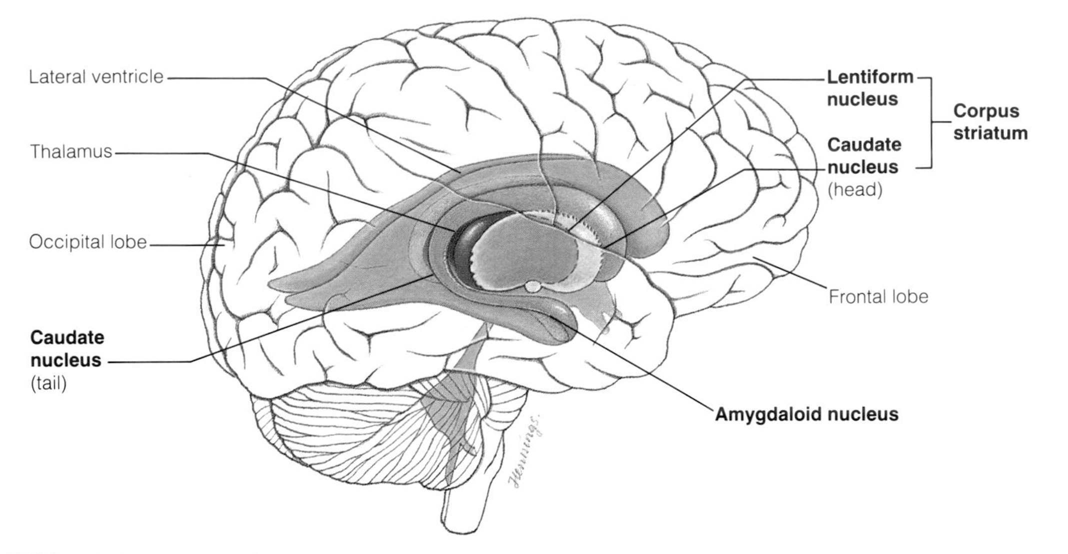

COLOR PLATE 3 The basal ganglia of the brain. [Source: Gaudin, A. J., and Jones, K. C. (1989). "Human Anatomy and Physiology." Harcourt Brace Jovanovich, San Diego, p. 299. Reproduced with permission.] [*See* Nervous System, Anatomy.]

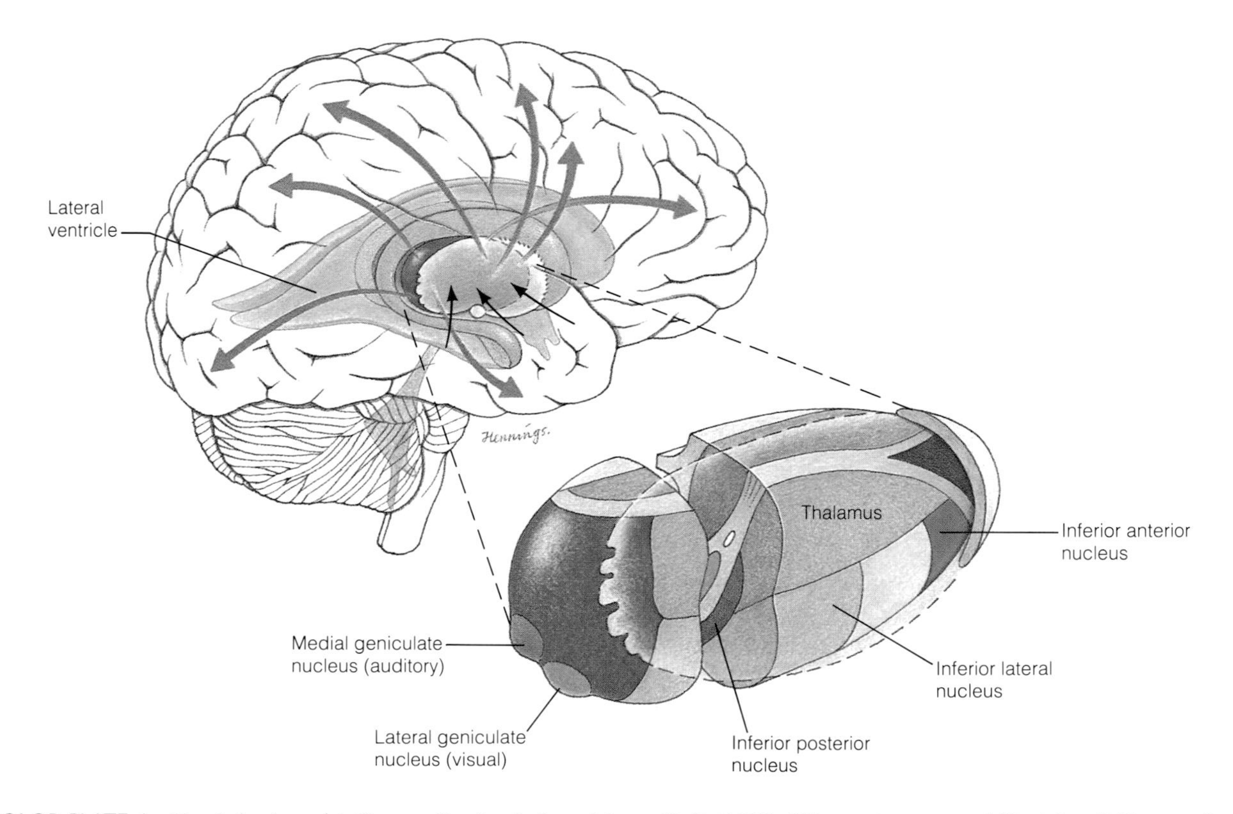

COLOR PLATE 4 The thalamic nuclei. [Source: Gaudin, A. J., and Jones, K. C. (1989). "Human Anatomy and Physiology." Harcourt Brace Jovanovich, San Diego, p. 300. Reproduced with permission.] [*See* Nervous System, Anatomy.

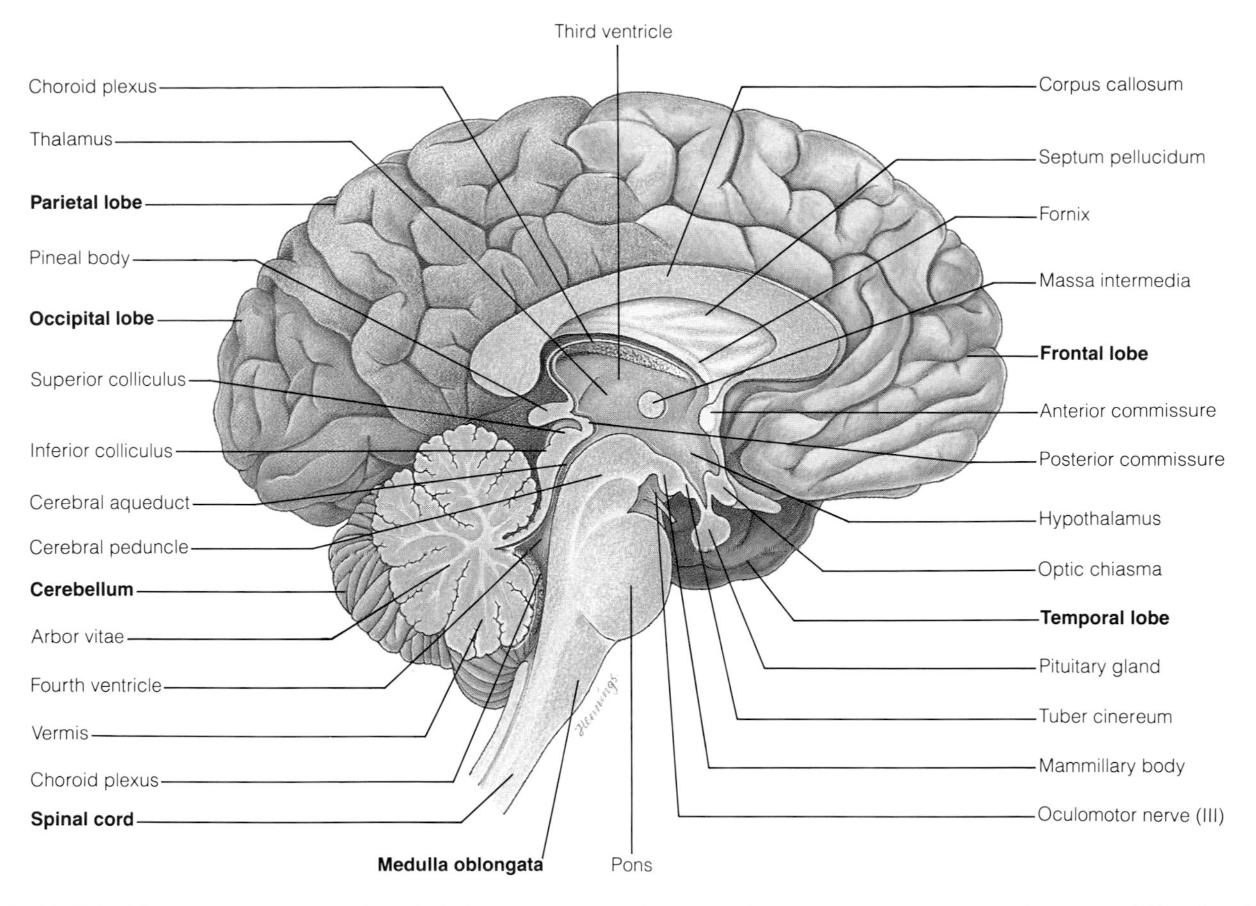

COLOR PLATE 5 Longitudinal section through the brain. [Source: Gaudin, A. J., and Jones, K. C. (1989). "Human Anatomy and Physiology." Harcourt Brace Jovanovich, San Diego, p. 302. Reproduced with permission.] [*See* Nervous System, Anatomy.]

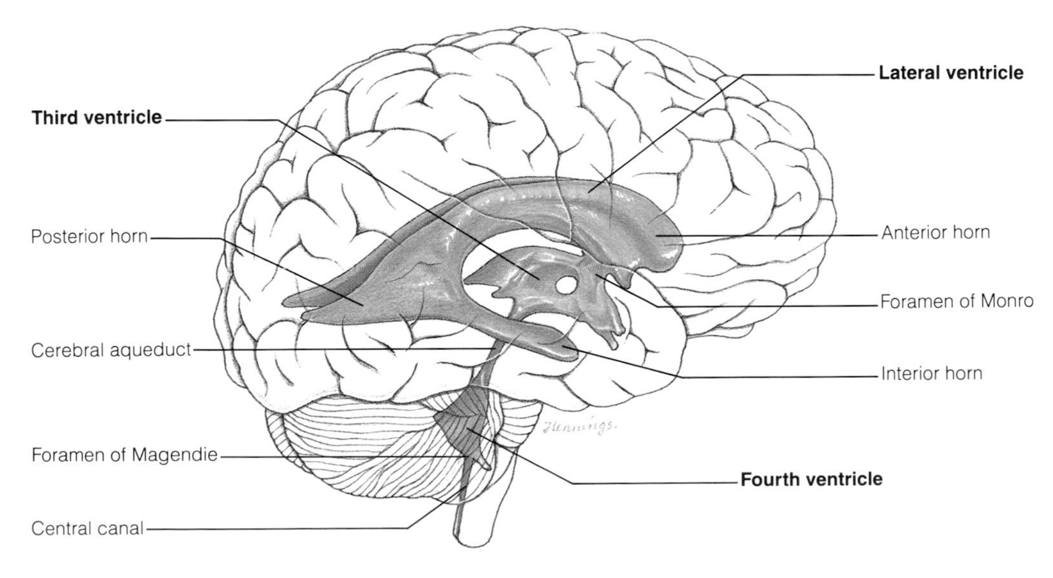

COLOR PLATE 6 Ventricles of the brain. [Source: Gaudin, A. J., and Jones, K. C. (1989). "Human Anatomy and Physiology." Harcourt Brace Jovanovich, San Diego, p. 309. Reprinted with permission.] [*See* Nervous System, Anatomy.]

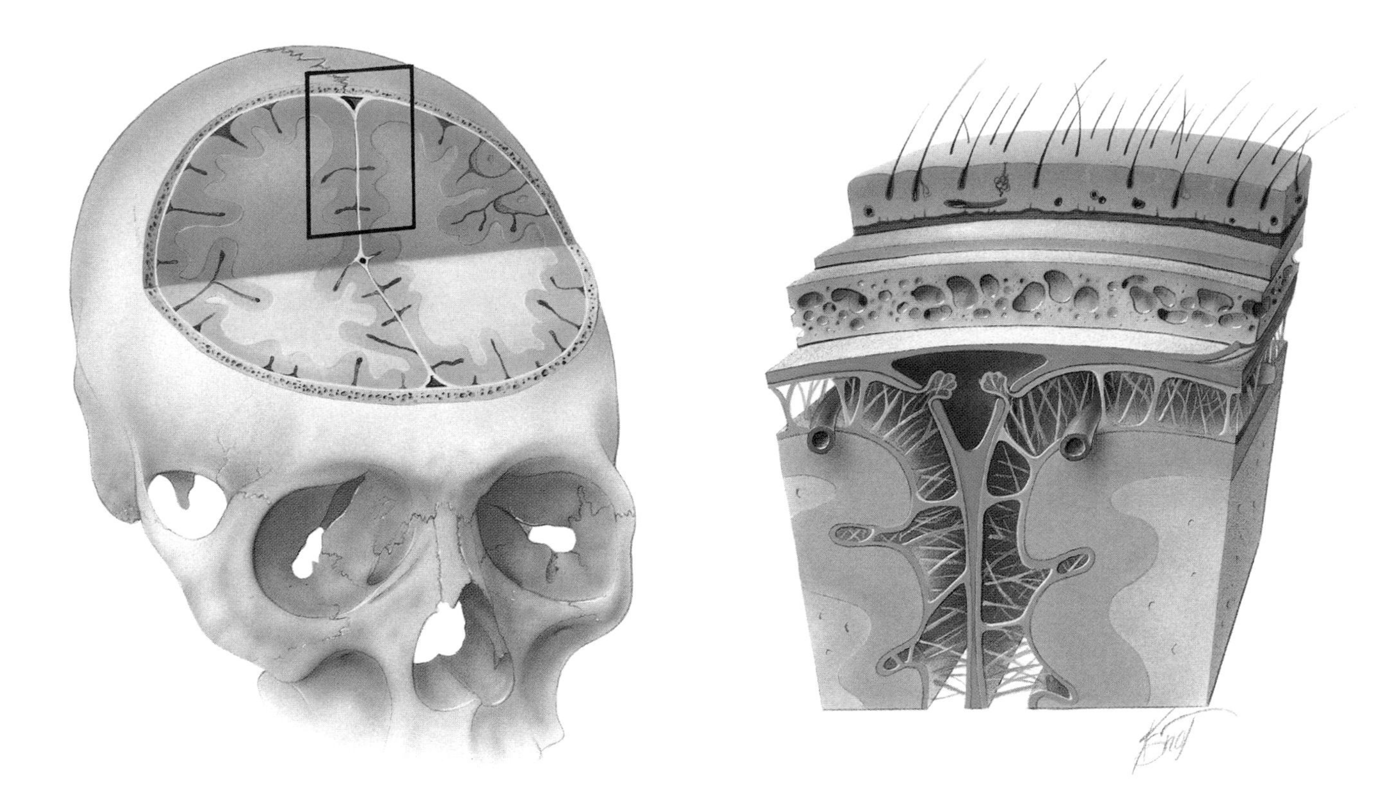

COLOR PLATE 7 Meninges of the brain. [Source: Gaudin, A. J., and Jones, K. C. (1989). "Human Anatomy and Physiology." Harcourt Brace Jovanovich, San Diego, p. 310. Reprinted with permission.] [*See* Nervous System, Anatomy.]

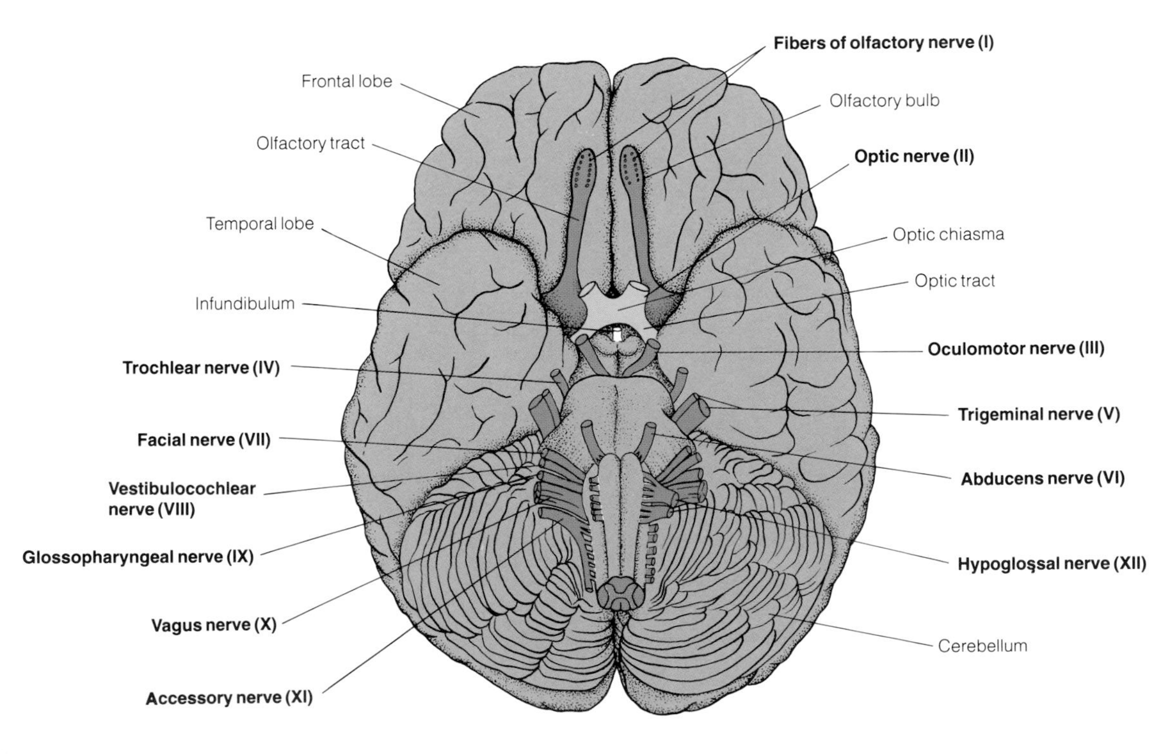

COLOR PLATE 8 The cranial nerves. [Source: Gaudin, A. J., and Jones, K. C. (1989). "Human Anatomy and Physiology." Harcourt Brace Jovanovich, San Diego, p. 323. Reprinted with permission.] [*See* Nervous System, Anatomy.]

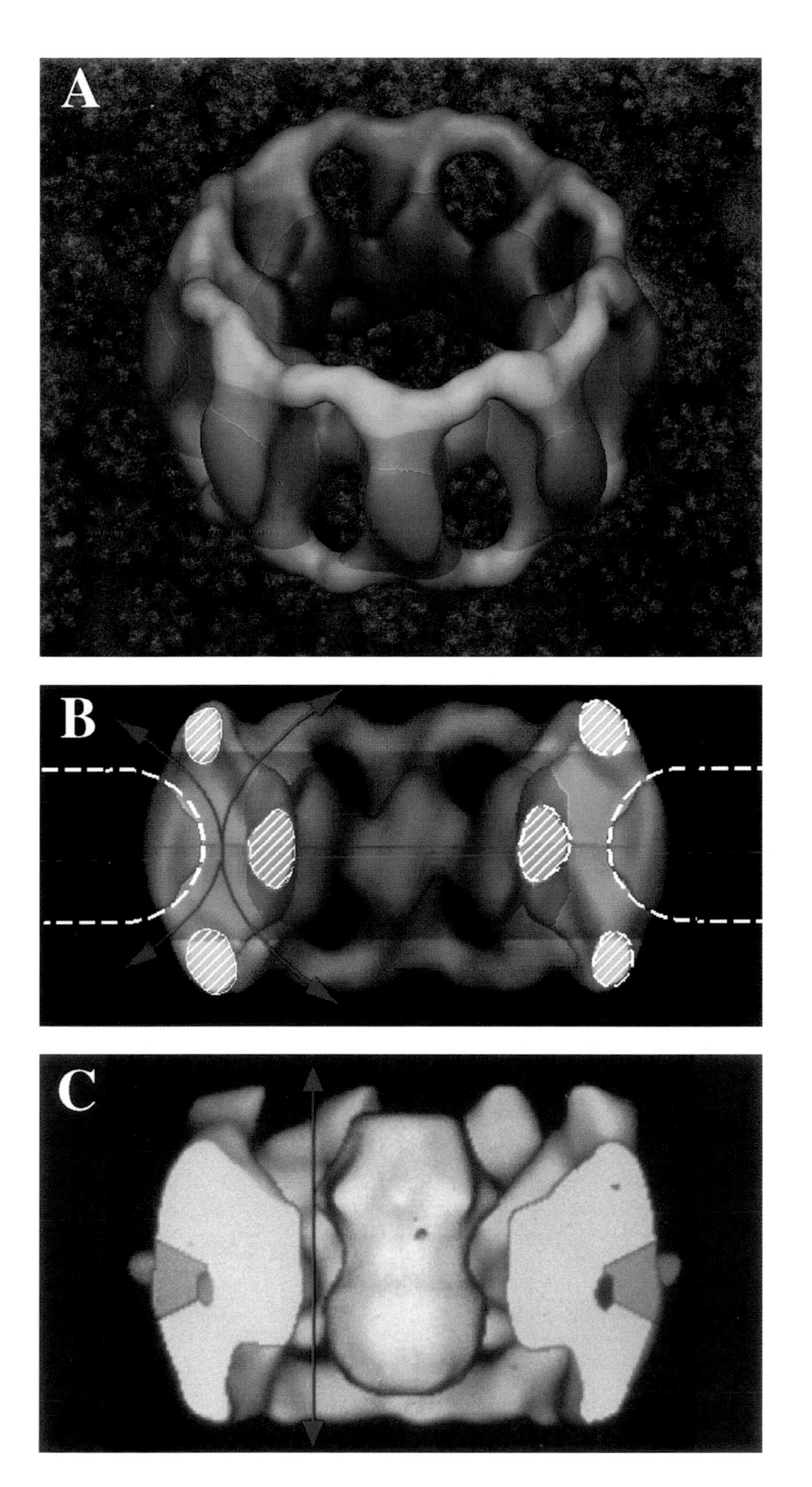

COLOR PLATE 9 Three-dimensional maps of the nuclear pore complex calculated by the random conical tilt procedure from samples negatively stained (A and B) or frozen in vitreous ice (C). (A) Tilted view of the pore complex, rendered to fit its molecular weight. In the background are negatively stained pore complexes that were obtained from *Xenopus* oocytes and used to calculate the three-dimensional map (see Hinshaw *et al.*, 1992). (B) Cut-away side view of the pore complex. The dotted line represents the predicted positioning of the nuclear membrane and the double-headed arrows indicate peripheral channels that may provide an alternative route for passive diffusion. Four major domains are color coded: yellow = ring, tan = column, green = annular, and blue = lumenal. (C) Three-dimensional map of the pore complex with the central plug or transporter (see Akey and Radermacher, 1993). [*See* Nuclear Pore, Structure and Function.]

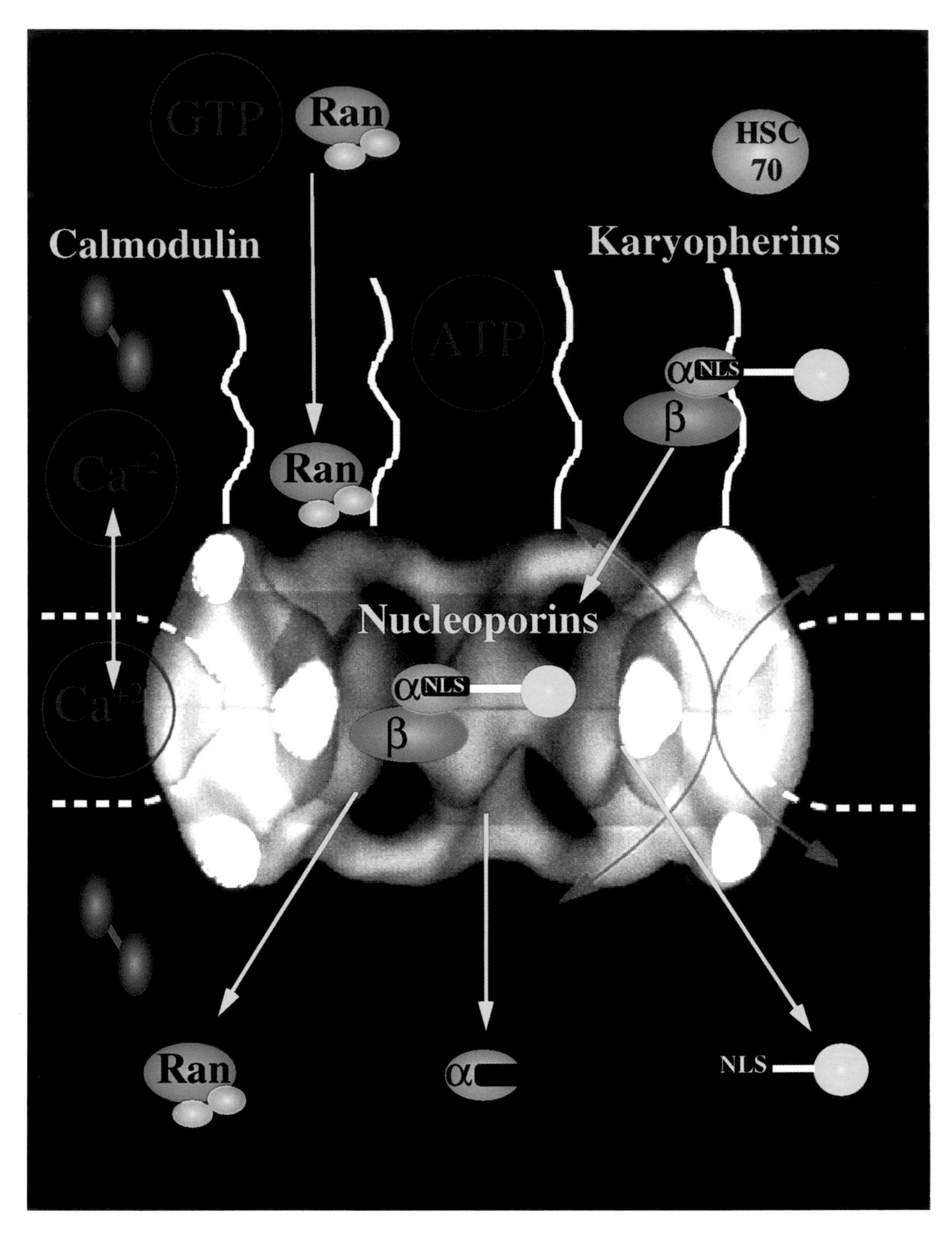

COLOR PLATE 10 The mechanism of nuclear protein import. The molecules thought to be involved in nuclear protein transport are indicated, as is their association with the nuclear pore. GTP acting through Ran/TC4 is involved in activation of nuclear transport. A similar role is envisioned for Ca^{2+}/calmodulin in the cytoplasm. Ca^{2+} is also required in the lumen of the nuclear membrane/endoplasmic reticulum. The transport process also requires ATP. Other protein components involved are the karyopherins involved in nuclear localization sequence binding and the nucleoporins. One potential model for nuclear protein import involves the generation of a series of successive association–dissociation reactions involving the karyopherins and the repetitive regions of the nucleoporins (GFSFG and GLFG motifs). This "thermal ratchet" mechanism would drive the inward movement of molecules after they interact with the filaments on the outer surface of the nuclear pore. [*See* Nuclear Pore, Structure and Function.]

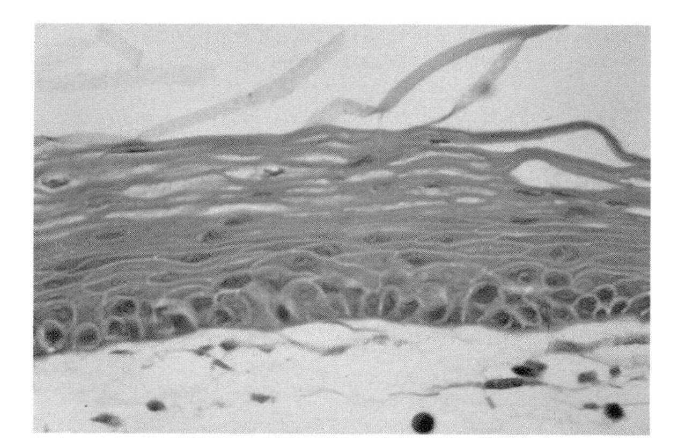

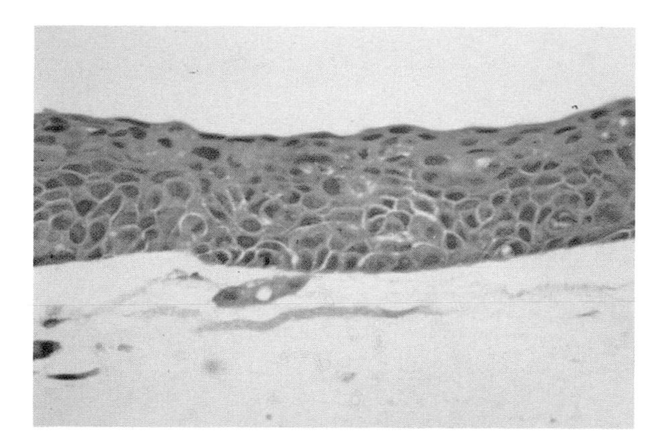

COLOR PLATE 11 Effect of HPV-16 DNA on *in vitro* differentiation of primary human keratinocytes as assayed in the organotypic culture system. *Top panel*, A control that received no HPV DNA and shows a normal pattern of keratinocyte differentiation; *Bottom panel*, The effect of the HPV-16 E6 and E7 genes on keratinocyte differentiation. Dividing cells are observed through the full thickness of the epithelium, and the gradient of differentiation is totally disrupted. Severe cases of cervical dysplasia show similar histological features. [*See* Papillomaviruses and Neoplastic Transformation.]

myelocytic cells. A proposed model suggests that the PML–RAR fusion may dimerize with normal PML. RA treatment would result in RA binding to the PML–RAR fusion protein and disruption of the dimer complex. PML could then function normally and allow cells to differentiate. Although the mechanism of transformation is as yet unclear, PML–RAR is an example of a novel oncogene generated by the fusion of two nononcogenic proteins.

VIII. CONCLUSION

In summary, a gene that encodes a protein that is involved in regulating cell growth has the potential to become oncogenic and to transform cells. Proteins involved in any step of a growth cascade, from extracellular signaling molecules (*v-sis*) to cell-surface receptors (*erbB*) to secondary messengers (*v-src*, H-*ras-1*) to transcription factors (*myc*), have this potential. Alterations that result in oncogenic transformation range from single point mutations that alter a protein's enzymatic activity to deregulated overexpression. In addition, gene fusions may also occur that result in the expression of a novel gene with oncogenic activity (PML–RAR).

BIBLIOGRAPHY

Hekseth, R. (1994). "The Oncogene Handbook." Academic Press, San Diego.

Hesketh, R. (1995). "The Oncogene Facts Book." Academic Press, San Diego.

Macara, I. G., Lounsbury, K. M., Richards, S. A., McKiernan, C., and Bsr-Sagi, D. (1995). The Ras superfamily of GTPases. *FASEB J.* **10,** 625.

Ryan, K. M., and Birnie, G. D. (1996). *Myc* oncogenes: The enigmatic family. *Biochem. J.* **314,** 713.

Sporn, M. B., Roberts, A. B., and Goodman, D. S. (eds.) (1994). "The Retinoids: Biology, Chemistry, and Medicine," 2nd Ed. Raven, New York.

Superti-Furga, G. (1995). Regulation of the Src protein tyrosine kinase. *FEBS Lett.* **369,** 62.

Watson, J. D., Hopkins, N. H., Roberts, J. W., Steitz, J. A., and Weiner, A. M. (eds.) (1987). "Molecular Biology of the Gene," 4th Ed. Benjamin/Cummings, Menlo Park, California.

Oogenesis

TONI G. PARMER
GEULA GIBORI
University of Illinois at Chicago

DAVID F. ALBERTINI
Tufts University School of Medicine

GLOSSARY

Dictyate stage Diplotene stage of the first meiotic division in which the meiotic process of the oocyte has been arrested

Germinal vesicle Nucleus of the primary oocyte

Germinal vesicle breakdown Breakdown or dissolution of the nucleus of the preovulatory dictyate oocyte, initiated when the oocyte resumes meiosis

Meiotic competence Ability of the primary oocyte to resume meiosis *in vitro*

Maturation promoting factor Protein phosphorylating enzyme activity that is required for the reinitiation and completion of meiosis

Oogonia Primordial germ cells that have entered the female gonad and continue their mitotic proliferation

Primary oocyte Oogonia that have entered into the first meiotic division

Primordial germ cells Undifferentiated, mitotically dividing stem cells found in both sexes, which are destined to give rise to the gametes

Secondary oocyte Oocyte that has completed the first meiotic division and contains the haploid number of chromosomes

OOGENESIS HAS BEEN DEFINED AS "THE ENTIRE period of female germ cell differentiation, from the oogonial stage to the mature gamete." The following discussion is mainly concerned with mammalian oogenesis with particular reference to humans. However, when necessary, reference to studies involving lower vertebrates will be made, as knowledge of the processes in mammals is incomplete.

I. GAMETOGENESIS: PRIMORDIAL GERM CELLS AND THE FORMATION OF PRIMARY AND SECONDARY OOCYTES

In vertebrate embryos, cells destined to give rise to the gametes are known as the primordial germ cells. In the mouse embryo, the primordial germ cells are first seen at the posterior end of the primitive streak about 8 days postcoitium. From this location, the cells migrate to the allantois, to the hindgut epithelium, up the dorsal mesentary, and, by 10–11 days postcoitum, they enter the germinal ridges (from which the gonads arise). The migration of these cells is believed to occur by amoeboid movement using lytic enzymes to penetrate cell membranes.

At the time that they can first be recognized, the primordial germ cells number approximately 40–50 in the human embryo. During migration to the germinal ridges, they undergo numerous mitotic divisions and increase to 600–1300 cells. Once they reach the germinal ridges, they become embedded in the epithelium, where they are known as oogonia. The oogonia continue to proliferate at a high rate markedly increasing in number. In the human, this stage of rapid proliferation begins in the fifth or sixth week of embryonic development. From the 10th week on, an increasing proportion of these oogonia enter meiotic prophase while the remainder continue to proliferate by mitosis.

The epithelium of the germinal ridge forms the surface of the gonad and is known as the cortex. The primitive sex cords, strands of compactly arranged cells in the interior of the gonad, form the medulla. In the female embryo, the medulla becomes reduced, the primary sex cords are resorbed, and the interior of the gonad becomes filled with loose mesenchyme permeated by blood vessels. The oogonia remain embedded in the cortex, close together in groups. They are connected by intercellular bridges, which are thought to result from incomplete separation of daughter cells during mitotic divisions. These bridges allow for the free exchange of cytoplasmic material between the connected cells and probably play a role in the synchronization of germ-cell differentiation. During the growth of the gonad, the cortical cells on the inner surface of the cortex split up into groups; strands of these cells then surround one or several of the oogonia that have entered meiosis. These oogonia are now called primary oocytes. The strands surrounding them become a single layer of flattened cells called follicular cells. The whole structure is called a primordial follicle. Oocytes at this stage of development can be seen at 4 months of gestation but usually remain few in number until about the fifth or sixth month.

By the fifth month of intrauterine life, the oogonia reach a maximum number of approximately 7 million and consist of both "resting" and dividing oogonia together with primary oocytes at all stages of meiotic prophase up to the diplotene stage. At birth the number of oogonia and oocytes declines to approximately 2 million. The reason for this loss is not known but is due mainly to a large decrease in the number of small oocytes. Oocytes in the pachytene and diplotene stages of meiosis also degenerate.

Once the primary oocyte reaches the diplotene stage of meiotic prophase, it enters a prolonged resting period until shortly before ovulation (Fig. 1). Before being released from the follicle, the oocyte undergoes its first meiotic division, producing a secondary oocyte and the first polar body. Nearly all of the cytoplasm is retained by the secondary oocyte. The second meoitic division begins almost immediately but is arrested at metaphase. At this stage, the secondary oocyte, with its first polar body still contained within the zona pellucida, is released at ovulation. The completion of meiosis takes place only at the time of fertilization. [*See* Meiosis.]

II. GROWTH OF THE OOCYTE

A. Ultrastructure of the Mammalian Oocyte and Changes in Cellular Organelles during Oocyte Growth

The fine morphology of oogonia is essentially very similar to that of the primordial germ cells. Like all mitotic cells, oogonia exhibit a pair of centrioles at either pole of their mitotic spindles. Shortly after the end of mitotic proliferation, the centrioles disappear, so that from the commencement of the meiotic divisions until the time of fertilization the oocyte is devoid of these structures.

The fine morphology of meiotic oocytes has been studied extensively in the rat, rabbit, cow, and human and has been shown to be essentially identical in these species. During early meiosis, prominent changes occur in the nucleus or germinal vesicle of the oocyte with little change in the cytoplasm. From the preleptotene to pachytene stages of early meiosis, chromatin condenses into isolated chromosomes; later, homologous chromosomes pair with each other. During diplotene, unpaired chromosomes reappear—similar to those seen during leptotene. In the human, this chromosomal pattern persists throughout the length of the resting period. In the rat and rabbit, it disappears and the chromatin assumes a net-like form in which it it distributed uniformly throughout the nucleoplasm.

At the end of meiotic prophase, the intercellular bridges between oocytes disappear, causing isolation of individual oocytes surrounded by several layers of follicle cells.

The life of the oocyte can essentially be divided into three main stages: the quiescent stage, the maturative stage, and the preovulatory stage.

1. Quiescent Stage

The quiescent stage refers to the period of meiotic arrest. This period varies among species of mammals. For example, in the human, it can vary between about 12 years and several decades. During this period, the oocyte has a spheroidal shape, with a diameter of approximately 50–70 μm, and is still metabolically active; for instance, it continues to synthesize RNA. Close to the nucleus, histological preparations show a crescent-shaped region densely packed with basophilic structures, known as the Balbiani vitelline body, or yolk nucleus complex. This area contains the majority of the cell's organelles; a prominent Golgi apparatus with few mitochondrial and endoplasmic reticulum elements. The rest of the cytoplasm consists almost exclusively of free ribosomes.

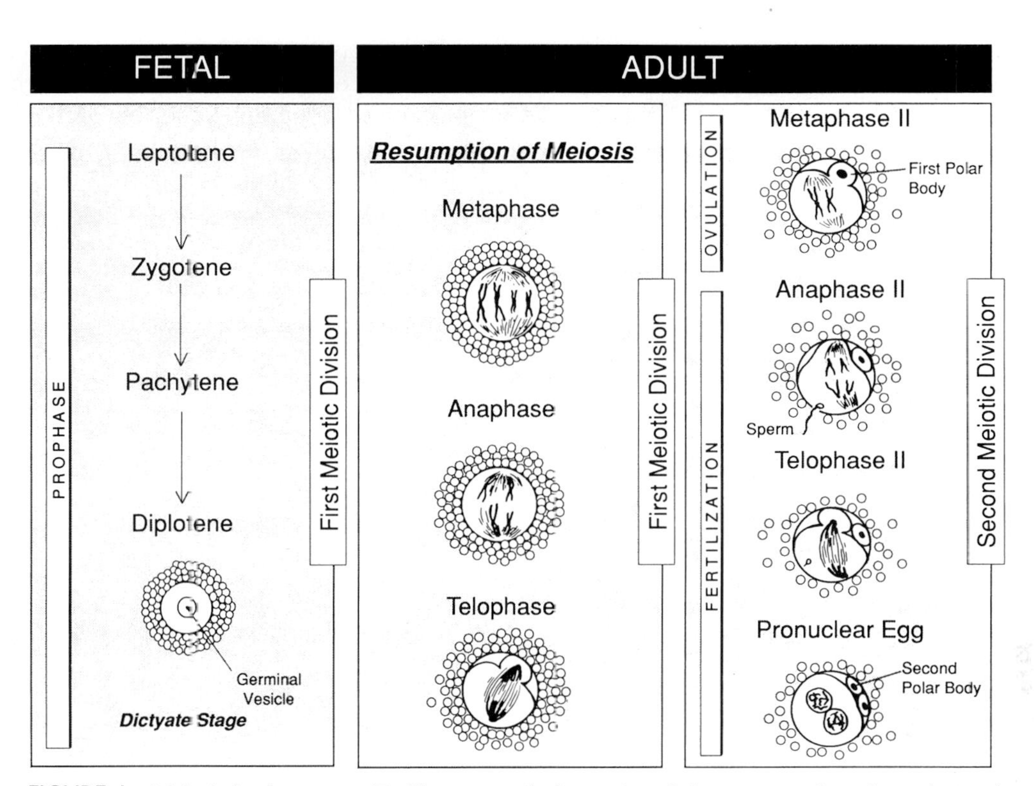

FIGURE 1 Meiosis in the oocyte. To illustrate, only four pairs of chromosomes have been depicted. During fetal development, the oocyte enters the prophase stage of the first meiotic division. At diplotene, the primary oocyte enters a prolonged resting period and is said to be in the dictyate stage. Shortly before ovulation meiosis is resumed. The oocyte completes the first meiotic division, resulting in a secondary oocyte and the first polar body. The second meiotic division begins almost immediately but is arrested at metaphase II. At this stage ovulation occurs. The completion of the second meiotic division takes place only at the time of fertilization.

The most prominent of the juxtanuclear organelles are the annulate lamellae. These structures have been identified in most mammalian species. They consist of stacks of up to 100 paired membranes that contain nuclear pore complexes. The amount of annulate lamellae varies considerably between species and their function is not known. However, their resemblance to nuclear envelope has led some investigators to suggest that they represent a storage pool of nuclear membranes that might be used in early embryonic development.

2. Maturative Stage

The beginning of the maturative stage is marked by the resumption of mitotic activity of the follicle cells surrounding the oocyte, presumably as a result of the actions of locally produced growth factors. The follicle cells exhibit morphological signs of enhanced secretory activity, such as large numbers of cytoplasmic ribosomes and prominent Golgi complexes. The secretory activity brings about the deposition of liquor folliculi, which in turn causes the formation of a fluid-filled space, the follicular antrum. As the volume of the follicular antrum increases, the cells that immediately surround the oocyte, which comprise the cumulus oophorus, separate from the remaining follicle cells. As a result, the oocyte and the surrounding cumulus come to occupy an eccentric position in the follicular cavity.

Before antrum formation, the oocyte increases in volume until it reaches a diameter of approximately 70–85 μm. Then its growth ceases, while the follicle continues to grow. This oocyte is said to be meiotically competent (i.e., it has the ability to resume meiosis *in vitro*) after it has completed it's growth. Only oocytes that have completed growth and meiotic maturation can be fertilized and undergo normal development.

During the period of active growth, the oocyte develops the zona pellucida, a thick, translucent coat that surrounds the plasma membrane of the oocyte and is made up of three glycoproteins with molecular weights of 83,000, 120,000, and 200,000. These proteins are produced by the oocyte; in fact, in culture, growing mouse oocytes, which have been denuded of follicle cells, still secrete these glycoproteins. The zona pellucida is not formed as a continuous extracellular membrane around the oocyte but, rather, is laid down during oocyte growth in discontinuous portions that eventually fuse.

In addition to the formation of the zona pellucida, the growth phase of the oocyte involves the proliferation of mitochondria, dispersal of multiple Golgi complexes, and the elaboration of rough endoplasmic reticulum. Secretory granules appear in association with the Golgi complexes. Because these granules assume a position in the oocyte cortex, they have been called cortical granules. Cortical granules are vital to the process of fertilization as their contents modify the zona pellucida and limit the number of sperm that can penetrate the oocyte.

During the formation of the zona pellucida, the cells of the cumulus oophorus exhibit an increase in surface activity and form cytoplasmic protrusions. These traverse the zona to reach the surface of the oocyte, to which they are firmly anchored by means of intercellular junctions. Also during this time, microvilli develop over the surface of the oocyte.

In most mammals, the oocyte nucleus and cytoplasm undergo changes in structure in preparation for the resumption of meiosis. The germinal vesicle exhibits a more condensed chromatin pattern tht is correlated with the acquisition of meiotic competence. Microtubules decrease in number and become localized in the cortex at this time in association with multiple microtubule organizing centers or centrosomes. These centrosomes are structures that organize the chromosomes on the meiotic spindle during the process of meiotic maturation, which takes place during the preovulatory stage in response to the ovulation inducing hormones.

3. Preovulatory Stage

At the formation of the first polar body, the oocyte enters the preovulatory stage. In this stage, the most prominent change is the relationship between the oocyte and the cumulus cells. The cumulus cells separate from the zona pellucida and from one another.

One of the most noticeable changes during this stage is the resumption and completion of the first meiotic division. It is highlighted by the breakdown of the germinal vesicle, referred to as germinal vesicle breakdown (GVBD). This is actually the condensation of the nucleolus, followed by the diappearance of the nuclear membrane and liberation of the chromosomes in a cytoplasmic area devoid of organelles.

Metaphase I soon follows and the chromosomes become arranged at the equatorial plate of the meiotic spindle. The first meiotic division results in the liberation of the first polar body into the perivitelline space.

The first polar body is morphologically similar to the oocyte: it contains microvilli and cortical granules. The absence of these components in the second polar body provides a means of distinguishing the two. Another difference is that the nucleus of the first polar body consists of isolated chromosomes, without a nuclear membrane, whereas the second polar body contains a nucleus complete with a double membrane.

Following extrusion of the first polar body the chromosomes of the now secondary oocyte forms a compact, crescent-shaped mass that moves into a metaphase II arrangement. In this stage the oocyte is ovulated.

B. Interactions between Oocytes and Follicular Cells

The cumulus cells, which surround the follicle, are believed to regulate oocyte growth by providing nutrients. This role is supported by the observation that occytes will grow *in vitro* only if cumulus cells are in contact with the oocyte. Gap junctions between cumulus cell projections and the oocyte provide channels for the passage of ions and small molecules (molecular mass <1000 Da). Such metabolic coupling (or metabolic cooperactivity) between the two cells is essential for oocyte function. [*See* Cell Junctions.]

In vitro studies with oocytes enclosed in cumulus cells, in fact, show that >85% of the metabolites derived from the culture medium and ending up in the oocyte are first taken up by and metabolized in the cumulus cells. They reach the oocyte via these gap junctions. In addition, a positive correlation exists between oocyte growth and the extent of ionic coupling between the oocyte and the cumulus cells.

The gap junctions may be important in maintaining meiotic arrest. The cumulus cells around an immature oocyte are closely apposed to the zona pellucida, which is studded with numerous cytoplasmic projections, whereas around a mature oocyte the cells are loosely organized and few projections are seen in the zona pellucida. Electrophysiological measurements indicate that ionic coupling between cumulus cells and oocyte decrease progressively as the time of ovulation approaches. Thus, the release of the oocyte from meiotic arrest may be the result of metabolic uncoupling of the cumulus cells from the oocyte, possibly induced by the luteinizing hormone (LH) surge. This will be further discussed in Section V.

III. INHIBITION OF OOCYTE MATURATION

A. Role of Cyclic Adenosine Monophosphate

A great deal of evidence indicates that cyclic adenosine monophosphate (cAMP) plays a fundamental role in the inhibition of oocyte maturation in mammals. Analogs of cAMP, such as dibutyryl cAMP, or inhibitors of cAMP degradation, such as the methylxanthines, have been shown to inhibit oocyte maturation. The resumption of meiosis in follicle-enclosed oocytes is inhibited by both dibutyryl cAMP and isobutylmethylxanthine. These effects on oocyte maturation probably reflect physiological events because the levels of cAMP in the follicle significantly decrease shortly before resumption of meiosis, both under *in vitro* and *in vivo* conditions.

Cyclic AMP produces a greater inhibition of maturation in the cumulus-enclosed oocyte than in oocytes denuded of their cumulus cells. Exposure to cholera toxin or forskolin, which activates adenylate cyclase and thus increases the levels of cAMP, inhibits spontaneous maturation in rat cumulus–oocyte complexes but not in denuded oocytes. These results led to the suggestion that the cells of the cumulus oophorus are the source of the inhibitory cAMP, which may be transferred to the oocyte via gap junctions. This idea was supported by the observation that a parallel increase in cAMP was seen in both oocyte and cumulus cells during the incubation of hamster cumulus-enclosed oocytes in the presence of forskolin. However, several investigators were unable to demonstrate the transfer of cAMP from the cumulus cells to the oocyte. Although they were able to demonstrate cumulus cell-dependent inhibition of oocyte maturation by cAMP modulators, they failed to find any difference in cAMP levels between oocytes cultured in the presence of forskolin and controls.

Thus, although compelling evidence indicates that cAMP can be transferred from cumulus cells to the oocyte and prevent maturation, it cannot be ruled out that, physiologically, the inhibitory effect is mediated by another agent, which requires cAMP for its action.

B. Role of Gonadotropins

Follicle-stimulating hormone (FSH) has been shown to inhibit the maturation of oocytes of several species, including rat, pig, and mouse. Compounds that increase levels of cAMP also inhibit oocyte maturation in isolated oocyte cumulus complexes, suggesting that cAMP mediates the FSH action. As shown later, LH also increases the level of cAMP in isolated oocyte cumulus complexes; however, FSH is significantly more active and is probably the major gonadotropin that stimulates cAMP accumulation in the oocyte inhibiting maturation. FSH and cholera toxin, however, do not inhibit oocyte maturation or stimulate cAMP accumulation in the absence of the cumulus cells, whereas forskolin has its effect in the presence or the absence of these cells. Thus, the oocyte appears to have the catalytic subunit of adenylate cyclase but seems to lack the guanosine triphosphate-binding protein, which is necessary for activation of the enzyme by either FSH or cholera toxin.

C. Role of Steroids

The involvement of steroid hormones in the inhibition of oocyte maturation is the subject of much controversy. Some researchers have reported that steroid hormones stimulate the rate of meiosis, although LH can still initiate meiosis in follicles even after the total inhibition of follicular steroidogenesis. Others argue that steroids are involved in the arrest of oocyte maturation. When mouse oocytes were cultured with testosterone, progesterone, or androstenedione, GVBD was partially inhibited, whereas estradiol, estrone, and dihydrotestosterone had no effect. In studies on pig oocytes, however, estradiol reduced the maturation of denuded oocytes, whereas progesterone inhibited this effect. A similar effect of estradiol was seen in denuded, cumulus-enclosed hamster oocytes, in which estradiol enhanced forskolin-dependent meiotic arrest. In other studies, estradiol was found to inhibit spontaneous maturation in denuded mouse oocytes. Thus, the precise role of follicular estrogen on oocyte maturation is obscure. The differences between studies may be due to differences in the nature of the culture media used. Reportedly, estradiol has the capacity to arrest meiosis of either intact or denuded oocytes *in vitro* but only if no exogenous protein(s) is present in the media. Because *in vivo* the oocyte cumulus complex is bathed by a highly proteinaceous fluid, estrogens may not play a physiological role in the maintenance of meiotic arrest. The preovulatory surge of LH induces the resumption of meiosis and a rise in follicular steroidogenesis, including a transient rise in estrogen production. Estrogen remains elevated in some species when GVBD takes place, in other species when the oocyte becomes committed to mature. Even though GVBD takes place later on, these findings argue against the suggested role of estrogen as a follicular inhibitor of the resumption of meiosis. However, estradiol may be involved in the maintenance of the structural integrity of the oocyte–cumulus complex. Thus, estrogen has been shown to regulate the appearance and maintenance of gap junctions between cells of the membrane granulosa and of the myometrium, and it may perform a similar role in the maintenance of junctions between cumulus cells or between cumulus cells and the oocyte.

The role of androgens and progestins in the maturation of mammalian oocytes is also unclear. Progesterone has been shown to facilitate the maturation of denuded rabbit and bovine oocytes, but a similar effect has not been seen on the maturation of human and porcine oocytes. The addition of steroids to the culture medium does not induce maturation of bovine, porcine, and rat oocyte–cumulus complexes. Only at high concentrations (100 μM) does progesterone block GVBD and polar-body formation from oocytes in medium- and large-sized follicles. Reportedly, large quantities of progesterone are present in media incubated with oocyte–cumulus complexes, indicating that cumulus cells can secrete steroid hormones. Thus, although the evidence for a primary regulatory role of steroids in oocyte maturation is unconvincing, what role these molecules have in modulating the meiotic process remains to be determined. [*See* Steroids.]

D. Oocyte Maturation Inhibitor

In 1935, Pincus and Enzmann showed that when rabbit oocytes are liberated from their follicles they undergo spontaneous maturation *in vitro*, without hormonal stimulation. This was found to be the case for all mammalian species studied, leading to the suggestion that the follicle cells "supply to the ovum a substance or substances that directly inhibit nuclear maturation."

Subsequent studies found that resumption of meiosis was inhibited when porcine oocytes were cocultured with porcine granulosa cells or with medium conditioned by granulosa cells, but not with the theca layer or the ovarian bursa. These results suggested that within the follicle the granulosa cells produce an oocyte maturation inhibitor (OMI), which keeps the oocyte in a state of meiotic arrest. Follicular fluid from various species, such as the rabbit, bovine, porcine, and human, exerts this inhibitory effect on the oocyte. This effect is not species specific: human follicular fluid, for example, can inhibit the maturation of rat oocytes. Oocyte maturation inhibitor activity appears to decline as follicular growth increases. Follicular fluid collected from humans indicated that OMI activity was much lower in follicles yielding mature and fertilizable oocytes as compared with follicles yielding immature or atretic oocytes.

Rat or porcine oocytes isolated from either medium or large follicles have been used in culture as a bioassay system to partially characterize and purify porcine OMI. The oocyte maturation inhibitor has been reported to have a molecular weight of less than 2000 with two major peaks of activity. It is thought to be a peptide, as OMI cannot be extracted by ether or charcoal but is trypsin sensitive and heat stable. OMI isolated from porcine granulosa cell extracts has been found to be similar in properties to OMI isolated from porcine follicular fluid, which is similar to OMI from bovine follicular fluid. Characterization and isolation of the putative inhibitor, however, have yet to be accomplished. Problems in the adequacies of the assay systems employed, lack of a standard OMI preparation, and the variability of the experimental approaches have hampered the complete isolation of this molecule.

It has recently been postulated that OMI may be the purine hypoxanthine. This conclusion was based on findings that the fractions of porcine follicular fluid with the greatest inhibitory effect on oocyte maturation also have an absorption maximum identical to that of hypoxanthine, have a retention time on high-pressure liquid chromatography of pure hypoxanthine, and exert inhibition of oocyte maturation identical to that exhibited by a commercial preparation of hypoxanthine. Hypoxanthine also augmented the effect of FSH on inhibition of oocyte maturation as well as that of dibutyryl cAMP. At high concentrations, hypoxanthine is known to inhibit cAMP phosphodiesterase activity, and this may be the mechanism of its action in the inhibition of oocyte maturation.

Another purine, adenosine, has also been shown to inhibit oocyte maturation at physiological concentrations (0.3–10 μM). Adenosine was also found to increase cAMP and progesterone production in response to FSH and LH in preovulatory oocyte-cumulus complexes, with little response in postovulatory complexes. In addition, adenosine augmented oocyte maturation inhibition by FSH. This response was specific for adenosine: no effect was seen with inosine, hypoxanthine, or guanine. Little effect of adenosine on oocyte maturation is seen in the absence of FSH.

The site of action of adenosine is probably cumulus cell dependent because adenosine itself has no effect on the rate of spontaneous maturation of denuded oocytes. This conclusion is supported by the fact that FSH action on the oocyte–cumulus complex is the cumulus cell and the action of adenosine is FSH dependent.

The mechanism of adenosine action is unknown. Adenosine action is associated with an enhancement of the accumulation of cAMP in the presence of FSH; several theories on the biochemical processes involved in this effect have been suggested. For example, adenosine may act on receptors to increase the accumulation of cAMP. Adenosine has been shown to increase adenosine triphosphate (ATP) levels in luteal, granulosa, and cumulus cells, whereas FSH depletes concentrations of ATP in the cumulus cells. Thus, adenosine may buffer the depletion of ATP by FSH, thereby maintaining elevated levels of cAMP, which is then shunted via gap junctions into the oocyte to inhibit maturation. Adenosine may also act on cumulus cells by decreasing intracellular levels of calcium, which has been shown to inhibit FSH-stimulated accumulation of cAMP.

IV. INDUCTION OF OOCYTE MATURATION

A. Role of Gonadotropins

Follicle-enclosed oocytes have been used to study the mechanism of induction of maturation because isolated oocytes resume meiosis spontaneously. The factor responsible for the initiation of meiosis is LH. However, neither the specific target cell(s) for LH nor the underlying biochemical events are known. Inhibition of steroidogenesis does not block the induced maturation by LH, and administration of progesterone, prolactin, or estradiol does not mimic its effect. FSH may be responsible for preparing the follicle for ovulation. Thus, at ovulation the processes that link the oocyte with the cumulus cells retract, and the resulting spaces now present between the cumulus cells are filled with a hyaluronidase-sensitive material. FSH *in vitro* has been shown to specifically stimulate the synthesis of hyaluronic acid by the mouse cumulus; therefore, the cumulus is thought to expand as the result of the gonadotropin-induced deposition of glycosaminoglycans, possibly hyaluronic acid between the cumulus cells. This effect is specific for FSH—it does not occur in resposne to LH or human chorionic gonadotropin (hCG). The action of FSH is essential for follicular growth prior to the LH surge; the cumulus expansion is seen only *in vitro,* no *in vivo,* suggesting that some follicular component inhibits expansion of the cumulus before the LH surge. An increase in proteinase activities (aminopeptidase, endopeptidase, trypsin-like, and elastase-like activities) in the cumulus cells after the LH surge may play a role in oocyte–cumulus changes.

Cyclic AMP, which appears to inhibit oocyte maturation in the preovulatory follicle, can also induce maturation. In fact, LH elevates cAMP levels in follicular fluid and treatment of follicles with dibutyryl cAMP induces maturation. Therefore, LH may induce oocyte maturation by a cAMP-mediated response. Indeed, forskolin, which enhances cAMP levels, has been shown to mimic the effect of LH on the ovarian follicle.

The seemingly contradictory effects of cAMP can be explained by the changes of the follicle before ovulation, which modulate the cAMP effect. Clearly maturation of the mammalian oocyte in the preovulatory follicle is suppressed by agents present within the follicle and is removed by the LH surge. By changing the metabolism of the oocyte–cumulus complex, LH would induce maturation in a manner similar to the spontaneous maturation that occurs in the isolated oocyte complex. Among the metabolic changes induced by LH, oxygen consumption is decreased with increased lactate production in the preovulatory cumulus at about the time the oocyte increases its respiration. This metabolic shift occurs before any alteration in the morphology of the oocyte–cumulus complex is seen, yet inhibition of the stimulatory effect of LH on follicular lactate production does not prevent its action on the resumption of meiosis.

Another change induced by LH is blocked communication in the oocyte–cumulus complex by causing an uncoupling among cumulus cells and between cumulus cells and the oocyte. This uncoupling reaction is also induced by cAMP and phosphodiesterase inhibitors. Thus, in the preovulatory phase, cAMP generated by the cumulus cells in response to FSH is transferred via the junctional processes to the oocyte to keep it in a state of meiotic arrest, but once the LH surge takes place the interruption of communication between the oocyte and cumulus cells stops the transfer of cAMP, and the oocyte can now resume meiosis.

However, not all laboratories have been able to show that termination of communication between the oocyte and cumulus cells takes place prior to the resumption of meiosis. For rat oocytes, some researchers have suggested that prior to uncoupling and the resumption of meiosis LH-induced maturation may be due to a desensitization of FSH receptor-activation of adenylate cyclase in cumulus cells. Thus, further investigations are required to determine more precisely the role of cAMP in the LH-induced maturation.

B. Role of Peptides and Growth Factors

Gonadotropin-releasing hormone (GnRH) or its agonist analog, the role of which is to cause FSH and LH release from the pituitary, can induce the resumption of meiosis in the hypophysectomized rat and in follicle-enclosed oocytes *in vitro*. This response appears to be specific for GnRH because a GnRH antagonist abolished the meiosis-inducing action. The antagonist, however, did not block the effect of LH, showing that GnRH does not appear to mediate the action of LH in the induction of oocyte maturation.

Epidermal growth factor (EGF), transforming growth factor-β (TGFβ), and insulin-like growth factors I and II (IGF-I and IGF-II) have been reported to stimulate oocyte maturation, whereas insulin and platelet-derived growth factor do not. TGFβ, EGF, and IGFs have minimal effects on cAMP production by oocyte–cumulus complexes, yet their effects are inhibited by cAMP analogs, phosphodiesterase inhibitors, and hypoxanthine. Their mechanisms of action require further study. [*See* Growth Factors and Tissue Repair.]

V. REGULATION OF METABOLISM DURING OOCYTE GROWTH

A. Changes in Enzyme Activity

Knowledge of the enzyme systems that exist in the growing oocyte in mammals is minimal. The activities of several enzymes, however, have been reported to increase during the growth period of the oocyte. Thus, mitochondrial succinate dehydrogenase reaches a peak in preovulatory oocytes, glucose-6-phosphate dehydrogenase (G-6-PD) and lactate dehydrogenase (LDH) reach maximum activity when the oocyte has a diameter of approximately 85 μm. In the mouse, LDH comprises about 5% of the total protein content of the oocyte, whereas the human and the rabbit have much lower levels.

RNA polymerase has also been shown to be present in the oocyte. RNA polymerase I, which is responsible for the synthesis of ribosomal RNA, increases during the initial phase of oocyte growth and then declines. Creatine kinase has been reported in mouse and rat oocytes and may play an important role in maintaining a high ATP–ADP ratio during oogenesis. These and other results indicate that the synthesis of many enzymes is completed in the oocyte before the onset of meiotic maturation.

B. DNA and RNA Synthesis

1. DNA

The DNA content of the growing oocyte has been examined in mammals by Feulgen microspectrophotometry. Oocytes contain more than the diploid amount of DNA; the "extra" DNA is associated with the mitochondria. The fully grown oocyte contains approximately 11 pg of genomic and 2–3 pg of mitochondrial DNA. The last synthesis of DNA occurs at meiotic prophase. [*See* DNA Synthesis.]

2. RNA

Most of the knowledge on this subject has been obtained from *in vivo* mouse studies. All the main RNA species have been shown to be present; ribosomal RNA (28S + 18S), 5 S-ribosomal RNA, transfer RNA, messenger RNA (mRNA), and heterogenous nuclear RNA. In the fully grown oocyte, 90–95% of total RNA is ribosomal and transfer RNA; the remainder is mostly mRNA. Total RNA content increases during oocyte growth and reaches a peak in fully grown oocytes. Its synthesis stops with antrum formation. Small, growing oocytes are estimated to contain approximately 0.29 ng of RNA, whereas fully grown oocytes contain about 0.57 ng. RNA of growing oocytes is unusually stable; in the mouse, for example, at least 80% of labeled RNA found 2 days after label administration is retained 10–20 days later.

During meiotic maturation, the fraction of mRNA in total RNA decreases from 19 to 10% and almost 20% of total RNA is degraded or deadenylated [i.e., loses the terminal poly(A) tail]. The mechanism(s) involved in the reduction of mRNA and in the degradation–deadenylation processes is unknown, although RNA polymerase activity does decrease in the mature oocyte.

Mammalian oocytes store a large excess of maternal mRNA that is translated in the early postfertilization period. The synthesis of RNA apparently is not necessary for the resumption of meiosis because inhibitors of RNA synthesis do not inhibit nuclear progression.

C. Protein Synthesis

As the oocyte grows, the cells volume increases tremendously. In the mouse, for example, its diameter increases from 15–20 μm to approximately 80–85 μm; the total protein content increases linearly with volume, reaching 28–30 ng in the fully grown oocyte. The absolute rate of protein synthesis increases from 1.1 pg/hr in a quiescent mouse oocyte to 42 pg/hr in a fully grown oocyte. Ribosomal proteins are synthesized throughout oocyte growth, even though ribosomal RNA synthesis is not detectable. The cytoskeletal proteins, actin and tubulin, are synthesized both during growth and in the fully grown oocyte. Cytoskeletal and ribosomal proteins are the most abundant and most stable proteins in the growing oocyte.

As the oocyte progresses from the dictyate to metaphase II stage of meiosis, the pattern of proteins synthesized changes markedly in a stage-specific way. This program is supported by preformed mRNA because mRNA synthesis at the resumption of meiotic maturation is undetectable at this time. The changes in protein synthesis do not take place in oocytes that fail to undergo GVBD, suggesting that they are triggered by the contact of the oocyte's nucleoplasm and cytoplasm at the time of GVBD.

The precise nature and role of the proteins synthesized during meiotic maturation are not known. In oocytes liberated from follicles, inhibitors of protein synthesis do not prevent GVBD but stop meiosis; they do, however, block the LH-induced maturation of follicle-enclosed rat oocytes. Thus, new protein synthesis must take place for nuclear progression to proceed beyond GVBD.

These changes in protein synthesis may be caused by a cytoplasmic factor produced or unmasked independently of the nucleus, which is also involved in GVBD and nuclear maturation. In fact, nucleate oocyte fragments of mouse oocytes resume protein synthesis and meiosis *in vitro*, and anucleate fragments undergo some of the changes in protein synthesis. These findings have led to the suggestion that meiotic competence of the oocyte is due to the synthesis of new proteins directed by mRNA already present in the cytoplasm of the oocyte. The nature of the functions of these proteins is unknown but recent studies suggest they are involved in meiotic cell cycle control.

VI. REGULATION OF MEIOTIC CELL CYCLE

Meiotic maturation represents the transition of the oocyte from the arrested G2 phase of the cell cycle to M phase. In most eukaryotic cells, this cell cycle transition is associated with the activation of an M-phase kinase, a protein phosphorylating enzyme known as $p34^{cdc2}$ whose activity is regulated by its interaction with cyclin proteins.

The cyclin–$p34^{cdc2}$ protein complex, also known as maturation promoting factor (MPF), is activated upon the resumption of meiosis in mammalian oocytes. By phosphorylating protein targets, chromosome condensation and spindle assembly occur in response to MPF activation during both meiosis I and meiosis II.

The degradation of cyclins is believed to occur at anaphase onset and serves to regulate the initiation of chromosome movement in both meiotic and mitotic cells. As mentioned earlier, what makes the meiotic cell cycle of oocytes unique is the arrest of meiosis at metaphase II, the state of cell cycle arrest of the ovulated oocyte. It is now known that the c-*mos* protooncogene, itself a kinase, is at least partly responsible for arresting the oocyte at metaphase II. While the details remain to be worked out regarding the metabolic regulation of the oocyte cell cycle, it is likely that the expression and modification of cell cycle proteins are precisely regulated during the acquisition of meiotic competence and resumption of meiosis during oogenesis.

BIBLIOGRAPHY

Albertini, D F. (1992). Regulation of meiotic maturation in the mammalian oocyte: Interplay between exogenous cues and the microtubule cytoskeleton. *BioEssays* **14**, 97.

Albertini, D. F., and Rider, V. (1994). Patterns of intercellular connectivity in the mammalian cumulus-oocyte complex. *Microsc. Res. Techn.* **17**, 125.

Behrman, H. R., Preston, S. L., Pellicer, A., and Parmer, T. G. (1988). Oocyte maturation is regulated by modulation of the action of FSH in cumulus cells. *In* "Meiotic Inhibition: Molecular Control of Meiosis" (F. Haseltine, D. J. Patanelli, and N. First, eds.). A. R. Liss, New York.

Brower, P. T., Gizang, E., Boreen, S. M., and Schultz, R. M. (1981). Biochemical studies of mammalian oogenesis: Synthesis and stability of various classes of RNA during growth of the mouse oocyte in vitro. *Dev. Biol.* **86,** 373.

Dekel, N., Lawrence, T., Gilula, N., and Beers, W. (1981). Modulation of cell-to-cell communication in the cumulus-oocyte complex and the regulation of oocyte maturation by LH. *Dev. Biol.* **80,** 356.

Dong, J., Albertini, D. F., Nishimori, K., Kumar, T. R., Lu, N., and Matzuk, M. M. (1996). Growth differentiation factor-9 is required during early ovarian folliculogenesis. *Nature* **383,** 531.

Ducibella, T. (1996). The cortical reaction and development of activation competence in mammalian oocytes. *Hum. Reprod. Update* **2,** 29.

Eppig, J. J., and Downs, S. M. (1984). Chemical signals that regulate mammalian oocyte maturation. *Biol. Reprod.* **30,** 1.

Guraya, S. S. (ed.) (1985). "Biology of Ovarian Follicles in Mammals." Springer-Verlag, Berlin.

Siracusa, G., De Felici, M., and Salustri, A. (1985). The proliferative and meiotic history of mammalian female germ cells. *In* "Biology of Fertilization" (C. B. Metz and A. Monroy, eds.). Academic Press, New York.

Tsafuiri, A., and Adashi, E. (1994). Local non-steroidal regulators of ovarian function. *In* "The Physiology of Reproduction (E. Knobil and J. D. Neill, eds.), Vol. I. Raven Press, New York.

Wassarman, P. M., and Albertini, D. F. (1994). The mammalian ovum. *In* "The Physiology of Reproduction." (E. Knobil and J. D. Neill, eds.), Chap. 3, p. 79. Raven Press, New York.

Opioids

LEI YU
Indiana University School of Medicine

GLOSSARY

Addiction Drug-seeking behavior, partially motivated to alleviate the adverse withdrawal symptoms

Agonist and antagonist Drug that produces a response is an agonist, and a drug that opposes or blocks the action of an agonist is an antagonist

Alkaloids Nitrogen-containing organic compounds with alkaline properties, found mostly in plants, but also in fungi and vertebrates

Analgesia Pain relief through pharmacological intervention, without a loss of consciousness

Dependence Condition often seen with prolonged or repeated use of a drug such as an opioid, in which abrupt discontinuation (withdrawal) of the drug leads to adverse physical and psychological symptoms

Pain Behavioral response to a noxious stimulus, whether mechanical, thermal, or chemical, is called nociception. In humans, nociception is interpreted as pain

Tolerance Condition in which the effectiveness of a drug is diminished with prolonged or repeated use, and higher and higher doses of the drug are needed to achieve the desired effect

HUMANS HAVE A LONG HISTORY OF USING OPIUM, of which morphine is a major active ingredient, to control pain and diarrhea and to experience feelings of euphoria. The term *opiates* was originally used to indicate alkaloid chemicals that have morphine-like effects. Nowadays, the term *opioids* has a wider meaning and includes opiates and functionally related drugs, whether alkaloids or peptides, naturally existing or synthetic, agonists or antagonists. The concept of "opioids" has also evolved, from that restricted to only the drugs with morphine-like activity, to the current definition—agonists whose action can be antagonized by naloxone (a morphine analogue) and antagonists that block the action of opioid agonists.

The major clinical usage of opioid agonists is for pain relief, or analgesia. Commonly used opioids such as morphine produce powerful analgesic effect to blunt the pain sensation, without strongly affecting other sensory modalities such as vision or touch. Prolonged or repeated usage of opioids, however, often results in tolerance, a condition in which the effectiveness of the drug is diminished and higher and higher doses are needed to achieve the desired effect. Dependence to the drug may also develop, in which case abrupt discontinuation (withdrawal) of the drug leads to adverse physical and psychological symptoms. When opioids are used for their euphoric effect, drug dependence is the major cause for addiction, the drug-seeking behavior to alleviate the adverse withdrawal symptoms. Opioids such as codeine also have antitussive effects, that is, the ability to suppress cough. Peripherally acting opioids such as loperamide are effective in controlling diarrhea by reducing gastrointestinal motility; such peripheral opioids do not appear to elicit tolerance or dependence.

I. ALKALOID AND PEPTIDE OPIOIDS

A. Alkaloid Opioids

Humans have been using opioids for several thousand years. In ancient times, people used the juice from the poppy plant (*Papaver somniferum*), opium, for medicinal purposes and in religious rituals. Since the sixteenth century, consumption of opium for euphoric feelings became more widespread around the world, especially in Asia, and caused significant social problems.

Opium contains a number of alkaloids, some of which are active opioids. In the beginning of the nineteenth century, the first active component was extracted from opium and was named "morphine" after the Greek god of dreams, Morpheus. This reflected the thinking at the time about the chief effect of opium—producing a dream-like state of mind, that is, feelings of euphoria. Subsequently, other alkaloid components were extracted from opium, including codeine (an opioid) and thebaine (not an opioid). To date, morphine and codeine are still widely used in medicine, underscoring the potent biological effects of these opioids.

Because of the potent pain-relieving effect of the naturally existing opioids, considerable effect has been devoted to making synthetic analogues of these natural compounds. Notable examples are naloxone, an antagonist that is often used to define whether an agonist has opioid activity; heroin, a morphine derivative with high abuse potential because it gives quick onset of euphoria; fentanyl, a potent opioid alkaloid widely used for clinical management of pain such as during childbirth; and loperamide, a peripherally active opioid alkaloid used to relieve diarrhea symptoms. [*See* Alkaloids in Medicine.]

B. Peptide Opioids

Many drugs act by interacting with a receptor, the specific protein in the cell membrane that binds to the drug to mediate the drug effects. It was assumed long ago that alkaloid opioids from the poppy exert their biological effects by acting upon opioid receptors in the brain, and that these opioid receptors are there not just for the chemicals from plants but for endogenous compounds made in the body that act on these receptors in a similar fashion as the plant alkaloids.

Sure enough, it was discovered 20 years ago that the mammalian brain makes its own opioids in the form of peptides. The three families of endogenous opioid peptides, enkephalins, endorphins, and dynorphins, are made from distinct genes and are processed from polypeptide precursors: pro-opiomelanocortin (POMC) is cleaved by peptidase activity to give β-endorphin, as well as other hormones such as γ-melanocyte-stimulating hormone (γ-MSH), adrenocorticotropin (ACTH), and β-lipotropin (β-LPH); proenkephalin is processed to give both met- and leu-enkaphalins; and prodynorphin can yield dynorphin A, in addition to leu-enkephalin and α- and β-neoendorphins. Around the same time as the discovery of endogenous opioid peptides, three classes of opioid receptors were also identified: mu (μ), delta (δ), and kappa (κ) receptors. It should be noted that endogenous peptide opioids do not have a high degree of selectivity for specific opioid receptors (see Section II), although enkephalins and endorphins can activate both μ and δ receptors, whereas dynorphins are more effective for κ receptors. Endogenous peptide opioids are thought to function during stress, because stress, or electrical or chemical stimulation of certain brain sites in animals, can produce analgesia, which is antagonized by naloxone. Pain relief through acupuncture is also believed to involve the endogenous opioids.

Aside from these endogenous peptide agonists, many synthetic peptides have been made. Some of these peptide opioids are found to be highly selective for a particular opioid receptor type. For example, DAMGO, an agonist, and CTOP, an antagonist, are selective for the μ opioid receptors, whereas DPDPE, an agonist, is selective for δ opioid receptors. These peptides are not used in clinical settings, because as peptides they do not cross the blood–brain barrier well to activate the receptors in the brain, and are quickly degraded in the liver and in the circulation. However, the availability of these synthetic peptide opioids greatly facilitates research, because they serve as important tools in animal experimentation to identify the specific functions of each opioid receptor type. [*See* Peptide Hormones and Their Convertases.]

II. OPIOID RECEPTORS

A. Pharmacological Classification

Opioids exert their physiological effects by binding to membrane receptors. Based on pharmacological properties, opioid receptors are broadly classified into three major classes. μ Receptors are named after morphine and are the major cellular target for most medically relevant opioid drugs, including both naturally existing drugs such as morphine and codeine as well as synthetic compounds such as fentanyl and methadone. In addition to these μ-selective alkaloids, μ opioid receptors are also activated by peptide opioids such as β-endorphin and enkephalins (endogenous) and DAMGO (synthetic). The two other classes of opioid receptors are δ and κ. δ Receptors bind to the endogenous peptide opioids β-endorphin and enkephalins as do μ receptors. There are also δ-selective opioids such as DPDPE. κ Receptors bind very tightly to the endogenous dynorphins, and there are a number of κ-selective alkaloid agonists such as U50488 and U69593.

To some degree, all three classes of opioid receptors share a common pharmacology. The prototypical opioid antagonist, naloxone, binds to all three receptors and antagonizes their effects equally well. Also, there is only limited selectivity for endogenous opioid peptides. β-Endorphin and enkephalins bind tightly to both μ and δ receptors with high affinity; dynorphins show better selectivity for κ receptors with very tight binding, but also bind to μ and δ receptors, albeit with lower affinity values.

B. Molecular Cloning

With recent advancements in molecular biology, efforts to clone the opioid receptor genes have been successful. The first opioid receptor to be cloned was a δ receptor, and the molecular cloning of μ and κ receptors quickly followed. Since pharmacological studies indicated multiple receptor subtypes within each of the μ, δ, and κ receptor families, it was somewhat surprising that so far there has been only one gene identified for each receptor type. It remains to be seen whether or not, for example, a single δ receptor gene encodes some or all of the pharmacologically defined δ receptor subtypes, or whether other δ receptor genes exist but are not identified yet.

Extensive research prior to receptor cloning showed that opioid receptors function through GTP binding proteins (G proteins). Based on the cloned receptor genes, the protein polypeptides for these receptors indeed belong to the supergene family of G protein-coupled receptors and possess the structure of seven hydrophobic peptide domains thought to form transmembrane regions. Underscoring the fact that all three opioid receptors share some degree of pharmacology, the genes, and thus polypeptides, of the cloned receptors also share sequence similarity.

Interestingly, molecular cloning has turned up a fourth, and novel, gene. The protein made from this novel gene does not bind to naloxone or other opioids; thus it does not fit the classic definition of an opioid receptor. However, this receptor gene contains a high degree of sequence similarity with the three cloned opioid receptor genes, thus it is definitely a new member of the opioid receptor gene family. Using this novel receptor as a detection assay, a novel endogenous peptide agonist has been purified that has some resemblance at the sequence level to all three families of endogenous opioid peptides, that is, endorphins, enkephalins, and dynorphins. Hence this novel peptide also appears to be related to the endogenous opioids. Animal studies indicate, however, that this peptide does not produce analgesia. Thus, it is still unclear how the novel receptor and the peptide should be classified. It is possible that this ligand–receptor system represents a yet to be identified biological function.

III. PAIN AND OPIOID-INDUCED ANALGESIA

A. Pain as a Sensation

The human body feels pain when there is a noxious stimulus, either from external sources such as a needle prick or from internal sources such as stomach pain. Nociception, and the resulting pain sensation, is a useful mechanism for survival. It serves as a warning signal, alerting the person to either remove from the noxious stimulus or seek help in dealing with the cause of the pain. Thus, pain sensation provides a protective mechanism for the body. However, sustained pain can be harmful, causing physiological and psychological suffering. Therefore, pain relief is sought by physical means or pharmacological intervention such as drug treatment.

As a type of sensation, pain is quite different from other sensations in that most sensations are able to adapt. For example, when one enters a dimly lit room, one's visual acuity is increased to adjust to the lighting situation. Similarly, one may smell the fragrance of flowers when first entering a room, but the sensation fades over time till one is no longer aware of the fragrance anymore. Such a phenomenon is variably termed adaptation, habituation, or desensitization.

Pain sensation, however, does not adapt much and may even intensify when the source of pain continues to be present. This is thought to be important for pain to serve as a warning sign until the source of pain is removed. On the other hand, the nonadaptive nature of pain sensation can be undesirable; for example, after a knife cut wound is treated, pain still lingers on. Also, during terminal illnesses such as late-stage cancer, sustained pain causes unbearable suffering.

There are two types of pain. Cutaneous pain is a sharp, pricking pain that is easily localized, such as a needle prick or a knife cut. This type of pain occurs mostly on skin and tissues close to the body surface. Visceral pain, on the other hand, is perceived as a dull, aching, and throbbing pain that is diffuse and difficult to localize. This type of pain includes pain from internal organ diseases as well as referred pain, is associated with tissue damage or destruction, and can be intense and unbearable. [*See* Pain.]

B. Cellular Basis of Pain

Like all the sensory signals, pain sensation is also conducted by nerve fibers. There are two types of fibers for nociceptive conduction. Larger Aδ fibers are lightly myelinated, with fast conduction velocity (usually 15–30 m/sec). They have a high threshold and require a strong noxious stimulus to activate. Smaller C fibers are unmyelinated, smaller fibers with a slower conduction velocity (1 m/sec). They are usually polymodal and capable of conducting nociceptive signals from mechanical, thermal, and chemical stimuli.

Nociceptive stimuli in the body are detected by nerve endings of sensory neurons that lie in the dorsal root ganglia, just outside the spinal cord. These signals are relayed to neurons in the dorsal horn of the spinal cord. From here, these signals are both processed by neuronal circuitries in the spinal cord and sent to the brain via ascending paths. Therefore, pain sensation is modulated by two mechanisms, a local one within the spinal cord and a supraspinal one from the brain via descending paths.

Some chemicals are involved in pain sensation. They include those that directly induce pain, such as bradykinin and substance P, two endogenous neurotransmitters that are released in the spinal cord when sensory neurons are stimulated, and capsaicin, a chemical that stimulates sensory neurons and makes chili pepper "hot." Other chemicals, such as prostaglandins, do not induce pain by themselves, but are capable of enhancing the pain sensation. They do so by increasing the sensitivity of pain nerve endings, thereby potentiating the nociceptive sensation. [*See* Neurotransmitters and Neuropeptide Receptors in the Brain.]

C. Cellular Effect of Opioid Action

Pain relief can be achieved in a number of ways. Opioids alleviate pain sensation by inhibiting brain and spinal cord neurons that carry the pain signal. Specifically, activation of opioid receptors regulates the activity of ion channels on these neurons. Because neurons rely on their membrane ion channels to carry out the cellular functions, regulation by opioid receptors thus controls how active a neuron is. Two classes of ion channels are regulated by opioid receptors: potassium (K^+) and calcium (Ca^{2+}) channels. Opioids, through the action of opioid receptors and G proteins, can activate a special type of K^+ channel. When these K^+ channels open, K^+ ions inside the cell tend to rush out, making the inside more negative than the outside, thus rendering the neuron electrically less excitable and less likely to fire action potentials as pain signals. Ca^{2+} channels are normally activated by membrane depolarization, and Ca^{2+} ions rushing into the cell make the inside of the cell more positive and more electrically excitable. In nerve terminals that contain neurotransmitters (such as substance P) packed inside small lipid vesicles called synaptic vesicles, the influx of Ca^{2+} ions triggers the fusion of synaptic vesicles with the membrane of the nerve terminal to release the neurotransmitters into the extracellular space between this nerve terminal (presynaptic) and the next neuron (postsynaptic). Activation of opioid receptors suppresses Ca^{2+} channel activity, therefore reducing the amount of the neurotransmitters released and thus the pain "signal."

D. Opioid-Induced Analgesia

The most common clinical use of opioid agonists is for pain relief, that is, analgesia. Opioids are very powerful analgesics, capable of relieving even the most severe pain. In addition, the analgesic effects of opioids, when used at proper dosages, are relatively selective for the pain sensation, without strongly affecting other sensory modalities such as vision or touch. For these reasons, opioids are often the drugs of choice in treating severe pain, such as those after an operation, during childbirth, and in terminal cancer patients. Because of the potential development of tolerance and dependence (see Section IV,C), however, opioids are not used when treating milder forms of pain.

Several μ-selective agonists are the most commonly used opioids for pain relief. Morphine, the first opioid ever identified, remains a popular analgesic in the clinical management of pain and serves as the standard against which the effectiveness of other opioid analgesics is compared. Frequently used opioids such as morphine, fentanyl, and related compounds are mostly alkaloids. Thus, once administered into the body, whatever the route, they can cross the blood–brain barrier and act on sites both in the central nervous system (CNS) and in the periphery (primarily in the bowel, see Section IV,A). The major sites that mediate opioid analgesia are in the CNS, including both the spinal cord (spinal analgesia) and the brain (supraspinal analgesia). Administration of fentanyl or opioid agonists into the spinal cord produces effective analgesia with a rather quick onset, because of the ready access of opioids to the primary nociceptive neurons in the dorsal horn of the spinal cord. Studies using animal models have also shown that profound analgesia can also be induced by administering opioids into a number of sites in the brain, notably the third ventricle, the periaqueductal gray, and several other sites in the midbrain and medulla. These supraspinal sites are effective because they modulate the spinal nociceptive system via descending paths from the brain to the spinal cord. When opioids are administered into both the spinal and supraspinal sites, synergistic effects occur, with a 10-fold reduction in the total dosage of the drug to produce a desirable analgesia. [*See* Brain, Central Gray Area.]

IV. OTHER PHYSIOLOGICAL EFFECTS OF OPIOIDS

A. Gastrointestinal Effects

Even at low doses, opioids are effective at decreasing gastrointestinal tract motility. In fact, this is one use of opium since ancient times to control diarrhea. The gastrointestinal effects of opioids are due in part to diminished propulsive peristaltic movements of the colon. Many opioid alkaloids, because of their ability to cross the blood–brain barrier, can induce constipation as a side effect when their intended effect is analgesia. Loperamide, a peripherally acting μ-selective agonist that does not cross the blood–brain barrier, provides effective treatment for the symptoms of diarrhea. It can also be used to treat irritable bowel syndrome, a condition for which the cause is not yet known. Unlike centrally acting opioids, prolonged use of loperamide does not appear to induce tolerance development, and certainly not physical or psychological dependence; therefore, it is considered a rather "safe" opioid drug in this regard. It is possible that the lack of tolerance/dependence liability of loperamide is due to its restricted access to only peripheral sites, although this remains to be determined.

B. Respiratory Effects

Respiratory functions can be affected by opioids. Morphine, codeine, and related opioids are used as antitussive medicine to suppress the cough reflex. In addition, high doses of opioids can induce respiratory depression, the most common cause of overdose-related death by opioids such as heroin. The respiratory effects by opioids are due in part to the drug's action in a CNS site, the cough center in the medulla. Chest muscular rigidity with high doses of opioids can also compromise respiration. It should be noted, however, that opioid effects on cough suppression and respiratory depression are not necessarily linked, because there are effective opioid cough suppressants that do not depress respiration.

C. Opioid Tolerance, Dependence, and Addiction

Although opioids are effective pain killers, they are reserved for treating more severe pain. This is because prolonged or repeated use of opioids may lead to the development of tolerance, a condition in which the effectiveness of the drug is diminished and higher and higher doses are needed to achieve the desired effect. Though this is usually not a major problem when opioid usage in clinical settings is limited to short (days to a week) periods of time and the signs of tolerance development are carefully monitored, it nonetheless represents a concern. Tolerance can be a major problem in some instances, especially during prolonged usage when treating terminal illnesses such as late-stage cancer, where sustained pain causes unbearable suffering.

Opioid dependence may also develop with repeated use. Dependence to a drug is defined by the appearance of adverse physical and psychological symptoms when the drug is abruptly discontinued. Withdrawal signs are particularly noticeable in opioid addicts. Thus, addiction is drug-seeking behavior to alleviate the adverse withdrawal symptoms as well as to seek the feelings of euphoria. The most commonly abused opioid is heroin. It is a morphine derivative, diacetylmorphine, which has a higher lipid solubility, thus enabling it to cross the blood–brain barrier more easily than morphine. Once in the brain, heroin is readily hydrolyzed to morphine, and it is the morphine that produces the heroin effects. Heroin is more widely abused than morphine because it has a quicker onset of action. Methadone, another μ-selective agonist, is currently used in the United States and other countries to combat heroin addiction. Methadone has a longer half-life than morphine and can be used effectively in maintenance programs to keep former heroin addicts from experiencing withdrawal symptoms.

BIBLIOGRAPHY

Brownstein, M. J. (1993). A brief history of opiates, opioid peptides, and opioid receptors. *Proc. Natl. Acad. Sci. U.S.A.*, **90**, 5391–5393.

Di Chiara, G., and North, R. A. (1992). Neurobiology of opiate abuse. *Trends Pharmacol. Sci.* **13**, 185–193.

Herz, A. (1993). "Opioids," Vols. I and II. Springer-Verlag, Berlin.

Kieffer, B. L. (1995). Recent advances in molecular recognition and signal transduction of active peptides: Receptors for opioid peptides. *Cell. Mol. Neurobiol.* **15**, 615–635.

Reisine, T., and Pasternak, G. W. (1996). Opioid analgesics and antagonists. *In* "Goodman & Gilman's The Pharmacological Basis of Therapeutics" (J. G. Hardman, L. E. Limbird, P. B. Molinoff, R. W. Ruddon, and A. G. Gilman, eds.), 9th Ed., pp. 521–554. McGraw–Hill, New York.

Optical Properties of Tissue

BRIAN C. WILSON
Ontario Cancer Institute

GLOSSARY

Bilirubin Yellow bile pigment formed in the breakdown of heme

Choroid Thin, pigmented vascular coating in the eyeball extending from the anterior margin of the retina to the optic nerve

Chromophore Chemical that absorbs light with a characteristic spectral pattern

Fluorescence Property of emitting light of a longer wavelength on absorption of light energy

Light fluence rate Measure of light "intensity," defined in terms of the optical power per unit area

Light scattering Change in direction of propagation of light in a turbid medium caused by reflection and refraction by microscopic internal structures

PROPAGATION OF ULTRAVIOLET (UV), VISIBLE, AND infrared (IR) radiation ("light") in tissues is governed by the absorption and scattering of light photons. At normal power densities, the absorption is caused by specific molecules ("chromophores"), which have different concentrations in different tissues and may have a strong wavelength dependence. Light energy absorption leads to reversible or irreversible photochemical, thermal, photomechanical, or other changes in the tissue. The scattering of light in tissue, resulting from microscopic variations in the refractive index, does not deposit energy in the tissue, but rather alters the spatial distribution of the light. This and the location of chromophores then determine the absorbed energy distribution and the resultant photobiological response. Transmission through the structures of the eye and the skin are important in the proper functioning of these organs. Transmission through or scattering from other organs is increasingly studied as lasers and optical technologies are applied to human biology and medicine.

I. FUNDAMENTALS OF LIGHT ABSORPTION AND SCATTERING

A. Background

The optical region of the electromagnetic spectrum includes UV (wavelength range 100–400 nm), visible (400–780 nm), and IR (>780 nm) radiation, as shown in Fig. 1. At shorter wavelengths lie X and γ rays, in which the photon energy is high enough to cause ionization of atoms in tissue, whereas at longer wavelengths lie the microwave and radiofrequency bands. The visible spectrum is divided into different color bands, the UV into three regions (i.e., UV-A, UV-B, and UV-C), and the IR into IR-A, IR-B, and IR-C. These divisions are related to particular physical and/or photobiological characteristics of the radiation, and other ways of dividing the spectrum are also used. For example, reference will be made to "near" and "far" UV or IR, corresponding to those parts of the spectrum lying closest to or farthest from the visible band, respectively.

A fundamental property of electromagnetic radiation is that the energy of each photon varies inversely with the wavelength. The photon energy in the optical spectrum corresponds roughly to molecular excitation energies, so that these photons can produce specific biophysical and biochemical effects in tissue, and these are strongly energy (i.e., wavelength) dependent.

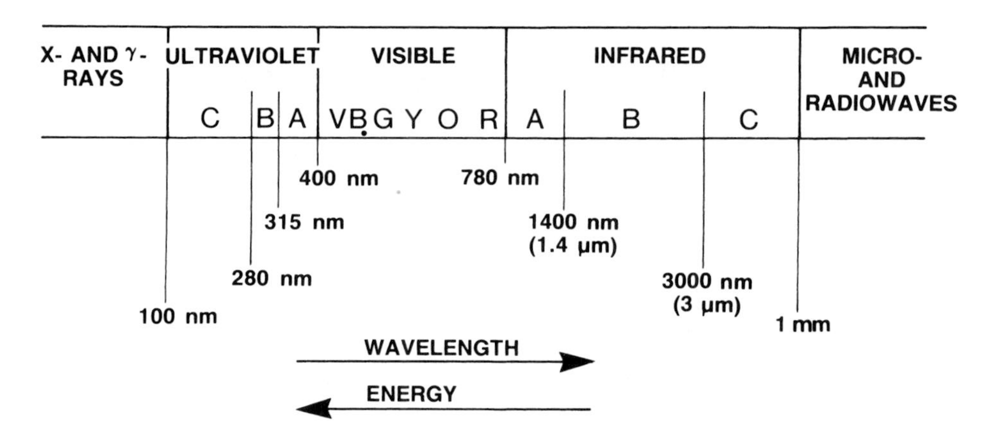

FIGURE 1 Electromagnetic spectrum showing division of the optical region of ultraviolet, visible, and infrared radiation. Note nonlinear wavelength scale. Some of the boundaries are variously defined: UVA/UVB may be 315 or 320 nm; UVA/visible, 380–400 nm; visible/IR, 760–780 nm. Normal colors [violet (V)–red (R)] are indicated in the visible region.

The optical region is of particular biological interest. It encompasses the natural spectrum of radiation reaching the earth's surface from the sun. A minimum exposure to at least parts of the optical spectrum is essential to human health and well-being, whereas overexposure may lead to tissue damage and disease. An example of the former is ricketts resulting from insufficient vitamin D, which is produced in the skin by exposure to UV-B. By contrast, chronic high exposure to UV-B is associated with induction of skin cancers. Treatment of diseases by light has been known since the time of the ancient Egyptians (the word *radiation* itself stems from Aton Ra, the Egyptian sun god). Herodotus, for example, in 425 BC. related the strength of the skull to sunlight exposure, whereas sunbathing was prescribed by early physicians for conditions as diverse as obesity, epilepsy, and jaundice. Vision itself depends on the transmission of visible light through the anterior structures of the eye and on the photochemical changes induced in the retina by absorption of the photon energy. Other wavelengths are harmful to the retina or lens and must be absorbed before reaching them.

These examples serve to illustrate fundamental principles in photobiology, namely, that

1. the photon energy must be absorbed by the particular tissue target before any biological effects can occur;
2. the biological effects resulting from energy absorption depend on the target and on the wavelength; and
3. the amount of light reaching a tissue target depends on the tissue through which the light must first travel, and hence on the light absorption and scattering properties of this tissue.

B. Light Absorption in Tissue

At high light power densities, which can be achieved using short-pulse lasers, light absorption in tissue is a complex, nonlinear process, in which molecules or atoms are excited to high energy levels, or even ionized. Although there is great interest in understanding the photophysics and photobiology in this regime, the applications are for the future and will not be discussed further.

At more normal power levels, as found with natural sources, lamps, and most lasers, light absorption is a linear process in which the rate of energy absorption is simply proportional to the incident light fluence rate [i.e., power density (W/cm^2)]. The absorption is due to specific molecules, known as chromophores, each of which has a distinct absorption spectrum. At any wavelength, the absorption coefficient of a particular tissue, which is a measure of the probability that a photon will be absorbed per unit distance traveled through the tissue, is the sum of the absorption coefficients of all the chromophores present. The absorption spectra of some important chromophores in soft tissues are shown in Fig. 2. In the visible region, hemoglobin is a significant absorber in the intravascular component of all tissues. It has a complex spectrum, which changes significantly between the oxy-

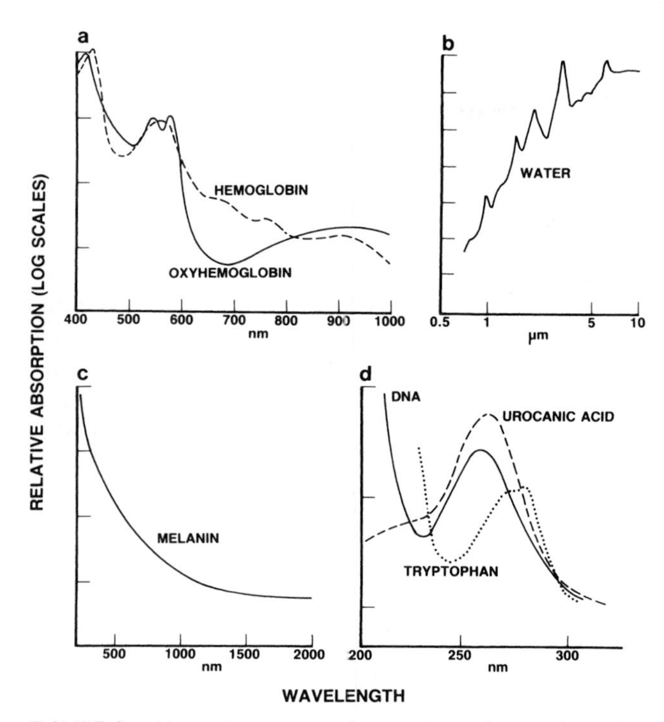

FIGURE 2 Absorption spectra of some tissue chromophores in the UV, visible, and IR. Note different wavelength ranges. Vertical scales are in arbitrary units, but in each case one division represents a factor of 10 in the absorption coefficient.

genated and reduced states. The double-peaked structure in the green region of the spectrum at 542/577 nm is particularly characteristic for oxyhemoglobin. There are several wavelengths where the absorption coefficient of hemoglobin is independent of the oxygenation ("isobestic points," e.g., ~815 nm). Comparisons of the ratio of absorption at isobestic and nonisobestic wavelengths may be used to determine the oxygenation status of tissues. Although its absorption is less in the near-IR than in the visible, hemoglobin is still a significant chromophore in this region.

The pigment melanin, a protein–polymer complex, is an important visible chromophore in some tissues. It has a relatively simple spectrum, with the absorption decreasing steadily with increasing wavelength. Melanin absorption plays a key role in the photoprotection and photoresponse of skin (see Section II,A). In the eye, the choroid also has a melanin layer, which absorbs light transmitted through the retina, thus preventing light being scattered back to the retina from the sclera. Melanin absorption is negligible above the near-IR range. Other visible pigments are present in different concentrations in various tissues. For example, bilirubin and beta-carotene are also found in skin and, together with hemoglobin and melanin, are responsbile for skin color.

In the UV range, absorption in tissue is mainly due to proteins and nucleic acids. The latter have a main absorption peak close to 265 nm because of the pyrimidine structure, whereas the aromatic amino acids are the absorbing sites in protein, with tyrosine peaking at 275 nm and tryptophan at 289 nm (see Fig. 2d). Numerous other small aromatic molecules also absorb in the UV. At wavelengths less than about 240 nm, high absorption results from peptide bonds. Absorption by water also becomes high at less than about 190 nm.

These various absorption characteristics contribute to the form of the "action spectrum" for photobiological effects in tissue (i.e., the wavelength dependence of biological responses to irradiation). This follows the first principle, namely, that energy absorption must occur before biological change can result. The correspondence is not exact, however, both because of the wavelength-dependent photobiological sensitivity of the targets and because of the influence of light scattering and the distribution of the chromophores. Examples are the action spectrum for UV-induced erythema (i.e., reddening of the skin as in sunburn), which peaks between about 240 and 290 nm, whereas skin cancer induction is greatest in the UV-B region between 280 and 325 nm.

As the wavelength increases beyond about 600 nm, and particularly in the IR-B and IR-C regions, the absorption of specific chromophores falls off, and water becomes the dominant IR absorber in soft tissues. The absorption coefficient of any tissue then depends largely on its water content. The IR absorption spectrum of water is complex (see Fig. 2b), with numerous narrow peaks. The absolute value of the water absorption can be large. At the highest peak around 2.9 μm, the coefficient is about 13,000/cm, so that, in a tissue with a 70% water content, such radiation is attenuated by 99% in less than 5 μm.

The total absorption in tissue is lowest between about 600 and 1300 nm, with the extinction of specific pigments having fallen off and that of water still being small. This red/near-IR "optical window" permits the greatest penetration of light into tissue and is, therefore, a region of particular interest in therapeutic and diagnostic applications.

An example of the total absorption spectrum in tissue is shown in Fig. 3. Even in highly pigmented liver, the total absorption is less than about 5/cm (0.5/mm) in the optical window. Thus, on average,

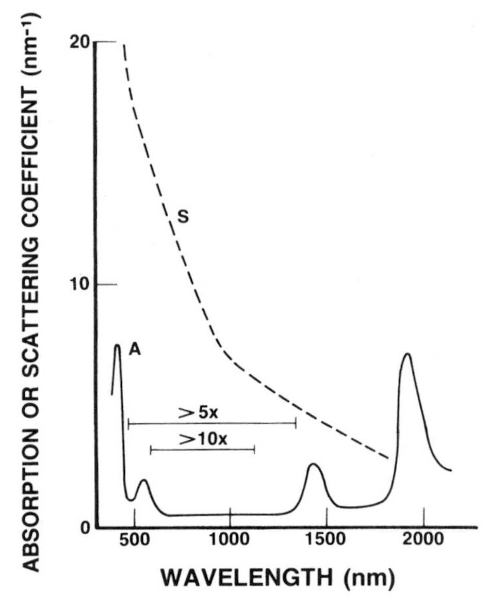

FIGURE 3 Example of the absorption (A) and scattering (S) spectra of tissue (liver *in vitro*). Bars show the wavelength ranges where the scattering is greater than 5 times or 10 times the absorption for this tissue, illustrating the concept of the optical window. [From P. Parsa *et al.* (1989). *Appl. Optics* **28**, 2325–2330, with permission.]

such photons travel more than 2 mm before being absorbed: however, because absorption and scatter are random processes, the distribution of these interaction distances is exponential, and 1% of the photons will travel more than 1 cm before being absorbed. For lightly pigmented tissues (e.g., brain) the average distance traveled before absorption may be several centimeters. At shorter visible wavelengths, hemoglobin and other pigments cause a marked increase in the total tissue absorption. In the example of Fig. 3, the 555-nm single absorption peak of deoxyhemoglobin is apparent, as is the so-called Soret band just above 400 nm, which corresponds to excitation of the heme ring structure by the light. Strong absorption peaks in the near-IR caused by water are also seen.

C. Light Scattering in Tissue

With the exception of the transparent structures of the eye, tissues are optically turbid, because photons may be scattered as well as absorbed. Scattering is normally elastic, so that the direction of the photons is changed but there is no loss of energy, and the wavelength is unaltered. The scattering is due to microscopic fluctuations in the refractive index of the tissue, corresponding to physical inhomogeneities. The angles through which the light is scattered and the scatter coefficient depend on the size and shape of the optical inhomogeneities relative to the wavelength and differences in the refractive index between the inhomogeneities and the surrounding medium. In scattering from molecules or structures whose size is much less than the wavelength (Rayleigh scattering), the scattering is weak, is nearly equal in all directions, and rapidly decreases with wavelength. When the wavelength and inhomogeneities are about the same size, the scattering is stronger, is more forward-directed, and decreases roughly inversely as the wavelength. Scattering from large inhomogeneities is highly forward-directed and only weakly dependent on wavelength (Mie scattering).

The inter- or intracellular structures causing light scattering in tissues have not been clearly identified and may well be different for different tissues. In most soft tissues, scattering is very forward-directed in the visible and near-IR. This suggests that the optical inhomogeneities are of the same order as or larger than the wavelength (i.e., on a micron scale). The scatter spectrum for liver is shown in Fig. 3 and is typical of soft tissues in this spectral range. The scattering decreases approximately linearly with wavelength, with little spectral structure. In the optical window, the scatter coefficient is much greater than the absorption coefficient. Also, over the wavelength range in Fig. 3, the scattering angular distribution is roughly constant, with an average scattering angle of about 20–30° (data not shown). At a given wavelength in the optical window, the scattering coefficients vary between tissues by a factor of about 5–10.

In the dermis of skin, scattering has been associated with collagen fibers, and optical changes seen in aging or photodamage may result from changes in collagen content or structure. In the brain, laying down of myelin with maturation is thought to be important in causing the increased light scattering seen in adult tissue compared with neonatal tissue. [*See* Collagen, Structure and Function.]

Although scattering is the result of microscopic variations in refractive index, the average refractive index of tissue determines the speed of light in the tissue and also the regular (specular) reflection at the tissue surface. The latter causes the "shining" appearance of skin or freshly exposed tissue surfaces. In the red part of the spectrum where measurements have been made, the average refractive index of most soft tissues is in the range 1.37–1.45, so that, when a visible light beam strikes an air–tissue interface, typically a few percent is specularly reflected. By contrast, a large fraction (up to ~60%) may be diffusely reflected after multiple scattering within the tissue.

On a microscopic scale, light photons crossing cell membranes between relatively aqueous and lipid environments are subject to both refraction (bending of light waves) and specular reflection, which result in scattering as seen on a macroscopic scale. This can be seen by passing a light beam through a cell suspension. In an aqueous environment, the beam is strongly scattered, but this can be markedly reduced by increasing the refractive index of the suspension medium to match that of the cell membranes.

D. Combined Absorption and Scattering in Tissue

For soft tissues in the optical window, the scattering coefficient is high and also much larger than the absorption coefficient. Typically, the average distance between scattering interactions is only tens of microns. However, many more scattering interactions than absorption events take place, and the scattering is forward-peaked. These properties account for the existence of the optical window: although the photons interact with the tissue tens or hundreds of times for each millimeter of their travel, in most cases the photons survive with no loss of energy and are deflected only slightly (and randomly) from the initial direction. Thus, a fraction of the photons may penetrate to considerable depth. For example, in near-IR transmission spectroscopy of the neonatal brain to measure the tissue oxygenation, light is applied to one side of the head, and detectors on the other side measure the transmitted fraction. Virtually no photons pass through the head without interacting, but a large transmitted signal can be detected nevertheless. This comes from the scattered light, whose average distance traveled in crossing the head is on the order of meters rather than centimeters.

The forward scattering of tissue effectively reduces the strength of the scattering coefficient. Photons may scatter many times through a small angle and travel an equivalent depth in tissue as they would if the number of scatters was less but each scatter was through a large angle. The angular dependence of the scatter and the scattering coefficient may be combined in the concept of a "reduced" or "transport" scattering coefficient, which is typically 10 or 20 times less than the true scattering coefficient. Then, unless the detailed scattering angles are specifically important, the transport coefficient is a convenient measure of effective scattering.

The absorption and scatter in tissue combine to determine how light entering the tissue is distributed. An example is shown in Fig. 4, where different tissue pigmentation and blood content lead to different and wavelength-dependent light distributions. In the extreme case of the absorption being much greater than the scattering (UV and far-IR), a light beam incident on tissue is attenuated exponentially, with little lateral spreading of the light or backscattering (diffuse reflectance). Conversely, where the tissue is very scattering, there is wide spreading of the beam; the decrease of fluence with depth below the surface is complex and may even show a subsurface peak several times higher than the incident value, and the diffuse reflectance is high.

An important consequence of high light scattering in tissue is that any simple spectral measurement (e.g., light transmitted through or diffusely reflected from a certain thickness of tissue) does not reveal the true absorption spectrum of the tissue. Rather, the spectrum is distorted by the scattering. Thus, in Fig. 4, the ratio of the intensities of light reaching the surface

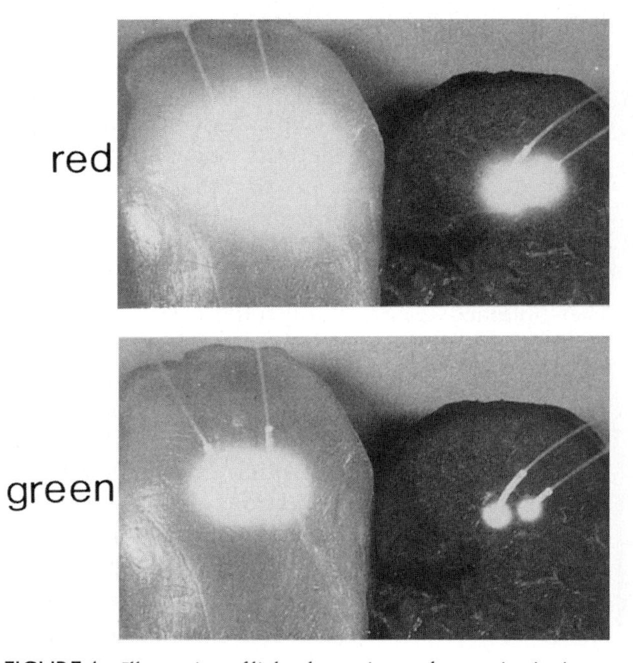

FIGURE 4 Illustration of light absorption and scattering in tissue. Each panel shows two different tissues (lightly pigmented 'white" avian muscle on the left and "red" bovine muscle on the right) with implanted optical fibers delivering either red (630 nm) or green (514.4 nm) light. Photographs show distribution of unabsorbed light reaching the surface of tissues after scattering. The optical power delivered to tissues is the same in all cases. [From J. A. Marijnissen and W. M. Star (1987). *Lasers Med. Sci.* **3**, 235–242, with permission.]

at the two wavelengths in each tissue is not the same as the inverse ratio of the absorption coefficients at these wavelengths. In general, scattering tends to "flatten out" peaks and valleys in the absorption spectrum.

In many practical situations, it is not necessary to consider separately the contributions of absorption and scattering to the light attenuation. The exponential decrease in fluence as light penetrates through tissue may be described adequately by a single "effective attenuation coefficient." This depends on factors such as the light beam size and shape as well as on the tissue optical properties and so varies from case to case. The effective attenuation coefficient is directly proportional to the tissue optical density [OD = $\log_{10}(I_0/I)$, where I_0/I is the ratio of the incident to transmitted fluence rate of a light beam passing through 1 cm of tissue].

E. Measurement Techniques

Measurement of tissue optical properties is not simple. In addition to the problems caused by light scattering, the tissue handling and preparation are important. Clearly, embedding techniques or histological stains cannot be used. Freezing may change the scattering properties, whereas keeping tissues at room temperature even for a limited period can alter the absorption spectrum and possibly the scattering. Loss of components during excision or sectioning, particularly blood, is an obvious source of error. With solid tissues, the condition of the surface can affect how light is scattered from it. The most direct information has been obtained using optically thin tissue sections (typically $<100\ \mu m$), but these are most subject to preparation artifacts. Thicker sections ($\sim$mm) may be used, but the determination of the true scattering and absorption values is complicated by the interplay between scattering and absorption in the overall light transmission through the section.

Comparisons of light distributions measured in intact tissues immediately postmortem with those in living tissue suggest that the scattering properties are not generally much altered but that the absorption in the visible range can change markedly because of drainage of blood from the tissues and reduced blood oxygenation.

Measurements *in vivo* have generally been done by inserting optical fibers into the tissue. The other end of the fiber is connected to a light detector, so that the distribution of input light in the tissue may be "mapped." The optical properties may be deduced if this mapping is sufficiently detailed. Recently, noninvasive tissue spectroscopy has become feasible, as illustrated by the case of neonatal brain measurements cited earlier. The general principle is to deduce the optical properties from the (spectrum of) light that is diffusely reflected from or transmitted through the tissue. An area of particularly active study is the use of time-of-flight spectroscopy. In this, short laser pulses are used, and the time spectrum of diffusely reflected or transmitted light is measured using fast detectors. The time taken for a photon to travel between the source and detector depends on the path that it follows through the tissue by multiple scattering. The fraction of the light that survives a particular path length depends on the tissue absorption. Thus, the time spectrum carries information on both the scattering and absorption properties of the tissue. Because the speed of light in tissues is so high (20 cm in 1 nsec), these techniques are at the leading edge of laser and optical detector technology.

F. Data on Human Tissue Optical Properties

The characteristics of tissues discussed here are based primarily on measurements in nonhuman tissues. With the exception of skin and eye, which are discussed next, there have been few systematic studies of human tissues. This situation is rapidly changing with the interest in laser applications in medicine. Although clearly there are differences in the detailed spectral characteristics of given tissues between human and other mammalian species, the general characteristics described are expected to apply to the human case also, because chromophore content (absorption) and tissue/cellular structure (scattering) are generally comparable, even if not identical. This has been borne out by the few measurements that have been made.

II. OPTICAL PROPERTIES OF THE SKIN, EYE, AND BLOOD

The skin and eye are unique organs in that they are naturally and routinely exposed to optical radiation. They have evolved to take advantage of parts of the optical spectrum but also to protect themselves from components that are harmful. Unlike most other tissues or organs that are, optically, relatively homogeneous, the skin and eyes are heterogeneous and optically complex. The skin has several distinct layers with widely varying absorption and scattering properties. These properties are also dynamic, changing with alterations in local vasodilation, vascular permeability, blood oxygenation, pigmentation, and thickness of the skin. The anterior structures of the eye are nonscattering, which is unique in itself, but have absorptions and hence transmittances, which are strongly wavelength-dependent in the UV, visible, and IR. There is a large literature on the optics of skin and eye, of which only the major features can be presented.

A. The Skin

In considering the skin, the critical issue is the vastly different penetration of light of different wavelengths into the various layers, namely, the stratum corneum, epidermis, and dermis. This is illustrated in Fig. 5. The stratum corneum is a thin (10–20 μm) proteinaceous layer of flattened, keratinized dead cells. It acts as a specularly reflecting surface with a refractive index of about 1.55, which gives 5–7% specular reflectance over the optical range, independent of pigmentation. It is relatively translucent (i.e., not highly scattering). The absorption is largely due to melanin, except at short UV wavelengths.

The epidermis, which varies in thickness between about 50–150 μm, absorbs most of the UV-C and far-IR striking the skin. The UV absorption is due to aromatic amino acids, urocyanic acid, and nucleic acids (see Fig. 2d). The epidermal absorption in the 250- to 300-nm range is high, regardless of the melanin content. However, in the UV-A, visible, and near-IR, melanin absorption in the epidermis has a profound influence on the penetration of light to the dermis and the spectrum of light diffusely reflected from the skin, which determines the skin color. Melanin granules are produced in the melanocytes of the basal layer and transferred to keratinocytes, which migrate outward (although in normal skin, the majority of melanin remains located in the basal cell layer). The dermis is also optically highly scattering, being a semisolid mixture of collagen (and reticulum and elastin) fibers embedded in a viscous ground substance of water and polysaccharides. It contains the blood capillaries, sweat and sebaceous glands, nerves, and lymphatics.

Both the production of melanin and its migration can be increased by UV exposure, especially UV-B, in the process of delayed tanning. (Short-term immediate tanning is the result of photochemically induced changes in preexisting melanin granules.) Once pres-

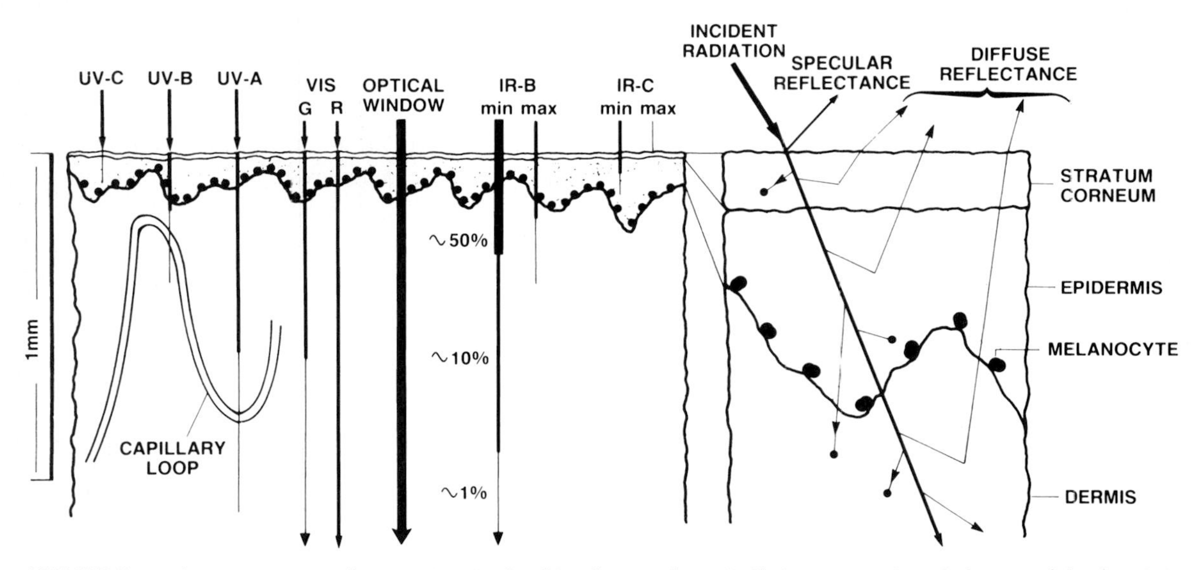

FIGURE 5 Light penetration and interactions in the skin, shown schematically in cross section. Only part of the dermis is indicated: its total thickness may be as much as about 4 mm, and below it (not shown) lies fatty subcutaneous tissue that acts as a heat insulator and shock absorber. On the left side, each line indicates approximate depth of penetration of different fractions of the incident light. For IR-B and IR-C, "max" and "min" represent wavelengths corresponding to local peaks and valleys in the water absorption spectrum. On the right of the diagram, the expanded cross section shows absorption and scattering processes in each layer, specular reflection from the stratum corneum, and contributions of light backscattered from various layers to the total diffuse reflectance of the skin. Most of the diffuse reflection comes from the dermis.

ent, the melanin protects the deeper layers. In the absence of this protection or in the case of acute overexposure to UV, damage may be caused to the dermal tissue (e.g., in sunburn), which results in capillary dilation and inflammatory response.

The effect of melanin content (and distribution) on skin color is seen in Fig. 6. In black skin, the melanin in the epidermis markedly reduces the reflectance of visible and near-IV light. At longer wavelengths, above about 1100 nm, the skin "color" is indistinguishable, because the melanin absorption in stratum corneum and epidermis becomes negligible (see Fig. 2c). High melanin content also reduces or eliminates the characteristic "signature" of other chromophores (e.g., hemoglobin). Thus, although in Caucasian skin the oxyhemoglobin band is clearly seen in Fig. 6, this is lost in Negroid skin because of high absorption by melanin of both the incident visible light and of light that is backscattered from the dermis.

In the dermis, the main chromophores in the visible region are hemoglobin (oxy- and reduced), beta-carotene, and bilirubin. Hemoglobin is entirely vascular, whereas beta-carotene is sequestered in dermal lipid and subcutaneous fat. Bilirubin may be both intra- and extravascular.

Although the concentration and distribution of chromophores play important roles in the optics of skin, it must be emphasized that the scattering of light, particularly in the dermis, also profoundly influences the spectral distribution of light in the skin. Much theoretical and experimental work has been directed at understanding the complex interplay of absorption and scattering in this multilayered structure. The details are particularly important in therapeutic applications of light in dermatology, where targeting of light energy to the correct structure is critical. Thus, in the laser treatment of portwine stain, the congenitally abnormal blood vessels causing this disfigurement are thermally occluded by pulses of light at 577 nm, corresponding to an oxyhemoglobin absorption peak, with little effect on overlying normal epidermis in which the absorption at this wavelength is low. Use of short pulses also ensures that the effect is produced in the blood vessels before the induced heating has time to spread and damage adjacent structures.

As a result of the complex optics of the skin, the action spectra for particular photobiological effects depend strongly on the depth of the specific targets and on the absorption and scattering of overlying layer(s). The sensitivity to different spectral regions can also be increased by deliberate or accidental exposure to photosensitizers, of which there are many man-made and natural species. Deliberate medical uses range from psoralen-UVA (PUVA) treatment of psoriasis to the treatment of skin cancers by porphyrins and other red- or near-IR-activated photosensitizers. [*See* Skin; Skin, Effects of Ultraviolet Radiation.]

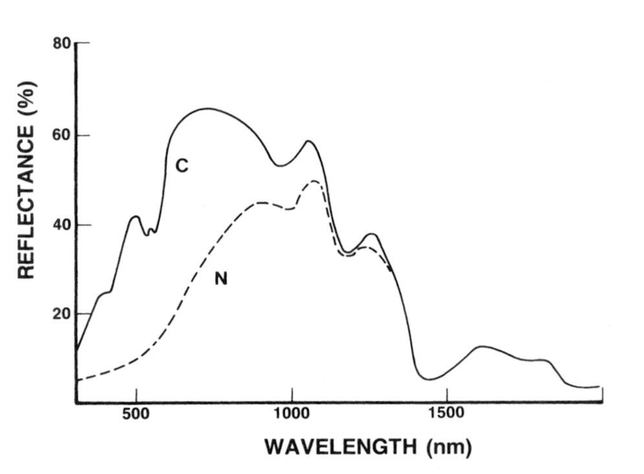

FIGURE 6 Reflectance spectrum of dark Negroid (N) and fair Caucasian (C) skin *in vivo*. Measurements include regular (specular) reflectance of a few percent. [From Anderson and J. A. Parrish (1982). "The Science of Photomedicine" (J. D. Regan and J. A. Parrish, eds.). Plenum, New York, with permission.]

B. The Eye

The focus in this section is not on the functioning of the eye as a light-imaging system. Rather, the transmission of optical radiation through the structures of the eye will be considered, as a function of wavelength from UV to IR. The essential characteristics are illustrated in Fig. 7. Visible light is transmitted with little absorption loss to the retina. Near-IR is partially ab-

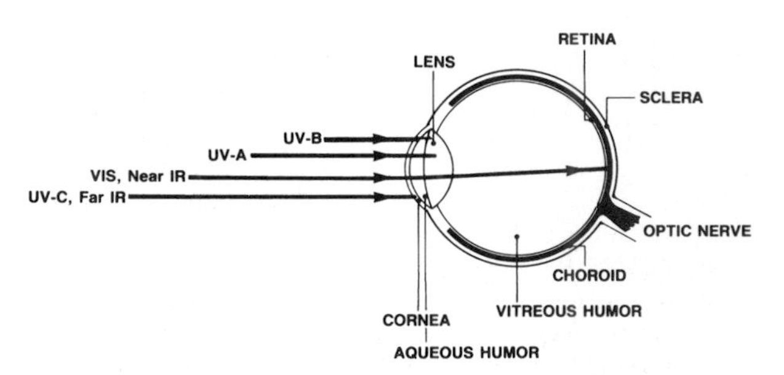

FIGURE 7 Schematic of transmission of different spectral regions through the occular media.

sorbed in the aqueous and vitreous humors. UV and far-IR are strongly absorbed by the anterior elements. UV-C and far-IR are completely blocked by the cornea. UV-B is mainly absorbed by the cornea with some reaching the lens, whereas UV-A passes through the cornea and aqueous with little loss but is completely absorbed by the lens. These various absorption losses are shown quantitatively in Fig. 8 for each component in the eye and for the whole structure through the retina. Note that the overall transmission is significant only in the visible and near-IR range. All the anterior elements of the normal eye are essentially nonscattering, thus allowing their functioning as high-quality optical elements.

As would be expected, the action spectra for damage to the occular structures are the consequence of the location of the main energy absorptions at each wavelength. Acute exposure to UV-B and UV-C primarily produces effects in the cornea and conjunctiva, which is the thin membrane lining the eyelids and covering the cornea. Examples are conjunctivitis (inflammation of the membrane) and photokeratitis (inflammation of the cornea), as in the effects known as *welders flash* or *snow blindness*. Cataract formation, which is a partial or total loss of the optical transparency of the lens (or lens capsule), is associated with chronic exposure to UV-A. The loss of transparency involves induction of both visible pigmentation and optical scattering.

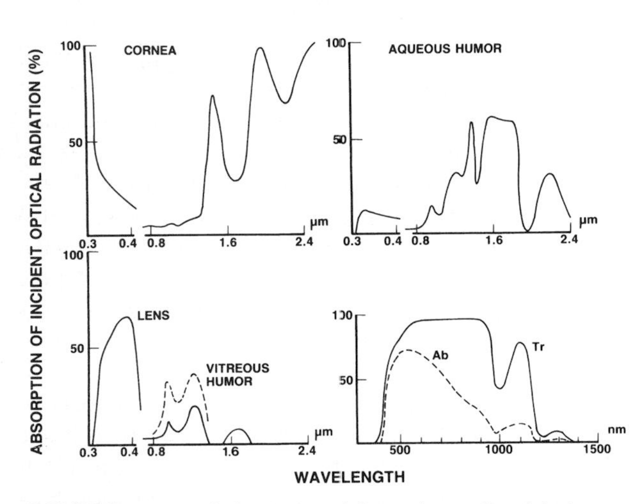

FIGURE 8 Spectral absorption of the ocular media of the human eye (first three graphs), expressed as percentage of the incident optical radiation absorbed by each structure. Note the break in curves in the visible, where absorption of all structures is low. The lower right graph shows percentage transmission of the whole ocular structure in the visible and near-IR and percentage absorption of light energy in the retina and choroid in this spectral region. Ab, absorbed; Tr, transmitted. [From D. Sliney and M. Wolbarsht (1980). "Safety with Lasers and Other Optical Sources." Plenum, New York, with permission.]

The sclera is highly scattering for the light that is transmitted through the retina and choroid. The effect of its backscattering into the eye is greatly reduced by the pigmented epithelial layer between the retina and choroid, as is evident in the case of albino individuals in whom this protective effect does not occur. [*See* Eye, Anatomy.]

C. Blood

The optical properties of blood are important not only because the absorption contributes significantly to the absorption of tissues in the visible part of the spectrum, but also because optical techniques are used in a number of *in vitro* and *in vivo* assay techniques to measure quantities such as oxyhemoglobin saturation, hemoglobin concentration, hematocrit (volume percentage of red blood cells), and arteriovenous oxygen difference. The absorption and scattering of visible and near-IR light by whole blood is determined primarily by those of the red blood cells. The optical coefficients thus depend on the hematocrit value, H: The absorption coefficient is proportional to H, whereas the scattering coefficient increases with H at small values but falls again at high hematocrit.

Because the absorption spectrum is due to hemoglobin, whose spectrum varies with oxygenation (see Fig. 2a), the absorption spectrum of blood has a complex shape that depends on the oxygen saturation, S. For example, at 660 nm, the absorption coefficient for $H = 50\%$ and $S = 0\%$ is about 4/mm, whereas for $H = 50\%$ and $S = 100\%$ (fully oxygenated), the value is about 0.4/mm (i.e., 10 times smaller). However, the corresponding values at 820 nm are roughly equal to those at about 1/mm. Thus, the effect of blood oxygenation on the light absorption is wavelength-dependent. The corresponding scattering coefficients do not depend on the oxygenation and change only slowly with wavelength in this range. As with complete tissues, the scattering is much higher than the absorption (for $H = 50\%$, the scattering coefficients at 660 nm and 820 nm are about 700/mm and 500/mm, respectively). In the visible range where measurements have been done, red blood cells are highly forward scattering, with characteristics consistent with their size and refractive properties. [*See* Hemoglobin.]

III. EFFECTS OF LIGHT ON TISSUE

The foregoing discussion has been concerned mainly with how tissue affects light, by absorbing and scattering it. The question of how light affects tissue is of primary concern in photobiology and photomedicine. Only some major aspects of this can be touched on. A central principle has been expounded already (i.e., that the light must be absorbed in the tissue before any biological effects can occur). Further, the spectrum and distribution of light in the tissue determine the chromophores affected and the location of energy deposition. Absorption results initially in raising of molecules to a higher energy state. A variety of mechanisms exists for de-excitation, depending on the chromophore, the degree of excitation, and the physicochemical environment of the chromophore. De-excitation pathways may be radiative or nonradiative.

In radiative processes, part of the energy may be reemitted as light. Fluorescence is the commonest mechanism, in which light of a longer wavelength (lower energy) than the incident light is given off, usually on a time scale on the order of nanoseconds. Not all tissue chromophores fluoresce. As with the absorption spectrum, the fluorescence emission spectrum is distinctive for particular fluorophores (fluorescent chromophores). At any given emission wavelength, there are a range of wavelengths that can excite fluorescence. The excitation spectrum is usually similar but not identical to the absorption spectrum. Examples of tissue autofluorescence spectra (i.e., the spectra of naturally occurring fluorophores in the tissue) are shown in Fig. 9. This illustrates the distinctiveness of such spectra and the fact that significant changes may occur with disease, caused by presence or absence of specific molecules. The plots also demonstrate the effect of reabsorption of excitation and emitted light by hemoglobin in the tissue, which accounts for some of the contour valleys (at 420, 540, and 580 nm; compare with Fig. 2a). Such fluorescence spectroscopy holds exciting potential for medical diagnosis. Fluorescence imaging is also widely used in tissue microscopy. Deliberately administered (exogenous) fluorophores may be used, both *in vitro* and *in vivo*.

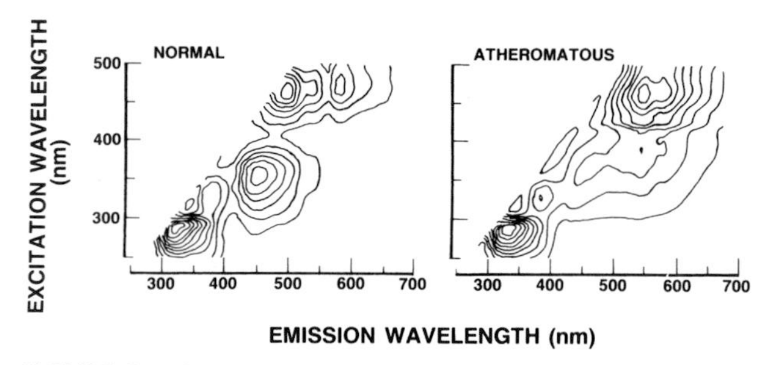

FIGURE 9 Fluorescence contour maps in normal and abnormal human aorta (*in vitro*). Note that the wavelength of emitted fluorescence is always greater than the excitation wavelength. (Data kindly provided by Drs. M. Feld and R. Richards-Kortum, Massachusetts Institute of Technology.)

Although radiative de-excitation is of great practical interest, it normally accounts for only a small fraction of the energy absorbed in tissue. Nonradiative mechanisms lead to the essential biological effects observed. Except in some laser techniques, the main results of energy absorption by tissue are either photochemical changes or heating of tissue. The former may be complex and depend critically on the chromophores present, on the irradiation wavelength, and on the biophysical and biochemical environment. The photochemical events lead to specific biological changes, depending on the chromophores and their location in the cell or tissue. This is described in the action spectrum for the particular photobiological effect.

Tissue heating is usually nonspecific. The biological effects depend on the temperatures produced and the nature of the tissue. This can range from physiological changes (e.g., altered blood flow) to biochemical changes (e.g., protein or collagen denaturation) to vaporization or even burning of tissues. Deliberate, localized heating is the basis for much of laser surgery. Heating of tissues can be localized to some degree by using wavelengths where particular chromophores absorb strongly (e.g., hemoglobin in blood vessels). Thermal effects may also be confined to some degree by using pulses of (laser) light on a time scale that is short compared with the time for heat to diffuse in the tissue, as illustrated earlier.

In recent years, there have been many reports of rather nonspecific effects of low light levels in tissue (typically ~mW/cm^2). Effects have ranged from accelerated wound healing to pain control. The photobiological basis for this "biostimulation," in which the wavelength appears to be critical, is not well understood but is likely related to triggering or stimulation of particular metabolic pathways.

At the other extreme of high optical power densities and short exposure times, photomechanical effects can be produced, which can result in cutting or ablation (physical removal of tissue from an irradiated surface). With intense short pulses of UV radiation, direct photochemical breaking of molecular bonds can occur, again resulting in cutting or ablation of tissue.

BIBLIOGRAPHY

Ben-Hur, E., and Rosenthal, I. (eds.) (1987). "Photomedicine," Vols. I–III. CRC Press, Boca Raton, Florida.

Carruth, J. A. S., and McKenzie, A. L. (1986). "Medical Lasers: Science and Clinical Practice." Adam Hilger, Bristol, England.

Kohen, E., Santus, R., and Hirschberg, J. G. (1995). "Photobiology." Academic Press, San Diego.

Preuss, L. E., and Profio, A. E. (eds.) (1989). Tissue optics. *Appl. Optics* **28**, 2207–2357 (multiple papers).

Regan, J. D., and Parrish, J. A. (eds.) (1982). "The Science of Photomedicine." Plenum, New York.

Sliney, D., and Wolbarsht, M. (1980). "Safety with Lasers and Other Optical Sources." Plenum, New York.

Smith, K. C. (ed.) (1989). "The Science of Photobiology," 2nd Ed. Plenum, New York.

Suess, M. J. (ed.) (1982). "Nonionizing Radiation Protection," Chaps. 1–3, Regional Publications, European Series 10. World Health Organization, Copenhagen.

Wilson, B. C., and Jacques, S. L. (1990). Optical reflectance and transmittance of tissues: Principles and applications. *J. Quant. Electr.* **26**, 2186–2199.

Oral Contraceptives and Disease Risk

ZHONG XU
LUE PING ZHAO
KATHLEEN MALONE
ROSS L. PRENTICE
Fred Hutchinson Cancer Research Center

GLOSSARY

Combined oral contraceptives Pills containing both an estrogen and a progestin, which prevent conception by suppressing ovulation. In each cycle, such pills are taken daily for 3 weeks followed by cessation for 1 week

Meta-analysis Formal means of combining the intrastudy information from two or more reports

Relative risk Ratio of diseases rates for "exposed" as compared with "unexposed" persons

STEROID CONTRACEPTIVES CONTAIN AN ESTROgen or a progestin, or a combination thereof. By far the most commonly used steroid contraceptive is the combined oral contraceptive (OC), in which each pill contains both an estrogen and a progestin. Such pills are taken daily for 3 weeks, followed by cessation for 1 week, during which withdrawal bleeding occurs, and the cycle is resumed. These preparations have proven effective in preventing pregnancy, primarily by suppressing ovulation. However, intensive study during the past 25 years has identified a corresponding impressive range of risks and benefits. The most important risks include elevations in cardiovascular disease rates, including coronary heart disease and stroke, whereas benefits include reduced rates of cancers of the ovary and endometrium. The possibility that combined OCs increase the risk of cancer of the breast among young women continues to be a topic of intensive controversy and study. Increased rates of cervical cancer and liver cancer among OC users have also been reported. Much of the available epidemiologic data involve older OC preparations in use during the 1960s and 1970s. The newer low-dose pills, containing less than 50 μg of estrogen and substantially less progestin than the early pills, now account for the vast majority of usage, particularly in developed countries. Although it is reasonable to anticipate an improved risk–benefit profile and a diminution of cardiovascular disease risk in particular with the low-dose preparations, additional years of experience will be necessary to establish such results.

I. PREPARATIONS AND PATTERN OF USE

The combined OC preparations suppress ovulation and, in the unlikely event that ovulation occurs, evidently reduce conception rates by affecting ovum transport through the fallopian tubes and by altering cervical mucus so that reduced spermoidal penetration occurs. These products were first marketed in the United States in the early 1960s. Their use rapidly spread throughout the world, and by 1980 approximately 80 million women had used combined OCs. The original Enovid pill contained 10 mg of norethynodrel (a progestin) and 150 μg of mestranol (an estrogen), but most current usage involves 1 mg or less of progestogen and less than 50 μg of an estrogen. The specific estrogen and progestin used vary among a considerable number of preparations. In the United States, norethindrone is a commonly used progestin and ethinyl estradiol is the most commonly used estrogen.

Sequential oral contraceptives were marketed widely in the United States up to about 1980 and are still in use in some countries. These preparations involve exposure to estrogen alone for approximately 15 days of a woman's menstrual cycle, approximately 5 days of an estrogen plus a progestogen, followed by cessation for a week with resultant withdrawal bleeding. Because of the substantial period of "unopposed" estrogen with such sequential preparations, one may gain valuable insights into the hormonal aspects of disease occurrence by comparing the epidemiologic effects of sequential and combined OCs. Such comparison might also lead to hypotheses concerning the effects of other steroid contraceptives, such as the minipill, for which few epidemiologic data yet exist.

The minipill has been sold in the United States since 1973. Such pills contain only a progestin, at a considerably lower dosage than is found in the combined pill. This pill does not suppress ovulation but appears to act by creating the thickened cervical mucus and by inhibiting ovum transport and implantation. This pill is not currently widely used. Some other strongly progestational contraceptives, particularly long-acting injectable preparations [e.g., norethindrone enanthate and, especially, depomedroxyprogesterone acetate (DMPA)], have been used extensively in a number of countries. DMPA, for example, is usually administered every 3 months in a dose of 150 mg. These products were licensed only recently for use as contraceptives in the United States.

The newest combined OCs are the multiphasics. These preparations, which are widely used in some developed countries, vary the doses of the two hormones during the pill cycle. Such variation is designed to keep hormone doses low while preventing ovulation and maintaining menstrual cycle control.

There has been a gradual increase in the use of OCs during the 1980s. In 1988 an estimated 14% of married women of reproductive age in developed countries and 8% of such women in developing countries, for a combined total of more than 60 million married women worldwide, use such preparations. By 1987, pills with estrogen doses less than 50 μg accounted for about 85% of all pharmacy purchases of combined OCs in developed countries.

From the beginning it has been clear that combined OCs offer almost perfect, reversible, contraceptive effectiveness when used properly. However, a range of noncontraceptive side effects have also been identified. Early clinical trials identified headaches, nausea, irregular menstruation, breast tenderness, and weight gain as possible short-term side effects of OC use. By the late 1960s and 1970s, more serious adverse effects, including thromboembolism, stroke, and heart disease, were identified, stimulating the dosage changes noted. Since the late 1970s there has been much study of OC use in relation to the risk of several cancers. Combined OC use has been found to convey noteworthy protection against ovarian and endometrial cancer, while possibly increasing the risk of certain other cancers. The following sections discuss the health risks and benefits of OC use. Much of the material updates the earlier reviews by R. L. Prentice and coworkers.

II. RISKS AND BENEFITS FOR CARDIOVASCULAR DISEASES

A. Background and Methods

Since 1968, four substantial cohort studies have been conducted to study the relation between oral contraceptive use and subsequent disease occurrence. The Royal College of General Practitioners' (RCGP) Oral Contraceptive Study enrolled 46,000 women aged 15 or older, about one-half of whom were never-users of OCs. The Oxford Family Planning Association Contraceptive Study recruited 17,032 women aged 25–39, 56% of whom were current OC users. The Walnut Creek Contraceptive Drug Study enrolled 16,638 women aged 18–54, of whom 28% were OC users, 33% were former users, and about 40% were 40 years of age or older. More recently, a cohort of 121,700 nurses aged 30–55 was assembled in the Nurses' Health Study with completed questionnaires. At baseline, 55% of participants reported that they had ever used oral contraceptives and 6% reported that they were current OC users. These cohort studies have provided much valuable information on the risk and benefits of OC use.

Case–control studies have also been carried out using a variety of study populations to relate OC use to specific diseases. These studies compared patterns of OC use among the "cases" developing a certain disease to the usage patterns of selected "control" subjects who had not developed this disease, thereby allowing the dependence of the disease odds ratio on OC use to be estimated. In the present context, the odds ratio is always sufficiently close to the relative risk that the two are interchangeable and the terminology "relative risk" is used throughout. The relative risk is simply the ratio of the disease occurrence rate among OC users to that among nonusers. Of course, the magnitude of the relative risk for a given disease can depend on aspects of OC use such as duration of use, time since last use, timing of use, or hormone dosage and on study subject characteristics such as age, reproductive factors, and the presence of disease risk factors. In addition, many studies have confined their analysis of disease risk to the index of ever/never use of OCs, limiting interpretability to some degree. Further, the diseases of interest here vary in terms of the duration of time following use that may be required before disease manifestation. Diseases with long "induction" periods, as has been hypothesized for breast cancer, may need follow-up periods that are so long as to make the case–control study design more practical. Cardiovascular disease increases probably with increasing current and/or recent use of OCs and thus may be more conducive to evaluation using a cohort study design. The size and quality of the studies relating OC use to a disease are often quite variable. Hence the tables provide separate summary relative risk estimates from cohort and from case–control studies, the results from which have been reported in the medical literature and have been included in the MEDLINE database. The logarithm of a summary relative risk estimates is calculated as a variance-weighted linear combination of the logarithms of the relative risk estimates from two or more specific studies.

B. Cerebrovascular Disease

As shown in Table I, the first three cohort studies (Royal College, Oxford, and Walnut Creek) give a summary cerebrovascular disease relative risk estimate of 2.9, with an approximate 95% confidence interval of (2,0, 4.1) for current use of OCs versus never-use. The confidence interval is denoted as 95% C.I. = 2.0–4.1 hereafter for simplicity. This estimate was based on reports in the early 1980s and hence primarily reflects the risk associated with the higher-dose pills in use in the 1960s and the 1970s. Most of these cerebrovascular disease events were classified as subarachnoid hemorrhage, for which the cohort study summary relative risk estimate is 2.0 (95% C.I. = 1.1–3.6), or nonhemorrhagic (thrombotic, embolic) stroke, for which the relative risk estimate is 3.8 (95% C.I. = 2.4–6.1).

As shown on the right side of Table I, the same three cohort data sources combine to yield evidence of elevated stroke risk among former users of OCs as well, with a relative risk estimate of 1.8 (95% C.I. = 1.3–2.6) for total cerebrovascular disease, 2.1 (95% C.I. = 1.1–3.4) for subarachnoid hemorrhage, and 1.9 (95% C.I. = 1.2–3.0) for nonhemorrhagic stroke. There was no evidence of disagreement among the three cohort studies in respect to the magnitude of the relative risks for either former or current OC use.

Cerebrovascular disease risk has also been studied in relation to OC use in case–control studies. By far the largest such study was carried out by the Collaborative Group for the Study of Stroke in Young Women, which identified 598 nonpregnant women aged 15–44 with various types of cerebrovascular disease and compared their OC use history with that of matched neighborhood and hospital controls. The case–control results for current OC use in Table I arise from the 12 studies reported after the mid-1970s. [*See* Stroke.] The summary relative risk estimate of current OC use from case–control studies for nonhemorrhagic stroke, in relation to current OC use, which have been estimated, is generally consistent with that from cohort studies. The summary risk estimate from case–control studies for hemorrhagic stroke in relation to current OC use is smaller than that from cohort studies. In general, however, these results from case–control and cohort studies are similar. Listed on the right-hand side of Table I are summary relative risk estimates for cerebrovascular diseases in relation to former OC use. The estimates from these case–control studies generally agree with those from cohort studies, although the estimates for all cerebrovascular diseases and for hemorrhagic diseases are slightly lower.

The RCGP cohort study also examined the relationship of dependence of the relative risk for total cerebrovascular disease to duration of OC use and time from cessation of OC use. There was a suggestion of a higher relative risk if the duration of current use exceeded 4 years, whereas there was no evidence that the elevated risk among former users declined with increasing years since cessation of use. However, a study based on the follow-up of 119,000 women in the Nurse's Health Study from 1976 to 1984, including 282 cerebrovascular disease events, did not detect any relation between stroke risk and former use of OCs. The reasons for the lack of agreement with the longer-standing cohort studies previously mentioned are unclear, although differences in hormone dosage may be a contributing factor.

Various studies also examined the dependence of the magnitude of the cerebrovascular relative risk associated with OC use on disease risk factors. The Collaborative Group study found little evidence of a

TABLE I
Summary Relative Risk Estimates and Confidence Intervals for Various Cardiovascular Diseases in Relation to Combined OC Use

		Current OC use			Former OC use		
Disease category	ICD code	Cohort	Case–control	No. of studies	Cohort	Case–control	No. of studies
All cerebrovascular diseases	430–438	2.9 (2.0–4.1)[a] *	1.6 (0.9–2.5)	12	1.8 (1.3–2.6)*	1.3 (1.1–1.4)*	10
Hemorrhagic		2.0 (1.1–3.6)*	1.2 (0.9–1.5)	6	2.1 (1.1–3.4)*	1.3 (1.2–1.5)*	5
Nonhemorrhagic	432–438	3.8 (2.4–6.1)*	4.2 (4.0–4.5)*	2	1.9 (1.2–3.0)*	3.0 (0.7–13.7)	1
Myocardial infarction	410	1.9 (1.1–3.3)*	1.7 (1.3–2.2)*	14	1.1 (0.6–1.9)	0.9 (0.7–1.1)	12
Peripheral arterial disease	440–448	1.6 (1.3–2.1)*			1.1 (0.8–1.5)		
Venous thromboembolism	450–453	3.2 (2.4–4.3)*	2.6 (1.6–4.1)*	8	1.0 (0.6–1.8)	0.9 (0.7–1.3)	6

[a]95% confidence interval in parentheses.
*Statistical significance at 95% level.

dependence of the relative risk associated with current OC use on cigarette-smoking histories, although the hemorrhagic stroke relative risk estimate was particularly elevated among "regular" cigarette smokers. The RCGP study found similar relative risks for current and former OC use among smokers and nonsmokers. This study also observed somewhat larger stroke relative risks among women aged 35 years or older compared with that among younger women.

The dependence of the stroke relative risk associated with OC use on hormone dosages is of particular interest in view of the lower dosages in most OC preparations in current use. The RCGP study included experience with three preparations having 50 μg of the estrogen ethinyl estradiol and varying doses (1, 3, or 4 mg) of the progestin norethindrone acetate. Based on 26 cerebrovascular events among never-users and 6, 14, and 5 events at dosages of 1, 3, and 4 mg of norethindrone acetate, the stroke risks in relation to current OC use were estimated as 2.0 (95% C.I. = 0.8–4.9), 3.8 (95% C.I. = 2.0–7.3), and 6.6 (95% C.I. = 2.5–17.1), respectively, at the three progestin dosages. Although based on small numbers of events, these data suggest a strong dependence of stroke relative risks on progestin dosage, at least for this particular progestin. Valuable support for such a dose response was provided by a 1980 study of reports to the Committee on Safety of Medicine in the United Kingdom, which included reports on 191 occurrences of stroke among current users of the same three preparations. This report also suggested somewhat lower stroke relative risk associated with OC preparations having 30 μg versus 50 μg of ethinyl estradiol. Collectively, however, these data sources do not provide a precise estimate of cerebrovascular disease risks associated with current or former use of low-dose OC preparations. Further observational study will be required to determine the extent to which the elevations in hemorrhagic and thrombotic stroke risks associated with the older OCs have been obviated by the newer preparations.

Hypertension is the major risk factor for hemorrhagic stroke. In fact, when taking account of random measurement errors in blood pressure measurement, one can note that the modest average elevations in blood pressure that have been observed among current OC users provide an explanation for the increased hemorrhagic stroke risks associated with both current and former OC use. Such average elevations for the older OC preparations were in the vicinity of 4–5 mm Hg systolic and 1–2 mm Hg diastolic blood pressure. The magnitude of the increase in blood pressure has been shown to vary directly with the progestin dosage, at least for certain OC preparations. [*See* Hypertension.]

Blood pressure elevations and consequent structural vascular changes also provide a plausible explanation for the elevated thrombotic stroke risk among former OC users, as well as for a portion of the increase among current users. Other factors, presumably including OC effects on blood coagulability, are likely to be involved in a full explanation of the noteworthy thrombotic stroke elevations among current users of the older OCs.

C. Ischemic Heart Disease

The cohort studies mentioned earlier and a number of case–control studies have been used to examine the relation between current and former OC use and the incidence of ischemic heart disease. For total ischemic heart disease occurrence, reports in the early 1980s from the RCGP and Walnut Creek studies combine to yield a relative risk estimate of 1.4 for current OC use, a value that is just significant at the 0.05 level, and a nonsignificant relative risk estimate of 1.2 for former OC use. When myocardial infarction events were excluded, these relative risk estimates became 1.3 and 1.2, respectively, neither of which was significantly different from zero. However, the myocardial infarction relative risk associated with current OC use is estimated as 1.9 (95% C.I. = 1.1–3.3) based on these two cohort studies, whereas the corresponding relative risk estimate and confidence interval are 1.1 (95% C.I. = 0.6–1.9) for former OC use, as shown in Table I. Fourteen case–control studies of nonfatal myocardial infarction combine to yield a similar summary relative risk estimate of 1.7 (95% C.I. = 1.2–2.2) for current OC use. There is no evidence of heterogeneity among these studies. A subset of these studies also examined former OC use, giving a summary relative risk estimate of 0.9 (95% C.I. = 0.7–1.1), similar to that from the cohort studies. Furthermore, the case–control study relative risk estimates for fatal myocardial infarction appear to be quite consistent with those for nonfatal infarction.

In summary, current use of the older OCs is associated with noteworthy elevation of myocardial infarction incidence based on a substantial series of studies that generally provide risk factors such as cigarette-smoking habits, serum cholesterol concentration, and blood pressure level. Control of these factors is essential because they have evidently played a role in decisions regarding contraceptive choices. However, standard methods of confounding factor control may lead to some underestimation of relative risks associated with OC use, in view of well-documented effects of OCs on blood pressure and on blood lipids and lipoproteins.

The studies summarized here suggest that any increase in myocardial infarction incidence among former OC users must be small, which is supported by the largest case–control study conducted in New England, from 1985 to 1988. This case–control study included 910 first myocardial infarctions and 1760 controls and shows a mild association between the duration of past OC use and the relative risk of myocardial infarction. The age-adjusted relative risk estimate was somewhat elevated and equaled 1.5 for women who had used for less than 11 months. This possible association is important, however, in view of the prevalence of this disease and the rapid increase in myocardial infarction incidence rates as women age. Furthermore, the second largest case–control study, involving 556 premenopausal myocardial infarction cases admitted to the coronary care units of 155 hospitals in Boston, New York, and the Delaware Valley during 1976–1978 and 2036 control subjects, did report a significant relative risk trend with duration of former OC use. Specifically, relative risk estimates were 1.0 (95% C.I. = 0.7–1.2), 1.4 (95% C.I. = 1.0–2.0), and 2.3 (95% C.I. = 1.5–3.8) for less than 5, 5–9, and 10 or more years of former OC use, respectively, and the increased risk was evident as long as 5–10 years after discontinuation of use. These findings need to be interpreted with some skepticism in view of Table I and particularly in view of a lack of any association between former OC use and coronary disease in the cohort study of U.S. nurses, which included 485 cases of nonfatal myocardial infarction or fatal coronary heart disease.

Two large case–control studies, reported in 1980 and 1981, examined the dependence of myocardial infarction among current OC users on duration of current use. Although the data cannot be said to be strong, when results from the two studies are combined, a significant trend is evident with relative risk estimates of 1.5 (95% C.I. = 0.9–2.5), 2.6 (95% C.I. = 1.6–4.2), and 4.4 (95% C.I. = 2.6–7.4) for less than 5, 5–9, and 10 or more years of current use, respectively.

The RCGP study also examined in a 1982 report the dependence of the relative risk among current OC users on the dose of the progestin norethindrone acetate among preparations having 50 μg ethinyl estradiol. Compared with never-users, the relative risk estimates and 95% confidence intervals of 1, 3, and 4 mg of this progestin are 0.8 (95% C.I. = 0.3–2.0), 1.8 (95% C.I. = 1.0–3.0), and 1.5 (95% C.I. = 0.5–4.9), based on rather limited numbers of cases. Also relative risk estimates for the 3-mg and 4-mg doses compared with the 1-mg dose were 2.2 and 2.2, based on a larger number of reports to the U.K. Committee on the Safety of Medicine, and the trend was highly significant. These reports imply a myocardial infarction incidence at 1 mg of norethindrone acetate that is about one-half that at 3 or 4 mg, but they do not provide a precise estimate of relative risk at the 1-mg dose as compared with nonuse.

Based on a total of only 25 cases in a 1987 report, ever-use of OCs was associated with a twofold increase in myocardial infarction risk in the Oxford cohort, whereas there was no evidence of increased risk among current users. Experience with preparations having less than 50 μg estrogen was limited in this analysis. A 1989 report from the RCGP study involved 158 myocardial infarction cases. Current OC use was associated with a relative risk estimate of 1.8, which is evidently of borderline significance, and former OC use was associated with a crude relative risk estimate of 1.3 that reduced to 1.0 after adjusting for cigarette smoking and social class. To the extent that the newer low-dose preparations prevail among current users, these results are somewhat disappointing. The authors argue that the excess risk associated with current OC use is confined to cigarette smokers. Collectively, however, the cohort and case–control studies suggest a relative risk for current OC use that is as large among nonsmokers as smokers. Of course, OC use has a greater impact on absolute myocardial infarction risk among smokers as compared with nonsmokers, because cigarette smoking is a strong risk factor for this disease. Similarly, there does not appear to be a strong dependence of the relative risk associated with OC use on study subject age, based on the collective observational data.

Oral contraceptives have been found to have substantial influences on lipid and lipoprotein metabolism. Serum high-density-lipoprotein (HDL) cholesterol concentration has been found to correlate negatively with progestin dose at a fixed estrogen dose, with 10–20% reductions in HDL cholesterol-associated certain OC preparations. Similarly noteworthy, 10–30% increases in low-density-lipoprotein (LDL) cholesterol concentrations have been shown to follow the use of OCs that are low in estrogen dose and high in progestin dose, whereas preparations that are relatively high in estrogen and low in progestin give rise to elevated HDL concentrations. These results, along with results showing a favorable impact of menopausal estrogens on HDL concentrations, suggest that effects of OCs on cholesterol fractions may explain much of the elevation in myocardial infarction incidence among current OC users. The collective observational data provide little evidence that myocardial infarction risk is affected with former OC use. [*See* Cholesterol.]

D. Peripheral Vascular Disease

As noted in Table I, the RCGP and Walnut Creek cohort study reports in the eary 1980s combine to yield a relative risk estimate of 1.6 (95% C.I. = 1.3–2.1) for peripheral arterial disease associated with current OC use. There was little evidence of increased risk among former OC users based on those same data sources.

Venous thromboembolism was one of the earliest adverse effects of OC use to be identified. The three long-standing cohort studies each reported data on idiopathic venous thromboembolism in the late 1970s. These studies combine to yield a relative risk estimate of 3.2 (95% C.I. = 2.4–4.3), although there is evidence ($P = 0.04$) of discrepancy among the estimates, with the RCGP estimate of 2.7 (95% C.I. = 2.0–3.8) being smaller than the other two. There was no evidence of elevated venous thromboembolism risk among former OC users based on data from the RCGP cohort.

The corresponding current OC use relative risk estimate for idiopathic venous thromboembolism was 2.6 (95% C.I. = 1.6–4.1) based on eight case–control studies, which is consistent with that obtained in cohort studies. Also consistent with the result from the two cohort studies was the estimated relative risk of 0.9 (95% C.I. = 0.7–1.3) for the former OC use on the basis of six case–control studies. The relative risks associated with current OC use for pulmonary embolism, the most threatening of the venous thromboembolic events, were similar to those for other such events, and OC relative risks were similar for fatal and nonfatal events.

In summary, idiopathic venous thromboembolism incidence is elevated about twofold among current users of the older OCs, as compared with a relative risk of about 2.5 for predisposed venous thromboembolic disease. OC use has been found to bring about a reversible increase in the coagulability of blood involving changes in platelet, coagulation, and fibrolytic systems, and changes in the structure of veins and arteries. The RCGP data and reports to the British Committee on the Safety of Medicine have been used to examine the dependence of risk on estrogen and progestin dosages. These sources do not allow one to disentangle the roles of the two hormones in a satisfactory manner, although it is reasonable to assume that the higher estrogen dose, if not higher doses of both agents, is associated with increased venous thromboembolism risk. Once again, the relative risk associated with OC use does not appear to depend much on the age or cigarette-smoking habits of the study subject or on the duration of OC use.

III. RISKS AND BENEFITS FOR CANCER

A. Endometrial Cancer

Women treated at the menopause with conjugated estrogens exhibit highly elevated rates of endometrial hyperplasia and of endometrial cancer. There is correspondingly strong evidence that sequential OCs, which exert a net estrogenic effect on the endometrium, also increase endometrial cancer risk. In fact, the summary endometrial cancer relative risk estimated associated with ever-use of sequential OCs is 2.0 (95% C.I. = 1.1–3.8) based on three case–control studies reported in the early 1980s.

Estrogens are presumed to increase endometrial cancer risk by enhancing the mitotic activity of endometrial cells, whereas progestins reduce DNA synthesis in the endometrium and lead to endometrial atrophy when given alone. Hence combined OCs, which contain both estrogen and progestin throughout the pill cycle, can be expected to give rise to lower endometrial cancer rates than do sequential OCs, which are now seldom used. Based on 13 substantial case–control studies reported since the mid-1970s, ever-use of combined OCs is associated with a relative risk of 0.4 (95% C.I. = 0.3–0.5). Hence rather than a twofold increase there is a twofold reduction in endometrial cancer risk associated with combined OC use. These studies also suggest somewhat greater endometrial cancer risk reductions with increasing duration of OC use. One of the studies examined risk as a function of time since cessation of use and reported a relative risk of 0.8 (95% C.I. = 0.5–1.3) after 10 or more years since exposure to combined OCs. Three long-standing cohort studies also suggest a negative association between combined OC ever-use and endometrial cancer, with the summary relative risk estimate of 0.8 (95% C.I. = 0.4–1.4).

B. Breast Cancer

Of the various diseases that may be associated with the use of combined OCs, none has been studied more extensively or intensively than breast cancer. Fully 47 case–control studies involving a total of more than 36,200 breast cancer cases have examined this relation, and the four cohort studies previously mentioned (RCGP, Oxford, Walnut Creek, and Nurses Health), in addition to a smaller cohort study conducted by the Group Health Cooperation of Puget Sound, have all reported their results on OC use and breast cancer on one or more occasions.

In addition to a number of past reviews of this topic, both a meta-analysis of all epidemiologic studies conducted between 1966 and 1989 and a detailed summary review of the literature have provided summary relative risk estimates for various indices of oral contraceptive exposure in relation to breast cancer risk. As noted in Table II, the 47 case–control studies combine in a meta-analysis to yield a summary relative risk estimate of 1.0 (95% C.I. = 0.9–1.1) for ever-use of combined OCs, while the most recent data from the five cohort studies combine to give an identical summary relative risk estimate of 1.0 (95% C.I. = 0.9–1.1). Overall there is no evidence of an increased risk of breast cancer in women who meet the criterion of ever having used oral contraceptives. Among those studies examining breast cancer risk in relation to oral contraceptive use among women of all ages combined, the majority found little evidence of increased risk related to prolonged duration of oral contraceptive use.

Individual relative risk estimates, however, still exhibit significant heterogeneity across various risk factors such as menopausal status and age at diagnosis. The fact that breast cancer risk is enhanced by a number of hormone-dependent events, including earlier age at menarche, late age at menopause, few or no children, and history of little or no lactation along with the high incidence of this disease in developed countries, where blood hormone concentrations are known to be comparatively high, has caused the OC and breast cancer relation to be the subject of considerable attention and controversy. In respect to exposure to OCs at young ages, six case–control studies of breast cancer before age 45 have reported relative risk estimates associated with ever-use of OCs before age 25. These studies combine to give a summary relative risk estimate of 1.3 (95% C.I. = 1.1–1.4). However, there is strong evidence ($P < 0.001$) of disagreement among the specific studies. A large number of case–control studies have examined the relationship between oral contraceptives and breast cancer risk among young women (under age 35 or under 45). The majority of these reported modestly increased risk estimates for long-term oral contraceptive use. Meta-analysis results and summary relative risk estimates suggest that the relative risk for long-term use among young women involves at least a 50% increased risk. Analysis by D. B. Thomas gave a summary risk estimate for duration of oral contraceptive use before first full-term pregnancy among all studies of 1.4 (95% C.I. = 1.2–1.7). I. Romieu *et al.* calculated a summary estimate of risk among women under age 45 in relation to four or more years of use before first full-term pregnancy of 1.7 (95% C.I. = 1.4–2.2). Findings are inconsistent with regard to current use of oral contraceptives with reported relative risks between 1.0 and 2.9, and no consistent patterns of risk have been observed in relation to time since last use.

TABLE II
Summary Relative Risk Estimates and Approximate 95% Confidence Intervals for Several Cancers in Relation to OC Use

Disease category		No. of studies	Ever-use of combined OCs
Endometrial cancer	Case–control	13	0.4 (0.3–0.5)*
	Cohort	3	0.8 (0.4–1.4)
Breast cancer	Case–control	47	1.0 (0.9–1.1)
	Cohort	5	1.0 (0.9–1.1)
Ovarian cancer	Case–control	19	0.7 (0.6–0.8)*
	Cohort	4	0.7 (0.5–0.9)*
Invasive cervical cancer	Case–control	15	1.1 (0.9–1.3)
	Cohort	3	2.1 (1.7–2.8)*
Liver cancer	Case–control	11	1.3 (0.9–1.8)
In developing countries		4	0.9 (0.5–1.5)
In developed countries		7	1.6 (1.1–2.2)*
Malignant melanoma	Case–control	10	1.0 (0.9–1.1)
	Cohort	3	1.1 (0.7–1.9)
Colorectal cancer	Case–control	6	1.1 (0.9–1.5)
Colon cancer	Case–control	3	0.9 (0.7–1.1)
	Cohort	1	0.8 (0.6–1.1)
Rectal cancer	Case–control	2	1.2 (0.8–1.7)
	Cohort	1	1.4 (0.7–2.5)

*Statistical significance at 95% level.

Little is known about the physiology of the growth and differentiation of normal human breast tissue and the joint and separate effects of endogenous and exogenous estrogen and progesterone on breast tissue. Estrogen apparently contributes to breast tissue proliferation and is postulated to affect breast cancer risk by stimulating growth of stem cells, intermediate cells, and established cancer cells. Progesterone is thought to increase alveolar cell growth and differentiation in the estrogen-primed breast, although there is also controversy concerning its proliferative influence. Both may act directly on breast tissue or indirectly through locally secreted growth factors, which regulate cell function. Despite the extensive study that has already taken place, there may be a need for further study of OC use and breast cancer, with a particular emphasis on preparation dosages and on pertinent hormone concentrations, for example, in blood or in breast fluid, to understand the etiology of OC use in relation to breast cancer as well as the biology of breast cancer. [*See* Breast Cancer Biology.]

C. Ovarian Cancer

Nineteen case–control studies have been reported since the mid-1970s that combine to yield a relative risk estimate of 0.7 (95% C.I. = 0.6–0.8) for ever-use of combined OCs in relation to all ovarian malignancies (see Table II), of which epithelial cancers predominate. The four cohort studies presented an almost identical relative risk estimate of 0.7 (95% C.I. = 0.5–0.9) for the ever-use of combined OCs in relation to ovarian cancer. There was no evidence of disagreement among the studies concerning the magnitude of this risk reduction, which is similar among parous and nulliparous women, arguing against the possibility of confounding by infertility. A recent case–control study has noted that a relative risk reduction of a similar magnitude applies to ovarian tumors of low malignancy potential. Another case–control study found similar risk reductions for epithelial ovarian tumors among women diagnosed in 1983–1985 as compared with those diagnosed in 1974–1977, perhaps suggesting that this OC benefit is retained by the newer preparations.

Various epidemiologic studies suggest that ovarian cancer risk is decreased by late menarche, pregnancy, and early menopause, all of which correspond to a reduction in a women's total number of ovulations, as does OC use. It may be that ovulation enhances risk by subjecting the ovarian epithelium to steroid-risk follicular fluid or to elevated levels of pituitary gonadotropins. The fact that menopausal steroids also reduce gonadotropin concentrations in the blood, without a corresponding reduction in ovarian cancer rates, suggests that mechanisms other than gonadotropin reduction may be required to explain the protective effect of OC use on ovarian cancer.

D. Cervical Cancer

Study of the relation of OC use to cervical cancer is hampered by the need to account carefully for sexual practices and pap smear screening activities. There have now been 15 case–control studies that have attempted to control for these as well as other potential confounding factors. These studies combine to yield a summary relative risk of 1.1 (95% C.I. = 0.9–1.3) for ever-use of combined OCs (see Table II), thereby providing no evidence of risk enhancement. However, this finding is not supported by cohort studies. Three cohort studies show the relative risk estimate of 2.1 (95% C.I. = 1.7–2.8) for OC ever-use supported by the association of OC and cervical cancers. Specifically, in two of these studies, with a small number of invasive cervical cancers, all disease events occurred among women who had used OCs, whereas the RCGP study reported an unadjusted relative risk estimate of 2.1 (95% C.I. = 1.1–4.3) for the ever-use of OCs.

There have also been six case–control studies that have examined OC use in relation to cervical carcinoma *in situ*. These studies yield a summary relative risk estimate of 1.2 (95% C.I. = 1.1–1.4) for ever-use of OCs, and there is evidence of increasing relative risk with longer duration of use. However, there is evidence of inconsistency ($P < 0.01$) among the study results, possibly because several of those studies made little attempt to control sexual practices. The three cohort study reports on this topic give a somewhat higher relative risk estimate of 1.5 (95% C.I. = 1.1–2.0), again with some evidence ($P = 0.06$) of heterogeneity among studies. Cervical squamous dysplasia occurrence has also been studied in relation to combined OC use. Cohort and case–control studies suggest significant relative risks, in the vicinity of 1.4, with evidence of higher relative risk among longer-term OC users. Once again, the individual relative risk estimates are more variable than would be expected on the basis of chance alone, and only limited efforts were made to control confounding.

The studies summarized here provide limited evidence of association between OC use and squamous cell carcinoma of the cervix. Based on modest sample sizes, these studies also provide arguable evidence of association between such use and adenocarcinoma of the cervix. There has been little attempt to date to investigate mechanisms for such association. Hormonal mechanisms would appear to merit exploration, as would be consistent with the association of this disease with reproductive factors such as multiparity. However, the strong associations of cervical cancer rates with early age at first intercourse and number of sexual partners leave open the possibility that the observed association with OC use could be affected by inadequate confounding factor control. Further studies that include biological measures of exposure to various sexually transmitted viruses would seem to be particularly worthwhile.

E. Liver Cancer

Seven case–control studies have examined the association between OC use and liver carcinoma in developed countries. These seven studies combine to give a summary relative risk estimate of 1.6 (95% C.I. = 1.1–2.2) for ever-use of OCs (see Table II) and 9.6 (95% C.I. = 4.0–22.8) for long-term use of OCs defined variously as more than 5 or more than 8 years. Although these studies are small and have methodological limitations, there is considerable consistency among their results, suggesting that primary hepatocellular carcinoma can be a complication of OC use, particularly long-term OC use. In developing countries, the association between OC ever-use and cervical cancer has not been proved, the relative risk estimate being 0.9 (95% C.I. = 0.5–1.5). This apparent discrepancy may be explained by the difference of relative risks of OC with hepatocellular carcinoma and with cholangiocarcinoma, incidences of which are different in developing and developed countries. In developing countries, hepatocellular carcinoma and cholangiocarcinoma relative risks with OC ever-use are 1.1 (95% C.I. = 0.6–2.0) and 1.3 (95% C.I. = 0.6–2.9), respectively. In developed countries, on the other hand, these two relative risks are 1.7 (95% C.I. = 1.2–2.4) and 0.3 (95% C.I. = 0.1–1.3).

There have also been two small case–control studies of OC use in relation to hepatic cell adenomas. These benign tumors sometimes result in intraperitoneal or intramural bleeding, which may be fatal. These studies indicate that OC users have elevated risk for this tumor and that the relative risk increases dramatically with duration of use. However, these are rare tumors, with the older OCs increasing risk by about three cases per 100,000 users per year.

Fortunately, primary hepatocellular carcinoma

among young women is also extremely rare in developed countries, where it has been estimated that about one liver cancer fatality per 100,000 long-term users per year can be expected. However, there would be some concern if elevated risks persisted for many years beyond cessation of OC use, and so further monitoring and evaluation are indicated.

Of much greater concern is the possibility that relative risks of the magnitudes indicated here would prevail in areas of Asia and Africa in which hepatitis B virus is endemic and liver cancer rates are high. However, a recent sizable case–control study among women in eight generally high-risk countries provides evidence that any elevation of hepatocellular carcinoma rates in the presence of prior hepatitis B virus infection must be small. Specifically, this study gives a liver cancer relative risk estimate of 0.7 (95% C.I. = 0.4–1.2) for ever-use of combined OCs, and there was no evidence of heterogeneity of relative risk estimate among the participating centers in this collaborative study.

F. Malignant Melanoma

Ten case–control studies and three cohort studies have reported malignant melanoma relative risk estimates associated with the ever-use of combined OCs. These studies combine to give a summary relative risk estimate of 1.0 (95% C.I. = 0.9–1.1), and the cohort studies give a summary estimate of 1.1 (95% C.I. = 0.7–1.9). This latter estimate is based on the report of a substantially increased risk in the Walnut Creek cohort, and the three cohort study relative risk estimates are not in agreement ($P = 0.002$). However, the case–control studies do suggest some modest elevation in risk associated with long-term OC use, defined variously as more than 5 or more than 10 years, with relative risk estimate of 1.3 (95% C.I. = 1.0–1.7). Somewhat larger relative risk estimates were reported among women who were 10 or more years from initial exposure in two of the case–control studies. In summary, these data do not provide convincing evidence that OCs enhance the risk of malignant melanoma, but do suggest that risk may eventually be elevated among long-term users. [*See* Melanoma, Malignant.]

G. Colorectal Cancer

Six case–control studies reported a somewhat elevated colorectal cancer relative risk associated with OC ever-use, with a risk estimate of 1.1 (95% C.I. = 0.9–1.5). For colon cancer, the risk estimate is 0.9 (95% C.I. = 0.7–1.1), and for rectal cancer it is 1.2 (95% C.I. = 0.8–1.7). Only one cohort study presented a relative risk estimate, giving 0.8 (95% C.I. = 0.6–1.1) for colon cancer and 1.4 (95% C.I. = 0.7–2.5) for rectal cancer. An Australian case–control study reported a reduced risk of colorectal carcinoma among OC users, although none of the three long-standing cohort studies has reported such an association.

H. Other Neoplasms

A protective effect of OC use for benign breast disease, including both fibroadenoma and fibrocystic disease, has been well documented. However, such protection evidently does not convey a reduced breast cancer risk. Reduced rates of retention cysts of the ovary have been observed among the OC users in several studies. Such reduction is not unexpected because OCs inhibit ovulation. Uterine leiomyomas (fibroids) have been variably reported as increased, unchanged, or decreased among OC users in the three long-standing cohort studies. The recent study using the Oxford cohort found that risk consistently decreased with increasing duration of OC use. Initial reports of an elevation in pituitary adenoma rate among women of childbearing age and of an elevation in thyroid cancer among OC users were not confirmed by subsequent studies.

IV. RISKS AND BENEFITS FOR OTHER DISEASES

The two British cohort studies have reported reductions in risk for several menstrual disorders but an increased risk of amenorrhea or scanty menstrual flow among OC users. The amenorrhea can persist for 2 years or more, but no increased risk of permanent infertility has been identified. Reduction in the incidence of iron deficiency anemia has been observed in all three cohort studies.

Elevated incidence of hay fever and allergic rhinitis, with some evidence of a dose response with the estrogen content of the OC preparation, has been reported, as have elevated rates of eczema, conjunctivitis, and diseases of the eyelid. Whether such associations reflect an OC effect on the immune system or whether reporting bias contributes to such associations has not been determined.

All three cohort studies have reported an increased rate of migraine headaches among OC users. There have also been variable reports of increased rates of referral for depression and an increased rate of hospitalization for mental or emotional illness among OC users. Somewhat increased rates of suicide and self-poisoning among OC users have also been suggested by the three long-standing cohort studies. That women with a predisposition for such events more frequently choose OCs as their method of contraception provides a possible explanation for these associations.

A number of other associations with OC use have been reported but have not been confirmed by other studies. These include reduction in risk of rheumatoid arthritis and in certain thyroid diseases (euthyroid swelling, thyrotoxicosis, and myxoedema).

Certain other conditions, including increased rates of regional enteritis and ulcerative colitis, and small increases in the rates of gastritis, duodenitis, and appendicitis have been reported in two or more of the three cohort studies. Gallbladder disease (cholecystitis and cholelithiasis) has also been related to short-term OC use in several studies. Also OC users have been shown in all three cohort studies to be at increased risk for cervical erosion, vaginal discharge, and pelvic infection. These last associations may simply reflect differing sexual practices between users and nonusers of OCs. However, OC use evidently reduces the risk of pelvic inflammatory disease, perhaps by thickening cervical mucous.

OCs are known to affect carbohydrate metabolism. Evidently estrogen increases glucose levels and suppresses the corresponding insulin response, whereas progestin stimulates insulin production. Hence, OC use may be contraindicated for women with a history of diabetes.

Finally, because OCs provide effective contraception and prevent ovulation, they reduce the risk of various complications of pregnancy and childbirth, including the risk of ectopic pregnancy.

V. IMPLICATIONS FOR TOTAL MORTALITY

It is evident that combined OCs convey a range of risks and benefits with respect to human disease. An analysis of total mortality rate provides one method of bringing together the risks and benefits in a manner that acknowledges aspects of the lethality and incidence rates for the life-threatening diseases. Table III gives summary relative risk estimates for total mortality from four cohort studies; in 1981 RCGP reported 249 deaths, 238 deaths through 1989 in the Oxford cohort, 138 deaths in the Walnut Creek cohort, and the Nurses' Health Study reported 2879 deaths between 1976 and 1988, excluding the Group Health Cohort since it observed only 78 deaths between 1971 and 1981. Results from the first three studies yield combined estimates of relative risks, because these studies were conducted during the same period of

TABLE III
Summary Mortality Relative Risk Estimates and Approximate 95% Confidence Intervals for OC Use

Disease category	Royal College[a] (1981)	Walnut Creek[b] (1981)	Oxford FPA[c] (1989)	Combined relative risk estimates	Nurses' Health[d] (1994)
Total mortality	1.4 (1.1–1.8)*	1.1 (0.7–2.4)	0.9 (0.7–1.2)	1.3 (1.2–1.5)*	1.0 (0.9–1.0)
Mortality from circulatory system diseases	4.2 (2.3–7.7)*	0.8 (0.3–1.9)	1.5 (0.7–3.0)	2.1 (1.0–4.5)*	0.9 (0.7–1.1)
All cancer mortality	1.0 (0.5–1.9)	1.4 (0.8–2.3)	0.8 (0.6–1.1)	1.0 (0.7–1.3)	1.0 (0.8–1.0)
Accidents and violence	1.5 (0.8–3.0)				1.3 (0.5–3.0)
Suicide			1.1 (0.3–3.6)		1.3 (0.9–2.0)
Other trauma			1.1 (0.2–12.4)		1.3 (0.9–1.8)
All other mortality	0.7 (0.3–1.6)	1.2 (0.7–2.1)	0.8 (0.4–1.9)	1.0 (0.7–1.3)	0.9 (0.8–1.1)

[a]Adjusted for age, parity at birth, cigarette consumption, and social class.
[b]Standardized by age, parity, smoking, education, and estrogen use.
[c]Standardized by age, parity, social class, and smoking habit.
[d]Adjusted for age, cigarette smoking, body mass index, and follow-up interval.
*Statistical significance at 0.05 level.

time. The results from the Nurses' Health Study are listed separately in Table III because the study has observed a relatively large number of deaths during a different period of time. The combined total mortality relative risk of the first three cohort studies was 1.3 (95% C.I. = 1.2–1.5) for ever-use versus never-use of the older OCs, which is not consistent with the estimate of 1.0 (95% C.I. = 0.9–1.0) from the Nurses' Health Study. Twenty-one percent of the deaths were from disease of the circulatory system. Whereas the first three studies combine to give a relative risk estimate of 2.1 (95% C.I. = 1.0–4.5), the Nurses' Health Study does not support this positive finding with an estimate of 0.9 (95% C.I. = 0.7–1.1). Forty-four percent of the deaths were from cancer, for which the combined relative risk of the first three studies is the same as that from the Nurses' Health Study, namely, 1.0. The remaining 35% of the deaths were attributable to accident and violence or other causes; for accident and violence the relative risks from both RCGP and the Nurses' Health Study are comparable, in the neighborhood of 1.4. Finally, the estimated relative risks for all other causes of deaths are around 1.0. Hence, the older OCs may increase mortality from accident and violence with little effect on the mortality rates for all other diseases combined, at least within the first decade or so after initiation of use. Additional studies are probably needed to resolve the controversies of total mortality and circulatory mortality relative risk estimates between the Nurses' Health Study and the three other cohort studies. Further, additional observation and analysis will be required to determine any long-term mortality consequences of the older OCs. Importantly, such additional observations will also be necessary to determine the extent to which the newer OCs have reduced or eliminated the elevation in cardiovascular disease mortality risk and the extent to which such OCs have altered the risk–benefit mortality profile for other diseases.

BIBLIOGRAPHY

Colditz, G. A. (1994). Oral contraceptive use and mortality during 12 years of follow-up: The Nurses' Health Study. *Ann. Intern. Med.* **120,** 821–826.

Harlap, S. (1991). Oral contraceptives and breast cancer: Cause and effect. *J. Reprod. Med.* **36,** 374–394.

Malone, K. E., Daling, J. R., and Weiss, N. S. (1993). Oral contraceptives in relation to breast cancer. *Epidemiol. Rev.* **15,** 80–97.

Olsson, H. (1989). Oral contraceptives and breast cancer. *Acta Oncol.* **18,** 849–863.

Porter, J. B., Walker, A. M., and Jick, H. (1987). Mortality among oral contraceptive users. *Obstet. Gynecol.* **70**(1), 29–32.

Prentice, R. L., and Thomas, D. B. (1987). On the epidemiology of oral contraceptives and disease. *Adv. Cancer Res.* **49,** 285–401.

Ramcharan, S., Pellegrin, F. A., Ray, R., and Hsu, J-P. (1981). "A Prospective Study of the Side Effects of Oral Contraceptives," The Walnut Creek Contraceptive Drug Study, Vol. III, NIH Pub. No. 81-564. National Institutes of Health, Washington, D.C.

Romieu, I., Berlin, J. A., and Colditz, G. (1990). Oral contraceptives and breast cancer: Review and meta-analysis. *Cancer* **66,** 2253–2263.

Royal College of General Practitioners' Oral Contraception Study. (1981). Further analysis of mortality in oral contraceptive users. *Lancet* **7,** 541–546.

Schlesselman, J. J. (1989). Cancer of the breast and reproductive tract in relation to use of oral contraceptives. *Contraception* **40,** 1–39.

Thomas, D. B. (1991). Oral contraceptives and breast cancer: Review of the epidemiologic literature. *Contraception* **43,** 597–642.

Vessey, M. P., Villard-Mackintosh, L., McPerson, K., and Yeates, D. (1989). Mortality among oral contraceptive users: 20 year follow up of women in a cohort study. *Br. Med. J.* **299,** 1487–1491.

Organ Transplantation

MARIA T. MILLAN
ROGER JENKINS
Beth Israel Deaconess Medical Center, Harvard Medical School

GLOSSARY

Allograft Organ transplanted between genetically nonidentical members of the same species

Antibody Proteins present in serum and on the surface of B cells that are induced by, recognize, and bind to antigen

Antigen Any molecule recognized by the immune system

B cell Immune cells that develops in the fetal liver and eventually in bone marrow and resides in lymph nodes and the spleen; it responds to antigen by dividing, differentiating, and producing antibody

Cytokine Substance released by white blood cells that modulates differentiation and division of cells; it can also act directly as a toxin to cells

Isograft Organ transplanted between genetically identical individuals

Macrophage White blood cell that, among other functions, phagocytoses antigen, processes it, and presents it to T cells

Rejection Immune reaction leading to damage of the transplanted organ

T cell Immune cell that develops in the thymus and recognizes antigens expressed on the host's cells

Xenograft Organ transplanted between different species

ANCIENT EGYPTIAN AND GRECO-ROMAN CIVILIZATIONS depicted the exchange of organs between humans and animals. In these legends, there were images of animals and humans joined at the waist and neck, such as Homer's chimera, a triple fusion of a lion's head, a goat's body, and a dragon's tail. During the Christian era, there were also accounts of miracles involving transplanting human organs. The most famous miracle is described by Jacobus de Voragine, Archbishop of Genoa, in the thirteenth century. In "Lives of the Saints" he recounted Saints Cosmas and Damian replacing a sexton's leg, consumed by cancer, with a leg taken from the cadaver of a Moor.

Today, the transplantation of human organs has become reality. Current medical technology and scientific advances have allowed the replacement of nonfunctioning or severely malfunctioning organs with organs from other individuals. This is termed allotransplantation—the transplantation of an organ (allograft) within the same species. Early experiments carried out in animals in the eighteenth century and continued work through the twentieth century led to clinical transplantation in the late 1950s and early 1960s. Advances in several areas made transplantation possible. Reliable surgical techniques were perfected for sewing in the transplanted organ and preservation solutions were created so that the donor organ can be safely maintained outside the body for a period of time prior to being transplanted. Despite these advances, clinical transplantation would not have been possible without the scientific work that defined the immunological process of transplant rejection. Once the process of rejection was better elucidated, immunosuppressive agents could be developed that permit long-term survival of transplanted organs.

I. HISTORY OF TRANSPLANTATION

The technique of "grafting" was first performed in the sixteenth century in plants. The advent of the natural sciences in the eighteenth century led to the application of this technique to animals. In 1749, Henri-Louis Duhamel du Monceau successfully transplanted spurs removed from a young chicken onto the comb of the same or another animal. John Hunter, father of British "scientific surgery," confirmed these experiments in 1767 by implanting teeth into this same comb before finally transplanting teeth from a human cadaver to a live human recipient. During the nineteenth century, experimental transplantation was in full swing. In 1869, a Geneva surgeon, Jacques-Louis Reverdin, performed the first successful epidermal grafts on human beings. It was observed that these grafts failed early on and this was attributed to technical imperfections as well as biological incompatibility.

During the twentieth century, dramatic progress was made in the field. It became apparent that organs, under the proper conditions, can survive outside of the body for a period of time. Alexis Carrel described a reliable method of joining vessels together (1906), and, among others, studied the importance of infection control and aseptic technique to the success of these procedures. In 1946, Ray Owen described an immunological tolerant state in vascular connections between dizygotic twin cattle. This substantiated the finding in the 1930s by Brown that skin grafts between identical twins were permanently accepted. These observations encouraged ongoing efforts at successful organ transplantation and encouraged attempts to overcome the immunological barriers to transplantation.

In 1954, Joseph Murray performed the first successful kidney transplant between identical twin brothers. In 1962, he performed the first unrelated kidney allograft. Murray went on to receive the Nobel Prize for this work. T. E. Starzl performed the first successful liver transplant in 1967. Around the same time, Christiaan Barnard performed the first human cardiac transplant. The initial successes in the clinical arena were tempered by failure due to acute rejection of the transplanted organ. Development of immunosuppressive agents, as will be discussed in more detail, has made long-term graft survival possible. Nevertheless, there is still considerable research directed at developing agents and methods to avoid rejection episodes and to achieve a state of "tolerance." Tolerance is a state of immunological nonreactivity whereby the recipient would accept a transplanted organ as if it was not "foreign." This concept was first introduced by Sir Peter Medawar and his colleagues in a paper published in 1943, in which they systematically studied the phenomenon of actively acquired tolerance between inbred strains of mice. They were able to overcome the barrier between "self" and "non-self" by exposing animals to certain antigens during the neonatal period.

II. REJECTION

Several types of rejection responses may occur when organs are transplanted into genetically nonidentical hosts. Hyperacute rejection, a phenomenon rarely seen today, occurs within minutes of implantation of the organ. It is the result of circulating preformed antibodies in the recipient that recognize and induce damage to the donor organ. These preformed antibodies may be derived from repeated pretransplant blood transfusions, pregnancy, previous transplantation, or bacterial presensitization. Acute rejection, the usual form of rejection encountered in the clinical realm, usually occurs within 3 months after transplantation, and this form of rejection is the main subject of this section. Chronic rejection is a late response, which is still being characterized, and seems to be a result of sensitization of the recipient to graft antigens.

When tissue is transplanted into a nonidentical recipient, antigens present on the cell surface of the transplanted tissues are recognized by the recipient as foreign. This so-called "non-self-recognition" initiates the rejection response. The strength of rejection is directly related to the degree of genetic disparity between donor and recipient, and the T cell is the pivotal element in this rejection response. Through a complex recognition system involving interaction with B cells and macrophages and their products, T cells respond to transplantation antigens expressed on the graft and are largely responsible for damage to the transplanted organ.

The major histocompatibility complex (MHC) is the chromosomal region (chromosome 6 in humans) encoding proteins that take part in rejection of the transplanted organ. The human leukocyte antigen (HLA) is the main MHC in humans and it was defined by D. B. Amos' group and others in the 1960s. The HLA is divided into three classes: class I, II, and III. Class I genes encode cell-surface transplantation antigens, are expressed on all tissues, and serve as the primary targets for T lymphocytes in graft rejection. Class II genes encode virtually all of the genetic material for the immune response and encode antigens present on the immune cells, macrophages, B cells, and activated T cells.

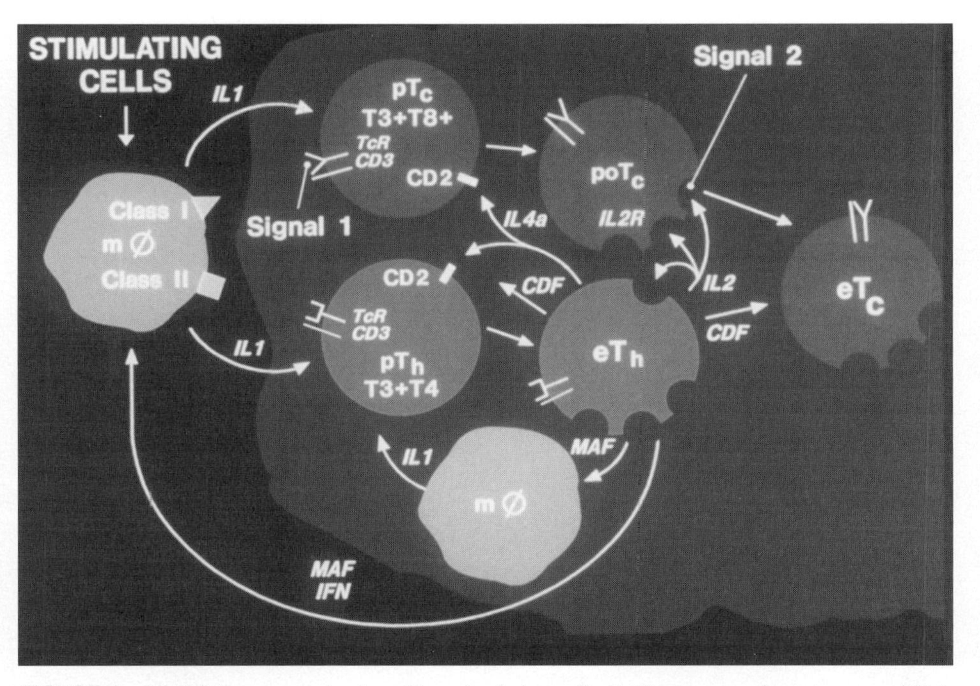

FIGURE 1 Model for acute rejection. The stimulating cells (MØ or macrophages) present HLA antigens (class I and II) to T cells (pT_h and pT_c). These cells then differentiate, become activated, and secrete cytokines such as IL-2. The "second signal," in this case IL-2, induces these T cells to become cytotoxic to the organ. The cytokines depicted, for example, MAF (macrophage-activating factor) and IL-1 (interleukin-1), are among a host of factors that also play a role in this cascade. (Figure borrowed with permission from F. H. Bach, M.D., SCI, Harvard Medical School.)

Acute rejection is most prevalent during the first 3 months after transplantation. This process is mediated primarily by T cells that express on their cell membrane a specific T-cell receptor that bears the antigen binding site in association with other proteins (CD3 complex). Recognition begins with the presentation of antigen by B cells and macrophages to the T cell, which recognizes the antigen in association with particular MHC molecules. These T cells, when given another stimulus (second signal), become activated and produce cytokines, such as interleukin-2 (IL-2), which promote further activation and differentiation of these T cells (Fig. 1). The immune cells and their products promote damage to the organ that may result in transient or permanent dysfunction. A variety of monocyte, B-cell, and T-cell cytokines, as well as complex cell–cell interactions and signaling events not addressed here, are responsible for modifying and augmenting this response. [*See* Major Histocompatibility Complex (MHC); T-Cell Activation.]

Based on these immunological principles, several screening tools have been developed. Tissue typing is the technique used to determine the MHC specificity carried by an individual's tissue. This can be performed serologically or by using the mixed lymphocyte culture (MLC), a method developed by F. H. Bach. By matching recipient and donor according to their MHC profiles, there is theoretically a better chance of long-term graft survival with a decreased chance of rejection. Hyperacute rejection, the result of preformed antibodies against the graft, can be averted using *in vitro* crossmatch. This method provides for the selection of recipients whose pretransplant sera do not lyse donor cells.

III. IMMUNOSUPPRESSION

To overcome the rejection process as described here, a multitude of agents have been developed (Fig. 2). Various combinations of these agents are used today. The regimens vary with the type of organ transplanted, the patient's clinical profile, and to a certain extent on the center's own experience.

The agents employed in the early 1900s were responsible for the initial successes in clinical transplantation. Radiation and chemical reagents were employed to nonselectively destroy rapidly dividing cells.

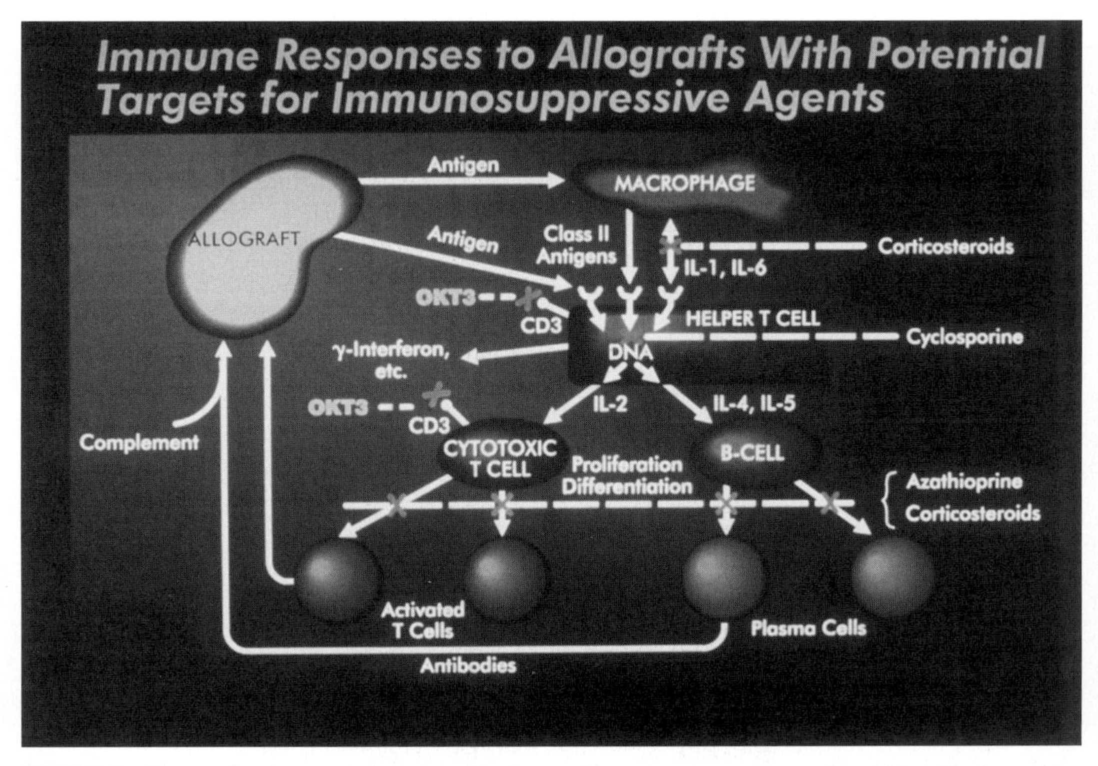

FIGURE 2 Targets for immunosuppression. Corticosteroids, among a spectrum of activities, block cytokine secretion, T-cell differentiation, and macrophage activation. Azathioprine, an antimetabolite, blocks DNA synthesis and therefore cell division and differentiation. OKT3, a monoclonal antibody targeted to the T-cell receptor, blocks T-cell activity via a mechanism that may involve cell lysis, membrane receptor cross-linking, and decreased receptor expression. Cyclosporine (and FK506) inhibits the gene expression and production of IL-2, the cytokine largely responsible for T-cell differentiation and activation. (Figure borrowed with permission from F. H. Bach, M.D., SCI, Harvard Medical School.)

The most common regimen at that time included azathioprine (Imuran) and steroids. Imuran, a purine analogue that interferes with DNA synthesis, was shown to prolong the survival of kidneys transplanted in dogs. Steroids, among other actions, interfere with antigen presentation, inhibit primary antibody response, reduce the number of circulating T cells, and inhibit secretion of cytokines. These agents are still extensively used today in immunosuppressive transplant regimens.

In the mid-1900s, agents directly targeted to T cells were developed. Antibodies that reacted to the T-cell surface were used as a means of inhibiting T-cell function. Antilymphocyte sera was produced in 1899 and was reapplied 70 years later in rodent models by P. Russell, A. P. Monaco, and P. Medawar. Clinical application by Starzl and subsequent refinement by J. S. Najarian led to powerful polyclonal reagents, antibodies that recognize a broad spectrum of antigens. More recently, monoclonal antibodies, homogeneous antibodies produced by a single clone directed to a single target, have been developed to target the T-cell receptor or its accessory molecules (e.g., anti-CD3 or OKT3).

In 1969, cyclosporine was isolated from the fungus *Tolypocladium inflatum* Gams. Proliferation of successful liver transplantation as well as kidney transplantation in the 1980s has been attributed to the use of this agent. Survival data from Pittsburgh in the first 67 patients receiving this immunosuppressive agent showed a dramatic improvement in 1-year graft survival rate from 30% under azathioprine and steroids to over 65% for patients receiving cyclosporine and steroids. In early clinical trials with kidney transplants, Sir Roy Calne showed that 1-year allograft survival was improved by 20 to 30% as compared with that obtained with azathioprine. Cyclosporine acts on the molecular level to inhibit T-cell cytokine synthesis (IL-2) and inhibits generation of the T-cell population responsible for organ destruction. Other agents with similar activity have been developed, including tacrolimus (FK506). Still other agents have been developed that block other steps in the maturation of T cells (e.g., RAPA, leflunomide, morpholinethyl-mycophenolic acid, and brequinar). Investigation in developing and perfecting immunosuppressive agents and protocols is ongoing. The eventual goal is to achieve permanent acceptance of the organ with minimal to no immunosuppression. [*See* Interleukin-2 and IL-2 Receptors; T-Cell Receptors.]

A major concern in the use of these potent immunosuppressants and an issue that complicates the course of many transplanted patients is the phenomenon of "opportunistic" infections. The immunosuppressed transplant patient is apt to incur a life-threatening illness that would not normally affect the immunocompetent host. They are at greater risk for infection with certain organisms (e.g., cytomegalovirus, herpes simplex virus, and *Pneumocystis carinii*). For this reason also, it would be desirable to achieve a regimen whereby the patient becomes tolerant and does not require life-long high-dose immunosuppression.

IV. CURRENT AND FUTURE DIRECTIONS

In addition to the standardization of surgical technique and the development of immunosuppression, other factors have contributed to the improved results in transplantation. These include improved anesthesia, critical care, and nursing support. A strong emphasis has also developed on the selection of appropriate patients for transplantation, that is, those who are more likely to survive and benefit from transplantation of a given organ. There are now agencies dedicated to the management of the donor and to the efficient care and delivery of the donor organ. In the late 1980s, the development and introduction of the University of Wisconsin preservation solution extended the safe cold preservation time of the donor liver from 8 hours to 24 hours. This dramatically improved the performance of the recipient operation. Results of transplantation of the various organs are listed in Fig. 3.

After the fine-tuning that occurred in the 1980s, the major limiting factor became the availability of suitable donor organs. Adjustments were made for more effective allocation of donor organs and criteria were modified so that "marginal" organs became acceptable for implantation. Additionally, the use of living-related and living-unrelated donors (in kidney transplantation) has increased the number of available organs. Nevertheless, organ shortage continues to limit the clinical application of transplantation. The mortality rate for transplant recipients on the waiting list is rising as there continues to be a steady increase in the number of potential recipients with a limited number of organ donors.

The organ shortage problem is especially prominent in the pediatric population. These patients are often unable to accommodate full-size adult organs (specifically the liver) and there is a very limited number of

Organ	Number of organs transplanted	% functioning at 3 yr.
Kidney (cadaveric donor)	47,010	69.0
Kidney (live donor)	13,881	83.9
Liver	16,370	61.1
Pancreas	3,084	61.6
Heart	12,421	72.9
Lung	1,936	52.7
Heart-Lung	360	48.1

FIGURE 3 Success rates for organ transplantation. These survival rates were computed using the Kaplan–Meier method and are based on the United Network for Organ Sharing Scientific Registry data as of October 7, 1995. These are the graft and patient survival rates from October 1987 to December 1993.

pediatric donor organs. In the past decade, techniques have been developed for transplanting pediatric patients with "reduced-size" livers. Using these techniques, a portion of liver from a live relative is retrieved and transplanted into the recipient pediatric patient. This is of particular import as there is a severe shortage of donors for infants and small children and the mortality rate for pediatric patients on the waiting list is very high, approaching up to 50% in some centers. With development of reduced-size living-related operations, the mortality rate for pediatric patients on the waiting list has decreased to as low as 10% in a number of centers.

The development of the field of transplantation is marked by dramatic breakthroughs in science and medicine. As we approach the next era, much effort is being invested into methods of creating reliable sources of donor organs. There is ongoing work in the development of mechanical organs, most notably the artificial heart. Another area of active investigation is that of xenotransplantation, the transplantation of organs between different species. The limited clinical experiences with xenografts, most recently in Pittsburgh and Los Angeles, have generated considerable interest and controversy. Although this experience did not result in long-term survival, research continues to investigate the rejection process that occurs in xenotransplantation, a process that appears to be distinct from that seen with allotransplantation. Recently reported is the birth of viable lambs, products of the transfer of DNA derived from a mammary gland cell. This development along with advances in molecular biology and gene therapy may afford a means of producing organs from donor animals which possess favorable characteristics for transplantation.

BIBLIOGRAPHY

Amos, D. B., and Sanfilippo, F. (1991). The immunology of transplant antigens. *In* "Textbook of Surgery" (Sabiston, ed.). Saunders, Philadelphia.

Kahan, B. D., and Ghobrial, R. (1994). Immunosuppressive agents. *Surg. Clinics of North America* **74**(5), 1133.

Kuss, R., and Bourget, P. (1992). "An Illustrated History of Organ Transplantation. The Great Adventure of the Century." Laboratoires Sandoz, Rueil Malmaison, France.

Sanfilippo, F., and Amos, D. B. (1991). Mechanisms and characteristics of allograft rejection. *In* "Textbook of Surgery" (D. C. Sabiston, Jr., ed.). Saunders, Philadelphia.

Simmons, R. L., Ildstad, S. T., Smith, C. R., *et al.* (1994). Transplantation. *In* "Principles of Surgery" (S. I. Schwartz, G. T. Shires, and F. C. Spencer, eds.), 6th Ed. McGraw–Hill, New York.

Starzl, T. E., *et al.* (1991). Liver transplantation: A 31-year perspective. *Curr. Probl. Surg.* **27**, 205.

Sullivan, K., and Leech, S. H. (1993). Concepts and background of histocompatibility testing. *In* "SEOPF Tissue Typing Reference Manual" (G. Tardif and J. M. MacQueen, eds.). SEOPF, Richmond, Virginia.

Wilmut, I., Schnieke, A. E., McWhic, J., Kind, A. S., and Campbell, K. H. S. (1997). Viable offspring derived from fetal and adult mammalian cells. *Nature* **385**, 816.

Wood, R. P., Ozaki, C. F., Katz, S. M., Monsour, H. P., Jr., Dyer, C. H., and Johnston, T. D. (1994). Liver transplantation. *Surg. Clinics of North America* **74**(5), 1133.

Oxygen Transport to Tissue

I. S. LONGMUIR
North Carolina State University

GLOSSARY

Anoxia Total lack of oxygen

Henry's law When a liquid is equilibrated with a gas, the amount of gas that goes into solution in the liquid is proportional to the partial pressure of that gas in the gas phase. Thus, the concentration of a gas in solution is $CO_2 = \alpha\, PO_2$, where α is the Bunsen solubility coefficient and PO_2 is the partial pressure of oxygen in the gas phase in equilibrium with the liquid. The activity of a gas in solution is a function of its partial pressure rather than its concentration as is the case with solid solutes

Hyperoxia and hypoxia Values substantially higher or lower, respectively, than the normoxic values

Normoxia Partial pressure of oxygen in various parts of the healthy human body at rest, breathing air at 760 mm Hg. These values are as follows: PI_{O_2} = 150 mm Hg, PA_{O_2} = 100 mm Hg, Pa_{O_2} = 96 mm Hg, $P\overline{V}_{O_2}$ = 40 mm Hg

Partial pressure Partial pressure of a gas in a given volume of a mixture of gases is the pressure that gas would exert if it alone were present in that volume. The sum of the partial pressures of each gas in the mixture is equal to the pressure exerted by the mixture. By convention all pressures in physiology are expressed in millimeters of mercury (mm Hg)

Symbols By convention the following symbols are used: FI_{O_2} is the fraction of oxygen in the inspired air; PI_{O_2} is the partial pressure of oxygen in inspired air; PA_{O_2} is the partial pressure of oxygen in the alveolar air; Pa_{O_2} is the partial pressure of oxygen in the arterial blood, that is, the blood leaving the lung capillaries; $P\overline{V}_{O_2}$ is the mean partial pressure of oxygen in the mixed venous blood, that is, the blood entering the lung capillaries; PT_{O_2} is the partial pressure of oxygen in the tissues

ALL HUMAN CELLS REQUIRE A CONTINUOUS SUPply of oxygen, but the urgency with which different cells need oxygen varies. The cells of the retina are the most demanding in this respect. After only a few seconds without oxygen (anoxia) they stop responding to light; the subject "blacks out." Muscle cells, on the other hand, can continue to contract for several minutes in the absence of oxygen. Oxygen is also needed for certain metabolic activities necessary for the maintenance of cell integrity. Cell function can be restored after brief periods of anoxia by restoring the oxygen supply, but if the anoxia is prolonged the cells will die. As little as 4 min of cerebral anoxia causes brain death.

The body's source of oxygen is the ambient air. Some surface cells, such as those of the cornea, can obtain all the oxygen they require directly, but all other cells depend on a complex distribution system. Air is inspired through the nose and mouth and passes down the trachea to the lungs. The main bronchus in each lung branches many times, and the small airways terminate in sacs called the alveoli. Each alveolus is about 0.25 mm in diameter and is surrounded by a mesh of small blood vessels called capillaries. Oxygen diffuses from the air into the capillary blood and reacts with the hemoglobin in the red blood cells. The blood drains into the left heart and is pumped through the aorta and its branches to all the tissues of the body. The smallest branches and the tissue capillaries are vessels about 5.0 μm across and separated from one another by variable distances averaging about 0.1 mm. The oxygen leaves the blood in the tissue

ENCYCLOPEDIA OF HUMAN BIOLOGY, Second Edition, VOLUME 6.

capillaries and diffuses through the cell to the sites of utilization.

I. THE OXYGEN CASCADE

The pathway of oxygen from ambient air to the intracellular sites of oxygen utilization is down the "oxygen cascade," so named because at each step there is a fall in partial pressure. The first step is the admission of air to each alveolus. Since air cannot be completely expelled from the lungs, even by forced expiration, the fresh air entering will be mixed with residual air from which some oxygen has been removed. As a result, the average PA_{O_2} is about 50 mm Hg lower than the PI_{O_2}. The second step is the diffusion across the alveolar membrane into the blood, where the oxygen binds to hemoglobin. Since the lungs are efficient gas exchangers, the fall in PO_2 at this stage is quite small. However, the difference between PA_{O_2} and PA_{O_2}, the A—a difference, is increased by two other events, shunting and ventilation/perfusion mismatch, as will be explained later. During the third step, the circulation of the blood from lung capillaries to those in the tissue, there is in general a very small fall in PO_2 due to oxygen consumption by the blood itself. In certain specialized structures, such as small arterioles invested by venae comites, there can be an appreciable diffusion of oxygen from the arterial blood to the venous blood. The final step of the cascade occurs in the transfer to tissues, where the PO_2 at points near the capillary wall may be nearly as high as the PA_{O_2}, while at points farthest from open capillaries it may be near zero.

II. VENTILATION OF THE LUNG

The lungs consist of a series of branching tubes, which terminate in small air sacs called alveoli. The human lung has 300 million such sacs, each about 0.25 mm in diameter. The sacs are lined with a thin layer of fluid and are therefore essentially bubbles. If the fluid were pure water the hydrostatic pressure inside each bubble would be sufficient to collapse the alveoli. However, this collapsing pressure is opposed by a surface monolayer of detergent (dipalmityllecithin), which efficiently resists surface pressures. Thus, the alveoli do not collapse. At the same time, if the alveoli are made to expand, there comes a point where the surface film is incomplete, and the surface tension of fluid free of monolayer opposes further inflation. Thus by the unique properties of dipalmithyllecithin, also called pulmonary surfactant, the volume range of each alveolus is restricted. Pulmonary surfactant is continuously secreted by cells in the alveolus. If the supply is disrupted by certain toxins, such as paraquat or a PI_{O_2} in excess of 500 mm for a day or more, the alveoli collapse.

If the lungs are removed from the chest cavity, they will collapse, partly because of the small excess of the surface tension of water over the collapse resistance of surfactant and partly because of the large amount of elastic tissue in the solid part of the lung. Normally, the lungs hang in a closed box, the thoracic cavity, which prevents collapse. The volume of the thoracic cavity is increased when the intercostal muscles and the diaphragm contract. This results in the inspiration of air into the lungs. When these muscles relax, the thoracic cavity decreases in volume and air is expired. Virtually all the change in volume of the lungs occurs in the alveoli.

Respiration, the alternative contraction and relaxation of the intercostals and the diaphragm, is controlled by the respiratory centers in the floor of the fourth ventricle of the brain. Although this action is involuntary, it is possible to override it to some extent by voluntary effort. The respiratory center is not an oxygen sensor but responds to changes in arterial blood pH and the partial pressure of carbon dioxide so that the air in the alveoli is kept at a PCO_2 of about 40 mm Hg. The lungs have two main functions: the uptake of oxygen and the excretion of carbon dioxide. The latter, although it is an excretion product, is not completely purged from the blood but is kept at a partial pressure of about 40 mm Hg, which keeps the pH of arterial blood at about 7.4, that is, essentially neutral. The effect of this control of carbon dioxide content in the alveolar air is to maintain the PA_{O_2} at about 100 mm Hg when normal air at sea-level barometric pressure is inspired. However, when the PI_{O_2} drops below 150 mm Hg the PA_{O_2} falls below 95 mm. At 70 mm Hg, certain peripheral chemoreceptors such as the carotid bodies are stimulated to override the respiratory center. As a result, the rate and depth of lung ventilation increase. This increases the Pa_{O_2}. There appears to be a wide variation in the Pa_{O_2} at which the override mechanism begins to operate and the degree to which hypoxia stimulates ventilation. The only physiological event that gives rise to a lowered PI_{O_2} is ascent to high altitudes, and subjects with the greatest hypoxic response are best able to ascend to great heights. In addition, living at altitude for some weeks increases the hypoxic drive and enables climbers to ascend further. [*See* Respiratory System, Physiology and Biochemistry.]

The air in the lungs is contained in the airways: the main bronchi, the secondary bronchi, the bronchioles, the alveolar ducts, and the alveoli. Oxygen diffuses from the air into the blood only in the latter structures. This air is called the alveolar air and the rest the "dead space" air. It is possible to calculate the voluem of the dead space by analyzing an expired breath. At the end of the breath the air will have come from the alveoli only, whereas the whole volume of the breath will contain a mixture of the whole of dead space air plus some alveolar air. Thus, from measurement of the oxygen content of an "end tidal" sample and of the mixed breath, the dead space air volume can be calculated. From this value the amount of air entering the alveoli per minute can be calculated by subtracting the product of the dead space volume and the breaths per minute from the volume of air inspired per minute. In a normal man at rest the minute alveolar volume ($\dot{V}_A$) is about 5 liters. This volume is increased many times on exercise.

The entire blood output of the right heart is delivered to the capillary meshes surrounding each alveolus. At rest this volume is again about 5 liters per minute ($\dot{Q} = 5$). It is clear that these two fluids, air and blood, need to be delivered equally to all parts of the lungs. In alveoli that are ventilated but not perfused with blood and in those that are perfused but not ventilated there can be no oxygen uptake. Ideally, each alveolus should receive the same volume of both blood and air. Every alveolus in the healthy lung tends to fill with air to about the same extent as a result of the surfactant mechanism previously described. The uniform perfusion of the pulmonary circulation is, however, more complex. The pressure in the pulmonary trunk varies between 20 mm Hg when the heart ventricles contract (systole) and 8 mm Hg when they are at rest (diastole). As a result, circulation in capillaries in those parts of the lung that lie above the heart when the subject is erect stops during diastole, whereas the circulation at the base of the lung is continuous. Such a mismatch of ventilation with air and perfusion with blood leads to inefficient gas exchange. There is a mechanism, however, which goes some way to making the distribution of blood in the lung more uniform. It is "oxygen autoregulation." When the oxygen tension in a part of the lung is low, the blood vessels constrict, reducing the blood flow to that region. If a volume of lung is not ventilated for any reason, the pulmonary blood flow ceases altogether. Since the lungs are obliged to accept all the blood ejected from the right ventricle, any reduction in flow in any part of the lungs must result in increased flow in the rest. When the $P_{I_{O_2}}$ falls, as in ascent to high altitudes, all the pulmonary vessels constrict. This results in a rise in pressure in the whole pulmonary circulation and so improves perfusion of the apices of the lung. Thus, the response of the lung blood vessels to hypoxia has the effect of altering the value of $\dot{V}a/\dot{Q}$ toward unity in all parts of the lung.

III. DIFFUSION ACROSS THE ALVEOLAR WALL

Since the $P\overline{v}_{O_2}$, the partial pressure of oxygen in the blood entering the capillary nets surrounding each alveolus, is less than that in the alveolar air, oxygen diffuses from air to blood. The rate at which it does so is proportional to the product of the diffusivity of oxygen (D), the area of the alveolar wall (A), and the partial pressure difference between air and blood (ΔP_{O_2}) and inversely proportional to the thickness of the barrier between air and blood (δ). As a result, oxygen flux $= (AD/\delta)\Delta P_{O_2}$. The total alveolar area in humans is greater than 100 m^2, about the size of a tennis court. The thickness of the alveolar wall averages about 1.4 μm. It is clear that the lungs constitute a very efficient gas exchange and in severe exercise can take up several liters of oxygen per minute. At the alveolar level the blood traverses the capillary net in less than a second, about 0.7 sec at rest, and half this during maximal exertion. During this time each 100 ml of blood takes up 5 or more ml of oxygen, and the P_{O_2} rises from about 40 to 95 mm Hg. As the P_{O_2} rises toward that in the alveolar air, the rate of uptake will fall. During expiration no oxygen enters the alveoli, so there will be a fall in Pa_{O_2} during this phase of the ventilatory cycle. However, since the volume of air in the alveoli is about 100 times that of the blood in the pulmonary capillaries, this fall will be negligible. During inspiration there will be a rise. These small fluctuations result in a just detectible variation in Pa_{O_2}. Although the Pa_{O_2} in the blood draining each alveolus is very nearly equal to the P_{O_2} in the gas phase in that alveolus, the mean Pa_{O_2} will be a few millimeters of mercury lower than the mean $P_{A_{O_2}}$ because of three factors: (1) the nonuniform distribution of blood and air, (2) shunting, whereby some of the output of the right heart bypasses the alveolar capillaries, and (3) some oxygen consumption by the lung cells. As a consequence, the so-called A—a gradient is about 5–6 mm Hg. The diffusing capacity of the lungs can be measured using the rate of oxygen uptake. However, this is difficult because of the uncertainty of the back pressure of oxygen in the blood. One can use another gas, carbon monoxide, which has about the same diffusivity as oxygen. The great merit of this gas is the avidity with which it binds to hemoglobin. As a result, the back pressure problem does not arise and the mass transfer equation can be written $(DA/\delta)/Pa_{CO}$.

In this way the diffusing capacity, DA/δ, can be measured from the uptake of carbon monoxide from a single breath.

IV. TRANSPORT OF OXYGEN IN THE BLOOD

A. The Equilibrium with Hemoglobin

Oxygen diffuses into the blood plasma first and then passes across the red cell membrane and reacts reversibly with hemoglobin. The red cells contain a high concentration of hemoglobin, about 30% w/v. The reaction of hemoglobin and oxygen is quite fast, and equilibrium is reached while the red cells are in the pulmonary capillaries, a period of less than a second. Each hemoglobin molecule can bind reversibly to four oxygen molecules, which, however, are not completely independent of one another. When one molecule of oxygen binds with one site, the shape of the molecule changes, and as a result the affinity of the other three sites for oxygen is increased. Attempts to develop equations relating the amount of oxygen bound to hemoglobin and the Po_2 in solution have been only partially successful. However, to understand the role of hemoglobin in oxygen transport it is convenient to consider the oxyhemoglobin dissociation curve. This is a plot of the percentage saturation of the binding sites against the Po_2 (Fig. 1, curves 2, 3, and 4). It will be seen that the curves are S-shaped and that at a Po_2 of 95 mm, the normal value for arterial blood, the hemoglobin is nearly fully saturated. Blood contains about 15 g of hemoglobin per 100 ml, and each gram of hemoglobin can bind to 1.3 ml of oxygen. Thus fully saturated blood contains about 20 ml of oxygen bound to hemoglobin. There is also some oxygen in solution, amounting to 0.25 ml at 95 mm Hg. At normal Pa_{O_2} values this is a negligible quantity, but if oxygen is inspired at a pressure of several atmospheres, the amount in physical solution can be sufficient to meet all the tissues' needs for oxygen. However, this partial pressure of oxygen

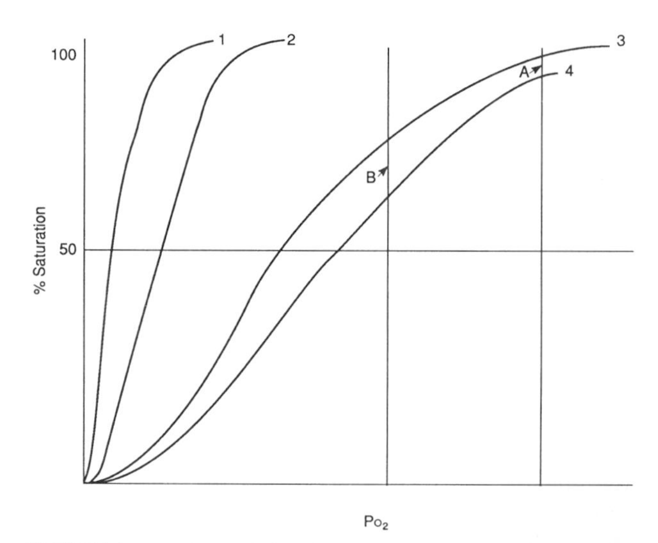

FIGURE 1 Oxyhemoglobin dissociation curves. (1) Hemoglobin monomer, one binding site; (2) hemoglobin solution; (3) red cells—P_{CO_2} that for arterial blood; (4) red cells—P_{CO_2} that for venous blood. The differences between the curves have been exaggerated to clarify the various points made in the text.

in the blood would be low enough to damage the brain and lungs and so could not be sustained for more than a brief period. Thus, the major function of the hemoglobin is to raise the oxygen capacity of blood severalfold without increasing the Pa_{O_2} to toxic levels. [*See* Hemoglobin; Red Cell Membrane.]

The shape of the oxyhemoglobin dissociation curve plays an important role in oxygen loading and unloading. As mentioned before, the curve is S-shaped since the effect of the addition of an oxygen molecule is to increase the affinity of the remaining three sites for oxygen. If the hemoglobin molecule is split into four parts, each with one binding site, this effect is lost and the curve becomes a simple hyperbole (see Fig. 1, curve 1). The midpoint of the dissociation curve of blood, the P_{50}, is at a Po_2 of about 26 mm. A pure solution of hemoglobin has a midpoint of less than 6 mm. With such a low value it can be seen the tissue oxygen tension would have to fall to very low values before a significant amount of oxygen could dissociate. All the hemoglobin in blood is inside the red cells, which also contain 2 : 3 bisphosphoglycerate (BPG) in an equimolar concentrate to that of hemoglobin. This molecule binds to deoxygenated hemoglobin, reducing its affinity for oxygen and shifting the P_{50} to a higher value. A number of other substances also shift the P_{50}. The most important ones, physiologically, are CO_2 and pH change. When the blood takes up CO_2 in the tissues, the P_{CO_2} rises and the pH falls (i.e., acidity increases), shifting the curve to the right (see Fig. 1, curves 3 and 4). In the lungs the reverse changes occur. As can be seen from the figure, the change increases the amount of oxygen taken up in the lungs and the amount given up in the tissues. Thus, the concomitant transport of carbon dioxide increases the efficiency of oxygen transport. A number of congenital and acquired factors influence the dissociation curve. More than a hundred abnormal hemoglobins have been discovered, many of which show reduced efficiency in oxygen transport. The conversion of some hemoglobin to carbonmonoxyhemoglobin or methemoglobin, both present to a small extent in human blood, causes a reduction in P_{50}, which can become significant if either is present in excess of 10%.

B. Blood Perfusion of Tissue

The oxygenated blood drains from the alveolar capillaries through the pulmonary veins into the left side of the heart. From the left ventricle the blood is pumped into the aorta. Arteries branch from this vessel and repeatedly divide into smaller arteries and then arterioles to deliver blood to all the organs and tissues of the body. The smallest vessels are the capillaries, vessels whose walls are one cell thick and have a lumen of 5.0 μm, less than the diameter of the red cell. Red cells, therefore, have to be distorted to some extent in order to pass through the capillaries. It is while the red cells are traversing the capillaries that they give up oxygen. They can release oxygen only when the P_{O_2} in the plasma falls by the diffusion of oxygen into the surrounding tissue, where the P_{O_2} is lower still.

The blood is distributed to the tissue in response to local oxygen need. This distribution is regulated in two ways. In the long term, the size of the arteries appears to be related to the oxygen need of a particular organ. In the embryo and in healing tissue angiogenesis, the formation of new blood vessels is stimulated by tissue hypoxia. When the oxygen consumption of tissues such as skeletal muscle or the heart is raised by a substantial increase in physical activity, the major vessels in those organs increase in diameter. Because the resistance to blood flow is inversely proportional to the fourth power of the radius, this change is associated with a greatly increased blood flow. The increase in flow due to vessel enlargement occurs in a matter of weeks, but in the short term, on the order of a few seconds, there is another effect, known as "oxygen autoregulation." This term has already been used to describe the reduced blood flow to hypoxic regions of the lung. In the peripheral tissues the response to hypoxia is inverted, and hypoxia increases the blood flow.

All three sets of vessels appear to be involved in the increased tissue perfusion. Arterioles and venules dilate by a relaxation of the smooth muscle in their walls. Capillaries lack this muscle coat; they increase perfusion by recruitment, the opening up of previously closed capillaries. Red cells traverse each capillary at a fairly constant rate. An open capillary has a diameter of about 5.0 μm, and this lumen is further restricted when the endothelial cells that make up the capillary wall swell. Then the red cells can no longer squeeze through, and oxygen delivery from that capillary ceases. Thus, the mechanism of "recruitment" may be the reversal of this reduction by a slight shrinkage of endothelial cells to the point where red cells can flow again. The tissue response to mild hypoxia may be a purely capillary reaction, which reduces the resistance to blood flow, resulting in increased arteriolar flow without any increase in diameter of the latter vessels. In addition, recruitment would also permit more efficient tissue oxygenation because it reduces the intercapillary distances. However, more severe hypoxia results in simultaneous dilatation of both anterioles and venules in addition to capillary recruitment. When the oxygen tension rises, the vessels constrict. Although this constriction is part of the process of "oxygen autoregulation," it is not clear whether or not a separate mechanism is involved.

Although the effects of oxygen tension on tissue perfusion with blood have been known for more than 70 years, the mechanism is still unclear. The most cogent theories are based on the metabolic consequences of oxygen starvation of the oxygen-consuming enzymes. When the cytochrome oxidase system, for lack of oxygen, is unable to produce adenosine triphosphate as fast as it is consumed, adenosine is released. However, arterioles that dilate in hypoxia do so at oxygen tensions above that at which cytochrome oxidase begins to become reduced. Aspirin and similar inhibitors of cyclooxygenase partially inhibit "oxygen autoregulation." This enzyme, which is essential for prostaglandin synthesis, requires a higher P_{O_2} for optimal activity. Its inhibition by lack of oxygen or by drugs may reduce the production of vasoconstrictor molecules. More complex models have been proposed, involving specific "reporter" enzymes whose sole function is in "oxygen autoregulation." It has also been proposed that the vasoconstriction that occurs when the Po_2 is high is due to oxygen-free radicals, which are produced in greater amounts in hyperoxia.

C. Transport of Oxygen from Blood to Tissue Site of Utilization

The capillary wall is highly permeable to oxygen, which can diffuse readily through all parts of this structure. Since the cells surrounding the capillary are continuously consuming oxygen, the Po_2 in them is lower than in the blood, so oxygen diffuses into the tissue. As the Po_2 in the blood falls, oxygen dissociates from oxyhemoglobin as described previously. The first attempt to model the transport of oxygen in tissue was based on two assumptions: uniform oxygen consumption and uniform passive diffusion. Since oxygen can diffuse for only a limited distance before being used completely, this leads to a picture of bounded cylinders of falling oxygen tension surrounding each capillary. The radius of this cylinder is a function of capillary Po_2, the diffusivity of oxygen in tissue, and the local rate of oxygen consumption. However, the first assumption of uniform oxygen consumption is modified by the finding that the oxygen-consuming enzymes are situated in highly localized structures in the cell (mainly the mitochondria). Attempts to measure the diffusivity of oxygen in cells showed anomalies in intracellular oxygen transport. In red muscle it is clear that myoglobin (a muscle protein) facilitates the transport of oxygen. In all cells, oxygen diffuses along the hydrophobic core of membranes 10 times as fast as through cytosol. Since most of the respiratory enzymes are localized within organelles associated with membranes, these membranes might constitute an intracellular oxygen transport system. There is some evidence that the endoplasmic reticulum, a double-membrane structure, is continuous with both the plasma membrane and the outer mitochondrial membrane. The major respiratory enzyme, cytochrome oxidase, is located in the inner mitochondrial membrane. Some oxidases are situated in the endoplasmic reticulum and the outer mitochondrial membrane and the rest within peroxysomes attached to the endoplasmic reticulum. Thus, all the oxygen-consuming enzymes are associated with a membrane along which oxygen can diffuse more rapidly than through the rest of the cellular constituents. A further complexity has been introduced by the observation that even in the presence of enough oxygen, cells switch off their respiration. It seems that individual cells are either respiring maximally or not at all. Thus, the original cylinder has become greatly distorted both spatially and temporally.

V. CLASSIFICATION OF RESPIRATORY ENZYMES

Respiratory enzymes can be classified as oxidases, oxygenases, and mixed function oxidases. Oxidases use oxygen as an electron acceptor. Cytochrome oxidase is the terminal enzyme of the electron transport chain in mitochondria and reduces oxygen to water by the addition of four electrons. The principal function of this enzyme system is to generate adenosine triphosphate, which is the fuel for muscular contraction and many other metabolic events. It is by far the major consumer of oxygen. Other oxidases donate two or one electron to produce hydrogen peroxide or superoxide free radical, respectively, which are utilized for killing foreign organisms.

The oxygenases introduce both atoms of the oxygen molecule into a second molecule. The mixed function oxidases reduce one atom of oxygen to water while incorporating the second atom into another substrate. An alternative classification of these enzymes is on the basis of their affinity for oxygen. The enzyme with the highest affinity is cytochrome oxidase, which can consume oxygen at its maximum rate until only a trace remains. At the other end of the scale are the flavoprotein oxidases, which have much lower affinity. Clearly these enzymes will require much higher oxygen concentrations, but since their major role is synthetic, the limiting factor is either the tissue demand for their products or the availability of substrates on which they act. Other enzymes have an intermediate range of affinities, as demanded by their functions; some of them play a role as oxygen sensors for oxygen autoregulation.

VI. TISSUE OXYGEN TRANSPORT IN THE FETUS

Tissue oxygen transport in the fetus appears to be qualitatively the same as in the adult but with a striking quantitative difference. The placenta, which is the source of oxygen before birth, is a remarkably inefficient oxygen exchanger compared with the lung. The Po_2 in the blood in the umbilical vein in the human fetus is only 29 mm Hg. This figure is further reduced by admixture with fetal venous blood before entering the arterial system. The low fetal Pa_{O_2} is partially compensated for by a low P_{50} in the fetal blood. Nevertheless, the "oxygen cascade" in the fetus has a gradient only one-third that of the adult. This gradient changes dramatically at birth.

BIBLIOGRAPHY

"Oxygen Transport to Tissue," Vols. I–XVII (1996). Advances in Experimental Biology and Medicine. Plenum New York/London.

Weibel, E. R. (1986). "The Pathway for Oxygen." Harvard Univ. Press, Cambridge, Massachusetts/London.

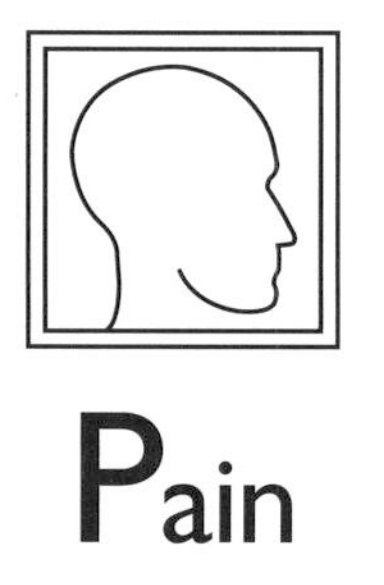

Pain

FRAN LANG PORTER
Washington University School of Medicine

GLOSSARY

Myelination Process in which a phospholipid sheath forms around nerve fibers. Myelination speeds conduction of the nerve impulse

Nociception Response induced by a painful, injurious stimulus

Noxious stimulus Stimulus that is painful, injurious, or harmful

Psychophysiology Study of the interactions between mental and physiological processes

PAIN REPRESENTS A COMPLEX, PSYCHOPHYSIOlogical event that involves the interaction of sensory, neurochemical, emotional, motivational, and cognitive systems. Pain provides important information by conveying the message that "something is wrong." It encourages action to prevent further injury. From an evolutionary point of view, pain is one of the most powerful ways to ensure the survival of an organism in a dangerous or potentially dangerous world.

I. DEFINITION

Pain is typically described along dimensions of quality, location, duration, and intensity, yet it is more than the total of these. Perceptions of pain and reactions to it can vary dramatically among individuals and even within the same individual over time, conditions, health, and psychological states. It is, therefore, difficult to quantify pain precisely.

The most widely accepted definition of pain was offered by the International Association for the Study of Pain (IASP) in 1979. This definition states that pain is:

> An unpleasant sensory and emotional experience associated with actual or potential tissue damage, or described in terms of such damage.

This definition emphasizes not only the sensory aspects of pain but also the importance of the emotional components of the pain experience. It further stresses the individualized nature of pain:

> Pain is always subjective. Each individual learns the application of the word through experiences related to injury in early life.

Thus, exhausting structural or physiological explanations may not be sufficient to understand the existence, absence, or persistence of pain. Pain represents an enormous challenge for clinicians and scientists alike.

II. PHYSIOLOGY OF PAIN

The physiological mechanism by which noxious stimuli produce a subjective correlate of pain is far from completely understood. Because pain is fundamentally a subjective experience, there are inherent physiological limitations to understanding it. However, a series of complex electrical and chemical events between tissue injury and the subjective experience of pain have been identified. These include transduction, transmission, and modulation.

A. Ascending Pain System

Transduction is the process by which noxious stimuli produce electrical activity in specific sensory nerve endings (Fig. 1, left). *Transmission* is the process by which this electrical activity is then transmitted over the peripheral nerve to the spinal cord. The pain message then synapses with cells of origin of the two major ascending pain pathways, the *spinothalamic* and the *reticulothalamic* pathways. These pathways make up a network of relay neurons that ascend from the spinal cord to the brain stem and the thalamus. The message is then transmitted to higher brain centers via reciprocal connections between the thalamus and the frontal and somatosensory cortices. [*See* Thalamus.]

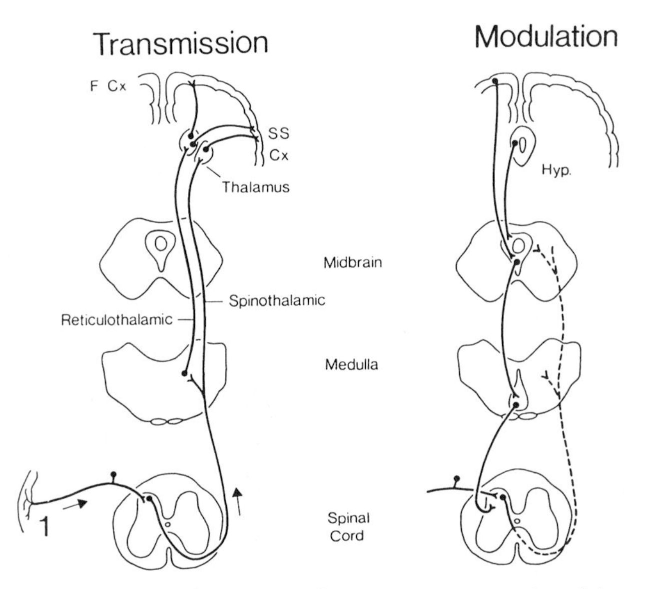

FIGURE 1 Neural pathways of pain transmission and modulation. (Left) Noxious stimuli activate the primary afferent nociceptor's peripheral endings by the process of transduction (1). The neural impulse is then transmitted to the spinal cord, where it synapses with the spinothalamic and spinoreticulothalamic pain pathways. The message is then relayed from the thalamus to the frontal cortex and the somatosensory cortices. (Right) Inputs from the frontal cortex and hypothalamus activate cells in the midbrain that control spinal pain transmission cells via cells in the medulla. [Reproduced, with permission, from H. L. Fields (1987). "Pain," p. 6. McGraw–Hill, New York.

1. Pain Transmission Fibers

There are believed to be two major types of fibers involved in the transduction and transmission of pain signals. These are described by their diameter, impulse conduction rate, extent of myelination, and ability to respond to noxious stimuli. A-delta fibers, the larger (2–5 m) of the two, conduct impulses relatively rapidly (12–80 m/sec), are thinly myelinated, and are specialized to give several types of information. Some respond primarily to mechanical stimulation and probably relay information regarding the site of an injury, whereas others respond to thermal or chemical stimulation. Still others may be polymodal in responsivity so that they will respond to any type of stimulation (mechanical, thermal, or chemical) but only after a certain threshold has been reached. Some A-delta fibers increase their discharge as stimulus intensity is increased so that the greater the stimulation, the more frequent the impulses are fired. Stimulation of A-delta fibers produces a sharp, pricking sensation.

The smaller (0.3–1.0 m) C pain fibers are unmyelinated, conduct impulses relatively slowly (0.4–1.0 m/sec), and respond to noxious stimulation. C fibers are almost all polymodal responders that are activated only by high threshold stimulation. C fiber stimulation is associated with a dull, diffuse, deeply perceived burning sensation.

The receptive field of each nociceptive neuron overlaps with those of other neurons so that every point on the skin surface, for example, lies within the receptive field of two to four neurons. At this peripheral level then, a sophisticated warning system reports that injury is occurring at a specific site via both A-delta and C fibers and provides information about the quality and intensity of the pain via A-delta fibers.

Evidence that both A-delta and C fibers contribute to the perception of pain comes from nerve block studies. Adults report two sensations in response to brief, intense stimuli. These include an early sharp and relatively brief pricking sensation (first pain) and a later dull, prolonged burning sensation (second pain). First pain disappears when myelinated fibers are selectively blocked by pressure, and the second pain is eliminated when unmyelinated fibers are selectively blocked with local anesthesia. It has been proposed that the first pain is elicited by A-delta fiber activity and the second by stimulation of C fibers so that each may contribute uniquely to the quality of the pain sensation.

It has been suggested, however, that this is an oversimplified view. Most naturally occurring skin stimuli will activate a broad range of receptors, so that the size of the area stimulated, the frequency with which the stimulus is applied, and the duration and location of the stimulus will influence the perceived sensation. In addition, it is probably incorrect to assume that A-delta and C fibers are solely responsible for pain transmission. When both of these fibers are blocked, activation of A-beta fibers produces a sensation of light pressure or tickling but not specifically pain. Moreover, both A-delta and C fibers can be activated by nonnoxious stimuli as well, suggesting that their function is not solely pain related. The possibility remains that there are subcategories of pain fibers, some activated by nonnoxious stimuli, others by noxious stimuli, and still others by both noxious and nonnoxious stimuli. One could hypothesize that each type would derive from a particular receptor type, but this has yet to be demonstrated.

2. Spinal Cord Projections

Once a pain message reaches the spinal cord, it ascends to the higher brain centers along the laminar organization of the gray matter of the spinal cord. This laminar organization is often used to describe the entry points of the pain fibers and is summarized in Fig. 2. The primary ascending nociceptor input is via A-delta and C fibers. The A-delta thermoreceptors terminate primarily in lamina I and the outermost part of lamina II. C fiber thermoreceptors and mechanoreceptors terminate mainly in lamina II. There are some nonnociceptive A-delta and C primary afferent fibers that also terminate in laminae I and II. The deep part of lamina II and laminae III and IV receive input only from nonnociceptive myelinated primary afferents. The larger-diameter myelinated primary afferent (A-alpha and A-beta) fibers that enter the spinal cord in the medial division of the dorsal root terminate in lamina III or deeper. These are primarily low-threshold mechanoreceptors. [*See* Spinal Cord.]

B. Descending Pain System

In addition to the ascending pathways, there are descending neural pathways that arise primarily in the midbrain and terminate in the dorsal horn, where they modulate the transmission and perception of pain signals. *Modulation* refers to the regulation of pain transmission neurons by means of selective inhibition at the level of the spinal cord. This descending pathway can be activated by stress or by certain analgesic

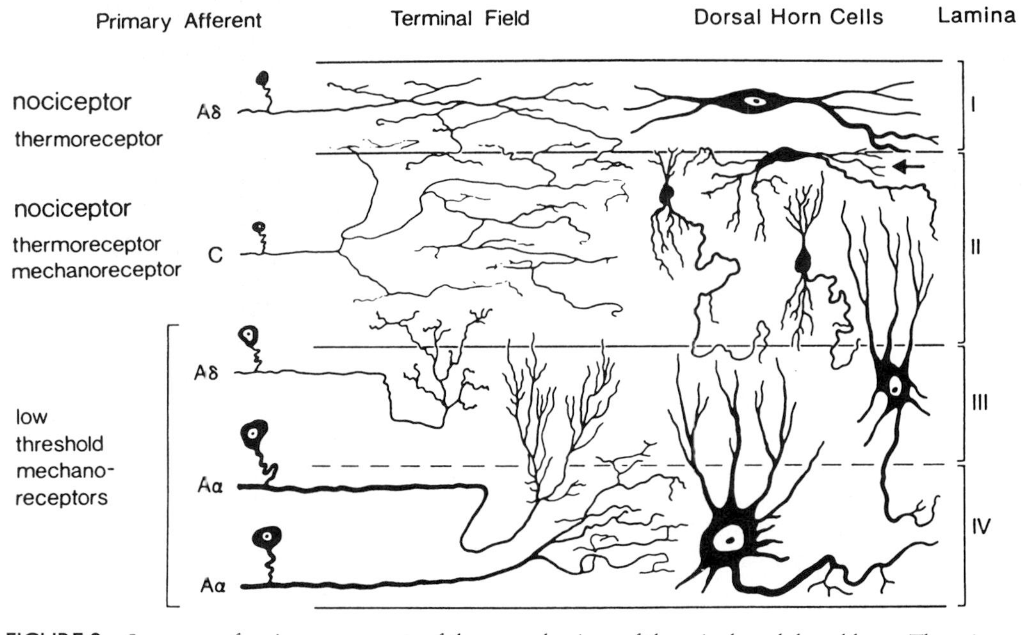

FIGURE 2 Summary of major components of the upper laminae of the spinal cord dorsal horn. The primary afferent nociceptor input is via A-delta and C fibers. A-delta nociceptors terminate primarily in laminae I and II. C fiber nociceptors terminate primarily in lamina II. [Reproduced, with permission, from H. L. Fields (1987). "Pain," p. 49. McGraw–Hill, New York.]

drugs like morphine. When the pain modulation system is active, noxious stimuli produce less activity in the pain transmission pathway. This, in turn, reduces the perceived intensity of the sensation produced by a noxious stimulus.

The descending pathways are a major component of the brain's own system of analgesia. The close anatomic association of endogenous opioid peptides with the pain-modulating networks has provided powerful support for the idea that these peptides actually contribute to analgesia under physiological conditions. Analgesia can be most reliably activated by noxious stimulation. For example, noxious stimuli that are relatively intense and long-lasting can reduce the responses of animals to subsequent noxious stimuli. Such analgesia can be reversed by naloxone, an opioid antagonist, or by lesions of the pain-modulating network.

Analgesia can also be produced by a variety of stressors that are not painful. These include restraint, hypoglycemia, and fear. This analgesia response can also be conditioned, as has been demonstrated in animals that were exposed to a pairing of a light or tone with a noxious stimulus. The precise source and identity of the opioid peptides involved in this analgesic response, however, are not currently known. Nor is it certain whether the anatomy of pain-modulating circuits is the same in humans as in animals. Despite the limitations of current data, it seems probable that this pain-modulating system contributes to the well-described differences in perceived pain among individuals.

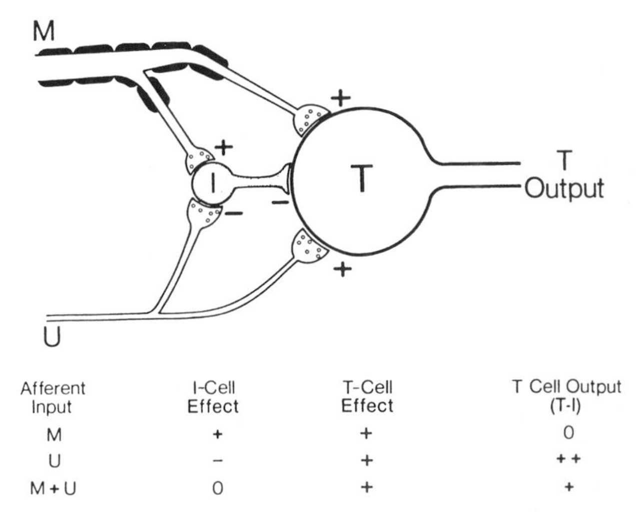

Afferent Input	I-Cell Effect	T-Cell Effect	T Cell Output (T-I)
M	+	+	0
U	–	+	++
M + U	0	+	+

FIGURE 3 Revised version of the gate control theory. (Top) Four neural components of pain: the unmyelinated, nociceptive primary afferent (U), the myelinated nonnociceptive primary afferent (M), the transmission cell (T), whose activity usually results in the sensation of pain, and an inhibitory interneuron (I), which is spontaneously active and whose activity inhibits the T cell, thus reducing perceived pain intensity. (Bottom) How perceived pain (T cell output) is the result of a balance of input from myelinated (M) and unmyelinated (U) primary afferents. [Reproduced, with permission, from H. L. Fields (1987). "Pain," p. 139. McGraw–Hill, New York.]

III. PAIN THEORIES

Several theories have been proposed to explain the neural mechanisms of pain. An intensity theory first suggested that pain was the result of overstimulation of sensory receptors in any modality. Then a specificity theory proposed that highly specialized modality-specific receptors were responsible for pain. These early theories were replaced by a gate control theory proposed by R. Melzack and P. D. Wall, a revised version of which is illustrated in Fig. 3. This version more completely accounts for both the excitatory–inhibitory and sensory–psychological influences associated with pain than did the previous theories.

According to the gate control theory, a neural mechanism, located in the substantia gelatinosa layer in the dorsal horn, acts as a synaptic gate to increase or decrease the flow of nerve impulses from the receptors via the peripheral nerve fibers to the brain. It suggests that the perceived intensity of pain is the result of a balance of input from myelinated fibers (M), unmyelinated fibers (U), and inhibitory interneurons (I) on pain transmission (T) cells. The activity of the pain transmission cell usually results in the sensation of pain, whereas the spontaneously active inhibitory interneuron inhibits the same cell, thus reducing perceived pain intensity. The perception of pain also can be inhibited by the excitatory effect of myelinated afferents on the inhibitory interneuron. Unmyelinated nociceptors, in contrast, inhibit the inhibitory interneuron, thus secondarily exciting the pain transmission cell. The unmyelinated primary afferent thus has both direct and indirect excitatory effects on this cell. A balance between these various inputs to that cell produces intermediate levels of pain intensity. The table at the bottom of Fig. 3 shows how perceived pain (pain transmission cell output) is the result of this balance of input from myelinated and unmyelinated primary afferents.

Psychosocial influences play important roles in the gate control system. Cognitive factors and past experiences can dramatically influence reactions to pain. Among pain theories, the gate control theory is the most relevant for understanding the cognitive aspects of pain. Additionally, it provides a theoretical basis for the use of psychological techniques in pain management, as well as sensory, peripheral procedures such as transcutaneous electrical nerve stimulation.

IV. PSYCHOLOGY OF PAIN

Twentieth-century medical science has emphasized the physiologically based sensory qualities of pain. Originally, however, pain was construed more in terms of inherent individual characteristics such as goodness or evil. Ancient civilizations, for example, believed that gods and demons caused pain to punish individuals lacking spiritual goodness. The ancient Greeks and Romans believed that pain was caused by the interaction of earth, fire, air, and water with the individual's soul. Aristotle suggested that pain was an emotion, as pervasive as anger, joy, or terror, an affect distinct from the classic five senses. This conceptualization would allow for pain to exist in the absence of any tissue injury. Later, Descartes described pain with an exclusive sensory-based model that did not incorporate emotional–psychological components. According to the IASP definition, contemporary thought holds that both affective and sensory qualities are components of the pain experience. In fact, painful experiences have been so intimately associated with affective distress that behavior that is lacking this emotional component may be dismissed as not being pain because it lacks face validity.

Pain is an intensely personal experience. The subjective nature of pain makes it difficult to quantify its emotional and psychological components. Recently, three distinct temporal forms of pain have been associated with certain affective states. *Phasic pain* is of short duration and occurs at the onset of injury. Behaviorally, it is characterized by efforts to withdraw from the source of injury accompanied by verbal and nonverbal expressions that are recognized as distress behaviors. *Acute pain* is provoked by tissue damage and includes both phasic pain and a tonic phase, which persists until healing takes place. Acute pain can be associated with fear and anxiety. There is controversy as to whether increased anxiety increases the reaction to painful stimuli. *Chronic pain* persists beyond the period of time required for healing and can be associated with emotional and behavioral disturbances such as depression, irritability, fear, hopelessness, and serious debilitation.

Because pain is a personal experience, there is wide variability in the psychological and emotional reactions to pain. It is clear that the pain experience is potentially affected by childhood experiences, ethnic and cultural variables, socioeconomic factors, genetic predisposition, birth order, gender, and a host of other physical, perceptual, cognitive, and emotional influences. For example, social expectations can strongly influence the way individuals perceive and respond to pain. Thus, wounds may be disregarded by soldiers on the battlefield, athletes on the playing field, and religious zealots during ceremonial rituals, whereas in other contexts the same tissue damage would be perceived as disruptive and painful. A study contrasting pain endured by wounded soldiers and civilians provided a major challenge to the specificity theory by demonstrating that the context or meaning of pain was a critical variable in the perception of pain. The influence of social expectations has been documented further with respect to cultural–ethnic and gender influences. Early studies suggested that distinct attitudes toward pain and the expression of responses to pain may be linked with cultural heritages. Other studies have indicated gender differences. Although men may tolerate greater degrees of pain than women, these differences may be attributed to gender-influenced social expectations rather than to gender-related physiological differences. For example, most studies have found no difference in pain *threshold* response (the point at which a stimulus is reported to be painful) between men and women, but pain *tolerance* (the point at which a stimulus is no longer tolerable) does appear to be influenced by gender-specific expectations. It is important to note that these social influences apply not only to the patient or recipient of pain, but to clinicians as well, whose own experiences and beliefs regarding pain can have significant diagnostic and therapeutic implications for the patient.

Individual personality styles have also been shown to influence pain perception. Much work has been done to identify the pain-prone patient, the individual who, because of personality, appears to be more vulnerable to pain. Other psychological factors such as anxiety and depression have been shown to heighten pain reactions. However, whether pain is "real" or "imaginary," or "physiological" (organic) or "psychological," to the patient pain is real, even if a doctor can find no physical reason for the pain.

V. DEVELOPMENTAL ASPECTS OF PAIN

It has generally been assumed that the ability to perceive and experience pain increases as a child develops; thus, pain is less commonly considered a problem in children than in adults. Traditionally, in fact, it was held that newborns were not even capable of experiencing pain. This belief was founded on several assumptions and the existing but limited knowledge about newborns.

Through the nineteenth century it was held that complete myelination of pain fibers was necessary for the transmission of pain impulses, contributing to the belief that the newborn was protected from pain. We now know that complete myelination is not essential to the transduction and transmission of pain signals. In fact, in the adult, pain signals are transmitted along nonmyelinated or thinly myelinated fibers. Furthermore, the neuroanatomic basis of pain begins to develop very early in fetal life, long before myelination is complete. As early as the 7th week of gestation, sensory receptors begin to develop, and they spread to all cutaneous and sensory surfaces by the 20th week of gestation. During a period from 6 to 30 weeks of gestation, afferent fibers are established linking the sensory receptors and interneurons in the dorsal horn. Cholinesterase staining studies have further revealed that by 24 weeks, thalamocortical synapses between these afferent nerves and cortical areas are established. Although these histological studies have supported the concept that the necessary structural components for pain are present in the fetus, studies to determine their physiological function have not yet been carried out. By 30 weeks of gestation, pain fibers to the brain stem and thalamus are completely myelinated and, by 37 weeks, the thalamocortical fibers are myelinated. Functional maturity of the cerebral cortex is observable by 30 weeks; pain is generally considered to be a cortical process.

A number of methodological problems are responsible for the previous assumptions regarding pain in infants and young children. Because infants lack an ability to describe pain to adults, the precise degree to which the infant experiences pain is difficult, if not impossible, to assess. This is problematic and emphasizes the need for indices that are independent of verbal skills.

The absence of discernible responses to presumably painful events has also been interpreted as meaning that infants do not experience pain. Individual differences in response magnitude may reflect differences in developmental stage, clinical conditions, therapies employed for those clinical conditions, endogenous pain-protective mechanisms, or previous experience with pain. Alternatively, the absence of response may simply reflect the inadequacy or inappropriateness of the instrument of assessment.

Finally, the assumption that infants do not remember early postnatal experiences has contributed to the assumption that early pain is neither remembered nor capable of exerting lasting effects on child behavior or development. Although memory for pain has not been systematically studied, it has been shown that the newborn and the young infant can learn and demonstrate memories of sensory events for relatively long time spans. Important animal studies have further demonstrated that even the fetus can remember and respond differentially to sensory stimuli depending on prior experience. A better understanding of how and to what extent early aversive experiences are encoded and affect subsequent behavior is needed.

VI. ASSESSMENT OF PAIN IN CHILDREN

A. Neonates

The accurate detection of acute pain is a prerequisite for providing maximally effective preventive and palliative efforts. In neonates this is especially challenging because of the absence of self-report. Yet the accurate measurement of pain will facilitate decisions regarding the utility and effectiveness of pain-relieving interventions. Further, pain assessment is essential to furthering our understanding of the nature of pain in infants and children and how it may change throughout the life span.

Scientific investigations of infant pain, however, are appropriately limited by ethical constraints. Infant pain is, therefore, experimentally evaluated predominantly in the context of clinically required or parentally requested invasive hospital procedures. Among healthy full-term infants, these include blood sampling (clinically required) and circumcision procedures (parentally requested). Among premature and sick infants, a broader range of invasive procedures includes blood sampling, lumbar puncture, tracheal intubation and aspiration, peripheral intravenous line placement, chest tube placement, umbilical artery catheterization, and arterial and venous cutdowns. Thus, infants' responses to stimuli that have been found to be unequivocally painful to adults are used as a means of studying pain in preverbal infants. Current assessment techniques are limited to evaluating how infants behave in response to pain (behavioral assessments) and how their bodies react to pain (physiological assessments).

Behavioral assessments include measures of facial expression, body movement, behavioral state, and crying. Charles Darwin was among the first to report that facial expressions were linked with specific emotions and that these were widely recognized, both cross-culturally and across species. Recent research has confirmed that adults can indeed reliably identify emotional states, including pain, in infants only minutes old using a system to code eye, eyebrow, nose, and mouth positions. Characteristics of the infant and contextual cues, however, contribute to large individual differences in the judgment of the intensity of pain from facial expressions. Facial expressions, therefore, are not standardized indices of infant pain. Shortly after birth, infants also demonstrate avoidance of pain by body movements, once thought to be only reflexive and now believed to represent precise and active attempts to withdraw from painful stimuli. Changes in behavioral state after pain also suggest that infants are not merely responding reflexively to pain. Furthermore, striking individual differences in behavioral state after painful unanesthetized hospital procedures, such as circumcision, indicate that even shortly after birth infants have individual styles of coping with pain. Among neonatal responses to pain, however, the cry is the most commonly studied and widely accepted index of pain. Variations in the fundamental frequency (perceived as pitch), duration, harmonic structure, and other acoustic characteristics are associated with increases in noxious stimulation. These acoustic variations are recognized by adults, confirming their communicative value.

Although behavioral responses to pain appear to have high face validity and are universally interpreted as meaningful, they are unsuitable as the sole criteria for neonatal pain indices. They are often not applicable to premature or sick infants who, due to their immaturity, clinical condition, or therapeutic program, may not demonstrate behavioral responses to pain. Further, behavioral measures have not been standardized and can reflect more the subjective biases of the observer than the infant's pain experience. Physiological measures, therefore, offer a potentially more objective, quantitative assessment of neonatal pain.

Physiological measures of pain include heart rate, respiratory rate, transcutaneously measured circulating oxygen and carbon dioxide levels, blood pressure, intracranial pressure, and palmar sweating. Among healthy, full-term infants undergoing noxious stimulation, marked changes in a broad range of these physiological parameters have been demonstrated, many of which are attenuated with anesthesia. Although many of these physiological changes are dramatic, they are often transitory, demonstrating prompt recovery in this population. Sharp physiological changes have also been widely reported among premature and sick infants in response to noxious stimuli but, more importantly, these changes are not without potential adverse consequences. Bradycardia, cyanosis, hypoxia, increased requirements for oxygenation and ventilation, and alterations in circulatory hemodynamics can occur in response to painful stimulation and contribute to the development of clinical conditions that are associated with significant neurodevelopmental morbidity and increased risk of death.

In addition to physiological measures of pain, hormonal and neurochemical indices are believed to reflect painful events. Changes in serum cortisol, for example, have been documented after routine unanesthetized circumcision procedures in newborns. Endogenously produced opioids, such as the endorphins extracted from cerebrospinal fluid, also increase after painful disturbances in neonates. It is not clear, however, whether these increased endorphin levels reflect a distinct response or a more generalized response to illness or environmental disturbances.

Currently there is no single, reliable index with which to detect pain in newborns. Although physiological measures are more readily quantified and less subjective than behavioral parameters, they require further clarification. Distinctions between responses to pain and responses to nonpainful stressors, and the relation between behavioral and physiological responses, need to be systematically investigated.

B. Children

Self-report measures, widely used with adults, may be utilized in children but with attention to several areas of concern. A major problem with children's self-report measures is the bias introduced by the specific demand characteristics of the situation. Children may describe their pain differently to their mothers than to a doctor, for example. Children also may lack the cognitive skills to understand questions regarding pain or to respond in an interpretable fashion to those questions. Finally the open-ended nature of self-report is not amenable to quantification and accurate measurement.

For very young children, happy–sad children's faces scales have been shown to be easily understood and inexpensive and appear to have excellent measurement characteristics. Faces depicting different degrees of pain are presented to the child and the child chooses the one that most closely approximates his or her own pain. Children over 7 years of age can also use a "pain thermometer," in which intensity of pain is graphically and numerically represented as a higher degree on the visual thermometer, or other visual analogue scales. The Visual Analogue Scale (VAS), for example, usually consists of a 10-cm line labeled at the ends with "no pain" and "most intense pain imaginable" or similar descriptions. The child appraises his or her pain with respect to where it falls on the scale. The McGill Pain Questionnaire is a verbal assessment scale that is suitable for older adolescents who are able to understand adjectives and can choose which words best describe their pain. This tool is unique among pain rating scales in that it assesses both sensory and emotional responses to pain and, thus, offers some evaluation of the anxiety or apprehension that may accompany pain experiences. The validity and reliability of many of these assessment scales have been well established. Primarily nonverbal methods have been used to measure the cognitive component of children's pain. These have included asking children to describe the color of their pain or to draw pictures of their pain. Unfortunately, there has been little or no validation of these projective techniques or systematic investigation of their utility in clinical settings.

VII. MANAGEMENT OF PAIN IN INFANTS AND CHILDREN

The management of pain in children is problematic because of the difficulties associated with detecting their pain, and because only limited research has been done in either area. Questions concerning the underutilization of medication for pain in infants and children have been raised. It is not surprising that newborns and young infants have not been aggressively treated for pain since it has traditionally been held that they were protected from pain owing to their developmental stage. There is clear evidence that infants and young children are undermedicated for pain, both within the hospital setting and by parents, when compared with similarly induced pain in adults. This approach is believed to stem largely from concern regarding the potential toxic or addictive effects of pain medications in children.

In 1992, the Agency for Health Care Policy and Research released a clinical practice guideline calling for "aggressive pain management before, during, and after surgery," noting that failure to control pain can cause unnecessary suffering and can delay recovery and prolong hospitalization. Undermedication and unnecessary pain are noted to occur particularly in infants, children, and the elderly, who often cannot express their needs. Although the risk–benefit ratio of anesthesia in those populations have not yet been adequately investigated, preliminary results suggest significant enhancement of recovery with adequate pain management.

VIII. TESTING PAIN IN NORMAL ADULTS

Studies of pain sensation in normal adults began with the assessment of pain thresholds and the goal of duplicating clinical pain sensation to objectively assess the efficacy of analgesia. Mechanical pressures, chemical stimuli, tourniquet pressure, cold, heat, and electrical stimuli were used in these early studies. The methodology and goals of these studies have now evolved to the use of sophisticated psychophysical procedures, behavioral assessments, and advanced physiological techniques to assess the mechanisms of both pain transmission and pain modulation. As part of this process, pain research investigators have attempted to establish standard criteria for experimental pain research stimuli. Gracely's criteria (1984 and 1985) include the following categories:

1. rapid onset;
2. rapid termination;
3. natural;
4. repeatable with minimal temporal effects;
5. objectivity: similar sensitivities in different individuals;
6. excite a restricted group of primary afferents.

It is clear that specific pain-production methods satisfy some but not all criteria. Thus, an ideal stimulus for one experiment may be inappropriate for others.

Similarly, criteria for pain measures have been advanced. Gracely's (1983) characterization of an ideal pain measure includes that it

1. be bias-free;
2. provide immediate information about accuracy and reliability;
3. separate sensory and emotional aspects;
4. assess experimental and clinical pain with the same scale to permit comparisons;
5. provide absolute assessments to allow for intergroup comparisons or comparisons over time.

A variety of verbal (e.g., stimulus threshold procedures, visual analogue scales, magnitude estimation) and nonverbal (e.g., behavioral, physiological) methods have been extensively used in experimental pain research. These studies have extended beyond demonstration of analgesic efficacy to studies of the mechanisms of pain and analgesia and the use of experimental methods to augment clinical pain report. These studies have also moved from the laboratory to the clinic, providing information about clinical pain syndromes and assisting in the measurement of clinical pain magnitude.

IX. PAIN IN OLDER ADULTS

The elderly represent a group of individuals, much like neonates, who are at risk for pain but who may not report pain. Because of their advancing age, the elderly have an increased incidence of physical illness and associated pain yet they report less pain than their younger counterparts. This phenomenon may be a function of the normal aging process, in which sensitivity to pain stimuli may be decreased. The additional contribution of various diseases, including dementia, to this failure to report pain has not been documented. However, minimal reporting of pain has been interpreted to mean that old people do not feel pain as intensely as younger people, yet laboratory studies have shown no age change in pain sensation. These observations have obvious and important implications for both the diagnosis of illness and the clinical management of pain in the elderly. For example, because there is no standardized method of detecting pain or assessing the effectiveness of pain-relieving techniques in elderly individuals, they may be inappropriately medicated for pain. Therefore, the elderly represent another group for whom an alternative index of pain is required, especially in the context of clinically required medical or care-giving procedures that may induce pain.

BIBLIOGRAPHY

Anand, K. J. S., and Hickey, P. R. (1987). Pain and its effects in the human neonate and fetus. *N. Eng. J. Med.* **317**, 1321.

Fields, H. L. (1987). "Pain." McGraw Hill, New York.

McGrath, P. J. and Unruh, A. M. (1987). "Pain in Children and Adolescents." Elsevier, Amsterdam.

Porter, F. (1989). Pain in the newborn. *In* "Clinics in Perinatology," (J. J. Volpe, (ed.) **16**, 549.

Porter, F. (1993). Assessment of pain in infants. *In* "Pain Management in Children and Adolescents" (N. L. Schechter, C. Berde, and M. Yaster, eds.), pp. 87–96. Williams & Wilkins, Baltimore.

Wall, P. D., and Melzack, R. (1989). "Textbook of Pain." 2nd ed. Churchill Livingstone, Edinburgh.

Pancreas, Physiology and Disease

CATHERINE CHEN
Beth Israel Deaconess Medical Center and Harvard Medical School

GLOSSARY

Cholinergic Describing nerve fibers that release acetylcholine as a neurotransmitter

Cirrhosis Distortion of liver structure by fibrosis and nodular regeneration

Endocrine Gland that manufactures and secretes hormones directly into the bloodstream

Endoderm Inner of the three germ layers of the early embryo

Exocrine Gland that discharges its contents by means of a duct onto an epithelial surface

Gluconeogenesis Biochemical process in which glucose is synthesized from noncarbohydrate sources such as amino acids

Glycogenolysis Biochemical process by which glycogen is converted into glucose-1-phosphate

Golgi complex Collection of membrane-bound vesicles in cells used to store and transport proteins

Jaundice Yellowing of skin and eyes due to excess bilirubin in blood

Meckel's diverticulum Congenital anomaly of gastrointestinal tract resulting from persistence of the embryonic yolk stalk

Parasympathetic nervous system One of two divisions of the autonomic nervous system with fibers from the brain and spinal cord that innervate organs, glands, and blood vessels

Proenzyme Inactive form of certain enzymes

Pseudocyst Fluid-filled space in an organ without a proper lining

Steatorrhea Passage of increased amounts of fat in feces

Sympathetic nervous system One of two divisions of the autonomic nervous system with fibers arranged in chains of ganglia that then innervate organs, glands, and blood vessels

THE PANCREAS IS AN ORGAN INVOLVED IN THE major functions of the digestive system in humans. The organ lies behind the abdominal cavity lined by the peritoneal membrane and is therefore retroperitoneal at the level of the second lumbar vertebra. The organ weighs 75 to 100 g, is 15 to 20 cm long, and is divided into three portions: the head, the neck, and the body/tail. The pancreas secretes enzymes needed for digestion of carbohydrates, proteins, and fats, and is involved in glucose homeostasis. Disease states include multiple causes of acute and chronic inflammation and benign and malignant neoplasms.

I. ANATOMY

A. Embryology

The pancreas begins to form during the fourth and fifth week of development of a human embryo. As part of the foregut, or cephalic portion of the primitive gut, the pancreas develops alongside the esophagus, stomach, duodenum, liver, and gallbladder. Two buds, one ventral and one dorsal, emerge from the endodermal lining of the duodenum. As the duodenum rotates to the right, the ventral pancreatic bud migrates clockwise behind the duodenum to lie immediately behind and below the dorsal pancreas bud.

Eventually, the pancreatic buds fuse to form the parenchyma and the ductal system of the pancreas (see Section I,B).

Two abnormalities may result from altered formation of the pancreas. An annular pancreas may form if the right portion of the ventral pancreatic bud migrates normally but the left portion migrates in the opposite direction. The duodenum then becomes surrounded by pancreas tissue and may lead to complete obstruction. In addition, heterotopic pancreas tissue may be found anywhere in the gut from the distal end of the esophagus to the tip of the primary intestinal loop that forms the majority of the small intestine. Pathologic pancreatic changes most commonly may be found in the stomach mucosa and in a Meckel's diverticulum.

B. Structural Anatomy

The pancreas lies nearly completely retroperitoneal surrounded by many organs and vascular structures of the upper abdomen (Fig. 1). The ventral bud becomes the uncinate process and inferior head of the mature pancreas, whereas the dorsal bud becomes the neck, body and tail, and the superior head of the pancreas. The ventral and dorsal ducts fuse to become the main pancreatic duct, or duct of Wirsung, which drains most of the pancreas. The duct of Wirsung runs the entire length of the pancreas, joins the common bile duct, and empties into the duodenum at the ampulla of Vater. A lesser duct, the duct of Santorini, often drains the superior portion of the head of the pancreas and empties separately into the duodenum through the lesser papilla. Variations in pancreatic ductal anatomy are common and occur in 40% of patients. The main duct and the lesser duct usually communicate but the lesser duct may end blindly and not into the duodenum. Pancreas divisum may result in 5–10% of patients where an incomplete fusion of ventral and dorsal ducts leads to a lesser duct draining the entire pancreas and an atrophic duct of Wirsung.

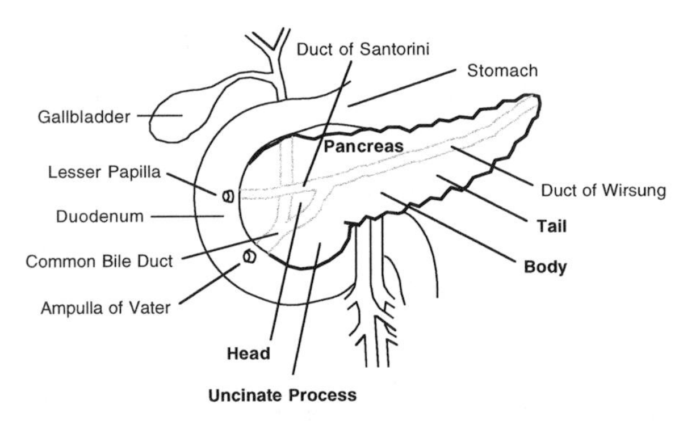

FIGURE 1 Anatomic relation of the pancreas to the stomach, duodenum, and extrahepatic biliary system.

C. Vascular and Lymphatic Anatomy

The pancreas receives blood supply from branches of the celiac and superior mesenteric arteries. The head of the pancreas is particularly vascular with arcades of branches from the gastroduodenal and superior mesenteric arteries forming collaterals in both anterior and posterior surfaces. The body and tail of the pancreas are supplied by at least 10 branches of the splenic artery. Variations in the arterial anatomy are common.

The venous drainage of the pancreas follows the arterial supply and usually lies superficial to the arteries. Eventually, all venous effluent empties into the portal vein.

The lymphatic drainage of the pancreas follows regional lymph node basins and is abundant. The major drainage from the head of the pancreas and the duodenum flows into celiac and superior mesenteric lymph nodes. The body and tail of the pancreas drain into nodes along the splenic vessels.

D. Innervation

The secretory functions of the pancreas described here are coordinated by a rich supply of nerves. Parasympathetic nerve fibers from the vagus nerve generally stimulate both exocrine and endocrine secretion from pancreatic islet cells. Sympathetic nerve fibers from splanchnic nerves tend to predominantly inhibit secretion. Peptidergic neurons secrete many hormones and amines but their precise physiologic role is unknown. Finally, the pancreas is innervated by a rich supply of afferent sensory fibers that may be involved in the severe pain experienced by patients with pancreatic cancer or chronic pancreatitis.

II. STRUCTURE AND HISTOLOGY

A. Exocrine Structure

The majority of the pancreatic parenchyma is composed of acinar cells and the ductal system, both of which form the backbone of the exocrine pancreas. Acinar cells are pyramidal and secrete digestive enzymes that are packaged into zymogen granules. The granules are released from the apex of the cell, which faces the lumen of the pancreatic ducts. Twenty to 40 acinar cells form a unit, the acinus, which also contains centroacinar cells responsible for the secretion of bicarbonate, fluid, and electrolytes. Exocrine enzymes and fluid secreted into the small pancreatic ducts eventually collect in the main pancreatic duct before emptying into the duodenum. [*See* Digestive Systems, Physiology and Biochemistry.]

B. Endocrine Structure

The endocrine pancreas controls glucose homeostasis and comprises 2% of the total pancreatic tissue. The four major cell types form nests of cells scattered throughout the organ called islets of Langerhans. Blood flow to the islets is distributed in a distinctive pattern and constitutes 20–30% of the total pancreatic arteriolar flow. Alpha (A), beta (B), delta (D), and PP or F cells secrete glucagon, insulin, somatostatin, and pancreatic polypeptide, respectively. These cells are distributed in varying quantities and locations within each islet. Other less abundant cells in the islets secrete a host of other peptides, such as vasoactive intestinal polypeptide (VIP), serotonin, pancreastatin, calcitonin gene-related peptide (CGRP), neuropeptide Y, and bombesin. [*See* Endocrine System.]

III. PHYSIOLOGY

A. Exocrine Function

The pancreas secretes 500 to 800 ml/day of clear, odorless, isosmotic fluid containing bicarbonate and digestive enzymes. The concentration of bicarbonate secreted by the centroacinar cells and the ductal epithelial cells varies from 20 mmol/liter in the basal state to 150 mmol/liter during maximum stimulation. The bicarbonate is formed from carbonic acid by the enzyme carbonic anhydrase. The fluid pH varies from 7.6 to 9.0, an alkaline environment that can neutralize gastric juices and carry inactive proteolytic enzymes to the duodenum. The fluid also contains sodium and potassium in concentrations similar to that of plasma, as well as chloride. Bicarbonate secretion is regulated predominantly by secretin, a protein released by the duodenal mucosa when the duodenal luminal pH drops below 3.0. Cholecystokinin (CCK), gastrin, and acetylcholine also weakly stimulate bicarbonate secretion, but CCK potentiates the effects of secretin. Atropine has an inhibitory effect on bicarbonate secretion.

Enzyme secretion by acinar cells is primarily regulated via hormonal and neural stimuli. Enteric cholecystokinin binds to acinar cells through membrane-bound receptors, releasing the second messengers calcium and diacylglycerol. Inactive enzymes are stored in zymogen granules, released from the apex of the acinar cells into the lumen of the acinus, and transported into the duodenum, where they are activated. Other stimulatory influences on enzyme secretion include acetylcholine from postganglionic fibers of the pancreatic plexus, secretin, VIP, and islet hormones.

Three major enzyme groups are secreted by the acinar cells, including amylases, lipases, and proteases. Amylase hydrolyzes glycogen and starch to glucose, maltose, and dextrins. Lipase emulsifies and hydrolyzes fats with the aid of bile salts. Proteases necessary for digestion of proteins include trypsinogen, chymotrypsinogen, procarboxypeptidases A and B, and ribonuclease.

B. Endocrine Function

Two hormonal secretory products of the endocrine pancreas, insulin and glucagon, are paramount in glucose homeostasis. Insulin is synthesized by B cells of the islet of Langerhans. It is a 56-amino-acid polypeptide (molecular weight 6000) comprising two chains, A and B, joined by two disulfide bridges. Insulin secretion is regulated by a complex interplay of nutrient, neural, and hormonal factors, especially glucose concentration. Maximal stimulation of the B cell occurs at a glucose concentration of 400–500 mg/dl, resulting in the secretion of insulin in two phases. Enteric glucose appears to be a stronger stimulus than intravenous insulin via an effect named the enteroinsular axis. Phase 1 releases a burst of insulin within 4–6 min, followed by a more sustained regulated release in phase 2. The stimulated B cell initially synthesizes proinsulin, which is cleaved into insulin and a residual C peptide in the Golgi complex of the cell. Insulin molecules are packaged into granules and released into the intervascular space. Insulin then binds to specific receptors on all cells except B cells, hepatocytes, and cells in the central nervous system, to enhance glucose transport. The half-life of secreted insulin is 7–10 min before metabolism by the liver, kidney, and skeletal muscle.

Other enteric hormones that stimulate insulin secretion include gastric inhibitory polypeptide (GIP) and glucagon-like peptide 1 (GLP-1). Amino acids such as arginine, lysine, and leucine and free fatty acids also promote insulin secretion. Stimulatory hormones include glucagon, GIP, and CCK, whereas inhibitory hormones include somatostatin, amylin, and pancreastatin. Neural regulation of insulin secretion occurs via cholinergic and β-sympathetic fibers, which are stimulatory, and α-sympathetic fibers, which inhibit insulin secretion.

Abnormalities in various endocrine functions of the pancreas may lead to diabetes mellitus, a disorder of carbohydrate metabolism. Type I, or insulin-dependent diabetes, occurs when B cells are destroyed, resulting in insulin deficiency. Insulin resistance in type II diabetes may occur with defects in insulin receptors. Rarer forms of diabetes may result from defects in the synthesis and cleavage of insulin. [*See* Diabetes Mellitus; Insulin-Dependent Diabetes Mellitus.]

Glucagon is secreted by A cells of the islet and enhances hepatic glycogenolysis. The control of glucagon secretion is counterregulatory to that of insulin secretion, thereby allowing both hormones to participate in the maintenance of basal glucose levels. A similar series of hormonal, nutrient, and neural factors participate in glucagon secretion. Elevated glucose concentrations, insulin, and somatostatin suppress glucagon secretion, whereas amino acids such as arginine and alanine are stimulatory. Neural stimuli of glucagon secretion parallel those of insulin secretion. Upon stimulation of the A cell, a single-chain, 29-amino-acid glucagon polypeptide (molecular weight 3500) is secreted, allowing glycogenolysis and gluconeogenesis. During physiologic stress, glucagon elevates blood levels of glucose to provide metabolic fuel. Glucagon is metabolized primarily in the kidney. [*See* Insulin and Glucagon; Peptide Hormones of the Gut.]

IV. ACUTE PANCREATITIS

Pancreatic inflammatory disease may be classified as acute or chronic, based primarily on clinical criteria. In addition, acute pancreatitis can resolve with restoration of normal pancreatic function, whereas chronic pancreatitis often leads to progressive and permanent pancreatic insufficiency. Acute pancreatitis is a complex disorder of the exocrine pancreas with a wide pathologic and clinical spectrum. The pathologic hallmark is acute acinar cell injury. Depending on the extent of regional and systemic inflammatory responses, a mild, self-limited, edematous pancreatitis usually occurs, or necrotizing pancreatitis may result in 5–10% of patients if pancreatic necrosis evolves.

Multiple causative factors have been associated with the development of acute pancreatitis, but actual pathologic mechanisms remain unclear (Table I). The incidence of pancreatitis varies in different countries depending on etiologic factors; in the United States, ethanol ingestion is the most common cause, whereas gallstones appear to be the more common trigger in England.

After initiation of acute pancreatitis, the normal sequence of enzyme secretion by acinar cells appears to be disrupted. Premature release of proteases may occur, leading to continued acinar cell injury. There is evidence for loss of normal acinar cell polarity with disordered release of enzymes through the basolateral cell membrane, leading to further injury. Acute inflammatory responses occur locally within minutes with the infiltration of polymorphonuclear leukocytes followed by monocytes and lymphocytes. Activated products of these inflammatory cells may cause a systemic response, resulting in pancreatitis-induced multiorgan system failure.

TABLE I
Causes of Acute Pancreatitis

Causes
Ethanol ingestion
Biliary tract stone disease
Trauma
Postoperative
Postendoscopic retrograde cholangiopancreatography (ERCP)
Direct
Metabolic
Hyperlipidemia
Hyperparathyroidism
Pregnancy
Mechanical obstruction of the ampulla of Vater
Tumor
Duodenal obstruction
Pancreas divisum
Infection
Viral (mumps; hepatitis; Coxsackie B)
Bacterial (*Mycoplasma*)
Fungal
Parasitic (*Ascaris*)
Drugs
Steroids
Diuretics
Azathioprine
Sulfonamides
Tetracycline
Valproic acid
Idiopathic

Acute pancreatitis most commonly occurs in adults between 30 and 70 years of age, but may affect all ages depending on the etiology. The cardinal clinical symptom is epigastric pain often radiating to the back. Several grading systems exist that estimate risks and outcomes based on specific features of the clinical presentation, such as the Ranson criteria or the APACHE score. Diagnosis centers on a clinical judgment aided by laboratory findings such as hyperamylasemia and imaging studies, including ultrasound and computed tomography (CT).

Management of acute pancreatitis is dependent on the severity of the clinical course and the possible development of late complications such as pseudocyst, abscess, or diabetes. Current approaches include general supportive care with a focus on resuscitation, ventilation, nutrition, and removal of etiologic stimuli. Surgical therapy is usually reserved for specific complications or correctable anatomic lesions.

Mortality rates for acute pancreatitis range from 6 to 21%, rising to 50% or more for necrotizing pancreatitis. With improvements in critical care, nutritional support, and specific pharmacologic therapies, mortality rates should improve.

V. CHRONIC PANCREATITIS

Chronic pancreatitis is a progressive disorder leading to permanent destruction of the pancreatic parenchyma and fibrosis of the gland. Eventually, clinical symptoms may reflect both exocrine and endocrine pancreatic insufficiency. The major cause of chronic pancreatitis in 70% of cases is ethanol consumption; however, only 10% of alcoholics develop the disease. The underlying mechanisms by which alcohol causes chronic pancreatitis are unknown. Recent theories include a direct toxic effect of alcohol on exocrine cells, or altered pancreatic protein and calcium secretion. Pathologic findings on microscopy include abnormal pancreatic lobules with dilated ductules, loss of ductal epithelium and acinar cells, localized ductal obstruction with proteinaceous plugs, and eventually fibrosis.

Clinically, most patients present with dull, constant, epigastric pain with radiation to the back. The pain can be a result of direct inflammation of the gland, intrapancreatic neural inflammation, associated pseudocysts, bile duct obstruction or duodenal obstruction, or elevated pancreatic ductal pressures. Loss of greater than 90% of pancreatic exocrine function leads to malabsorption, with abdominal bloating, steatorrhea, weight loss, and fat-soluble vitamin deficiencies. Exocrine deficiencies usually precede endocrine insufficiencies and pancreatic diabetes.

Diagnosis is again determined from the clinical presentation with the aid of tests of exocrine function. Routine blood tests are not helpful, and in fact serum amylase levels are often normal or subnormal. Collection of pancreatic secretions is invasive and has been largely replaced by radiologic studies to image pancreatic structure. Calcification of the pancreas on a plain abdominal film is pathognomonic of chronic pancreatitis. Other useful studies include ultrasound, CT, and endoscopic retrograde cholangiopancreatography (ERCP) to image pancreatic ductal pathology.

Treatment includes abstinence, enzyme replacements for malabsorption, analgesics, and occasionally celiac ganglia blocks for chronic pain management. Intractable pain is an indication for surgical treatment, which may involve pancreatic ductal drainage procedures or pancreatic resection.

Patients with chronic pancreatitis have an excess mortality close to 40% over 20 years. The most common causes of death include cancers of the aerodigestive tracts, diabetic complications, and cirrhosis.

VI. PANCREATIC ADENOCARCINOMA

Carcinoma of the pancreas is the eighth most common cancer and the fourth leading cause of adult cancer death in the United States. It accounts for 25,000 deaths per year and its aggressive nature is reflected by only 3% of patients expecting to survive 5 years from the time of diagnosis. The etiology in most cases is unknown, although epidemiologic associations have been shown with cigarette smoking, coffee, alcohol, organic solvents, and petroleum products. Pathologically, the tumors are most often adenocarcinomas arising from ductal epithelium. The head of the pancreas is involved in 65% of cases, the body and tail in 30%, and the tail alone in 5%. Unfortunately, 25% of patients demonstrate local invasion and regional lymph node spread at the time of diagnosis, and 60% of patients have distinct metastases.

Clinically, patients present more commonly with weight loss, pain, and jaundice. The sudden onset of diabetes mellitus without a predisposing cause should prompt an evaluation for pancreatic cancer. Diagnostic procedures often include ultrasound, CT, and ERCP, with an aim to evaluate the extent of disease and tumor resectability.

Surgical resection of carcinoma of the head of the pancreas remains the only potentially curable treatment. The standard operative approach begins with a systematic evaluation of resectability followed by a pancreaticoduodenectomy. In most series, only 15–20% of cases are resectable. The 5-year survival rate following curative surgery remains less than 25%, with a median survival of 20 months. Palliation may be accomplished by operative biliary bypass, gastric bypass, and splanchnicectomy. Multicenter trials are ongoing to evaluate the role of adjuvant chemotherapy and/or radiation therapy in the treatment of pancreatic cancer.

VII. PANCREATIC ENDOCRINE TUMORS

In clinical studies, pancreatic endocrine tumors have a prevalence of one case per 200,000 population. They arise most commonly in the islet cells of the pancreas but may also be found in the duodenum and adrenal glands. Over 75% of islet cell tumors secrete functioning hormones. Entopic tumors, such as insulinomas, glucagonomas, and somatostatinomas, produce hormones or peptides usually found in the pancreas. Ectopic tumors, such as gastrinomas and VIPomas, produce hormones or peptides not normally found in the pancreas. Recognition of the clinical symptoms caused by excess hormone production allow diagnosis of the functioning pancreatic endocrine tumors. Tumor localization often requires CT scans, magnetic resonance imaging (MRI), selective visceral angiography, and selective venous sampling for hormone detection. Resection is possible if no metastatic disease or invasion of vascular structures is found. In general, islet cell tumors have a better prognosis than pancreatic ductal adenocarcinoma. In addition, gastrinomas are found at a higher frequency in patients with multiple endocrine neoplasia-I (MEN-I), an inherited syndrome characterized by a familial predisposition to develop pancreatic islet cell tumors, pituitary tumors, and parathyroid hyperplasia.

BIBLIOGRAPHY

Berger, D. H., Feig, B. W., and Fuhrman, G. M. (eds.) (1995). "The M. D. Anderson Surgical Oncology Handbook." Little, Brown, Boston.

Greenfield, L. J. (ed.) (1997). "Surgery: Scientific Principles and Practice," 2nd Ed. Lippincott–Raven, Philadelphia.

Isselbacher, K. J., Braunwald, E., Wilson, J. D., Martin, J. B., Fauci, A. S., and Kasper, D. L. (eds.) (1994). "Harrison's Principles of Internal Medicine," 13th Ed. McGraw–Hill, New York.

Langman, J. (1995). "Langman's Medical Embryology," 7th Ed. Williams & Wilkins, Baltimore.

Pancreatic Toxicology

DANIEL S. LONGNECKER
Dartmouth Medical School

GLENN L. WILSON
University of South Alabama College of Medicine

GLOSSARY

Acute pancreatitis Disease characterized by inflammation, abnormal destruction of cells, and hemorrhage in the pancreas

Carcinoma of the pancreas Cancer of the exocrine pancreas

Chronic pancreatitis Disease characterized by atrophy (shrinkage) and fibrosis (scarring) of the pancreas

Cystic fibrosis Disease caused by a genetically determined metabolic abnormality that alters pancreatic secretory function and leads to a form of chronic pancreatitis

Diabetes mellitus Group of diseases generally characterized by lack of insulin secretion and/or increased cellular resistance to insulin resulting in hyperglycemia and other metabolic disturbances

Endocrine Related to internal secretion (i.e., into the blood)

Exocrine Related to external secretion (e.g., into the intestine)

Insulin-dependent diabetes mellitus A form of diabetes characterized by insulin deficiency due to the destruction of insulin-secreting cells. Individuals with this disease require insulin therapy to maintain life

Islet (pancreatic) Small group of endocrine glandular cells that is the functional unit of the endocrine pancreas

Noninsulin-dependent diabetes mellitus A form of diabetes characterized by hyperglycemia resulting from insulin secretory defects and/or increased resistance to the actions of insulin. Individuals with this disease do not require insulin therapy to maintain life

Xenobiotic Chemical from the environment (i.e., outside of the body) such as a drug or insecticide

Zymogen granule Storage form of digestive enzymes within the pancreatic acinar cells

THE PANCREAS IS USUALLY A "SILENT" INTERNAL organ that can sustain injury to its cells as a result of exposure to xenobiotics that are potentially poisonous, even when not involved in a major way with their metabolism, detoxification, or excretion. The recognized frequency of xenobiotic toxicity for the pancreas is low, but when toxicity occurs, death or permanent impairment of health may result.

Five major diseases of the pancreas contribute significantly to ill health in the United States. These diseases are acute pancreatitis, chronic pancreatitis, carcinoma of the pancreas, diabetes mellitus, and cystic fibrosis. All but the last of these can be induced in laboratory animals by acute or chronic administration of chemical agents or carcinogens. Pancreatitis and diabetes in humans have been attributed to toxic chemicals in some cases. Separate sections of this article focus on toxic injury (1) to the exocrine pancreas, which secretes digestive enzymes into the intestine, and (2) to the endocrine pancreas (islets), which secretes hormones such as insulin into the blood primarily for the regulation of carbohydrate metabolism.

I. TOXICITY IN THE EXOCRINE PANCREAS

A. Structure and Function of the Exocrine Pancreas

The pancreas constitutes about 0.1% of adult body mass in humans (Fig. 1). The exocrine pancreas is

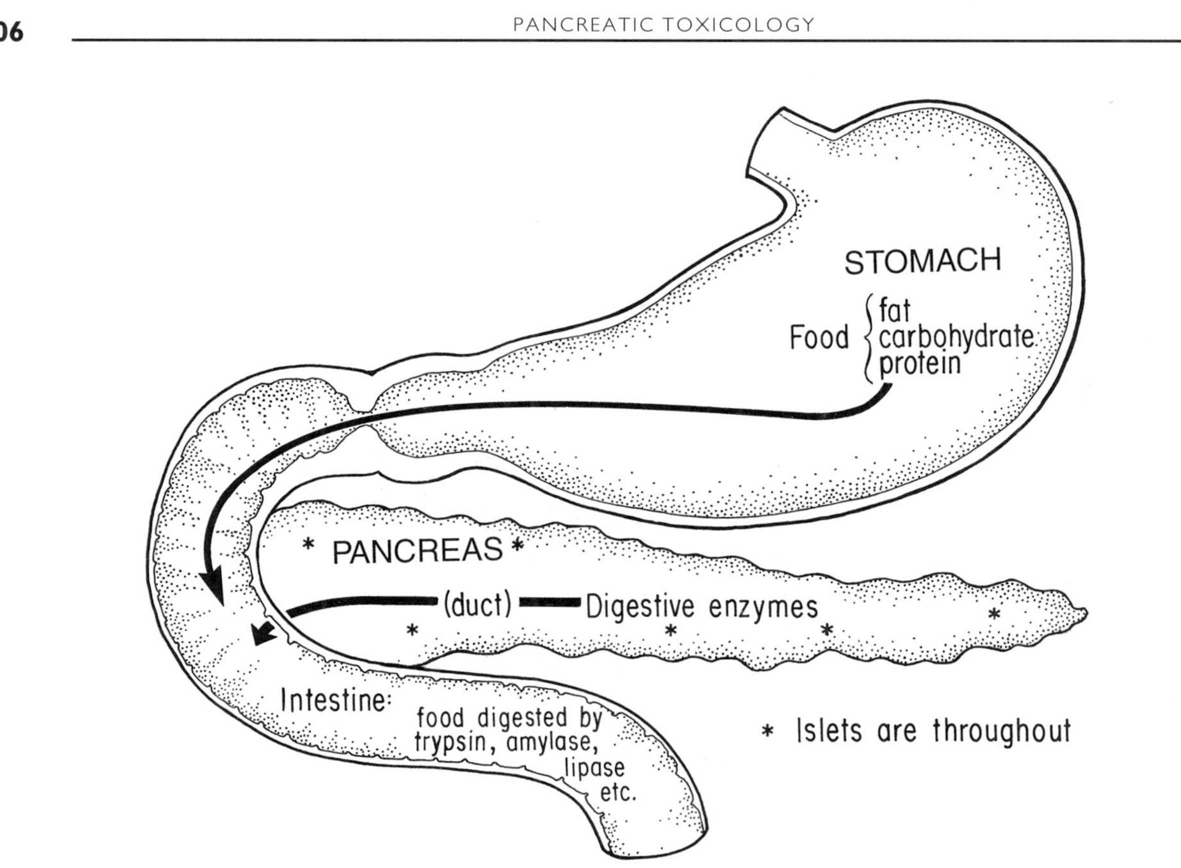

FIGURE 1 General scheme of the pancreas location and function. The pancreas lies in the upper abdomen partially behind the stomach. The pancreatic duct delivers digestive enzymes into the first part of the small intestine (duodenum).* Islets are found throughout. Drawing by Joan E. Thomson.

composed of two basic functional cell types: acinar and ductal. More than 80% of the gland consists of acinar cells. These cells secrete digestive enzymes into the pancreatic duct system. The duct system comprises 2–4% of the pancreas. Cells of the smallest terminal ducts that contact acinar cells are commonly designated as centroacinar cells.

Acinar cells secrete enzymes by releasing zymogen granules into the duct system in response to nerve stimuli or in response to specific circulating gastrointestinal hormones such as cholecystokinin (CCK), for which the cells have receptors. Several of the exocrine digestive enzymes are synthesized as inactive enzyme precursors that are activated in the intestine.

Pancreatic enzymes are required for normal digestion and absorption of food in the gastrointestinal tract. The exocrine pancreas appears to possess great functional reserve. In humans with chronic pancreatitis, apparently atrophy and scarring must progress to the point that enzyme secretion is reduced <10% of normal levels before malabsorption occurs, causing fat and proteins to increase in the feces.

B. Mechanisms of Toxicity in the Exocrine Pancreas

For the most part, xenobiotics reach the pancreas through the bloodstream after absorption and systemic distribution. Thus, toxic agents may reach the pancreas after exposure by skin contact, inhalation, injection, or ingestion. The possibility also exists that ingested xenobiotics in the duodenum may reflux into the pancreatic duct and reach the ductal cells. Such reflux has been documented in experimental animals when duodenal luminal pressure is increased, but this is not regarded as a regular or predictable route of exposure.

The pancreas contains drug-metabolizing enzymes, although the level of these enzymes is perhaps 1% of that found in liver cells. Induction of such enzymes by xenobiotics (e.g., betanaphthaflavone) has been demonstrated in animals. Both acinar and ductal cells contain such enzymes, but differences in the distribution of enzymes between these cell types have been noted. The activation of toxins and procarcinogens by metabolism in the pancreas has been documented in animals.

C. Response of the Exocrine Pancreas to Injury

Numerous chemical agents are known or suspected to induce acinar cell injury or death, to alter acinar cell or duct cell function, or to induce neoplastic change in the pancreas. These agents include ethanol, naturally occurring antibiotics such as puromycin, chemotherapeutic agents such as 5-fluorouracil, and a variety of other compounds (Table I).

1. Cytotoxicity

Water or fat may accumulate in the cytoplasm of acinar cells as a result of toxic injury. One hallmark of sublethal injury to acinar cells is the presence of dense cytoplasmic bodies (autophagic vacuoles) that provide evidence of autodigestion of part of the cell. These are best seen by electron microscopy but can be identified by light microscopy when they are large (i.e., in the size range of the nucelus or larger). In electron micrographs, autophagy is manifest as the segregation of cytoplasmic organelles within a membrane that is probably derived from the endoplasmic reticulum of the cell. Mitochondria, zymogen granules, and even nuclei may be recognizable in such vacuoles. Digestion of the contents of an autophagic vacuole ensues promptly after its formation so that only lipid, membranous, or granular debris may remain. Activation of digestive enzymes of acinar cells within the vacuoles probably contributes to the degradation of the contents.

Groups of lethally injured acinar cells are quickly destroyed after cell death occurs because of their high content of hydrolytic digestive enzymes. This process is called necrosis. The release of the enzymes lipase and phospholipase into the tissue results in the destruction of adjacent fat cells. The release of enzymes and necrotic debris incites inflammation and the changes of acute pancreatitis. At this stage of injury, significant degrees of necrosis should be detectable by assay of serum for the level of activity of pancreatic exocrine enzymes, such as amylase or lipase, which leak into the blood.

The digestive enzymes, when activated within the pancreas, can digest protein, RNA, DNA, phospholipids, carbohydrates, and lipids. Varying degrees of acute pancreatitis can be induced in animals by giving them chemicals that are cytotoxic for acinar cells. Some poisons induce damage that closely mimics the most severe forms of acute hemorrhagic pancreatitis. Intermediate degrees of pancreatic necrosis simulating nonfatal degrees of pancreatitis have been induced by the naturally occurring antibiotics puromycin and azaserine, whereas lower doses of these and other agents or treatment with less cytotoxic agents may induce milder degrees of damage.

Experimental studies have shown that acinar and ductal cells can divide, and regeneration of the exocrine pancreas can occur following toxic injury. When the injury is mild or moderate, the pancreas may be restored to normal size and structure. When injury is more severe, regeneration may be incomplete, and a portion of the pancreatic tissue may be replaced by fat. Peptide hormones such as CCK that stimulate pancreatic secretion also seem to stimulate pancreatic growth. They have been demonstrated to cause enlargement of the pancreas when present in excess. [*See* Peptide Hormones of the Gut.]

Changes in ductal and centroacinar cells due to toxic injury are less distinctive than those described in acinar cells. They may show swelling or undergo necrosis. Ductal changes often appear to be secondary to inflammation or necrosis in adjacent acinar tissue. The lumen of ducts may contain condensates of the proteins of the pancreatic secretions. Metaplasia of ductal epithelium into mucous or squamous cells has been described. Hyperplastic and neoplastic changes are seen in carcinogen-treated rodents.

TABLE I
Agents and Drugs Accepted to Cause Pancreatitis in Humans

Azathiaprine	Metronidazole
Chlorothiazide	Pentamidine
Erythromycin	Sodium valproate
Ethanol	Sulfonamides
Furosemide (Lasix)	Tetracycline
Hydrochlorothiazide	

2. Carcinogenesis

Nonlethal mutations appear to occur in acinar and ductal cells exposed to mutagens. The induction of DNA damage in both acinar and ductal cells by such potential carcinogens has been documented in hamsters and rats. Focal changes in the pattern of cell growth that appear to be clonal (i.e., resulting from the division of a single abnormal cell) can be identified following exposure to pancreatic carcinogens. Chemical carcinogens of diverse types have been shown to affect the pancreas of animals, although nitrosamines are the most common type. Several models for the induction of carcinomas in the pancreas of experimental animals by carcinogenic chemicals have been developed. Two of these have been extensively characterized and studied: (1) azaserine is used as a pancreatic carcinogen in rats, and (2) *N*-nitrosobis(2-oxopropyl)-amine is used in the Syrian golden hamster. Because of these studies in animals, it is increasingly accepted that certain chemicals may be the cause of carcinomas in the human pancreas.

3. Specific Toxic Agents

A wide variety of chemicals have been reported to cause injury or alteration of the exocrine pancreas of experimental animals. These include a sufficient number and diversity of agents to document the vulnerability of the pancreas to xenobiotic toxins. The list includes several amino acid analogs. This group of compounds has a specific propensity to affect the acinar cells, probably because α-amino acids readily enter the acinar cells by amino acid transport pathways. The pancreas has been shown to concentrate both normal and abnormal amino acids to a greater degree than most other tissues.

Medical literature contains reports of acute pancreatitis in association with 25–30 drugs, and this list continues to grow. The risk of pancreatitis, although low, seems well established for some of these agents, such as azathiaprine, chlorothiazide, and furosemide. Firm linkage is accepted when evidence indicates pancreatitis during treatment of the patient with the drug in the absence of other known causes of pancreatitis if the pancreatitis resolves on withdrawal of the drug, and pancreatitis recurs when the drug is again given. In well-studied series of patients given azathioprine, the incidence of pancreatitis has been as high as 6.2%. For other drugs, only rare sporadic cases have been reported, and for about two-thirds of the drugs, the proof that the agent was the cause of pancreatitis is circumstantial or weak.

Alcohol (ethanol) appears to have a direct cytotoxic effect on the pancreas and to alter pancreatic acinar cell function. An excessive intake of alcoholic beverages is regarded as a major cause of both acute and chronic pancreatitis in humans in the United States. In Europe, alcoholism is primarily recognized as a risk factor for chronic pancreatitis. The dose of alcohol required to induce chronic pancreatitis has been estimated to be in the range of 150–175 g of ethanol per day for several years. The dose of alcohol required to induce acute pancreatitis appears to be less predictable and may reflect individual genetic variability in susceptibility. The precise mechanism by which alcohol induces pancreatic injury is not understood. [*See* Alcohol Toxicology].

Agents that stimulate pancreatic secretion above the physiologic level have been shown to cause acute pancreatitis. One natural toxin, an insecticide, and a peptide analog of CCK serve as examples. The toxin is scorpion venom, which stimulates cholinergic discharge in the pancreas. Acute pancreatitis is frequently reported to follow the sting of scorpions in Trinidad. The insecticide that has induced acute pancreatitis clinically and experimentally contains an anticholinesterase, which can apparently reach the pancreas and amplify cholinergic stimulation of pancreatic secretion. The last example of this group is caerulein, a 10-amino acid peptide analog of CCK, which is more potent than the natural hormone in stimulation of pancreatic acinar cell secretion. Continuous infusion of caerulein for 8–10 hr induces pancreatitis in experimental animals.

D. Evaluation of Exocrine Pancreatic Toxicity

Evaluation of the toxicity of agents for the exocrine pancreas must be done almost entirely *in vivo* for practical reasons. Culture of acinar cells is difficult because of their propensity to self-destruct. No normal acinar cell lines are available. Isolation of significant numbers of ductal cells is technically difficult, although they can be cultured for longer periods and immortal cell lines are available. Primary cultures of acinar and ductal cells have been used for studies of toxicity and DNA damage and for measuring drug-metabolizing enzymes.

Determination of urine and/or serum amylase levels can also be applied *in vivo* and seems to offer the most practical initial screening approach at the present time for the detection of acute exocrine pancreatic injury. This technique lacks specificity for the pancreas unless isoenzyme studies are done to allow the recognition of amylase of pancreatic origin.

II. TOXICITY IN THE ENDOCRINE PANCREAS

A. Structure and Function of the Endocrine Pancreas

In the adult pancreas, the endocrine portion comprises only about 1–2% of the cells. These endocrine cells are contained in small ovoid clusters (100–200 μm in diameter), called the islets of Langerhans, which are distributed throughout the pancreas. A normal human pancreas contains approximately 500,000 islets.

Three major cell types have been identified in the islets using differential histochemical staining. These cells have been termed α, β, and δ cells. Each makes and stores a peptide hormone in small secretory granules. α cells contain the hormone glucagon, β cells contain insulin, and δ cells contain somatostatin. It is now possible to identify these three cell types using immunocytochemical staining with antibodies directed toward the hormonal products of the cells. More recent studies have identified other hormone-containing cells in the islets. The most numerous of these cells contain a 36 amino acid polypeptide. These cells have been termed pancreatic polypeptide (PP) cells.

β cells are the most abundant in the endocrine pancreas. Although the number varies from islet to islet, approximately 80% of endocrine cells are β cells. α cells comprise approximately 15% of the endocrine cells, and δ cells around 4%. The remaining 1% consists mainly of PP cells. In humans, β cells occupy the central portion of the islets, while the other cell types are found around the periphery. Penetrating the β-cell core of human islets are large vascular channels lined with α and δ cells.

Islets can be viewed as an endocrine organ whose major function is to manage energy production. Regulation of glucose metabolism is key because glucose is the primary fuel used to provide energy in humans. Energy metabolism is regulated by the balance of insulin and glucagon secretion by the islets. When islet function is altered as a result of toxic injury, loss of control of glucose metabolism and energy production is the most frequent and important result. The clinical disease state is diabetes mellitus. [*See* Diabetes Mellitus; Endocrine System; Insulin and Glucagon.]

B. Mechanisms of Endocrine Pancreatic Toxicity

As in the exocrine pancreas, xenobiotics reach the islets through the bloodstream. Islet cells also contain drug-metabolizing enzymes and thus seem able to metabolize some chemicals to more toxic forms. By far the most prevalent evidence of toxic injury seen in the endocrine pancreas involves insulin-secreting β cells, and injury of these cells may lead to diabetes. It appears that α cells and δ cells are less often affected by toxic agents than β cells. However, it may also be true that deficiency of somatostatin or glucagon is better tolerated than the lack of insulin so that toxicity for α and δ cells may sometimes escape notice.

C. Responses of the Endocrine Pancreas to Injury

1. Cytotoxicity in Insulin-Dependent Diabetes Mellitus

Chemical toxins such as streptozotocin have been shown to cause the abrupt onset of diabetes in a variety of animals, including rat, mouse, Chinese hamster, dog, sheep, rabbit, and monkey. Damaged β cells initially exhibit a vacuolated cytoplasm, followed by shrinkage of the nuclei as the cells detach from one another. As cellular destruction proceeds, nuclear damage and disintegration of the nuclear membrane become obvious. Generally, β-cell destruction is complete by 24 hr after the toxin is given. Following significant β-cell loss, the islets in diabetic animals are small. Islet cells are arranged in small cords that are surrounded by fibrous tissue. Using immunocytochemical techniques, it has been demonstrated that these cords are composed of two-thirds glucagon-containing α cells and one-third somatostatin-containing δ cells. Insulin-containing β cells are almost totally absent in these islets.

Inflammation of the islets, termed insulitis, is a feature observed in humans at the time of onset of the insulin-dependent form of diabetes and in some animal models of diabetes. This condition is characterized by an infiltration into the islets of leukocytes, predominantly lymphocytes. The exact pathogenetic consequences of insulitis in chemically induced diabetes are as yet not fully defined; however, this inflammation has been found to be an integral part of an autoimmune response directed against β cells in some animal models of spontaneous diabetes.

Direct evidence that xenobiotics can destroy β cells and cause insulin deficiency and diabetes in humans has been provided by case reports of individuals who ingested the rat poison Vacor, *N*-(4-nitrophenyl)-*N'*-(3-pyridinylmethyl)urea. Many of these people developed severe insulin deficiency and diabetes. Autopsy examination in some fatal cases has demonstrated a loss of β cells. In toxicity studies in animals, Vacor was either nontoxic (dogs, cats) or so lethal that the islet toxicity was not detected (rats). Thus, species differences precluded the detection of β-cell toxicity in animals.

Additional support for a role involving environmental toxins in some forms of insulin-dependent diabetes mellitus has been provided by studies in laboratory animals. These have shown that specific *N*-nitroso compounds (streptozotocin, chlorozotocin) and other complex amines (alloxan) can cause diseases very similar to those seen in humans. These and other structurally similar compounds pose a potential threat to humans, either through formation of these agents in the body or through trace contamination in the environment. The *N*-nitroso compound streptozotocin is a naturally occurring antibiotic.

An example of the potential threat of ingested toxins has been provided by epidemiologic studies in Iceland. These studies suggest that insulin deficiency may occur in some male offspring of mothers who have ingested smoked mutton. Analysis of this meat has revealed a significant content of several *N*-nitroso compounds. Additional recent evidence linking these compounds to diabetes has been supplied by analysis of population-based registries of childhood diabetes. These show a correlation between *N*-nitroso compounds in the diet and precursor nitrates in drinking water with the onset of diabetes.

There is no effective regeneration of β cells in the islets of adult animals treated with islet cell toxins; however, in neonates there is marked regeneration of β cells following a toxic insult. Neonatal rats treated with a high dose of streptozotocin show a dramatic increase in β-cell mass following initial β-cell destruction. This results from both replication of surviving β cells and the budding of new islets from ducts. However, the regenerative process is not complete because these rats are glucose intolerant in adult life.

Growing evidence indicates that autoimmune mechanisms play a role in the destruction of β cells, resulting in the insulin-dependent form of diabetes in humans. Genetic factors also play some role in determining susceptibility to this process, and it is uncertain what triggers the autoimmune destruction of β cells. There has been a conceptual change with regard to the role of toxic agents. Rather than precipitating diabetes abruptly by massive β-cell destruction, toxic chemicals or viruses possibly may trigger the development of autoimmune processes directed against β cells in genetically susceptible individuals. Alternatively, it is conceivable that agents toxic for β cells can augment ongoing immunological processes and hasten the onset of clinical manifestations of diabetes. Streptozotocin apparently does trigger an autoimmune response against β cells in mice. Repeated subdiabetogenic doses of streptozotocin, administered to certain strains of mice, can cause diabetes associated with islet inflammation and β-cell destruction. This diabetes can be prevented only by combined treatment with nicotinamide to block the overt toxic effects of the chemical and antilymphocyte serum to negate cell-mediated immunity directed against β cells.

Streptozotocin also can augment the diabetogenic activity of viruses. Low subdiabetogenic doses of streptozotocin, in combination with Coxsackie B3 or B4 virus or the B variant of encephalomyocarditis virus, can cause overt diabetes in mice. In the absence of streptozotocin, none of these viruses precipitate diabetes. In humans, where some forms of diabetes are strongly associated with autoimmunity, xenobiotics may interact with ongoing immunological phenomena, in a similar manner, to hasten the destruction of a critical mass of β cells and produce insulin deficiency.

2. Functional Alterations in Noninsulin-Dependent Diabetes Mellitus

In noninsulin-dependent diabetes mellitus, there are alterations at the level of the β cell in cell number, which can be reduced by as much as 40% of that of nonobese control subjects, and defects in insulin release in response to glucose stimulation. Based on studies with identical twins, generally it has been believed that genetic factors were primarily responsible for the pathogenesis of this form of diabetes. However, with the discovery that a disease very similar to that in humans could be produced in adult rats by treating them as neonates with streptozotocin, it is plausible that xenobiotics can play a part in the etiology of this disease. As adults, these animals exhibit almost no elevation in fasting plasma glucose concentrations compared to controls and have only mildly elevated nonfasting glucose levels. However, streptozotocin-treated animals are markedly glucose intolerant when challenged with a bolus of glucose. Rats with this chemically induced disease have a defective first-phase insulin release. Although the insulin response to glucose is defective, these rats exhibit normal responses to other stimulants of insulin release. As the rats age, they develop a resistance to the actions of insulin in fat cells after 4 months and in the heart by 9 months. This cardiomyopathy is similar to that seen in humans. It has been shown that the development of the cardiomyopathy correlates with the advancement of the secretory defect in these animals. The secretory defect has been demonstrated to be the result of a progressive insensitivity to the actions of glucose, in the stimulation of both insulin secretion and synthesis, and this defect cannot be reversed by normoglycemia. Interestingly, this defect has been associated with a decrease in mitochondrial DNA gene expression. From these experiments, it is apparent that an environmental insult early in life can have dire consequences much later. Identical twins could be concordant for the development of noninsulin dependent diabetes resulting from this type of insult, as they would be exposed to the same environment as neonates.

3. Induction of Glucose Intolerance

A variety of drugs have been identified as causing mild hyperglycemia and glucose intolerance. For the most part, these effects are temporary, and when the chemical is removed, the hyperglycemia resolves. Thus, it appears that a temporary state of β-cell dysfunction is induced rather than death of β cells. β cells can apparently return to normal function when exposure to the offending agent is stopped. A description of the best characterized of these chemicals follows. The benzothiodiazines comprise a group of diuretic and antihypertensive drugs. The diuretic chlorothiazide is the prototype of this class. In some of the individuals treated with chlorothiazide, hyperglycemia occurs. This chemical also exacerbates glucose intolerance in pre-existing diabetics. Other members of this class (e.g., diazoxide, furosemide, trichlorothiazide, hydrochlorothiazide) have similar effects. As noted earlier, this group of drugs is also toxic for the exocrine pancreas. Benzothiodiazines apparently act through two specific mechanisms. These drugs inhibit insulin release, most likely by interrupting the calcium mobilization in the β cells so that secretory granules cannot be released. It is also apparent that these chemicals have extra-islet effects because they increase hyperglycemia in severely diabetic animals, which are essentially insulin deficient. This is probably due to the release of epinephrine and other catecholamines, which promote the breakdown of glycogen and elevate circulating glucose levels.

The only obvious morphological alteration in β cells from animals treated with benzothiodiazines is an increased accumulation of secretory granules. This change is not permanent. When the drugs are withdrawn, their diabetogenic effects are reversed and β cells regain their normal structure.

Cyproheptadine is an antiserotonin, antihistaminic compound that has been used clinically to stimulate weight gain in adults and children. When given to adult rats in repeated high doses, this chemical produces an inhibition of proinsulin synthesis, a reduction in pancreatic insulin, and glucose intolerance. Morphologically, extensive vacuolization can be seen in β cells by light microscopy, and a progressive degranulation of β cells, dilation of endoplasmic reticulum, and loss of ribosomes from the surface of the rough endoplasmic reticulum can be observed by electron microscopy. All these effects are reversed when treatment with the drug is stopped; however, permanent changes in β cells can be induced in the offspring of pregnant rats given repeated low doses of cyproheptadine during the last 8 days of gestation. This treatment initially causes a 50% reduction in fetal pancreatic and serum insulin concentrations. In contrast, mothers exhibit no changes in insulin levels. By 50 days after birth, progeny of the drug-treated dams are glucose intolerant, have twofold increased levels of pancreatic insulin, and have an accentuated response to the insulin-lowering action of cyproheptadine on the β cells. These results demonstrate that a permanent postnatal defect in β-cell function can be produced by prenatal exposure to cyproheptadine in rats.

Cyclosporin is an 11 amino acid cyclic polypeptide commonly used as an immunosuppressive agent in patients receiving tissue and organ transplants. This chemical inhibits insulin secretion in mice, rats, and humans *in vivo* and *in vitro*. β cells in cyclosporin-treated animals exhibit severe degranulation, cytoplasmic vacuolization, and dilation of the endoplasmic reticulum. When cyclosporin is removed, both functional and morphological abnormalities disappear. The mechanism of cyclosporin-induced effects on β cells is not yet known but may be related to a general destabilization of mRNA, which is especially severe in β cells because of selective uptake.

A variety of other agents have been reported to cause hyperglycemia in certain individuals. These include nicotine acid, β-adrenergic-blocking agents, and high doses of some hormones such as estrogen.

C. Carcinogenesis

Tumors in the endocrine pancreas have been reported after exposure to a small number of xenobiotics. Alloxan and streptozotocin cause islet cell tumors in rats. The number of animals having tumors can be increased by treatment with nicotinamide concomitantly with the toxin to minimize lethal cell injury. Other chemicals that have been reported to induce neoplasms are 6-diethylaminomethyl-4-hydroxyaminoquinoline-1-oxide, heliotrine, and azinophosmethyl. All these chemicals cause tumors in rats, but mice seem to be resistant to the carcinogenic effects of these xenobiotics on the endocrine pancreas. β cells are the predominant type of cell found in islet cell tumors, and most of the tumors are benign. The number of secretory granules within the cytoplasm of these cells is variable. Neoplasms may contain scattered α or δ cells.

D. Evaluation of Endocrine Pancreatic Toxicity

1. Evaluation of Endocrine Toxicity *in Vivo*

Tests of β-cell function include the determination of serum glucose and blood glycohemoglobin levels and the determination of serum insulin and c-peptide levels. Glucose tolerance tests can be performed by drawing serial blood samples for 2 hr following oral administration of a bolus of glucose. Serum insulin levels measured by radioimmunoassay also can be followed after a glucose challenge.

Elevated fasting blood glucose or glucose intolerance and deficient serum insulin levels in response to glucose are likely to be the most sensitive parameters for the evaluation of β-cell function. Reversibility of β-cell injury can be assessed by reevaluating these parameters at intervals after exposure to toxic agents. α- and δ-cell function are not usually tested *in vivo*, although levels of both glucagon and somatostatin can be determined by a radioimmunoassay.

The integrity of all islet cell types can be evaluated histologically and ultrastructurally in animals at autopsy following fatal exposure to a toxic agent. A few victims have died following the ingestion of Vacor, providing evidence of its toxicity for β cells.

2. Evaluation of Endocrine Toxicity *in Vitro*

Because of the complexities of chemical metabolism and the difficulties of determining mechanisms of cytotoxicity in animals, *in vitro* techniques may be more suitable for the systematic analysis of toxin-induced β-cell injury. Several cell lines that secrete insulin can be used to study potentially diabetogenic chemicals in cell cultures. The most definitive way to investigate the effects of a given toxin *in vitro* is to use primary cultures of islets obtained via enzymatic digestion of the pancreas. Techniques were originally described to isolate islets from the pancreas of the rat. These techniques have been modified to enhance the yield and viability of the islets obtained and have been adapted for the isolation of islets from the human pancreas. Monolayers of β cells remain viable and diploid in culture for months; however, they do not grow to confluency. When given a glucose challenge, they respond by increasing their output of insulin in a normal biphasic manner. Upon removal of the stimulus, the insulin secretion returns to basal levels. These cells also respond in a similar manner to nonglucose secretagogues. In toxicity tests, cultured β cells have been found to be selectively sensitive to the toxic effects of the known diabetogens streptozotocin and alloxan. Although these cells must be prepared from fresh pancreas tissue for every experiment, they closely approximate the function of β cells *in vivo* and are ideally suited for screening and studying the mechanisms of action for specific β-cell toxins.

III. CONCLUSIONS

Toxicologic studies of new chemicals to be utilized in ways that will result in significant human exposure should include acute and chronic tests focused on the following: (1) acinar cell cytotoxicity, (2) effects on the duct system or pancreatic secretions that will lead to duct obstruction, (3) cytotoxicity for islet cells, and (4) carcinogenicity. Because a significant fraction of pancreatic disease is of unknown etiology, the recognition and control of pancreatotoxic agents may reduce the incidence of pancreatitis, diabetes, and pancreatic carcinoma.

Because diabetes is a heterogeneous collection of diseases with more than one distinct etiological pathway, any one of several scenarios could be operative for the effects of toxic agents in individual cases. These possibilities include direct destruction of a critical mass of β cells, triggering of autoimmune processes against β cells, or augmentation of separate ongoing diabetogenic factors (e.g., autoimmune processes, viruses) to hasten inevitable clinical manifestations or functional alterations of β cells or target cells for the actions of insulin.

BIBLIOGRAPHY

Bateman, D. N. (1991). Gastrointestinal disorders. *In* "Textbook of Adverse Drug Reactions" (D. M. Davies, ed.), pp. 237–238. Oxford University Press, Oxford.

Eizirik, D. L., Sandler, S., and Palmer, J. P. (1993). Repair of pancreatic beta cells: A relevant phenomenon in early IDDM? *Diabetes* **42**, 1383–1391.

Go, V. L. W., DiMagno, E. P., Gardner, J. D., Lebenthal, E., Reber, H. A., and Scheele, G. A. (eds.) (1993). "The Pancreas: Biology, Pathobiology, and Disease" Raven Press, New York.

Lefebvre, P. J., and Pipeleers, D. G. (eds.) (1988). "The Pathology of the Endocrine Pancreas in Diabetes." Springer-Verlag, Berlin.

Longnecker, D. S., and Wilson, G. L. (1991). Pancreas. *In* "Handbook of Toxicologic Pathology" (W. M. Haschek-Hock and C. G. Rousseaux, eds.), pp. 253–278. Academic Press, San Diego.

Mallory, A., and Kern, F., Jr. (1980). Drug-induced pancreatitis: A critical review. *Gastroenterology* **78**, 813–820.

Rossini, A. A., Greiner, D. L., Friedman, H. P., and Mordes, J. P. (1993). Immunopathogenesis of diabetes mellitus. *Diabetes Rev.* **1**, 43–75.

Taylor, S. I., Accili, D., and Imai, Y. (1994). Insulin resistance or insulin deficiency: Which is the primary cause of noninsulin dependent diabetes mellitus? *Diabetes* **43**, 735–740.

Wilson, G. L., and Leiter, E. H. (1990). Streptozotocin interactions with pancreatic $\beta 1$ cells and the induction of insulin-dependent diabetes. *In* "Current Topics in Microbiology and Immunology" (T. Dyberg, ed.). Springer-Verlag, Berlin.

Wilson, G. L., and LeDoux, S. P. (1993). Interactions of chemicals with pancreatic $\beta 1$-cells. *In* "Lessons from Animal Diabetes IV" (E. Shafrir, ed.). Smith-Gordon and Company Limited, London.

Papillomaviruses and Neoplastic Transformation

KARL MÜNGER[1]
Harvard Medical School

BRUCE WERNESS
Roswell Park Cancer Center

LEX M. COWSERT
ISIS Pharmaceuticals

WILLIAM C. PHELPS
Glaxo Wellcome Co.

GLOSSARY

Colposcope Magnification device used in the gynecological examination of the uterine cervix

Enhancer DNA element that positively affects gene expression in a position- and orientation-independent manner

Episome Genetic element that exists in an extrachromosomal state

Koilocyte Cervical epithelial cell with marked cytoplasmic clearing; thought to represent a cytopathic effect of human papillomavirus infection

Oncogene Gene whose expression leads to cellular transformation in cultures or tumor formation in an animal

Open reading frame Portion of a genome with uninterrupted amino acid coding potential

Tumor suppressor gene Gene whose expression is vital for normal cell growth and differentiation. Failure of a cell to produce the functionally active gene product leads to cellular transformation. Such genes are also referred to as "antionocogenes" or "recessive oncogenes"

THE PAPILLOMAVIRUSES ARE A CLASS OF SMALL DNA tumor viruses clinically associated with epithelial lesions. They have been isolated from many higher eukaryotic species and contain a double-stranded circular genome approximately 8000 base pairs (bp) in size. Roughly 70 different human papillomavirus types (HPVs) have been characterized. A subgroup of 20 HPVs have been associated with lesions of the anogenital tract. They are further classified into "low-risk" and "high-risk" types according to the clinical lesions these viruses cause. The low-risk HPVs, such as HPV-6 and HPV-11, are mostly detected in genital warts and other benign lesions that have a very low probability for malignant progression. The high-risk

[1]Corresponding author: Pathology Department, Harvard Medical School, Goldenson Building, Room 113, 200 Longwood Avenue, Boston, MA 02115.

HPVs, such as HPV-16 and HPV-18, however, are often detected in precancerous lesions that can progress to anogenital cancers. Although tissue culture systems for the propagation of a few papillomaviruses have now been established, most of the present knowledge on their biology has been obtained from studies with molecularly cloned viral genomes. A number of *in vitro* systems have been established and have allowed the genetic dissection of the viral functions involved in cellular immortalization and transformation. The transforming and immortalizing functions of the HPVs correlate with the classification of high risk and low risk; high-risk HPVs transform and immortalize much more efficiently than do low-risk HPVs. A variety of studies have indicated that the high-risk human papillomaviruses encode two oncoproteins: E6 and E7. [*See* Viruses and Cancer.]

The E7 gene encodes a multifunctional protein that is structurally and functionally related to other oncoproteins, the adenovirus E1A proteins, and the large tumor antigen of simian virus 40 (SV40 TAg). The E7 protein can activate expression of the adenovirus E2 gene by acting on its promoter and is able to cooperate with the *ras* oncogene to transform primary baby rat cells. Like the adenovirus E1A proteins, E7 is able to form a complex with the retinoblastoma tumor suppressor gene product, pRB. The E7 gene is sufficient for transformation of established rodent fibroblast cell lines (e.g., NIH 3T3 cells).

Studies with human skin epithelial cells (keratinocytes) have shown that the E6 gene also encodes an oncoprotein and that combined expression of both the E6 and E7 genes is required to induce keratinocyte immortalization.

I. MOLECULAR CHARACTERIZATION AND NATURAL OCCURRENCE OF PAPILLOMAVIRUSES

The human papillomavirus genome consists of a double-stranded, covalently closed DNA circle of approximately 8000 bp. A schematic map of HPV-16 is presented in Fig. 1. All of the viral mRNAs are transcribed from one DNA strand of the papillomavirus genome. The coding strand contains approximately 10 major open reading frames (ORFs). These have been classified as either "early" (E) or "late" (L), based on their position in the papillomavirus genome and their expression pattern in nonproductively versus productively infected cells (Table I). The structural organization of the genome is well conserved among

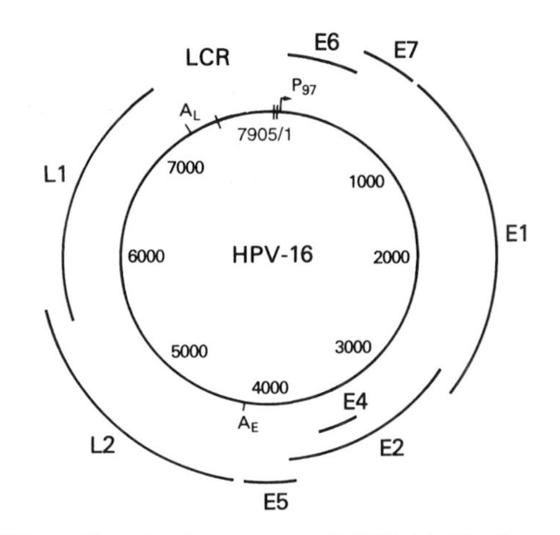

FIGURE 1 The circular genome of HPV-16. The long control region (LCR) contains the only mapped promoter (P_{97}) of HPV-16. E, L, early and late open reading frames, respectively. A_E, A_L, early and late polyadenylation sites, respectively.

the different papillomaviruses. A region about 1000 bp in length situated between nucleotides 7100 and 100 with no extensive coding potential is referred to as the "upstream regulatory region" (URR) or "long control region" (LCR). This region of the genome contains multiple cis elements that are critical for the regulation of viral transcription and replication.

Papillomavirus genomes are packaged in an icosahedral capsid 55 nm in diameter. They have been isolated from a wide range of higher vertebrates, including humans. In all of these species, they generally include warts and benign papillomas and/or fibroepithelial tumors (i.e., fibropapillomas).

From studies with the cottontail rabbit papillomavirus, it became clear that in the presence of certain chemical compounds (cocarcinogens), some papillomavirus-induced lesions had the potential to progress to malignant carcinomas. This is one of the few animal systems that permits study of progression of experimentally induced tumors in the natural host organism.

The papillomaviruses are highly host specific and exhibit a high degree of tissue specificity. Regulation of papillomavirus gene expression is thought to be intimately linked to the differentiation program of the epithelial host cells. Differentiation of keratinizing epithelium is a dynamic process in which individual cells move from the most internal layer of the epithelium (the basal layer) to the most superficial layer (the corneum). Based on cellular morphology and biochemical differentiation markers, normally stratified keratinizing epithelium can be divided into four distinct layers: basal, spinous, granular, and cornified.

The basal layer is a single sheet of undifferentiated

TABLE I
Papillomavirus Open Reading Frames and Their Assigned Functions

ORF	Assigned function
E1	Replication; binding to viral origin, DNA helicase, ATPase, E2 binding
E2	Transcriptional modulation of HPV promoters, DNA binding (ACCN6GGT), replication, E1 binding
E4	Highly phosphorylated cytoplasmic protein; associated with cytokeratin network
E5	Transforming transmembrane protein, localized to Golgi and endoplasmatic reticulum membranes; associates with growth factor receptors and a 16-kDa subunit of the vacuolar H(+)-ATPase; interferes with growth factor receptor signaling
E6	Transforming nuclear protein, associates with host cellular proteins. A complex between E6 associated protein (E6-AP) and E6 interacts with p53 and induces its rapid degradation through the ubiquitin-mediated proteolysis pathway. Inactivates the functions of p53 and induces genomic instability. E6/E6-AP complex functions as a ubiquitin ligase with p53 as a substrate. Interacts with E6-BP, an putative Ca-binding protein
E7	Transforming nuclear phosphoprotein, associates with host cellular proteins including the retinoblastoma tumor suppressor protein pRB and the related proteins p107 and p130. Interferes with the regulation of the transcriptional activities of E2F transcription factors leading to activation of E2F-dependent promoters. Associated with cell cycle machinery including a serine/threonine protein kinase activity. Induces genomic instability. Substrate for phosphorylation by casein kinase II
L1	Minor capsid protein
L2	Major capsid protein

small cells of uniform morphology. In normal epithelium, dividing cells are confined to the basal cell layer. As cells migrate from the basal layer to the cornified layer, they undergo a program of terminal differentiation, resulting in changes in morphology and in the expression pattern of a number of epithelium-specific proteins (e.g., the keratins).

One common feature of papillomavirus infections is the perturbation of the normal differentiation program. In an infected epithelium there is delayed differentiation and disorganization of the granular layer. Mitotically active cells can be observed in suprabasal layers, whereas in normal squamous epithelial cells they are limited to the basal layer. As cells move toward the more superficial layers, the cytoplasm of the infected cells becomes progressively clearer and the cells become larger and more irregularly shaped. It is thought that the virus particles enter through an abrasion in the epithelium and infect the basal cells, where the papillomavirus DNA is maintained at a low copy number as a stable nuclear episome. The full vegetative functions required for the replication and propagation of the papillomaviruses, including the expression of the late structural proteins of the virus particles, are expressed only in the terminally differentiated outermost layers of the epithelium.

So far, only a very few papillomaviruses have been propagated in the laboratory and, as a result, molecular studies with these viruses have been severely hampered.

Early insights on the molecular biology and genetics of the papillomaviruses were obtained from studies with bovine papillomavirus type 1 (BPV-1), which infects cattle and causes benign fibropapillomas. The cloned genome of BPV-1 DNA readily transforms certain established rodent fibroblast cells *in vitro*. In these transformed cell lines, the majority of the BPV-1 DNA is maintained as a autonomously replicating multicopy plasmid with an average copy number of 10–200 per cell nucleus. The viral functions involved in stable plasmid replication in transformed cells could be manifestations of the normal biology of the virus, and therefore they could represent a model of the latent nonproductive stage of a papillomavirus infection.

II. CLINICAL MANIFESTATIONS OF HUMAN PAPILLOMAVIRUS INFECTIONS

A. General

Most papillomavirus infections cause benign epithelial hyperplastic lesions commonly referred to as warts. Because most warts spontaneously regress, early concepts regarding the origin and treatment of warts have been diverse and colorful. That such infections were viral was inferred from experiments by the Italian researcher G. Ciuffo, which described the transmission of common human warts by cell-free sterile filtrates of homogenized warts.

In the 1960s, virus particles were isolated from such lesions and studied by electron microscopy. It was at first believed that only a single HPV type was responsible for the whole spectrum of warts in humans and that the observed histopathological differences among these lesions were related to the nature of the particular type of epithelium at the site of infection. Only with the application of molecular biological techniques did the heterogeneity of the many HPV types become apparent.

B. Benign Skin Warts

Most warts of the skin typically contain HPV types 1, 2, 3, 4, or 10. Common skin warts are most often found in children and young adults. Later in life the incidence of these lesions decreases, presumably due to immunological and physiological changes. Given the ubiquity of skin warts, the reported incidence of malignant progression is remarkably small. Moreover, the sites at which these warts are commonly found (i.e., hands or feet) are often exposed to mechanical trauma and possibly to cocarcinogenic compounds. It therefore seems likely that the potential of malignant progression of an HPV-associated lesion must at least be partially determined by the HPV type.

C. Epidermodysplasia Verruciformis and Nonmelanoma Skin Cancer

Epidermodysplasia verruciformis (EV) is a rare, genetically transmitted disease that is clinically characterized by chronic skin lesions that appear as reddish macules. HPV-5 and HPV-8 are often associated with EV. Approximately one-third of these patients eventually develop multiple skin carcinomas. In general, the carcinomas arise in sun-exposed areas of the skin and thus it is suspected that ultraviolet radiation could play a cocarcinogenic role with the specific HPV types in their development. HPVs that are related to EV-associated types have also been detected in and isolated from a large number of nonmelanoma skin cancers of immunosuppressed patients. [*See* Skin, Effects of Ultraviolet Radiation.]

D. Laryngeal Papillomas and Oral Cancers

Laryngeal papillomas are benign epithelial tumors of the larynx. Most of these lesions contain either HPV-6 or HPV-11. These two HPV types are generally associated with genital condyloma acuminatum (see Section II,E). Spontaneous malignant conversion of laryngeal papillomas is rare; however, they are persistent and tend to expand. It has been observed that therapeutic X irradiation of juvenile laryngeal papillomas led to an increased conversion to carcinomas 5–40 years after treatment. HPVs have also been detected in some oral cancers. The oral cancer-associated HPVs are similar but may not be identical to the high-risk HPVs isolated from anogenital tract cancers.

E. Genital Warts and Cancers of the Anogenital Tract

Genital warts are venereally transmitted. Clinically, they can be categorized into two major groups: condyloma acuminatum and flat genital warts. Molecular studies have demonstrated that greater than 90% of condylomas contain either HPV-6 or HPV-11 DNA. They generally occur on the penis, on the vulva, or in the perianal region. They can spontaneously regress or persist for years. Progression to an invasive carcinoma occurs only at an extremely low frequency.

Unlike other genital warts, those occurring on the uterine cervix usually exhibit a flat, rather than acuminate, morphology; they are usually recognizable only after the application of dilute acetic acid and with the aid of a colposcope. A papillomavirus etiology for cervical dysplasia was suggested in the late 1970s by cytologists who recognized that the cytological changes characteristic of cervical dysplasia observed on a Pap smear were due to HPV infections. This association was important, because previous clinical studies had established that cervical dysplasia [also referred to as cervical intraepithelial neoplasia (CIN)] was a precursor to carcinoma *in situ*, which in turn could give rise to invasive squamous epithelial cell carcinoma. HPV-16 and -18, as well as other high-risk types, were cloned directly from cervical carcinoma.

Epidemiological studies show that more than 90% of the human cervical carcinomas and the deprived cell lines harbor high-risk HPVs. The koilocytic cells found in Pap smears and tissue sections, often in association with moderate cervical dysplasia, are now recognized to be manifestations of an HPV infection. Cervical lesions classified by histological criteria as mild to moderate CINs spontaneously regress in about 40% of all cases, but progression to carcinoma *in situ* or invasive cervical carcinoma can occur in a low percentage of cases.

III. *In Vitro* TRANSFORMATION BY HUMAN PAPILLOMAVIRUSES

A. Transformation of Established Rodent Fibroblasts

Cell transformation was obtained with HPV-16 using a recombinant plasmid encoding the full-length genome of HPV-16, which was transfected into the murine fibroblast cell line NIH 3T3. After 4–6 weeks, cell clones exhibiting a transformed morphology were obtained. These were tumorigenic when injected into nude mice. It is interesting that the low-risk HPV types (i.e., HPV-6 and HPV-11; see Section II) also fail to transform these cells *in vitro*.

B. Transformation of Primary Cells

1. Transformation of Primary Rodent Cells

Although high-risk HPVs are unable to transform primary rodent cells, they encode a function that allows the indefinite growth (i.e., immortalization) of these cells. As with other immortalizing oncogenes, they can cooperate with an activated *Ras* oncogene to transform primary rat cells. Only high-risk HPV types scored positive in this assay.

2. Immortalization of Primary Human Epithelial Cells

Potentially more relevant to human carcinogenesis is the observation that HPV-16 and HPV-18 can immortalize primary human fibroblasts and skin epithelial cells (i.e., keratinocytes). The keratinocyte cell lines derived from these experiments lack the characteristics of transformed cells, but they exhibit altered growth and differentiation properties. [*See* Keratinocyte Transformation.]

These altered differentiation properties are most clearly recognizable when HPV-expressing keratinocytes are grown in raft culture. These are prepared by growing the cell on a collagen matrix containing feeder cells to provide required growth and differentiation factors. At the liquid–air interface they undergo terminal differentiation to fully differentiated keratinizing epithelium, made up of several layers imitating regular epidermis (Color Plate 11, top panel).

HPV-16-expressing human foreskin keratinocytes fail to differentiate properly under these conditions and show many of the histological features characteristic of naturally occurring HPV lesions.

IV. ONCOGENIC ACTIVITIES OF HIGH-RISK HPVs EXPRESSED IN CERVICAL CANCERS

A. HPV Integration and Transcription in Cervical Carcinoma Cells

In most HPV-associated cervical carcinomas and derived cell lines, the majority of the viral DNA is stably inserted (i.e., integrated) in the host DNA. This is in contrast to precancerous lesions, in which most of the HPV DNA is maintained as autonomously replicating plasmid DNA. There is no specific site of integration of the viral DNA into the host genome, although integration has been noted in some cases in proximity to cellular protooncogenes. Integration is more specific with respect to the viral genome. For integration, the viral DNA ring must be opened, and this frequently occurs in the E1 or E2 ORFs. Integration is often also accompanied by deletions of the viral genome. Therefore, integration of HPV DNA into the host genome is a terminal event for the life cycle of the virus.

The viral transcripts of several high-risk HPV-expressing cervical carcinoma cell lines have been described in some detail. The major viral mRNA species that were detected are all derived from a single promoter located in the viral LCR just upstream from the E6 ORF. These mRNAs encode either a full-length E6 gene product or truncated versions (due to internal splicing), termed E6*, as well as the full-length E7 protein. The E6–E7 regions of both HPV-16 and HPV-18 have been implicated in cellular transformation in all of the *in vitro* systems discussed.

The expression of the HPV E2 ORF is frequently disrupted as a consequence of viral integration. The HPV-16 and HPV-18 E2 ORFs encode trans-acting factors that modulate viral transcription by direct binding to DNA sequences located in the LCR. In HPV-16 and HPV-18, the predominant activity of E2 is to repress transcription. Therefore, disruption of deletion of E2 caused by the integration of HPV DNA has profound consequences on the viral transcriptional regulatory circuits. One of the consequences of HPV DNA integration is the deregulated expression of the E6 and E7 genes, which might contribute directly to malignant progression.

B. E6 Protein

The HPV E6 genes encode zinc-binding proteins of approximately 150 amino acids in length. They have been detected in nuclear and nonnuclear membranes and are rather basic. The high-risk HPV E6 proteins exert their transforming properties at least in part by forming complexes with and functionally inactivating the tumor suppressor protein p53. As a consequence of complex formation, the high-risk HPV E6 proteins accelerate the degradation of p53 via the ubiquitin-dependent proteolysis system, resulting in lower steady-state p53 levels in high-risk HPV E6-expressing cells compared to the parental cells. Since the low-risk HPV E6 proteins do not form specific complexes with p53, they have no direct effect on its stability. The biochemical pathway of E6-mediated p53 degradation has been delineated in detail. It was recognized that an additional cellular protein of 100 kDa, designated E6 associated protein (E6-AP), is necessary for binding of E6 to p53 and for the ubiquitination of p53. The E6 and E6-AP complex can act as an enzyme (ubiquitin ligase) and directly participate in the final step of covalently attaching the ubiquitin polypeptide to the p53 tumor suppressor protein.

Mutations in p53 constitute a frequent genetic abnormality in human cancers. The p53 tumor suppressor protein is a DNA-binding transcriptional activator. The p53 tumor suppressor acts as a sensor at a checkpoint of the cell division cycle where positive and negative signals are integrated. As part of this sensor function, the p53 levels increase in response to damaged DNA. This increase of transcriptionally active p 53 induces a growth arrest of the insulted cells in the G1 phase of the cell division cycle. Alternatively, in some cell types the damaged cells are eliminated by programmed cell death ("apoptosis"). Elevated levels of p53 enhance the expression of several p53-responsive cellular genes including the cyclin-dependent kinase (cdk) inhibitor p21$^{cip1/waf1/sdi1}$. This protein can directly bind to and inhibit the enzymatic activities of cyclin/cdk complexes. As a direct consequence of inhibiting these motors of the cell division cycle, the cells cease to divide and growth arrest in the G1 phase. High-risk HPV E6-expressing cells fail to induce p21$^{cip1/waf1/sdi1}$ and do not growth arrest after treatment with DNA-damaging agents since the function of p53 is annulled by high-risk HPV E6-mediated degradation. The resulting loss of p53-mediated cell cycle checkpoint control may in part account for the observation that cervical carcinoma cells often have abnormal karyotypes.

Several findings indicate that there are additional biological functions of E6 that are not mediated by its interaction with E6-AP and p53. One such protein, named E6-BP1 (for E6 binding protein 1), has been isolated and encodes a previously unidentified putative calcium-binding protein. The isolation of additional targets of E6 will shed some lights on the molecular mechanisms that are targeted by E6.

C. E7 Protein

The HPV E7 proteins are acidic zinc-binding nuclear phosphoproteins of approximately 100 amino acids in length. They are functionally similar to the Ad E1A proteins and the SV40 TAg and can transcriptionally activate the Ad E2 promoter and cooperate with *ras* to transform primary rat cells.

As suggested by their functional similarity, the three oncoproteins HPV E7, Ad E1A, and SV40 TAg are structurally related (Fig. 2). The amino acid sequence similarity extends to a short portion of the conserved region (CR) 1 and the entire CR2 of Ad E1A. In each of the three viral oncoproteins, these conserved regions are critical to the transforming function and participate in the binding of several cellular proteins. These include the retinoblastoma tumor suppressor gene product pRB and the structurally related p107 and p130 proteins. Because pRB, p107, and p130 share two conserved sequence motives in the carboxyl terminus, which serve as the binding "pocket" for the viral oncoproteins, they are referred to as the "pocket proteins."

The biological activity of pRB is regulated by cell division cycle-dependent phosphorylation/dephosphorylation. In the G0 or G1 phases of the cell division cycle, pRB is present in a hypophosphorylated form. Since pRB acts as a negative regulator of cell growth at the G1/S boundary, the hypophosphorylated pRB may have growth-suppressive activities. Upon progression of the cell division cycle from G1, pRB is phosphorylated at multiple amino acid residues by one or more cyclin/cdk complexes and the protein is hyperphosphorylated during the S, G2, and early M phases. In late M phase, hyperphosphorylated pRB is reconverted to the hypophosphorylated form through a cell cycle-specific activity of a phosphatase. Like SV40 TAg, the HPV-16 E7 oncoprotein preferentially binds to hypophosphorylated pRB. This provides an attractive model for the functional inactivation of the

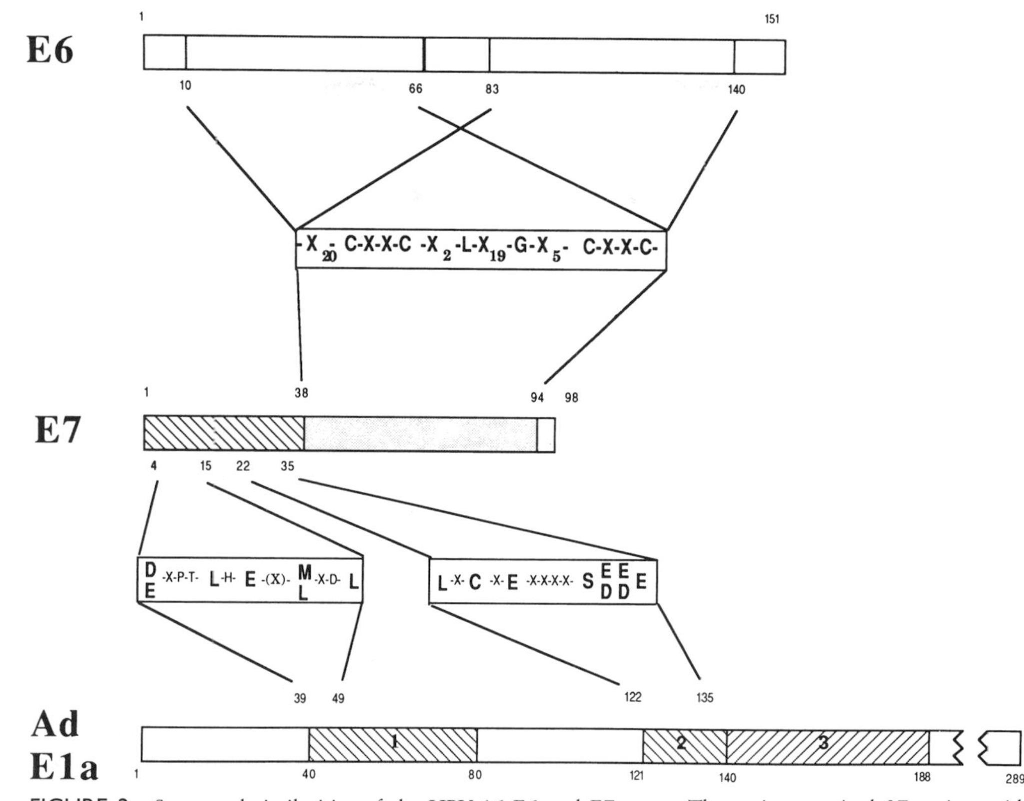

FIGURE 2 Structural similarities of the HPV-16 E6 and E7 genes. The amino-terminal 37 amino acid residues are strikingly similar to portions of conserved domains 1 and 2 of the adenovirus (Ad) E1A protein. In HPV-16 and Ad E1A these sequences are required for cellular transformation properties and contain the sequences necessary for interaction with the retinoblastoma tumor suppressor gene product p105-RB. The carboxy-terminal part of E7 contains two copies of a Cys-X-X-Cys sequence motif that is also conserved in E6. These sequences have been implicated in the binding of Zn^{2+}. Based on these structural similarities, it has been hypothesized that the E6 and E7 genes might have evolved from a common ancestor. The standard single-letter abbreviations for amino acid residues have been used (C, Cys; D, Asp; E, Glu; G. Gly; L, Leu; M, Met; P. Pro; T, Thr; X, any amino acid residue).

growth-suppressive activity of pRB. The high-risk HPV E7 protein interferes with the regulatory pathways that are governed by the pocket proteins in normal cells by displacing and/or abrogating the functions of cellular proteins that are normally bound to the pocket proteins pRB, p107, and p130. A number of cellular proteins that can interact with and may be regulated by the pocket proteins have been described. These include members of the E2F transcription factor family. Their transcriptional activities are governed by interactions with pocket proteins: when bound to pRB, E2F can function as a transcriptional repressor; when not bound to a pocket protein, E2F can activate the transcription of target genes. The E2F transcription factors have been shown to contribute to the temporal regulation of expression of a number of genes involved in cell division cycle control and in nucleotide biosynthesis. By interacting with the pocket proteins, the HPV E7 protein subverts the normal regulation of the E2F transcription factor activity and certain E2F/pocket protein complexes are significantly reduced in HPV E7-expressing cells.

A short sequence in the CR2 homology domain of E7 is necessary for complex formation with the pocket proteins, and this sequence is shared by the low-risk and high-risk HPV E7 proteins. The low-risk HPV E7 proteins interact less efficiently with pRB than do high-risk HPV E7 proteins. The ability of the HPV E7 protein to efficiently interact with pRB has been closely linked to the cellular transformation activities in several assay systems. The difference of the high-risk and low-risk HPV E7 proteins in transforming rodent cells correlates with the difference in pRB binding efficiency. There is a single consistent amino acid sequence difference in the pRB binding sites between the high-risk and the low-risk HPV E7 proteins that largely accounts for the differences in pRB binding and cellular transformation.

As with the high-risk HPV E6 proteins, there is compelling evidence that there are biological functions of E7 that are not mediated by interactions with the pocket proteins. Additional cellular targets of the HPV E7 proteins may therefore exist. The E7 protein can interact with members of the cell division cycle machinery, and a serine/threonine protein kinase activity as well as cyclins have been detected in complex with E7. It is not known, however, if these are direct interactions or whether they are mediated by the pocket proteins pRB, p107, and/or p130. Possibly as a consequence of such protein–protein interactions, the high-risk HPV E7 proteins can overcome the p53-mediated cell division cycle checkpoint in G1. This may contribute to the ability of E7 to induce genomic instability.

V. DISRUPTION OF THE pRB AND p53 PATHWAYS IN CERVICAL CARCINOGENESIS

Studies with a panel of cervical cancer cell lines have suggested that the cellular regulatory pathways that are governed by p53 and pRB may be physiologically relevant targets of the HPV E6 and E7 oncoproteins during cervical carcinogenesis. The premise for these experiments was that HPV-positive and HPV-negative cervical carcinoma cell lines should differ in their status of p53 and pRB if these tumor suppressors were relevant and essential targets of the HPV E6 and E7 oncoproteins. In accord with these predictions, the HPV-positive cells lines all had wild-type p53 and normal pRB, whereas both HPV-negative cervical carcinoma cell lines contained mutated p53 and RB genes. Similar analyses with primary tumor tissue yielded more ambiguous results and this simple correlation was not observed in all studies. With the definition of additional downstream targets and upstream regulators in the p53 and pRB pathways (such as certain cyclins, cdk's, and cdk inhibitors) and the availability of more sensitive and reliable methods for the detection of HPV in tumors, it will be possible to further test the validity of these predictions.

VI. ADDITIONAL FACTORS INVOLVED IN CERVICAL CARCINOGENESIS

Epidemiological studies suggest that only a relatively small percentage of women infected with high-risk HPVs develop any clinical lesions and only a small number of these, if left untreated, will progress to cervical cancer. The immunological response to an HPV infection is critical for progression or regression of these precursor lesions. The immune system, however, is not the only determinant for malignant progression and additional cellular mutations such as the inactivation of cellular tumor suppressors and/or the activation of oncogenes may also be important. This is illustrated by the previously summarized results that HPV-immortalized human keratinocyte cell lines develop a fully transformed phenotype only after prolonged passage in tissue culture or after transfection with an additional oncogene. Studies with somatic cell hybrids of the HeLa or SiHa cervical carcinoma cell lines and normal human cells have suggested that a tumor suppressor gene located on chromosome 11 may be inactivated in these cervical cancer cell lines. Functional disruption of this locus during carcinogenic progression, in addition to the disruption of the viral negative regulatory circuits governed by E1 and E2, is thought to further contribute to the disruption of cell cycle checkpoint control and result in elevated and uncontrolled cellular replication and chromosomal instability, possibly followed by clonal selection of malignant cells and tumorigenesis. With the availability of suitable animal models that recapitulate the multistep process of cervical carcinogenesis, it will be possible to further study these critical molecular events.

BIBLIOGRAPHY

Bosch, F. X., Manos, M. M., Munoz, N., Sherman, M., Jansen, A. M., Peto, J., Schiffman, M. H., Moreno, V., Kuman, R., Shah, K. V., Alihonou, E., Bayo, S., Mokhtar, H. C., Chicareon, S., Daudt, A., Delosrios, E., Ghadirian, P., Kitinya, J. N., Koulibaly, M., Ngelangel, C., Tintore, L. M. P., Riosdalenz, J. L., Sarjadi-Schneider, A., Tafur, L., Teyssie, A. R., Rolon, P. A., Torroella, M., Tapia, A. V., Wabinga, H. R., Zatonski, W., Sylla, B., Vizcaino, P., Magnin, D., Kaldor, J., Greer, C., and Wheeler, C. (1995). Prevalence of human papillomavirus in cervical cancer: A worldwide perspective. *J. Natl. Cancer Inst.* 87, 796–802.

Dyson, N., Howley, P. M., Münger, K., and Harlow, E. (1989). The human papillomavirus-16 E7 oncoprotein is able to bind to the retinoblastoma gene product. *Science* **243**, 934–937.

Laimins, L. A. (1993). The biology of human papillomaviruses—From warts to cancer. *Infect. Agents and Disease* **2**, 74–86.

Lechner, M. S., Mack, D. H., Finicle, A. B., Crook, T., Vousden, K. H., and Laimins, L. A. (1992). Human papillomavirus E6 proteins bind p53 *in vivo* and abrogate p53-mediated repression of transcription. *EMBO J.* **11**, 3045–3052.

McCance, D. J., Kopan, R., Fuchs, E., and Laimins, L. A. (1988). Human papillomavirus type 16 alters human epithelial cell differentiation *in vitro*. *Proc. Natl. Acad. Sci. USA* **85**, 7169–7173.

Münger, K., and Phelps, W. C. (1993). The human papillomavirus-E7 protein as a transforming and transactivating factor. *Biochem. Biophys. Acta* **1155**, 111–123.

Scheffner, M., Werness, B. A., Huibregtse, J. M., Levine, A. J., and Howley, P. M. (1990). The E6 oncoprotein encoded by human papillomavirus types 16 and 18 promotes the degradation of p53. *Cell* **63**, 1129–1136.

Scheffner, M., Münger, K., Byrne, J. C., and Howley, P. M. (1991). The state of the p53 and retinoblastoma genes in human cervical carcinoma cell lines. *Proc. Natl. Acad. Sci. USA* **88**, 5523–5527.

Vousden, K. H. (1994). Interactions between papillomavirus proteins and tumor suppressor gene products. *Adv. Cancer Res.* **64**, 1–24.

Werness, B. A., Levine, A. J., and Howley, P. M. (1990). Association of human papillomavirus types 16 and 18 E6 proteins with p53. *Science* **248**, 76–79.

White, A. E., Livanos, E. M., and Tlsty, T. D. (1994). Differential disruption of genomic integrity and cell cycle regulation in normal human fibroblasts by the HPV oncoproteins. *Genes & Develop.* **8**, 666–667.

Zur Hausen, H., and De Villiers, E. M. (1994). Human papillomavirus. *Ann. Rev. Microbiol.* **48**, 427–447.

Parathyroid Gland and Hormone

PAUL L. MUNSON
University of North Carolina

GLOSSARY

Branchial or pharyngeal pouch One of a paired set of four embryonic endodermal structures that give rise to the thyroid and parathyroid glands, the thymus, and other important tissues

Osteoblast Bone cell responsible for bone formation

Osteoclast Large, multinucleated cell responsible for bone resorption, both matrix and mineral

Secretory granules Discrete bundles of granules, enclosed in a membrane, of products of a secretory cell, such as hormones, destined for extrusion from the cell into the extracellular fluid

THE PARATHYROID GLAND IS IMPORTANT FOR good health because it produces, secretes, and regulates the secretion of a hormone known simply as parathyroid hormone. The major function of the parathyroid gland and parathyroid hormone is to maintain within rather narrow limits a certain concentration of calcium (approximately 10 mg/dl) in the blood plasma that is ideal for efficient neuromuscular function and for the development and maintenance of a healthy skeleton.

I. PARATHYROID GLAND

A. Anatomy

The parathyroid glands, small tan to reddish-brown structures, usually four in number in the human species, are located in close association with the thyroid gland. Each of the upper two glands, the "external glands," is located on the upper surface of the right or left pole of the thyroid gland. The lower two glands, the "internal glands," are usually located within the lower right and left lobes of the thyroid gland but can also occur behind the esophagus or in the mediastinum. Occasionally, the total number of parathyroid glands in humans is five to eight. Each normal human gland weighs, on the average, approximately 40 mg and measures 4–6 mm in its greatest dimension. In humans, the blood supply for all four glands is from a branch of the inferior thyroid artery, arising from the subclavian artery, and the blood drainage is into the thyroid vein, but there are differences in other species.

B. Histology

The major cells of the parathyroid gland are the "chief cells." These cells, each measuring 6–8 μm in diameter, occur in closely packed sheets. The chief cell is the site of biosynthesis, storage, and secretion of parathyroid hormone. The parathyroid hormone is contained in and secreted from secretory granules in the chief cell. The number of these granules is relatively small compared with other endocrine glands; little parathyroid hormone is stored. Other cells seen on histological sections of the parathyroid gland include fat cells, which constitute 10–50% of the gland in adults, and "oxyphil" cells, less than 5% of the gland, which make their first appearance at puberty and increase in number with age. Neither fat cells nor oxyphil cells have any known function. A separate cell type is responsible for production and secretion of parathyroid hypertensive factor.

C. Embryology

The parathyroid glands originate in the endodermal layers of the third and fourth pairs of branchial pouches. During embryogenesis, they migrate to the thyroid gland.

D. Species Distribution on the Evolutionary Scale

The parathyroid glands occur in all terrestrial vertebrates; they first appeared in amphibians and are not present in fish. The glands are ordinarily four in number in most species, as in the human, except for the rat (a widely used species for experiments on parathyroid hormone), in which there are normally only two parathyroid glands, located on the upper poles of the thyroid gland.

II. PARATHYROID HORMONE: THE POLYPEPTIDE CHAIN

Parathyroid hormone is a single-chain polypeptide made up of 84 amino acid residues. It lacks cysteine and substituted amino acid residues but contains two methionines. Neither terminus of the chain is blocked. The molecule does not contain carbohydrates or other non-amino acid cofactors. There is a preponderance of basic amino acid residues; overall, the molecule is basic in character.

The amino acid sequences of parathyroid hormone from several different species—ox, pig, human, rat, and chicken—have been determined. They are quite similar but not identical. Figure 1 shows the structure of the human hormone and the differences in sequence between it and the bovine, porcine, and rat hormones. Human parathyroid hormone has been efficiently synthesized in *Escherichia coli,* using a synthetic gene, and purified and characterized.

III. BIOSYNTHESIS OF PARATHYROID HORMONE

The human gene for parathyroid hormone is located on the short arm of chromosome 11. After transcription of the DNA into mRNA in the nucleus of the chief cell, mature mRNA moves into the cytoplasm, and parathyroid hormone is biosynthesized as part of a larger 115-amino acid polypeptide chain known as

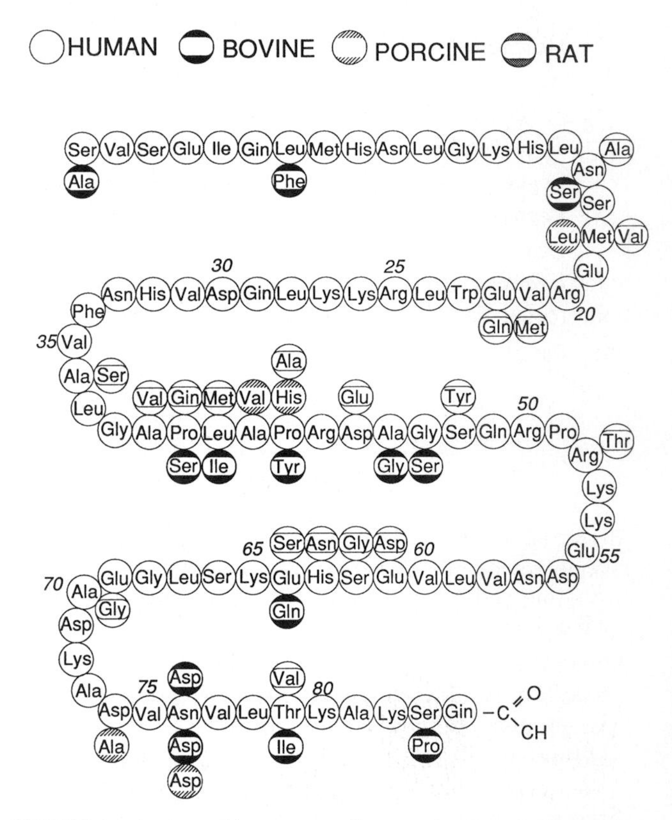

FIGURE 1 Amino acid sequence of human, bovine, porcine, and rat parathyroid hormone. Points of difference between the sequences of the four molecules are indicated by the symbols for species. [Reprinted, with permission from M. Rosenblatt, H. M. Kronenberg, and J. T. Potts, Jr. (1989). Parathyroid Hormone: Physiology, Chemistry, Biosynthesis, Secretion, Metabolism, and Mode of Action. *In* "Endocrinology" (L. J. DeGroot, ed.), 2nd Ed., p. 852. Saunders, Philadelphia.]

"preproparathyroid hormone." The synthesis of this larger molecule takes place on polyribosomes bound to membranes of the rough endoplasmic reticulum of the chief cells. The next stage consists of two steps. In the first step, 25 amino acids at the amino terminus (known as the signal or leader sequence) are removed by proteolytic enzymes in or near the rough endoplasmic reticulum. (The function of the signal sequence is to direct the protein across the membrane of the endoplasmic reticulum and into the secretory pathway.) The product, preproparathyroid hormone, contains 6 more amino acids at the amino terminus of parathyroid hormone (the function of the prosequence has not been determined). The proparathyroid hormone is then moved to the Golgi apparatus, where it is converted by another proteolytic enzyme, furin, to parathyroid hormone. Parathyroid hormone (1–84) itself is the major form of the hormone contained in the mammalian glands; only 7% of the total is proparathyroid hormone and there is an even lower percentage of preproparathyroid hormone. After biosynthesis, the parathyroid hormone is transported to secretory granules, from which it is secreted, along with a variable amount of carboxy-terminal parathyroid hormone fragments.

Neither preproparathyroid hormone nor proparathyroid hormone is secreted and neither has any significant biological activity. Their biological significance is confined to their role as biosynthetic precursors of parathyroid hormone.

IV. TERMINAL METABOLISM AND DISPOSITION OF PARATHYROID HORMONE

Most of the parathyroid hormone in the plasma is taken up by the liver and kidney. Parathyroid hormone is also bound to receptors on the osteoclasts, but the amount bound, less than 1%, is insignificant compared with that in liver and kidney.

Proteolytic enzymes of the Kupffer cells in the liver blood capillaries and in other cells at various sites in the kidney split off large carboxy-terminal fragments from parathyroid hormone. Although these fragments are biologically inactive, they may have been measured as active intact parathyroid hormone by the older simple radioimmunoassay methods and thereby confuse interpretation of the results. However, this is not a problem with the newer two-site radioimmunometric assays, which are more specific for intact parathyroid hormone.

Both the intact hormone and the carboxy-terminal fragments are excreted by the normal kidney. The clearance half-time for intact hormone is less than 5 min, but for carboxy-terminal fragments it is at least 20–40 min. In uremia, the half-time of the fragments may be prolonged as much as 100 times. The half-time of intact hormone is prolonged also, but much less so than for the carboxy-terminal fragments because it is taken up and broken down by the liver and the carboxy-terminal fragments are not. Meprin, an integral membrane protein, is predominantly involved in the degradation of parathyroid hormone by the renal microvillar membranes.

V. STRUCTURE–ACTIVITY RELATIONSHIPS FOR PARATHYROID HORMONE

There is considerable cross-reactivity between the parathyroid hormones from different species, but the relative potencies of the hormones from the four species shown in Fig. 1 differ considerably, depending on what assay method is used for the comparison. In an *in vivo* method (chick hypercalcemia), more meaningful for human biology, the ratio of bovine to porcine to human was 9 : 3 : 1.

Synthetic polypeptides that contain the amino acid sequence 1–34 of the human and bovine parathyroid hormones are about as active on a molar basis as the entire 1–84 sequence in most assay methods. In the adenylate cyclase assays, both *in vitro* and *in vivo,* and the chick *in vivo* assay, the sequence 1–25 was the minimum for retaining any detectable biological activity. Removal of the two amino-terminal amino acids resulted in complete loss of activity.

A substance similar to parathyroid hormone is the "parathyroid hormone-like protein," which has been isolated and characterized from extracts of tumors and found to have homology in structure with the first 13 amino acids of parathyroid hormone. It also shares most of its biological activities with parathyroid hormone. Furthermore, it appears to be responsible for the hypercalcemia of malignancy and is elaborated by several different classes of tumors (squamous, bladder, and ovarian carcinomas) not in the parathyroid gland.

Research toward development of potent antagonists for parathyroid hormone is in progress. A synthetic peptide patterned after a partial sequence (7–34) of the parathyroid hormone-like protein was 6–8 times more potent as an antagonist than any comparable peptide from bovine parathyroid hormone, but it is still not potent enough to be practical as a treatment for hyperparathyroidism.

Along with parathyroid hormone, the parathyroid gland also secretes a larger polypeptide, originally known as "parathyroid secretory protein" (molecular weight 70,000), renamed "parathyroid chromogranin A." It is a glycosylated protein similar or identical to chromogranin A found in secretory granules of the adrenal medulla. Current evidence suggests that it is a precursor of pancreastatin, which may affect the secretion of parathyroid hormone. (Pancreastatin was discovered as a strong inhibitor of glucose-stimulated insulin release.)

VI. ASSAY OF PARATHYROID HORMONE IN PLASMA

The concentration of parathyroid hormone in blood plasma is too low to be measured by any simple bioassay, such as by increase in serum calcium after injection into rats or chicks.

In preference, radioimmunoassay methods are widely used in clinical diagnosis as well as in research. The older, simple methods had the disadvantage of lack of specificity, measuring parathyroid hormone fragments as well as intact parathyroid hormone. The newer two-site immunoradiometric assays are highly specific for intact parathyroid hormone 1–84 and can distinguish quite well between normals and patients with hyperparathyroidism or hypoparathyroidism. The assays can also distinguish between patients with nonparathyroid hypercalcemia and can be used to monitor the course of parathyroid surgery.

The actual range of parathyroid hormone concentration in the plasma of normal healthy persons is rather broad, ranging from 10 to 55 pg/ml using an immunoradiometric assay.

VII. FUNCTIONS OF PARATHYROID HORMONE AND HOW THEY ARE PERFORMED

A. Calcium Homeostasis

The major function of parathyroid hormone is to raise the concentration of calcium in the blood plasma to an optimum level of about 10 mg/dl and to keep it there. The rapid fall in plasma calcium after removal of the parathyroid glands is illustrated in Fig. 2. In the absence of parathyroid hormone, as shown in the figure, the plasma calcium may fall to as low as 5 mg/dl, a concentration that reflects the apparent equilibrium between bone mineral and body fluids. To avoid an excess of parathyroid hormone that would result in hypercalcemia, the rate of secretion is regulated by negative feedback, as described further in the following.

The calcium in blood is essentially all in the plasma, where it is divided about equally between protein-bound and ionized calcium. (A small percentage, about 3%, is complexed with organic ions such as citrate.) Only the ionized calcium fraction is physiologically active, but because in most situations the forms of plasma calcium are in rapid equilibrium, any change in total calcium is reflected immediately in a corresponding change in the concentration of ionized calcium. Exceptions to this statement are: in humans with serum protein abnormalities, total serum calcium can change considerably without much change in ionized calcium, and in birds, during the normal egg-laying cycle, the total serum calcium may even double (by binding to phosvitin) without any change in ionized calcium.

Three organ systems affected by parathyroid hormone contribute to the ability of the hormone to maintain the normal serum calcium concentration: bone, kidney, and intestine.

1. Bone

The mineral phase of bone is predominantly hydroxyapatite, $Ca_{10}(PO_4)_6(OH)_2$. A part of the calcium in

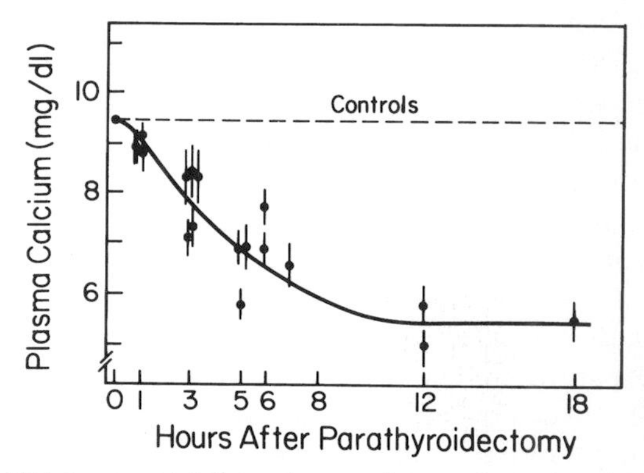

FIGURE 2 Rapid fall in plasma calcium after removal of the parathyroid glands in young male rats. Each point and vertical line represent the mean plus standard error of four to six rats. [Redrawn with permission, from data in Fig. 2 in A. H. Tashjian, Jr. (1966). Effects of parathyroidectomy and cautery of the thyroid gland on the plasma calcium level of rats with autotransplanted parathyroid glands. *Endocrinology* 78, 1144–1153.]

bone is adsorbed to the surface of bone in dynamic equilibrium with the calcium in plasma. Parathyroid hormone favors outflow of calcium from bone over inflow into bone and thereby prevents a fall in the plasma calcium when the calcium balance is negative. In the absence of the hormone, inflow is favored, with the result being a lowering of the plasma calcium. The action of parathyroid hormone on a responsive subpopulation of osteoblasts, including those that line bone surfaces, appears to be principally responsible for the minute-to-minute and hour-to-hour regulation of plasma calcium.

A second parathyroid hormone-mediated process that affects plasma calcium is "bone remodeling," in which areas of bone are being "resorbed" (matrix as well as mineral removed) and then replaced by new bone formation. The balance between the two activities of bone formation and bone resorption affects the plasma calcium concentration over the long term, but the changes are too slow to explain the rapid plasma calcium-raising activity of an injection of parathyroid hormone under experimental situations. [*See* Bone Remodeling.]

Bone resorption is performed by the osteoclast, a large, multinucleated bone cell. Although parathyroid hormone has a strong effect to increase the number of osteoclasts and their level of activity, no receptors for parathyroid hormone have been found on these cells. On the other hand, osteoblasts (the bone-forming cells) are rich in parathyroid hormone receptors. It has now been shown in *in vitro* experiments that parathyroid hormone stimulates osteoclastic bone resorption most vigorously in the presence of osteoblasts. Parathyroid hormone acts on osteoclasts by stimulating osteoblasts to elaborate osteoclast-stimulating factors, and it is in this way that parathyroid hormone effectively but indirectly stimulates bone resorption. Some evidence for parathyroid hormone receptors on osteoclasts has been obtained, however, and it seems likely that parathyroid hormone affects osteoclasts by direct as well as indirect stimulation.

The processes of bone resorption and bone formation are tightly coupled, so that when parathyroid hormone stimulates bone resorption, bone formation occurs also. At normal concentrations of plasma calcium and parathyroid hormone, the two processes tend to be in balance, but during hypocalcemia and at high concentrations of parathyroid hormone, the effect on resorption outweighs formation.

Parathyroid hormone is best known as a hypercalcemic hormone but, interestingly, it can be an anabolic hormone under certain circumstances. When given intermittently, it can produce a lowering of blood calcium and promote bone formation. This property of parathyroid hormone is being exploited with encouraging results in the treatment of osteoporosis with parathyroid hormone alone and in combination with other agents, such as estrogen, calcitonin, vitamin D, and bisphosphonates.

2. Kidney

Parathyroid hormone also acts on the kidney to promote calcium homeostasis. It does so by (1) reducing the renal excretion of calcium and (2) stimulating the production of calcitriol, which increases intestinal absorption of calcium and acts on bone to help maintain the plasma calcium concentration.

Most of the calcium in the glomerular filtrate is not excreted in the urine but is reabsorbed in the proximal tubule independently of parathyroid hormone. Nevertheless, the added action of parathyroid hormone to enhance reabsorption of calcium, which is on the thick ascending and granular portions of the distal tubule, is extremely important. The additional calcium reabsorbed under the influence of parathyroid hormone could account for as much as one-fifth to one-third of the total extracellular fluid calcium.

3. Intestine: Calcitriol

The third major organ system by which parathyroid hormone affects calcium metabolism is the small intestine. Most, if not all, of this effect is mediated through the hormonal metabolite of vitamin D, calcitriol (1α,25-dihydroxycholecalciferol), the production of which is increased by parathyroid hormone.

Calcitriol is produced in the kidney by the action of the enzyme renal 25-hydroxyvitamin D 1α-hydroxylase on the precursor of calcitriol, calcidiol (25-hydroxycholecalciferol), produced by the liver from cholecalciferol. Cholecalciferol (vitamin D_3) originates in the skin by the action of an ultraviolet part of sunlight (290–315 nm) on 7-dehydrocholesterol. Vitamin D, either vitamin D_2 (ergocalciferol) or vitamin D_3, may also be supplied as dietary supplements in commercial foods (mostly milk). Except for fish with a high fat content there is very little vitamin D in natural foods. Parathyroid hormone affects the production of calcitriol by stimulating the renal 1α-hydroxylase, which results in the supply of an adequate quantity of calcitriol or its analog, 1α,25-dihydroxyergocalciferol, to promote a healthy rate of absorption of calcium by the small intestine. Although all segments of the small intestine—duodenum, jejunum, and ileum—are involved in the absorption of calcium, it is the duodenum that is most sensitive to the effect of calcitriol. [*See* Vitamin D.]

B. Effect on Metabolism of Inorganic Phosphate

Parathyroid hormone also has an effect on the concentration of inorganic phosphate in the blood, which, like calcium, is located almost entirely in the plasma. Again, the effect is produced by actions of parathyroid hormone on three organ systems—bone, kidney, and intestine—of which the kidney is predominant.

1. Kidney

About 75% of filtered phosphate is reabsorbed by the proximal tubule, the remainder by the distal tubule and cortical collecting loop. Parathyroid hormone depresses reabsorption of inorganic phosphate in both the proximal and distal renal tubules, thereby increasing the quantity of inorganic phosphate excreted in the urine and decreasing the concentration of inorganic phosphate in the plasma.

2. Bone

The effect of parathyroid hormone to increase outflow of mineral from bone in the short term and during bone resorption in the long term tends to increase the concentration of inorganic phosphate as well as of calcium in the plasma, but the renal effect on phosphate predominates, so that, overall, the result is a decrease in plasma inorganic phosphate, as can be seen in hyperparathyroidism.

3. Intestine

Inorganic phosphate, unlike calcium, is readily absorbed from the small intestine. Nevertheless, calcitriol increases its absorption, and therefore, indirectly, by its effect on calcitriol production, parathyroid hormone favors the absorption of phosphate as well as calcium from the intestine.

4. Importance

The effect of parathyroid hormone on phosphate metabolism is secondary in importance to its effect on calcium metabolism. The supply of phosphate in the food and its absorption from the intestine are ample, unlike the situation for calcium. Therefore, the loss of phosphate due to the action of parathyroid hormone is not usually detrimental. On the other hand, excretion of phosphate protects the body from hyperphosphatemia, which tends to lower plasma calcium by several mechanisms.

C. The Adenylate Cyclase–Cyclic AMP System in the Mechanism of Action of Parathyroid Hormone

Much experimental evidence indicates that the increase in plasma calcium after an injection of parathyroid hormone is mediated by an increase in cyclic AMP. The same is true for the effect of parathyroid hormone on the decrease in reabsorption of inorganic phosphate by the renal tubule. It is thought that after parathyroid hormone binds to its receptor, the receptor interacts with and activates a guanyl nucleotide-binding protein "G," which in turn activates adenylate cyclase and increases the hydrolysis of ATP to cyclic AMP. G proteins are membrane-associated proteins that facilitate activation or inhibition of second-messenger effector systems in response to receptor activation. Although the concept that parathyroid hormone acts through cAMP is well supported, there is also strong evidence for another second messenger, namely, inositol 1,4,5-triphosphate.

VIII. REGULATION OF SECRETION OF PARATHYROID HORMONE

The most important factor regulating the rate of secretion of parathyroid hormone is the plasma concentration of ionized calcium. (This is unlike the situation in most endocrine cells, in which calcium is required for hormone secretion.) An increase in calcium inhibits secretion of parathyroid hormone, and a decrease "stimulates" it by releasing the gland from inhibition. The interaction between ionized calcium and secretion of parathyroid hormone constitutes a valuable negative feedback system that works to regulate the plasma calcium concentration within narrow limits between 8.8 and 10.5 mg/dl. Figure 3 illustrates changes in the plasma concentration of parathyroid hormone in response to changes in plasma calcium.

The reaction of the feedback system is quite rapid. Observations in experimental animals indicate that the gland responds within 1 min of an induced fall in serum ionized calcium to increase hormone secretion.

The first parathyroid hormone released is that which has been stored in "mature" secretory granules. Later, if the calcium concentration stays low for a long time, with continuous release of parathyroid hormone, a greater proportion of the secreted hormone

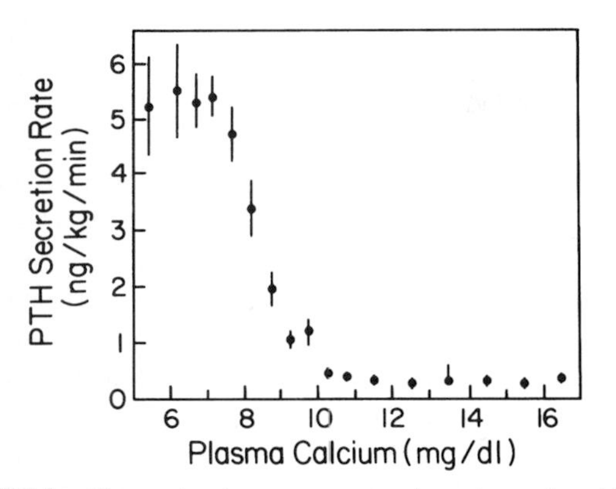

FIGURE 3 Change in plasma concentration of parathyroid hormone in relation to induced changes in plasma calcium concentration. Each point and vertical line represent the mean ± S.E. of repeated measurements in 2–12 calves. [Redrawn with permission, from data in Fig. 2 in G. P. Mayer and J. G. Horst (1978). Sigmoidal relationship between parathyroid hormone secretion rate and plasma calcium concentration in calves. *Endocrinology* **102,** 1036–1042.]

is newly synthesized hormone. It has been calculated that there is enough parathyroid hormone in the gland to last 7 hr at normal serum calcium concentrations but only 2.5 hr under protracted hypocalcemic conditions.

The fact that calcium and other agents that inhibit parathyroid hormone secretion also inhibit accumulation of cyclic AMP within parathyroid cells suggests that there is an intimate relationship between secretion of parathyroid hormone and the adenyl cyclase–cyclic AMP system. On the other hand, cyclic AMP affects secretion from a preformed hormone pool, whereas calcium controls secretion of newly synthesized as well as stored parathyroid hormone, suggesting a certain independence of the two factors. We must conclude that, in spite of intensive investigation, the exact mechanisms whereby plasma ionized calcium regulates secretion of parathyroid hormone are not well known.

Recently, E. R. Brown *et al.* reported cloning of a protein from a cDNA library derived from bovine parathyroid cell mRNA with characteristics of a calcium sensor. The protein derived from this cDNA has a predicted topological structure that is similar to the seven membrane-spanning domains of the G-protein-coupled superfamily of receptors. The receptor responds to extracellular calcium and other polyvalent cations with increase in intracellular phosphatidylinositol turnover. Using the bovine calcium receptor cDNA, a similar nucleotide sequence was identified in human parathyroid adenoma cells. Thus, the transducing mechanism for hypocalcemic stimulation of PTH secretion is beginning to be elucidated.

There are other factors that have been shown to affect the secretion of parathyroid hormone: an increase in the magnesium concentration inhibits the secretion of parathyroid hormone in a manner similar to that of calcium, but the parathyroid is less responsive to magnesium than to calcium. Paradoxically, severe and prolonged hypomagnesemia may lead to hypoparathyroidism. High concentrations of potassium, on the other hand, stimulate secretion of parathyroid hormone. Catecholamines also can increase the secretion of parathyroid hormone, as can various other factors. Nevertheless, the feedback relationship between ionized calcium and parathyroid hormone secretion is the dominating influence on hormone secretion and plasma calcium concentration.

There appears to be a modest circadian rhythm in the rate of secretion of parathyroid hormone, with the rate during the night about twice that in the daytime. In rats, the plasma parathyroid concentration is considerably elevated during lactation. This phenomenon has also been seen in women secreting large amounts of milk, such as mothers nursing twins. Both the circadian changes and the lactation-associated increase in parathyroid hormone secretion appear to be related to factors other than plasma calcium. [*See* Circadian Rhythms and Periodic Processes.]

There is clear evidence that increasing the plasma concentration of calcitriol decreases secretion of parathyroid hormone. The findings suggest some sort of feedback relationship between parathyroid hormone and calcitriol, with parathyroid hormone increasing the production of calcitriol and calcitriol decreasing the secretion of parathyroid hormone. The physiological significance of the interactions of the two feedback systems on each other, if any, has not yet been worked out. In fact, in lactating rats, the concentrations in the plasma of both plasma calcitriol and plasma parathyroid hormone are elevated.

Recently, V. L. Schultz *et al.* observed that some but not all anesthetics tested, namely, pentobarbital, methoxyflurane, and xylazine combined with ketamine, increased serum parathyroid hormone markedly (as much as threefold). On the other hand, diethyl ether and ketamine alone were negative. The comparisons were made with blood from decapitated, unanesthetized rats. It is not yet known whether similar effects occur in humans. The effects in rats suggest that there may be some complex neuronal effects on the secretion of parathyroid hormone.

IX. REGULATION OF THE BIOSYNTHESIS OF PARATHYROID HORMONE

To supply enough extra parathyroid hormone to reverse prolonged hypocalcemia, hormone production is increased. Evidence suggests that during hypocalcemia, the parathyroid hormone gene is nearly maximally active and that during hypercalcemia it is inhibited. Furthermore, during hypocalcemia most of the biosynthetic product is intact (1–84) nonfragmented parathyroid hormone, whereas during hypercalcemia there is a high percentage of parathyroid hormone fragments due to intraglandular cleavage. Finally, after prolonged hypocalcemia, the number of chief cells is considerably increased. Taken together, this evidence supports the idea that a low concentration of plasma calcium increases net synthesis of parathyroid hormone by stimulating both formation of new chief cells and hormone production within each cell. In contrast, a high concentration of calcium decreases the parathyroid hormone supply. The biosynthesis of parathyroid hormone is, however, poorly understood.

X. INTERACTIONS WITH OTHER HORMONES

A. Calcitriol

In addition to the important interactions of parathyroid hormone with calcitriol that have already been discussed, parathyroid hormone and calcitriol work together in a poorly understood manner to promote the normal growth, development, and maintenance of the skeleton, as well as to increase net outflow of calcium from bone. In vitamin D-deficient animals, a larger amount of parathyroid hormone is needed to produce its usual effect. Excessive treatment with calcitriol or related compounds can, like excessive parathyroid hormone, result in extensive bone demineralization with resulting bone fragility and soft tissue calcification.

B. Calcitonin

A second hormone that interacts with parathyroid hormone is calcitonin, produced in humans and other mammals in the parafollicular or C cells of the thyroid gland.

Adequate doses of calcitonin can antagonize the action of parathyroid hormone by counteracting the effect of parathyroid hormone on outflow of calcium from bone and by inhibiting resorption of bone by osteoclasts. Overall, the effect of calcitonin is to decrease the concentration of calcium in plasma.

Calcitonin is used to treat hypercalcemias due to excessive osteoclastic activity, hyperparathyroidism, vitamin D toxicity, myelomas, Paget's disease of bone, and bone malignancies. Furthermore, long-term treatment with calcitonin has been shown to be beneficial in osteoporosis. To avoid the inconvenience of frequent injections of calcitonin, preparations for administration of calcitonin intranasally have been developed and shown to be efficacious.

C. Glucocorticoids

In a variety of experimental animals (dogs, cats, rats, and mice), it has been found that the hypocalcemia that occurs after parathyroidectomy is greatly reduced by adrenalectomy. According to recent experiments in rats, this effect of adrenalectomy is due to removal of the source of supply of glucocorticoid, which, in rats, is corticosterone. When corticosterone was given to adrenalectomized, parathyroidectomized rats at physiological concentrations, the plasma calcium was reduced to the level after parathyroidectomy alone.

In contrast, glucocorticoids did not have a serum calcium-lowering effect in parathyroid-intact rats. Since substantial doses of parathyroid hormone did not reverse the serum calcium-lowering effect of glucocorticoid, other possible parathyroid-related factors that are responsible for the lack of the effect of glucocorticoid (and of adrenalectomy) in intact rats are being investigated.

XI. DEFICIENCY AND EXCESS OF PARATHYROID HORMONE

A. Hypoparathyroidism

Hypoparathyroidism may occur because of congenital absence or deficiency of the parathyroid glands (rare) or, more frequently, because of deliberate or inadvertent parathyroidectomy. Deliberate parathyroidectomy is performed to remove a parathyroid adenoma or parathyroid carcinoma or to relieve secondary hyperparathyroidism (such as may occur as the result of renal failure), but ordinarily not all of the parathyroid tissue is removed. Usually, the remaining gland(s) will increase its function and eventually provide sufficient hormone so that hypoparathyroidism, if it occurs, will be only temporary. Inadvertently, one or more parathyroid glands may be removed during surgical thyroidectomy, resulting in temporary or permanent hypoparathyroidism, depending on how much functional tissue has been left intact.

The immediate result of total parathyroidectomy is a fall in serum calcium, which, if it is severe and not treated adequately, may result in tetany and even death due to respiratory failure (neuromuscular inadequacy). A second immediate result is a decrease in urine inorganic phosphate and the hyperphosphatemia. These results are not as serious as the hypocalcemia and are not life-threatening. An additional effect that is slower in onset is a fall in the blood concentration of calcitriol (because of diminished production) and the eventual decrease in intestinal absorption of calcium, contributing further to the low level of calcium in the circulation. Over the longer term, the severity of hypoparathyroid hypocalcemia depends on the amount of parathyroid tissue left in the patient, the level of function of this tissue, and the efficacy of supportive treatment.

The immediate treatment of acute hypoparathyroidism is administration of calcium salts intravenously and orally, but this gives only temporary relief. For chronic treatment, parathyroid hormone administration would seem to be appropriate but is not used. Instead, vitamin D_2 or D_3 or a related compound (dihydrotachysterol), or calcitriol, is used. They are all active orally and effective. Initially, the dose is varied, with periodic monitoring of the plasma calcium, toward the objective of maintaining the plasma calcium in the low normal range. However, unexpected hypercalcemia may develop. When this occurs, treatment must be stopped immediately until the serum calcium falls to normal, which takes a variable amount of time, depending on which vitamin D preparation has been used. This time is longer (4–6 weeks) with vitamin D_2 and D_3 because they are slowly released from stores in fat depots. A synthetic compound, closely related in chemical structure to vitamin D, dihydrotachysterol, has a much shorter duration of action (about 2 weeks) and is therefore preferred. Calcitriol is even shorter in its duration of action (2–10 days).

Pseudohypoparathyroidism is the name given to rare types of hypoparathyroidism that are caused not by a deficient supply of parathyroid hormone but by end organ resistance (bone and kidney) to the action of parathyroid hormone. The major symptom is hypocalcemia with all its attendant problems. One result of the hypocalcemia is an increase in the production, secretion, and plasma concentration of parathyroid hormone, which, however, do not correct the hypocalcemia in these patients with end organ resistance to the hormone. Pseudohypoparathyroidism, like hormone-deficient hypoparathyroidism, is effectively treated with a vitamin D preparation.

B. Hyperparathyroidism

Primary hyperparathyroidism may occur because of hyperplasia of the chief cells or development of an adenoma (more rarely, parathyroid carcinoma). The most prominent symptom is hypercalcemia caused by supernormal effects of excessive parathyroid hormone on bone, kidney, and intestine. Important secondary effects may include osteopenia and ectopic calcification in the kidney, aorta, and lung. For reasons unknown, parathyroid hormone secretion by hyperplastic glandular tissue and adenomas is inadequately inhibited by the elevated plasma calcium concentration.

Secondary hyperparathyroidism occurs in association with advanced renal failure and renal osteodystrophy. The latter term includes all the defects in bone and calcium metabolism that appear with the decline in renal function. The increased production and secretion of parathyroid hormone in secondary hyperparathyroidism is assumed to be the result of hypocalcemia, but this has not been well documented. Eventually, during the progression of renal failure, the hypocalcemia is replaced by poorly suppressible hypercalcemia due to the hyperparathyroidism. Factors in renal failure that tend to encourage hypocalcemia are decreased production of calcitriol (with poor intestinal absorption of calcium), hyperphosphatemia, and skeletal resistance to parathyroid hormone.

Primary hyperparathyroidism is treated by surgical removal of the abnormal tissue, whether it is adenoma, carcinoma, or hyperplastic glands. In the latter case, there are two alternative preferred procedures:

1. subtotal parathyroidectomy (removal of three and one-half glands), or
2. removal of all four glands with an immediate autograft of one of the glands into an arm.

Hypocalcemia may be a problem in the immediate period after surgery until the remaining parathyroid tissue has been restored to adequate function and may require treatment similar to that used for hypoparathyroidism from other causes. Development of synthetic antagonists to parathyroid hormone is being attempted but none of them that have been synthesized up to this time is potent enough to be useful in the treatment of hyperparathyroidism.

Nonsurgical approaches to the treatment of secondary hyperparathyroidism, such as combined oral calcitriol and calcium carbonate, are now being used increasingly with considerable success. In cases that require surgery, the recommended procedures are the same as for primary hyperparathyroidism due to hyperplasia.

XII. PARATHYROID HYPERTENSIVE FACTOR

The recently discovered parathyroid hypertensive factor (PHF) has been implicated in some forms of hypertension, predominantly the salt-sensitive and low-renin forms. It was first detected in plasma of hypertensive rats and human hypertensive patients. It has a delayed hypertensive effect (45–90 min) after injection into normotensive rats.

The structure of PHF has not yet been determined but it appears to include a peptidic part that is essential for activity. The molecular weight is in the neighborhood of 3000. Antisera to PHF have shown no cross-reactivity with bPTH 1–84, bPTH 1–34, other shorter PTH fragments, and several other vasoactive substances.

The origin of PHF in the parathyroid gland has been clearly established by observations in parathyroidectomized and parathyroid transplant rats, by cross-transplants, and by its production from cultured parathyroid cells from spontaneously hypertensive rats (SHR) and not from normotensive rats. A novel parathyroid cell type has been identified in SHR but not in normotensive rats.

The mechanism of action of PHF involves an increase in calcium channel activity in vascular smooth muscle cells. The beneficial effects of PHF, if any, may be related to its modulation of intracellular calcium concentration in target cells of various tissues.

BIBLIOGRAPHY

Aurbach, G. D., Marx, S. J., and Spiegel, A. M. (1992). Parathyroid hormone, calcitonin, and the calciferols. *In* "Williams Textbook of Endocrinology" (J. D. Wilson and D. W. Foster, eds.), 8th Ed., pp. 1397–1476. Saunders, Philadelphia.

Aurbach, G. D., Marx, S. J., and Spiegel, A. M. (1992). Metabolic bone disease. *In* "Williams Textbook of Endocrinology" (J. D. Wilson and D. W. Foster, eds.), 8th Ed., pp. 1477–1517. Saunders, Philadelphia.

Bikle, D. D., and Negro-Vilar, A. (eds.) (1995). "Hormonal Regulation of Bone Mineral Metabolism." Endocrine Society, Bethesda, Maryland.

Bilezikian, J. P. (ed.) (1993). "The Parathyroids." Raven, New York.

Hirsch, P. F. (1995). Calcitonin. *In* "Principles of Pharmacology" (P. L. Munson, R.A. Mueller, and G. R. Breese, eds.), pp. 737–748. Chapman & Hall, New York.

Munson, P. L. (1988). Parathyroid hormone and calcitonin. *In* "Endocrinology: People and Ideas" (S. M. McCann, ed.), pp. 239–284. American Physiological Society, Bethesda, Maryland.

Munson, P. L., Mueller, R. A., and Breese, G. R. (eds.) (1995). "Principles of Pharmacology." Chapman & Hall, New York.

Nishino, K., Hirsch, P. F., Mahgoub, A., and Munson, P. L. (1991). Hypocalcemic effect of physiological concentrations of corticosterone in adrenalectomized-parathyroidectomized rats. *Endocrinology* **128**, 2259–2265.

Pang, P. K. T., *et al.* (1994). Parathyroid hypertensive factor: A new circulating substance in essential hypertension. Proceedings of a symposium. *J. Cardiovasc. Pharmacol.* **23**(suppl. 2), s1–s49.

Potts, J. T., Jr., *et al.* (1995). Parathyroids. *In* "Endocrinology" (L. J. DeGroot, ed.), 2nd Ed., pp. 917–1273. Saunders, Philadelphia.

Schultz, V. L., Boass, A., Garner, S. C., and Toverud, S. U. (1995). Several anesthetics, but not diethyl ether, cause marked elevation of serum parathyroid hormone concentration in rats. *J. Bone Mineral Res.* **10**, 1298–1302.

Parkinson's Disease

EDITH G. McGEER
University of British Columbia

GLOSSARY

Blood/brain barrier Membranous "wall" between blood vessel and brain tissue that limits the entry of materials into the brain and thus acts as a protective mechanism

Dementia Loss of mental faculties, particularly memory

En cascade Neuronal changes occurring subsequently to, and as a consequence of, the degeneration of other neurons that connect with them

Extrapyramidal system Group of subcortical nuclei, also called the basal ganglia, having to do with movement control

Free radicals Unstable oxidation products

Inclusion bodies Variety of abnormal swellings visible by histological techniques in degenerating neurons

Neurotransmitter One of a group of specific chemicals used by the nervous system to carry messages from one neuron to another

Neurotrophic factors Materials contributing to the development and/or maintenance of neurons; the best-known example is nerve growth factor

Peptide Compound made up of a chain (polymer) of amino acids

Peroxides Unstable, highly oxidized compounds

Substantia nigra Major nucleus in the extrapyramidal system

Superoxides Unstable, highly oxidized compounds

PARKINSON'S DISEASE IS ONE OF THE MOST COMmon movement disorders, affecting an estimated 2.5% of people over the age of 65, but rarely occurring before 40. It seems to be more common among men than among women. It is an extrapyramidal disorder, with the principal pathology being within the basal ganglia, one of the three systems above the level of the spinal cord that act to control movement. Pathology in these systems, as well as in the spinal cord or muscles, leads to a variety of diseases with motor symptoms (Fig. 1). The chief symptoms of Parkinson's disease are rigidity, tremor, and difficulty in initiating movement. Ameliorative treatments have been found, based on the major identified neurochemical deficit, but they do not halt the progressive degeneration. Transplantation has been tried, but so far benefit has been seen in relatively few cases. The rate of progression can vary greatly from case to case or from time to time in a single case, but the life expectancy in treated parkinsonism is now near normal. The basic cause is unknown, although some cases appear to follow a virus infection, and parkinsonian-like symptoms can be produced in humans or monkeys following exposure to certain toxins. Oxidative stress or loss of neurotrophic factors are other hypothesized causes, and much basic, and some clinical, research is now directed at those possibilities. Parkinson's disease is not necessarily accompanied by any mental change, although about 66% of cases show depression and the tendency to develop dementia is much greater than that in the age-matched spouses of parkinsonian patients. The occurrence of a familial parkinsonian-dementia complex among the native population on the island of Guam also suggests that the types of neurons whose losses lead to these two distinct types of symptoms may be vulnerable to similar degenerative processes.

I. GENERAL DESCRIPTION

The classic triad of symptoms are rigidity, tremor, and flexion. The rigidity results in a general absence

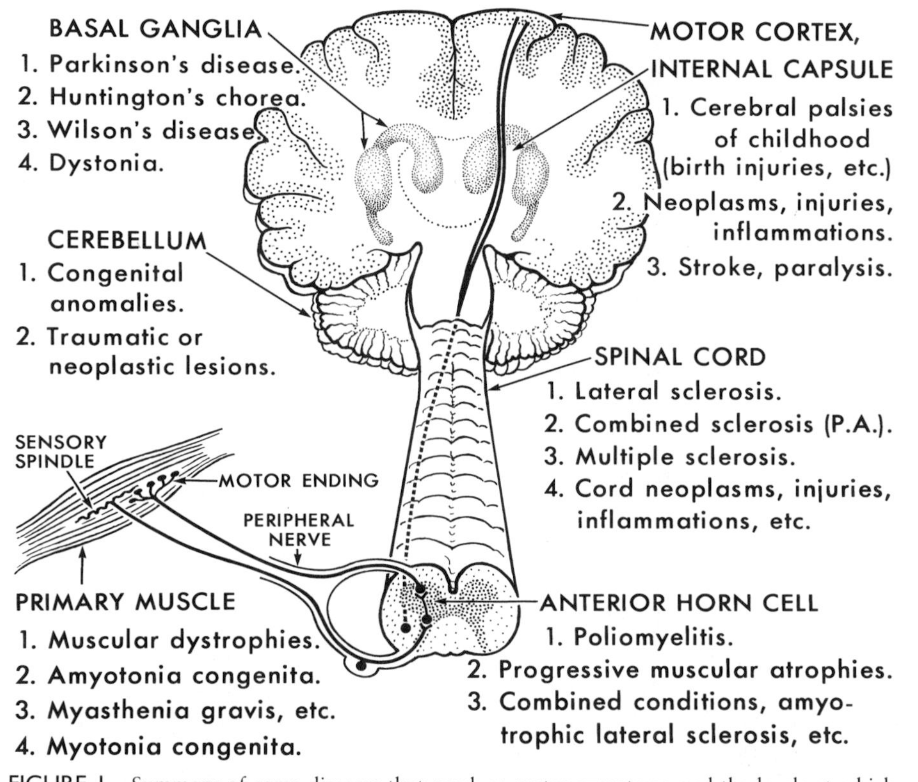

FIGURE 1 Summary of some diseases that produce motor symptoms and the levels at which they attack the nervous system.

of motor activity and involves all the voluntary muscles. The unnatural immobility of facial muscles leads to the typical parkinsonian "mask." The tremor affects primarily the fingers and hands, giving rise to a cigarette-rolling movement. It is present at rest but disappears during movement. The flexion is of the whole body. The head is flexed on the chest, the body is bowed, and the arms, wrist, and knees are bent. An advanced example of the untreated disease can be diagnosed at a glance, but is rarely seen now with the advent of drugs that commonly relieve all but the final stages. A person with Parkinson's disease has problems in initiating or changing movements. Getting up from a chair may be a major struggle. They tend to walk with short, rapid steps and, once started, have difficulty in stopping or changing direction.

II. CENTRAL NERVOUS SYSTEM PATHOLOGY

A. Histological Pathology

The most important and consistent pathological change seen in the brain in Parkinson's disease is loss of the large, pigmented neurons of the substantia nigra (Fig. 2). There are lesser losses of neurons in many other regions of brain in most cases of Parkinson's disease. However, both the results of drug treatment (Section V) and the appearance of parkinsonian symptoms in animals whose only neuronal loss is in the substantia nigra (Sections III and IV) have led to the supposition that the loss of these pigmented neurons is the crucial pathological factor. It may be relevant that these pigmented cells are gradually lost during normal human aging (see Fig. 4A); this loss helps to explain the difficulties of movement seen in aged individuals.

A common feature of the histology of Parkinson's disease is the appearance of inclusions called Lewy bodies in the substantia nigral neurons. These do, however, appear in various types of degenerating neurons in other neurological diseases.

B. Chemical Pathology

Substantia nigra means black body and this name was given to the structure because, unlike most of the gray matter of the brain, the substantia nigra of an adult

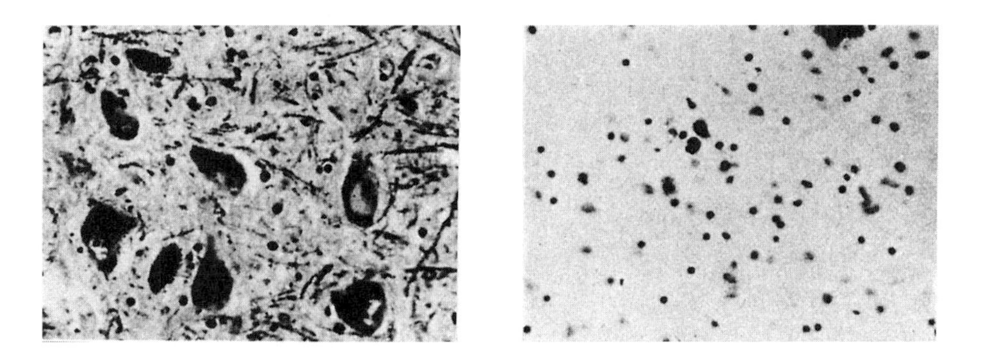

FIGURE 2 Histological changes in the substantia nigra in Parkinson's disease. Note the dramatic loss of the large pigmented neurons.

human appears black on gross inspection. The reason is that the large neurons in this nucleus make a neurotransmitter called dopamine, which can be polymerized to melanin, the pigment that is also responsible for the dark coloration of suntanned or black skin. Thus, the major chemical defect in Parkinson's disease is believed to be in the dopamine systems. Dopamine concentrations are significantly decreased all over the brain, but the largest losses are in the caudate and putamen, two other large nuclei of the basal ganglia and the major areas served by the substantia nigra neurons that use dopamine as a neurotransmitter. There is no visible neuronal degeneration in these target areas.

Changes in many other neurotransmitter systems have been reported from postmortem studies on parkinsonian brains, particularly in the basal ganglia. Some of these may be functional and secondary to changes in the dopamine system or represent degeneration "en cascade." In any case, the relatively minor decreases in the concentrations of these other neurotransmitters does not seem to be of primary importance to the disease.

III. THEORIES OF DISEASE CAUSATION

A minor proportion (15% or less) of parkinsonian cases appear to be familial, suggesting that a genetic defect may play some role in the etiology. But the general lack of concordance for Parkinson's disease in identical twins indicates that genetics is not the primary factor in most cases. Parkinson's disease can occur after, and as a result of, an infection of the brain; such "postencephalitic" cases are becoming rarer owing to better methods of rapid and effective treatment of the infections. The cause of the majority of cases (so-called idiopathic parkinsonism) is unknown and may be multiple.

The fundamental question in this and other neurodegenerative disorders is "Why do these particular neurons die?" Various theories have been advanced, including viruses or the action of specific neurotoxins. What the toxin or toxins may be remains unknown. It is known that victims of sublethal carbon monoxide poisoning or workers in manganese mines often develop a movement disorder. The latter is now believed to be closer to dystonia than to parkinsonism. Great excitement has been generated by the finding that there is a selective destruction of dopaminergic neurons in the substantia nigra in humans and primates treated systemically with *N*-methyl-4-phenyl-1,2,3,6-tetrahydropyridine (MPTP). The neurotoxic potential of this compound was discovered accidentally and tragically when a number of cases of advanced parkinsonism began appearing, about 1980, among Californians in their late teens or early twenties; it was established that a common factor was their use of a street drug that contained MPTP as a minor impurity.

The mechanism by which the destruction of dopamine neurons occurs is not yet completely known but it involves an oxidation product of MPTP, formed in the brain by the action of an enzyme called monoamine oxidase B (MAO-B), with the oxidation product (MPP^+, Fig. 3) being taken up into dopamine neurons by the specific dopamine uptake mechanisms that are normally active in recapturing the neurotrans-

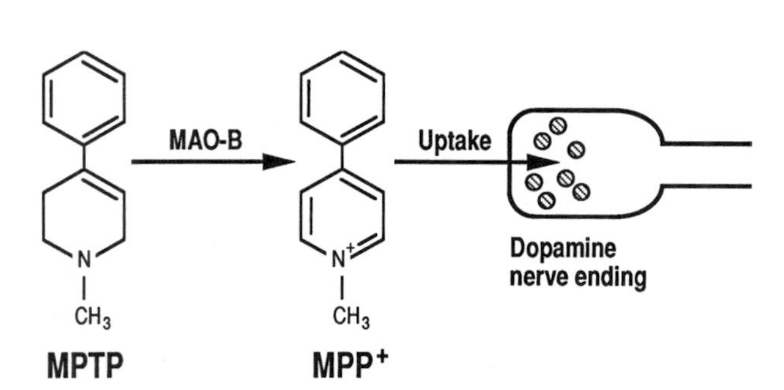

FIGURE 3 Conversion of MPTP to MPP^+ and selective uptake of the latter by dopamine nerve endings.

mitter released at the synapse. Inhibitors of MAO-B or of the specific uptake process block the toxicity of MPTP. Some believe an environmental toxin similar to MPTP may be responsible for idiopathic parkinsonism. No firm evidence exists, however. Perhaps the most significant finding so far from the MPTP work is that all the classic signs of parkinsonism can occur in humans or monkeys in which the only neurotransmitter affected is dopamine.

One general possibility lies in the vulnerability of these neurons to superoxides, peroxides, or free radical compounds. A related suggestion is that the melanin that accumulates in such neurons in humans eventually poisons them and that a shortage of antioxidants may favor melanin formation. When free melanin occurs, it tends to bind iron and it has been argued that this iron–melanin complex can catalyze the formation of further free radicals and thus continue the destructive process.

Another general hypothesis suggests that the problem may be loss or genetic lack of a neurotrophic factor necessary for survival of the dopamine neurons. A number of such factors have been shown to protect dopamine neurons in culture and at least one, basic fibroblast growth factor, appears to be lost early in the parkinsonian process (Fig. 4B). Such a loss does not occur during normal aging (Fig. 4B).

Some have argued that idiopathic parkinsonism may result from an early insult (infectious or toxic) that kills some of the dopamine neurons but that symptoms do not occur until the normal age-related attrition takes the number below a "parkinsonian threshold" (Fig. 4A). A technique to test this hypothesis is now available in positron emission tomography (PET), which allows study of brain chemistry in humans during life. PET studies with a chemical called 6-^{18}F-fluorodopa can give a picture of the concentration of dopamine nerve endings in the caudate and putamen. Clear decreases can be seen in parkinsonians and even in some MPTP users who do not as yet show clinical symptoms (Fig. 5). Sequential PET studies of such individuals should show whether there is a progressive loss in aging with eventual development of the disease.

IV. ANIMAL MODELS

Primates treated with MPTP undoubtedly yield the animal model that duplicates human parkinsonism most exactly. Treatment of mice, rats, or other small laboratory animals with MPTP is being widely explored, but the effects are generally transient so that an accurate model of dopamine cell loss is not obtained. Permanent loss of dopamine neurons is seen in monkeys after MPTP treatment and such animals are an excellent model in which to study possible treatments for parkinsonism. However, work with such primates is limited.

Another approach, which has been used for many years in exploring possible therapies for parkinsonism, is the injection, under anesthesia, of a trace amount of a chemical called 6-hydroxydopamine into rat brain. This close structural relative of dopamine is easily taken up by dopamine neurons and is oxidized to a toxic derivative that selectively destroys them. One good example of the use to which this model was put was the demonstration that transplantation of dopamine-producing cells into the brain might be useful in alleviating parkinsonism.

If 6-hydroxydopamine is injected into only one side of the brain (unilaterally) of adult rats, and the rats are challenged with the dopamine-releasing drug amphetamine, they will rotate because of the unequal release of dopamine in the undamaged and lesioned brain hemispheres. This rotation can be quantitated. The transplantation of substantia nigra cells from fetal rats into such lesioned adults can restore the movement to near normal (Fig. 6A). The transplant is put into the brain just above the caudate (Fig. 6B), so clearly the dopaminergic cells are not involved in the extensive normal circuitry, and yet just getting the dopamine to the caudate putamen seems to help the symptoms. Transplants into the brain seem much less subject to rejection than transplants into the periphery—a fact that has been used to argue that the brain is "immunologically privileged." The "privilege" is a limited one, however, and only a fraction of the transplanted cells generally survive. Such transplants have also been reported to "normalize" the behavior of MPTP-treated monkeys and these experiments

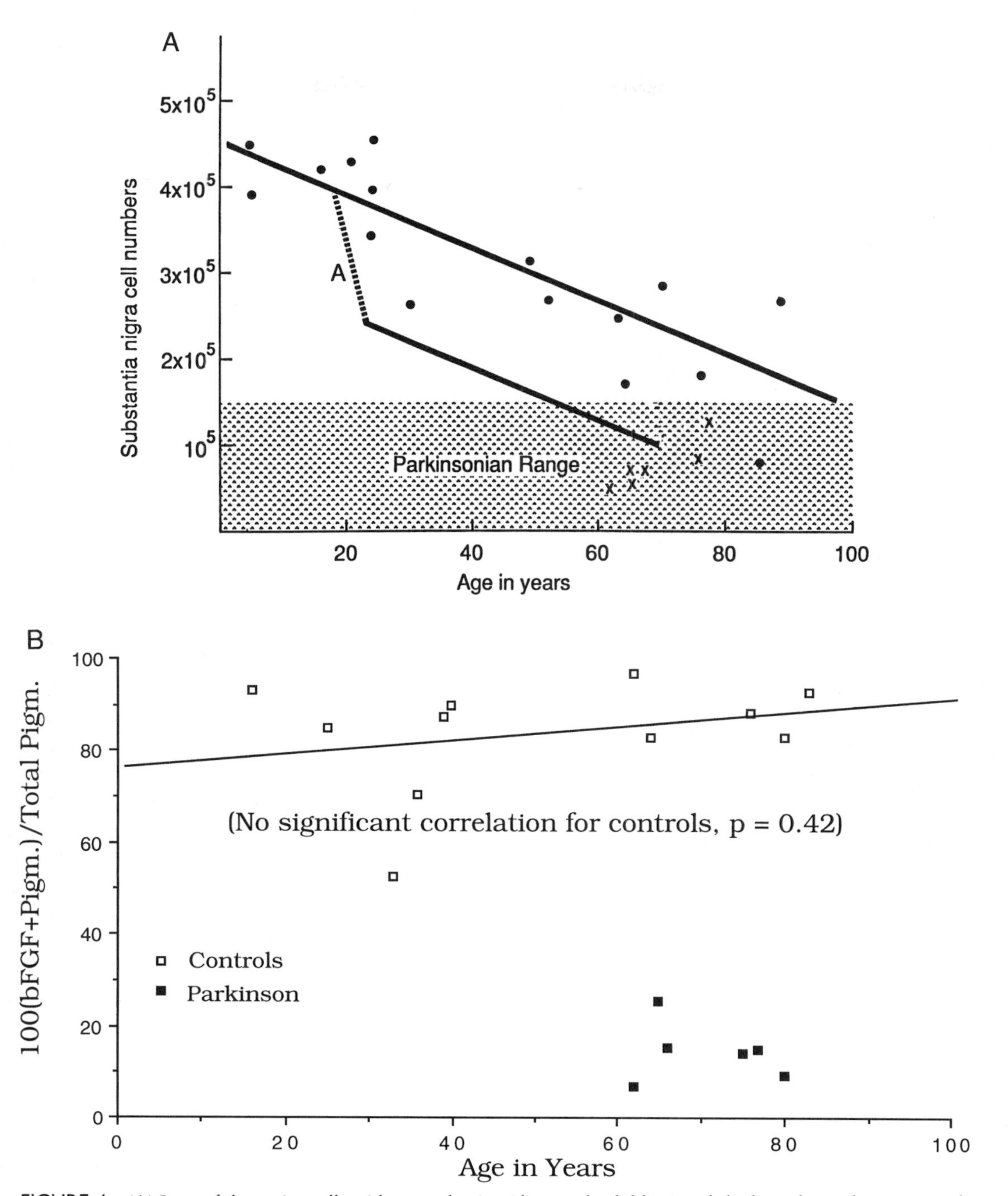

FIGURE 4 (A) Loss of dopamine cells with normal aging (dots, and solid line) and the hypothesized remote insult plus normal aging (line A) leading to numbers of cells similar to those found in parkinsonian brains (crosses). (B) Percentage of remaining dopamine cells that contain basic fibroblast growth factor. This percentage is high and unchanged with age in normal individuals but is very low in Parkinson's disease, indicating that loss of this neurotrophic factor occurs early in the pathology.

have formed the basis for transplant work in human parkinsonian patients (see Section V).

Most of the drugs used in the treatment of schizophrenia either block the action of dopamine in brain or deplete its levels. Such drugs can produce parkinsonian symptoms in both humans and animals but the symptomatology is generally reversed on removal of the drug. There is no permanent damage to the dopa-

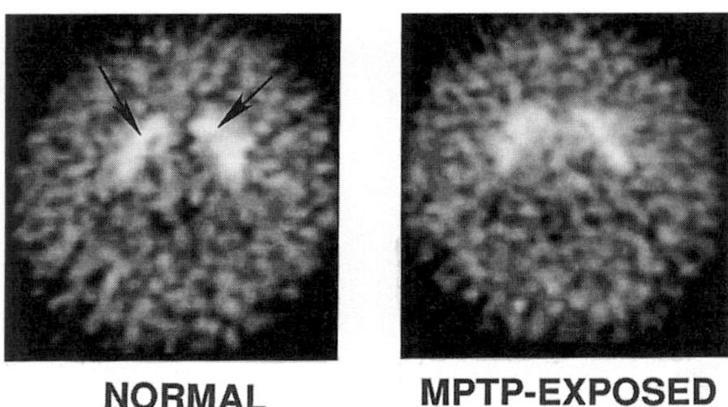

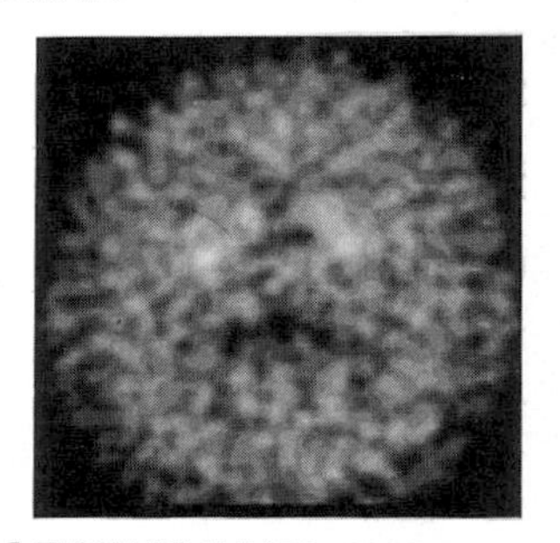

FIGURE 5 PET pictures after 6-^{18}F-fluorodopa administration to a normal control, an MPTP-exposed, asymptomatic individual, and a Parkinson's disease patient. The light areas indicated by arrows in the normal correspond to the caudate putamen; the lighter these regions, the greater the number of dopamine nerve endings. There is clear loss in the parkinsonian individual, with a lesser but significant loss in the MPTP-exposed individual. (Photograph courtesy of Dr. D. B. Calne and the UBC/TRIUMF PET program.)

mine system and thus, for this and other reasons, animals treated with such drugs are not a good model of human parkinsonism.

V. TREATMENT

The treatment of choice for many years in Parkinson's disease has been L-dopa (also called levadopa). Dopa is the normal precursor of dopamine and is effective because it can cross the blood/brain barrier—which dopamine cannot. Dopa is readily converted in the brain into dopamine by an enzyme called dopa decarboxylase, which normally occurs in excess and is found in both dopamine neurons and other brain cells. Hence, even when many of the dopamine neurons have been lost, there is generally enough dopa decarboxylase in the brain to effect this conversion. If dopa is given by itself, much of it is metabolized in the body before it reaches the brain. Hence, it is usually given together with an inhibitor of dopa decarboxylase so chosen that the inhibitor cannot cross the blood/brain barrier but will block peripheral decarboxylation. [*See* Parkinson's Disease, Biological Basis and Therapy.]

A major problem in the treatment of parkinsonism with dopa is the occurrence of an on/off phenomena, that is, that the dopa may suddenly and transiently lose its effectiveness for no apparent reason. The mechanism is still unknown, although changes in receptor sensitivity or in other neurons in the extrapyramidal system have been suggested.

Another treatment approach is the use of synthetic compounds that are direct dopamine agonists, that is, that act like dopamine at dopamine receptors. Several types of dopamine receptors are known, but the most common seem to be those called D1 and D2. Action of dopamine at these receptors results in two different types of changes in the chemistry of the membranes and hence different modulation of the excitability of the neurons that carry these receptors. There are far more D1 than D2 sites but, so far, it appears that only drugs that act primarily as D2 agonists are effective in parkinsonism. A well-known example is bromocriptine.

A somewhat different approach is the use of the MAO inhibitor deprenyl. Since dopamine is also metabolized by MAO-B, deprenyl might have some beneficial effect by increasing the life of dopamine in the brain. Such an effect would probably be minor, however, because there are other brain enzymes that also destroy dopamine. The major reason for trying deprenyl is the hypothesis that dopamine cell death in Parkinson's disease may be due to some chemical that, like MPTP, requires oxidation by MAO-B before it becomes toxic. On this hypothesis, deprenyl would not be expected to reverse established symptoms but rather might slow the progression of the disease. Such slowing of a disease by a drug is much harder to establish than alleviation of symptoms, and reports on the efficacy of deprenyl are controversial.

Another, more debatable approach to treatment of human parkinsonism is transplants, such as have been shown effective in animal models (see Section IV). A major problem is the source of the tissue. Most centers have used part of the patient's own adrenal medulla. This gland, which lies near the kidneys, normally makes dopamine only as a precursor to noradrenaline and adrenaline and is therefore probably not the ideal tissue to use. Only limited success has been reported. The use of human aborted

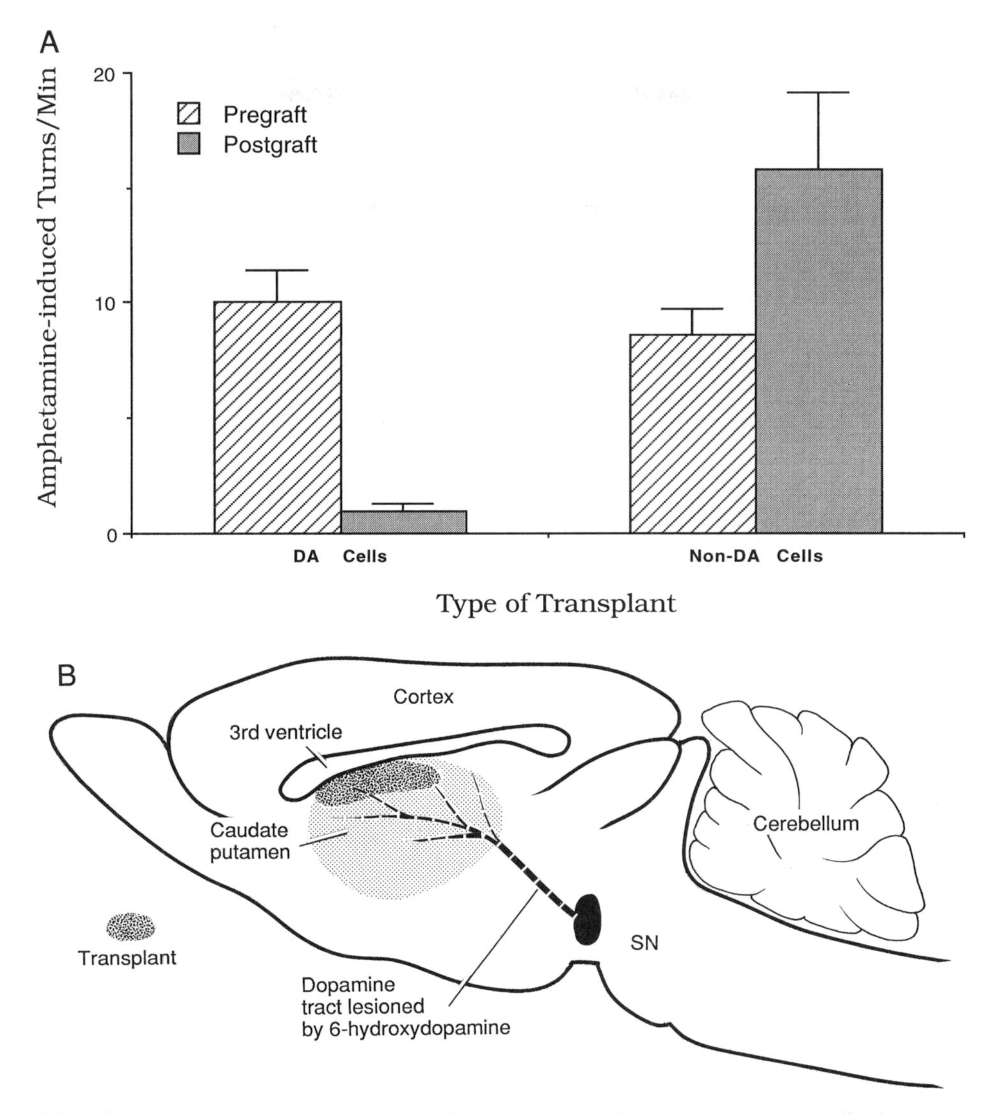

FIGURE 6 (A) Bar graph showing typical effect of transplanted fetal substantia nigra cells (dopamine cells) compared with transplanted non-DA cells (sciatic nerve) on rotation in rats with unilateral 6-hydroxydopamine lesions. Rotation was studied immediately before and 4 months after the transplants. (B) Sagittal section of rat brain showing general position of the dopamine projection from the substantia nigra to the caudate putamen, as well as the position of the transplanted cells in experiments such as used in (A).

fetal tissue may have technical advantages but raises serious ethical questions. Some evidence suggests that what benefit is seen after some transplants comes from neurotrophic stimulation of the patient's own dopamine neurons rather than survival of such neurons in the transplant. An approach that offers some promise for the future in this and other neurodegenerative diseases is to develop tissue culture cell lines expressing the enzyme(s) necessary for synthesis of the desired chemical. In theory, such cell lines can be banked and grown to give an unlimited supply of reproducible cells that could be treated before implantation with maturation factors that would cause their differentiation into neuron-like cells incapable of cancerous multiplication. One difficulty may be that experiments in animals suggest that the percentage of such cells that survive after transplantation is much less than the percentage of surviving cells in a transplant of fetal tissue. However, considerable research along these lines is being done and may provide solutions to this problem.

The possibility of using neurotrophic factors to protect and stimulate the patient's own dopamine cells is also being actively explored. The problem here is that most such factors are peptides that are rapidly metabolized in the peripheral body and do not easily cross the blood/brain barrier. The active research along these lines suggests, however, that some treatment that can halt the progression of the disease rather than simply alleviate the symptoms will eventually be found.

BIBLIOGRAPHY

Fisher, L. J., and Gage, F. H. (1994). Intracerebral transplantation—Basic and clinical applications to the neostriatum. *FASEB J.* **8**, 489–496.

Gerlach, M., Ben-Shachar, D., Riederer, P., and Youdim, M. B. (1994). Altered brain metabolism of iron as a cause of neurodegenerative diseases? *J. Neurochem.* **63**, 793–807.

Jenner, P. (1994). Oxidative damage in neurodegenerative disease. *Lancet* **344**, 796–798.

Marsden, C. D. (1994). Parkinson's disease. *J. Neurol. Neurosurg. Psychiatry* **57**, 672–681.

Youdim, M. B. H., and Riederer, P. (1997). Understanding Parkinson's disease. *Scientific American* **276**(1), 52–59.

Parkinson's Disease, Biological Basis and Therapy

TIMOTHY J. COLLIER
Rush Presbyterian-St. Luke's Medical Center

JOHN R. SLADEK, JR.
Finch University of the Health Sciences, The Chicago Medical School

GLOSSARY

Blood–brain barrier Anatomical arrangement in which modifications of cells associated with the walls of blood vessels and within the brain function to keep some components of the circulating blood separate from the tissue fluid surrounding brain cells

Neuron Basic functional unit of the nervous system; a cell specialized to transmit information via electrical impulses and chemical signals

Neurotransmitter Chemical substance released from nerve endings to transmit information from one cell to another, for example, dopamine and glutamate

Oxygen radical Form of unstable molecule that can react with biological membranes causing damage

Receptor antagonist A receptor is a region on the surface of a nerve cell specialized for responding to neurotransmitters and participating in information transmission. A receptor antagonist is a molecule or drug that prevents the functional interaction between neurotransmitter and receptor, blocking transmission of information

Striatum Brain region involved in modulation of body movement

PARKINSON'S DISEASE IS A NEURODEGENERATIVE syndrome affecting portions of the brain involved in expression of body movements. Symptoms of the disease include a slowing and decrease in movement, tremor in the limbs, and disturbance of posture and balance. The behavioral signs have been linked to death of nerve cells in a brain region called the substantia nigra, which utilizes the chemical messenger dopamine to transmit information to a target brain region known as the striatum. This neural system is referred to as the nigrostriatal dopamine pathway (Fig. 1). It is the progressive loss of dopamine in the striatum that leads to worsening behavioral signs. Though use of the drugs levodopa and carbidopa can result in dramatic symptomatic relief during the early stages of the disease, these drugs begin to produce severe unwanted side effects with prolonged use, and lose efficacy late in the disease process. A variety of new therapies are being tested to delay nerve cell degeneration, replace lost dopamine cells, and restore the chemical balance of the brain circuitry that generates body movements.

I. THE BIOLOGY OF PARKINSON'S DISEASE AND LEVODOPA THERAPY

The behavioral signs of Parkinson's disease were first described by James Parkinson in "An Essay on the

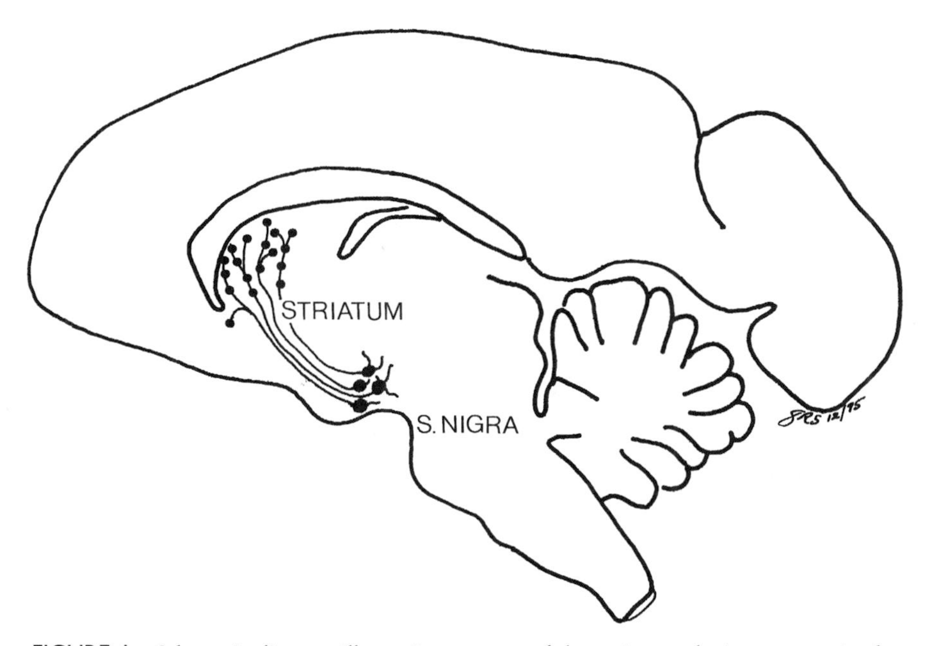

FIGURE 1 Schematic diagram illustrating a group of dopamine-producing neurons in the substantia nigra (S. Nigra) in the primate brain that project their axons into an area of the brain called the striatum, which helps regulate body movement. This so-called "nigrostriatal pathway" loses neurons in Parkinson's disease, with a resultant depletion in the amount of the neurotransmitter dopamine that is available to influence the regulation of motor activity by the striatum.

Shaking Palsy" published in 1817. The disease involves degeneration of brain cells that modulate body movement and is characterized by four clinical signs: bradykinesia (slowing and poverty of movement), resting tremor, cogwheel rigidity (ratchet-like movement when passively moving a limb), and postural instability. The disease generally has an onset after age 40, increases in frequency with advancing age, and has a protracted time course of 10 years or more. It was not until 1960, when H. Ehringer and O. Hornykiewicz published their findings demonstrating a severe depletion of the neurotransmitter dopamine in the striatum of the brains of Parkinson's disease patients, that we began to understand the neurochemical basis for this disorder. Over the next 5 years, experiments in rats and monkeys that were designed to damage the cells of the substantia nigra (a group of neurons in the ventral midbrain) implicated these brain cells as the source of striatal dopamine and yielded the concept that pathology of the nigrostriatal dopamine system was the causal factor for Parkinson's disease. [*See* Parkinson's Disease.]

Until 1961, there was no reliable therapy for the symptoms of Parkinson's disease. With the revelation that parkinsonian symptoms were a product of nigrostriatal system dopamine depletion, it was postulated that Parkinson's disease patients might benefit from administration of levodopa, the naturally occurring amino acid precursor of dopamine that is derived from beans, including the fava bean, and known to cross the blood–brain barrier. The actual transmitter, dopamine, was not considered a candidate for administration since it does not cross the blood–brain barrier. The conversion of levodopa to dopamine in the brain, particularly in the striatum, would increase levels of dopamine in this brain region that is critical for the control of body movement and ameliorate the symptoms of Parkinson's disease. For the next 6 years both oral and intravenous administration of low doses of levodopa (usually less than 1/2 g per day) provided sporadic, brief symptomatic relief for some Parkinson's disease patients, but enthusiasm for levodopa therapy among physicians was waning. Then in 1967, G. C. Cotzias and colleagues reported that high doses of orally administered levodopa (3–16 g per day) produced pronounced improvement in the symptoms of Parkinson's disease. Soon after, it was established that coadministration of carbidopa, a drug that selectively prevents the conversion of levodopa to dopamine in peripheral tissues, but not in brain, allowed significant decreases in the oral dose of levodopa administered and prevented unwanted peripheral side effects such as changes in blood pressure. This combination of levodopa and carbidopa comprises the principal current treatment of Parkinson's disease. Although the combination revolutionized treatment, these drugs did not represent a cure, and with long-term use new disease-related side effects emerged. Levodopa/carbidopa therapy also caused on–off periods: daily, alternating periods during which the drugs were maximally effective and patients enjoyed relatively normal movement ("on" periods) and periods of waning drug effect during which patients were immobile ("off" periods). The increasing doses of levodopa required as the disease progressed yielded periods of overactivity of the dopamine system marked by unwanted extra bursts of involuntary movement, termed dyskinesias. Finding a drug dose that is effective on parkinsonian symptoms but does not produce unwanted side effects becomes increasingly problematic as the disease progresses. Eventually, late in the disease process, any significant therapeutic effect of levodopa may be lost. Thus, the search for other therapies remains a priority. [*See* Blood-Brain Barrier.]

II. AN EXPERIMENTAL MODEL FOR PARKINSON'S DISEASE

In 1977, a 23-year-old chemistry graduate student was referred to the National Institutes of Mental Health for evaluation. He arrived mute, unable to move, and unable to swallow. His symptoms were ameliorated by administration of levodopa. During the next 5 years, several more young adults were referred for treatment to Stanford University, where they too were diagnosed with the signs of Parkinson's disease. It was a mystery why these young individuals appeared to have the symptoms of a disease generally associated with old age. Further investigation revealed that all of these young people were drug abusers, and all had taken forms of synthetic narcotics that were found to be contaminated with a by-product of their synthesis: the molecule MPTP (1-methyl-4-phenyl-1,2,3,6-tetrahydropyridine). Subsequently, the first patient died of a drug overdose, and autopsy revealed that he had severe degeneration of substantia nigra dopamine neurons. This was consistent with his behavioral symptoms and mimicked the pathology present in Parkinson's disease. The discovery of the unique selective toxicity of MPTP for the nigrostriatal dopamine system represented a tragedy for those afflicted, yet it offered an opportunity for scientists to explore the use of MPTP to create an animal model of Parkinson's disease that could yield insights into the causes of the disease and an important testing ground for new therapies.

Initial attempts to create parkinsonian animals by administration of MPTP to rats, cats, and guinea pigs produced no effects. However, when MPTP was administered to monkeys, it was found that these animals were as sensitive to the drug as were humans. They rapidly developed a parkinsonian syndrome that mimicked the disease in its behavioral features, neuronal cell loss, depletion of the neurotransmitter dopamine, and responsiveness to the standard therapy of levodopa/carbidopa administration. The excitement generated by these findings was accentuated by the fact that all of these features were observed in a species closely related to humans. It is through study of these MPTP-treated parkinsonian monkeys that a variety of recent insights into potential risk factors and causes of Parkinson's disease have been identified, and new therapies have been brought to clinical trials.

III. DELAYING DISABILITY: MONOAMINE OXIDASE TYPE B INHIBITORS AND ANTIOXIDANTS

Early studies of the metabolism of MPTP indicated that this molecule was a preferred substrate of the enzyme monoamine oxidase type B (MAO-B) with an affinity equal to that of its natural substrate dopamine. MAO-B acted on MPTP to generate the metabolite MPP^+. The conversion of MPTP to MPP^+, and the accumulation of MPP^+ in substantia nigra dopamine neurons, led to dopamine neuron death. A next logical step was to determine whether blockade of the action of MAO-B would prevent the toxicity of MPTP. Two groups of researchers independently demonstrated that giving monkeys a relatively nonspecific MAO inhibitor drug (pargyline) prior to MPTP treatment prevented the development of parkinsonian behavior, dopamine depletion, and substantia nigra cell loss normally produced by MPTP administration. One of these studies additionally showed that the specific MAO-B inhibitor drug deprenyl also prevented MPTP toxicity. Although it was not clear whether the chemical reaction that lent MPTP its dopamine cell toxicity was related to the events that produce Parkinson's disease, it suggested that if an endogenous toxin or environmental toxin contributed to Parkinson's disease, it might operate via some similar chemical mechanism as MPTP. Other lines of evidence were beginning to suggest that interfering with the action of MAO-B might be beneficial for Parkinson's disease patients. First, MAO-B normally was active in breaking down dopamine in the brain. Therefore, slowing this process might allow dopamine to exist for longer periods of time in the brain and provide for greater activity from the remaining dopamine available to the Parkinson's disease patient. Second, products of dopamine breakdown produced by MAO-B include hydrogen peroxide and reactive oxygen radicals that in high enough concentration are toxic to dopamine neurons. Taken together, all of this emerging information suggested that drugs that diminish the activity of MAO-B and inactivate oxygen radicals may be useful as therapeutic agents for Parkinson's disease.

Based on this rationale, a clinical trial was begun in 1987 to test the effects of the selective MAO-B inhibitor deprenyl and the antioxidant substance tocopherol (vitamin E) on the progression of behavioral symptoms of early Parkinson's disease. The so-called DATATOP (Deprenyl and Tocopherol Antioxidative Therapy of Parkinsonism) trial studied 800 patients in multiple research hospitals in North America. Patients were all in the early stages of Parkinson's disease and received one of four treatments: placebo (no drug), placebo + active deprenyl, placebo + active tocopherol, or both active drugs. The primary end point was the length of time to the onset of disability prompting the clinical decision to begin administration of levodopa. After approximately 2 years of study, these investigators concluded that deprenyl, but not tocopherol, had a significant impact on the progression of the symptoms of Parkinson's disease. This effect consisted of a delay in the onset of disability by about 9 months and a slight but sustained improvement in the behavioral rating of symptoms; furthermore, it was tolerated well by patients, with minimal negative side effects. It remains unclear whether deprenyl acts via short-term improvement of behavioral symptoms (a symptomatic effect similar to that of levodopa), by slowing degeneration of dopamine neurons (a protective effect), or both. The study group recommended that deprenyl be considered as a therapeutic option for initial treatment of Parkinson's disease.

IV. RESTORING BALANCE TO THE PARKINSONIAN BRAIN: SURGICAL PALLIDOTOMY AND GLUTAMATE RECEPTOR ANTAGONISTS

Studies of MPTP-treated parkinsonian monkeys have contributed greatly to our present understanding of how dopamine loss in the nigrostriatal system influences the activity of neurons in regions connected to the striatum. Indeed, recordings of the electrical activity of neurons and measurement of the metabolic activity of brain regions in MPTP-treated monkeys indicate that the behavioral symptoms of parkinsonism are in part due to overactivity in particular brain regions as a consequence of dopamine cell death. Thus, dopamine loss, due to either MPTP or Parkinson's disease, creates an imbalance in brain circuits subserving body movement, and only by restoring balance to the system can the symptoms diminish. Levodopa therapy and deprenyl attempt to replace dopamine and slow dopamine loss, respectively, in pursuit of restoring balance. However, another approach is to intervene at the level of brain cells made overactive by dopamine loss and restore balance by decreasing this overactivity. One brain region that becomes overactive in MPTP-treated monkeys is the subthalamic nucleus, which in turn overexcites a secondary region, the globus pallidus. In MPTP-treated monkeys, surgical damage to the subthalamic nucleus improves symptoms of parkinsonism. However, surgical ablation of this region in humans is extremely difficult without creating serious unwanted side effects. Alternatively, surgical damage to the globus pallidus is safer and could achieve the same effect. Indeed, surgical pallidotomy currently is receiving a great deal of attention. In its most careful form, the operation employs recordings of the electrical activity of brain cells in the globus pallidus as they respond to movements of the patient's limbs, which allows localization of the site to be lesioned surgically. Following surgery, an average improvement of approximately 50% in the rating of behavioral symptoms has been reported, with this benefit being stable for up to 16 months. Improvement is more pronounced on the side of the body opposite to the surgical damage. Although questions remain as to the long-term benefit of the surgery, and risk always accompanies surgical procedures, this approach represents one way to restore balance to the brain circuits disorganized by Parkinson's disease.

The same rationale of balancing the activity of circuits has generated clinical trials that utilize drugs to silence the overactivity driven by the subthalamic nucleus. The overactive brain circuits utilize the excitatory chemical transmitter glutamate, and recent studies in MPTP-treated monkeys have administered molecules that block the activity of glutamate in an attempt to improve symptoms of parkinsonism. The experiments demonstrate that these glutamate receptor antagonist drugs can be effective when given alone, but are particularly potent in combination with levodopa. As with any drug, side effects are a concern, particularly since glutamate also functions in many of the brain circuits required for learning and memory. However, to date, side effects have not been significant in early testing. As more information is collected on the use of these drugs in patients with Parkinson's disease, their use in combination with levodopa may comprise an attractive, improved drug therapy for restoring chemical balance in the parkinsonian brain.

V. SURGICAL REPLACEMENT OF DOPAMINE NEURONS: NEURAL GRAFTING OF EMBRYONIC TISSUE

One of the most conceptually straightforward approaches among the modern therapies for Parkinson's disease is that provided by neural grafting strategies. Since the behavioral symptoms of Parkinson's disease can be attributed to death of dopamine neurons, one way to treat the disease is to attempt to replace dopamine neurons. Drawing upon a concept that had been tried as early as 1890, and the accumulating information concerning the growth properties of the developing nervous system, scientists in Sweden published a report in 1970 describing successful transplantation of embryonic neural tissue to the anterior eye chamber of experimental rats. They found that if embryonic brain cells were removed at a time prior to completing their connections with other brain regions, they could be transplanted into adult animals, survive this grafting, and establish connections with the host brain because they retained their capacity for further growth and development. In 1979, a team of American and Swedish scientists reported that this neural grafting strategy could be used with embryonic dopamine neurons, and that dopamine neurons grafted into rats with experimentally induced damage to the substantia nigra survived transplantation and replaced a supply of dopamine to the striatum, improving the symptoms of this rat model of Parkinson's disease.

Following years of important tests in rats, it was time to explore the use of brain cell transplantation in a species more closely related to humans. In 1986, we and a team of scientists from Yale University published an initial report of successful embryonic dopamine neuron transplantation in MPTP-treated parkinsonian monkeys. From 10 years of studies in monkeys, our own work and that of others have demonstrated that embryonic dopamine neurons derived from appropriately aged donors exhibit excellent survival and robust growth and integration with the dopamine-depleted host striatum (Fig. 2A). These grafted cells provide a long-term source of replacement dopamine that is effective in diminishing the behavioral signs of MPTP-induced parkinsonism in these monkeys. Though certain issues unique to the use of neural grafts in human patients remain unresolved, including whether Parkinson's disease will afflict the replacement cells and what effect continued use of levodopa may have on grafted cells, transplanted dopamine neurons theoretically could provide a cellular source of dopamine for the life span of the patient.

In 1989, a report by scientists in Sweden described their initial attempt at grafting human embryonic dopamine neurons into patients with Parkinson's disease. During the next 5 years, a handful of reports described variable, modest improvement of the symptoms of Parkinson's disease following neural grafting. Interestingly, the one report of marked improvement following dopamine neuron transplantation was in two of the young adults that had ingested MPTP and had been assessed at Stanford University almost 10 years previously. The marked success following grafting in these MPTP parkinsonian humans, in contrast to initial results from grafting in older individuals suffering from Parkinson's disease, raised a variety of questions about whether grafts could be effective in elderly individuals and whether the disease was less treatable than MPTP poisoning. Aside from the patients involved, the other critical difference in the successful grafting study was the amount of embryonic tissue grafted: 2–8 times more than had been utilized previously. This increase in the number of dopamine neurons grafted was key, and in 1995 Kordower and colleagues reported that dramatic improvement of the symptoms of Parkinson's disease was achieved following grafting of similar large numbers of embryonic dopamine neurons. To date, one of the Parkinson's disease patients that had received embryonic tissue grafts has come to autopsy. This patient died 18 months after transplant surgery from causes unrelated to tissue grafting. Study of his brain revealed robust survival of grafted dopamine neurons and clear signs of behaviorally significant dopamine replacement (Fig. 2B). Although the long-term efficacy of neural grafts remains to be determined, and the philosophical and practical issues associated with the use of embryonic tissue remain controversial, there is considered optimism that neural grafting may represent the most enduring therapy for Parkinson's disease yet developed.

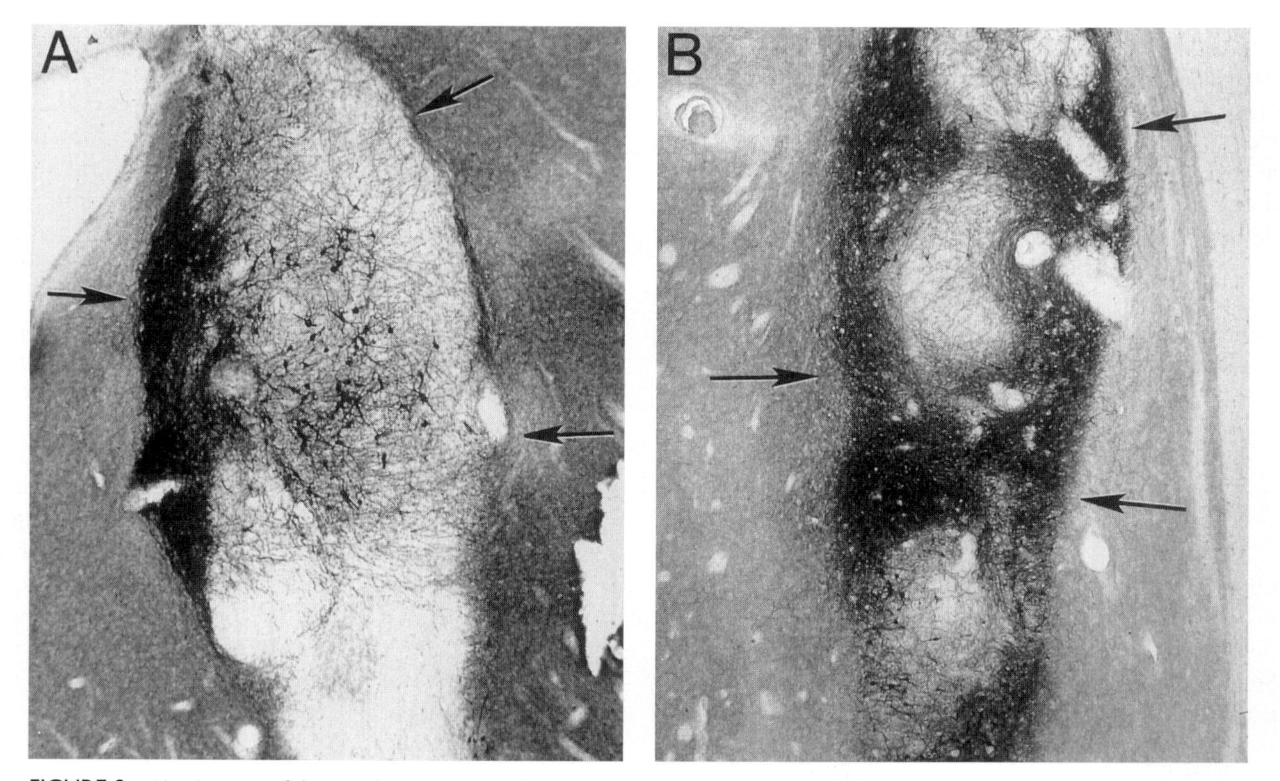

FIGURE 2 (A) A successful transplant (arrows) of embryonic dopamine neurons is depicted after several months of survival in the brain of a nonhuman primate that exhibited movement disorders similar to those in Parkinson's disease. This histological section was prepared with an immunohistochemical technique that marks dopamine neurons with a dark stain. The transplant is viewed at a low power and reveals the presence of considerable stained material representing dopamine neurons and fiber networks that extend into the recipient brain. Transplants like this are able to reverse the motor disability of debilitated animals that serve as models of the disease. This micrograph was derived from collaborative studies with Drs. Eugene Redmond, Robert Roth, Jane Taylor, and John Elsworth of the Yale University School of Medicine. (B) A graft similar to that seen in (A) reflects the comparable survival of grafted dopamine neurons in a parkinsonian human brain 18 months after surgery. The similarity in the size, shape, and distribution of dopaminergic neurons in each graft, as well as the pronounced behavioral improvement in each instance, also demonstrates the utility of animal models of human disease for attempts at development of new therapeutic strategies.

ACKNOWLEDGMENTS

Research on the potential therapeutic value of neural transplants in a nonhuman primate model of Parkinson's disease that formed the basis for discussion in part of this article was supported by PHS Grant POI NS 24032. The authors are grateful for the valuable input into the research concepts presented herein by their colleagues at Yale University (Drs. Eugene Redmond, Robert Roth, Jane Taylor, and John Elsworth), the Chicago Medical School (Dr. Kathy Steece-Collier), and the Rush School of Medicine (Dr. Jeffrey Kordower).

BIBLIOGRAPHY

Collier, T. J., and Sladek, J. R., Jr. (1988). Neural transplantation in animal models of neurodegenerative disease. *News Physiol. Sci.* **3**, 204.

DeLong, M. (1990). Primate models of movement disorders of basal ganglia origin. *Trends Neurosci.* **13**, 281.

Greenamyre, J. T., Klockgether, T., Turski, L., Zhang, Z., Kurlan, R., and Gash, D. M. (1992). Glutamate receptor antagonism as a novel therapeutic approach in Parkinson's disease. *In* "Excitatory Amino Acids" (R. Simon, ed.). Thieme Medical Publishers, New York.

Kolata, G. (1983). Monkey model of Parkinson's disease. *Science* **220**, 705.

Koller, W. C., and Paulson, G. (eds.) (1995). "Therapy of Parkinson's Disease," 2nd Ed. Marcel Dekker, New York.

Kordower, J. H., Freeman, T. B., Snow, B. J., Vingerhoets, F. J. G., Mufson, E. J., Sanberg, P. R., Hauser, R. A., Smith, D. A., Nauert, G. M., Perl, D. P., and Olanow, C. W. (1995). Neuropathological evidence of graft survival and striatal reinnervation after the transplantation of fetal mesencephalic tissue in a patient with Parkinson's disease. *N. Engl. J. Med.* **332**, 1118.

Langston, J. W., Ballard, P., Tetrud, J. W., and Irwin, I. (1983). Chronic parkinsonism in humans due to a product of meperidine-analog synthesis. *Science* **219**, 979.

The Parkinson Study Group (1993). Effects of tocopherol and deprenyl on the progression of disability in early Parkinson's disease. *N. Engl. J. Med.* **328**, 176.

Sladek, J. R., Jr., Redmond, D. E., Jr., Collier, T. J., Blount, J. P., Elsworth, J. D., Taylor, J. R., and Roth, R. H. (1988). Fetal dopamine neural grafts: Extended reversal of methylphenyltetrahydropyridine-induced parkinsonism in monkeys. *In* "Progress in Brain Research" (D. M. Gash and J. R. Sladek, Jr., eds.), Vol. 78. Elsevier Science Publishers, Amsterdam.

Pathology, Comparative

ROBERT LEADER
Michigan State University

JOHN GORHAM
Washington State University

GLOSSARY

Antigen Any sort of material such as an organism, toxin, or chemical capable of eliciting a specific immune response

Atherosclerosis Form of arteriosclerosis consisting of deposits of cholesterol-containing plaques in the arteries

Disease Any deviation from normal structure or function of any of the organs of the body

Epitope One specific antigenic site

Genome Complete set of genes, or hereditary substances possessed by an individual, contained in the nucleus of each cell

Germ cell Sperm cells in the male or ova in the female from which all other body cells are formed and which carry all hereditary characteristics

Immune system Cellular components, including lymphocytes, macrophages, and granulocytes, along with related biochemical reactions in the body having the function of defending the individual against foreign nonself substances and organisms

Lesion Any measurable pathological deviation in tissue resulting in abnormal appearance or function

Neutralizing antibody Immunoglobulin molecule capable of neutralizing a virus or other infectious agent

Retrovirus Group of viral agents, so named because they contain the enzyme reverse transcriptase, also known as RNA-dependent DNA polymerase

Somatic cell All cells of the body other than germ cells

PATHOLOGY IS THAT BRANCH OF MEDICINE DEALing with the essential mechanisms of disease, especially, but not limited to, the structural and functional changes in organs of the body that cause or are caused by disease. Comparative pathology is that branch of pathology that emphasizes comparisons of disease phenomena among various species, usually with the ultimate objective of learning more about the diseases of human beings, but at the same time with intrinsic interest in understanding the diseases of all animals, including humans.

I. BACKGROUND

Biomedical research and experimental pathology have grown logarithmically since the early 1940s, but comparative medicine goes much further back into history. The study of animals as models for human disease was recorded in Greece as early as the 4th century BC by Aristotle, who was funded by Alexander the Great, as later described by Pliny: "Fired by the desire to learn the natures of animals, he trusted the prosecution of this design to Aristotle. For this end he placed at his disposal some thousands of men in every part of Greece, among them hunters, scholars, fishers, park keepers, herdsmen, bee wards as well as keepers of fish ponds and aviaries in order that no creature might escape his notice." One of the most successful examples of the utility of comparative medicine comes from the work of Jenner, who noticed in 1796 that cowpox infection in milkmaids gave them subsequent immunity to smallpox. This eventually led to the total elimination of smallpox, one of the greatest scourges of mankind. This did not happen until more than 180 years after Jenner's first observation, and teaches a

ENCYCLOPEDIA OF HUMAN BIOLOGY, Second Edition, VOLUME 6.

lesson in patience as well as the value of keen observation.

II. DISEASES INDUCED IN ANIMALS

Comparative pathology can divide easily into two major compartments: studies of naturally occurring diseases of animals and experiments purposefully causing specific conditions in experimental animals. This division is important, both from a practical and an emotional viewpoint. Specific lesions produced in animals are usually intended to mimic as closely as possible some human functional or structural abnormality. Knowledge gained from this process is then extrapolated and applied to the primary condition in the human species with the goal of curing or at least ameliorating human pain and suffering. Advantage to humans is the main objective, although other animals may benefit.

There has been and continues to be much controversy about the appropriateness and morality of the use of animals in research to gain more knowledge about human disease. This article is a proper forum that addresses the substantive but not the ethical aspects of these questions. Animal studies have been essential to progress in every field of human biology. Without the use of animals we would not know of the Pavlovian physiological and behavioral responses to sounds and other external stimuli. Paralytic poliomyelitis would remain unconquered. Heart and artery surgery would be impossible. Antibiotics would not be available, and simple infections would still carry the risk of fatality. The impact of the acquired immune deficiency syndrome (AIDS) on the human population worldwide has been devastating and remains to be solved, but work with a similar disease in rhesus monkeys has given much knowledge about the biology of the virus and its impact on the human immune system.

One need only consider the Nobel Prizes awarded during this century to see the stellar contributions of animals in medical research to the benefit of all animals and human beings. Of the 135 recipients of the Nobel Prize in Physiology or Medicine from 1901 to 1984, most used mammals in their research. One-third were cited for work that involved no warm-blooded animals and an additional 17 were cited for work involving only humans. A wide range of species including vertebrates, invertebrates, cell cultures, and even higher plants were used as sources of model systems. Careful scrutiny of this series of awards reveals with the greatest clarity the true unity of biological phenomena. Humans are not an isolated species, but relate biologically, emotionally, and behaviorally to all the other passengers on our planet. We all fly together and will either survive or perish together.

III. NATURALLY OCCURRING CONDITIONS

Naturally occurring conditions of animals frequently reveal information that directly benefits the species being studied as well as humans. There are many illustrations of this, including the development of canine distemper vaccine many years ago by growing the virus in the membranes of embryonating eggs. Canine distemper is a member of the paramyxovirus group closely related to measles, and the successful canine distemper vaccine was later extended by development of a vaccine for measles in children, one of the more important steps in modern control of human infectious diseases. It is not to be claimed that one event led directly to the other, but there was a pattern. The adaptation of a dog virus to growth in eggs, resulting in its attenuation, and subsequent proof that this was an infectious agent capable of stimulating immunity, but without causing disease, was a major achievement. The primary goal was a vaccine that benefited dogs by controlling one of their most devastating infectious diseases; human medicine was later rewarded by the contribution of a significant block of knowledge about the behavior of viruses adapted to growth in nonprimary hosts. The entire chapter of the development of attenuated vaccines for such human diseases as poliomyelitis, measles, and other diseases was aided by the classical work of Haig in canine distemper.

Control of the human immunodeficiency virus (HIV), the cause of AIDS, is one of the most baffling medical problems facing modern society. The disease was first described and its cause, a retrovirus, determined in the early 1980s. Unfortunately there are currently no vaccines to prevent AIDS, and the treatments that kill the virus in infected cells within the human host have only a partial value. [*See* Acquired Immunodeficiency Syndrome, Virology.]

While the AIDS epidemic continues to spread, research workers are investigating animal diseases caused by related retroviruses, which provide convenient and meaningful models for study of host–virus interactions. Most animal species are infected with retroviruses. An understanding of the mechanisms by which these organisms survive in the host and produce chronic disease with highly variable incubation periods is important for our understanding of AIDS and the success of current efforts directed at prophylaxis, treatment, and improved diagnosis of retroviral infections.

Some of the currently known animal retroviral infections that have been intensively studied include maedi/visna of sheep, equine infectious anemia (EIA), simian immunodeficiency, and caprine arthritis–encephalitis. The characteristics of these retroviral diseases are similar to the diseases in humans caused by HIV, as shown in Table I. Most of the retroviruses causing this group of diseases are characterized by rapid changes in their antigenicity, which enable them to evade the defense mechanisms of the infected hosts. In the case of EIA, the first antigenic change occurs within a few weeks after the first infection and at least six epitope variants can appear within the first year. As in AIDS and other known retroviral diseases of animals, all horses once exposed remain infected for life. The simian immunodeficiency virus induces suppression of the immune mechanism, which makes it important as an animal model of HIV infection. Caprine arthritis–encephalitis virus is a persistent retrovirus infection of goats and is important to study because viral-neutralizing antibodies do not inhibit the progressive course of the disease. It is readily apparent that the interplay between this complex spectrum of viruses and their hosts resembles in many ways the relationship between HIV and the human host, and yet each has its own peculiar biological nature. Much can be learned by studying the entire group.

More recent examples of human and animal benefits from studying animal models can be found in the field of immunology. In the early 1970s, veterinarians reported that Arabian foals, which appeared normal at birth, died of a variety of respiratory infections before they reached 5 months of age. It was found that the foals were deficient in both arms of the immune system—antibody (B lymphocytes) and cellular immunity (T lymphocytes). This defect, which is known as severe combined immunodeficiency (SCID), is transmitted genetically as an autosomal recessive trait in both humans and horses. Approximately 25% of the Arabian horses in the United States are carriers of this disorder and have served as a laboratory for further studies of the condition.

TABLE I
Animal Models of Retroviral Infections

Characteristic of HIV (AIDS)	Most appropriate current animal model (1989)
Immunodeficiency	Feline leukemia Simian immunodeficiency virus
Antigenic variation	Maedi-visna in sheep Equine infectious anemia Caprine arthritis and encephalitis
Encephalitis	Maedi-visna in sheep
Viral persistence	All animal retroviruses

Perryman and his co-workers at Washington State University have successfully transplanted bone marrow from a full sibling donor in the treatment of equine human SCID. Because bone marrow transplantation is the only means of treating this poorly understood syndrome in children, the work on horses is particularly important. Another model of SCID in mice has been discovered. Both animals provide useful models in helping us understand this perplexing condition. [*See* Bone Marrow Transplantation.]

Animal models have also been intensively studied to gain better understanding of the biochemical nature of hereditary metabolic diseases.

Many important examples of well-recognized animal models, along with analyses of their value in contributing to the solution of human disease, have been published in the scientific literature. The subjects range from amyloidosis in the mouse to a systemic lupus erythematosus in the dog to pattern baldness in the chimpanzee. Genetically transmitted metabolic defects in animals can be very useful for investigations of the molecular pathology of the human counterparts. Characterization of the underlying genetic defect and its pathophysiological manifestations allows specific design and evaluation of therapeutic approaches, which could not be applied in clinical trials of the human condition. Furthermore, as our knowledge of the splicing of genes becomes more precise, it is possible that studies of these animal models may eventually lead to cures. Such cures would benefit not only human patients, but affected animals as well. Table II outlines examples of some specific metabolic defects that have been well characterized in animals.

IV. MAJOR CONTRIBUTIONS

The greatest killer in the populations of Western civilization is cardiovascular disease, particularly coronary artery atherosclerosis, which accounts for several hundred thousand deaths in the United States alone

TABLE II
Confirmed Animal Models of Human Lysosomal Storage Diseases[a]

Disease	Enzyme defect	Animal species
Pompe syndrome	Glucosidase	Cow
Mannosidosis	Mannosidase	Cow, cat, goat
Gangliosidosis	Gangliosidase	Cat, calf, dog
Gaucher	Glucosidase	Dog
Krabbe	Galactosidase	Dog, mouse
Niemann–Pick	Sphingomyelinase	Dog, cat
Hurler	Iduronidase	Cat
Maroteaux–Lamy	Arylsulfatase	Cat

[a]Adapted from Desnick, R. J., Patterson, D. F., and Scarpelli, D. G. "Animal Models of Metabolic Diseases." A. R. Liss, New York.

each year. Many naturally occurring and induced models in animals have been used in studies of heart disease, including those involving primates, rabbits, dogs, mice, rats, swine, pigeons, chickens, turkeys, and quail. All of these have contributed significantly in their own special way. Recent major advances about the knowledge of the metabolism and heredity of various types of lipids have focused attention on the metabolic and dietary aspects of lipids as risk factors. This has made the studies in primates loom even larger in importance, as these animals are metabolically closer to humans than others. Recent work in primates indicates strongly that atherosclerotic plaques produced over many years may be reversible and may regress significantly under rigorous medical and dietary regimens. This gives hope and encouragement to many millions of patients who have already reached the age where they are most likely to have significant levels of atherosclerosis. A rabbit strain with a genetically transmitted abnormality of lipid metabolism has also proven very important. Each of these models contributes to our knowledge, most because of their similarity to human disease, but often because of specific differences. Differences allow comparisons of definitive steps in mechanisms of disease, increasing the usefulness of the animal. [*See* Atherosclerosis.]

It is important not to overlook the many important contributions of behavioral research to human and animal welfare. The classic example is that of Pavlov (Nobel Prize, 1904), who demonstrated that distinct physiological responses could be elicited in the dog by external stimuli, such as ringing a bell. With the emphasis of modern medicine on stress, it is important to be aware of the basic work done in defining the stress syndrome by Selye. Weiss showed that rats subjected to very small electrical shocks developed classical stress manifestations, but were totally protected when warned of the shock by a buzzer. Many other important contributions can be found in research dealing with the processes of learning and language, particularly in gorillas and chimpanzees; in the rehabilitation of patients with neuromuscular disorders (Nobel Prize, C. S. Sherrington and E. D. Adrian, 1932); and in studies of social behavior patterns (Nobel Prize, K. von Frisch and K. Lorenz, 1973).

Considerable attention has been given to the interdependence of species, in recognition of the value of companion animals such as cats, dogs, and birds to the well-being of aged and handicapped persons.

Those who work with animal models for human disease often come to realize that there is more than a spatial, mechanical relationship between them and the animals they study. An animal is not merely a test tube. Animals, including mice and rats, and certainly dogs, cats, and primates, react to the attitudes of their keepers; the outcome of experiments can therefore be influenced in a major way by the quality of care given to the animals. Those activists who continuously press for the total elimination of animals from experimental medicine profess that biomedical laboratories are only a series of cruel torture chambers, where scientists subject animals to the crudest and most useless procedures. Certainly, there are abuses, but they are the exception. There are strict federal rules for the proper handling of animals, and all laboratories must abide by them in order to receive research support. Federal funding sources have demonstrated their concern by withholding research funds on a number of occasions.

The bond between humans and other animals is firmly rooted in our history from the time of cave dwellers. Many committed scientists are currently devoting careful study to this relationship, and without question, the presence of animals can benefit many patients. Specific examples include favorable responses in blood pressure on the part of hypertensive patients in the presence of pets. This relationship has been especially fruitful in homes for the aged, where the companionship of a cat or dog or bird can make the difference between mere existence and enjoyment of daily life. The relationship between people and animals is a complex one, which cannot be whisked away by wanton accusations of cruelty in laboratory procedures. We are all interdependent. As has been eloquently expressed by Bustad and others, the bond is strong and permanent. If we work to strengthen it, all species will benefit.

V. IMPACT OF MOLECULAR PATHOLOGY

Since the early 1980s there has been an explosive development of ideas that apply the techniques of molecular biology to pathologic studies in animals. One of the truly exciting areas is the production of transgenic animals. These animals will be useful for the study of gene expression and regulation in a wide variety of studies, including pathology, neurobiology, immunology, and in normal as well as abnormal development.

The term transgenic was first applied by Gordon and Ruddle in 1981 to animals that have integrated foreign DNA in their somatic or germ cells as a consequence of the introduction of DNA from another individual of the same or different species. In this technique, a fertilized egg is removed from an animal such as a mouse, pig, or cow, and foreign DNA is injected into the male pronucleus. The egg is then implanted into a foster mother. As the fertilized egg develops, the foreign genes are expressed in all new cells produced. Development of this technology in association with DNA splicing was followed by a cascade of many important results. Much attention was given by the lay press to the work of Brinster and Palmiter, who microinjected the gene for rat growth hormone into mouse embryos, resulting in mice that grew much faster and reached weights 70–80% greater than their noninoculation littermates.

Not long after this discovery, the same idea was applied to the production of domestic animals. Although the results in swine did not give the dramatic success seen in mice, the experiments are continuing. The sheep β-lactoglobulin gene injected into mice caused secretion of this protein in the milk of the mice. In the meantime, Chinese workers have successfully introduced rat growth hormone genes into carp and claim that some transgenic fish grow to twice their normal size.

It is now possible to use retroviruses, which can carry genomic material into mammalian cells, as vectors to incorporate desirable genes for such traits as growth, egg production, or disease resistance into recipient animals. The future possibilities for these kinds of experiments to impact on human health are enormous. [*See* Retroviruses as Vectors for Gene Transfer.]

All of these experiments have stimulated stormy discussions among scientists, theologians, ethicists, and politicians. The argument has just opened as this torrent of developments in the manipulation of genes has only begun. It cannot be stopped, indeed will gain momentum in the future, and will be a major future focus of research workers worldwide.

VI. FUTURE DEVELOPMENTS

Undoubtedly the application of the emerging knowledge of the transfer of genes will have a major impact on the field of comparative pathology. This is illustrated by a paper in *Science,* which is authored by an unlikely consortium of scientists from the U.S. Department of Agriculture Laboratory at Plum Island, the University of California at Davis, and a California biotechnology company, who joined forces to produce a vaccine by recombinant techniques similar to those previously used for the production of a smallpox vaccine. This involved insertion into vaccinia virus of specific genes programmed to manufacture immune stimulating factors against a specific disease. The disease was rinderpest, which affects mainly cattle and buffalo, is worldwide in distribution, and causes millions of deaths each year. A modified live virus vaccine grown in tissue culture or in other species is available but is very fragile and often deteriorates in the variable environmental conditions to which it is subjected in the field. The new vaccine provides the safety of a benign agent but with the ability to replicate and be amplified in the inoculated hosts. These advanced techniques resulted in a type of vaccine that may also be effective for immunization against canine distemper and measles, as there is a great deal of cross-immunity among the viruses of this closely related family.

An article in the *New York Times* looks at the commercial potential for gene-altered or transgenic animals. A wide range of as many as 1000 strains of transgenic mice have already been produced, in addition to 12 strains of transgenic pigs, several breeds of rabbits and fish, at least two breeds of rats, and one transgenic cow with another still under development.

These animals can act as "molecular factories" for production of a variety of proteins for medical uses such as immunoglobulins, thyroid, pancreatic, and growth hormones and so forth. But all is not perfect. Only a small proportion of attempts to transplant genes are successful. This weakness is no serious problem in mice, which have a very short generation time, and can be easily bred to large numbers, but poses serious problems in larger species such as cows and pigs. Unquestionably, this area of research will grow tremendously in the future, but there will continue to be controversy about its usefulness and its ethical and religious connotations. Some maintain that we must not tinker with God's work, but the age of the gene is here. The genie has emerged and will not be put back into the bottle.

BIBLIOGRAPHY

Bustad, L. K. (1988). Living together: People, animals, environment. A personal historical perspective. *Perspect. Biol. Med.* **31,** 2.

Cuthbertson, R. A., and Klintworth, G. K. (1988). Transgenic mice: A gold mine for furthering knowledge in pathology. *Lab. Invest.* **58,** 5.

Gorham, J. R. (1987). Biotechnology and veterinary medicine. *In* "Proceedings of the 91st Annual Meeting of the U.S. Animal Health Association."

Leader, R. W. (1967). The concept of unity in mechanisms of disease. *In* "The Rockefeller University Review" (Jan.–Feb.).

Leader, R. W. (1967). The kinship of animal and human disease. *Sci. Am.* **216,** 1.

Leader, R. W. (1969). The Chediak-Higashi anomaly: An evolutionary concept of disease. *NCI Monogr.* **32.**

Leader, R. W., and Padgett, G. A. (1980). The genesis and validation of animal models. *Am. J. Pathol.* **101,** 3s.

Leader, R. W., and Stark, D. (1987). The importance of animals in biomedical research. *Perspect. Biol. Med.* **30,** 4.

Narayan, O., Zink, M. C., Huso, D., Sheffer, D., and Crane, S. (1988). Lentiviruses of animals are biological models of the human immunodeficiency viruses. *Microb. Pathogen.* **5,** 149.

Perryman, L. E., and Magnuson, N. S. (1982). Immunodeficiency disease in animals. *In* "Animal Models of Inherited Metabolic Diseases." A. R. Liss, New York.

Schmeck, H. (1988). Gene-altered animals enter a commercial era. *New York Times,* Dec. 27, 1988.

Scott, G. B. D. (1993). "Comparative Primate Pathology." Oxford Univ. Press, London.

Slauson, D. O., and Cooper, B. J. (1990). "Mechanisms of Disease: A Textbook of Comparative General Pathology," 2nd Ed. Williams & Wilkins, Baltimore, MD.

Yilma, T., Hsu, D., Jones, L., Owens, S., Grubman, M., Mebus, C., Yamanaka, M., and Dale, B. (1988). Protection of cattle against rinderpest with vaccinia virus recombinants expressing the HA or F gene. *Science* **142,** 1058.

Pathology, Diagnostic

DIANE C. FARHI
MICHAEL E. LAMM
Emory University and Hospital and Case Western Reserve University

GLOSSARY

Anatomic pathology Study of disease by visual and microscopic examination

Autopsy Study of the body after death (postmortem examination)

Clinical pathology (laboratory medicine) Study of disease by laboratory methods

Cytology Microscopic study of cells

Diagnostic pathology Study of disease as applied to the diagnosis and care of patients

DNA Repository of genetic information in cells

Experimental pathology Study of disease in a research laboratory setting

Histology Microscopic study of tissue

Pathology Study of disease

PATHOLOGY IS THE BRANCH OF MEDICINE PARTICularly concerned with the study of disease per se, its origins, development, and manifestations. Although its historical orientation has been toward morphological evaluation by naked eye or microscope, pathology today embraces an array of methodologies, including chemical, physical, and immunological. Diagnostic pathology is the application of such techniques along with the pathologist's training and experience to medical diagnosis.

I. ANATOMIC PATHOLOGY

Anatomic pathology is the study of disease through the examination of tissues and cells by a variety of methods, ranging from visual inspection of the whole body and its organs to analysis of subcellular constituents. The means of obtaining tissue or cells for study is dictated by the nature and site of disease and the expertise and facilities available; in turn, the means used for obtaining tissue may have an impact on the diagnosis rendered.

The oldest technique for examining tissues is visual inspection of the entire body, body cavities, and organs. This method, used for centuries, was employed as the basis for the first reliable anatomic texts in the 15th century and landmark treatises on pathology in the 18th and 19th centuries. Histology, or the microscopic study of tissues rather than whole organs, began as a discipline in the early 19th century, gathering momentum with increasing use and refinement of the microscope and improved methods for preserving and preparing tissues for examination. By the early 20th century, microscopy became a mainstay of diagnostic techniques in pathology, which it remains today. A variation on microscopy was introduced with the advent of fluorescence technology, which permits the identification of tissue elements through the application of fluorescent dyes.

The theory and principle of microscopy as a diagnostic tool were carried further in the mid-20th century, when a beam of electrons rather than a beam of light was directed at tissues. Instead of an image formed by transmitted and absorbed light, an image is generated by transmitted and absorbed electrons. The power of this technique, known as electron microscopy, lies in its capacity to resolve images many orders of magnitude smaller than visible by light microscopy. Although not routinely used in diagnostic pathology, the study of cells at this level has contributed much to the understanding of disease processes. [*See* Electron Microscopy.]

To derive the maximum information from any microscopic method, tissues are first stained with any of a variety of reagents. The staining of tissues with biological dyes reached its greatest development in the late 19th and early 20th centuries, at which time the stains now used routinely were formulated. These are based on combinations of naturally occurring and synthetic chemicals with affinities for acidic, basic, and other substances. The cornerstone of diagnostic stains, a combination of hematoxylin and eosin, colors nucleic acids (and therefore nuclei) blue with hematoxylin, and proteins (and therefore cytoplasm) orange–red with eosin. Many other stains are in frequent use, several as direct applications of classical chemistry methods, such as the Prussian blue stain for iron. These stains are employed for the detection of minerals, sugars, pigments, enzymes, and many other substances.

Fluorescent microscopy and electron microscopy also depend on staining techniques for optimum visualization. In the former, a staining reagent may be chemically tagged with a fluorescent marker such as fluorescein, which emits a characteristic wavelength in the visible spectrum when excited by a light of shorter wavelength. The goal of staining in electron microscopy is to render some cellular elements electron dense, leaving others electron lucent, in a fashion analogous to light microscopy; this is usually accomplished by staining with heavy metals like osmium, uranium, and lead.

Modern advances in immunology, including the understanding of antigen–antibody interactions, have added a new dimension to the staining of tissues and individual cells. Antibodies can be produced in animals and by cells in tissue culture to a vast and growing array of cellular and extracellular constituents. An antibody directed against the substance of interest is incubated with a slice of tissue or cells in suspension, and the resultant binding of antibody to its target is observed. The visualization of this interaction is achieved by tagging the antibody with a fluorescent marker, a heavy metal, or an enzyme that generates a colored product. These techniques, known collectively as immunostaining, have greatly increased our understanding of disease and have been added to the standard diagnostic armamentarium.

The foregoing methods of diagnosis are all applied to the main fields of anatomic pathology: autopsy pathology, surgical pathology, and cytology. These are defined primarily by the way in which tissue is obtained and how it is processed.

Autopsy pathology is the study of the body and its tissues after death, or postmortem. Following medical determination of death and the obtaining of necessary legal permits, the body is examined externally. This may take place at the scene of death in the case of a forensic autopsy (see below) or at a hospital morgue in the case of death at home, in the hospital, or other facility. The body is then examined internally, with care taken not to affect the external appearance, and tissue samples are removed for microscopic study. Specimens may also be removed for study in other diagnostic pathology laboratories (see below). The autopsy is a detailed, comprehensive study of the cause of death and other pathologic processes and results in a report synthesizing the clinical and pathologic findings into a cohesive whole. It is the basis of all anatomic studies and provides opportunities for educating medical students and physicians and for quality control of medical care. Despite the many advances in clinical medicine, in a significant percentage of cases the autopsy continues to reveal major findings that were unsuspected or misinterpreted during life, even after state-of-the-art diagnostic evaluation. Findings revealed at autopsy may have far-reaching medicolegal implications regarding adequacy of diagnosis and care, aside from the autopsy's importance in forensic pathology (see below).

Autopsy findings are crucial in assessing the natural history of disease and the effects of medical care; in fact, modern medicine as a discipline and as an effective therapeutic endeavor may be said to have originated with the development and acceptance of the autopsy as standard practice. Serious concerns have been raised in recent years regarding the decline of autopsy rates in the United States. If this decline continues, it will undoubtedly prove to be a substantial impediment to progress in understanding human disease.

Surgical pathology is based on the examination of tissues removed at surgery. As surgical techniques improved in the 19th and 20th centuries, more specimens were available for pathologic examination antemortem. Interest shifted to detection and diagnosis of disease in the early stages and alterations brought about by treatment. The removal of whole organs or substantial parts of organs for treatment and study has been supplemented or replaced in many cases by tissue biopsy, or removal of a small portion for diagnosis. Improving biopsy techniques have broadened surgical pathology to include some hitherto relatively inaccessible tissues and have permitted repeated tissue studies in difficult or evolving cases. Thus, surgical pathology is a changing area in terms of both new techniques for the study of tissues and new types of specimens being handled. Increasingly sophisticated methods are being brought to bear on ever smaller samples. The questions asked of the surgical pathologist center on the nature of the diagnosis, its distinction from related diagnoses (differential diagnosis), the stage of the natural history at which the disease is seen, and the impact of therapy. Because the patient is alive at the time of pathologic examination, the implications of the pathologist's diagnosis for therapeutic decision-making are great.

An important aspect of surgical pathology involves the communication and interaction of the surgical pathologist and the surgeon while the patient is undergoing an operation. The pathologist may enter the operating room to inspect the abnormality and to render an opinion. Usually a small piece of tissue is given to the pathologist by the surgeon. The pathologist then performs a rapid frozen-section microscopic evaluation, which takes only a few minutes instead of the many hours required for a standard microscopic preparation. Based on the pathologist's frozen-section diagnosis, the surgeon decides on a course of action. The rapidity of frozen-section diagnosis thus allows further surgery or other measures to be taken during the same operation. Without this procedure, definitive surgical treatment might have to be delayed, mandating a separate, second operation. Frozen-section diagnosis also permits other forms of therapy such as antibiotics, anti-cancer agents, or radiation to be instituted immediately for life-threatening infections or malignancies.

Cytology is the microscopic study of individual cells obtained from a patient. It is similar to surgical pathology in the questions asked and implications for clinical care; it differs primarily in the type of specimen submitted for study. Cells are obtained from samples of body fluids and secretions by scraping from external and internal surfaces and by aspiration from internal tissue through a narrow-gauge needle. Cytologic samples are usually collected through less invasive means than surgical pathology specimens and are thus ideally suited for low-cost mass screening studies directed toward the detection of early, potentially curable malignancies and premalignant conditions. A noteworthy example is the contribution of cytopathology to the reduction in morbidity and mortality from carcinoma of the uterine cervix via the Papanicolaou (Pap) test.

II. CLINICAL PATHOLOGY (LABORATORY MEDICINE)

The study of tissues and cells has been augmented in the last century by the study of body fluid components. Organized as a distinct branch of pathology in the 1920s, clinical pathology, or laboratory medicine, is devoted to diagnosis through laboratory methods. This is a rapidly growing area of medicine in which technologic advances have repeatedly revolutionized the scope of laboratory diagnosis. In some areas, subcellular analysis as carried out in anatomic pathology and in clinical pathology merge, and the traditional divisions between these two disciplines disappear. In general, however, the laboratory activities of clinical pathology fall into the following broad diagnostic categories: hematology, blood banking, microbiology, clinical chemistry and toxicology, immunology, and cytogenetics.

Hematology, the study of blood, encompasses the examination of the cellular constituents of blood and blood-forming organs, principally the bone marrow, and the proteins and other substances peculiar to blood, such as the clotting factors. The main activities of a hematology laboratory include quantitative analysis and microscopic evaluation of red and white blood cells and platelets; preparation, staining, and evaluation of aspirated bone marrow samples; tests of coagulation, including quantitative and qualitative evaluation of clotting factors and platelet function tests; detection of abnormal hemoglobins; and specialized tests directed at red and white cell defects, fetal–maternal transfusion, serum viscosity, and others. Many quantitative hematologic tests originally done manually are now performed on automated equipment capable of processing minute amounts of blood rapidly and at low cost; however, skilled microscopic examination remains a very important aspect of this field.

Blood banking, or transfusion medicine, is a related area in that it is concerned with blood, but the emphasis is on blood as a tissue transplant or transfusion. The discipline of blood banking has received major impetus through the needs generated by trauma, surgery, organ transplantation, and treatment of blood disorders. Transfusion of blood may be accompanied by immunologic reactions in the recipient and the transmission of infectious disease. Thus, the activities of the blood bank comprise not only the major public health task of obtaining a continuous supply of fresh blood but assuring its immunologic compatibility and safety when transfused. The blood bank, or transfusion medicine laboratory, collects blood from suitable donors; determines the blood type; tests for infectious agents; separates blood as needed into components such as plasma, platelets, and red blood cells; irradiates or washes components as necessary; tests for donor–recipient compatibility; oversees transfusions; and analyzes adverse transfusion reactions if they occur. Many laboratories also remove blood or specific blood components from patients as a therapeutic maneuver. Improved methods of storing and handling blood are continually being developed to extend the life of this perishable product and permit its more efficient use. In addition, new methods are being rapidly implemented to eliminate the transfusion of potentially lethal organisms, such as the human immunodeficiency virus [the agent of acquired immunodeficiency syndrome (AIDS)]. [*See* Blood Supply Testing for Infectious Diseases.]

Microbiology, the study of infectious agents, is one of the oldest branches of clinical pathology. Microbiology laboratories examine a wide variety of specimens harboring diagnostic evidence of infectious disease, including tissues, fluids, secretions, and blood. Originally these laboratories were directed primarily toward the finding and identifying of intact organisms such as bacteria and fungi, but as the spectrum of infectious agents has widened, laboratory techniques have grown to include testing for viral nucleic acids and proteins and immunologic responses, among others. These approaches to the diagnosis of infectious disease are particularly suited to the detection of obscure, latent, or treated infections. The mainstays of diagnosis, however, remain culture in or on suitable media or cells for bacteria, fungi, and viruses and microscopic examination for parasites. Laboratory work in this area still, to a large degree, relies on skilled microscopic examination. The impact of molecular biology (see below) in microbiology has already been great and is increasing.

Clinical chemistry and toxicology are linked in that these disciplines center on identification and quantification of extracellular molecules in tissues and fluids. The clinical chemistry laboratory measures minerals, fats, proteins, sugars, vitamins, hormones, and a wide variety of other substances in small samples of blood, plasma or serum; in some cases, these tests are made on other specimens, such as urine, cerebrospinal fluid, amniotic fluid, or tissue. Many of these tests are performed in groups designed to identify dysfunction of a particular organ or to resolve a specific differential diagnosis. The toxicology laboratory identifies and quantifies therapeutic drugs in body fluids and tissues, as well as alcohol and nonmedically administered narcotics and poisons. Chemical determinations have proliferated dramatically with the advent of highly efficient automated methods and increasingly sensitive analytical techniques, such as radioimmunoassay and gas and liquid chromatography. [*See* Radioimmunoassays.]

Recent progress in the understanding of the immune system has resulted in the development of clinical immunology laboratories devoted to the identification and monitoring of immunologic disorders. Tests are designed to quantify and assess the function of the major classes of lymphocytes, identify and quantify immunoglobulins and certain other proteins, diagnose cancers of the immune system, identify antibodies of clinical interest, and carry out tissue typing needed for successful transplantation of organs. [*See* Lymphocytes.]

Cytogenetics is that branch of medicine concerned with the detection and analysis of chromosomes, cell elements that contain the DNA governing inherited traits. Laboratories devoted to cytogenetics are often found in departments of pathology, but because much of their work is related to pediatrics and obstetrics, cytogenetics laboratories are sometimes located in these or other departments. The field of cytogenetics has grown substantially in the past few decades and has gained added momentum with the application of molecular biologic techniques capable of identifying individual DNA fragments and even single genes. The cytogenetics laboratory is primarily occupied with growing human cells from various sources in tissue culture, inducing and arresting mitosis, and analyzing the number and configuration of the chromosomes revealed. Chromosomal identification is accomplished by means of sophisticated staining techniques in combination with high-resolution photomicroscopy and requires highly skilled and experienced personnel. The results have a great impact on prenatal diagnosis, the diagnosis of malignancies and premalignant conditions, and the management of patients undergoing bone marrow transplantation, among others. The increasing use of techniques designed to uncover genetic disorders is rapidly expanding medical options in the diagnosis and management of these diseases; thus, genetic counseling services and the cytogenetics laboratory are very closely allied. [*See* Genetic Counseling.]

III. FORENSIC PATHOLOGY

Forensic pathology, or legal medicine, refers to that branch of pathology concerned with determining the cause of death under unattended, violent, or suspicious circumstances. Under such conditions, the death must be reported to the coroner or medical examiner, who decides if an autopsy should be performed. It is often appropriate to include a toxicological study of body tissues and fluids to determine whether drugs or poisons are present. In a case of traumatic death, the forensic pathologist seeks to ascertain whether death was due to accident, suicide, or homicide. In the case of a bullet wound, at what distance was the gun fired and in which direction? If there were two bullet holes, which was the entrance and which the exit wound, or were there two bullets? Could the wound have been self-inflicted? In the case of a motor vehicle accident, did the driver's preceding death (e.g., from a heart attack) cause the accident, or did the accident cause the driver's death, and if so, was the driver under the influence of drugs or alcohol? When multiple people die in disasters such as earthquakes, explosions, or airplane crashes, or in individual deaths in which the body bears no identification papers, the forensic pathologist endeavors to establish the identity of the deceased.

In situations like those mentioned earlier, the implications of the pathologist's conclusions extend far beyond determining the cause and circumstances of death from a strictly medical viewpoint. Important legal and financial considerations may arise that bear on inheritances, insurance coverage, and lawsuits. The forensic pathologist regularly interacts with the police, law courts, and the press. Nor should one minimize the psychological aspects of dealing with bereaved families, for whom the sudden and unexpected death of a family member may induce strong feelings of guilt, especially when a child is involved; it may be comforting indeed if the pathologist can assure the family that the loved one died of natural causes and not as a result of carelessness or inattention. Identifying conditions relevant to genetic counseling is an important part of the pathologist's contribution in cases of stillbirth and deaths of infants.

IV. COEVOLUTION OF DIAGNOSTIC PATHOLOGY AND CLINICAL PRACTICE

The practice of diagnostic pathology is inextricably linked to the practice of clinical medicine. As advances are made in the clinical diagnosis of certain disorders by epidemiologic evidence, physical examination, or instrumentation, demand grows for new and better laboratory tests to confirm or establish the diagnosis. These tests may range from culture or immunostaining of a newly identified infectious agent to the development of rapid, efficient methods for detecting clinically important enzymes. Conversely, as new technology enters the pathology laboratory from advances in instrumentation or basic biomedical research, the possibility of more rapid and accurate clinical diagnosis increases, spurring interest in diseases that may have been rarely diagnosed, poorly understood, or even unknown. Expertise in clinical medicine and diagnostic pathology grow together, the impetus arising in each depending on the situation. This relationship permits planning of controlled studies, resulting in the careful accumulation of reliable medical knowledge.

V. PATHOLOGY AS A BRIDGE BETWEEN BASIC MEDICAL SCIENCE AND CLINICAL PRACTICE

Diagnostic pathology is uniquely situated in medicine between clinical practice and basic medical research. The relationship of pathology to basic science is no less important than its position vis-à-vis clinical work; in fact, diagnostic pathology is a crucial bridge between these areas. Through application in pathology laboratories, new techniques developed in academic and industrial research laboratories rapidly gain entry to the diagnostic and therapeutic medical world. In the last few decades, the movement of research tools into diagnostic laboratories has accelerated dramatically; the lag time in this transit is now often measured in months rather than years.

Two examples of this movement are the recent rise of the immunology laboratory, alluded to earlier, and the current interest in the development of molecular diagnostic laboratories. Clinical immunologists have quickly applied improvements in methods for analyzing proteins to immunoglobulin identification, moving into several generations of immunologic assays. Especially remarkable has been the rapid application of monoclonal antibodies and flow cytometry to diagnostic pathology. The use of such antibodies for cell-surface marker assessment via flow cytometry has changed status from an expensive, cumbersome research procedure to a standard feature of medium to large hospital laboratories; methods have even been developed for transporting specimens in culture media from small hospitals to larger laboratories for these sophisticated tests. The functions of the flow cytometer have grown to include chromosome analysis of malignant cells, and further applications appear limitless.

As a second example, molecular diagnostic laboratories, now at an early stage of development, show great promise. Until recently, this field was confined to the research laboratories in which it was developed. With technical improvements and commercial marketing, analysis for DNA and RNA, including specific segments, is now possible in well-equipped diagnostic laboratories. These procedures include the polymerase chain reaction (PCR); enzymatic digestion of nucleic acids; nucleic acid electrophoresis and hybridization with radioactive or enzyme-labeled nucleic acid probes; nucleic acid electrophoresis and immunostaining; and hybridization of nucleic acids in intact cells, tissue, or tissue digests. Such methods have proved immeasurably useful in protecting transfusion recipients from transmission of the viral agent of AIDS. They have also been used in identifying other transmissible agents, diagnosing malignancies of the immune system, and assessing genetic alterations in tumors. With these techniques, single genes may be located on individual chromosomes, and the specific genetic makeup of individual human beings may be identified from minute specimens, known as genetic "fingerprinting." These powerful tools hold great implications for the practice of clinical and forensic medicine. In summary, pathology functions as an intellectual bridge between medical research and medical practice. Further, as discussed elsewhere in this article, pathologists must function as a communication bridge between their own discipline and other medical specialists.

VI. TRAINING OF A PATHOLOGIST

Why might a young physician choose a career in pathology, a field which does not entail the practice of medicine in the usual sense of directly caring for a sick patient? Instead, the pathologist tends to function behind the scene, providing patient-related services indirectly. The pathologist is primarily concerned with diagnosis and has often been referred to as the "doctor's doctor," advising the clinician on the nature of the patient's condition and what needs to be done.

Pathologists are attracted to their profession by its breadth and intellectual challenges, particularly the need to keep one's knowledge current in order to apply new scientific advances to medical problem-solving. Many pathologists enjoy working with analytic instruments and computers. Some are attracted by the opportunity to organize and manage laboratories and the flow of information. Another positive feature is the opportunity to teach in a variety of settings involving medical colleagues, physicians in training, medical students, and medical technologists.

After graduation from medical school, the young physician intent on a career in pathology embarks on a period of specialty training to meet the requirements for certification as a pathologist. Training may be confined to either anatomic or clinical pathology (see earlier), but most pathologists-to-be choose combined training programs, as these best provide the broad range of skills needed for practice in a community hospital. Such programs entail a series of rotations through the different departmental laboratories, followed by more advanced training in the major areas. However, the opportunity to pursue an aspect of pathology in even greater depth may be especially attractive to an individual planning an academic career in a medical school or teaching hospital. Departments of pathology in such settings have more resources in personnel and facilities, thus offering considerable opportunities for each pathologist to develop and maintain expertise in a subspecialty of diagnostic pathology and/or research. Additional training in some subspecialty area of diagnostic pathology such as cytology, hematopathology, and blood banking is required for subspecialty certification.

The opportunity to pursue research into the causes, nature, and manifestations of disease has always been an attractive part of pathology. Traditionally, research in pathology emphasized morphological evaluation of the changes wrought by different disease processes in cells, tissues, and organs. This aspect continues to be important, but increasingly the pathologist of today uses a wide range of research disciplines available to the modern biomedical researcher, including biochemistry, molecular biology, and immunology to mention a few. There really are no limits to the kind of research a pathologist might do, and the experimental pathologist must be prepared to keep up with new developments in a number of related fields in order to apply them to a research problem. The distinguishing feature of the pathologist's research is not the technique being applied but rather the orientation to disease—its cause, appearance, and detection.

VII. ORGANIZATION OF PATHOLOGY IN DIFFERENT SETTINGS

Pathology is practiced differently in various settings. In an academic medical center embracing a medical school and teaching hospital, a department may easily have more than 20 pathologists. Because all the clinical departments in such a hospital are populated by medical specialists and subspecialists, it is incumbent on the pathology department to provide state-of-the-art diagnostic services in a variety of areas. Because no single pathologist or even a small group can be expected to acquire and maintain the enormous depth and variety of expertise needed in a large referral hospital, many diagnostic pathologists, each specializing in a different area, are necessary. For example, the pediatric pathologist primarily interacts with the pediatricians, the neuropathologist with the neurologists and neurosurgeons, and the microbiologist with the clinicians specializing in infectious diseases. The subspecialty pathologist, by definition, is expected to deal effectively with highly unusual or complicated cases. In contrast to the large teaching hospital, the smaller community hospital is usually staffed with relatively few pathologists, each of whom is likely to be a generalist in anatomic pathology with additional responsibility for one or two aspects of clinical pathology. Although most practice within a hospital setting, some pathologists practice in free-standing private reference laboratories, which provide diagnostic services on specimens obtained from physicians' offices or hospitals.

VIII. THE PATHOLOGIST AND THE PRACTICE OF MEDICINE AT LARGE

In a narrow sense, the pathologist functions to perform and interpret diagnostic procedures on specimens obtained from patients; however, more broadly, the pathologist serves as a true clinical consultant very much involved in the care of the individual patient. This aspect of pathology practice is particularly pronounced in smaller hospitals where the clinical services are less highly specialized. The pathologist may be asked to advise whether or not certain tests are appropriate for a given patient's problem, in addition to helping to interpret the results to follow. In addition to the role of advisor in the diagnosis and management of individual patients, the pathologist often serves as a general educational resource for the hospital staff. The pathologist is expected to keep abreast of advances in diagnostic medicine and biomedical research and is, thus, in an excellent position to bring new findings and concepts to the attention of clinical colleagues. The educational role of the pathologist is much in evidence when individual cases are discussed in departmental or hospital-wide conferences. In a common format, a case history is presented, followed by a discussion of the clinical and laboratory findings by the clinician and the pathologist. Such conferences provide stimulating interchanges of information and views from which all can learn.

The pathologist is also in a position to monitor the quality of medicine practiced in the hospital at large. Are too many tests being ordered? Are some tests ordered when others would be more useful or when others should be ordered first as a screening procedure? Are too many operations being performed in which pathological examination of removed tissues and organs discloses no abnormality? In short, the pathologist works closely with clinical colleagues and hospital administrators to make sure that laboratory tests are ordered appropriately and performed correctly, promptly, and at the least possible cost.

IX. CONCLUSION

In conclusion, diagnostic pathology is an intellectually challenging, diverse, and heterogeneous branch of medicine that links basic biomedical research to clinical medical practice. As such, the pathologist must be continually on the alert to adapt emerging medical and scientific knowledge to daily patient-related work.

BIBLIOGRAPHY

Cotran, R. S., Kumar, V., and Robbins, S. L. (1994). "Pathologic Basis of Disease," 5th Ed. Saunders, Philadelphia.

Henry, J. B. (ed.) (1996). "Clinical Diagnosis and Management by Laboratory Methods," 19th Ed. Saunders, Philadelphia.

Hill, R. B., and Anderson, R. E. (1988). "The Autopsy: Medical Practice and Public Policy." Butterworths, Boston.

Intersociety Committee on Pathology Information. (1996). "Pathology as a Career in Medicine," revised ed. Bethesda, MD.

Rubin, E., and Farber, J. L. (eds.) (1994) "Pathology," 2nd Ed. Lippincott, Philadelphia.

Smith, R. D. (1986). Medical education and its impact on the future of pathology. *Arch. Pathol. Lab. Med.* **110,** 296–298.

Smith, R. D., Anderson, R. E., and Benson, E. S. (1985). Manpower needs and supply in academic pathology. *Arch. Pathol. Lab. Med.* **109,** 889–893.

Pathophysiology of the Upper Respiratory Tract

KEVIN T. MORGAN
Glaxo Wellcome, Inc., North Carolina

DONALD A. LEOPOLD
Johns Hopkins Bayview Medical Center

GLOSSARY

Cribriform plate Thin plate of bone, separating the front of the brain from the olfactory or upper region of the nose, through which the olfactory nerves pass carrying the sense of smell

Lamina propria Layer of tissue supporting the nasal epithelium that contains many blood vessels, nerves, glands, and other structures

Pathophysiology Study of basic processes responsible for functional disorders of an organism

Reflex Involuntary action in response to a stimulus

Rhinorrhea Free discharge of thin nasal mucus

Serous secretion Clear fluid that moistens membranes

Sinusitus Inflammation of the sinuses

Transudation Passage of serum or other body fluid through a membrane or tissue surface

Viscoelastic Specific physical characteristics that give mucus both fluid and solid properties

MUCH CAN BE LEARNED ABOUT BIOLOGICAL processes from studies of the pathophysiology of disease. With the exception of inflammation and infections, injury induced by inhaled air pollutants and trauma, and the effects of cardiovascular disease and drugs, the respiratory tract is generally trouble free. However, the upper respiratory tract (nose, pharynx, and larynx) does present an important portal of entry for many pathogenic (disease causing) organisms, including viruses, bacteria, and protozoa. The inspired air is a source of irritants, such as toxic chemicals, that can damage the respiratory tract and possibly other organs. Allergens, such as pollen grains or household dust, also enter by this route to play havoc with our upper respiratory tract comfort, in addition to inducing both local and more widespread diseases. Examination of abnormalities associated with diseases and investigation of the mechanisms responsible for these changes can provide insight into the nature of normal body function, as well as reveal characteristics of many adaptive or defensive responses.

The upper respiratory tract carries out a number of activities that improve our chances of survival in a hostile world. Food can be detected using the sense of smell, which is dependent on the activity of olfactory structures in the nose. Olfaction can also provide warning of potentially dangerous airborne materials, such as toxic gases. In animals, and probably also in humans, the sense of smell plays a role in sexual activity. The upper respiratory tract, through its air conditioning activity, contributes to general body function by providing protection to the more delicate lining of the lower respiratory tract. These air conditioning functions include warming, cleaning, and humidifying the inspired air. The activity of this region of the respiratory tract is further complicated by its close relationship with the alimentary tract. The pharynx

ENCYCLOPEDIA OF HUMAN BIOLOGY, Second Edition, VOLUME 6.

and larynx work together as a mechanical unit to permit passage of food or water into the gastrointestinal tract and transport of vomit in the reverse direction. During swallowing and vomiting it is essential that respiration be only briefly interrupted and that harmful materials such as stomach acid not enter the lower respiratory tract, where they can cause severe damage and even death. The upper respiratory tract can carry out these many tasks only because it also is provided with a number of effective defenses of its own. The wide range of functions carried out by the upper respiratory tract is reflected in its structural, physiological, and biochemical complexity.

I. UPPER RESPIRATORY TRACT STRUCTURE AND FUNCTION

The upper respiratory tract provides a continuous though complex connecting passage between the ambient air and the lower respiratory tract. In the nasal passages, this airway is tortuous because of the presence of the nasal turbinates and channels that connect to the nasal sinuses. There are distinct differences between the anatomy of the nasal passages of mammalian species, with humans resembling in some respects certain nonhuman primates (Fig. 1). However, structurally and functionally the nasal passages of the majority of mammals have many features in common. In the pharynx and larynx, the airway is subject to many changes of shape associated with phonation and swallowing. The airway has a moist lining, or mucosa, which exhibits quite specific regional characteristics. The mucosa can be separated histologically and functionally into three major types, each with its own characteristic epithelium: squamous, respiratory, and olfactory. A fourth epithelial type, transitional epithelium, has been described in the nose of rats and nonhuman primates but not as yet in humans. The pharynx is lined by protective squamous epithelium, consistent with the need for this region to handle air, food, water, and occasionally corrosive stomach acid during vomiting. Much of the larynx is lined by squamous epithelium, though distally; where it connects with the trachea, it is lined by respiratory epithelium. [*See* Respiratory System, Anatomy.]

Each of these epithelia has different histological, physiological, and biochemical characteristics that influence its susceptibility to specific diseases. Squamous epithelium is much like skin and is not a common site of upper respiratory tract disease. The respiratory epithelium plays a major role in cleaning the inspired air through maintenance of the mucociliary apparatus (see Section II,B). Respiratory epithelium represents a common site of upper respiratory tract disease. Finally, the olfactory epithelium (Fig. 2), which is confined to a limited region of the upper nasal cavity, is a highly specialized extension of the nervous system that is metabolically quite active and exhibits an array of interesting maladies. The olfactory epithelium is responsible for the sense of smell. Underneath each

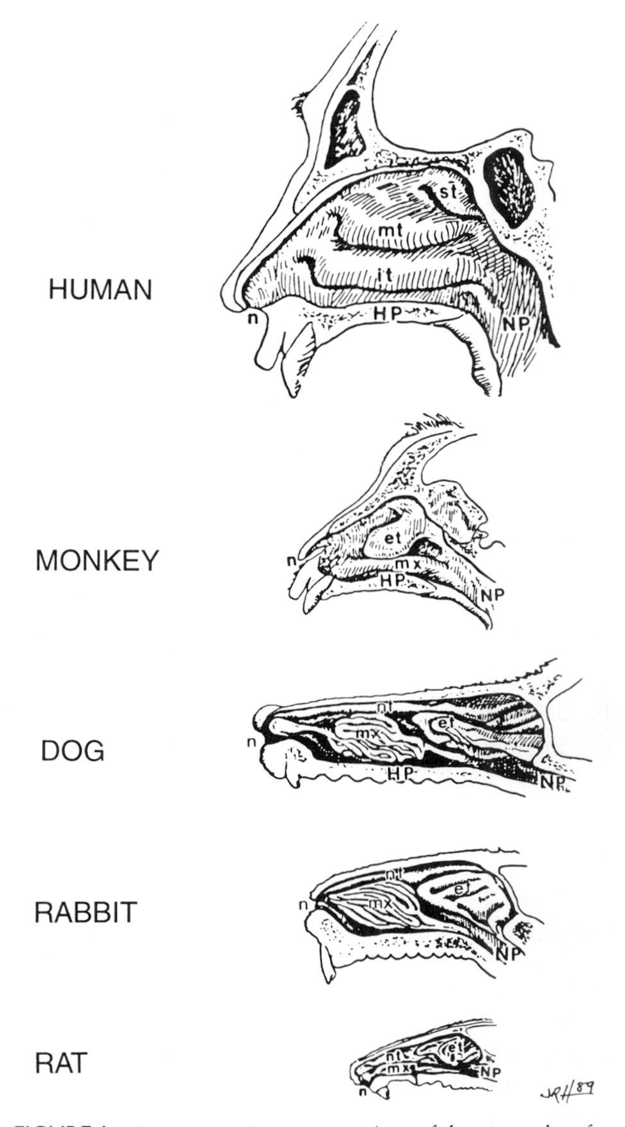

FIGURE 1 Diagrammatic representations of the mucosal surface of the nasal passages of the human, squirrel monkey, dog, rabbit, and rat to show internal anatomy and location of selected structures. HP, hard palate; n, naris; NP, nasopharynx; et, ethmoid turbinate; nt, nasoturbinate; mx, maxilloturbinate; mt, middle turbinate; it, inferior turbinate; st, superior turbinate. [From J. R. Harkema (1991). *Toxicol. Pathol.* **19**, 321–336.]

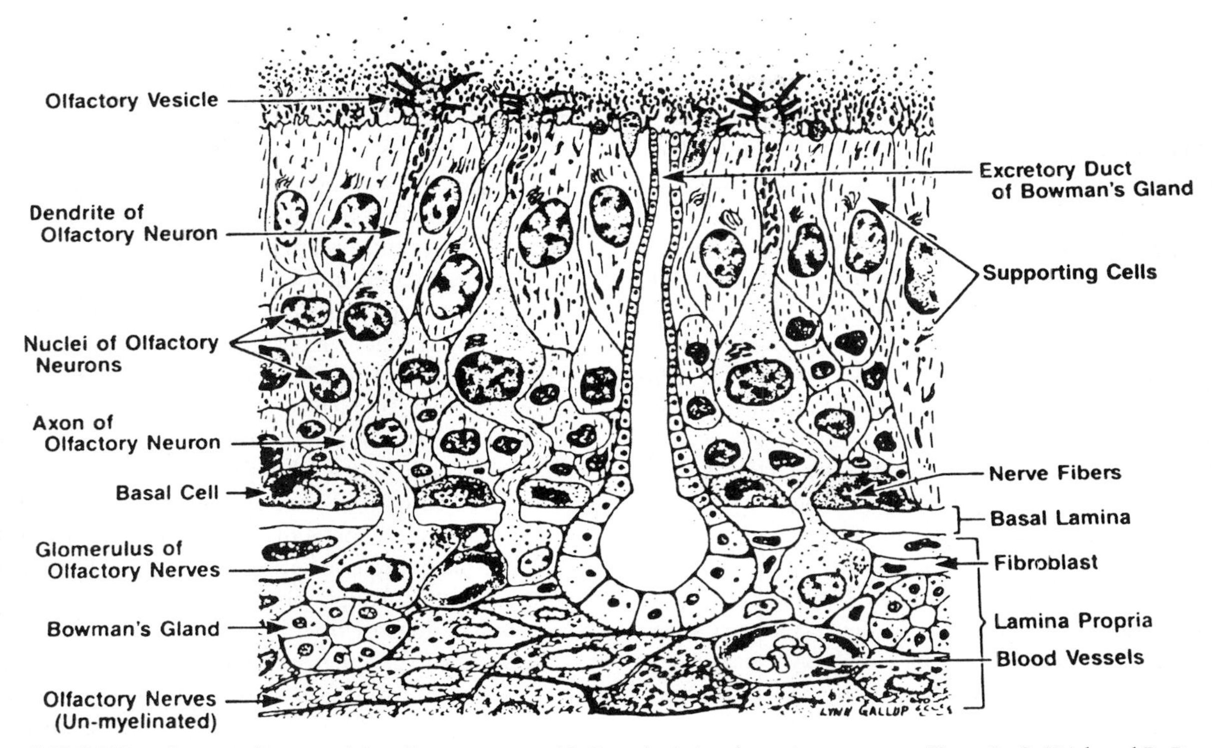

FIGURE 2 Schematic diagram of the olfactory neuroepithelium depicting the major structures. [From L. C. Uriah and R. R. Maronpot (1990). *Environ. Health Perspect.* 85, 187–208.]

of these epithelia lie many glands, nerves, and blood vessels that also have regional characteristics. During studies of respiratory disease it is important to be aware of these topographic features of the upper respiratory tract lining and to discern whether changes are localized to specific regions or mucosal types.

Diseases of the upper respiratory tract are frequently induced by inhaled materials, as the nostrils provide a direct portal of entry for many airborne gases or particles. The nose is extremely effective at removing inhaled materials from inspired air. If you blow your nose after exposure to a dusty environment, the material you see deposited in your handkerchief was removed by your nose from the inspired air, demonstrating the ability of this system to protect your lungs. In contrast to adult humans, who can breathe both nasally and oro-nasally, human infants and laboratory rats are obligate nose breathers. This represents an important distinction, as mouth breathing has been shown to permit deeper penetration of gases and aerosols into the lower respiratory tract, presumably as a result of the loss of the effective scrubbing action of the nose. The physicochemical properties of inspired materials play a major role in the effectiveness of the nose to scrub them from the inspired air. Smaller particles (2–5 μm in diameter) and certain gases and vapors (carbon monoxide, acetone, phosgene) readily penetrate this defensive system to enter the lungs. Large particles (over 10 μm in diameter) are almost entirely removed by the nose, as are highly water soluble gases, such as formaldehyde and sulfur dioxide, and possibly also ultrafine particles (<0.2 μm in diameter). It is the removal of these materials, which include potentially infectious agents such as bacteria and viruses, that puts the upper respiratory tract at risk of disease. The upper airways, however, are provided with elaborate defenses.

II. UPPER RESPIRATORY TRACT DEFENSES

A. Reflexes

The upper respiratory tract possesses a number of defensive or protective reflex responses. The normal function of these reflexes is to:

1. reduce or prevent further entry of invading agents, such as occurs in reflex apnea (cessation of breathing) on exposure to irritant gases;

2. remove the offending agent, as in coughing or sneezing; and
3. carry out adjustments of airway or vascular tone, or changes in quantity and nature of serous or mucus secretions as a component of complex protective responses, including inflammation.

Many of these normal reflexes are exhibited as disease symptoms. The common cold provides many examples of normal reflex defenses, such as coughing, sneezing, increased nasal serous secretions, and changes in vascular perfusion of the nasal mucosa, which are clinically manifested as partial obstruction of the nasal airway. Interestingly, the trigeminal reflex, which causes depression of respiratory rate in laboratory rodents, has been used extensively to assess the toxicity of inhaled irritants and to assess interspecies differences in the amount of potentially toxic airborne materials deposited in the respiratory tract. For researchers studying upper respiratory tract diseases, a wealth of information is to be gained from the way in which these reflexes respond to infectious agents and noxious air polllutants.

B. Mucociliary Clearance

The major routes of inspiratory airflow in the nose are lined by respiratory epithelium that provides an important airway defense system, the mucociliary apparatus. This apparatus plays a major role in cleansing the inspired air of inhaled gases and particulates. The nose filters about 10,000 liters of air per day, and the delicate, but very effective, mucociliary apparatus is readily perturbed by certain air pollutants. Altered mucociliary function can provide a sensitive indicator of early nasal damage.

The mucociliary apparatus derives its name from two of its major components, mucus and cilia. The mucociliary apparatus provides a continual "river" of slimy mucus that is driven over the mucosal surface by the combined action of millions of tiny, hair-like processes, the cilia. The human nose produces about a liter of mucus per day. A mucociliary apparatus is present in both the upper and lower respiratory tracts. In the nose the mucus flows posteriorly toward the nasopharynx, whereas in the trachea, bronchi, and other lower airways mucus flows upward toward the pharynx. This mucus, with any entrapped materials, is eventually swallowed. In the nose, mucus flow patterns are highly organized. In laboratory animals it has been found that not only mucus flow patterns but also mucus flow rates, ciliary length, and beat frequency are highly site specific. Mucus flow rates range from less than 1 mm/min to more than 2 cm/min. Understanding the pathophysiology of the mucociliary apparatus requires knowledge of the cellular and subcellular components of this system.

A wide range of materials has been shown to damage the mucociliary apparatus, including cigarette smoke, toxic gases, and viral and bacterial infections. In the upper respiratory tract, minor or even extensive damage to the mucociliary apparatus may be tolerated with little evident disease. However, in the lower respiratory tract, disturbance of this system can be life-threatening as a result of obstruction of the small airways, as can occur in patients with chronic bronchitis, a disease associated with cigarette smoking, or cystic fibrosis, a genetically controlled condition that inflicts considerable suffering on affected children and young adults through obstruction of the lower airways with plugs of thick mucus.

Because of the inherent complexity of the system, however, determination of the pathophysiology of these responses is difficult. Defects in mucus flow, and thus reduced clearance efficiency, may result from changes in the viscoelastic properties of mucus, modification of the depth or consistency of the liquid layer (hypophase) that lies beneath the flowing surface blanket of watery, sticky mucus, altered ciliary beat, cellular damage in the supporting epithelium, or disruption of normal function of glands, blood vessels, and nerves in the underlying lamina propria. Furthermore, each of these factors may have a multitude of causes.

It is evident from studies of abnormalities of the mucociliary apparatus that much remains to be learned about this highly effective airway defense system. If material deposited on the nasal airway surface penetrates the mucociliary apparatus, a number of other defenses are encountered, including both biochemical and immune systems designed to inactivate or detoxify invading agents or organisms. These defenses can contribute to as well as protect from disease processes in the upper respiratory tract.

C. Metabolism of Xenobiotics

Until recently there was little interest in the metabolism of xenobiotics (foreign chemicals) in the upper respiratory tract. However, with the increasing incidence of nasal cancer and other lesions in the nasal passages of laboratory animals used in toxicology studies designed to predict human risk, there has been increased interest in upper airway biochemistry. One important discovery was the finding that nasal tissues, especially the olfactory mucosa, have high concentrations of a number of enzymes that metabolize xenobiotics. These biochemical studies have been extended by the use of histochemistry, a technique that can localize enzymes to specific cell types. A picture of the complex pattern of enzymes that exist in the nose is gradually emerging. However, much remains to be learned about nasal metabolism, especially in humans.

The localization of enzymes that inactivate or activate xenobiotics plays an important role in interpretation of biochemical events that lead to tissue damage and subsequent disease processes. For certain materials, such as formaldehyde and acetaldehyde, metabolism by their respective dehydrogenases results in inactivation through conversion to a less toxic metabolite. It has recently become evident, however, that a number of chemicals are metabolized in the nasal mucosa to more toxic metabolites. These metabolites can produce degenerative changes and in some cases may result in nasal cancer. Examples of the latter mode of toxicity include effects of certain nasal carcinogens, including several nitrosamines and phenactin, which induce lesions in the olfactory mucosa. The site specificity of these responses has been attributed to the high levels of cytochromes P-450 in this region of the nose. In fact, recent immunocytochemical studies have demonstrated that the P-450 activity is specifically located in Bowman's glands of the olfactory mucosa and in olfactory epithelial sustentacular cells. Knowledge of enzyme activity and location, and correlation of this information with detailed study of the histopathology of nasal responses to certain chemicals, provides considerable insight into the role of metabolism in nasal disease and disease resistance. [*See* Cytochrome *P*-450.]

D. Antimicrobial and Immune Defenses

A multitude of host defense mechanisms serve to protect the upper respiratory tract from infection. These defenses include mechanical cleansing, humoral and cell-mediated (immune) responses, and interferon. A number of other substances in nasal secretions may also provide protection against infectious agents. Such substances include lactoferrin, an iron-binding protein with a broad-spectrum antibacterial action. Nasal secretions also contain high levels of lysozyme, an enzyme that kills certain bacteria.

Mechanical cleansing of the nose is carried out by the mucociliary apparatus described earlier, as well as by sneezing and blowing the nose. However, the efficacy of these activities in protecting against infections remains to be established. Humoral immunity involves both local secretory activity and systemic antibody responses. In fact, considerable amounts of the immunoglobulin IgA are present in nasal secretions, while there is much less IgG. IgA in these secretions plays a major role in protection of the nose against both bacterial and viral infections. During viral upper respiratory tract infections there may be considerable transudation of serum proteins, particularly IgG and albumin, which may provide an effective nonspecific inflammatory reaction to infection.

The role of cellular immunity in resisting infections of the upper respiratory tract has received much less attention. As in other tissues, however, infiltration by neutrophils, macrophages, and other cell types represents a consistent feature of nasal inflammation. These cellular responses, which are designed to inactivate invading organisms, represent an important component of nasal inflammation induced by infectious agents and noxious chemicals. Furthermore, the nose acts as an important reservoir for microorganisms that can cause human disease, including *Hemophylus influenzae* (a bacterium), common cold viruses, influenza virus, and respiratory syncytial virus. Evidently, a delicate balance exists between these infectious agents and the nasal defenses, about which much remains to be learned.

III. SELECTED UPPER RESPIRATORY TRACT DISEASES

There are many upper respiratory tract diseases. A few issues have been selected for discussion to illustrate the value of studies of disease processes for increasing our understanding of human biology. In many instances, the nature of the mechanisms responsible for upper respiratory tract diseases is still unknown. However, recent work in laboratory animals has shed considerable light on these complex processes. To discuss potential mechanisms of human diseases, it will be necessary to refer frequently to studies in laboratory animals.

Diseases are often named, classified, and ultimately investigated on the basis of the symptoms they evoke. Diseases of the upper respiratory tract are no exception. Through careful study of the symptoms of a disease process, and identification of the mechanisms responsible for them, a great deal can be learned about normal function and the nature and limits of the body's defense systems. Study of the common cold, for instance, has revealed many aspects of responses of the upper respiratory tract to viral infection, the role of nasal secretions and immune defenses, and resolution of the disease process. It is hoped that such studies will one day lead to the development of effective vaccines. Examination of laboratory animals exposed to toxic chemicals has demonstrated the susceptibility of specific cells types in the olfactory epithelium to each of these chemicals. Enzyme histochemical studies, combined with biochemistry, have revealed the nature of subcellular mechanisms responsible for these highly selective effects of certain chemicals. Furthermore, this work has resulted in increased understanding of the relationships between morphological and functional changes, and the kinetics of repair and associated cell replication rate. Studies of nasal cancer in laboratory rats have also identified important molecular events, such as the production of DNA adducts (i.e., chemical addition products) and activation of oncogenes in the nasal passages after exposure to certain carcinogens. This work contributes to understanding normal function through characterization of background levels of adducts and assessment of the role of DNA repair in upper respiratory tract defense. [*See* DNA Repair.]

Three areas of research are briefly discussed: allergic rhinitis, effects of chemical air pollutants, and olfactory dysfunction. Allergic rhinitis was chosen because of its general familiarity. Studies of chemical air pollutants were considered because this is a matter of increasing concern for the public and is an area of active research in both laboratory animals and humans. The issue of olfactory dysfunction was selected because the authors consider this to be of general interest for nonscientists and scientists alike, and it is an area of current research activity for both of the authors.

A. Allergic Rhinitis

Allergic diseases are a major source of illness or disability in people of all ages. Allergic rhinitis presents as a set of symptoms resulting from nasal inflammation induced by allergic or hypersensitivity reactions. Allergic reactions were first described in some of the earliest writings. In fact, the death of one pharaoh was attributed to an allergic reaction to a bee sting in an account provided in Egyptian hieroglyphs. Hippocrates and Galen described allergic reactions in their writings, and they even suggested that they may result from environmental factors. The name "allergy" was given to this condition in 1906 by von Pirquet, who created the term from "allol," the Greek word for "change in the original state." This example demonstrates the common process of disease classification based on a characteristic deviation from the "normal," or healthy, condition. Detailed study of mechanisms of diseases commonly reveals multiple mechanisms inducing a common set of symptoms. Thus, increasing knowledge of mechanisms of disease generally adds to the complexity of this classification process. [*See* Allergy.]

Allergic rhinitis is a common condition that may not be taken seriously by doctors because it is not life-threatening. However, a constantly itchy, excessively runny or stuffy nose disturbs sleep, is socially embarrassing, and can clearly impair the quality of life for affected individuals. It may even induce significant changes in patterns of facial growth. A better understanding of this form of rhinitis came after the discovery of immunoglobulin E (IgE) and its role in allergic disease in the mid-1960s. This work has led to considerable research on the mechanisms responsible for allergic conditions. The result of this work has been improved treatment regimens, which have recently included the development of immunotherapeutic approaches.

A number of conditions must be differentiated from allergic rhinitis, including viral and bacterial infections, septal deviations, sinusitis, Kartagener's syndrome, vasomotor rhinitis, and even responses to certain medications. Diagnosis is based on exclusion of other diseases, distinction between infectious and noninfectious disease, and separation of allergic from nonallergic patients on the basis of history and a number of allergy tests. Rhinitis is divided into three broad groups—allergic, nonallergic, and infectious. There is some overlap, however, as patients may have two or more of these conditions simultaneously.

Understanding the pathophysiology of allergy requires knowledge of the immune system. The immune system consists of a set of cellular and humoral components that interact with many molecular structures (antigens) in a manner that permits distinction of self from non-self. This process is required for the identification of foreign materials, such as invading bacteria or viruses, and activation of systems designed to eliminate the invaders. This system can also respond inappropriately and lead to hypersensitive (allergic) conditions, such as allergic rhinitis. Interestingly, the science of immunology that examines and defines these complex immune processes was originally an outgrowth of microbiology, which occupies itself with the study of infectious or disease-causing organisms. [*See* Immune System.]

Immunologic studies have demonstrated that hypersensitivity reactions occur as a number of distinct types of disease processes (Types I to IV). Allergic reactions may be very rapid, as in immediate hypersensitivity, such as acute allergies to pollen or household dust. This condition is characterized by sneezing, eye irritation, nasal congestion and obstruction, and rhinorrhea. The mechanisms involved are not entirely known, but they involve IgE, mast cells and basophils, and several physiologically active compounds, or mediators, such as histamine, leukotrienes, and prostaglandins. These mediators induce edema of the nasal mucosa, spasm of smooth muscles, attraction of inflammatory cells, activation of platelets, and increased production of mucus. These complex mechanisms are then expressed as the characteristic symptoms of the disease. In contrast, delayed hypersensitivities, such as contact dermatitis, have a slower onset and involve a completely different set of cellular and tissue reactions. Despite differing mechanisms, acute and delayed hypersensitivities do have common features, including release of histamine and consequent pruritis (itching).

B. Air Pollutants, Toxic Chemicals, and Upper Respiratory Tract Lesion Distribution

It is becoming increasingly evident, especially from inhalation toxicology studies in laboratory animals, that upper respiratory tract lesions induced by noxious chemicals or particulates generally occur in specific locations in the nose or larynx. This is true for lesions on the middle turbinate of nickel workers, formaldehyde-induced lesions in the lateral meatus of rats, and responses to cigarette smoke on the dorsal surface of the epiglottis of rats. Certain materials administered by a noninhalation route (ingestion or injection) have also been found to induce nasal lesions in specific regions of the nose, such as those produced by phenaceten and the tobacco-specific nitrosamine[4-(*N*-methyl-*N*-nitrosamino)-1-(3-pyridyl)-1-butanone (NNK).

The site-specific nature of these responses is attributable to (1) regional deposition patterns of inhaled materials, (2) regional tissue susceptibility, or (3) a combination of these factors. Regional deposition may account for the site specificity of lesions as a result of "hot spots" of deposition, resulting in areas that receive a high exposure compared with other regions. This site-specific distribution of deposition is attributable, at least in part, to regional airflow characteristics. For example, the anterior curvature of the middle turbinate is a preferential site of particle deposition in the human nose. This region of the nose is also a frequent site of nasal cancer in nickel workers, presumably as a result of high exposure of this site. Another common site of particle deposition is the region adjacent to the ostium internum, just inside the nasal entrance, where prominent air turbulence is developed, which would be expected to favor deposition. Regional deposition due to airflow characteristics probably also accounts for site specificity of lesions induced by certain toxic gases, such as formaldehyde in rats and rhesus monkeys.

Airflow patterns in the upper respiratory tract are extremely complex, and they can be difficult to study because of the inaccessible nature of these airways, which are encased in bone and cartilage. A number of approaches have been developed to study upper respiratory tract airflow, including the use of airway casts or models. These models, which have been developed in some cases specifically for the purpose of determining airflow patterns for interpretation of toxicology studies, have shed considerable light on the nature of nasal airflow.

The major routes of inspiratory nasal airflow pass over regions of the nose that are lined by respiratory epithelium, with smaller flow streams passing to the olfactory region. Differences in airflow patterns between species may account for differences in the distribution of responses to certain toxic materials, such as formaldehyde. Lesions also occur in specific sites in the larynx, though less work has been published on regional deposition patterns in this organ. Common sites of laryngeal lesions that may be attributable to airflow-related deposition patterns include the posterior (dorsal in rats) surface of the epiglottis and the vocal cords. Factors other than airflow may also account for the presence of specific laryngeal lesions, however. Secondary tuberculous laryngitis in humans, for instance, has been attributed to deposition of tubercle bacilli in the larynx as a result of their transport from lungs in the sputum. Study of deposition of materials in the upper airways has led to considerable research on airflow which in turn has shed light on the important anatomical and physiological features of these complex passageways.

In addition to site-specific deposition, lesions may occur in certain areas of the upper respiratory tract as a result of a regional tissue susceptibility. For instance, inflammation of the sinuses can be exacerbated by poor drainage of inflammatory exudate from these sites. This can result from mucosal swelling in the narrow passageways that normally permit drainage of the sinuses. Activities associated with marked changes in ambient pressure, such as scuba diving and airplane flights, can also initiate sinusitis. Thus the design of sinuses and their system of drainage renders them susceptible to infectious and inflammatory diseases. The more frequent presence of nodules on the vocal cords has been associated with singing or extensive use of the voice. If this association is truly causal in nature, the vibratory function of the vocal cords may account for this apparent susceptibility. Lesions of the olfactory mucosa in rats exposed to a number of chemicals, including nitrosamines, 3-methyl indole, and methyl bromide, have been attributed to site-specific metabolism of these materials, presenting another example of site-specific susceptibility of a tissue resulting in a disease process. In this case, the metabolic processes that may normally play a role in upper respiratory tract defense can induce disease through activation of these materials. The site-specific nature of squamous cell carcinoma of the larynx of tobacco smokers and of the throat of smokers and alcohol drinkers may also provide societally important examples of site-specific susceptibility to toxic chemicals.

C. Olfactory Dysfunction

The perception of chemical stimuli is accomplished almost entirely through the first (olfactory) and the fifth (trigeminal) cranial nerves. The olfactory nerve provides information on the quality and intensity of the odorant, whereas the trigeminal nerve assesses the odorant's pungency. When this system malfunctions, there can be either a change in the intensity of an odorant stimulus (hyperosmia, hyposmia, or anosmia) or a change in its quality (dysosmia). There is a natural range of olfactory ability that differs among individuals, with women generally having a better sense of smell than men. Olfactory function can also be adversely affected by a number of environmental contaminants, including tobacco smoke. Such variables should be considered when attempting to assess responses to experimental variables in olfactory function studies in human subjects. [*See* Olfactory Information Processing.]

The first event in the process of olfactory stimulation is the movement of odorous chemical molecules through the narrow, dark passageways of the nose. Any process that obstructs their flow to the olfactory region at the top of the nasal cavity will decrease olfactory ability. Thus, nasal airflow characteristics play an important role in this process. Some materials may naturally fail to reach the olfactory area in sufficient quantities because they are rapidly removed from the inspired air by the more anterior regions of the nose. One group of conditions that adversely influences the sense of smell through impairment of air passage to the dorsal nose are the *obstructing nasal diseases*. These conditions include the common cold, where the accompanying edema of the nasal membrane physically prevents airflow, and thus transport of odorant molecules, to the olfactory receptors. Nasal allergic conditions can also lead to decreased olfactory ability for the same reason—mucosal edema—although this condition generally has a much longer duration. Obstruction from chronic sinusitis, nasal polyps, or nasal tumors can similarly decrease olfactory ability. Medical or surgical therapy is generally useful for people in this category, since their olfactory ability can return if the obstruction is removed. Individuals who have had a laryngectomy and are forced to breathe through a tracheostome in their neck have poor olfactory ability simply because their respiratory airstream no longer flows past the olfactory receptors.

Changes in the nerves of the olfactory system, in the nose or in the brain, account for most other types of olfactory dysfunction. In laboratory animals exposed to gaseous irritants, such as chlorine or dimethylamine gas, there is extensive destruction of olfactory sensory cells. However, very limited information is available on the functional impact of these changes. In the case of methyl bromide, which induces extensive olfactory epithelial degeneration after a single inhalation exposure, olfactory function is also lost. However, the sense of smell returns to these animals during the subsequent recovery process. The ability of the olfactory epithelium to carry out this repair makes it unique as a neural system. In fact, olfactory sensory neurons are one of the few known postembryonic nerve cells capable of regeneration, as a result of which they have attracted considerable interest by neurobiologists. This is a good example of toxic responses being used for basic research on normal function. Interestingly, in laboratory animals there is frequently information on structural changes in the olfactory epithelium, whereas olfactory function data are generally not available. In human subjects the reverse situation is usually the case; in humans it is difficult to observe all the stages of an olfactory degenerative process because of ethical difficulties associated with such controlled experiments and the invasive nature of sample collection.

There is an unfortunate group of people who have lost some or all of their olfactory ability after an upper respiratory tract infection. This condition is categorized as *postinfectious olfactory dysfunction*. It is thought that the virus that caused the infection entered the olfactory nerve cells and induced modifications that resulted in their loss of function, or that the infection destroyed the sensory cells completely. This condition can occur after most minor upper respiratory tract infections and usually presents as a failure of olfactory ability to return after the nasal stuffiness has cleared. Fortunately, this condition is quite unusual. The loss can be partial or complete, is generally permanent with no known therapy, and no suitable animal model is available to study this condition.

After the odorant molecule has interacted with the primary olfactory neurons, using molecular mechanisms, including second-messenger systems, the chemical information is changed into electrical messages. These messages are produced in the form of action potentials in the sensory neurons. The information is relayed to the olfactory bulb in the brain, which is located on the cranial side of the cribriform plate, through which these nerve fibers pass as many small bundles. The small nerve fibers as they pass through the cribriform plate are susceptible to mechanical damage during head trauma, especially after occipital injury. *Posttraumatic olfactory loss* tends to be severe and shows little general improvement. A curious aspect of this condition is the lack of recovery, even though neurons in the normal individual are constantly regenerating.

Many body functions decline with age, and olfaction is no exception. During the sixth or seventh decades there is a precipitous loss of olfactory function that is thought to be due to loss of primary neurons. This *age-associated olfactory deficit* can be so severe that the average 80-year-old has only half the olfactory ability of a person in his or her thirties. As mentioned earlier, environmental toxins are another major cause of olfactory dysfunction. Individuals who persistently inhale the vapors of acids, aldehydes, heavy metals, solvents, and other sometimes unknown chemicals can suffer considerable olfactory epithelial damage and loss of their sense of smell. Furthermore, nasal cancer, which may be a consequence of environmental pollutants, can also lead to extensive loss of olfactory function through tissue destruction.

Each of the causes of olfactory dysfunction mentioned here can cause either a decrease or a total loss of olfactory ability. Sometimes, individuals will complain of a change in the quality of their sense of smell. This problem may present as either a distortion of the perception of an odorant, such as a rose smelling like vinegar (parosmia), or the perception of an odor when there is no evidence, in the opinion of adjacent observers, of such an odorant being present (phantosmia). The cause of these problems is all too frequently unknown, but they have been noted to occur in conjunction with many of the previously described causes of olfactory loss. People with these problems are greatly distressed, since much of their life is no longer perceived as pleasant. Such concern is frequently accompanied by loss of weight and possibly clinical depression. However, new therapies are becoming available to relieve these problems. When our olfactory function is normal, we are generally not aware of the pleasure and protection it provides. Dysfunction of our sense of smell, however, will quickly remind us of its importance for our quality of life.

IV. SUMMARY

Much can be learned about normal function from disease processes. Allergic rhinitis represents "inappropriate" responses of normal immune defenses to allergens to yield a disease process characterized by excessive activity of this system as symptomatic hypersensitivity. Studies of the sense of smell have revealed the complex nature of this sensory system, the considerable metabolic capacity of the olfactory mucosa, and the role that metabolism plays in responses of this mucosa to toxic chemicals. Toxicology studies in laboratory animals, designed to assess human risks from exposure to air pollutants, are shedding light on normal upper airway function through the identification of critical physiological, metabolic, and molecular biological features of the airway mucosa and its defenses. Research directed at understanding disease processes is continuing to reveal the exquisite design and overwhelming complexity of the upper respiratory tract and is contributing in a significant way to furthering the science of human biology.

BIBLIOGRAPHY

Feron, V. J., and Bosland, M. C. (eds.) (1989). "Nasal Carcinogenesis in Rodents: Relevance to Human Health Risk." PUDOC, Wageningen, The Netherlands.

Dahl, A. R. (1988). The effect of cytochrome P-450-dependent metabolism and other enzyme activities on olfaction. *In* "Molecular Neurobiology of the Olfactory System, Molecular, Membraneous, and Cytological Studies" (F. L. Margolis and T. V. Getchell, eds.), pp. 51–70. Plenum, New York.

Getchell, T. V., Doty, R. L., Bartoshuk, L. M., and Snow, J. B. J., Jr. (1991). "Smell and Taste in Health and Disease." Raven, New York.

Korenblat, P. E., and Wedner, H. J. (eds.) (1984). "Allergy, Theory and Practice." Harcourt Brace Jovanovich, Orlando, Florida.

Mygind, N., and Pipkorn, U. (eds.) (1987). "Allergic and Vasomotor Rhinitis: Pathophysiological Aspects." Munksgaard, Copenhagen, Denmark.

Negus, V. (1958). "The Comparative Anatomy and Physiology of the Nose and Paranasal Sinuses." Livingstone, Edinburgh/London.

Proctor, D. F., and Andersen, I. (eds.) (1982). "The Nose: Upper Airway Physiology and the Atmospheric Environment." Elsevier Biomedical Press, New York.

Reznik, G., and Stinson, S. F. (eds.) (1983). "Nasal Tumors in Animals and Man," Vols. I, II, and III. CRC Press, Boca Raton, Florida.

Widdicombe, J. G. (1986). The physiology of the nose. *Clinics in Chest Med.* 7, 159–170.

Pediatrics

MARY ELLEN AVERY
Harvard Medical School

GLOSSARY

Antibiotics Substances (e.g., penicillin) derived from molds or bacteria that are used as drugs to combat certain infections caused by other microorganisms

Gene therapy Insertion of a normal gene into an appropriate progenitor (pluripotential) cell, once the gene and its product (a missing enzyme) are isolated, when an individual is born with inborn errors of metabolism, which result from a missing or altered gene that is necessary for the synthesis of an enzyme

Hilar Pertains to hilum of the lung, or lung root, through which nerves and vessels enter the lung; lymph nodes clustered around the major bronchi are called hilar nodes

Neonatology Specialization in the care, development, and diseases of newborn infants

Oral rehydration therapy Oral administration of a solution of sodium chloride and glucose in a 1 : 1 molar ratio with additional potassium chloride to dehydrated individuals; the mixture is balanced to facilitate absorption from the intestine to restore lost liquid; when given slowly (by spoon) to dehydrated infants, it is well tolerated and obviates the need for more expensive (and often unavailable) intravenous medication

PEDIATRICS (GREEK *PAIES*, CHILD; *IATROS*, PHYSICIAN) is the medical specialty dedicated to the maintenance of health and diagnosis and treatment of diseases of children. Childhood is usually defined as the interval between birth and the completion of adolescence, or through puberty. For children with chronic, handicapping illnesses, a pediatrician may be the primary physician throughout life.

I. HISTORY

A. Seventeenth Century

Little evidence indicates that the illnesses of children were the subject of any serious study or the principal concern of physicians before the seventeenth century. For the most part, the care of young children was handled by nurses and mothers. Thomas Sydenham (1624–1689), an English physician, described many of the diseases common in children, such as scarlet fever, measles, and chorea. The most widely read prenatal book in the English language was by Walter Harris, "Acute Diseases of Infancy," published in 1689. It was considered the most popular pediatric book of its era until it was supplanted in 1784 by Michael Underwood's "A Treatise on the Diseases of Children."

B. Eighteenth Century

In colonial America, the care of children was provided by mothers, and only sick children were seen by physicians. Hospitals, in the modern sense of the word, did not exist until the late eighteenth century; the delivery of infants was entirely in the hands of midwives, who also served as pediatricians. Although we do not have accurate figures, the mortality among infants in colonial America was staggering. Illnesses rarely seen today, such as diphtheria, cholera, typhoid fever, and smallpox, were among the hazards of childhood. In the absence of understanding the causes of these disorders, let alone the principles of hydration and the use of antibiotics or immunization, treatment of most of the diseases was bloodletting, sweating, or purging.

In the eighteenth century in Salem, Massachusetts, the mortality rate for female children was as high as 313 per 1000 livebirths. The average number of children born per family was between 8 and 10, and only an average of 6 survived to adulthood.

In London, attention to the illnesses of children in England was highlighted by the writings of George Armstrong, who opened the first hospital and dispensary for children in 1769.

C. Nineteenth Century

In the nineteenth century, the population of the United States rose from only 5 million at the beginning of the century to approximately 23 million by midcentury. The first pediatric hospital in North America was established in 1855, when the Children's Hospital of Philadelphia opened for "the reception of children suffering from acute diseases and accidents"; this was more than half a century after Hôpital Des Enfants Malades opened in Paris in 1802. The Great Ormond Street Hospital in London was founded in 1852, and The Children's Hospital, Boston, became the second North American children's hospital when it was established in 1869. Vaccination against smallpox was introduced in approximately 1800, when Benjamin Waterhouse vaccinated his 5-year-old son. Widespread use of vaccination did not occur for another 100 years. In fact, the germ theory of infectious diseases was not widely accepted until the studies of Robert Koch and Louis Pasteur in the late nineteenth century.

Pediatrics as a medical discipline can be dated from the close of the eighteenth to the beginning of the nineteenth century in France. The textbook "Traité des Maladies des Enfants Nouveau-nés et à la Mamelle" was published by Billard in 1828 and is considered to be the first serious text in the field of infant diseases. Contributions of the French included the invention of the stethoscope by René Laennec in 1816, as well as the denouncement of the leading intervention of the times—bloodletting. Meanwhile, infant mortality rates in Boston between 1840 and 1845 showed no improvement over those of 1790. As urban centers developed, the death rate for infants actually increased.

Pediatrics did not emerge as a distinct branch of medicine until the second half of the nineteenth century, a period of enormous population growth in the United States. Before then, most hospitals did not admit infants under 2 years old. By the end of the century, 24 children's hospitals were scattered across the country. and about half of the medical schools had a chair in pediatrics.

The first pediatrician to devote himself full-time to the care of children in the United States was the German-born clinician Abraham Jacobi, who established the first children's clinic at the New York Medical School in 1860. The first full professorship in pediatrics was established by vote of the medical faculty at the Harvard Medical School in 1893 when Thomas Morgan Rotch was appointed full professor. That same year, he presented the first paper on prematurity ever given before the American Pediatric Society, in its fifth year of existence.

The first meeting of the American Pediatric Society was held in September 1889, when Dr. Jacobi delivered his presidential address on "The Relations of Pediatrics to General Medicine." Interestingly, while the first presidents of the Society were (at least part-time) pediatricians, the fourth president was the famous professor of medicine Sir William Osler, who was also greatly interested in diseases of children. During this time (the second half of the nineteenth century), more than one-quarter of the children born in the developed world died before their fifth birthday, and "summer diarrhea" was the cause of nearly half of these deaths.

D. Twentieth Century

The heralding of modern pediatrics in America dates from publication in 1896 of the textbook "The Diseases of Infancy and Childhood," by L. Emmett Holt, which had a major influence on the development of American pediatrics. In its first edition, this textbook recommended hypodermoclysis, the subcutaneous infusion of saline in the treatment of the dehydration from "cholera infantum," otherwise known as summer diarrhea. This was the first effort to infuse liquids into a dehydrated baby, heralding the more systematic understanding of fluid and electrolyte balance, one of the major achievements of the twientieth century.

A major emphasis in the first part of the twentieth century was infant feeding and the gradual movement toward artificial feeding of infants with cow milk. Before that time, throughout human history, infants had been fed human milk, as experience indicated it was the only safe and effective way of feeding babies. Once it became possible to pasteurize cow milk and to add carbohydrate to it and dilute it, various formulas were advocated by those who felt that they had embarked on a "scientific" approach to infant-feeding. However, some pediatricians opposed pasteurization for fear it would decrease the nutritional properties of cow milk.

The emergence of biochemical research in pediatrics, initiated by Holt at the Babies Hospital in New York, was given an enormous boost when John Howland and McKim Marriott demonstrated that the acidosis accompanying diarrhea and dehydration was a consequence of bicarbonate loss in the stools; meanwhile Howland assembled a talented group of scientists and clinicians at the Harriet Lane Home of the Johns Hopkins Hospital, many of whom would later head pediatric departments in other medical schools. Among the major advances in the first part of the twentieth century was awareness of the importance of vitamins in the prevention of rickets and scurvy, understanding of the microbial basis of many infectious diseases that regularly weakened and killed children, and the evolution of antibiotics and antisera, which made it possible to treat many of these diseases. Major advances in public health resulted in nearly universal immunization, thus removing many causes of childhood morbidity and mortality, such as diphtheria and tetanus.

II. THE WORLD'S CHILDREN

A. Developing Countries

Pediatrics, as it emerged in Europe and America, has been appreciated around the world and applied in some centers in most every country in the world. However, the tragic fact, as noted by James Grant in the UNICEF publication "The State of the World's Children, 1988," is that one death in every three in the world is the death of a child under 5 years old. In some developing countries, death rates for children are little different than they were in the nineteenth century in Europe and America. In other words, the progress of recent years has not been extended to more than a fraction of the world's children.

The leading causes of death in poorer countries are diarrheal diseases and malnutrition. Approximately 1 million children die of malaria each year and nearly 2 million die of measles, the latter of which can be prevented by a single vaccination. Acute respiratory infections, including whooping cough, which can be largely prevented by immunization, also contribute to an estimated 0.6 million deaths. Tetanus, which is also preventable, kills 0.8 million infants a year, and many other diseases are exacerbated in the presence of undernutrition, which is nearly universal in some developing countries.

Significant progress is being made in some areas with the advocacy of oral rehydration therapy and with efforts to achieve universal immunization. Smallpox, once a leading killer of children, no longer exists in the world today because of the eradication of the virus. Possibly, poliomyelitis and measles can also be eradicated, because the viruses that cause them have no animal reservoir, the diseases are readily recognizable, and effective immunization is available. If all susceptibles are immunized throughout the world, the viruses will disappear, and the diseases can be eliminated forever. [*See* Infectious Diseases, Pediatric.]

In the United States, infant mortality rates (deaths in the first year of life) were 165/1000 livebirths in 1900 and 29/1000 livebirths in 1950, and fell to approximately 10/1000 livebirths in 1987. In Japan, Scandinavian countries, and some other European countries, infant mortality rates are even lower—<6/1000 livebirths.

III. THE PRESENT ERA

A. The Development of Subspecialties

The last half of the twentieth century has seen the development of multiple subspecialties within pediatrics. The largest of these is neonatology (study of the care, development, and diseases of newborn infants), which has contributed to the improved outlook for survival of low-birth-weight infants in particular (Table I). Other subspecialties in pediatrics include neurology, cardiology, and endocrinology (three of the earliest), followed by the study of infectious diseases, genetics, hematology, oncology, nephrology, gynecology, gastroenterology, and pulmonology, to name a few of an ever-expanding list. In addition, behavioral pediatrics and child psychiatry are important subspecialties of pediatrics and psychiatry. The surgical specialties also have their pediatric counterparts in orthopedics, otolaryngology, and ophthalmology.

As subspecialties and further subdivisions of them evolve, the need for someone to translate the new knowledge into patient care has emerged, and we now recognize the increasing role of the generalist or the primary care pediatrician to do this.

In the sections that follow, the most common illnesses and injuries that affect children's health are discussed. All of these will present to the generalist but often require the services of the subspecialist. Only by working together can the generalist and subspecialist succeed in achieving the common goal of pedi-

TABLE I
Infant, Neonatal, and Postneonatal Mortality Rates in the United States for Selected Years (1993 Highlighted)[a]

Year	Total	Under 28 days	Postneonatal
1915	99.9	44.4	55.5
1920	85.8	41.5	44.3
1930	64.6	35.7	28.9
1940	47.0	28.8	18.2
1950	29.2	20.5	8.7
1960	26.0	18.7	7.3
1970	20.0	15.1	4.9
1975	16.1	11.6	4.5
1980	12.6	8.5	4.1
1981	11.9	8.0	3.9
1982	11.5	7.7	3.8
1983	11.2	7.3	3.9
1984	10.8	7.0	3.8
1985	10.6	7.0	3.7
1986	10.4	6.7	3.6
1987	10.1	6.5	3.6
1988	10.0	6.3	3.6
1989	9.8	6.2	3.6
1990	9.2	5.8	3.4
1991	8.9	5.6	3.4
1992[b]	8.5	5.4	3.1
1993[b]	**8.3**	**5.3**	**3.0**

[a]Adapted from the National Center for Health Statistics (1995). "Vital Statistics of the United States." U.S. Department of Health and Human Services, Hyattsville, MD.
[b]Provisional data.

atrics—the health and well-being of the world's children.

B. Diagnosis and Treatment of Diseases

1. Infectious Diseases and Host Defenses

Collectively, infections are the major causes of morbidity in children and contribute substantially to mortality. Infectious diseases are the consequence of invasion of the body by microorganisms that then induce inflammation. They are classified as viral, bacterial, fungal, protozoal, rickettsial, or parasitic.

The development of effective vaccines against poliomyelitis, measles, diphtheria, pertussis, tetanus, rubella (German measles), and mumps has resulted in the near elimination of these formerly common and sometimes lethal infectious diseases (Table II). In the late 1980s, new vaccines have become available to prevent invasive *Haemophilus* influenza type B infection, which is one of the most common causes of meningitis in infants. A vaccine to prevent varicella infection (chicken pox) is currently being developed. Despite years of sustained research, no vaccine is available for one of the leading causes of death in the world in the 1990s—malaria.

Another monumental achievement has been the development of effective antibiotics, starting with the pioneering observation of Alexander Fleming in the discovery of penicillin in 1928 at St. Mary's Hospital, London; shortly thereafter, Selman Waksman at Rutgers discovered streptomycin. Many major accomplishments evolved from here, including the discovery of isoniazid, which accelerated the decline of tuberculosis, for which no effective vaccine exists even now. Although bacteria quickly developed defenses against some of the antibiotics (antibiotic resistance), a number of new compounds were tested and found effective, and the armamentarium of drugs against not only bacteria but some viruses is continually increasing. Despite vaccines and antibiotics, many viral and parasitic illnesses remain lethal at the present time and one in particular, human immunodeficiency virus, which causes acquired immunodeficiency syndrome (AIDS), is in epidemic proportions in certain regions of the world and may become a pandemic if the organism continues to elude efforts to develop an effective vaccine or a drug to combat the infection. [*See* Antibiotics.]

The body's defenses against infection include the production of specific antibodies and stimulation of cells that can kill the infecting organism. This complex system is known as the immune system and is orchestrated in part by the central nervous system, lymphoid tissues, and cells derived from precursors in the bone marrow called macrophages. [*See* Immune System.]

2. Clinical Genetics

Surely the greatest scientific discovery of our times (1953) was that of the structure of DNA, the chemical of the genes, which inaugurated the era of molecular biology. Although this new science has provided complete insight into some aspects of growth and differentiation, clinical applications in pediatrics remain relatively few. They are, for the most part, most relevant for prenatal diagnosis of some hereditary diseases. Widely applied examples include sickle-cell anemia, thalassemia, and hemophilia A.

Gene therapy is on the horizon and appears to be possible, but as of 1995, sustained success has not

TABLE II

Recommended Childhood Immunization Schedule, United States, January 1995 (Vaccines are listed under the routinely recommended ages; shaded bars indicate range of acceptable ages for vaccination)[a]

Vaccine \ Age	Birth	2 mos	4 mos	6 mos	12[b] mos	15 mos	18 mos	4–6 yrs	11–12 yrs	14–16 yrs
Hepatitis B[c]	HB-1									
		HB-2		HB-3						
Diphtheria, tetanus, pertussis[d]		DTP	DTP	DTP	DTP or DTaP at 15+ m			DTP or DtaP	Td	
H. influenzae type b[e]		Hib	Hib	Hib	Hib					
Polio		OPV	OPV	OPV				OPV		
Measles, mumps, rubella[f]					MMR			MMR or	MMR	

[a]Approved by the Advisory Committee on Immunization Practices (ACIP), the American Academy of Pediatrics, and the American Academy of Family Physicians (AAFP).

[b]Vaccines recommended in the second year of life (12–15 months of age) may be given at either one or two visits.

[c]Infants born to HBsAg-negative mothers should receive the second dose of hepatitis B vaccine between 1 and 4 months of age, provided at least 1 month has elapsed since receipt of the first dose. The third dose is recommended between 6 and 18 months of age. Infants born to HBsAg-positive mothers should receive immunoprophylaxis for hepatitis B with 0.5 ml Hepatitis B Immune Globulin (HBIG) within 12 hr of birth, and 0.5 ml of either Merck Sharpe & Dohme vaccine (Recombivax HB) or of SmithKline Beecham vaccine (Engerix-B) at a separate site. In these infants, the second dose of vaccine is recommended at 1 month of age and the third dose at 6 months of age. All pregnant women should be screened for HBsAg in an early prenatal visit.

[d]The fourth dose of DTP may be administered as early as 12 months of age, provided at least 6 months have elapsed since DTP3. Combined DTP–Hib products may be used when these two vaccines are to be administered simultaneously. DTaP (diphtheria and tetanus toxoids and acellular pertussis vaccine) is licensed for use for the 4th and/or 5th dose of DTP vaccine in children 15 months of age or older and may be preferred for these doses in children in this age group.

[e]Three *H. influenzae* type b conjugate vaccines are available for use in infants: HbOC [HibTITER] (Lederle Praxis); PRP-T [ActHIB; OmniHIB] (Pasteur Mérieux, distributed by SmithKline Beecham; Connaught); and PRP-OMP [PedvaxHIB] (Merck Sharp & Dohme). Children who have received PRP-OMP at 2 and 4 months of age do not require a dose at 6 months of age. After the primary infant Hib conjugate vaccine series is completed, any licensed Hib conjugate vaccine may be used as a booster dose at 12–15 months.

[f]The second dose of MMR vaccine should be administered *either* at 4–6 years of age *or* at 11–12 years of age.

been achieved. Some success has been achieved in animals by insertion of a gene in bone marrow stem cells, which can then be reimplanted in the bone marrow.

The virtual explosion of information in molecular biology has brought important new insights into the pathogenesis of genetically determined disorders, of which over 4500 have been described, each related to a single gene locus. Prenatal diagnosis is feasible by either chromosomal identification or tests on fetal fibroblasts, obtained by amniocentesis before 20 weeks of pregnancy.

Many congenital malformations can also be diagnosed prenatally with the advent of ultrasonography. For example, most major malformations of the heart, nervous system, or urinary tract can be diagnosed before birth, so delivery can be planned in a setting where prompt surgical correction is possible.

3. Cardiology

Throughout the twentieth century, but most significantly in the last half of it, surgeons have pioneered in operative correction of congenital malformations of the heart. Initially described in detail by Maude Abbott in the famous "Atlas of Congenital Malformations," these cardiac lesions could not be treated until a pioneering surgeon, Robert Gross in Boston, tied the first ductus arteriosis. In addition, another surgical pioneer, Alfred Blalock, with his cardiologist colleague Helen Taussig, developed in 1945 the diagnosis and treatment of tetralogy of Fallot, a complex malformation that causes cyanotic congenital heart disease. In the 1980s, with the help of cardiopulmonary bypass, complete repair of the most common major malformations has been undertaken operatively, and a number of them recently have been treated through the use of therapeutic catheterization. Cardiac valves that are damaged and constricted can be dilated with balloons, and septal defects can be closed with plastic devices inserted through a cardiac catheter.

New technology has not only provided superior tools for diagnosis, such as modern imaging with computerized scans and magnetic resonance imaging, but has also brought to the bedside efficient, effective, mechanical ventilators and the capacity to monitor continuously the vital signs (temperature, heart rate, respiratory rate, blood pressure, and oxygen saturation of the blood). In pediatric cardiology in particular, ultrasonography has revolutionized noninvasive diagnosis. First the M-mode echo (motion mode) was used for a one-dimensional view of the heart and then a two-dimensional echo was devised to demonstrate spatial relations. More recently, studies based on the Doppler effect (which detects motion) permitted calculation of flows, and color Doppler adds the capacity to discriminate valvular stenosis and insufficiency as well as intracardiac shunts.

Surgical correction of major lesions, such as transposition of the great vessels, was not possible before cardiopulmonary bypass of blood through an extracorporeal heart–lung machine was achieved. Coupled with deep hypothermia, it is now possible to stop the heart for about 40 min, long enough to enter it and correct major lesions. Cardiac transplantation in children is now performed with constantly improving outcomes.

4. Malignancies

Cancers are the third leading cause of death in children between the ages of 1 and 4 years, lagging only behind injuries and congenital malformations, and are the second leading cause of death between the ages of 5 and 19 years. The most common types of tumors by race are shown in Fig. 1. Fewer than 10,000 cases are diagnosed each year in the United States. The mortality rate has declined by almost 50% in the past 30 years. The most dramatic improvement has been in the kidney tumor of infants, known as Wilm's tumor, and in certain types of leukemia. [*See* Leukemia.]

The most common malignancy of childhood, acute leukemia, had been assumed to be always fatal. The remarkable observations of Sidney Farber, Louis Dia-

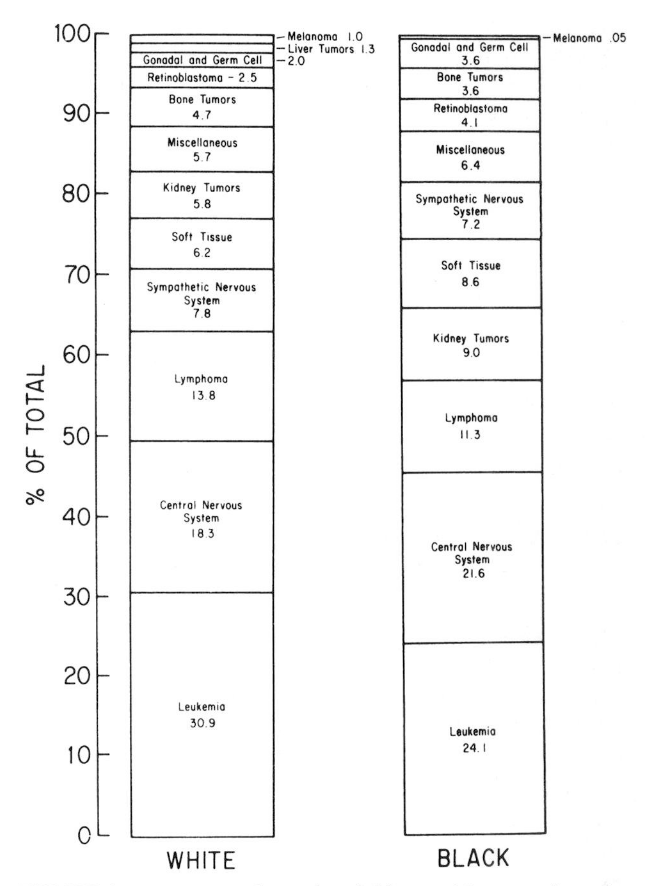

FIGURE 1 Cancer incidence for children ≤15 years of age by site and race. Data obtained from SEER program 1973–1976. [Reproduced, with permission, from C. Pratt (1985). *Pediatr. Clin. North. Am.* **32**, 541.]

mond, and others resulted in the effective use of a folic acid antagonist, aminopterin, in producing a remission (although no cures) in children with leukemia. This pioneering observation, published in 1948, opened the era of cancer chemotherapy. A combined approach with surgical excision, radiation, and chemotherapy was applied to many solid tumors and transformed the outlook from uniform fatality to a high probability of recovery.

Bone marrow transplantation is a drastic but sometimes lifesaving treatment for malignant disease and some immunodeficiency states that have failed chemotherapy, surgery, or radiation. Three types of transplantation are being used in the 1980s: autologous transplants involve the patient's own cells that have been collected and treated *in vitro;* syngeneic transplants are from one identical twin to another; and allogeneic transplants occur between donor and patient who are not identical twins but have similar histocompatibility antigens. (These antigens are responsible for regulating the immune response, which allows identification of self from nonself; the more nearly identical they are, the less the chance of rejection.)

Human bone marrow transplants were first successful in the 1950s, but only in the mid-1970s were they widely used. Worldwide, by the late 1980s, more than 10,000 bone marrow transplants have been done. [*See* Bone Marrow Transplantation.]

5. Neonatology

Neonatology is the study of events that occur in the first 28 days after birth, which is called the neonatal period. Many conditions are either unique to or most prominent during this critical period when the infant must make an adaptation to extrauterine existence.

The survival of newborn infants, especially those of very low birth weight, is dependent on understanding the course of growth and development *in utero,* some insight into placental function, and measurement of the cardiopulmonary adaptations at the time of birth. Essential to the maintenance of life to all infants is meeting their nutritional requirements. The mother's milk is usually optimal. Modification of cow milk formulas can also support growth if maternal milk for some reason is not available.

Many studies elucidated the main physiologic and biochemical events that occur in the perinatal period. Application of this knowledge made it possible to provide interventions that facilitated the establishment of respiration and its support through mechanical ventilation when the infants are unable to sustain lung function on their own. Essentially, the successful use of ventilators led to the establishment of neonatal intensive care units, where skilled nurses and trained neonatologists can produce the appropriate environmental, mechanical, and nutritional supports for growth after birth, even after birth at 25 weeks of gestation.

The advances in neonatal care are dependent on advances in technology, such as the miniaturization of equipment that allows measurements of oxygen and carbon dioxide tension (blood gases) on a drop of blood, so that serial evaluation is possible. Microchemical determinations permit assessment of fluid and electrolyte needs as well as caloric requirements. In incubators, temperature can be controlled by servo devices from the infant's skin. Thus, cooling of the skin will trigger more heat and, conversely, when the infant becomes overheated, the incubator temperature is lowered. This adjustment of the thermal environment is critical for very small infants who lose body heat easily and sometimes are unable to increase metabolism sufficiently to maintain a normal body temperature. The ability to individualize the environment to meet the infant's needs has been a central component of neonatal intensive care.

An aggressive approach to diagnosis and treatment has also greatly improved the outlook for a number of infants. The recognition of the multiple causes of hyperbilirubinemia (jaundice) in the newborn infant and the ability to prevent some and treat others has reduced the risk of one form of cerebral palsy associated with excess bilirubin in the blood during the newborn period.

One of the major causes of respiratory distress in the newborn infant is known as hyaline membrane disease. Advances in understanding this disorder and preventing it have taken place in the past few decades. Infants born before their lung has matured, with respect to the capacity to synthesize and secrete pulmonary surfactants, are predisposed to atelectasis (closure of the terminal airways at end expiration). This situation leads to severe mismatch of ventilation and blood circulation in the lung and often requires ventilatory support. Even with such support, some infants progress from hyaline membrane disease to a chronic lung disease of prematurity, sometimes called bronchopulmonary dysplasia. This latter disorder is presumably the consequence of a combination of injury in an immature lung from the high pressures needed to ventilate the surfactant-deficient lung and, in some instances, the injury produced by the higher inspired oxygen concentration necessary to keep the infant alive in the first days of life.

Other disorders of respiration that can produce major problems are persistent pulmonary hypertension, which can occur in term and postterm infants and, occasionally, those born prematurely. Aspiration of meconium (the material accumulated in the intestinal tract during gestation and passed, usually for the first time, after birth) can produce lethal airway obstruction.

Newborn infants can be born with major malformations, which need to be diagnosed and sometimes treated promptly. These include congenital malformations of the heart, obstruction of the intestine, or deficiencies in the abdominal wall that can be fatal if not treated surgically in the newborn period.

In North America, specialized services for very low birth weight infants, especially pediatric surgical services, cannot be available in all centers where babies are born. Consequently, so-called tertiary care centers, usually associated with major teaching hospitals, have been established, and regionalization of perinatal care is now widely practiced. Where possible, mothers are referred to centers for delivery of the baby when trouble is expected; if the trouble is unexpected and it occurs in a community hospital, transport teams are often available to bring the infant to a neonatal intensive care unit. Transport incubators have been designed so that intensive care can be carried out during the actual period of transport by ambulance or even plane. Secondary care institutions provide individuals trained in the needs of small infants once they have stabilized and no longer require ventilators. Thus, back transport can take place from the tertiary care center to the hospital where the child was born as a step toward discharge home.

6. Pulmonology (Lung Disorders)

The leading causes of admission to most pediatric services in North America in the 1990s are lung disorders in infants and children. In the infant group, hyaline membrane disease, as just discussed, affects approximately 40,000 infants per year in the United States.

In children in the first and second year of life, bronchiolitis, an inflammation of the small airways, occurs in epidemics usually between October and April in the Northern Hemisphere. It may be associated with respiratory syncytial virus but also occurs with parainfluenza and even adenoviruses. The infants characteristically have overinflated lungs and airway-narrowing that leads to prolonged expiration and wheezes, which are indistinguishable from those in asthma.

In older children, asthma is a major problem and is thought to affect 2.4 million children in the United States under the age of 17 years. Recurrent attacks of airway-narrowing from bronchospasm or edema can be relieved with bronchodilators, and in severe attacks with increased inspired oxygen and corticosteroids. Acute attacks can usually be modified by medications that can be given at home.

Pneumonias caused by either bacteria or viruses are common at any age but are most severe in immunocompromised individuals. Many infants have a relative depression in their gamma globulin (important in defense of bacterial infection) at about 3–5 months of age. During this period, they are especially vulnerable to pneumonias. Children are also relatively immunocompromised in the wake of any viral infection but notoriously following measles, which allows bacterial pathogens to produce secondary pneumonias and tracheitis.

The ever-increasing population of individuals on chemotherapy for malignant disease usually have some suppression of their immune system, and a major complication of their treatment is pneumonia, often caused by organisms that are not virulent to a normal host.

Most lower airway infectious disease is associated with upper airway infection. Involvement of tonsils and adenoids as well as sinuses is not uncommon as a precipitating event for attacks of asthma and sometimes pneumonia.

Sudden infant death syndrome is defined as a quiet death that occurs during sleep, usually without evidence of a struggle, in a circumstance in which no adequate explanation is available after an autopsy. It is reported in 1.5/1000 livebirths and is more common in blacks and also in males. The age at greatest risk is 2–8 months but it has occurred in the first month and rarely after 9 months. Other findings include an increased risk in preterm infants and a slightly increased risk in siblings of victims. The cause is unknown but is probably multifactorial. Sudden death has been reported more frequently in infants asleep in the prone position, in which suffocation is more likely than sleep in the lateral or supine (back) position.

The lung is the organ most commonly involved in tuberculosis, and the form of tuberculosis in childhood is strikingly different from that in adults. The primary lesion can occur at any site in the lung, and the disease is usually not localized to the apices of the lung, as it often is in adults. In children, the so-called primary complex is often asymptomatic, although the enlarged hilar nodes can erode a bronchus and produce endobronchial disease. In the first years of life, a danger exists in the spread of the disease through the bloodstream, so that tuberculous meningitis and osteomyelitis are complications of pulmonary infection. The disease is spread by the airborne route and, consequently, the lung is always the portal of entry for the tubercle bacillus. A number of antimicrobial drugs, including isoniazid and rifampin, are usually effective in treating tuberculosis.

Cystic fibrosis is a disorder of the exocrine glands that produces abnormalities in many secretions, including lung mucus. The initial symptoms may be referrable to pancreatic dysfunction in the form of absence of the digestive enzymes, trypsin, and lipases, which results in malabsorption of food. Failure to gain weight and bulky stools may be present. Pulmonary dysfunction may date from early childhood and may be mild or severe. Indeed, the manifestations of the disease vary greatly among individuals. The condition is so strongly associated with elevated concentrations of chloride in the sweat that a positive sweat test is confirmatory of the diagnosis. The inheritance is recessive, and both parents must be carriers of the abnormal genes. The common form of cystic fibrosis is at least related to a gene on the long arm of chromosome 7. Many mutations (over 200) associated with cystic fibrosis have been identified through linkage studies. Each gene probably specifies a protein associated with epithelial cell chloride channels. [*See* Cystic Fibrosis, Molecular Genetics.]

Affected children often have progressive colonization of their lungs with one or another *Pseudomonas* organism, which is often resistant to antibiotics. The predisposition to purulent pulmonary disease is clearly the consequence of some abnormality within the lung, because a few individuals with cystic fibrosis who have had lung transplants from normal individuals have not had recurrence of the disease in the transplanted lung.

Cystic fibrosis is one of the most common hereditary disorders of the white population. It has a prevalence of about 1/2000 births. The estimated prevalence in the black population is 1/17,000 births, and it is rarest among Asians.

Great strides have been made in prolonging life and improving the quality of life for affected individuals. Cystic fibrosis was formerly considered lethal in childhood, but most individuals with this disorder are now living into their adult years, but as they reach adulthood many of them require repeated hospitalizations for aggressive physical therapy to dislodge their impacted pulmonary secretions and intensive intravenous antibiotic therapy to try to reduce some of the bacterial flora in the lungs. Death is usually associated with pulmonary failure.

7. Gastroenterology

Among the most common symptoms of gastrointestinal disorders in children are vomiting and diarrhea. Diarrheal diseases remain a major cause of morbidity and mortality in many parts of the world and are a continuing reason for hospitalization of infants in the developed world. The cause is most often bacterial or viral contamination of food or water (which can be for a number of different reasons). Although oral rehydration therapy has had a major impact on reduction of deaths from dehydration, some infants sustain water losses through the inflamed intestinal tract to the extent that they are in a state of shock and require intravenous support. The ability to give appropriate concentrations of fluids and electrolytes intravenously has been one of the major advances of the twentieth century. The knowledge of the appropriate concentration of electrolytes and glucose in oral solutions is of lifesaving importance. The World Health Organization and UNICEF have distributed electrolyte and glucose packets, diluteable in water, at very low cost in most developing countries, greatly reducing deaths caused by dehydration that is diarrhea-related. An appropriate oral rehydration solution is usually well tolerated even if infants have been vomiting. It must be given slowly at brief intervals, usually by spoon.

Occasionally, infants do not respond to oral electrolyte solutions, and after a few days attention to caloric intake becomes very important. Intravenous alimentation has been a major advance in this context. Solutions of amino acids, lipids, and glucose can be given intravenously in situations where the intestinal tract must be put to rest to restore the integrity of the mucosa.

Surgical correction of major malformations, such as omphalocele (herniation of the intestinal contents through a defect in the abdominal wall), and relief of intestinal obstructions have been made possible by the ability to maintain the nutritional status of the infant through intravenous alimentation. Among older children, ulcerative colitis and Crohn disease are serious, chronic, inflammatory diseases of the bowel, and they too are occasionally treated by "resting the bowel" and maintaining nutrition with long-term intravenous alimentation.

8. Endocrinology and Metabolism

Endocrinology is the study of the chemical messengers (hormones) that control aspects of growth and metabolism. Pediatric endocrinologists are most often consulted about disorders of growth (too tall, too short) or excessive weight or failure to gain weight. Congenital abnormalities of the reproductive system and external genitalia are other disorders best understood in terms of hormonal imbalance during fetal life. [*See* Endocrine System.]

The endocrine glands of the body include the anterior pituitary, which secretes growth hormone and adrenocorticotropic hormone (ACTH), thyroid-stimulating hormone, gonadotropic hormone, and prolactin. Deficiency of growth hormone results in growth failure in childhood. Recombinant growth hormone produced by genetic engineering is available and effective. Excess growth hormone results in gigantism:

after closure of epiphyses of long bones, the excessive growth is in the face, hands, and feet, a condition called acromegaly. The hormones of the posterior pituitary are vasopressin, also called antidiuretic hormone (a deficiency of which will result in polyuria and a condition referred to as diabetes insipidus), and oxytocin (which promotes milk ejection in mothers who breast-feed). The thyroid gland synthesizes thyroxin, which regulates the level of metabolism in all the cells of the body. Deficient thyroxin results in cretinism, which was formerly an important cause of mental and physical retardation. Routine screening of blood of newborn infants for thyroid activity reveals that about 1/4000 infants are deficient. Early treatment is effective. Excessive thyroxin increases metabolism, and weight loss, tachycardia, and heat intolerance are among the symptoms of an overactive thyroid. [*See* Thyroid Gland and Its Hormones.]

The parathyroid glands synthesize parathyroid hormone, whose principal known target organs are bone and kidney. A lowered concentration of calcium in the blood is the most important factor promoting secretion of parathyroid hormone, which in turn causes increased reabsorption of calcium and decreased reabsorption of phosphate by the renal tubules. The hormone also increases the formation of an active form of vitamin D in the kidney and, hence, facilitates calcium phosphate absorption in the intestinal tract. Deficiency of parathyroid hormone results in low serum calcium levels, which can result in seizures. Loss of calcium from bone results in soft bones and curvature of growing bones, known as rickets. [*See* Parathyroid Gland and Hormone.]

The adrenal glands consist of a cortex that secretes cortisol, sex hormones (particularly androgens), and mineral corticoids and a medulla that produces catecholamines. Overactivity of the adrenal cortex in fetal life results in masculinization of the female from excessive androgens, a condition known as the adrenogenital syndrome. Gonads, both testes and ovaries, are capable of producing androgens and estrogens; however, under normal circumstances the testes produce predominantly androgens, and the ovaries produce primarily estrogen. The placenta should be viewed as another endocrine organ that secretes many hormones essential for a normal pregnancy. After the first few weeks of fertilization, progesterone becomes the major and indispensable hormone of pregnancy. Human chorionic gonadotrophin is another hormone of pregnancy that circulates in the mother's bloodstream and is the basis of a pregnancy test. [*See* Adrenal Gland.]

The pancreas contains islets of cells that produce insulin, glucagon, and somatostatin. Insulin and glucagon regulate circulating glucose levels; insulin deficiency results in diabetes mellitus, insulin excess in hypoglycemia. Diabetes onset in childhood is usually sporadic and may follow a viral illness such as mumps, which can affect pancreatic islet cells. Such individuals become insulin-dependent for life. Only pancreatic transplantation offers a possibility for cure but is not now recommended in children. [*See* Insulin and Glucagon.]

Metabolism refers to the biochemical processes within cells that maintain their integrity and function. In general, the field of clinical metabolism involves the study of various disorders of chemicals important for life. Most "inborn errors of metabolism" are now understood in terms of their mode of inheritance and their biochemical basis. Some are treated by withholding the substrate, which cannot be metabolized (e.g., a low-phenylalanine diet in individuals with phenylketonuria, or no galactose in the diet of an infant who cannot metabolize it, as in galactosemia). Research in the 1990s is focusing on gene replacement to permit synthesis of the missing gene product (usually an enzyme).

The past three decades have seen discovery of many systems of chemical messengers in addition to the classic ones just described. Cell-to-cell communication (called paracrine function) and within-cell communication by chemical messengers (autocrine function) are central to regulation of metabolism. The tools of molecular biology have made possible the field often identified as cell biology. The ability to disrupt a chemical messenger selectively by inactivating a gene is now doing for endocrinology what organ ablation in animals accomplished a century ago.

Nutrition of infants and children has had much attention through the ages and continues to be a topic of current interest. No serious student of the subject doubts the need for vitamins, minerals, and an appropriate balance of carbohydrate, protein, and fat. A prudent diet for children >2 years old contains 30–40% of calories in fat, with more emphasis on vegetable and fish protein than on meat protein, to lessen risks of atherosclerosis. Amounts of polyunsaturated fats should about equal the amount of saturated fats. Carbohydrates should supply about half the total calories, and protein should be 10–20% from varied sources. Salt intake should not be excessive. Total calories should be adjusted to achieve optimal weight (neither too thin nor too fat). If a family history of premature death from coronary disease is present, diets more restricted in salt and fat intake are recommended. [*See* Nutritional Quality of Foods and Diets.]

During the first 15 years of life, dietary iron intake should be 8–15 mg daily. Iron deficiency results in anemia, which is most common during rapid growth between 6 and 36 months of age. All children should receive the recommended daily vitamin intake in food or as a supplement. [*See* Copper, Iron, and Zinc in Human Metabolism.]

9. Nephrology (Kidney Disorders)

The principal problems confronting the pediatric nephrologist relate to congenital malformations of the kidney and urinary tract infections, which may or may not be associated with anatomic abnormalities. Other functions of the kidney that are the concern of nephrologists include disorders of acid–base balance, which are often associated with inborn errors of function of the renal tubule. Hypertension is controlled by the renin–angiotensin–aldosterone system, as well as extra renal factors such as atrial naturiuretic peptide, vasopressin, and catecholamines. In general, both systolic and diastolic blood pressure rise slowly with age. In children <2 years old, the upper limit of normal blood pressure is about 110/70; by 10 years old, it is about 130/90.

The kidney is sometimes the target for immunologic disorders, one of which is glomerulonephritis. This condition, associated with bloody urine, is the result of increased permeability of the capillaries in the glomeruli of the kidneys (and often a reduction of the level of circulating complement C3). Hypertension is common. The condition sometimes is preceded by a pharyngeal or cutaneous infection with group A beta hemolytic streptococci and usually resolves after 1–2 weeks.

Another glomerular lesion that leads to increased protein losses in the urine with consequent accumulation of fluid in the body is nephrosis. The condition may be recurrent and chronic, but in about 85% of the cases it is of gradual onset and self-limited. Serum complement C3 is normal in nephrosis, and the condition usually is responsive to corticosteroids. A more serious form, but also fortunately rarer, is the congenital nephrotic syndrome, which may require support of the infant for several months of life to permit adequate growth for kidney transplantation. Renal transplantation is now so successful that virtually any child with chronic renal failure should be considered a candidate regardless of age. Immunosuppression is necessary when the donor kidney is not of identical tissue type.

10. Bone Disorders

Bone disorders of children can be related to inborn errors of metabolism, disturbances in calcium and phosphorus balance, infection, and immunologic disorders, such as rheumatoid arthritis. Adolescents, particularly those who have had excessive stretching of the back ligaments during sports or dancing, for example, may develop curvature of the spine known as scoliosis. Early detection of scoliosis is important, as it is possible to insert a rod that will maintain the spin in a straight position until the vertebrae adequately fuse. Surgery is usually restricted to curves with angles $>40–50°$. Study of the natural history of scoliosis suggests that curves $<40°$ will not progress further at the end of the growth spurt, whereas those $>50°$ tend to be progressive.

11. Neurology and Behavioral Disorders

In pediatrics, neurology encompasses a wide variety of congenital defects: brain injury associated with asphyxia at birth, infections of the nervous system, and disorders associated with convulsions and coma.

Mental development in children proceeds in a series of successive stages. Within a few hours of a normal birth, an infant fixes on faces and has a generally alert appearance. Enhancement of memory continues during the first months of life. By the end of 1 year, the infant is formulating speech sounds, which during the second year gradually merge into connected words and sentences. During childhood, development increases to more complex levels of performance, much of which depends on appropriate stimulation and learning experiences.

The period of human development is relatively long compared to that of most animals and permits time for the brain to acquire and store information and achieve the capacity for reasoning, thus allowing the individual to participate in complex social and technological domains.

In general, the neurologist approaches the patient with the idea of localizing in space and in time the defect that manifests itself as dysfunction of the nervous system and ascertaining its cause. Pediatric neurologists are concerned with developmental delays in motor function, as well as cognitive function, and in speech and language. Infants and children are also subject to some hereditary disorders that lead to progression of neurologic dysfunction by virtue of accumulation of abnormal substances in neurons (Tay-Sachs disease) and demyelinating disorders such as multiple sclerosis and the neurologic lesion of AIDS.

In addition, some fixed disorders (i.e., not progressive) include all forms of motor disorders, collectively called cerebral palsy.

Seizures in childhood may be from many different causes. The most common are simple febrile seizures, which are usually generalized and occur most often from 6 months to 6 years of age and are, by definition, under 15 min in duration. They are usually single, and full recovery is the rule. More complex seizures require diagnostic investigation to explore the possible causes, such as infections, tumors, malformations, or strokes (vascular occlusions). When the cause cannot be defined, the condition is known as epilepsy. Most of the time, individuals prone to seizures can become symptom-free with appropriate medications. [*See* Epilepsies; Seizure Generation, Subcortical Mechanisms.]

Neurologists, psychiatrists, and psychologists are concerned with behavioral disorders, some of which are thought to have a biochemical basis. Major depressive disorders can occur in children and adolescents. Young people are also prone to a chronic eating disorder characterized by loss of weight and distortion of body image (anorexia nervosa) or excessive eating followed by induced vomiting (bulimia). [*See* Eating Disorders.]

An attention-deficit hyperactivity disorder is characterized by a short attention span, high distractibility, and impulsive behavior. The biologic basis of this relatively common dysfunction in childhood is not known.

A serious pervasive developmental disorder known as infantile autism can lead to a lifetime of impaired ability to communicate verbally with others. These individuals have abnormal motor behavior, emotional lability, and sometimes self-harming behavior. This condition has been associated with mental retardation, although this is not a consistent finding. Approximately one-third of affected individuals can learn to function in sheltered or supervised environments, but two-thirds of affected individuals need constant supervision and support throughout life.

12. Skin Disorders

A wide variety of diseases are manifest in the skin. In general, the most common disorders are simple infectious lesions (such as impetigo), contact dermatitis (such as poison ivy), insect bites, and hypersensitivity lesions (such as hives).

Acne is a common inflammatory condition of skin that appears predominantly over the face and shoulders, about 1–2 years before the onset of puberty. It results from androgen stimulation of sebaceous (oily) glands. Obstruction of the glands promotes accumulation of secretions (whiteheads) that in time become open and darken as blackheads; when infected they are pimples. Acne may be drug-induced at any age, most commonly by cortisone or similar medications.

A number of more serious congenital lesions are known to involve the skin. These include diseases such as neurofibromatosis (tumors of peripheral nerves that can appear as lumps under the skin) and epidermolysis bullosa, a serious inherited disorder of the epidermis, which leads to peeling of the skin and in its severe form can be fatal.

A group of scaly lesions are also disorders of the epidermis, one of which is psoriasis, which can be present in infancy and is likely to recur; in some instances, it persists throughout life. Another skin disorder is eczema, an inflammatory disorder most common in young children who have remission and exacerbations over a 3- to 5-year period. Intense itching and weeping of the lesions are characteristic. A family history of allergic disorders is common.

13. Injuries and Substance Abuse

a. Injuries

Each year, more than 22,000 children die in the United States of injury-related causes, most commonly related to motor-vehicle accidents. This statistic makes injuries the leading cause of death beyond the neonatal period. They exceed deaths from cancer, congenital anomalies, and heart disease all together. An estimated 20% of all children will sustain at least one injury that requires medical attention; 17% of all pediatric hospitalizations nationwide are injury-related. The morbidity from injuries is also severe, and they can impose lifelong serious disability.

b. Substance Abuse

Substance abuse is defined as the intentional introduction of a drug into the body for stimulation and pleasure. Some of the drugs are labeled "recreational," which is unfortunate because most of them, such as tobacco and alcohol, ingested over a period of time can result in addiction as well as injury to the lung (cigarette smoke) and to the brain or liver (alcohol).

More powerful stimulants that have come into wide use for the sake of artificial stimulation include amphetamines, cocaine, and the hallucinogens marijuana, lysergic acid diethylamide (LSD), and phencyclidine (PSP or angel dust); sedative hypnotics, such as barbiturates and benzodiazepines, especially valium, are also widely used. Among some children, glue-

and hydrocarbon-sniffing are practices that can produce very dangerous side effects with injury to the heart, brain, and liver. Cardiac arrythmias and sudden death have been reported. [*See* Cocaine and Stimulant Addiction.]

The rising use of these agents among high school students in particular is a cause of major concern. In a 1986 study, more than half of high school seniors reported using marijuana, and about 5% used it daily; cocaine was used by about 13% of high school students. The form of cocaine known as "crack" is particularly dangerous and has been associated with sudden death among its users. [*See* Marijuana and Canabinoids.]

c. Poisoning

Childhood poisonings continue to be an important public health problem. A major nationwide effort to respond promptly to poisonings has helped to establish poison control centers, which respond to over 1,500,000 calls each year. The approach must be prevention, which will require a major coordination of educators to inform school students of the dangers of exposure to poisons, even when they are labeled recreational drugs, and efforts to restrict availability of some of the more dangerous and potentially lethal ones, such as cocaine.

d. Lead Poisoning

Lead poisoning remains a significant problem for the fetus *in utero,* as well as for infants and children. Lead is ubiquitous in the environment, deriving primarily from automobile exhaust, produced by lead-containing gasoline. Also, paint in older houses being renovated produce a lead-containing dust that can be inhaled, and sometimes the water supply is affected when pipes are soldered with lead. Rarely, acid fruit juices kept in ceramic pitchers with lead paint glazes can leach enough lead into the juice to cause acute and even fatal poisoning. Burning car batteries can cause massive lead intoxication from the inhalation of fumes.

Elevated lead levels in blood have been reported in approximately 12% of black children and 2% of white children in the United States, mostly in those living in urban areas and in old housing with lead-containing plaster and paint. The main source of exposure for adults is the diet, with products such as bone meal (which may be contaminated with lead) and drinking water from lead leached out of pipes that are usually in older homes. Even low levels of exposure before birth are associated with significantly slower achievement of developmental milestones in the first years after birth. Acutely, the symptoms may relate to gastrointestinal problems, abdominal pain, anemia, and neurologic abnormalities.

Treatment requires immediate separation of the affected infant or child from the source of lead and decontamination of the environment. Often, hospitalization of the child is required for adequate chelation therapy to mobilize lead from bone and enhance secretion in the urine.

e. Child Abuse

Any form of physical maltreatment of children that inflicts injury, and even neglect, with its consequent emotional deprivation constitutes child abuse. Many state laws mandate reporting even suspicion of injury or neglectful situations so that the situation can be investigated for the protection of the child.

Child abuse and neglect are now recognized as medical conditions with complex aspects that often require a multidisciplinary approach to management. The pediatrician responds to the need to diagnose and treat the injuries and has an obligation to mobilize the services of social workers, psychiatrists, and sometimes lawyers to protect the child from recurrent assault.

As the extent of sexual abuse of small children increases, public awareness is increasing as well. In general, forced genital contact or rape are crimes against children and are considered such by the legal system.

f. Burns and Smoke Inhalation

Burns are second only to motor-vehicle-related injuries as a cause of mortality and morbidity among children. Sometimes the injury is flame-related and affects the skin; in other circumstances, it involves smoke inhalation, which accounts for nearly half of fire-related deaths. Toxicity of smoke is such that it causes massive pulmonary edema, sometimes after a delay of several hours. Individuals exposed to smoke inhalation must be immediately hospitalized because vigorous respiratory support can be lifesaving. After massive skin burns, loss of water through the skin is a great danger resulting in dehydration, which develops quickly. Rehydration should begin as soon as possible after a burn, using oral electrolyte solution to replace some of the extensive losses of fluid across the burned surface. Intravenous fluids are required with extensive burns. [*See* Thermal Injuries.]

Prevention involves surveillance on many fronts, including the reduction of the exposure of infants to water with a temperature over 50°C, creation of barriers around grates or radiators that may have hot surfaces, and the establishment of smoke detectors and fire extinguishers in all homes. The immediate response to a burn should be the application of cold water to delay the inflammatory response, and, in the event of smoke inhalation, oxygen should be provided to the victim as soon as respiratory distress is evident. Hospitalization is required for all serious burns.

IV. CONCLUSION

This brief discussion of diagnosis and treatment of diseases, injuries, and psychological disorders illustrates some of the problems present in childhood and emphasizes required sophisticated diagnostic and therapeutic interventions. Disorders of other organ systems are often referred to the specialist, and discussion of them can be found in other portions of this volume. For example, very few pediatricians treat eye diseases in infancy, although they are the ones that must be alert to detection of them. Likewise, problems of the ears, nose, and throat, other than straightforward infection, are referred to an otolaryngologist.

Pediatricians are child advocates, who point out the enormous discrepancy between what can be done and what is done for children of the world. Their central role is to maintain the health of the child, diagnose and treat illness, and at all times join forces with parents, educators, and other concerned citizens to work for equal access to preventive and therapeutic services for children everywhere.

ACKNOWLEDGMENTS

I am grateful to Dr. Hughes Evans for review of historical sections of this review and to Dr. Lewis First and Dr. Karen McAlmon for helpful discussion of which issues to include and advice on appropriate emphasis.

BIBLIOGRAPHY

Antonarakis, S. E. (1989). Diagnosis of genetic disorders at the DNA level. *N. Engl. J. Med.* **320**, 153.

Avery, M. E., and First, L. R. (eds.) (1994). "Pediatric Medicine." Williams & Wilkins, Baltimore.

Cone, T. E. (1979). "History of American Pediatrics." Little, Brown & Company, Boston.

Crick, F. (1988). "What Mad Pursuit: A Personal View of Scientific Discovery." Basic Books, New York.

Grant, J. P. (1994). "The State of the World's Children." Oxford Univ. Press, New York.

Greven, P. J. (1970). "Four Generations, Population, Land and Family in Colonial Andover (Mass.)." Cornell Univ. Press, Ithaca, New York.

Howland, J., and Marriott, W. McK. (1916). Acidosis occurring with diarrhea. *Am. J. Dis. Child.* **11**, 309.

McKusick, V. A. (1988). The morbid anatomy of the human genome, IV. *Medicine* **67**, 159.

Peter, G. (ed.) (1994). "Report of the Committee on Infectious Diseases" (The 1994 Red Book), 23rd Ed. American Academy of Pediatrics, Elk Grove Village, Illinois.

Peptide Hormones and Their Convertases

MICHEL CHRÉTIEN
NABIL G. SEIDAH
Clinical Research Institute of Montreal

GLOSSARY

Convertases Novel enzymes which have the unique property to cleave proproteins at motifs made of basic amino acid residues

Feedback control Regulation of the flux through a biological pathway such that a given point in the pathway is able to influence earlier points

Hormone specificity Uniqueness of fit between a hormone or any ligand and a receptor or other binding molecules

Receptor Cell protein able to strongly bind a signaling molecule

HORMONES BELONG TO A CLASS OF SUBSTANCES known to have a powerful action in minute amounts. They are important in biological processes as intracellular and intercellular messengers. The term hormone was coined from a Greek root to indicate an agent able to excite or activate. The concept, first expressed by Bayliss and Starling in 1902, was based on an active substance present in a glandular secretion that could be mediated by way of the blood to produce a physiological effect. Recent developments have extended this concept to the fact that a hormone can also act contiguously to adjacent cells or even within its own site of secretion. Thus the terms paracrine and autocrine have been introduced and used regularly. This has led to the current view of a hormone as an agent produced by any tissue of the body that upon secretion may act locally or be transported by the blood to various organs and tissues, where its reactions may affect the organism as a whole. This article considers only hormones that are protein or peptide in their structure, conventionally designated polypeptide hormones. A rapidly growing group of active polypeptide agents called growth factors and some viral envelope glycoproteins will be discussed briefly in relation to human diseases. The steroid prostanoid or catecholamine hormones will not be presented.

I. THE NATURE OF POLYPEPTIDE HORMONES AND THEIR BIOLOGICAL EFFECTS

Historically, polypeptide hormones were thought to belong to a single class of substances secreted by endocrine or ductless glands, but they are now recognized to be produced and released by a large variety of cells in numerous tissues, including the nervous system, in which they constitute a major source of neurotransmitters and/or neuromodulators. Polypeptide hormones are secreted at a variable rate in response to environmental or other stimuli. A crucial step in their release is the maturation needed to become active, including the cleavage of their high molecular weight precursors at specific motifs by recently discovered converting enzymes, named "convertases." A final key element in the cell receiving the chemical message is the receptor(s) specific for each hormone. Defects in processing, insufficient or excessive production, or a defect in the receptor or its means of signal transmission may cause clinical disorders, which, in many

ENCYCLOPEDIA OF HUMAN BIOLOGY, Second Edition, VOLUME 6.

cases, have helped in understanding the normal endocrine system functions while opening new avenues to better understand the physiopathology of a number of important human diseases and to develop novel modes of therapy. [*See* Diabetes Mellitus.]

Peptide hormones stimulate and control many cellular and metabolic processes of widely varying natures, but they have several properties in common.

1. They resemble other crucial biological agents such as enzymes or vitamins in that they are effective in tiny amounts. For example, their blood levels can be as low as 10^{-10} to 10^{-12} M and their half-lives ($t_{1/2}$) can be 5 to 60 min.
2. In many cases, they are produced in a tissue other than that in which they ultimately have their biological effects. They also act on adjacent cells or within their own secretory cells as metabolic regulators.
3. When secreted into the circulation, their concentration in the blood at any given time gives an indication of the activity of the gland in question and the degree of exposure to an interaction with the target cells.
4. Structurally, polypeptide hormones are rather small proteins, with molecular weights of 30,000 or less; they may also be as small as enkephalin and thyrotropin-releasing hormone, which contain five and three amino acids, respectively. They are water soluble and, hence, do not usually require carrier proteins in circulation. The variety in nature and the activity of peptide hormones are partially shown in Table I.

The role and the sources of peptide hormones are likely to continue to extend in this way because advanced techniques in protein chemistry and molecular biology permit the isolation and characterization of polypeptides in ever greater numbers, often before we know what their cellular role is or whether they have biological activity. [*See* Peptide Hormones of the Gut.]

A. Regulation of Hormone Levels

The effect of a peptide hormone mainly depends on its concentration in tissues and in circulation. The producing cells have a basal or steady secretion and respond to need by increasing from a basal to a stimulated level.

1. Basal Peptide Hormone Secretion

Basal secretory levels cannot be properly evaluated because of the difficulty of measuring secretion rates. The blood levels of most peptide hormones can now be measured by radioimmunoassays and they reflect quite accurately the state of the actual secretion. [*See* Radioimmunoassays.]

2. Stimulated Hormone Secretion

Different forms of stimulation can lead to enhanced secretion, e.g., the secretion of insulin by the β cell of the pancreas. The principal factor controlling insulin secretion is the concentration of glucose in the blood. A rise in the glucose level causes insulin to be released. Other agents, including arginine, glucagon, and other hormones, can also modulate insulin secretion but are less effective and are dependent on the presence of a glucose effect. Certain α-adrenergic agents and somatostatin can lower insulin release. Thus, stimulated secretion responds to a complex hierarchy of agents and effects. [*See* Insulin and Glucagon.]

B. Neurosecretion and the Hypothalamus

Peptide hormone secretion under neural regulation may be visualized by two general mechanisms. A nerve terminal may make direct contact with a secreting cell or a neuron may itself secrete a peptide, which may act directly to alter target cell function or pass via the circulation to act on a second secretory cell. Such neurons act as both nerve and endocrine cells and are called neurosecretory cells. An example are cells in the hypothalamus that, when stimulated by other nerve cells in the higher regions of the brain, secrete a specific peptide into the portal blood vessels of the pituitary stalk. Such agents have been called releasing factors and several have been characterized (Table I). In some respects, this portrays a hierarchy system of hormone action with the hypothalamus as the main regulatory tissue.

C. The Endocrine Cascade

The level of a particular hormone in the blood and its effect on target cells are the result of a complex set of dynamic interacting components.

1. The level of hormone available in the blood is dependent on its rate of synthesis and the secretion of the hormone from the tissue of origin.
2. The actual transport in the circulation is believed to take place in an unaided fashion, but certain hormones such as growth hormone, corticotropin-

TABLE I
Active Polypeptide Hormones

Polypeptide hormone[a]	Properties	Physiologic effects
1. Pituitary		
ACTH (adrenocorticotropin)	39 AA[b]	Regulates adrenal function Promotes lipolysis in fat cells Promotes an increase in pigmentation
β-END (β-endorphin)	31 AA	Produces analgesia and euphoria
α-MSH (α-melanocyte-stimulating hormone)	13 AA	Promotes an increase in pigmentation
β-MSH (β-melanocyte-stimulating hormone)	18 AA	Promotes an increase in pigmentation
TSH (thyroid-stimulating hormone)	201 AA glycosylated	Regulates thyroid function
FSH (follicle-stimulating hormone)	204 AA glycosylated	Stimulates follicular development in the ovary and gametogenesis in the testes
LH (luteinizing hormone)	204 AA glycosylated	Promotes luteinization of the ovary and stimulates Leydig cell function of the testes
GH (somatotropin)	191 AA	Regulates intermediary metabolism Increases protein synthesis, amino acid transport, and fat mobilization Decreases sensitivity to insulin
PRL (prolactin)	198 AA	Promotes lactation
AVP (vasopressin)	9 AA	Increases water resorption in kidney tubules
Oxytocin	9 AA	Promotes contraction of the uterus Stimulates mammary glands
2. Neurones		
TRH (TSH-releasing hormone)	33 AA	Stimulates TSH and PRL secretion
LHRH (LH-releasing hormone)	10 AA	Stimulates LH and FSH secretion
GRH (GH-releasing hormone)	40 AA and 44 AA	Stimulates GH secretion
CRF (corticotropin-releasing hormone)	41 AA	Stimulates secretion of POMC-related products
Somatostatin	14 AA and 28 AA Also found in pancreas and gut	Inhibits secretion of GH and TSH and many extrapituitary hormones
Enkephalins	5 AA	Inhibits the release of neurotransmitter Induces analgesia
BNP (brain natriuretic peptide)	26 AA	Natriuretic Diuretic Hypotensive
Endorphin	17 AA	Induces analgesia
Neuromedin B	10 AA Also found in gastrointestinal (GI) tract, pituitary, central nervous system (CNS)	Promotes muscle contraction
NPY (neuropeptide Y)	36 AA	Vasoregulator Inhibits secretion of LH and GH
CGRP (calcitonin gene-related peptide)	37 AA	Mediates smooth muscle contraction Alters blood pressure and heart rate
3. Pancreas and GI tract		
Insulin	51 AA Two chains	Promotes glucose uptake, glycogenesis, and lipogenesis Promotes protein synthesis
Glucagon	29 AA	Promotes glycogenolysis
PP (pancreatic polypeptide)	36 AA	Increases gut motility
Gastrin	17 AA and 34 AA	Promotes gastric secretion and motility Increases growth of gastric mucosa Increases secretion of pancreatic enzyme as well as insulin, somatostatin, and pancreatic polypeptide

(continues)

TABLE I (*Continued*)

Polypeptide hormone[a]	Properties	Physiologic effects
CCK (cholecystokinin)	33 AA, 12 AA, and 8 AA Also found in CNS	Regulates processes in the pancreas, the biliary system, and gut
Secretin	27 AA	Stimulates secretion of water and bicarbonates by pancreatic acinar cells Increases pancreatic enzymes release
VIP (vasoactive intestinal peptide)	28 AA Also found in CNS	Promotes smooth muscles relaxation Increases secretion of water and electrolytes from pancreas and gut
Bombesin	14 AA Also found in nerves	Induces hypothermia Stimulates pancreatic secretion and smooth muscle contraction
Motilin	22 AA	Increases GI tract motility Decreases acid secretion
Neurotensin	13 AA Also found in hypothalamus	Promotes vasodilation and vascular permeability Decreases plasma volume and causes hypotension
Substance P	11 A Also found in nerves	Acts as a sensory and excitatory neurotransmitter Analgesic Promotes contractions of GI smooth muscle Is a potent vasoactive agent
GIP (gastric inhibitory polypeptide)	42 AA	Inhibits gastric acid secretion Promotes insulin release
Galinin	29 AA Also found in CNS	Inhibits insulin release Promotes release of GH and prolactin
4. Thyroid and parathyroid		
CT (calcitonin)	32 AA	Lowers serum calcium
PTH (parathyroid hormone)	84 AA	Stimulates renal calcium resorption Promotes bone calcium resorption Enhances GI tract calcium absorption

[a]Hormones are listed in a more commonly used full or abbreviated name with the alternative form in parentheses. Hormones have been classed with the tissue from which they were first isolated or where they are found in the greatest amount, although many hormones have been found in more than one tissue.

[b]Amino acid(s).

releasing factor (CRF), vasopressin, and oxytocin may have carrier proteins, which act as specific transport systems.

3. In some cases, an enzymatic activation of a precursor can occur in the circulation, e.g., the conversion of angiotensinogen to angiotensin I and then angiotensin II.

4. Finally, hormone levels in the blood may be decreased by degradation to restore a proper balance. There is also a dynamic balance of the various hormones in the organism. For example, release by the hypothalamus of CRF acts on special cells (corticotrophs) in the pituitary, causing the release of adrenocorticotropin (ACTH), which in turn acts on the adrenal to induce the release of steroid hormones (Fig. 1). These hormones then feed back on the hypothalamus and the pituitary with a negative stimulus, which tends to balance the whole process. [*See* Hypothalamus.]

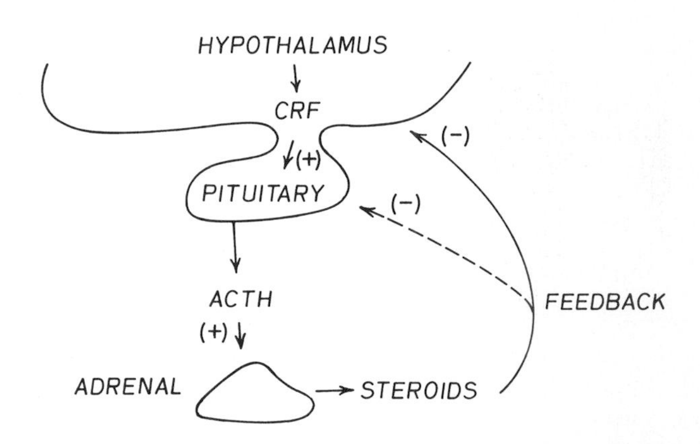

FIGURE 1 Feedback regulation. Regulation by the hypothalamus of the pituitary and, in turn, the adrenal is shown schematically. CRF secreted by the hypothalamic tissues stimulates the pituitary to release ACTH, which acts on the adrenal gland. The adrenal cells produce steroid hormones, which feed back negatively on the hypothalamus and, to a lesser extent, on the anterior pituitary, tending to decrease the secretion of CRF and ACTH, respectively.

D. Cellular Effects: A Multitargeted System

The range of responses of cells to a particular hormone is intrinsic to the target cell and largely depends on the level and nature of the differentiated state of the cell. A given peptide hormone may act on several different types of target cells and exhibit different responses in each of them. For instance, when insulin acts on fat cells, it is known to induce glucose transport and lipid biosynthesis, whereas in liver it stimulates amino acid transport and glycogen synthesis.

Essentially all cells in the body are a target for one or more hormones. A single hormone may induce several responses in target cells, and, conversely, several different hormones can provoke similar activities in a particular target cell. As a general rule, most hormones have a wide range of activities, which are recognized as our ability to perceive and measure subtle changes within the target cells and/or tissues.

The interaction of the hormone with a target cell is brought about by a receptor inserted in the plasma membrane, which has high affinity and specificity for the hormone. Each target cell has on its surface a limited number of receptors for a particular hormone and not all of them need be bound to the hormone to provide a maximum biological effect.

1. Receptors and Hormone Binding

A receptor provides the cell with a specific and unique ability to recognize and strongly bind one or more bioactive hormone(s) from among the many diverse substances in circulation. Even though the hormone may be present at a very low concentration among other proteins and peptides found at much higher concentrations, it can be distinguished by a specific receptor and specifically attracted to it. At the heart of this interaction is the strong binding of a region of the receptor with a complementary segment of the hormone. Thus, the tightness of fit or the affinity observed in the hormone–receptor complex is a crucial characteristic intrinsic to this form of intercellular communication. These events are schematically represented in Fig. 2.

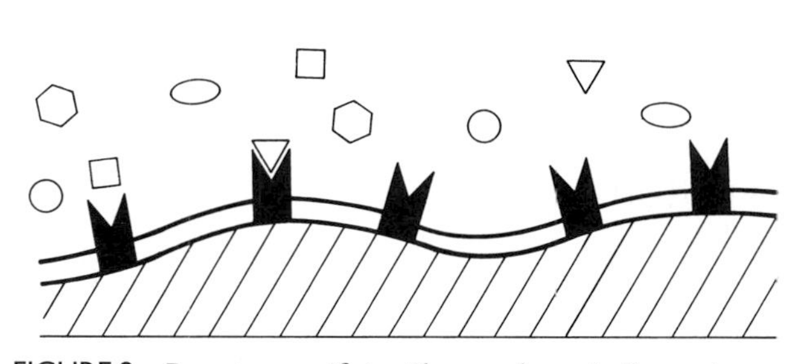

FIGURE 2 Receptor specificity. Shown schematically are the target cell receptors being exposed to the blood containing their corresponding hormone (triangles) and to other hormones represented by other geometric figures.

2. Hormone Binding and Biological Activity

It is necessary to understand the relationship between the hormone–receptor binding and the ability to produce biological effects. All active hormones or hormone analogs having activity are designated as agonists. Other analogs, called antagonists, bind to the receptor, sometimes with an affinity even greater than that of the native hormone, but do not produce any biological activity. Studies involving hormone analogs have shown that the binding region of the hormone is usually different from the region of the hormone that provides the biological response. Thus, antagonist molecules can readily attach to the hormone-binding site on the receptor molecule, but are unable to trigger the cellular activity; however, their presence on the receptor-binding site prevents the binding of any bioactive hormone, if it is present. This mechanism has helped produce some of the best drugs, such as β blockers and H2 antagonists. It is hoped that similar approaches will eventually bring about many other potent drugs in the polypeptide hormone field. [*See* Receptors, Biochemistry.]

II. BIOSYNTHESIS OF POLYPEPTIDE HORMONES

One of the most interesting features in the biosynthesis of peptide hormones is that the active form of the hormone found in the tissues and in circulation is a short fragment of a large precursor molecule.

A. The Origin of the Prohormone Theory

The prohormone theory stipulates that bioactive polypeptide hormones are generated by the posttranslational cleavage of larger inactive polypeptidic precursors. The theory was enunciated in 1967 by two independent research groups on the basis of experimental results derived from different approaches. Steiner and colleagues demonstrated with elegant *in vitro* studies that insulin is a processed form of the larger proinsulin. Chrétien and Li were led to the precursor model after they had determined the amino acid (aa) sequences of β-lipotropin (β-LPH), γ-LPH, and β-melanotropin (β-MSH). They observed that the β-MSH sequence was imbedded in the sequences of both β-LPH and γ-LPH (Fig. 3). They concluded that β-MSH was a product of proteolytic processing of the other two hormones. They also noted that this processing occurred after pairs of the basic amino acids lysine (K) or arginine (R) and presumed that these cleavages might take place within the pituitary. This insightful observation was later confirmed for proinsulin and, through the years, for a multitude of other precursors of regulatory proteins (Fig. 3).

B. The Discovery of Proprotein Convertases (PCs)

The prohormone theory entailed the existence, in the same cells as the precursors, of endoproteases which could process them preferentially after paired basic residues. For more than two decades, attempts to isolate enzymes with this cleavage specificity were unsuccessful. Progress was hampered by the lack of appropriate enzymatic assays, by the overwhelming presence of contaminating enzymes in extracts, and by the limited amounts of the physiologically relevant convertases in endocrine tissues.

The first molecular characterization of an authentic prohormone convertase was achieved through yeast genetics. Like insulin and ACTH in higher organisms, the yeast α-mating factor (α-MF) derives from the proteolytic processing of a larger inactive precursor after paired basic residues. Mutant yeasts were generated that were incapable of mating because they could not process pro-α-MF to α-MF. By complementing these mutant cells with gene fragments from mating-competent yeast cells, the *KEX2* gene was isolated. Its product, kexin, when expressed in mammalian cells that are normally devoid of this activity, could induce correct processing of pro-opiomelanocortin (POMC) into β-endorphin and adrenocorticotropin (ACTH). This prototypic enzyme was found to be structurally related to the subtilisin family of serine proteinases.

Initial attempts to clone the mammalian homolog of the *KEX2* gene using the yeast cDNA as a probe were unsuccessful. Meanwhile, an oncogene candidate molecule called "furin" was serendipidously found to have strong sequence homology with kexin and to have convertase activity on precursors such as the pro-von Willebrand factor. During the same period, the polymerase chain reaction was used in an attempt to identify endocrine homologs of kexin in mammals. It is a well-fated coincidence that the first two research groups who proposed the prohormone

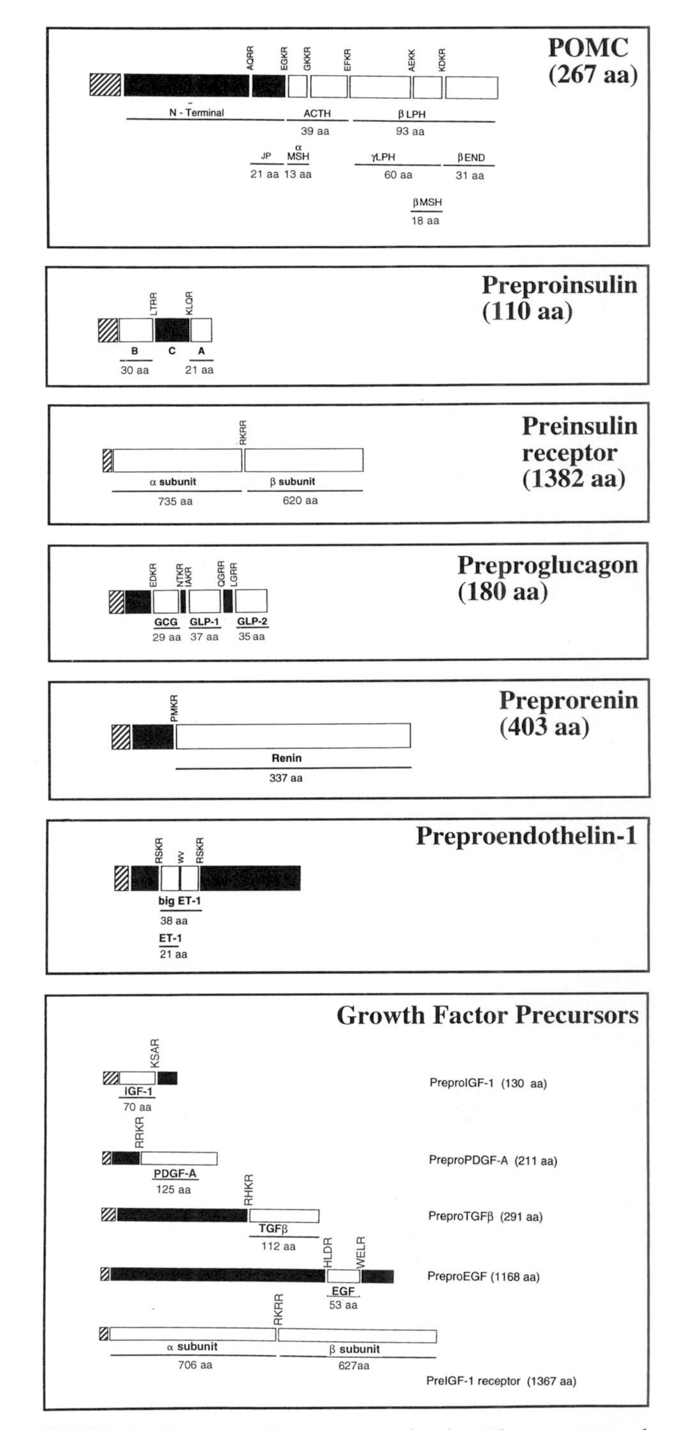

FIGURE 3 Structure of precursor molecules. The sequences of the convertase(s) cleavages sites are emphasized. ACTH, adrenocorticotropic hormone; β-END, β-endorphin; LPH, lipotropic hormone; MSH, melanocyte-stimulating hormone; GCG and GLP, glucagon and glucagon-like peptides, respectively; ET-1, endothelin-1; IGF-1, insulin growth factor-1; PDGF-A, platelet-derived growth factor-A; TGF-β, transforming growth factor-β; EGF, epidermal growth factor.

theory in 1967 were also, in 1990, the first to identify the first two endocrine and neural prohormone convertases known as PC1 and PC2.

The sequence homology among furin, PC1, and PC2 indicated that they are closely related members of a family of kexin-like enzymes. Additional members have been identified since. They are PACE4, PC4, and PC5 (Fig. 4).

C. Intracellular Processing

During synthesis, the signal sequence, which mediates the entry of the polypeptide in the endoplasmic reticulum, is removed, leaving a resultant precursor, which usually requires further processing to yield an active substance(s). The precursor penetrates into the cisternae of the endoplasmic reticulum (ER) where the signal sequence is removed and, in some cases, zymogen activation can take place. However, the final processing takes place in the Golgi complex and in one or more secretory vesicle-storage granules, whereupon the products are ultimately secreted from the cell. It is interesting to note that PC1 and PC2 have different cleavage site specificity as revealed by *in vivo* studies which showed that PC1 produces ACTH and β-LPH from POMC whereas PC2 induces the activation of smaller peptides such as α- and β-MSH and β-endorphin (Fig. 5).

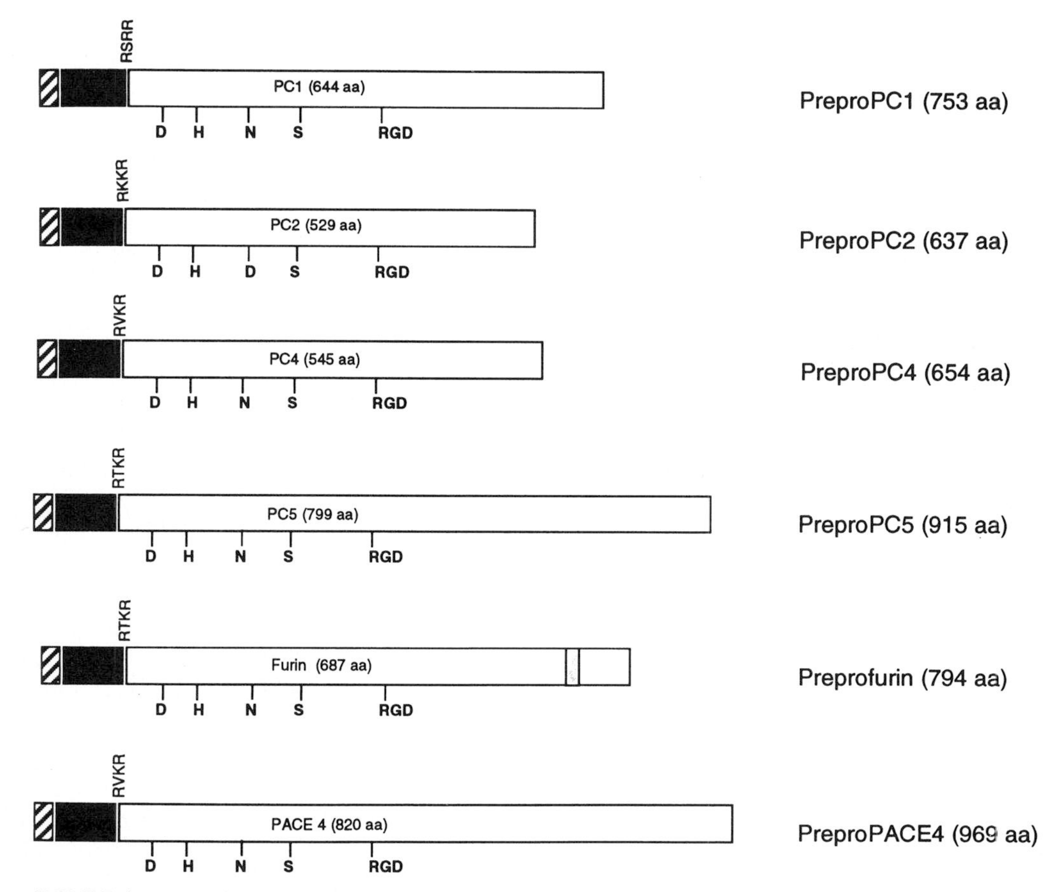

FIGURE 4 General structure of proprotein convertases. In this and subsequent figures, all sequences represented are from human, unless otherwise indicated. The signal peptide is represented by a hatched box, the prosegment by a black box, and the active molecule by an open box; the four residues preceding the convertase-processing sites are shown vertically. Amino acids are indicated by single letters (A, Ala; D, Asp; E, Glu; F, Phe; G, Gly; H, His; I, Ile; K, Lys; L, Leu; N, Asn; P, Pro; Q, Gln; R, Arg; S, Ser; T, Thr; V, Val; W, Trp; Y, Tyr). Note the conservation of the general organization of these proteins, particularly in the catalytic region that is highlighted by the active site residues D, H, and S; the N (D in PC2), which facilitates substrate binding; and the RDG motif, which may mediate an interaction with membranes. Also note that activation of these enzymes occurs at PC-like sites, suggesting autocatalysis or heterocatalysis by another enzyme of the same family. The shaded box in the furin sequence represents a transmembrane-spanning domain.

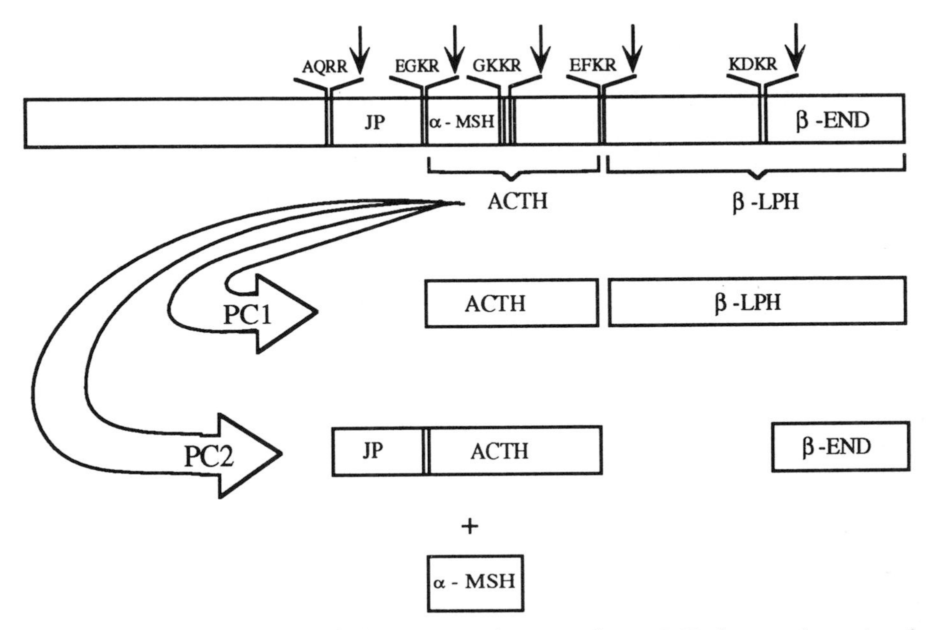

FIGURE 5 PC1 produces specifically ACTH and β-LPH whereas PC2 cleaves other pairs of basic amino acids to give α-MSH and β-endorphin.

Processing is generally understood to also include any transformation of the initial polypeptide, which may take place during the total course of biosynthesis before the final mature and active hormone is produced. Specific amino acid modifications are also involved, suchas glycosylation, amidation, acetylation, phosphorylation, sulfation, covalent attachment of lipid, disulfide bond formation, and even asymmetric center inversion, which ultimately lead to the secretable peptide form.

III. ROLE OF CONVERTASES IN HUMAN DISEASES

From the apparent variety of physiologically important substrates that may require proteolytic activation by PCs, it is presumed that reduced, excessive, or ectopic production of one or the other of these enzymes will cause serious pathological conditions in humans, manifesting either as generalized physiological imbalances or as localized anomalies. The clinical symptoms that such abnormal expression could induce will probably be very pleotropic due to the relatively broad distribution of many of these enzymes.

This section examines the relevance of PCs in clinical medicine. A selected number of diseases in which precursor-derived proteins or peptides have been shown or presumed to be determining or contributing factors, including endocrine, proliferative cardiovascular, central nervous system, and viral diseases, will be discussed.

A. Endocrinopathies

Excessive or deficient production or activation of hormones or their receptors has been implicated in the pathogenesis of endocrinopathic disorders. Considering the critical role of PCs in prohormone processing, these enzymes must contribute to the pathophysiology of several endocrine diseases. This possibility will be discussed here in the context of clinical endocrine disorders such as Cushing's disease and diabetes.

1. Cushing's Disease

Cushing's disease is an overproduction of cortisol, resulting from ACTH hypersecretion by the pituitary, and is most often caused by a corticotroph adenoma. Nelson syndrome, a variant of Cushing's disease, is characterized by the further development of a semiautonomous ACTH/MSH-producing tumor following surgical bilateral adrenalectomy with the possibility of an increased PC2 type of cleavage.

Without therapy, Cushing's disease leads to significant morbidity. Selective transphenoidal resection of ACTH-secreting pituitary adenomas is the initial treatment of choice. The efficiency of this procedure is approximately 85%. Alternate therapies currently in use (cyproheptadine, bromocriptine, sodium valproate) have been found effective only in a small number of patients. Moreover, most of the inhibitors of the adrenal function which can effectively suppress cortisol secretion can also cause very serious side effects.

Novel therapeutic approaches need to be developed. In this search, enzymes that process POMC to ACTH represent obvious targets. Therefore, specific inhibitors of these enzymes, when they become available, could be used to attenuate or even reverse the symptoms of this pathology. They could also be useful in the treatment of various carcinomatoses associated with ectopic ACTH secretion, particularly lung cancers.

2. Diabetes

Insulin, its receptor, and glucagon are critical regulatory proteins for glucose metabolism. Diabetes mellitus often arises from molecular defects in these molecules or from their abnormal expression. Their biosynthesis requires the processing of their respective precursors at PC-like motifs.

a. Insulin

Human insulin is made of two peptide chains, A and B, linked by two disulfide bonds. It is generated from a precursor polypeptide of 110 aa by the removal of an internal peptide C through cleavages after RR_{31-32} and KR_{64-65} (Fig. 3).

PC1 and PC2, which are found in pancreatic β cells, have been proven to be proinsulin convertases. PC1 preferentially cuts after RR_{31-32} and PC2 after KR_{64-65}. This processing is very efficient in normal subjects, as only small amounts of proinsulin enter circulation. There have been reports of familial hyperproinsulinemia due to point mutations in the insulin gene which resulted in the absence of processing, but the possibility of a β-cell deficiency in PC1 or PC2 warrants examination in diabetic patients, given their apparent role in proinsulin processing. A patient with PC1 deficiency has recently been described.

b. Insulin Receptor

Insulin cellular effects are mediated by its cell surface receptor. The receptor is a heterotetrameric membrane protein made up of two α and two β subunits. It is synthesized as a 1382 amino acid precursor which gets cleaved into α and β subunits after the sequence $KRKR_{732-735}$. The site of processing fits the consensus cleavage signal for furin, indicating that its deficiency or sequence changes at the cleavage site of the insulin proreceptor represent alternate pathogenic events in the development of insulin resistance and eventually of noninsulin-dependent diabetes mellitus (NIDDM). An $Arg_{735}Ser$ mutation at the processing site, which leads to a deficiency in the active insulin receptor and hence to insulin resistance, has been observed in one diabetic patient.

c. Glucagon

In patients with NIDDM, the glucagon reserve capacity of pancreatic α cells is impaired. Glucagon maintains glucose homeostasis. Its precursor is produced in pancreatic α cells as a 180 amino acid polypeptide containing the sequences of glucagon and two related peptides. These peptides are released by proteolytic cleavages at five pairs of basic residues (Fig. 3).

3. Hypertension

Hypertension is a hemodynamic disorder of complex etiology. Its pathophysiology could involve anomalies in the mechanical properties of the vascular walls as well as electrolytic alterations due to renal malfunctions. Renin and endothelin (ET) are two secretory molecules that have been implicated in the development of hypertension. They are products of processing of their respective precursors at PC-like sites.

a. Renin

Renin converts angiotensinogen to angiotensin I, which is matured in circulation to angiotensin II, a hormone that plays a critical role in sodium chloride homeostasis and vasoconstriction. The major site of renin production is the juxtaglomerular apparatus in the renal cortex.

Renin is a fragment of a larger inactive precursor made of 403 amino acids, generated by proteolytic processing at one of its seven pairs of basic residues. Gene transfer experiments have shown that of all the six convertases known, only PC1 and PC5 can process human prorenin into active renin.

b. Endothelin-1

Endothelin-1 (ET-1) is a member of a family of potent vasoconstrictive peptides produced by the vascular endothelium. Endothelin has been found in various nonvascular tissues, suggesting that it may have many different biological functions. ET-1 affects circulatory hemodynamics and the response to alterations in cardiovascular homeostasis in pathological conditions. Its association with various clinical conditions such as hypertension and renal and heart failure has been substantiated. Increased plasma levels of ET-1 in patients with essential hypertension have been reported.

The human preproET-1 is 212 amino acids long. Its cleavage after $RSKR_{49-52}$ and $RSKR_{90-93}$ produces ET1-38, also called big ET. Further processing of big ET between Trp_{73} and Val_{74} results in the formation of ET1-21 (Fig. 3). A neutral metalloprotease apparently responsible for this cleavage has been purified from porcine aortic endothelium. In humans, big ET1 is the major circulatory form of endothelin. It has been suggested that it may represent a storage form of the hormone.

On the one hand, furin and PC5 mRNA have been detected in endothelial cells. On the other hand, it has been found that furin is four to five times more effective at converting proET to ET1-38 than PC1 or PC2, further supporting a possible role of this convertase in proET processing.

B. Proliferative Disorders

Polypeptidic growth factors are physiological regulators of normal cell division. After binding to its receptors, a growth factor initiates a cascade of molecular signaling events that are relayed by second messengers to the nucleus, culminating in significant changes in the pattern of gene expression. The result of this process on cell division could be either stimulatory or inhibitory, depending on the factor, the responsive cell, and the tissue context. Stimulatory factors include, among others, platelet-derived growth factor (PDGF), epidermal growth factor (EGF), fibroblast growth factor (FGF), and insulin-like growth factors (IGF). Inhibitory factors include cytokines such as interferon, interleukin, and tumor necrosis factor (TNF). Other factors, such as transforming growth factors (TGF), may be stimulatory or inhibitory.

Most polypeptide growth factors are fragments of larger precursor proteins and many are generated by cleavage after sequences of basic residues that could be recognized by convertases.This is the case for EGF, IGF-I and II, PDGF-A, and most members of the extended family of TGF, including inhibin and activin. Figure 3 shows the structure of the precursors to some of these polypeptides.

Proliferative diseases, as consequences of unchecked cell division, often involve the immediate participation of growth factors and, more upstream in the expression pathway, of the convertases responsible for their maturation. Cancer and atherosclerosis are two diseases in which the contribution of growth factors appears to be crucial.

1. Cancer

Carcinogenesis is a progressive derangement of mechanisms that control cell proliferation and differentiation. The results of this multistep process are tumor establishment and malignant transformation. Although growth factors have been long suspected to be the culprit, the discovery that the normal counterpart of the oncogene v-*sis* of the simian sarcoma virus is the cellular gene encoding the β chain of PDGF has confirmed this suspicion.

There are many ways in which a cell can evolve into malignancy with the contribution of growth factors. The most direct mechanism involves autocrine stimulation, whereby a cell is constantly stimulated by an endogenous mitogenic factor. Another way is the excessive proliferative response of a cell to paracrine factors produced by adjoining cells. Another possibility is cell refractiveness to inhibitory factors. Finally, malignant cells can produce angiogenic factors which promote tumor vascularization and growth. In short, all anomalies of expression or activity of growth factors or their receptors are potentially oncogenic.

Many cancerous cells overproduce growth factors and their receptors. Some are even dependent on these factors for the maintenance of their transformed phenotype. Malignant gliomas, sarcomas, and anaplastic thyroid cancers usually express PDGF, whereas melanomas produce FGF. Neoplastic lung cells secrete mostly IGF-I, EGF, and TGF-α.

Initial studies with lung cancer cells have shown that non-small cell carcinomas of the lung produced high amounts of furin, whereas small cell carcinomas expressed increased amounts of PC1. Preliminary results revealed that PC1 and PC2 are present in human lung carcinoma containing POMC-derived peptides. It remains to be determined whether this increased PC expression is critical for the neoplastic phenotypes of the cells.

Current knowledge concerning the role of growth factors in cancer offers opportunities for imaginative therapeutic interventions.

2. Atherosclerosis

Myocardial and cerebral infarctions are the most prevalent causes of death in the Western world. They result from atherosclerosis, a proliferative response of vascular endothelial cells to chronic injuries. Numerous growth-regulatory molecules (stimulatory

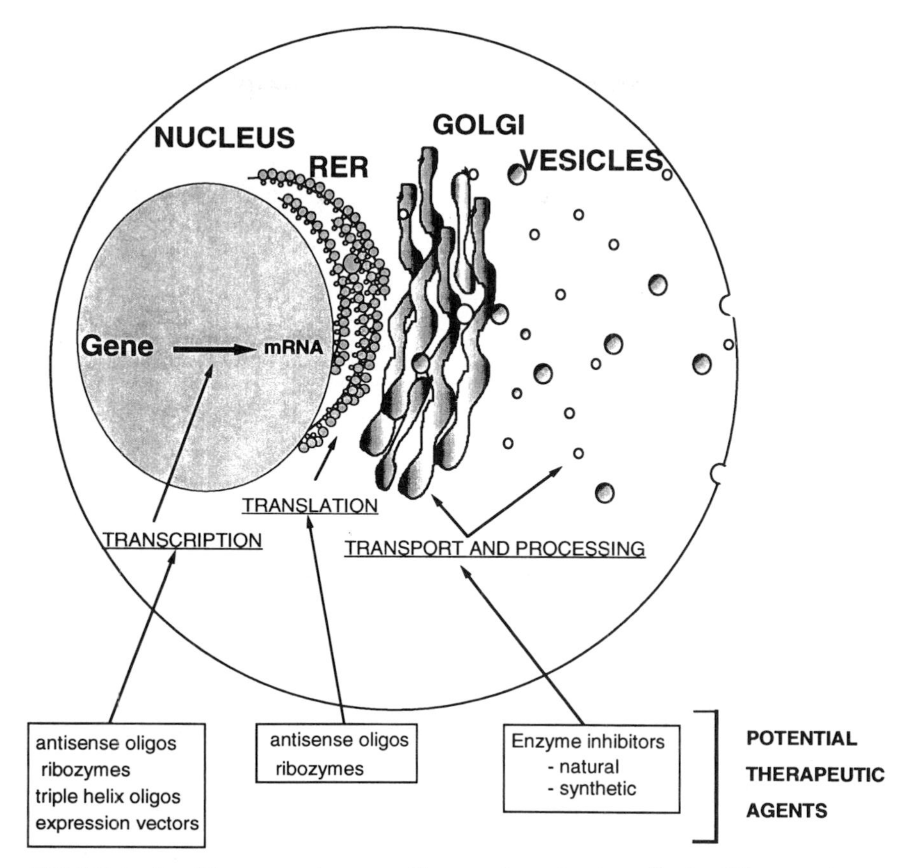

FIGURE 6 Possible steps and means of interventions aimed at affecting PC expression or activity. Within the cell, PC expression may be interferred with at the gene level by blocking transcription in the nucleus or at the mRNA level by arresting translation in the rough endoplasmic reticulum (RER). Its activity can be inhibited during transport through the Golgi apparatus or within secretory vesicles by cobiosynthesized peptidic inhibitors or by using exogenously administered synthetic inhibitors.

and inhibitory) play critical roles in inducing the fibroproliferative response that contributes to the formation of the advanced lesions of atherosclerosis. These molecules determine whether the potentially occlusive fibrous plaque lesions of atherosclerosis will progress, regress, or remain unchanged. The group includes PDGF, EGF, FGF, colony-stimulating factors (CSF), TGF-β, and TNF. PCs probably play a role in the activation of some of these growth factors, thus possibly contributing to the pathogenesis of atherosclerosis.

3. Viral Infections

Viral infection often requires the interaction of viral envelope glycoproteins with proteins at the plasma membrane of susceptible cells. Some of these glycoproteins must be proteolytically processed to efficiently mediate viral infectivity. Experimental evidence shows that this processing is sometimes mediated by cellular proteases at sites compatible with the specificity of PCs (Table II).

The glycoprotein hemagglutinin (HA) of the influenza virus is a typical example. Pro-HA is a 547 amino acid polypeptide that is activated by cleavage after the RKKR site. This posttranslational cleavage is a determinant of the virus infectivity and virulence. In pathogenic strains of influenza virus, proHA contains the K-X-K/R-R sequence at the cleavage site. Furin can process proHA at this site, and this cleavage could be inhibited by peptidyl chloromethylketones.

The envelope glycoprotein gp160 of the human immunodeficiency virus (HIV) is another example of a viral protein that needs to be cleaved at a PC-like site to become active. Furin has been shown to be able to

TABLE II
Examples of Proteolytic Activation Sites of Viral Envelope Gycoproteins[a]

Virus	Glycoprotein	Cleavage site
Mumps virus	F protein	S-R-**R**-H-**K**-**R**⌀F-A-G-I
Influenza virus H5	Hemagglutinin	R-K-**R**-K-**K**-**R**⌀G-L-F-G
Respiratory syncytial virus	F protein	K-K-**R**-K-**R**-**R**⌀F-L-G-F
Yellow fever virus	prM	S-R-**R**-S-**R**-**R**⌀A-I-D-L
Human cytomegalovirus	gB	H-N-**R**-T-**K**-**R**⌀S-T-D-G
Varicella zoster virus	gpII	N-T-**R**-S-**R**-**R**⌀S-V-P-V
Human immunodeficiency virus	gp160	V-Q-**R**-E-**K**-**R**⌀A-V-G-I

[a]The basic residues of the furin consensus cleavage site (R-X-R/K-R) are shown in bold letters.

catalyze this conversion. Furin-specific α-antitrypsin variants, which inhibit gp160 processing, also prevent the syncytium formation characteristic of HIV-infected cell populations.

The authors, in their own laboratory, have shown that both PC1 and furin, which are expressed in $CD4^+$ cells, can convert gp160 into gp120 and gp41 and that this activation can be blocked by a synthetic peptide inhibitor.

C. Possible Agents

1. Proprotein Convertases in Therapies?

The preceding sections described a sampling of pathologies that can arise either because of a lack of proteolytic activation of a precursor to a functional protein (i.e., diabetes) or because such a precursor, when produced in higher amounts, is too efficiently processed, leading to an excess of its active product (i.e., POMC in Cushing's disease). The authors have also pointed to the possible role of many other bioactive proteins in the pathophysiology of several diseases (i.e., growth factors in proliferative disorders and envelope glycoproteins in viral infections). Many of these active proteins are generated by single or multiple proteolytic cleavages at basic sites which may be recognized by proprotein convertases. The underlying theme has been that if PCs are physiologically involved in the processing of these factors, they would constitute as critical participants in pathogenic processes and would represent metabolic steps to modulate through enzyme supplementation or specific inhibition. For the moment, however, the direct implication of PCs in pathological states still awaits clear-cut evidence beyond quantitative and coincidental correlations. This is undoubtedly because these enzymes, suspected for decades, have been discovered only recently and their pathophysiological relevance has barely been examined.

2. Practical Considerations in the Design of PC-Based Therapies

The possibility of a role of PC also implies that novel therapies can be designed with these enzymes as pharmacological agents or targets. Enzymes and their inhibitors have been used previously in the treatment of a variety of diseases. Administration of polyethylene glycol-adenosine deaminase has been used in the treatment of severe forms of combined immunodeficiency. The symptoms of Gaucher disease can be alleviated by treatment with glucocerebrosidase. Various synthetic inhibitors of the angiotensin-converting enzyme are widely used to control hypertension. Inhibitors of hydroxymethyl glutaryl-CoA reductase are commonly used to lower blood cholesterol in patients with hypercholesterolemia.

The question therefore is not whether it makes pharmacological sense to use PCs and their inhibitors in treatments, but whether it is practically achievable. Although most PCs are secretory proteins, their pH and calcium concentration requirements are such that their normal sites of action will probably be within the intracellular secretory compartments. Any therapies using or aimed at these enzymes will involve the expression or the transfer of the therapeutic agents within these compartments. The emerging technologies of gene therapy may offer appropriate tools and vehicles to that end: (1) antisense oligonucleotides or ribozymes could be administered near the affected tissues to specifically inhibit the expression of a particular PC at transcription or translation and (2) vectors in which the genes for PCs or their proteic inhibitors are under the control of cell-specific promoters could be used to target expression to particular cells (Fig. 6). It is possible, however, that some PCs (e.g., furin, PACE4) may also act as ectoenzymes at the cell surface, in which case their inactivation with specific inhibitors could be achieved with relative ease. In addition, it is possible that the enzymology of proprotein convertases will find some immediate application in the production of peptidic pharmaceuticals through biotechnology. These enzymes should greatly facilitate the accurate processing of inactive precursors to bioactive forms during their large-scale biosynthesis in cellular bioreactors.

In conclusion, the authors propose that the newly discovered proprotein convertases could be causative or contributing factors in many human diseases. Clearly, the hypothesis needs to be fully verified by actual cases. The main goal at this stage is to suggest a new outlook and a new field of investigation for clinical research. The results of this investigation may lead to a new understanding of the role of these enzymes in normal and pathological physiology as well as to the development of a new arsenal of tools for the diagnosis, the prognosis, and, ultimately, the treatment of various diseases.

IV. SUMMARY

Multicellular organisms require communication to organize and coordinate the activities of development, cell division and growth, metabolism and maintenance, behavior, and emotional expression. The biochemical communication, which brings about such diverse cellular activities, is affected in part by many hormones, one class of which is polypeptide hormones. To completely describe the polypeptide hormone as a messenger, the characteristics of secreting cells and cells that are subject to the hormone effect—target cells—must be considered. Together they act as sending and receiving components in a biological communication system.

Although varying in size, polypeptide hormones are usually of low molecular weight and are generally water soluble so that once secreted from the producing cell, they are secreted to adjacent cells or transported in the circulation to their site of action. A critical entity in the communication system is the receptor situated in the plasma membrane of the target cell. The receptor has an extremely high specificity and affinity for its intrinsic hormone, which allows recognition and binding to the target even at very low concentrations.

The discovery of the convertases and their relationships with the biosynthesis of receptors, growth factors, and viral envelope glycoproteins adds new dimensions to the biological importance of the polypeptide biosynthetic system in disease processes and to eventual new types of therapies.

BIBLIOGRAPHY

Alberts, B., Bray, D., Lewis, J., Raff, M., Roberts, K., and Watson, J. D. (1983). "Molecular Biology of the Cell." Garland Publishing, New York.

Chrétien, M., and Li, C. H. (1967). Isolation, purification and characterization of γ-lipotropic hormone from sheep pituitary glands. *Can. J. Biochem.* **45**, 1163–1174.

Chrétien, M., Mbikay, M., Gaspar, L., and Seidah, N. G. (1995). Proprotein convertases and the pathophysiology of human diseases: prospective considerations. *Proc. Assoc. Am. Phys.* **107**, 47–66.

Imura, H., Shizume, K., and Yoshida, S. (eds.) (1988). "Progress in Endocrinology." Excerpta Medica, Amsterdam.

Krieger, D. T., Brownstein, M. J., and Martin, J. B. (eds.) (1983). "Brain Peptides." Wiley, New York.

Mbikay, M., Seidah, N. G., and Chrétien, M. (1993). From proopiomelanocortin to cancer: Possible role of convertases in neoplasia. *Ann. N.Y. Acad. Sci.* **680**, 13–19.

Seidah, N. G., and Chrétien, M. (1992). Pro-protein and prohormone convertases of the subtilisin family: Recent developments and future perspectives. *Trends Endocrinol. Metab.* **3**, 133–140.

Steiner, D. F., Cunningham, D., Spiegelman, L., and Aten, B. (1967). Insulin biosynthesis: Evidence for a precursor. *Science* **157**, 697–700.

Willians, R. H. (1981). "Textbook of Endocrinology." Saunders, Philadelphia.

Peptide Hormones of the Gut

TADATAKA YAMADA
CHUNG OWYANG
University of Michigan Medical Center

GLOSSARY

Antrum Distalmost portion of the stomach

Endocrine action Action that is mediated by transfer of a hormone from the effector cell to the target cell via the circulation

Hormone Chemical messenger from specific cells in one part of the body that traverses to a nearby cell or to a distant site, where it exerts an action on a target cell that recognizes it

Parietal cell Cell in the stomach lining that is responsible for gastric acid secretion

Peptide Molecule consisting of a relatively small number of amino acids linked together, amino-terminal end to carboxyl-terminal end; in this article, the focus will be on hormones of the gut that are peptides

Receptors Sites on a target cell that recognize and bind a specific hormone and initiate the transduction of the binding event to a cellular response

PEPTIDE HORMONES OF THE GUT SERVE AS THE chemical messengers that connect cells or organs in the body so that they can perform integrative tasks. The messengers can be delivered via the circulation or the interstitial space. The function of any organ in the gastrointestinal tract is tightly controlled by a variety of neural and hormonal regulators that interact in a temporally appropriate fashion to aid in the process of nutrient assimilation. There exists a diverse array of peptide hormones, many that can be grouped into families that may have arisen from a common ancestor by tandem gene duplication. Peptide hormones have been implicated in the pathogenesis of various gastrointestinal disorders but also hold promise as useful adjuncts in the diagnosis and treatment of other diseases.

I. PEPTIDE HORMONE PHYSIOLOGY

A. Introduction

The concept that regulatory substances could be released at one site of the body to control physiological functions at another site arose from the pioneering studies of Bayliss and Starling in 1902, who noted that the duodenum, on acidification, releases a substance that causes the pancreas to secrete bicarbonate-containing juice. This substance, which was known only by its *function* at the time, was called a *hormone* and labeled as *secretin*. Three years later, Edkins described another hormonal activity, labeled as *gastrin*, that mediated the secretion of gastric acid in response to alkalinization of the stomach. During the first half of this century, a number of putative hormonal substances were described by their functions in this fashion, although their structures were unknown.

In 1959 a dramatic improvement in the detection of these hormonal substances, many of which proved to be small polypeptides, was made possible by the development of the technique of radioimmunoassay by Yalow and Berson. In this technique, the binding of a radioactively labeled antigen (such as a hormone) to an antibody that recognizes it is competitively inhibited by unlabeled antigen in an assayed sample. As depicted in Fig. 1, the quantity of specific antigen in the assayed sample is a measurable function of the

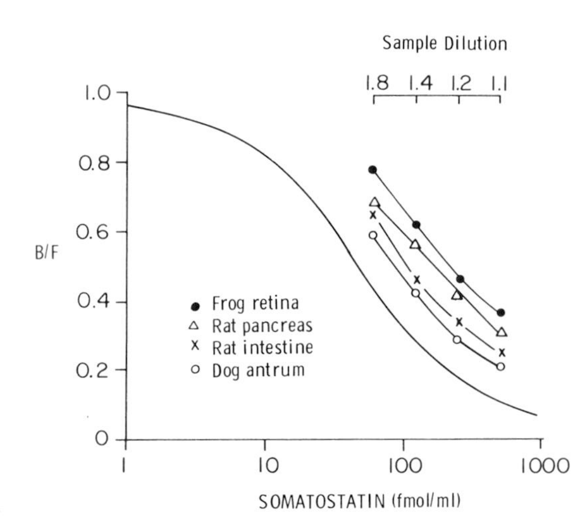

FIGURE 1 Standard curve for radioimmunoassay of somatostatin. The ratio of ^{125}I-Tyr^{1}-somatostatin bound (B) to antibody or free (F) is a semilogarithmic function of unlabeled somatostatin added to the incubation mixture. Extracts from various organs displace label binding in parallel to the standard curve, and thus the concentration of somatostatin in these samples can be estimated. [From J. DelValle and T. Yamada (1990). *Annu. Rev. Med.* **41**, 447–456.]

displacement of labeled antigen from the antibody. The development of the radioimmunoassay technique was paralleled by the evolution of advanced peptide purification techniques [e.g., gel filtration (separation of molecules by size), ion-exchange chromatography (separation by charge), immunoaffinity chromatography (separation by antigenic properties), reverse-phase high-pressure liquid chromatography (separation by hydrophobicity), and fast protein liquid chromatography (more rapid purification using combinations of the foregoing)]. Moreover, through peptide microsequencing techniques, it became possible to obtain the amino acid sequences of peptides that were purified in only minute amounts. Thus, for example, a peptide such as somatostatin (see Section II,F,1), which was first purified in 1973 from the hypothalami of 500,000 sheep by a team of scientists after several years of struggle, could be purified and structurally analyzed from a single human stomach in 2 weeks by an undergraduate student less than a decade later. The availability of peptide sequences permitted scientists to take an additioinal step in structural analysis through recombinant DNA technology. By fashioning synthetic oligodeoxynucleotide probes specific for known amino acid sequences (on the basis of the genetic code), it is possible to screen such a "library" of recombinant bacterial or viral clones containing DNA molecules complementary to messenger RNA extracted from a tissue known to contain the peptide hormone of interest. By determining the nucleotide sequence of the recombinant DNA molecule thus identified, one can deduce the amino acid sequence of the precursor molecule that is processed via a variety of steps to form the biologically active peptide. [*See* Radioimmunoassays.]

B. Mechanisms of Peptide Hormone Action

Modern advancements in research technology have reversed the problem faced by scientists studying peptide hormones earlier in this century. Now, instead of physiological functions in search of the hormones responsible for them, we are faced with a host of peptide hormones, the physiological functions of which are unknown. In the broadest sense, these small molecules serve as the means of communication between one part of the body and another, and between distant or adjacent cells. They permit the body to function as an integrated unit rather than as an amalgamation of unrelated parts. The mechanisms by which peptides exert these effects are multiple (Fig. 2). In the classic *endocrine* sense, a peptide hormone

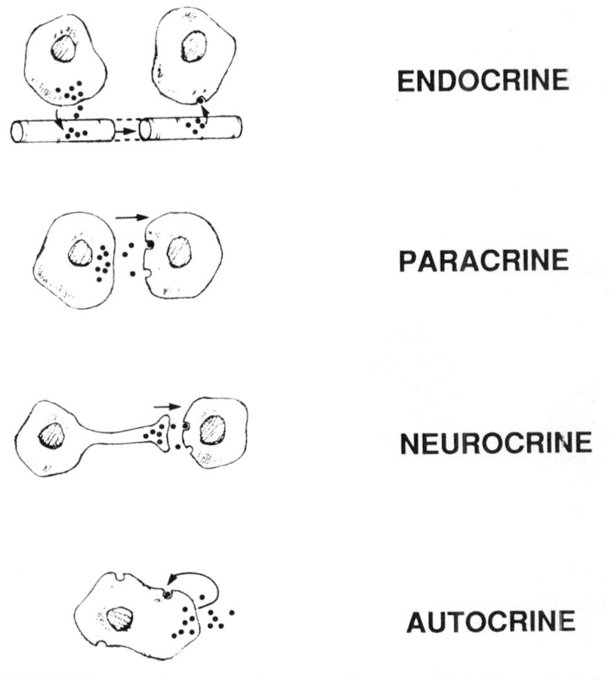

FIGURE 2 Mechanisms of action of gut peptides. [From J. DelValle and T. Yamada (1990). *Annu. Rev. Med.* **41**, 447–456.]

is released by cells at one site into the bloodstream and, via the circulation, delivered to a distant target cell possessing receptors that recognize the hormone. Under these circumstances, the physiological event caused by the hormone can be related to its concentration in the circulation. Another mechanism by which peptide hormones can act is called the *paracrine* effect. In this case, a peptide is released into the interstitial space and acts on an adjacent cell that has appropriate receptors. *Neurocrine* actions of peptides are specialized forms of paracrine action in which the interstitial space is the synaptic junction between a nerve and its target cell. Under other circumstances, a peptide released locally can directly influence the cell from which it originated. This feedback action is termed *autocrine*. The physiological function of peptide hormones as paracrine, neurocrine, and autocrine effectors is difficult to study because of the inability to sample their concentrations in the interstitial space.[1] Nevertheless, using a variety of means involving pharmacological, histochemical, and biochemical approaches, peptides have been implicated as functioning physiologically through a combination of these various routes. Some peptides, such as somatostatin, act through all four mechanisms. Others, such as vasoactive intestinal polypeptide, appear to function only through one (neurocrine in this case).

C. Physiological Functions of Hormones

Most studies of peptide hormone actions must be classified as *pharmacological* as opposed to *physiological,* for the latter implies proximity to the events that occur in real life. To demonstrate the physiological effect of a hormone, one must be able to produce an observed pharmacological effect of the substance at a target organ by reproducing concentrations of the hormone that the target organ is exposed to under normal or physiological circumstances. For example, gastrin is known to be a potent stimulant of gastric acid secretion (see Section II,A,1). When an animal eats a meal containing a substantial amount of protein, the concentration of gastrin in the circulation rises to a certain level, perhaps 50 fmol/ml. This, in turn, is followed by an increase in gastric acid secretion, perhaps to 25 mEq/hr. Most importantly, when the same animal under fasting conditions is given an intravenous infusion of gastrin to produce a circulating concentration of 50 fmol/ml, a level of gastric acid secretion the same as that observed after a meal is achieved. Moreover, when the gastrin that is released in response to meal ingestion is neutralized in the circulation with antibody to gastrin, the acid secretory response of the stomach is virtually abolished. Such gross *in vivo* studies are often required to ascertain the true physiological role of peptide hormones in integrated systems. Of course, it is possible to circumvent these cumbersome studies if specific receptor antagonists are available. By observing the alterations in physiological functions induced by single applications of a highly selective receptor antagonist for a peptide, it is possible to deduce a functional role for the peptide in question. Unfortunately, such useful antagonists are available for only a few peptides such as opioids or cholecystokinin. Future development of these antagonists will greatly facilitate studies to distinguish between the physiological versus pharmacological actions of peptide hormones.

D. Gastrointestinal Functions Under Hormonal Control

Virtually every function of the gut is tightly controlled by various hormonal regulators. These functions include exocrine and endocrine secretion, gut motility, intestinal fluid and electrolyte transport, nutrient assimilation, appetite regulation, and even pain perception. Although a description of the various peptide hormones and their actions is detailed in Section II, it is useful to examine their role in controlling one of the major functions of the gut, gastric acid secretion, to gain insight into the intricate interactions between various hormones and between hormones and other paracrine/neurocrine effectors involved in regulating physiological processes.

The stimulation of acid secretion is initiated by the simple contemplation of food. This so-called *cephalic* phase of gastric acid secretion can be induced by the thought, sight, taste, or smell of food without any of the food entering the stomach. Although under some ordinary conditions the cephalic phase accounts for only a fraction of the acid secretory response to a meal, it can account for as much as two-thirds under some circumstances. The vagus nerve is thought to carry the acid stimulatory signals from the brain to the stomach because vagotomy abolishes the cephalic stimulus. The primary mediator carried by the vagus

[1]This is an unusual use of the term "hormone," which is defined as an agent that is released from one organ or region into the systemic circulation to act on a target in another region. By tradition, peptide effectors are called "hormones" for lack of a better term, although they are endocrine peptide, paracrine peptide, neurocrine peptide, and autocrine peptide *effectors*.

is thought to be acetylcholine, although gastrin, a known stimulant of acid secretion, is also contained within the nerve. Acetylcholine may stimulate acid secretion by direct action on gastric parietal cells, but indirect actions via stimulation of gastrin secretion from G cells in the stomach or inhibition of the secretion of somatostatin, a hormone known to inhibit acid secretion, may also be important.

Once food enters the stomach, another important series of hormonally mediated events occurs. The proteins contained within a meal, more specifically the breakdown products of proteins (the amino acids and their decarboxylated derivative amines), stimulate gastrin secretion. The stretching of the stomach wall by the ingested meal contributes to gastrin release. As mentioned earlier, gastrin accounts for the major portion of the acid secretory response to meal ingestion. Gastrin by itself is not an efficacious direct stimulant of gastric parietal cells, which are responsible for acid secretion, but in combination with histamine induces virtually maximal stimulation. Histamine is produced by various cells in the stomach, depending on the animal species, and appears to act as a paracrine effector. Because the intracellular mediator for histamine action (adenylate cyclase/cyclic adenosine monophosphate) is different from that for gastrin (Ca^{2+}/phospholipid-dependent protein kinase), the two agents potentiate the actions of each other to produce an effect that exceeds the sum of the individual effects. At the same time that acid secretion is being stimulated by one set of events, another process is set forward to modulate the secretory response by inhibiting it. The ingested meal and the accompanying acid combine to stimulate somatostatin secretion. Somatostatin not only inhibits the secretion of acid by direct action on parietal cells but also acts in conjunction with acid to inhibit the secretion of gastrin.

When the ingested meal enters the small bowel, other events that affect acid secretion occur. Although there appears to be an ill-defined acid stimulatory hormone (termed *enterooxyntin*) released by the presence of nutrients in the intestine, this effect is small and overcome by the greater effects of a series of acid-inhibiting intestinal hormones, including somatostatin, neurotensin, gastric inhibitory polypeptide (GIP), secretin, and peptide YY. These inhibitory hormones are all considered to be candidates for the role of "enterogastrone," a hormonal activity described by its function earlier in the century.

The physiological regulation of the stomach makes intuitive sense from the functional requirements of the digestive process. Acid is needed for the primary digestive enzyme of the stomach (i.e., pepsin) to function optimally. Thus, the initial phase of digestion that occurs in the stomach requires acid. However, when food enters the small intestine, it is further digested by pancreatic enzymes that function optimally in a less acidic environment. Thus, gastric acid secretion must be turned off and the pancreas must be stimulated to secrete enzymes and neutralizing bicarbonate by other hormones such as cholecystokinin (CCK) and secretin, which are released by nutrients and acid, respectively, in the intestine. This elegant interplay of hormonal effectors and inhibitors acting in a regionally and temporally integrated fashion applies not only to the stomach but to all organs in the gut and is essential for the normal function of the gastrointestinal tract.

II. PEPTIDE HORMONE FAMILIES

Structural analysis of peptide hormones has led to the identification of great similarities between groups of them. Because of the similarities within a given family of peptides, its members are presumed to have arisen via tandem duplication of a common ancestral gene. After duplication, members of a gene family may diverge with respect to structural and functional characteristics. The hormone families consisting of peptides sharing substantial structural similarity are summarized in Table I. A number of peptides that do not fall into a large family group are referred to as orphan peptides.

A. Gastrin/CCK Family

1. Gastrin

The peptide hormone gastrin was postulated to exist as early as 1905 when Edkins first described a substance in the mucosa of the gastric antrum that induced the stomach to secrete acid. For many years thereafter, however, there was controversy over whether this substance was simply histamine or a new substance. This problem was resolved when Gregory and Tracy isolated gastrin from porcine antrum and confirmed its biological potency. The structural feature that is required for gastrin's biological activity is the presence of an amide moiety on the carboxyl-terminal phenylalanine residue. As indicated earlier gastrin is released into the circulation in response to meal ingestion. There are several discrete phases to this response, including those associated with the

TABLE I
Gastrointestinal Peptide Families

Family	Major members	Principal biological actions
Gastrin	Gastrin	↑ acid secretion, ↑ tissue proliferation
	Cholecystokinin (CCK)	↑ pancreatic secretion ↑ gallbladder contraction
Secretin	Secretin	↑ pancreatic secretion
	Glucagon	↓ sphincter of oddi pressure ↓ intestinal motility and absorption
	Enteroglucagon	↑ insulin release
	Vasoactive intestinal polypeptide (VIP)	↑ pancreatic and intestinal secretion
	Gastrin inhibitory polypeptide (GIP)	↓ gastric secretion, ↑ intestinal secretion
	Glicentin	↑ hepatic glucose, ↓ acid secretion
	Oxyntomodulin	↑ insulin release, ↓ acid secretion
	Growth hormone-releasing factor (GRF)	↑ release of growth hormone
	Peptide histidine isoleucine (PHI)	↑ pancreatic secretion
Pancreatic polypeptide	Pancreatic polypeptide (PP)	↓ pancreatic secretion
	Peptide YY (PYY)	↓ pancreatic secretion ↓ gallbladder contraction
	Neuropeptide Y (NPY)	↑ vasoconstriction
Opioids	Enkephalin	↓ intestinal transit
	Beta endorphin	↓ intestinal transit
	ACTH	↑ cortisol release
	α-Melanocyte stimulatory hormone (αMSH)	↑ melanin release
	Dynorphin	↓ intestinal transit
Tachykinin-bombesin	Substance P	Contraction gastrointestinal smooth muscle
	Substance K	Contraction gastrointestinal smooth muscle
	Neuromedin K	Contraction gastrointestinal smooth muscle
	Gastrin-releasing peptide (GRP)	↑ release gastrin
	Neuromedin B	↓ acid secretion
Orphan peptides	Somatostatin	↓ acid secretion, ↓ pancreatic function
	Neurotensin	↓ acid secretion, ↓ gastric motor activity
	Galanin	↑ plasma glucose, ↓ fundus contraction
	Pancreastatin	↓ islet somatostatin release
	Motilin	↑ motility

sight, thought, taste, or odor of the meal, with gastric distension that results from the presence of food in the stomach, with alkalinization of the gastric lumenal contents by the buffering action of the meal, and with the proteins and their breakdown products in the ingested food. Although gastrin is present in the stomach, it exerts its acid secretory action only after entering the peripheral circulation and then being transported to the basolateral surfaces of the parietal cells. There, specific gastrin receptors of roughly 74,000 Da in size bind the peptide and transduce the binding signal into a cellular response that appears to be mediated by mobilization of intracellular Ca^{2+} and activation of a Ca^{2+}/phospholipid-dependent protein kinase on the cell membrane. Aside from inducing acid secretion, gastrin has another major function to promote growth of gastrointestinal mucosa. This "trophic" action of gastrin is not well characterized in normal tissues but may have clinical significance in gastrointestinal neoplasms, particularly colon cancer.

2. Cholecystokinin

Cholecystokinin was initially isolated from hog intestine as a peptide hormone that was 33 amino acids long, but has since been found to exist as peptides that are 58, 39, 33, 22, 8, and 4 amino acids long. It has 100% structural homology with gastrin in the carboxyl-terminal 6 amino acids, which comprise the biologically active portion of the peptide. It is not surprising, therefore, that CCK can bind to gastrin receptors and exert the same pharmacological effects as gastrin. Normally, however, CCK circulates in the blood in concentrations that are only 5–10% as high as gastrin; thus, under physiological conditions, CCK does not contribute significantly to gastrin's effects. The feature of CCK structure that accounts for its unique function is the presence of a sulfated tyrosine residue 7 amino acids toward the amino-terminal end from the carboxyl-terminal phenylalanine-amide residue. CCK is produced in discrete endocrine cells scattered throughout the small intestine and in neurons of the cerebral cortex. It is released from the intestine in response to a meal stimulus, but the mechanisms for this release process are not known. It is hypothesized that ingested nutrients, particularly the fats and aromatic amino acids, stimulate the release of a yet unidentified CCK-releasing peptide into the lumen of the small intestine that stimulates CCK release. The pancreatic enzymes released by CCK, in turn, digest the CCK-releasing peptide and, thus, inhibit in a feedback fashion further stimulation by CCK. The actions of CCK appear to be mediated by two classes of receptors, peripheral-type receptors that require CCK to be sulfated, and central CCK receptors that do not distinguish between sulfated and nonsulfated CCK. The primary "peripheral-type" CCK effects are the stimulation of pancreatic enzyme secretion, the induction of pancreatic growth, and the contraction of the gallbladder. CCK action on the pancreas appears to be mediated by specific receptors that are linked, like gastrin, to mobilization of intracellular Ca^{2+}. The major "neural-type" effects of CCK may be its action on gastrointestinal smooth muscle to delay gastric emptying and enhance intestinal motility. A provocative function that has been ascribed to CCK is that of a satiety signal that causes animals to stop eating.

B. Secretin/Vasoactive Intestinal Polypeptide/Glucagon

1. Secretin

As noted earlier, secretin was the first hormone for which a biological function was described. It derives its name from its action as a stimulant of pancreatic bicarbonate secretion. It is found in the small intestine, primarily concentrated in the duodenum, as a 27-amino-acid peptide with a carboxyl-terminal amide residue. Secretin is released from the intestine on acidification by the passage of gastric contents into the duodenum. The threshold pH required for its release appears to be 4.5. Nutrients, particularly fats, can also stimulate secretion under experimental conditions, but the physiological significance of this observation is uncertain because no such stimulation is observed in the absence of acid. The major function of secretin is to stimulate the secretion of bicarbonate from the pancreatic ductular cells, thus accounting for the neutralization of the acid secreted by a meal and emptied into the intestine. Another function of secretin is to inhibit gastric emptying, perhaps thereby protecting the duodenum from exposure to excessive amounts of acid.

2. Vasoactive Intestinal Polypeptide

Vasoactive intestinal polypeptide (VIP) is a peptide with a primary structure similar to that of secretin. It consists of 28-amino-acid residues and derives from a gene that encodes still another related peptide called peptide histidine-isoleucine in animals (PHI) or peptide histidine-methionine in humans (PHM). In contrast to classic peptide hormones such as secretin, VIP is detectable only in neurons of the enteric and peripheral nervous systems. In the gut, VIP neurons are present primarily in the submucous plexus, and their fibers form a rich and dense network within the lamina propria. VIP neurons are innervated by preganglionic cholinergic fibers, thus they appear to function as intermediaries between central nervous system neurons and their target organs in the gut. The major function of VIP appears to be that of an inhibitor of smooth muscle contraction, thus it relaxes the lower esophageal and anal sphincters and promotes both gastric and intestinal relaxation. Another function of VIP, by virtue of its ability to relax vascular smooth muscle, is to enhance gastrointestinal blood flow during the period after ingestion of a meal. VIP may also be important in regulating the hydrated state of intestinal contents via its actions to promote fluid and electrolyte secretion into the lumen of the gut. Under some pathological conditions, the presence of excess VIP may induce diarrhea.

3. Glucagon

Pancreatic glucagon is a 29-amino-acid peptide that also shares a high degree of structural homology with other members of the secretin family. It derives from a larger precursor that may be cleaved to produce a peptide of 69 amino acids known as glicentin or another peptide of 37 amino acids known as enteroglucagon or oxyntomodulin, which is pancreatic glucagon with an extension of eight amino acids at the carboxyl terminus. The latter two peptides are found in endocrine cells of the intestinal mucosa, whereas the former is found in α cells in the pancreatic islets of Langerhans. Pancreatic glucagon is considered to be the counterpart to insulin in glucose homeostasis in that it mobilizes glucose from carbohydrate stores in response to hypoglycemia. The specific physiological function of enteroglucagon is not known for certain, but its secretion is stimulated by glucose and digested fats in the intestine. [*See* Insulin and Glucagon.]

C. Opioids

One of the most profound observations in human biology was the finding that the body makes endogenous peptides that act in the manner of alkaloid opioid analgesics. There are three main classes of these peptides: the enkephalins (methionine or leucine enkephalin) derived from proenkephalin, the dynorphins derived from prodynorphin, and β-endorphin derived from proopiomelanocortin. These peptides have a common structural feature in consisting of the pentapeptide Tyr-Gly-Gly-Phe-Met (or Leu). They are widely distributed in the central and peripheral nervous systems, and in the gut they are found primarily in myenteric plexus neurons that arborize into the circular muscle layers. These peptides act on three classes of receptors labeled mu, delta, and kappa to produce different actions, often with the same physiological effect. For example, mu and delta opioid agonist activity hyperpolarizes myenteric neurons by increasing a K^+ conductance, whereas kappa activity blocks Ca^{2+} channels. Although the mechanisms are different, both actions have the effect of inhibiting neuronal action. The net effect of opioid peptides on gut motility is to slow transit of ingested material through the intestinal lumen. This effect may be exerted either by influencing the activity of other enteric neurons or by direct action on smooth muscle cells. Opioids are also known to have the effect of inhibiting intestinal secretion by diminishing electrolyte transport into and promoting fluid absorption from the intestinal lumen. The combination of delayed transit and decreased secretion produced in the gut by opioids makes them useful in the therapy of diarrheal states.

D. Pancreatic Polypeptide Peptide YY/Neuropeptide Y

1. Pancreatic Polypeptide

Pancreatic polypeptide (PP) is a 36-amino-acid peptide that was initially discovered as an impurity in preparations of insulin. It is localized almost exclusively to the pancreas, where it is found in specific cells in the islets of Langerhans as well as other cells scattered throughout the acinar tissues. PP is released in response to ingestion of nutrients (primarily protein and fats), and this response appears to be mediated by cholinergic neurons because it can be abolished by antagonizing muscarinic cholinergic receptors with atropine. The physiological function of PP has not been established conclusively, but the peptide appears to be an inhibitor of pancreatic secretion. This action would imply a role for PP in the feedback regulation of pancreatic secretion stimulated by the enteropancreatic neural reflex. The ingestion of a meal stimulates a cholinergic neural response in the pancreas. The cholinergic neurons appear to act simultaneously both to stimulate pancreatic acinar cells and to stimulate the release of PP, an inhibitor of pancreatic secretion. In this fashion, PP may function to modulate or finely regulate pancreatic secretion.

2. Peptide YY

Peptide YY (PYY) is so named because it is a peptide with a tyrosine (Y) residue at the amino terminus and another tyrosine (Y) residue at the carboxyl terminus. It is one of a series of peptides that have been purified on the basis of their structural properties (specifically, the presence of a carboxyl-terminal amide moiety) without any clues as to their function. Like PP, PYY consists of 36 amino acid residues; however, its localization is very different. It is found mainly in distinct endocrine cells in the small intestine, primarily in the distal portion and in the proximal colon. PYY is released into the circulation by entry of fat into the small intestinal and, unlike release of PP, this response does not appear to be mediated via cholinergic neurons. The main function of PYY appears to be that of an enterogastrone (i.e., an inhibitor of gastric acid secretion derived from the intestine). As noted earlier, such substances are thought to be important feedback modulators of the acid secretion induced by a meal. Two other effects of PYY that may be important in processing ingested nutrients are its ability to inhibit pancreatic secretion and to delay gastric emptying.

3. Neuropeptide Y

Neuropeptide Y (NPY) derives its name from the fact that it is a neuropeptide with amino- and carboxyl-terminal tyrosine (Y) residues. It, too, is a 36-amino-acid peptide purified on the basis of its carboxyl-terminal amide moiety. Unlike PP and PYY, however, it is found strictly in neurons of the central and peripheral nervous systems. In the gut, NPY is located in both submucous and myenteric plexus neurons. A portion of the NPY fibers are associated with blood vessels in the gut. These fibers derive from extramural neurons containing norepinephrine.

It is not clear how NPY release is stimulated in the gut, but its action appears to produce sustained vasoconstriction and, in so doing, reduces enteric blood flow and motility.

E. Tachykinins

The tachykinin family consists of a number of amphibian, molluscan, and mammalian peptides including substance P, one of the most thoroughly studied of all the gut peptides. These peptides were termed "tachykinins" because of their characteristic rapid onset of action in gut tissues in contradistinction to the "bradykinins," which elicit much slower responses. Occasionally the tachykinins are referred to as "neurokinins." The carboxyl-terminal amino acid sequence Phe-X-Gly-Leu-Met-NH_2 is highly conserved in the members of the tachykinin family. The amino acid at position X defines much of the specificity of a given tachykinin. For example, the aromatic amino acid phenylalanine determines substance P specificity, whereas the aliphatic amino acid valine is characteristic for neurokinin A and B. Tachykinins are present in almost all layers of the intestinal wall and can be synthesized by neurons in the myenteric and submucous ganglia. Substance P can also be found in enterochromaffin cells of the gut. It is of particular note that tachykinins are found in the dorsal root and nodose ganglia. The fact that dorsal root ganglion neurons that innervate the gut may send processes to the skin may explain the referred pain noted in many gastrointestinal disorders. The receptors for the tachykinins have been characterized particularly well because their genes have been cloned and analyzed. They exhibit a structure characteristic of other receptors whose signals are mediated by guanine nucleotide binding proteins in that they traverse the cell membrane seven times with extracellular domains containing sites for addition of carbohydrates and intracellular domains containing sites for addition of phosphate residues. The primary effects of tachykinins appear to be stimulatory on gastrointestinal motility via excitatory neurons regulating peristalsis, although the physiological significance of these effects is not confirmed. Other potentially important effects of tachykinins include regulation of blood flow, intestinal secretion, and pancreatic secretion.

F. Orphan Peptides

1. Somatostatin

Somatostatin received its name from its initially described ability to inhibit pituitary growth hormone secretion. It was isolated from the hypothalamus almost by accident by investigators seeking to find the stimulant of growth hormone. The peptide exists in two primary molecular forms of 14 and 28 amino acids in length. Somatostatin is found in a wide variety of tissues in the body, including the central, peripheral, and enteric nervous systems and the gastroenteropancreatic systems. In the gut, it is found in intrinsic and extrinsic neurons of the submucosal and deep muscular layers as well as in specialized D cells in the mucosal lining of virtually the entire gastrointestinal tract. The release of somatostatin from the gut in response to a meal may result from a multiplicity of factors. Nutrients may act directly to stimulate D cells that communicate with the lumenal contents in the gastric antrum. Othere factors (e.g., postprandial gastric acid and gastrin secretion) may also contribute to somatostatin secretion. The major function of somatostatin seems to be to act as a natural brake for the body. It is characterized by virtually uniform inhibitory actions in the gut, on exocrine secretion, endocrine secretion, motility, fluid and electrolyte transport, blood flow, and tissue growth and proliferation. The mechanisms by which somatostatin exerts these effects on the target cell are uncertain but may involve inhibitory actions at multiple sites, including the cell's receptors for other neurohumoral messengers, the intracellular mechanisms that transduce the receptor signal to a cellular action, or even the terminal action of the cell itself (e.g., exocytosis). Given its widespread inhibitory effects, it is not surprising that somatostatin has potential usefulness in treating a variety of gastrointestinal disorders resulting from too much activity of one sort or another (e.g., diarrhea).

2. Motilin

Motilin is a 22-amino-acid peptide found in discrete endocrine cells in the small intestine. It derives its name from its action on gut motility. Unlike other peptide hormones of the gut, motilin concentrations in the blood decrease in response to a meal in seemingly paradoxical fashion. Instead, during the interdigestive period there are regular periodic increases in plasma motilin levels, which coincide with acceleration of gastric emptying and initiation of a wave of intestinal contraction that courses through the entire gut at regular intervals of 60–120 min. These contractions, called the migrating motor complexes, are known as the "housekeepers" of the gut that function to clean it out from time to time. Ingestion of a meal prevents the normal cyclic increases in plasma motilin and, thereby, the migrating motor complexes. The mechanism by which motilin exerts its effects has yet to be elucidated, but experimental studies suggest that the neural circuits responsible for programming the migrating motor complexes reside in the enteric nervous system and involve cholinergic pathways. Motilin presumably acts on enteric neurons as well as on gastric smooth muscle to initiate the intestinal housekeeping process.

3. Galanin

Galanin, a 29-amino-acid peptide, is another of the gut hormones purified on the basis of its carboxyl-terminal amide moiety. Its name derives from the presence of a glycine residue at its amino terminus and an alanine residue at its carboxyl terminus. In the gut it is found exclusively in enteric neurons concentrated mainly in the small intestine. Like somatostatin, galanin exerts primarily inhibitory action. For example, it appears to be an important regulator of glucose homeostasis by virtue of its inhibitory action on insulin secretion. Its function in the gut may result from its inhibitory actions on motility.

4. Neurotensin

As its name suggests, neurotensin was initially isolated on the basis of its ability to influence systemic blood pressure (i.e., it induces hypotension). It is a 13-amino-acid peptide that is related to two other peptides that share some structural homology at the carboxyl terminus, neuromodulin N and xenopsin. Neurotensin in the gut is primarily found in endocrine-type cells in the ileal mucosa, with smaller amounts being present in the rest of the gut. Small amounts are also found in enteric neurons. The peptide is most potently released by ingested fat. Although the physiological function of neurotensin has not been determined, it may play a role in a variety of gastrointestinal events activated by fat in the lumen, including inhibition of acid secretion, enhancement of intestinal fluid secretion, stimulation of pancreatic output, and increase in blood flow.

III. CLINICAL SIGNIFICANCE OF PEPTIDE HORMONES

A. Disease Pathogenesis

Peptide hormones of the gut have been implicated in the pathogenesis of a wide variety of human disorders, including peptic ulcer disease, hypergastrinemia syndromes, irritable bowel syndrome, gallstones, diarrheal disorders, reflux esophagitis, Hirschprung's disease, and even morbid obesity. The most striking correlation between peptide hormones and gastrointestinal disease is found in the many syndromes associated with hormone-secreting tumors of the pancreas. These interesting "accidents" of nature have provided clues as to the physiological function of some peptide hormones. Because the cells that produce hormones in the pancreas are thought to have an embryological origin common to other neuroendocrine tissues, the pancreatic endocrine neoplasms almost always involve more than a single hormone and often involve other neuroendocrine organs. Nevertheless, the specific syndromes that arise from these tumors generally can be ascribed to the effects of a single hormone.

The most common of the pancreatic endocrine neoplasm syndromes was described by Zollinger and Ellison in 1955 when they noted the association between hypersecretion of gastric acid, refractory peptic ulcer disease, and nonbeta cell tumors of the islets of Langerhans in the pancreas. Later the substance responsible for the bulk of clinical findings was identified as gastrin. One interesting effect of the uncontrolled acid hypersection induced in this disorder by gastrin is that many of the patients have diarrhea largely because of the inability of pancreatic enzymes to digest ingested nutrients in the presence of an acidic environment in the intestine. Gastric acid secretion has long been known to be a physiological action of gastrin, but another curious manifestation of Zollinger–Ellison syndrome, gastric mucosal hyperplasia, contributed to the discovery that gastrin has another important function as a promotor of tissue growth.

A syndrome complex of severe watery diarrhea, low serum potassium concentrations, and reduction in gastric acid secretion in association with a pancreatic islet cell tumor was described in 1958 by Verner and Morrison. The diarrhea was so severe that the disorder was called "pancreatic cholera" by some. Initially, this syndrome was thought to be a variant of Zollinger–Ellison syndrome; however, the offending agent was later identified to be VIP. A cardinal feature of the disease is the presence of diarrhea in the absence of nutrient ingestion, implying that it results from excess intestinal secretion rather than nutrient maldigestion or malabsorption. Like cholera toxin, VIP induces the activation of adenylate cyclase and subsequent production of cyclic adenosine monophosphate in intestinal epithelial cells. This induces the secretion of water and electrolytes into the intestinal lumen. In some patients, the diarrhea can exceed 8 liters/day; thus severe dehydration and internal electrolyte disturbances may result.

Another hormone, somatostatin, was implicated in the pathogenesis of a clinical syndrome in a somewhat circuitous fashion. Although the hormone was described in 1973, no clinical abnormality had been associated with its overproduction until 1977 when a diabetic patient with modest diarrhea underwent surgery for gallstones and was found to have an incidental pancreatic tumor. On histochemical analysis of the tumor, it was noted to be abundantly occupied by cells containing somatostatin. It was only in retrospect that the symptoms of the patient could be correlated to the actions of the hormone. The ability of somatostatin to inhibit gallbladder contraction led to stasis of bile and gallstone formation, the inhibition of digestive enzyme secretion from the pancreas by somatostatin led to nutrient maldigestion and diarrhea, and the inhibitory action of somatostatin on insulin secretion led to diabetes. Subsequently, a host of other patients with somatostatin-secreting tumors of the pancreas have been identified.

Aside from these disorders, there are numerous other syndromes associated with overproduction of insulin, glucagon, substance P, neurotensin, adrenocorticotrophic hormone (ACTH), and PP among others. The features of the various syndromes are summarized in Table II. In some patients, the features of one or more of these syndromes may be combined in a clinical entity known as multiple endocrine neoplasia Type I. The vast majority of patients with this syndrome have pancreatic endocrine neoplasms producing numerous peptide hormones in the pancreas; however, most of these patients also have hormone-secreting tumors in their pituitary and parathyroid glands. The disease is generally transmitted in an autosomal dominant pattern; thus the relatives of all patients with peptide-secreting endocrine neoplasms should be screened carefully for signs of endocrine abnormalities.

B. Disease Diagnosis

Peptide hormones, because of their many actions, are potentially useful as aids in diagnosis and treatment of a number of pathological conditions. Obviously, the measurement of peptide hormones in the plasma is an essential element of diagnosing endocrine neoplasms. Some endocrine neoplasms, because of their pancreatic ductular origin, can be induced to secrete hormones in response to stimulation with secretin, a hormone that generally induces bicarbonate secretion from pancreatic ducts. This "secretin test" is particularly useful in diagnosing patients with Zollinger–Ellison syndrome whose serum gastrin levels are only marginally elevated. Secretin and CCK infusions are also useful in evaluating patients with disorders of pancreatic exocrine function. Because these two peptides induce bicarbonate and enzyme secretion from the pancreas, respectively, by infusing the peptides and measuring the maximal secretory responses of the pancreas, it it possible to quantify the relative

TABLE II
Pancreatic Endocrine Neoplasm Syndromes

Peptide hormone	Regulatory action	Clinical manifestations
Gastrin	Acid secretion Cell proliferation	Gastric acid hypersecretion Peptic ulcer disease Diarrhea
Somatostatin	Widespread inhibitory actions	Gallstones Diabetes Diarrhea
Vasoactive-mediated polypeptide (VIP)	Intestinal fluid and electrolyte secretion Vasomotor actions	Watery diarrhea Electrolyte abnormalities Flushing
Insulin	Decreased blood sugar Anabolic state	Hypoglycemia Behavioral disturbances
Glucagon	Increased blood sugar Catabolic state	Skin rash Diabetes Anemia
Adrenocorticotrophic hormone (ACTH)	Cortisol secretion	Moon-like face Muscle wasting Weakness

degree of organ dysfunction. CCK's other primary action to stimulate gallbladder contraction can also be used diagnostically to evaluate patients with chronic abdominal pain that may originate from abnormalities in the motility of the biliary tree. Other hormones that have effects on the motility of the intestine are useful adjuncts in radiological and endoscopic evaluation of the gastrointestinal tract. Both glucagon and somatostatin in appropriate doses can relax the smooth muscles in the intestine. This effect "quiets" the movements of the intestine sufficiently to take more detailed X rays of the gut and also to permit cannulation of the biliary tree where it terminates in the duodenum so that it can be injected with radiological contrast dye for X-ray evaluation (i.e., endoscopic retrograde cholangiopancreatography or ERCP).

C. Disease Therapy

The application of hormones in therapy of pathological states has been restricted to only a few special circumstances. Obviously, the opioids have proved to be useful in treatment of pain. Opioids are also contained within many standard remedies for diarrhea. These substances are effective by virtue of their ability to inhibit intestinal motility as well as secretion of fluid and electrolytes. Erythromycin derivatives have been found, quite by accident, to substitute for motilin in inducing intestinal smooth muscle contraction. This property has been explored as a potential means for treatment of patients with intestinal dysmotility.

In the past, native peptide hormones have not been useful as therapeutic agents because they are small, easily degraded molecules and, thus, disappear quickly after being administered. Recently, several peptide analogues with greatly prolonged half-lives have been synthesized. Thus, for example, somatostatin analogues may have a plasma half-life of 2 hr or more, whereas the natural compound disappears within minutes. One new somatostatin analogue, octreotide, has been used effectively in treating patients with endocrine neoplasms. In a sense, somatostatin analogues are ideal for treatment of these disorders because the peptide has the capability of inhibiting both the secretion of peptides from tumors as well as the effects of the secreted peptides at their target organs. Because of somatostatin's widespread inhibitory actions, the octreotide analogue has potential usefulness in the treatment of a variety of other disorders, such as diarrhea, pancreatitis, motility disorders, and gastrointestinal bleeding.

Other applications of peptide hormone biology to clinical conditions involve the use of specific peptide receptor antagonists. Naloxone, an opioid receptor antagonist, has proved to be a useful antidote for narcotic overdoses. More recently, antagonists for CCK have been developed as potentially useful agents in treating disorders of intestinal motility or appetite regulation. Gastrin receptor antagonists with potential applications in treatment of peptic ulcer disease are being developed.

Given the multitude of effects that peptide hormones have in the body, the development of additional stable analogues of peptide hormones or selective peptide receptor antagonists holds great promise for treatment of many other human disease states.

BIBLIOGRAPHY

DelValle, J., and Yamada, T. (1989). Secretory tumors of the pancreas. *In* "Gastrointestinal Diseases" (M. H. Slesinger and J. S. Fordtran, eds.), 4th Ed. Saunders, Philadelphia.

Fordtran, J. S., and Sleisenger, M. H. (eds.) (1993). "Gastrointestinal Disease: Pathophysiology, Diagnosis, Management," 5th Ed. Saunders, Philadelphia.

Johnson, L. R. (ed.) (1996). "Gastrointestinal Physiology," 5th Ed. Mosby–Year Book, New York.

Makhlouf, G. M. (1989). "Neural and Endocrine Biology, Vol. 2: *Handbook of Physiology*, Sec. 6: *The Gastrointestinal System*." American Physiological Society, Bethesda, Maryland.

Walsh, J. H. (1987). Gastrointestinal hormones. *In* "Physiology of the Gastrointestinal Tract" (L. R. Johnson, ed.), 2nd Ed. Raven, New York.

Wolfe, M. M., and Jensen, R. T. (1987). Zollinger–Ellison syndrome: Current concepts in diagnosis and management. *N. Engl. J. Med.* **317**, 1200.

Yamada, T. (1987). Local regulatory actions of gut peptides. *In* "Physiology of the Gastrointestinal Tract" (L. R. Johnson, ed.), 2nd Ed. Raven, New York.

Peptides

TOMI K. SAWYER
Warner-Lambert Company

CLARK W. SMITH
Pharmacia and the Upjohn Company

GLOSSARY

Anabolic hormone Chemical messenger (e.g., peptide, steroid, biogenic amine) that stimulates directly or indirectly biochemical changes of intermediary metabolism, which result in the biosynthesis of macromolecules (e.g., protein, polysaccharide, lipid, nucleic acid)

Heterodimeric peptide Peptide (or protein) having two nonidentical subunits (e.g., α and β) that interact, usually in a noncovalent manner, to produce the biologically active form of the molecule

Hyperglycemic action Effect producing abnormally high concentrations of glucose in the blood

Juxtaglomerular cells Epithelial cells of the afferent arteriole located in the kidney and proximate to the point where the arteriole abruptly branches into a network of capillaries

Nociceptic Pain sensory responsive

Second messengers Chemical messengers (e.g., cyclic AMP, calcium, inositol triphosphate) that are produced intracellularly in response to extracellular "first" messengers (e.g., peptide, biogenic amine) acting through cell membrane receptors and transmembrane signaling mechanisms

Secretagogue Chemical messenger (e.g., peptide, steroid, biogenic amine) that stimulates the release (or secretion) of a specific cellular product

Secretory granules Cytoplasmic storage vesicles derived from the Golgi complex; they contain secretory products of a cell

Signal sequence Amino-terminal sequence of amino acids (primarily hydrophobic) that is requisite for transport of newly synthesized cytosolic peptides (or proteins) into the lumen of the endoplasmic reticulum, where the signal sequence is then enzymatically removed

Tropic growth effect Growth stimulatory effect that may be induced directly or indirectly by trophic chemical messengers at their target tissues (e.g., muscle, bone)

Tyrosine autophosphorylation Protein (e.g., receptor, enzyme) self-phosphorylation that may be induced by hormonal and/or allosteric mechanisms and results in functional modification due to the formation of phosphate ester linkage on the amino acid side-chain group of Tyr

PEPTIDES ARE CHEMICAL MESSENGERS THAT EXIST in all forms of life and serve as hormones, neurotransmitters/neuromodulators, and growth factors, regulating the cardiovascular, gastrointestinal, reproductive, and immune systems in higher organisms, including humans. All peptides are composed of amino acids (i.e., residues) covalently linked by amide (CONH) bonds formed between an α-carboxyl (CO_2H) group of one residue and an α-amino (NH_2) group of another residue. The specific ordered-sequence (Fig. 1) primary structure of a peptide determines both its three-dimensional properties (i.e., secondary structure, or conformation, and tertiary structure) and its biological properties. Specific examples of such properties for a few selected peptides are described in Section II.

Peptide hormones (e.g., adrenocorticotropin and insulin) normally exist in very low concentrations (10^{-6} to 10^{-12} M) in bodily fluids or tissues, in which they persist for short times (1–30 min), because they are rapidly inactivated by proteolytic enzymes and/or excreted (i.e., cleared) from the body. Their transitory actions are suitable for continuous regulation (i.e.,

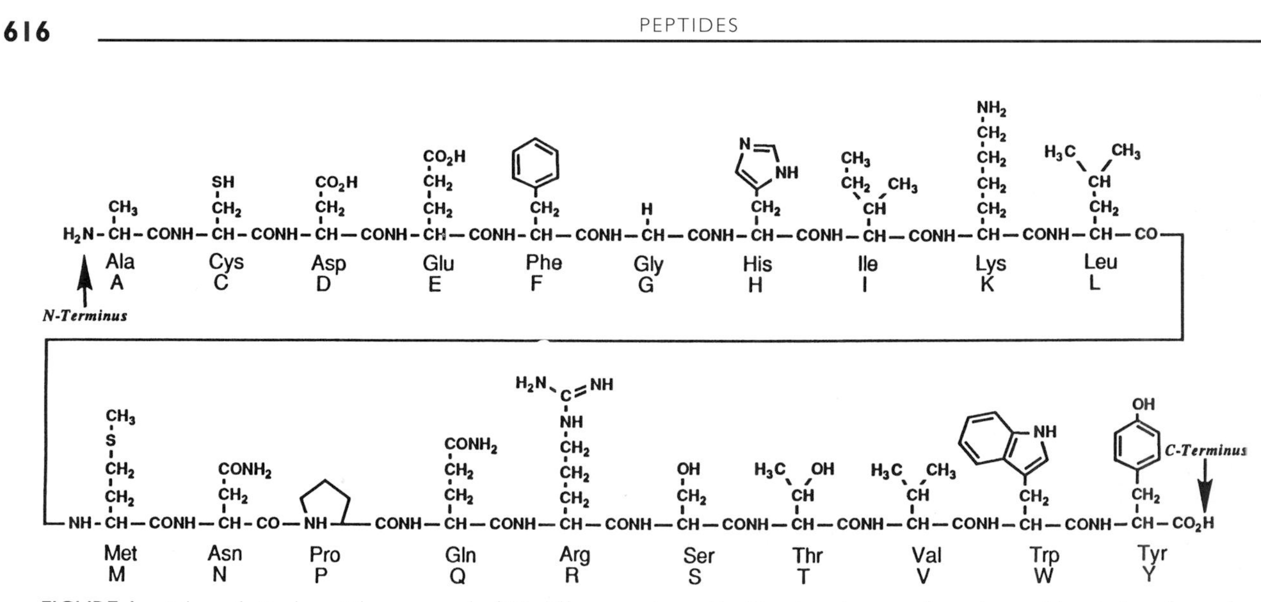

FIGURE 1 A hypothetical peptide composed of 20 different amino acids. The three-letter and one-letter abbreviations for each residue are given.

homeostasis). Another common trademark of peptides is their specificity at target cells (e.g., secretory or neuronal), which depends on the existence of receptors, or proteins typically found on the cell surface, each capable of recognizing and binding a particular peptide, forming with it biochemically active complexes. These peptide–receptor complexes are reversible, and dissociation ultimately results in an unstimulated state of the target tissue. However, the cellular response, once initiated, can proceed independently of the initial events occurring at the cell surface, until it is terminated by other signals or deactivation processes. As a result of these chemical and biological characteristics, research on peptides relies on many technological tools, such as (1) radioimmunoassay, to detect and quantitate peptides in various biological tissues or fluids; (2) radioreceptor assay, to determine peptide–receptor interactions and explore the mechanism of action of peptides; and (3) artificial synthesis, either direct or indirect, through DNA recombination technology, to prepare the quantities of peptides required for *in vitro* or *in vivo* studies.

I. NATURALLY OCCURRING PEPTIDES

A. Biosynthesis and Metabolism

Peptide hormones are synthesized, like all proteins, on ribosomes by translating specific mRNAs, which reproduce the sequences of the genes present in the DNA of chromosomes. Most peptides are made up of biologically inert and higher-molecular-weight precursor molecules (i.e., prepropeptides), which are later split, or processed, enzymatically. The "pre" portion of the precursor usually comprises an amino-terminal extension of 15–30 amino acids to the sequence of the main peptide, called the signal sequence. This sequence directs the transport of the peptide to the endoplasmic reticulum and is removed as the peptide is transported into its cavity.

The remaining peptide can still be an inactive precursor, a propeptide, which can be processed in various compartments (e.g., secretory granules or the blood). The peptide can be modified by sulfation, glycosylation, amidation, or acetylation, and, if it is to be secreted, it is packaged into secretory vesicles within the Golgi apparatus. Peptide neurotransmitters/neuromodulators are similarly biosynthesized and stored in secretory vesicles within neurons until they are released at synapses.

Degradation and inactivation of peptides can occur at a variety of physiological sites (e.g., the liver, kidneys, lungs, and brain) through the action of proteolytic enzymes (exopeptidases or endopeptidases) that recognize and cleave the bonds between certain amino acid residues.

B. Mechanisms of Action

The high specificity of peptides for their respective target tissues is achieved by release close to the site of action and rapid degradation, as well as by high specificity for their receptors. Receptors are cell-surface macromolecules that bind specific peptides, but with high affinity and noncovalently, according to their unique amino acid sequence and resultant three-dimensional conformation and topography. Peptide receptor subtypes have also been identified and pharmacologically defined (e.g., μ, δ, and κ opioid receptors for the endogenous opioid peptides endorphin, enkephalin, and dynorphin, respectively). The mechanisms of action of the majority of peptides are based on physiological regulation of peptide receptor number and various forms of information transfer (i.e., signal transduction).

As shown in Fig. 2, signal transduction from the peptide–receptor complex to the cellular genes can occur by transmembrane modulation of certain enzymes, such as adenylate cyclase (which elevates cAMP), guanylate cyclase (which elevates cGMP), calmodulin (which mobilizes Ca^{2+}), and/or enzymes that affect the metabolism of arachidonic acid (via prostaglandin synthesis) or phosphatidylinositol (via inositol triphosphate and diacyglycerol synthesis). Several key intracellular signals (i.e., secondary messengers) can mediate the biological roles—either stimulatory or inhibitory—of a particular peptide. In many cases GTP-binding proteins localized at the cell membrane can act as intermediaries of signal transduction between the peptide–receptor complex and the enzymes mentioned.

C. Peptide Biological Systems

A myriad of biological activities have been described for the more than 200 peptides that have been isolated and characterized in humans or other vertebrate species. Only a few peptides (Table I) are described here to exemplify some of their more fundamental biological properties. An important theme of this section is that many of the classic peptide hormones, which are chemical messengers of the endocrine system, have now been found within the brain, the gastrointestinal tract, and many other tissues. Thus, as an extension of their hormonal properties, which were first described, several of these peptides manifest neural, metabolic, growth, cardiovascular, gastrointestinal, reproductive, and immune regulatory functions (see Table I).

1. Adrenocorticotropic Hormone and Melanocyte-Stimulating Hormone

Adrenocorticotropic hormone (ACTH) and melanocyte-stimulating hormone (MSH) are structurally similar peptide hormones derived primarily from the pitu-

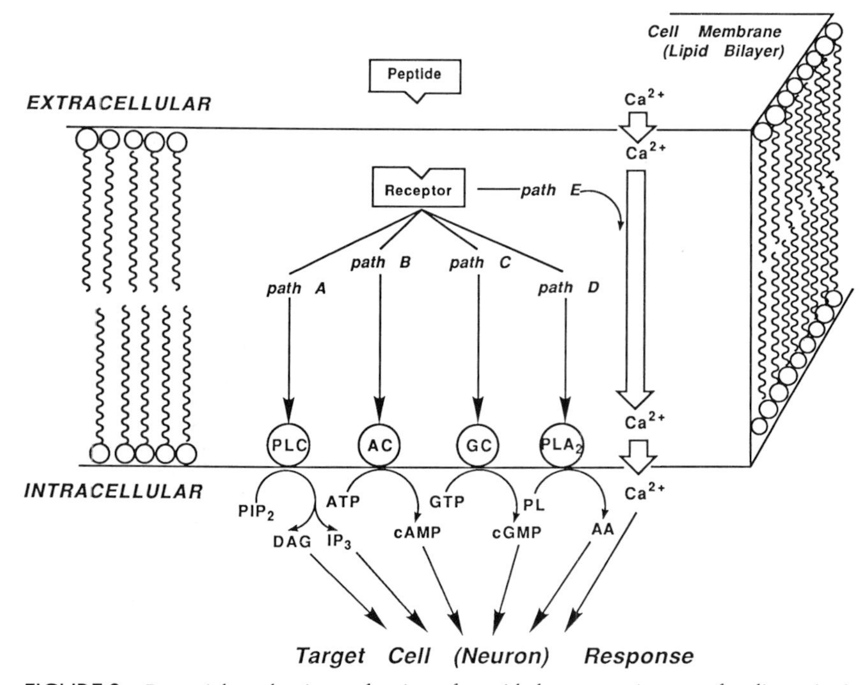

FIGURE 2 Potential mechanisms of action of peptide hormones (see text for discussion). PLC, phospholipase C; AC, adenylate cyclase; GC, guanylate cyclase; PLA_2, phospholipase A_2; PIP_2, phosphoinositol bisphosphate; DAG, diacylglycerol; IP_3, inositol triphosphate; PL, phospholipid; AA, arachidonic acid.

TABLE I
Some Known Biologically Active Peptides

Peptide	Amino acids[a]	Major biological actions[b]
Adrenocorticotropic hormone	39	Adrenal steroidogenesis increased
Angiotensin II	8	Vasoconstriction increased, aldosterone secretion increased, dipsogenia (i.e., thirst) increased
Bradykinin	9	Smooth muscle contraction increased, vasodilation increased, inflammation analgesia increased
Calcitonin	32	Blood Ca^{2+} decreased
Cholecystokinin 8	8	Pancreatic enzyme and electrolyte secretion increased, gallbladder contraction increased, satiety increased, secretion of insulin and glucagon increased
Corticotropin-releasing factor	41	Corticotropin secretion increased
Dynorphin	17	Sedative analgesia increased, feeding behavior increased
β-Endorphin	31	Opiate-like activity increased, central analgesia increased, respiratory depression increased, euphoria increased
Leu-Enkephalin	5	Spinal analgesia increased, emotional effects increased
Epidermal growth factor	53	Epidermal growth and keratinization increased, growth of corneal epithelium increased
Follicle-stimulating hormone	220	Spermatogenesis in males increased, ovarian follicle growth and estradiol synthesis in females increased
Gastrin 17	17	Pepsin, gastric acid–HCl, and pancreatic enzyme secretion increased; stomach, gallbladder, and intestine smooth muscle contraction increased
Glucagon	29	Blood glucose increased, gluconeogenesis increased, glycogenolysis increased
Gonadotropin-releasing hormone	10	Luteinizing hormone and follicle-stimulating hormone secretion increased
Insulin	51	Blood glucose decreased, protein synthesis and lipolysis increased
Luteinizing hormone	210	Testicular androgen synthesis in males increased, ovarian estradiol and progesterone synthesis in females increased
α-Melanotropin	13	Melanin biosynthesis and secretion increased
Motilin	22	Gastrointestinal villous motility increased
Nerve growth factor	118	Sympathetic neurite development increased
Oxytocin	9	Milk secretion increased, uterine contraction increased
Parathyroid hormone	84	Blood calcium increased
Prolactin	199	Milk synthesis increased, corpus luteum progesterone synthesis increased
Secretin	27	Pancreatic secretion of H_2O, bicarbonate, electrolyte, and protein increased
Somatomedins	50–80	Peripheral nervous system growth and development increased
Somatostatin 14	14	Somatotropin, thyrotropin, parathyroid hormone, calcitonin, renin, and gastric acid secretion decreased
Somataotropin	191	Hepatic somatomedin synthesis increased
Somatotropin-releasing hormone	44	Somatotropin secretion increased
Substance P	11	Peripheral nervous system pain transmission increased, central nervous system pain transmission decreased
α_1-Thymosin	28	Lymphocyte proliferation and differentiation increased
Thyrotropin	201	Thyroid hormone T_3 and T_4 synthesis and secretion increased
Thyrotropin-releasing hormone	3	Thyrotropin and prolactin secretion increased
Vasoactive intestinal polypeptide	28	Vasodilation increased, bronchodilation increased, glucagon and insulin secretion increased, glycogenolysis increased, lipolysis increased
Vasopressin	9	Renal H_2O absorption increased, vasoconstriction increased

[a]Prevalent form of the peptide in humans.
[b]Major actions (some peptides affect multiple biological actions) as related to either an increased/stimulated or decreased/inhibited physiological response.

itary gland. Proopiomelanocortin is a precursor protein (31,000 molecular weight) for both ACTH and several molecular variants of MSH (i.e., α-MSH, β-MSH, and γ-MSH). α-MSH is identical to the first 13-amino-acid sequence of ACTH, but, as a result of posttranslational modifications, is N-acetylated and C-amidated (Fig. 3). The biogenetic regulation and enzymatic processing of proopiomelanocortin have been intensively studied and are a classic example of prepropeptide hormone biology. [*See* Pituitary.]

ACTH secretion from the pituitary gland is stimulated by the hypothalamic peptide corticotropin-releasing factor. ACTH activates the adrenal gland to secrete the corticosteroids (i.e., aldosterone and cortisol), which in turn regulate the metabolism of carbohydrates, lipids, and proteins, as well as Na^+ resorption at various target tissues. ACTH is also active on melanocytes (MSH-like stimulation of melanin synthesis/secretion) and adipose tissue (lipolysis). [*See* Adrenal Gland.]

The release of MSH from the pituitary gland is controlled by mechanisms that are more poorly defined, but may include the inhibition of secretion by hypothalamic catecholamines, which, if interrupted, results in MSH release. α-MSH activates melanin pigmentation in most vertebrate species, and it is significantly more potent (i.e., more than 100-fold) than ACTH in this effect. The structural similarity between MSH and ACTH results in some overlap of biological properties. Other examples of activities attributable to both MSH and ACTH include certain central nervous system–behavioral effects (i.e., arousal, attention, memory retention, and learning) and antipyretic or pyretic thermoregulation. Interestingly, the melanotropic activity of MSH is uniquely enhanced by heat–alkali treatment, which results in chemical racemization (i.e., the stereochemical conversion of an L- to a D-amino acid) of specific residues within its sequence. Racemized MSH effects highly potent and sustained (prolonged) melanotropic activity *in vitro* and *in vivo*. [*See* Catecholamines and Behavior.]

ACTH and MSH appear to act through a common molecular mechanism—the stimulation of adenylate cyclase—which then produces cAMP (the secondary messenger), increasing its concentration within the target cell.

2. Angiotensin and Bradykinin

The regulation of blood pressure involves the biological activities of several peptide and nonpeptide factors. Angiotensinogen-derived peptides, more specifically angiotensins II (ANG II) and III (ANG III), are physiologically active in increasing blood pressure (i.e., hypertensive). Both ANG II (see Fig. 3) and ANG

ACTH	H-Ser-Tyr-Ser-Met-Glu-His-Phe-Arg-Trp-Gly-Lys-Pro-Val-Gly-Lys-Lys-Arg-Arg-Pro-Val-Lys-Val-Tyr-Pro-Asn-Gly-Ala-Glu-Ser-Ala-Glu-Ala-Phe-Pro-Leu-Glu-Phe-OH
ANG II	H-Asp-Arg-Val-Tyr-Ile-His-Pro-Phe-OH
Bradykinin	H-Lys-Arg-Pro-Pro-Gly-Phe-Ser-Pro-Phe-Arg-OH
CCK 33	H-Lys-Ala-Pro-Ser-Gly-Arg-Val-Ser-Met-Ile-Lys-Asn-Leu-Gln-Ser-Leu-Asp-Pro-Ser-His-Arg-Ile-Ser-Asp-Arg-Asp-Tyr(SO_3H)-Met-Gly-Trp-Met-Asp-Phe-NH_2
Enkephalin-Met^5	H-Tyr-Gly-Gly-Phe-Met-OH
β-Endorphin	H-Tyr-Gly-Gly-Phe-Met-Thr-Ser-Glu-Lys-Ser-Glu-Thr-Pro-Leu-Val-Thr-Leu-Phe-Lys-Asn-Ala-Ile-Val-Lys-Asn-Ala-His-Lys-Lys-Gly-Gln-OH
Dynorphin	H-Tyr-Gly-Gly-Phe-Leu-Arg-Arg-Ile-Arg-Pro-Lys-Leu-Lys-Trp-Asp-Asn-Gln-OH
Gastrin 17	(pyro)Glu-Gly-Pro-Trp-Leu-Glu-Glu-Glu-Glu-Glu-Ala-Tyr(SO_3H)-Gly-Trp-Met-Asp-Phe-NH_2
Glucagon	H-His-Ser-Gln-Gly-Thr-Phe-Thr-Ser-Asp-Tyr-Ser-Lys-Tyr-Leu-Asp-Ser-Arg-Arg-Ala-Gln-Asp-Phe-Val-Gln-Trp-Leu-Met-Asp-Thr-OH
GnRH	(pyro)Glu-His-Trp-Ser-Tyr-Gly-Leu-Arg-Pro-Gly-NH_2
α-MSH	Ac-Ser-Tyr-Ser-Met-Glu-His-Phe-Arg-Trp-Gly–Lys-Pro-Val-NH_2
Somatostatin 14	H-Ala-Gly-Cys-Lys-Asn-Phe-Phe-Trp-Lys-Thr-Phe-Thr-Ser-Cys-OH

FIGURE 3 Primary structures of selected peptides of human origin (see text for discussion). ACTH, adrenocorticotropic hormone; ANG, angiotensin; CCK, cholecystokinin; GnRH, gonadotropin-releasing hormone; α-MSH, α-melanocyte-stimulating hormone.

III are small linear peptides ultimately generated from ANG, a biologically inactive precursor glycoprotein secreted from the liver. ANG is processed to give another biologically inactive precursor, ANG I, which is a decapeptide corresponding to the amino-terminal sequence of ANG. Conversion of ANG to ANG I is specifically effected by circulating renin, an enzyme secreted from the juxtaglomerular cells of the kidney in response to hypovolemia (i.e., a reduction in blood volume), an increase in the sodium concentration in the blood, and/or a central nervous system stimulus (catecholamine related). ANG I is converted to ANG II by the angiotensin-converting enzyme, in lung tissue. Finally, removal of the first amino acid at the amino terminus of ANG II by aminopeptidase yields ANG III.

ANG II is an extremely potent vasoconstrictor on arterioles and effects an immediate elevation of blood pressure. Long-range effects of ANG II and/or ANG III include the stimulation of aldosterone release from the adrenal cortex, which then effects Na^+ retention and a fluid volume increase (i.e., antidiuresis). All of the components of the ANG–renin–angiotensin-converting enzyme system have been identified in the brain, and two biological actions that are known in this regard include the stimulation of drinking behavior and the stimulation of pituitary vasopressin (i.e., antidiuretic hormone) secretion.

ANG II acts via specific receptors, but the mechanism is not known in detail. On endothelial (i.e., blood vessel) tissue the signaling pathway apparently involves the regulation of a Ca^{2+} channel, rather than the activation of adenylate cyclase. The resultant changes in cytosolic Ca^{2+} levels seem to provide the molecular basis for muscular contraction in the blood vessel wall.

Kinins are another family of peptides that, in contrast to the ANGs, are physiologically important in effecting vasodilation (i.e., hypotensive), among many other activities. In response to localized tissue injury, the kinins are derived from plasma protein precursors, called kininogens, via the enzymatic processing activities of plasma and/or tissue kallikreins. Specifically, bradykinin (See Fig. 3) and kallikin are the biologically active kinins and, in contrast to the ANGs, are inactivated by the angiotensin-converting enzyme (or kininase II). A bradykinin-like peptidergic system also exists in the brain. [*See* Kinins: Chemistry, Biology, and Functions.]

Bradykinin contracts venous smooth muscle to provide increased blood flow and vascular permeability to damaged tissue. Included in this biological action are the activation of sensory pain fibers and stimulation of the synthesis of both prostacyclin and endothelium-derived relaxing factor. As a mediator of inflammatory pain transmission, bradykinin has recently been shown to act on neurons that respond to pain.

Bradykinin receptors apparently are pharmacologically distinct in different target tissues. Receptor-mediated signal transduction for bradykinin appears to include phosphatidylinositol hydrolyis, guanylate cyclase activation (i.e., increased cGMP), and arachidonic acid metabolism via phospholipase C and/or phospholipase A_2 mechanisms. It is believed that each signal transduction pathway is associated with a different receptor.

3. Cholecystokinin and Gastrin

As exemplified earlier for ACTH and MSH and for the opioid peptides, chemical similarity exists for two families of gastrointestinal peptides, namely, cholecystokinin and gastrin (see Fig. 3), and secretin-related peptides (see Section I,A,7). In the cases of both cholecystokinin (CCK) and gastrin, many molecular variants of each peptide have been identified (e.g., CCK 39, CCK 33, CCK 12, CCK 8, gastrin 34, and gastrin 17). Nevertheless, usually only one of these molecular forms constitutes the main form stored in the secretory granules of the cell in which the peptide is synthesized. The carboxy-terminal pentapeptide sequences of the CCK and gastrin peptides are identical. Both CCK and gastrin peptides could possess a tyrosine residue that is sulfated on the side chain, but the sites at which this amino acid exists in the peptides are different. The difference might play a major role in effecting their different biological functions and specificities. Recently, the common carboxy-terminal tetrapeptide sequence of CCK or gastrin (i.e., CCK 4 or gastrin 4) has been discovered in the pituitary gland and the hypothalamus. [*See* Hypothalamus.]

CCK 33 is secreted from the duodenol mucosa when food is present (particularly amino acids, HCl, and certain free fatty acids) and causes contraction and emptying of the gallbladder, as well as stimulation of the release of pancreatic enzyme. Other known physiological roles of the CCK peptides include gastric emptying and potentiation of secretin-induced pancreatic bicarbonate secretion. CCK peptides might also regulate the growth of the exocrine pancreas; however, in contrast to the gastrins, CCK peptides do not exhibit a trophic growth effect on the gastrointestinal mucosa. Recently, it has been proposed that CCK 8 could function as a satiety hormone, and within the central nervous system CCK might possess neurotransmitter/neuromodulator (e.g., dopaminergic) activities.

The mechanism of action of CCK is initiated by binding to its receptors, which differ according to target tissue. Recent studies show that brain and pancratic CCK receptors are structurally and functionally distinct. CCK binding to its pancreatic receptors effects Ca^{2+} outflux and increases cytosolic cGMP. Concomitant increased metabolism of phosphatidylinositol has also been determined.

Gastrin is produced in the antral mucosa of the stomach and in the G cells of the duodenal mucosa. Similar to CCK, food is the primary physiological stimulus of gastrin secretion, and its release is further regulated by the autonomic nervous system and, possibly, by other peptides, somatostatin, and vasoactive intestinal polypeptide.

Gastrin is the most potent known secretagogue of gastric acid. It also stimulates pepsinogen secretion (perhaps indirectly via HCl release), relaxes the pyloric sphincter, stimulates pancreatic enzyme secretion, and increases intestinal motility.

4. Cytokines

Only recently, the biological responses of the immune system have been shown to be governed by soluble peptides, known as cytokines, which regulate the proliferative and differentiative responses of the respective immune lineage cells. [*See* Cytokines and the Immune Response.]

These peptides have multiple biological functions, formerly attributed to separate factors. For example, the 17,500 molecular weight interleukin 1 (IL-1) possesses the biological activities of lymphocyte-activating factor, endogenous pyrogen, leukocytic endogenous mediator, mononuclear cell factor, catabolite, osteoclast-activating factor, and hemopoietin 1. Two forms of IL-1 (α and β) have been identified. Both forms are unique in that they are synthesized as precursors (31,000 molecular weight) that do not have signal cleavage sequences, despite the fact that IL-1 is found extracellularly (see Section I,A). Although IL-1α and IL-1β have only 26% amino acid homology, they interact with the same receptor. The pair serves as a prime example of how different ligands with identical receptor interactions can be made through the assembly of similar overall three-dimensional structure using different building blocks.

IL-1 facilitates the release of another cytokine, IL-2 (14,700 molecular weight), from T lymphocytes, formerly known as T-cell growth factor, killer cell helper factor, or thymocyte mitogenic factor. In turn, IL-2 stimulates the release of another cytokine, γ-interferon, which is believed to act as a powerful differentiation signal, inducing macrophage killer activity, cytotoxic T-cell differentiation, and B-cell immunoglobulin secretion. This cascade of cytokines is only one example of the broader observation that, during activation of cellular or humoral immunity, some cytokines are dependent on the action of others that precede them in a precise sequence of production and site of action. Thus, not only do the cytokines regulate the biological response of cells in a primary manner, they further control the secondary response of the cells to secrete other cytokines needed to expand or support the differentiation of antigen-specific cells. Furthermore, it is now known that many biological systems, in addition to the immune system, are modified by the actions of secondary cytokines synthesized and secreted in response to an initiating cytokine. [*See* Interferons; Interleukin-2 and the IL-2 Receptor.]

5. Enkephalin, Endorphin, and Dynorphin

The endogenous opioid peptides were discovered after the elucidation of receptors within the central nervous system that were pharmacologically characterized as receptors for opiates (i.e., morphine, an agonist; naloxone, an antagonist). Because there was no evidence to suggest that morphine-like alkaloids occur naturally in the animal species, it was proposed that there might exist an endogenous opiate-like substance that would serve as natural ligands at these receptors. Several opioid peptides were subsequently discovered and now include endorphin, enkephalin, and dynorphin (see Fig. 3). Several molecular variants of each of these peptides exist (e.g., Met- and Leu-enkephalin), and each peptide is derived from a prepropeptide. Proopiomelanocortin is a precursor for endorphin. The biological sources of the opioid peptides include the brain and the pituitary gland, and they have also been localized to peripheral neural nerves of the gastrointestinal tract and other tissues. Discrete neuronal localization of endorphins, determined by immunocytochemistry, has shown that high concentrations are found in brain regions involved with pain transmission, respiration, motor activity, pituitary hormone secretion, and mood.

The primary role of the opioid peptides is analgesic, as a result of enkephalin-containing spinal cord neurons connected to peripheral sensory nerve cells that convey pain information. These primary afferent neurons utilize the neurotransmitter peptide, substance P, to transmit the pain stimuli, and the release of substance P is directly suppressed by enkephalin, or opiate agonist. Interestingly, acupuncture-induced analgesia is believed to occur via indirect stimulation of such endorphin receptors.

The opioid receptors have been classified into three groups by both pharmacological and biochemical methods: μ, δ, and κ. These receptors appear to mediate such actions as central analgesia (i.e., via β-endorphin), emotional behavior and spinal analgesia (i.e., via Leu-enkephalin), and sedative analgesia and appetite stimulation (i.e., via dynorphin). The principal receptor-mediated mechanism of action for the opioid peptides is by the inhibition of adenylate cyclase, and the biochemical basis of tolerance could be related to decreased receptor signaling, and that of withdrawal symptoms is connected to overproduction of cAMP, due to a compensatory mechanism of opioid-induced chronic suppression of adenylate cyclase activities.

6. Gonadoliberin

Hypothalamic gonadoliberin, or gonadotropin-releasing hormone [GnRH, or leuteinizing hormone (LH)/follicle-stimulating hormone (FSH) RH] (See Fig. 3), is a decapeptide that stimulates the secretion of both gonadotropins, lutropin (i.e., LH) and follitropin (i.e., FSH), from the pituitary gland. The action of GnRH on FSH release is selectively inhibited, in turn, by the antisecretagogue known as inhibin. Gonadal steroids regulate the action of GnRH by negative feedback inhibition of LH/FSH. The biosynthesis of GnRH is modulated by the neurotransmitter γ-aminobutyric acid. Extrahypothalamic GnRH also exists and regulates the biological activity of the testes, ovary, and placenta and exhibits central neurotransmitter effects related to sexual behavior. In addition, such GnRH is involved in tumor growth (e.g., prostate cancer).

The mechanism of action of GnRH is apparently mediated primarily via specific receptors that stimulate adenylate cyclase. As a result of subsequent increased intracellular Ca^{2+}, gonadotropin secretory granules release their contents to their extracellular space. GnRH is released in a pulsatile manner, and receptor down-regulation (i.e., decreased activity) is observed if exogenous GnRH is administered chronically, with resulting tissue refractoriness and, in the case of females, ovulation inhibition.

7. Insulin and Glucagon

Insulin is a heterodimeric peptide consisting of an A chain (21 amino acids) and a B chain (30 amino acids), which are covalently linked by two disulfide links between cysteine residues. Proinsulin contains an interconnecting sequence of 23 amino acids between the amino terminus of the A chain and the carboxy terminus of the B chain. The chemical complexity of insulin has prevented its total synthesis by standard chemical methods. Therefore, insulin research and insulin therapy (i.e., in diabetes mellitus) have relied primarily on insulins isolated from various species or on the biosynthetic production of recombinant insulin. The three-dimensional structure of insulin, determined by X-ray crystallography, suggests that a specific conformation of the peptide backbone and the spatial proximity of amino acids derived from both the A and B chains are required for biological activities.

Insulin is secreted from the β cells of the pancreatic islets in response to an elevated blood glucose concentration. The major physiological role of insulin is to lower blood glucose (i.e., hypoglycemic action) by enhancing the cellular uptake of glucose and stimulating the subsequent conversion of glucose to glycogen, protein, or fat in liver, muscle, and adipose tissues, respectively. Insulin stimulates the active transport of glucose, amino acids, K^+, Na^+, and PO_4^{2-} into its target tissues. Insulin secretion is inhibited by normal levels of blood glucose, as well as by somatostatin derived from pancreatic D cells.

The mechanism of action of insulin is complex. The insulin receptor (450,000 molecular weight) is a membrane glycoprotein with an α_2–β_2 subunit structure, linked by disulfide bonds. Insulin binds to the α subunits, causing the β subunits to undergo tyrosine autophosphorylation and to become an active receptor kinase. It is important to note that, in contrast to other peptide hormones that affect intracellular protein phosphorylations via, for example, cAMP as a secondary messenger, insulin causes a direct phosphorylation event. This biological mechanism is also observed for growth factors. Insulin-bound receptors rapidly become clustered; invaginate, forming coated pits, lined with clathrin molecules at the cytoplasmic side; and, finally, internalize by vesicular endocytosis. The immediate intracellular events associated with insulin action could include modification of membrane protease activity, activation of phosphokinase C or phospholipase C, and/or Ca^{2+} mobilization.

Glucagon (See Fig. 3) is derived primarily from the α cells of the pancreas. It is chemically similar to other peptide hormones (e.g., secretin, vasoactive intestinal peptide, and gastric inhibitory peptide). Glucagon affects glucose homeostasis in a manner opposite that relative to insulin. It produces hyperglycemic action by directly stimulating hepatic glucose release by glycogenolysis (i.e., the conversion of glycogen to glucose) and gluconeogenesis (i.e., the conversion of amino acids and glycerol to glucose). The gluconeogenic action of glucagon is particularly important in the physiological events associated with prolonged fasting and exercise, whereas its glycogenolytic activities are rapid and short term to elevate blood glucose levels. Glucagon is lipolytic, thereby liberating free fatty acids and glycerol.

The mechanism of action of glucagon at its target tissues is initiated by receptor-specific binding, followed by transmembrane signaling to adenylate cyclase and an elevation of cytosolic cAMP. The glycogenolytic action is mediated by cAMP-dependent intracellular conversion of inactive phosphorylase *b* to active phosphorylase *a*. The glucagon receptor (63,000 molecular weight) is a dimeric protein, with binding sites for both glucagon and a guanine nucleotide triphosphate-binding signal-transducing protein. [*See* Insulin and Glucagon.]

8. Somatotropin and Somatostatin

Somatotropin, or growth hormone (GH), is primarily produced by the pituitary gland, and its secretion is regulated by hypothalamic somatotropin-releasing factor and somatotropin release-inhibiting factor (i.e., somatostatin). Human GH is chemically similar to prolactin: it consists of 191 amino acids and has two intramolecular disulfide bonds.

GH is an anabolic hormone (i.e., it promotes the accumulation of reserves) in terms of its direct or indirect (via somatomedins) effects on metabolism, cellular proliferation, and growth. It stimulates amino acid uptake into muscle protein and stimulates the extracellular deposition of collagen. GH stimulation of hepatocytes effects glycogenolysis (i.e., a hyperglycemic response, as described for glucagon) and secretion of somatomedins, which in turn stimulate cellular growth in many target tissues.

The mechanism of action of GH is not well understood. It apparently binds to specific receptors on its target tissues, but the subsequent steps of intracellular signaling are complex. The cloning of cDNA encoding GH receptors has shown a structural similarity between cell-surface GH receptors and circulating (i.e., serum-derived) GH-binding proteins. It has been proposed that the BH receptor is a transmembrane protein with an extracellular domain and a cytosolic domain, connected by a short intramembrane sequence, and that signal transduction is direct (as opposed to, e.g., signaling mechanisms based on adenylate cyclase).

Somatostatin is perhaps the most studied and important peptide inhibitory factor known. It contains 14 amino acids and is cyclic, by virtue of an intramolecular disulfide bridge between two cysteine residues (See Fig. 3). Somatostatin secretion from the hypothalamus functions primarily to inhibit pituitary GH release. Somatostatin is widely distributed in the central nervous system and is known to potentiate L-dopa and affect sedation and hypothermia. Pancreatic somatostatin inhibits the production of insulin and glucagon and attenuates gastric acid secretion.

Two types of somatostatin receptor have been characterized in pituitary, brain, and pancreatic tissues. Binding of somatostatin to its receptors in believed to cause inhibition of adenylate cyclase activity, which is stimulated by the agonists (e.g., somatotropin-releasing factor). However, other signaling mechanisms (e.g., Ca^{2+} related) might also be pertinent.

II. BIOTECHNOLOGY AND BIOMEDICAL RESEARCH

A. Recombinant Peptides

It is no coincidence that our progress in understanding the biochemical properties of certain peptides corresponds with advances in modern molecular biology and the ability to clone and express the genes in prokaryotic or eukaryotic cell lines. Some of the peptide hormones are present in such small quantities that their isolation and complete identification by protein sequencing techniques are extremely laborious. The advent of recombinant DNA technology has permitted peptide sequence determination through use of the corresponding cDNA. Moreover, the expression of the cloned genes has yielded peptides in sufficient quantities to characterize and explore their biological activities. For example, genes of about 15 human cytokines have been cloned. They include three colony-stimulating factors (macrophage, granulocyte, and granulocyte–macrophage), interferons α and γ, interleukin-1 (α and β) through -6, lymphotoxin, tumor necrosis factor, and transforming growth factor β. The molecular weight of these cytokines (i.e., 17,000–30,000) spans the gray zone of an arbitrary definition of where peptide ends and protein begins, and the smallest cytokines reach the absolute size limits (i.e., about 150 residues) for chemical synthesis. Thus, all peptides can be obtained by either chemical synthesis or recombinant DNA technology. Chemical synthesis is important for the lower-molecular-weight peptides, which are usually unstable to the proteases present in bacterial expression systems.

The list of biologically important human peptides produced by recombinant DNA technology continues to expand. Human growth hormone (i.e., somatotropin), important for treating dwarfism of hypopituitary origin (Table II), was first obtained only from cadaver brains. It is now readily available through bacterial culture. As mentioned earlier (Section I,C,7), the chemical complexity of insulin makes its synthesis impractical, so replacement therapy for diabetes uses porcine insulin isolated from the pig pancreas. However, porcine insulin does not have the same sequence as human insulin, and the risk of immune response is higher than with the human sequence. Again, recombinant DNA technology now allows the production of human insulin. The A and B chains of recombinant human insulin are produced separately, not together, as in natural proinsulin. Extensive research is required to discover conditions under which biologically active insulin could be obtained with its two disulfide bonds correctly formed.

Recombinant DNA technology has been important not only for the production of high-molecular-weight peptides, but even more for conducting research into the mechanism of peptide–receptor interactions. It is now possible to systematically alter peptide sequences by site-directed mutagenesis of the gene DNA and study structure–activity relationships.

TABLE II
Abnormal Conditions of Excessive or Deficient Physiological Levels of Naturally Occurring Peptides

Peptide	Relationship	Symptom (disease)
Adrenocorticotropic hormone	Deficiency	Abnormal carbohydrate metabolism (Addison's disease)
	Excess	Protein catabolism increased (Cushing's disease)
Gonadotropin-releasing hormone	Deficiency	Secondary hypogonadism
	Excess	Secondary hypergonadism
Insulin	Deficiency	Hyperglycemia and glucosuria (diabetes mellitus or insulin-dependent diabetes, if resulting from cytotoxic autoantibodies to β cells)
Parathyroid hormone	Excess	Abnormal blood Ca^{2+} increased, hyperparathyroidism
Somatomedin	Deficiency	Abnormal growth decreased (Larin-type dwarfism)
Somatotropin	Deficiency	Abnormal growth decreased (hypopituitary dwarfism)
	Excess	Abnormal growth increased (giantism in children)
	Excess	Abnormal growth increased (acromegaly in adults)
Thyrotropin	Excess	Abnormal metabolism increased (Graves' disease)
Vasopressin	Deficiency	Hypovalemia–dehydration increased (pituitary-type diabetes insipidus)

B. Peptide Therapeutics

For their important physiological actions, peptides are at the basis of several human pathophysiological (disease) states (See Table II). Replacement therapy using naturally occurring, recombinant, or synthetic peptides is frequently used to offset abnormally low levels of the endogenous peptide (e.g., diabetes mellitus, Addison's disease, and dwarfism). Medicinal chemistry and/or biotechnology strategies to design more effective (i.e., in terms of biological potency, metabolic stability, duration of action, and oral absorption) peptide-based drugs have become a major challenge and effort for the pharmaceutical industry.

Synthetic peptide derivatives might simply contain D-amino acid substitutions or side-chain or backbone amide modifications, or might be changed with respect to their three-dimensional properties (i.e., conformation and structural flexibility) via the use of intramolecular bridging (i.e., cyclization). Such synthetic peptide analogs might act as superagonists or competitive antagonists. In addition, some exogenous (i.e., not found in the host animal) naturally occurring peptides could possess some of these chemical or pharmacological properties when given *in*

TABLE III
Some Known Synthetic Peptides Demonstrating Noteworthy Biological Properties[a]

Synthetic peptide	Structure–activity relationship
H-Tyr-D-Ala-Gly-*N*-Me-Phe-Gly-ol	Potent μ-selective opioid agonist
H-D-Phe-Cys-Tyr-D-Trp-Orn-Thr-Cys-Thr-NH_2	Potent μ-selective opioid antagonist
H-Tyr-D-Pen-Gly-Phe-D-Pen-OH	Potent δ-selective opioid agonist
H-D-Phe-Cys-Phe-D-Trp-Lys-Thr-Cys-Thr-ol	Potent somatostatin agonist
H-Sar-Arg-Val-Tyr-Val-His-Pro-Ala-OH	Potent ANG II antagonist
*p*Glu-His-Trp-Ser-Tyr-D-Ser(*t*-Bu)-Leu-Arg-Pro-NHEt	Potent GnRH agonist
H-D-Arg-Arg-Pro-Hyp-Gly-Thi-Ser-D-Phe-Thi-Arg-OH	Potent bradykinin antagonist
Ac-Ser-Tyr-Ser-Nle-Glu-His-D-Phe-Arg-Trp-Gly-Lys-Pro-Val-NH_2	Potent MSH agonist

[a] *N*-Me-Phe, N^{α}-methyl-L-phenylalanine; Gly-ol, glycinol; Orn, L-ornithine; Pen, L-penicillamine; Sar, sarcosine; D-Ser(*t*-Bu), D-serine; Pro-NHEt, L-proline ethylamide; Hyp, 4-hydroxy-L-proline; Thi, L-2-thienylalanine; Nle, L-norleucine.

vivo. Peptide antagonists are useful for examining the precise physiological roles of peptides, and perhaps as drugs to block the undesired effects of inappropriate (or excessive) levels of an endogenous peptide. A few selected peptide derivatives relevant to this discussion are shown in Table III. Finally, the search for nonpeptide compounds that bind to endogenous peptide receptors (e.g., morphine at the opioid receptor) will undoubtedly advance our understanding of peptide biology and, possibly, provide new leads in the design, discovery, and development of peptide-related drugs.

BIBLIOGRAPHY

Farrar, W. L., Ferris, D. K., and Harel-Bellan, A. (1989). The molecular basis of immune cytokine action. *CRC Crit. Rev. Ther. Drug Carrier Syst.* 5, 229.

Hadley, M. (1995). "Endocrinology," 4th Ed. Prentice–Hall, Englewood Cliffs, New Jersey.

Nogrady, T. (1988). "Medicinal Chemistry: A Biochemical Approach," 2nd Ed. Oxford Univ. Press, New York.

Taylor, M. D., and Amidon, G. L. (1995). "Peptide-Based Drug Design." American Chemical Society, Washington, D.C.

Wallis, M., Howell, S. L., and Taylor, K. W. (1985). "The Biochemistry of the Polypeptide Hormones." John Wiley & Sons, New York.

Perception

H. BOUMA and D. G. BOUWHUIS
Institute for Perception Research/IPO in Eindhoven

E. P. KÖSTER and J. H. A. KROEZE
Utrecht University

E. G. G. EIJKMAN
Nijmegen University

F. L. ENGEL, J. 'T HART, A. J. M. HOUTSMA, F. L. VAN NES, and J. A. J. ROUFS
Institute for Perception Research/IPO in Eindhoven

GLOSSARY

Attention Selective perception of certain objects or events out of many possible ones

Hearing Perception mediated by certain acoustical vibration impinging upon the ear

Kinesthetic perception Perception mediated by mechanical and thermal stimulation outside or inside the body

Perception Interface between the outside world and the human brain, as regards the intake of sensory data, the processing into sensory attributes, and the recognition as objects or events

Smell Perception mediated by certain volatile chemical substances impinging upon the sensory surfaces of the nose

Taste Perception mediated by certain chemical substances impinging upon the tongue

Vision Perception mediated by certain electromagnetic radiation impinging upon the eye

HUMAN PERCEPTION IS TAKEN HERE AS THE physicochemical interface between the outside world and the human brain, plus the sensory and cognitive processing of the data transforming these into information. Each sense organ is generally activated by its specialized type of stimulation and gives rise to its specific set of sensory attributes. Senses that take in stimulation from distant sources are hearing, through certain acoustical vibration, vision, through certain electromagnetic radiation, and smell, by a number of volatile chemical substances. Sense organs that react to stimulation from close sources are taste, by certain chemical substances in the mouth, and the skin senses (and kinesthetic senses), from mechanical stimulation at the outside or inside of the body. Next, the sensory attributes and their dynamic variations are subjected to processes of recognition, that is, interpreted within a framework of evolved experience and recent context. An important special case is the recognition of spoken language (speech) through the ear or written language (reading) through the eye, where the language elements are symbolic carriers of information between humans. In the overwhelmingly rich sensory stimulation to which humans are exposed, powerful perceptual selection processes make the chaotic input amenable. All perceptual functions are far from constant because they are in active development over life from one's earliest to final moment.

I. GENERAL

A. Transduction

Transducers are located in the sense organs, which, after initial filtering, map the physical data to neural data. So in the eye, imaging is an optical filtering process and the rod and cone receptors of the retina are the transducers proper. In the ear, the ear canal, the chain of middle ear bones, and basilar membrane provide filtering, and the hair cells act as transducers. When physical parameters are outside the range of filters and transducers, no perception results, even if perceptual terms are commonly used (infrared, ultrasound).

B. Sensory Attributes

After transducers, neural processing takes over. This is quite complex and leads to (1) largely or partly automatic reflexes in muscles and glands and (2) impressions and interpretations about the outside world (perception). The two categories are interdependent, and active perception, in which movements of the observer occur, may be different from passive perception. Sensory attributes are coupled to physical stimuli and reflect direct processing, which may be of a complex nature; examples in vision are brightness and color as coupled to spatial luminance distribution and wavelength composition; examples in hearing are loudness and sound direction as dependent on physical spectral intensities and differences in intensity and time of the signals for the two ears, respectively. For the sense organs, with the possible exception of smell, the number of sensory attributes is relatively small. Constant stimulation need not produce constant sensation (adaptation). The field of study that relates sensory attributes to physical stimuli is called psychophysics.

C. Recognition

Another type of processing is the interpretation of the sensory attributes of one or more senses by mapping them onto internal concepts of the outer world that have developed earlier. The central term here is recognition, because learning has been involved, perceptual experience is relevant, and interpretation may well differ among observers. The activated concepts imply meaning to the observer, which relates to networks of relevant associations. Examples are recognition of faces and of musical sounds. Associations may sometimes be described in physical terms, that is, in terms of physical stimuli, which would produce similar activations. The recognition and associations form a basis for further cognitive evaluation and for intended movements, which bring about changes in the physical world. Natural language is concerned with the manipulation of symbolic information. The symbols themselves, speech or print, are coupled to meaning by language conventions rather than by physical attributes. Units are called "words," and their ordering rules are described by the syntax. Natural language permits an incredibly rich gamut of expressions from a similarly rich gamut of inner worlds. Subtlety and ambiguity lie close together. Speech perception and reading are the two perceptual components.

D. Selection

Perception is usually quite robust and stable in the presence of wide variations of physical stimuli (perceptual constancies). This should perhaps be attributed to the types of processing involved and to priorities in processing (selection). Priorities in processing are dependent both on sensory attributes coupled to external physical stimuli (conspicuity) and on internal factors relating to the search of specific information by the subject (directed attention). As the relevant distribution of sensory attributes becomes more complex, priorities in processing are more pertinent. Conspicuity and directed attention may be in mutual conflict; however, the rule seems to be that the two processes are cooperative in a sense that the desired information is coded in terms of its most conspicuous attributes.

Robustness and stability of perception are prerequisites for integrating the perceiving subject in each of the many physical environments, thus preventing perceptual chaos. The basis for this seems to be that the subjects act on an internal model of the physical world and incoming data are interpreted in terms of shifts and changes in this internal model. Components of such internal models reside, as we say, in long-term memory and are revived as required in the internal model of the world "here and now." Language data are coded as internal speech in a short-term buffer. It holds some five to nine words or word combinations. Several other memory components are short term as well, but properties of the former language type are best known.

E. Development

Because it is brought about by physical stimuli, perception depends on the physical environment. However, the great amount of learning involved in perception makes perceptual processes themselves also very much dependent on the world around us. In this sense, human perception bears the marks of the environment that we live in. Much of the present environment has been brought about by technology and consequently both our internal concepts as built up in the preceptual process and the perceptual processes themselves are geared to this technological world.

II. HEARING

A. Transduction

The human hearing organ consists of an acoustical–mechanical part comprising outer, middle, and inner ear and an electrochemical part, which includes a variety of neural structures that run from the 8th (auditory) nerve to the temporal lobe of the brain's cortex. The outer ear (pinna, ear canal) mainly transmits sound waves from the outside air to the eardrum. In the middle ear, three small bones (malleus, incus, stapes) with attached ligaments and muscles connect outer ear (eardrum) and inner ear (oval window). The lever action of these bones and the area reduction of eardrum to oval window form a mechanical transformer between the low impedance of air in the ear canal and the much higher impedance of the fluid inside the inner ear. Automatic stiffening of the middle ear muscles when the ear is exposed to intense sound (middle ear reflex) causes sound attenuation, which protects the inner ear. The inner ear (cochlea) is a rolled-up tube partitioned almost over its entire length (35 mm) by the basilar membrane. This membrane assumes different resonance patterns, dependent on the vibration frequency of the stapes. Thousands of (inner) hair cells, situated along this membrane and innervated by fibers of the 8th nerve, convert the mechanical motion of the membrane into electrical pulses, which propagate along 8th nerve fibers to other nuclei of the auditory nervous system. The actual processing of those nerve pulses, which results in the sensation of hearing, may occur at many neural levels anywhere between the termination of the 8th nerve (cochlear nucleus) and the auditory cortex. [*See* Cochlear Chemical Neurotransmission; Ears and Hearing.]

The sensitivity range of the human ear is on the one hand limited by the hearing threshold, which is shown in Fig. 1. On the other hand, our hearing range is limited by the discomfort or pain threshold, which is typically around 130 dB, independent of frequency. For long-duration sound exposures, a hearing-damage threshold can be as low as 90 dB.

B. Auditory Attributes

Relationships between perceptual, subjective attributes of sound and its physically measurable, objective characteristics are well established. Perceived loudness of a tone is a power function of its sound pressure (exponent 0.6). Perceived pitch has a simple monotonic relationship with frequency (tone height) but also a cyclic one (tone chroma). This causes pitches of successive octaves to sound similar. Perceived timber depends primarily on the spectral profile of a sound but also on temporal features such as attack and decay. Perceived location of sound depends systematically on interaural arrival-time differences (low frequencies) and interaural intensity differences (high frequencies). There are also many cross-relationships. Loudness and timber both depend on frequency, and the pitch of a tone can change when only its intensity changes.

C. Recognition

What we hear is determined not only by the absolute or differential sensitivity of our ears but also by our

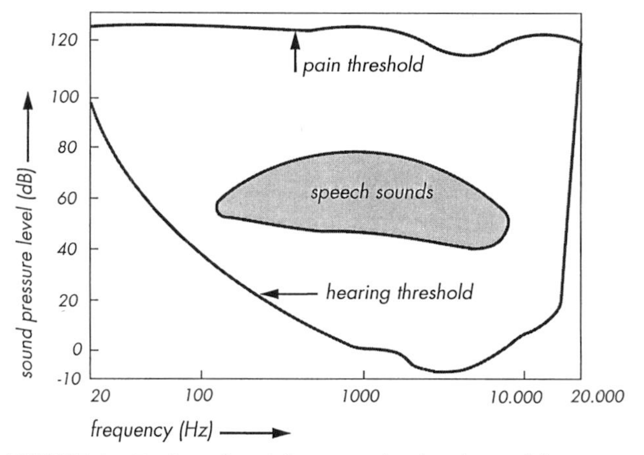

FIGURE 1 Region of sound pressure level and sound frequency in which normal hearing occurs.

memory capacity for sound. Roughly speaking, we can barely distinguish tones that differ by 1 dB in intensity (i.e., 12% in sound pressure) or 0.2% in frequency. Integrating this over the entire intensity and frequency ranges of our ear, we find that we can distinguish about 340,000 different tones. However, if we have to recognize or identify every one of those tones, we end up making mistakes. In fact, the largest number of different tones that we can reliably identify is about seven. Identification performance is primarily limited by memory capacity and not so much by differential sensitivity of our ears. Memory capacity for sound can grow with training. During infancy and childhood, we learn to recognize and handle many different speech sounds of the language (categorical perception). Trained musicians, especially those who have absolute pitch, can recognize considerably more than seven notes.

D. Hearing Impairments

Hearing impairments are first of all those associated with upward threshold shifts. They can be broadly divided into conductive and sensori-neural hearing losses. The former are broadband and exhibit simultaneous upward shifts of hearing and pain thresholds. The latter are often frequency-specific and generally show a mere increase in hearing threshold without an equivalent rise in pain threshold. Reduction of dynamic range by a sensori-neural disorder often causes an abnormal growth of perceived loudness with sound intensity, called recruitment. Some other hearing disorders, not necessarily connected with hearing loss, are tinnitus (hearing a constant tone when there is no sound) and monaural diplacusis (hearing several tones or noise when the sound is a pure tone). Binaural diplacusis (hearing slightly different pitches in each ear when they receive the same tone) is a common phenomenon but can become excessive in pathological cases. Noise-induced temporary threshold shifts recover with time. However, when exposure levels are too high (gun shots, disco) or the ears have been exposed for many years to sound pressure levels of 90 dB or higher (factory noise), permanent threshold shifts may result.

Current technology has given us virtually unlimited control over the sounds we can produce (computer music) and over means of encoding or decoding sound (telephone communication). Often, one can save communication costs or sound storage space by simply not encoding those portions of a sound that our ears would not perceive anyway (perceptual entropy). For the hearing-impaired it is sometimes useful to process the sound of a hearing aid so that it will match the person's residual hearing. Finally, for those who suffer from total bilateral cochlear hearing loss, various cochlear electrode implant techniques are available through which the 8th nerve can be stimulated by electrical transformations of sound, resulting in sensations that resemble hearing.

III. SPEECH

Speech is the sound produced by the vocal mechanism of humans as a carrier of meaningful messages that are coded according to the rules of language.

A. Production

The energy is provided by a stream of air from the lungs; the various speech organs convert it into a source sound and shape it into the many different speech sounds. For most speech sounds, the source consists of vocal cord vibration, which gives rise to a quasi-periodic sound with many overtones. Varying the repetition frequency results in changes in perceived pitch.

B. Vowels and Consonants

The source sound is shaped in the vocal tract. This consists of the pharyngeal and oral cavities (and the nasal cavity, which is only relevant for nasal sounds, such as the first and last sounds in "man"). The vocal tract has a flexible form, mainly determined by the movability of the tongue body and tip. At many different places in the oral cavity, a reduction of its cross-sectional area can be made, which gives rise to as many differences in timber, noticeable in, for example, the 12 vowels of English in the words "heat," "hit," "head," "hat," "hard," "hot," "hod," "hook," "hoot," "hut," "herd," and "the."

If somewhere in the vocal tract a sufficiently narrow constriction is made, or a closure followed by a release, the airstream from the lungs may become turbulent. This causes noise and, in these cases, the sound source is situated in the vocal tract itself. Examples are "s" and "t." Again, the position of the constriction determines the nature of the speech sound [e.g., "k" (back) and "t" (front)].

The articulatory characteristics mentioned thus far are reflected in acoustic properties: Figs. 2a and 2b show the waveforms of the vowels in "heat" and "hoot." The energy distribution as a function of frequency (the spectrum) appears to be specific for each

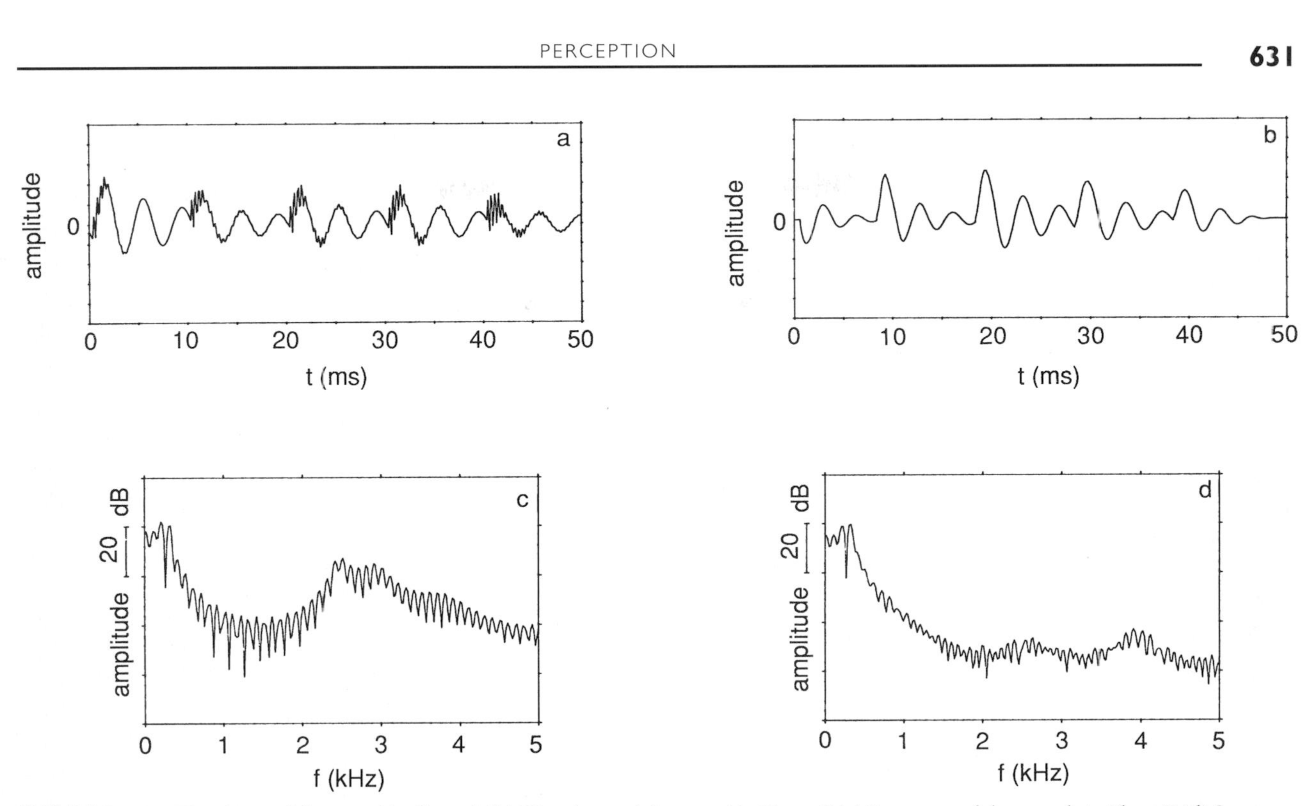

FIGURE 2 (a) Waveform of the vowel in "h<u>ea</u>t." (b) Waveform of the vowel in "h<u>oo</u>t." (c) Spectrum of the vowels in "h<u>ea</u>t." (d) Spectrum of the vowel in "h<u>oo</u>t."

speech sound. Figures 2c and 2d show examples for the same speech sounds as in Figs. 2a and 2b.

Many of the acoustic features give rise to distinct auditory attributes, such as hissing sounds ("s") versus sounds with pitch ("m"), high ("h<u>ea</u>t") versus low ("h<u>oo</u>t") timber, and short ("t") versus long ("s") sounds. But this does not imply that normal speech perception is based on the identification of all the auditory attributes for each of the successively incoming speech sounds. Indeed, isolated speech sounds can only be recognized with about 70% certainty.

C. Recognition

Cognitive attributes must also be taken into account: we should consider that the listener largely bases his or her interpretation on stored knowledge. An appeal to cognition can take place at different levels of abstraction. For instance, it has become apparent that in a word recognition task, the number of perceived syllables together with the location of the word stress strongly reduces the number of candidate words. Nevertheless, only the low level of speech sound recognition is decisive in distinguishing between, for example, "elegant," "element," and "elephant." On the other hand, if the word is presented in a syntactically correct sentence, its status as either a noun or an adjective may be revealed, and this will help to separate the first word from the other two (e.g., the context "That's an . . . solution" only allows "elegant").

Clearly, in normal discourse the contextual information is of great help in these situations. But a truly complicating factor is that in fluent speech there are practically no pauses between words (like spacings in print). Our impression is that we can nevertheless process the continuous stream of speech as if it were segmented into words. This can hardly be understood without the assumption that a constant, intensive appeal is made to an internally represented dictionary. Thus, speech perception is a very complicated activity, in which knowledge of the language plays an important part, on levels ranging from phonology (sound structure) to semantics (meaning). There is an enormous gap between the abundant information available in the speech signal and the restricted amount of it the listener can cope with. Therefore, speech perception must be considered to be a highly selective process. [*See* Language.]

D. Speech Technology

Speech perception can be studied by means of measuring the acceptability of artificial speech. An interesting by-product of this method is that, in the end, it may lead to reliable acoustic specifications for speech synthesis-by-rule. This is, in fact, one of the objectives of speech technology, which furthermore is concerned with—parsimoniously coded—speech transmission, with automatic speech recognition, and with speaker identification. The products of these efforts are beginning to find their way in a world that becomes increasingly technical: computer systems that respond to spoken commands or queries, in artificial speech rather than in orthographic form on a screen, are already one of the possibilities. But also communicatively disabled persons may benefit from these facilities: a reading machine for the visually impaired or a keyboard-to-speech device for vocally handicapped people. Mention may be made of semiautomatic devices as well, such as the intonable electrolarynx for laryngectomees or those who lack the use of their voice temporarily.

IV. VISION

A. The Eye and Visual Pathways

The eye consists of a bulb-like transparent body (vitreous body) that is enveloped by the sclera. At the front side of the eye bulb, a clear window enables light to enter the eye, through a hole in the iris diaphragm (pupil), and fall onto the retina. The optics of the eye (cornea and lens) create an image of the outside world on the receptor mosaic within the retina. There are two kinds of photoreceptors: rods and cones, with rods operating at night and cones at daylight. They transduce the light distribution into a stream of nervous activity, which is modified at different retinal cell layers and guided into the optic tract. Eventually, the information stream of the left half of the retinas reaches the left part of the visual cortex of the brain, the right retinas feeding the right part of the cortex, both passing a crossover (optic chiasm) and a neural nucleus (lateral geniculate nucleus). [*See* Eye, Anatomy; Vision, Physiology; Visual System.]

B. Physical Stimuli

Light, emitted by a natural or artificial light source, is reflected by object surfaces and projected onto the retina. It enables a human being to obtain information from the outside world and to orient and act accordingly. The relevant physical magnitude of the reflected light is luminance, which can be characterized as light density. The illumination pattern at the retina caused by the properties of the visible surround is proportional to its luminance pattern. Spectral components of this light stimulate three kinds of cone differently, which causes color vision. [*See* Retina.]

C. Visual Attributes

Variations in luminance and spectral content evoke the visual attributes brightness, brightness contrast, sharpness, and color. Details can be detected by differences in brightness and color. Visual acuity characterizes the ability to see small details. Generally, the visual attributes are not trivially linked to the physical stimuli. The luminance distribution evokes, for instance, an internal image of the outside world that is largely determined by object properties such as surface reflectance, whereas the light source properties and observer position are relatively unimportant. Other visual attributes that also mainly reflect object properties are depth, movement, and texture.

D. Cognitive Aspects

It is commonly acknowledged that the main goal of vision is to derive three-dimensional object shape from the information contained in the retinal luminance distribution. In other words, the visual system aims at reconstructing the outside world from the light density variations.

Visual attributes such as brightness, color, movement, and depth can be regarded as lower-level interpretations that form the input to higher-level interpretations. These involve the grouping of related items within the visual field to larger wholes, and exchanges with memory items. For instance, features of body shape, color of hair, and movement enable us to identify an individual.

In visual recognition, in search, and in reading, the eyes normally move over the text in irregular jumps (saccades), each eye pause giving rise to recognition of objects in the fixation area. Integration across eye saccades occurs smoothly in an unknown way.

E. Visual Disorders

Defects may occur anywhere in the visual pathways. Optical defects may result in a retinal image that is not optimal. Lens errors, for instance, may evoke blurred imaging on the retina, whereas light straying at particles in the optic media diminishes contrast. Neural defects may already occur at lower processing levels, such as color weakness. A typical example of malfunctioning at higher levels is word blindness (dyslexia), better referred to as reading weakness.

F. Vision in a Technical World

Luminance distributions of pictures displayed on film and television are intended to evoke the same perceptual sensations as the corresponding luminance distributions that occur in real scenes. This does not imply that they should be identical. In fact, the displayed images are often far more parsimonious than the real ones. Generally, it is sufficient if the differences cannot be seen. A typical example is the apparent continuity of a television image, which in fact is a line-type image that is periodically refreshed.

V. READING

Reading concerns the visual perception of language signs. The signs are coupled to meaning by convention, either directly as in ideographs or by alphabetic or syllabic code (characters), which relates to speech sounds. [*See* Reading Processes and Comprehension.]

A. Eye Movement Control

Generally, the process of reading is preceded by searching for the desired part of a text on a page. During search, the eyes skim over the page, their motion being guided by certain text attributes. Just as in other static visual situations, the eyes move stepwise so that in a series of fixations of about 250 msec the successive characters can be imaged on the central part of both retinas. Saccades between fixations are 8 ± 4 letter spaces in alphabetic languages. Other saccades are necessary from the end of a line or column to the beginning of the next one, and back to previous parts of the text. [*See* Eye Movements.]

B. Word Recognition

During fixation pauses, information is extracted from the visual reading field, which measures 10–20 letter positions of alphabetic text or 1–3 ideographs. Both character recognition and word contour recognition contribute to this process. Luminance contrast and discriminability between characters of similar configuration are important.

C. Temporal Integration

Time is a crucial factor in reading: if the reading speed is too low, it takes too long to absorb the full content of a sentence and, therefore, its context, thus preventing integration. There is a minimum reading rate of 20 words per minute. Reading rate is the result of a complex interplay between necessary recognition time and the prevention of backward masking, saccade length, and accruing comprehension.

D. Reading and Language

A reader's knowledge of the language interacts with all components of reading. Visual patterns are decoded as characters and words, in turn representing sounds of speech and semantic units. Characters in a prescribed order and orientation represent words of the language, with a different sound pattern, and a specific grammatical function and meaning. Fluent reading requires extensive use of redundancy of the printed text at graphical, orthographic, lexical, syntactic, and semantic levels.

E. Reading Disorders

In many cultures that depend so much on the printed text, a reading disorder is a considerable handicap. Reading disorders may be due to insufficient intellectual development or a visual impairment. A specific reading disorder—dyslexia—occurs in about 6% of the male population. Its origin is now sought in the chain of cognitive processes following vision. [*See* Dyslexia–Dysgraphia.]

F. Reading in a Technical World

An increasing part of what people read now is presented with electronic means on electronic displays. Reading proceeds basically in the same way for paper-based as for screen-based texts, but negative effects may result from display properties. The presentation of bright text on a dark background, for instance, together with the presence of a reflecting glass sheet in front of the screen, may hamper reading if the surrounding luminance is high. With proper care, many such negative effects may be avoided. Also sharpness, color, contrast, and character configuration are relevant.

VI. SMELL

A. Transduction

The olfactory epithelia, two pigmented areas of 2–4 cm^2 in the olfactory clefts on both sides of the nose, consist of receptor cells, sustentacular cells, and Bowman's glands, which secrete the watery olfactory mucus (composition unknown) in which the odor-receptive cilia (80) sprouting from the protruding knobs of the bipolar receptor cells are bathed. The axons of these cells (neurons that are replaced by new ones from a basal cell layer during life) reach the olfactory bulb, where they form synapses. In this layered structure, many interconnections are found. The mitral cells (secondary neurons) form the olfactory tract to central parts of the brain. Granular cells receive afferent and centrifugal input and exert inhibitive influences on other cell types in the bulb. Odorous molecules can reach the receptors via ortho-nasal (sniffing) or retro-nasal (exhalation of vapors from the mouth) stimulation. The nature of the receptive mechanism is unknown, but receptor proteins are involved. On average, the olfactory receptor responds to 30% of the odors presented, but different receptors respond to different sets of odors and correlations between the sensitivities to odors are low (average $r = \pm 0.30$), suggesting a fair degree of independence and specificity. [*See* Olfactory Information Processing.]

Human olfactory sensitivity strongly varies among odors and among individuals. Adaptation (reduction of sensitivity during stimulation) is strong and recovery after adaptation slow. Odors often suppress each other in mixtures.

B. Chemical Stimuli

The physicochemical properties determining the odor of a molecule are unknown. Volatility is a prerequisite, and water and lipid solubility, molecular shape, and functional groups are important. Enantiomers can have distinguishable odors.

C. Sensory and Cognitive Attributes

Odors warn (fire, cadavers), convey pleasure (food), and play a role in sex (perfume, body odors). Odorous quality is quite varied: many thousands of odors can be distinguished, but not verbally labeled. Odor memory is predominantly episodic. Odors have (inborn) or acquire strong affective values. Pleasant odors become unpleasant when strong.

D. Olfactory Disorders

Specific anosmia (i.e., insensitivity to one odor or a group of closely related odors in otherwise normally sensitive persons) is not exceptional. General anosmia (complete insensitivity), hyposmia (reduced sensitivity), parosmia (disturbed perception of the nature of the stimulus), and kakosmia (perceiving all odors as putrid) can be caused by head traumata, viral infections, hormone or neurotransmitter deficiencies, or obstructions in the nasal pathways. Recovery prospects depend on the nature of the cause. Parosmia can occur as a transient stage in the regrowth of olfactory receptors after destruction of the epithelium. Olfactory hallucinations are sometimes caused by brain tumors.

E. Odors in a Technical World

Industry, intensive agriculture, waste treatment, and traffic frequently cause malodors. Direct scaling methods for odor nuisance have been developed. Malodors may cause social problems, sleeplessness, and nausea.

VII. TASTE

A. Transduction

Taste is a part of complex oral sensations, to which touch, pain, and temperature may also contribute. Transductors are the taste buds in the lingual papilla or in the soft palate and epiglottis. They consist of 50–150 elongated epithelial cells with microvilli projecting outward into a "taste pore" in the epithelial layer. Taste cells on the anterior tongue are innervated by the chorda tympani (CT), those on the tongue's edges by the CT and glossopharyngeal nerve (GN), and taste buds on the back of the tongue by the GN only. The vagal nerve (VN) innervates the epiglottis. Gustatory nerve targets include the neocortex, required for associations such as the retention of learned taste aversions, and the limbic system, serving hedonic aspects. Stimulus intensity, reflected by neural response magnitude, predicts sensation strength, whereas qualities (sweet, salty, sour, bitter) are associated with four nerve fiber groups. Most stimuli activate fibers of all groups, leading to distinct across-fiber patterns. In suprathreshold mixtures, mutual suppression may occur. Weak stimuli sometimes show mixture-enhancement. With prolonged stimulation, taste declines and water may take on a different aftertaste (adaptation). Recovery occurs after stimulus cessation. Cross-adaptation between different taste substances occurs mainly within and not between taste qualities. Only small regional sensitivity differences between qualities exist on the tongue. Bitterness is stronger on the back of the tongue than on the tip; however, on the tip, bitterness recognition is superior. [*See* Tongue and Taste.]

B. Taste Stimuli and Sensory Attributes

Sweet is initiated by binding to complementary proteinaceous sites. In addition to sugars, many sweeteners are known. One is the protein thaumatin, equisweet to a 3000-fold sucrose weight. Aspartame, equisweet to a 200-fold sucrose weight, is widely used in foods and drinks. Sensitivity to bitter-tasting urea is bimodally distributed in the population, suggesting a genetically controlled receptor substance. Acids taste sour. Many salts taste salty, but they often are predominantly bitter. [*See* Sweetness, Stereochemistry.]

C. Taste Disorders

Taste complaints, in most cases, can be ascribed to misperceived olfactory disturbances. Except for slow decline in the elderly, taste rarely causes complaints. The decline of taste in the elderly is not uniform across compounds, so the receptor composition on the tongue may not be uniformly altered. Localized loss or change of taste may be associated with destruction of tongue tissue, nerve damage (e.g., a tumor), or central pathology.

D. Taste Technology

Taste substances are used in many products. Often sugars are replaced by noncaloric or noncariogenic sweeteners. Low-sodium diets and low-energy drinks require carefully balanced substitutes so as to maintain their hedonic value. Industrial and consumer taste panels assist in psychophysical assessment of most foods and beverages.

VIII. SKIN SENSES

A. Transduction

The skin senses (or somatic senses) subserve the perceptions of touch, pressure, warmth, cold, pain, and also electric current. Their main purpose is to explore form and roughness of surfaces, to search for comfort with respect to temperature, to avoid painful contacts that could do harm, and to inform about movement and position of the body and the limbs. In the evolutionary process, they are probably among the oldest senses whereby animals tried to probe their environment. In two different layers of the skin, the dermis and the epidermis, a manifold of nervous structures can be discovered, which support the skin sensitivity. Their supposedly specialized functions are still a matter of debate. In the dermis, there are rather large so-called Pacinian corpuscles. They are internally shaped like an onion and are almost certainly subserving the sense of touch. Also in the dermis, fluid-filled compartments containing collagen fibrils with nerve endings are found, called Ruffini organs. In the epidermis, one finds Meissner's corpuscles, consisting of laminar cells interwoven with nerve endings, and also disc-like Merkel's corpuscles. Some of these corpuscles are believed to function in a network of touch and pressure-sensitive transducers. There is also an abundance of free nerve endings in the skin, the function of which is most difficult to establish. Somehow, their responses elicit different sensations of warmth, cold, and pain, which are very well discernible. Finally, in the hairy skin one finds hair follicle nerve end organs responding to the bending of hairs. [*See* Proprioceptors and Proprioception; Skin and Touch.]

The large number (on the order of 10^6) of nerve fibers ending in the skin layers are distributed very unevenly over the body surface. The sparse distribution on the back in contrast to the dense distribution on the fingertips, mouth, and tongue reflects the different sensitivities in these areas.

One form of perception, located deeper in the body, is kinesthesis, the sense of movement and of posture; this will not be treated here.

B. Sensitivity to Physical Stimuli

Because the sensitivity over the skin area varies considerably, only approximate indications can be given here. Sensations of the sense of touch (also often referred to as the sense of vibration) requires an amplitude of indentation of the skin of $>0.1-10$ μm in the frequency range of 20–1000 Hz. A feeling of pressure can be elicited by placing weights of $>10-100$ mg on the skin. Warmth can be felt by temperature rises of $>0.1-0.2$°C over a sufficiently large area within a sufficiently short time interval. Cold sensations begin to arise at a smaller temperature drop (-0.05°C) under approximately similar conditions. Pain can be evoked under a variety of conditions: by damaging the skin mechanically, by applying an aggressive chemical to the skin surface, by exposing the skin to temperatures of >45°C, or by sending an electric current in the milliamp range through the skin.

Sensations of itch and itchy skin ("alloknesis") can be evoked by intracutaneous injection of chemicals such as histamine. For instance, a dose of 1 μg histamine applied intracutaneously produces alloknesis with a latency of about 10 sec and a duration of about 25 min.

C. Sensory Attributes

A conspicuous property of the skin senses is the fact that, except for pain, they all show a marked decrease in perceived "strength of sensation" when a stimulus remains constant over time. For instance, the pressure of a coin placed on the skin is felt for only a short time. Plunging in cold or warm water gives initially an overwhelming sensation of cold or warmth, but this sensation wears off rather quickly. The sense of vibration escapes the prevailing adaptation by the nature of its stimulus, which is a rapid change of indentation of the skin. A further important aspect of skin sensation is the fact that its magnitude increases as the stimulus area is enlarged up to an upper bound. This is called the summation area. The summation area of touch, pressure, pain, cold, and warmth gets larger in this order. The summation area is also approximately inversely related to the density of nerve fibers and is therefore smallest for the tongue and the fingertips. Finally, the summation area is related to the so-called two-point threshold, which is the minimum distance of two point-like stimuli that are just perceived separately.

D. Cognitive Attributes

In exploring the environment, touch among the skin senses is dominant. It is amazing how many different textures of solid surfaces can be recognized. Even the nature of fluids may be felt by moving the hand in the fluid.

The touch sense also can have an important signaling function in interpersonal relationships as the way of touching one another can have an emotional content. The senses of warmth and cold do not seem to carry more specific conscious information than is related to comfort. The alarm function of the sense of pain is evident; many fast retraction reflexes depend on this sense. Intractable pain causes patients suffering from this pathological condition to experience an unbearable pain in the skin (often in the face) for which no adequate external reason can be found.

E. Technical Use of the Skin Senses

Of the human's sensory systems, the skin senses play a restricted role in communication. It is not until the visual or the auditive sense is impaired or overloaded that the tactile sense is called upon for extra communication. In the past, many ingenious devices employing the sense of touch have been put to use for the visually handicapped. The temperature senses are not commonly thought of as useful for communication purposes owing to their relatively slower response, their poorer localizability, and their fast adaptation. Attempts have been made to enlarge the information capacity of the human operator by recruiting the skin senses in a so-called "skin vision" system, whereby information from the outside world is "projected" onto the skin. The interpretation of such "projections" will have to be learned. Up until the present, only blind people have been applying such projection systems with limited success.

IX. PERCEPTUAL SELECTION

A. Reduction of Complex Data

A drastic reduction occurs in the amount of perceptual data impinging on the observer. This occurs through body, head, and eye movements and through sensory and cognitively controlled selective processes.

B. Object and Subject Factors

We classify the factors that control the selective processes into object and subject factors. The object factors are of external origin and relate to properties of the stimuli (conspicuity). The subject factors are of internal cognitive origin and derive from the subject's motives, desires, and expectations (directed attention). The two factors are also named involuntary and voluntary determinants of selective attention, respectively.

Voluntary selection may be precategorical (i.e., directed at expected stimulus location) or postcategorical (i.e., depending on stimulus recognition). Benefit primarily exists in the neglect of nonattended stimuli. Also, some direct benefit of selective attention can be demonstrated. An example of precategorical voluntary attention is the neglect of messages to one ear when the two ears receive different speech messages. The "cocktail party effect" is related in fact to both precategorical selection (sound direction) and postcategorical selection (speech content). Visually, only certain attributes can be effectively selected among other attributes present. Selection also occurs among signals coming simultaneously from different senses. So pain may be sometimes reduced by shifting attention toward other perceptual information.

Figure 3a demonstrates how the conspicuity of a target (center) depends on differences with neighboring distractors. Figure 3b indicates that a high conspicuity corresponds to a large visual field in which the stimulus will be detected. Selection develops with aging in a subtle way.

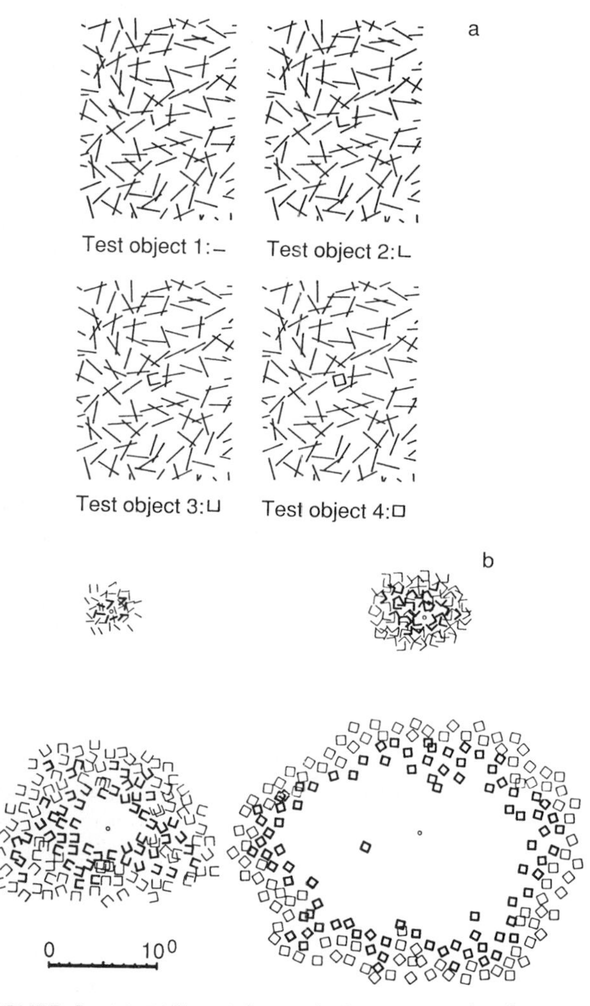

FIGURE 3 (a) Differential conspicuity as a result of target–background differences for test objects 1–4. In each case, the target object is in the center. (b) Areas around the fixation point in which targets 1–4 can be detected in a single glance in the same background as in (a). Target objects in bold were detected; the others were not discovered.

C. Selection in a Technical World

Research on auditory attention was initiated in the early 1950s through a need to understand the performance difficulties faced by air traffic controllers and pilots who were required to respond quickly and accurately to a range of visual and auditory inputs. However, how to catch and hold attention is also a very practical problem for the advertiser, the road sign designer, the newspaper make-up editor, and, nowadays, the designer of multimodal human–computer interfaces.

X. DEVELOPMENT OF PERCEPTUAL FUNCTIONS

When during a person's lifetime unchanged stimuli gradually give rise to different sensations and cognitive percepts, we speak of perceptual development. Changes in perception are brought about by maturation, sensory development, perceptual learning, and physiological changes, which ultimately may also lead to deterioration of perceptual functions.

A. Infancy

In infants, the components of the visual system mature in the order in which visual information is processed. With the retina maturing first and the cortex last, elementary stimulus variables can be handled better than complex patterns in infancy. A corollary of this is that during the first 2 months, both foveas are not consistently aligned for binocular fixation. A sensory limit on blur detection amounts to ±1.4 diopters, while convergence may vary in a 3° range in 1-month-old infants. Stereoacuity develops rapidly from 3 months of age onward, and near space perception after 4–5 months is similar to that of adults, but visual space seems to be confined to some 1.5 m.

Although newborns might be color-deficient, as the lateral geniculate nucleus and the prestriate cortex are not yet mature then, no evidence indicates this. At 4 months, infants can perceive hues and do so categorically, in much the same fashion as normal trichromatic adults. By then, myelinization of the optic radiations is complete.

Visual acuity is moderate at birth; at <1 month of age, resolution is just <1°, but at 2 months, acuity is good for distances at least between 30 cm and 1.5 m. At 6 months, acuity is about 5 min of arc, and at 1 year it is comparable with that of adults (at least 1 min of arc). Shape perception shows a gradual development too. At 1 month, infants prefer to look at a face-like pattern rather than a plain geometrical one, but they scan details only at 2 months. Whether or not infants up to 6 months of age have the same kind of size and shape constancy as adults is unknown; stimulus invariance seems to involve much perceptual learning and not to be a direct consequence of maturation.

Hearing, coupled with localization, can be shown to exist at birth but is more sensitive to complex sounds than to pure tones. Recognition of the mother's voice may happen early but is probably fully evolved at 1 or 2 months. Ample evidence indicates that 1- to 4-month-olds can easily distinguish between many different speech sounds and categorize them.

One-year-old infants can recognize objects by touch alone, but little is known about earlier touch perception, which is strongly linked to unconditioned reflexes.

A surprising ability, present at 3 weeks, is to imitate adults' facial gestures. Full control of facial gestures also provides evidence of the ability to recognize smells (6 weeks) or tastes. Apparently, preference for sweets is not learned or modified by experience.

B. Childhood

Although young children can distinguish detailed visual features quite well, at 5 years of age they are still unable to grasp fully the relation between the whole and the parts. Only well after 10 years of age are they able to complete an embedded figure test at adult speed. They are also less sensitive to figure orientation. Size constancy is much better than that of infants but is restricted largely to distances of 3 m. When growing up, children are increasingly better able to confine their attention to specific information-bearing parts of the stimulus pattern as a whole and use their attention strategically for perceptual tasks.

C. Learning

The simplest kinds of learning consist of habituation and sensitization. Habituation occurs when repeated presentations of a stimulus evoke increasingly lower sensation levels. Sensitization, on the contrary, increases sensation level after sufficient stimulus repetition but, in general, requires cognitive effort. Both concepts are to be distinguished from adaptation, which may encompass both effects. Adaptation is operative in a short time range (minutes or less), and its effects are completely reversible. Habituation and sensitization, however, have enduring effects, up to weeks or months or even longer, and may not be reversible at all. Practice is a common means to induce sensitization and is effective in a large range of psychophysical tasks. [*See* Learning and Memory.]

The human perceptual world derives a great deal from extensive experience with the objective physical environment (e.g., resulting in the perception of the world as upright). Perceptual learning is said to occur when basic stimulus features are interpreted on a higher abstract level to represent a different unit. This procedure may be repeated many times, whereby the stimulus is residing over a hierarchy of ever more detailed subordinate feature categories. Thus, abstraction becomes a means for perceptual efficiency when dealing with masses of environmental stimulation. Conversely, attention is increasingly employed to select wanted information and to cancel or suppress irrelevant perceptual information. Reading is a good example of cumulative and abstract perceptual learning; from visual features to characters, symbols or words, phrases, sentences, and thematic and semantic units. The ability to recognize words routinely takes years of practice; at 10 years of age, normal readers are some 150 msec slower than adults to name shortly presented words and about 10% less accurate. Text redundancy (e.g., residing at the orthographic, lexical, and syntactic level) can be employed only after extensive practice; both sensitization and perceptual learning, as well as selective attention, must take place.

Most stimulus patterns, for reasons of efficiency, tend to be perceived as instances of categories; the latter are obtained by continuous exposure and learning and depend heavily on human perceptual experience. Perceptual categories form the framework of the recognition process.

D. Physiological Changes

One of the best-documented cases of physiological change is that of decreased accommodation range of the eye, known as presbyopia. Due to continuous growth of fibrous tissue within the lens capsule, it becomes less elastic and transparent. Whereas the accommodation range at 8 years of age may span up to 19 diopters, it decreases almost linearly until the age of 50 years, after which it stabilizes at a range of 1 diopter on average. The presence of fibers and particles within the lens and the eyeball fluid will scatter incoming light and cause glare, resulting in reduced subjective contrast and acuity. Yellowing of the lens causes a slight reduction of sensitivity to green, blue, and violet for older people. In addition, they have more difficulty in ignoring irrelevant visual stimulation, thus needing more time to process information.

A common aging form in hearing is presbyacusis, involving a progressive hearing loss for high frequencies. This goes together with a decrease in speech intelligibility from the 30th year onward.

During a person's lifetime, changes occur in the anatomical structure, the neural pathways, and physiological functioning of all sensory systems; in general, this implies reduced sensitivity and performance levels. In many cases, however, sensory sensitivity of older people is hardly or not at all impaired compared with that of younger people, the elderly generally employ a stricter criterion for perception and recognition, which means their sensitivity only seems affected, but it is actually equally effective. This appears also to be the case for touch sensitivity and pain. Little is known about age-related changes in taste and smell.

BIBLIOGRAPHY

Aslin, R. N., Alberts, J. R., and Peterson, M. R. (eds.) (1981). "Development of Perception," Vol. 2. Academic Press, New York.

Bartoshuk, L. M. (1988). Taste. *In* "Stevens' Handbook of Experimental Psychology" (R. C. Atkinson, R. J. Herrnstein, and G. Lindzey, eds.), 2nd Ed., Vol. 1, Chap. 9, "Perception and Motivation" (R. D. Luce), pp. 461–499. John Wiley & Sons, New York.

Boff, K. R., Kaufman, L., and Thomas, J. P. (1986). "Handbook of Perception and Human Performance." John Wiley & Sons, New York.

Borden, G. J., Harris, K. S., and Raphael L. J. (1994). "Speech Science Primer; Physiology; Acoustics and Perception of Speech," 3rd Ed. Williams & Wilkins, Baltimore.

Charness, N. (ed.) (1985). "Aging and Human Performance." John Wiley & Sons, Chichester, England.

Singer, H., and Ruddell, R. B. (eds.) (1985). "Theoretical Models and Processes of Reading," 3rd Ed. Lawrence Erlbaum Associates, International Reading Association, Newark, Delaware.

Thomas, E. F., and Silver, W. L. (eds.) (1987). "Neurobiology of Taste and Smell." John Wiley & Sons, New York.

Tinker, M. A. (1964). "Legibility of Print." Iowa State Univ. Press, Ames.

Periodontics

NIGEL G. CLARKE
University of Adelaide

GLOSSARY

Caries Process of tooth decay resulting from the production of acid by bacteria from fermentable food residues. However, not all people are equally vulnerable and not all teeth in susceptible people are equally at risk. The factors responsible for a susceptible tooth surface are not well understood but are probably similar to those described here for periodontitis

Continuous tooth eruption Natural minute daily vertical movement of teeth thought to be the mechanism compensating for the loss of tooth structure from the grinding surfaces of the teeth. This also compensates for the continuous growth that occurs in the lower jaw adjacent to the joints with the base of the skull

Dental abscess Inflammation occurring at any location within the periodontal ligament due to infection within the pulp in the core of a tooth. The link between the internal and external parts of the tooth may be via communication channels (lateral or accessory canals) or via the permeable tubular dentin and root cementum

Gingivitis Inflammation of the gingivae (gums) around the necks of teeth resulting from the presence of bacteria on the tooth surface in the crevice between the tooth and gum

Periodontal disease An imprecise term that may commonly include gingivitis and an extension of inflammation within the gingivae to alveolar bone (periodontitis). An older term was pyorrhoea (pus from the gums)

Periodontal ligament Connective tissue that links the cemental covering of a tooth to the adjacent supporting alveolar bone. It provides formative, nutritive, sensory, protective, and defensive functions

Periodontitis Loss of alveolar bone supporting the teeth when gingivitis extends into the underlying bone. This was thought to be a common event but now appears to be very unusual

Pulpitis Inflammation of the dental pulp, the tissue found in the internal cavity that occupies the core region of the tooth. There are several pathways available to bacteria to gain access to the dental pulp

PERIODONTICS IS THE STUDY OF THE ANATOMY, physiology, and function of the supporting structures of the teeth, including their maintenance and management of disease. The object of periodontics is to keep the natural teeth for life.

I. ANATOMY AND PHYSIOLOGY OF THE PERIODONTIUM

The anatomical and physiological properties of an organ are linked to the function that the organ serves. The teeth are principally required to masticate food and they function as tools in hunter-gatherer cultures. They play a role in the formation of words and sounds. The anatomical components of the periodontium are formed from two hard and two soft tissues. The hard structures are specialized bone and form the surface of the tooth socket (the alveolar bone) and the covering of the root surface (the cementum). These two hard tissues provide anchorage for the fibers of the periodontal ligament.

These structures are collectively described as the attachment apparatus of the teeth. The second soft tissue component of the periodontium is the visible and superficial gingival covering of the bone, cementum, and periodontal ligament, commonly referred to as the gums. The periodontal ligament has five functions in performing the demands placed upon it: formative, nutritive, supportive, protective, and sensory. [*See* Dental and Oral Biology, Anatomy.]

A. Functions of the Periodontal Ligament

1. Formative

Specialized connective tissue cells on both sides of the periodontal ligament form new bone and cementum throughout life. The activity repairs the wear and tear that can occur to the hard tissues and enables progressive reformation of the fiber attachment to tooth and bone. New fibers are embedded into the bone on the outer surface and into new cementum on the inner surface of the periodontium. Fibroblasts within the ligament itself form new fibers and remove old or damaged collagen tissue.

2. Nutritive

Blood vessels are present throughout the periodontal ligament and supply the basic building blocks required by the specialized cells forming the bone, cementum, and periodontal ligament. Fibroblasts also produce a matrix (ground substance) in the periodontal ligament that provides a pathway for the spread of nutrients through the periodontal ligament.

3. Supportive

Considerable forces are applied to teeth and four effective systems distribute the forces and prevent damage to the periodontal tissues.

a. Fibers

Straightening of the collagen fiber bundles of the periodontal ligament absorbs light loads applied along the tooth crown and resists the extraction of a tooth.

b. Fluid Displacement

Ground substance present between the fiber bundles cannot be compressed under load and is displaced laterally into bone through fine holes (foramina) in the surface. The periodontal ligament fiber bundles control the fluid displacement, thus preventing vertical movement. The fluid ground substance returns to the periodontal ligament on release of the tooth from load. Blood is the second fluid displaced from the periodontal ligament to bone when the tooth is loaded.

c. Bone Elasticity

Loads applied laterally to a tooth crown are resisted by the strength and elasticity of the bone supporting the teeth. The supporting alveolar process absorbs heavy loads by bending and fractures only when its high value Young's Modulus is exceeded. This can occur in the extraction of multirooted teeth.

d. Tooth Contact Points

Loads applied to the biting surfaces of the teeth are distributed to adjacent teeth through their contacting points as a direct consequence of the forward angulation of the roots of opposing teeth.

4. Protective

The periodontal ligament is a most highly innervated tissue with several different types of nerve endings responding to all directionally applied forces. Output from these nerve endings when a status of overload approaches is interpreted as pain by the brain. Outflow of impulses from the brain to the muscles of mastication reduces the load and maintains the masticatory system in balance.

5. Sensory

A tooth can detect being stroked by a strand of cotton wool. This ability of the periodontal ligament sensors and the brain to detect very light touch is an indication of the efficiency of the sensory components located within the periodontal ligament. These sensory systems are used in chewing, speech, and swallowing. The teeth and the mechanisms of tooth support play important roles for the human organism. It is appropriate that when disease occurs all efforts be made to retain their function for the longest possible time by the best methods available.

B. Continuous Tooth Eruption

Continuous tooth eruption is a biological mechanism that occurs within the periodontium at approximately 0.1 mm per year. The process is needed to compensate for the loss of the structure on the biting surfaces of the teeth as a result of continuous wear. Attrition in the upper tooth structure was especially severe in hunter-gatherer cultures and could even result in the exfoliation of teeth in extreme instances.

Continuous tooth eruption is also needed to compensate for the increasing distance between the upper and lower jaws as a result of continuing growth of the mandible adjacent to the joints with the floor of the skull. This growth is appproximately 1 mm between the age of 20 and 60 years. When compounded with tooth wear, this natural growth makes it imperative for continuous tooth eruption to maintain the lower face height.

II. ANCIENT CONCEPTS AND PRACTICES OF PERIODONTICS

The ancient cultures of Greece, Rome, Egypt, India, and China had concepts for the cause (etiology) of periodontal disease and its treatment. Various medicines were developed for the purpose of reducing painful inflammation of the gums and localized gingival swellings were cauterized with a red-hot metal rod. The Middle Ages in Europe brought considerable advances in the treatment of the periodontal diseases and the basic dental anatomy began to be understood. The multirooted teeth (molars) were identified and the presence of a circulation within the teeth was discovered. The invention of the microscope resulted in the investigation of scrapings from the surfaces of the teeth, which revealed an enormous amount of bacteriological activity. This discovery led to the hypothesis that bacteria caused periodontitis. This hypothesis has been maintained since that time but is still unproven.

III. MODERN CONCEPTS OF PERIODONTICS

John Hunter in England and Pierre Fauchard in France were the fathers of modern dentistry. They developed appropriate treatment methods for many aspects of dentistry, including tools for the scaling of tooth surfaces. The concepts of tooth scaling were incorporated into the curricula of the first dental schools when they began in Europe in the late 1700s.

The twentieth century witnessed a great expansion in the practice of periodontics. The contribution emanating from several American Dental Schools in the 1950s and 1960s dramatically extended the knowledge and understanding of the anatomy, physiology, and diseases of the tissues supporting the teeth. Numerous controversies developed in periodontics, many of which were resolved by examination with the electron microscope.

The historical record of the occurrence of periodontal diseases, coupled with the modern writings of anthropologists, led to the opinion that periodontal disease was an ancient problem that has afflicted all cultures throughout history. Early epidemiological studies of periodontal disease in the 1950s and 1960s appeared to confirm that view by finding that periodontitis was present and progressive in all of the cultural groups that were surveyed. Recent analysis of both modern and ancient skulls and epidemiological studies using improved but still imperfect methods fail to support the concept of a widespread progressive disease. The skulls held in many museums around the world do have disease of the supporting bone around the teeth, but, contrary to the earlier anthropological findings, the damage had a dental (tooth) origin rather than a gingival origin (periodontitis).

The incidence of periodontitis was also grossly overestimated in earlier studies of skulls because continuous tooth eruption was not factored into the assessment. Both continuous tooth eruption and periodontitis result in an increase in the distance between the tooth neck and the alveolar crest of bone. In the assessment of periodontitis in ancient skulls, simple measurements of the distance between the tooth neck and bone crest were made. Therefore, allowance for continuous tooth eruption in these skulls was overlooked; any increase above the baseline was attributed to periodontitis. When the effects of dental abscesses and continuous tooth eruption are factored into the assessment of periodontitis in skulls, it is difficult to demonstrate the presence of periodontitis.

Perhaps the most important finding of the 1990s is that a person's health and environment (host factors) play an important role in the initiation of periodontal breakdown. It has become increasingly evident that smoking plays an important role in the development of periodontitis. Stress is also an important factor but is more difficult to measure than smoking. Diet has an impact on the health of the periodontium but its effect is also difficult to assess. Periodontitis appears to conform with most chronic diseases in having a multifactorial cause.

IV. HUMAN CHRONIC DISEASES

Chronic diseases are the result of long-term interactions between a person and their environment. These interactions are subtle and the link between cause and effect is often difficult to make, ensuring that the various contributing factors of many illnesses have to be inferred because they cannot be modeled. The causes of chronic diseases are multifactorial, including factors that arise from cellular dysfunction, aging, environmental, social, and genetic components. A necessary factor must be present if the disease is to occur but that factor alone is not sufficient to cause disease. The necessary cause may be some form of bacteria or virus that will not produce symptoms until other essential causes (often psychosocial) compromise the health of the person. Whether their good health is maintained or an illness develops depends on the environment and the ability of the host to withstand environmental stressors. Three components form an etiological matrix for chronic disease: the virulence of the microorganism or agent, host resistance or susceptibility, and the environmental conditions (social structure, emotions, diet, health care, and conflict).

Tuberculosis (TB) is an excellent example of a chronic disease once thought to have a specific bacterial cause but that is now recognized to have a multifactorial etiology. It occurs when sufficient psychosocial factors operate in conjunction with the necessary bacteria. White blood cell (phagocytic) activity, the main defense against the tuberculosis bacteria (tubercle bacillus) in TB patients, is diminished during periods of emotional excitement. In the human population, the distribution of the tubercle bacillus is much wider than the distribution of tuberculosis, demonstrating that other factors are essential for the initiation of disease than just the presence of the bacteria.

Similarly, the initiation of periodontitis requires that bacteria be present to initiate gingivitis but some people develop disease and others do not, even though they all harbor the same bacteria. Therefore, it is probable that additional factors are important in the etiology of the condition; it is arguable that it is more important to determine who gets the disease and in what circumstances than which bacteria were present.

V. ETIOLOGY OF PERIODONTITIS

Epidemiological studies over the last decade have dramatically modified earlier beliefs concerning the widespread nature of the periodontal diseases. Instead of virtually everyone having the disease, it has become apparent that only a very small percentage of a population demonstrates evidence of the disease. This is true for populations that practice high levels of oral hygiene and in cultures where there is no opportunity or little value has been placed upon the cleaning of teeth. Furthermore, even in the cases where periodontitis does occur, only a small number of teeth are found to be seriously affected.

In studies of periodontitis (as in other chronic diseases) it has been difficult to establish by observation the relative importance of various local and systemic factors that operate over many years to result in the disease. Periodontitis was believed to result from gingivitis induced by a combination of all of the bacteria that were present in the region of the tooth neck (a nonspecific infection). Over the years, the classification of bacteria has been greatly refined and many new species have been cultured. As more species of bacteria were found, each new bacterial type was proposed to be the cause of periodontitis. Stained bacterial smears reveal that there could be more specimens to uncover by improved collection and culturing methods. However, it is a fundamental microbiological principle that indigenous bacteria are incapable of inducing disease in their usual environment. Indigenous bacteria can cause disease if either the host's resistance is lowered or bacteria are relocated from their usual habitat. Examples of these principles are found in the poor resistance to infections of diabetics and other chronically ill patients. The classic example of normal bacteria causing trouble on relocation within the host is cystitis caused by *Escherichia coli* from the rectum.

A. Generalized Periodontitis

It is significant that it has not been possible, under test conditions, to induce periodontitis in healthy sites by infecting them with bacteria from diseased sites. This is singularly fortunate as the relocation of bacteria occurs during dental examinations. If periodontitis could be spread in this way, the dental profession would be the principal agents in the spread of the disease. There is no evidence to support that periodontitis can be spread by either professional or social contact with a specific or generalized mixing of oral bacteria. The distribution of the bacteria is much wider than the distribution of the disease; they are found in healthy mouths and in the healthy sites of diseased mouths. The association between these bacteria and the breakdown of the periodontium is weak, but periodontitis is still commonly referred to as an infection.

There is no relationship between the bacteria commonly promoted as periodontopathogens and the distribution of periodontitis. All of the bacteria that have been proposed to be periodontal pathogens are normal indigenous human bacteria of the gingival crevice. These bacteria are ubiquitous, normally harmless, and make up part of the natural oral flora that serve a protective function. The occupation of all of the oral sites by indigenous bacteria that are suited to the conditions makes it difficult for foreign bacteria to gain a foothold in competition with the normal flora. Generalized periodontitis probably represents the consequences of poor personal defense to widespread gingivitis in compromised hosts. It is paradoxical to have the normal flora cast in both defensive and aggressive roles.

Support for the specific cause of periodontitis was provided by the findings that different forms of the disease were associated with different bacterial species (a specific infection). Juvenile periodontitis was found to be associated with *Haemophilus actinomycetumcomitans* and adult periodontitis with *Bacteroides gingivalis*. These findings were confirmed by several laboratories and the site-specific concept of periodontitis was established, but until recently little progress has been made in understanding who is at risk and at which periodontal location. In addition to finding specific bacteria in specific sites, the evidence for the association of bacteria with periodontitis is based on the presence of antibodies in plasma to specific bacteria. Also cited is the *in vitro* production of bacterial toxins that are capable of inhibiting human defense cells. However, not all antibody-forming agents promote disease and the ability to ward off defense cells might equally be essential for the survival of bacteria in a hostile (to bacteria) environment. But it is also arguable that the broken-down sites in the periodontium provide the conditions enjoyed by some bacteria.

B. Localized Periodontitis

Evidence for the pathogenesis of localized periodontitis may be found from anthropological, epidemiological, microbiological, and clinical sources. It was mentioned earlier that recent anthropological studies of skulls do not support the concept that there is an attack on the crestal alveolar bone from the inevitable presence of gingivitis in cultures where little or no oral hygiene was possible. These studies demonstrate that damage to the alveolar bone was restricted to individual teeth, the damage arising from severe tooth wear that perforated the central pulpal space of the tooth or as a result of caries, which has the same effect as severe attrition. Where adjacent teeth were similarly perforated, careful observations indicate that the generalized damage is actually the collective consequences of localized damage around several individual teeth. This process has the clinical name of an endo/perio defect and, although historically it has not been considered to be a significant cause of localized deep pockets, it now appears that it may be the primary cause.

Localized periodontitis probably represents the consequences of relocating bacteria from their normal habitat to the dental pulp. The bacteria's presence in that tissue challenges the host's defenses and the consequences of that fight, if sustained, result in the production of highly inflammatory peptides that have access to the supporting periodontal ligament via the natural pathways between the pulp and the periodontal ligament.

The lack of evidence for a direct bacterial cause of human periodontitis, localized or general, together with the principle of microbiology, suggests that the heavy emphasis placed on specific bacterial causes for the various periodontal disorders should be viewed with caution. Intense search for a bacterial cause for periodontitis for more than 30 years has failed to provide satisfactory evidence.

VI. RISK FACTORS AND INDICATORS (MARKERS) FOR PERIODONTITIS

The search for risk indicators for periodontitis reflects the need to be able to predict who is at risk of developing periodontitis and which sites are periodontally active. Systemic disease has been suspected in the establishment of periodontitis. However, studies of groups of periodontal patients have failed to find the presence of systemic disease. Host factors commonly associated with chronic diseases are not themselves disorders but are the life-style factors and psychosocial consequences of stress. Identifying risk factors for periodontal disorders may require a closer examination of the factors normally attributed to initiating other chronic diseases. It may become clear why one individual develops periodontitis, another heart disease, while a third develops no chronic illness at all.

A. Personal Risk Factors for Periodontitis

The personal risk factors for periodontitis are probably the bacterial agent, the social environment, and psychosocial, systemic, and life-style factors.

1. Necessary Cause: The Agent

The necessary factor for periodontitis is the bacterial agent, which must be present if disease is to occur but which does not alone produce periodontitis. The bacterial agent might be any or all of the microbial species that combine to form dental plaque. The agent may be assisted by factors that encourage plaque accumulation and hinder its removal, such as gingivally placed restorations, rough tooth surfaces, tooth anatomy (grooves), caries, and the diet. However, bacteria colonize the gingival crevice in the absence of any assisting factors and independent of culture, race, or social class. [*See* Dental Caries.]

2. The Social Environment

The social environment includes an individual's family, community, industrial, and political relationships, influenced by educational, cultural, socioeconomic, and health status. These factors can impact on the agent and psychosocial, life-style, and systemic components. It is not necessary for all of the factors that can be associated with disease to be present for it to become established. The agent must be present, and since for periodontitis the agent is derived from the normal commensal flora, it is inevitably found in all mouths. The greater number of factors present increases the chance of the initiation of generalized periodontitis.

It has been observed that social stress in a community results in the establishment of high levels of chronic disease. Psychosocial disease probably results from an individual's failure to adapt to social relationships that begin to fail, when bereavement occurs, when industrial change and redundancy intrude, or when political turmoil and poverty are inescapable. The effects of stress are mediated by the autonomic nervous system (ANS). The gingivae and dental pulp are both stimulated by the ANS and the mouth has been described as an important site for observing the effects of psychosocial stress. Stress has a clearly established harmful impact on the defensive systems of the host by increasing the susceptibility to disease. Psychosocial stressors that affect the immune system have been implicated in the provocation and exacerbation of infectious and autoimmune diseases, including allergies and cancer.

The pharmacoactive amines released in states of stress have a vasoconstrictive effect with the potential to constrict blood flow in the peripheral circulation. Sustained neuroendocrinal and neurohormonal stimulation can have a negative effect on target cells to initiate organic damage. Activation of the ANS by stress can result in the initiation of psychosocial disease. Stress may be significant in determining the intensity of gingival inflammation by modulating the immune response and precipitating the extension of gingivitis. Periodontal lesions have been observed to be severe and widespread as stressors increased.

3. Life-style Factors

Several life-style choices that are associated with other chronic diseases also contribute to periodontitis. Cigarette smoking contributes to the genesis of periodontitis probably due to the vasoconstrictive property of nicotine. Diet and the abuse of alcohol may also have a role in the initiation of periodontitis. These factors, together with the physiological consequences of stress, have the potential to reduce a host's resistance and increase the potential for disease. Other chronic diseases (e.g., diabetes) increase the susceptibility to periodontitis. Periodontitis appears to be a multifactorial chronic disease, where reduced host performance can result in the loss of alveolar crestal bone as a result of inadequately defended gingivitis.

VII. TREATMENT

Throughout the ages, instruments have been used to clean teeth, the fundamental approach to periodontal therapy. The rationale for any treatment is based on two principles: that the cause of the problem should be removed and that the damaged structure should be repaired as well as possible.

Prior to the development of microscopy, microbiology, and radiography, appreciation for problems of the gingivae relied on observations made by the naked eye. It was simple to detect the presence of debris that naturally collected on the surfaces of teeth and to see the resultant redness and swelling of the gingivae. It was also simple to observe that many teeth became loose in middle life with an increasing detachment of the periodontal ligament from the tooth to form a "pocket." It was reasonable to assume that the changes early in life to the gingivae from local debris led to the loss of support and ultimately the loss of teeth. These were reasonable assumptions, but the evidence no longer supports them. However, the old concepts still govern the delivery of most periodontal services around the world in the mid-1990s.

A. Hygiene Phase

Treatment has been directed toward the removal of calcified layers of bacteria from the surfaces of teeth at the tooth/gum interface. Two methods have commonly been employed: hand instrumentation using scalers and/or curettes that scraped the root surfaces of the pocket, removing attached calculus and also a portion of the root covering (cementum). This procedure is usually completed without the use of local anesthetic. In some instances, pain control methods are used when desired by the operator or the patient. The second method and probably the one most frequently employed uses an ultrasonic machine to achieve the desired clinical cleaning of the teeth. In association with this approach, patients are encouraged to maintain very high standards of bacterial removal (plaque control) using toothbrushes for the accessible surfaces and some other approach (usually floss) to clean between the teeth. This hygiene phase of periodontal therapy is usually completed by the general dental practitioner. Patients with complex problems are often referred to specialist practitioners where they are available. These hygiene phases of treatment address the first reason for therapy but they do not re-form the lost tissues (the second objective of treatment).

B. Surgical Phase

Consistent with the earlier hypothesis of a bacterial cause for the periodontal diseases, procedures were designed to eliminate the pockets to facilitate hygiene for the patient. Where deep pockets have formed around teeth it is difficult to remove the plaque from the pockets. The first surgical approach, which was practiced until the 1960s, was a simple cut through the gingivae around the necks of the teeth to the level of the base of the pocket (gingivectomy). Some gingivae was left to maintain a desired level of protection for the underlying periodontal structures. The postoperative period was made comfortable by covering the wound with a special dressing. The main disadvantage of the technique was that it significantly lengthened the clinical crown, an unsightly result when practiced in the anterior part of the mouth. Some regrowth of the tissue occurred over time and the pockets gradually returned. The practice of gingivectomy is now unusual and limited to the occasional removal of overgrowth of gingival tissue often as a result of drug therapy (e.g., dilantin in epilepsy).

An improved procedure was developed in the 1960s that eliminated pockets and facilitated hygiene, but avoided the disadvantages of the gingivectomy. It was referred to as the Widman flap procedure and required that the gingivae and oral mucosa be lifted from the underlying bone to provide excellent access to the root surfaces. This approach provided optimal conditions for scaling the root surfaces and allowed choices to be made of where the elevated tissue should be replaced at the conclusion of the procedure.

Long-term studies, over 5 years, revealed that there was no difference in the results achieved from simple scaling procedures or the Widman flap approach. The appearance of the gingivae could be much improved, but none of the tissue that was lost was re-formed. Attempts to replace lost tissue were made using the transplantation of bone marrow from the hip. The clinical results were often very good, with the bone surviving in the new location and often filling the defect very adequately. However, though the bone survived in close proximity to the tooth, there was no rebuilding of the normal periodontal structures between the bone and the tooth and no new attachment resulted.

Other materials have been tried over the years to fill in localized bony defects, but the problem has always been that no new periodontal ligament tissue formed between the implanted material and the tooth surface. A completely new approach emerged in the 1980s—guided tissue regeneration—after it became apparent that the formative cell of the gingivae could not rebuild the basic structures of the periodontal attachment. The formative cell (fibroblast) of the periodontal ligament, however, has the potential and a technique was devised to optimize the regeneration by placing a membrane between the gingival tissues and tooth extending below the crest of bone. Regeneration of the periodontal ligament tissue could then occur from the surface of the ligament and extend along the surface of the tooth without competition from gingival fibroblasts. The clinical results were encouraging, but microscopy showed that the new attachment was minimal (approximately 0.5 mm).

The high cost of the membranes used, the need for a second-stage procedure to remove the membranes, and the minimal improvement in attachment indicate that the search for an effective surgical solution to tissue loss remains. It is a particularly difficult problem because nature did not provide a model that could be copied. The periodontal tissues are originally formed by the vertical movement of the tooth dragging bone along by the periodontal ligamental connection. This process is not available to the operator in attempts to repair damaged tissue. However, advances in biochemical methods may uncover growth hormones and chemical messengers that are capable of stimulating formative cells into the production of the periodontal tissues that have been damaged by the disease process.

Long-term studies of patients maintained at the highest possible levels of care show that a small number of teeth will continue to break down. It would appear that treatment cannot prevent breakdown from occurring in those few locations where disease will become active in those small numbers of people who will experience disease. Other long-term studies demonstrate that similar numbers of sites break down in groups offered no treatment at all. It is doubtful whether treatment prevents breakdown or is capable of predictably regenerating tissue once it is lost.

The essential role that host factors, stress, smoking, diet, and chronic diseases have in the etiology of periodontitis has not yet been incorporated into the philosophy of treatment. Hippocrates noted that in treating disease it was essential to "Look at the person who has the disease, not the disease that the person has." When this sage advice is followed, it becomes much less important to worry about which bacteria is predominant in a pocket and more necessary to concentrate on improving the overall health and the defensive performance of the host. Periodontics has had a long history, but there is some way to go before major breakdowns to the tissue can be prevented and restored.

BIBLIOGRAPHY

Baelum, V., *et al.* (1988). Tooth mortality and periodontal condition in 60–80-year-old Chinese. *Scand. J. Dental Res.* **96**, 99.

Clarke, N. G. (1990). Periodontal defects of pulpal origin: Evidence in early man. *Am. J. Phys. Anthropol.* **82**, 371.

Clarke, N. G. (1992). Some anatomical factors that influence the pathways followed by dental inflammatory exudates. *In* "Craniofacial Variation in Pacific Populations" (S. Molnar and T. Brown, eds.). Department of Dentistry, The University of Adelaide, Adelaide, Australia.

Clarke, N. G., and Carey, S. E. (1985). Etiology of chronic periodontal disease: An alternative perspective. *Am. Dental Assoc. J.* **110**, 689.

Clarke, N. G., and Hirsch, R. S. (1990). Periodontitis and angular alveolar lesions: A critical distinction. *Oral Surg. Oral Med. Oral Pathol.* **69**, 564.

Clarke, N. G., and Hirsch, R. S. (1991). Physiological, pulpal and periodontal factors influencing alveolar bone. *In* "Advances in Dental Anthropology" (M. Kelly and C. Larsen, eds.). Alan R. Liss, New York.

Clarke, N. G., and Hirsch, R. S. (1992). Two critical confounding factors in periodontal epidemiology. *Community Dental Health* **9**, 133.

Clarke, N. G., and Hirsch, R. S. (1995). Personal risk factors for generalized periodontitis. *J. Clin. Periodontol.* **22**, 136.

Clarke, N. G., *et al.* (1986). Periodontal disease in ancient populations. *Am. J. Phys. Anthropol.* **71**, 173.

Personality

NATHAN BRODY
Wesleyan University

Revised by
JONATHAN CHEEK
Wellesley College

GLOSSARY

Behavioral inhibition system System influenced by septal and hippocampal regions of the brain that inhibit action

Correlation Statistical measure of the relationship among a set of paired scores; it varies from +1.00 to −1.00. Zero correlation indicates lack of relationship; positive or negative correlations measure the magnitude of a positive or negative relationship among the scores

Extraversion Personality dimension related to the tendency to be with other people rather than alone

Factor analysis Statistical procedures used to determine the dimensionality of a correlation matrix

Neuroticism Personality dimension related to adjustment and the tendency to develop mild psychopathologies

Temperament Relatively enduring emotional and behavioral reactions that appear in infancy and are influenced by genetic constitution

Trait Tendency of an individual to exhibit consistent behavior in diverse situations over time

Within-family environment Variations in the environment encountered by individuals reared in the same family

PERSONALITY PSYCHOLOGY MAY BE DEFINED AS the study of individual differences in the psychophysical systems that determine people's characteristic thoughts, feelings, and actions. There are approximately 30,000 words in English that in one way or another refer to differences among individuals. It is possible to use this extensive linguistic resource to develop a scientifically viable descriptive system for individual differences in personality through the use of a statistical procedure called factor analysis.

I. DESCRIPTION OF PERSONALITY

A. Factor Analysis

The use of factor analysis can be explained by reference to a hypothetical study that is assumed to be representative of a large class of such investigations. Assume that a group of individuals is presented with a relatively comprehensive set of terms that may be used to describe themselves. The terms are arrayed in the form of scales anchored at each end by terms that are opposite in meaning, for example, neurotic and normal. Each bipolar scale permits the individual to rate himself or herself with a number reflecting position on the scale. If there are 100 scales, each person's self-description on these scales may be exhaustively described by his or her 100 ratings. If there are 100 subjects in this study, an exhaustive description of the data set would contain 10,000 numbers. It is possible to develop a more economical desciption of the data set, and it is entirely likely that there are significant redundancies among the set of personality descriptors selected for investigation. For example, the description "neurotic" is usually assigned the description "poorly adjusted" rather than "well adjusted." The redundancies among all possible pairs of descriptive scales may be noted by obtaining a correlation for all possible pairs or scales indicating the degree of relationship between each pair. There are 4950 such correlations in our hypothetical sample. It is likely that there will be subsets of scales that exhibit relatively high correlations among the members of the set but relatively low correlations with other scales. The set of clustered scales may be said to measure a common hypothetical factor. Factor analysis may be used to discover and define the underlying dimensions or number of factors needed to summarize and account for the redundant relationships that are obtained among the set of correlations.

In countless investigations of what is assumed to be a relatively exhaustive set of personality descriptors, it has been found that the matrix of correlations may usually be explained by assuming that there are a relatively limited number of factors present—typically varying from three to six. Among the best-defined and constantly recurring factors is *extraversion*–a tendency to prefer to be with people rather than alone. The opposite or low end of the factor is called *introversion*. Other factors that are constantly found include *neuroticism* (sometimes called *adjustment*) and *impulsivity*, which includes descriptive scales such as persistence and conscientiousness. Among other factors that have been found are *culture, agreeableness,* and *aggression*. The exact subset of factors needed to define self-report ratings of personality is still a matter of dispute. However, there is virtually unanimous agreement that the subset of factors must include extraversion, neuroticism, and impulsivity. There is also some agreement, if not unanimous accord, that the set of factors is relatively limited, and that something like five or six dimensions are relatively exhaustive.

A conceptually analogous experiment may be repeated using characterizations made by someone who is acquainted with the person being rated, instead of self-report ratings of one's own personality. Such ratings may be factor analyzed and it is usually found that the dimensional structure of these acquaintance ratings corresponds to the dimensional structure of the self-report ratings. Moreover, factor scores derived from self-report measures of personality tend to correlate with factor scores derived from personality ratings made by individuals who are acquainted with the person. Thus, one's self-description of personality tends to be in partial agreement with the description of personality obtained from one's acquaintances.

B. Trait Theory

The discovery that descriptions of personality may be summarized in terms of a limited number of dimensions is, strictly speaking, a discovery about the language used to describe personality. It is possible to argue that the dimensions of personality derived from factor analysis do in fact provide us with an adequate descriptive account of personality. Each of the factorially derived dimensions of personality may be assumed to be a measure of a personality trait. A trait is defined as a consistency in behavior exhibited by an individual in diverse settings over relatively long periods of time. Historically, there have been two views about trait descriptions of personality. One group of psychologists holds that the descriptions are linguistic fictions that do not in fact describe consistencies in the social behavior of individuals. Another group argues that the descriptions are accuate reflections of consistent behavior patterns. The former group argues that correlations between trait measures derived from self-reports or ratings and actual measures of behavior are quite low; they also claim that individuals tend to exhibit characteristically different behaviors in different social setings. Thus, an individual might be aggressive in job settings involving subordinates, and nonaggressive and meek in the home with his or her spouse. Such an individual's characteristic behaviors would not be adequately summarized by assigning a score on a dimension of aggressiveness to the individual. These psychologists tend to argue that personality may be optimally understood in terms of the idiosyncratic pattern of responses of each individual to the characteristic social situations that he or she encounters.

It is increasingly clear that the debate between trait theorists and their critics is subject to empirical resolution. The resolution may be accomplished by reference to the use of aggregated measures of behavior. Any single observation of behavior is likely to be unreliable as an index of the characteristic behavior of an individual. Consider an example. To predict punctuality from trait ratings of conscientiousness, a psychologist might observe whether or not individuals for whom conscientiousness ratings had been obtained arrive at class on time. If the correlation between the ratings and the behavioral index of being on time were low, the psychologist might conclude that there is little relationship between the trait rating and this particular index of social behavior. The conclusion may, however, be in error because a single observation of the punctuality of an individual is not likely to be a reliable index of the characteristic behavior of that individual. A person who is characteristically punctual might be late on a single occasion owing to unusual and unrepeatable circumstances. [*See* Behavior Measurement in Psychobiological Research.]

The use of aggregated measures of behavior in which individuals are observed in the same settings on several occasions is likely to increase the reliability of the measurements and result in more accurate predictions of behavior from trait ratings. It is generally conceded that individuals are quite consistent in their behavior in the same social situation. Individuals tend to be temporally consistent in their behavior in the same setting. They are also consistent in their characteristic trait ratings. Longitudinal studies of adult personality with decade or longer time lags between measurements find that individuals rate themselves in a similar way on different occasions.

Critics of trait theory generally concede that individuals are temporally consistent in their behavior. They have, however, argued that this does not provide support for trait theory. The critical issue for them is the extent to which individuals are consistent in their behavior in different settings. Consider again our punctuality example. Critics would argue that aggregated measures of punctuality measured by observing time of arrival at a class would not necessarily predict aggregated measures of time of arrival at a party. Therefore they would argue that personality is temporarily consistent and situationally inconsistent.

Trait theorists can again rescue their position by the use of aggregated measures. Assume that aggregated measures of behavior in different situations that are assumed to be indicative of a trait are obtained for a group of individuals. It is possible to randomly divide the set of measures into two subsets. An aggregated score for each of the subsets can be obtained for each individual. Each score represents the average level of behavior reflecting a particular trait exhibited in a particular subset of situations. The pairs of scores will have a substantial positive correlation, indicating that an individual's tendency to behave in a manner that is reflective of a particular trait in one set of situations will predict with substantial accuracy the individual's tendency to exhibit behavior indicative of that trait in a different set of situations. This analysis indicates that the dispute between trait theorists and their critics is not really substantive but rather reflects a preference for different levels of analysis. Trait theorists are interested in the broadest level of aggregated consistency in behavior that is exhibited by an individual. They would want to know what an individual's characteristic behavior is in a variety of situations. Their critics are more likely to argue that two individuals with the same aggregated trait score are likely to exhibit characteristically different patterns of behavior in different situations. Therefore, they prefer to describe personality in terms of an analysis that emphasizes the pattern of differential responsiveness of individuals to different situations.

Although trait descriptions may not be entirely accurate descriptions of the fine-grained dimensions of an individual's social behavior, there is some evidence that they do relate to important social outcomes. The utility of a trait descriptive analysis of personality can be found in longitudinal studies in which trait measures are predictive of measures of behavior obtained years later. Consider some examples. Ratings of aggressiveness based on peer ratings of a child's classmates in elementary school have been found to predict criminality and indices of antisocial behavior obtained in young adulthood. Measures of temperament, those emotional and behavioral reactions that appear early in life and are influenced by genetic constitution, can also be predictive of later social behavior. For example, in one study, ratings of difficult temperament in infants during the first year of life were predictive of reading problems and behavioral difficulties in adjusting to elementary schools. Trait ratings of neuroticism and impulsivity based on the aggregated ratings of five acquaintances at the start of a marriage have been found to be more predictive of marital unhappiness and divorce than social background measures of the partners or knowledge of life crises. In particular, if the male partner in a prospective marriage was described as nonneurotic and nonimpulsive and the female partner was described as nonneurotic, the marriage was more likely to be described as successful, and was less likely to lead to divorce than if the male partner was described as neurotic and impulsive and the female partner was described as neurotic. Individuals who rate themselves as impulsive or who are rated as impulsive by their acquaintances are likely to have academic difficulties in college even if they have high scores on tests of academic ability. These and many other findings indicate that trait ratings predict important social outcomes. These findings suggest that trait ratings capture characteristic differences in the behavior of individuals that lead, over time, to different social outcomes.

II. BEHAVIORAL GENETICS OF PERSONALITY

A. Twin Studies and Family Studies

Why do individuals differ in traits? It is a truism to assert that heredity and environment jointly determine individual differences in trait ratings. Behavior–genetic methods may be used to provide a more precise understanding of the nature of genetic and environmental influences on traits. Behavior–genetic analyses are variants and elaborations of two fundamental research designs introduced by Francis Galton in the latter part of the nineteenth century—the twin method and the family method. The twin method capitalizes on an experiment of nature. There are two kinds of twins—monozygotic (MZ) and dizygotic (DZ). The former derive from the splitting of a single fertilized egg and are, as a result, genetically identical. The latter derive from the fertilization of two separate eggs and are as genetically similar as siblings.

The family method in its more refined versions involves comparisons between children in adoptive families and biological families. The resemblance between adopted children and their biological and adoptive parents provides evidence about the relative importance of genetic influences and differences attributable to variations in the environment associated with being reared in different families. In recent years there has been an increase in both the number and sophistication of studies of the behavior genetics of personality traits. There are, however, major lacunae in the existing literature. For example, virtually no studies of personality traits are based on ratings by acquaintances for adults. Most of the studies are based on self-report trait measures. Nevertheless, a large body of evidence supports several tentative generalizations about the role of heredity and environment in the development of individual differences in traits. Twin studies indicate that MZ twins reared in the same home exhibit correlations of about .5 on self-report measures of personality traits. If trait scores were determined solely by genotypes, the correlation would be 1.00. This implies that personality traits are influenced by the environment. Since the correlation of .5 is based on MZ twins reared in the same home, this environmental influence must be attributable to differences experienced by MZ twins reared in the same home. These are called within-family environmental influences.

Twin studies almost invariably find that the correlation among same-sex DZ twins is less than that for MZ twins. Depending on the trait being measured, the correlations are generally .25 or lower. The differences in the magnitude of the correlations of MZ and DZ twin pairs is usually interpretable as a measure of the influence of genetic factors on the trait. Since MZ twin pairs are genetically identical, and DZ twin pairs are not, it can be argued that the greater similarity in genetic characteristics of MZ twins leads to greater similarity on the phenotypic characteristic. Note that this conclusion assumes that MZ and DZ twin pairs experience environments of equal variability within the pair. Although there is little persuasive reason to doubt this assumption, we shall directly examine its validity when we consider studies of separated twins reared in different families. A crude measure of the degree of genetic influence on a trait, ignoring several complexities, may be obtained by doubling the difference between within-pair MZ and DZ correlations for the trait. This analysis suggests that approximately 50% of the phenotypic variance on personality traits is attributable to genetic influences.

In behavior–genetic analyses it is traditional to partition the total phenotypic variability in a trait into component sources of variance that may be thought of as the characteristic influences on trait variation. Our analyses of twin studies of personality have established that personality traits are influenced both by genotypes and by variations in the within-family environment. A quantitative analysis of these two sources of variance shows that they account for all of the phenotypic variability of personality traits. This analysis leads to a somewhat counterintuitive result—that personality traits are not influenced by variations attributable to being reared in different homes. The family that rears a child does not, on this analysis, influence personality trait scores. This conclusion is also supported by studies of twins reared in different families.

Several modern studies have supplied data on personality resemblance for twins reared in separate homes. Though the samples are relatively small, the results indicate that MZ twins reared in separate families tend to be as similar to each other as MZ twins reared in the same family. These studies support the conclusion that there are no between-family environmental influences on personality traits.

Twin studies also provide information about the nature of genetic influences on different personality traits. It is possible to distinguish between additive and nonadditive genetic influences. Nonadditive genetic influences tend to decrease the phenotypic similarity of individuals who are not genetically identical. Nonadditive influences are attributable to dominance and to epistasis—the interactive influence of genes. In the limiting case, nonadditive influences can lead to zero phenotypic resemblance among individuals who are not genetically identical. Evidence of relatively high correlations for MZ twin pairs combined with relatively low correlations for same-sex DZ twin pairs, and low correlations between parents and children, provide support for the existence of nonadditive genetic influences. Extraversion appears to be a trait that is subject to nonadditive genetic influence. Twin studies indicate that the correlation among same-sex DZ twin pairs for this trait is less than half the value of the MZ correlation. Since DZ twins have a genetic correlation of at least .5, a phenotypic correlation for DZ twin pairs of less than half that of the corresponding MZ pairs suggests that there are nonadditive genetic influences on extraversion.

Family studies provide data that are in some respects congruent with the results of twin studies. Correlations between parents and children and between biological siblings reared in the same family for self-report measures are relatively low—generally below .3. In adopted families, the correlations between adopted parents and their adopted children and between biologically unrelated siblings reared in the same family are close to zero. These results imply that the relatively low degree of resemblance between parents and children and between siblings reared together in biological families are primarily attributable to the genetic relationship among family members. The absence of significant relationships between adopted parents and their adopted children, as well as between biologically unrelated children reared in the same family, provides additional support for the assertion that personality traits are not influenced by variations attributable to being reared in different families.

B. Genetic and Environmental Influences

Although genetic analyses provide us with some insights into the sources of individual differences in personality traits, they do not provide a clear understanding of the ways in which genetic influences and within-family environmental influences combine to determine a person's personality traits. At this stage of our knowledge, we can only speculate about the influence of the environment on personality. Traditionally, personality theorists tended to believe that personality was determined by socialization practices associated with variations in the environment of different families. The personality of one's parents, the atmosphere of the home, and the ways in which children were fed, toilet trained, and disciplined were among the influences stressed by psychologists as critical for the development of personality. Behavior–genetic research indicates that all of these influences can be eliminated as determinants of individual differences in personality traits. We can only speculate about possible within-family environmental influences. Among the variables that might be important are prenatal influences, birth traumas, and illnesses. There may be important variations in the actual or perceived socialization experience encountered by children in the same family. One child may be favored or treated more leniently than another. It should also be noted that the term "within-family environmental influence" is a misnomer, because it encompasses events that occur outside of the context of the family. These might include the influence of friends, lovers, books, religious experiences, and a huge array of potential environmental events that occur outside the home. We have virtually no knowledge about the potential influence of any of the above types of within-family environmental events on individual differences in personality traits.

Our knowledge of the way in which genotypes influence personality is equally speculative. One account suggests that genotypic influences develop from passive to active over time. Passive influences refer to genotypically influenced variations in response to events encountered by an individual. Such variations might lead individuals to be treated differently and to experience different environments. In the most active influence, an individual might learn to select environmental niches that are compatible with his or her genotypically influenced personality characteristics. For example, a child might be born with a genetic disposition to be fearful. Such a child is likely to find the world threatening. His or her peers, observing the child's fearfulness, may become threatening and select the child as a victim. In this example the genotypes would influence the environmental events encountered by the child. The child might eventually learn to structure his or her environment to avoid threats and fear-inducing events. This attitude in turn might influence such things as choice of friends, profession, and recreational activities. This example shows that we cannot understand the impact of the environment on individuals without taking into account the differences in the characteristics that individuals bring to the environment.

III. BIOLOGICAL BASIS OF PERSONALITY

A. Eysenck's Theory

Evidence that personality traits have a genetic basis suggests that traits must have a biological basis. That is, genes exert their influence on personality by influencing the structure and functioning of the nervous system. Among the more influential accounts of the biological basis of personality are those advanced by Hans Eysenck and Jeffrey Gray. Eysenck's theory was substantially revised in 1968 and it is that version that shall be considered here. Eysenck's theory provides a biological account of extraversion and neuroticism. We shall consider his theory of extraversion because it has received more research support. Eysenck assumes that individual differences in extraversion are related to individual differences in the arousability of a diffuse cortically arousing ascending reticular system in the brain. Introverts relative to extroverts would have a hyperarousable reticular system. Eysenck, borrowing a Pavlovian concept, assumes that there is a limit to cortical arousal. At this upper threshold of arousal, transmarginal inhibition occurs and is likely to reduce or prevent further cortical arousal. Since introverts have hyperaroused nervous systems, they are likely to encounter transmarginal inhibitions at lower levels of stimulation than extroverts. These assumptions imply that introverts are likely to be more aroused than extroverts when they encounter stimuli that are low in intensity and that extroverts are likely to be more aroused than introverts when they encounter stimuli that are high in intensity.

A number of studies using the galvanic skin response—a psychophysiological index of arousal—have provided support for Eysenck's theory. Thus, in studies in which individuals were exposed to tones of different loudness, introverts were found to have more intense galvanic skin responses than extraverts when loudness was low, and less intense responses when the loudness was high. In another study, in which stimulation was varied by manipulating caffeine dosage, introverts were found to exhibit more intense levels of galvanic skin response than extroverts when they were exposed to placebos or low- or medium-dose levels of caffeine. At high-dose levels of caffeine, extroverts had slightly more intense galvanic skin responses than introverts.

Extraverts and introverts, when exposed to stimuli that differ in arousal potential, not only exhibit different levels of psychophysiological arousal, but are also likely to respond at different levels of efficiency. For example, in a study in which auditory thresholds for the detection of sounds were measured under different conditions of light intensity, extroverts had slightly higher auditory thresholds at low levels of light intensity and lower auditory thresholds at high levels. As light intensity increased, extroverts tended to improve their performance (i.e., lower their thresholds), whereas introverts tended to decline in performance. Another experiment employed a vigilance task in which individuals were required to maintain vigilant attention and detect the occurrence of a randomly occurring visual stimulus. It was found that introverts were more vigilant and alert than extroverts when performing the task under relatively low noise levels, whereas extroverts were more vigilant and alert under conditions of relatively high noise levels. There is also evidence that introverts prefer situations that are lower in stimulation and arousal potential than extroverts.

This brief review of research indicates that there is some evidence supporting Eysenck's theory of extraversion with respect to psychophysiological indices, performance measures, and preferences for different situations. How do these differences relate to the fundamental definition of extraversion in terms of a preference to be with other people or to be alone? There is research suggesting that the mere presence of other individuals leads to the development of a hyperaroused state in the human organism. The higher level of arousal induced in introverts may be excessive, leading to inefficient performance. Parties and gatherings of large numbers of individuals are likely to be noisy and high in arousal potential, and therefore are likely to be more aversive to introverts than to extroverts. Thus, on this analysis, differences in the arousability of the nervous system in different situations may be used to explain the characteristic social behaviors of individuals who differ in extraversion.

B. Gray's Theory

Jeffrey Gray has proposed a biological theory of anxiety. The potential to experience anxiety is assumed to be related to neuroticism and extraversion. Individuals who are high on neuroticism and low on extraversion (i.e., neurotic introverts) are assumed to be subject to the development of anxiety. Gray assumes that anxiety is related to the hyperarousal of a behavioral inhibition system that is related to activity in the septohippocampal regions of the brain. Antianxiety drugs such as the benzodiazepines (commercially available as Librium and Valium), barbituates, and alcohol are all assumed to decrease the activity of the behavioral inhibition system. Similarly, lesions in the septohippocampal regions of the brain are also assumed to decrease the activity of the behavioral inhibition system. [*See* Hippocampal Formation.]

The behavioral inhibition system is activated by threats of punishment and loss of reward as well as by certain innate triggers. The system, as is implied by its name, is assumed to inhibit motor action, alert the organism, and induce a state of hyperattentional arousal that leads the organism to scan the environment for potential danger or potential aversive events. The system is not assumed to be directly related to learning or performance. Rather, its sole function is to stop action in order to permit an appraisal of potential dangers in the environment. Gray assumes that introverted neurotics, who have hyperarousable behavior inhibition systems, are more responsive than stable extroverts to potential negative effects in the environment.

Gray has attempted to extend his theory to an explanation of several neurotic conditions. He assumes that introverted neurotics are susceptible to anxiety attacks, phobias, and obsessive–compulsive behaviors. Each of these forms of neurotic behavior is related to the activity of the behavioral inhibition system. Anxiety states are defined as the phenomenological representation of the arousal of the behavioral inhibition system. Gray assumes that most phobias are triggered by a genetically programmed tendency for potentially dangerous stimuli to elicit activation of the behavioral inhibition system. He indicates that the stimuli that elicit phobias appear to be the same in most cultures. Introverted neurotics, by virtue of their hyperarousable behavioral inhibition system, are likely to respond with inhibitory responses and withdrawal when encountering stimuli that trigger potential phobic responses. Obsessive–compulsive behaviors are also assumed to be a consequence of arousal of the behavioral inhibition system. These responses are also assumed to be cross-culturally similar and to serve a common functions tendency to focus attention on potentially dangerous components of the enrvironment. Obsessive and compulsive thoughts and actions are assumed to reflect a preoccupation with potential dangers in the environment. Compulsions may be understood as a tendency to remove environmental contaminations. Obsessive thoughts may be understood as deriving from a heightened attention to potential dangers attributable to the arousal of the behavioral inhibition system.

It should be noted that Gray's theory of personality differs from Eysenck's in two principal respects. First, Gray places affect rather than generalized cortical arousal at the center of his analysis of differences between introverts and extroverts. Second, Gray assumes that the most important dimensions of personality are defined by the combination of extraversion and neuroticism. He contrasts neurotic introverts with stable extroverts and neurotic extroverts with stable introverts. Eysenck tends to treat neuroticism and extraversion as independent dimensions of personality.

C. New Directions

Marvin Zuckerman has attempted to update the theories of Eysenck and Gray by integrating them with the latest research on the functioning of brain mechanisms. For example, Zuckerman believes that Eysenck's neuroticism dimension and Gray's behavioral inhibition system are both part of a more general mechanism, which is located in the amygdala, that regulates emotionality. Zuckerman also emphasizes the significance of new research on the biochemistry of the human nervous system. The apparent effectiveness of the selective serotonin reuptake inhibitor fluoxetine in psychiatric treatments for hypersensitivity to rejection and other aspects of depression has stimulated interest in the relationship between neurotransmitters and personality traits.

C. Robert Cloninger has suggested that a new psychobiological model should be developed to link individual differences in genetically influenced personality dimensions with levels of specific neurotransmitters. Cloninger links dopamine to the personality dimension of novelty seeking, serotonin to harm avoidance, and norepinephrine to reward dependence. One implication of this new proposal is that the strategy previously used for identifying important personality dimensions should be reversed. Rather than starting with factor analysis of personality descriptions and then investigating the biological foundations of those dimensions, Cloninger's approach starts with basic biological findings about neurotransmitters and genetics and then links these biological factors to individual differences in personality and social behavior. Although drugs that psychiatrists use for treatment of various personality disorders do have some effects that provide support for Cloninger's model, further research is needed before his approach can be evaluated objectively.

Another new approach to the biological basis of personality goes further in challenging the use of personality dimensions derived from factor analysis. Jerome Kagan argues that studies of infants and young children reveal coherent profiles of emotion, behavior, and physiology that form temperamental categories. His approach focuses on types of people rather than on personality dimensions. For example, a profile of fearful and avoidant responses to unfamiliar events defines an inhibited category of children who make up about 15% of the population. Inhibited children tend to have highly reactive sympathetic nervous systems and high levels of the neurotransmitter norepinephrine and the stress hormone cortisol. Kagan speculates that the amygdala is one part of the brain that plays an important role in the physiology of temperamental inhibition, which is consistent with Zuckerman's suggestion that the amygdala deserves more attention than it received in Gray's theory. A novel point in Kagan's argument is his suggestion that inhibited and uninhibited children do not form the end points of a single quantitative continuum that defines a personality dimension. Instead, he proposes that these two types of children are qualitatively different in their physiological profiles and form two discontinuous temperamental categories. The idea of qualitative types as a way of categorizing people has been out of favor in personality psychology, but Kagan's suggestion that many more temperamental categories could be identified if researchers would look for them might well be proven correct in the future. In the meantime, Kagan has succeeded so far in demonstrating that the inhibited temperamental category identified during infancy is related to tendencies toward the personality traits shyness and introversion in later childhood, adolescence, and adulthood.

IV. CONCLUSION

This overview of personality has presented a conception of personality based on trait theory. This theory assumes that there are a limited number of basic personality dimensions that are influenced by genotypes, which in turn are assumed to influence the structure and function of the nervous sytem. It should be noted that other approaches to understanding personality do not assign a central role to biological concepts. The challenge currently facing personality psychologists is the task of integrating theories that emphasize biology with those that emphasize social learning and self-concept.

BIBLIOGRAPHY

Buss, A. H. (1995). "Personality: Temperament, Social Behavior, and the Self." Allyn & Bacon, Needham Heights, Massachusetts.

Cloninger, S. C. (1996). "Personality: Description, Dynamics, and Development." Freeman, New York.

Funder, D. C. (1997). "The Personality Puzzle." Norton, New York.

Kagan, J. (1994). "Galen's Prophecy: Temperament in Human Nature." Basic Books, New York.

Kramer, P. D. (1993). "Listening to Prozac." Penguin, New York.

Zuckerman, M. (1991). "Psychobiology of Personality." Cambridge Univ. Press, Cambridge, England.

Personality Disorders

SUSAN VAN METER
ALLEN J. FRANCES
Duke University Medical Center

GLOSSARY

Affective instability Rapid fluctuations of mood that last several hours to a few days

Cognition A person's quality of perceiving, recognizing, judging, reasoning, and imagining

DSM IV "The Diagnostic and Statistical Manual of Mental Disorders," 4th ed., is the currently used system to diagnose personality disorders

Electroencephalography Recording of electric potentials of the brain that is derived from leads placed on the scalp

Serotonin Specific chemical agent found in both the central nervous system and the peripheral tissues

PERSONALITY TRAITS ARE A CONSTELLATION OF enduring patterns of perceiving, thinking, and relating to the environment and oneself across a wide range of social and personal situations. When personality traits are inflexible with maladaptive patterns that result in subjective distress or impairment in social or occupational functioning, they constitute a personality disorder. The core feature of a personality disorder is a consistent pattern of inner experience and behavior that deviates significantly from accepted cultural norms and is manifested in at least two of the following: cognition, affective regulation, interpersonal functioning, or impulse control. This maladaptive pattern is stable and of long duration, with onset in late adolescence or early adulthood, and occurs across all aspects of the individual's life. Importantly, the observed behavioral pattern is not a consequence of another mental disorder or a general medical condition. Research has focused on determining the most appropriate ways to define personality disorders as well as elucidating their development, treatment, and impact on other illnesses.

I. HISTORICAL BACKGROUND

Personality disorders develop from the same factors that contribute to the development of normal personality. Biological, social, and psychological factors all contribute in varying degrees to different personality traits and disorders. Genotypic variability is adaptive from an evolutionary standpoint for the species, but at times can be maladaptive to the individual when some genetic variants are expressed. Data support genetic contributions to at least some personality disorders. Personality is also quite responsive to the environment, and maladaptive personality traits will at times be a result of pathologic familial, social, and/or economic factors. For example, it can be adaptive to be submissive and dependent in one's family of origin but not after one leaves home. Given the amount of time spent with and the importance of the family of origin, this behavior pattern will be slow to respond to changes in environment that occur with separation from the family unit. Some personality traits will be adaptive in one situation but not in others (e.g., aggressive competitiveness can contribute to success in a career but not in a marriage). It is then inaccurate to say that personality disorders are simply maladaptive, for they will often represent extreme or

ENCYCLOPEDIA OF HUMAN BIOLOGY, Second Edition, VOLUME 6.

atypical variants of adaptive genetic variation and/or behavior patterns that are adaptive in one situation or time but not in others.

Personality has been a major focus of theory and research for psychology since its founding as a science in the late 19th century. Psychologists have been attempting to identify and measure the basic dimensions of personality, which is not easy given that there are over 2700 different trait terms in the English language. Some have suggested that only two basic dimensions of control (dominance versus submission) and affiliation (love versus hate) are necessary to describe all variation in personality style, whereas others have distinguished between 810 different character types.

II. EPIDEMIOLOGY

The prevalence of personality disorders has been assessed in inpatient and outpatient psychiatric, medically ill, and community populations. These different subject samples as well as varied assessment techniques have yielded a wide range of rates for personality disorders. Most studies have been conducted on psychiatric inpatients who have additional psychiatric diagnoses. In these samples, rates as high as 67% have been reported, with many patients having two or more personality disorder diagnoses. Personality disorders are common in psychiatric outpatients, with rates ranging from 12 to 100%. In the medical setting, outpatients who are at risk for human immunodeficiency virus (HIV) infection have a 10% rate of personality disorders. There are no systematic studies using DSM-IV criteria that assess all personality disorders in a randomly selected sample of the general population. Earlier studies demonstrate that up to 30% of community samples have maladaptive personality traits. The overall rates of personality disorders range from 5 to 10% in the community. Antisocial personality disorder, one of the most rigorously studied personality disorders, has a prevalence of about 3% in the community.

III. DIAGNOSIS

Personality disorders differ quantitatively, not qualitatively, from normality because they frequently encompass behaviors which occur in diminished number or intensity in the normal population. The distinction between health and illness also varies among cultures that have different societal norms.

Personality traits represent a repertoire of reactions to the environment across many situations and at different points in time. *State* conditions reflect a person's condition at a given point in time. Many psychiatric conditions, e.g., depression, can significantly alter a person's interpersonal coping style. Upon remission from such a state condition, the individual's personality may be quite different from that observed when he or she was ill. It is crucial to determine if a patient's behavior represents a change from his or her baseline or is an enduring personality style. [*See* Depression.]

When determining if a personality style is maladaptive enough to be consistent with a personality disorder, attention must be paid to the context in which the behavior occurs. An observed behavior may be in response to the demands of a situation or social role instead of an enduring personality style. Cultural norms should also be considered when making a diagnosis of personality disorder.

DSM-IV utilizes a multiaxial approach in assessing mental disorders and the highest level of functioning of an individual. Axis I diagnoses include clinical syndromes such as schizophrenia, depression, substance abuse, and anxiety disorders. Axis II include personality disorders. Diagnoses on both axes are made if a person has coexisting conditions. There are 10 personality disorders which are grouped into three clusters.

Cluster A comprises the odd and eccentric personality disorders that are characterized by a pronounced difficulty in forming close attachments to others. This cluster includes schizotypal, schizoid, and paranoid personality disorders.

The schizotypal personality disorder is typified by strikingly odd behavior with cognitive or perceptual distortions. Cognitive distortions can include magical thinking such as the belief that they possess special powers or abilities. Perceptual changes include ideas of reference and perceptual miscues such as the sense of someone calling their name. Speech is usually digressive, vague, and idiosyncratic, but without true derailment. Appearance is often strange. Social situations are problematic and frequently a source of anxiety; interpersonal relationships are, by their own choosing, limited in number. Unlike schizophrenia, a break with reality with frank hallucinations is rare, usually occurring during periods of stress, with rapid recompensation.

An individual with schizoid personality disorder

tends to be a loner who does not form relationships or respond to others in a meaningful manner. A person with this disorder neither seeks nor enjoys close relationships with either friends or family. The constricted range of emotional expression in the schizoid patient is notable. Such a person frequently appears aloof and indifferent to the criticism or praise of others.

The patient with paranoid personality disorder has the pervasive and unwarranted tendency to interpret the actions of others as being purposefully demeaning or threatening. Without justification, such a person will question the loyalty of friends or peers. He or she will be easily slighted, tend to react with excessive anger, and bear grudges for a long time. Paranoid beliefs may result in hypervigilance, secretiveness, hostility, and anxiety. It may be difficult to distinguish this disorder from schizotypal personality disorder because they share a mistrust of others and social anxiety. However, schizotypal personality disorder tends to be associated with both perceptual and cognitive distortions.

The dramatic cluster includes histrionic, borderline, narcissistic, and antisocial personality disorders. Patients with these disorders frequently are flamboyant, impulsive, and seeking social contact. An individual with one of these disorders will frequently have traits, if not diagnoses, of another disorder in this group.

The person with histrionic personality disorder tends to be attention seeking, self-dramatizing, and excessively gregarious. Typically, there is excessive concern with physical appearance and a constant need for reassurance, approval, or praise from others. An individual with this disorder is usually very self-centered and becomes angry when they are not the focus of attention. Such an individual often behaves in an inappropriately sexual or provocative manner regardless of the situation. Emotional expression is shallow and mercurial. Speech tends to be dramatic but superficial.

Narcissistic personality disorder entails a pattern of an inflated sense of self-importance, a need for excessive admiration, and a lack of empathy. These individuals are preoccupied with fantasies of their own beauty, fame, power, wealth, and achievements. Their self-esteem is very fragile, which results in a need for constant attention from others and a heightened sensitivity to criticism. Interpersonal relationships are shallow and tend to be exploitive.

Antisocial personality disorder is characterized by a failure to conform to social norms with a disregard for the rights of others. Also called sociopathy and psychopathy, this diagnosis is given to individuals who are at least 18 years old and have evidence of conduct disorder in childhood. Conduct disorder occurs in childhood and encompasses a pattern of aggression, property destruction, theft, lying, school truancy, and fire setting. The pattern of antisocial behavior persists into adulthood with engagement in criminal activity with a profound lack of regret or empathy for others harmed by their actions. They can appear charming or seductive, usually in the service of initiating or maintaining a manipulative interpersonal relationship. They are often impulsive, with frequent abuse of illicit substances and alcohol, and precipitous changes in jobs, residences, or relationships.

Borderline personality disorder encompasses a behavior pattern of intense and chaotic relationships with fluctuating and extreme attitudes toward others. There is a pervasive fear of abandonment with actual or threatened loss promulgating affective instability, a distorted sense of self, and engagement in impulsive behaviors. The affective instability results in relatively rapid shifts in mood from a characteristic baseline dysphoria to anxiety, irritability, or depression. The disturbance in identity is reflected in frequent changes in self-image, long-term goals, and values. Impulsivity falls into one of two categories and is usually self-destructive. Activities such as substance abuse, binge eating, casual sexual relations, and gambling are frequently seen. The other type of impulsivity is seen in repeated suicidal behavior, suicidal threats, and self-mutilation.

The avoidant or anxious cluster includes dependent, obsessive-compulsive, and avoidant personality disorders. Dependent personality disorder is composed of a pattern of excessive reliance on others which is reflected in the tendency to permit others to make decisions, to feel helpless when alone, to subjugate one's needs and values to others, and to tolerate mistreatment. Such an individual requires excessive advice and reassurance with even everyday decisions. Submissive behaviors are used to secure caregiving from others. Individuals are plagued with self-doubt and tend to minimize or deny their own abilities. He or she will commonly live with a domineering, overprotective, and infantalizing person.

The avoidant patient is inhibited, introverted, and anxious. Additional features include low self-esteem, hypersensitivity to rejection, apprehension, social awkwardness, and a fear of being embarrassed. The criterion regarding social withdrawal overlaps with schizoid personality disorder; however, the schizoid patient is indifferent to others whereas the avoidant person desires relationships but is too shy to pursue them. The avoidant and dependent personality disorders share characteristics of interpersonal insecurity with the desire for relationships and low self-esteem.

The person with obsessive-compulsive personality disorder tends to be perfectionistic, constricted, and excessively disciplined. Behavior is rigid, formal, and intellectualizing. A preoccupation with trivial details or procedures often exists that hinders the individual from having a broad view of a situation. Decision making is very difficult and is often postponed or avoided to the point of interfering with task completion. The need for order and perfection pervades the individual's life and can be seen in their approach to leisure activities and interpersonal relationships. An individual with this disorder resists the authority of others while unreasonably insisting that others agree with their views.

DSM-IV encourages the listing of personality traits if an individual does not meet the criteria necessary for a personality disorder. In addition, an individual can have more than one axis II disorder. These provisions reflect the fact that all individuals have certain personality traits and that patients with personality disorders frequently possess more than one type of axis II pathology.

IV. ETIOLOGIC MODELS

Different theories exist that attempt to explain the development of personality and personality disorders. Much of the supporting evidence for these hypotheses is theoretical because it is often very difficult to test them in a controlled experimental situation. At this time, there is no definitive model that has been proved to be more correct that the others. Earlier writings emphasized either environmental or biologic factors as causative, but more recent literature proposes that personality development depends on the interaction of an individual's constitution and experiences within and without the home.

A. Psychoanalytic Theory

Psychoanalytic theory proposes that an infant is born with instinctual attachments and sexual and aggressive drives. Personality development is a process in which the child gains increasing control of his instinctual impulses by learning and developing identifications with important caretakers. A successful effort for adaptation results in normal behavioral tendencies. If a child's needs are too intense to be satisfied adequately or if expected needs are not satisfied because of a lack of response from the environment, the resulting experiences may serve as a trauma with resultant "fixation" points which impede subsequent maturation. Disturbances in development at particular developmental phases will result in specific characteristic personality formations. The quality of interactions between the individual and their environment results in unconscious conflicts which influence behavior. [*See* Psychoanalytic Theory.]

B. Social Learning Theory

Social learning theory posits that social experiences, particularly those occurring early in one's development, are the major determinants of behavior and personality. This theory places less emphasis on one's constitution and more on the environment. Specific social experiences result in an individual's behavior, cognition, and affect. For learning theorists, attention is directed to the careful identification of conditions that determine how behavior is learned and unlearned. These early experiences have great impact in early life and are more difficult to modify than those experiences occurring later in life. The developing child learns not only from exposure to and instruction from his parents, but also from events observed in their environment. For example, styles of interpersonal communications are obtained from observation of everyday events. Excessive, insufficient, erratic, or inconsistent methods used by parents to modulate a child's behavior will result in maladaptive personality patterns. For example, the child will become socially avoidant and withdrawn if he or she cannot meet excessive parental demands and harassments.

C. Temperament

Very young infants tend to differ from one another in biologic functions, reactivity to stimuli, adaptability to change, characteristic moods, distractibility, and persistence. These innate endowments constitute each person's temperament, which interact with the environment in ways that accentuate initial behaviors. An individual's temperament appears to endure during the course of development. A good match between a child's temperament and environment fosters a healthier personality development.

Genetic studies are very useful ways to determine the relative contribution of environment and heredity to personality. The literature investigating the genetic influences on personality has used different methods and patient populations, including twins reared together and apart, and first-degree relatives of patients with personality disorders. Family studies suggest that schizotypal and paranoid personality disorder patients have a familial association with schizophrenia. Borderline personality disorder appears to run in families and to have an association with mood disorders, alcoholism, and other substance abuse disorders.

Twin studies have determined that both environment and genetics contribute to the development of personality traits. The genetic influence appears to account for approximately 50% of measured personality traits. Most of the environmental contribution to personality is due to an individual's specific environment rather than to a common familial milieu.

D. Biology

Several advances in biological techniques have helped to elucidate the biological substrates of personality. Many of these techniques were initially used to investigate the biological correlates of depression and schizophrenia. Patients with personality disorders and maladaptive extremes of certain traits differ from control groups in a number of different biologic parameters. Several studies have shown an inverse correlation between central serotonin levels and impulsivity and physical aggression. These behaviors underlie both certain personality dimensions as well as personality disorders, particularly borderline and antisocial personality disorders. Borderline and antisocial patients exhibit abnormal brain functioning as measured by electroencephalography. Schizotypal patients demonstrate abnormal results in tests measuring information processing and eye movements. Neuroendocrine abnormalities have been noted in patients with borderline personality disorder. [*See* Depression: Neurotransmitters and Receptors.]

Although most of these studies are being used for research purposes, biologic tests such as those described earlier may in the future aid in personality evaluation and lend credence to the role of biologic substrates in personality characteristics.

V. PERSONALITY DISORDERS AND OTHER DIAGNOSES

By definition, the presence of a personality disorder impairs one's ability to adjust to stress. Therefore it is to be expected that personality disorders will have an impact on both psychiatric and medical conditions. Personality pathology influences the predisposition, presentation, course, and treatment response of several psychiatric illnesses, including schizophrenia, mood, anxiety, eating, and substance abuse disorders. The exact nature of the relationship between personality disorders and the major syndromes can be conceptualized in different ways. Certain personality traits may (1) predispose, (2) modify, (3) represent a complication of, (4) represent an attentuation of, or (5) coexist independently with specific axis I disorders.

Borderline, histrionic, dependent, and avoidant personality disorders frequently occur in patients with major depression. Conversely, individuals with borderline, histrionic, dependent, avoidant, antisocial, schizotypal, and obsessive-compulsive disorders have an increased incidence of mood disturbances. The presence of character pathology in depressed individuals may be associated with major depression alone. Specifically, the presence of borderline character disorder is associated with an increased incidence of anxiety and substance abuse, as well as attempted and completed suicide. The presence of personality pathology in general has been correlated with earlier onset and poorer treatment outcome of depression.

It is difficult to ascertain if certain personality disorders predispose to or represent a complication of schizophrenia because of the chronic course and usual long prodrome of this illness. Schizotypal and paranoid personality disorders have a familial association with schizophrenia.

Borderline, antisocial, and histrionic patients often exhibit impulsive behavior as evidenced by their increased use of alcohol and other psychoactive substances. In the histrionic or borderline patient, substance abuse may represent an attempt to alleviate the symptom of affective instability. For the antisocial patient, substance abuse may result from or initiate an antisocial lifestyle. Complications of substance abuse in such patients include increased impulsivity and self-destructiveness, particularly attempted and completed suicide.

Individuals with avoidant personality disorder may develop simple phobias, generalized anxiety disorder, or panic disorder. Social hypersensitivity and perception of threat predispose schizotypal and paranoid patients to develop anxiety disorders.

Debate exists regarding the relationship between personality disorder and the onset and course of medical illnesses. Nonetheless, the occurrence of serious medical illness interacts with and may aggravate the personality traits possessed by any patient. Medical illness represents a particularly difficult event for patients with personality disorders because of their impaired ability to deal adaptively with stress. A patient's personality traits also affect the quality of the doctor–patient relationship, acceptance of a treatment plan, and treatment outcome. Assessment of the patient's personality optimizes the ability to cope with a medical condition. The compulsive patient will do best with a physician who invites him or her to be an active participant in the treatment plan. Conversely, the dependent patient may benefit best from a relationship in which the physician makes the decision.

VI. TREATMENTS

The various different treatments of personality disorders include psychodynamic, supportive, interpersonal, behavioral, cognitive, and pharmacologic therapies.

There is very little empirical basis to the choice of a particular treatment option, and frequently pharmacotherapy is combined with a particular psychotherapy. The interpretation of results is also difficult because these patients frequently have more than one personality diagnosis and a coexisting axis I disorder.

Unlike the other therapies, which are more target symptom oriented, psychodynamic psychotherapy focuses on changing the patient's personality structure. Therefore, this therapy usually strives to decrease the inflexible quality of the maladaptive traits and to increase the self-awareness of behaviors in order to improve interpersonal functioning. This treatment also attempts to reduce the automatic quality of behavioral responses that result from unconscious conflicts. Patients with dependent, obsessive-compulsive, avoidant, and histrionic personality disorders are most likely to benefit from this form of therapy. A more specific psychodynamic psychotherapeutic treatment targeting specific personality disorders, primarily borderline, narcissistic and schizoid disorders, has been developed, with the use of specific interventions depending on the personality pathology. [*See* Psychotherapy.]

Supportive psychotherapy attempts to aid the patient's coping skills without changing his or her personality structure. Supportive treatment can help the patient through periods of medical, interpersonal, or occupational stresses. For instance, admiring the narcissistic patient, providing information to the obsessive-compulsive patient, or maintaining distance from the schizoid patient can obtain these goals.

Interpersonal psychotherapy attempts to improve the patient's skills in establishing and maintaining relationships. Such an approach depends on the fact that personality styles often elicit complementary responses in others. Such an approach may be particularly suitable for personality disorders as these conditions often represent maladaptive interpersonal styles. In this treatment, the therapist assumes an interpersonal style that encourages more adaptive and flexible functioning in the patient in order to diminish the usual pattern of mutually debilitating relations.

Behavior therapy promotes the patient's involvement in more adaptive and pleasurable activities. Behavior techniques involving assertiveness training and graded exposures for social anxiety may be useful for dependent and avoidant patients, respectively. The schizoid patient may benefit from shaping of his or her social behavior.

Cognitive therapy focuses on diminishing central, irrational assumptions causing patient's beliefs and behaviors. Such a therapeutic approach may be particularly useful for personality disorders which frequently include specific cognitive styles. The borderline patient may benefit from addressing the exaggerated attitudes of others in order to develop more realistic perceptions. Cognitive techniques that focus on assumptions of threat or inadequacy may be helpful with the avoidant and dependent patient. The obsessive-compulsive patient may be aided by addressing his or her irrationally rigid, severe beliefs and moral standards.

Dialectal behavioral therapy is a highly specialized form of therapy involving a systematic and integrated approach of both cognitive and behavioral techniques. This therapy targets specific symptoms of borderline personality disorder, including suicidality, with an emphasis on a therapeutic balance between acceptance and change.

Pharmacotherapy may be useful when it focuses on such features as affective dysregulation in borderline personality disorder, perceptual disturbances of schizotypal disorder, or the anxiety of avoidant personality disorder. Low-dose neuroleptics have been found to be useful for the anxiety and cognitive disturbances of schizotypal and borderline personality disorders. Antidepressants and occasionally mood stabilizers such as lithium have been useful for the mood symptoms frequently seen in individuals with axis II disorders. Care in the choice of drug treatment is needed as there is a possibility of impulsive misuse or use of the prescribed medication in a suicide attempt. This is especially of concern in individuals with borderline and antisocial personality disorders.

VII. CONCLUSIONS

Everyone has a personality that imparts a particular and somewhat predictable quality to their relationships, responses to the environment, and cognitive style. A personality disorder exists when a specific constellation of personality traits causes pervasive and enduring maladaptive behaviors across many situations. These disorders rarely occur in isolation and impact on both other psychiatric diagnoses and medical illness. A complex interaction between the individual and their environment results in these conditions. A growing literature is elucidating the underpinnings of these conditions as well as treatment strategies.

BIBLIOGRAPHY

American Psychiatric Association (1994). "Diagnostic and Statistical Manual of Mental Disorders," 4th Ed. American Psychiatric Association, Washington, D.C.

Dahl, A. A. (1994). Hereditary in personality disorders: An overview. *Clin. Genet.* **46**, 138–143.

Fisher, F. D., and Leigh, H. (1986). Models of the doctor–patient relationship. *In* "Psychiatry" (R. Michels, J. O. Cavenar, H. K. H. Brodie, A. M. Cooper, S. B. Guze, L. L. Judd, and G. L. Klerman, eds.). Lippincott, Philadelphia.

Frances, A. J., and Widiger, T. (1986). The classification of personality disorders: An overview of the problems and solutions. *Psych. Assoc. Annu. Rev.* **5**, 240–258.

Liebowitz, M. R., Stone, M. H., and Turkat, I. D. (1986), Treatment of personality disorders. *Am. Psych. Assoc. Annu. Rev.* **5**, 356–393.

Livesley, W. J., Jang, K. L., Jackson, D. N., and Vernon, P. A. (1993). Genetic and environmental contributions to dimensions of personality disorder. *Am. J. Psych.* **150**, 1826–1831.

Marin, D. B., Frances, A. J., and Widiger, T. A. (1990). Personality disorders. *In* "Clinical Psychiatry for Medical Students" (A. Stoudemire, ed.). Lippincott, Philadelphia.

Merikangas, K. R., and Weissman, M. W. (1986). Epidemiology of DSM-III axis II personality disorders. *Am. Psych. Assoc. Annu. Rev.* **5**, 258–277.

Oldham, J. M. (1994). Personality disorders: Current perspectives. *J. Am. Med. Assoc.* **272**, 1770–1776.

Siever, L. J., Friedman, L., Moskowitz, J., Mitropoulou, V., Keefe, R., Roitman, S. L., Merhige, D., Trestman, R., Silverman, J., and Mohs, R. (1994). Eye movement and schizotypal psychopathology. *Am. J. Psych.* **151**, 1209–1215.

Siever, L. J., Klar, H., and Coccaro, E. (1985). Psychobiologic substrates of personality. *In* "Biologic Response Styles" (H. Klar and L. Siever, eds.). American Psychiatric Association, Washington, D.C.

Widiger, T. A., and Costa, P. T., Jr. (1994). Personality and personality disorders. *J. Abnorm. Psychol.* **103**, 78–91.

Widiger, T. A., and Hyler, S. E. (1986). Axis I/II interactions. *In* "Psychiatry" (R. Michels, J. O. Cavenar, H. K. H. Brodie, A. M. Cooper, S. B. Guze, L. L. Judd, and G. L. Klerman, eds.) Lippincott, Philadelphia.

Winston, A., Laikin, M., Pollack, J., Samstag, L. W., McCullough, L., and Muran, J. C. (1994). Short-term psychotherapy of personality disorders. *Am. J. Psych.* **151**, 190–194.

Person Perception

JOHN N. BASSILI
University of Toronto

GLOSSARY

Actor–observer effect The tendency to attribute other people's behaviors to their personalities while attributing our own to features of the situation

Associative networks Metaphor used to characterize the mental representation of information in terms of *nodes* and *links*

Attributions The inferences people make to explain behavior

Construct accessibility The state of readiness of our social knowledge structures for processing incoming information

Discounting principle The notion that confidence in the role of a cause in producing an effect decreases if other plausible causes are also present

Display rules Rules about the appropriateness of emotional expressions in particular situations

Implicit personality theory The informal assumptions people make about relationships between character traits

Impression management The processes people use to control self-relevant information

Information integration The notion that impressions require the integration of information about others and the models of how this integration is effected

Memory-based judgments Judgment processes that require the recall of information stored in memory

On-line judgments Judgment processes that occur during the encoding of information

RESEARCH ON PERSON PERCEPTION EXPLORES how we form impressions of people and how we use these impressions to make judgments about them. At times, impressions are subjective, involving assumptions based on the way people look or on hear-say information. At other times, impressions involve scrupulous efforts at objectivity, such as when a jury has to decide the guilt or innocence of an accused. However impressions are arrived at, they play a central role in social functioning because they represent an important aspect of people's understanding of their social world.

I. THE CHALLENGE OF PERSON PERCEPTION

The aim of the perceiver is usually to form an accurate impression of another person. Two elements of person perception pose a challenge in reaching this goal. The first is that unlike inanimate objects, people care about the impressions others form of them and take an active role in controlling the information about themselves that others are permitted to see. This process is known as *impression management.* To form accurate impressions of others, the perceiver must see through the impression management efforts of others, a task that is not always easy or successful. The second challenge for the perceiver is created by the indirect nature of person perception. The elements of the impressions we form of others include character traits, intentions, interests, abilities, moods, and attitudes. None of these properties can be perceived directly. There is no look, sound, or feel to the trait of kindness, the intention to help, or the state of empathy. These properties have to be inferred from outward manifestations such as actions, demeanors, verbalizations, and expressions.

ENCYCLOPEDIA OF HUMAN BIOLOGY, Second Edition, VOLUME 6.

II. SURFACE INFORMATION: APPEARANCES, EXPRESSIONS, AND DEMEANORS

A. Assumptions Based on Appearance: The Case of Physical Attractiveness

Appearance often provides the first information we have about another person. A person's sex, height, weight, and age, as well as the way he or she is dressed and groomed, are readily apparent. This information can lead the perceiver to make strong assumptions about the person.

Several studies have demonstrated that we make positive assumptions about physically attractive people. In one study, subjects were given pictures of physically attractive people, some who were average in attractiveness, or some who were unattractive. The subject then rated the kindness, sensitivity, interestedness, strength, poise, modesty, sociability, and sexual warmth of the person depicted in the picture. Averaging across these characteristics, it was found that attractive persons were rated highest and unattractive persons, lowest. This bias in favor of attractive people extended to ratings of assumed occupational status, marital competence, and happiness.

The tendency to generalize from one positive characteristic, such as attractiveness, to the whole of the person is called the *halo effect*. Early research on the effects of physical attractiveness on person perception uncovered so many characteristics favoring attractive people that some researchers concluded that attractiveness surrounded the person with a positive aura. It was found that physically attractive people are more likely to have their written work evaluated positively, to be recommended for hiring following a job interview, to receive lighter sentences from mock juries, and to be seen as effective psychological counselors. The bias was found to affect the perception of men and women equally and to extend to the perception of children. Adults, for example, judge transgressions committed by attractive children less severely than the same acts committed by unattractive children. Together, these findings suggest that in the eyes of the perceiver, what is beautiful is good.

An accumulation of research findings on the effects of attractiveness on perception, however, suggests that the notion that what is beautiful is good is too general and undifferentiated. With the many positively biased assumptions made about attractive people, there are negative assumptions and evaluative characteristics on which no bias is shown. Attractive people are dealt with harshly by mock juries if it appears that they exploited their looks in the commission of their crime, as in a swindlle or con game. Moreover, studies analyzing ratings of personality characteristics have found that physical attractiveness has its impact primarily on ratings of sociability, social competence, and popularity. The impact of physical attractiveness on strength, adjustment, and intelligence is only moderate and it is negligible on integrity and concern for others.

The heterogeneity of perceptions of attractive people suggests that the halo effect metaphor is inappropriate. Instead, the assumptions we make about attractive people stem from implicit personality theories. These are informal theories we hold about the kinds of personal characteristics that tend to go together in people. In the same way that we assume intellectuals are introverted and idealistic, we also assume that attractive people are sociable and popular. These assumptions are fairly specific and reflect the links that exist in our implicit personality theories between attractiveness and other trait attributes.

The assumptions we make about physically attractive people are not necessarily inaccurate. *Self-fulfilling prophecies* describe the tendency to act on assumptions in such a way as to elicit confirmatory behavior from other people. It has been shown, for example, that college men who phoned a college woman they assumed to be attractive exhibited more socially skilled verbal behavior and elicited more socially skilled verbal behavior from her than when they assumed that she was unattractive. It is possible that the positive attention attractive people receive increases their self-confidence and provides them with opportunities to hone their social skills.

B. The Perception of Emotions and the Detection of Lies

Facial expressions are particularly rich in their potential for expressing emotions. Consider the faces shown in the top panel of Fig. 1. Even if these pictures were

FIGURE 1 Posed expressions of six basic emotions. The bottom panel shows time exposures of blackened faces covered with white dots. The camera shutter was opened just before the beginning of an expression and was closed just after the emotion reached its apex. [The bottom panel is from Bassili, J. B. (1979). Emotion recognition: The role of facial movement and the relative importance of upper and lower areas of the face. *J. Pers. Soc. Psychol.* 37, 2049–2058. American Psychological Association, reprinted by permission of the publisher.]

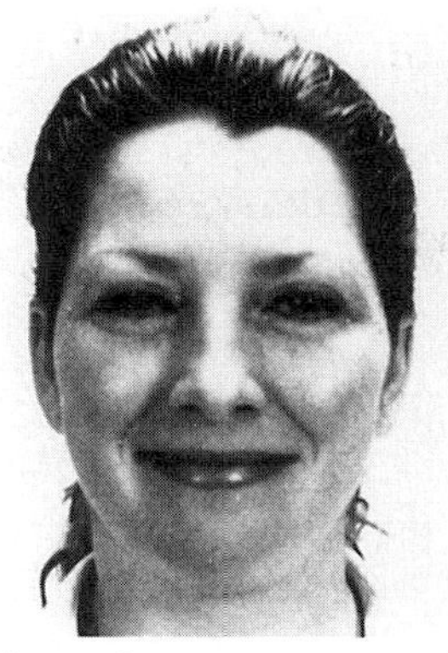

Happiness

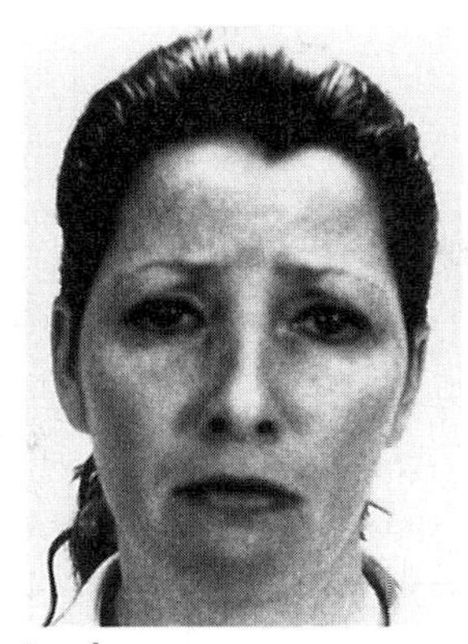

Sadness

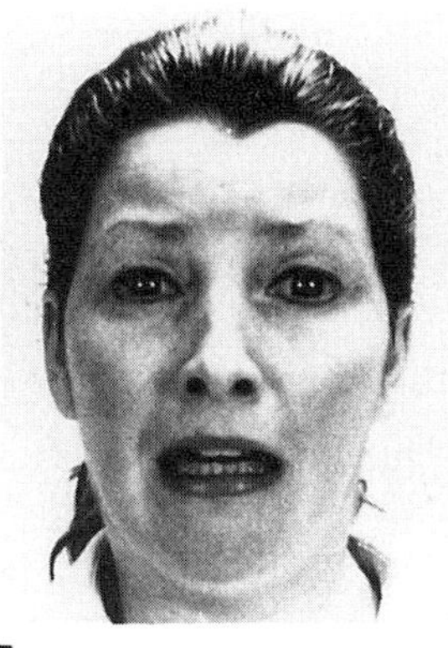

Fear

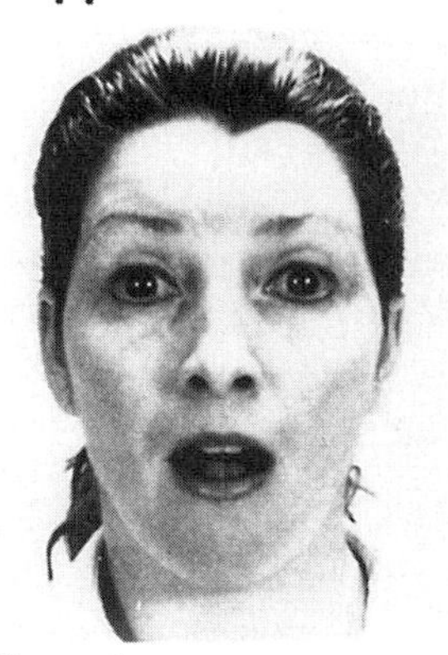

Surprise

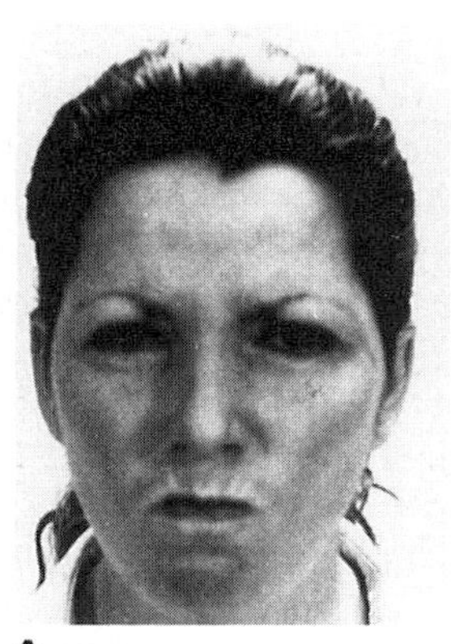

Anger

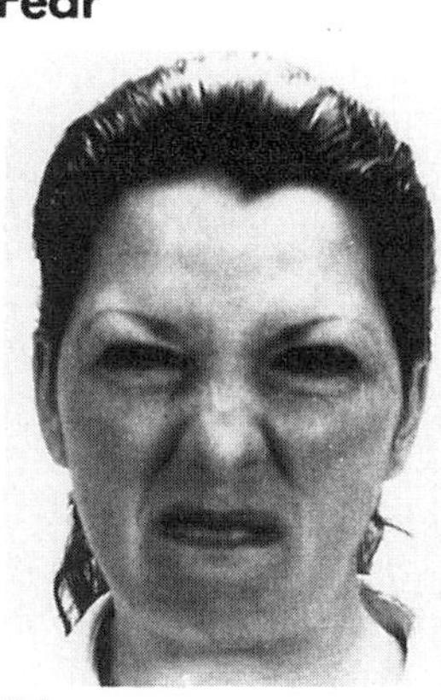

Disgust

Happiness

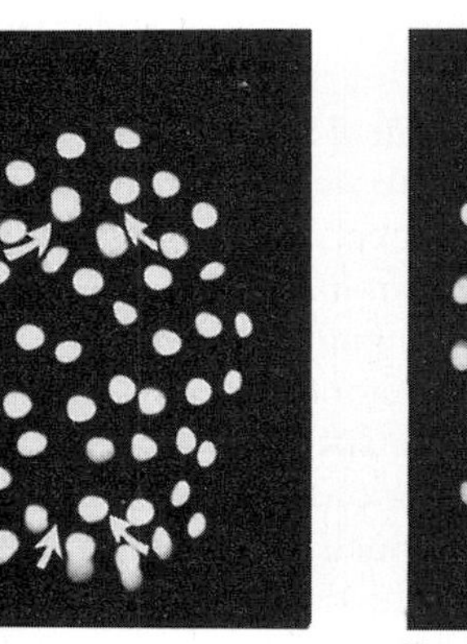

Sadness

Fear

Surprise

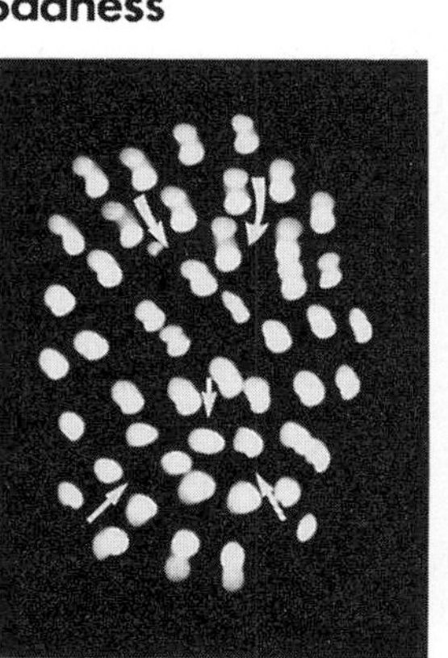

Anger

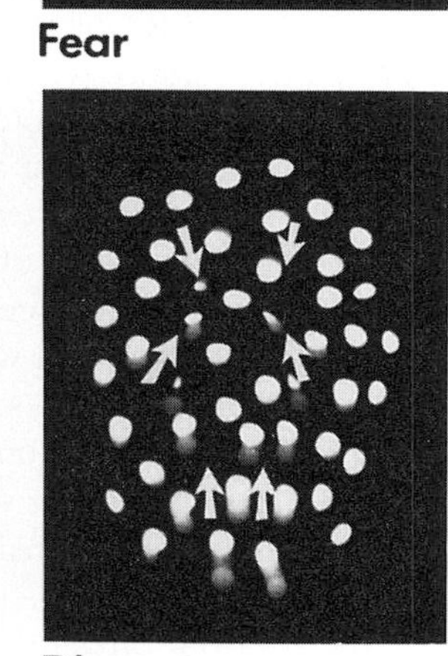

Disgust

not labeled, you would have no difficulty identifying each of the emotions expressed. Research has demonstrated that people are highly accurate at recognizing the posed expressions of happiness, sadness, fear, surprise, anger, and disgust. It is not difficult to understand why this is the case. The expressions shown in Fig. 1 are very rich in information. Each expression contains a number of distinctive elements involving the shape of facial features (e.g., the smiling mouth in happiness) and the presence of wrinkles on the forehead in fear.

In addition to static information in these pictures, research has shown that dynamic motion information is also distinctive in each of the expressions and provides further cues for their identification. The white dots shown in the bottom panel of Fig. 1 are randomly placed on blackened faces. When stationary, these dots do not reveal much about the surface on which they are placed. Observers cannot even guess that they are placed on faces. This indicates that all feature information in normal photographs is eliminated in the blackened faces. Yet, when videotapes of these faces are shown to viewers, they can immediately tell that they are looking at faces and are remarkably accurate at recognizing each of the six emotions. Static and dynamic facial information together provide rich cues for the recognition of emotions.

In his theory of evolution, Darwin argued that there was great survival value to the accurate signaling of emotional states. Darwin proposed that because there is a link between the signals used by subhuman primates and those used by humans, these signals are universal. Research on the universality of emotional expressions has produced findings consistent with Darwin's position. In fact, the six emotional expressions shown in Fig. 1 are recognized much the same way across cultures. The possibility that this consistency is caused by exposure to western facial expressions is ruled out by a study conducted in a remote part of New Guinea with people who had no contact with western culture. Brief stories describing an emotion-inducing situation were read to the New Guineans, who were asked to select from a number of pictures the one that depicted the emotion. On average, the choices made by members of this group agreed with those of western subjects in over 80% of the cases. Only in the case of fear and surprise was the confusion rate high in the New Guinean group, a finding that suggests surprising situations may generally have a fearful component in this culture.

The impressive recognition rates of expressions such as those shown in Fig. 1 should not be taken as evidence that people are always accurate at perceiving the emotional states of others. Matching six expressions with six corresponding emotion labels is not nearly so challenging as identifying emotions in everyday life, for several reasons. First, the emotional expressions used in research, eliciting impressive recognition rates, are usually posed by highly trained actors. Second, these expressions are pure in that they convey a single emotion rather than complex blends (such as fearful surprise in reaction to a disgusting stimulus). Third, and most important, emotional expressions are often influenced by learned *display rules* about the appropriateness of facial expressions in various situations.

The effect of display rules on emotional expressions was demonstrated in research with bowlers. The facial expressions of bowlers were observed to discover whether smiling resulted from happiness with a successful performance or from social engagement with partners. If a bowler is happy following a good roll, then the frequency of smiles should be higher following a strike or a spare than following a poor roll. In fact, the association between performance and smiling was weak. Smiling was found to depend primarily on whether the bowler was engaged in social contact with his or her companions.

Perhaps the most instructive situation with respect to the accurate identification of emotions involves the detection of deceit in situations in which the liar is anxious about being apprehended.The criminal under investigation and the smuggler questioned by a customs officer usually experience fear. Yet the facial expression presented in such situations seldom resembles the expression of fear shown in Fig. 1. Under such circumstances, people mask the fear they actually feel and often simulate other feelings such as relaxation or joviality.

How successful is the falsification of emotional expressions? In a realistic and well-controlled study, passengers about to board a plane were given fake contraband to smuggle past a customs inspector. A randomly selected control group of passengers were given no contraband. The "smugglers" and the "innocent" participants were videotaped during the interview with the customs inspectors. These interviews were later shown to undergraduate students, who coded the behavior of the travelers and also tried to pick out the smugglers among them.

The study yielded several interesting results. The verbal and nonverbal behavior of the smugglers did not differ systematically from those of the innocent travelers. This indicated that smugglers were successful in masking the cues that would have given away their guilt. Customs inspectors were no more likely to search smugglers than innocent travelers, and undergraduate students who later viewed the videotapes were also unable to pick them out. Despite this lack of accuracy, customs inspectors and undergraduate observers agreed on who they thought was smuggling contraband. Both groups relied on the same cues to decided who was guilty. Specifically, travelers who paused before giving their answers, who gave short answers, who avoided eye contact with the customs inspector, and who shifted their bodies a lot were deemed to be hiding something, even though these cues were not reliably connected with actual guilt.

This study illustrates that people can mask their true feelings with success and that perceivers face a serious challenge in trying to see through the mask. In fairness, the passengers who served as subjects in this study did not experience the level of anxiety a real smuggler would experience. The subjects, therefore, may have had an easier time controlling their self-presentations. It is likely that intense emotions tend to leak, despite efforts at masking them. Research suggests that the lower body is more likely than the face to carry cues of leakage. A cautious assumption about accuracy in the recognition of emotions in general and of deceit in particular is that people differ in their skills at falsifying the information they present to others and perceivers differ in their skills at seeing through falsification.

III. UNDERSTANDING THE CAUSES OF BEHAVIOR

The cues we gather from appearances, expressions, and demeanors form only a small part of the information we use in person perception. When we go beyond the snap judgments we form on the basis of these cues, we usually pay increasing attention to a person's actions. The inferences we make to explain behavior are called *attributions*. In 1958 Fritz Heider introduced the field of *attribution theory* by arguing that to function successfully in the social environment, a person has to relate the actions of others to their personal characteristics and to characteristics of the situation. It matters, for example, whether a boss attributes an employee's lateness to exceptional traffic conditions or to the employee's lack of commitment to the job, or whether a husband attributes his wife's frequent trips to a city to the presence of a large business account or to that of a lover.

A. The Covariation Principle

At the heart of attribution theory is the notion that people tend to attribute effects to causes with which they covary. The method is analogous to John Stuart Mill's method of difference, namely that the cause of an event resides in the conditions present when the event occurs and absent when it does not.

In 1967 Harold Kelley formulated an important model of attribution based on the principle of covariation. According to this approach, there are three causes for social behavior: the *person* who performed the behavior, the *entity* to which the person responded, and the *circumstances* that surrounded the behavior. Consider the case of Paul, who laughs uproariously at a comedian at an office party. An observer could attribute Paul's behavior to something about him (a giddy personality), something about the entity (the comedian being funny), and/or something about the circumstances (the festive mood of the party).

Kelley suggested that perceivers examine three types of information in their efforts to explain behavior: *distinctiveness* (does the person respond to other entities in the same way?), *consensus* (do most other people respond to this entity in the same way?), and *consistency* (does the person respond to this entity in the same way in different circumstances?). In our example, Paul's behavior would be low in distinctiveness if he responds with uproarious laughter to other comedians, whereas it would be high in distinctiveness if he does not easily laugh at other comedians. Consensus would be low if few other people laughed at this comedian, whereas it would be high if most other people laughed at this comedian. Finally, consistency would be low if Paul did not laugh at the comedian at other times (such as when the comedian appeared on television or at a nightclub), and it would be high if Paul laughed at the comedian at other times.

To make an attribution, people look at the pattern of information formed by distinctiveness, consensus, and consistency. Three patterns are particularly important. First, high distinctiveness, high consensus, and high consistency tend to yield an entity attribution. That is, if Paul laughs at few other comedians, if most other people laugh at this comedian, and if Paul laughs at this comedian every time he sees him, then the preceiver will attribute Paul's laughter at the office party to the talent of the comedian. Second, low distinctiveness, low consensus, and high consistency tend to yield a person attribution. That is, if Paul laughs at all comedians, if most other people do not laugh at this comedian, and if Paul laughs at this comedian every time he sees him, then the perceiver will attribute Paul's laughter to something about Paul (his giddiness). Third, high distinctiveness, low consensus, and low consistency tend to yield a circumstance attribution. That is, if Paul laughs at few other comedians, if few other people laugh at this comedian, and if Paul does not laugh at this comedian on other occasions, then the perceiver will attribute Paul's laughter to something about the particular circumstances at the office party.

It is instructive to relate the information contained in distinctiveness, consensus, and consistency to the covariation principle. Distinctiveness represents variation in the entity (the comedian), consensus represents variation in the person (Paul), and consistency represents variation in the circumstances (the office party). According to the covariation principle, effects are attributed to the causes with which they covary. High distinctiveness indicates that the laughter disappears when the comedian is different, thus contributing to the impressions that it is something about the comedian that caused Paul's laughter. Low consensus indicates that the laughter disappears when the person is different, thus contributing to the impression that it is something about Paul that caused his laughter. Low consistency indicates that the laughter disappears when the circumstances are changed, thus contributing to the impression that it is something about the office party that caused Paul's laughter.

The covariation principle of attribution, as it is embodied in Kelley's model, has received strong empirical support. A number of studies have presented subjects with scenarios describing a person's reaction to an entity in a particular situation and have manipulated distinctiveness, consensus, and consistency information. In general, subjects' attributions to each of the three causes have been consistent with the predictions of the model.

B. Correspondent Inference

Kelley's model provides a comprehensive analysis of causal inference. There is one class of attribution, however, that is particularly important in person perception—attributions to people's personality characteristics. The theory of correspondent inference, postulated by Edward Jones and his colleagues, focuses on these attributions. Correspondent inference refers to the perceived congruence between a behavior and an underlying personality characteristic. Suppose you saw a person behaving in a friendly manner. Under what conditions would you infer that the person has a friendly disposition?

The first step in making correspondent inferences is to infer whether the behavior was intended. Under most circumstances, we do not learn much about other people from behaviors that they do not intend. The case of accidental behavior is particularly relevant here. An intended shove may lead to the inference of an aggressive disposition, but a shove resulting from a loss of balance normally does not. There are exceptions to this rule. Traits such as clumsiness, forgetfulness, and insensitivity result in behavior that is largely unintended. Thus, while intention is necessary to the inference of most dispositions, there are cases in which the disposition itself implies that the person does not have control over the behavior.

Having established the presence or absence of intention, the perceiver next attempts to determine what trait caused the behavior. A common way for the perceiver to make trait inferences is to focus on the effects of behavior. This is illustrated best in situations where a person chooses between paths of action that have different effects. Imagine a person contemplating the purchase of one of two cars. Both cars are known for their performance and reliability, but one has attractive, conservative lines, while the other is bedecked with wings and scoops that give it an eye-catching appearance. The person decides to purchase the showy car. According to correspondent inference theory, a perceiver would infer that the person likes getting attention. This is because the attention-getting property of the car is the only effect that is not common to the two alternatives. The fact that the chosen car has good performance and reliability is not particularly informative because the other car also has these characteristics. The theory of correspondent inference also predicts that the perceiver will make more confident inferences when there are few rather than many effects that are not common to the alternatives. If, in our example, the showy car was also cheaper, smaller, and more economical than the conservative car, then the perceiver could not have attributed its purchase specifically to its eye-catching quality.

Predictions of the model have received good empirical support. Subjects in some studies read descriptions of situations where actors made choices between several alternatives. In general, the greater the number of effects that were not common to the alternatives, the less confident subjects were of their attributions.

C. The Discounting Principle and the Correspondence Bias

The idea that confidence in the role of a particular cause in producing an effect decreases as other plausible causes are also present has been embodied in a general principle called *discounting*. The principle applies to situations in which a behavior may have been caused by a personality disposition or by any number of causes external to the person. Imagine Peter being courteous to Paul. This behavior may have been caused by Peter's trait of courtesy or by something else. For example, courtesy may have been normative in that interaction, or Peter may have been ingratiating himself to Paul in preparation for asking of him a large favor, or perhaps Peter's wife implored him to show courtesy to her friend Paul. According to the discounting principle, the more the perceiver is aware of alternative plausible causes, the less confident he or she will be that any one of the causes was responsible for a behavior.

The discounting principle was illustrated in a series of experiments on the perception of other people's attitudes. Subjects in these experiments read essays and were asked to judge the attitudes of the authors. The contents of the essays, as well as the conditions under which they were purportedly written, were varied. Some essays were favorable to Fidel Castro, the Cuban leader, and the rest were unfavorable to him. Moreover, some subjects were told that the position taken by the author was freely chosen, but others were told that the author was assigned the position by the experimenter.

Subjects who thought that the authors had chosen the direction of their essays inferred that the opinions expressed by the authors reflected their true attitudes. Discounting, however, occurred in cases where subjects thought that the position taken by the essay was assigned by the experimenter. From the point of view of the subject, there are two plausible causes for the opinions expressed in such cases: the authors' attitudes and the demands made by the experimenter. As a result, subjects were less likely to infer that the authors' personal attitudes were consistent with those expressed in the essays in these conditions.

Something intriguing emerged in these experiments on the perception of attitudes. Despite the fact that subjects in these experiments were told explicitly that the positions presented in the essays had been assigned to the authors, they still inferred that the written positions reflected the authors' true attitudes. In other words, while discounting occurred in these experiments, its effect was far from complete. This phenomenon has become known as the correspondence bias. The bias refers to people's tendency to make attributions that give too much weight to personal dispositions and not enough weight to situational forces. So pervasive has this bias proven in attribution research that it has also become known as *the fundamental attribution error*.

D. Salience and the Actor–Observer Effect

Fritz Heider's pioneering work in attribution anticipated the correspondence bias. According to Heider, other people's behavior is usually so salient to us that we tend to focus on it rather than on the situation. Taken a step further, this idea suggests that any salient stimulus is likely to be seen as playing an exaggerated causal role.

Several experiments have supported this notion. For example, if the attention of the perceiver is focused on one member of a group during a conversation, that person will be perceived as playing a more central causal role in directing the conservation than other members of the group. Perceivers have also been found to attribute more causality to a person who stood out by being the only member of a visible minority in the group, or even by being dressed in a way that was different from the rest of the group. These effects are not limited to situations in which the perceiver has little information relevant to causality other than salience.

There is one important exception to the correspondence bias. It involves situations in which people make attributions about their own behavior rather than that of others. Instead of attributing their own behavior internally, people tend to attribute it to external causes. For example, students who are asked why they chose their major tend to give answers that describe properties of the field (e.g., computer science is an important technological field). When asked about their best friend's choice of majors, however, they give dispositional reasons (he has always been a computer whiz). This asymmetry in attribution for a person's own behavior versus that of others is called the actor–observer effect.

The effects of salience on causal attributions can help explain the actor–observer effect. As observers, we tend to focus on behavior. As actors, however, we do not focus on our own behavior. Instead, our attention is directed at the features of the environment to which we are reacting. What is salient to actors and to observers is very different, and this is why they tend to make attributions to different causal factors.

E. Construct Accessibility and the Interpretation of Behavior

Salience is not entirely a property of the information available to a perceiver. Another important determinant of how people attend to and interpret social information is the level of activation of their social knowledge structures. Abstract knowledge structures that are relevant to the interpretation of information about people are known as trait constructs. Research has shown that trait constructs become more accessible immediately after their use. This phenomenon is known as priming. As a result of priming, a trait construct is more likely to be used to interpret new information and to affect relevant judgments. To illustrate the effect of construct accessibility on the interpretation of social behavior, consider having lunch with a colleague at work. During lunch you talk about shy people. A little later you are joined by an employee who has only been with the firm a few days. He asks if he can join you, sits quietly, and eats his lunch without making much eye contact. According to the notion of construct accessibility, you will be more likely to attend to these apparently shy behaviors and more likely to interpret them as reflecting a shy disposition after having been engaged in conversation about shyness than if your construct of shyness had not been primed by the prior conversation. This is particularly true with ambiguous behaviors. If the behaviors were obviously shy, then you would perceive them as such whether or not your construct for shyness were primed. If the behaviors were not at all shy, then the construct of shyness would not be relevant to the behaviors and would not be likely to be invoked in interpreting them.

The level of activation of social constructs is affected not only by priming. Each of us tends to pay attention to particular aspects of social behavior more than to others. Some of us are particularly attuned to the competitiveness of behavior; others, to its friendliness. Chronic differences in levels of construct activation account for these differences. Together, priming and chronic effects play an important role in the interpretation of ambiguous social information.

F. Self-Serving Biases in Attribution

Classical theories of attribution focus on the rationality of the perceiver. The covariation and the discounting principle, as well as the theory of correspondent inference, all paint a picture of a quasi-scientific perceiver eager to determine the causes of behavior rationally. The impact of salience, a nonrational basis for causal inference, on attribution demonstrates that there are limits to the metaphor of the scientific perceiver. In addition, research has revealed other biases demonstrating that perceivers sometimes make defensive rather than rational attributions.

A well-documented bias in attribution is the tendency to take personal responsibility for successes while attributing failures externally. There are two motivational reasons for this. First, such attributions serve to protect and enhance the way people see themselves. In one study, subjects were led to believe that they had failed at a task and were then induced to attribute their failure either to the difficulty of the task or to their own lack of ability. Subjects were more upset and had lower self-esteem when they attributed failure to lack of ability rather than to the difficulty of the task. Second, these attributions serve an impression-management function by bolstering the image the person presents to others. People who expect to fail at a task tend to exert less effort than people who expect to succeed, presumably because this provides them with an excuse for their poor performance that does not reflect on their own ability.

Another well-documented bias involves the attribution of responsibility for misfortunes such as accidents and crimes. In such situations people tend to blame the victim, even with little evidence. This finding suggests that people make attributions that promote their belief in the fairness of the world. There is an important motivational reason for maintaining a belief in a just world. If the world is just and victims of mishaps deserve their misfortune, then such misfortunes will not happen to blameless people like us. This interpretation is supported by data that show that our propensity for blaming the victim decreases when we feel similar to him or her. If a person who is similar to us deserves an awful calamity, then perhaps the same could happen to us!

The self-serving biases documented by attribution researchers serve to make a broader point about person perception. Motivation often plays an important role in our interpretation of information about people. Our sentiments about others, about ourselves, and our particular goals in social interactions influence the way we perceive events around us.

G. When Are Attributions Made?

Heider conceived of attributions as a necessary step in understanding social reality. He believed that the motivation to engage in attribution is natural and spontaneous. A number of approaches have been taken to test this assumption. One approach has been to analyze naturally occurring responses to social events to identify the factors conducive to attributional thinking.

A particularly interesting example of this approach is a study of commentary given by members of winning and losing teams following baseball and football games. The explanations in newspaper articles for the outcome of a number of games were coded for attributional content. This coding revealed a tendency to attribute wins internally and losses externally. More to the point, the sheer number of attributions was greater following an unexpected outcome than following an expected one. An unexpected outcome in this context involves a loss by the favored team to the underdog. One reason for attributions, therefore, is to make sense of unexpected events. Other research involving the coding of responses has demonstrated that people are also more likely to make attributions following failure than following success at a task, possibly because such outcomes also tend to be unexpected.

Trait attributions are particularly important in person perception. Studies have explored the spontaneous attribution of traits to others. These studies present subjects with sentences of the form: "The secretary solved the mystery halfway through the book," and then use indirect memory techniques to test whether subjects made the trait attribution suggested by the sentence (e.g., the secretary is clever). The results of these studies suggest that trait attributions are made spontaneously with only moderate frequency. In general, people have to be intent on forming impressions of others before attributing traits to them.

IV. ORGANIZING AND USING OUR MENTAL REPRESENTATIONS OF OTHERS

A. Impression Formation

In his pioneering work on how people form impressions of others, Solomon Asch presented subjects with lists of traits that described other persons and asked them about the impressions they had formed. Asch was particularly interested in the capacity to take disparate and even contradictory information about others and to make the information fit in a way that revealed the fabric of others' characters. To illustrate, consider a person who is both brilliant and silly. A perceiver may resolve the apparent contradiction in this description by thinking of a person who is brilliant intellectually but silly with practical day-to-day chores.

According to Asch, people can form coherent impressions of others because social information can usually be given a number of interpretations. Brilliance can be interpreted as meaning great depth of intelligence or dazzling astuteness. Silliness can be interpreted as shallowness or mindlessness. The perceiver can successfully juxtapose the trait of brilliance and that of silliness because each of these traits has an interpretation compatible with an interpretation of the other (depth of intelligence and mindlessness).

Several phenomena of person perception derive from this theoretical stance. *Central traits* play a particularly important role in guiding the interpretation of other traits in an impression. Consider the following two descriptions of another person: (1) intelligent, skillful, industrious, warm, determined, practical, and cautious; and (2) intelligent, skillful, industrious, cold, determined, practical, and cautious. The impressions engendered by these two descriptions differ radically despite the fact that the descriptions differ by only a single trait (warm versus cold). The traits warm and cold are central in this context because they influence the interpretation of each of the other traits. Warm and cold are the only traits in the list that convey information about the interpersonal motives of the target person. These traits allow the perceiver to read different meanings, which are primarily relevant to talents and abilities, into each of the other elements of the description.

Other phenomena compatible with this approach are the *primacy effect* and the *context effect*. The primacy effect is the tendency for early information to have more impact on impressions than later information. According to Asch, early information sets an expectation about the person. This expectation becomes relevant to interpreting subsequent information. The context effect is the tendency for a trait to be perceived more positively in the context of positive traits and more negatively in the context of negative traits than it is perceived in a neutral context. This results from the impact of context on the interpretation of the target trait.

B. Information Integration

The notion that our impressions of others represent a complex interpretive process has not been accepted universally. The best-articulated alternative to this approach has been presented by Norman Anderson, who noted that evaluations constitute the most powerful element of impression formation. Anderson proposed an averaging model of evaluations in impression formation. This is the model in its simplest form. Suppose a personnel manager meets a job candidate who appears intelligent and efficient but who has poor social skills. These traits differ on how positive and negative they are, and this is what the personnel manager focuses on. Assume that on a scale ranging from -10 to $+10$, intelligence is a $+9$, efficiency is a $+8$, and poor social skills is a -5. According to the averaging model, the personnel manager will form an overall impression of the job candidate that represents the average of the three traits, i.e., $9 + 8 - 5 / 3 = 4$.

The averaging model is noteworthy for its simplicity and its openness to precise experimental testing. Norman Anderson and his colleagues have conducted an extensive research program to test the model and have garnered impressive evidence for it. One particularly interesting test involves the comparison of the averaging model with its obvious counterpart, the additive model. According to the additive model, an impression is the sum rather than the average of the scale values of the component traits. Thus, in our example, the overall evaluation of the job candidate would be $9 + 8 - 5 = 12$. The fact that 12 is larger than 4 is not important because these numbers are meaningful only in the context of each model. However, the comparison between the two models becomes quite revealing if we assume that the candidate is also athletic and that this quality has a scale value of $+2$. If we incorporate this value into the impression according to the averaging model, the overall evaluation becomes $9 + 8 + 2 - 5 / 4 = 3.5$. However, if we incorporate this value into the impression according to the additive model, the overall evaluation becomes $9 + 8 + 2 - 5 = 14$. Averaging has resulted in a less positive impression, and adding has resulted in a more positive one. Clearly, the two models make very different predictions about impression formation. The research that has tested these predictions has generally favored the averaging model. The practical lesson for us is that if we have succeeded in creating a positive impression with our best traits, we should resist the temptation to cram our self-presentation with mildly positive information about ourselves.

C. Associative Networks and the Structure of Our Impressions

With the advent of the information-processing revolution in psychology, researchers have become interested in specifying the exact nature of the mental representations formed in person perception. Reid Hastie as well as Thomas Srull and Robert Wyer proposed that these representations are composed of idea *nodes* and of associative *links*. Figure 2 shows such an associative network. At the top level of the structure is a node identifying features of the person such as his or her proper name. At the second level are concepts that describe the person and that organize more specific information about him or her. Trait concepts are assumed to play a dominant role at this level because they describe others and organize information about them. At the bottom level are specific facts about the person, such as behaviors.

Most of the evidence for these structures comes from research on memory for information about others. Such research is based on the notion that the retrieval of information from an associative network involves movement from one node to another along existing links. For example, in recalling information about Peter, the retrieval process begins at the top node and then descends toward either the generous node or the intelligent node. For simplicity we will assume that the links to the two nodes are of equal strength and that the selection between them is random. Suppose that the search proceeded down to the generous node. The information contained in that node would be recalled (i.e., Peter is generous). The retrieval process would then continue randomly to one of the three behavioral nodes subsumed under the trait generous. Upon retrieving that information, the process would return to the higher node and select the next link randomly from the four emanating from that node. It is possible for the retrieval process to return to the Peter node or even for it to retrace its steps to the behavioral node just retrieved. In the latter case, the process would return to the generous node and select a pathway again. The retrieval process would stop upon reaching a point where already recalled information is encountered repeatedly.

D. Evidence for Trait-Behavior Associations in Impressions

One of the most important assumptions in the network structure is that behaviors are organized in terms of the trait concepts they exemplify. This as-

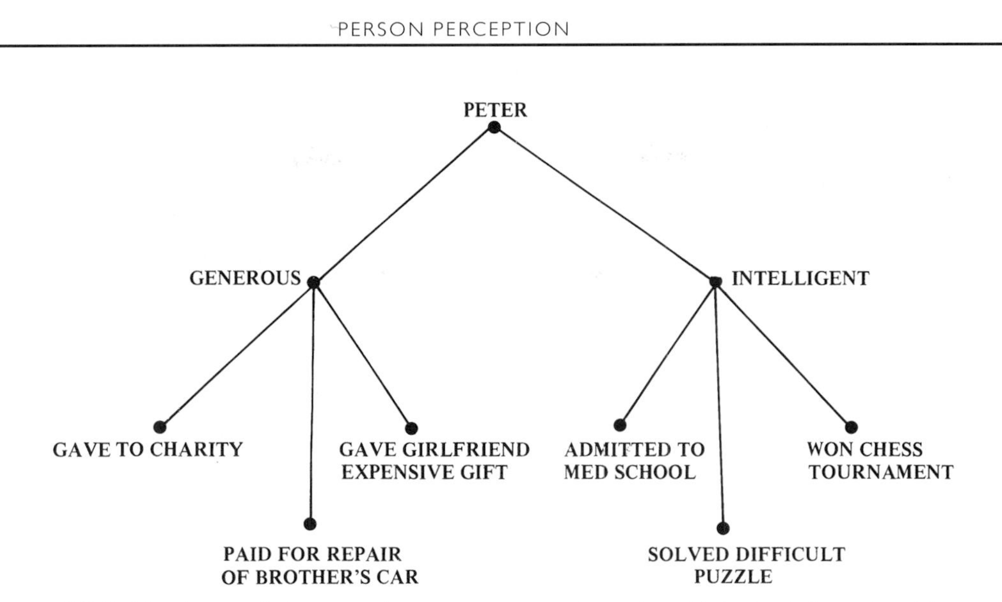

FIGURE 2 An associative network structure representing trait and behavior information. [Reprinted with permission from Bassili (1989).]

sumption has been supported by research showing that when we recall the behaviors of a person about whom we have formed an impression, we tend to recall behaviors that exemplify a particular trait as a group. In recalling information about Peter, a subject is likely to report the generous behaviors and the intelligent behaviors in clusters.

The fact that clustering results from links with trait concepts is supported by a number of other findings. There is little clustering by trait category when the information about a person is learned without intention to form an impression. As seen earlier, trait concepts are not normally applied to the interpretation of behavior under such conditions. It is interesting that with a drop in clustering, these conditions also result in much poorer recall than conditions in which information is learned with the intention of forming an impression. Research has shown that subjects are more likely to recall a behavior exemplifying a trait if they had also recalled the trait than if they had not. In other words, trait concepts provide powerful organizing elements in our memory structures and act as midway cueing stations for the recall of others' behaviors.

E. Memory for Behaviors that Confirm or Disconfirm our Expectations

So far we have considered associations between behaviors and trait concepts without considering direct associations between behaviors. Interbehavior associations have been explored in one of the most extensive research programs in person memory. This research is based on the phenomenon that people remember behaviors inconsistent with their expectations about another person better than behaviors that fit their expectations. This is not to say that we do not remember consistent behaviors well. Such behaviors are remembered much better than behaviors irrelevant to an expectation. Still, the question arises as to why inconsistent behaviors are recalled best.

The answer is that inconsistent behaviors are surprising and puzzling, and this makes us think about them in relation to other behaviors of the person. According to the associative approach, the act of thinking about two behaviors in relation to each other creates an association between them. The result of such thinking is that inconsistent behaviors are associated more with other behaviors than with consistent behaviors. This is illustrated in Fig. 3. Applying the search procedures discussed earlier to this network, it becomes apparent that nodes representing inconsistent behaviors are more likely to be reached than nodes representing consistent behaviors. The latter nodes, in turn, are more likely to be reached than nodes representing irrelevant behaviors.

F. On-Line Versus Memory-Based Evaluations

It was shown earlier that evaluations are of primary importance in impression formation. The mental rep-

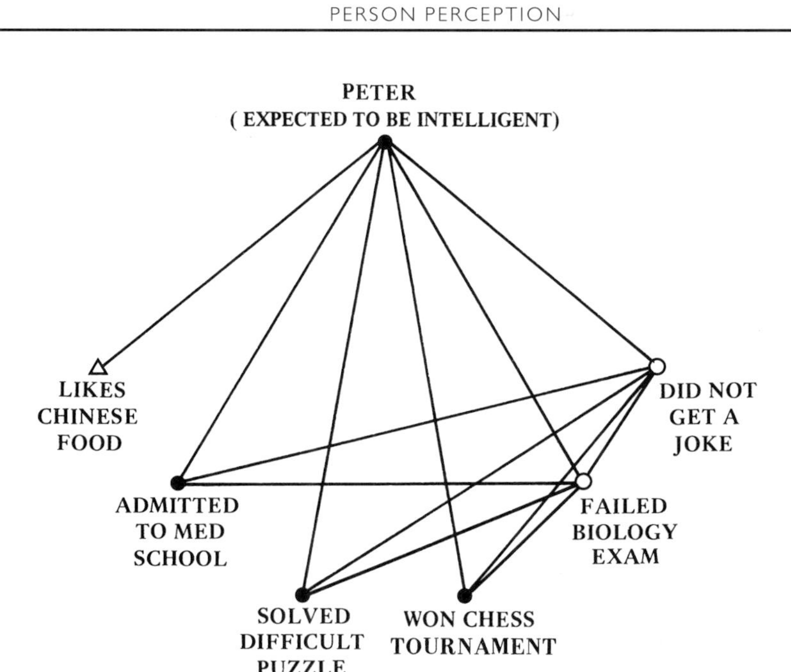

FIGURE 3 Example of associations formed between behaviors that are consistent with (black circles), inconsistent with (white circles), or irrelevant to (triangle) an expectation.

resentations we have of others provide the basis for the evaluative judgments we make about them. There are two situations in which this can happen: those in which we evaluate people immediately as we require information about them, and those in which our evaluations are based on information we recall from memory. In other words, the former type of evaluations are performed *on-line,* whereas the latter are *memory based.* As it turns out, there are important differences between the cognitive representations that are used in on-line versus memory-based evaluations.

On-line evaluations are by far the more common. Suppose that having just been introduced to Peter, you learn that he was admitted to medical school, that he gave to a charity, etc. Two things will happen as you acquire each piece of information about Peter. The information will have some impact on your general liking for him, and the information will be stored in memory in a network such as the one shown earlier. The important thing to note here is that your liking for Peter will be based on an evaluative representation, updated simultaneously with the appearance of each new piece of information. This representation is independent of the associative network in which the behavioral information is stored.

Suppose that a few weeks later, a mutual acquaintance asks you about your feelings toward Peter. Research suggests that your response will rely on the latest update of your general evaluation of Peter rather than on your memory of the information that led to your evaluation. You may answer that you like Peter very much, without remembering why you feel this way. In fact, research has often demonstrated that there is little connection between feelings and the informational basis for these feelings. With the passage of time, the recall of information deteriorates and is independent of the evaluative representation of a person.

There is one situation in which the relationship between evaluations and memory for relevant information is good. This is when the evaluations are memory based. Suppose that you had no interest in forming an impression of Peter when you learned behavioral information about him. In this case, your mental representation would consist of a simple list of behaviors linked with Peter. Upon having to answer your friend's query a few weeks later, you would have to determine your feelings on the basis of information you could retrieve from memory. What is forgotten will have no impact on your evaluation. Naturally, what you feel toward Peter and what you remember of him will be closely connected in this situation.

V. SUMMARY

The processes through which we get to know people are collectively known as person perception. These processes can be triggered by superficial information such as appearance or nonverbal behaviors. We often form preliminary impressions about people on the basis of such information. These impressions can be elaborate because of assumptions we make on the basis of implicit personality theories. One example of this is the assumption that physically attractive people have better social skills and are generally more outgoing than physically unattractive people.

When we observe others' behavior, we are often motivated to understand its causes. We usually do so by attributing the behavior to properties of the person or of the situation. It is through this process of attribution that we infer the intentions and character traits of others. Because behaviors are often more salient to us than the situational context within which these behaviors occur, our attributions are usually biased toward the person. The opposite is true of attributions about our own behavior. The accessibility of our social constructs also plays a role in determining the salience of information and its interpretation. Construct accessibility is determined by recent prior activation and by chronic predisposition. Attributions are generally motivated by a desire to form an impression about a person or to understand why an unexpected event occurred.

The information we acquire about others is organized into meaningful impressions. We are able to form coherent impressions of others despite the disparate information we have about them because social information can usually be interpreted in a number of ways. Our overall evaluations of others are often based on the average value of the information we have about them. Models of how we store information about others assume that our mental representations consist of nodes and of associations between them. These models have been successful in explaining why we tend to remember behaviors that violate our expectations better than behaviors that confirm them. They have also been instrumental in delineating the differences between evaluations formed during acquisition of information about a person and evaluations based on memory.

BIBLIOGRAPHY

Bassili, J. N., ed. (1989). "On-line Cognition in Person Perception." Lawrence Erlbaum Associates, Hillsdale, New Jersey.

Ekman, P. (1985). "Telling Lies: Cues to Deceit in the Marketplace, Politics, and Marriage." W. W. Norton Co., New York, N.Y.

Fiske, S. T. (1993). Social cognition and social perception. *In* "Annual Review of Psychology," (M. R. Rosenzwieg and L. W. Porter, eds.). Annual Reviews, Palo Alto, CA.

Fiske, S. T., and Taylor, S. E. (1991). "Social Cognition," 2nd Ed. Addison-Wesley, Reading, MA.

Ross, M., and Fletcher, G. J. O. (1985). Attribution and social perception. *In* "Handbook of Social Psychology." (G. Lindzey and E. Aronson, eds.). Random House, New York.

Persuasion

JOHN T. CACIOPPO
GARY G. BERNTSON
RICHARD E. PETTY
Ohio State University

GLOSSARY

Attitude General evaluative response or disposition to respond

Influence Effect of events and others on behavior

Persuasion Power of persons to alter attitudes and behavior through information

ALL ORGANISMS HAVE BIOLOGICAL MECHANISMS for approaching, acquiring, or ingesting certain classes of stimuli and withdrawing from, avoiding, or rejecting others. For simple organisms, the stimuli that potentiate approach or withdrawal, the form of the response, and the mediating mechanisms are relatively fixed. For more complex organisms such as humans, multiple mechanisms contribute to approach and withdrawal tendencies; this potentiation can manifest consciously (cognitively, emotionally) as well as behaviorally and can be stored in memory in the form of attitudes, and both the eliciting stimulus and the form of the response are subject to generalization and modification. Persuasion refers to the power of persons to alter attitudes and behavior through information. Attitude and behavior change resulting from a communication constitute persuasion regardless of the communicator's intent or the recipient's awareness that an attitude has been changed. Not included under the rubric of persuasion are changes in knowledge or skill (i.e., education) or changes in behavior that require another's surveillance or sanctions (i.e., compliance). Persuasion, therefore, represents a form of self-control and social control that does not rely on coercion. In addition, innate and relatively inflexible predilections to approach or withdraw—such as reflexes or fixed action patterns—and irreversible changes in parameters of approach or withdrawal—such as diminished response vigor due to aging—may be related to attitudes but are not considered instances of persuasion. Traditionally, studies of the antecedents of persuasion have focused on characteristics of the source of a recommendation and on rational or emotional forms of argumentation linking a particular recommendation to a person's beliefs and values. Any information that changes an individual's predilection to react to a class of persons, objects, or issues in a consistently positive or negative fashion could be included under the rubric of persuasion, however. Studies of the consequences of persuasion have tended to focus on changes in attitudes and cognition and the persistence of these changes, but the physiological and behavioral effects have also been investigated.

I. ATTITUDES AND PERSUASION

People's perceptions of events and people in their world are organized in part in terms of long-term evaluative responses to stimuli. This organization is reflected in the structure of language; people's conceptual organization of motivation, emotions, and moods; facial expressions of emotion; and everyday behaviors. The evaluative (attitudinal) categorization of stimuli has even been found to emerge in some circumstances prior to an individual recognizing the eliciting stimulus.

Attitudes, in turn, influence what people perceive and feel about their world and can have direct and indirect effects on behavior across a wide range of situations. The direct effect of attitudes on behavior represents the tendency for people to approach, acquire, support, protect, and promote liked, in contrast to disliked, objects, persons, and issues. Although there may be intervening psychological operations between attitudes and behavior, such as accessing the relevant attitude and formulating a behavioral intention, these operations pertain primarily to the response side of the information-processing sequence (e.g., response execution). Promotional campaigns for products in stores, for instance, are designed simply to mobilize people to try the products. Finally, the causal relationship between attitudes and behavior is reciprocal rather than unidirectional. The feedback from actions or inactions toward some target constitutes a powerful source of information that can shape subsequent beliefs and attitudes about the target. Product trials can provide compelling reasons to individuals for purchasing, or not purchasing, particular products again. [*See* Attitude and Attitude Change.]

The indirect effect of attitudes on behavior stems from their influence on individuals' selective attention to, interpretation of, and recollection of people and events in their world and, subsequently, on their behavior and on the behavior of others toward them. In addition to the direct behavioral effects of attitudes, therefore, attitudes can also shape an individual's experience and representation of the world. If information favoring both sides of an issue is presented to groups with opposing attitudes, the discrepancy between the groups' attitudes might intuitively be expected to diminish somewhat. The opposite result has been found to be the case, however. Individuals often accept the evidence that supports their initial attitude and are critical of the evidence contradicting their initial attitude. The result is that both groups find reasons to strengthen their initial attitude, and disagreement between the two groups can actually be heightened by information favoring both positions.

In addition to having a directive function that channels activity into specific types of responses toward certain stimuli, attitudes have also been viewed by some as having the dynamic function of energizing people to act. Thus, individuals who come to feel strongly about another person, topic, or issue may not only channel their thoughts and behavior toward the target accordingly when given the opportunity, but may also be excited to create opportunities to act in a positive or negative fashion toward the target.

In sum, attitudes are central to people's conceptual organization of their world and to the organization of their behavior. Because both attitudes and behavior are multiply determined, the correspondence between attitudes and behavior tends to be modest but significant. The correspondence between attitudes and behavior can be enhanced by: (1) general rather than specific measures of attitudes and behaviors, (2) direct experience with the attitude target, (3) prior knowledge and thought about the target, (4) a highly accessible attitude toward the target, (5) weak social norms governing behavior toward the target, (6) personal control over one's behavior toward the target, (7) strong attitudes, and (8) chronic tendencies by individuals to think about the reasons for their attitudes. By influencing attitudes and factors that moderate attitude–behavior correspondence, persuasion can potentiate a broad class of positive or negative behaviors toward a person, object, or issue.

II. TWO ROUTES TO PERSUASION

A. Background

The resolution of conflicts and the mobilization of effort to serve the goals of a collective are cornerstones of civilization. The history of humankind reveals physical force and intimidation—not persuasive skill—to be the key mode of achieving political, social, and economic control. William McGuire has noted that persuasion has played a central role in social control in only four epochs: Athens from 427 to 338 B.C. (during which time Plato and Aristotle considered the processes underlying persuasion), Rome from approximately 150 to 43 B.C. (during which time Cicero wrote about oration and persuasion), in Europe from approximately 1470 to 1572 (during the Italian Renaissance), and the present period of the mass media, which began to take form in the eighteenth century. The scientific study of what and how factors affect persuasion gained momentum during World War II, when the mass media played an important role in recruiting and indoctrinating troops, maintaining the morale of the Allied forces and residents, and assaulting the morale of the Axis troops.

This early research was organized by the question "who said what to whom, how, and with what effect"—that is, in terms of the external stimulus factors of source (e.g., expertise, trustworthiness), message (e.g., one-sided, two-sided), recipient (e.g., sex, intelligence), and modality (e.g., print, auditory). Two assumptions underlying much of this early work were that each of these factors had general and independent effects on persuasion and that a close correspondence existed between attitude change and behavior change across situations. Both assumptions proved to be oversimplifications. After accumulating a vast quantity of data and a large number of theories, there was surprisingly little agreement concerning if, when, and how the traditional source, message, recipient, and modality variables affected persuasion. Existing literature supported the view that nearly every independent variable studied increased persuasion in some situations, had no effect in others, and decreased persuasion in still other contexts. This diversity of results was even apparent for variables that on the surface, at least, appeared to be quite simple. For example, although it might seem reasonable to propose that by associating a message with an expert source, agreement could be increased (e.g., see Aristotle's "Rhetoric"), experimental research suggested that expertise effects were considerably more complicated. Sometimes expert sources had the expected effects, sometimes no effects were obtained, and sometimes reverse effects were noted.

B. The Central and Peripheral Routes to Persuasion

Empirical and theoretical advances over the past decade have led to a more comprehensive framework for organizing, categorizing, and understanding the basic processes underlying these diverse effects. Specifically, the many different empirical findings and theories in the field have been viewed as emphasizing one of two relatively distinct *routes to persuasion.* The first is attitude change that occurs as a result of a person's careful and thoughtful consideration of the merits of the information presented in support of an advocacy (*central route*). The second is that occurring as a result of some simple cue in the persuasion context (e.g., an attractive source) that induces change without necessitating scrutiny of the merits of issue-relevant information (*peripheral route*). This model of the psychological operations underlying persuasion, which is depicted in Fig. 1, highlights that attitudes are multiply determined and that attitudes whose verbal expression is similar may have different antecedents and consequences. For instance, the issue-relevant thinking that characterizes the central route to persuasion can result in the integration of new arguments, or one's personal translations of them, into one's underlying belief structure for the attitude object. In addition, by scrutinizing the strengths and weaknesses of a recommendation, the information and the consequent attitude are rendered more coherent, accessible, and cross-situational. Attitudes formed through the central route, therefore, are relatively persistent, resistant to counterpersuasion, and predictive of behavior.

If people were to try to adopt only those attitude positions about which they had thought carefully, they either would be unable to venture into novel situations or would be unable to respond to the myriad stimuli to which they are exposed each day. This is true even though individuals might be motivated generally to hold "correct" attitudes. The numerous stimuli to which individuals must respond daily, coupled with individuals' limited time and cognitive resources, make it adaptive to also be capable of using cognitively less demanding shortcuts (e.g., simple cues, habits, rules-of-thumbs) to guide attitudinal reactions in some situations. Although such a mechanism for attitude change (peripheral route) can guide responses to a wide variety of stimuli while minimizing the demands on individuals' limited cognitive resources, the resultant attitudes and behavior are based on information that is only superficially or peripherally related to the actual merits of the position. Hence, some responses potentiated by this generally adaptive mechanism may be unreasonable and maladaptive. These maladaptive features of attitudes derived through the peripheral route are diminished somewhat by their relatively short persistence, susceptibility to change, and weak influence on behavior.

C. The Elaboration Likelihood Continuum

The model outlined in Fig. 1 has provided a general framework for understanding how a variety of factors, such as speed of speech and source credibility, can increase, decrease, or have no effect on attitude change. If the central route is followed, the perceived cogency of the message arguments and factors that may bias argument processing (e.g., prior knowledge, initial opinion) are predicted to be important determinants of the individual's acceptance or rejection of

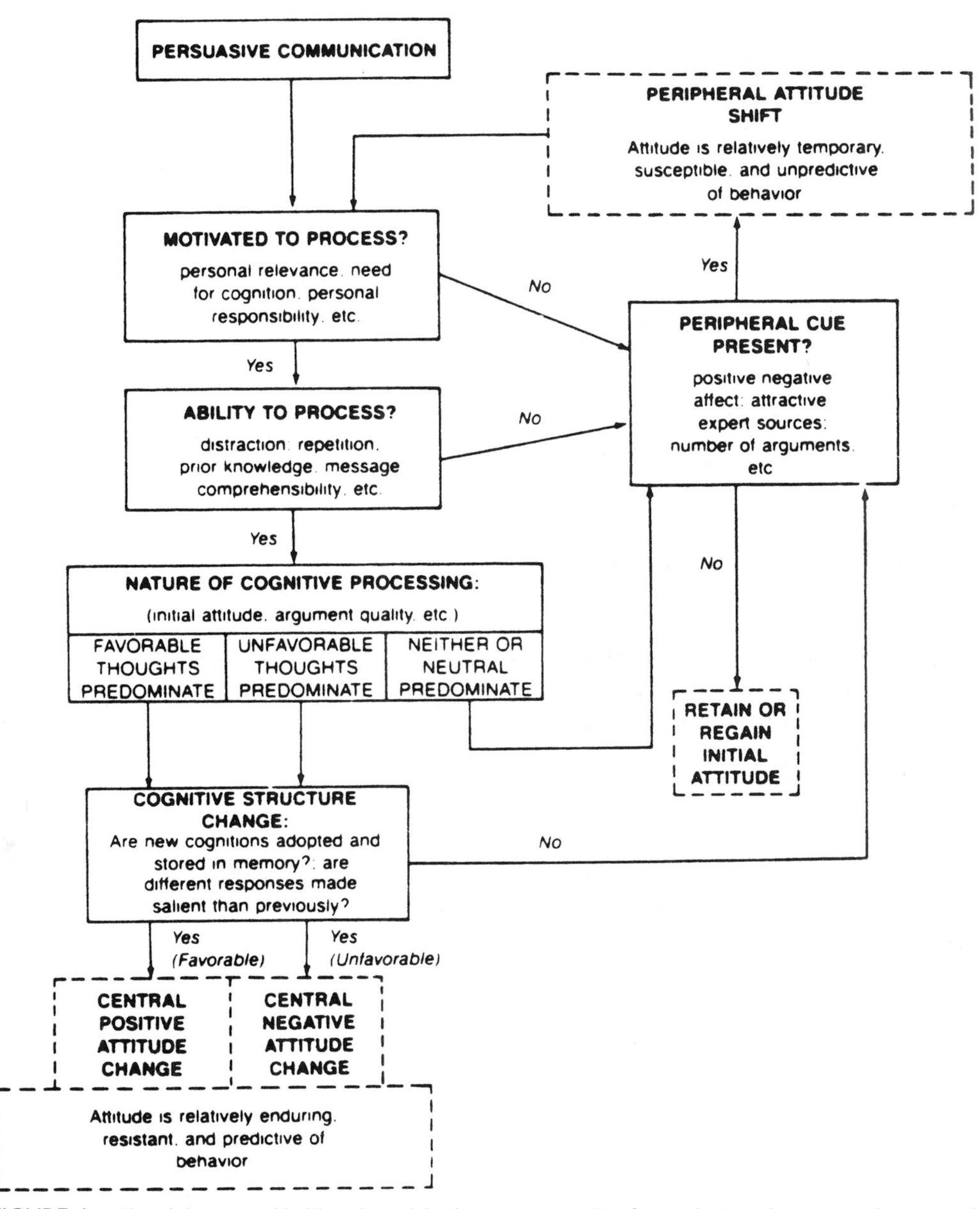

FIGURE 1 The elaboration likelihood model of persuasion. This figure depicts the two anchoring end points on the elaboration likelihood continuum: the central and peripheral routes to persuasion. [From R. E. Petty and J. T. Cacioppo (1986). "Communication and Persuasion: Central and Peripheral Routes to Attitude Change." Springer-Verlag, New York.]

the recommendation, and factors in the persuasion setting that might serve as peripheral cues are relatively unimportant determinants of attitudes. If, on the other hand, the peripheral route is followed, then the strength of the message arguments and factors that bias argument processing become less important and peripheral cues become more important determinants of attitudes. That is, there is a trade-off between the central and the peripheral routes to persuasion.

Importantly, the conditions that lead to influence through the central versus the peripheral route have also been specified. For instance, many attitudes and decisions are either perceived to be personally inconsequential or involve matters about which people are uninformed. In these situations, people may still want to be correct in their attitudes and actions, but they are not willing or able to think a great deal about the arguments for or against a particular position. Peripheral cues provide a means of maximizing the likelihood that one's position is correct while minimizing the cognitive requirements for achieving this position.

Implicit in the central route, on the other hand, is that people must relate the incoming message arguments to their prior knowledge in such a way as to evaluate the cogency and scope of the arguments—that is, they expend cognitive effort to examine the information they perceive to be relevant to the central merits of the advocacy. When conditions foster people's motivation and ability to engage in this issue-relevant thinking, the *elaboration likelihood* is said to be high. This means that people are likely to attend to the appeal, attempt to access relevant information from both external and internal sources, and scrutinize or make inferences about the message arguments in light of any other pertinent information available. Consequently, they draw conclusions about the merits of the arguments for the recommendation based on their analyses and derive an overall evaluation of, or attitude toward, the recommendation. Thus, the central and the peripheral routes to persuasion can be viewed as anchors on a continuum ranging from minimal to extensive message elaboration or issue-relevant thinking, and issue-relevant thinking may be relatively objective or biased. Factors governing an individual's motivation and ability to scrutinize the truthfulness of various attitude positions determine whether the central or the peripheral route operates.

D. Determinants of the Likelihood That Issue-Relevant Thinking Mediates Persuasion

Motivational variables are those that propel and guide people's information processing and give it purposive character. There are a number of variables that have been found to affect a person's motivation to elaborate on the content of a message. These include: (1) task variables such as the personal relevance of the recommendation, (2) individual difference variables such as need for cognition, and (3) contextual variables such as the number of sources advocating a position. These kinds of variables act on a directive, goal-oriented component that might be termed *intention* and a nondirective, energizing component that might be termed *effort* or *exertion*.

Intention is not sufficient for motivation, for instance, since one can want to think about a message or issue but not exert the necessary effort to move from intention to thought and action. If both intention and effort are present, then motivation to think about the advocacy may exist, but message elaboration may still be low because the individual does not have the ability to scrutinize the message arguments. There are a number of variables that can affect an individual's ability to engage in message elaboration, including *task variables* such as message comprehensibility, *individual difference* variables such as intelligence, and *contextual variables* such as distraction and message repetition. Contextual variables that affect a person's ability to elaborate cognitively on issue-relevant argumentation can also be characterized as factors affecting a person's opportunity to process the message arguments.

Experiments have demonstrated that if task, individual, and contextual variables in the influence setting combine to promote motivation and ability to process, then the arguments presented in support of a change in attitudes or behavior are thought about carefully. If the person generates predominantly favorable thoughts toward the message, then the likelihood of acceptance is enhanced; if the person generates predominantly unfavorable thoughts (e.g., counterarguments), then the likelihood of resistance or boomerang (attitude change opposite to the direction advocated) is enhanced. The nature of this elaboration (i.e., whether favorable or unfavorable issue-relevant thinking) is determined by whether the motivational and ability factors combine to yield relatively objective or relatively biased information processing and by the nature of the message arguments. If elaboration likelihood is low, however, the nature of the issue-relevant thinking is less important, and peripheral cues become more important determinants of attitude change (see Fig. 1).

A number of experiments have explored ways to stimulate or impair thinking about the message arguments in a persuasive appeal. Distraction, for instance, can interfere with a person's scrutiny of the arguments in a message and thereby alter persuasive impact. In an illustrative experiment on distraction and persuasion, students listened to a persuasive message over headphones while monitoring in which of the four quadrants of a screen a visual image was projected (a distractor task). In the low-distraction condition, images were presented once very 15 seconds, whereas in the high-distraction condition, images were presented once every 5 seconds. Importantly, neither rate of presentation was so fast as to interfere with the students' comprehension of the simultaneously presented persuasive message, but the students' argument elaboration was much more disrupted in the high- than in the low-distraction condition. The results revealed that the students' postcommunication attitudes were a function of message processing when distraction was low but not when distraction was high.

Numerous task, contextual, and individual difference variables have been identified that enhance or impair argument elaboration by affecting a person's motivation or ability. Moderate levels of repetition of a complicated message can provide individuals with additional opportunities to think about the arguments and, thereby, enhance argument processing. Messages worded to underscore the self-relevance of the arguments enhance individuals' motivation to think about the arguments. Being singly responsible rather than one of many assigned to evaluate the recommendation can induce more issue-relevant thinking, as individuals are unable to diffuse their responsibility for determining the veracity of the recommendation.

III. ARGUMENT ELABORATION VERSUS PERIPHERAL CUES AS DETERMINANTS OF PERSUASION

The hypothesis that there is a trade-off between argument scrutiny and peripheral cues as determinants of a person's susceptibility or resistance to persuasion has also been supported by more than a decade of research. In an illustrative study, two kinds of persuasion contexts were established: one in which the likelihood of relatively objective argument elaboration was high and one in which the elaboration likelihood was low. This was accomplished by varying the personal relevance of the recommendation: students were exposed to an editorial favoring the institution of senior comprehensive exams at their university, but some students were led to believe that these comprehensive exams would be instituted next year (high elaboration likelihood) whereas others were led to believe that the exams would be instituted in 10 years (low elaboration likelihood).

To investigate the extent to which students' argument scrutiny determined attitudes, half of the students heard eight cogent message arguments favoring comprehensive exams, and the remaining students heard eight specious message arguments favoring the exams. Finally, to examine the extent to which peripheral cues were important determinants of attitudes, half of the students were told that the recommendation they would hear was based on a report prepared by a local high school class (low expertise), whereas half were told that the tape was based on a report prepared by the Carnegie Commission on Higher Education (high expertise). Following the presentation of the message, students rated their attitudes concerning comprehensive exams and completed ancillary measures. Results indicated that argument quality was the most important determinant of the students' attitudes toward comprehensive exams when they believed that the recommendation was consequential for them personally, but that the status or expertise of the source was the most important determinant of the students' attitudes when they believed that the recommendation would not affect them personally. These results held even though comprehension of the message arguments, and judgments of the expertise of the source, were equal across the experimental groups.

IV. OBJECTIVE VERSUS BIASED ARGUMENT PROCESSING

Message processing in persuasion research was traditionally thought to imply objective processing. This, too, proved to be an oversimplification. When an individual is motivated to scrutinize arguments for a position, there are no assurances that the information processing will be objective or rational. *Objective argument processing* means that a person is trying to seek the truth wherever that may lead. When a variable enhances argument scrutiny in a relatively objective manner, the strengths of cogent arguments and the flaws in specious arguments become more apparent. Conversely, when a variable reduces argument scrutiny in a relatively objective fashion, the strengths of cogent arguments and the flaws of specious arguments become less apparent. Objective processing, therefore, has much in common with the concept of "bottom-up" processing in cognitive psychology because elaboration is postulated to be relatively impartial and guided by data (in this case, message arguments).

In contrast, *biased argument processing* means that there is an asymmetry in the activation thresholds for eliciting favorable or unfavorable thoughts about the advocacy. Consequently, the encoding, interpretation, and recall of the message arguments are distorted to make it more likely that one side will be supported over another. Biased processing has more in common with "top-down" than "bottom-up" information processing, because the interpretation and elaboration of the arguments are governed by existing cognitive structures, such as a relevant knowledge or attitude schema, which guide processing in a manner favoring the maintenance or strengthening of the original schema. Research on factors such as the role of initial attitudes described in Section I has demonstrated that people are sometimes motivated and able to augment even specious arguments to arrive at a more cogent line of reasoning for their desired position.

V. PERSUASION VARIABLES HAVE MULTIPLE AND INTERACTIVE EFFECTS

Another reason why the processes underlying persuasion have appeared enigmatic is that some variables may increase argument processing at one level of the factor, but may actually bias or decrease argument processing at a different level of that factor. For instance, repeating a long or complicated persuasive message can provide individuals with additional opportunities to think about the message arguments and, therefore, enhance relatively objective argument scrutiny. Excessive exposures to a persuasive message can become tedious, however, and can actually motivate a person to reject the recommendation. Hence, the same stimulus factor—message repetition—had quite different effects on issue-relevant thinking as the amount of this factor increased.

Factors previously thought to have simple effects on information processing and persuasion have also been found to have quite different effects depending on the presence or absence of other factors. For instance, presenting a persuasive message on a noninvolving issue in rhetorical rather than declarative form can increase an individual's propensity to think about the message arguments. When the recommendation is already personally involving, however, the insertion of rhetorical questions in the message arguments can actually interfere with the individual's ongoing idiosyncratic argument scrutiny.

In sum, the introduction of new factors (e.g., arguments presented in rhetorical rather than declarative form) can have striking but explicable effects on people's cognitive processes and attitudes. Current models of persuasion are now able to account for rather complicated patterns of data even though the intervening processes are fairly straightforward.

VI. BIOLOGICAL ASPECTS OF ATTITUDES AND PERSUASION

Despite early conceptualizations of attitudes as postural orientations and neural predispositions to respond, programmatic research on the biological mechanisms underlying attitudes and persuasion is fairly recent. This gap in theory and research on persuasion is due in part to the methodological approaches championed by the pioneers in the field, the interest by early theorists in applying persuasion research to address social problems (e.g., wartime propaganda), and the relative ease for governments and institutions to manipulate environmental rather than biological factors to achieve social control. Occasionally, study of the biological aspects of attitudes and persuasion has been dismissed because attitudes are said to be acquired through experience, as if environmental influences were antithetical to biological mechanisms. More recently, attention is being paid to the physiological manifestations by which the elementary psychological operations underlying persuasion can be indexed and to the physiological mechanisms through which attitudes and persuasion are accomplished.

A. Genetic Factors

The achievement of strong behavioral proclivities in animals through selective breeding (e.g., herding behavior in border collies, aggressiveness in the pit bull) raises the possibility that some attitudes, and the manner in which individuals respond to information pertinent to attitude change, may be partially determined genetically. The existing data, though sparse, support both possibilities. Most studies bearing on the genetic contributions to attitudes and persuasion are based on the similarities observed between monozygotic twins reared apart from an early age. These studies have documented genetic contributions to individual differences in general intelligence, positive and negative affective dispositions, interests, general social attitudes (e.g., liberalism/conservativism), and job satisfaction. Although some of the individual differences shaped by genetic factors may influence attitudes directly, most of these dispositional factors (e.g., intelligence, affective disposition, interests, values) would likely influence attitudes and persuasion by increasing objective or biased message processing or by affecting what constitutes a compelling argument or peripheral cue to a particular individual.

The contribution of genetic factors tends to be modest, so that environmental factors are also major determinants of attitudinal reactions. Perhaps a more important finding to emerge from research on behavioral genetics is that genetic and environmental factors are not as separable as once thought. Environmental factors can inhibit or trigger the expression of genetic influences, and genetic factors can lead individuals to seek and remain in certain environments. Monozygotic twins reared apart, for instance, have been found not only to express similar levels of satisfaction with their jobs, but to hold jobs that are similar in terms of complexity, motor skills, and physical demands. Possible mechanisms of heritability for this finding range from affective disposition to cognitive and physical capacity.

B. *In Utero* Factors

Based on the billions of dollars spent annually each year on advertising, it has been estimated that the average person in the United States has the potential to be exposed to over 1400 persuasive appeals per day. Even if only a small fraction of these appeals are effective, this deluge of appeals suggests that an individual's attitudes are under nearly constant challenge. As noted earlier, attitudes based on little prior knowledge are particularly susceptible to change. One of the more surprising findings is that repeated, unreinforced exposures to a novel or unfamiliar stimulus result in a positive attitude toward the stimulus. That is, repeated exposure to a novel stimulus that results in neither reward nor punishment breeds preference for this stimulus over a similar stimulus to which an individual has not been exposed. This *mere exposure* effect has been demonstrated using stimuli as diverse as nonsense words, ideographs, polygons, and faces, and the mere exposure effect is enhanced by factors such as a heterogeneous exposure sequence, a moderate number of presentations of the target stimulus (e.g., less than 100), brief exposure durations (e.g., less than 5 sec), and a delay between the stimulus presentations and attitude measurement.

Attitude change due to information emanating from the environment has also been documented *in utero* and appears to be a variation on the mere exposure effect. In an illustrative study, pregnant women recited a speech passage aloud each day during their last 6 weeks of pregnancy. Their newborns were tested within a few days following birth to determine whether the sounds of the recited passage were more reinforcing (i.e., preferred) than the sounds of a novel passage. A matched group of newborns who had not been exposed previously to either passage were also tested to determine whether one passage was simply more likeable. Results revealed that the passage to which the fetuses had been exposed during the third trimester was preferred over the comparison passage, whereas the matched group of newborns exhibited no preference for one over the other passage. *In utero* recordings indicate that the auditory frequencies to which fetuses are exposed range between approximately 125 and 1000 Hz. The fundamental frequency of the speech of women, but not men, tends to fall within this frequency range. As would be expected, therefore, this "mere exposure" effect is not found for fathers' voices, but instead a preference for paternal voices over less familiar male voices develops postnatally once paternal voices have become perceptually salient among other male voices.

C. Autonomic Factors

The activity of the autonomic division of the peripheral nervous system has been of interest in studies of attitudes and persuasion because it was thought to reflect, if not represent, the emotional substrate of attitudes and attitude change. Reports of the autonomic differentiation of attitudes and emotions appear occasionally, but the autonomic responses found to differentiate hedonic states have tended to differ across studies, as factors such as the implications of the attitude for an individual's action vary. A more robust finding to emerge from psychophysiological research is that autonomic activity varies as a function of the intensity of the emotion (regardless of valence), cognitive effort, and behavioral effort aroused by an attitude object or persuasive appeal. In an illustrative study, students indicated their agreement or disagreement with 20 controversial statements. Several weeks later students were tested individually while a measure of autonomic activation—palmar skin resistance—was recorded. The 20 statements were again read to each student, but in addition students were told that a fictitious majority of their peers held a similar or dissimilar attitude. Students were then asked to restate their original attitude toward each statement, at which time skin resistance was measured. The major finding was that autonomic activity was higher the greater the discrepancy between the attitudes of the student and the fictitious majority, except when students were absolutely confident of their original attitude.

General autonomic arousal, as might be achieved by exercise or by watching a sexually explicit film, has also been found to enhance an individual's normal affective reaction to an unrelated but evocative stimulus presented several minutes after the initial evocative stimulus. The specific mechanism responsible for this effect is not entirely clear, although the effect appears to be limited to instances in which (1) the second stimulus itself elicits a clear and dominant affective response and (2) the second stimulus is presented while individuals' autonomic activity is still elevated from the initial evocative stimulus but the individuals no longer feel aroused. More recent studies that have examined the effect of exercise-induced arousal on argument scrutiny and persuasion have found that peripheral cues (e.g., celebrity status) have a greater influence on attitudes at high than at moderate levels of arousal, whereas the quality of the message arguments has a greater influence at moderate than at high levels of arousal.

D. Somatic Factors

Attitudes were defined initially in terms of a postural orientation and the resulting disposition for an ensuing action. Subsequent theory and experimentation has focused on the latter dynamic component evoked by a stimulus. Somatic factors have been of interest in these inquiries for two reasons: (1) measures of skeletomuscular activity have provided information regarding cognitive and emotional responses to attitude stimuli, and (2) skeletomuscular activity has been found to influence attitudes and persuasion.

As noted earlier, the initial studies of bodily responses and attitudes focused on autonomic activity. Over the past quarter century, however, somatic responses to emotionally evocative stimuli have been found to be highly differentiated. Research on facial efference (measured using electromyography) and observable facial actions has revealed that: (1) individuals perform at better-than-chance levels when categorizing facial expressions of happiness, sadness, fear, anger, disgust, surprise, and contempt; (2) the inductions of what individuals report as being positive and negative emotional states are associated with distinctive patterns of facial actions (emotional expressions); (3) distinctive expressions of emotion are displayed by neonates and the blind as well as sighted adults; (4) cultural influences can, but do not necessarily, alter these expressions significantly; (5) the variability in emotional expressions that can be observed across individuals and cultures is attributable to factors such as the individual or cultural significance of a given evocative stimulus and to cultural prescriptions regarding the display of emotions; and (6) although many subtle emotional processes or events are not accompanied by visually perceptible expressive facial actions, the valence and intensity of these emotions are accompanied by distinctive patterns of facial efference that are measurable using surface electromyography. Heightened facial electromyographic activity can also be recorded over the perioral (speech muscle) regions during silent language processing, and this has provided a method of examining gross differences in message processing in persuasion.

Manipulations of somatic activity such as body posture, arm flexion/extension, and facial expression also appear capable of modulating affective reactions, biasing message processing, and influencing persuasion. Early research, for instance, suggested that people have difficulty feeling a particular emotion (e.g., joy) when they are posed in a contrasting stance (e.g., anger). In another study, students were led to believe that they were participating in a study on the comfort and sound quality of stereo headphones when listeners were engaged in movement (e.g., dancing, jogging). Some students were told that they should move their heads up and down (vertical movements condition) about once per second to test the headphones, whereas others were told to move their heads from side to side. A final group of students heard no specific statements about head movements. Head movements were chosen because of their strong association with agreeing and disagreeing responses in a wide variety of cultures. Students heard muscial selections and either an editorial in favor of raising tuition at their university or one in favor of reducing tuition. Following the broadcast, the students answered questions, including what they thought tuition should be. Results revealed that students who heard the editorial favoring an increase in tuition supported a higher level of tuition than did students who heard the editorial favoring a reduction in tuition. More interestingly, this effect was modulated by head movements. Vertical head movements (as if nodding in agreement) led to the most attitude change to both editorials, and horizontal head movements (as if shaking in disagreement) led to the least attitude change to the editorials.

In conceptually related research, stimuli evaluatively categorized while performing isometric upper arm flexion (a somatic act associated with approach) are subsequently preferred to stimuli evaluatively categorized while performing isometric upper arm extension (a somatic act associated with withdrawal). The attitudinal effects of motor processes (arm flexion or extension) have also been found to be more evident when the stimuli had few associations (e.g., pronounceable nonwords) than when the stimuli had many associations in memory (e.g., familiar words). Somatic manipulations have been less effective in inducing affective responses, but a growing number of studies suggest that somatic events can modulate affect and persuasion.

VII. CENTRAL NERVOUS SYSTEM SUBSTRATES OF DISPOSITIONS TO RESPOND

From a biological perspective, attitudes and persuasion are ultimately products of the operation of the nervous system. In this regard, the features of the conceptual central and peripheral routes to persuasion evidence striking parallels with functional levels of organization in the nervous system. Although both the central and peripheral routes to persuasion can involve the highest levels of the central nervous system (CNS) (e.g., the cerebral cortex), it is probable that the elaboration likelihood continuum has its ultimate origin in fundamental ontogenetic and phylogenetic trends in CNS development. Historical and recent findings from the neurosciences support this view.

The highest levels of the CNS show the greatest expansion and elaboration through both the development of the individual (ontogeny) and that of the species (phylogeny), and serve to differentiate the adult human from the infant, and from other animals. In contrast, lower levels of the CNS (e.g., the spinal cord) evidence a more common, primordial oragnization throughout ontogeny and across phylogeny. Basic approach/withdrawal dispositions, however, are intrinsic to all levels of CNS organization, as documented by both experimental studies in animals and clinical findings in humans with spinal cord injuries. Cord transections isolate spinal networks from higher neural influences, leaving the lower regions of the body (e.g., the legs and trunk) under the exclusive control of spinal mechanisms. In spite of this loss of higher neural controls, the spinal cord is intrinsically capable of supporting reflexive limb-withdrawal to a noxious stimulus. This response is mediated, in part, by a relatively simple three-neuron circuit. The simplicity of this basic reflex circuit, however, does not imply an immutability of spinal networks. Just as attitudinal dispositions are subject to change, so are spinal dispositions. For instance, spinal networks can learn to withdraw from innocuous stimuli that, although not painful themselves, come to predict the occurrence of pain stimuli through Pavlovian conditioning.

The isolated spinal cord can also evidence primitive approach responses, as indicated by basic genital reflexes to tactile stimulation (erection, pelvic thrusting, and ejaculation). Thus, basic approach/withdrawal behaviors can be seen not only in the simplest of organisms, but also at the lowest level of organization in the mammalian central nervous system. Moreover, these basic approach/withdrawal dispositions do not differ from attitudinal dispositions in modifiability—spinal reflexes are clearly subject to learned modification with experience. At the same time, no one would mistake the primitive responses of the isolated spinal cord for the richness of the reactions of an intact organism to aversive or sexual contexts. The fundamental distinctions between these classes of reaction are two: (1) stimulus dependence and response variability and (2) the level of contextual control.

The first striking difference between the spinal organism (i.e., an organism with a lesion separating the spinal cord from the brain) and an organism with an intact connection between the brain and spinal cord is the repertoire of responses to, for example, an aversive stimulus. Although a spinal organism may show limb-withdrawal, the intact organism also exhibits more global escape and avoidance responses, aggression, vocalization, or instrumental responses that serve to eliminate or diminish the aversive stimulus. A related difference is apparent in the persistence of behavioral responses. Although the spinal withdrawal is highly stimulus bound, the intact organism may evidence behavioral activation, agitation, and escape attempts that persist well after the pain stimulus is withdrawn. That is, the aversive reaction of the intact organism is less dependent on the immediate sensory environment. Indeed, the stress that frequently characterizes human existence in contemporary societies is seldom directly related to pain stimuli.

This latter feature anticipates the second major distinction in the approach/withdrawal responses of spinal and intact organisms. That difference is in the complexity of contextual controls over behavior. The approach/withdrawal responses of the spinal organism appear to be sensitive only to relatively simple dimensions of a stimulus, such as its modality, intensity, or body location. In contrast, the aversive or sexual reactions of an intact organism frequently depend on highly complex relational features of the social/environmental context. In the intact organism, sexual arousal may not require direct tactile stimulation, but may be manifest in the presence of a specific individual, or by the thought of a specific individual, in an appropriate social context.

Studies of decerebrate organisms further illustrate the increase behavioral flexibility and the expansion of relational contextual controls over behavior that result from higher-order neural organizations. The decerebrate organism, deprived of the cerebral hemispheres, can display highly complex orofacial consummatory responses (orientation, chewing, swallowing) to palatable items placed within the mouth, and vigorous rejection responses to nonpalatable items. These approach/withdrawal behaviors parallel the basic capacities of the spinal preparation, but evidence a degree of complexity and integration beyond that characteristic of spinal systems. Moreover, these complex reflexive responses are not only influenced by the immediate sensory features of the stimulus (palatability), but are also sensitive to internal motivational conditions (e.g., food deprivation, metabolic need). Thus, even though mechanisms for reflex ingestive responses to palatable foods may be relatively hard-wired, their ultimate expression is further controlled by an additional class of internal stimuli.

Although decerebrates are sensitive to metabolic needs and are capable of competent ingestive (i.e., approach/withdrawal) responses in the presence of a suitable goal object, they still fail to adequately regulate food intake or maintain body weight in typical environments. What appears to be lacking in these organisms is the ability to anticipate metabolic need, to evidence goal searching behavior in the absence of an immediate stimulus for ingestion, or to respond to the normal contextual controls over food intake (e.g., social convention, passage of day, consideration of caloric need). These controls require a contextual representation that transcends simple dimensions of environmental stimuli (e.g., presence of food, level of food deprivation) and entails relational aspects among stimuli (e.g., passage of time together with the presence of food, or hunger together with the memory of the location of food). It is this latter class of contextual controls, entailing relational features among stimuli, or transcendent representations of the environment, that liberates an organism from the immediate dictates of the sensory environment and confers what has been interpreted as deliberative or goal-directed action.

The progression of increased behavioral flexibility and the expansion of relational contextual controls over behavior constitute hallmarks of higher-level neural organizations. These higher-level systems appear to be organized partly in a hierarchical fashion, and both extend the sensory processing of lower systems and expand on the motor repertoire and flexibility of lower mechanisms. Moreover, this hierarchical organization permits multiple levels of analysis and control over behavioral processes. In response to a pain stimulus, for example, lower-level processing may predominate initially, resulting in a rather stereotyped but highly adaptive, short-latency limb-withdrawal. The significant advantage of lower-level processing is that, although somewhat inflexible, it is highly efficient and places minimal burdens on higher-level processing substrates. Indeed, for the initial protective response, elaborate processing of the stimulus is not necessary, and in fact may be maladaptive. Lower-level processing, however, does not preclude further analysis at higher perceptual levels. In the case of a pain stimulus, this further analysis may be manifest in subsequent emotional reactions (e.g., fear, anxiety), which may motivate subsequent behavior (e.g., avoidance, aggression).

This pattern of multiple-level analysis and control confers significant advantages. By their nature, higher-level organizations must integrate information from varied modalities and sources and exert control over diverse aspects of behavior. This convergence of sensory information, the need for integration with prior memories, and the divergence of output control can create a processing bottleneck that taxes the information-processing capacity of neural networks. Consequently, these higher-level systems may require active attentional focus and may have limited capacity for multiple concurrent activities. Those stimuli or conditions that do not effectively compete for attentional resources may be subject to only lower-level processing, or to fairly elementary processing by higher-level neural networks. An important question, and an actively researched area, relates to the determinants of which stimuli are selected for further processing. This is also a fundamental question in the area of attitudes and persuasion, since it addresses the distinction between peripheral and central routes to persuasion and their underlying mechanisms.

Parallels can be noted in the distinctions between central and peripheral routes to persuasion and the continuum between hierarchical and lower-level information processing. The peripheral processing route, lower-level neural processing, and elementary information processing within a higher level of neuraxis are characterized by: (1) minimal cognitive elaboration, (2) limited flexibility, (3) stimulus dependency or relative lack of persistence, and (4) reliance on rather simple, nonrelational features of the stimulus context. In contrast, the central route to persuasion as well as higher-level neural processing are characterized by: (1) elaborate, integrative analysis of multiple stimuli, (2) maximal flexibility, (3) persistence and resilience, and (4) reliance on complex relational features of the stimulus context (i.e., abstractions). In both neural processing and persuasion, analysis may frequently take place across multiple levels, within a level to varying degrees, or may shift between one level or another, depending on the context and competing demands.

With rapidly developing electrocortical and neuroimaging techniques, views are available of apsects of brain function in conscious individuals during a variety of cognitive and affective activities. These techniques, together with more basic studies of cellular processes, information transformations among neurons, and behavioral analyses of attitudes and persuasion, should provide the tools for a truly interdisciplinary study of social psychological phenomena such as attitudes and persuasion. For instance, evaluative categorization processes have been found to be measurable using event-related brain potential, and these measures have been found to be relatively insensitive to attitudinal response (e.g., selection/execution) processes. In a typical study, subjects are exposed to sequences of six traits and perform a dichotomous evaluative categorization task. Evaluative inconsistency may be varied by embedding, for instance, very positive, moderately positive, moderately negative, and very negative traits in sequences containing predominantly very positive traits. Results have shown that highly and moderately evaluatively inconsistent traits, compared to mildly inconsistent or consistent traits, evoke a larger-amplitude late positive potential (LPP) that is maximal over centroparietal scalp regions. Furthermore, extremely evaluatively inconsistent traits evoke a larger-amplitude LPP than moderately evaluatively inconsistent traits even though both sets of traits are judged to be negative. Research has also shown that the amplitude of the LPP is larger when subjects evaluatively categorize inconsistent rather than consistent attitude stimuli even when they intentionally misreport their attitudes. Finally, analyses of the spatial distribution of the LPP in these studies have revealed that, consistent with the relative involvement of the right hemispheres in hedonic processes, evaluative categorizations evoke a more asymmetrical (right lateralized) LPP than nonevaluative categorizations of the same stimuli.

In sum, all species must be capable of discriminating positive from negative stimuli and events in their environment. For simple organisms with limited environmental niches, relatively fixed (reflex) mechanisms can effectively process environmental stimuli to support life. All species have this evolutionary legacy, but the development of more complex stimulus representation, appraisal, and reasoning processes can greatly expand the adaptability and flexibility of a species. In humans, attitudes and evaluations can be socially and culturally transmitted and can help people understand events and guide their actions in their world. Because attitudes are so central to people's conceptual organization of their world, attitudes are subject to modification based on vicarious experience and information (persuasion). Social scientists have provided a rich corpus of effects, principles, and theories of persuasion, whereas neuroscientists have provided basic knowledge of neural organization and motivation processes. Social scientists and neuroscientists initially worked in isolation, but common interests have fostered interdisciplinary research and intertheoretic translations that are providing new insights into the processes of attitude formation and persuasion.

BIBLIOGRAPHY

Arvey, R. D., Bouchard, T. J., Jr., Segal, N. L., and Abraham, L. M. (1989). Job satisfaction: Environmental and genetic components. *J. Appl. Psychol.* **74,** 187–192.

Berntson, G. G., Boysen, S. T., and Cacioppo, J. T. (1993). Neurobehavioral organization and the cardinal principle of evaluative bivalence. *Ann. N. Y. Acad. Sci.* **702,** 75–102.

Cacioppo, J. T., Crites, S. L., Jr., Gardner, W. L., and Berntson, G. G. (1994). Bioelectrical echoes from evaluative categorizations. I. A late positive brain potential that varies as a function of trait negativity and extremity. *J. Personality Social Psychol.* **67,** 115–125.

DeCasper, A. J., and Spence, M. J. (1986). Prenatal maternal speech influences newborns' perception of speech sounds. *Infant Behav. Dev.* **9,** 133–150.

Eagly, A., and Chaiken, S. (1993). "The Psychology of Attitudes." Harcourt Brace Jovanovich, Fort Worth, Texas.

McGuire, W. (1985). Attitudes and attitude change. *In* "Handbook of Social Psychology" (G. Lindzey and E. Aronson, eds.). Random House, New York.

Petty, R. E., and Cacioppo, J. T. (1986). "Communication and Persuasion: Central and Peripheral Routes to Attitude Change." Springer-Verlag, New York.

Tesser, A. (1993). The importance of heritability in psychological research: The case of attitudes. *Psychol. Rev.* **100,** 129–142.

Pesticide Residues in Foods

ROBERT A. HOWD
ANNA M. FAN
California Environmental Protection Agency[1]

GLOSSARY

De minimus risk Legal term defining a risk level too small to be considered of health significance

Margin of safety Ratio between the highest dose of a chemical that has no adverse effect in an exposed organism and the actual or estimated exposure dose

Pesticide Product used to kill or control a pest, including insecticides, rodenticides, herbicides, fungicides, antimicrobials, plant and animal growth regulators, and other biocontrol agents (except antibiotics)

Reference dose (RfD) U.S. EPA estimate of the maximum daily dose of a chemical that is likely to be without appreciable risk of deleterious effects over a lifetime

Risk assessment Scientific process of estimating the probability of an adverse effect in an organism or population following exposure to a chemical (or other potential hazard)

Tolerance Legally enforceable, maximum allowable level of a pesticide residue in a food commodity

TO UNDERSTAND THE ISSUES CONCERNING PESTIcide residues in food, it is necessary to consider uses of pesticides, how they are regulated to avoid excessive residues in foods, and how scientists evaluate the potential risks of food residues of the chemicals to help ensure a wholesome and safe food supply. The safety of pesticide residues continues to attract wide public attention, although according to some opinions, the degree of concern has greatly exceeded actual risks. It is argued that, compared to other risks encountered in our daily lives, including those caused by excessive food consumption or failure to maintain a balanced diet, pesticide residues do not present a significant adverse impact. Responsible growers, distributors, and regulators work together to ensure a wholesome food supply. However, maintaining a safe and abundant food supply depends in practice on the adequacy and availability of information on the chemicals used, good agricultural practices, and enlightened public policies. By design, pesticides are used to kill or control pests, and thus in most cases they are expected to be toxic to living organisms. The regulations of the United States Environmental Protection Agency (U.S. EPA) promulgated under the Federal Insecticide, Fungicide and Rodenticide Act (FIFRA) are intended to avoid or minimize risk to all nontarget species. Data submitted to register pesticide products enable the U.S. EPA and related responsible agencies to make appropriate regulatory decisions to ensure safety of the products. These decisions are based on information on the efficacy, environmental persistence, persistence in treated plants or animals, toxicological effects, and potential exposure to the pesticides proposed for registration.

Pesticides can be broadly categorized as agricultural versus nonagricultural based on the use pattern (e.g., crop or noncrop use). The greatest volumes of pesticides are used in agriculture, followed by use in industry, commerce and government, and then home and garden use. Pesticides applied to foods or used near foods to prevent infestation by animals, plants, or

[1]The views expressed are those of the authors and do not necessarily represent those of the Office of Environmental Health Hazard Assessment, the California Environmental Protection Agency, or the State of California.

ENCYCLOPEDIA OF HUMAN BIOLOGY, Second Edition, VOLUME 6.

microorganisms may reasonably be anticipated to result in trace residues in foods. However, the chemical nature of the products and their patterns of use are very diverse, providing great differences in the likelihood of contamination problems. Factors important in the evaluation of pesticide residues include the amount or concentration of the pesticide applied in each use, the total amount of a chemical and all similar chemicals used, the time(s) of application (preplant, preharvest, postharvest), and the persistence of the chemical(s) in the environment. Amounts of pesticides used vary from hundreds to millions of pounds annually. Some chemicals are unique, whereas others are members of a common chemical/structural group like the organophosphates. Persistence varies greatly by chemical class and structural variations within a class and is related to properties such as fat and water solubility, volatility, and chemical reactivity. Chemical stability and fat solubility are particularly relevant because they can result in bioaccumulation, as has been observed for many organochlorine pesticides. All of these factors contribute to residual traces of pesticides or their metabolites in foods; the details and significance are discussed in the following sections.

I. AGRICULTURAL USES OF PESTICIDES

A. Pesticide Types

1. Insecticides

Among the insecticides, important categories of chemicals used for their acute toxic effects include chlorinated hydrocarbons, cholinesterase inhibitors (organophosphates and carbamates), pyrethrins, formamidines, avermectins, and various broad-spectrum fumigants.

Chlorinated hydrocarbons (e.g., DDT, chlordane, lindane) cause stimulation or depression of the central nervous system. Clinical signs and symptoms include nausea, mental confusion or semiconsciousness, jerking of limbs, dizziness, lethargy, weakness, and anorexia. In general, organophosphates (e.g., malathion, diazinon, chlorpyrifos) are irreversible inhibitors of the enzyme acetylcholinesterase, which is required for nervous system function. Clinical manifestations of moderate inhibition include headache, visual disturbances, pupillary abnormalities, and increased secretions. Severe poisoning results in nausea, vomiting, pulmonary edema, changes in heart rate, muscle weakness, respiratory paralysis, confusion, convulsion or coma, and death. Carbamates (e.g., aldicarb, carbaryl) are reversible cholinesterase inhibitors. Signs and symptoms of intoxication are similar to those of organophosphate poisoning. Pyrethrins are also nervous system stimulants, with effects similar to those of DDT. They may be organic (e.g., pyrethrum, derived from chrysanthemums) or synthetic (e.g., permethrin, cypermethrin, fluvalinate). The synthetic forms are more stable in sunlight and usually more toxic and effective than the original plant extracts. Formamidines such as amitraz are stimulants that inhibit the enzyme monoamine oxidase. Avermectins (abamectin) are microbially derived antibiotics with insecticidal actions. The fumigant pesticides are broad spectrum in their actions and are of several different chemical types.

2. Herbicides

Herbicides may be either broad spectrum or selective in their actions. There are several chemical classes—the old traditional ones such as copper salts (nonselective); and newer types including nonselective postemergent chemicals such as glyphosate (Roundup); selective plant hormone-type pesticides such as 2,4-D, usually used postemergent; contact desiccants such as paraquat and diquat; and defoliants such as DEF (tribufos). They can have many different toxic effects in mammals and other nontarget organisms (see Section II,C).

3. Fungicides

Fungicide types include the traditional sulfur and copper products; non-cholinesterase-inhibiting carbamates or thiocarbamates such as benomyl and ziram, thiazoles such as etridiazole, triazines such as anilazine, substituted organics such as chlorothalonil, dicarboximides such as folpet or captan, fumigants such as methylisothiocyanate (MITC) and methyl bromide, and other miscellaneous compounds. Though the fungicides generally have low mammalian toxicity, many of them have been shown to cause cancer or reproductive toxicity in experimental animal studies.

4. Other Agents

Other agricultural pest control agents are significant for special purposes. Dozens of product types can be used to kill, attract, or repel various pests.

The dry powder insecticides such as boric acid, sodium borate, and diatomaceous earth are popular low-toxicity products, approved for use in food preparation and storage areas. However, they are not without risk. Acute poisonings with boric acid have injured and killed infants. Diatomaceous earth can contain crystalline silica, which is harmless when taken by mouth (so food residues are of no public health significance) but is a potential carcinogen by inhalation.

Rat poisons have resulted in many cases of poisoning due to food contamination, when the chemicals were accidentally mixed into grain or processed foods. Sodium fluoroacetate (also known as Compound 1080) was withdrawn from general agricultural use in the United States because of its extraordinarily high toxicity and hazard. Residues from the other chemicals such as warfarin have not been a problem in foods.

Ultraviolet light-reactive dyes represent a new category of insecticide, currently being tested against fruit flies. When consumed by a transparent-bodied insect, dyes such as phloxine B apparently form free radicals in sunlight, which poison the insect. The dyes are of low toxicity to mammals and birds.

Insect growth regulators work by a variety of mechanisms, but often disrupt maturation. They are generally of low toxicity to nontarget organisms and are used in low amounts so there is not a residue problem.

Pheromones are male and female insect attractants, which may be used to trap pests or disrupt aggregation and mating. These specific uses do not result in food residue problems.

Other attractants and repellents are used in traps (e.g., fermented protein and sugar solutions) and sprayed on plants or soil areas. Most products of this type, such as hot pepper sprays and mountain lion urine, are not of toxicological concern for food residues.

Biological products such as *Bacillus thuringiensis* (BT, a bacterial product that inhibits insect molting), insect viruses, and fungi that are harmless to plants but inhibit growth of other fungi are under active development by many manufacturers. Live insects that attack other insects (e.g., ladybugs for aphids and parasitic wasps for whiteflies) are also becoming popular. No associated toxicological or food residue problems are presently known.

Genetically engineered products, such as corn or soybeans modified to produce a BT toxin, may or may not become a health concern, depending on the properties of the added gene product. For example, introduction of a natural alkaloid toxin to discourage predation, or a potentially allergenic protein, could lead to adverse effects in humans or animals.

5. Food Handling or Processing Chemicals

Postharvest chemical applications to prevent infestations or microbial growth are a major category of pesticide use. The chemicals involved include soaps and detergents, quaternary surfactants, phenols, chemicals that release free chlorine or bromine, peroxides, and sulfites, plus some of the pesticides mentioned earlier. Foods may contact their residues on food processing machinery, or be sprayed or dipped with such chemicals for sanitizing or storage. Traces of the late-treatment antimicrobials and postharvest fumigants are common on food as consumed, and thus only very low toxicity chemicals are appropriate for these purposes.

B. Pesticide Tolerances in Food

A series of field tests are currently required of pesticide manufacturers to determine how to use their new proposed pesticides for maximum efficacy in crop protection. Crop samples are analyzed to determine the concentrations of the chemical and its metabolites in the edible portions. The pesticide doses are estimated, taking into consideration food consumption patterns. These dose estimates are compared to the safe dose determined from the animal testing (the highest no observed effect level, or NOEL). If available, the EPA's Reference Dose (RfD) or the United Nations Food and Agriculture Organization's (FAO) Acceptable Daily Intake (ADI) levels for the pesticide are used for comparison. A tolerance, or maximum allowable level for the pesticide in the food, is then proposed that is far enough below the NOEL, RfD, and/or ADI to ensure that a toxic dose is unlikely to be achieved, including a margin for uncertainty. The tolerance is also proposed high enough so the food residue levels should not exceed the tolerance when the pesticide is used according to good agricultural practice.

The U.S. EPA then sets the tolerances based on a review of the available data and the applicable regulations. To limit the possibility of excessive exposures to the pesticide residues in food, or to combined levels of similarly acting chemicals in food, the EPA may adopt a tolerance lower than suggested in the manufacturer's tolerance petition. The EPA may also seek to limit the use of chemicals that are particularly hazardous to the applicators or to the environment in setting tolerances and use conditions. In balancing the various considerations, the EPA may adopt different tolerances for different foods. Food items that are consumed in small quantities, such as spices, could be allowed larger tolerances than foods consumed in larger amounts, such as most fruits and vegetables. If the food is a staple of some population group, like apples are for children (including apple juice, apple sauce, and raw apples), the tolerance could be even lower. A less restrictive tolerance might be adopted for livestock feed, if the pesticide does not accumulate in the animal tissues. The assumptions used to set tolerances in specific foods are not published, however, which has led to concerns that potentially sensitive populations such as infants and the elderly may not have been considered in some cases as needed. In response to this concern, an additional safety factor of up to 10-fold to account for potential greater susceptibility of infants to pesticide residues has been specified in the recently signed Food Quality Protection Act of 1996.

To evaluate the adequacy of tolerances for acute exposures, the variability in pesticide residue levels must be considered. That is, what is the likelihood of a consumer being exposed to a residue at or above the legal tolerance? Food monitoring consistently shows less than 1% of above-tolerance samples, and the average pesticide residue level at less than 10% of the tolerance. Interindividual variability in food consumption is also studied. Modeling of food consumption patterns and actual residue levels, as discussed in Section III, can be used to evaluate the potential for acute adverse effects from pesticide exposures.

To estimate the health risks associated with repeated daily exposures to pesticides in foods, chronic effects such as cancer must be considered. Exposure is estimated based on average lifetime consumption of all foods that may contain the pesticide residues. The exposure estimates are also based on average levels of the pesticides found in the foods, rather than on the tolerances; the average residue level is always well below the tolerance. The determination of appropriate tolerance levels must take into account any accumulation of the residues in human tissues (as well as in plant and animal tissues). However, chronic consumption of all chemicals with a similar mechanism of toxicity has not, in the past, been considered in setting individual pesticide residue tolerances. The 1996 Food Quality Protection Act requires the EPA to consider exposure to pesticides with a common mechanism of toxicity in adopting tolerances. [*See* Food Toxicology.]

II. PESTICIDE TOXICITY TESTING AND RESIDUE MONITORING

A. Pesticide Laws

Pesticides are regulated in the United States under the Federal Insecticide, Fungicide, and Rodenticide Act and the Federal Food, Drug and Cosmetic Act (FFDCA). FIFRA specifies requirements for pesticide registration and use; the FFDCA regulates tolerances for pesticides in food. Both laws have been significantly revised by the Food Quality Protection Act of 1996. Revisions to FFDCA have, among other things, replaced the Delaney Clause (regulating carcinogens in processed foods) with a new general safety standard for both raw commodities and processed foods, specified as "a reasonable certainty of no harm." The FFDCA previously authorized pesticide residue tolerances for raw commodities under Section 408, applying a risk–benefit standard to pesticide residues, including those shown to cause cancer in laboratory animals. Under this standard, residues of carcinogenic chemicals were legal if the chemical caused a *de minimus* risk (often defined as less than a 1 in 1 million chance of causing cancer) and provided benefits to balance any risks. The FFDCA Section 409 tolerances for processed foods were subject to a zero tolerance for carcinogens IF the pesticide concentrated in the processed food, compared to the raw commodity (a provision inserted by Congressman Delaney in 1958). Accordingly, pesticides shown to be carcinogenic in a relevant animal study would be prohibited in many processed foods and, by extension, in the original raw commodities. If the processed food residue levels were equal to or lower than those in the raw commodity, then Section 408 applied, and the carcinogens were allowed.

The Delaney Clause has been the subject of controversy in the pesticide and food safety area for years. It presented regulatory agencies with a dilemma based on conflicting requirements for raw commodities versus processed foods, resulting in a situation in which presumably safer, negligible-risk pesticides would be prohibited in favor of others that could pose higher total risks, based on noncancer effects, which was known as the "Delaney Paradox." However, the Delaney Clause had little impact on pesticide regulatory practices until the late 1970s, when advances in toxicity testing, updated risk assessment methodology, and the associated guidelines led to the classification of certain pesticides as "possible" or "probable" human carcinogens. Thereafter (until the late 1980s) the U.S. EPA did not register a new pesticide on food crops if the chemical was considered a potential carcinogen and would require a Section 409 tolerance. In 1985, the Agency asked the Board of Agriculture of the National Research Council to study the Agency's methods for adopting tolerances for pesticide residues in food, requesting specifically an examination of the current and likely future impacts of the Delaney Clause on the tolerance-setting process. In its 1987 report, the Board recommended that pesticide residues in food be regulated on the basis of consistent standards. U.S. EPA was advised to apply a negligible risk standard (*de minimus*) when considering use of a carcinogenic pesticide on both raw and processed foods. For the next few years, tolerances under Sections 408 and 409 were approved based on a working definition of "negligible," which is "in the range of one in a million." This practice was challenged in court, and in July 1992, the U.S. Ninth Circuit Court of Appeals ruled that the negligible risk policy violated the Delaney Clause. The effect of this was to require several dozen pesticides or specific crop tolerances to be canceled.

The impending cancellations caused renewed interest in changing the law. After various debates and attempts, the Federal Food Quality Protection Act of 1996 was passed with several requirements to provide increased protection from potential adverse effects of pesticide residues in foods and to eliminate the Delaney Clause. Under this new law, the EPA must determine that tolerances are "safe," defined as "a reasonable certainty that no harm will result from aggregate exposure" to the pesticide. Tolerances for pesticide residues will be set by the same process in all types of food, raw or processed; furthermore, the standards apply to all human health risks, not just cancer risks. New hazard evaluations, and new tolerances, will be phased in over the next several years, based on these standards.

B. Toxicity Test Requirements

To register a pesticide in the United States, an extensive series of data requirements must be met. In general, pesticides must be shown to be both safe and effective. A series of animal tests must be conducted to evaluate the potential toxicity to nontarget organisms and to document the chemical's absorption and metabolism in mammalian test species. Limited environmental degradation and transport studies must also be carried out. A sensitive assay for the chemical and any toxic metabolites in plant and animal tissues must be provided. The concentration of pesticide residues in the edible plant tissues at harvest is determined, and the safety of those levels is evaluated, both to humans and to animals, if applicable (e.g., for alfalfa and feed grains).

The toxicity test procedures include acute, subchronic, and chronic studies, including dermal and eye irritation, reproductive, and developmental studies (Table I), plus selected *in vitro* tests (Table II). Various dosing routes, including oral, dermal, and inhalation, are evaluated. The *in vivo* studies require about 6 to 50 animals of both sexes at multiple doses for most tests. Rats and mice are the suggested test species for most of the general toxicity tests, with guinea pigs, rabbits, and dogs being used for several of the more specialized studies such as eye irritation, dermal sensitization, and teratology. Not all of the tests are necessarily required for every pesticide. Antimicrobial and biochemical pesticides, for example, are subject to a tiered testing protocol that requires a minimal list of studies if acute toxicity is low and there would be little exposure. For agricultural pesticides, the acute toxicity to fish, gamebirds, and honey-

TABLE I
Standard U.S. EPA *in Vivo* Toxicity Tests for Pesticides

Acute tests
LD_{50} by oral, dermal, and inhalation routes (two or more species)
Eye and skin irritation
Skin sensitization
Subchronic tests
90-day oral toxicity
Repeated dose dermal toxicity (21, 28, and/or 90 day)
Subchronic inhalation toxicity
Prenatal developmental toxicity
Reproduction and fertility study
Chronic tests
Chronic toxicity/carcinogenicity (two species)
Special tests
Neurotoxicity study (for organophosphates)
Metabolism and pharmacokinetics
Dermal penetration
Genetic/mutation studies—*Drosophila* sex-linked recessive test, rodent specific locus test, chromosomal cytogenetic studies (sperm, bone marrow, red blood cell), and dominant lethal, heritable translocation, and sister chromatid exchange assays

TABLE II
Standard U.S. EPA *in Vitro* Tests for Pesticides

Escherichia coli reverse mutation test
Salmonella reverse mutation test (Ames assays)
Gene mutation in *Aspergillus nidulans* and *Neurospora crassa*
In vitro mammalian cytogenetics
Bacterial DNA damage or repair
Unscheduled DNA synthesis
Mitotic gene conversion in *Saccharomyces cerevisiae*
Sister chromatid exchange

bees is also usually determined. No human toxicity tests are required, although pesticide manufacturers may choose to determine the acute toxicity of their products in limited human studies.

In practical applications, the acute toxicity values for pesticides based on LD_{50} in rats and skin and eye effects have been used to catalog and label pesticides, as shown in Table III.

C. Major Adverse Effects of Pesticides

Because of the intricate relationship between pesticide use and human exposures, it is important to have an understanding of the toxicity of pesticides and the associated implications for human health. The most significant health effects or toxicity issues are summarized in the following sections.

1. Pesticides as Carcinogens

The potential for human carcinogenicity of pesticides is evaluated based on both human epidemiological and experimental animal data. In correlating human and animal data, species and dose differences result in many inconsistencies and uncertainties. For humans, simultaneous exposure to multiple chemicals and the undefined duration and extent of exposure make it difficult to establish cause and effect relationships. In case–control studies, excess incidences of soft-tissue sarcoma, leukemia, and lymphoma have been noted among farmers and foresters exposed to organochlorine insecticides, particularly DDT, chlordane, and lindane. Animal cancer bioassays for DDT, chlordane, and many other halogenated hydrocarbons of this type are usually positive. In several case–control and cohort studies of herbicide manufacturers and applicators, soft-tissue sarcoma and non-Hodgkin's lymphoma have been associated with phenoxyacetic acid herbicides like 2,4-D and 2,4,5-T, but these chemicals have not been shown to be carcinogenic in animals.

Under the current pesticide registration requirements, long-term carcinogenicity studies in two species are required if a food or crop tolerance is involved. One can be a combined chronic and carcinogenicity study and it is usually carried out in rats. Pathological lesions are diagnosed and reported as hyperplasia, nodules, and benign or malignant tumors, and by organ and tumor type. Interpretation involves consideration of many factors, including pattern of tumors (type, sites, time of appearance), concurrent toxicity, current and historical control rates, study design, the presence of contaminants, conduct of study, high mortality or early deaths, and other factors. U.S. EPA has identified over 50 pesticides as having carcinogenic potential from among the several hundred active ingredients in current use. [*See* Chemical Carcinogenesis.]

2. Pesticides as Reproductive and Developmental Toxicants

There has been increasing concern regarding reproductive and developmental effects of chemicals, including pesticides, in humans. Pesticides are generally

TABLE III
Acute Toxicity Classifications for Label and Use Restrictions

	Hazard indicators: Signal word and toxicity categories			
Toxicity measures	Danger I	Warning II	Caution III	Caution IV
Inhalation LD_{50} (mg/liter)	0–0.2	0.2–2	2–20	>20
Oral LD_{50} (mg/kg)	0–50	50–500	500–5,000	>5,000
Dermal LD_{50} (mg/kg)	0–200	200–2,000	2,000–20,000	>20,000
Eye effects	Corneal opacity >7 days	Corneal opacity <7 days, irritation >7 days	No corneal opacity, irritation <7 days	No irritation
Skin effects	Corrosive	Severe irritation at 72 hr	Moderate irritation at 72 hr	Mild or slight irritation at 72 hr

evaluated in rats and rabbits for potential adverse effects in both parents and offspring.

Benomyl produced developmental and male reproductive toxicity. Molinate produced sperm abnormality in rats. The organochlorines such as DDT, hexachlorobenzene, lindane, and toxaphene have been implicated in abnormal menses and impaired fertility; such effects may be mediated through direct interactions with estrogen receptors or through metabolism to estrogenic analogues. Some organophosphates may also produce abnormal menses and premature menopause but the available information is very limited.

Dibromochloropropane (DBCP) is one of the most extensively studied pesticides for adverse reproductive effects. It was used as a nematocide mostly along the southern Atlantic and Pacific coasts and Hawaii from the mid-1950s until its ban in the late 1970s on crops such as citrus, grapes, peaches, pineapples, soybeans, and tomatoes. Testicular atrophy was first reported in 1961 in rats, guinea pigs, and rabbits; in 1977, low sperm counts and abnormal sperm were discovered in DBCP production workers. Widespread use of the chemical resulted in persisting groundwater contamination in some areas, but it is not a significant food contaminant.

Another human reproductive toxicant is chlordecone, or Kepone. It was used as an agricultural insecticide and fungicide from 1958 until its ban in 1975 because of neurotoxicity in production workers. The chemical primarily affects sperm motility and viability as seen in humans and in rat studies. It has been a persistent problem in fish and shellfish in the James River in Virginia, but is not a significant contaminant of produce. [*See* Teratology.]

3. Pesticides as Genotoxic Agents

Testing for genotoxic effects of pesticides using a battery of *in vitro* short-term tests and *in vivo* assays identifies chemicals that produce effects such as mutation, chromosome translocations and breakage, and formation of DNA adducts. The finding of mutagenicity of many carcinogens has been interpreted as implying that carcinogens are mutagens or that mutagens are presumed carcinogens. However, several different mechanisms can alter cell division to produce tumors (such as with DDT, chlordane, and heptachlor) without a mutagenic effect. Overall, genotoxicity should be evaluated as a distinct biological end point in addition to supportive evidence for carcinogenicity.

4. Pesticides as Immunomodulators

Immunotoxicity is a relatively new area of concern in pesticide evaluation. Testing for such effects has not been required for pesticide registration, except for skin sensitization, but there is a current effort to include evaluation of more parameters. Pesticides can increase or decrease immune system responsiveness. Contact or respiratory hypersensitivity (allergic responses) is among the most striking effects. Immunosuppression can decrease body defense mechanisms leading to increased susceptibility to infection or tumor development. Guinea pigs and mice are the most common test animals, but the results of a battery of tests are not necessarily a good predictor of human responses because of the complexity of immune systems and the wide variability.

Pesticides reported to have immunological effects in animals include methylparathion, DDT, mirex, malathion, pentachlorophenol, aldicarb, hexachlorobenzene, chlordane, and tributyltin oxide. Sensitivity to malathion was discussed intensively when the chemical was used for Mediterranean fruit fly eradication in southern California in 1990. The effects of pentachlorophenol (technical grade) and 2,4,5-T (manufactured between 1958 and 1969) were shown to be mediated by 2,3,7,8-tetrachlorodibenzo-*p*-dioxin (TCDD) present as a contaminant. Persons exposed to methylisocyanate (used as an intermediate in the production of carbamate pesticides) following the accident in Bhopal, India, were reported to have effects on immune responses, primarily as effects on lymphocytes, but little effect of the chemical was seen in animals. Pesticides showing hypersensitivity reactions, which are of particular importance to occupational exposure, include malathion, captan, benomyl, and naled in the guinea pig skin sensitization test, and dinitrochlorobenzene (DNCB), benomyl, maneb, and naled in the human diagnostic patch test.

5. Pesticides as Neurotoxic Agents

Although most types of pesticides can have acute effects on nervous system function, more long-lasting or cumulative adverse effects caused by chronic exposure are also possible. Current toxicity tests may not reveal such effects, particularly on the developing nervous system. Single high-level exposures to organophosphates can cause persistent nerve or muscle damage, for example. Mercurials can also cause permanent brain damage, but these are no longer used as pesticides since their cumulative hazards have been identified. The concern is that other such effects may be occurring but are as yet unrecognized owing to the lack of sensitive animal tests and the limitations of epidemiology studies. The recent development of U.S. EPA guidelines for a battery of sensitive behavioral, neurological, and electrophysiological tests helps address the need for test methods, but such tests are not required for most pesticides.

D. Use of Toxicity Data in Risk Assessment

Animal tests commonly utilize high doses, including the "maximum tolerable dose" (MTD), which is defined as the maximum dose that causes no severe toxicity or weight loss, up to a maximum test level (usually 1 or 2 g/kg body weight, depending on the test). From the series of tests, the lowest observed effect level (LOEL) and no observed effect level (NOEL) are also estimated for noncarcinogens. For carcinogens, one of various mathematical models is used to estimate risk of tumors at different (chronic) exposure levels. No specific extrapolation to estimate risk of an adverse effect has been common for noncarcinogens, because of the presumption that at a sufficiently low dose, there will be no risk of an adverse effect (i.e., that there exists a dose threshold). A presumed no risk level is estimated by applying safety or uncertainty factors to the LOEL or NOEL dose. Thus, the NOEL for kidney cytotoxicity in a rat chronic study might be divided by 100 to obtain a dose estimated to avoid all risk of kidney toxicity to humans.

More specifically, an uncertainty factor of 10 is generally used to estimate safe levels to humans from data in another mammalian species. A factor of 10 is also used to account for the potential variability among humans, including sensitive subgroups. Another factor of 10 is often used in risk assessment to account for the uncertainty in estimating chronic toxicity level, if only acute tests are available (although for pesticides, chronic studies are generally required). This does not assume that humans are 100 times as sensitive as rats, but merely that, in the absence of more specific knowledge, it is prudent to acknowledge the uncertainty by setting a lower maximum allowed exposure. This is known as the margin of safety (MOS) or margin of exposure (MOE). Actual human data, if available, might show humans to be less sensitive to kidney effects and more sensitive to cardiovascular effects than rats, for instance. Extra sensitivity might be found or assumed in some specific group, such as infants or people with kidney disease.

More accuracy in dose extrapolation within and between species for both carcinogens and noncarcinogens may be accomplished with better understanding of toxic mechanisms and comparative physiology. Mathematical models of the distribution and metabolism of chemicals in tissues based on measured biochemical and physiological parameters, known as physiologically based pharmacokinetic (PBPK) models, have been shown to produce excellent cross-species extrapolations. When the toxic action and potency of a chemical are known from studies in isolated tissues or cells, the *in vivo* pharmocodynamics can be modeled very accurately. This can greatly decrease the uncertainty in toxicity predictions and justify lower uncertainty factors. For carcinogens, it may be possible to demonstrate that tumors are occurring by a threshold mechanism (e.g., stimulation of cell division) rather than by a potentially nonthreshold mechanism such as DNA alkylation. The new draft U.S. EPA cancer guidelines (1996) acknowledge this potential, but present knowledge is inadequate to make the determination for most carcinogens.

As has been discussed at length in many contexts, tests in animals are limited in their ability to predict adverse effects in humans. However, based on the many physiological similarities, a panoply of different tests in multiple species can help to identify disruptions of critical physiological systems, which will provide warning of a potential effect in humans. With experience, the design and interpretation of animal tests have continued to improve, although perhaps the limitations still get more attention than the successes. Effects for which there is still much uncertainty include cancer, certain types of neurotoxic effects, immunotoxicity, endocrine effects, and developmental effects. Improvements in cross-species modeling are not sufficient to improve extrapolations when the toxic mechanisms are unknown or in dispute.

Other important questions for providing acceptable margins of safety include the potential combined effects of many compounds in the environment and the level or type of effect that should be considered significant (or "adverse") for risk assessment. Chemicals that interact with endocrine receptors such as DDT and chlordane, for example, could be interacting to produce effects at lower levels than either alone. Another pesticide might form a complex with a blood protein that is inconsequential in untreated individuals, but that increases the proportion of free hormone in the blood to potentiate the effects of DDT and chlordane. All potential effects and interactions cannot possibly be predicted or tested for. Prudence requires application of a safety factor to acknowledge this overall uncertainty in estimating safe levels of chemicals in foods to establish the pesticide tolerances.

III. ESTIMATING PESTICIDE EXPOSURE

A. Commodity Sampling for Pesticide Residues

The U.S. Food and Drug Administration and U.S. Department of Agriculture (USDA) sample and analyze foods to determine pesticide residue levels. The State of California also has a large food sampling program. The FDA sampling programs follow three different approaches, referred to as regulatory monitoring, incidence/level monitoring, and market basket surveys for the Total Diet Study. The USDA samples meat and poultry, and also tests produce under its Pesticide Data Program to help support minor use pesticide registration.

Under FDA's regulatory monitoring program, which is intended to evaluate and enforce adherence to pesticide tolerances, about 1500 to 2000 samples per year are analyzed around the country, including imported foods. Less than 1% of the samples are found with illegal residues, most of which are trace levels of chemicals for which there is no tolerance in the particular commodity. A much smaller proportion have residues over the tolerances. In the corresponding and complementary California program ("marketplace surveillance"), about 5000 samples are taken of locally grown or imported foods. In 1994, the most recent year for which records are available, no residues were detected in 66% of samples. An illegal residue was found in 1.5% of the foods, of which 0.4% were overtolerance violations (about 20 samples). The proportion of samples that contained residues at 50 to 100% of tolerance was 1.34%. This is not a completely random sampling program, so results may not be applicable to all commodities.

The FDA incidence/level monitoring program is designed to provide more information on specific commodities or pesticides. Infant foods and commodities treated with benomyl/thiabendazole, ethylenethiourea (ETU), aldicarb, or daminozide (Alar) were recently selected categories. Several hundred samples per year are analyzed in this program to support priority investigations. Foods are collected in specific regions, not across all growing areas. For the infant food study, eight categories of food were analyzed for the foregoing chemicals plus several organochlorine and organophosphate pesticides. All detected pesticide residues were well below tolerances.

The FDA's Total Diet Study (TDS), initiated in 1961, is a yearly program that determines levels of various pesticide residues, contaminants, and nutrients in a typical "market basket" of foods and estimates intakes of these substances for specific age/sex groups in the U.S. population. The studies are conducted by the FDA's Center for Food Safety and Applied Nutrition (CFSAN). Food samples are collected three or four times a year from different regions (east, west, north, and south) of the United States, from local grocery stores and restaurants. The TDS has gone through three major revisions; the most recent one, in 1991, incorporated the food consumption data from the 1987–1988 USDA National Food Consumption Survey to select the market basket items. There are now 260 foods in the program, representing the main ingredients in the U.S. food supply. Traditional items include fruits, vegetables, grain products, eggs, meat, and dairy products, as well as mixed dishes, fast foods, desserts, beverages, fats, and sweeteners. These have been analyzed individually since 1982, not as composites, to evaluate the variations in residue levels. The coverage has also been expanded to include 14 age/sex groups, including 6- to 11-month infants, 2-year-old children, 6-year-old children, 10-year-old children, and separate female and male groups of ages 14–16, 25–30, 40–45, 60–65, and 70+. The pesticides assayed for include organohalogens, organophosphates, *N*-methyl carbamates, chlorphenoxy acids, pyrethroids, ethylenethiourea, substituted urea herbicides, and benomyl/thiabendazole.

The most recent results on the TDS provided dietary intakes of pesticides for the period of July 1986–April 1991. The 16 most frequently detected pesticides are shown in Table IV in order of decreasing frequency of occurrence.

Except for carbaryl, for which only 1512 total items were analyzed for *N*-methyl carbamates, the findings were based on analyses of 4914 items. The estimated daily intakes for males aged 14–16 years (a relatively high exposure group) were compared to the FAO/WHO ADIs (acceptable daily intakes) for chemicals for which an ADI exists. The results did not show any intake exceeding the ADI. The organochlorinated pesticides including DDT are continuing a slow decline in residue levels and incidence of positive samples since most of them were banned in the 1970s and 1980s. Dieldrin is still found at one of the highest percentages of the ADI, in both fresh produce and fish, from earlier use. Of the contaminants resulting from current use, malathion, chlorpyrifos, and diazinon are especially notable not only for their high incidence in foods but also for causing toxicity to invertebrate organisms in agricultural runoff.

TABLE IV
Pesticides Most Commonly Found in the July 1986–April 1991 Total Dietary Study

Pesticide	Percentage of total positive samples	U.S. EPA RfD (μg/kg/day)	Rat oral LD_{50} (mg/kg)
Malathion	20	20[a]	885
p,p'-DDE	16	0.5[b]	115[b]
Diazinon	11	0.9	300
Chlorpyrifos	9	3.0	135
Chlorpyrifos-methyl	9	10	941
Dieldrin	8	0.05	40
Endosulfan	7[c]	6[a]	18
Methamidophos	6	0.05	13
Hexachlorobenzene	6	0.8	10,000
Dicloran	5	—	4,000
Chlorpropham	4	200[a]	3,800
Heptachlor epoxide	4	0.01	100[d]
Lindane	4	0.3	76
Carbaryl	3	100	307
Acephate	3	4	866
Dimethoate	3	0.2	250
Ethion	2	0.5[a]	27

[a]Includes only the parent chemical.
[b]Value for DDT.
[c]Includes endosulfan I, endosulfan II, and endosulfan sulfate.
[d]Value for heptachlor.

B. Food Consumption Surveys

Several national surveys have been conducted on the amounts of different foods eaten by Americans. These usually involve recall over 1-day or 3-day periods rather than actual measurement of food consumed. The primary purpose of such surveys is to determine the nutritional quality of people's diets to help direct consumer education efforts. The surveys are also useful for estimating exposures to toxic substances in our diets. In the largest of the surveys, the USDA Nationwide Food Consumption Survey (NFCS), food consumption has been tabulated by age, sex, region of the United States, season, and ethnic group in the years 1977–1978 and 1987–1988. Other food consumption surveys, such as the National Health and Nutrition Examination Surveys (NHANES), conducted by the National Center for Health Statistics, and the Continuing Surveys of Food Intakes of Individuals (CSFII), conducted by USDA, help fill in the coverage gaps from NFCS and provide independent estimates using different methods and population samples.

Ideally, the number of people questioned would be large enough to determine not only average consumption of different foods but also the statistical distribution of consumption. For instance, it is useful to know the average grams/day of apples, bananas, and orange juice consumed by children 1–6 years old, and also the amount at the upper 90% or 95% percentiles of consumption. However, in some cases, particularly for smaller groups of people such as women who are breast-feeding children, and for foods eaten by relatively few people during the study period, the consumption values cannot be accurately calculated. Other surveys may oversample from smaller population groups to correct for such statistical deficiencies.

C. Exposure

To estimate exposure to a particular pesticide in food, one must have consumption estimates for all the pesticide-treated foods in the American diet, as well as the actual or potential residue levels in all commodities likely to contain that pesticide. Consumption estimates were collected in the 1977–1978 and 1987–1988 NFCS for hundreds of raw and processed food types, which were combined into crop groups and about 200 individual commodities by the U.S. EPA for data analysis. The EPA uses a custom software program called the Dietary Residue Evaluation System to calculate exposure, based on the NFCS food surveys and estimated food residue levels. With this system, estimates of either annual (chronic) exposures or acute (single) exposures to a pesticide can be obtained. Contributions of each of the individual food commodities to exposures can also be determined. A food consumption computer program is also available from a private company, Technical Assessment Systems, Inc., which utilizes the NFCS and NHANES data bases. These programs allow population exposure estimates based on age, sex, region of the country, and season.

Exposure estimates can account for the fraction of the total population that consumes a particular food. That is, to determine the likelihood of an individual consuming an excessive dose of aldicarb in potatoes (acute exposures), only the people who ate potatoes would be considered. However, to calculate the average daily dose of aldicarb in the diet, the population exposure estimates would include all members of the population, both consumers and nonconsumers, for all commodities. This could be used to determine if total average exposure to a chemical residue is acceptable, or if a high pesticide residue in a particular crop would produce excessive exposures.

Another relevant analysis, estimating the combined average exposure to all chemicals of a particular type (such as cholinesterase inhibitors), can be assembled from multiple runs. This is adequate for determining exposure to specific measured or assumed concentrations of pesticides (e.g., the tolerance levels) for the average consumer in each of the population groups. As will be discussed in the next section, more extensive data analyses are also possible in the context of a detailed risk assessment.

D. Pesticide Risk Assessments

Risk of exposure to pesticide residues in food is estimated by utilizing the food consumption values described in Section III,B and the safe pesticide exposure levels described in Section II,C. For assessment of exposure to a carcinogen, this process provides a numerical estimate of risk, such as a lifetime cancer risk of 1 in 10 million for consuming dieldrin in kohlrabi. New draft U.S. EPA cancer risk assessment guidelines may change the estimated cancer potencies and risks for many pesticides and commodities, when the guidelines are finalized and begin to be used in risk assessments.

For noncarcinogenic effects, the result of the risk assessment is an estimate of the ratio of the safe level to the consumption level, which is expressed as a large margin of safety. For example, combined exposures to cypermethrin might result in a daily average dose to children 7 to 18 years old of 1.5 μg/kg-day (compared to the RfD of 10 μg/kg-day and a NOEL of 1 mg/kg-day, based on loose stools in a 1-year dog feeding study). The MOS from the NOEL would be $1000/1.5 = 667$. The conclusion would be that the risk of significant adverse effects is negligible. Attempts are under way to estimate actual human risk based on measured variability in the two species, pharmacokinetics of the pesticide, and known or extrapolated species sensitivity for certain end points, but the more quantitative approaches are not yet widely used for regulatory purposes.

IV. PESTICIDE CONTAMINATION INCIDENTS AND ANALYSIS

A. Contamination Incidents

There have been relatively few reported cases in the United States involving illnesses caused by legal use of pesticides on food. Situations or incidents that have received nationwide attention involve sulfites in wine, dried fruit, and fresh produce; ethylene dibromide (EDB) in cereal grains and bakery products (1984); illegal use of aldicarb in cucumbers and watermelons (1985, 1987); daminozide in apples (1987); procymidone in wine (1990); aldicarb in bananas (1991); and pesticides in fish (ongoing).

Although deaths have occurred from the use of sulfites as a preservative in various foods, this problem has decreased since the prohibition of spraying of sulfite solutions on fresh produce in grocery stores and restaurants to preserve freshness. The problem is acute bronchoconstriction associated with inhalation of sulfur dioxide released by the action of salivary enzymes on sulfite in foods. Asthmatics can be especially sensitive to the bronchoconstrictive effects.

In the case of EDB, the fumigant was first thought not to leave a residue because of its high volatility. Subsequent findings of its residue in grains and grain products led to a reevaluation, the establishment of new tolerances, and subsequent cancellation for food use. The toxicological concern was the potent carcinogenicity observed in animal tests. No human illness had been associated with this contamination, but excessive estimated risks, especially to infants and children, because of their potential high exposures through cereals and cereal grain products in baby food and other baked goods, was the rationale for cancellation.

Aldicarb has high acute toxicity. Its residues from illegal use on watermelons led to about 300 reports of symptoms of varying severity in California and Oregon during a July 4 holiday period. As little as one bite of watermelon was associated with illness. The toxicity was attributed mostly to the metabolite aldicarb sulfone, the primary toxic residue found in foods. An inspection program was rapidly established that traced the contamination back to the illegally treated watermelon fields.

Daminozide (Alar) is a plant growth regulator that was used to control ripening of apples and prevent premature fruit drop. The health concern was related to potential carcinogenicity in humans based on studies of both daminozide and its breakdown product, unsymmetrical 1,1-dimethylhydrazine (UDMH), in animals. UDMH tends to concentrate during processing to make apple juice. The potential high risk to children because of children's high consumption of apple products caused considerable controversy, which eventually led the manufacturer to withdraw the product from use.

Procymidone is a fungicide widely used in Europe, especially on grapes, but not on crops in the United States. Residues detected in imported wines led to detainment of the shipments. At that time, the chemical had never been reviewed by the U.S. EPA and there was no established tolerance for any domestic commodity. In response to a petition from the manufacturer, the Agency expedited its review of the chemical. A maximum residue level (MRL) had already been set for procymidone in wine in the Codex Alimentarius (the European list of standards). Although the Agency was asked to adopt the MRL on an interim basis, U.S. law does not allow EPA to defer to another authority for food safety judgments, so importation of wine made from grapes treated with the chemical was delayed until a tolerance was established.

The aldicarb in banana situation involved the finding of elevated residue levels in individual bananas, rather than in composites (as usually measured). The concern was acute toxicity, particularly since this is a high-consumption food in infants. Another concern was that the nonspecific symptoms could lead to unrecognized illnesses in infants and children, who might be given more bananas as soft food by their unsuspecting parents, which would compound any problems. Shipments were allowed to continue under close monitoring for aldicarb until the suspect crops were cleared from the market. There were no reports of illnesses associated with this episode.

Fish contamination with pesticides is a result of widespread environmental contamination with lipophilic organochlorine pesticides. Most of the problem pesticides, such as the DDT/DDE family, dieldrin, heptachlor, and chlordane, have already been banned. Commercial fish are monitored for pesticide residues by the U.S. FDA, and sport fish by various state agencies. Health advisories have been released by many states advising citizens to limit their consumption of fish caught in various bodies of water. Most of these are based on excessive levels of methyl mercury, rather than pesticides.

B. Analysis and Discussion

The recent passage of the Food Quality Protection Act of 1996 provides a stimulus for reevaluation of the safety of pesticide residues in food. It removes the paradoxical situation in which relatively toxic compounds that do not concentrate in processed foods could remain on the market while safer, lower-toxicity compounds subject to the Delaney Clause must be removed. As a result it allows U.S. EPA and related programs and agencies to devote resources to higher-priority, higher-risk issues.

The relatively few reported problems with exposures to pesticides in U.S. food demonstrate the overall success of the present regulatory system. Though some problems result from illegal pesticide usage, most of the incidents involved the recognition of a health concern with the availability of new toxicological and monitoring data, rather than frank injuries. It should be noted that health concerns may arise from either acute or chronic toxicity. The former might be readily recognized but the latter is not. Continued residue monitoring may reveal contamination and accumulation, providing an early warning of a potential problem. An ongoing effort to update our knowledge of pesticide toxicology and of the residue levels in the nation's food supply also seems necessary because of the continued introduction of new compounds and changes in pesticide use patterns.

It should be understood that in some cases risk estimates are based on theoretical risks rather than actuarial risk, and therefore an elevated risk does not mean that people are immediately or irreversibly harmed from consuming the foods. Successful avoidance of excess risk associated with long-term or lifetime exposures is the goal. Appropriate public health measures can include providing consumers with information on competing risks, which involves the consumption of natural food toxins as well as chemicals used to prevent predation and spoilage. Encouragement of integrated pest management practices, whilch utilize a variety of approaches to pest control to minimize pesticide use, is also desirable.

Effective risk communication is particularly important for pesticide residue issues. Experience shows that the public wants a simple answer—Is the residue "safe" or "unsafe?" People tend not to be interested in the concepts of negligible or *de minimus* risk. When a risk/benefit ratio is mentioned, the public thinks of benefits to farmers with risks borne by consumers, and may have a feeling of helplessness about involuntary exposures to pesticide residues in food. The regulatory community has been sensitive to the values of society, which is reflected in the protective nature of the risk assessment process. Maintaining a balance between the public's concern for absolute food safety and the acknowledgment of competing risks and benefits of pesticide use is the regulatory challenge. This requires the best scientific approaches, frank discussions of data limitations, and an acknowledgment of the uncertainties. The goal is to communicate honestly with the public, so as to produce a public that is informed, involved, interested, reasonable, solution oriented, and collaborative.

BIBLIOGRAPHY

Ballantyne, B., and Marrs, T. C. (eds.) (1992). "Clinical and Experimental Toxicology of Organophosphates and Carbamates." Butterworth Heinemann, Oxford, England.

Benbrook, C. M. (1996). "Pest Management at the Crossroads." Consumers Union, Yonkers, New York.

Fan, A., Howd, R., and Davis, B. (1995). Risk assessment of environmental chemicals. *Annu. Rev. Pharmacol. Toxicol.* **35**, 341–368.

Fan, A. M., and Chang, L. W. (eds.) (1996). "Toxicology and Risk Assessment; Principles, Methods, and Applications." Marcel Dekker, New York.

Fenner-Crisp, P. A. (1995). The NAS report: How can the recommendations be implemented? *Environ. Health Persp.* **103**, 159–162.

Hayes, W. J., Jr., and Laws, E. R., Jr. (eds.) (1991). "Handbook of Pesticide Toxicology." Academic Press, San Diego.

Landrigan, P. J., *et al.* (1993). "Pesticides in the Diets of Infants and Children." National Research Council, National Academy Press, Washington, D.C.

Levine, T. E. (1992). Assessment and communication of risks from pesticide residues in food. *Food Drug Law J.* **47**(2), 207–214.

Winter, C. K. (1992). Dietary pesticide risk assessment. *Rev. Environ. Contam. Toxicol.* **127**, 23–67.

Yess, N. J., Gunderson, E. L., and Roy, R. R. (1993). Food and Drug Administration monitoring of pesticide residues in infant foods and adult foods eaten by infants/children. *JAOAC Int.* **76**, 492–507.

Phagocytes

SUZANNE GARTNER
Henry M. Jackson Foundation Research Laboratory

GLOSSARY

Complement A group of serum proteins activated in a cascading fashion following interaction with antibody–antigen complexes or certain bacterial components

Cytokine A molecule secreted by a cell that can mediate and regulate the functions of that cell or those of other cells

Fc fragment A portion of an immunoglobulin molecule generated by papain digestion. It is composed of two heavy chains and has binding sites for activation of complement, but no binding sites for antigen

Granuloma A small nodular lesion associated with chronic inflammation and characterized by the presence of numerous macrophages, which can fuse to form multinucleated giant cells

Major histocompatibility complex (MHC) Four closely linked loci on chromosome 6 (in humans) that code for strong transplantation antigens. These antigens are present on the surface of all cells except mature erythrocytes. The MHC is also called the human leukocyte antigen complex

Opsonization An enhancement of phagocytosis mediated by the coating of the antigen with a specific antibody

Phagocytosis Process whereby phagocytes ingest and degrade material

Phagolysosome An intracellular vesicle in phagocytes created by the fusion of a primary lysosome with a membrane-bound vacuole containing engulfed material (a phagosome); called secondary lysosomes

Receptor A protein structure, located on the surface of a cell, that serves as a specific recognition and binding site for other molecules

HUMAN PHAGOCYTES ARE BONE MARROW-DErived cells of two categories, polymorphonuclear leukocytes and mononuclear phagocytes, whose specialized function is the engulfment and digestion of particulate matter. Together these groups of cells serve as a first line of defense against invading microbes. Mononuclear phagocytes also perform a number of other diverse functions. As scavengers, they continuously remove and degrade debris and aging cells from the tissues. As participants in the immune response, they process and present antigen, act as cytotoxic effector cells, and produce and secrete a number of immunoregulatory molecules. Phagocytes also play a role in certain disease conditions, most notably chronic diseases and diseases in which macrophages serve as host cells for the pathogen.

I. DEFINITION, DESCRIPTION, AND ORIGIN

Phagocytes are defined functionally by their ability to engulf matter such as particulate material or other cells from their environment. This ability is characteristic of many species within the plant and animal kingdoms and is a major means by which unicellular organisms obtain nutrition. It has been suggested that this ability also played a role in the evolutionary process from unicellular to multicellular organisms. For example, it is conceivable that mitochondria, the energy-generating organelles present in higher cells, arose from ingested bacteria. In humans, essentially only two kinds of cells, polymorphonuclear leukocytes and mononuclear phagocytes, have the capacity to phagocytize. Both kinds of cells originate from a common precursor in the bone marrow and are present in blood and tissues, and both play key roles in defending the body from invading potential pathogens.

A. Polymorphonuclear Leukocytes

Polymorphonuclear leukocytes (commonly called polys), as the name suggests, are characterized by the irregularity of their nuclear structure. This category includes two kinds of phagocytic cells, the neutrophil and the eosinophil, along with the nonphagocytic basophil.

1. Eosinophil

This cell is named for its most striking feature, eosinophilia, the bright red color its numerous large granules take on as a consequence of being stained with the acid eosin. Ultrastructurally these membrane-bound granules are ovoid, and each contains a single, irregular, electron-dense crystalloid core. The core is composed of a poorly characterized insoluble basic protein, whereas the matrix of the granule contains a variety of hydrolytic enzymes, including histaminase, arylsulfatase-B, and phospholipase-D, as well as a peroxidase distinct from that present in neutrophils. They account for 1 to 6% of the white cells in circulating blood, but more than 99% of their total numbers are resident in the bone marrow and other tissues. Predominant sites include the skin, lungs, gastrointestinal tract, and vagina—tissues frequently exposed to the external environment. The half-life in blood of an eosinophil has been estimated to be approximately 5 hr. Once in the tissues, the eosinophil does not reenter the circulation except under certain pathological conditions. Like neutrophils, they are mobile and respond to chemotactic factors. Eosinophils are highly phagocytic for antigen–antibody complexes but are generally less phagocytic than neutrophils. They can ingest bacteria and fungi. However, despite the fact that they produce more peroxide than neutrophils, they are less effective in killing bacteria, presumably because they lack lysozyme and phagocytin, and also because they cannot utilize the peroxidase-H_2O_2-Cl pathway.

Two primary functions have been ascribed thus far to the eosinophil: a role in the termination of allergic reactions and a role in host defense against parasites. Type I (anaphylactic) hypersensitivity is a systemic or local, rapidly developing immunologic reaction resulting from the release of potent vasoactive amines from basophils and mast cells. This release is triggered by the binding of a previously seen antigen to specific IgE molecules attached to basophils and mast cells. Also released is eosinophil chemotactic factor of anaphylaxis (ECF-A), which attracts eosinophils to the area. The eosinophils undergo degranulation, which results in the release of the granule enzymes, enzymes that degrade those released by the basophils and mast cells. The final result is the termination of the anaphylactic reaction. In parasite-associated granulomas, as many as 50% of the cells present are eosinophils. Antibody-dependent, eosinophil-mediated destruction of the parasite is thought to be accomplished as follows: Antiparasite IgG and IgE antibodies bind to specific receptors for the Fc part of antibodies present on the eosinophil, which leads to degranulation of the cell. The released basic proteins and peroxidase then act to kill the target parasite. The activation of the eosinophil by cytokines has also been suggested to have a role in this process.

2. Neutrophil

Neutrophils (so called because they do not stain with acidic dyes such as eosinophils or with basic dyes such as basophils) are the most abundant of the leukocytes in blood, where they usually constitute 55–65% of the total population. The majority of neutrophils, however, reside in the bone marrow where they serve as a rapidly mobilizable defense reserve. Their total life span is thought to be about 8 days, approximately 10 hr in the blood, 1 to 2 days in the tissues, and the remainder in the marrow. The mature neutrophil possesses a highly lobulated nucleus, usually consisting of five lobes joined together by thin strands of nuclear material. A more immature form of the neutrophil with a bilobular nucleus can also be seen in blood. An increase in the numbers of these so-called bands in blood is indicative of their recruitment from the marrow, in response to injury or infection.

The cytoplasm of neutrophils contains azurophilic granules, designated primary granules, as well as numerous smaller granules, designated specific granules. The azurophilic granules are large, spherical, and electron dense and are actually lysosomes. They contain a number of enzymes such as acid hydrolases and neutral proteases (collagenase, elastase, cathepsin) as well as potent antibacterial enzymes such as lysozyme, myeloperoxidase, and D-amino-oxidase. The specific granules are small, more numerous, and irregular in shape and density. They contain the bactericidal enzymes lysozyme and lactoferrin, as well as alkaline phosphatase, whose function is not really known.

Neutrophils play a primary role in the acute inflammatory response. At the site of injury or infection, they phagocytize and destroy invading microorganisms and cell debris. Neutrophils also destroy microbes and degrade debris by releasing their enzymes directly into their environment. They have little capacity for protein synthesis, and hence are incapable of any prolonged, continuous function. Neutrophils, along with mononuclear phagocytes and K cells, can also participate in antibody-dependent, cell-mediated cytotoxicity (ADCC). This is a process that has been observed *in vitro* whereby an IgG-coated target cell is lysed on contact with the neutrophil. It involves the binding of the Fc receptor on the neutrophil to the Fc portion of the exposed IgG molecule. Low concentrations of antibody are sufficient, phagocytosis is not involved, and the phenomenon is complement independent. [*See* Neutrophils.]

B. Mononuclear Phagocytes

The mononuclear phagocyte system, previously termed the reticuloendothelial system, includes the monocytoid components of bone marrow, monocytes of the peripheral blood, and tissue macrophages. These cells are specialized in phagocytosis and intracellular digestion. Unlike neutrophils, mononuclear phagocytes are general scavengers, in that they continually engulf and degrade aging and decaying cells from the tissues. They also perform a number of other functions. In particular, macrophages function as mediators of the immune response. Many tissue macrophages have specialized names. For example, Kupffer cells are macrophages in the liver, and microglial cells are presumed to be a kind of brain macrophage. Tissue macrophages are thought to develop from blood monocytes following their migration from blood into the tissues. [*See* Macrophages.]

In conventionally fixed and stained preparations, monocytes are comparatively large and possess a large, eccentrically located nucleus, which is often indented or horseshoe shaped. They have small azurophilic granules scattered throughout their cytoplasm. These granules are primary lysosomes containing a number of enzymes functional at acid pH including peroxidase, acid phosphatase and aryl-sulfatase, and acid nucleases. (Secondary lysosomes present in macrophages are vacuolar structures that are the sites of current or past digestion.) Monocytes compose 2–10% of the blood leukocytes. Their stores in the bone marrow are small, and their transit time in the blood is estimated to be about 32 hr. As tissue macrophages, they appear to live at least several weeks, and perhaps several months. Unlike neutrophils, monocytes are capable of continuous lysosomal activity and regeneration, and they can utilize the aerobic, in addition to the anaerobic, metabolic pathway.

The microscopic appearance of the macrophage varies considerably depending on the activation state of the cell and its tissue microenvironment. Resting cells appear rounder, and small, active ones are larger and possess several cytoplasmic projections. Like monocytes, quiescent macrophages have many primary lysosomes. During phagocytosis, lysosomes fuse with phagosomes (vacuoles containing the ingested material), creating phagolysosomes. As a consequence, the number of lysosomes decreases in the active phagocytic macrophage. However, because lysosomes can be regenerated in these cells, their declining numbers can be later increased. Phagolysosomes are the sites of enzymatic digestion of the phagocytized material. Residual matter may remain in the cell indefinitely or be released into the surrounding tissue by exocytosis. Mature macrophages often contain a number of residual bodies—storehouses of incompletely digested material. These features are responsible for the name macrophage, which is derived from the Latin meaning large eater, and contribute to the scavenger appearance of the macrophage, irregular with respect to shape and inclusions. The tremendous degradative ability of the macrophage is essential for the disposition of metabolic breakdown products, a prime example being the ingestion and breakdown of aged red blood cells by spleen macrophages.

II. PHAGOCYTOSIS

Phagocytosis, also called endocytosis, is defined as the process of internalization and degradation of extracellular material. Essentially, three distinct steps—recognition, engulfment, and degradation—characterize the phenomenon. Recognition is mediated through specific receptors located on the surface of the phagocyte. Some receptors can bind the foreign material directly. For example, human phagocytes have receptors (mannose–fucose receptors) that directly bind to sugars present on the walls of yeast cells. More frequently, however, phagocytes use the receptor for the Fc part of antibodies and for proteins of the complement system, which are present in blood. In these cases the foreign material (antigen) is first bound by immunoglobulin (IgG) or immunoglobulin and the C3 component of complement. The antigen is then indirectly bound to the phagocyte either by the binding of the Fc portion of the antigen-complexed immunoglobulin molecule to the Fc receptor on the phagocyte or by the binding of complement-coated, antigen–antibody complexes to the phagocyte via the complement receptor. The distribution of the receptor–ligand complexes on the surface of the phagocyte plays a role in the efficiency of phagocytosis. Clustered antibody or complement molecules are much more effective. Thus erythrocytes coated with clusters rather than monomers of IgG are phagocytized more avidly.

It should be mentioned that most cells (one exception being the mature erythrocyte) are capable of ingesting material from their environment. This is a receptor-mediated process called pinocytosis and it differs from phagocytosis in three principal ways: (1) pinocytosis, but not phagocytosis, can occur at temperatures below 18°C; (2) phagocytosis, but not pinocytosis, requires the participation of actin filaments within the cell; and (3) particles taken up via pinocytosis are generally less than 1 μm in diameter, whereas phagocytized particles can be considerably larger (e.g., other cells). The target material for pinocytosis is soluble molecules, whereas the target material for phagocytosis is usually particulate matter. Also, pinocytosed material is taken up in special clathrin-coated vesicles.

Certain features of the receptor–ligand interaction merit further attention. The binding of the targeted material to the receptor of the phagocyte initiates a series of events. However, binding alone is not sufficient, at least in some cases. For example, quiescent macrophages can bind but not phagocytize complement-coated particles. It has been shown, however, that the addition of T-cell lymphokines or other activators of macrophages (e.g., phorbol myristate acetate) can drive the process of phagocytosis to completion. Another important feature is that the phagocytosis receptors, the Fc, complement, and oligosaccharide receptors, all function independently of each other, although Fc and complement receptors can work cooperatively in the phagocytosis of a particle. Furthermore, the binding of a ligand to one kind of receptor does not appear to signal the other kinds of receptors.

Following binding of the target particle to the phagocyte, the membrane of the cell begins to advance, ultimately enclosing the particle within a membrane-bound vesicle. This process is accomplished via the polymerization and depolymerization of actin filaments located in the cytoplasm beneath the membrane. The movement of the cell membrane leads to additional ligand–receptor interactions which, in turn, lead to further advancement of the membrane. This process has been likened to the closing of a zipper. A single membrane-to-membrane fusion event closes the phagosome. The phagosome may then move deeper into the cytoplasm of the cell. Engulfment can be a relatively rapid process. A neutrophil can ingest a chain of 10 to 15 bacteria within a few minutes, and a macrophage can ingest even more. Both neutrophils and macrophages contain cytoplasmic stores of membrane components; it is presumed that they are used to replace membrane consumed during phagocytosis. Indeed, recycling of membrane components has been documented for phagocytosing macrophages. Obviously phagocytosis is an energy-requiring process. Macrophages and neutrophils have considerable stores of glycogen and obtain most of their energy from glycolysis. Macrophages also have large stores of creatine phosphate, which they tap during phagocytosis and other times of high energy demand.

The initial event of degradation is the acidification of the phagosome, which results in the dissociation of the bound material from its receptor. Acidification is a consequence of the pumping of protons into the phagosome and is triggered when the particle becomes bound to its membrane receptor. The dissociated receptor is often recycled back to the plasma membrane. The acidified phagosome then fuses with a lysosome whose potent degradative enzymes function optimally at pH 4.0. Enzymatic digestion of the engulfed material follows. Two categories of bactericidal mechanisms have been identified, one oxygen dependent and the other oxygen independent. The term respiratory burst refers to the increase in oxygen uptake by macrophages and neutrophils during phagocytosis. This process is not actually activated by the ingestion of material, nor is it a necessary part of phagocytosis. However, much of the oxygen is reduced to superoxide anion, which presumably is toxic for the ingested microorganisms. The other major oxygen-dependent pathway is the H_2O_2–myeloperoxidase–halide system. Activation of an oxidase for reduced pyridine nucleotide during phagocytosis leads to the liberation of H_2O_2 within the phagolysosome. Although the details of the process are not known, it is clear that the bactericidal effect of H_2O_2 is tremendously enhanced in the presence of the enzyme myeloperoxidase plus a halide ion, chloride ion being the most physiologically relevant. This mechanism appears to be effective against viruses and fungi as well.

Two oxygen-independent bactericidal mechanisms have been proposed: (1) the inhibition of bacterial growth by acid pH and (2) the direct digestion of cell wall and membranes of the bacteria via specific enzymes. The pH of the phagolysosome continues to remain low (at approximately 4.0) as a consequence of the production of hydrogen ion, which is derived from the action of carbonic anhydrase or the production of lactic acid. Few bacteria can proliferate at this pH; many can be directly killed by the lactic acid. In addition, the enzyme lysozyme can digest, to at least some degree, the bacterial cell walls of all bacteria but especially those of gram-positive cocci, by hydrolyzing the muramic acid–*N*-acetylglucosamine bonds. In the neutrophil, this digestion can be followed by the action of the arginine-rich cationic protein, phagocytin, which serves to lyse bacterial membranes. Under certain *in vitro* circumstances, neutrophils can be induced to release their lysosomal enzymes into their environment. The *in vivo* relevance of this phenomenon is unclear, but it may play a role in certain pathological situations such as immune complex-associated glomerulonephritis. Not surprising is the fact that some pathogens have developed ways to circumvent their destruction by modifying their host and adapting themselves so that they can survive and even proliferate within macrophages. Examples include the tuberculosis bacillus and certain retroviruses. Such pathogens employ a number of approaches to accomplish this, and clearly this phenomenon represents a major mechanism of disease generation.

A final point regarding phagocytosis is the movement of phagocytes to a site of inflammation, which is mediated by chemotactic factors, a process called chemotaxis. Neutrophils, eosinophils, and macrophages are all susceptible to chemotaxis, although they do not all respond to the same attractants. The response is a sensitive one, in that these cells can recognize a gradient of chemoattractant as small as 1% across the dimension of the cell. The two most important chemoattractants for neutrophils are bacterial products and the components of the complement system. Eosinophils also respond to complement (C5 fragments), and at sites of anaphylactic allergy, they respond to ECF-A released from IgE-sensitized basophils or mast cells when these cells contact the specific antigen. Chemotactic factors for monocytes include the C3 and C5 fragments of complement, various bacterial factors, denatured proteins arising from dead cells, and uncharacterized factors from neutrophils and lymphocytes. Genetic or acquired deficiencies in chemotaxis are associated with an increase in susceptibility to microbial infection.

III. MACROPHAGE SURFACE RECEPTORS

Cells of the mononuclear phagocyte lineage display a number of important receptor proteins on their surfaces. Those involved in phagocytosis, the mannose–fucose receptor and the Fc and complement receptors, were mentioned earlier, but the latter two require further discussion. In addition, mononuclear phagocytes possess receptors for certain growth factors and cytokines, most importantly receptors for (1) interferons α, β, and γ; (2) interleukin-1 (IL-1); (3) tumor necrosis factor (TNF), and (4) colony-stimulating factor (CSF). Macrophages also produce and secrete many of these factors.

Human macrophages have three distinct Fc receptors. FcRI can bind both monomeric and polymeric IgG with high affinity, and its expression can be increased by interferon γ. FcRI manifests a IgG subclass preference IgG1 > IgG3 > IgG4 > IgG2. The two other Fc receptors, FcRII and FcRIII, bind IgG with a lower affinity. The FcRII receptors are also found on other types of cells, including B lymphocytes. Several receptors for the different complement fragments have been detected on the surface of human macrophages. The best characterized are the CRI and CR3 receptors, but also present are the CR4, C5a, C1q and factor H receptors. CR1, previously named the immune adherence receptor, is present on a number of different kinds of cells (neutrophils also) and is the receptor for the C3b component of complement. CR3 is present on mononuclear phagocytes and granulocytes (some lymphocytes also) and binds to the iC3b component of complement. The CR1 and CR3 receptors participate not only in phagocytosis but also play a role in ADCC. The C5a receptor on the macrophage membrane plays a key role in arachidonic acid metabolism by mediating the release of arachidonate derivatives. The lipid arachidonic acid is a major constituent of the macrophage membrane, which is released into the cytoplasm of the cell following appropriate signaling. The arachidonic acid is then metabolized to form a family of compounds called eicosanoids. These compounds are potent, short-lived, short-range hormones. Representatives are the prostaglandins, leukotrienes, and thromboxanes. [*See* Lymphocytes.]

By a process of differentiation and maturation called hematopoiesis, cells of the blood derive from a single precursor—the pluripotent stem cell—that resides in the bone marrow. This pluripotent cell gives rise to a number of other progenitor cells, each of which becomes committed to only one of the specific differentiation pathways. Assays have been developed that permit the clonal growth of the various progenitor cells in soft agar. Specific factors, referred to as CSFs, are required to drive the clonal proliferation of the cells and also appear to regulate hematopoiesis *in vivo*. Hematopoietic cells of varying degrees of differentiation display receptors for these factors, which are produced by several kinds of cells located in various sites throughout the body.

Macrophage CSF (M-CSF, CSF-1) is produced by macrophages and fibroblasts. Its receptor is present only on macrophages. Granulocyte–macrophage CSF (GM-CSF) is produced by macrophages, bone marrow stromal cells, and activated lymphocytes; its receptor is present on hematopoietic stem cells and myeloid cells. Granulocyte CSF (G-CSF) is produced only by macrophages and stromal cells. Its receptor is present on both granulocytes and a progenitor cell which can give rise to either macrophages or granulocytes. Finally, interleukin-3 (multi-CSF) is produced by activated lymphocytes, and its receptor is present on a progenitor cell capable of giving rise to granulocytes, mast cells, and lymphoid cells. In all cases, the binding of these regulatory molecules to their appropriate target cells induces cell maturation. Of particular interest is that the CSF-1 receptor, a tyrosine kinase, is identical to the product of the proto-oncogene c-*fms*. This receptor is present on monocytes and macrophages at all stages of differentiation, with the most mature cells bearing the greatest number of receptors.

The IL-1 receptor is an 80-kDa glycoprotein with a high affinity for its ligand. Both IL-1α and IL-1β bind with high affinity and compete with each other for binding sites. The TNF receptor is a protein with a molecular mass in the range of 60 to 80 kDa. Like IL-1, it exhibits a high affinity for its ligand. Similarly, the receptors for interferon α and β are high-affinity protein receptors with molecular masses in the 100- to 120-kDa range. Although many of the postbinding events regarding these regulatory molecules have not been delineated, it is thought that the receptor–factor complexes are endocytosed.

IV. MACROPHAGE-DERIVED CYTOKINES

As alluded to earlier, macrophages produce a number of soluble factors with a variety of functions. Those to be noted here include interleukin-1, interleukin-6, interferons α and β, and tumor necrosis factor. These mediators act locally, have multiple activities, and are synthesized *de novo* following stimulation of the cell. Although macrophages are a major source of these factors, other kinds of cells can also produce them. Certain features of these cytokines are shown in Table I. Two additional cytokines produced by macrophages have been identified. One, monocyte-derived neutrophil chemotactic factor (MDNCF), is produced in response to endotoxin, tumor necrosis factor, and interleukin-1, and promotes degranulation and activation of neutrophils as well as chemotaxis. The other, designated IP10, is 42% homologous with MDNCF and is produced by macrophages following interferon γ stimulation. Its functions await elucidation. [*See* Cytokines and the Immune Response.]

Macrophages also produce and secrete a considerable number of other molecules. An important example is apolipoprotein E, which participates in the transport of lipids.

V. MACROPHAGE ACTIVATION AND CYTOTOXICITY

A. Activation

Activated macrophages are characterized by morphologic alterations as well as by changes in the secretory status of certain products. Activated macrophages become larger and their cytoplasm more ruffled, and their overall protein content increases, along with increases in the production of the lysosomal enzymes. Their surface receptor repertoire changes (e.g., the number of Fc receptors increases) and, as a consequence of continuous endocytic sampling of their surroundings, they turn over the equivalent of their entire membrane approximately every 30 min. They also have an enhanced killing capacity for both endocytosed bacteria and tumor cells. The activation state can be induced by either an immunologic or a nonimmunologic interaction. In the former case, the macrophage is activated as a consequence of its interaction with a sensitized T lymphocyte, whereas in the later case, the activator is either a bacterial component or another soluble mediator. Unfortunately, con-

TABLE I
Macrophage-Derived Cytokines

Characteristic	Interleukin 1 (IL-1α, IL-1β)	Interleukin 6 (IL-6)	Tumor necrosis factor (TNF)	Interferon (IFNα, IFNβ)
Gene	Chromosome 2	Chromosome 7	Chromosome 6	Chromosome 9
Protein (mature form)	159 aa[a] 153 aa	190 aa (glycosylated)	157 aa	IFNα I 165–166 aa II 172 aa (nonglycosyl.) IFNβ 166 aa (glycosylated)
Receptor	High affinity	High affinity	High affinity	High affinity
Effector function	↑ T- and B-cell growth; regulates production of cytokines by other cells	↑ Ig secretion from B lymphoblasts; ↑ growth of thymocytes and some T-cell clones	Lyses some tumor cells; ↑ IL-2R on T cells; activates macrophages; activates neutrophils	Inhibition of cell and viral growth; enhances accessory cell activity of macrophages
In vivo effects	Primary mediator of systemic inflammatory reaction (e.g., fever, neutrophilia)	Participates in acute-phase inflammatory reaction	Cachexia, regression of some tumors; mediator of acute-phase inflammatory reaction	Modulation of antibody production; graft rejection; antiviral effect

[a]Amino acids.

troversy exists with regard to lymphocyte-derived macrophage-activating factors. Many immunologists believe that the previously described macrophage-activation factor (MAF) is indistinguishable from interferon γ. However, most would agree that both GM-CSF and interleukin-4 (previously called B-cell stimulatory factor 1) are potent inducers of macrophage cytotoxicity.

Clearly T lymphocytes play a primary role in the activation of macrophages at sites of inflammation. The microbes or tumor cells provide antigens, which stimulate the proliferation of lymphokine-secreting, sensitized T cells. Interactions with these T cells activate the macrophages. However, the capacity of macrophages to be activated by the lymphokine-producing T cells is limited. Likewise, the cytotoxic activity of these macrophages is short-lived; it peaks within 8 to 12 hr following lymphokine stimulation and is totally gone within 24 hr. These processes are controlled by means of a negative feedback mechanism, presumably because strict regulation serves to depress the activation response as soon as it is no longer required.

B. Cytotoxic Effector Function

Many of the known mechanisms whereby macrophages kill microbes were mentioned earlier. With regard to tumor cell killing, tumor necrosis factor can be directly toxic to tumor cells. Also, it is thought that cytolytic proteases can act synergistically with reactive oxygen intermediates to kill tumor cells. However, some tumor-targeted killing is oxygen independent, as immunologically activated macrophages have been shown to kill tumor cells under anaerobic conditions. The killing of tumor cells by activated macrophages appears to require intimate contact between the two cells because soluble toxic factors cannot be demonstrated. Although activated but not nonactivated or inflammatory macrophages can bind selectively to a number of different tumor cells, no receptor associated with this binding has been identified.

C. Delayed-Type Hypersensitivity Response

The delayed-type hypersensitivity response (also known as type IV hypersensitivity) is initiated by antigen-sensitized T lymphocytes at the site of antigen deposition. A classic example is the tuberculin skin reaction, which is characterized by reddening and a hardening of the tissue at the site of injection. The response peaks about 24 to 72 hr following injection and then slowly resolves. An important feature of the reaction, which accounts for the hardening of the tissue, is the deposition of fibrin at the site. Although sensitized T cells are the initiators of the response, their numbers account for only approximately 10% of the cells in the lesion. The predominant cell in the reaction is the monocyte–macrophage. The hypersensitivity response is amplified by the sensitized T-cell-driven recruitment of monocyte–macrophages and additional unsensitized T lymphocytes; the recruitment is lymphokine mediated. The ultimate goal is the rapid accumulation of a population of nonspecific, cytotoxic macrophages. These effector macrophages can destroy a wide variety of targets, including bacteria, viruses, fungi, and some parasites and tumor cells. The response is governed by a number of complex interactions between the T cell and the macrophage, and between their secreted lymphokines.

VI. ANTIGEN PRESENTATION

A primary event in the initiation of an immune response is the proper presentation of the antigen to lymphocytes. Cells capable of this function are called antigen-presenting cells (APC). They consist of mononuclear phagocytes. Langerhans' cells of the skin, follicular dendritic cells of lymph nodes, and B lymphocytes. Proper recognition requires that the antigen be codisplayed with molecules of the class II major histocompatibility complex (MHC class II). CD4+ helper T lymphocytes recognize antigen in the context of MHC class II, whereas CD8+ cytotoxic T cells recognize antigen in the context of MHC class I molecules. Obviously, appropriate expression of MHC class II molecules on macrophages is crucial for optimal antigen presentation. The expression of the MHC molecules is both positively and negatively regulated in many different ways. For example, interferon γ produced by activated T cells or by tumor necrosis factor-activated natural killer cells can act to increase MHC class II expression on macrophages. In contrast, prostaglandins can decrease the expression. [*See* CD8 and CD4: Structure, Function, and Molecular Biology.]

It is clear that macrophages in some way process antigen, but the precise steps involved have not been elucidated. Unlike the B-cell response, the T-cell proliferative response to antigen presented by APC can be induced by denatured or fragmented proteins. Macrophage processing of antigen involves internalization of the material followed by intravesicular chemical processing at low pH. (These vesicles are endocytic vacuoles that have been acidified but have not fused with primary lysosomes, so they lack hydrolases.) Approximately 30 to 60 min after its native form is internalized, fragments of the processed antigen appear as an immunogenic determinant on the surface of the macrophage. How the determinant reaches the cell surface is unclear, but evidence from the murine system, indicating that the mouse equivalent of MHC class II proteins can bind peptides, has led to the hypothesis that these proteins act as transporters of the immunogenic peptides.

VII. PHAGOCYTES IN INFLAMMATION

Acute inflammation is a short-lived response to tissue injury and is characterized by the emigration of leukocytes, particularly neutrophils, to the site of the injury and by the exudation of fluid and plasma proteins. Chronic inflammation is a more diverse process and is usually of longer duration. Chronic inflammatory lesions are characterized by the presence of lymphocytes and macrophages and by the proliferation of fibroblasts and small blood vessels (angiogenesis). These are dynamic processes, and phagocytes are key participants. [*See* Inflammation.]

Neutrophils and monocyte–macrophages are primarily responsible for the destruction of invading microbes present in acute inflammatory lesions. A few additional points concerning their emigration should be mentioned. A phenomenon called margination, in which leukocytes moving within the bloodstream orient themselves along the periphery of the flow, has been observed. This occurs more frequently when the flow slows or becomes stagnant, as at sites of inflammation. Ultimately, many leukocytes stick to the endothelial cell surface. Although the mechanisms responsible for the sticking have not been thoroughly delineated, chemotactic factors are thought to play a role. Following adhesion, the leukocytes migrate along the endothelium until they locate a hole, usually the junction between neighboring endothelial cells. The leukocyte then moves itself through the hole, traverses the basement membrane, and ultimately journeys into the tissue.

The predominant cell type within an acute inflammatory lesion is a function of the nature of the initiating stimulus and the age of the lesion. Within the first 6 to 24 hr, neutrophils are more frequently represented. After that time, monocytes begin to predominate. The production of a monocyte chemoattractant by neutrophils and the longer life span of the monocyte–

macrophage are thought to account for this situation. Some examples of the nature of the stimulus include (1) acute *Pseudomonas* infections are characterized by a preponderance of neutrophils for up to 48 hr; (2) eosinophils predominate in certain hypersensitivity reactions; and (3) in some viral infections, lymphocytes are the first emigrating cells.

The accumulation and persistance of macrophages in chronic inflammatory lesions occur as a result of the continued recruitment of monocytes from the blood, the local proliferation of macrophages, and the prolonged survival and immobilization of macrophages. These cells continue to attempt to destroy and/or digest the targeted material. If the material is too large to be engulfed or resists degradation, the macrophages may form a granuloma. Granulomas are collections of inflammatory cells, mostly macrophages, rimmed by lymphocytes. They are usually 1–2 mm in diameter. Ultimately the macrophages can fuse to one another to form a multinucleated foreign body giant cell. (Foreign body giant cells in which the nuclei are located on the periphery of the cell are called Langhans' giant cells.) Classic examples of diseases associated with chronic inflammation include tuberculosis and leprosy (both caused by mycobacteria), schistosomiasis, and cat-scratch fever. Also, chronic inflammation resulting from the presence of large, toxic, indigestible material such as asbestos fibers is involved in the development of certain lung diseases. Finally, macrophages may play another role in chronic inflammation. Macrophage-derived factors have been implicated in the angiogenesis and fibroblast proliferation characteristic of chronic lesions.

BIBLIOGRAPHY

Dean, R. T., and Jessup, W. (eds.) (1986). "Mononuclear Phagocytes: Physiology and Pathology." Elsevier, Amsterdam, The Netherlands/New York.

Gallin, J. I., Goldstein, I. M., and Snyderman, R. (eds.) (1992). "Inflammation: Basic Principles and Clinical Correlates," 2nd Ed. Raven Press, New York.

Paul, W. (ed.) (1993). "Fundamental Immunology," 3rd Ed. Raven Press, New York.

Powandi, M., Oppenheim, J. J., Kluger, M. J., and Dinarello, C. A. (eds.) (1988). "Monokines and Other Nonlymphocytic Cytokines." A. R. Liss, New York.

van Furth, R. (ed.) (1985). "Mononuclear Phagocytes: Characteristics, Physiology and Function." Nijhoff, Amsterdam, The Netherlands (available from Kluwer Academic, Norwell, MA).

van Furth, R. (ed.) (1992). "Mononuclear Phagocytes: Biology of Monocytes and Macrophages." Kluwer Academic, Norwell, MA.

pH and Carbohydrate Metabolism

ROBERT D. COHEN
University of London

GLOSSARY

Acidosis Lowering of pH in a body fluid compartment (e.g., extracellular, intracellular); acidosis may be "respiratory" because of CO_2 retention or "metabolic" because of other causes

Gluconeogenesis Synthesis of new glucose from precursors (e.g., amino acids) and intermediates of carbohydrate catabolism (e.g., lactate, pyruvate, 2-oxoglutarate)

Glycolysis Conversion of monosaccharides (e.g., glucose and fructose to pyruvate and lactate)

pH Negative logarithm (base 10) of the hydrogen ion (proton) activity. Thus if hydrogen activity is 10^{-7} molar, pH is 7, which is neutral. Lower pH is acid, higher is basic

ANIMALS POSSESS POWERFUL MECHANISMS FOR maintaining the pH of extra- and intracellular fluids within relatively narrow ranges. This is because hydrogen ion activity is a determinant of the structural conformation of proteins (e.g., enzymes) and the state of ionization of weak acids and bases. Changes in these properties influence fluxes through biological membranes and through metabolic pathways. Accordingly, there are important effects of pH changes on several major pathways of carbohydrate metabolism, including the conversion of glucose and fructose to pyruvate and lactate, and subsequent oxidation of pyruvate to carbon dioxide and water; these catabolic pathways are major sources of energy production. In addition, the anabolic conversion of precursors such as lactate and pyruvate and certain amino acids to glucose (gluconeogenesis) is also substantially influenced by pH. The situation is complicated by the fact that these reaction pathways are often H^+ (proton) producing or consuming, and their rates are therefore themselves determinants of pH, both in the cells where they occur and elsewhere. These interactions between pH and carbohydrate metabolism are considered here, beginning with an outline of basic mechanisms, followed by an attempt at physiological integration.

I. BASIC MECHANISMS

A. Effects of pH on Individual Reactions of Carbohydrate Metabolism

Both catabolic (degradation) and anabolic (synthetic) pathways of carbohydrate metabolism are characterized by bottlenecks, i.e., nonequilibrium (rate-controlling) steps that are markedly affected by pH in the cellular compartment in which they take place. For the glycolytic pathway, which is located entirely within the cell cytosol, the main bottleneck is the conversion of fructose 6-phosphate to fructose 1,6-bisphosphate, catalyzed by the enzyme phosphofructokinase (PFK) (Fig. 1). PFK is markedly inhibited by the lowering of cytosol pH, probably by increasing the inhibitory action of ATP on the enzyme. The gluconeogenic pathway from three-carbon precursors such as lactate and pyruvate occurs mainly in the liver and kidney and has several potentially rate-controlling steps. Those most obviously affected by pH are the conversion of pyruvate to oxaloacetate by pyruvate carboxylase, which takes place in mitochondria, and the subsequent conversion of oxaloacetate to phosphoenolpyruvate by the enzyme phosphoenolpyruvate carboxykinase (PEPCK). The latter reaction

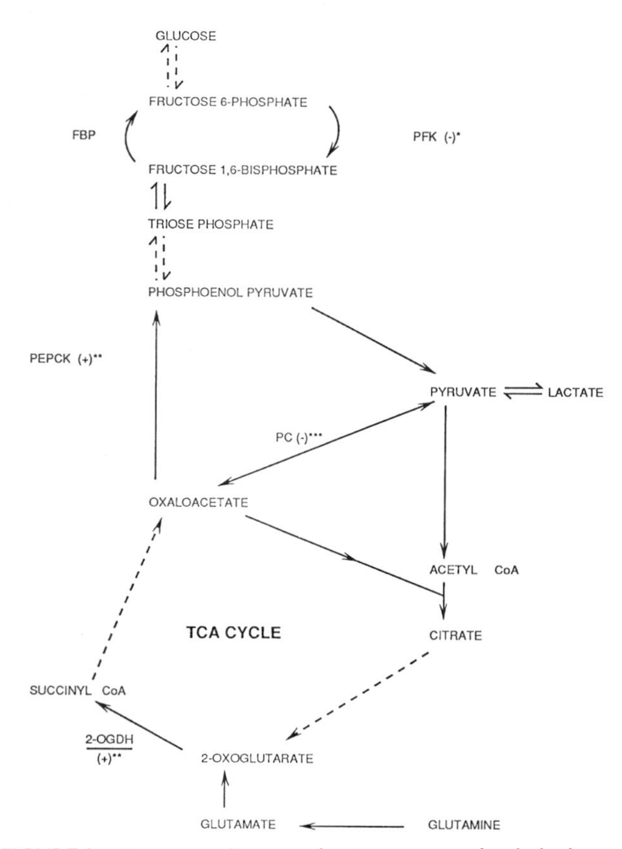

FIGURE 1 Composite diagram of main reactions of carbohydrate metabolism indicating (*underlined*) the catalytic steps at which physiologically relevant effects of pH change have been clearly demonstrated. Glycolysis (glucose → pyruvate) occurs in all tissues. Gluconeogenesis (pyruvate → glucose, or glycogen) is confined to liver, kidney, and, under some circumstances, skeletal muscle. The tricarboxylic acid (TCA) cycle occurs in all tissues demonstrating oxidative metabolism. (+) or (−) indicates overall stimulatory or inhibitory effects of lowering pH. *Asterisks* indicate those tissues in which these effects may be important: *most tissues; **kidney; ***liver. PFK, phosphofructokinase I; PC, pyruvate carboxylase; PEPCK, phosphoenolpyruvate carboxykinase; FBP, fructose 1,6-bisphosphatase; 2-OGDH-2-oxoglutarate dehydrogenase. The *interrupted arrows* represent transitions involving multiple enzyme steps. See text for further explanation.

is entirely cytosolic in some species but, in others including humans, takes place both in the cytosol and in the mitochondria.

There are quite different physiologically relevant modes of action of pH on these two enzymes of gluconeogenesis. In the case of hepatic pyruvate carboxylase *in vitro,* a fall of pH within the physiological range inhibits the obligatory allosteric activation by acetyl coenzyme A. In the case of (renal) PEPCK, lowering the blood pH of the animal (i.e., induction of acidosis) induces its synthesis, with consequent acceleration of gluconeogenesis via an effect on the regulatory region of the PEPCK gene.

A further mode of pH action on enzymes is exemplified by the marked activation of renal 2-oxoglutarate dehydrogenase observed in acid conditions both *in vitro* and *in vivo;* this effect is achieved by an increase in affinity of the enzyme for its substrate. 2-Oxoglutarate dehydrogenase catalyzes the conversion of 2-oxoglutarate to succinyl coenzyme A in the tricarboxylic acid cycle; it is also concerned in the pathway of glutamine metabolism, and its role in this respect appears to be an important regulator of the renal adaptation to acidosis (see next section).

A final important effect of pH changes is in the posture of "redox" reactions. Using the cytosolic conversion of lactate to pyruvate as an example, this reaction is represented by the equation

$$\text{lactate} + \text{NAD}^+ \rightarrow \text{pyruvate} + \text{NADH} + \text{H}^+$$

in which NAD^+ is reduced to NADH. This reaction, catalyzed by lactate dehydrogenase, is close to equilibrium under most circumstances and the law of mass action may therefore be applied:

$$\frac{[\text{lactate}]}{[\text{pyruvate}]} = \frac{K[\text{NADH}][\text{H}^+]}{[\text{NAD}^+]}.$$

A change in the left-hand side (usually denoted as L/P) may therefore be due to either or both a change in $[NADH]/[NAD^+]$ or in $[H^+]$. Thus a fall in cytosolic pH of 0.3 units (corresponding to a twofold change of $[H^+]$) would result in a doubling of L/P without a change in the redox status as determined by $[NADH]/[NAD^+]$. Similar considerations apply to other redox reactions (e.g., those catalyzed by 3-hydroxybutyrate dehydrogenase and by glutamate dehydrogenase, both of which are intramitochondrial).

In all of the just-described examples the effect is substantial within the range of pH that might be expected in the relevant organs (and cell compartments within those organs) under physiological and pathophysiological conditions.

B. Effect of pH on the Supply of Substrate to the Reactions of Carbohydrate Metabolism

1. Membrane Transport Effects

pH has major effects on the transport of substrates across cell membranes. The most well-characterized example in carbohydrate metabolism is that of lactate, in relation both to its exit from cells in organs producing lactate (e.g., skeletal muscle and erythrocytes) and to its entry into cells in organs responsible for lactate disposal (e.g., the liver and kidneys). Because of its considerable physiological and clinical importance, the lactate example will be discussed in some detail.

Lactate (and almost certainly pyruvate) traverses cell membranes by at least two mechanisms, passive nonionic diffusion not mediated by a carrier and a transporter-mediated process, both of which are pH dependent. Nonionic diffusion is a very general mode of distribution of weak acids and bases across biological membranes; the molecular species actually traversing the membrane is the uncharged form, which is much more soluble in the membrane lipid bilayer than is the charged form. For a weak acid, the distribution of the total acid (i.e., dissociated plus undissociated) between two compartments (1) and (2) (separated by a lipid membrane), in which the hydrogen ion activity is $[H^+]_1$ and $[H^+]_2$, respectively, is given by

$$\frac{R_1}{R_2} = \frac{[H^+]_2 ([H^+]_1 + K)}{[H^+]_1 ([H^+]_2 + K)}, \qquad (1)$$

where R is the total concentration of acid as defined earlier, K is the acid dissociation constant, and the subscripts refer to the two compartments. In deriving this equation, it is assumed that the permeability to the charged moiety is negligible compared with that of the undissociated acid, that transit of the undissociated acid across the membrane is by passive diffusion alone, and that equilibrium has been reached. The main implication of Eq. (1) is that weak acids will tend to be concentrated in the more alkaline of the two compartments, and that if $[H^+]_1$ and $[H^+]_2 \ll K$ then $R_1/R_2 = [H^+]_2/[H^+]_1$. Under these conditions at equilibrium, a gradient of 1 pH unit between the two compartments results in a 10-fold concentration difference of the total weak acid. Departures from the theoretical distribution may result from (a) nonequilibrium conditions (e.g., removal of the weak acid by metabolism after traversing the membrane) or (b) significant permeability of the ionized form, in which case the electrical potential across the membrane becomes an additional determinant of R_1/R_2. It should be noted that for some weak acids such as lactate (with pK_a 3.8), the concentration of the undissociated moiety is several orders of magnitude smaller than that of the lactate ion in the physiological range of pH. Under some circumstances, the nonionic diffusion mechanism is the principal mode of transit across cell membranes because the low concentration of the undissociated form is more than compensated by the much greater ability to cross the cell membrane. It is important to appreciate that this mechanism is always kinetically first order (i.e., it does not saturate) in contrast to the transporter mechanism described in the next paragraph.

The transporter mechanism for the entry of lactate (ionized) is semispecific; certain other monocarboxylates, notably pyruvate, are also transported. In the majority of mammalian species examined, it is largely stereospecific for the natural L-isomer of lactate. It is widely distributed but has been best characterized in erythrocytes and in hepatocytes. The erythrocyte lactate transporter has been reconstituted in active form into artificial liposomes. The gene for the mammalian erythrocyte transporter has now been cloned and sequenced. This gene is also expressed in cardiac and some skeletal muscle cells and in gut epithelium. It appears to be distinct from that coding for monocarboxylate transport in liver. In hepatocytes, the lactate transporter activity is markedly increased in the starved and diabetic states as opposed to the fed state. Transport is notably pH dependent, a downhill proton gradient across the membrane being associated with high transporter activity and vice versa. The mechanism of the pH dependency is at least partly explained by the transporter acting as a "symport" for both lactate ions and protons. Although the stoichiometry for lactate and H^+ appears to be 1 : 1 in some cells, in others (e.g., the hepatocyte) the ratio of lactate/H^+ transported is in excess of unity. Another possibility is that the pH effect is mediated by proton-directed conformational changes in the transporter. The distinct pyruvate transporter of the inner mitochondrial membrane is similarly pH dependent. [*See* Cell Membrane Transport.]

2. Supply of Substrate to Organs of Carbohydrate Metabolism

Complete metabolism of glucose to CO_2 and water requires oxygen. A lack of oxygen diverts pyruvate to lactate, instead of the tricarboxylic acid cycle, to reconvert NADH to NAD^+, thus allowing anaerobic glycolysis and some ATP production to continue. Oxygen is also required on the synthetic side (i.e., gluconeogenesis), which has a particularly heavy requirement for ATP.

pH affects oxygen delivery to the various organs in several ways. The first is through a complex series of changes in the affinity of hemoglobin for oxygen.

Acute falls in blood pH produce an immediate change resulting in the more ready release of oxygen from hemoglobin to the tissues. However, in more chronic acidosis the affinity returns toward its original value because of a pH-induced decline in the erythrocyte content of 2,3-diphosphoglycerate (2,3-DPG), which binds to hemoglobin and causes decreased affinity. 2,3-DPG synthesis is unique to erythrocytes and takes place in a "side-loop" of the glycolytic pathway arising as an alternative route of 1,3-diphosphoglycerate metabolism. A fall in erythrocyte pH lowers 2,3-DPG by a complex mechanism.

The second major effect of pH on substrate supply for carbohydrate metabolism is on blood flow through the organs concerned. Acidosis has markedly negative effects on myocardial contractility, leading to a diminished myocardial output and a consequent fall in individual organ blood supply. This has the potential for increasing anaerobic glycolysis and the production of lactic acid in peripheral organs, and of diminishing lactate disposal in organs such as the liver. Lactate uptake by the perfused rat liver falls slowly at first with diminishing blood flow but declines rapidly when flow is below 25–30% of normal, and the liver may switch to lactic acid production. A third important effect of lowered pH on carbohydrate metabolism is on the entry of glucose into cells, which in many cases is dependent on the presence of insulin. It has long been apparent in diabetic ketoacidosis that there is a marked resistance to the blood glucose lowering effect of insulin. At least part of this phenomenon is related to an acidosis-induced decrease in the number of insulin receptors in insulin-sensitive cells, with consequent impairment of insulin-dependent glucose entry. [*See* Insulin and Glucagon.]

3. Proton Production and Consumption during the Metabolism of Carbohydrate

When an electrically neutral carbohydrate such as glucose is fully converted to neutral products such as carbon dioxide and water, there is no net production or consumption of protons. In contrast, if the metabolic reaction gives rise to charged anionic products such as lactate or pyruvate, protons are released in equivalent amounts:

$$\underset{\text{glucose}}{C_6H_{12}O} \rightarrow \underset{\text{lactate}}{2CH_3\,CHOHCOO^-} + 2H^+.$$

Similarly, if as in gluconeogenesis, anionic precursors are converted to glucose, protons are consumed in the overall reaction:

$$2CH_3\ CHOHCOO^- + 2H^+ \rightarrow C_6H_{12}O_6.$$

There is therefore *in the steady state* a close stoichiometry of proton production or consumption as indicated in the just-mentioned equations. Thus in glycolysis, the production of two protons would be expected for each molecule of glucose converted to lactate. That this stoichiometry is disrupted in nonsteady state conditions can be illustrated by writing down the equations of the individual reactions of glycolysis, having full regard to the ambient pH, the acid dissociation constants of phosphate, and the adenine nucleotides ATP and ADP and the charge on the reactants. Summation of these equations indicates that the process of glycolysis of one glucose molecule results in the production of two lactate ions, the conversion of two molecules of each of ADP and inorganic phosphate to two molecules of ATP, and an amount of protons considerably less than the two expected. However, this situation does not represent a steady state because, for instance, the ATP concentration has been increased. If one then takes into account the proton production accompanying the full use of that additional ATP (for energy-driven reactions, e.g., muscle contraction) so that the true steady state is maintained, it is simple to show the expected stoichiometry of two protons produced per one molecule of glucose converted to lactate. [*See* Adenosine Triphosphate (ATP).]

These proton producing and consuming reactions alter the pH at the site where they occur and thus modify the rates of those reactions by feedback. Intracellular acidification in skeletal muscle has been clearly demonstrated in skeletal muscle during anaerobic exercise in humans using noninvasive ^{31}P magnetic resonance spectroscopy. In isolated rat liver perfused with medium containing lactate as the sole substrate, the hepatic venous effluent is always slightly more alkaline than the inflowing perfusate, despite a higher partial pressure of carbon dioxide (P_{CO_2}) than in the portal vein. This could in theory be due to entry of the lactate solely as the undissociated acid, protons being thus continuously abstracted from the perfusate. However, this cannot be more than a part of the process because it has been shown that hepatic intracellular pH rises with increasing lactate uptake; the simplest explanation is that some fraction of the lactate entering the hepatocyte does so in the ionized form, and intracellular protons are consumed during its subsequent conversion to glucose. Evidence shows that the lactate transporter transfers lactate at least partly in the ionized form.

II. PHYSIOLOGICAL INTERACTIONS OF CARBOHYDRATE AND ACID–BASE METABOLISM

In normal resting adult humans, about 1300 mmol lactate enters the circulation per day, principally from skeletal muscle, skin, gut, brain, and erythrocytes. The similar quantity of protons produced with this lactate titrates tissue and blood bicarbonate. In health, the bicarbonate deficit thus created is corrected when lactate is taken up and converted to neutral products (e.g., glucose, or CO_2 and water), principally by the liver, but also by the kidney and heart and under some circumstances by the skeletal muscle itself. The process of regeneration of bicarbonate may be represented as:

$$2CH_3\ CHOHCOO^- + 2H^+ \rightarrow C_6H_{12}O_6$$

$$2H_2O \rightarrow 2H^{+\nearrow} + 2OH^-$$

$$2OH^- + 2CO_2 \rightarrow 2HCO_3^-$$

As just indicated, the first and probably all of these reactions take place within the liver cell, and virtually all the regenerated bicarbonate is then exported to the circulation to restore the deficit. Lactate oxidation to CO_2 and water has the same overall effect and is more prominent in kidney, heart, and skeletal muscle than in the liver. Even in resting humans, the proton turnover associated in this way with lactate production and removal forms a major fraction of proton turnover through the circulation.

This cycle of lactate production and disposal, with its accompanying proton turnover, is affected by pH at the many stages detailed in the first part of this account. The marked drop in intracellular pH (pH_i) in skeletal muscle accompanying glycolysis has been suggested as a mechanism of fatigue, related to the feedback inhibition of glycolysis at the PFK step by low pH (see Fig. 2). The declining force during fatigue is negatively and linearly related to pH_i. Furthermore, exercise endurance in humans is diminished by the induction of acidosis with ammonium chloride and enhanced by the administration of sodium bicarbonate. Measurements of intermediate metabolism of glycolysis in skeletal muscle during these maneuvers are consistent with an effect at the PFK step. Yet if this mechanism were the principal explanation of fatigue, a fall in muscle ATP would be expected. In fact, the fall during fatigue is relatively slight, and some additional factor such as compartmentation of ATP would have

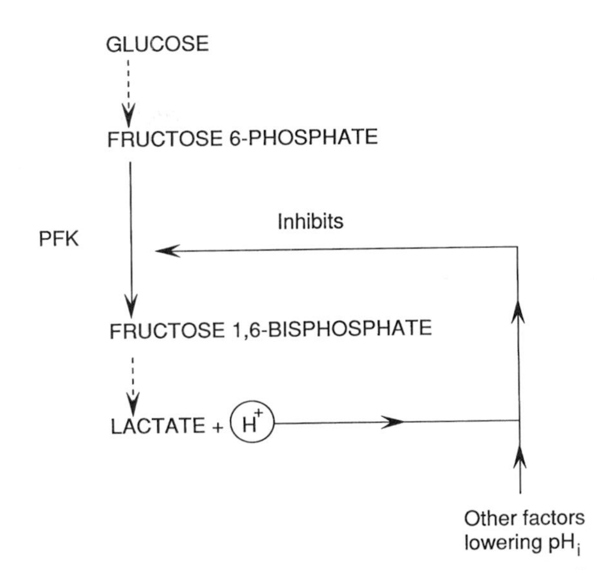

FIGURE 2 Skeleton of glycolysis demonstrating negative feedback of generated hydrogen ions on the phosphofructokinase (PFK) step. Other factors that lower cell pH (pH_i) have the same effect. The interrupted arrows represent transitions involving a number of steps.

to be present. The matter thus remains unresolved. It has been suggested that the inhibition of PFK by lowered pH has been evolved to protect the animal against the likely fatal acidosis that would be caused by totally unbridled glycolysis. [*See* Glycolysis.]

A further example of a pH effect on glycolysis probably exerted at the PFK step is seen during respiratory alkalosis, in which both extra- and intracellular pH has been elevated by lowering of the partial pressure of CO_2 due to increased pulmonary excretion of CO_2. Glycolysis appears to be stimulated in many tissues, including erythrocytes, and blood lactate is thereby increased. The extent of this effect is, however, variable, depending *inter alia* on the response of the lactate removal mechanism. In normal humans, spontaneous hyperventilation, or mechanical ventilation under anesthesia, produces only small increments in blood lactate; in contrast, more substantial elevations are seen in mechanically hyperventilated dogs. The production of lactic acid in response to low P_{CO_2} goes some way toward correcting the alkalosis.

Lactate produced by glycolysis and entering the circulation is removed principally by the liver; any tendency to mild lactic acidosis is counteracted by an increased hepatic lactate uptake, due to both the elevation of blood lactate concentration and the increased pH gradient across the hepatocyte plasma membrane, which stimulates both the transporter and diffusion mechanisms of lactate entry. At normal or slightly elevated blood concentrations of lactate (i.e., between 0.5 and 1.5 mmol/liter), the transporter pathway seems to be the dominant entry mechanism (Fig. 3). Consistent with these mechanisms is the observation that hepatic lactate disposal increases in humans undertaking moderate exercise, despite a 50% fall in hepatic blood flow. Observations both in humans and in animals suggest that the hepatic lactate uptake increases with blood concentration up to a level of 3–5 m*M* and then remains constant despite further elevation. At 3–5 m*M* the lactate transporter is already saturated, and the bulk of entry must be via the nonionic diffusion pathway, which cannot saturate. The plateau effect above 3–5 m*M* must therefore be due to saturation of one of the nonequilibrium reactions of gluconeogenesis. The role of the liver in correcting the very marked degree of lactic acidosis (blood lactate up to 20–25 m*M*), which may be achieved during severe exercise, is therefore limited, and other mechanisms of the correction seen on cessation of exercise must be involved. A striking observation is that the rate of decline of lactic acidosis after cessation of maximal exercise is greater if the subject persists with submaximal exercise than if he or she rests completely. This phenomenon is due to the use of lactate as a fuel of oxidation in submaximally exercising skeletal muscle. This cannot happen in maximum exercise when the demand for oxygen outstrips the supply and anaerobic glycolysis therefore produces rather than consumes lactate.

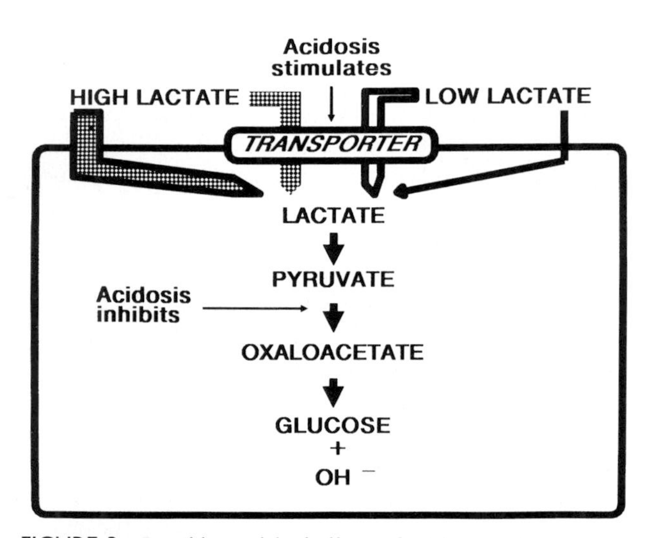

FIGURE 3 Possible model of effects of acidosis on hepatic disposal of lactate. A schematic hepatocyte is depicted with a lactate transporter at the perisinusoidal pole. The top right-hand corner illustrates events at normal or only minimal raised blood lactate concentration. A minor fraction of lactate entry takes place through the nontransporter diffusion pathway. The major fraction is transporter mediated. Both pathways are stimulated by an increased proton gradient downhill from the outside to the inside of the cell, with an increase in overall lactate disposal. The top left-hand corner depicts events occurring at high blood lactate concentrations (>2 m*M*). The transporter is saturated so no further increase can take place via this pathway. The main entry is through the diffusion pathway. Although this may potentially increase, as shown, the point of rate limitation has shifted to the pyruvate → oxaloacetate step of gluconeogenesis, which is inhibited by acidosis, so overall lactate disposal falls.

Severe lactic acidosis also occurs in circulatory insufficiency (shock) due to both increased production of lactate and protons (because of tissue hypoxia) and major impairment of lactate removal. Observations in perfused rat liver have demonstrated that provided perfusate lactate levels are substantially elevated, thereby saturing the transporter, acidosis itself is a powerful inhibitor of lactate disposal via gluconeogenesis. This contrasts with the stimulatory effect of acidosis seen at lower lactate levels, presumably related to stimulation of the transporter. At the higher levels of lactate, the site of the inhibitory action of acidosis on gluconeogenesis is at a step between pyruvate and oxaloacetate (Fig. 3). The degree of inhibition is closely related to hepatic pH_i, whether this is achieved by raising P_{CO_2} ("respiratory acidosis") or lowering bicarbonate ("metabolic acidosis"). Several stages in the pyruvate → oxalocetate step are possible sites of the inhibitory effect of low pH; an obvious but unproved possibility is the effect of low pH on the allosteric activation of pyruvate carboxylase, referred to earlier.

The pathological significance of this inhibitory effect is considerable. In lactic acidosis the prime homeostatic requirement is to dispose of lactate at an accelerated rate so that bicarbonate may be rapidly regenerated. This inhibitory effect of acidosis on gluconeogenesis is precisely the opposite of what is required and, in fact, exacerbates the lactic acidosis. Several types of lactic acidosis have a particularly fulminating course, and it has been suggested that this may in part be due to a vicious circle set up by acidotic inhibition of gluconeogenesis from lactate. In clinical shock, the prospect for survival becomes increasingly poor when blood lactate concentration exceeds 5 mmol/liter; the mechanism just described may be one component in this poor prognosis. The antidiabetic biguanide phenformin was responsible for a large number of fatalities because of fulminating lactic acidosis in the 1970s, before this drug was abandoned. Phenformin has multiple effects on biological membranes because of its insertion into them and consequent disturbances of surface charge. One of its actions is to inhibit the lactate transporter; it may be speculated that for this reason the homeostatic defenses against mild lactic acidosis are removed, allowing lactic acidosis to develop to the degree in which the inhibitory effect on gluconeogenesis is evident. Unlike in recovery from exercise, there is no substantial alternative disposal route for lactate and hydrogen ions in either shock or phenformin toxicity.

The kidney has a smaller role in lactate disposal than the liver, but one that animal studies suggest could be significant. The kidneys use lactate both as a fuel of oxidation and for conversion into glucose. In normal rats, the kidneys account for about 20% of the disposal of an intravenously injected sodium lactate load. If rats are made progressively acidotic by feeding ammonium chloride, the rate of extrarenal removal of lactate falls. However, this is partially counteracted by an increased rate of renal lactate disposal, so that in severely acidotic rats the kidneys may account for up to 40% of removal of the lactate load. Whether this effect is seen in humans is unknown. It seems likely to be due to the induction of PEPCK as described in Section I. This effect does not occur in liver, thus accounting for the difference in behavior of the liver and kidneys toward lactate in acidosis. It should be noted that lactate loss in the urine is insignificant until blood lactate reaches 8–10 m*M* and urinary excretion of buffered protons is small compared with proton removal by the liver during lactate metabolism.

The complex interactions of pH and carbohydrate metabolism are well-illustrated in diabetic ketoacidosis. In this condition there is gross overproduction of glucose by the liver, accounting for the major part of the hyperglycemia. The substrates for glucose formation include both amino acids and three-carbon precursors such as lactate and pyruvate. There may be severe extracellular acidosis, usually attributed to the production of 3-hydroxybutyrate and acetoacetic acids ("ketone bodies") by the liver, and it is at first sight anomalous that diabetic ketoacidosis is only rather infrequently accompanied by severe lactic acidosis; the latter would be expected because, *inter alia,* of the acidotic inhibition of gluconeogenesis from lactate described earlier. Some observations provide a possible solution to this paradox. First, it is not possible to inhibit gluconeogenesis from lactate by perfusing livers taken from diabetic ketoacidotic rats with acidotic medium, whereas, as indicated earlier, this inhibition is readily demonstrated in livers taken from normal rats. Surprisingly, the intracellular pH in perfused livers from diabetic ketoacidotic rats is higher than that in the "normal" perfused livers, although both sets of livers are perfused with similarly acidotic media. Second, if rats are made progressively acidotic by inducing diabetic ketoacidosis, by ammonium chloride ingestion, or by hydrochloric acid infusion, hepatic cell pH (pH_i) measurements *in vivo* by ^{31}P nuclear magnetic resonance spectroscopy indicate that although hepatic pH_i falls in ammonium chloride and hydrochloric acid acidosis, there is almost no effect of progressive blood acidosis on hepatic pH_i in the diabetic ketoacidotic livers. It may be that this unique protection of hepatic pH_i in livers from diabetic ketoacidotic animals permits normal or increased gluconeogenesis from lactate to proceed despite severe blood acidosis. The basis of this protection is speculative but may be due to stimulation of gluconeogenesis from lactate by the ketogenic state, with consequent increased generation of bicarbonate. This bicarbonate could be sufficient to neutralize protons produced during ketogenesis, thus preventing a fall in hepatic pH_i. Because acidosis tends to inhibit ketogenesis, another effect of maintenance of hepatic pH_i in diabetic ketoacidosis may be that the partial oxidation of fatty acid substrate to ketone bodies may be encouraged to continue despite the severe extracellular acidosis.

The acid–base consequences of fructose administration have been of considerable interest because this monosaccharide was included as a component of intravenous feeding solutions for clinical use. It was noted that subjects to whom these solutions were administered at a substantial rate developed lactic acidosis. Unlike glucose, fructose is rapidly converted in part to lactate by the liver. It has been shown in both humans and animals that this is due to the brisk phosphorylation of fructose to fructose 1-phosphate by fructose 1-kinase; this reaction requires ATP, which is rapidly depleted in liver cells during fructose infusion. Studies in rat perfused livers in which hepatic cytosol pH has been followed either by ^{31}P magnetic resonance spectroscopy or by intracellular pH-sensitive microelectrodes have shown striking falls during fructose infusion. This fall in cytosol pH is initially partially due to the H^+ released in the fructokinase reaction during the phase of fall in ATP concentration; later, low cytosol pH is maintained by the substantial quantities of lactate and protons released during fructolysis. Fructose has now been abandoned as an intravenous feeding agent. It should be noted that the ordinary absorption of fructose from the hydrolysis of dietary sucrose does not proceed at rates sufficient to produce the effects just described.

The kidney has classically been regarded as having a major role in acid–base homeostasis. Although this view has been subjected to substantial reinterpretation, it is undisputed that acidosis results in a substantial increase in the rate of urinary excretion of ammonium ions. [*See* pH Homeostasis.] This has probably been wrongly interpreted as representing the excretion of protons buffered as NH_4^+ but is more correctly regarded as diverting nitrogen to urinary excretion as NH_4^+ rather than to urea synthesis; urea synthesis incurs the undesirable penalty of generating two protons for every molecule of urea synthesized in the liver, so some switch from urea synthesis to NH_4^+ exertion in acidosis would be advantageous for acid–base homeostasis. The mechanism of increased ammoniagenesis is now fairly well understood and involves well-defined interactions of pH, carbohydrate, and amino acid metabolism. NH_4^+ is derived from the renal hydrolysis of glutamine, followed by oxidative deamination of the resulting glutamate to 2-oxoglutarate, with two molecules of NH_4^+ being generated. The latter reaction, catalyzed by glutamate dehydrogenase, is at near equilibrium, and it is necessary to remove the products to allow the reaction to proceed. NH_4^+ is removed by nonionic diffusion into the acidified urine. 2-Oxoglutarate is removed by gluconeogenesis, the flux-controlling enzyme in this process being PEPCK, which, as indicated earlier, is increased in quantity by chronic acidosis. Inhibition of this enzyme prevents the increase in NH_4^+ production seen in acidosis. In addition, 2-oxoglutarate dehydrogenase is stimulated by acute acidosis; this effect diminishes renal 2-oxoglutarate concentration even when PEPCK is inhibited.

III. CONCLUDING REMARKS

Cells possess three main categories of defense against tendencies to alter their internal pH: (1) physicochemical buffering, (2) movements of protons (or other relevant ions, e.g., OH^-, HCO_3^-) across the plasma membrane, and (3) alteration in the internal production or consumption of protons. This account has been substantially concerned with the last of these three mechanisms and has reviewed the manner in which changes in flux through the pathways of carbohydrate metabolism may alter proton production or consumption. It has been shown that these changes are sometimes homeostatic and at others antihomeostatic for the maintenance or correction of acid–base status. Furthermore, they are frequently themselves caused by pH alteration.

Such emphasis as has been given has been in consideration of perturbations in the acidotic direction. The reason is that in terms strictly of carbohydrate metabolism, this is the direction in which most of the major disturbances occur—exercise, shock, and ketoacidosis being prime examples. However, there is a view that the threat of disturbance in the alkalotic direction has been much underestimated, particularly in biological and evolutionary terms rather than from the clinical perspective. But because this threat derives from the metabolism of the carbon skeletons of amino acids, it has not been considered here. The reader is referred to the appended bibliography for source material on this topic.

BIBLIOGRAPHY

Atkinson, D. E., and Bourke, E. (1984). The role of ureagenesis in pH homeostasis. *Trends Biochem. Sci.* **9**, 297–300.

Cohen, R. D. (1990). *In* "The Metabolic and Molecular Basis of Acquired Disease" (R. D. Cohen, B. Lewis, K. G. M. M. Alberti, and A. M. Denman, eds.). Bailliere Tindall & Cox, London.

Cohen, R. D. (1994). Lactic acidosis: New perspectives on origins and treatment. *Diabetes Rev.* **2**, 86–97.

Cohen, R. D., and Guder, W. G. (1988). Carbohydrate metabolism and pH. *In* "pH Homeostasis" (D. Häussinger, ed.). Academic Press, London.

Madias, N. E. (1986). Lactic acidosis. *Kidney Int.* **29**, 752–774.

Iles, R. A., Stevens, A. N., and Griffiths, J. R. (1982). NMR studies of metabolites in living tissue. *Prog. Nucl. Magn. Reson. Spectroscopy* **15**, 49–200.

Pharmacogenetics

BERT N. LA DU, JR.
University of Michigan

GLOSSARY

Ecogenetics Essentially, the same meaning as pharmacogenetics but used in reference to chemicals of the environment rather than chemicals that are used as therapeutic agents

Polymorphic enzymes Multiple forms of an enzyme, differing slightly in their structure and catalytic activities

PHARMACOGENETICS IS THE STUDY OF GENETIcally transmitted traits in humans and animals that confer exaggerated or unusual responses to therapeutic drugs and to chemicals of our environment. These genetic traits may affect pharmacokinetics, defined as the processes by which drugs are absorbed from their site of administration, distributed throughout the body, metabolized, and eliminated from the body. They can also alter the pharmacodynamics (i.e., the tissue responsiveness to the drug at the organ or cellular level) because of some structural alterations in receptor proteins at the sites inside the cell or the cellular membranes, where critical interactions take place between drug molecules and receptors.

The objectives of pharmacogenetic investigations are listed in Table I. Apparently, pharmacogenetics is a very pragmatic specialty that seeks to understand the basis of unusual reactions to drugs and to improve the effectiveness and safety of drugs by avoiding these adverse reactions.

I. EXAMPLES

About 60 different pharmacogenetic conditions are known, and some representative examples are briefly characterized in Table II. They range from the striking differences among people in their ability to taste phenylthiourea (PTC), to the serious, life-threatening hereditary condition called malignant hyperthermia with muscular rigidity. The former represents an inherited difference in the bitter taste threshold for PTC, which has not yet been explained at the molecular level but appears to represent no significant clinical vulnerability; the latter is a very serious and, fortunately, rare condition, which is most likely to occur during surgery in young, healthy adults who have often been given two commonly used drugs during anesthesia: halothane and succinylcholine. Until quite recently, nearly half of the subjects who developed a pronounced rise in body temperature and a generalized rigidity of skeletal muscles died from the resulting heart or kidney failure. Because the unusual susceptibility is inherited as a dominant trait, many family members of affected people are also at risk, even though the condition is rare. A pharmacological test, based on the contractile response of a skeletal muscle biopsy preparation to caffeine, has been used successfully to identify susceptible relatives. As more is learned about the genetic basis of this condition, developing simpler diagnostic tests should be possible.

II. UNDERLYING CAUSES

Many pharmacogenetic conditions are relatively rare, and the majority are of the pharmacokinetic type, characterized by a marked reduction in the rate of drug metabolism. Slow metabolizers with a deficiency of one of the drug metabolism enzymes are likely

TABLE I
Objectives of Pharmacogenetic Studies

1. Identify the genetic trait associated with the pharmacogenetic condition.
2. Determine the mechanism by which the trait causes the unusual response to the drug, or drugs.
3. Determine the mode of inheritance of the trait (dominant or recessive; autosomal or sex-linked).
4. Estimate the gene frequencies of the trait in the general population and selected ethnic and geographic areas.
5. Evaluate the clinical importance of the trait.
6. Develop a simple test to detect the trait, without exposure to the drug, if a diagnostic test is needed.

to develop a much higher drug concentration than normal in the blood and tissues. If repeated, standard doses of the drug are given to such people, the drug may accumulate and reach toxic concentrations unless appropriate adjustments are made in the dose and dosing interval to compensate for the individual's reduced rate of metabolism. [*See* Pharmacokinetics.]

New pharmacogenetic conditions have been discovered primarily by investigating patients who have shown an exaggerated response to a drug or experienced some unusual adverse reaction to a drug. The investigation may establish that the patient is unusually slow in metabolizing the drug or that the patient transformed the drug to a different pattern of drug metabolites, some of which produce toxic reactions. If close relatives of the patient can also be tested without undue risk, the slow metabolism may be duplicated or the usual metabolic pattern may be obtained in these family members. Another family member with the reduced rate of metabolism would increase the likelihood that genetic factors are responsible for the unusual findings. The first patient showing an unusual resistance to a coumarin anticoagulant drug required 20 times the usual dose of the drug to obtain the expected anticoagulant effect. By remarkable coincidence, this patient had an identical twin brother, who also received a coumarin-type anticoagulant within a few weeks of his brother's treatment. He displayed exactly the same unusual type of drug resistance. A survey of additional family members revealed that several others were also resistant to the drug and that resistance was inherited as a dominant trait.

Most pharmacogenetic conditions are completely unsuspected. The affected people have no indication of their unusual genetic trait and show no disturbance in their metabolism of sugars, proteins, fats, or other ordinary chemical compounds. From this observation, it is clear that pharmacogenetic traits selectively involve different metabolic systems and a different set of enzymes from those concerned with the metabolism of endogenous chemicals that arise during the course of normal intermediary metabolism.

An interesting biological problem is posed by these distinctions in metabolic systems. What is the nature of the special enzymatic systems utilized by mammalian species to inactivate and eliminate drugs and other foreign chemicals, and how did these drug metabolism enzyme systems evolve? We know that there are two major pathways by which drug molecules are transformed in the body: (1) direct attack by special enzymes such as the P450 enzymes of the liver endoplasmic reticulum, which catalyze drug oxidations, and reductions to produce primary changes in the drug molecule; and (2) the conjugation reactions, catalyzed by transferase enzymes that utilize endogenous donor compounds such as activated sulfate or glucuronate. These conjugation reactions generally yield products of the drug that are inactive pharmacologically because important functional groups are blocked in the course of forming the glucuronides, acetylated derivatives, or sulfate conjugates (Fig. 1). Surprisingly, these inactivating systems for drug molecules show extremely high variability from animal species to species, and even within the human population, from one person to another. Half-lives of drugs (the time required for half of the drug to be removed from the body) can vary as much as 10- to 20-fold. This wide variation explains why a standard dose of a drug will be too high for some people but too low for many others, even though it is calculated to be the right amount for the "average" person. Individualization of the drug dose is often necessary to obtain optimal therapeutic results in most patients.

The other important observation is that whether a person is a fast or slow metabolizer of a particular drug, he or she will persist in his or her individual rate, if rechallenged, with little change over many years. The logical conclusion to be drawn from these two findings is that there is considerable genetic heterogeneity in the drug metabolism enzyme systems, and these characteristics are set for each individual primarily by genetic rather than environmental factors. It is not unexpected, then, that a large number of genetic traits in humans have been identified that are responsible for the wide range of activities of the drug metabolism enzymatic systems. Some of these genetic traits

TABLE II
Some Examples of Pharmacogenetic Conditions

Defective protein or polymorphism	Susceptible phenotype	Drugs or foreign compound	Disorder	Incidence
N-Acetyltransferase	Slow acetylator	Isoniazid	Polyneuritis	60% Caucasian; 10–20% Orientals
Serum cholineserase	Atypical and other variants	Succinylcholine	Prolonged paralysis	1:2000 Caucasians
Serum paraoxonase polymorphism	Slow metabolizer	Paraoxon	Pesticide poisoning	50% Caucasians
Debrisoquine hydroxylation deficiency	Slow metabolizers	Debrisoquine, sparteine	Speciic P450 deficiency, overdose toxicity	6–9% Caucasians
Mephenytoin hydroxylation deficiency	Slow metabolism	Mephenytoin, mephobarbital	Specific P450 deficiency, overdose toxicity	5% Caucasian; 23% Japanese
Aldehyde dehydrogenase deficiency	Poor metabolizer	Ethanol	ALDH isozyme I deficiency, overdose toxicity	30–50% Orientals
Phenyloin hydroxylation deficiency	Slow metabolizers	Phenytoin	Specific P450 deficiency, overdose toxicity	Very rare
Glucose-6-phosphate-dehydrogenase deficiency	Deficient red blood cells	Primaquine, many drugs	Specific enzymatic defect, hemolysis	Frequent in American Blacks and in tropical countries
Malignant hyperthermia	Defective binding of calcium in muscle membrane	Halothane, succinylcholine	Hyperthermia, hyperrigidity (often fatal)	About 0.005% in Caucasians
Warfarin resistance	Resistant phenotype; requires 20 times as much drug as normal	Dicoumarol-type anticoagulants	Receptor defect; increased affinity for vitamin K	Very rare
Porphobilinogen synthetase inducibility	Abnormal inducibility of enzyme	Barbiturates	Drug induces δ-amino-levulinate dehydratase, porphyria	Rare
Hypoxanthine-guanine phosphoribosyl transferase	Enzymatic deficiency	Allopurinol	Drug relatively ineffective for treating gout in patients with this deficiency	Rare
Catechol-*o*-methyl transferase	Enzymatic deficiency	Isoproterenol, L-dopa	Uncertain	Rare
Unknown	Bitter taste in "tasters"	Phenylthiourea	Marked variation in taste threshold	Nontasters, 30% Caucasians

cause the individuals carrying them to react in very unusual ways, even when the drug is given in the generally recommended dose.

Pharmacogenetic studies attempt to identify the biochemical basis for the conditions and to determine which enzymatic system is altered. Some pharmacogenetic conditions can now be diagnosed by a simple blood test, and the drug sensitivity can be accurately predicted without exposing the subject to the drug. For example, several of the serum cholinesterase vari-

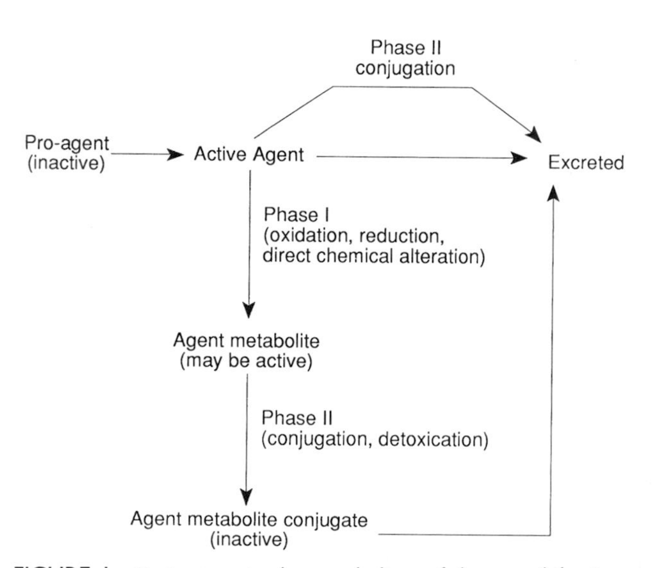

FIGURE 1 Basic steps in the metabolism of drugs and foreign chemicals.

ants that cause an exaggerated response to the muscle-relaxant agent succinylcholine can be accurately diagnosed by a simple *in vitro* test with serum (or plasma). The sample is incubated with its substrate benzoylcholine and certain inhibitors, dibucaine and fluoride, to determine the status of the enzyme in question, cholinesterase. A number of the cholinesterase variants have been analyzed at the DNA level and shown to represent point mutations or frameshift mutations that affect the amount or quality of the enzymatic structure. Although the succinylcholine sensitivity condition is rare, it is inherited as a simple Mendelian trait. Close relatives (siblings, parents, and children) are much more likely to share the same type of drug sensitivity than nonrelatives. Accordingly, these "high risk" individuals may want to have their blood typed before receiving the drug to find if they would also be sensitive to the drug.

The isoniazid acetylation polymorphism can be used to illustrate a topic of current interest in pharmacogenetic investigation. About half the population in this country are slow and half rapid, so the slow acetylation phenotype, determind by a double dose of a recessive gene, is not rare, as is so often the situation with genetic disorders. Also, the consequence of being either a rapid or a slow metabolizer is probably related directly to the level of *N*-acetyltransferase activity of the liver. Unfortunately, this is a difficult organ to evaluate by direct tests, but information about the clinical importance of the acetylation polymorphism can also be obtained indirectly. For example, it can be determined whether certain adverse drug reactions or complications occur with greater frequency in people who are rapid or slow acetylator phenotypes. Up until the early 1990s, however, it was necessary to administer test drugs to people, such as isoniazid, to identify their status, by measuring the proportion of the free drug and the acetylated drug in a urine sample.

The *in vivo* typing method has limited the number of studies that could be done, but an alternative method has been developed that permits acetylator phenotyping by measuring the ratios of certain urinary metabolites of caffeine after ingestion of a test dose of the drug or after drinking a cup of regular coffee. No doubt, the next advance will be a diagnostic test that can be carried out on DNA samples from blood leukocytes when a peerson's acetylator genotype can be deduced from specific DNA sequences. As it becomes practical to study a larger number of subjects, there will be better opportunities to find correlations between rapid or slow acetylator phenotype and adverse reactions to isoniazid and other drugs. Slow acetylators of isoniazid seem to be more likely to have severe adverse reactions to several other drugs, develop lupus erythematosis either spontaneously or as an adverse drug reaction, and seem to have bladder tumors more often after chronic exposure to some industrial chemicals.

Because many more people are fast or slow acetylators than the number who develop the previously mentioned complications, additionial genetic and environmental factors must contribute to the development of the toxic reaction or protect against its development. A more complete study requires that additional analyses be undertaken to evaluate what other genetic markers and environmental components occur regularly in the subgroup that did develop the toxic drug reactions.

Liver enzymes such as the P450 cytochrome systems, flavin-monooxygenases, xanthine oxidase, and several alcohol and aldehyde dehydrogenases that catalyze various types of drug oxidation reactions are particularly important in drug metabolism. Of these, the cytochrome P450 systems have received particular attention in recent years and are the target of much current pharmacogenetic interest because of their unique roles in so many types of drug metabolism. Some genetic variants or deficiencies of specific P450s are included in Table II, but many more types are suspected as being associated with slow metabolism or with adverse reactions from toxic metabolites of other drugs and environmental chemicals. Again, it has been very difficult to obtain enough human liver tissue to characterize the individual P450 proteins and to identify the variations in their structure that result from genetic mutations. [*See* Cytochrome P450.]

III. FUTURE DIRECTIONS

Drug metabolism and toxicity studies have been done with rats and mice of defined, highly inbred genetic strains and with congenic animals produced by selective breeding to carry particular genetic traits on a specified genetic background. These pharmacogenetic experiments in the animals have identified certain forms of P450 that seem to be of particular importance in the metabolism of selected types of drugs and foreign chemicals, such as the hydroxylation of polycyclic hydrocarbons, the hydroxylation of debrisoquine and sparteine, the hydroxylation of mephenytoin, and the oxidation of ethanol. Cytochrome P450 enzymes in human liver seem to be analogous with similar special substrate preferences. A number of the genes coding for the P450 proteins have been cloned, and some genetic defects that modify the structure of the proteins or affect the rate of synthesis of the enzyme have been identified. These advances utilizing the methods of molecular biology have revolutionized the study and diagnosis and molecular disorders caused by genetic defects, and these methods also allow investigators to assess the status of a number of the drug metabolism enzymes, including the liver P450 enzymes, by analyzing the persons' blood cell genomic DNA. The new DNA analytic and diagnostic methods are overcoming the major obstacles in human pharmacogenetic studies of the past: problems of safety in studying drug effects in patients, if these studies may provoke toxic reactions, and the difficulty in obtaining for detailed study the drug metabolism enzymes that are present primarily in liver or other tissues that are very difficult to sample directly. These new techniques will result in the rapid advance of pharmacogenetic investigations toward the objectives listed in Table I, particularly in the identification of new traits that cause or contribute to adverse drug effects, and the development of simple diagnostic tests to identify susceptible people.

Only a few pharmacogenetic examples have been discovered that represent pharmacodynamic alterations in tissue-intrinsic sensitivity to a drug. Resistance to coumarin-type anticoagulants is one example. These anticoagulants decrease the rate of synthesis of several liver proteins required in blood clotting. The resistant individuals show a normal response to such drugs, but it requires about 20 times as much drug than a normal person. Because these resistant people also have an exaggerated response to the antidotal effect of vitamin K, it is believed that essential receptors in the liver have been altered because of some genetic mutation that has increased the binding affinity for the vitamin and decreased it for the anticoagulant drug. This example serves as a prototype, and we can anticipate that there will be many more pharmacogenetic conditions of this type. Analyses are currently being made using the techniques of molecular biology of many specific drug receptor proteins, and genetics variants can be expected.

BIBLIOGRAPHY

Calabrese, E. J. (1984). "Ecogenetics: Genetic Variation in Susceptibility to Environmental Agents." Wiley-Interscience, New York.

Kalow, W., ed. (1992). "Pharmogenetics of Drug Metabolism." Pergamon Press, New York.

Kalow, W., Goedde, H. W., and Agarwal, D. P. (1986). "Ethnic Differences in Reactions to Drugs and Xenobiotics." Alan R. Liss, New York.

Omenn, G. S., and Gelboin, H. V. (1984). "Genetic Variability in Response to Chemical Exposure." Banbury Report #16, Cold Spring Harbor Laboratory, New York.

Vogel, F., and Motulsky, A. G. (1986). "Human Genetics: Problems and Approaches," 2nd Ed. Springer, New Yor.

Pharmacokinetics

BRUCE CAMPBELL
Servier Research and Development Limited, Slough, United Kingdom

GLOSSARY

Bioavailability Amount of orally administered drug that becomes available to the systemic circulation following formulation dissolution, absorption, and hepatic first pass, all processes that may reduce the total or absolute bioavailability

Clearance Proportionality constant, measured in units of flow (liters · hr^{-1}), that provides a measure of the extraction or removal of a drug from a given volume of blood as it passes through a particular eliminating organ (e.g., liver, kidney, or whole body) at equilibrium over a given time

Drug metabolism Mechanistic study of how the body chemically modifies a foreign chemical (xenobiotic) to produce metabolites, in its attempt to remove the compound from the body. Often, but not always, this detoxifies the compound by increasing polarity to enable more rapid renal excretion, but metabolites can be more active or toxic than the parent drug

Ecogenetics Essentially the same as pharmacogenetics, but used in reference to chemicals of the environment, rather than chemicals used as therapeutic agents

Half-life Time taken, measured in minutes or hours, for the drug concentration (blood or tissue) to fall by one-half

Pharmacodynamics Action of a drug in terms of beneficial (pharmacology) or detrimental effects (toxicology) through its interaction with enzymes, ion transport, or receptors; it can be thought of as what the drug does to the body

Pharmacogenetics Study of the inherited variation in drug response; in the context of this article, it is the genetic variation in pharmacokinetics or metabolism

Pharmacokinetics Study of the time course of the movement of a chemical through the body; it represents what the body does to the drug

Polymorphic enzymes Multiple allele forms of an enzyme, differing slightly in their structure and catalytic activities

Steady state Blood equilibrium plateau or plasma level reached following repeated administration of a drug when the amount absorbed (rate in) equals the amount eliminated (rate out), normally taken to occur 3–5 half-lives after dosing commences

Therapeutic window Range of drug levels between a minimal effective level (MEL) when desirable activity starts, and a minimal toxic level (MTL) when toxicity or unacceptable side effects occur

Toxicokinetics Generation of pharmacokinetic data within animal toxicological studies in order to assess the comparative exposure of the chemical in humans

Volume of distribution Proportionality constant measured as the volume (liters) that a drug distributes in the body at equilibrium, relating the drug concentration in plasma (blood) to the amount of drug in the tissues

PHARMACOKINETICS COMES FROM THE GREEK words *pharmakon* (drug or poison) and *kinesis* (movement) and is the study of the time course of movement of a chemical within the body. Pharmacokinetics, also called *biodisposition,* can be used to examine the fate of any natural or synthetic chemical, drug, pesticide, industrial food additive, protein, and so on, in any animal species or plant. However, most recent studies have been undertaken with drugs in animals humans, so for convenience, pharmacokinetics will be discussed in terms of this association. It is measured by the analysis of the drug or its metabolite in body fluids, tissues, excreta, or expired air at various times after administration. In simple terms, it describes what the body does to the drug and examines *absorption, distribution, metabolism,* and *elimination* (ADME), normally in mathematical terms that reduce complex physiological processes into relatively simple equations. Parameters derived from such equations are specific for the compound and study conditions (age, disease, race, smoking), but may also be used predictively in association with pharmacodynamics (see Section IV) or toxicity to characterize the likelihood of therapeutic or adverse effects of compounds when the conditions are changed.

I. HISTORY AND INTRODUCTION

Although many scientists, since the advent of modern medicines, have described the time course of drug action, it is generally accepted that it was Axel Theorell, a Swedish physiologist, who in 1937 first proposed physiologically based kinetic models. He divided the body into discrete compartments, each with its own volume, representing the blood, a drug depot, intra- and extracellular fluid, kidney elimination, and an organ for degradation. Using rates of exchange from each compartment, he was able to describe the movement of drugs within the body making certain simple assumptions and using differential rate equations. However, it was not for a further 16 years that the actual word *pharmacokinetics* was first coined by F. H. Dost in his book "Der Blutspregel-Kinetic der Konzentration Sabläufe in der Krieslauffussigkeit." Nine years later, the first symposium was held on the subject in Borstel, Germany. Over the subsequent years up until 1970, there was a growing awareness of the importance of this subject, and the fundamental concepts and mathematical equations were derived and tested by various researchers, including Wagner, Gillette, Nelson, Levy, Gibaldi, and Riegelman in the United States, and Kruger-Thiemer, Rowland, and Dettli in Europe. Meanwhile, Williams in London was painstakingly compiling, by review and research, the basis for the understanding of drug metabolism. Unfortunately, analytical methods were relatively crude and unable to measure less than $\mu g \cdot ml^{-1}$ concentrations, and were often nonspecific, measuring both unchanged drug and metabolites. The full use of the newly derived kinetic formula, therefore, had to wait another 10 years for the development of specific analytical techniques capable of routinely measuring nanogram (10^{-9} g) or lower quantities in body fluids.

The large amounts of data generated by such work led to another problem, that of computer power to store and analyze the results. Although some programs were available in the early 1970s (e.g., NONLIN, AUTOAN), these needed relatively expensive hardware and some expertise. In the 1980s, there has been a proliferation of simpler programs that can run on less expensive microcomputers and can be used by expert and student alike. These include TOPFIT, SIPHAR, ELSFIT, MK Model, PC NONLIN, and others.

Although the youngest of the disciplines in drug development, pharmacokinetics has now become one of the most important. It alone provides a synthetic framework for all the other disciplines of *drug discovery, pharmacology, toxicology, biopharmacy, formulation development,* and *clinical studies.* For example, it can provide a greater insight into which animals to choose for preclinical studies, what dosage changes are needed for aged or diseased patients, the choice of coadministered drug and food, the frequency of drug administration required, what level of compliance is achieved, why some patients do not respond or have side effects, and so on. Lack of pharmacokinetic information is also one of the most important reasons why new drugs are not passed for marketing by the U.K. regulatory authorities, the Committee on Safety of Medicines (CSM). This pivotal position for pharmacokinetics has led to a plethora of papers and journals, and many drug-related reports now make some reference to this ubiquitous science.

II. ANALYTICAL METHODS

Over the last 20 years, there have been dramatic changes in analytical techniques for the measurement of chemicals in biological fluids, with sensitivities improving by more than a millionfold. This has been achieved by a greater efficiency of extraction, by improvement in the detectors of chromatographic systems, and by the increasing use of analytical mass spectrometry.

A. Extraction

Before analysis, the compound is extracted from its biological matrix (Table I), the contaminating impuri-

TABLE I
Major Body Fluids That Can Be Used for the Study of Pharmacokinetics

Plasma	Urine	Expired air	Cerebrospinal fluid
Red cells	Bile	Tears	Saliva
Hair	Feces	Milk	Sperm
Serum	Sweat		Vaginal fluid

ties removed, and the solution finally concentrated. Extraction can be achieved by either organic solvents not miscible with water or column chromatography.

(1) Solvents (ether, chloroform, hexane, etc.) are shaken with the aqueous body fluid adjusted to a pH that maximizes the formulation of a lipid-soluble form of the drug. The latter preferentially dissolves in the organic layer, leaving the biological contaminants in the aqueous layer.
(2) For column extraction, the biological fluid is poured into a glass or plastic column containing a solid-phase separator, which traps the drug by either ion exchange or lipophilicity (see Section III,A,2), allowing unwanted contaminants to pass through and be washed away. The drug is ultimately released by use of mild acids, bases, or solvents, and the extract is concentrated before analysis.

B. Thin-Layer Chromatography

Separation and semiquantitation of compounds can be achieved by applying the concentrated organic extract onto the bottom of a thin layer of silica, cellulose, or alumina attached to glass plates, and allowing a solvent mixture to run up the absorbent layer. The differential affinity of the drug to the layer and solvent produces a separation of the components. Individual spots or bands are visualized by placing the plates under ultraviolet or fluorescent light, or by spraying with coloring reagents. When radioactivity is used (see Section II,F), the plates are covered by an X-ray plate, and after several days or weeks, depending on the activity, the photographic plates are developed. The superficial identity can be made by comparing the relative distance run from the origin (RF value) to known standards. Semiquantitation (10^{-4} to 10^{-7} $g \cdot ml^{-1}$) is achieved by (1) *visual comparison* with standards, (2) *cutting the paper* and solubilizing the drug, (3) *densitometry*, using a measure of the amount of light absorbed when a particular wavelength is passed through the drug spot, either when sprayed to form a color or through an X-ray plate after contact with the area of radioactive, or (4) a *linear analyzer*, which counts the radioactivity coming directly from the active compound on the plate.

C. Gas Liquid Chromatography

Gas liquid chromatography (GLC), developed in the 1960s, revolutionized pharmacokinetics, because it could separate and accurately measure compounds down to 10^{-9} $g \cdot ml^{-1}$ simply and reproducibly. The concentrated mixture is injected into a long, coiled glass column (2 mm $\times$ 2 m) containing a coated solid support held in a temperature-programmable oven and purged by a low-pressure gas system (nitrogen or helium). The drug, metabolite, or solvent is volatilized at an optimal temperature; the rate at which it passes through the column is proportional to its binding to the coating on the solid phase and to a lesser extent its molecular weight. The coating and temperature are variable, so optimal separations can be achieved. The method has been improved by using very thin but longer (0.5 mm $\times$ 30 m) capillary columns, allowing small amounts to be quantified with better peak resolution. The separated compounds are measured by various types of detectors: (1) nonspecific flame ionization detector (FID); (2) nitrogen specific; (3) phosphorous specific; (4) electron capture (EC); and (5) mass spectrometry (MS). The amount detected is compared to a known amount of an added internal standard using calibration curves. The GLC technique, however, has the drawback of not being able to analyze temperature-labile molecules and those with large molecular weights, and is thus being used less.

D. High-Performance (Pressure) Liquid Chromatography

The most commonly used analytical technique, high-performance liquid chromatography (HPLC), has now taken over from GLC as it combines sensitivity (10^{-9} $g \cdot ml^{-1}$), stability, and a greater adaptability to automation. It is an analogous technique to GLC with the concentrated extract applied to a stainless-steel column packed with a solid phase, but used at room temperature. The compounds are forced through under high pressure (2000 psi) with either a simple single-solvent (isocratic) or a multiple-solvent (gradient) mixture, and the compounds either differentially bind to the solid phase (normal phase: NP-HPLC) or, more commonly, differentially solubilize in a gradient elution (reverse phase: RP-HPLC). Altering the amount of polar to nonpolar constituents in the solvent allows the compounds to move through the column at different speeds to achieve separation. Detection devices are (1) ultraviolet absorbance; (2) fluorescence, normal and laser; (3) radioactivity; (4) electrochemical; and (5) mass spectrometry.

E. Radioimmunoassay and Radioreceptor Assay

1. Radioimmunoassay

Radioimmunoassay analytical techniques developed in recent years have provided a breakthrough in the routine and rapid measurement of drugs and hormones at low levels in the body. They were used initially in 1960 by Yalow and Berson in New York to measure insulin and by Ekins in London to measure thyroxine. A competition assay method can be outlined as follows and divided into *antibody production, synthesis of radiolabel,* and *assay:* (1) the drug (or hapten) to be analyzed, unless of high molecular weight (>5000), is reacted with a protein (normally albumin) to produce an *immunogen* capable of being antigenic; (2) the immunogen ($\sim$100 μg) is injected intradermally in an oily *adjuvant,* for example, Freunds to enhance the immune response, normally into guinea pigs, rabbits, or sheep, at monthly intervals for periods up to 6 months. This stimulates the production of *antibodies* (*immunoglobulins*) specific to the injected immunogen and harvested from the animal for use; (3) in a competitive radioimmunoassay (RIA), the original drug is radiolabeled with either ^{125}I or tritium to obtain a label of high specific activity and this is added to the antibody at a concentration for all the radioactivity to be bound; (4) when an unknown amount of the drug to be analyzed is added to a solution containing the radioactive drug–antibody complex, an amount of radioactive drug is displaced proportional to the added cold drug; (5) the remaining radioactive drug–antibody complex is removed by charcoal, centrifugation, or second antibody precipitation. The supernatant-free radioactive drug is counted by scintillation spectrophotometry for tritiated compounds and by gamma counting for ^{125}I. Newer techniques of *scintillation proximity assay* obviate the need for the removal of the drug–antibody complex. By constructing calibration curves with known quantities of the cold drug, the amount of displaced radioactive ligand proportional to the concentration of the drug in the sample relates to the concentration of the unlabeled drug in the sample. [*See* Radioimmunoassays.]

RIA is rapid (300 samples/day), sensitive (<10–100 pg·ml^{-1}), and simple (little expertise required); automation is easy and is cheap for many samples; sample volumes are low (100–300 μl); and no extraction is necessary.

Among its disadvantages are a long development time (up to 6 months); initial development is expensive, particularly for few samples; and specificity may be poor with cross-reactivity to metabolites or compounds of similar structures.

Another type of RIA is the *solid-phase assay,* which may be either direct or indirect. In the direct assay, the antigen is bound to the plastic surface of the wells in microtiter plates. ^{125}I-labeled, antigen-specific antibodies are then applied in solution for approximately an hour at 37°C. After this, the plates are washed, and the amount of bound radioactivity is counted in a gamma counter. For greater sensitivity, the indirect assay may be more appropriate. In this case, the antigen is bound to the plate as before and a nonradiolabeled, antigen-specific antibody is applied. After incubation and washing, a second radiolabeled antibody that recognizes the first antibody as an antigen is added. After incubation and washing, the plate is counted as before. This has the advantage of magnifying the amount of radiolabel that will be attached (albeit indirectly) to the original antigen.

A variation on this procedure is ELISA (enzyme-linked immunosorbent assay). It is similar in principle to the solid-phase RIA with the exception that an enzyme (e.g., alkaline phosphatase) is coupled to the antibody rather than a radiolabel. Subsequent incubation of the antibody-bound enzyme with its substrate produces a colored product in proportion to the concentration of the original antigen. Such a reaction may be simply quantified by colorimetric analysis.

2. Radioreceptor Assay

The radioreceptor assay is in many ways similar to RIA and was first described for the assay of adrenocorticotropic hormone (ACTH) by Lefkowitz in 1970 using homogenates of adrenal cortical tissue. In radioreceptor assay (RRA), the binding of the radioactive ligand is, in contrast to RIA, to a specific receptor, normally at which the drug or hormone may elicit its pharmacological activity; this is not always the case. The receptors are prepared directly from tissues of lung, brain, kidney, platelets, and so on, and may be crude homogenates or purified and solubilized preparations. All the advantages of RIA are applicable to RRA, with the additional advantage that only the active biological component is measured. Thus, where a racemate is administered, only the active eutomer that has a high specificity for the active receptor will be measured, and therefore RRA can be thought of as a biological activity measurement. It is a technique that can be relatively more quickly developed than RIA. However, the disadvantages include lack of specificity, since an active metabolite or other similarly active drug will be measured. Multiple receptor binding (e.g., β blockers to $\beta 1$ and $\beta 2$ receptors) makes interpretation difficult. Frequent production of receptors with each batch need to be tested and characterized before use. Because of these disadvantages, neither RIA nor RRA has been used extensively for pharmacokinetic analysis, mainly because of lack of specificity of analysis.

F. Radioactive Measurement

As part of the development of a new drug, it is important to completely follow the biodisposition of the parent compound and its metabolites, particularly if they are active or toxic. Although this can be done using the analytical techniques described previously, it is made considerably easier if the molecule is labeled with a radioactive marker, usually ^{14}C or ^{3}H, although sometimes ^{35}S or ^{33}P can be used. The radioactive compound is synthesized at or near the active nucleus away from metabolic or chemical degradation, and must be highly purified before use. Once administered to an animal, all parts of the molecule can be measured in all tissues by excision or in urine, feces, and expired air as $^{14}CO_2$ by collection, and a total balance over a period of time can be made to show no specific retention in the body. The measurement of radioactivity is undertaken in a *scintillation counter* after suitable aqueous solubilization of the tissue or excreta in a solution that contains a scintillant. These low-energy radioactive molecules produce β particles that interact with the scintillant in solution, emitting a flash of light that is counted by multiple photoelectric cells. The amount of radioactivity present is therefore proportional to the total amount of light produced over a given time when compared to a standard of known radioactivity. A more recent development is the use of *positron emission tomography* (PET), in which the molecule is labeled with a higher-energy ^{11}C or ^{18}F isotope and the distribution throughout the living body can be made by external monitoring. This technique cannot only be used to measure the pharmacokinetics of a drug or a receptor ligand, but can also show the internal energy requirements of a small area of a tissue, and has been used to investigate various diseases such as Alzheimer's, Parkinson's, dementia, cancer, and cardiovascular problems.

G. Metabolite Identification

Foreign chemicals are often broken down or metabolized to smaller molecules (see later discussion), and it is necessary to identify these metabolites to see if they are active or toxic. A number of techniques are available, but the most frequently used is *mass spectrometry* (MS), which specifically identifies an unknown compound by distinguishing the component parts or masses once the molecule has been fractured. The sample is introduced into a source either directly, as a solid, or after separation by GLC or HPLC. It is bombarded by *electron impact* or by *fast atoms,* or *undergoes chemical reaction* that ionizes the molecule into fractional parts. The charged ions pass through a focusing system, either magnetic or voltage depending on the type of MS, and are focused according to their mass and charge onto an electron multiplier detector to produce a computerized printout of all the masses of the parent drug and the parts that have been fractured. The extent of the breakdown of the molecule can be controlled and depends on the stability of the compound. There are different types of MS, and each is used for specific purposes. The original systems use a magnet for focusing, providing high resolution and good absolute identification, but are slow and lack control of the fragmented ions. More recently, *quadrapole* focusing has been introduced using voltage scanning, which provides more versatility and speed but lacks the precision of measuring molecular weights to several decimal places. These machines are particularly suitable for analysis of samples from HPLC and make more specific detectors. But perhaps the most useful advance has been the use of MS/MS that allows more rapid identification of metabolites. Samples that do not need to be purified are introduced into the system and the first MS focuses only a single ion of interest into a collision cell. This reacts under controlled conditions with an inert nobel gas (argon). The single ion is fragmented into smaller parts that are subsequently focused again by another quadrapole or electrostatic application before detection. The advantage is that because there is no cleanup and no loss of material, and very little of the metabolite is needed, identification is speeded from days to hours. More recently, the technique of accelerated MS which was previously used for carbon dating has been used for radioactive ADME studies. Very low levels of ^{14}C can be measured, allowing a several thousand-fold reduction in dose. Other techniques for metabolite characterization include *nuclear magnetic resonance* (NMR), a technique in which the drug is placed within a magnetic flow and bombarded with radio frequencies that resonate the odd-numbered nucleons of ^{1}H or ^{15}N, inducing a transition energy state within the molecule and which can interact with protons in neighboring atoms. The resultant spectrum is a fingerprint for the molecule under investigation.

III. PHYSICOCHEMISTRY AND MEMBRANE TRANSPORT

For a drug to produce an activity, it should normally be absorbed, pass through various tissues and body fluids, and finally interact with its site of action. To do this, it must be able to dissolve in aqueous fluids and pass through the various lipoprotein membranes in the body. This is known as *drug transport* and the rates and extent of those processes are dependent on the physicochemical properties of the drug and the tissues of the body involved (Table II).

A. Membrane Transport

Membranes throughout the body are a lipoprotein sandwich composed of an inner, mainly lipid core, and an outside surface layer of protein (Fig. 1). This is due to the configuration of the phospholipid with the *hydrophobic* or *lipophilic* (water hating or *lipid*) portion of the molecule facing inward and the *hydrophilic* (water liking or polar) layer facing the outside, more aqueous environment. In addition, there are narrow aqueous-filled channels and larger aqueous pores crossing the membrane. Chemicals can be transported by two principal mechanisms: (1) *passive diffusion* and (2) *active transport*. Some larger molecules (e.g., polypeptides) can also be internalized into the cell by *phagocytosis*. [*See* Cell Membrane Transport; Membranes, Biological.]

Passive diffusion is the movement of a molecule down a concentration gradient until equilibrium occurs, and is the predominant transporting mechanism of most xenobiotic chemicals. The rate of penetration of such molecules by diffusion is dependent on (a) the *permeability* (P), (b) the *surface area* (S), and (c) the *concentration difference* ($Cn_1 - Cn_2$) of the molecules on either side of the membrane according to Fick's law of diffusion:

$$\text{Rate of penetration} = (Cn_1 - Cn_2) \cdot (S) \cdot (P) \quad (1)$$

Although the concentration gradient is the driving force for molecular transport under normal conditions, it is the *permeability* that determines and controls the rate and extent of transport. The four major determinants of permeability are *membrane composition and thickness, lipid solubility,* and *ionization,* and the latter two are the most important physicochemical factors influencing both drug transport and binding.

1. Membrane Composition

The membrane composition is relatively constant within the body and for most drugs this does not

TABLE II
Major Factors That Can Influence the Biodisposition of a Chemical

Absorption	Distribution	Urinary excretion	Metabolism
Aqueous solubility[a] (ionized form)	Lipid solubility[a]	Lipid solubility[a]	Lipid solubility[a]
Gut mobility	—	Renal function[a]	Hepatic function[a]
pKa	pKa	pKa	—
pH GI tract	pH of tissue[a]	pH of urine[a]	—
Blood flow (depot injec.)	Blood flow	Renal blood flow	Hepatic blood flow[a]
Coadministered compounds	—	Coadministered compounds	Coadministered compounds[a]
Disease[a] (e.g., Crohn's)	Body composition[a]	Molecular weight	Chemical structure[a]
Metabolism	—	—	—
Dissolution of formulation[a]	—	—	Genetics[a]

[a]Most important.

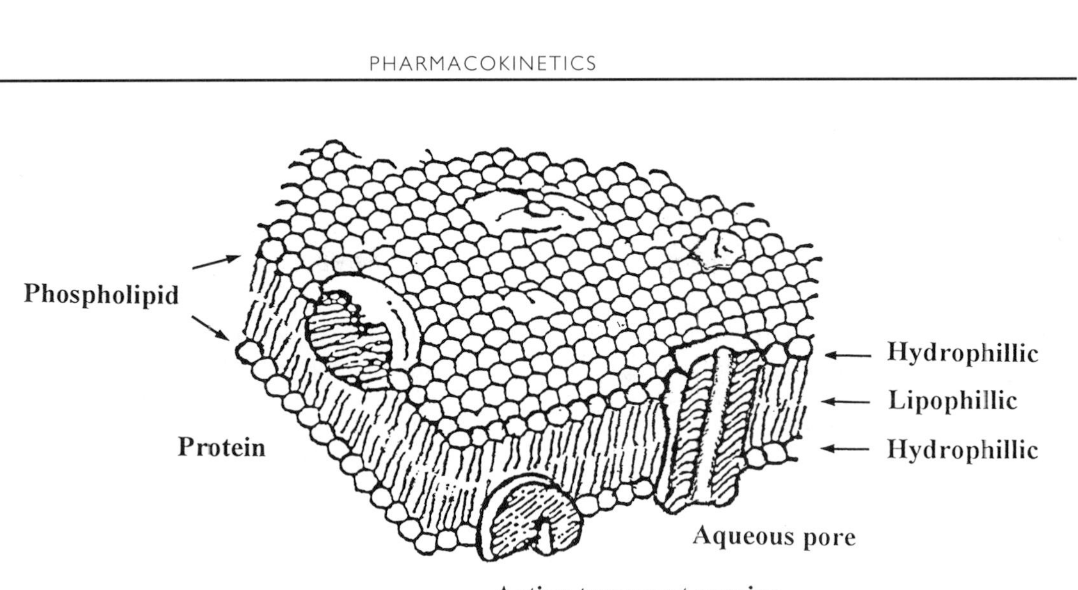

FIGURE 1 Schematic representation of a membrane showing the lipid layer with lipophilic (lipid) center, the aqueous pore that transports small water-soluble compounds, an enzyme carrier-mediated transport, and protein molecules that have various functions including cell recognition and adhesion. Most drugs will pass through the phospholipid sandwich by dissolving in the lipophilic layer.

present any difficulty for penetration. This is not the case for high-molecular-weight compounds (>5000), particularly for filtration in the Bowman's capsules of the kidney. The brain is an exception, with a *blood–brain barrier* composed of tight junctions that specifically exclude certain foreign compounds that may be toxic. The skin is another membrane, with *epidermal layers* that protect the body from toxic insult.

2. Lipid Solubility

Lipid solubility is a measure of the extent to which a drug can dissolve in a water-immiscible organic solvent and can be measured by equilibrating a drug between an aqueous buffer of controlled pH (7.4) and an organic solvent, normally octanol. The ratio of drug concentrations in the solvent and aqueous phase is called the *distribution coefficient* (D), but since this is dependent on the pKa of the drug and pH of the solution, the real distribution or *partition coefficient* (P) of the unionized drug is given by Eqs. (2) and (3) depending on whether the compound is an acid or base:

$$\text{Log P (acid)} = \log \text{D} + \log(1 + 10^{\text{pH}-\text{pKa}}) \quad (2)$$

$$\text{Log P (base)} = \log \text{D} + \log(1 + 10^{\text{pKa}-\text{pH}}) \quad (3)$$

3. Ionization and pH Partition Hypothesis

Most drugs are either *weak acids* (A) or *weak bases* (B) and therefore exist in solution, either as their polar ionized form (A^-, acid; BH^+, base) or in lipid-soluble unionized form (HA, acid; B, base) depending on the pH of the solution of body fluid. When the pH equals the pKa, the ionization constant of the compound, concentrations of both are equal:

$$\begin{array}{lcc} & \text{Unionized} & \text{Ionized} \\ \text{Acid} & [HA] + [H_2O] \Leftrightarrow & [HA^-] + [H_3O^+] \quad (4) \\ \text{Base} & [B] + [H_2O] \Leftrightarrow & [BH^+] + [OH^-] \quad (5) \end{array}$$

When the pH becomes more acidic, the equilibrium shifts to more unionized for acids or less unionized for bases and vice versa when the pH is basic. Since it is probable that only the unionized lipid form is transported through the lipid membranes, both the pKa of drug and the pH are important for passive diffusion.

Since it is the unionized form that is transported, the relative amount of drug on each side of a membrane at equilibrium is governed by the relative ratio of ionized to unionized drug on both sides. This can be calculated from the pH difference of tissue (pH_t) and plasma (pH_p) and the pKa of the drug according to the Henderson–Hasselbalch equations:

$$\text{Weak acid} = \frac{\text{Conc in tissue}}{\text{Conc in plasma}} = \frac{1 + 10^{pH_t - pKa}}{1 + 10^{pH_p - pKa}} \quad (6)$$

$$\text{Weak base} = \frac{\text{Conc in tissue}}{\text{Conc in plasma}} = \frac{1 + 10^{pKa - pH_t}}{1 + 10^{pKa - pH_p}} \quad (7)$$

Thus, weak bases will concentrate in more acidic solutions compared to plasma (pH_p 7.4; e.g., stomach, pH 1–2, red cell, pH 7.2, milk, pH 7.2, urine, pH 6–7), whereas the opposite is true for weak acids.

As early as 1897, Overton and Meyer suggested that drugs and chemicals cross lipid membranes by dissolving in the lipid layer and that the rate and extent of this diffusion were linearly related to the lipophilicity (log P) of the unionized form since only this would be transported:

$$\text{Log diffusion} = b \log P + a \tag{8}$$

Although for some systems this relationship may be true (e.g., the greater the lipophilicity, the faster and more extensive the transport), it has become apparent that it was unable to explain the observation that at values of high log P, transport becomes less extensive. More recently, Hansch and others have defined a parabolic relationship [Eq. (9)] between diffusion and lipophilicity, suggesting that there is a maximum value of log P above which a very lipid drug stays associated with the lipoprotein membrane rather than passing through:

$$\text{Log diffusion} = -a \log P - b(\log)^2 + C \tag{9}$$

It is of interest that the optimum value of log P is approximately 2 (octanol: buffer pH 7.4) regardless of whether transport is across the skin, stomach, intestine, blood–brain barrier, kidney tubule, or even plant leaves. Using this relationship, it is possible to design drugs according to whether inclusion or exclusion to a particular organ (e.g., the central nervous system) is required.

In general, a drug with a high lipid solubility is more likely to have (a) high protein binding and may displace similarly bound drugs; (b) high distribution volume leading to low blood levels, long half-life, infrequent dosing, and slow buildup to steady state; and (c) low renal clearance due to low filtration (high protein binding) and high tubular reabsorption leading to high hepatic clearance and first-pass metabolism and circulatory metabolites.

4. Blood Flow

Blood flow is another important consideration for transport. If a drug can easily pass through membranes, there is no barrier for its diffusion into tissues. Under these circumstances, the rate of uptake can be reduced if blood flow or perfusion is slow, producing a *perfusion rate limitation* to transport. On the contrary, if a drug cannot easily diffuse into the tissue, blood flow changes will have little effect and the drug has a *diffusion rate limitation.*

B. Active Transport

Active transport normally carries water-soluble endogenous compounds like glucose and amino acids across and up a concentration gradient through water channels. Unlike passive transport, this process is compound specific, usually carrier-*mediated* or enzyme-*linked* and is not normally associated with drugs, except for some that are structurally similar to natural chemicals and that share the same transporter or are *biliary* or *renally secreted.*

IV. BIODISPOSITION

A. Absorption

The process by which a drug moves from its site of administration into the body to where it acts is called *absorption.* Normally this is following oral administration but many other routes may be used, where the drug must cross membranes before entering the systemic circulation (Fig. 2). Before a drug is absorbed from any part of the *gastrointestinal* (GI) *tract,* it

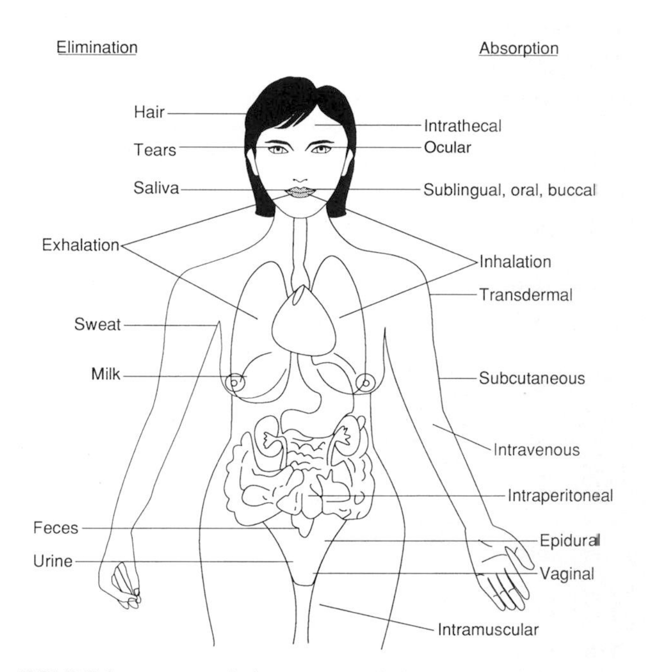

FIGURE 2 Routes of absorption and elimination of drugs and chemicals from the human body.

must be in solution so that the unionized part can be transported by passive diffusion. In a few cases, highly ionized compounds like the quaternary ammonium drugs are slowly and incompletely absorbed. The important factors that control absorption and other biodisposition processes are shown in Table II.

1. Formulation

Formulations are developed to control the delivery of the drug into the body. Thus solutions, syrups, suspensions, soft gelatin capsules, and sublingual lozenges are rapid delivery systems with early peak blood levels (15 min–1 hr), whereas compressed tablets and hard gelatin capsules are slower delivery systems yielding peak levels of drugs at 1 to 4 hr, depending on the rate of dissolution of the formulation. *Dissolution* is a process by which a formulation disintegrates into granules, deaggregates into particles, and dissolves into solution. It can be controlled and slowed down by *drug delivery systems* that include *enteric coating, sustained release,* and *osmotic pumps. Enteric coating* is used to protect an acid-labile drug from stomach degradation in order to be released and absorbed lower in the GI tract. It can also protect against drug-induced irritation. *Sustained release* slows the release of a drug to achieve constant or controlled levels so that one dose can be administered less frequently, usually over 24 hr. This is particularly useful for rapidly eliminated drugs. *Osmotic pump systems* achieve the same effect by allowing water to pass through a semipermeable membrane and slowly pumping the drug out as a solution at a constant rate.

Recently, there have been important advances in drug delivery systems that allow drugs to be administered by many routes, in particular by systemic administration to bypass hepatic metabolism (e.g., rectal, buccal) or to speed up absorption (sublingual, inhalation), or through the skin by patches (nicotine) or cream (hormone replacement therapy) or subcutaneously with depot injections or implants (contraceptive), intravenous pumps (insulin), and systems that target the drug to particular sites such as in cancer treatment.

2. Solubility

Although the lipid-soluble form passes through membranes for GI tract absorption, the drug must initially be in aqueous solution. Thus, most drugs, unless they have an inherent water solubility, are administered as their salts, hydrochloride or sulfate for weak bases, sodium or potassium for weak acids.

3. pKa

The rate of absorption from the GI tract, buccal cavity, vagina, or skin will to some degree be dependent on the pKa of the drug and pH of the body fluid. Once into solution, the ease of membrane transport is proportional to the amount of unionized drug in equilibrium and its lipid solubility. Thus, in the acid pH (1–2) of the stomach, weak acids are unionized and some absorption can take place, whereas weak bases remain ionized and are not absorbed. However, few drugs are absorbed to any great extent in the stomach.

4. Surface Area

The stomach is a physiological vat with a relatively smooth surface, secreting acids and enzymes to degrade food. The combination of a small surface area (1 m^2) and low blood flow (150 $ml \cdot min^{-1}$), when compared with the intestine with its highly convoluted microvilli that increase the surface area and blood flow to 200 m^2 and 1000 $ml \cdot min^{-1}$, respectively, means that most drugs are not absorbed in the stomach (except alcohol) but are absorbed from the intestine regardless of their pKa, although the rates may be different.

5. Other Factors

A number of other factors influence drug absorption, including (a) food may have little effect for some drugs, but for others it can enhance or inhibit absorption by altering dissolution and transport or delaying gastric emptying; (b) fluids will normally enhance the rate of absorption; (c) supine posture or no physical activity can slow down absorption owing to delayed gastric emptying and reduced splanchnic (GI tract) blood flow, whereas increased activity will increase absorption; (d) coadministration of drugs may enhance or inhibit total absorption because of their effects on gastric motility, blood flow, and hepatic metabolism; and (e) diseases such as cancer, Crohn's disease, and cardiac failure can also reduce absorption.

B. Bioavailability

When drugs are administered, they are conveniently packaged in formulations such as tablets, capsules, suppositories, sprays, or patches to aid in the absorption of the product. Bioavailability measures the factors that influence the rate and extent of arrival of a drug at its site of action, but in practical terms, it is defined as the fraction (F) of the dose reaching the systemic circulation as unchanged (not metabolized) drug following administration by any route. *Absolute bioavailability* compares the amount of drug available from the test formulation with an intravenous (IV) dose, which is 100% available, whereas *relative bioavailability* compares the availability of two different formulations that may be used clinically, and if they produce similar blood levels, can be said to be bioequivalent. This is of particular importance with the increasing use of generic copies.

Different methods of measurement, design, and statistical analysis are used. Various kinetic parameters are used to measure bioavailability in terms of rate (a) peak concentrations (C_{max}) and extent, (b) time to C_{max} (T_{max}), and (c) infinite *area under* the plasma (blood) time *curve* (AUC) after single and repeated dosing. Experiments are undertaken on a sufficiently large number of healthy volunteers (n = 12–50) so that the data will provide an 80% probability of detecting a 20% difference in these parameters (80/20 power rule), particularly AUC, as the measure of extent of absorption. Various study designs are used, including (a) *crossover design,* in which each subject takes two formulations to obtain a within-subject comparison, (b) a *latin-square design* for more than 2 to 4 formulations, or (c) *incomplete block* for more than 4 formulations, in which subjects do not take all the formulations. Statistical analysis of the parameters is under constant discussion but includes (a) the standard *power approach* or hypothesis testing using analysis of variance (ANOVA) to test the null hypothesis for no difference; and (b) *confidence intervals* (symmetrical and nonsymmetrical) about a mean difference, which provides a more meaningful basis for clinical decisions.

In general, formulation differences in effect are likely to be more important for drugs with a narrow therapeutic window, with a steep slope of the kinetic dynamic relationship. There is also a growing awareness that clinical end points can and should also form the basis of more meaningful comparative bioavailability testing whenever possible.

C. Distribution

Once a drug is absorbed into plasma, it is rapidly circulated around the body by the blood once every 2–3 min. The process by which a drug is reversibly transferred from one part of the body to another is called distribution and is usually calculated from the site of measurement (plasma) and expressed as a dilution volume. A drug will pass from the circulating blood, through membranes, binding to tissue macromolecules before interacting with its locus of activity. The extent of distribution for any drug is largely dependent on its lipid solubility and tissue permeability, its relative binding to blood and tissue protein, and the blood flow to the tissue. Thus, a drug may be highly bound to plasma protein but still distribute extensively into the rest of the body if its affinity to tissues is greater.

1. Volume of Distribution

Once a compound leaves the blood circulation, it is distributed around the body into a certain volume and the concentration in blood is thereby reduced. This apparent dilution of the blood can be measured and is called the *apparent volume of distribution.* In Fig. 3, the initial rapid decline in plasma levels is mainly due to this uptake into tissues. Thus, after a rapid intravenous injection, the initial volume of distribution (V_i) is given by the dose (D) and the blood concentration at the time of administration (C_o):

$$V_i = D/C_o$$

Normally this value will be approximately 5–10 liters, the volume of blood, but at later times after administration more of the drug will have been distributed until an equilibrium occurs when the amount going into the tissues equals the amount coming out. The volume of distribution is then at steady state (V_{SS}) and is maximal. In general, if the V_{SS} is approximately 5 liters, it has been retained in the circulation, and if 100 liters, it has been distributed equally throughout the body. For some compounds the V_{SS} may be several thousand liters, indicating considerable uptake by the tissues in a volume that is very much greater than the body volume. Thus, the volume of distribution has little anatomical meaning but is a convenient mathematical way of measuring how much has disappeared from the blood circulation.

2. Blood Protein Binding

Most drugs bind to blood components to some degree. In the red cell, the major proteins for binding are hemoglobin and the enzyme carbonic anhydrase, whereas in plasma, the two main proteins are albumin, to which many acidic and neutral drugs bind,

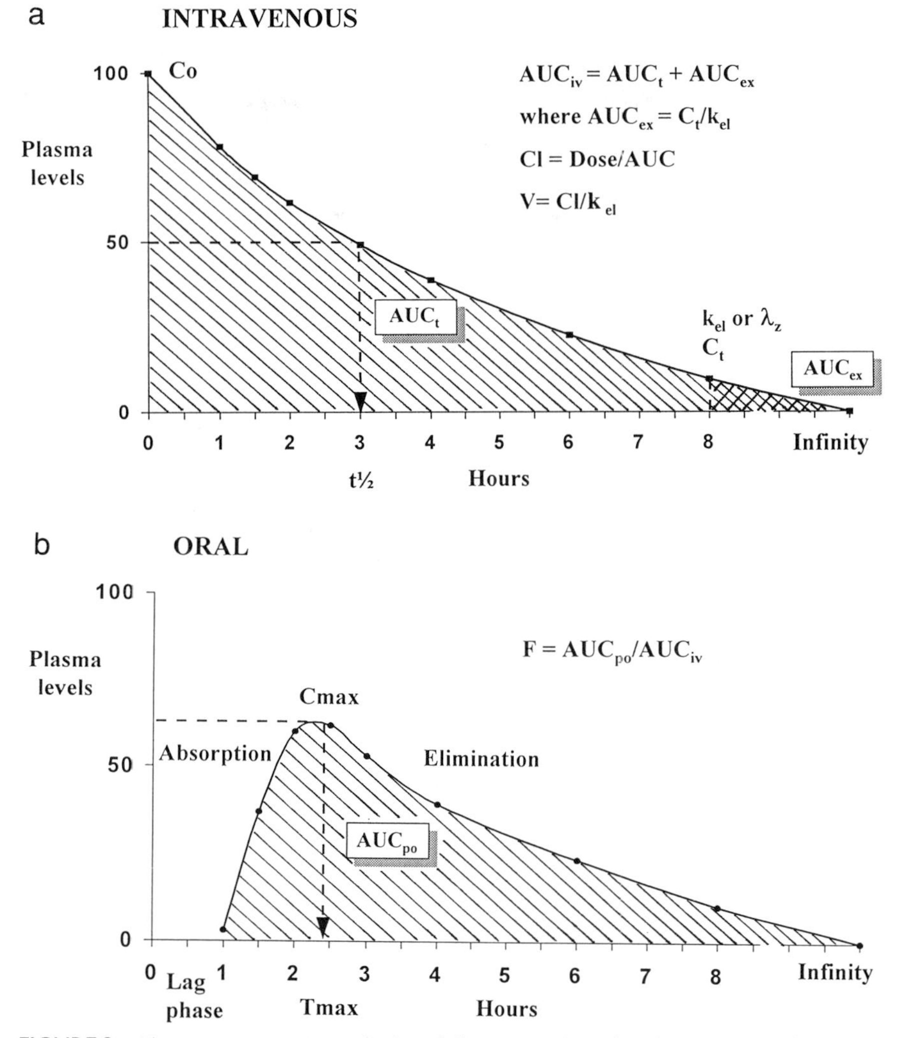

FIGURE 3 Plasma concentrations of a drug following single oral and intravenous administration showing derivation of some kinetic parameters. The area under the plasma time curve (AUC_t) can be measured by summation of each area under sequential time passes.

and α-glycoproteins, to which basic drugs tend to bind. Some binding can also occur to globulins and lipoproteins. Protein binding can be measured *in vitro* by incubating the drug (C) with plasma or specific protein and then separating and measuring the free unbound drug (C_u) from that bound ($C - C_u$). Percentage binding (PB) is given by

$$PB = \frac{C - C_u}{C} \cdot 100 \qquad (10)$$

Separation of C_u is achieved by (a) *equilibrium dialysis* using passive diffusion into a protein-free buffer (pH 7·4) through a semipermeable membrane (SPM); (b) *ultrafiltration,* which forces the aqueous solution of free drug through a semipermeable filter; (c) *gel filtration,* which separates protein-bound drug from free drug by molecular weight sieving; or (d) *electrophoresis,* which achieves separation by differential migration in an electrical field.

In practice, however, the percentage unbound (PB-100) or fraction unbound (f_u) is clinically most useful. Although all binding will affect kinetics, it only becomes important when it exceeds 90%. A change from 90 to 95% bound may seem little but actually reduces f_u by 50%. Binding can be reduced in disease states because of low albumin levels (hepatic and cardiac failure, nephrotic syndrome), whereas binding to α-glycoprotein (a stress protein) is increased owing to increased levels in inflammatory arthritis, Crohn's disease, myocardial infarction, burns, and other conditions. It has long been thought that changes in protein binding would alter the therapeutic activity of the drug, as more free drug would be available at the site of action. With the exception of high-clearance drugs administered intravenously, this is not the case, as more drug is at the same time available for elimination and at steady state; although the total levels may alter, the free active drug levels remain constant if there is no change in free clearance. Theoretically, it can be argued that steady-state levels (C_{uss}) are dependent on free clearance (CL_u) and not volume; if CL_u does not alter, then neither will C_{uss}. Protein binding is therefore of less importance than once considered.

D. Elimination

If the body was unable to eliminate drugs, one dose would last a lifetime. Nonspecific mechanisms must therefore exist to remove the 5000 foreign substances estimated to be taken into the human body each day. There are many routes possible (see Fig. 2), but the most important are *excretion, metabolism, biliary, exhalation,* and *lactation.* The kinetic parameters associated with elimination are clearance and half-life.

a. Clearance

Clearance is a proportionality factor relating the rate of drug elimination (k_{el}) to plasma drug concentration (C):

$$k_{el} = CL \cdot C \qquad (11)$$

Since concentration is measured as an amount of drug per unit volume, clearance can also be expressed as the volume of blood cleared of drug in unit time:

$$CL = V \cdot k_{el} \qquad (12)$$

The total body clearance (CL) is the sum of the individual clearances from each eliminating organ:

$$CL = CL_{metab} + CL_{kidney} + CL_{others} \qquad (13)$$

Thus, clearance can also be thought of physiologically as the loss or extraction of drug by that organ as blood passes through. It is therefore related to the blood flow (Q) and the degree of extraction (E) for the particular drug:

$$CL = Q \cdot E \qquad (14)$$

For highly extracted or hepatically metabolized drugs where $E \simeq 1$, the clearance approaches hepatic blood flow (Q_H) (1500 ml · min^{-1}) [see Eq. (11)], and elimination is limited only by perfusion through the organ and not by events within the liver (perfusion-rate limited). These drugs have high first-pass metabolism and low systemic bioavailability, and the clearance can be markedly changed by alterations in portal blood flow (disease, food, concomitant drugs).

In contrast, for low-extraction drugs, $E < 0.3$, extraction is not controlled by blood flow but by other processes: (a) enzymatic reactions; (b) biliary excretion; (c) the degree of protein binding within blood; or (d) diffusion into tissue (diffusion-rate limited).

Clearance is important to kinetics, as estimates of availability or the amount entering the body (F) can be made from intravenous data assuming only hepatic metabolism:

$$F = 1 - \frac{CL}{Q_H} \qquad (15)$$

In addition to the total amount absorbed (F), the most important determinant of average steady-state drug levels ($C_{ss} \cdot av$) after repeated dosing is clearance:

$$C_{ss} \cdot av = \frac{F \cdot D/\tau}{CL} \qquad (16)$$

where F · D is the absorbed dose administered over a certain dosing interval, τ. Clearance can be simply measured by calculating the area under the infinite plasma concentration time curve (AUC) (Fig. 3) after intravenous administration. It cannot be measured after oral dosing unless the extent of bioavailability is known:

$$CL = \frac{Dose}{AUC_{iv}} \qquad (17)$$

b. Half-life

The time required for a drug to decline to half its concentration has long been considered to be an important kinetic parameter, perhaps because it can be easily measured (see Fig. 3). In fact, it is a hybrid constant dependent on clearance and the distribution volume,

$$t\tfrac{1}{2} = \frac{0.693 \cdot CL}{V} \qquad (18)$$

and a long half-life may be due to a high distribution volume or a slow clearance. Similarly, both volume and clearance could alter without changing the half-life. Although a poor indicator of elimination, it gives information on (a) time of drug in the body; (b) time to reach steady state on continual drug administration ($\approx 4\text{–}5 \times t\frac{1}{2}$); and (c) time for dosing interval, approximately every half-life. Drugs may have several half-lives, often dependent on different rates of elimination from different volumes, but the last or terminal half-life ($t\frac{1}{2}_z$) is the most important and is used for most kinetic calculations. Half-life is related to the elimination constant k_{el} (λ_z) according to

$$t\frac{1}{2} = \frac{0.693}{k_{el}} \quad (19)$$

1. Renal Excretion

The major route of elimination of the more polar drugs and metabolites is by renal excretion:

$$\text{Rate of urinary excretion} = CL_R \cdot C \quad (20)$$

where CL_R = renal clearance and C = plasma drug concentration. The appearance of the drug in urine is the net result of (a) filtration, (b) secretion, and (c) reabsorption:

$$\text{Rate of excretion} = \text{Rate of filtration} + \text{Rate of secretion} - \text{Rate of reabsorption} \quad (21)$$

Filtration, a passive process of the kidney nephron, occurs across the glomerular membranes in the Bowman's capsule. Filtration rate at any instant is dependent on plasma concentration, protein binding (only free drug C_u is filtered), molecular weight (exclusion above 1–2000), and glomerular filtration rate of plasma water (GFR). In healthy individuals, GFR is approximately 125 $ml \cdot min^{-1}$ measured by inulin or creatinine, compounds that are not protein bound, secreted, or reabsorbed. For a drug with a filtration rate based on unbound drug different from this value, other processes must be occurring:

$$\text{Rate of filtration} = GFR \cdot C_u \quad (22)$$

Secretion, an active process, transports drug from blood into the proximal tubule of the nephron if renal clearance is greater than $f_u \cdot GFR$. There are separate mechanisms for acids and bases, but each lacks specificity, which can lead to dangerous drug interactions of coadministered drugs showing the same secretory mechanisms (e.g., quinidine and cimetidine):

$$CL_R = f_u \cdot GFR + \left[\frac{\text{secretion rate} - \text{reabsorption rate}}{C}\right] \quad (23)$$

Reabsorption is a passive process that transports the drug back into the circulating blood from urine all along the nephron, starting at the proximal tubule where water is absorbed. It occurs to some extent for many drugs and is dependent on urine flow, urinary pH, pKa, and lipid solubility. Urine flow is important for drugs that are extensively reabsorbed ($CL_R < 20\ ml \cdot min^{-1}$). At high flow rates, because the system is never in equilibrium, the urine will wash the drug through the nephron without allowing adequate time for diffusion. The rate of reabsorption is dependent on the concentration gradient and on the ability of the unionized drug to diffuse through tubular membranes. The extent of reabsorption is greater for highly lipid-soluble drugs. According to the Henderson–Hasselbach equations (6) and (7), it is dependent on the pKa and the pH difference between urine and plasma. Urinary pH can fluctuate between 5·0 to 7·5 (average 6·3), depending on diurnal variation, exercise, food, and drugs, so the pKa of the drug can play an important role in renal excretion. Thus excretion of a weak base (pKa 10) can vary from 1% of the dose in alkaline to 80% in acid urine. In general, highly polar drugs are not reabsorbed; strong bases are not reabsorbed when the pKa is >12 nor for acids with pKa <2 irrespective of urinary pH, as they exist mostly in the ionized form. Urinary pH fluctuations have a pronounced effect on reabsorption for lipid-weak bases with pKa between 7·0 (high reabsorption) and 11 (low), whereas for weak acids, the pKa needs to be between 3·0 (low) and 7·5 (high).

2. Metabolism

For those drugs with a greater lipid solubility and subsequent higher protein binding, the body cannot excrete them into the urine since they will not be filtered, and the small amount that does pass through the Bowman's capsule will be extensively reabsorbed. The body therefore attempts to make the compound more water soluble, either by breaking down the xenobiotic foreign chemical by enzymatic biodegradation or by adding a more polar group. This metabolism or biotransformation primarily occurs in the liver, since this is the first point of entry into the body via the portal vein when a compound is taken orally. However, metabolism, albeit to a lesser extent, can occur in the kidney, lung, intestinal wall and contents (microflora), and blood. When a drug is metabolized during absorption through the liver it is called *first-pass hepatic metabolism,* and this becomes important for those drugs that are extensively metabolized and have a high clearance. Blood flow to the liver is approximately 1500 $ml \cdot min^{-1}$ and therefore there is a limit to the rate of metabolism. Those with high clearances (>1000 $ml \cdot min^{-1}$) are extensively metabolized during absorption and little unchanged drug gets into the circulation. This is undesirable since there is (a) less available drug, (b) extensive production of metabolites that may be inactive, of different activity, or cause toxicity, (c) larger individual subject variation, (d) increased likelihood of drug interactions, and (e) greater effect of diseases. The majority of drugs have hepatic clearances (CL) ranging from 100 to 800 $ml \cdot min^{-1}$, but they can be as low as <5 $ml \cdot min^{-1}$ or as high as >1400 $ml \cdot min^{-1}$.

Metabolism is catalyzed by a variety of different enzymes and is able to modify almost all chemicals that are ingested into the body. These have been categorized into two types of reactions.

a. Phase I

Phase I reactions occur in the microsomal fraction of hepatic cells to produce a variety of products: (a) cleaving the molecule, (b) oxidation (-OH, -NO, -SO_2), (c) reduction (-COOH, -SH, -OH), and (d) dealkylation (-NH_2). These reactions have been shown to be catalyzed by a family of mixed function enzymes called *cytochrome P450,* since in solution they give a characteristic UV-specific maximum at this wavelength. At least 19 of those isoenzymes have been identified in humans. The most important, CYP3A4, constituting approximately 60% of the total cytochrome content in the human liver, is relatively nonspecific and will break down a variety of compounds. Others include the CYP1A family, which catalyzes aromatic carcinogens, and CYP2D6, which oxidizes functional groups that are 5–7° distant from a nitrogen.

Some enzymes are increased with repeated administration of a chemical (*enzyme induction*) that increases metabolic clearance and lowers blood levels, whereas others are inhibited either by the same compound or by the coadministration of others (*enzyme inhibition*). This can also occur with food (Brussels sprouts, grapefruit juice, barbecued meat, etc.). Similar problems exist if too much drug is administered, for example, in overdosage, when *enzyme saturation* occurs and clearance is reduced. In these cases, there is a potential change in the activity of a drug blood level and subsequently the dosage may need to be altered.

b. Phase II

Phase II reactions occur in the cytoplasm of the hepatic cell and these enzymes add large polar groups to the molecule, and with the exception of morphine all have been found to be inactive. These addition products are called *conjugates* and involve a diverse group of transferase enzymes that react with endogenous compounds such as glucoronides, sulfates, glutathione, glycine, and acetyl, methyl and mercapturic. Many other enzyme systems are also present, but they are less well studied (monoamine oxidases, flavin monooxygenases, alcohol dehydrogenase, etc.), confirming the wide spectrum of metabolizing capability of the liver to deal with all xenobiotics.

c. Species Differences

Although other animals will also metabolize ingested chemicals, they do not have exactly the same enzyme systems found in humans and the quantitative and qualitative metabolic profiles are often different. Some species completely lack certain pathways [e.g., dog (acetylation), cat (glucuronidation), guinea pig (mercapturic acid)], and this makes it difficult to compare the pharmacology or toxicology undertaken in animals with that expected in humans.

d. Polymorphism

Drug biodisposition and response are, to some degree, controlled by inheritance, but the differences between individuals are normally diffuse and indistinguishable. When specific feature differences of genetic origin are measurable, this is known as *genetic polymorphism.* In drug metabolism, genetic polymorphism can be detected in population studies by polymodal frequencies of rates of metabolism: clearance or steady-state drug levels showing two or more frequency peaks for slow and fast eliminators. This is caused by small changes at the active site involving only one or two amino acid substitutions, but these different *alleles* can cause important changes in the enzyme activity, either increasing or more normally reducing activity. There are three well-recognized metabolic polymorphisms: (a) *acetylation,* due to differences in the enzyme *n*-acetyltransferase, exists in 60% of North Americans, 40% of British, and 10% of Eskimos, for drugs like isoniazid, procainamide, hydralazine, and many sulfonamides, and levels can differ more than fourfold in these patient populations receiving the same dosage regimen, leading to side effects and toxicity or ineffective therapy; (b) *hydroxylation* and *dealkylation,* due to a deficiency of cytochrome 2D6, exist in 8% of Caucasians but are higher for Asians, for drugs like debrisoquin, nortriptyline, phenformin, perhexiline, and metoprolol; and (c) *hydrolysis,* due to differences in plasma cholinesterase, occurs in 1 in 2500 of the general population, in particular for succinylcholine. An increasing number of polymorphisms are being found for many of the metabolizing enzymes, providing explanation for much of the individual variations in patient blood levels.

Because of the potential side effects associated with slow metabolism, attempts have been made to identify or *phenotype* patients before therapy by administering single or multiple markers, or *genotype* them by measuring their specific DNA sequences from blood or tissue samples. However, the clinical value of this approach has yet to be proven.

3. Biliary Excretion

Biliary excretion is a complex and poorly understood process in which drugs are cleared from circulation by being actively secreted against a concentration gradient from hepatocytes into bile. It is a universal but relatively slow process. Biliary flow (BF) is only 0.5 ml $\cdot$ min^{-1}, and it is thought to be a safety overflow valve for compounds not easily metabolized or renally excreted:

$$\text{Biliary clearance} = \text{BF} \cdot \left(\frac{C_{\text{bile}}}{C}\right) \tag{24}$$

where C_{bile} = drug concentration in bile and C = plasma drug concentration.

There are separate mechanisms for biliary secretion of acids, bases, and unionized compounds, with two main controlling factors, molecular weight and polarity. The structure of the drug is also important because the process is active, but little is known about this mechanism.

a. Molecular Weight

Molecular weight is a major determinant, as many drugs or metabolites with a MW between 250 and several thousand are excreted in the bile. However, the lower-end threshold is species dependent, so that for large animals, such as humans and monkeys, secretion will not occur below approximately 500, whereas biliary excretion for small animals like rat and mouse occurs above 250.

b. Polarity

In addition to high MW, high polarity will encourage biliary excretion. Phase II conjugated metabolites (e.g., glucuronides, which have both an increase in the MW of 176 and an increase in polarity) are preferentially eliminated via the bile.

Once excreted into the bile, the drug passes down the biliary tract into the gallbladder (absent in rats), the contents of which (bile acids) are emptied into the small intestine during meals to aid the digestive process. The drug may be directly absorbed again or the conjugate may be hydrolyzed by gut flora back to the unconjugated drug (aglycone), which may also be reabsorbed. This process can continue in a cycle: absorption, biliary excretion, absorption, and so on, and is known as enterohepatic circulation. Under extreme conditions, it can cause secondary *absorption* peak oscillations as small amounts of drug escape the cycle and enter the systemic circulation. Similarly, escape may occur down the gastrointestinal tract, causing fecal excretion, but because enterohepatic circulation can occur after IV administration, high fecal drug levels do not always equate with poor absorption. This process is probably of limited importance in man.

4. Lactation

Although lactation is not a major route of elimination, excretion into milk is important because during weaning, neonates may be exposed to unacceptable large doses of the drug via ingestion of the mother's milk. This is of particular concern if additionally neonatal clearance is reduced. The amount of drug in milk is, according to the Henderson–Hasselbach equation, dependent on (a) pH of milk, (b) pKa, (c) lipid and protein binding in milk and plasma, and (d) time of lactation. The pH partition hypothesis enables calculation of the free drug ratio in milk and plasma, and as milk pH is slightly more acid (7$\cdot$2) than that of plasma, basic drugs will be concentrated in milk. However, the protein content of milk is relatively low, and therefore the degree of uptake is particularly related to the lipid solubility of the drug. However, the binding to milk fat and protein is dependent on the time of lactation, as early milk (colostrum) is low in lipid, but subsequently increases during the first month postpartum. The infant dose from milk can be calculated from maternal steady-state levels (C_{ss}), the ratio of the areas under the milk/plasma time curves over the same period (AUC_M/AUC), and the volume of milk ingested (V_{milk}), on average about 150 ml $\cdot$ kg^{-1} $\cdot$ day^{-1}:

$$\text{Infant dose} = C_{ss} \cdot \left(\frac{AUC_M}{AUC}\right) \cdot V_{milk} \qquad (25)$$

E. Repeat-Dose Kinetics

When a drug is repeatedly dosed, plasma levels will build up or accumulate to steady-state levels (C_{ss}) when the rate of drug entering the body equals the rate of elimination (Fig. 4). The time taken to reach steady state is usually between 3 and 5 half-lives, but may be longer if small but significant deep compartments with longer unrecognized terminal elimination rates are present. For drugs with short half-lives, much less than the dosing interval, steady state is reached after one dose. As stated previously, the magnitude of these plateau levels is dependent on only clearance and dosage, and not on body weight. However, the volume of distribution will influence the magnitude of the maxium and minimum steady-state levels (C_{ss} maximum and C_{ss} minimum) and half-life (see Fig. 4), but not average levels. Thus, obese subjects with a potentially larger volume will reach C_{ss} more slowly and with longer half-lives and the oscillations between maximal and minimal levels will be smaller at a fixed dosing interval, but the C_{ss} levels will normally be the same as in nonobese subjects given the same dose. The infinite area under the plasma time curve (AUC) is theoretically equal to the area under the plasma time curve during a dosing interval at steady state. If this is not found, then the kinetics of the drug may have changed (*inhibition or induction*) during repeated dosing with a corresponding change in k_{el} at the end of dosing, assuming no change in distribution. Similarly, the accumulation of the drug can be calculated by comparing the AUC during the dosing interval after a single dose and at steady state.

F. Stereoselectivity

Most drugs have one or more carbon atoms, each one of which has four bonds. If all the chemical substitu-

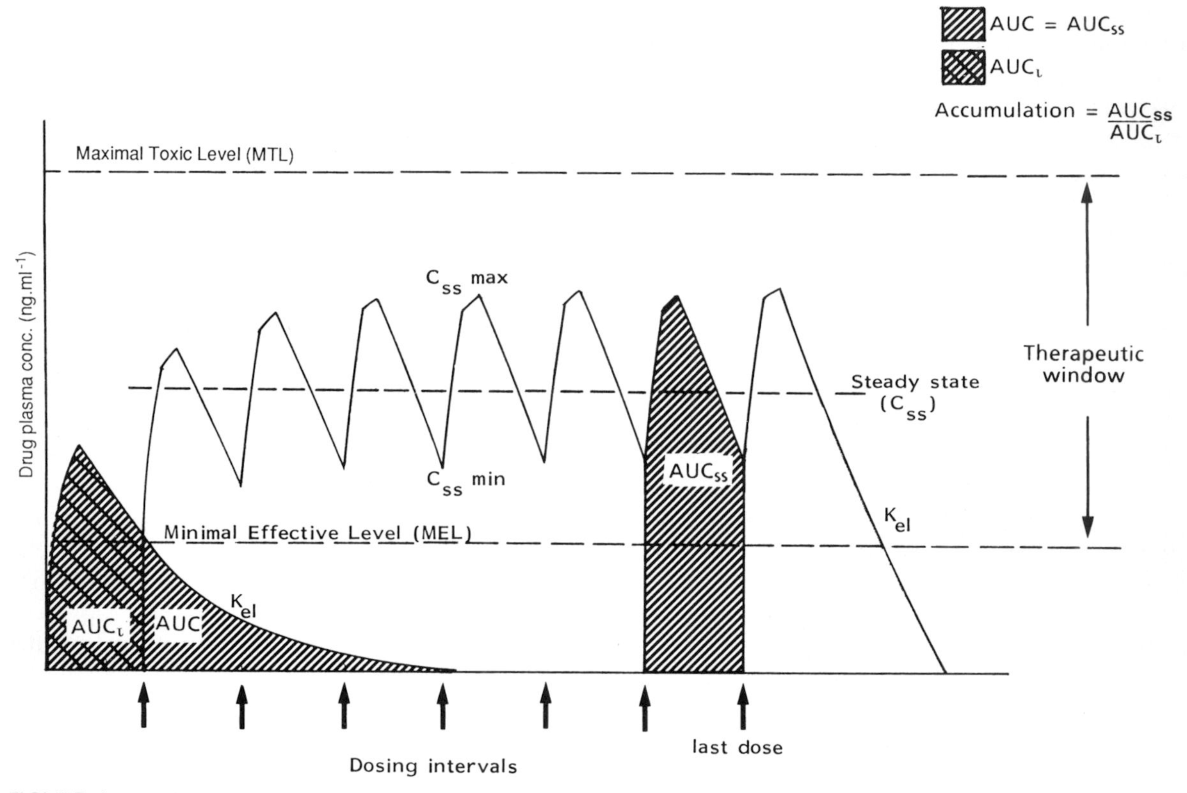

FIGURE 4 The buildup of plasma levels to steady state within an ideal therapeutic window during repeat dosing of a drug.

ents associated with these bonds are different, then the carbon atom can be said to be *asymmetric,* and the molecule will have a *chiral* center. Chirality comes from the Greek word χειρ, meaning "hand." A molecule with an asymmetric carbon atom can exist as two separate but spatially different (stereo) and noninterchangeable compounds, and just like our own hands, are nonsuperimposable mirror images of each other but with the same elementary chemical composition. These compounds are called *enantiomers* (Greek εναντιο, for "opposite shape"), and there may be two or more existing for one formula, depending on the number (n) of asymmetric carbon atoms, according to the formula

$$\text{Number of isomeric pairs} = 2^n. \qquad (26)$$

If isomerism occurs at only one asymmetric atom in a molecule with more than one center of asymmetry, then the result is a stereoisomer, not a mirror image, and is called a *diastereoisomer.* Unlike enantiomers, they have different physicochemical properties. Enantiomers are said to be optically active; when they are in solution, they bend polarized light to either the right [*dextro* (Greek) rotary] or left [*laevo* (Greek) rotary] and are symbolized, respectively, by the prefix D- or (+) and L- or (−). Confusion can arise by the additional use of prefixes R- [*recto* (Latin) = right] and S- [*sinister* (Latin) = left], which are designations of an absolute configuration. In nature, dominance of one or the other of the isomers is due to a preferential stability of one form (e.g., D-glucose L-amino acids), but chemical synthesis usually produces an equal mixture of both, called a *racemate* (prefix rac-). Approximately 25% of all drugs have asymmetry, and more than 80% are administered as the racemate of two or more separate compounds. Because the body can "perceive" these spatial differences, there is now increasing awareness that isomers may be stereospecific with different pharmacology, toxicology, pharmacokinetic, and metabolic profiles.

To show stereoselective differences in kinetics, it is necessary to separate and measure the isomers. This can be achieved by GLC or HPLC using recent progress in chiral chromatographic discrimination using (a) chiral derivatization by reagents, (b) chiral elements in HPLC, or (c) chiral stationary phases. Although relatively few chiral drugs have been investigated (<10%), it has been shown that only in those processes in which the drug binds to a *receptor* or a specific site-recognition protein will there be stereoselectivity in kinetics. Thus, physical processes of passive diffusion or absorption, distribution, and renal elimination (filtration and reabsorption) do not normally show selectivity, whereas active processes such as renal and biliary secretion and metabolism do show enantiomeric differences. Protein binding is also stereoselective but the difference in free-drug concentration rarely exceeds twofold, and volumes of distribution are rarely different.

The most important aspect of stereoselective kinetics is the differences in the rates and sometimes routes of metabolism, which have been shown to occur for all the pathways given in Table III. Some species are different from others: thus deamination is faster for the D-isomers of phenylethylamines in humans but not in rats, and ring hydroxylation is faster for L-isomers in humans but not in other animals. This may be due to the species differences in the isoforms of the cytochrome P450 involved in these reactions. For some of the nonsteroidal anti-inflammatory drugs, there is metabolic *enantiomeric conversion* from the inactive R-isomer to the active S-isomer. The combination of different isomeric clearances leading to different steady-state levels of compounds with sometimes markedly different activities does lead to problems in interpreting kinetic–dynamic interaction and the use of therapeutic drug monitoring when racemates are administered. It is now accepted that isomers, when administered together as the raemate, should be thought of as two or more separate compounds, which should be analyzed separately and the pharmacokinetic analysis under-

TABLE III
Common Metabolic Pathways

Phase I	Phase II
Oxidation	Conjugation
Deamination	Glucuronide
Dealkylation	Peptide
Epoxidation	Sulfate
Hydroxylation	Glutathione
Sulfoxidation	Methylation
Reduction	Acetylation
Alcohol dehydrogenase	Acylations
Azoreduction	Mercapturic acid
Nitro reduction	
Dehalogenation	
Hydrolysis	
Amide hydrolysis	
Ester hydrolysis	
Aromatization	

taken accordingly. Calculation of kinetic parameters from combined levels can lead to conclusions that are not simply an average of the two isomers. Whenever racemic mixtures are used clinically, these different isomeric activities should be taken into account.

V. PHARMACOKINETIC ANALYSIS

Pharmacokinetic analysis is the most difficult and complex aspect of pharmacokinetics, and it is not possible within the scope of this article to go into great detail. If there is further interest, the books of Rowlands and Tozer (1995) and Gibaldi and Perrier (1982) will provide a more complete description. The aim of pharmacokinetic analysis is to convert large quantities of drug-level data, normally from plasma, into a small number of parameters that can be used predictably to define the biodisposition of the drug or chemical under investigation. Although there are three types of such data analysis—*direct, empirical,* and *theoretical*—in practice, their use overlaps depending on the analytical objectives.

A. Direct Analysis

For direct analysis, the parameters are taken directly from the data either with no further manipulation (e.g., for C_{max}, T_{max}, etc., see Fig. 3), or by relatively simple empirical computation of the data, for example, the area under the concentration time curve to time t (AUC_t) is the sum of

$$AUC_t = \frac{(t_1 - 0) \cdot (C_0 + C_1)}{2} + \frac{(t_2 - t_1) \cdot (C_1 + C_2)}{2} + \cdots + \frac{(t_t - t_{t-1}) \cdot (C_{t-1} - C_t)}{2} \quad (27)$$

where C_0, C_1, and C_t and plasma drug concentrations are from zero time up until any time ι.

B. Empirical Analysis

Empirical analysis is a *model-independent* approach that makes no assumptions about the physiological processes involved. For most drugs, the concentration–time profiles after an IV injection can be described by the polyexponential equation

$$C_t = C_1 e^{-\lambda_1 \cdot t} + C_2 e^{-\lambda_2 \cdot t} + \cdots + C_n e^{-\lambda n \cdot t} \quad (28)$$

where C is the drug concentration and λ the exponential coefficient for each in phase of the curve. In practice, n is rarely more than 3, because phases greater than this cannot easily be separated. When the data are fitted by these equations, then the same kinetic parameters as derived from direct analysis can be calculated, for example,

$$AUC = \frac{C_1}{\lambda_1} + \frac{C_2}{\lambda_2} \cdots \frac{C_N}{\lambda_N} \quad (29)$$

Simple graphical estimates of compartment models can also be achieved using the method of residuals (Fig. 5).

In general, estimates of the parameters by empirical analysis are better than those by direct analysis, particularly half-life and computed maximal blood levels. Empirical analysis also describes mathematically the complete curve even at time points when no sampling has been made, so it also has the advantage of enabling better prediction or simulation of blood levels after single or repeated administration, and can be combined with simultaneous pharmacodynamic modeling. Unfortunately, empirical analysis has the disadvantage that it cannot be used to predict the effect on pharmacokinetics of physiological pathological changes such as renal or hepatic disease.

C. Theoretical Analysis

Theoretical analysis is a *model-dependent* approach that assumes that some underlying process is contributing to the observed changes in body-fluid levels. There are three different types of these models: compartment, physiological, and systems.

1. Compartment Model

The representation of pharmacokinetics as if the body consists of a number of discrete compartments has been an extensively used model. Although these compartments do not need to have any physiological identification, they are generally considered to correspond approximately to (a) *central:* blood and rapidly perfused tissues (lung, heart, kidneys, etc.), (b) *tissue:* more slowly perfused tissue (skin, muscle, etc.), (c) *deep:* very slowly perfused tissue (fat, bone, etc.), and (d) *elimination* (urine, feces, etc.) (Fig. 6).

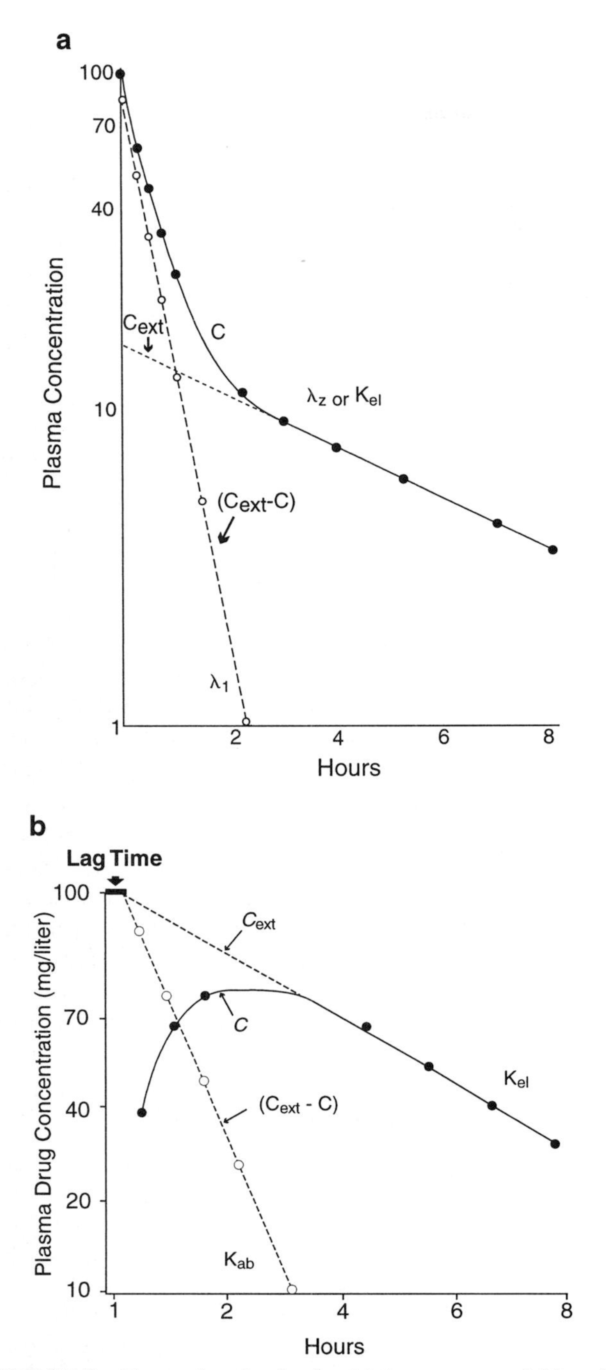

FIGURE 5 Plasma drug levels after (a) intravenous and (b) oral administration. The method of residuals is used to graphically calculate (a) the initial elimination rate exponent (λ_1) and (b) the absorption rate (K_{ab}). The terminal log linear elimination is back extrapolated (---) and any drug concentrations along the line are called C_{ext}. The observed concentration C is subtracted from C_{ext} at the same time and plotting these residuals (C_{ext}-C) against this time will give a straight line (——) with slopes (first-order rates only) equal to (a) absorption or (b) initial elimination rates.

The concepts behind this modeling are simple. The drug is assumed to go in and out of the compartments at rates directly proportional to the amount in each, but which may be different for each compartment. Unfortunately, although the concept is easy, the calculation of the rate constants of exchange (k) and the amount of drug in each compartment at any one time is not, particularly for multiple compartments. Thus, for a simple one-compartment IV bolus model, the rate constant for elimination k_{el} can be determined by integrating the differential equations

$$dA/dt = -k_{el} \cdot A \text{ and } A = \text{Dose}_e{}^{-k_{el}t} \qquad (30)$$
$$dE/dt = +k_{el} \cdot A$$

where A is the amount of drug in compartment 1.

However, as the number of compartments and differential equations increases, the arithmetical calculation gets so complex that results of the required numerical integration are not always reliable. The main advantage of this modeling is that, once solved, the equations allow computation of the amount of drug in any compartment at any time after dosing and are conceptually easy to understand, particularly by the nonpharmacokineticist. However, in recent years, such compartmentalization has come under criticism because it uses an abstract *black box* approach with little real anatomical basis, is prone to error, and can be overinterpreted, particularly if the model is incorrect.

2. Physiological Model

Although based on compartmental concepts, physiological or blood flow modeling attempts to overcome some of their disadvantages by defining the kinetic processes in terms of actual blood flow, protein binding, membrane permeability, real tissue volumes, and metabolic clearances. An example of a physiological model is shown in Fig. 7. The drug is taken up into the various tissues of the body from arterial blood at a rate that is dependent on the organ blood flow (Q_o) and the relative binding to the tissue. This can be either calculated theoretically from tissue and blood protein binding, permeability constants, and so on or obtained experimentally in animals by measuring tissue-to-blood ratios. Such an approach leads to a more realistic and understandable handling of kinetic results since it relies on known anatomical and physiological data. Subsequently it is of potentially greater clinical significance, particularly when kinetic predictions of the effects of disease or age are needed. In addition, since the comparative physiology in animals

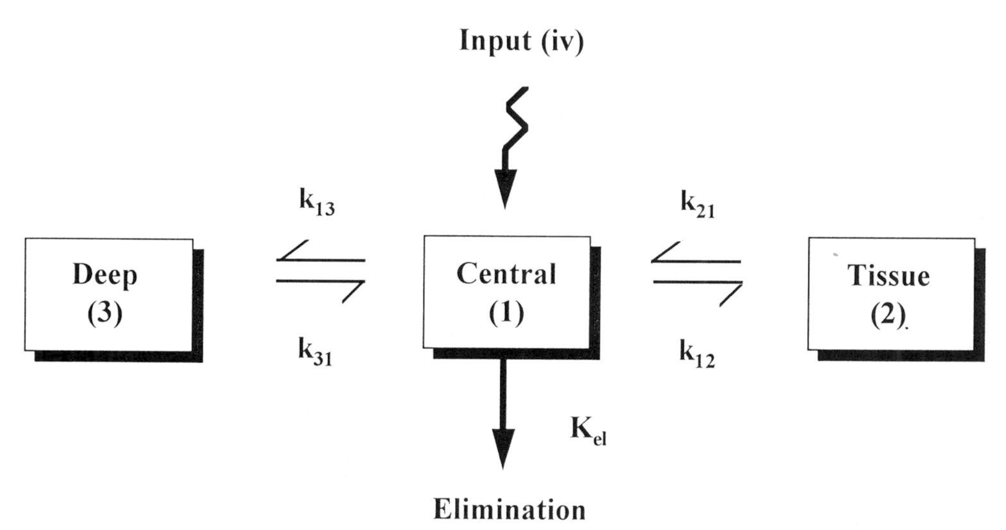

FIGURE 6 Three-compartment model.

is known, such models allow a better basis for comparing species differences in biodisposition. Despite the obvious advantages of such modeling, the technique has not gained widespread acceptance for the following reasons: (a) difficulty of measurement of the necessary physiological variables, particularly in humans, (b) models tend to be overambitious while body fluid analysis is insufficient for such overcomplexity, and (c) the statistical validity of these models is difficult to check. Because of this, together with the inherent variability for such parameters, the final results can sometimes be rendered meaningless. Despite these drawbacks, some success has been achieved with anticancer agents and anesthetics, where tissue concentrations may be particularly critical. Future research will see more investigators attempting to overcome these problems by using simpler and more verifiable models.

3. Systems Dynamics Model

The systems dynamics model, the newest of pharmacokinetic approaches, gets away from the highly structured and restrictive compartmental models previously described. Instead, it uses a mathematical approach to quantify general properties of physiological systems without the necessity to describe preconceived kinetic models. It therefore uses fewer but more verifiable assumptions and parameters. This model-independent procedure considers drug transport as a stochastic or random behavior of drug molecules in the body and uses mathematical operations such as *superposition* for multiple-dose predictions, *deconvolution* for curve fitting and disposition, and *decomposition* for absorption, distribution, and elimination. In the main, it assumes linear processes to be operational, but nonlinearity may also be modeled. Perhaps its greatest advantage will come from the use of systems analysis in pharmacodynamics, as it can be theoretically based on the current understanding of the mechanisms of action involved. Systems analysis, however, does have some disadvantages, the main being that this more generalized approach does not provide all the kinetic parameters of conventional approaches, and that the mathematics at the moment is unfamiliar to many scientists.

4. Summary

Kinetic models should (1) simply explain the biodisposition of a drug in terms of a small number of parameters from biofluid analysis; (2) be able to extrapolate or predict other situations; and (3) relate to pharmacodynamic processes. However, kinetic modeling is only as good as the data obtained and the assumptions used. No one model is the only answer, and sometimes the model can be wrong, even though the data fit is adequate. The following questions should always be asked: Is the model too complex or too specific to answer the objective of the study, based on the quantity or quality of the data, and are other models more useful alone or in combination?

VI. POPULATION KINETICS

An important use of kinetics is the prediction of dosage changes for certain patients groups: the elderly,

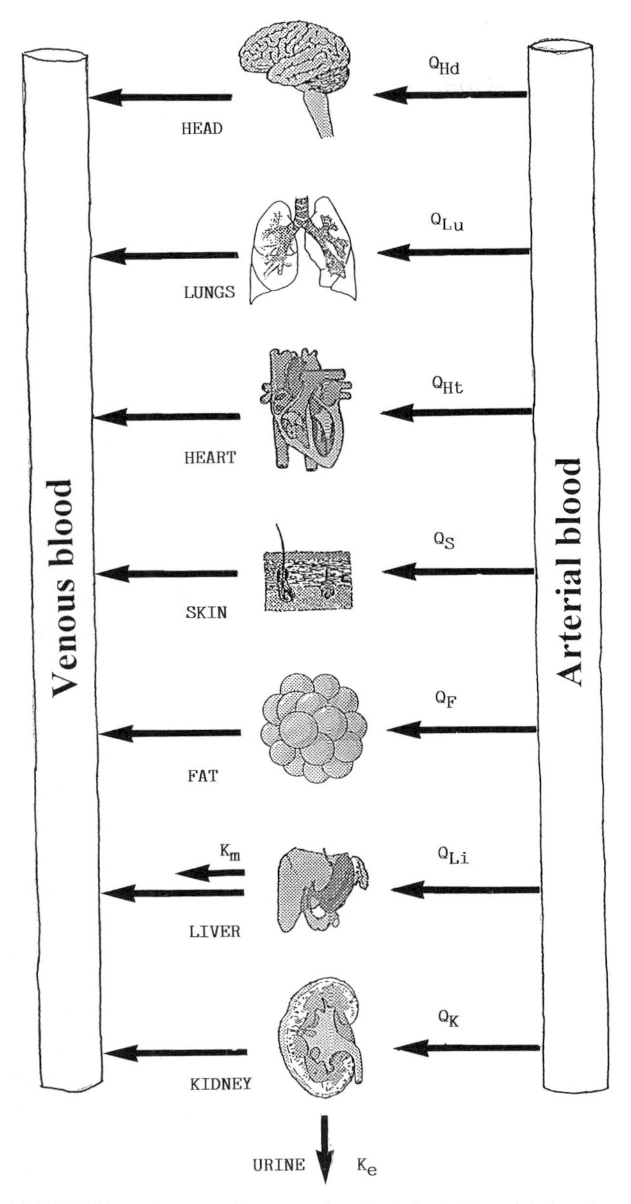

FIGURE 7 Schematic diagram of a physiological model showing that the clearance to each tissue is calculated as a function of blood flow (Q) and uptake and elimination of liver (K_m) and kidney (K_e).

the obese, for renal or hepatic failure, and so on. In the past this has been undertaken by comparing the kinetics in small groups (≈11–12) of healthy volunteers with the group that is to be investigated. This has disadvantages since the group size is small and may not be representative of a larger population, and it can require many separate studies. Population kinetics can overcome these problems. This can be achieved by measurement of kinetic parameters in a large population where individual estimates are statistically combined to produce overall population parameters that can be simply related to certain demographic details (age, weight, renal clearance, etc.). The errors of these estimates can also provide confidence in the accuracy of these population parameters. Three approaches can be used to relate response (steady-state drug levels) to the covariate (time): the naive averaging of data, the standard two-stage method, and the mixed-effect model or population approach.

A. Naive Averaging of Data

The naive averaging of data (NAD) approach averages all the plasma levels from all individuals studied at each observational time and analyzes the data for kinetics as if they came from one individual. This method is commonly used, but should be avoided because it smooths out the data, providing no information about the individual or the error, and often the apparent population estimates are incorrect. The naive pool method (NPD) is a more general variant of the NAD, fitting all the data in one step. It is used for initial population estimates, but suffers from the same problems.

B. Standard Two-Stage Method

The standard two-stage (STS) method allows, in the first stage, for an evaluation of kinetics from each individual and, in the second stage, combines these to produce mean population parameters and relates them to various pathological factors to show the influence, say, of body weight on steady-state levels. There are several variations and improvements to the two-stage approach, including global two-stage (GTS), iterative two-stage (ITS), and generalized iterative two-stage (GITS). They use either the initial estimates of population kinetics to feed back and improve individual values to obtain better population results (Bayesian approach), or more improved algorithms with more appropriate data weighting to compute the basic kinetics.

These procedures suffer in that they use traditional designed studies comparing data from specific illness groups with those from healthy volunteers. The disadvantages include (1) all population groups cannot be studied; (2) large amounts of controlled data are needed from small groups ($n = 10$–20); (3) sparse data cannot be used; (4) highly controlled hospitalized studies are needed; and (5) data may be polarized using extreme differences, for example, healthy or severe renal failure patients.

C. Population Kinetics

To overcome the foregoing problems and to obtain kinetic parameters that are more representative for all types of patients, Sheiner and Beal have introduced the concept of population kinetics. This is based on the principle of extended least squares and is computed on a program called NONMEM (nonlinear mixed-effect model). Instead of using a small number of individuals to define the population, the emphasis is reversed so that a small number of random, but well-defined, concentrations from a large number of individuals are used to make up a population from which individual kinetics can be defined. The advantages of this approach are that (1) it can use sparse data, only two or three points per individual at any time, but well documented; (2) data can come from any study where drug levels are measured; (3) data from specific studies (e.g., smoking) can be added at any time to improve estimates; (4) it provides good average population estimates; and (5) population standard deviations and estimates of residual error are all obtained, providing statistical confidence in the data. This analysis can be extended to important pathophysiological variables related to disease states. Thus, it is possible to say that clearance (CL) of a particular drug and therefore steady-state levels are related to body weight (W), age (Ag), and renal clearance (CL_R) as shown in

$$CL = (W \cdot 0.2) \cdot \left(\frac{CL_R}{70}\right)^{-0.7} \cdot \left(\frac{Ag}{65}\right)^{-0.5} \qquad (31)$$

The disadvantages are that (1) the software is not easy to use and few people are trained to use this approach; (2) the randomness of study design is a difficult concept to accept; and (3) the applicability and acceptability of the data generated have not yet been fully tested, either clinically or with regulatory authorities. However, this population approach linked with Bayesian feedback, incremental learning, and improvement of estimates can be used for all aspects of drug development including pharmacodynamics and toxicokinetics, and most importantly it enables the prescribing doctor to potentially tailor the dosage of a medication to fit.

VII. EFFECT OF AGE AND DISEASE

The most important kinetic changes that may occur in disease, smoking, or pregnancy, compared to those in healthy volunteers, are summarized in Table IV. The extent of the changes, if they occur at all, is very much dependent on the drug and the individual.

A. Elderly

Although many physiological changes occur in elderly patients, the most important, in terms of kinetics, is the reduction in renal blood flow (50% of normal at 65), which has important consequences for renally cleared drugs. Other changes include reduction in cardiac index (distribution and elimination), hepatic efficiency (metabolism), splanchnic blood flow (absorption), breathing capacity (elimination of anesthetics), and protein binding (distribution). There are large individual differences, and age alone is not a good index of functional deterioration.

B. Neonates

The newborn has an increased gastric pH, blood flow, and circulating endogenous fetal compounds (proteins, bilirubin), but decreased gastric emptying, intestinal blood flow, hepatic metabolism (drug dependent), α-glycoprotein levels, and particularly renal clearance (large intersubject variation). However, within the first few weeks and months after birth, maturation can dramatically alter physiological processes, and it is very difficult to predict dosage changes, as each drug and each subject behaves differently. Drugs have to be used very carefully in these infants. [*See* Antimicrobial Agents, Impact on Newborn Infants.]

C. Renal

Renal failure can occur directly or indirectly as a consequence of hypertension, diabetes, infection, and other conditions, and in elderly patients it can be estimated by creatinine clearance. The main change is a reduced renal clearance, which substantially reduces the clearance for those drugs primarily eliminated in the urine. Several nomograms are available relating renal clearance to dosage changes for such drugs. It should also be remembered that although the parent drug may not be renally eliminated, the more polar metabolites will be, and if active or toxic, then dosage changes may be necessary. The acidosis caused by

TABLE IV
Major Changes in Kinetic Parameters That Can Occur in Age and Disease Compared with Healthy Volunteers[a]

	Elderly age > 65	Young age < 5	Hepatic	Renal CL_{CR} < 50	Cardiac failure	Pregnancy	Burns	Intestinal	Smoking
Absorption (F)	⇩	⇩	↑ L	⇩↔	⇩	⇧	⇧ ⇩	↓	↔
Peak levels (C_{max})	⇧↔	↑	↑	↑↓↔	↓	↔	⇩	↓	↔
Fraction unbound (f_u)	↑ ↓ G A	⇧	↑	⇩ ⇧ G A	⇩ G (myocardial infarcation)	⇧	↑ G	⇧ ⇩ A G (Crohn's)	⇩ G
Distribution (V)	⇧ ⇩ Li P	⇧	↓	⇧ P	⇩ ⇧ L P	⇧	↑	⇩↔ G	↔
Half-life ($t_{\frac{1}{2}}$)	↑	⇧	↑	↑ P	↑	↔	⇧ ⇩	⇧↔ G	↔
Hepatic clearance (CL_H)	⇩	⇩	↓	⇩↔	↓	⇧	⇩ ↑ H L	↔	↑↔
Renal clearance (CL_R)	↓	⇧ N	↔	↓ P	↓	⇧	⇧ ⇩	↔	↔
Steady-state levels (C_{SS})	↑	⇧	↑ L	↑ P	↑	⇩↔	⇩ ⇧	↓	↑↔
Dosage changes	⇣	⇣	↓ L	↓ P	↓	↔⇡	↔	↑	⇣↔

[a]Li, lipid-soluble drugs; G, α-glycoprotein binding; N, neonates; P, polar drugs; A, albumin binding; H, high-clearance drug; L, low-clearance drug; ---, for some drugs; ↑, major change; ⇧, minor change; ↔, no change

retention of uric acid may reduce absorption and alters tissue distribution and protein binding. Hepatic clearance can also be indirectly reduced by circulating inhibitors of metabolism.

D. Hepatic

Hepatic disease can include cirrhosis, acute and chronic, hepatitis, and obstructive jaundice. Unlike renal disease, there are few reliable markers for hepatic function, and simple correlations with kinetic parameters are difficult. The major pathological changes are a reduction of metabolism, hepatic blood flow, and albumin concentrations, and an increase in portacaval shunting. Hepatic clearance can be substantially reduced (≈50%) for highly extracted drugs attributable to reduced blood flow and bypass shunting, whereas for poorly extracted drugs, clearance may be reduced, increased, or remain the same.

E. Cardiac Failure

Cardiac disease can include cardiac failure, hypovolemic shock and myocardial infarction, but the main physiological change is a reduced blood perfusion to all organs, resulting in reductions in rates of absorption, distribution and elimination. Although in some cases these factors cancel each other out, the net overall effect is longer half-lives and higher steady-state levels for many drugs.

F. Gastrointestinal Tract

GI tract disorders can include cancer, Crohn's disease, irritable bowel, achlorhydria and, in general, they significantly reduce the rate and extent of absorption of drugs. The stress protein, α-glycoprotein, is increased, which may reduce the volume of distribution of certain basic drugs.

G. Pregnancy

There are many physiological changes associated with pregnancy and these will differ according to fetal growth and hormonal balance. There are reductions in gastric activity, intestinal motility, and albumin, and there are increases in tidal volume, cardiac output, blood flow and volume, fat deposition, edema, volume (fetus, placenta, breasts, etc.), and particularly renal clearance (50%), but hepatic clearance remains unchanged. For most drugs, steady-state levels do not change, but increased dosage is needed for some drugs that are extensively renally cleared.

VIII. KINETIC–DYNAMIC RELATIONSHIPS

Unfortunately, the relative ease of measuring drug concentrations in blood compared to objective dynamic measurements has led medicinal scientists to assume that blood levels are synonymous or simply and linearly related to activity. Thus, the assumption is that a doubling of blood levels leads to a doubling of activity. Often this is not the case, and large changes in levels may produce only small changes in effect or vice versa. It is therefore necessary to relate or model drug levels with activity in order to maximize the use of kinetics.

A. Receptor Binding

In the 1930s, Clark first proposed from chemical reactions that for a drug to illicit an activity, it must *interact* in some way with a *receptor substance* and that this could be expressed mathematically by the *law of mass action,* thus

$$[\text{Drug} + \text{Receptor}] \overset{K_D}{\rightleftharpoons} [\text{Drug} - \text{Receptor}] \Rightarrow \text{effect} \quad (32)$$

Rearrangement produces the Michaelis–Menten-type equation, which relates the effect to the maximal effect (E_{max}), the dissociation constant (K_D), and the free-drug concentration at the receptor site (D_u):

$$\text{Effect} = \frac{E_{max} \cdot D_u}{K_D + D_u} \quad (33)$$

This produces a sigmoidal curve when effect is plotted against drug concentration with a small effect at low levels and greater response linear portion, which then attains a maximal effect when all the receptors are filled.

Unfortunately, for most drugs, it is not possible to sample at the site of action, and the assumption must be made that the free drug in the blood (C_u) is readily accessible to the receptor. Thus Eq. (31) becomes

$$\overset{\text{Blood}}{C} \rightleftharpoons C_u \rightleftharpoons D_u \rightleftharpoons \overset{\text{Receptor}}{DR} \rightarrow \text{effect} \quad (34)$$

B. Dynamic Models

Assuming that the drug has reached the site of activity and is in equilibrium with the drug level in blood, it is possible to relate these levels to the measured pharmacological or clinical effect in tissues using the Michaelis–Menten-type equation

$$\text{Effect E} = \frac{E_{max} \cdot C^N}{EC_{50} + C^N} + E_0 \quad (35)$$

where E_{max} is the maximum obtainable effect, EC_{50} is the drug concentration in blood at 50% maximum effect and is a measure of potency, E_0 is a baseline figure, C is the drug concentration, and N is the slope or Hill coefficient, a measure of the steepness of the drug level response curve. This relationship is not a straight line, and from Fig. 8a, it can be seen that when this relationship is graphed on Cartesian paper, it follows a curve reaching an asymptote. However, it is best represented graphically with the drug levels plotted on a logarithmic scale (Fig. 8b) (*sigmoidal E_{max} curve*), where it can be seen that at low drug levels, the effect increases slowly as the number of molecules starts to fill the available receptors. This is followed at 20% effect by a sharper linear increase in activity over a smaller range of drug levels. Finally, after 80% of the maximum activity, the curve flattens off as the receptors fill up, and at the E_{max}, large increases in drug levels do not increase the effectiveness of the drug. The slope of this curve (N) is important for the activity of a drug and they normally vary between 0.5 and 4.0. Thus, for those drugs with shallow slopes (<1), when there are large increases in drug levels, there is only a correspondingly small increase in activity. These compounds show fewer side effects, have a *wide therapeutic margin,* and normally have a longer duration of activity and may be administered only once or twice a day. They are therefore relatively safe, and this type of profile is ideal for drugs used

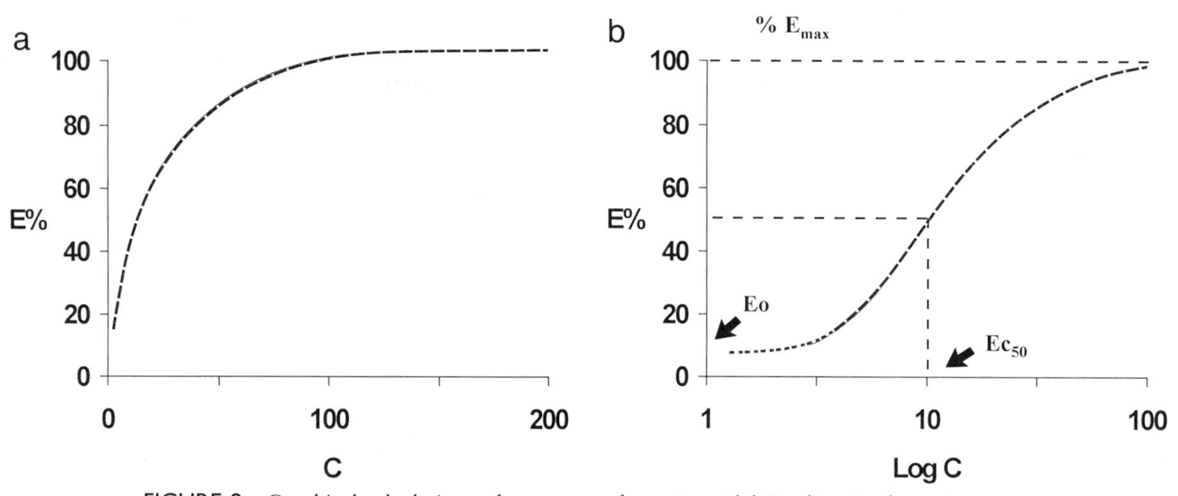

FIGURE 8 Graphical calculations of parameters from sigmoidal E_{max} kinetic–dynamic curves.

long term and where side effects are not wanted in diseases such as hypertension or diabetes. Compounds with steep slopes (>2) exhibit a more rapid and intense activity for only small changes in drug levels, and therefore have a *narrow therapeutic margin* and are less easily controlled. Such drugs are used in hospital settings for pain relief or induction of anesthesia, and generally have a shorter duration but more intense activity.

C. Kinetic–Dynamic Modeling

It has been assumed so far that the drug in the blood is in instantaneous equilibrium with the drug at the receptor site, but for most drugs, this is not true, and the pharmacokinetic changes may have to be considered. To overcome this, both the dynamic and kinetic modeling should be done simultaneously. The resultant equations are complex and may compound any errors inherent in either system. In addition, it is necessary to give the active site a size or compartment. To overcome this problem, researchers have used a so-called *effect compartment model,* a minimalistic conceptualization in that the active site is assumed to have a negligible volume. Various software programs are available to undertake this analysis, including ELSMOS, SIPHAR, and NONLIN. One of the problems that these programs can overcome is hysteresis.

D. Hysteresis (Anticlockwise Hysteresis)

Hysteresis comes from the Greek word Υστερησις, meaning "to come late." In this context, it has been coined for the delay in obtaining maximal activity compared with time of maximal plasma drug levels. It is often caused by the time needed for a drug to be transported from the systemic circulation to the site of action. This is graphically shown when plasma levels are plotted against activity and the points are joined up in time order to produce an anticlockwise directional curve (Fig. 9).

Thus, at early times after an IV administration with initial high plasma levels, but nothing in the brain, there is an increasing effect with time for equivalent drug levels. Inhalation produces less hysteresis, whereas the slower oral absorption shows none, as there is more time for equilibrium between plasma and brain. Note that after 15 min, all routes of administration show the same kinetic–dynamic curve, and thus for modeling, only those latter time points should be used. Removal of hysteresis can also be achieved mathematically by using a *link-effect model,* which collapses the hysteresis and provides estimates of the rate of uptake into the active site.

Hysteresis can also occur in three ways. (1) *Active agonist metabolites* produce maximal blood levels later than the parent drug. This can occur for many drugs, particularly those with which first-pass metabolism is important. (2) In *delayed or cascade activity,* for some drugs (e.g., antihypertensives and antidepressants) there is a delay of several weeks before maximal activity is seen. Similarly, the anticoagulant action of warfarin takes several days owing to the cascade activity of the various clotting factors in blood. (3) In *sensitization,* the activity increases with time because of the action of the drug itself, for example, an increased number of receptors (up-regulation).

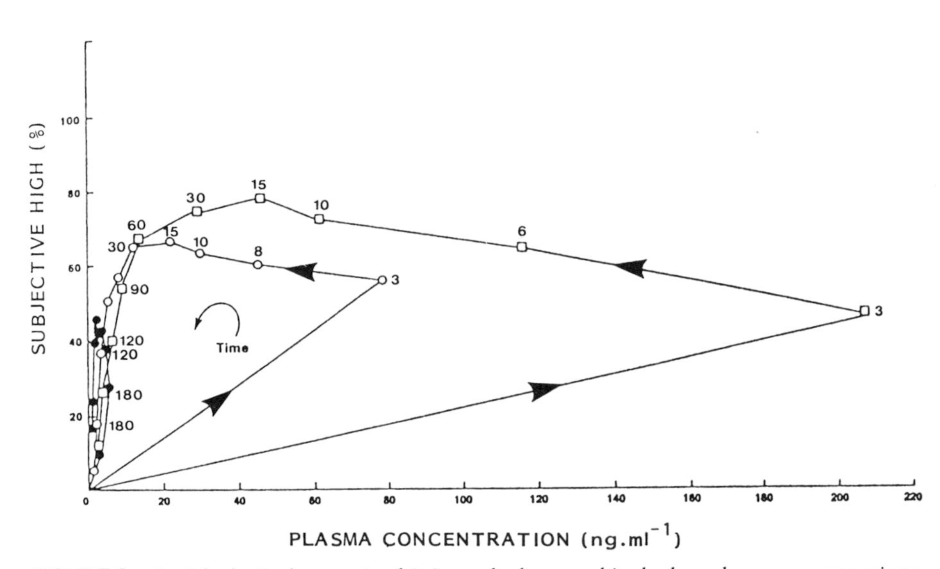

FIGURE 9 Anticlockwise hysteresis of Δ-9-tetrahydrocannabinol when plasma concentrations are related to a subjective high, after different routes of administration. □, IV; ○, smoking; ●, PO. [From Chiang *et al.* (1984).]

E. Proteresis (Clockwise Hysteresis)

Proteresis can be graphically seen as a decreasing effect with time for equivalent drug concentrations. It occurs when there is (1) *tolerance* due to a reduction in the number of receptors (down-regulation) or an endogenous transmitter, and occurs with addictive drugs but also with the nitrate vasodilators and antibiotics; (2) *active but antagonistic metabolites* that can inhibit the action of the administered drug; (3) *feedback regulation,* in which a physiological mechanism, or enzymatic, biochemical, or nerve function, comes into play to counteract and dampen the action of the drug with time; and (4) *learning* owing to insufficient time for subject familiarization with psychometric or behavioral tests for drugs that depress the central nervous system. This often occurs with drugs of addiction such as cocaine and benzodiazepines.

IX. THERAPEUTIC DRUG MONITORING

For many drugs, there are large intersubject variations in blood levels, for certain drugs up to 40-fold, attributable to race, age, food, disease, coadministered drugs, and gender, so it has long been the goal of pharmacokinetics to be able to tailor a particular dosage for a particular patient so that steady-state levels would be within the therapeutic window or targeted concentration range. The measurement of drug levels to provide this prediction, either as single-dose kinetics or more routinely at steady state, is called *therapeutic drug monitoring* (TDM) and is being used in hospitals for drugs with a narrow therapeutic window like phenytoin, digoxin, theophylline, lithium, gentamicin, and warfarin. When there is little difference (≃ twofold) between therapeutic and toxic side effects for this service to be of benefit, several assumptions have been made: (a) steady-state levels can be predicted from a few blood levels; (b) all subjects produce a similar response from comparable drug levels; and (c) the procedure is cost-effective. Steady-state levels can now be relatively well predicted, using initial population kinetic values and Bayesian feedback methods to impose the fit for the individual patient from only three or four drug levels; however, most up-to-date evidence now shows that few drugs have a drug-level response curve that is the same between or within individuals. Even for those drugs with which TDM is extensively used, data do not support clear definitions of a therapeutic window, often because of the lack of carefully controlled studies. Because of the lack of reliability of TDM, the cost–benefit ratio of such expensive and time-consuming practices is now being questioned by the payers of health care. Many clinicians still use clinical end points to decide on dosage changes, with drug levels considered only as backup information.

TABLE V
Symbols and Derivations[a]

Symbol	Definition	Derivation
Ae (Au)	Total amount of unchanged drug excreted into the urine [mg]	Data
AUC (AUC_∞)	Area under the plasma drug concentration time curve to infinity [ng · hr · ml^{-1}]	(i) $AUC_t + C_t/k_{el}$ (ii) $\frac{D}{V \cdot k_{el}}$
AUC_t	Area under the plasma drug concentration time curve to the last measured time point, t [ng · hr · ml^{-1}]	$(t_1 - t_0) \cdot (C_0 + C_1)/2 + (t_2 - t_1) \cdot (C_1 + C_2)/2 + \cdots (t_t - t_{t-1}) \cdot (C_t + C_{t-1})/2$
AUMC	Infinite area under the first moment versus time curve [ng · hr^2 · ml^{-1}]	$C_1/\lambda_1^2 + C_2/\lambda_2^2 + \cdots + C_n/\lambda_n^2$
C (Cp)	Total concentration of drug in the plasma [ng · ml^{-1}]	Data
C_b	Total concentration of drug in blood [ng · ml^{-1}]	Data
C_0	The initial or back extrapolated to zero time, plasma concentration [ng · ml^{-1}]	Data
CL (Clp)	Total plasma clearance of drug [ml · min^{-1}]	(i) $\frac{F \cdot D}{AUC}$ (ii) $k_{el} \cdot V$
CL_B	Total blood clearance [ml · min^{-1}]	$\frac{F \cdot D}{AUC\ (blood)}$
CL_{CR}	Renal clearance of creatinine [ml · min^{-1}]	$\frac{\text{Urinary excretion rate}}{\text{Plasma concentration}}$
CL_H	Hepatic clearance [ml · min^{-1}]	(i) $E_H \cdot Q_H$ (ii) $CL - C_R$
CL_{int}	Intrinsic clearance of drug from eliminating organ [ml · min^{-1}]	(i) $\frac{CL_H}{Cu}$ (ii) $\frac{CL_R}{Cu}$
CL_R	Renal clearance of drug [ml · min^{-1}]	$\frac{Ae}{AUC}$
$C_{max,ss}$	The maximum concentration at steady state [ng · ml^{-1}]	$C_{max} + (C_{min} \cdot e^{-k_{el}} \cdot t_{max})$
C_{min}	Minimum plasma concentration reaching during a dosing interval [ng · ml^{-1}]	Data
$C_{min,ss}$	The minimum trough drug level at steady state [ng · ml^{-1}]	(i) Data (ii) $\frac{C_{min}}{1 - e^{-k_{el} \cdot \iota}(i.v.)}$
$C_{av,ss}$ (C_{ss})	Average concentration at steady state during repeated dosing [ng · ml^{-1}]	$\frac{AUC}{\iota}$
C_t	Plasma concentration at last measured time point	Data
C_u	Unbound concentration in plasma [ng · ml^{-1}]	Data
D	Dose administered [mg]	Data
D_L	Loading dose to achieve desired steady-state levels [mg]	$V \cdot C_{ss}$
D_M	Fixed maintenance dose to keep steady-state levels constant [mg]	$D_L \cdot (1 - e^{k_{el} \cdot \iota})$
E	Effect (unit of effect or %)	$\frac{E_{max} \cdot C^Y}{EC_{50}^Y + C^Y}$
EC_{50} (E_{50})	Concentration of drug producing 50% of maximal effect [ng · ml^{-1}]	Data
E_H	Hepatic extraction ratio (no units)	$\frac{CL}{Q_H}$
E_{max}	Maximum effect (unit of effect or 100%)	Data
E_0	Baseline effect (unit of effect or %)	Data
ER (E)	Extraction ratio (no units)	$\frac{CL}{Q}$

continues

TABLE V (*Continued*)

Symbol	Definition	Derivation
F	Fraction of drug entering the body intact or absolute availability (no units)	$\frac{AUCp.o. \cdot Di.v.}{AUCi.v. \cdot Dp.o.}$
f_e	Fraction of absorbed drug extracted unchanged into the urine (no units)	(i) $\frac{F \cdot A_e}{D}$ (ii) $\frac{CL_R}{CL}$
F_{max}	Theoretical maximum oral availability for a drug assuming first-order first-pass hepatic metabolism	(i) $1 - ER$ (ii) $1 - (CL_H/Q_H)$
f_{up} f_{ub}	Fraction unbound of total drug concentrations in plasma/blood (no units)	$\frac{C_u}{C}$ or $\frac{C_u}{C_b}$
GFR	Glomerular filtration rate [$ml \cdot min^{-1}$]	$\approx CL_{CR}$ normally 120 $ml \cdot min^{-1}$
k_{12}	Transfer rate constant from central (1) to peripheral (2) compartment [hr^{-1}]	from integration (1) $\frac{dC_1}{dt} = -k_{12} \cdot C_1 + k_{21} \cdot C_2$
k_{21}	Transfer rate constant from peripheral (2) to central (1) compartment [hr^{-1}]	$\frac{dC_2}{dt} = k_{12} \cdot C_1 - k_{21} \cdot C_2$
k_a	Absorption rate constant [hr^{-1}]	Data
K_D	Dissociation constant of a drug from its receptor [$ng \cdot ml^{-1}$]	Data
k_e	Urinary excretion rate constant [hr^{-1}]	$C_R \cdot C$
k_{el}	Elimination rate constant [hr^{-1}]	$\frac{0.693}{t\frac{1}{2}}$
MRT	Mean residence time within the body [hr]	$\frac{AUMC}{AUC}$
Q_H (HBF)	Hepatic blood flow (hepatic artery + portal vein) [$ml \cdot min^{-1}$]	On average = 1050 + 300 $ml \cdot min^{-1}$
Q_R (RBF)	Renal blood flow ($ml \cdot min^{-1}$)	On average = 1100 $ml \cdot min^{-1}$
Rac (Ra)	Accumulation ratio (no units)	(i) $\frac{AUC}{AUC_t}$ (ii) $\frac{1}{1 - e^{-k_{el} \cdot \iota}}$
S	Slope of linear models of kinetic–dynamic relationships	Data
t	Time after drug administration [hr]	Data
t_{lag}	Lag time [hr]	Data
T_{max}	Time of maximal plasma concentration during a dosing interval [hr]	Data
$t\frac{1}{2}$ ($t\frac{1}{2}\beta$)	Half-life of the terminal slope [hr]	(i) $\frac{0.693}{\lambda_z}$ (ii) $\frac{0.693 \cdot V}{CL}$
V (Vd or Vd_{area})	Apparent volume of distribution derived from total plasma concentration after distribution equilibrium has been achieved [liters]	(i) $\frac{D}{AUC \cdot \lambda_z}$; (ii) $\frac{t\frac{1}{2} \cdot Cl}{0.693}$
V_1 (V_0)	Initial volume of distribution at zero time [liters]	$\frac{D}{C_o}$ (for i.v. bolus)
V_p; V_T	Plasma volume; tissue volume	Data
V_{ss}	Apparent volume of distribution under steady state conditions derived from total plasma concentrations [liters]	$MRT \cdot CL$
Y (N)	Slope factor or Hill coefficient in concentration–response relationship (no units)	Data
λ_i	Exponents of the ith exponential term of a polyexponential equation [hr^{-1}]	Data
λ_1, λ_2	Exponential coefficients [hr^{-1}]	Data
λ_2	Terminal exponential coefficient	Data
ι	Dosing interval [hr]	Data

[a]For example, for a two-compartment model. Because of the mathematical nature of pharmacokinetics, symbols are frequently used to simplify equations. This list is not exhaustive but defines those that are most frequently used together with alternatives ( ) and frequently used units [ ] and derivation. Data means that the parameter can be obtained directly without computation. (Note: To convert $ml \cdot min^{-1}$ to $liters \cdot hr^{-1}$ divide by 16.6.)

X. FUTURE DIRECTIONS

The importance of pharmacokinetics in drug development and understanding drug action cannot be overemphasized. As a science, it has matured and come a long way from the highly mathematical derivations that few nonkineticists understood, to more physiological interpretations of the parameters that provide comprehensible rationales for drug usage (Table V). The future will see (a) improvements in analytical methods as lower doses of more specifically acting drugs are used, and a better separation of asymmetric molecules; (b) use of noninvasive techniques such as MRI, PET (positron emission tomography), and bioelectrical impedance, which can measure drugs within the body by scanning and not by taking samples, allowing human distribution studies to be undertaken; (c) greater awareness of chronopharmacokinetics, in which the kinetics parameters will change depending on when the drug is administered; (d) greater understanding of metabolic differences between animals, and between individuals differing by race, environment, way of life, and genetic polymorphism linked to an understanding of the specific metabolizing enzymes involved; and (e) better indices of disease states, such as hepatic, renal, respiration, and cardiac, so that drug dosage can be adjusted accordingly. There is now the realization that kinetics is only a probe; in isolation, without relating it to dynamic activity, it is of little interest. More emphasis will be put on dynamic modeling, using the newer population approaches and systems analyses. Even the newest of mathematical theories, neural networks and fuzzy logic are now being used in kinetic dynamic modeling to more fully cope with the complexities of the human body. These techniques will form the basis for prediction kinetics, replacing the more classic kinetic methods and naive therapeutic or compliance monitoring, and lead to a greater understanding of the disease processes themselves.

ACKNOWLEDGMENTS

I would like to thank Kerrie Campbell for her artistic acumen and Danielle Bowra for her perseverance in proofreading this paper.

BIBLIOGRAPHY

Benet, L. Z., Massond, M. S., and Gambertoglio, J. G. (1984). "Pharmacokinetic Basis for Drug Treatment." Raven, New York.

Campbell, D. B. (1990). The use of kinetic–dynamic interactions in the evaluation of drugs. *Psychopharmacology,* **100,** 433–450.

Gibaldi, M., and Perrier, D. (1982). "Pharmacokinetics." Marcel Dekker, New York.

Gibson, G. G., and Skett, P. (1993). "Introduction to Drug Metabolism," 2nd Ed. Chapman & Hall, London/New York.

Rowland, M., and Tozer, T. N. (1995). "Clinical Pharmacokinetics: Concepts and Application," 3rd Ed. Williams & Wilkins, Baltimore.

Phenylketonuria, Molecular Genetics

RANDY C. EISENSMITH
SAVIO L. C. WOO
Howard Hughes Medical Institute, Baylor College of Medicine

GLOSSARY

cDNA Complementary DNA. A DNA molecule prepared from messenger RNA using the enzyme reverse transcriptase

Guthrie test Semiquantitative bacterial inhibition assay for the determination of serum phenylalanine levels in newborns. The Guthrie test requires only a small volume of blood, is specific, inexpensive, and well suited to handle large numbers of samples

Haplotype Genetic constitution of an individual with respect to one member of a pair of allelic genes

Polymerase chain reaction Amplification of specific regions of DNA through the repeated annealing and subsequent extension of oligonucleotide primers by the heat-stable enzyme *Taq* polymerase

Restriction fragment-length polymorphism Benign nucleotide substitutions within the genomic DNA of individuals that have no phenotypic manifestations and detected only by restriction endonucleases

Short tandem repeat Short nucleotide sequence, usually 2–5 bp in length, that is tandemly repeated several times within the genome

Variable number tandem repeat Nucleotide sequence of variable length that is tandemly repeated several times within the genome

CLASSICAL PHENYLKETONURIA (PKU) IS AN AUTOsomal recessive genetic disorder caused by a deficiency of hepatic phenylalanine hydroxylase (PAH). The disorder is characterized by an accumulation of phenylalanine in the serum, resulting in hyperphenylalaninemia (HPA) and associated abnormalities in aromatic amino acid metabolism. Untreated PKU patients develop severe postnatal brain damage and irreversible mental retardation. PKU is among the most common inborn errors of amino acid metabolism in humans, with an average incidence of approximately 1 in 10,000 Caucasian births.

I. BIOCHEMICAL BASIS OF CLASSICAL PHENYLKETONURIA

Classical phenylketonuria was first reported in 1934 by Følling, who observed the presence of increased levels of phenylpyruvate in the urine of a group of mentally retarded patients. A year after the discovery of this disease, Penrose confirmed Følling's earlier suspicion that PKU is an autosomally transmitted recessive genetic disorder. The biochemical defect present in PKU patients was established by Jervis in 1947, when he observed that the administration of phenylalanine produced a prompt elevation in serum tyrosine levels in normal individuals, but not in individuals with PKU. From these studies, he concluded that PKU patients lack the ability to convert phenylalanine into tyrosine. This conclusion was supported by the subsequent demonstration of the conversion of phenylalanine to tyrosine in postmortem liver samples from normal individuals but not in those from PKU patients.

In studies inspired by the prior success of dietary management in the treatment of galactosemia, several investigators, including Louis Woolf, Horst Bickel, and others, began the administration of low-phenylalanine diets to young PKU patients in the early 1950s. This dietary treatment led to the reduction of serum phenylalanine and urinary phenylpyruvate levels in these children. In addition, some improvement in behavioral performance has been reported in PKU patients receiving dietary therapy.

As these initial findings were extended in additional studies, it became apparent that there is an inverse relationship between the age of onset of dietary therapy and the ultimate IQ level attained by treated patients. This observation provided a strong incentive for the implementation of neonatal screening programs for PKU. With the development of the Guthrie test, such neonatal screening programs became feasible. From the collective results of mass screening programs in Western countries, the incidence of PKU has been estimated at approximately 1 in 10,000. This value corresponds to a carrier frequency of approximately 1 in 50 for this autosomally transmitted recessive disorder.

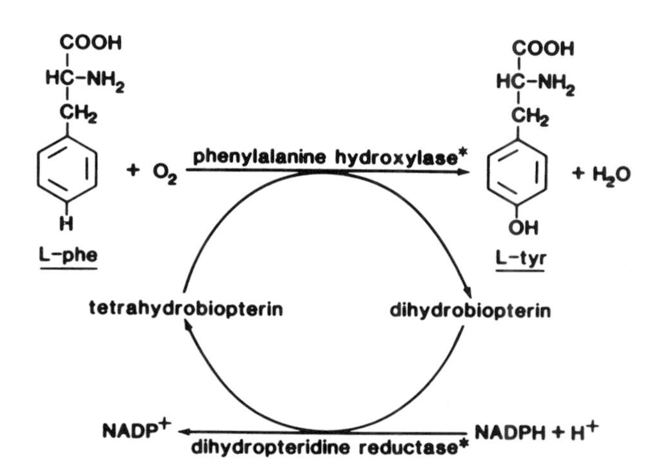

FIGURE 1 The phenylalanine hydroxylating system in humans. Asterisks denote gene products implicated in the most common genetic disorders of phenylalanine metabolism.

II. PHENYLALANINE HYDROXYLATING SYSTEM IN HUMANS

When Jervis first demonstrated that liver samples from PKU patients were unable to convert phenylalanine to tyrosine, relatively little was known about the enzyme or enzymes responsible for this process. This situation improved dramatically throughout the 1960s and the early 1970s, in large part due to the work performed in the laboratory of Seymour Kaufman. The results of this work are summarized in Fig. 1.

Phenylalanine hydroxylase, a mixed-function oxygenase, catalyzes the hydroxylation of the essential amino acid L-phenylalanine to L-tyrosine. This enzyme, expressed exclusively in the liver in humans, utilizes L-phenylalanine and molecular oxygen as substrates and requires the cofactor L-erythrotetrahydrobioptrerin (BH_4) as an electron donor. Although the cofactor is oxidized to the dihydro- form during each reaction cycle, adequate cofactor levels are maintained in the liver through the reduction of dihydrobiopterin to tetrahydrobiopterin by the enzyme quinonoid dihydrobiopterin reductase (QDPR).

Most cases of PKU are associated with a deficiency of PAH activity (classical PKU). However, deficits in QDPR activity or other enzymes involved in the biosynthesis and metabolism of cofactor BH_4 can also produce elevations in serum phenylalanine levels which lead to PKU. Although these atypical forms of PKU represent only about 1 to 2% of all PKU cases, the existence of atypical PKU complicates somewhat the diagnosis of PKU patients. Furthermore, patients with atypical PKU require significantly different regimens of treatment and counselling and may face significantly different prognoses than patients with PAH deficiencies. The characterization of this multienzyme phenylalanine hydroxylating system marked the end of the classical era of PKU research.

III. MOLECULAR BASIS OF PHENYLKETONURIA

A. Molecular Cloning of Human PAH and Determination of Its Primary Structure

The molecular era of PKU research began with the cloning of the rat PAH cDNA in the laboratory of Dr. Savio Woo. First, PAH mRNA was purified from rat liver by polysome immunoprecipitation and was used to prepare a rat PAH cDNA. The authenticity of this clone was established by hybrid-selected translation and was confirmed by matching the nucleotide sequence with the partial amino acid sequence of the purified rat enzyme. Second, the rat PAH cDNA clone was used as a specific hybridization probe to isolate several human PAH cDNA clones from a human liver cDNA library. The longest of these clones contained an open reading frame which encoded a protein composed of 451 amino acids. The predicted amino acid sequence of human PAH, as deduced from the nucleotide sequence of this clone, was shown to be over 90% homologous to the amino acid sequence of rat PAH.

Because native human PAH initially appeared to be a polymeric enzyme composed of multiple subunits, it was not clear whether human PAH is a heteropolymer or a homopolymer. This issue was critical to the understanding of the genetics of PKU. If the enzyme is a heteropolymer, multiple genetic loci may be expected. To resolve this issue, an expression vector was produced that contained the human PAH cDNA driven by an appropriate promoter. Introduction of this expression vector into cultured mammalian cells resulted in the expression of PAH mRNA, the production of immunoreactive PAH protein, and the presence of pterin-dependent enzymatic activity similar to authentic human PAH. These gene transfer experiments demonstrated that human PAH is a single gene product, permitting the use of the human PAH cDNA as a probe to perform molecular analysis of the PAH gene and the PKU locus in humans.

B. Molecular Structure of the Human PAH Gene and Its Chromosomal Location

Chromosomal assignments for human genetic loci can be made by using cloned genes in molecular hybridization studies to probe genomic DNA isolated from human/rodent cell hybrids that contain different combinations of human chromosomes. A panel of such hybrid cell lines was analyzed by Southern hybridization using the human PAH cDNA clone as a hybridization probe. The results of these studies indicated that the human PAH locus is on chromosome 12. This result was confirmed in subsequent experiments using both deletion chromosome mapping and *in situ* hybridization. Detailed examination of the results of the *in situ* hybridization experiments further localized the human PAH loci to the q22-q24.1 region of chromosome 12. The PAH cDNA had previously been shown to contain all of the genetic information necessary for the expression of PAH, which is the enzyme deficient in PKU, so the PKU locus in humans must also lie within this region.

The structural organization of the PAH gene was subsequently established by Southern analysis of normal human genomic DNA using the human PAH cDNA as a hybridization probe. Preliminary observations suggested that the chromosomal PAH gene is over 65 kb long and contains multiple intervening sequences. Because of the large size of this gene, a human genomic DNA library was constructed using cosmid vectors, which permitted the analysis of relatively large fragments. Screening of this cosmid library with the human PAH cDNA resulted in the isolation of four cosmids that contain overlapping PAH genomic sequences. Detailed analysis of these four cosmid clones indicated that the human PAH gene is approximately 90 kb in length, contains 13 exons with intron sizes ranging from 1 to 23 kb, and encodes a mature messenger RNA of approximately 2.4 kb.

C. Restriction Fragment-Length or Repeat Polymorphism (RFLPs) in the Human PAH Gene

An examination of the PAH locus in normal and PKU individuals by Southern analysis using the human PAH cDNA as a hybridization probe identifies several RFLPs in or near the human PAH gene. These RFLPs are the result of benign nucleotide substitutions within the genomic DNA of individuals that have no phenotypic manifestations and can be detected only by restriction endonucleases. The relationship of eight RFLP sites to the 13 exons which constitute the complete coding region of the 90-kb human PAH gene is shown in Fig. 2. Each different combination of RFLP sites defines a unique haplotype. Seven of these RFLPs are diallelic by Southern analysis, while the *Hin*dIII polymorphism is triallelic. Thus, these eight RFLPs define 384 specific haplotypes, more than 70 of which have been observed thus far. The 6 most common of the more than 80 haplotypes identified thus far in the PAH locus are defined at the bottom of Fig. 2. [*See* DNA Markers as Diagnostic Tools.]

Detailed examination of the multiallelic *Hin*dIII system demonstrated that the polymorphism is not the result of the presence or absence of a single restriction site, as in traditional RFLPs, but rather is caused by the presence of a variable number tandem repeat (VNTR) located between two constant *Hin*dIII sites. This VNTR system can contain 3–13 copies of a 30-bp repeat (Fig. 2). Alleles of different repeat sizes have been observed on normal and mutant PAH chromosomes. The incorporation of this VNTR into the existing RFLP haplotype system increases the potential number of PAH haplotypes to 1280.

A second, highly polymorphic repeat, located within intron 3, has also been identified in the PAH gene (Fig. 2). This short tandem repeat (STR) contains at least nine alleles, each differing from the next by

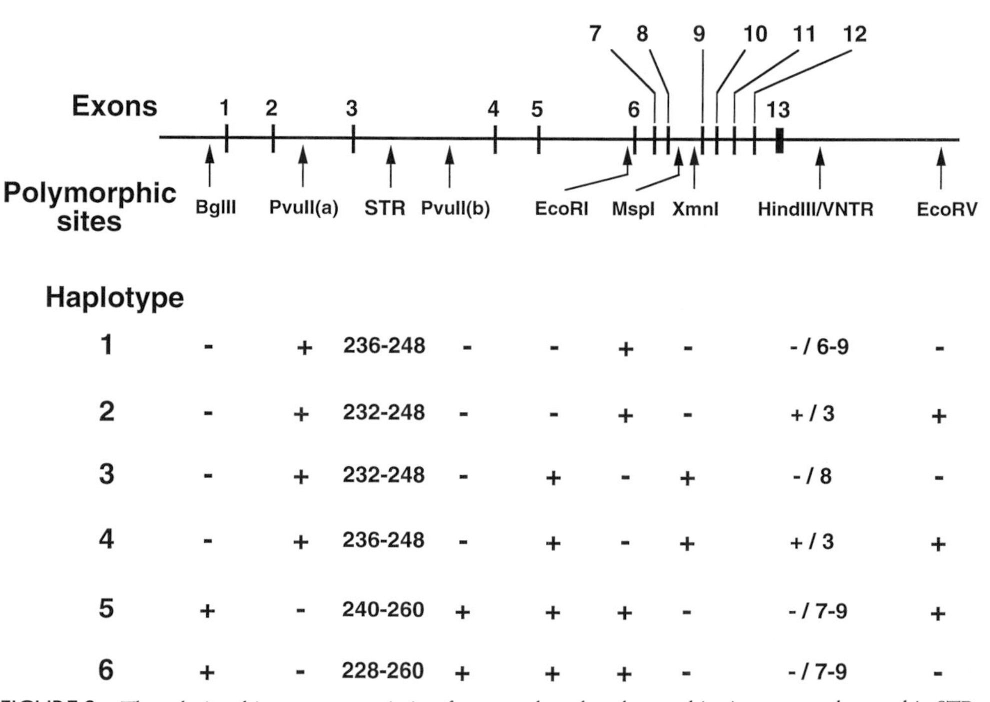

FIGURE 2 The relationship among restriction fragment-length polymorphic sites, one polymorphic STR, one polymorphic VNTR, and the 13 exons which constitute the complete coding region of the human PAH gene (top). Each specific combination of RFLP, STR, and VNTR sites constitutes a unique haplotype. The six most common PAH haplotypes among Caucasians in the Northern European population are shown (bottom). Pluses and minuses denote the presence or absence of a particular sized fragment following the digestion of genomic DNA with a given restriction endonuclease. Numbers under the STR denote the range of size variation of this system in base pairs. Numbers under the VNTR denote the range of size variation of this system, given as multiples of the 30-bp repeat.

a single 4-bp TCTA repeat. With the inclusion of this STR polymorphism, the possible number of unique PAH haplotypes increases to 11,520. The finding of only a small fraction of this number of potential haplotypes suggests that many of these polymorphisms are in linkage disequilibrium with one another.

D. Linkage-Based Tests for Prenatal Diagnosis and Carrier Screening

Because these restriction and repeat polymorphisms are tightly linked to the PAH gene, they can be used to follow the transmission of normal or mutant chromosomes in PKU families. For example, by comparison of the hybridization patterns of parental and proband DNA samples following digestion with one or more restriction endonucleases, the inheritance of normal or mutant alleles can be easily determined in a PKU family. Figure 3 shows the first practical application of RFLP haplotype analysis to the prenatal diagnosis of PKU. In Fig. 3, genomic DNA from the father, the mother, the proband, and an unborn child were isolated, digested with the restriction endonuclease *Hin*dIII, and analyzed by Southern hybridization using the human PAH cDNA as a probe. Both

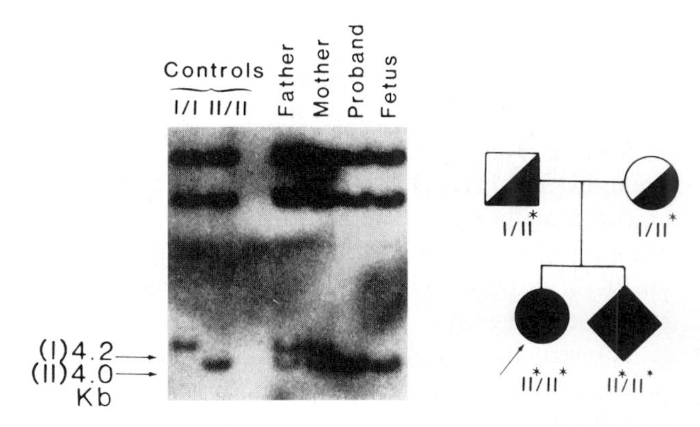

FIGURE 3 The application of RFLP haplotype analysis for the detection of mutant PAH alleles in a family with a prior history of phenylketonuria (see text for details).

parents are heterozygous for the *Hind*III polymorphism, as evidenced by the presence of both 4.2- and 4.0-kb bands, while the proband contains only the 4.0-kb band. Thus, the mutant PAH gene is associated with the 4.0-kb allele in this family. Because this type of analysis was informative in this family, fetal DNA was obtained and similarly analyzed. The fetus was also homozygous for the 4.0-kb fragment, thus permitting a prenatal diagnosis of PKU. This diagnosis was confirmed shortly after birth.

In a sample derived from several European countries, haplotype analysis based on all eight RFLP sites is informative in about 86% of all PKU families. However, because of the presence of only a relatively small number of RFLP haplotypes among east Asians, similar haplotype analysis is informative in only about 32% of all Chinese and Japanese PKU families. The degree of informativity of linkage-based tests can be significantly increased through the inclusion of the highly polymorphic repeat systems. Due to its high degree of polymorphism, the STR system alone is informative in over 80% of European Caucasians and over 70% of east Asians. Combining the STR and VNTR systems with one or more of the diallelic RFLPs present in the PAH gene results in a further increase in the informativity of linkage-based tests to over 95% in Caucasians and to over 75% in east Asians.

Although the application of haplotype analysis permitted the development of prenatal diagnosis of PKU, this technique is not without limitations. Most importantly, prenatal diagnosis using RFLP analysis can only be provided to those families with a prior incidence of PKU. Unfortunately, the vast majority of new cases of PKU (over 95%) are the result of random mating events, and thus are undetectable by traditional haplotype analysis. To overcome these limitations, the specific molecular lesions responsible for PKU must be isolated and characterized.

E. Mutations in the Human PAH Gene

1. Mutation Analyses

Haplotype analysis of a number of PKU families from many European countries indicated that two PAH haplotypes (haplotypes 2 and 3) are much more prevalent among mutant chromosomes than among normal chromosomes. Since close associations between RFLP haplotypes and specific mutations had previously been observed in other genetic disorders such as β-thalassemia, the relationship between PAH haplotypes 2 and 3 and PKU chromosomes was examined by direct molecular analysis.

The PAH genes from PKU patients bearing PAH haplotype 2 or 3 chromosomes were isolated by molecular cloning and sequence analysis was performed. The results of these analyses from an individual homozygous for mutant haplotype 2 revealed the presence of a C $\rightarrow$ T transition in exon 12, which resulted in the substitution of the amino acid tryptophan for arginine at amino acid residue 408 of the PAH protein. Similar analyses from an individual homozygous for mutant haplotype 3 revealed the presence of a G $\rightarrow$ A transition at the consensus splice-donor site at the exon 12/intron 12 boundary region. In the mutation terminology currently employed by the PAH Gene Mutation Analysis Consortium, the c $\rightarrow$ t missense mutation associated with haplotype 2 is termed R408W, where "R" is the single-letter amino acid code for the original arginine residue, 408 is the number of the codon containing the mutation, and "W" is the single-letter amino acid code for the mutant amino acid tryptophan. The mutation in the intron 12 splicing signal is termed IVS12nt1g $\rightarrow$ a, indicating that it is a single base g $\rightarrow$ a substitution occurring at the first nucelotide of the 12th intronic or intervening sequence in the PAH gene.

The introduction of the polymerase chain reaction (PCR) and its combined application with single-strand conformational polymorphism (SSCP) analysis, denaturing gradient gel electrophoresis (DGGE) analysis, or direct automated sequencing greatly facilitated the detection of PAH mutations. To date, more than 250 PAH mutations have been identified. Approximately half of these are missense mutations. Nonsense mutations account for another 10%. The remaining mutations are deletions or insertions of various types. The most common deletions are those of single exons caused by mutations in or near the splice donor or acceptor sites. Larger deletions are relatively rare. Most of the insertions involve a few base pairs inserted within the exonic regions of the gene, but at least two are induced by missense mutations within introns that produce cryptic splice sites. [*See* Mutation Rates.]

2. Expression Analyses

Many sequence alterations can be found in the PAH genes of normal and phenylketonuric individuals. What evidence is necessary to distinguish benign substitutions from disease-causing mutations? The first line of evidence is usually genetic: the putative mutation should segregate with the disease phenotype in PKU families, should never be detected in a screen of a large number of normal chromosomes, and should be the only substitution apart from proven polymorphisms observed in the entire gene. Such genetic evidence is usually adequate and easily obtained for common mutations. However, in some cases the genetic evidence is inconclusive. In all instances, an examination of the effect of the putative mutation on protein function is desirable. These examinations can be performed on proteins produced by the *in vitro* expression of mutant cDNAs created by site-directed mutagenesis.

In vitro expression analysis has now been performed on about 30 missense, splicing, or nonsense mutations and two single codon deletions. At least 15 of these mutations result in the expression of mutant PAH proteins with significant amounts of residual enzyme activity. The remaining characterized mutations result in the expression of mutant PAH proteins with little or no residual activity.

3. Genotype/Phenotype Correlations

The finding of substantial amounts of residual activity associated with some mutant PAH proteins led to the examination of the relationship between *in vitro* expression levels of mutant PAH proteins and the severity of the disease phenotype in patients bearing these mutations. Such studies have demonstrated statistically significant relationships between the level of PAH activity predicted from genotype and the biochemical and intellectual phenotypes of patients with varying degrees of PKU severity. From these studies, it is clear that the heterogeneity of PKU and HPA phenotypes can be largely explained by variability within the PAH locus.

IV. POPULATION GENETICS OF PHENYLKETONURIA

Although homozygosity for mutant PAH alleles should impair the ability to reproduce in the absence of some dietary treatment, PKU exists at a relatively high frequency in many populations. Mechanisms that could account for the continued prevalence of PKU include founder effect/genetic drift, heterozygote selection, reproductive compensation, elevated mutation rate, and the involvement of multiple loci that confer similar disease phenotypes.

A number of studies have suggested that the strong linkage disequilibrium observed between PAH mutations and specific haplotypes in different human populations is the result of founder effect and genetic drift. The most well-documented example of founder effect at the PAH locus is a deletion mutation involving exon 3. This mutation is the sole cause of PKU in the Yemenite Jewish population of Israel, and genealogical studies have traced this mutation back to a common ancestor who lived in San'a, the capital of Yemen, in the mid-18th century.

Unlike this relatively localized mutation, genealogical records alone cannot establish whether the present distribution of the major mutant PAH alleles in Europe and Asia may also be due, at least in part, to founder effect and genetic drift. Evidence supporting the existence of multiple founding populations for PKU in Europe and Asia comes primarily from the strong associations observed between mutations prevalent in these areas and specific RFLP haplotypes and the strong gradients in the relative frequencies of these mutations within Europe or Asia. For example, the frequency distribution of the common European PAH mutation R408W shows a strong east–west gradient. This mutation accounts for 50% or more of all mutant alleles in eastern European populations, where it is strongly associated with RFLP haplotype 2. This high frequency and strong haplotype association has led several investigators to propose a Slavic origin for this allele. However, R408W is also relatively common in Ireland and Scotland. In these populations, R408W is largely associated with RFLP haplotype 1. These data and others suggest a second, independent origin for R408W in Europe.

Although the existing evidence suggests a number of different founding populations in diverse ethnic and racial groups, the presence of so many mutant PAH alleles in different populations implies that founder effect and genetic drift alone cannot be responsible for the high incidence of PKU. Some form of heterozygote selection is likely. Although no definitive evidence has been presented, several ideas have been proposed. One form of selection might come from pressure exerted on a genetic locus closely linked to PAH. The proximity of the interferon-γ locus in region 12q24.1 provides some speculative basis for this hypothesis, but there is no direct evidence. A second form of selection might come from some compensatory effects brought about by the slight elevations in serum phenylalanine levels present in heterozygotes. A third possible mechanism of selection could involve reproductive compensation, mediated by either a higher rate of reproduction among heterozygotes or a higher survival rate among their offspring. If selection has acted to maintain mutant PAH alleles in European, Asian, and Middle Eastern populations where PKU is prevalent, then the mechanism of selection should be common to the different climatic, cultural, and dietary conditions that existed in these regions. Resolving the possible contributions of founder effect, genetic drift, and heterozygote advantage in the origins and subsequent distribution of mutant PAH alleles in human populations remains one of the largest challenges in PKU research. [*See* Population Genetics.]

ACKNOWLEDGMENT

This work was supported in part by NIH Grant HD-17711 to R. C. Eisensmith and S. L. C. Woo. S. L. C. Woo is also an investigator with the Howard Hughes Medical Institute.

BIBLIOGRAPHY

Eisensmith, R. C., and Woo, S. L. C. (1992). Molecular basis of phenylketonuria and related hyperphenylalaninemias: Mutations and polymorphisms in the human phenylalanine hydroxylase gene. *Hum. Mut.* **1**, 13.

Eisensmith, R. C., and Woo, S. L. C. (1995). Molecular genetics of phenylketonuria: From molecular anthropology to gene therapy. *In* "Advances in Genetics" (J. C. Hall, J. C. Dunlap, T. Friedmann, and F. Gianelli, eds.), Vol. 32. Academic Press, San Diego.

Güttler, F. (1980). Hyperphenylalaninemia: Diagnosis and classification of the various types of phenylalanine hydroxylase deficiency in childhood. *Acta Paediatr. Scand. Suppl.* **280**, 7.

Scriver, C. R., Eisensmith, R. C., Woo, S. L. C., and Kaufman, S. (1994). The hyperphenylalaninemias of man and mouse. *Annu. Rev. Genet.* **28**, 141.

Scriver, C. R., Hoang, L., Byck, S., and Prevost, L. (eds.), (1995). "PAH Mutation Analysis Consortium Database," McGill University Press, Montreal.

Scriver, C. R., John S. W. M., Rozen, R., Eisensmith, R., and Woo, S. L. C. (1993). Associations between populations, PKU mutations and RFLP haplotypes at the PAH locus: An overview. *Dev. Brain Dys.* **6**, 11.

Scriver, C. R., Kaufman, S., Eisensmith, R. C., and Woo, S. L. C. (1995). The hyperphenylalaninemias. *In* "The Metabolic and Molecular Bases of Inherited Disease" (C. R. Scriver, A. L. Beaudet, W. S. Sly, and D. Valle, eds.), 7th Ed., Vol. 1. McGraw-Hill, New York.

pH Homeostasis

DIETER HÄUSSINGER
Heinrich-Heine University of Düsseldorf

GLOSSARY

Acids, bases Most commonly, an acid is a proton donor, and a base is a proton acceptor

Buffer Solution of a conjugate acid–base pair capable of ameliorating the effect of added acid or base on pH in that solution

Buffer strength In its differential form, defined as the ratio $\frac{dB}{dpH}$, with dpH reflecting the increase of pH following the addition of a small amount of base (dB); the highest buffer strength of a buffer system is found when the pH of the solution equals the pK value of the buffering conjugate acid–base pair

Conjugate acid–base pair Acids dissociate into a proton and their conjugate base, whereas proton binding to a base leads to the formation of its conjugate acid

pH Initially defined as the negative decadic logarithm of the free proton concentration; the modern pH scale is only a relative measure of the proton's chemical potential; thus, the inverse antilog of pH represents the proton concentration in a first approximation. The term "free proton concentration" is convenient and widely used; however, physicochemically the free H^+ has a short lifetime and is present predominantly as the hydronium ion:

$$H^+ + H_2O \rightleftharpoons H_3O^+$$

BIOLOGICAL COMPOUNDS SUCH AS PROTEINS have multiple acid or base groups; the degree of their protonation is critical for their function. The different intra- and subcellular compartments exhibit a characteristic and remarkably constant proton concentration, i.e., pH, within narrow limits, although numerous reactions taking place in these compartments will consume or produce protons. Such compartmental homeostasis of pH is achieved by a variety of H^+ or HCO_3^- pumps and exchangers, which are located in the bordering membranes. These transport proteins not only stabilize the compartmental pH but also can build up differences of proton concentrations at the two sides of a biological membrane, generating electrochemical proton gradients. The pH in the various compartments differs, and the functions of the proton concentration and the transmembrane proton concentration gradients are variable. For example, the low lysosomal pH allows enzymatic hydrolysis, and pH gradients are used in endosomes for the sorting of receptor–ligand complexes and serve as the driving force for the transport of biogenic amines in chromaffin granules. The so-called proton motive force, i.e., the electrochemical proton gradient across the inner mitochondrial membrane, drives adenosine triphosphate (ATP) synthesis. However, the free proton concentration is a physiologically important regulator of biological processes. Small, controlled changes of the cytoplasmic pH under the influence of growth factors and hormones may be important in normal and malignant growth. Mechanisms responsible for the maintenance of physiological proton concentrations in cellular and subcellular compartments (intracellular pH homeostasis) depend on the proton concentration in the external environment. For mammals, control of the proton concentration in the extracellular fluid (extracellular pH homeostasis) has become systemic, i.e., affects the whole body. Specialized cell types have evolved for the regulation of extracellular systemic pH by the interplay of different organs, mainly the lung, liver, and kidney. Thus, homeostasis of pH evidently does not mean an absolute constancy of pH but can be defined as the integration of events allowing the free proton concentration in the various compartments to play its role as a physiological coordinator of biological processes.

ENCYCLOPEDIA OF HUMAN BIOLOGY, Second Edition, VOLUME 6.

I. INTRACELLULAR pH HOMEOSTASIS

A. General Considerations

Because cells consist of different compartments with different free proton concentrations, the term intracellular pH (pH_i) describes more or less an average pH of many compartments. Accordingly, pH_i depends on the technique employed for its determination and represents an experimentally determined cellular pH, which will be an average of the pH of those compartments reached by the respective pH_i-determining technique. In many cases, pH_i will come close to the cytosolic pH; then intra- and extracellular pH differences will approximate the pH gradient across the plasma membrane. All techniques for pH_i assessment yielded transmembrane pH gradients far below those predicted theoretically by the Nernst equation, indicating that the distribution of H^+ across the plasma membrane is not in electrochemical equilibrium.

$$V_m = R \cdot T \cdot F^{-1} \ln([H^+]_e/[H^+]_i), \quad \text{(Nernst equation)}$$

where V_m is the voltage difference across the membrane in volt, R is the gas constant, T is the absolute temperature, F is Faraday's constant, and subscripts $_i$ and $_e$ refer to intracellular and extracellular, respectively. At 37°C, this equation becomes, in terms of pH,

$$V_m = 0.061(pH_i - pH_e).$$

Thus, with a membrane potential of −60 mV (inside negative) and pH_e of 7.4, the electrochemical equilibrium would predict a pH_i of 6.4, whereas actually pH_i values of about 7.0 are measured. Thus, cells must have mechanisms for H^+ extrusion, which keep the pH_i at values above equilibrium. Many of these pH_i-controlling systems are biochemically or functionally characterized; they underly a complex regulation that provides the basis for a control and coordination of biological processes by the proton concentration.

B. Methods for Assessment of pH_i and Subcellular pH

1. Weak Acid or Base Distribution

This technique for the determination of pH_i relies on the permeability of biological membranes for the uncharged, but not for the charged, species of a metabolically inert weak acid–base pair. After equilibration between the intra- and extracellular space, the concentrations of the ideally freely diffusible, uncharged species are equal in both compartments, whereas the concentrations of the charged corresponding acid or base are determined by the proton concentrations on either side of the membrane. For the weak acid HA, which dissociates $A^- + H^+$, the distribution of A^- between the intra- and extracellular space is inversely related to that of H^+:

$$[H^+]_i/[H^+]_e = [A^-]_e/[A^-]_i.$$

Employing the weak acid HA as an indicator (i.e., at a concentration sufficiently low to avoid significant pH_i poisoning), pH_i can be calculated from pH_e, the pK value of the indicator system HA/A^-, and the concentrations of dissociated and undissociated indicator $[HA + A^-]$ in the intra- and extracellular space:

$$pH_i = pK + \log[(10^{pH_e - pK} + 1) \times ([A^-]_i + [HA]_i)([A^-]_e + [HA]_e)^{-1}]$$

The most commonly used pH_i indicator is the weak acid 5,5-dimethyloxazolidine-2,4-dione (DMO). DMO is frequently employed as a ^{14}C-labeled compound, and the distribution of radioactivity between the extracellular space and tissue allows the determination of pH_i after correction for intra- and extracellular water spaces. The technique is comparatively easy to handle; however, repetitive or continuous determinations of pH_i are not possible. pH_i values determined by means of weak acid–base indicators will represent some average of the subcellular compartments among which the indicator is distributed. Weak acid–base pairs can also be used to determine the pH in subcellular organelles such as mitochondria, isolated by suitable cell subfractionation techniques.

2. Fluorescent Dyes

This technique employs an indicator whose spectrofluorometric properties are a function of the environmental pH. This indicator is loaded into the cells, which are excited by light of a suitable wavelength, while monitoring pH-dependent absorbance or fluorescence at another wavelength. Because the spectrofluorometric characteristics of a dye are different in extra- and intracellular environments, the system requires calibration. Indicator compounds frequently used are fluorescein derivatives. Biscarboxyethylfluorescein (BCECF), which is not electrically charged, rapidly permeates the plasma membrane and is trapped inside the cells after hydrolysis by intracellular esterases, yielding the active pH-sensitive compound carboxyfluoresceine. This technique allows continuous pH_i registration and is applicable to single and/or small cells and isolated organelles. Time-dependent bleaching and leakage of the dye out of the cells are the drawbacks.

3. pH-Sensitive Microelectrodes

The difference of electrical potential between a pH-sensitive (glass or liquid ion exchange) microelectrode and a reference microelectrode with very small tips ($<0.2\ \mu m$), both introduced into a single cell, is used to measure pH_i. The electrodes can be placed separately into adjacent, electrically coupled cells or can be combined to one double-barreled microelectrode. pH_i values determined by this technique most closely approach the cytosolic pH and allow the continuous measurement of pH_i. This technique is limited to cells large enough to sustain microelectrode puncture.

4. Nuclear Magnetic Resonance Spectroscopy

Because the nuclear magnetic resonance (NMR) frequency of endogenous inorganic phosphate is dependent on the $H_2PO_4^-/HPO_4^{--}$ ratio (so-called chemical shift), the inorganic phosphate peak of the ^{31}P-NMR spectrum can be used to determine pH_i. "Pure" inorganic phosphate signals from the intracellular compartment can be obtained. How much subcellular compartments can be distinguished is not fully resolved and depends on the tissue: whereas in liver ^{31}P-NMR probably gives whole cell, rather than cytosolic pH, in heart muscle it largely reflects the cytosolic pH. In addition to ^{31}P-NMR, ^{1}H-NMR and ^{19}F-NMR also have been employed to study pH_i, using the pH-dependent ^{1}H chemical shift on histidine and the pH-dependent distance of the two ^{19}F peaks of difluoromethylalanine. A major advantage of the NMR technique is that the relationship between pH_i and metabolic processes can be studied *in vivo*. [*See* Magnetic Resonance Imaging.]

C. Mechanisms and Regulation of Proton Translocation Across Biological Membranes

In principle, two mechanisms for proton transport across biological membranes can be distinguished: electroneutral exchange mechanisms (Na^+/H^+ antiport, HCO_3^-/Cl^- exchange, K^+/H^+ ATPase) and electrogenic proton pumps (H^+ ATPases). These transporters are responsible for the maintenance of pH in the respective compartments and the generation of electrochemical proton gradients. There is also nonionic diffusion. Many weak acids and bases rapidly equilibrate across the membrane by nonionic diffusion of their uncharged species. In the case of weak acids, such as many monocarboxylic acids, diffusion into the cell is accompanied by an import of protons. Another example is CO_2, which rapidly diffuses across membranes into cells, where it is hydrated to H_2CO_3, which rapidly dissociates to H^+ and HCO_3^-, leading to a fall of pH_i. When uncharged species of weak bases such as NH_3 enter the cell, a rise of pH_i occurs due to protonation of NH_3 yielding NH_4^+. [*See* Ion Pumps.]

D. How Cells Respond to an Intracellular Acid Load

When the interior of a cell is subjected to an acid load, pH_i initially drops and recovers with time to the initial pH_i value. Several mechanisms contribute to this pH_i homeostatic response: physicochemical buffering, adaptations of cellular metabolism, stimulation of proton extrusion out of and bicarbonate import into the cell, and probably, to a minor extent, a concentrative shift of protons from the cytosol into subcellular organelles.

Physicochemical buffering is the most rapid response to an intracellular acid load and represents proton binding to various intracellular bases according the mass action law with formation of the corresponding acids. Physicochemical buffering depends on the buffer strength of the cell interior (normally about 70 mmol/liter) and determines the maximal deviation of pH_i following an intracellular acid or alkali load. Part of the intracellular acid load is also removed from the cytosol by H^+ uptake into various organelles (H^+ ATPases of the vacuolar system). Simultaneously, H^+ extrusion from other organelles such as mitochondria decreases. Such net proton shifts into subcellular organelles represent another buffering mechanism with respect to cytosolic pH (organellar buffering); however, its quantitative importance is yet unclear. Buffering mechanisms ameliorate pH changes but can neither prevent the pH_i change nor return pH_i to its original level. Return to normality is brought about by H^+ extrusion from the cell via Na^+/H^+ exchange and/or HCO_3^- import into the cell via Cl^-/HCO_3^- exchange mechanisms in the plasma membrane. These transporters are activated when pH_i

falls. During this process, H^+ previously bound to intracellular buffers is released again according to the mass action law; the entire acid load is finally extruded until pH_i and the cellular buffering mechanisms are returned to their initial state.

As long as pH_i is below its normal value, alterations of cellular metabolism accompany the pH_i recovery phase. They can be seen as metabolic adaptations to the need of proton removal during an intracellular acidic challenge; the opposite happens in an alkaline challenge. Because protons are formed or consumed in a variety of metabolic reactions, an inhibition of H^+ formation and/or a stimulation of H^+ consumption provides a pH_i homeostatic response. One example is the inhibition of H^+ producing glycolysis in acidosis due to a 90% inhibition of phosphofructokinase activity when the pH_i falls from 7.2 to 7.1. Further, lowering of the steady-state concentrations of organic mono-, di-, and tricarboxylates consumes H^+: their tissue levels drop following a decrease of pH_i. Metabolic adaptations for pH_i homeostasis may exhibit organ-specificity: in the liver, a decrease of pH_i inhibits urea synthesis, a liver-specific pathway consuming HCO_3^- (or producing H^+), which plays a major role in whole-body pH homeostasis (see also Section II,C).

E. Biological Significance of pH_i and Subcellular pH Homeostasis

The free proton concentration in subcellular compartments is a physiologically important regulator of a variety of biological processes. pH values of different subcellular compartments are given in Table I. Only a few examples are given below.

TABLE I
pH Values in Different Compartments

Compartment	pH
Extracellular	7.36–7.44
Cytosol	6.9–7.2
Mitochondria	7.4–7.7
Lysosomes	4.5–5.4
Endosomes	about 5
Secretory vesicles	
Parotid gland	6.8
Pancreatic islet cells	5.0

1. The pH_i Signaling Hypothesis

Activation of Na^+/H^+ exchange under the influence of a variety of mitogens and growth factors was found to increase cytosolic pH from its normal value of 6.9–7.2 by up to 0.4 pH units. This persistent rise in pH_i was seen as an early signal, which triggers proliferation and cell growth. Indeed, inhibitors of Na^+/H^+ exchange blocked the proliferative response to many growth factors. Furthermore, mutant fibroblasts, lacking Na^+/H^+ exchange activity, which do not show an increase of pH_i following mitogen treatment, have an impaired activation of DNA synthesis. However, in many cases a proliferative response to mitogens and growth factors does not seem necessarily dependent on Na^+/H^+ exchange activity or a rise in pH_i; this is especially true if bicarbonate is present in the experimental systems. Thus, a rise in pH_i has probably only a permissive role for the initiation of cell growth and differentiation in conjunction with other important signals. Stimulus activation of Na^+/H^+ antiport under these conditions may not only trigger but also act to counter an increased intracellular proton formation accompanying metabolic hyperactivity during proliferative responses. Elevations of pH_i may also play a role in malignant transformation of cells: tumor-promoting phorbol esters stimulate Na^+/H^+ exchange probably via activation of protein kinase C, growth factor-like substances are produced by a number of transformed cell lines, and the oncogene product of the simian sarcoma virus shows marked homology with platelet-derived growth factor. Some evidence indicates that activation of Na^+/H^+ exchange is involved in tissue regeneration and organ hypertrophy.

A rise in pH_i following hormonal activation of Na^+/H^+ antiport will stimulate phosphofructokinase and could, in part, explain the stimulation of glycolysis by Ca^{2+}-mobilizing hormones such as α-adrenergic agonists or vasopressin. [*See* Glycolysis.]

In neutrophils, the action of chemotactic peptides is blocked by inhibition of the Na^+/H^+ exchanger. Furthermore, manipulation of pH_i in neutrophils modulates their chemotactic responsiveness. The stimulus-activated production of partially reduced reactive oxygen species by neutrophils and macrophages, such as O_2^- or O_2^{--}, is accompanied by a considerable intracellular acidification, which is counteracted by Na^+/H^+ exchange. Also, some evidence supports a regulatory role of the Na^+/H^+ exchanger on the activation of O_2^--producing membrane-bound reduced nicotinamide adenine dinucleotide phosphate oxidase in neutrophils. [*See* Neutrophils.]

2. pH_i and Membrane Potential

In a variety of tissues, a rise of pH_i (intracellular alkalosis) leads to a hyperpolarization of the cell membrane potential due to an increase of K^+ conductance by opening pH-sensitive, Ca^{2+}-activated K^+ channels. Conversely, intracellular acidosis leads to a depolarization. The effects of pH_i on membrane potential in turn affect Na^+-coupled metabolite transport into cells, as well as electrogenic bicarbonate transport. Thus, in the basolateral membrane of the proximal kidney tubule depolarization leads to an intracellular alkalinization, and hyperpolarization acidifies the cell interior. Intracellular acidification affects the electrical coupling between epithelial cells through gap junctions, causing uncoupling due to pH sensitivity of gap-junction channels.

3. pH_i and Regulation of Cell Volume

When cells are exposed to anisotonic media, osmotic water fluxes lead to cell swelling or shrinkage. Subsequent activation of volume-regulatory mechanisms restore the initial cell volume within minutes. A regulatory volume decrease following osmotic cell swelling is due to an increased K^+ conductance in most cell types. A regulatory volume increase following cell shrinkage in hypertonic media is amiloride-sensitive in many cell types such as rat liver cells, dog erythrocytes, or human lymphocytes. Parallel activation of Na^+/H^+ and of Cl^-/HCO_3^- exchange in these cells mediates regulatory volume increase resulting in a net gain of Na^+, Cl^-, and H_2O inside the cell.

4. Regulation of Intracellular Membrane Flow

The vacuolar system comprises the endoplasmic reticulum, the Golgi apparatus, endosomes, lysosomes, secretory vacuoles, and other membrane-bordered structures participating in transport and processing of soluble or membrane-bound macromolecules. This system is dynamic and presents a constant flow of membrane among the different structures. Many components of the vacuolar system exhibit an acidic internal pH, created by H^+ pumps, which is an important prerequisite for proper functioning of the vacuolar system.

The low intralysosomal pH provides not only optimal conditions for the activity of lysosomal hydrolases but may also serve as a signal for the activation of several lysosomal proteases. For example, cathepsin D is transported to the lysosome as a catalytically inactive precursor but is autocatalytically activated by the low intralysosomal pH through proteolytic cleavage. Endosomes represent an intermediary, prelysosomal compartment involved in the transport of internalized insoluble or soluble material or receptor–ligand complexes from the plasma membrane to the lysosomes. In these vesicles, sorting of receptor–ligand complexes takes place under low pH, which favors receptor–ligand dissociation; this determines the fate of the receptor and the ligand (either recycling or lysosomal degradation). The acidic endosomal pH, however, is only one signal for sorting and routing; other factors include Ca^{2+} and ligand valency. The low endosomal pH also plays a role in targeting lysosomal proteins from the Golgi apparatus to the lysosomes. Newly synthesized acid hydrolases bear phosphomannosyl groups that bind to the membrane-bound mannose-6-phosphate receptors at neutral pH but rapidly dissociate at acidic pH. Such dissociation occurs in the acidic compartment of uncoupling of receptors and ligands (CURL); the mannose-6-phosphate receptor recycles to the Golgi apparatus, whereas the hydrolases reach lysosomes. Acidotropic agents, such as chloroquin or methylamine, perturb the intravesicular acidic pH and divert precursors of lysosomal enzymes to a secretory pathway out of the cell. Also, some evidence indicates that acidic pH is involved in the accumulation of proteins to be secreted in secretory vesicles.

II. EXTRACELLULAR pH HOMEOSTASIS

A. General Considerations

The pH in various intracellular compartments critically depends on the pH in the cellular environment, i.e., the extracellular fluid (pH_e) in higher organisms. This is because intra- and extracellular pHs are linked by several processes that transport protons or bicarbonate either actively (H^+ pumps and Na^+/H^+ and HCO_3^-/Cl^- exchange) or passively (nonionic diffusion, passive H^+, and HCO_3^- movements due to electrochemical gradients). By these mechanisms, changes in pH_3 will affect pH_i in the same direction.

The bicarbonate buffer is of special importance for homeostasis of pH_e. In the extracellular space, the bicarbonate buffer is in chemical equilibrium with other buffer systems (collectively referred to as nonbicarbonate buffers, such as phosphate and proteins). According to the Henderson–Hasselbalch equation, the pH of a bicarbonate buffer is determined by the ratio of concentration of bicarbonate to that of physically dissolved CO_2, which is the product of the physical solubility of CO_2 and the partial CO_2 pressure:

$$pH_e = pK' + \log([HCO_3^-]_e/[CO_2]_e).$$

In organisms, the concentrations of HCO_3^- and CO_2 can be effectively regulated by excretion or metabolic disposal. The open system maintaining a steady state of $[CO_2]$ substantially increases operational buffer capacity. A physiological pH_e of about 7.4 is maintained by a coordinate interplay of specialized cells, mainly in lung, kidney, and liver. These cells act to stabilize not only the extracellular $[HCO_3^-]/[CO_2]$ ratio (and accordingly pH_e) but also the absolute concentrations of CO_2 and HCO_3^- in the extracellular space, despite the fact that CO_2 and HCO_3^- are continuously generated during the respiratory oxidation of foodstuff; thus, constancy of pH_e is largely brought about by finely tuned mechanisms disposing of CO_2 and HCO_3^- from the organism at the same rate as they are generated. Carbonic anhydrases accelerate CO_2/HCO_3^- equilibria in the various compartments and guarantee sufficiently rapid transfer of CO_2 between tissues and extracellular fluid on the one hand and the extracellular fluid and the external environments on the other.

B. CO_2/HCO_3^- Equilibria and Transport between Tissues: The Role of Carbonic Anhydrases

The following reactions are involved in the equilibrium of CO_2 and the HCO_3^- system:

$$CO_2 + H_2O \rightleftharpoons H_2CO_3 \qquad \text{(reaction I)}$$

$$H_2CO_3 \rightleftharpoons H^+ + HCO_3^- \qquad \text{(reaction II)}$$

Reaction II is very fast and reaches chemical equilibrium almost instantaneously, whereas reaction I is slow: at 37°C and a physiological pH of 7.4, the uncatalyzed hydration–dehydration reactions will be half complete within 3.6 sec after a step concentration change of one of the reaction components. Because the blood transit time in capillaries (where gas exchange takes place) is <1 sec, carbonic anhydrates (CA) activities are required to accelerate formation of HCO_3^- from CO_2 in peripheral capillaries and of CO_2 from HCO_3^- in lung capillaries.

Five different CA isoenzymes (CA I–V) (Table II) facilitate CO_2/HCO_3^- equilibration and transport across membranes or supply HCO_3^- for biosynthetic processes. They are distinguished with respect to their sensitivity to sulfonamide inhibition, serological characteristics, and tissue distribution. Some examples on the physiological roles of CA will be given. CO_2 produced in peripheral tissues rapidly diffuses to the blood and equilibrates with the plasma and the red cell interior. High CA activities in the erythrocytes guarantee rapid equilibration of the CO_2/HCO_3^- system inside them. The accompanying intracellular acidification produces an allosteric change of hemoglobin, which favors tissue oxygenation (Bohr effect). Electroneutral HCO_3^-/Cl^- exchange mediates the exit of HCO_3^- to the plasma. Whereas inside the erythrocytes, CO_2/HCO_3^- equilibria are reached during capillary transit in the presence of CA, this does not necessarily apply for the plasma CO_2/HCO_3^- system, especially in tissues lacking extracellular membrane-bound CA, such as liver. Membrane-bound CA present in capillaries of skeletal muscle accelerates CO_2/HCO_3^- equilibration in plasma so that equilibrium is reached already in end capillary blood. The processes occurring during gas exchange among blood, plasma, and muscle tissue are a mirror image of those taking place during the gas exchange in alveolar capillaries. In the lung, besides a membrane-bound CA on the luminal surface of lung capillaries, an intracellular CA has been described that participates in alveolar CO_2 exchange by facilitating CO_2 diffusion.

C. Organ-Specific Contributions to pH_e Homeostasis

pH homeostasis in the extracellular space is brought about by a complex interplay between specialized organs, mainly lung, liver, and kidney. Their coordinate action adjusts pH_e close to 7.4 by regulating the extracellular CO_2 and HCO_3^- concentrations. These organs have sizeable excretory or metabolic pathways for irreversible elimination of CO_2 or HCO_3^- (or of constituents of other extracellular buffers in chemical equilibrium with the HCO_3^-/CO_2 system). These pathways, in turn, are controlled by pH_e, $[CO_2]_e$, and $[HCO_3^-]_e$ so that there is a feedback control circuit between pH_e and CO_2/HCO_3^- elimination.

Clearly, any organ can influence pH_e by either release or uptake of acids or bases into or from the extracellular space. These processes, however, must be viewed as an integral part of overall metabolic fluxes. Their potential effects on pH_e are usually bal-

TABLE II
Carbonic Anhydrases (CA)

Isoenzyme	Localization	Tissue	K_i (acetazolamide)
CAI	Cytoplasmic	Lung, erythrocytes, brain, colon, cornea	2×10^{-7} *M*
CA II	Cytoplasmic or membrane bound	Lung, erythrocytes, white skeletal muscle, liver, kidney, bone, brain, stomach, small intestine, lung	10^{-8} *M*
CA III	Cytoplasmic	Red skeletal muscle, liver, erythrocytes	3×10^{-4} *M*
CA IV	Membrane bound	Skeletal muscle, kidney, lung, brain	10^{-6}–10^{-7} *M*
CA V	Mitochondrial	Liver, kidney	10^{-7} *M*

anced by proton fluxes in opposite directions. One example is gastric acid secretion, which transiently renders the blood somewhat alkaline (alkaline tide) until it is compensated by pancreatic HCO_3^- secretion. Similarly, during severe exercise, lactic acid production by the muscle may result in transient extracellular acidosis due to the accumulation of lactate plus H^+ in the extracellular space but spontaneously resumes following the removal of lactate via gluconeogenesis or oxidation, i.e., processes consuming protons. Normally, the steady-state concentrations of acidic or basic metabolic intermediates in the extracellular space remain constant because flux rates through reactions generating protons are matched by proton-consuming reactions of equal magnitude (biochemical steady state). Under pathological conditions only, imbalances of these metabolic proton fluxes give rise to acid–base disturbances (e.g., when the oxidative pathway from glucose to CO_2 is interrupted during hypoxia, whereas glycolytic pyruvic and lactic acid formation continues). Although normally intermediate metabolite concentrations remain remarkably stable despite wide variations of flux through the whole metabolic pathway, the amount of acidic or basic end products formed during the metabolism of carbohydrate, fat, and protein is variable and provides a continuous threat to acid–base homeostasis. In man, these end products include CO_2 (12–20 mol/day), HCO_3^- (about 1 mol/day), NH_4^+ (about 1 mol/day), H_2SO_4 (10–20 mmol/day), and $H_2PO_4^-$ (10–20 mmol/day). Elimination of these metabolic end products at the same rate as they are generated to achieve a biochemical steady state is the hallmark of pH_e homeostasis. Interestingly, major pH homeostatic mechanisms are frequently linked to other vital functions, explaining at least in part a limitation of pH homeostatic responses. For example, respiration links CO_2 elimination to oxygen uptake. Under extreme conditions the need for sufficient oxygen uptake may conflict with hyperventilatory CO_2 exhalation. Similarly, bicarbonate homeostasis is closely linked to ammonium homeostasis, and under certain conditions ammonium toxicity can only be avoided for the price of acid–base disturbances.

1. Role of the Lungs

The lungs are the major organ for irreversible elimination of CO_2; this process underlies a complex feedback control by the arterial pH and PCO_2. Pulmonary CO_2 excretion is regulated by alveolar ventilation ($\dot{V}_A$), which may rise in humans from 5 liters/min at rest to 50 liters/min during heavy exercise. The relationship between $\dot{V}_A$, the arterial partial pressure of CO_2 ($PaCO_2$), that in inspired air ($PiCO_2$) and metabolic CO_2 production $\dot{M}_{CO_2}$ is described by the following equation:

$$PaCO_2 - PiCO_2 = \dot{M}_{CO_2}(\dot{V}_A \cdot \beta)^{-1},$$

where β is the capacitance coefficient (β = 0.0517 mmol/1 torr). In resting humans, $PiCO_2$ is close to 0, $\dot{M}_{CO_2}$ = 10 mmol/min, and $\dot{V}_A$ = 5 liters/min. Accordingly, a normal arterial PCO_2 = 40 torr results. Increases in $\dot{M}_{CO_2}$ or $PiCO_2$ lead to an increase of $PaCO_2$ and a decrease in pH_e when $\dot{V}_A$ remains unchanged. Changes of arterial pH and PCO_2 are sensed by central (medulla oblongata) or peripheral (carotid and aortic bodies) chemoreceptors, resulting in an increased impulse rate of afferent nerve fibers from the chemoreceptors to the respiratory centers in the brain stem. This signal stimulates the efferent output of motoneurons to respiratory muscles and increases alveolar ventilation until arterial PCO_2 levels and pH are returned to normal. Thus, a control circuit with a negative feedback loop exists for the stabilization of arterial PCO_2. The adequate stimulus of the chemoreceptors appears to be the pH rather than the PCO_2;

in addition, arterial (peripheral) chemoreceptors are sensitive to changes in the arterial oxygen concentration. The pH sensitivity explains why $PaCO_2$ levels become subnormal when the blood bicarbonate concentration is lowered, stabilizing the pH_e close to 7.4. This situation exists, for example, during severe exercise when plasma bicarbonate decreases due to net lactic acid production by the muscle and the arterial PCO_2 are adjusted by ventilation to values below 40 torr. Good evidence indicates that ventilation control by signals from chemoreceptors is supplemented by afferent impulses from muscle or joint mechanoreceptors and from collaterals from cortical motor centers. [*See* Respiratory System, Physiology and Biochemistry.]

2. Role of the Kidneys

The mechanisms by which renal function contributes to homeostasis of pH_e include the reabsorption of filtered bicarbonate from the tubular fluid, net excretion of protons or bicarbonate, which can be assessed by titration of the final urine back to the pH of the extracellular fluid and participation in the coordinate regulation of bicarbonate and ammonium homeostasis in an interorgan team effort with the liver.

The glomerular filtrate contains bicarbonate at a concentration equal to that in plasma, i.e., normally about 25 m*M*. The reabsorption of filtered bicarbonate largely occurs in the proximal tubule involving luminal Na^+/H^+ exchange, and ATP-driven H^+ secretion, membrane-bound luminal and cytosolic CAs, and a basolateral Na^+ coupled HCO_3^- transport. *In vitro* proximal tubular H^+ secretion underlies complex control by hormones (parathyroid and thyroid hormones, glucocorticoids) and the acid–base status. In acidosis, proximal tubular bicarbonate reabsorption is increased by activation of Na^+/H^+ exchange and insertion into the luminal membrane of ATP-driven H^+ pumps usually stored in endocytotic vesicles. The distal tubule largely determines the final urinary pH by increasing acidity through an electrogenic H^+ ATPase located in the luminal membrane. Proton secretion into the lumen is accompanied by the intracellular formation of HCO_3^-, which leaves the cell across the basolateral membrane via a Cl^-/HCO_3^- exchange. The cortical collecting duct can exhibit bicarbonate absorption or secretion depending on the acid–base status. Alkalosis stimulates bicarbonate secretion, whereas acidosis stimulates bicarbonate reabsorption. Two types of intercalated cells are involved in this process. Type A cells absorb HCO_3^- and contain an apical H^+ ATPase and a basolateral Cl^-/HCO_3^- exchanger. An opposite polarity of these transport systems in type B cells allows bicarbonate secretion. Upon an acute decrease of pH in the cell, new H^+ pumps are inserted into the apical cell membrane of type A cells by migration from internal membranes via an exocytotic mechanism; whether or not an increase of cell pH induces endocytotic removal of the H^+ pumps is yet unclear. Chronic adaptation to acidosis includes an increase of H^+ secreting type A cells. Some evidence indicates that the acid–base status may regulate the interconversion of type A and B cells, and both cell types may represent a different functional state of the same cell. As a pH homeostatic response, interconversion of type A and B cells could then change from bicarbonate secretion to absorption or vice versa.

The pH in urine normally varies between 5 and 8, and the net amount of protons excreted or retained by the kidney can be estimated by adding measured amounts of acid or alkali to the urine (titration) until its pH reaches the same value as plasma pH. This titratable acid excretion is normally about 30–50 mmol/day and largely reflects protons bound to urinary buffers such as phosphate. This is the direct net effect of the kidneys on pH_e homeostasis; this seems a small contribution, especially when comparison is made with CO_2 excretion by the lungs (10–20 mol/day) or with HCO_3^- disposal during hepatic urea synthesis (about 1 mol/day; vide infra). When bicarbonate is infused, however, urinary bicarbonate excretion can provide a spillover for excess bicarbonate, which cannot be eliminated metabolically (see later). Above a certain threshold concentration of bicarbonate in plasma, further increases result in increased urinary bicarbonate excretion proportional to the amount of filtered bicarbonate. This bicarbonate threshold concentration varies with the PCO_2; its increase in respiratory acidosis (increased steady-state PCO_2 in chronic lung disease to maintain normal daily CO_2 exhalation) allows an increase of the plasma $[HCO_3^-]$ to maintain normal pH.

There has been some confusion on the role of the kidneys in pH homeostasis regarding their role in ammonium excretion (so-called renal ammoniagenesis). In physiology, NH_4^+ excretion into urine was viewed as net proton excretion (so-called nontitratable acid). Although NH_4^+ is an acid according to the Bronstedt–Lowry definition, it is a very weak acid ($pK = 9.1$), which at physiological pH values dissociates protons in only minimal amounts. Thus, NH_4^+ movements from plasma to urine are per se without a relevant effect on plasma pH. The role of renal ammoniagenesis, therefore, required reinterpretation on the basis of a close link of ammonium and bicarbonate homeostasis in the body.

3. Coordinate Regulation of Bicarbonate and Ammonium Homeostasis by Liver and Kidneys

Whereas CO_2 and H_2O are the only end products of fat and carbohydrate metabolism, the complete oxidation of proteins or bipolar amino acids at a physiological pH also yields HCO_3^- and NH_4^+ as well as small amounts of H_2SO_4 (derived from sulfur-containing amino acids). Due to the chemical structure of an "average" amino acid, HCO_3^- and NH_4^+ are formed in almost equimolar amounts, i.e., about 1 mol/day each in a person oxidizing 100 g protein/day. The major route for elimination of this metabolically generated HCO_3^- is urea synthesis. This liver-specific pathway can be viewed chemically as an irreversible, energy-driven neutralization of the strong base HCO_3^- (pK = 6.1 for the CO_2/HCO_3^- system) by the weak acid NH_4^+ (pK = 9.1 for the NH_4^+/NH_3 system:

$$2HCO_3^- + 2NH_4^+ \xrightarrow{\text{energy}} H_2NCONH_2 + CO_2 + 3H_2O$$

Accordingly, urea synthesis consumes HCO_3^- and NH_4^+ in the same stoichiometry as these compounds arise during protein breakdown. The daily average excretion of 30 g urea in humans corresponds to an irreversible bicarbonate elimination of about 1 mol/day. Hepatic urea synthesis (and accordingly the rate of irreversible HCO_3^- consumption in the liver) reveals a complex and sensitive feedback control by the extracellular pH, [HCO_3^-] and [CO_2]. Acid–base control of urea synthesis occurs at the level of mitochondrion: the formation of NH_4^+ and HCO_3^- inside mitochondria by glutaminase and CA V is under pH control and determines the rate of synthesis of carbamoylphosphate, a major rate-limiting step of the urea cycle. Thus, when the urea cycle is inhibited in acidosis, the sparing of bicarbonate is a direct pH–homeostatic response of the liver. A sophisticated structural and functional organization of the pathways of urea synthesis and glutamine metabolism in the liver acinus allows modulation of the urea cycle according to the requirements of acid–base homeostasis without threat of hyperammonemia. This is because in the liver acinus ammonia-rich blood from the viscera first gets into contact with cells capable of urea synthesis (but not glutamine synthesis). These cells take up as much NH_4^+ as is needed for acid–base-controlled HCO_3^- disposal via the urea cycle. NH_4^+ that is not utilized by ureogenic cells is washed to a more downstream, small hepatocyte population, which is located just before the venous outflow and is devoid of urea cycle enzymes. These cells take up NH_4^+ with high affinity and synthesize glutamine. Thus, an increase in hepatic glutamine formation maintains nontoxic ammonium levels when urea synthesis is switched off in acidosis to spare bicarbonate: hepatic NH_4^+ detoxication is shifted from urea to glutamine synthesis. The final elimination of surplus NH_4^+ is provided by renal ammoniagenesis. In this process, glutaminolysis in the kidney liberates NH_4^+, which is excreted into the urine. Thus, renal ammoniagenesis provides a spillover for elimination of NH_4^+, without concomitant HCO_3^- consumption, and may be viewed primarily as an ammonium homeostatic mechanism. Similar to hepatic urea synthesis, renal ammoniagenesis is sensitively regulated by the acid–base status: renal NH_4^+ excretion is stimulated in acidosis in parallel with the inhibition of bicarbonate-consuming urea synthesis in the liver. Thus, liver and kidney act as a team for maintaining ammonium and bicarbonate homeostasis. These relationships are schematically depicted in Fig. 1: a pH-controlled shift of NH_4^+ elimination from urea synthesis to urinary excretion maintains ammonium homeostasis in the whole body and provides the basis for an undisturbed feedback circuit between acid–base and hepatic HCO_3^- consumption. Failure of the renal ammoniagenesis indirectly leads to metabolic acidosis: the pH-sensitive control of urea synthesis is overridden by the need to eliminate potentially toxic NH_4^+, resulting in excessive stimulation of urea synthesis and concomitant HCO_3^- consumption. Under normal conditions, the production of NH_4^+ during protein oxidation exceeds HCO_3^- generation by about 40 mmol/day due to the neutralization of H_2SO_4, which is simultaneously formed from sulfur-containing amino acids. This explains the normal urinary NH_4^+ excretion of about 40 mmol/day, which can rise up to 400 mmol/day in severe metabolic acidosis, when the bicarbonate-consuming urea synthesis is strongly inhibited.

D. Disturbances of pH_e Homeostasis and Compensation

Disturbances of pH_e homeostasis occur when acid or alkali loads (or losses) overwhelm mechanisms of pH_e homeostasis or when pH_e homeostatic mechanisms

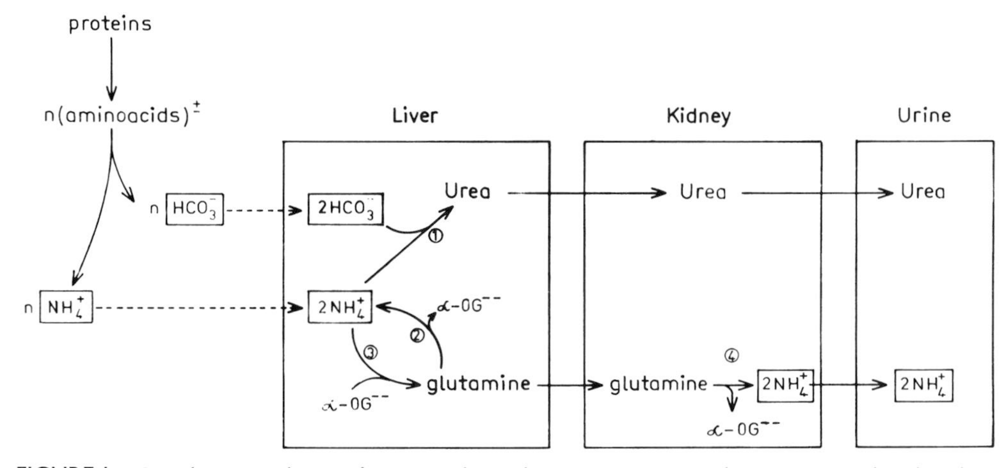

FIGURE 1 Coordinate regulation of HCO_3^- and NH_4^+ homeostasis. NH_4^+ and HCO_3^- are produced in almost equivalent amounts during protein breakdown. Whereas urea synthesis irreversibly consumes HCO_3^- and NH_4^+ in equal proportions, HCO_3^- is spared when urea synthesis is switched off in acidosis (bicarbonate homeostatic response of the liver). Under these conditions, NH_4^+ homeostasis is maintained by NH_4^+ excretion into urine (ammonium homeostatic response of the kidney). A feedback control loop between bicarbonate-consuming urea synthesis and the extracellular pH, $[CO_2]$ and $[HCO_3^-]$, adjusts hepatic HCO_3^- consumption to the acid–base status. Numbers in circles refer to major points of flux control by the extracellular acid–base status. In metabolic acidosis, flux through the urea cycle (reaction 1) and hepatic glutaminase (reaction 2) is decreased, whereas flux through hepatic glutamine synthetase (reaction 3) and renal glutaminase (reaction 4) is increased. (From Häussinger *et al.* (1986). *Biochem. J.* **236**, 261–265.)

are impaired due to pulmonary, hepatic, or renal dysfunction. pH_e disturbances are classified as acidosis (plasma pH <7.36) and alkalosis (plasma pH >7.44), both of which can be of metabolic, respiratory, or mixed respiratory–metabolic origin. In pure metabolic pH derangements, the primary disturbance is a fall or rise of plasma $[HCO_3^-]$. Respiratory compensation occurs by adjusting the P_{CO_2} to values restoring a near-physiological pH_e of 7.4, according to the Henderson–Hasselbalch equation. In pure respiratory pH derangements, the primary disturbance is a deviation of the arterial P_{CO_2} from its normal value (38–42 mm Hg), which is secondarily compensated by lowering or increasing the plasma $[HCO_3^-]$ (metabolic compensation). Respiratory disturbances can be compensated metabolically, and, vice versa, metabolic disturbances are compensated through respiration, with restoration of the pH_e to almost normal values; the compensation, however, will cause abnormalities of plasma $[HCO_3^-]$ and $[CO_2]$. Their restoration to the normal $[HCO_3^-]$ value of 25 mmol/liter is only achieved when the underlying primary disturbance is corrected. Compensation is achieved by an inhibition of urea synthesis (sparing of HCO_3^-) together with an increased renal ammoniagenesis (ammonium homeostatic response) and is quantitatively less important via increased net H^+ excretion by the kidneys (titratable acid). When compensation by these mechanisms fails, persistent severe metabolic acidosis results in compensatory bone resorption, which supplies base equivalents (carbonate, phosphate).

E. Functional Significance of pH_e Homeostasis

Apart from its relevance for pH_i homeostasis, homeostasis of pH_e is crucial for a variety of events occurring in the extracellular space, as shown by the following examples. pH_e affects hormone-binding to cell-surface receptors, explaining relative insulin resistance and a decreased catecholamine activity in acidosis. Carrier-mediated lactate uptake into hepatocytes is stimulated in acidosis, whereas amino acid uptake decreases. In many organs, a fall in pH_e leads to vasodilatation. An acidic pH_e in capillary blood decreases the affinity of hemoglobin for oxygen, facilitating oxygen delivery to peripheral tissues.

BIBLIOGRAPHY

Astrup, P., and Severinghaus, J. W. (1986). "The History of Blood Gases, Acids and Bases." Munksgaard, Copenhagen.

Atkinson, D. E., and Camien, N. M. (1982). The role of urea synthesis in the removal of metabolic bicarbonate and the regulation of blood pH. *Curr. Top. Cell. Reg.* **21**, 261–302.

Bourke, E., and Häussinger, D. (1992). pH homeostasis: The conceptual change. *Contrib. Nephrol.* **100**, 58–88.

Grinstein, S. (ed.) (1989). "Na^+/H^+ Exchange." CRC Press, Oxford.

Häussinger, D. (ed.) (1988). "pH Homeostasis." Academic Press, London.

Lidofsky, S. D., Fitz, J. G., and Scharschmidt, B. F. (1993). Mechanisms and functional role of intracellular pH regulation in hepatocytes. *Prog. Liver Dis.* **11**, 69–83.

Mellman, I., Fuchs, R., and Helenius, A. (1986). Acidification of the endocytic and exocytic pathways. *Annu. Rev. Biochem.* **55**, 663–700.

Noel, J., and Pouyssegur, J. (1995). Hormonal regulation, pharmacology and membrane sorting of vertebrate Na^+/H^+ exchanger isoforms. *Am. J. Physiol.* **268**, C283–C296.

Tashian, R. E., and Hewett-Emmett, D. (eds.) (1984). Biology and chemistry of the carbonic anhydrases. *Ann. N. Y. Acad. Sci.* **429**.

Phospholipid Metabolism

DENNIS E. VANCE
University of Alberta

GLOSSARY

Diacylglycerol Intermediate in phospholipid biosynthesis and an activator of protein kinase C

Phosphatidylcholine Quantitatively, the most important phospholipid found in human cells

Phosphatidylethanolamine Quantitatively, the second most important phospholipid found in human cells

Phospholipases Enzymes that degrade phospholipids

Phospholipid transfer protein Protein that transfers phospholipids between cellular membranes

PHOSPHOLIPIDS ARE BIOLOGICAL COMPOUNDS that contain phosphate and have both hydrophobic and hydrophilic components. Phospholipids are apparently critical for life, because all living organisms and cells have phospholipids in their membranes. Phospholipids provide the basic structure and the permeability barriers of cellular membranes. The central role of phospholipids in biology is underscored by the nearly complete lack of genetic alterations in their metabolism in humans and other animals, suggesting that such alterations are lethal. The metabolism of phospholipids includes the pathways for synthesis (anabolism) and degradation (catabolism). In addition, the pathways and mechanisms by which phospholipids move among the organelles of cells is an important aspect of phospholipid metabolism. How cells regulate metabolism of phospholipids is also discussed. Phospholipids are a major source of cellular second messengers in the transduction of signals from the cell membrane (examples are diacylglycerols, inositol-trisphosphate, and platelet-activating factor). How these second messengers arise from phospholipids and how the cell regulates the metabolism of these compounds are considered. The structures of phospholipids and their roles in membranes were discussed in the article on Lipids. [*See* Lipids.]

I. PHOSPHOLIPID SYNTHESIS IN PROKARYOTES

Although the emphasis of this article is on metabolism of phospholipids in animal systems, it is instructive to consider briefly the prokaryotes, chiefly *Escherichia coli.* The metabolic pathways of phospholipids in this organism have been clearly defined, and many of the enzymes involved in biosynthesis have been purified. In *E. coli,* phospholipid biosynthesis can be considered to begin with the synthesis of fatty acids from acetyl-ACP (acyl carrier protein) and malonyl-ACP, as diagrammed in Fig. 1. Almost exclusively, the purpose of fatty acid synthesis in this bacterium is for phospholipid synthesis. The major fatty acid derivatives formed are palmitoyl-ACP, a saturated fatty acid with 16 carbons in a thioester linkage to acyl carrier protein, and vaccenoyl-ACP, an ACP derivative of a monounsaturated fatty acid with 18 carbons (see Fig. 1). These fatty acyl-ACP derivatives react sequentially with glycerol-3-phosphate (G-3-P) to yield phosphatidic acid, which is present as a trace component in *E. coli* and immediately reacts with cytidine triphosphate (CTP) to give cytidine diphosphate CDP-diacylglycerol. It is noteworthy that CTP is the only high-energy

ENCYCLOPEDIA OF HUMAN BIOLOGY, Second Edition, VOLUME 6.

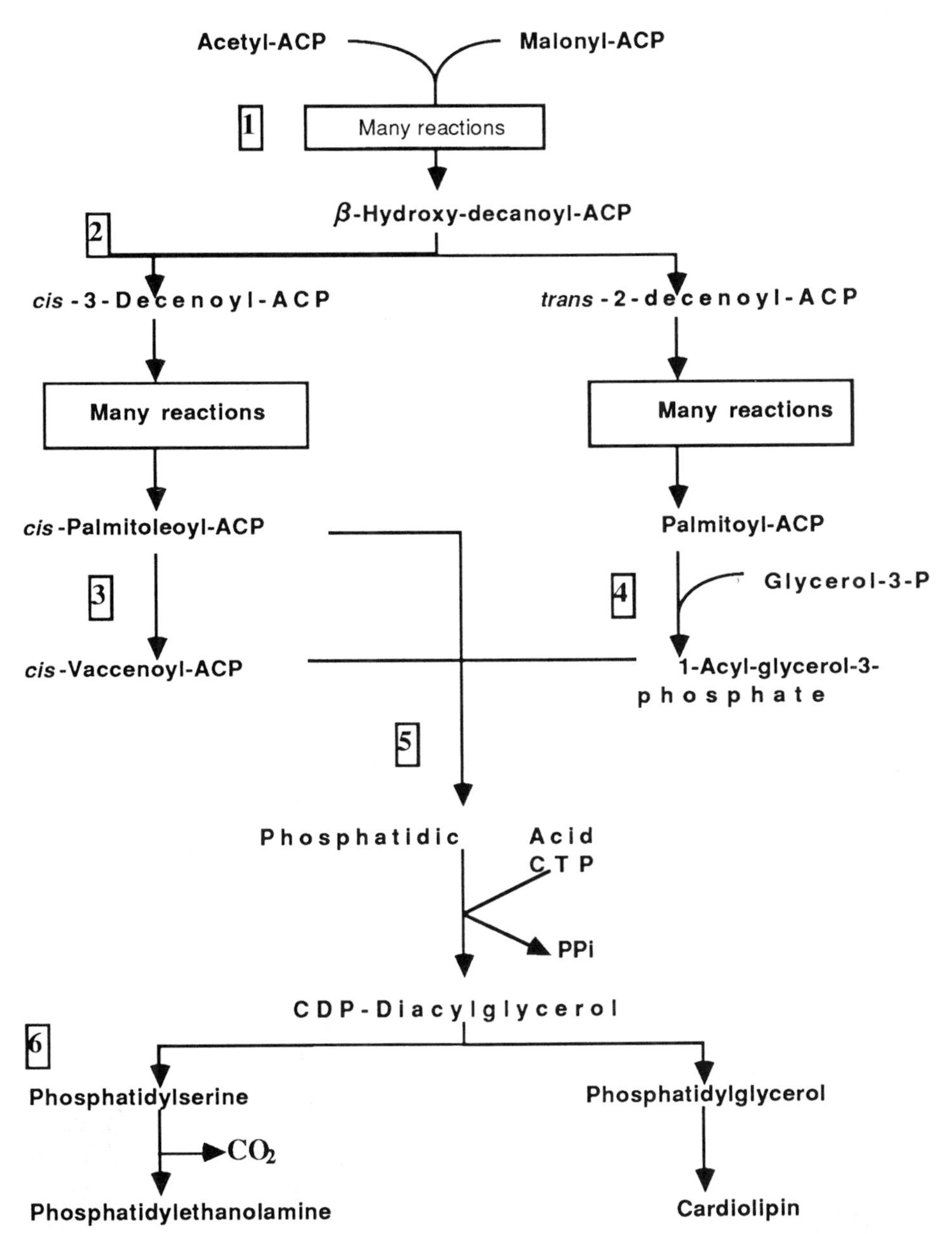

FIGURE 1 Biosynthetic pathway for phospholipids in *E. coli*. The synthesis of fatty acids begins with the condensation of acetyl-ACP and malonyl-ACP (1). Many additional reactions occur until a branch point is reached (2). At this juncture the cell commits itself to make either a saturated or monounsaturated fatty acyl-ACP derivative. The first unsaturated fatty acyl-ACP formed is *cis*-palmitoleoyl-ACP, and this is elongated to the 18-carbon monounsaturated fatty acyl derivative *cis*-vaccenoyl-ACP (3). The synthesis of the phospholipids begins with the acylation of glycerol-3-P with a 16-carbon acyl group, palmitoyl-ACP (4). Usually, an unsaturated fatty acid is inserted into the 2 position to give phosphatidic acid (5). This compound reacts with CTP to yield CDP-diacylglycerol, which is then converted to the major phospholipids in *E. coli* (6). ACP, acyl carrier protein; CDP, cytidine diphosphate; CTP, cytidine triphosphate; PPi, inorganic pyrophosphate.

nucleotide used for phospholipid biosynthesis in all organisms. The pathways branch to yield either phosphatidylethanolamine (PE; the major phospholipid in *E. coli,* comprising 75–85% of the total) or phosphatidylglycerol and cardiolipin (which account for the other 15–25% of the phospholipids in the organism). The synthesis of the acyl-ACP derivatives takes place in the cytosol, whereas the acylation of G-3-P and subsequent reactions occur at the inner surface of the inner membrane of *E. coli.*

Regulation of phospholipid synthesis in *E. coli* apparently occurs at an early but unknown step of fatty acid synthesis. This seems reasonable because the fatty acids in this organism (in contrast to mammals) are primarily used for phospholipid synthesis. Such regulation is also energy-efficient, as 94% of the energy consumed in phospholipid synthesis is used for the manufacture of the fatty acyl components. The growth temperature of the organism determines the proportion of unsaturated fatty acids made and incorporated into phospholipids by modulating the synthesis of *cis*-3-decenoyl-ACP (see Fig. 1). How the cell decides whether to partition CDP-diacylglycerol toward PE or phosphatidylglycerol and cardiolipin is not known. The commitment to make one of the phospholipids is apparently not based on the amounts of enzymes, as overexpression of phosphatidylserine synthase by 10-fold had no effect on the amount of phosphatidylserine and PE made. Similarly, an increase in the amount of phosphatidylglycerolphosphate synthase did not preferentially direct CDP-diacylglycerol toward phosphatidylglycerol or cardiolipin.

In the presence of an appropriate culture medium, bacteria continue to grow and divide; hence, phospholipid turnover (i.e., degradation and resynthesis) apparently is not required. Thus, the major phospholipid of *E. coli,* PE, is relatively stable and is not degraded or metabolized further. In contrast, the phospholipids of animal cells are in a constant state of synthesis and turnover.

II. PHOSPHOLIPID SYNTHESIS IN ANIMALS

A. Metabolic Pathways

The biosynthesis of phospholipids is far more complex in animal cells than in *E. coli.* Six major phospholipid species are found in animal cells compared with only three in the bacterium. The fatty acid substituents are much more varied, particularly in that animal cells have polyunsaturated fatty acids, which are generally absent in bacteria. In addition, in animal cells, phospholipid biosynthesis occurs on several different organelles.

The general scheme for phospholipid synthesis in animal cells is presented in Fig. 2. In animal cells, the biosynthesis of fatty acids is not directly linked with phospholipid synthesis as in *E. coli.* Although fatty acids are required for phospholipid synthesis, fatty acids are also a major source of energy and are stored in triacylglycerols for future energy requirements. As in *E. coli,* phospholipid synthesis begins with the acylation (attachment of a fatty acid) of G-3-P, usually with a saturated fatty acyl-CoA. A second acylation, usually with an unsaturated fatty acyl-CoA, yields phosphatidic acid. An alternative pathway beginning with dihydroxyacetone-phosphate (an intermediate in glycolysis) is equally important for the biosynthesis of phosphatidic acid (see Fig. 2). Phosphatidic acid is at a branch point in metabolism leading to either CDP-diacylglycerol or diacylglycerol.

As in *E. coli,* CDP-diacylglycerol is the precursor of phosphatidylglycerol and cardiolipin. One difference is that in animal cells, cardiolipin arises from the condensation of phosphatidylglycerol with CDP-diacylglycerol, whereas in *E. coli,* two molecules of phosphatidylglycerol condense to form cardiolipin and glycerol. The other phospholipid made from CDP-diacylglycerol is phosphatidylinositol (see Fig. 2). This lipid is further metabolized to phosphatidylinositol-4,5-P_2 and is discussed later (see Section V). Unlike in *E. coli,* CDP-diacylglycerol is not a precursor of phosphatidylserine or PE in animal cells.

The major metabolic fates for diacylglycerol are the synthesis of phosphatidylcholine (PC), PE, and triacylglycerol (see Fig. 2). The synthesis of triacylglycerol is important for the storage of energy in liver, adipose, and other cells but will not be discussed further in this article. PC is formed via the transfer of phosphocholine from CDP-choline to diacylglycerol by an enzyme that occurs on the endoplasmic reticulum and Golgi membranes and possibly at lower concentrations on other cellular membranes. CDP-choline is made from choline via phosphocholine as shown in Fig. 2. Choline is a dietary requirement for humans and other animals. Choline deficiency results in reduced levels of PC in liver, serum, and other tissues with an accumulation of triacylglycerol in liver. Such a deficiency is largely a laboratory phenomenon owing to the ubiquitous occurrence of PC, the major source of dietary choline, in animal and vegetable

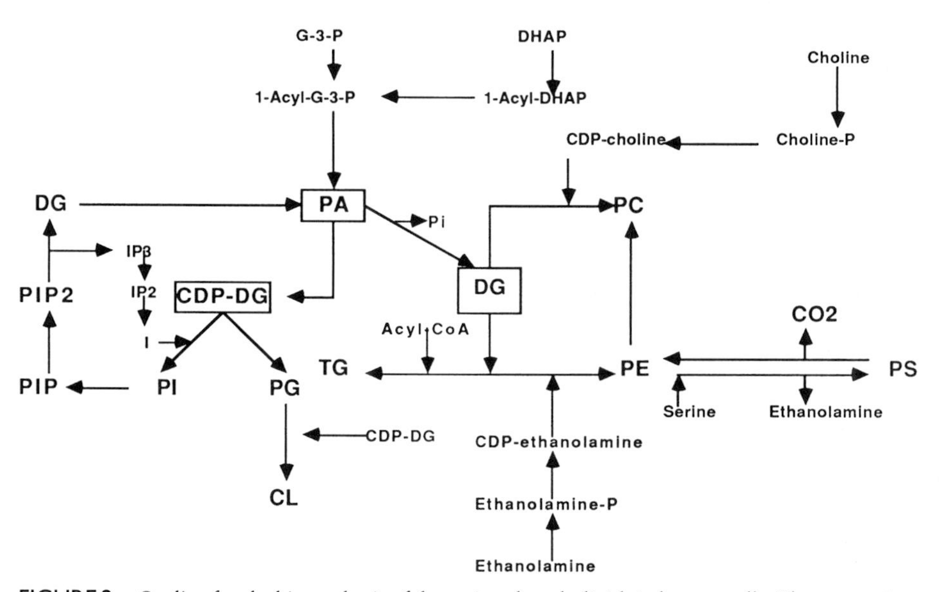

FIGURE 2 Outline for the biosynthesis of the major phospholipids in human cells. The conversion of PE to PC appears to be quantitatively significant only in the liver. CDP-DG, cytidine-diphosphodiacylglycerol; CL, cardiolipin; DG, diacylglycerol; DHAP, dihydroxyacetone phosphate; G-3-P, glycerol-3-phosphate; I, inositol; IP2, inositol-diphosphate; IP3, inositol-triphosphate; PA, phosphatidic acid; PC, phosphatidylcholine; PE, phosphatidylethanolamine; PG, phosphatidylglycerol; Pi, inorganic phosphate; PI, phosphatidylinositol; PIP, phosphatidylinositol-4-phosphate; PIP2, phosphatidylinositol-4,5-bisphosphate; PS, phosphatidylserine; TG, triacylglycerol.

sources in our daily diet. The major mechanism in animals for the biosynthesis of choline is via the conversion of PE to PC (and the subsequent catabolism of PC to choline) via the addition of three methyl groups by a single methyltransferase enzyme found principally on the endoplasmic reticulum. In animals, PE methyltransferase is recovered in significant amounts only from the liver. Approximately 30% of PC made in the liver has been estimated to derive from methylation of PE.

Quantitatively, PE is the second major phospholipid found in animals. Ethanolamine is phosphorylated by a cytosolic kinase, which is also responsible for the phosphorylation of choline. The other reactions of PE biosynthesis are depicted in Fig. 2 and seem to be catalyzed by different enzymes from those used to make PC. PE can also be made from the decarboxylation of phosphatidylserine in mitochondria. The relative contributions of the two major pathways for PE biosynthesis are not established.

The biosynthetic pathway for phosphatidylserine in animals is quite different from that in *E. coli*. Phosphatidylserine, which comprises 5–10% of animal cell membranes, is made via an exchange of serine for ethanolamine from PE on the endoplasmic reticulum as shown in Fig. 2. This appears to be the sole pathway for the biosynthesis of ethanolamine in nature.

The fatty acid substituents of PC, PE, and other phospholipids can be remodeled by a deacylation–reacylation pathway (Fig. 3). This seems to be an important mechanism for the enrichment of polyunsaturated fatty acids into phospholipids, particularly

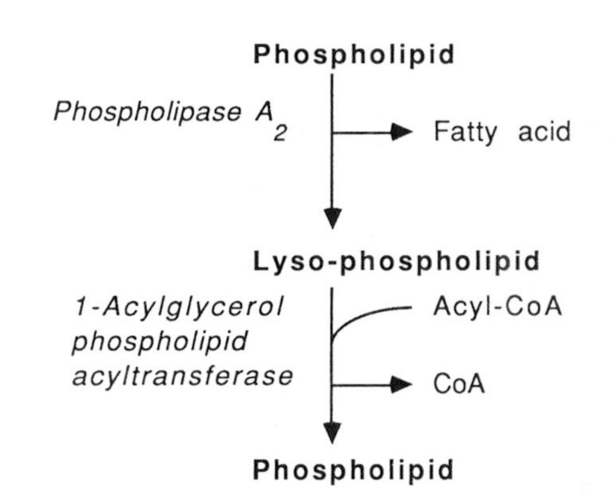

FIGURE 3 Deacylation–reacylation of a phospholipid occurs in most human cells and provides a mechanism for the cell to alter the fatty acid composition of a phospholipid. This is a major pathway for the introduction of arachidonic acid into phospholipids.

for the introduction of arachidonic acid (the precursor of prostaglandins and other eicosanoids). In the type II cells that line lung alveoli, this pathway is used for the enrichment of a saturated fatty acid (palmitic acid) into PC. The resulting dipalmitoyl-PC is a major component of lung surfactant secreted by these type II cells. The surfactant provides a surface tension, which prevents lung collapse when a person breathes out air. In premature infants with respiratory distress syndrome, a defect appears in the secretion of lung surfactant.

Another major phospholipid found in animal cells is sphingomyelin. PC is the donor of phosphocholine for the synthesis of sphingomyelin (Fig. 4). This reaction is believed to occur on either Golgi and/or plasma membranes of cells. The plasma membrane is enriched in sphingomyelin compared with other membranes of the cell. As its name suggests, sphingomyelin is abundant in the myelin membrane.

B. Enzymes of PE and PC Biosynthesis

A major problem that has severely impeded progress in the enzymology of phospholipid biosynthesis is that many of the enzymes are membrane proteins that contain transmembrane sequences. This makes very difficult the process of purification from other proteins and membrane lipids. The solution to this problem is to use detergents such as the bile acid sodium deoxycholate to dissolve the membrane lipids. However, it is tricky to solubilize the enzyme without denaturing and inactivating it. Nevertheless, careful and dedicated work has resulted in the purification of several enzymes that catalyze the reactions indicated in Fig. 2. This has facilitated the cloning and sequencing of the cDNAs for many of these enzymes.

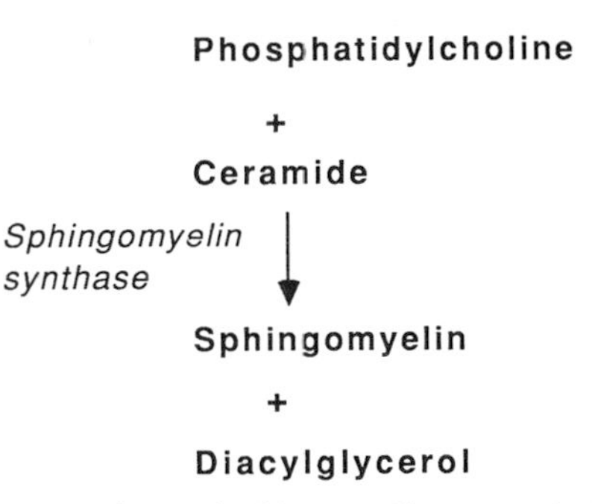

FIGURE 4 Biosynthesis of sphingomyelin occurs via the transfer of phosphocholine from phosphatidylcholine to ceramide. Ceramide is a sphingolipid, which has a different structure but a similar shape to diacylglycerol.

An alternative approach to obtain information on the structure and function of an enzyme, which avoids purification, is to identify the cDNA for a particular enzyme by complementation of a strain of yeast that is missing this enzyme activity. This strategy was recently used to isolate a human cDNA that encodes choline kinase, the enzyme that converts choline to choline phosphate (see Fig. 2). The yeast used in these experiments would not grow because they lacked choline kinase activity and, therefore, could not make PC. When transfected with the cDNA encoding choline kinase, the yeast were able to grow. This approach allowed the investigators to identify which cDNA from a library encoded choline kinase. Human choline kinase is a cytosolic protein with a molecular weight of 52,065. It also catalyzes the conversion of ethanolamine to ethanolamine phosphate (see Fig. 2).

The next enzyme in the CDP-choline pathway for PC biosynthesis, CTP : phosphocholine cytidylyltransferase, is found in both a soluble and membrane-associated form. The interaction with the membrane occurs via an amphipathic helix rather than transmembrane sequences and membrane binding is important in the activation of the cytidylyltransferase as discussed in the next section. This enzyme has also been localized to the nucleus of certain cells, the function of which is currently unknown. The cDNA for the human enzyme has been cloned and encodes an enzyme with a molecular weight of 42,000. The last enzyme in the CDP-choline pathway, CDP-choline : 1,2-diacylglycerol cholinephosphotransferase, has neither been purified nor cloned from a mammalian source.

The enzyme that converts PE to PC, PE methyltransferase, is an integral membrane protein and is largely localized to the endoplasmic reticulum of liver. It was purified from rat liver and has a molecular weight of approximately 20,000. The cDNA was subsequently cloned from a rat liver library. When an antibody to the cloned enzyme was used to identify the subcellular location of this enzyme, a major surprise was discovered. There was no reactivity with proteins on the endoplasmic reticulum. Instead, the antibody reacted specifically with a recently discovered subcellular fraction of rat liver, the mitochondria-associated membrane. Thus, the cloned cDNA encodes an isoenzyme of PE methyltransferase that is a specific marker for this novel subcellular fraction. Why there is a unique isoenzyme of PE methyltransferase localized to the mitochondria-associated membrane is not known. Nor is the function of this unique subcellular membrane known.

As mentioned earlier, the enzyme that phosphorylates ethanolamine is choline kinase. The second enzyme in the CDP-ethanolamine pathway, CTP : phosphoethanolamine cytidylyltransferase, has been purified from rat liver and exists as a dimer with a molecular weight of 100,000 to 120,000. The last enzyme in this pathway, CDP-ethanolamine : 1,2-diacylglycerol ethanolaminephosphotransferase, is an intrinsic membrane protein that has neither been purified nor cloned from a mammalian source.

As indicated in Fig. 2, PE can also be formed from decarboxylation of phosphatidylserine. This enzyme, localized to mitochondria, is an intrinsic protein that has also not been purified nor the cDNA cloned from a mammalian source. The cDNA for the yeast enzyme has been cloned and studies with strains of yeast in which the mitochondrial enzyme had been "knocked out" demonstrated the existence of a second, previously unknown phosphatidylserine decarboxylase located outside of the mitochondria.

For additional information on these enzymes and others involved in phospholipid biosynthesis, see the volume on "Phospholipid Biosynthesis" in the Methods in Enzymology series cited in the Bibliography.

C. Regulation

The understanding of how animal cells regulate the rate of phospholipid synthesis is at a relatively primordial level compared with that of other lipid metabolic pathways, such as fatty acid or cholesterol biosynthesis. One major obstacle has been the relatively modest variations in the rate of phospholipid synthesis as a result of different perturbations, such as fasting. Second, most of the enzymes have not been purified or only recently have been obtained in pure form, as these enzymes are usually tightly integrated into cellular membranes. A third difficulty is that many of the substrates are water-insoluble.

The regulation of PC biosynthesis is best understood. Control of this pathway appears to be centered on the enzyme CTP-phosphocholine cytidylyltransferase, which converts phosphocholine to CDP-choline (Fig. 5). The cytidylyltransferase is a dimer of identical subunits that is recovered from the cytosol in an inactive form and from cellular membranes (endoplasmic reticulum and Golgi), where the enzyme is activated by phospholipids. Considerable evidence indicates that the rate of PC biosynthesis in animal cells is governed by the reversible translocation of the cytidylyltransferase between a soluble form and cellular membranes. Current studies are directed toward elucidation of the mechanism, which regulates the subcellular distribution of the cytidylyltransferase.

One governing feature appears to be the concentration of PC in the membranes (see Fig. 5). As the concentration goes down, for example, when animals

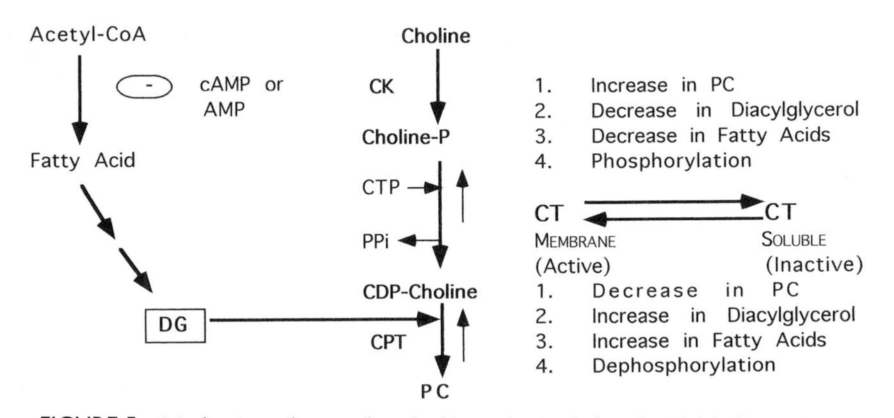

FIGURE 5 Mechanisms that regulate the biosynthesis of phosphatidylcholine (PC) via the CDP-choline pathway. The enzymes involved in the pathway are choline kinase (CK), CTP : phosphocholine cytidylyltransferase (CT), and CDP-choline : 1,2-diacylglycerol cholinephosphotransferase (CPT). The normally rate-limiting reaction is catalyzed by CT, which is found on cellular membranes in an active form and in a soluble, inactive form. The mechanisms that regulate the translocation of CT between the soluble and membrane-bound forms are indicated in the figure and discussed in the text. The supply of diacylglycerol (DG), one of the substrates for CPT, can also regulate the rate of PC biosynthesis as discussed in the text.

or cells are made choline deficient, there is increased association of the cytidylyltransferase with cellular membranes. The decrease in PC levels is due to a lack of the substrate choline for biosynthesis. When choline is supplied, the amount of substrate is no longer limiting, the PC levels rise, and the cytidylyltransferase is released into a soluble form. Thus, as in other metabolic pathways, the concentration of the end product is regulating its biosynthesis by altering the activity of the rate-limiting enzyme in the sequence.

The supply of diacylglycerol also appears to have regulatory importance in the CDP-choline pathway. First, the supply of diacylglycerol as a substrate for the reaction catalyzed by the cholinephosphotransferase appears to be important (see Fig. 5). Diacylglycerol levels can be influenced by the rate of fatty acid biosynthesis, which can be regulated by the production of cAMP or AMP. Alternatively, influx of fatty acids from the diet will boost the synthesis of diacylglycerol. In addition to acting as a substrate, increased diacylglycerol levels enhance the binding of the cytidylyltransferase to membranes where the enzyme is activated, thereby promoting the biosynthesis of the other substrate for the cholinephosphotransferase reaction, CDP-choline (see Fig. 5). Thus, the concentration of diacylglycerol acts as a feedforward modulator of PC biosynthesis. Fatty acids can also directly regulate the binding of the cytidylyltransferase to membranes in cells in culture, but the physiological relevance of this observation remains to be established. [*See* Fatty Acids.]

The cytidylyltransferase is highly phosphorylated on serine residues when in its soluble, inactive form and markedly dephosphorylated when bound to membranes in its active form. It has been shown that a decrease in the level of PC or an increase in the concentration of fatty acids in cultured hepatocytes initially causes translocation of the cytidylyltransferase to membranes with subsequent dephosphorylation. Thus, the dephosphorylation may function to stabilize the membrane–protein interaction. There may be other examples where the phosphorylation reactions govern the binding of the cytidylyltransferase to membranes without a change in the lipid composition. One example of this might be in the regulation of PC biosynthesis in the cell cycle. When cytidylyltransferase is most active in the cell cycle, the enzyme is in a dephosphorylated state. The enzyme is highly phosphorylated when the cytidylyltransferase is less active. Moreover, the enzyme is a substrate for serine phosphorylation by cyclin B/cdc2 kinase, an enzyme implicated in the regulation of the cell cycle.

Much less is known about the regulation in mammals of the other pathways for phospholipid biosynthesis indicated in Fig. 2. The regulation of PE biosynthesis via the CDP-ethanolamine pathway is governed by the activity of the phosphoethanolamine cytidylyltransferase and the supply of diacylglycerol. Interestingly, this cytidylyltransferase is completely different from the cytidylyltransferase in the CDP-choline pathway discussed earlier. The phosphoethanolamine cytidylyltransferase is active in a soluble form.

As stated previously, PC can also be made in liver by the methylation of PE. The PE methyltransferase appears to be constitutively expressed in liver and present at very low levels in nonhepatic tissues. Studies in intact animals provide evidence that the supply of methionine, the precursor of *S*-adenosylmethionine (the methyl donor), regulates the rate of PE methylation. Similarly, in cultured cells the supply of PE has also been demonstrated to alter the rate of methylation. Thus, biosynthesis of PC via this pathway seems to be regulated by the supply of substrates.

III. PHOSPHOLIPID CATABOLISM—PHOSPHOLIPASES

A. Background

Phospholipases, which are present in virtually all cells, are enzymes that degrade phospholipids, as shown in Fig. 6. The products are further catabolized, reincorporated into phospholipids, or serve as second messengers, as discussed in Section V. These enzymes can be classified into functional categories—digestive and regulatory. The digestive phospholipases have been studied most extensively, because they are proteins that are usually soluble in aqueous buffers. This category includes phospholipases, isolated from pancreas, that are secreted into the intestinal lumen and digest dietary phospholipids. Snake venom is another important source for phospholipases. A third source of digestive phospholipases is the lysosomes of cells, which degrade phospholipids that enter the cell via phagocytosis. The regulatory phospholipases are also discussed in Section V.

B. Phospholipases A_2

The phospholipases A_2 cleave the fatty acyl group from the 2 position from phospholipids. The pancre-

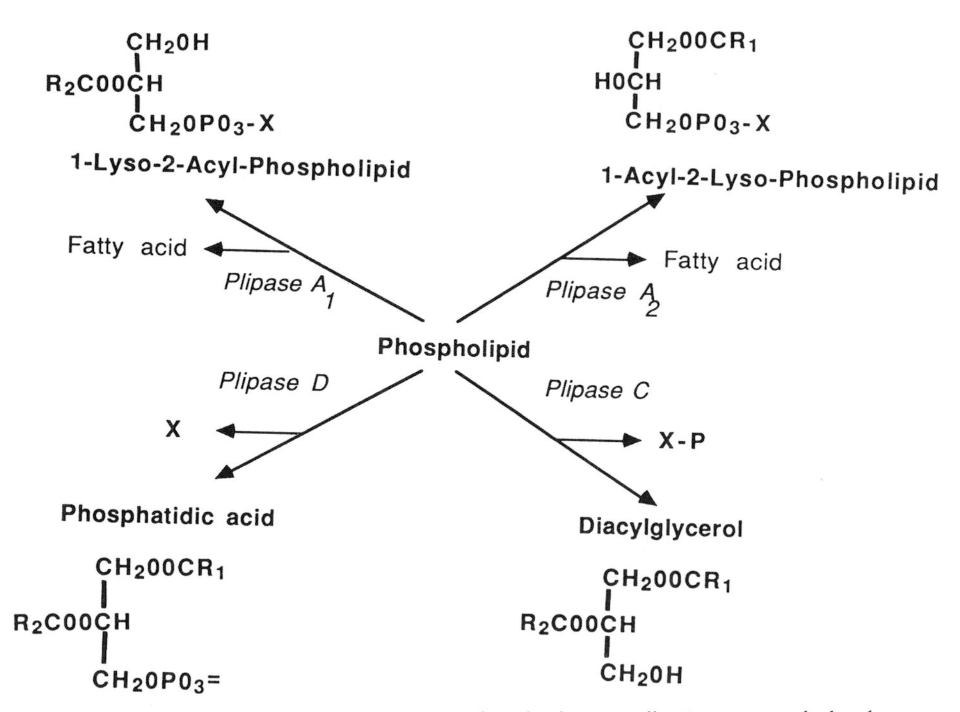

FIGURE 6 Four major phospholipase reactions found in human cells. X represents the head group of the phospholipid, such as choline or inositol, and X-P represents the phosphorylated head group. Plipase is an abbreviation for phospholipase. The designation Lyso indicates the absence of a fatty acyl group at the indicated position on the phospholipid.

atic phospholipase A_2 is secreted from the pancreas into the intestine as a proenzyme, where it is activated by proteolytic cleavage to yield an enzyme with a molecular weight of approximately 14,000. This processing allows the enzyme to bind to phospholipid aggregates and catalyze the hydrolysis of the fatty acid from the 2 position of the phospholipid, as shown in Fig. 6. The binding and catalytic sites appear to be at different positions on the phospholipase. Phospholipase A_2 purified from cobra venom is homologous to the pancreatic enzyme but is active as a dimer of two identical subunits, whereas the pancreatic enzyme acts as a monomer. Both enzymes have an absolute requirement for calcium.

Unlike the phospholipases A_2 that are secreted, a cytosolic enzyme has been purified and cloned that has a molecular weight of approximately 85,000. Calcium does not appear to be involved in the catalytic mechanism but rather in mediation of the binding of the enzyme to cellular membranes where the phospholipid substrates are located. This high-molecular-weight phospholipase A_2 has been implicated in the specific release of arachidonic acid for eicosanoid biosynthesis. However, low-molecular-weight enzymes have also been implicated in this process. More research will be required to understand the functions of the various phospholipases A_2.

C. Phospholipases A_1, C, and D

Phospholipases A_1 are widely distributed in nature, usually have a broad specificity for phospholipid substrates, and often catalyze the hydrolysis of lysophospholipids. A soluble enzyme has been studied from rat liver lysosomes that has optimal activity at pH 4.0 and does not require calcium. Two phospholipases A_1 have been purified from *E. coli*. Curiously, bacterial mutants defective in either or both of these enzymes have normal growth characteristics and phospholipid turnover. Hence, the function of these enzymes in *E. coli* remains uncertain.

Phospholipases C have also been most intensely studied in bacteria, particularly *Bacillus cereus;* two have been isolated from this organism. One enzyme with broad specificity has a requirement for zinc. A second does not require zinc but, curiously, is specific for phosphatidylinositol, which is not a phospholipid found in bacteria. The mammalian phospholipases C that have been studied are also generally specific for phosphatidylinositol and related inositol-containing phospholipids and require calcium. However, one described lysosomal phospholipase C does not need calcium and has a pH optimum of 4.5.

Plants were originally identified as the only source of phospholipases D; however, this enzyme activity has been recently described in several animal cells and tissues.

IV. TRANSFER OF PHOSPHOLIPIDS IN CELLS

In mammalian cells, the bulk of the phospholipids are made on the cytosolic face of the endoplasmic reticulum yet are distributed on every membrane in the cell. The phospholipid composition of membranes within cells differs, and there is an asymmetry in the distribution of phospholipids between the two leaflets of some, if not all, membranes. How the specific phospholipid composition of individual membranes is established and maintained is not known.

The first question is how do the phospholipids move from the site of synthesis on the cytosolic face to the inner leaflet of the endoplasmic reticulum bilayer? In pure phospholipid vesicles, phospholipids do not spontaneously cross the lipid bilayer. Because phospholipids do cross the bilayer in biological membranes, a protein catalyst of lipid transbilayer movement (flip-flop) might be involved. There are now several reports of proteins called flippases, which accelerate the movement of phospholipids from one side of the endoplasmic reticulum to the other.

The best described flippase is mdr2. The name was originally derived because the flippase has high homology to several multidrug-resistant proteins. The function of the flippase was elucidated when it was "knocked out" in a mouse and the investigators discovered the total absence of PC, normally a major constituent, in the bile. Subsequent studies demonstrated that the flippase functions to move PC from the cytosolic side to the outside of the bile canalicular plasma membrane in liver. A flippase in the red cell membrane specific for PE and phosphatidylserine has also been described.

The mechanism of the movement of phospholipids from their sites of synthesis on the endoplasmic reticulum to other cellular membranes is better understood than the transbilayer movement of phospholipids. Phospholipids are poorly soluble in an aqueous environment; thus, solubilization and diffusion seems an unlikely mechanism for movement of these lipids between membranes. A likely possibility is the transfer of phospholipids by binding to transport proteins. Proteins that catalyze the exchange or transfer of phospholipids between membranes *in vitro* have been isolated and studied intensely. However, what function, if any, these proteins perform *in vivo* remains a moot question over 20 years after their discovery. Possibly, these proteins are involved only in the exchange of phospholipids between membranes and not in the net transfer from the endoplasmic reticulum to a target membrane. Another likely mechanism for lipid transport is via vesicles that bud from the endoplasmic reticulum and transport lipids and proteins to Golgi and other organelles. These vesicles would contain phospholipids and fuse with target membranes. Thus, in addition to the transport of proteins, these vesicles would transport phospholipids. A third possible transport mechanism is continuity of the membranes between adjacent organelles and movement of phospholipids between organelles by diffusion within the bilayer. Future studies should establish which of these possible mechanisms function in eukaryotic cells and the quantitative contribution of the various transport processes.

V. FUNCTIONAL ASPECTS OF PHOSPHOLIPIDS

A. Sources of Second Messengers

It has been known for many decades that the precursor of the eicosanoids, arachidonic acid, was present at the *sn*-2 position of phospholipids. This fatty acid is probably released by the action of a phospholipase A_2 and is rapidly converted into prostaglandins, thromboxanes, leukotrienes, or one of the hydroxyeicosanoic acids. Eicosanoids are 20-carbon derivatives of fatty acids that have hormone-like actions near their site of synthesis. Among many different functions, they have been shown to be important in the modulation of adenylate cyclase, facilitation of ion transport, platelet aggregation, bronchoconstriction, increase of vascular permeability in skin, and chemotaxis of polymorphonuclear leukocytes.

Only in the last decade, scientists have recognized the importance of other species of phospholipids as storage forms of second messengers, which are active in various biological processes. Foremost is the lipid phosphatidylinositol-4,5-bisphosphate, the metabolism of which is shown in Fig. 2. This lipid is present in small quantities (<1% of the total cellular phospholipids) in the plasma membrane of many different types of cells. When a hormone or other agonist binds to a receptor on the membrane, a specific phospholipase C is activated and degrades the lipid to inositol-1,4,5-triphosphate and diacylglycerol, both of which are second messengers. The inositol derivative binds to the endoplasmic reticulum and causes the release of calcium into the cytosol of the cell, where it activates many different cellular processes. For example, the binding of the hormone vasopressin to liver cells results in the breakdown of phosphatidylinositol-4,5-bisphosphate to inositol-1,4,5-trisphosphate. This molecule mobilizes calcium, which activates the enzyme phosphorylase, which in turn degrades glycogen. The other product, diacylglycerol, activates an enzyme called protein kinase C, which phosphorylates a large variety of different proteins involved in stimulus–response couplings. Thus, protein kinase C has been implicated in catecholamine secretion, insulin release, steroid synthesis, arachidonic acid release, and many other cellular responses. Considerable evidence indicates that guanine triphosphate-binding proteins act as a bridge between the receptor and the activation of phospholipase C.

Other phospholipids, particularly PC, have been implicated in the production of diacylglycerol for the activation of protein kinase C. This diacylglycerol might be generated directly via the action of a phospholipase C or via the action of a phospholipase D to yield phosphatidic acid (see Fig. 6), which is degraded to diacylglycerol and phosphate by phosphatidic acid phosphatase.

A surprise discovery in 1979 was platelet-activating factor (1-alkyl-2-acetyl-*sn*-glycerol-3-phosphocholine). This compound aggregates platelets at a concentration of 10^{-11} M and induces an antihypertensive response when 60 ng are injected intravenously into rats. A variety of blood cells, healthy tissues, and tumor cells synthesize and degrade platelet-activating factor. It is made in cells stimulated by agents such as the calcium ionophore A23187. The biosynthesis involves the acetylation at the *sn*-2 position of 1-alkyl-2-lysoglycerophosphocholine, which arises from 1-alkyl-2-acylglycerophosphocholine via the action of a phospholipase A_2. Alternatively, platelet-activating factor can be made by the transfer of phosphocholine to 1-alkyl-2-acetyl-glycerol.

B. Anchoring of Proteins to Membranes

In recent years, some proteins have been discovered to link covalently to phosphatidylinositol in the plasma membrane. In this linkage, the COOH-terminal amino acid is bonded in an amide linkage to the amino function of ethanolamine phosphate, which is linked via a series of carbohydrates (terminating in glucosamine) to the inositol moiety of phosphatidylinositol (Fig. 7). Among the proteins found in such a linkage are cell-surface hydrolases (e.g., alkaline phosphatase), protozoa coat proteins, proteins on the surface of lymphocytes, and proteins involved in the adhesion of cells to other cells. The protein attached to the carbohydrate can be released from the membrane by phospholipase C cleavage of the phosphatidylinositol moiety. It is not known if this degradation is simply a normal event in membrane recycling or if the cleavage might be important in regulation of interactions at the cell surface. Presumably, this particular linkage of proteins to cell membranes is somewhat advantageous and will be elucidated in future studies.

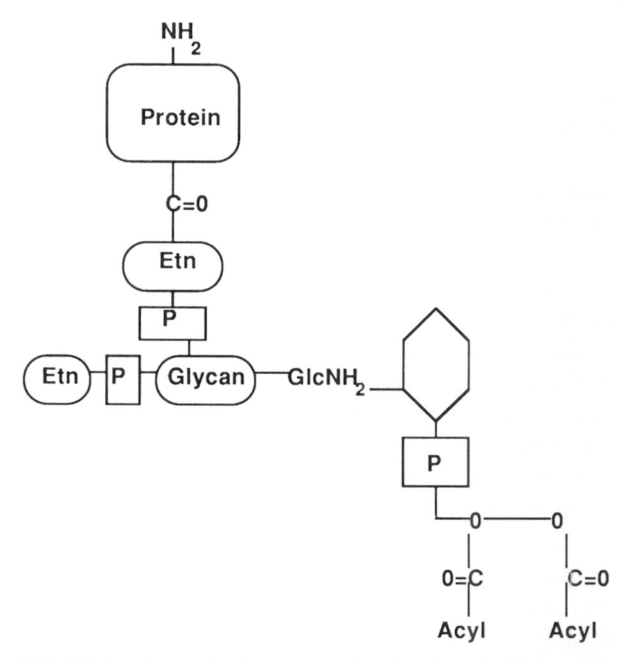

FIGURE 7 Generalized structure for the phosphatidylinositol–carbohydrate–ethanolamine–phosphate structure, which anchors certain proteins to the plasma membrane of a few types of cells. The remaining part of the structure represents phosphatidylinositol. Etn, ethanolamine; $GlcNH_2$, glucosamine; Glycan, an oligosaccharide; P, phosphate.

VI. CONCLUSION

Ten years ago, few scientists would have predicted the large number of functions that are now ascribed to phospholipids. Because the structure in the phospholipids is so diverse, we can only wonder what lies ahead in the next decade. However, it is clear that the functions of phospholipids and the enzymes involved in their metabolism will be best established in experiments where a particular enzyme activity has been deleted. This has been a traditional approach in bacteria and yeast. With the advent of gene targeting in mice, this approach will be fundamental in the elucidation of phospholipid function in mammals.

BIBLIOGRAPHY

Dennis, E. A. (1994). Diversity of group types, regulation and function of phospholipase A_2. *J. Biol. Chem.* **269,** 13057.

Dennis, E. A., and Vance, D. E. (eds.) (1992). "Phospholipid Biosynthesis," Methods in Enzymology, Vol. 209. Academic Press, San Diego.

Englund, P. T. (1993). The structure and biosynthesis of glycosyl phosphatidylinositol protein anchors. *Annu. Rev. Biochem.* **62,** 121.

Trotter, P. J., and Voelker, D. R. (1994). Lipid transport processes in eukaryotic cells. *Biochim. Biophys. Acta* **1213,** 241.

Vance, D. E., and Vance, J. E. (1996). "Biochemistry of Lipids, Lipoproteins and Membranes." Elsevier, Amsterdam.

Phosphorylation of Microtubule Protein

JESÚS ÁVILA
JAVIER DÍAZ-NIDO
Centro de Biología Molecular "Severo Ochoa," Universidad Autónoma de Madrid

GLOSSARY

Microtubule protein Protein components of the microtubule fraction obtained from cytosol through cycles of polymerization–depolymerization (assembly–disassembly); it contains tubulin and associated proteins

Neuronal plasticity Ability of nerve cells (neurons) to alter their morphology and physiology (especially their extent of response to other neurons) depending on their immediate environment, which consists of hormones, trophic factors, stimulatory and inhibitory neurotransmitters, and neuromodulators

Phosphoprotein phosphatase Enzyme that hydrolyzes monophosphoric esters on proteins, with release of inorganic phosphate

Phosphorylation Metabolic process of introducing a phosphate group into an organic molecule; protein phosphorylation usually refers to the formation of monophosphoric esters with hydroxy amino acids present in proteins

Posttranslational modification Any chemical modification of a protein that occurs after the completion of the polypeptide chain; it comprises both the irreversible splitting off of certain peptide fragments and the reversible modification of various amino acids by addition or removal of various chemical moieties (phosphate, adenosine phosphate-ribose, acetate, fatty acids, sugars, sulfate, etc.)

Protein kinase Enzyme that catalyzes the transfer of a phosphate group from a high-energy donor, usually adenosine triphosphate, to a protein to form a phosphoprotein

PROTEIN PHOSPHORYLATION IS A CHEMICAL modification consisting of the formation of monophosphoric esters, mainly at hydroxy amino acids (serine, threonine, and tyrosine), which usually results in changes in the properties of the modified proteins. Both the major component of microtubules, tubulin, and their associated proteins appear to be modified in this way. Because microtubule functions within cells depend on the ability of microtubules to polymerize (assemble) and depolymerize (disassemble) and to interact with other cell structures, the modification of microtubule proteins by phosphorylation affects important aspects of cell physiology and pathology that depend on these properties.

I. INTRODUCTION

Microtubules are long, slender polymers built of the protein tubulin and are present in the cytoplasm of all eukaryotic cells, forming part of the cytoskeleton, the system of various filament types that spans the cell. Microtubules seem to perform a multitude of essential roles within the cells, including the generation and maintenance of cell shape, the intracellular transport and sorting of cytoplasmic organelles, the assembly of the mitotic (and meiotic) spindle at cell division, the subsequent segregation of chromosomes, and the formation of microtubule arrays involved in cell motility such as cilia and flagella. This variety of functions is based on the ability of microtubule protein to polymerize and depolymerize reversibly and to interact with other subcellular structures.

Protein phosphorylation, in the same way as other posttranslational modifications of proteins, has been found to control many metabolic processes through the modification of some key enzymes, and it is also thought to regulate the ability of certain structural proteins to associate with other proteins. According to this view, phosphorylation of microtubule protein might provide a mechanism for regulating the assembly and disassembly of microtubules, as well as their interaction with other structures, thus controlling the cellular functions in which microtubules are involved. [*See* Microtubules; Protein Phosphorylation.]

Although they are found in all eukaryotic cells, microtubules are especially abundant in neurons, where they constitute the inner scaffolding of their appendages (neurites)—axons and dendrites—and also serve as tracks for organelle transport between cell bodies and neurite endings. For this reason, mammalian brain extracts have been the primary source of microtubule protein through cycles of *in vitro* polymerization–depolymerization at appropriate temperatures. These preparations contain tubulin as the major component, in addition to a group of proteins originally referred to as microtubule-associated proteins (MAPs) (Table I). MAPs are thought to promote microtubule assembly, stabilize the resulting microtubules, and interact with other cytoskeletal elements and cell organelles, thus being largely responsible for the functional diversity of microtubules. It has now become clear that the term "microtubule-associated protein" is no longer adequate for all of these proteins, as some of them interact only transiently with tubulin. Thus the term microtubule-associated proteins should be used only to refer to those tubulin-binding proteins that are able to stimulate the assembly of tubulin and remain bound to the resulting microtubules. All other proteins that interact with tubulin should be named microtubule-binding proteins (MTBPs). However, many authors still use the term "microtubule-associated proteins" for both types of tubulin-binding proteins. The largest group of MTBPs correspond to "motor" proteins. These are nucleotide-sensitive microtubule-binding proteins with ATPase activity that allows them either to carry specific organelles along microtubules or to catalyze the sliding of one microtubule on another microtubule.

Brain microtubule protein preparations also contain very minor components, among which a variety of enzymes involved in posttranslational modifica-

TABLE I
Brain Microtubule Proteins

Protein	Mass (10^3)[a]	Main localization	Suggested function
MAP1A	350	Neuron (dendrite/axon), glia	Microtubule stabilization Dendrite maturation
MAP1B (MAP5)	325	Neuron (axon/dendrite), glia	Microtubule assembly promotion Initial neurite outgrowth
MAP1C (dynein)	300	Ubiquitous	Organelle transport Microtubule sliding
MAP2A/B	270	Neuron (dendrite)	Microtubule bundling Dendritic microtubule organization Dendritic plasticity regulation
MAP3/4	180–220	Glia, nonneural cells	Mitotic spindle organization
Kinesin	120	Ubiquitous	Organelle transport Microtubule sliding
MAP2C/D	70	Neuron (axon/dendrite), glia	Microtubule bundling Initial neurite outgrowth
Tau	50–65	Neuron (axon)	Microtubule bundling Microtubule stabilization Axonal maturation
α-tubulin	52	Ubiquitous	
β-tubulin	50	Ubiquitous	

[a]Molecular masses correspond to the values determined by gel electrophoresis.

tions of microtubule proteins have been the best characterized. These enzymes are not considered MAPs nor MTBPs, as they usually bind to MAPs or MTBPs rather than to tubulin. Phosphorylation is the main modification affecting both tubulin and associated proteins and, consequently, it is one of the most important putative regulatory mechanisms of microtubule functions. Thus, a great deal of attention has been paid to the protein kinases found in microtubule protein preparations. These include a cyclic adenosine monophosphate (cAMP)-dependent protein kinase, also called protein kinase A (PKA), a type II calcium/calmodulin-dependent protein kinase (CaM kinase II), protein kinase C, glycogen synthase kinase 3, casein kinase 2, and some MAP kinases and cyclin-dependent kinases. Table II summarizes the main protein kinases that have been described to modify microtubule components. Recently the presence of phosphoprotein phosphatases (mainly phosphatase 2B, also named calcineurin, and phosphatase 2A) in microtubule protein preparations has been demonstrated. Thus microtubules contain the enzymes that are necessary for the reversible phosphorylation and dephosphorylation of microtubule components (Fig. 1).

The study of microtubule protein phosphorylation is carried out through two complementary approaches. One is the *in vitro* approach; it analyzes the phosphorylation of isolated microtubule proteins by purified protein kinases. Using this approach, the localization of some phosphorylation sites on the molecules and the functional properties of the phosphorylated protein in *in vitro* assays (for instance, tubulin-binding assays) have been determined. The second approach attempts to correlate the phosphorylation of certain microtubule components with cellular events in which cytoskeletal changes take place. This is mainly based on data obtained from appropriate *in situ* model systems (cell cultures, tissue slices, *in vivo* phospholabeling) and allows one to ascertain the physiological relevance of the phosphorylation reactions observed *in vitro*. The use of immunocytochemistry with antibodies specific for certain phosphorylation sites is particularly relevant in this respect.

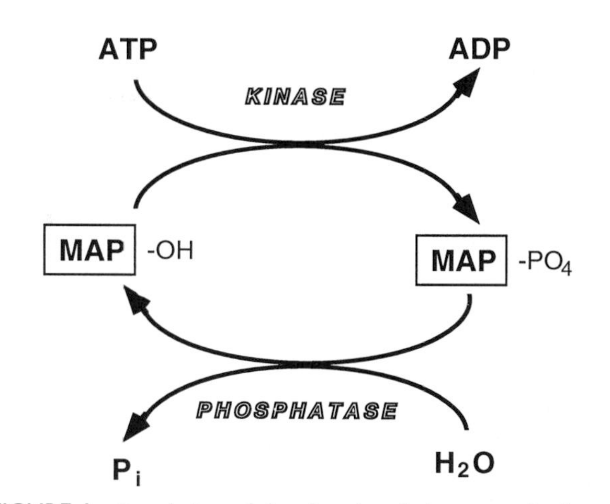

FIGURE 1 Regulation of the phosphorylation state of microtubule-associated proteins (MAP) by protein kinases and phosphatases.

II. TUBULIN PHOSPHORYLATION

Both α- and β-tubulin can be *in vitro* phosphorylated by a variety of protein kinases, including casein kinase

TABLE II
Protein Kinases That Modify Microtubule Proteins

Protein kinase	Major substrates
Second-messenger-dependent protein kinases	MAP2, tau
Cyclic AMP-dependent protein kinase (PKA)	
Cyclic GMP-dependent protein kinase (PKG)	
Ca^{2+}/calmodulin-dependent protein kinase (CaM-kinase II)	
Protein kinase C (PKC)	
Proline-directed protein kinases (PDPKs)	MAP1A, MAP1B, MAP2, MAP4, tau
Cyclin-dependent kinases (CDKs)	
MAP kinases (MAPK) or extracellular-regulated kinases (ERK)	
Glycogen synthase kinase-3 (GSK-3)	
Independent kinases	
Casein kinase 2 (CK2)	MAP1A, MAP1B, tau
Casein kinase 1 (CK1)	

2, type II Ca^{2+}/calmodulin-dependent protein kinase (CaM kinase II), and protein tyrosine kinases.

Tubulin phosphorylation by casein kinase 2 occurs predominantly at serine 444 of the minor neuronal-specific βIII-tubulin isoform. Phosphorylated βIII-tubulin is readily dephosphorylated by protein phosphatase 2A. Assembled tubulin is a better substrate for casein kinase 2 than unpolymerized tubulin, and dephosphorylation hampers the polymerization ability of βIII-tubulin.

In vivo phospholabeling studies have demonstrated that brain βIII-tubulin is indeed phosphorylated at serine 444. This phosphorylation is also observed when neuronal cells are induced to extend neurites. Cell fractionation analyses indicate that phosphorylated βIII-tubulin is largely restricted to the cytoskeletal fraction and is nearly absent from the soluble fraction. Furthermore, βIII-tubulin phosphorylation is inhibited by microtubule-depolymerizing drugs, suggesting that phosphorylation occurs mainly on previously assembled tubulin. It is therefore tempting to speculate that βIII-tubulin phosphorylation may stabilize assembled tubulin inside neurites.

Tubulin phosphorylation by either CaM-kinase II or protein tyrosine kinases takes place at residues on both α- and β-tubulin molecules, which have not been characterized in detail. In both cases, phosphorylation notably decreases the ability of tubulin to polymerize and bind to MAPs *in vitro*. Interestingly, complexes between tubulin and these kinases have been identified. However, at the moment there is no clear evidence that brain tubulin is *in vivo* phosphorylated by any of these protein kinases.

Tubulin phosphorylation in tissues other than brain has not been studied because of the relative paucity of microtubules in nonneuronal cells. An exception are platelets, in which microtubules play an essential role in maintaining cell shape. Phosphorylation of α- and β-tubulin correlates well with the changes in cell shape that accompany platelet aggregation, but the implicated kinases are still unknown.

III. PHOSPHORYLATION OF BRAIN MICROTUBULE–ASSOCIATED PROTEINS

Microtubules are necessary for the extension and maintenance of the neurites (axon and dendrites) that endow neurons with their characteristic shapes. Microtubule-associated proteins are notably abundant in brain and have been implicated in the regulation of microtubule functions during neuronal morphogenesis and plasticity. In particular, brain MAPs are thought to control the promotion of tubulin polymerization inside growing neurites, the stabilization of microtubules within maturing neurites, the development and maintenance of the distinct morphologies of dendrites and axons, and the cytoskeletal rearrangements underlying neuronal plasticity. The effects of MAPs on neuronal microtubules can be modulated by phosphorylation and dephosphorylation of the MAP molecules in response to the extracellular signals (neurotrophic factors, hormones, neurotransmitters, neuromodulators, adhesion molecules) that govern the activities of different protein kinases and phosphatases. Indeed, several identified MAPs are known to be good substrates for phosphorylation and dephosphorylation, both *in vitro* and *in vivo*.

A. MAP1 Phosphorylation

MAP1A and MAP1B are encoded by two different genes but show extensive amino acid sequence similarities. MAP1B (also named MAP5) is abundant in developing neurons, being particularly concentrated within their growing axons, and is present at a lower level in both axons and dendrites of mature neurons. MAP1A is more prominent in dendrites of mature neurons.

Both MAP1A and MAP1B are heavily phosphorylated in the living rat brain. MAP1B phosphorylation has also been observed during neurite growth in cell lines of neuronal origin and has been more thoroughly studied thanks to the availability of specific antibodies to phosphorylation-sensitive epitopes.

There appear to be at least two major modes of MAP1B phosphorylation. One induces an important upward shift in the electrophoretic mobility of the protein and may be catalyzed by proline-directed protein kinases (PDPKs). These phosphorylation sites can be readily dephosphorylated by protein phosphatases 2B (calcineurin) and 2A. MAP1B phosphorylated at these sites is predominantly localized in the distal growing segments of developing axons, and axonal maturation is in most cases accompanied by dephosphorylation. The functional consequences of this type of MAP1B phosphorylation are not yet understood, although it can be speculated that they might contribute to the dynamic configuration of microtubules that has been observed in growing axon terminals. Thus, this specific mode of MAP1B phosphorylation may be correlated with the ability of an axon to grow. The dephosphorylation of MAP1B at this kind of sites may lead to more stabilized and tightly packed microtubule bundles during axonal maturation. Axons in which no dephosphorylation occurs might have a higher regenerative potential.

The other type of MAP1B phosphorylation hardly modifies the electrophoretic mobility of the protein and is catalyzed by casein kinase 2. These phosphorylation sites can be easily dephosphorylated by protein phosphatases 2A and 1. Evidence obtained form neuroblastoma cells suggests that MAP1B phosphorylation by casein kinase 2 may favor its binding to microtubules and may be essential for neurite growth. This mode of MAP1B phosphorylation is maintained in the adult mammalian brain and is present in both axons and dendrites.

B. MAP2 Phosphorylation

There are several forms of MAP2 that arise from alternative splicing, including high-molecular-mass proteins known as MAP2A and MAP2B and low-molecular-mass forms named MAP2C and MAP2D. MAP2C is expressed in the developing mammalian brain and is down-regulated during brain maturation, whereas MAP2A/B is more abundant in the adult brain. MAP2C has a widespread distribution in brain, being present in neuronal cell bodies, dendrites, and axons, as well as in glial cells. In contrast, MAP2A/B is selectively localized in dendrites and neuronal cell bodies.

MAP2A/B (simply referred to as MAP2) is one of the best *in vitro* substrates for cAMP-dependent protein kinase, type II Ca^{2+}/calmodulin-dependent protein kinase, protein kinase C, and proline-directed kinases such as MAP kinases, cyclin-dependent kinases (cdk1—also named cdc2—and cdk5), and glycogen synthase kinase 3. MAP2 can also be *in vitro* dephosphorylated by protein phosphatases 1, 2A, and 2B (calcineurin).

Current evidence suggests that some of these phosphorylation and dephosphorylation events also occur *in vivo*. Furthermore, the functionality of MAP2 seems to be modulated by its degree of phosphorylation. Highly phosphorylated MAP2 containing up to 46 phosphates per molecule binds less efficiently to tubulin than underphosphorylated MAP2 containing up to 16 phosphates per molecule. However, completely dephosphorylated MAP2 seems to be the least efficient in tubulin binding. *In vitro* studies have shown that extensive phosphorylation of MAP2 with several purified protein kinases decreases its association with tubulin. At least in the cases of protein kinase C and MAP kinase, this has been correlated with the phosphorylation of sites on the microtubule-binding domain of the MAP2 molecule. However, nothing is known about the protein kinases responsible for the phosphorylation at other sites that can stimulate the binding of MAP2 to microtubules.

In vivo phosphorylation of MAP2C has been observed during early brain development at a time of axonal growth. Part of this phosphorylation seems to be catalyzed by PDPKs.

In adult brain, high-molecular-mass MAP2 is heavily phosphorylated. A physiological role for MAP2 phosphorylation and dephosphorylation in neuronal plasticity in the adult brain has been proposed. [*See* Nervous System, Plasticity.] It is thought that the alteration of the MAP2 phosphorylation state in response to certain neurotransmitters may trigger cytoskeletal changes that contribute to synaptic remodeling. In fact, a rapid and selective MAP2 dephosphorylation that is catalyzed by the Ca^{2+}/calmodulin-dependent phosphatase calcineurin after stimulation of NMDA (*N*-methyl-D-aspartate)-type glutamate receptors has been described in rat hippocampus. This suggests that the phosphorylation state of MAP2 in hippocampal neurons may be controlled by signal transduction cascades activated by different neurotransmitters, as indicated in Fig. 2.

Furthermore, some preliminary evidence has shown a correlation between some modifications in MAP2 phosphorylation and certain examples of synaptic plasticity. Thus, MAP2 becomes more dephosphorylated at the end of the critical period for plasticity in the developing visual cortex of the cat. Interestingly, the prolongation of the critical period attained in dark-reared cats is paralleled by a delay in MAP2 dephosphorylation. Moreover, both the end of the critical period and MAP2 dephosphorylation can be induced by exposure of the dark-reared cats to light. It consequently appears plausible that MAP2 phosphorylation may perform a general role in neuronal plasticity, allowing cytoskeletal rearrangements within dendrites in response to changes in the synaptic input that such dendrites receive.

C. Tau Phosphorylation

Several tau proteins with apparent molecular weights ranging from 55,000 to 65,000 (as determined by gel electrophoresis) are found in the mammalian central nervous system. Additional tau isoforms with an apparent molecular weight of 110,000 have been identi-

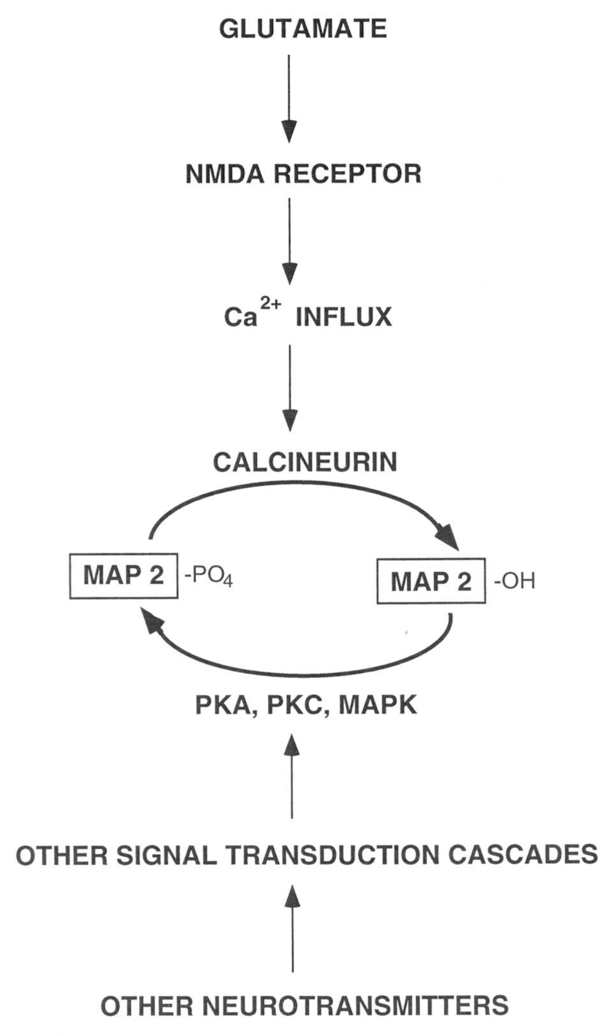

FIGURE 2 Role of neurotransmitters in the modulation of MAP2 phosphorylation and dephosphorylation.

fied in the peripheral nervous system. All of these tau protein isoforms are generated by alternative splicing. In brain, tau proteins are mainly localized within axons, although neuronal cell bodies and dendrites also contain some tau.

Tau proteins are *in vitro* phosphorylated by several protein kinases, including cyclic AMP-dependent protein kinase, type II Ca^{2+}/calmodulin-dependent protein kinase, protein kinase C, casein kinase 1, casein kinase 2, and PDPKs such as MAP kinases, cyclin-dependent kinases (cdk1, cdk2 and cdk5), and glycogen synthase kinase 3 (α and β).

Some of the residues modified by these protein kinases have been identified (Fig. 3 and Table III). Several serine and threonine residues corresponding to (Ser/Thr)-Pro motifs can be phosphorylated by PDPKs. Phosphorylation at some of these sites (which are located in the flanking regions of the tubulin-binding domain of the tau molecule as shown in Fig. 3) decreases the affinity of tau for tubulin, thus reducing the ability of tau to stabilize microtubules. Both protein phosphatase-2A and 2B (calcineurin) can readily dephosphorylate these sites *in vitro*. Phosphorylation of some serine residues located on the tubulin-binding domain (especially Ser^{262}) has a larger effect in diminishing the binding of tau to tubulin.

Phosphorylation at (Ser/Thr)-Pro motifs induces both an upward shift in the electrophoretic mobilities of tau proteins and a changed immunoreactivity toward different antibodies to phosphorylation-sensitive epitopes. Data obtained from transfected cells indicate that GSK-3 is one of the most effective PDPKs in this respect. *In vivo,* this type of phosphorylation is especially prominent in fetal tau as compared with adult tau, presumably as a consequence of a high activity of certain PDPKs (for instance, CDK-5 and GSK-3β) and a low activity of certain phosphatases (for instance, PP-2B) in the embryonic mammalian brain. This mode of tau phosphorylation may favor microtubule dynamics during axonal growth in developing neurons. Consequently, the dephosphorylation of these sites may contribute to microtubule stabilization during axonal maturation. Accordingly, fetal-type tau phosphorylation has some similarities with the MAP1B and MAP2C phosphorylations catalyzed by PDPK as discussed earlier. It is important to note that the hyperphosphorylation of tau proteins at these sites may contribute to the microtubule dysfunction that is found in Alzheimer's disease (see Section V).

IV. PHOSPHORYLATION OF MOTOR PROTEINS

Microtubules serve as tracks for the conveyance of organelles within cells, with the cooperation of a subclass of MTBPs referred to as "motor" proteins or microtubule-interacting ATPases. The best-characterized motor protein families include MAP1C (also named cytoplasmic dynein, which is completely unrelated to either MAP1A or MAP1B) and kinesin.

The study of fish melanophores has provided evidence for the involvement of protein phosphorylation–dephosphorylation in the regulation of bidirectional transport of pigment granules along microtubules. Movement away from the cell center requires cAMP, is accompanied by the phosphorylation of a M_r 57,000 protein, and is prevented by the

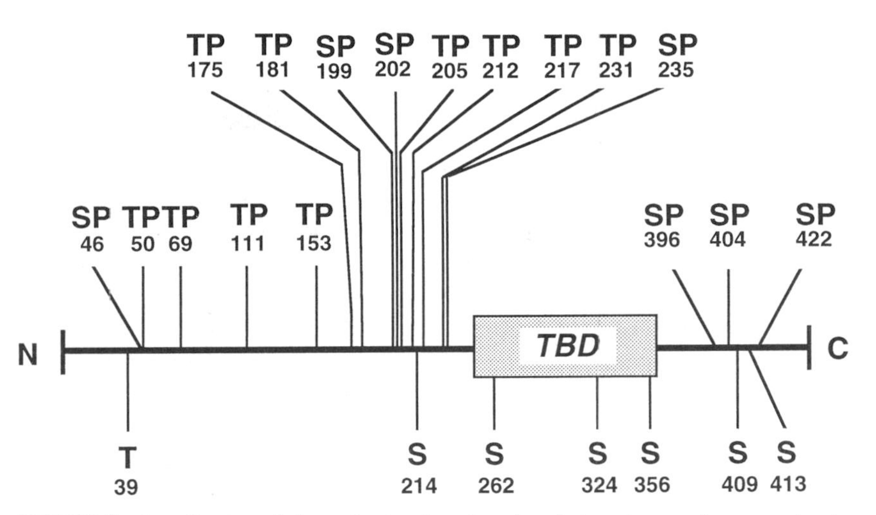

FIGURE 3 Localization of the major *in vitro* phosphorylation sites on the tau molecule. Numbers refer to the longest human tau isoform with 441 residues. TBD represents the tubulin-binding domain of the molecule.

protein inhibitor of the cAMP-dependent protein kinase. Furthermore, phosphatase inhibitors block both the movement of pigment granules toward the cell center and the concomitant dephosphorylation of the M_r 57,000 protein. These findings correlate well with the fact that pigment aggregation and dispersion are regulated by hormones that alter intracellular cAMP levels. However, the identity of the M_r 57,000 protein (possibly a motor protein-associated polypeptide) and its role in pigment granule transport are still unknown.

There is also evidence that cAMP levels modulate the movement of secretory vesicles in *Aplysia* bag cell neurons, possibly through the modification of the linkage between organelles, motor proteins, and microtubules. Major changes in microtubule-dependent vesicle motility have also been observed in cultured mammalian cells by drug treatments affecting intracellular protein kinases (particularly the cAMP-dependent protein kinase) and phosphatases.

Dynein, kinesin, and kinesin-associated proteins have been shown to be phosphorylated both *in vitro* and *in vivo*. The precise localization of the modified residues remains to be established. Current evidence suggests that the phosphorylation of kinesin and certain kinesin-associated proteins may modulate both motor–organelle and motor–microtubule interactions.

It thus appears that protein phosphorylation may control organelle movement. However, most studies have not yet correlated the changes in organelle motility with alterations in the phosphorylation state of motor proteins or associated polypeptides.

TABLE III
In Vitro Phosphorylation of Tau Protein[a]

Protein kinase	Major phosphorylation sites
PKA	Ser^{214}, Ser^{324}, Ser^{356}, Ser^{409}, Ser^{416}
CaM-kinase II	Ser^{416}
PKC	Ser^{324}
CDK-1 (cdc2)	Ser^{202}, Thr^{205}, Thr^{231}, Ser^{235}
CDK-5	Ser^{195}, Ser^{202}, Thr^{205}, Thr^{231}, Ser^{235}, Ser^{396}, Ser^{404}
MAP kinase	Ser^{46}, Thr^{181}, Ser^{199}, Ser^{202}, Thr^{212}, Ser^{235}, Ser^{396}, Ser^{404}, Ser^{422}
GSK-3α	Thr^{212}, Thr^{231}, Ser^{235}, Ser^{262}, Ser^{324}, Ser^{356}, Ser^{404}
GSK-3β	Ser^{199}, Thr^{231}, Ser^{396}, Ser^{413}
CK2	Thr^{39}

[a]All residue numbers refer to the longest human tau isoform with 441 residues. Abbreviations are as in Table II.

V. POSSIBLE IMPLICATIONS OF MICROTUBULE PROTEIN PHOSPHORYLATION IN NEUROLOGICAL DISORDERS

Because microtubule protein phosphorylation may be critically involved in the regulation of nerve cell morphology and in intraneuronal transport mechanisms, its alterations may lead to changes in the organization of microtubules, their associated proteins, and their interactions with other cell organelles, giving rise to a variety of neuropathies.

A clear case in point is provided by the experimental neuropathy caused by okadaic acid, a potent phosphatase inhibitor that induces a hyperphosphorylation of cytoskeletal proteins and neurodegeneration.

Likewise, abnormal cytoskeletal protein phosphorylation is thought to underlie some toxic neuropathies like the one produced after aluminum salt intoxication. Intracranial injections of aluminum in rabbits produce a severe neurodegenerative syndrome, characterized by the intraneuronal accumulation of hyperphosphorylated neurofilament proteins. In addition, chronic oral administration of aluminum to rats increases the phosphorylation of MAP2 and the heavy neurofilament subunit in the brain stem and cerebral cortex. This increased phosphorylation may constitute the first step in the deposition of hyperphosphorylated cytoskeletal proteins, including MAPs and neurofilaments, causing their irreversible aggregation.

Accumulations of abnormally phosphorylated cytoskeletal proteins are also found in several naturally occurring neurodegenerative disorders such as Alzheimer's disease, Parkinson's disease, Pick's disease, and amyotrophic lateral sclerosis. The cytoskeletal changes may constitute a response to insults due to metabolic disturbances, reductions in neurotrophic factor availability, or changes in synaptic interactions. The resulting disorganization of the cytoskeleton often leads to the formation of intraneuronal inclusions, alteration of neuronal morphology, and interference with axonal transport systems that may contribute to the accelerated degeneration of affected neurons.

The most frequent neurodegenerative disease is senile dementia of Alzheimer's type, which has therefore been most thoroughly studied. Alzheimer's disease is characterized by the degeneration of neurons in various cortical areas, which results in memory failure and other intellectual deficits. Two types of aberrant structures, senile amyloid plaques and neurofibrillary tangles, appearing within the brains of the patients serve as histopathological hallmarks of the disease. [*See* Alzheimer's Disease.]

Senile plaques are made up of extracellular amyloid deposits that form initial amorphous aggregates ("diffuse" plaques) that become increasingly fibrillar ("compacted" or "mature" plaques). Amyloid deposits are constituted by the β-amyloid peptide (β-AP), which is a fragment of an integral membrane protein referred to as the β-amyloid precursor protein (β-APP or simply APP). There are two major routes of APP processing. The normal (secretory) proteolytic processing of APP occurs through cleavage within the β-AP portion of APP, which generates a secreted form of APP and a nonamyloidogenic peptide fragment. The alternative route is a cleavage that produces intact β-AP and is consequently amyloidogenic.

Neurofibrillary tangles (NFT) are derived from intraneuronal inclusions, constituted by the aggregation of paired helical filaments (PHF), which become extracellularly located after neuronal cell death. In addition to NFT, PHF are also found within dystrophic neurites ("neuropile threads"). A modified form of tau protein is the major if not the exclusive component of PHF.

There is ample evidence supporting the view that abnormalities in signal transduction and protein phosphorylation are basic to the pathogenesis of Alzheimer's disease (Table IV). Thus, PKC, PDPKs, CK2, and protein tyrosine kinases seem to be altered in Alzheimer's disease, with a relative sparing of PKA and CaM-kinase II. Likewise, phosphoprotein phosphatases are defective in the brains of Alzheimer's disease patients.

The presence of PKC (particularly the βII isoform) in "diffuse" amyloid plaques suggests that PKC alteration might be an early biochemical marker of Alzheimer's disease. Interestingly, PKC controls APP processing, possibly through the phosphorylation of enzymes involved in APP cleavage. Reduced PKC activity can contribute to a diminished secretory processing and an augmented amyloidogenic cleavage of APP. This latter aberrant processing may also be favored after phosphorylation of APP by PDPK.

Abnormal hyperphosphorylation of tau protein may be one of the mechanisms driving the self-assembly of tau into the PHFs constituting Alzheimer's NFTs. In fact, highly phosphorylated tau is the major constituent of PHFs isolated from the brains of Alzheimer's disease patients. Hyperphosphorylated tau is also seen in the cytoplasm of some neurons containing delicate fibrillary inclusions, which may be precursors of NFTs. Thus, the accumulation of highly phosphorylated tau appears to be one of the earliest cytoskeletal changes in the process of NFT formation.

Perhaps there are sequences on the tau molecule that normally prevent aggregation into PHF, and phosphorylation at these sites may trigger PHF formation. Additionally, PHF-tau is completely unable to bind to tubulin and to stabilize microtubules. Because of their negligible affinity for tubulin, abnormally hyperphosphorylated tau molecules may easily interact with each other, probably at the microtubule-binding domain, to form dimers and finally PHF.

TABLE IV
Protein Phosphorylation Dysfunction in Alzheimer's Disease[a]

	Alterations	Putative substrates
Protein kinases		
PKC (βII)	Reduced activity Translocated to cytosol Associated with amyloid plaques	APP-processing enzymes
PDPKs	Increased activity	APP, tau, MAP1B
CK2	Reduced activity Associated with NFTs	MAP1B, tau
PTK	Reduced overall activity Maintained cytosolic activity	PDPK
Phosphatases		
PP2A	Decreased activity (?)	Tau, MAP1B
PP2B	Decreased activity (?)	Tau, MAP1B
Phosphoproteins		
APP	Abnormally proteolyzed	
Tau	Hyperphosphorylated (by PDPK)	
MAP1B	Hyperphosphorylated (by PDPK)	

[a]Abbreviations: APP, amyloid precursor protein; PKC, protein kinase C; PDPK, proline-directed protein kinase; CK2, casein kinase 2; NFT, neurofibrillary tangle; PP2A, phosphatase 2A; PP2B, phosphatase 2B, PTK, protein tyrosine kinase.

Thus, the abnormal hyperphosphorylation of tau may lead to both microtubule disorganization and PHF development, which might result in neuronal cell death.

The abnormal hyperphosphorylation of PHF-tau has been analyzed in great detail (Table V). The phosphorylation of PHF-tau consists of fetal-type phosphorylation (which is mainly due to PDPKs), additional proline-directed phosphorylation sites, and other non-proline-directed phosphorylation sites. It is interesting to note that fetal tau retains a low but still significant binding to tubulin, whereas PHF-tau is completely unable to associate with tubulin. This underscores the importance of the non-fetal-type phosphorylation that is specific for PHF-tau. Possibly PHF-tau is phosphorylated to a high degree because of the synergistic action of multiple protein kinases. Of particular relevance in this regard is glycogen synthase kinase 3 (GSK-3), as it is the most effective kinase able to modify tau in transfected cells. GSK-3 phosphorylates both proline-directed and non-proline-directed sites on tau protein. Phosphorylation of some sites by GSK-3 is dependent on prior phosphorylation by other kinases.

In addition to tau, hyperphosphorylated MAP1B has been found associated with NFTs.

An intense research effort is now under way to elucidate whether increased kinase activity or de-

TABLE V
Major Phosphorylation Sites Identified in PHF-Tau

Site[a]	Phosphorylated in fetal tau	Phosphorylated in PHF-tau
Ser^{148}	Yes	Yes
Ser^{199}-Pro	Yes	Yes
Ser^{202}-Pro	Yes	Yes
Ser^{208}	No	Yes
Ser^{210}	No	Yes
Thr^{212}-Pro	No	Yes
Ser^{214}	No	Yes
Thr^{217}-Pro	Yes	Yes
Thr^{231}-Pro	Yes	Yes
Ser^{235}-Pro	Yes	Yes
Ser^{262}	Yes	Yes
Ser^{396}-Pro	Yes	Yes
Ser^{400}	Yes	Yes
Thr^{403}	No	Yes
Ser^{404}-Pro	Yes	Yes
Ser^{409}	Yes	Yes
Ser^{412}	No	Yes
Ser^{413}	Yes	Yes
Ser^{422}-Pro	No	Yes

[a]All residue numbers refer to the longest human tau isoform with 441 residues.

creased phosphatase activity is responsible for the hyperphosphorylation of tau and MAP1B in Alzheimer's disease. Preliminary evidence suggests that some phosphatases are underproduced in affected neurons or that their activity is somehow inhibited.

Interestingly, proteolytic products of APP constitute extracellular signals that may affect tau protein phosphorylation within neurons. Amyloid peptide can activate GSK-3 and tau phosphorylation in cultured neurons.

Clearly a more thorough understanding of the abnormalities in protein phosphorylation in Alzheimer's disease may contribute to the development of novel strategies for therapeutic intervention.

VI. MICROTUBULE PROTEIN PHOSPHORYLATION IN THE CELL CYCLE

Microtubule protein phosphorylation is also implicated in the regulation of cell division in all eukaryotic cells at two key points: the interphase to mitosis transition (prometaphase) and the metaphase to anaphase transition.

The few long and relatively stable cytoplasmic microtubules nucleated by the centrosome in interphase disappear at the beginning of mitosis and are converted into the numerous, short, and highly dynamic microtubules nucleated by the mitotic poles that result from the division of the centrosome. The molecular mechanisms responsible for the increased microtubule-nucleating ability of mitotic poles and for the increased microtubule dynamic instability occurring at the beginning of mitosis are not yet entirely known, although their dependence on protein phosphorylation is clear. Indeed, activation of cdk1 (cdc2) kinase is thought to trigger the onset of mitosis. Addition of purified mitotic cdc2 kinase to *Xenopus* egg extracts results in the activation of the mitotic regime of microtubule dynamics and of the nucleating activity of centrosomes.

cdc2 kinase is bound to microtubules not only in *Xenopus* eggs but also in the mitotic spindles of cultured mammalian cells. Mitotic MAPs are therefore possible substrates for this mitotic kinase. In particular, cdc2 seems to associate with MAP4 through cyclin B, the regulatory subunit of mitotic cdc2.

A prometaphase-specific increase in the phosphorylation of MAP4 has been observed in cultured mammalian cells. The phosphorylation of MAP4 during mitosis may therefore be catalyzed by cdc2 and may contribute to the enhanced dynamics of mitotic microtubules. However, additional factors of MAP4 phosphorylation may be required to explain the dynamics of mitotic microtubules and the increase in the nucleating activity of centrosomes.

Finally, protein phosphatases are involved in the metaphase to anaphase transition. For instance, the kinesin-like motor protein CENP-E, which is phosphorylated by cdc2 kinase at prometaphase, is dephosphorylated at the onset of anaphase. This dephosphorylation activates the microtubule cross-linking ability of CENP-E, which results in the association of the protein with microtubules at the spindle midzone. Presumably the dephosphorylation of other cdc2 substrates also contributes to the completion of mitosis.

BIBLIOGRAPHY

Aoki, C., and Siekevitz, P. (1988). Plasticity in brain development. *Sci. Am.* **259**, 34–42.

Avila, J., Domínguez, J., and Díaz-Nido, J. (1994). Regulation of microtubule protein dynamics by microtubule-associated protein expression and phosphorylation during neuronal development. *Int. J. Dev. Biol.* **38**, 13–25.

Buendia, B., Draetta, G., and Karsenti, E. (1992). Regulation of the microtubule-nucleating activity of centrosomes in *Xenopus* egg extracts. *J. Cell Biol.* **116**, 1431–1442.

Goedert, M. (1993). Tau protein and the neurofibrillary pathology of Alzheimer's disease. *Trends Neurosci.* **16**, 460–465.

Liao, H., Li, G., and Yen, T. J. (1994). Mitotic regulation of microtubule cross-linking activity of CENP-E kinetochore protein. *Science* **265**, 394–398.

Lovestone, S., Reynolds, C. H., Latimer, D., Davis, D. R., Anderton, B. H., Gallo, J. M., Hanger, D., Mulot, S., Marquardt, B., Stabel, S., Woodgett, J. R., and Miller, C. C. J. (1994). Alzheimer's disease-like phosphorylation of the microtubule-associated protein tau by glycogen synthase kinase-3 in transfected mammalian cells. *Curr. Biol.* **4**, 1077–1086.

McIlvain, J. M., Burkhardt, J. K., Hamm-Alvarez, S., Argon, Y., and Sheetz, M. P. (1994). Regulation of kinesin activity by phosphorylation of kinesin-associated proteins. *J. Biol. Chem.* **269**, 19176–19182.

Morishima-Kawashima, M., Hasegawa, M., Takio, K., Suzuki, M., Yoshida, H., Titani, K., and Ihara, Y. (1995). Proline-directed and non-proline-directed phosphorylation of PHF-tau. *J. Biol. Chem.* **270**, 823–829.

Roush, W. (1995). Protein studies try to puzzle out Alzheimer's tangles. *Science* **276**, 793–794.

Saitoh, T., Masliah, E., Jin, L-W., Cole, G. M., Wieloch, T., and Shapiro, J. P. (1991). Protein kinases and phosphorylation in neurologic disorders and cell death. *Lab. Invest.* **64**, 596–616.

Sim, A. T. R. (1992). The regulation and function of protein phosphatases in the brain. *Mol. Neurobiol.* **5**, 229–246.

Walaas, S. I., and Greengard, P. (1991). Protein phosphorylation and neuronal function. *Pharmacol. Rev.* **43**, 299–349.

Photoelectric Effect and Photoelectron Microscopy

O. HAYES GRIFFITH
KAREN K. HEDBERG
University of Oregon

GLOSSARY

Fluorescence microscopy Type of optical microscopy in which the image is formed by light emitted by the specimen (e.g., fluorescence). The microscope is operated in the dark-field mode with a blocking filter so that only the fluorescence light, and not the exciting light, reaches the ocular. In combination with a fluorescent dye-labeled antibody to label specific sites, this is called immunofluorescence microscopy

Ionization potential Minimum voltage required to remove an electron completely from a molecule (or atom or ion). The ionization energy is the potential times the charge of the electron. By convention, ionization potentials are usually reported in units of energy (electron volts)

Photoelectric effect Ejection of electrons from the surface of a solid, a liquid, or a gas when it is illuminated by light

Photoelectron imaging Any form of imaging in which the source of information is the distribution of points from which electrons are ejected from the specimen by the action of ultraviolet light (i.e., the photoelectric effect). The highest-resolution technique of photoelectron imaging is presently photoelectron microscopy, also known as photoemission electron microscopy

Photoelectron microscopy Type of microscopy in which a specimen is exposed to light and the resulting emitted electrons are used to form an image of the exposed surface. The electrons are first accelerated and then focused by means of an electron lens system similar to that of a transmission electron microscope. In physics, photoelectron microscopy is often referred to as photoemission electron microscopy, abbreviated PEM or PEEM

Photoelectron spectroscopy Analysis of the kinetic energies of electrons ejected from gases, liquids, or solids by electromagnetic radiation, usually ultraviolet light, X rays, or synchrotron radiation. The ionization potentials identified from the photoelectron spectra are used to determine the electronic structure of molecules and surfaces

Work function In plots of the electron emission from a conductive metal surface versus wavelength of incident light, the work function corresponds to the threshold of photoemission. The onset of photoemission is generally not sharp. The work function is determined from a fit of the experimental curve with an analytical equation (e.g., Fowler equation)

PHOTOELECTRON MICROSCOPY IS THE ELECTRON optical analog of fluorescence microscopy. Light is focused on a specimen, and the emitted electrons are accelerated and imaged by means of an electron optical system. The enlarged image is captured on film and yields a photograph of the exposed surface of the specimen. Contrast is provided by intrinsic differences in the ionization potentials of cell components, by extrinsic labeling with photoemissive markers, and by surface topography. Photoelectron microscopy can be used in the study of a variety of biological specimens, including well-spread cells in culture, cell organelles, cytoskeletal structures, membranes, and DNA. The traditional method is to use ultraviolet (UV) light from a mercury arc lamp or laser that is just capable of ionizing the most weakly held outer electron (i.e., the threshold region). A recent variation is to use synchrotron radiation as the light source. With the much higher energies available from synchrotron radiation, it is possible to probe core electron levels and thus to obtain information regarding elemental distributions. Because of the possibility of obtaining analytical information, this is referred to as photoemission spectromicroscopy. The trade-off is that the resolution is lower than in conventional photoelectron microscopy, which uses threshold illumination.

I. THE PHOTOELECTRIC EFFECT

The basic concept of the photoelectric effect is illustrated in Figure 1. Historically, the photoelectric effect has played important roles in establishing that electrons are subparticles of atoms and in developing the quantum theory of radiation (e.g., the concept that despite its wave nature, light nevertheless has some properties akin to those of particles). Albert Einstein received the Nobel Prize in 1921 for his seminal 1905 paper, which introduced the concept of units of light energy called photons or quanta. Any material will exhibit a photoelectric effect if the incident light quanta are of sufficiently high energy (e.g., UV light) to eject electrons. All three states of matter—solid surfaces, liquids, and gases—can exhibit a photoelectric effect.

There are several types of information available in the photoelectric effect. These include: (1) the number of electrons released per incident photon (this number, which is usually much less than one, is also called the efficiency of photoemission or the photoelectron quantum yield), (2) the kinetic energy of the emitted electrons, (3) the angular distribution of emitted electrons, and, for solids, (4) the positions from which the electrons leave the surface. Different instruments have been developed over the years that utilize each of these sources of information.

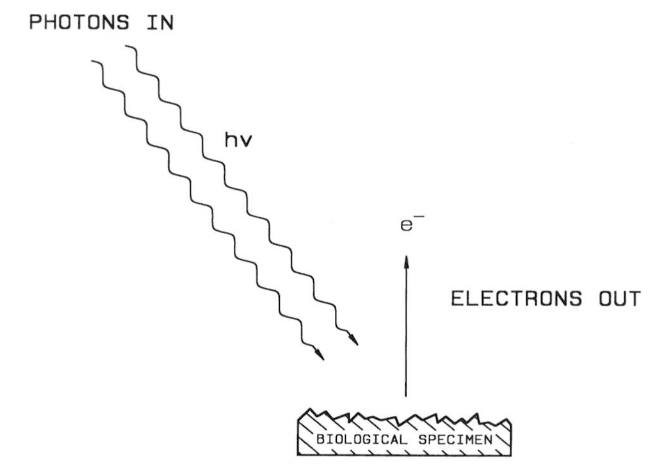

FIGURE 1 Diagram illustrating the photoelectric effect. Light ($h\nu$) striking a surface will cause the ejection of electrons providing that the energy of the photons is sufficiently great. For most materials, ultraviolet light is required.

Today, the photoelectric effect is used in many ways. Photocells and photomultipliers take advantage of the fact that light is converted into an electron current (e.g., Fig. 1). Photocathodes are surfaces that are specially prepared to have a high efficiency for emitting electrons. The best photocathodes generally contain the element cesium (e.g., the cesium–antimony or silver–oxygen–cesium photocathodes). Most optical spectrophotometers in chemistry and medical laboratories contain a photocell to detect changes in transmitted light by recording changes in the electric current. Image intensifiers and television cameras contain photocathodes to convert the light image into an electron image, which can then be amplified and scanned electronically.

Closely related to the photoelectric effect are the primary processes of photosynthesis. The chief difference is that in the (external) photoelectric effect the electrons are removed essentially to infinity, whereas in the light reactions of photosynthesis the photoejected electrons are trapped near the reaction centers.

Photoelectron spectrometers are instruments that shine a monochromatic beam of light on a specimen and measure the kinetic energy, number, and sometimes the angular distribution of emitted photoelectrons. Gases, in particular, exhibit very sharp and well-resolved photoelectron spectra, from which information about the electronic structure of the molecule is derived. UV light removes only the more loosely bound or valence electrons, including those that are involved in chemical reactions. Hence UV photoelectron spectroscopy, or UPS, is used to study the molecular orbitals of atoms and molecules. If the energy source is a beam of X rays, the technique is called X-ray photoelectron spectroscopy (XPS) or electron spectroscopy for chemical analysis (ESCA). X-ray photons have sufficient energy to remove core electrons as well as valence electrons, and XPS is therefore applicable for chemical analysis (e.g., detecting elements and oxidation states).

II. THE PHOTOELECTRON MICROSCOPE

In photoelectron microscopy, instead of collecting the total current (as in a photocell) or analyzing the kinetic energies (as in a photoelectron spectrometer), the positions at which the electrons leave the specimen are preserved by means of a series of electron lenses. Historically, simple versions of photoelectron microscopes were built in Germany as early as 1932 for the study of metals. The first use of photoelectron microscopy to image organic and biological surfaces was in 1972. This remains a relatively small field compared to other forms of electron microscopy. [*See* Electron Microscopy.]

There is a strong analogy between photoelectron microscopy and the widely used technique of fluorescence microscopy. The essential features of these two methods are compared in Figure 2. Fluorescence involves the absorption of light by the specimen followed by emission of light at longer wavelengths. A glass objective lens focused on the specimen collects a portion of this fluorescent light and forms an enlarged image of the pattern of emission. The image is further enlarged by an ocular (projector lens), resulting in an image of the biological specimen in which bright areas correspond to regions rich in fluorescence emission. The idea in photoelectron microscopy is to retain the contrast inherent in an emission experiment, that is, imaging bright (strongly photoemissive) objects against a dark background, while greatly improving the resolution compared to optical methods. UV radiation is used to eject electrons, and the electrons are accelerated and focused by electron objective, intermediate, and projector lenses. Thus, photoelectron microscopy is the electron optical analog of fluorescence microscopy.

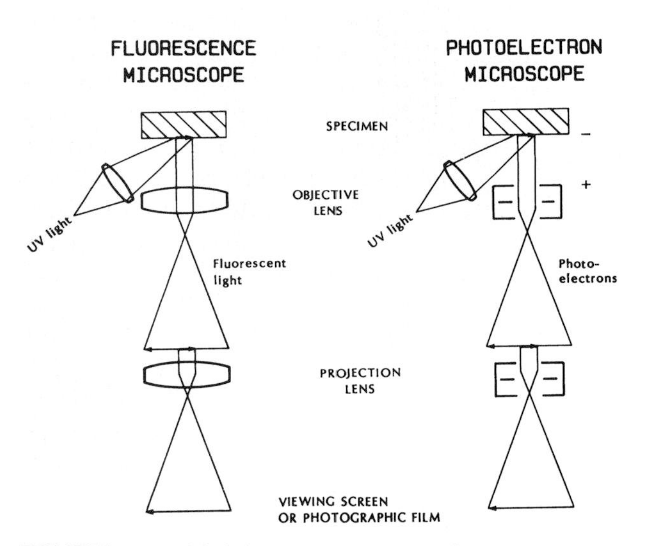

FIGURE 2 Simplified diagrams comparing a fluorescence microscope (left) with a photoelectron microscope (right). Both microscopes rely on light as the source of excitation and both magnify the image in the same general way. The fluorescence microscope uses glass optical lenses and the photoelectron microscope uses electron lenses.

III. ADVANTAGES AND LIMITATIONS

Photoelectron microscopy is still a developing technique and instrumentation is not yet widely available. The resolution of current photoelectron microscopes is between 5 and 10 nm. The main advantage of photoelectron microscopy is the source of contrast. Small differences in the bonding of the outermost electrons can cause significant contrast between molecules that would be otherwise difficult to detect, for example, hemes, chlorophylls, and aromatic hydrocarbons. A second advantage is the extremely high sensitivity to surface relief as encountered in studies of fine fibers, membrane proteins, cytoskeletal elements, and DNA. A third advantage is lower specimen damage compared with electron beam methods. The main limitation is that, owing to the extreme sensitivity to topography, the specimens must be relatively flat. Another limitation is caused by the fact that light travels unimpeded through water but eletrons do not. It follows that, for example, fluorescence microscopy is performed on wet and often living specimens, whereas the specimens for photoelectron microscopy must be frozen or dehydrated. This is a limitation of essentially all electron microscopes. Thus, a frequent strategy is to use a combination of optical microscopy and electron microscopy. Photoelectron microscopy can be viewed as one specialized type of electron microscopy. It differs from the better-known types such as transmission electron microscopy and scanning electron microscopy in that there is no electron gun. The specimens are the source of electrons in photoelectron microscopy.

IV. SELECTED APPLICATIONS

Photoelectron microscopy, like scanning electron microscopy, is a technique for studying the exposed surfaces of specimens. It follows that the study of cell surfaces is an appropriate area for its application. An example is a study in which photoelectron microscopy was used to examine the effects of phorbol ester tumor promoters on cell-surface fibronectin. Fibronectin is a large glycoprotein that plays a role in adhesion, migration, and differentiation. It binds to the surfaces of fibroblastic cells (cells of mesenchymal origin, generally destined to be connective tissue) at specific

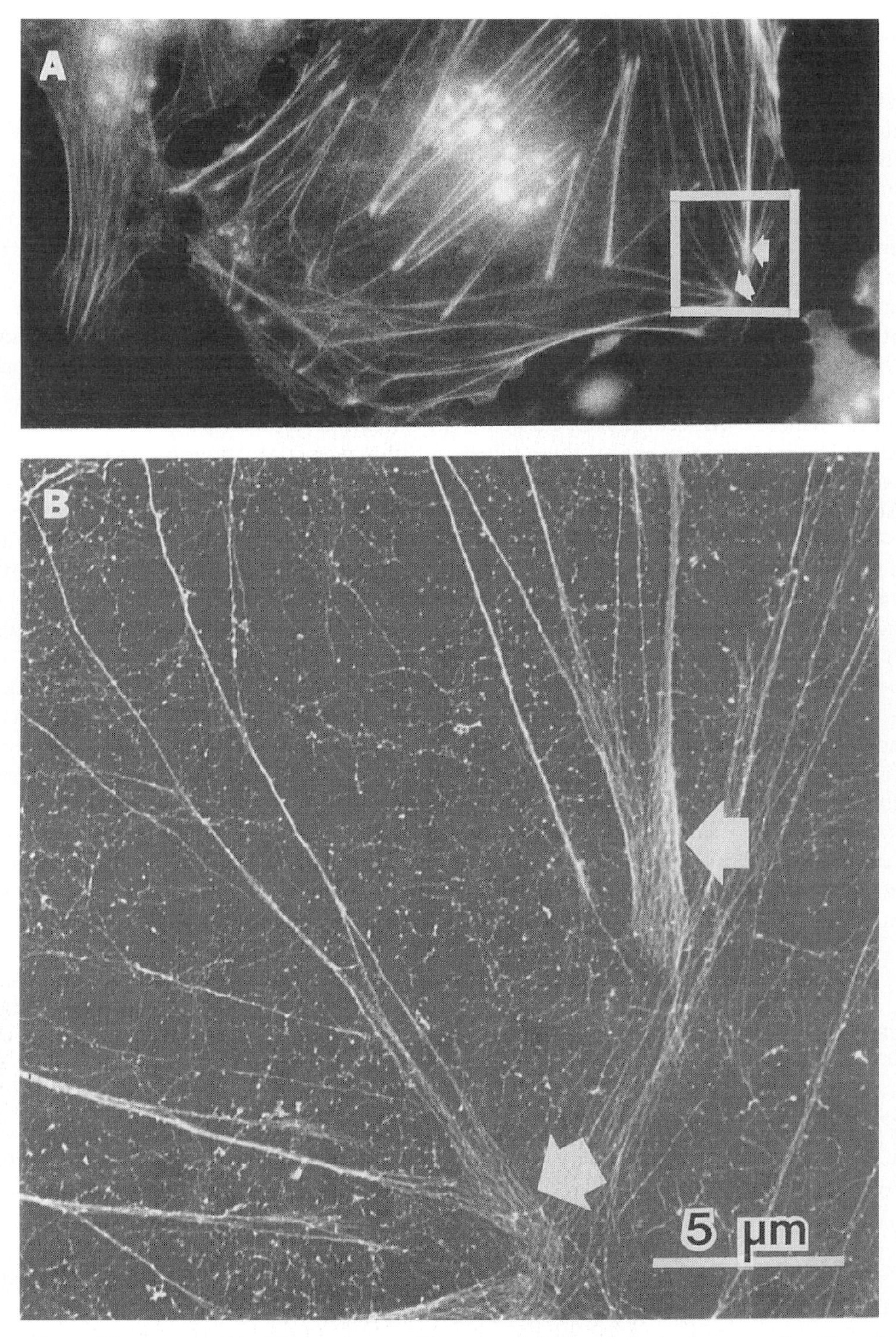

FIGURE 3 Comparison of fluorescence (A) and photoelectron (B) images of the cytoskeleton of a mouse fibroblast (Swiss 3T3 cell line) cultured on a sample mount. The box in (A) indicates the boundaries of the region shown in the higher-magnification photoelectron micrograph (B). The surface membranes of the cells were removed with neutral detergent, and then the cells were selectively labeled for fluorescence visualization of filamentous actin by means of rhodamine-conjugated phalloidin. The bright streaks in (A) indicate the presence of actin microfilament bundles crisscrossing the main cell body (the bright spots in the center are nonspecific fluorescence from the nuclear region of the cell). The fluorescent label seen in (A) does not serve as a photoelectron label. The major source of contrast in the photoelectron image (B) is the topography of the exposed cellular structures. This topographical contrast allows the actin microfilament bundles present in the boxed region to be easily identifiable. Many of the microfilament bundles in the fluorescence micrograph appear to have blunt, slightly bulbous ends. In the photoelectron micrograph, these ends are resolved to be splayed meshworks of filaments terminating on the lower cell surface (arrows). These regions, known as adhesion plaques or focal contacts, are believed to contain transmembrane proteins that link the actin cytoskeleton inside the cell to adhesive proteins on the substrate beneath the cell.

transmembrane receptors. To distinguish it from other proteins on the cell surface, fibronectin on human fibroblasts was labeled by an indirect antibody method with 6-nm-diameter colloidal gold as the marker (e.g., immunophotoelectron microscopy). Colloidal gold is much more photoemissive than cell-surface components, and the small gold spheres are readily recognized as patterns of bright dots against the darker cell-surface background. The contrast provided by this label can be boosted by silver enhancement, which provides a coating of metallic silver on the gold particles. This approach made it possible to observe the change in patterns and the release of cell-surface fibronectin induced by the tumor promoter. The images of cell surfaces are well resolved because photoelectron emission from within or beneath the cell does not reach the sample surface and therefore does not contribute to the image. For this kind of study, cells are generally unstained and uncoated. Even without labeling, the cell surfaces are visible by the topographic contrast inherent in photoelectron microscopy. Thus, photoelectron microscopy can be used to simultaneously image both labeling patterns and the topography of the cell surface.

Components inside cells may be studied after removal of the cell membrane with detergents or by mechanical shearing. The cytoskeleton of cultured cells, including microtubules, intermediate filaments, and actin-containing filament bundles, are within the useful imaging range of photoelectron microscopy. These photoelectron images are especially detailed and informative. One representative application is the study of effects of activators of protein kinase C on the actin cytoskeleton of cultured cells. Colloidal gold, or silver-enhanced colloidal gold, is the most frequently used marker, although much smaller (e.g., molecular) markers are possible. Double labeling, with one fluorescent marker (e.g., a dye) and one photoemissive marker, can also be used so that both fluorescence and photoelectron images can be obtained from the same specimen. Another variation is to use a fluorescence marker but no photoemissive marker, and to compare the fluorescence (labeled) image with the photoelectron (unlabeled) image of the same region. In this case, the photoelectron image is primarily due to the topography of the cytoskeletal elements. An example is shown in Figure 3, with a fluorescence image (top) and an enlarged photoelectron image of the boxed area (bottom). Note the much higher resolution of the photoelectron image.

As a final example, it has recently been shown that nucleic acids can be imaged by photoelectron microscopy. Figure 4A illustrates the photoemission of DNA. This experiment is reminiscent of autoradiography, in which electrons are released by radioactive decay. However, in photoelectron microscopy the DNA is

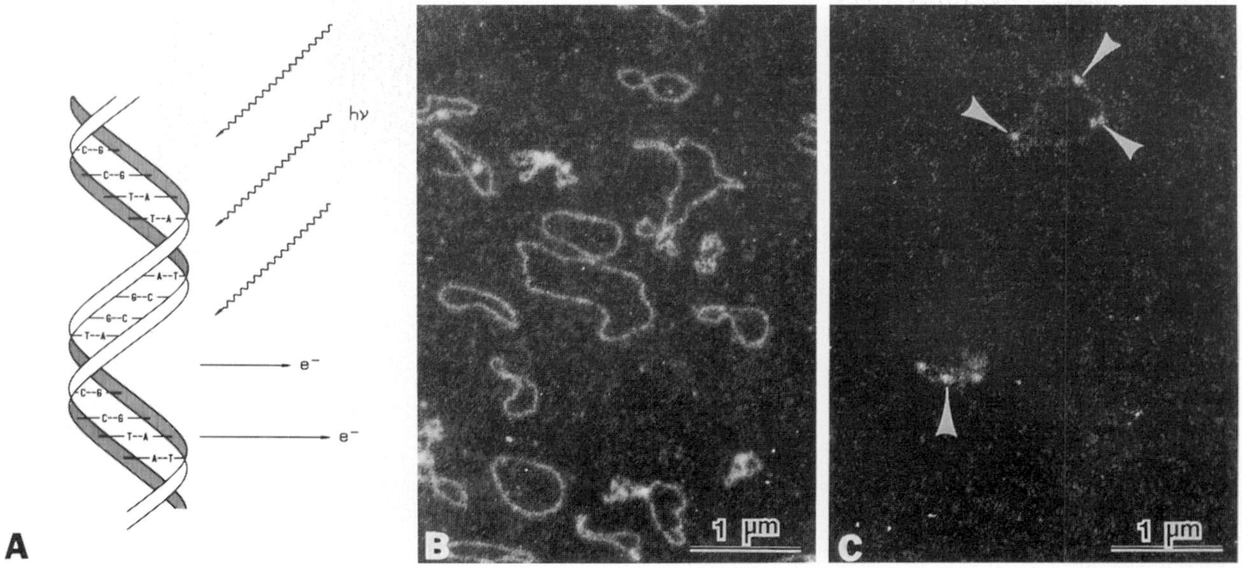

FIGURE 4 Photoelectron imaging of DNA. (A) Diagram illustrating photoemission of DNA. (B) Photoelectron images of DNA plasmids (pBR322) prepared by the cytochrome *c* spreading technique. (C) *Escherichia coli* RNA polymerase protein complexes (arrows) bound to naked DNA plasmids (no cytochrome *c*). [Adapted, with permission, from O. H. Griffith *et al.* (1990). *Biopolymers* 29, 1491–1493.]

not radiolabeled and the electrons are released by the action of UV light. This makes possible a higher resolution because the number of electrons released is larger and their kinetic energies are much smaller than with autoradiography. Figure 4B is a photoelectron image of pBR322 DNA plasmids spread by the standard cytochrome *c* method used in transmission electron microscopy. The cytochrome *c* coats the DNA so the effective diameter is probably 7 to 10 nm, but still much less than that of conventional electron microscopy preparations that utilize metal coating or staining. Figure 4C is a photoelectron image of naked DNA plasmids (no cytochrome *c*) with RNA polymerase complexes bound to them. The DNA is only 2 nm in diameter. The significance of photoelectron imaging of DNA lies in the information content. The photoelectron image is formed by valence electrons, which are sensitive to molecular structure. It may prove possible, for example, to use this technique in studies of protein–DNA interactions and small molecule (carcinogen)–DNA interactions. A rapid method of physical mapping of chromosomes (but not base sequencing at present instrument resolution) is another possibility if the differences in ionization potentials of the bases provide sufficient contrast to produce a modulation in brightness along the images of the DNA duplex.

BIBLIOGRAPHY

Eland, J. H. D. (1984). "Photoelectron Spectroscopy." Butterworths, London.

Ghosh, P. K. (1984). "Introduction to Photoelectron Spectroscopy." Wiley–Interscience, New York.

Griffith, O. H. and Engel, W. (eds.) (1991). Proceedings of the Second International Symposium and Workshop on Emission Microscopy and Related Techniques. *Ultramicroscopy* **36** (1–3), 1–274.

Griffith, O. H., and Rempfer, G. F. (1987). Photoelectron imaging: Photoelectron microscopy and related techniques. *In* "Advances in Optical and Electron Microscopy" (R. Barer and V. E. Cosslett, eds.), Vol. 10, pp. 269–337. Academic Press, London.

Habliston, D. L., Birrell, G. B., Griffith, O. H., and Rempfer, G. F. (1993). Photoelectron imaging of DNA: A study of substrates and contrast. *J. Phys. Chem.* **97**, 3022.

Hedberg, K. K., Birrell, G. B., Mobley, P. L., and Griffith, O. H. (1993). Transition metal chelator TPEN counteracts phorbol ester-induced actin cytoskeletal disruption in C6 rat glioma cells without inhibiting activation or translocation of protein kinase C. *J. Cell Phys.* **158**, 337.

Hedberg, K. K., Birrell, G. B., Habliston, D. L., and Griffith, O. H. (1995). Tunable label contrast on the cell surface: Photoelectron imaging with multiple wavelength excitation. *J. Microsc. Soc. Am.* **1**, 253.

Picornaviruses/Poliovirus

ANIKO V. PAUL
MATTHIAS GROMEIER
ECKARD WIMMER
State University of New York at Stony Brook

GLOSSARY

Complementation Process in which defects of mutant genes are corrected by the compensatory action of two homologous genetic systems

***De novo*, cell-free synthesis** Demonstration that poliovirus can be grown in a cell-free extract of mammalian cells

Homologous recombination Process in which incomplete nascent RNA strands switch templates at homologous nucleotide sequences without addition or deletion of nucleotides

Neurovirulence Ability of a virus to cause damage to the central nervous system and neurological symptoms

Picornavirus Small (pico) RNA (rna) virus

Recombination Exchange of genetic elements between two viruses during replication

PICORNAVIRUSES SUCH AS POLIOVIRUS ARE SMALL RNA-containing animal viruses causing a variety of diseases.

I. INTRODUCTION

Poliovirus, discovered in 1909 by K. Landsteiner and E. Popper in Vienna, has terrorized humankind for millennia. It causes a disease that destroys motor neurons and, consequently, muscle function. The disease, called poliomyelitis, may kill or it may leave victims paralyzed for the rest of their lives.

In an unprotected (nonvaccinated) population, poliovirus infects a vast number of humans, replicating almost unnoticed in the gastrointestinal tract. In only a small percentage of the infections (~1%), the virus finds its way into the central nervous system (CNS), where it targets the motor neurons (see Section III). Selection for CNS invasion is irrespective of a person's age, gender, race, or socioeconomic stratum. Why poliovirus chooses to inflict one child with poliomyelitis but spares his or her siblings or friends remains to be explained. It is a selection made seemingly at random.

Particularly mysterious was the observation that in developing countries, the incidence of poliomyelitis increased with an increasing standard of hygiene. Indeed, poliomyelitis has occurred only during this century in epidemic proportions. This conundrum, which defied commonly held perceptions of controlling infectious diseases, has been partially explained after closely monitoring the epidemiological behavior of poliovirus. In unvaccinated populations, the ubiquitous virus infects very young infants while they are still protected from neuroinvasion by maternal antibodies. The two events, "passive" immunization with maternal antibodies and infection within the first months of life, lead to life-long protection against the disease. However, raised standards of hygiene may interfere with either process. Thus, a child may be infected without carrying maternal antibodies (as the mother may never have had a poliovirus infection) or after protection by maternal antibodies has waned. In any case, a population without early exposure to poliovirus remains unprotected and is thus a fertile ground for poliomyelitis epidemics caused by poliovirus that has been imported from elsewhere.

Although poliovirus is highly infectious, the infections are remarkably seasonal. In North America and Europe, poliovirus strikes during the summer months, a phenomenon that has been linked to the virus' preference for high climatic humidity.

In industrialized countries, the terror of poliomyelitis came to an end in the 1950s through the development and application of two highly effective vaccines: the killed vaccine developed by Jonas Salk, and the live, attenuated vaccine developed by Albert Sabin. By the year 2000, almost one hundred years after its discovery, poliovirus may cease to be a curse on human life altogether. By then, poliovirus may be eradicated worldwide (see Section VIII).

Poliovirus is a relatively simple molecular entity consisting of only five types of macromolecules (60 copies each of four capsid proteins plus one copy of vRNA). Its organic matter can be described in the most basic chemical terms ($C_{332,662}$; $H_{492,388}$; $N_{98,245}$; $O_{131,196}$; $P_{7,501}$; $S_{2,340}$; for the consensus sequence of a quasi-species, see Section VI). Therefore, poliovirus can be considered a chemical with a life cycle and with properties that are pathogenic for humans. Viewed in the electron microscope, poliovirus has the appearance of a golf ball.

Because of its devastating impact on society, poliovirus has become one of the most thoroughly studied infectious human pathogens. This has led to a number of achievements important to the medical and biological sciences in general. Among these are: (a) development of the plaque assay for animal viruses; (b) demonstration of polyribosomes; (c) description of the polyprotein; (d) description of an RNA-dependent RNA polymerase and of replication intermediates; (e) discovery of a protein-linked 5′ terminus of genome RNA characteristic for numerous RNA viruses; (f) development of "reverse genetics" of RNA viruses by generating infectious cDNA (reverse transcription of viral genomic RNA, genetic engineering of the virus-specific cDNAs, and reversion of the cDNA to genomic RNA); and (g) discovery of a new genetic element called "internal ribosomal entry site" (IRES) that mediates cap-independent translation.

Poliovirus is the first mammalian RNA virus whose genome structure and genetic organization have been elucidated. Moreover, poliovirus has been synthesized *de novo* in a cell-free medium of uninfected cells, an achievement that has nullified the dictum that viruses can proliferate only in intact cells.

II. CLASSIFICATION

Poliovirus belongs to the *Picornaviridae,* a very large virus family that has been divided into the genera of *Enterovirus, Rhinovirus, Hepatovirus, Cardiovirus,* and *Aphthovirus*. These small plus-strand RNA viruses, which depending on the species infect humans, animals, and insects, can cause numerous diseases ranging from the serious (poliomyelitis, hepatitis, myocarditis, foot-and-mouth disease) to the benign (common cold). Properties of poliovirus, the prototype of *Picornaviridae,* will be discussed in some detail.

III. TISSUE TROPISM AND PATHOGENICITY

Poliovirus is characterized by a highly restricted host and tissue tropism. Only primates develop disease following poliovirus infection, and disease symptoms are predominantly neurological. Poliovirus infections of the CNS are characterized by specific clinical manifestations and pathohistological signs.

Infection with poliovirus follows ingestion. It leads to first rounds of viral replication in tonsular tissue and at an unknown site within the gastrointestinal tract. As a result, large amounts of infectious particles are excreted with stool by infected individuals during the initial stages of the infection. Following the "enteric phase," virus can be found in lymphatic structures associated with the gastrointestinal tract (lymphatic phase) and the bloodstream (viremic phase). There are few, if any, symptoms of infection at this stage. However, in a minor proportion of infections (ca. 1%) the viremic phase leads to invasion of virus into the central nervous system. Predominantly a distinctive subset of neurons within the CNS is susceptible to poliovirus replication. Infection and, ultimately, destruction of motor neurons of the spinal cord and the brain stem produce the characteristic clinical signs of poliomyelitis: flaccid paralysis.

The molecular determinants of the unusually restricted cell tropism of poliovirus have been the subject of intense investigations over many years. It became evident that the interactions of poliovirus with its host are manyfold and complex. Viral factors as well as host components contribute equally to produce the specific pathogenic features of poliomyelitis.

The first insight into the molecular structures underlying the ability of poliovirus to destroy motor neurons (neurovirulence) was gained from the study of attenuated variants of the naturally occurring poliovirus strains. These variants (named after Sabin), attenuated viral strains used as oral vaccines, attained a defect in the ability to infect cells of the CNS. The growth properties in the gastrointestinal tract, however, were largely unaltered. This property made the attenuated Sabin strains ideal candidates for use as immunizing agents for the prevention of poliomyelitis. Sequence comparison between the Sabin strains and their wild-type progenitors, which was possible for the type 1 and type 3 strains, revealed genetic alterations that accounted for the loss of neuropathogenicity. Various parts of the viral genome were found to contribute to the attenuation phenotype. Particularly the 5′ nontranslated region (5′ NTR) harboring the IRES (see the following) has been identified to be an important determinant of neuropathogenicity. All three Sabin strains contain a single mutation within their respective IRESes that contributes critically to the attenuation phenotype. However, the exact mechanism of how IRES elements can influence the clinical outcome of a poliovirus infection remains a matter of debate. Apart from this single, attenuating point mutation in the IRES, the genomes of the Sabin vaccine strains harbor other mutations that equally strongly contribute to the attenuation phenotype. For example, poliovirus type 1, Sabin [PV1(S)], carries 51 mutations compared to the wild-type progenitor strain, poliovirus type 1, Mahoney [PV1(M)]. It is the cumulative effect of these many mutations, particularly in the IRES and the coding regions for the capsid and the viral RNA polymerase, that makes PV1(S) the safest of all three types of the vaccine strains, even though PV1(M) is the most virulent of the wild-type poliovirus strains.

Apart from viral components influencing the pathogenic features of poliovirus, host factors play an important role in the determination of the neurovirulent phenotype. Host factors needed for replication or expression of the viral genome may be distributed in a cell-specific manner and thus determine sites of susceptibility toward poliovirus infection. Little is known about this aspect of infections of humans. Various attempts to isolate host proteins that are required for virus growth have not identified candidates that could explain the restrictive pattern of viral infection within the CNS. The lymphocyte homing protein CD44, however, may be such a cellular protein. The single most important determinant for poliovirus tissue tropism and pathogenesis is the human poliovirus receptor (hPVR) (see Section V,A).

IV. STRUCTURE OF THE VIRION

Poliovirus is a nonenveloped, icosahedral particle (Fig. 1) with a diameter of 28 nm and a molecular weight of 8.5×10^6 daltons, of which about 30% is RNA and the rest consists of protein. The capsid consists of 60 copies each of four virus-specific coat polypeptides (VP1, VP2, VP3, and VP4). Of these VP1,

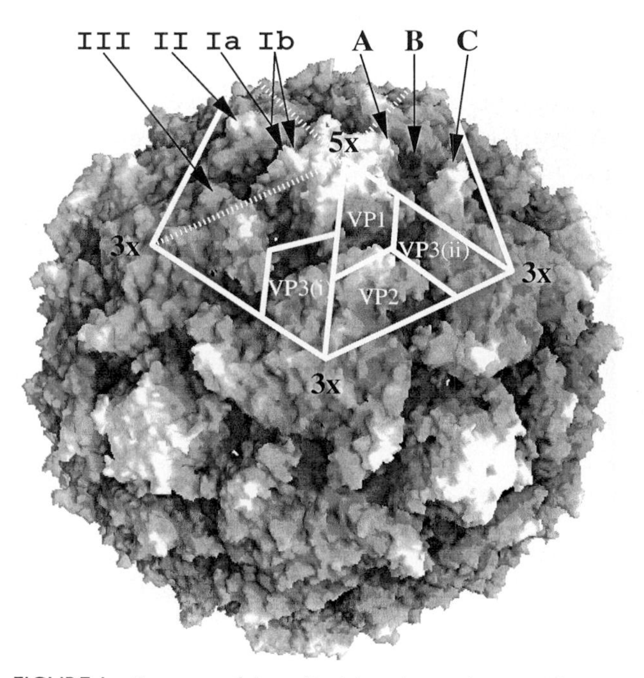

FIGURE 1 Structure of the poliovirion. A complete capsid structure of PV1(M) is illustrated as a water-accessible molecular surface. One of the 12 pentameric subunits of the capsid and its 5 constituent triangular pseudoprotomeric subunits are illustrated. The 5X and 3X labels indicate the locations of the fivefold and the threefold axes of this pentamer. The twofold axes occur at the intersection of the three adjacent pentamers. The central pseudoprotomer illustrates the subunit geometry of VP1, VP2, and VP3(ii). The biologically relevant protomer (to viral assembly) is pear-shaped and consists of VP1, VP2, and VP3(i). The internal VP4 protein is not visible from the surface. The canyon's north wall (A), south wall (C), and bottom (B) are indicated. The major antigenic sites to which neutralizing antibodies bind are labeled Ia, Ib, II, and III on an adjacent pseudoprotomer.

VP2, and VP3 each form eight-stranded, antiparallel beta barrels that assemble to cast the surface of the capsid, whereas the small VP4 lines the inner surface. No posttranslational modifications of the viral capsid proteins have been identified, with the exception of the N termini of VP4, which are myristoylated. Inside the protein shell resides the single-stranded plus-sense genomic RNA, approximately 7500 nucleotides (nt) long, which functions as mRNA. The 5′ end of the RNA is uncapped, and instead the terminal uridine nucleotide is covalently linked to a tyrosine residue of the virus-encoded, small peptide VPg (Fig. 2). No modified nucleotides other than the unique 5′ terminus have been found in the genome. Poliovirus genomic RNA possesses a 5′ NTR that is surprisingly long (742 nt), a characteristic common to all picornaviruses. This region contains two cis-acting elements (cloverleaf and IRES) that are important for viral proliferation. The 5′-terminal cloverleaf-like structure (nt 1–88) functions as a recognition signal in plus-strand RNA synthesis. The second genetic element within the 5′ NTR, the highly structured IRES (nt 320–631), is essential for the initiation of protein synthesis. At the 3′ end of the viral RNA, a short (72 nt) heteropolymeric NTR is followed by poly(A) (about 60 nt long; the length of this homopolymer can vary from 40 to 200 nt), which together form an ordered structure important for the initiation of minus-strand RNA synthesis.

V. LIFE CYCLE

A. Uptake

Virus particles enter susceptible cells via docking to the human poliovirus receptor (see Section V,B), a

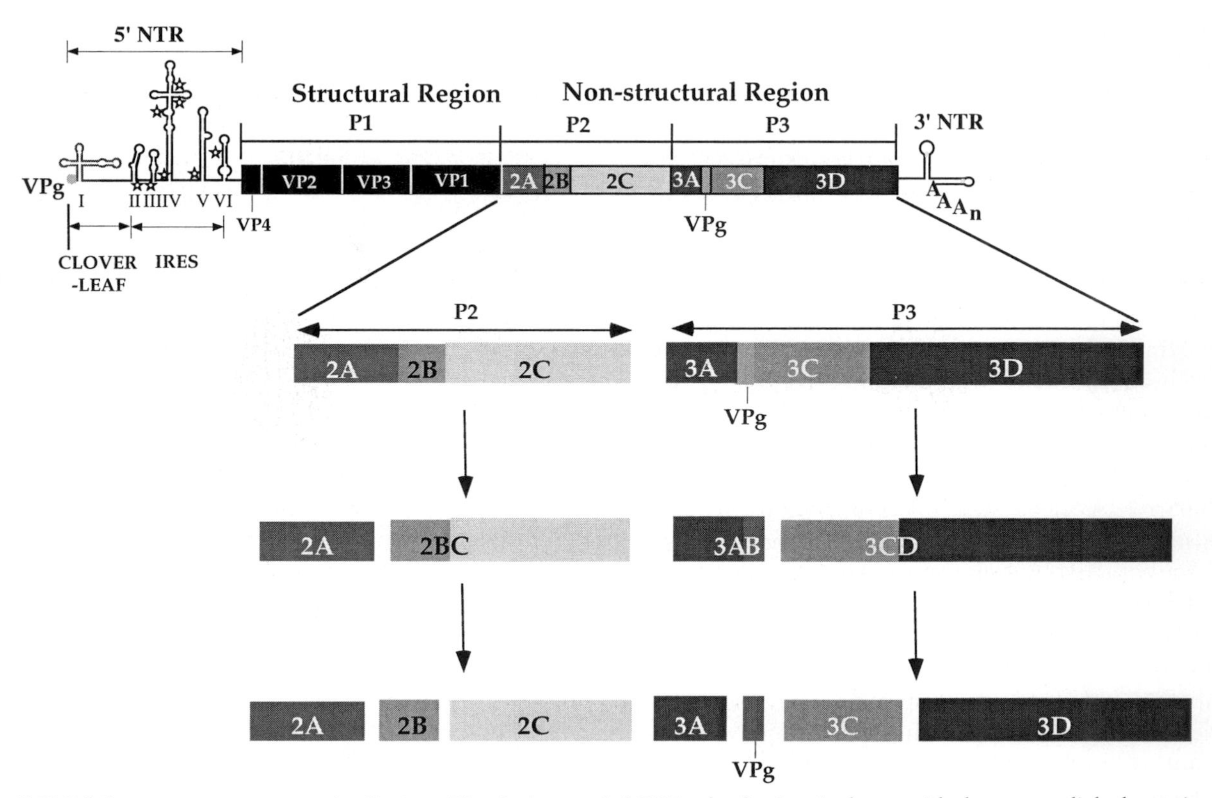

FIGURE 2 Genomic structure of poliovirus. The single-stranded RNA of poliovirus is shown with the genome-linked protein VPg (3B) at the 5′ end of the nontranslated region (single line) and the 3′ nontranslated region (single line) with the poly(A) tail. The boxed region shows the polyprotein, and the vertical lines within the box indicate proteinase cleavage sites. The locations of the structural (P1) and nonstructural (P2, P3) region proteins are shown on top. RNA structural domains within the 5′ nontranslated region are shown by Roman numerals, cloverleaf (I) and IRES (II–VI). Stars indicate the positions of noninitiating AUG triplets. Steps in the processing of the P2 and P3 domains of the polyprotein are shown enlarged below. Proteinase $2A^{pro}$ separates the structural and nonstructural domains, and all other cleavages (except between VP2 and VP4) are carried out by proteinase $3C^{pro}$ or $3CD^{pro}$.

molecule sufficient to promote virus entry and uncoating (removal of the protein shell of the virus). During this process, virions undergo a series of changes that ultimately result in the release of the viral RNA into the cellular cytoplasm from its confinement within the capsid. The exact sequence of structural changes remains to be solved, but a variety of subviral structures on their way toward exposure of the viral genome have been isolated and described. Initially, binding of viral particles to the receptor leads to a profound rearrangement of the capsid. The smallest of the four structural proteins, the internal VP4, is expelled from the capsid and the N termini of VP1 are moved outward. The resulting "A particle" is hydrophobic and sensitive to proteolytic enzymes. These changes facilitate escape of the viral genome from a shell that, under physiological conditions, is very stable. The subcellular compartments involved in the uptake of poliovirus remain unknown. Whether entry occurs by "receptor-mediated endocytosis" or via the interaction of the "A particle" with the plasma membrane remains a matter of controversy.

Poliovirus entry into susceptible cells is a highly inefficient process. Only a minor proportion of infectious particles ever reach the inside of the cell. The majority of "A particles," which can no longer interact with the receptor, are sloughed off the cell surface and are lost by degradation. Successful transportation and release of the genome trigger intracellular events of replication.

B. Receptor

The cardinal molecular mediating uptake of poliovirus into host cells is the human poliovirus receptor. hPVR is a highly glycosylated cell-surface protein (molecular weight >80,000) belonging to the immunoglobulin superfamily. It has the general structure ⟨H_2N–V–C2–C2–(membrane-spanning domain)-(intracellular domain)–COOH⟩, where V and C2 refer to immunoglobulin domains of different size. The gene for hPVR is expressed in the form of four splice variants of which two are secreted forms (because they have lost the membrane-spanning domain). The membrane-associated receptors differ only in the structure of their C-terminal domains. Proteins homologous to hPVR have been identified in cells of monkeys and mice, but only the monkey glycoproteins (referred to as mPVR) have the propensity to serve as poliovirus receptor. The mouse homologue cannot bind poliovirus, an observation correlating with the host range of poliovirus. However, mouse-adapted strains of poliovirus have been isolated that, on injection into the animals, cause neurological disease and death in mice distinct from poliomyelitis. The receptor of normal mice for the mouse-adapted poliovirus strains is obscure. The nonpathogenic function of hPVR and its homologues in mammals remains to be determined.

Available evidence suggests that hPVR is the only receptor for poliovirus. It has two functions: binding the virion (a virion is a single virus particle) and converting it to "A particles." Transformation of mouse cells with either hPVR-specific cDNA or with the hPVR gene renders the mouse cells susceptible to poliovirus, an observation suggesting that under these conditions hPVR is essential and sufficient for infection. Similarly, mice transgenic for hPVR are susceptible for poliovirus. On injection of inoculum (either intracutaneous, intramuscular, or intravenous), these animals develop a neurological disease that is clinically and pathologically indistinguishable from human poliomyelitis. In contrast to humans, however, these hPVR *tg* mice cannot be infected with poliovirus orally by conventional feeding schedules.

Genetic and biochemical evidence identified the V domain of hPVR as the virus binding and destabilizing domain; fusing the V domain onto the N terminus of other cellular receptors (such as CD4 or ICAM-1) yielded functional poliovirus hybrid receptors.

Since hPVR is considered to be the main determinant of tissue tropism and pathogenicity of poliovirus, it was expected that the pattern of its expression resembles the preference of poliovirus for its target organs. Surprisingly, analyses for hPVR-specific mRNA revealed expression of the gene in many human tissue cells albeit in very low amounts. Because of technical difficulties (small quantities of hPVR protein, exquisite sensitivity of the important epitopes to denaturation, lack of monoclonal antibodies recognizing denatured protein), this apparent widespread expression of hPVR mRNA has not been followed up by immunohistochemical procedures that would identify the functional receptor.

C. Translation and Proteolytic Processing

Following entry into the host cell, the viral RNA uses the machinery of cellular protein synthesis to express its genetic information. Poliovirus mRNA is identical to genomic RNA except that the 5′-terminal VPg has been removed by an unidentified cellular protein yielding a pU terminus. Translation of the uncapped mRNA by a cap-independent mechanism, and promoted by the IRES element of the 5′ NTR (See Fig. 2), is initiated at the eighth AUG codon at nt 743. This process probably involves direct binding of the 40*S* ribosomal subunits to the IRES, but the mechanism is as yet unknown. Apart from canonical translation factors operating in cellular protein synthesis, IRES elements also recruit nuclear proteins (e.g., polypyrimidine tract binding protein) for function. Genetic analysis revealed that the intact sequence of nearly the entire IRES (about 400 nt) is required for function. Even minor alterations may render the IRES inefficient or even inactive.

Poliovirus RNA is typical of picornaviruses in that it contains a single large open reading frame that is translated into a polyprotein (See Fig. 2). This polyprotein has three distinct domains: the P1 region encodes the structural proteins that form the capsid, and the P2 and P3 regions specify proteins required for RNA replication. Proteolytic processing of the polyprotein into functional units, which commences already during protein synthesis, is carried out by two viral proteinases. The primary cleavage event of the polyprotein, separation of the capsid from noncapsid precursors, is carried out cotranslationally by proteinase $2A^{pro}$ at a tyrosine/glycine amino acid pair between VP1 and $2A^{pro}$. Processing of the large precursors to yield a variety of products is carried out by proteinase $3C^{pro}$ and its precursor $3CD^{pro}$. These enzymes cleave between glutamine/glycine pairs but residues upstream of the scissile bond are also important determinants of the cleavage signal. The amino acid in the P4 position, usually an alanine, modulates the kinetics of cleavage. Thus alanineXXglutamine/glycine is a preferred cleavage site for $3C^{pro}$ and $3CD^{pro}$. Interestingly, $3CD^{pro}$ is required for cleavage of the viral capsid precursor.

The viral proteinases $2A^{pro}$ and $3C^{pro}$ are cysteine proteinases with a catalytic triad and three-dimensional structure reminiscent of serine proteinases (trypsin, chymotrypsin). This has been deduced from sequence comparisons and genetic and biochemical analyses. It has been ultimately proven by the solution of the crystal structures of $3C^{pro}$ of hepatitis A virus, human rhinovirus 14, and poliovirus.

D. RNA Replication

Replication of the viral RNA follows the general pathway of lytic RNA viruses but details of the mechanism are still poorly understood. Following entry of the viral RNA into the cytoplasm, the plus-strand RNA is transcribed into complementary RNA (minus strand), yielding a double-stranded RNA, the so-called replicative form (RF). Available evidence suggests that the RF provides the template (minus strands) for progeny virion RNA synthesis. Genetic and biochemical evidence has shown that all of the nonstructural proteins of the P2 and P3 regions and even some of their precursor polypeptides are required for RNA replication. Proteins of the P2 domain are predominantly responsible for the structural and metabolic changes that occur in the infected cell. For example, proteinase $2A^{pro}$ is important for the shutoff of host cell protein synthesis, whereas 2C and 2BC are important in the formation of membraneous vesicles that are anchors for the machinery of viral RNA replication. The P3 proteins are more directly involved in the process of RNA synthesis. The viral RNA polymerase $3D^{pol}$, the cleavage product of $3CD^{pro}$, catalyzes the elongation of both plus and minus strands. Since $3D^{pol}$ is strictly primer dependent *in vitro*, it has been proposed that an uridylylated form of VPg (VPg-pU) serves as a primer to yield VPg-linked progeny RNA. Polypeptide 3AB, a small basic protein, has multiple important functions in RNA replication owing to its ability to bind to RNA and to form complexes with either $3D^{pol}$ or $3CD^{pro}$. [*See* RNA Replication.]

E. Encapsidation

The precursor for the viral capsid (P1) is cleaved to three polypeptides (VP0, VP3, VP1) that remain associated to form a "protomer" (see Fig. 1). Five "protomers" assemble to a "pentamer" [(VP0, VP3, $VP1)_5$], and twelve pentamers associate to yield a procapsid. During this process virion RNA is encapsidated, but the encapsidation signal residing in the RNA is obscure. Finally, the shell of the procapsid undergoes a maturation cleavage resulting in the processing of VP0 to VP2 and VP4. The mature virions, about 25,000 to 50,000 per infected cell, are released by cell lysis. The entire replication cycle in tissue culture in human HeLa cells is completed in about 7 hr. [*See* Virology, Medical.]

VI. GENETICS

The genetics of poliovirus is dominated by an average mutation rate per base pair per replication of 6.3 $\times 10^{-4}$. This high rate of mutation, a phenomenon common to all RNA viruses, results from the fact that, during evolution, RNA viruses have made the choice not to develop proofreading and editing functions (DNA and RNA polymerases operate with a similar error frequency of roughly 10^{-4}). Consequently, virions of an RNA virus isolate constitute a vast population of different genotypes replicating at the threshold of error catastrophe—the hallmark of a quasi-species. For quasi-species, a wild-type isolate is defined as (a) a population of viruses not having a single sequence and (b) a population of viruses having a consensus sequence.

The high error rate in genome replication may be the main reason for the small size of the genome. An advantage of small ($<10^4$ bases) over large genome size is that more progeny RNAs can be generated from input RNA per unit time and thus more replicas are likely to survive in a hostile environment. This, combined with rampant genetic variation (mutation and recombination), allows for the rapid genetic adaptation to a new environment (host cell adaptation, drug resistance, etc.). A harsh consequence of the small genome size, however, is that poliovirus must live under conditions of severe genetic austerity. That is, the virus encodes just enough proteins to survive. No viral gene products, and hardly any viral genome sequence, can be deleted without lethal consequence to the virus.

The extent of genome compression due to the high error frequency in replication has led to the elimination of sequence motifs that could effect temporal expression of poliovirus genes. This in turn has led to the evolution of the viral polyprotein, the only translation product of the viral genome-length mRNA. This polyprotein, however, is not very stable as functional polypeptides are generated rapidly by proteolytic processing of the polyprotein, catalyzed in *cis* and *trans* by virus-encoded proteinases (see Section V, C). As mentioned before, some precursor polypeptides in the processing cascade have functions distinct from that of their end products, and several processing products may have multiple distinct functions (e.g., 3AB, $3C^{pro}$, $3CD^{pro}$, and $3D^{pol}$). This strategy has provided the means to expand the number of tools available for virus replication. It also serves to explain why experiments of genetic complementation, though well documented for some viral mutants but not for others, have not allowed scientists to deduce genetic complementation groups. Therefore, it must be concluded that the poliovirus genome is monocistronic.

Genetic variation of the poliovirus genome, through misincorporation of nucleotides, is augmented by genetic recombination occurring with astoundingly high frequencies ranging from 10^{-2} (intratypic) to 10^{-5} (between different serotypes). Homologous recombination takes place by a mechanism of copy choice during minus-strand synthesis, and it occurs between heterologous genomes as well as, preferentially, between sibling genomes. Using the HeLa cell-free system of *de novo* poliovirus growth *in vitro* (see Section VII), genetic recombination has also been achieved *in vitro* by simultaneously growing genetically different poliovirus strains (starting with isolated viral RNAs) in extracts of uninfected HeLa cells. [*See* HeLa Cells.]

Considering the genetic "plasticity" of the poliovirus genome, the restriction of poliovirus to only three serotypes is surprising. A hypothesis has been developed that suggests that the serotype restriction is related to the specific manner in which poliovirus capsid interacts with the cognate cellular receptor.

The study of the genetics of poliovirus was revolutionized by the developments in genetic engineering that have led to the generation of infectious poliovirus cDNA. The viral RNA genome that was converted to double-stranded DNAs was suddenly available for genetic manipulations hitherto unheard of. As a result of the cellular transcription machinery, transfection of suitable cells with the virus-specific double-stranded DNA then yielded poliovirus at low frequency. The efficiency of this process was greatly enhanced when it was shown that the virus-specific cDNA could be transcribed with phage T7 RNA polymerase to yield unlimited amounts of viral RNA that was nearly as infectious as RNA isolated from virions.

VII. *DE NOVO*, CELL-FREE SYNTHESIS OF POLIOVIRUS

Viruses, lacking the genetic information as well as the tools to provide most of the essential components to replicate, are obligatory intracellular parasites. The complexity of viral proliferation—macromolecular synthesis of polypeptides and genomic nucleic acid, and encapsidation—has led to the textbook wisdom that viruses can replicate only in intact cells.

Seeding a cellular extract prepared from uninfected HeLa cells that had been freed from nuclei, mitochondria, and endogenous mRNA with full-length poliovirus RNA leads to the production of poliovirus with properties that are indistinguishable from those of virus grown in intact cells. Analysis of this reaction revealed that (1) the viral polyprotein is synthesized and processed, followed (2) by replication of the viral RNA added to the extract, and (3) by encapsidation of newly made genomic RNA.

The observation of cell-free, *de novo* synthesis has revised the perception that viral proliferation is absolutely dependent on living cells. The cell-free synthesis of poliovirus has led to further insight into mechanisms governing replication. Although as yet an isolated example, it can be envisioned that cell-free synthesis of other viruses, even enveloped viruses, will be achieved in the future.

VIII. ERADICATION

Following the success in the global eradication of human smallpox virus, the 42nd World Health Assembly of the World Health Organization adopted an action plan (WHA42.22) to eradicate poliovirus by the year 2000. This bold decision, in effect since May 19, 1988, was based on the observations that (1) humans are the only natural hosts of the poliovirus (there is no known animal reservoir), (2) poliovirus exists in only three stable serotypes, and (3) two excellent vaccines are available, of which the live attenuated vaccine of Albert Sabin is highly suitable for the coverage of large proportions of a population by mass vaccination. Indeed, this latter point has been proven to be true particularly in vaccination campaigns in South America that converted areas endemic for poliomyelitis to those virtually free of circulating wild-type polioviruses. [*See* Infectious Diseases.]

Progress toward global eradication of poliovirus at this time (May 1997) is impressive. Indeed, eradication may be accomplished in the year of 2000. Vaccination against poliomyelitis then may stop in 2005. One can only hope that in a world marred with large areas of poverty or with political upheavals leading to violent conflicts, the action plan WHA42.22 will be completed as expected.

IX. CONCLUSIONS

Numerous problems of the tropism and replication of poliovirus and of picornaviruses remain unsolved. These include the function of several viral polypeptides, the mechanism of initiation of protein synthesis, the mechanism of genome replication, the steps of virus penetration and assembly, and the possible involvement of cellular factors in these processes.

BIBLIOGRAPHY

Jang, S. K., Pestova, T., Hellen, C. U. T., Witherell, G., and Wimmer, E. (1990). Cap-independent translation of the picornavirus RNAs: Structure and function of the internal ribosomal entry site. *Enzyme* **44,** 292–309.

Harber, J., Bernhardt, G., Lu, H.-H., Sgro, J.-Y., and Wimmer, E. (1995). Serotype-dependent binding of poliovirus to mutants of the receptor and identification of an uncoating-defective-receptor complex. *Virology* **214,** 559–570.

Hogle, J. M., Chow, M., and Filman, D. J. (1985). The three dimensional structure of poliovirus at 2.9 Å resolution. *Science* **229,** 1358–1365.

Kirkegaard, K., and Baltimore, E. (1986). The mechanism of RNA recombination in poliovirus. *Cell* **47,** 433–443.

Kitamura, N., Semler, B. L., Rothberg, P. G., Larsen, G. R., Adler, C. J., Dorner, A. J., Emini, E. A., Hanecak, R., Lee, J. J., Lee, S., van der Werf, S., Anderson, C. W., and Wimmer, E. (1981). Primary structure, gene organization and polypeptide expression of poliovirus RNA. *Nature* **291,** 547–553.

Koike, S., Taya, C., Kurata, T., Abe, S., Ise, I., Yonekawa, H., Nomoto, A. (1991). Transgenic mice susceptible to poliovirus. *Proc. Natl. Acad. Sci. USA* **88,** 951–955.

Kuhn, R. J., and wimmer, E. (1987). The replication of picornaviruses. *In* "Molecular Biology of Positive Strand RNA Viruses" (D. J. Rowlands, M. A. Mayo, and B. W. J. Mahy, eds.), pp. 17–51. Academic Press, London.

Lawson, M. A., and Semler, B. L. (1990). Picornavirus protein processing: Enzymes, substrates and genetic regulation. *Curr. Top. Microbiol. Immunol.* **161,** 49–87.

Mendelsohn, C. L., Wimmer, E., and Racaniello, V. R. (1989). Cellular receptor for poliovirus: Molecular cloning, nucleotide sequence, and expression of a new member of the immunoglobulin superfamily. *Cell* **56,** 855–865.

Molla, A., Paul, A. V., and Wimmer, E. (1991). Cell-free *de novo* synthesis of poliovirus. *Science* **254,** 1647–1651.

Molla, A., Jang, S. K., Paul, A. V., Reuer, Q., and Wimmer, E. (1992). Cardioviral internal ribosomal entry site is functional in a genetically engineered dicistronic poliovirus, *Nature* **356,** 255–257.

Nomoto, A., Morgan-Detjen, B., Pozzatti, R., and Wimmer, E. (1977). The location of the polio genome protein in viral RNAs and its implication for RNA synthesis. *Nature* **268,** 208–213.

Racaniello, V. R., and Baltimore, D. (1981). Cloned poliovirus complementary DNA is infectious in mammalian cells. *Science* **214,** 916–919.

van der Werf, S., Bradley, J., Wimmer, E., Studier, W. F., and Dunn J. J. (1986). Synthesis of infectious poliovirus RNA by purified T7 RNA polymerase. *Proc. Natl. Acad. Sci. USA* **83,** 2330–2334.

Wimmer, E., Hellen, C. U. T., and Cao, X. M. (1993). Genetics of poliovirus. *Annu. Rev. Gen.* **27,** 353–436.

Wimmer, E., Harbor, J. J., Bibb, J., Gromeier, M., Lu, H.-H., and Berhard, G. (1994). The poliovirus receptors. *In* "Cellular Receptors for Animal Viruses" (E. Wimmer, ed.). Cold Spring Harbor Press, Cold Spring Harbor, New York.

Pituitary

WILLIS K. SAMSON
University of North Dakota School of Medicine

GLOSSARY

ACTH Adrenocorticotropic hormone, major posttranslational product of pro-opiomelanocortin gene transcription

AVP Vasopressin or antidiuretic hormone, a neuropeptide produced in unique neurons in the hypothalamic paraventricular and supraoptic nuclei and released from nerve terminals in the posterior pituitary (neural lobe)

FSH Follicle-stimulating hormone, produced in gonadotrophs

GH Growth hormone, produced in somatotrophs

LH Luteinizing hormone, produced in gonadotrophs

OT Oxytocin, a neuropeptide produced in unique hypothalamic neurons in the paraventricular and supraoptic nuclei, released from nerve terminals in the posterior pituitary (neural lobe)

Portal vessels Venous plexus, which connects the hypothalamus and anterior lobe of the pituitary gland; it delivers humoral agents of neural origin, which control hormone production and release

PRL Prolactin, produced in lactotrophs

TSH Thyroid-stimulating hormone, produced in thyrotrophs

THE PITUITARY IS AN AGGREGATION OF NEURAL and endocrine tissue located on the ventral surface of the brain on the midline, lying beneath and in part continuous with the hypothalamus (Fig. 1). The posterior (neural) lobe is a ventral extension of the axons of hypothalamic neurons. From these terminals are released the neuropeptide hormones oxytocin (OT) and vasopressin (AVP). Therefore, the neural lobe is important for the hormonal control of fluid and electrolyte homeostasis and lactation. The intermediate lobe of the pituitary is distinguishable only during embryological development in the human; its cells merge with both the neural and anterior lobe cells after birth. The largest portion of the pituitary gland is the anterior lobe, which originates from ectodermal derivatives on the roof of the developing oral cavity. Cells migrate dorsally toward the expanding neural tube and establish a position in front of and partially surrounding the neural lobe and its connection with the hypothalamus, the infundibular stem. A peculiar vascular connection, the hypophyseal portal vessels, develops to connect the anterior lobe with the hypothalamus, through which trophic factors gain access to the hormone-producing cells of the gland. Hormones released from these cells control a wide variety of physiologic functions, including growth, metabolism, and reproduction.

I. DEVELOPMENT AND STRUCTURE

A dorsal evagination of the ectodermal lining of the stomodeum just anterior to the endoderm of the pharynx initiates during Week 4 of gestation in humans. Cells separate from the roof of the presumptive buccal cavity to form a spherical mass called *Rathke's pouch.* These cells will develop into the hormone-secreting elements of the anterior pituitary gland or adenohypophysis. By Week 6 they have approximated the ventral outgrowth of neural tissue, which will constitute the neurohypophysis or neural lobe. At the time

ENCYCLOPEDIA OF HUMAN BIOLOGY, Second Edition, VOLUME 6.

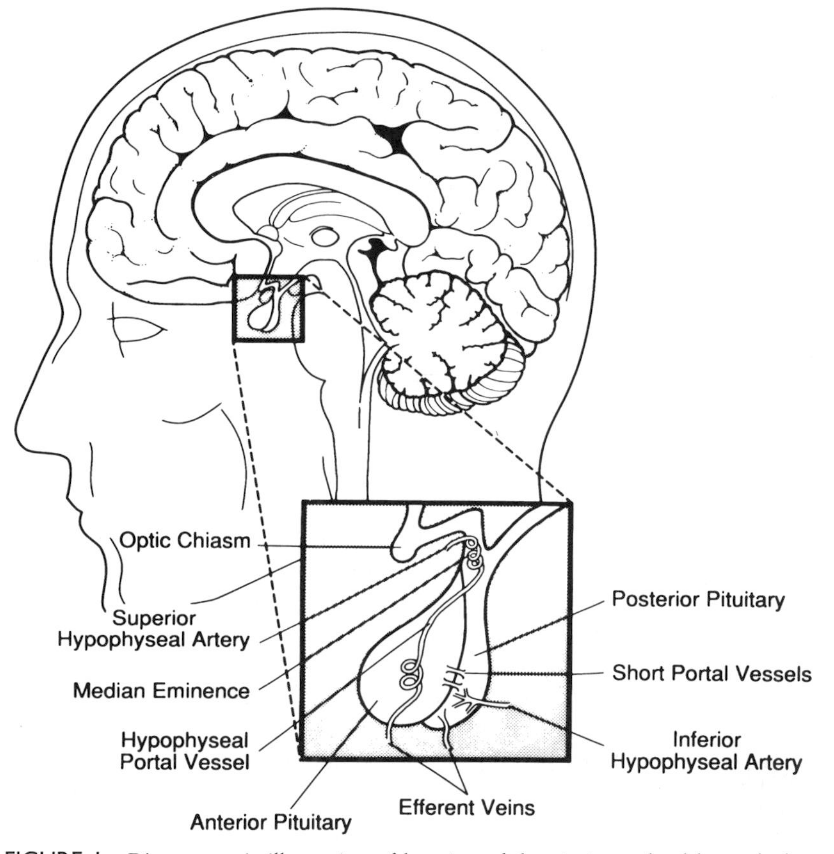

FIGURE 1 Diagrammatic illustration of location of the pituitary gland beneath the hypothalamus. Positions of the adenohypophysis (anterior pituitary) and the neurohypophysis (posterior pituitary) are demonstrated, as well as the arterial supply of these tissues. The hypophysial portal vasculature is schematized for simplicity.

of apposition, Rathke's pouch undergoes considerable structural rearrangement with the collapse of its hollow center. The anterior wall proliferates to become the glandular tissue characterizing the bulk of the anterior lobe, the pars distalis. The thin posterior wall contacts the developing neurohypophysis and fuses with it. In the embryo this tissue remains distinguishable as the pars intermedia. Lateral extensions of the developing pouch surround the infundibular stem as the pars tuberalis. The neurohypophysis or neural lobe has been formed by the ventral extension of axons from nerve cells located in the developing hypothalamic paraventricular and supraoptic nuclei. As these axons reach the base of the hypothalamus in the midline, they form a ridge of tissue in the median eminence and, progressing ventrally as the infundibular stem around which the pars tuberalis adheres, end in secretory terminals in the infundibular process or neural lobe.

During the sixth week of gestation, mesenchymal elements have aggregated into the portal vessel system, which will provide the vascular link between the hypothalamus and the developing adenohypophysis. The superior hypophysial artery, most commonly a branch of the internal carotid artery, supplies tissue at the base of the hypothalamus surrounding the separation of the infundibular stem, and the stem itself. This region of the hypothalamus is the median eminence, which appears as a vascularized ridge of tissue. Capillaries emanating from the superior hypophysial arteries in the median eminence form a dense network and are characterized by fenestrated endothelial elements through which neural factors released locally diffuse. These capillary loops give rise to long portal veins that descend along the infundibular stem and ramify into the sinusoids of the adenohypophysis. Via this unique portal vessel system, neurally derived agents gain access to the hormone-producing cells of the anterior lobe. The infundibular process (neurohypophysis) receives arterial supply via the inferior hypophysial artery. In addition to the venous drainage of this tissue, a short portal system of vessels connects the sinusoids of the neurohypophysis with those of the anterior lobe, providing another potential pathway for the delivery of trophic (stimulatory and inhibitory) factors to the endocrine cells of the adenohypophysis.

The adenohypophysis, or anterior lobe, comprises then the large pars distalis, the pars tuberalis, and the indistinct pars intermedia. The neurohypophysis is mainly the infundibular process, or neural lobe, and the infundibular stem. Some authors consider the median eminence part of the neurohypophysis, and although the internal layer of the median eminence does contain hypothalamo–neurohypophyseal fibers of the OT and AVP neural systems projecting toward the infundibular stem, the majority of the tissue in the median eminence is composed of nerve terminals terminating in the vicinity of the hypothalamo–hypophyseal portal vasculature; therefore this tissue is more related to anterior pituitary lobe function. For the sake of this review, the median eminence will be considered hypothalamic tissue. In the adult human, the pituitary gland (both adenohypophysis and neurohypophysis) measures 1.0–1.5 cm in length and width and 0.5 cm in depth. It weighs approximately 0.5 g, except during pregnancy when the gland may weigh as much as 1.0 g. The gland sits protected in a bony cavity of the sphenoid bone, the sella turcica. It is separated from the overlying nervous tissue of the hypothalamus by a layer of dura mater, the diaphragma sella, through which the infundibular stem passes. The major venous drainage of the pituitary gland is via the cavernous and transverse sinuses and the internal jugular veins.

II. ANTERIOR LOBE FUNCTION

A. Morphology

Adenohypophyseal cells are encapsulated in a dense collagenous matrix and are arranged in sinusoidal fashion in close apposition to the thin-walled vascular elements of the gland. Until recently the glandular cells of the adenohypophysis were classified on the basis of their affinity for routine histologic dyes. Two classes of cells were identified: chromphils and chromophobes. Modern techniques such as immunocytochemistry using antibodies directed against specific proteins produced by the cells have enabled more accurate identification of cell types. Six distinct, glandular cell types have been identified, and their characterization is described in Table I.

B. Control of Hormone Release

Anterior pituitary function is primarily controlled by the action of trophic substances, releasing and inhibiting factors, delivered to the gland by the long and short portal vessels. These factors originate for the most part in nerves located in the hypothalamus and rostral diencephalon. The majority of the trophic factors are small peptides, although other factors (e.g., more classic neurotransmitters) do exert potent effects on adenohypophyseal hormone synthesis and secre-

TABLE I
Adenohypophyseal Morphology: Cell Types and Hormone Content[a]

Cell type	Staining affinity	Hormone content	Secretory granule size	% of total
Somatotroph	Acidophilic	Growth hormone	150–600 nm	50%
Lactotroph	Acidophilic	Prolactin	150–350 nm or 300–800 nm	10–25%
Corticotroph	Basophilic	Pro-opiomelanocortins (ACTH, lipotropins, endorphins)	250–400 nm	15–20%
Thyrotroph	Basophilic	Thyroid-stimulating hormone	100–200 nm	<10%
Gonadotroph	Basophilic	Luteinizing hormone and follicle-stimulating hormone	150–250 nm or 350–450 nm	10–15%
Mammosomatotroph	Acidophilic	Prolactin and growth hormone	400–1200 nm	

[a]Other cell types have been identified on the basis of secretory granule content and affinity or avidity for acidic and basic dyes; however, the identity of their secretory product(s) has not yet been elucidated.

tion. The primary controller of pituitary function then is the hypothalamus. This tissue, which lies below the thalamus on the base of the diencephalon, is a heterogeneous population of neurons receiving abundant afferent input from other brain areas. Because of the absence of the blood–brain barrier at some locations, the hypothalamus also receives afferent stimuli that are humoral in nature. The hypothalamic cell groups, called *nuclei,* process incoming signals, integrating this information before sending out efferent neural messages controlling a wide variety of essential functions, including cardiovascular, renal, and gastrointestinal function, as well as autonomic nervous system activity and brain-derived behaviors. In coordination with these neural functions, the hypothalamic integrators also control such varied functions as metabolism, growth, and reproduction via their delivery of trophic factors to the median eminence, where on release they diffuse into the portal vessels and gain access to the pituitary gland. Thus the pituitary sees a constant variety of humoral cues, neuroendocrine in nature, which control the release of its hormones into the peripheral circulation. [*See* Hypothalamus.] In addition to the classic long, hypophyseal portal vessels, a similar, yet smaller portal vasculature connects the neurointermediate lobe of the pituitary gland with the anterior lobe. It is thought that these vessels may be responsible for the modulation of anterior lobe hormone secretion by interaction with factors delivered by the long portal vessels. There is also evidence that factors delivered by the short portal vessels may play a major role in the hypothalamic control of prolactin secretion.

A secondary level of control of adenohypophyseal function is exerted directly at the level of the anterior pituitary gland. Hormones released from the anterior lobe exert a variety of actions in distant tissues, some of which result in the elaboration and secretion of target tissue-specific hormones into the blood (e.g., estrogens from the ovary in response to gonadotropin stimulation) (Table II). These peripherally derived humoral agents not only reach the integrative centers of the hypothalamus where they can exert positive (stimulatory) or negative (inhibitory) effects on the hypothalamic control of adenohypophyseal function, but also exert selective effects on the anterior lobe

TABLE II
Hormones of the Adenohypophysis

Hormone	Origin	Structural identity	Major trophic factors
Growth hormone	Somatotrophs and mammosomatotrophs	191 amino acids (21,000 MW)	Somatostatin (−) Growth hormone-releasing hormone (GHRH, +)
Prolactin	Lactotrophs and mammosomatotrophs	198 amino acids (22,000 MW)	Dopamine (−) Thyrotropin-releasing hormone (TRH, +) Vasoactive intestinal peptide (VIP, +) Oxytocin (+)
Pro-opiomelanocortins	Corticotrophs		Corticotropin-releasing hormone (CRH, +)
ACTH		39 amino acids (4500 MW)	Vasopressin (+)
Alpha-MSH		13 amino acids (1800 MW)	Norepinephrine (+)
Beta-lipotropin peptides			
Gamma-lipotropin		58 amino acids (6000 MW)	
Beta-endorphin		31 amino acids (3500 MW)	
Thyrotropin (TSH)[a]	Thyrotrophs	Alpha subunit (92 amino acids) Beta subunit (112 amino acids)	Thyroid-stimulating hormone (TRH, +)
Luteinizing hormone (LH)[b]	Gonadotrophs	Alpha subunit (92 amino acids) Beta subunit (121 amino acids)	Luteinizing hormone–releasing hormone (GnRH)
Follicle-stimulating hormone (FSH)[c]	Gonadotrophs	Alpha subunit (92 amino acids) Beta subunit (118 amino acids)	GnRH (+) Activin (+) Inhibin (−)

[a]TSH is a glycoprotein with about 14% by weight carbohydrate (28,000 total MW).
[b]LH is a glycoprotein with about 16% by weight carbohydrate (30,000 MW).
[c]FSH is a glycoprotein with >20% by weight carbohydrate (32,000 MW).

cells themselves. These long-loop feedback actions of target tissue-specific factors play an important role in fine-tuning the ability of the anterior pituitary to respond to hypothalamic signaling. In this manner the synthetic and secretory activity of the gland can be accelerated, limited, or synchronized in an appropriate physiologic fashion.

Two other levels of control involve feedback effects of the anterior lobe hormones themselves. There is evidence that these hormones may gain access to the hypothalamus and exert direct effects on the hypothalamic neurons, elaborating the releasing and inhibiting factors controlling their secretion. This is termed *short-loop feedback* and is exemplified by the ability of prolactin to gain access to the brain and affect the activity of dopaminergic neurons. Hormones have also been implicated in a form of autofeedback in which they control the activity of other pituitary cell types (paracrine control) or of their own cells of origin (autocrine control). This phenomenon has been termed *ultrashort-loop feedback,* and although these feedback loops have yet to be conclusively demonstrated in humans, abundant data derived from animal studies suggest the presence of such control mechanisms in humans as well.

C. Cell Types: Hormone Production, Actions, and Control of Release

1. Somatotrophs

Growth hormone (GH)-producing cells are located mainly in the lateral wings of the pars distalis and are medium in size with a spherical or oval shape. They possess a centrally located, spherical nucleus, well-developed rough endoplasmic reticulum, and extensive Golgi complexes. These profiles, together with the presence of abundant secretory granules averaging about 300 nm in diameter, reveal the endocrine nature of these cells. The abundance of these somatotrophs accounts for the fact that the human pituitary contains about 10% by weight GH. Surprisingly, the number or activity of the somatotrophs is not influenced by age, suggesting then that the increased effectiveness of GH during development is due not to more hormone being secreted but instead to a more receptive tissue response to GH. GH has a plasma half-life of approximately 20 min and circulates in adults at normal levels of 2–4 ng/ml (5–8 ng/ml in adolescents). Secretion occurs in a pulsatile fashion with more exaggerated pulsatility present during puberty. Levels of GH remain fairly constant throughout the day, with the exception of minor surges 3–4 hr after meals and a major (often 10-fold) surge during the initial period of deep sleep (at the onset of stages III and IV, before REM sleep).

The actions of GH are best described by conditions present during hyper- or hyposecretion. Hyposecretion is characterized by dwarfism, a condition in which a symmetrical reduction in growth occurs. If hypersecretion occurs early in life, the resulting gigantism is reflected in a symmetrical overgrowth of all tissues, including bones, muscles, connective tissues, and visceral organs. Hypersecretion after fusion of epiphyseal plates cannot stimulate bone growth, and as a result only the soft tissues of the body (connective tissue, skin, visceral tissues) and the cancellous bone (hands, jaw) respond, a condition known as acromegaly. GH receptors are present in the liver and on fibroblasts, adipocytes, and lymphocytes. GH administration results in decreased plasma amino acid composition, reflecting a stimulation of amino acid transport into cells, particularly liver and muscle, and increased protein synthesis (anabolism). The effects on protein synthesis are thought to occur at a pretranslational level. GH infusion also results in positive nitrogen balance as well as the retention of calcium, phosphorus, potassium, and magnesium. The hormone antagonizes the action of insulin, resulting in a decreased use of carbohydrates and the promotion of hepatic and muscular glycogenesis. With the diversion of amino acids for protein synthesis and carbohydrates for glycogen formation, lipolysis and fatty acid oxidation are stimulated, resulting in ketogenesis. Mitosis increases in the hematopoietic system.

The most striking effect of GH is growth of cartilage and bone. To exert these effects, the hormone must first act in the liver to stimulate the production and release of factors called *somatomedins.* These peptides (approximately 7500 MW) share significant structural similarity with proinsulin and therefore have been termed *insulin-like growth factors* (IGFs). IGFs circulate in plasma associated with a binding protein or proteins, and several subclasses of IGF receptors have been characterized. Many of, if not all, the biological activities of GH can be ascribed to the action of the somatomedins. They stimulate collagen synthesis, sulfate incorporation into cartilage, and mitosis. Long bones (e.g., the tibia and femur) grow by increases in ossification of the diaphyseal and epiphyseal plates. The IGFs stimulate thickening of the plates, resulting in increased ossification and long bone growth. This cannot occur even in the presence of excess GH after epiphyseal plate closure has occurred in adolescence. [*See* Growth, Anatomical; Peptides.]

Metabolic, hormonal, and neural factors control somatotroph activity. Insulin-induced GH release is caused not by a direct effect of insulin per se, but instead by the hypoglycemic action of insulin. Delayed postprandial GH secretion, particularly after a high-protein meal, is caused by the direct stimulatory effects of certain amino acids (e.g., arginine and leucine). Falling levels of plasma free fatty acids also result in GH secretion. Metabolic inhibition of GH secretion occurs in hyperglycemia, in obese individuals, and when free fatty acid levels are high. Hormonal effects on GH secretion include stimulatory effects of estrogens, glucagon, and vasopressin and inhibitory effect of GH itself by an autocrine action. Plasma levels of GH are also low in hypothyroidism. By far the most significant regulation of somatotroph function is exerted by the hypothalamus and other central nervous system structures.

Two peptides of hypothalamic origin are the primary regulators of secretion. Somatostatin, a peptide that exists in both a 14- and an N-terminally extended 28-amino-acid form, inhibits basal and stimulated GH secretion *in vitro*. *In vivo* administration of the peptide in humans results in little effect on the already low basal levels of hormone; however, the sleep-related surge can be suppressed, and hypersecretion of GH in acromegalics can be reversed. Somatostatin-producing neurons are found adjacent to the walls of the third cerebroventricle in the periventricular region of the hypothalamus. Fibers containing the peptide project to the median eminence, where somatostatin is released into the vicinity of the capillary loops of the portal vasculature. The mechanism of action of somatostatin at the level of the somatotroph is thought to be via an inhibition of stimulated adenylate cyclase activity and to an action downstream from the formation of cAMP, perhaps caused by an action on calcium influx. GH-releasing hormone (GHRH) is a 44-amino-acid peptide produced in cells of the infundibular region of the hypothalamus. Short axonal processes deliver the peptide to the adjacent median eminence. It is active *in vitro* at doses similar to those present in portal blood of rats and is an effective stimulator of GH secretion when infused in humans. GHRH stimulates secretion of GH via adenylate cyclase-dependent and -independent [probably via increases in calcium influx and changes in phosphatidylinositol (PI) turnover] mechanisms. It synergizes with other agents known to stimulate these two signal transduction pathways (e.g., prostaglandins). The releases of both somatostatin and GHRH are controlled by numerous neural and hormonal factors, including the somatomedins, which have been hypothesized to feedback in a negative fashion at both the pituitary and hypothalamic levels to dampen GH secretion. Interactive effects of somatostatin and GHRH on their own release have been described. Alpha-adrenergic agents act centrally to stimulate GH secretion, perhaps explaining stress-induced elevations in plasma GH levels. Beta-adrenergic agents exert inhibitory control. Finally, in lower species there is evidence for short-loop negative feedback of GH on the hypothalamic cells responsible to GHRH release or on those neurons per se.

2. Lactotrophs

Prolactin (PRL)-producing cells are located mainly in the lateral posterior aspect of the pars distalis, adjacent to the posterior lobe. The adenohypophysis contains approximately 0.1 mg PRL, although the hormone undergoes rapid postmortem degradation, so this might be a low estimate. Two types of lactotrophs exist. The most abundant cells are small and angular, often arranged in contact with each other. Although they possess well-developed endoplasmic reticulum and Golgi complexes, they have sparse, small granules (150–350 nm). The Golgi complexes are stacked near the cell surface, giving the appearance that these cells are highly active and that the newly synthesized PRL is rapidly secreted. The second class of lactotrophs represents those cells in which stored PRL accumulates. They appear in isolation throughout the pars distalis, are large and often irregular in shape, and possess abundant, large secretory granules (300–800 nm). [*See* Golgi Apparatus.]

The number of lactotrophs varies greatly during life. The fetal pituitary is producing PRL already by the tenth week of gestation, and under the influence of maternal estrogens, both fetal and maternal pituitaries demonstrate a hyperplastic population of lactotrophs. Numbers of cells fall then in childhood as they also do postpartum in the absence of lactation. In the nonpregnant state, adult female pituitaries have approximately the same number of lactotrophs as found in males. Plasma PRL levels in adult, nonpregnant, nonlactating females are about 10 ng/ml and are slightly lower in males and children. Levels in plasma progressively rise during pregnancy to reach third-trimester levels of as high as 300 ng/ml. The plasma half-life of PRL is 20–30 min, and its high metabolic clearance rate suggests a secretion rate of about 0.2 mg/day, indicating the high synthetic activity of the lactotroph. Other than in pregnancy, the most remarkable secretion of PRL occurs in response to suckling, with plasma levels rapidly increasing to as high as 10 times basal with mechanical stimulation of the nipples. This mechanoreceptor-mediated stimulation of PRL secretion also occurs in the nonlactating female but not in males, indicating a sexual dimorphism with regard to the control of PRL secretion. Under basal conditions, PRL secretion is pulsatile in nature, with only minor fluctuations except during sleep when one or more major surges occur, resulting in a doubling or tripling of plasma levels. Unlike nocturnal GH surges, these PRL surges are not synchronized with the sleep cycle and normally occur toward the end of the period of sleep. Virtually any form of stress, whether caused by physical intervention or a manifestation of a perceived threat (psychogenic stress), results in PRL secretion. Additionally, stimulation of mechanoreceptors in the vagina and on the uterine cervix during intercourse results in PRL secretion. The pathways involved in these reflex releases of PRL are thought to involve brain opiate, serotonergic, dopaminergic, and peptidergic systems. Although present in rodents, the existence of a midcycle PRL surge in human females is still controversial.

The mammary gland is the primary site of action of PRL. The breasts rapidly develop during both puberty and pregnancy. The growth and differentiation is called *mammogenesis,* and although PRL is thought to play an important role, its actions appear to be secondary to those played by estrogen and progesterone. Ductile development is stimulated by increasing estrogens, and lobulo-alveolar growth in the estrogen-primed gland is induced by progesterone. Permissive effects of cortisol, GH, and insulin are also involved. PRL's effects are thought to be mediated via specific membrane receptors, which result in an activation of phospholipase A activity and eventually increased transcription and translation. The proliferative effect then is not only one involving mitosis, but also the *de novo* synthesis of enzymes such as protein kinases and milk proteins. During pregnancy, the high levels of estrogens and progesterone not only stimulate mammogenesis, but also actually inhibit milk production. After parturition, levels of these steroids fall, and if suckling ensues, the lactogenic effects of PRL are uncovered. These effects appear to require the permissive actions of adrenal steroids, as well as GH, insulin, thyroxine, and parathyroid hormone. [*See* Steroids.]

Prolactin acts in the adrenal gland and ovary to increase the availability of steroid precursors and in the presence of luteinizing hormone (LH) stimulates the secretion of progesterone. Effects on the growth and development of accessory sex glands have been described. Receptors are present in the prostate and testis, suggesting a role for PRL in growth and development of these tissues. In situations of hypersecretion, an antigonadotrophic effect is expressed, probably at the level of the hypothalamus, which in part explains the relative infertility associated with lactation. An exciting recent development has been the implication of PRL in the control of immune system function. Lymphocytes express PRL receptors and indeed production of the hormone has been reported in these cells as well. There is strong evidence that PRL acts in these tissues as a growth factor, which is supported by the observations that the PRL receptors are structurally homologous to those of other hematopoietic growth factors, including the cytokines. The clinical relevance of these observations is suggested by the recent finding that suppression of elevated PRL levels in patients suffering from systemic lupus erythematosus resulted in resolution of disease state. Thus a potential pathologic role for PRL in states of autoimmunity must be considered clinically relevant. Finally, although controversial, a renotropic action of PRL is suggested by clear natriuretic effects demonstrated in lower species and the presence of PRL receptors in the kidney.

Neural and hormonal factors control the release of PRL. As mentioned previously, estrogen stimulates hormone secretion, probably by both a direct effect on the lactotroph and effects exerted on hypothalamic control mechanisms. The major, recognized, neural factor regulating factor PRL release is dopamine, which exerts potent inhibitory effects on the lactotroph. Dopaminergic neurons originating in the infundibular region project to either the median eminence or neural lobe, delivering the catecholamine to the long and short portal vessels. Much of the magnitude of PRL secretion observed under a variety of physiologic circumstances can be explained on the basis of reduction of dopaminergic neuronal activity and the ensuing removal of inhibitory tone. Inactivation of adenylate cyclase appears to be one of the cellular mechanisms for dopamine's PRL-inhibiting effect. However, dopamine withdrawal alone cannot completely describe the timing or extent of PRL secretion, leading to the hypothesis that PRL-releasing factors must also be present, which either act in the absence of dopamine's inhibitory tone or can stimulate PRL release even in the presence of dopamine. Several peptides of hypothalamic origin have been demonstrated to act as PRL-releasing factors in humans. Thyrotropin-releasing hormone (TRH), vasoactive intestinal peptide (VIP), and oxytocin all are capable of stimulating PRL secretion; however, the exact role played by each of these peptides in the physiologic regulation of PRL release is still unclear. In the case of TRH, an involvement of phospholipase C and PI hydrolysis resulting in inositol triphosphate and diacylglycerol formation has been reported. The resulting increase in intracellular calcium and in levels of protein kinase C are thought to be the mechanisms by which the peptide stimulates PRL release. VIP's PRL stimulatory effect is thought to be caused by a stimulation of adenylate cyclase activity. Recently, evidence has been presented for the presence of novel peptides with potent PRL-releasing and -inhibiting activities, suggesting that the identities of the primary factors responsible for the physiologic control of lactotroph function remain unidentified.

3. Corticotrophs

Located primarily in the center of the pars distalis and at the border adjacent to the neural lobe, the corticotrophs are medium to large cells with a prominent perinuclear body, thought to be lysosomal in origin. The rough endoplasmic reticulum is well developed, Golgi complexes are abundant, and typical secretory granules are 250–400 nm in diameter. Corticotrophs appear early in fetal life, and the number of these cells does not seem to be affected by age or sex. Hyperplasia is observed in certain disease states. Corticotrophs that were originally part of the fetal intermediate lobe migrate during development into both the posterior aspect of the pars distalis and into the neural lobe. These cells differ not only in size from those of the adenohypophysis, being smaller and more cuboidal, but also in the manner in which they modify the major posttranslational secretory products.

The human pituitary contains about 0.2 mg of adrenocorticotropic hormone (ACTH), which is one of the primary secretory products of the corticotroph. ACTH is posttranslationally derived from a large glycoprotein, pro-opiomelanocortin (POMC), which is the primary product of the POMC gene transcription and translation. Posttranslational processing of the POMC gene product results in the formation of a variety of bioactive peptides, which may undergo further posttranslational processing, including glycosylation and phosphorylation, although the significance of these final two steps in the human remains obscure. The POMC gene encodes a large protein, which in addition to ACTH contains the sequences of beta-lipotropin (LPH) and, at the N terminus, a signal peptide followed by a large N-terminal fragment, for which no clear bioactivity has been demonstrated. Along the peptide backbone of the POMC gene product exist numerous arginine and lysine doublets, which provide cleavage sites for the generation of multiple-peptide fragments. In the human adenohypophysis, the two major, bioactive fragments are the 39-amino-acid peptide ACTH and the 91-amino-acid LPH at the carboxy terminus. ACTH can undergo further processing to yield alpha-melanocyte-stimulating hormone (MSH), a 13-amino-acid peptide, and the 22-amino-acid corticotropin-like intermediate peptide (CLIP). This process probably does not occur in the pars distalis, but instead in the cells comprising the vestiges of the intermediate lobe. Within the beta-LPH peptide reside the sequences for gamma-LPH and beta-endorphin. These 58- and 31-amino-acid peptides, respectively, are liberated from the beta-LPH molecule within the corticotroph. Cells remaining from the fetal intermediate lobe, now present at the boundary between the anterior and posterior lobes and dispersed within the neural lobe, have the capacity to process gamma-LPH further to form the 18-amino-acid peptide beta-MSH and to generate the 12-amino-acid peptide gamma-MSH from the N-terminal fragment of POMC.

The plasma half-life of ACTH is approximately 20 min, and only 10–20% of total content is secreted each day. Under normal, nonstressed conditions, levels of ACTH in plasma are low, >10 pg/ml. ACTH is secreted in a pulsatile fashion, with pulses of only minor amplitude occurring at intervals of about 20 min throughout the day. A major increase in pulse amplitude is responsible for elevated, surge-like levels present at night. Peak levels of ACTH are present in plasma between 2 and 8 A.M. In addition to this circadian fluctuation in hormone secretion, ACTH release is stimulated by external factors, including stress (pain or anxiety), pyrogens, and hypoglycemia.

The primary action of ACTH is exerted in the adrenal gland, where the peptide stimulates the conversion of cholesterol to pregnenolone. After binding to its specific adrenal receptor, ACTH increases PI turnover, resulting in an activation of adenylate cyclase. The ensuing increase in phosphorylation of key enzymes eventuates both a stimulation of cholesterol esterase activity, releasing free cholesterol, and an activation of the rate-limiting 20–22 desmolase reaction, thus initiating the enzymatic pathway for the biosynthesis of adrenal steroids in general and, in particular, cortisol. Because ACTH assays have proven laborious, plasma cortisol determinations are often used to access ACTH secretion. Alpha-MSH stimulates melanocytes, and the hyperpigmentation of ACTH excess might be caused by this action. Beta-endorphin exerts profound effects on neuronal activity, yet it is unclear whether peptide of pituitary origin accesses the brain. Potentially significant effects of beta-endorphin on cardiac and renal function are now being examined in animal and human studies. [*See* Cholesterol.]

The 41-amino-acid peptide corticotropin-releasing hormone (CRH) is the primary hypothalamic factor regulating corticotroph activity. Specific CRH receptors have been localized on the corticotroph, and activation of these receptors results in increased adenylate cyclase activity, which can be correlated with ACTH secretion. A downstream effect mediated via calcium influx has also been described. Most of the neural pathways involved in the central control of ACTH release converge on the CRH-producing neurons of the hypothalamus, located predominantly in the periventricular regions. Cholinergic, serotonergic, adrenergic, GABAergic, and histaminergic influences on CRH release in the median eminence have been described. Glucocorticoids are thought to exert long-loop negative feedback on ACTH secretion by actions within the hypothalamus and directly at the level of the corticotroph. Other neuropeptides have been described to have ACTH-releasing activity as has norepinephrine. Vasopressin exerts CRH actions via unique receptors, not characteristic of either the V-1 (vascular) or V-2 (renal) binding subclasses. OT and VIP have been implicated in the control of corticotroph activity; however, their roles in the control of ACTH secretion in the human are unclear.

4. Gonadotrophs

Gonadotrophs are medium-sized cells found throughout the pars distalis, often adjacent to capillaries. Their frequent close apposition to lactotrophs has suggested paracrine interactions between these two cell types. Although most gonadotrophs, which make up 10–15% of the cells in the adenohypophysis, produce both LH and follicle-stimulating hormone (FSH), ultrastructural evidence for subpopulations of gonadotrophs producing only LH or FSH exists. Physiologic and pharmacologic studies also support the concept of distinct subpopulations. The rough endoplasmic reticulum and Golgi complexes are well developed, and the spherical nucleus is often eccentric in location. Secretory granules containing FSH have been reported to be larger (350–450 nm) than those containing LH (150–250 nm). In the absence of circulating gonadal steroids, large gonadotrophs called *castration cells* are present. These cells are characterized by dilated and enlarged cytoplasmic organelles, indicating enhanced synthetic activity.

The gonadotropins are glycoproteins composed of alpha and beta subunits. They share a common alpha subunit, which is also similar to the alpha subunit of thyrotropin. Biologic specificity is contributed by their unique beta subunits, which do share some structural homology. Owing to the wide differences in purity of the available standard reference preparations, levels of LH and FSH in plasma and pituitary gland have been expressed in terms of an immunologic preparation and quantitated as International Units of activity (IU). Levels present in the pituitaries of adult men and menstruating women are approximately 700 IU for LH and 200 IU for FSH. These levels would be approximately 100–300 ng of hormone. Plasma levels are in the low ng/ml range and, with the exception of during the periovulatory period, are stable in adult men and women. Plasma gonadotropins are low in prepubertal children. Gonadotropins are secreted in a pulsatile fashion with peaks of LH occurring every 1–2 hr. This type of secretion is called *circhorial* because of its approximately hourly nature. Peaks of FSH are less evident because of its greater half-life in plasma than that of LH. Diurnal, or daily, fluctuations in LH and FSH secretion are minor in the adult; however, such changes do appear during sleep in pubertal females. FSH appears to be preferentially secreted before puberty. A third mode of gonadotropin secretion is related to the ovarian cycle of adult, premenopausal women. These surges are the gonadotropin response to positive gonadal steroid feedback. In postmenopausal women, the low circulating levels of estrogen and to some degree inhibin result in a rise in gonadotropin secretion, often to concentrations 10–15 times greater than those present before ovarian failure.

In the testes, FSH stimulates spermatogenesis via an action on the Sertoli cells. It also induces LH receptors on the Leydig cells, thereby promoting the action of LH to stimulate testosterone production. In the ovary, FSH stimulates follicular growth past the early antrum stage. It also acts on the granulosa cells to cause the conversion of androgens to estrogen, via an action on aromatase activity, and synergizes with estrogen to increase LH receptors on the granulosa cells. Follicular growth is also stimulated by LH. Additionally, LH stimulates androgen production by the thecal cells, thereby synergizing with FSH to stimulate estrogen production. Follicular rupture during ovulation and corpus luteum formation is stimulated by unknown mechanisms by LH, which thereafter exerts a luteotropic effect stimulating estrogen and progesterone formation.

Neural and hormonal factors control gonadotroph activity. The decapeptide hormone gonadotropin-releasing hormone (LHRH) is produced in neurons located in the infundibular region and the rostral hypothalamus. These neurons project to extrahypothalamic sites, where they are thought to be involved in the generation of stereotypic sexual behaviors and to the median eminence for release of gonadotropin-releasing hormone (GnRH) into the vicinity of the capillaries of the portal vessels. GnRH binds to specific receptors on the gonadotrophs and stimulates cAMP formation; however, it is not clear that this is the second-messenger system involved in the stimulatory effect of the decapeptide. Instead the role of extracellular calcium has been supported by cell culture studies, which revealed a calmodulin-sensitive pathway. The mechanism of action of the calcium effect of GnRH seems to involve phospholipid turnover via activation of phospholipase A-2 and arachidonic acid formation with subsequent protein phosphorylation. Other by-products of PI breakdown might also be involved in GnRH activation of the gonadotroph because diacylglycerol formed by PI hydrolysis activates protein kinase C, which synergizes with calmodulin.

Although the existence of a separate hypothalamic, FSH-releasing hormone has been speculated on for many years, it has failed to be identified. Instead a glycoprotein of gonadal origin, originally named *inhibin,* acts at both the hypothalamic and pituitary levels to inhibit FSH secretion. Two forms of inhibin have been isolated from porcine follicular fluid, each composed of alpha and beta subunits linked by disulfide bridges. Inhibin A differs from inhibin B only in the structure of the beta subunit. Recent studies in humans point to the stimulatory effect of FSH on inhibin release, suggesting a physiologically significant long-loop negative feedback from gonads to the hypothalamo–pituitary unit. Recently it has been reported that a protein formed by disulfide linkage of the respective beta subunits of inhibin A and inhibin B has potent FSH-releasing activity. This protein has been named *activin.*

In addition to the novel glycoprotein-mediated gonadal influences on the gonadotroph, gonadal steroids exert profound negative and positive effects on the release of the gonadotropins. Mainly exerted by estradiol, the negative feedback effects of gonadal steroids are a result of dampening of the hypothalamic pulse generator, which controls episodic GnRH secretion, and a result of a diminution of pituitary responsiveness to GnRH. Thus, in the follicular phase of the human menstrual cycle when estradiol and progesterone levels are low, the frequency of LH pulses is higher than during the luteal phase, when progesterone and estradiol secretion is maintained. At midcycle in the human female, rising estrogen levels from the developing follicle begin exerting a positive feedback effect by increasing GnRH pulsatility and sensitizing the gonadotroph to the action of the decapeptide. The midcycle surges of LH and FSH are then terminated by the high secretion rates of estrogen and progesterone from the corpus luteum.

5. Thyrotrophs

Thyrotrophs make up only a small proportion of adenohypophyseal cells (5–10%), are located in the center of the pars distalis, and are characterized by their large size and the presence of small secretory granules. These cells produce the glycoprotein hormone thyroid-stimulating hormone (TSH), which like the gonadotropins consists of alpha and beta subunits. These subunits are products of separate genes and are assembled only after addition of the carbohydrate side chains. The alpha subunit is structurally homologous to that of LH and FSH. The adenohypophysis contains approximately 0.1 mg TSH. The plasma half-life of the glycoprotein is about 50 min. Normal plasma levels are 1–4 ng/ml, which are often still referred to in terms of International Units (normal mean, 1.8 μU/ml). These low levels often complicate accurate determination of secretory activity of the thyrotroph, necessitating measurement of plasma thyroid hormones as an indicator of thyrotroph function.

Adenylate cyclase activity of the thyroid is stimulated as a consequence of TSH binding to a specific receptor. The increased cAMP levels result in protein phosphorylation and numerous changes in thyroid cell activity. Morphologic changes occur, the cell becoming cylindrical in shape and losing its cuboidal profile, and the amount of colloid in the follicles declines. TSH stimulates iodide transport, thyroglobulin synthesis, iodotyrosine and iodothyronine formation, thyroglobulin proteolysis, and thyroxine (T4) and triiodothyronine (T3) release.

Neural and hormonal factors control the release of TSH. A small (3-amino-acid) peptide known as TRH is produced in neurons of the preoptic and parvocellular paraventricular hypothalamic areas and delivered to the median eminence. TRH binding results in increased cytosolic calcium, probably caused by mobilization from mitochondrial stores. This free calcium binds to calmodulin and activates protein kinases, leading to eventual protein phosphorylation and granule extrusion. Continued TRH exposure, however, down-regulates the number of TRH receptors at the cell surface, resulting in decreased thyrotroph responsiveness. Estrogens up-regulate the TRH receptor, although T4 causes a decrease in the number but not affinity of TRH receptors. T4 can be converted within the thyrotroph to T3 by the action of 5′-monodeiodinase. In addition to down-regulating the number of TRH receptors, thyroid hormones decrease TSH synthesis. Finally, although there long has been speculation that thyroid hormones can exert long-loop negative feedback effects on hypothalamic mechanisms controlling TSH release, only recently have studies in monkeys demonstrated this possibility and also the transcription of the thyroid hormone receptor in the hypothalamus.

III. POSTERIOR LOBE FUNCTION

A. Morphology

The posterior lobe, often called the *neurohypophysis*, weighs approximately 0.10–0.15 g in the adult. As discussed earlier, the lobe is made up primarily of axonal processes that originate in the supraoptic and paraventricular hypothalamic nuclei. These fibers are unmyelinated and possess terminal swellings characteristic of neurosecretory endings. The lobe also contains corticotrophs that were originally part of the fetal pars intermedia and glial elements called *pituicytes.* The hypothalamo–neurohypophyseal fibers deliver OT and AVP to the neural lobe in association with specific proteins, the neurophysins, which were once thought to function as carrier molecules but are now known to be posttranslational products of the OT and AVP prohormones. Their biologic functions remain undetermined. Neurons producing OT and AVP also project to brain sites other than the neural lobe, including regions associated with stereotypic behaviors that have been associated with the peptides, and to the median eminence. Therefore, these peptides are communicated via either the long portal vessels from the median eminence or the short portal vessels from the neural lobe to the adenohypophysis. Releasing factor activities for OT at the levels of the corticotroph and lactotroph and for AVP at the levels of the corticotroph and thyrotroph have been described, although their role as trophic agents in human physiology has yet to be convincingly established.

Both hormones are peptides consisting of nine amino acids, containing an internal disulfide linkage between cystine residues at positions 1 and 6. The presence of this disulfide bridge is necessary for the peptides to maintain full bioactivity. They differ in structure from each other only at the 3 and 8 positions. As mentioned earlier, each hormone is synthesized as part of a larger prohormone, the neurophysin molecule (approximately 10,000 MW) being cleaved off during posttranslational processing. Estrogen treatment stimulates the production of OT and its associated neurophysin. Thus this neurophysin has been termed the *estrogen-stimulated neurophysin.* AVP-associated neurophysin production is stimulated by nicotine, so this glycoprotein is known as the *nicotine-stimulated neurophysin.* Individual neurons of the magnocellular supraoptic and paraventricular nuclei produce either OT or AVP, never both. The prohormones are transferred to the Golgi apparatus after synthesis on the ribosomes and then packaged into neurosecretory granules, in which posttranslational cleavage of the hormone from its associated neurophysin takes place. Delivery down the axon to the neurohypophysis occurs with a flow rate of 8 mm/hr (faster than normal axoplasmic flow). These hypothalamo–neurohypophyseal cells are neurons capable of generating and propagating action potentials that on arrival at the terminals result in depolarization and exocytosis of the neurosecretory granules. This release phenomenon is called *stimulus-secretion coupling.* Both hormones circulate in blood unbound to carrier proteins and are rapidly removed from the circulation by the kidney, liver, and brain. Although the older literature suggests a plasma half-life for OT and AVP of only 5 min, recent studies using physiologic dose levels of the hormone suggest half-lives of less than 1 min. Plasma levels in adults under resting conditions are approximately 1–2 pg/ml for AVP and 10–100 pg/ml for OT. OT levels are slightly higher in adult, nonpregnant, nonlactating females than in males and are known to fluctuate during the menstrual cycle in correlation with estrogen levels.

B. Posterior Pituitary Hormones

1. Vasopressin

Two classes of AVP receptors have been established. V-2 receptors are present on the peritubular (serosal) surface of cells in the distal convoluted tubules and medullary collecting ducts in the kidney. Adenylate cyclase activity is stimulated in these cells by AVP, and the cAMP formed activates a protein kinase on the luminal membrane, which results in protein phosphorylation and an enhancement of permeability of the cell to water. The change in permeability stimulated by AVP permits back diffusion of solute-free water down an osmotic gradient from hypotonic urine to the hypertonic renal medullary interstitium. The end result is an increase in urine osmolality (relative to glomerular filtrate or plasma) and a decrease in urine flow. A second class of AVP receptor exists on vascular smooth muscle. Activation of these V-1 receptors results in vasoconstriction, shunting of blood away from the periphery, and elevation of central venous pressure. This is demonstrated by the observation that total peripheral resistance increases in a linear fashion with increases in plasma AVP levels. Normally, the increase in total peripheral resistance caused by AVP is buffered by baroreceptor mechanisms, which reflexively cause a decrease in cardiac output and sympathetic tone, so that other than regional perfusion effects, the increase in arterial pressures is not great.

Release of AVP can be stimulated by osmotic and nonosmotic factors. These are best tpyified by water deprivation and hemorrhage-induced release of AVP, respectively. Water restriction results in increased plasma osmolality as the water lost in urine and via evaporation and respiration is not replaced. This increase in plasma osmolality is sensed by specialized cells called *osmoreceptors*, located in two vascular regions of the central nervous system where the blood–brain barrier is missing: the vascular organ of the lamina terminalis (OVLT) and the subfornical organ (SFO). Increases in plasma osmolality result in a loss of cellular water from these osmoreceptors and, via neuronal relays probably involving acetylcholine-containing neurons, a stimulation of AVP release. The sensitivity of the osmoreceptor to changes in plasma osmolality can be enhanced by circulating angiotensin II. This mode of stimulation of AVP release is exquisitely sensitive, AVP being released in response to changes in plasma osmolality as small as 1%. Water loading, however, results in decreased plasma osmolality, a net gain of water in the osmoreceptors, and an inhibition of AVP release.

Any stimulus resulting in a decrease in central blood volume (hemorrhage of greater than 8%, orthostatic hypotension induced by quiet standing, positive pressure breathing) is a potent stimulus for AVP secretion. However, maneuvers that increase total blood volume (isotonic saline or whole blood infusion, cold water immersion) suppress AVP release via an alpha-adrenergic mechanism. Low (left atrial) and high (carotid and aortic) pressure baroreceptors detect small alterations in blood volume (pressure) and, via afferents in the ninth and tenth cranial nerves, communicate volume–pressure changes to the central nervous system. Medullary structures receiving this input send efferents to the hypothalamus, which normally inhibit AVP release. Decreases in blood pressure result in a lowering (unloading) of neuronal activity in the afferents from the baroreceptors, less inhibition of AVP release, and increased circulating levels of the hormone. Renin released from the juxtaglomerular apparatus of the kidney in response to hypovolemia initiates the formation of angiotensin II and synergizes with it to sensitize the osmoreceptors leading to additional stimuli for AVP release. This phenomenon explains the increased sensitivity of the osmoreceptors to changes in plasma osmolality in volume-depleted states. This resetting of the osmoreceptor is thought to play an important role in the maintenance of normal fluid and electrolyte balance. A variety of other factors can modulate or alter AVP release. Secretion is induced by increased $PaCO_2$ or reduced PaO_2, pain, stress, heat, beta-adrenergic agents, gonadal steroids, opiates, barbiturates, nicotine, and prostaglandins. Hormone release is inhibited by cold, alpha-adrenergic agents, ethanol, and the cardiac hormones.

Secondary actions of AVP have been described. AVP appears to act within the central nervous system to facilitate the consolidation and retrieval of memory. Studies in humans have indicated a beneficial effect of AVP on short-term memory. As mentioned earlier, under certain circumstances, AVP can act as a CRH within the adenohypophysis. This effect is thought to represent in physiologic terms a potentiative effect to that of CRH during stress. The most common example of altered AVP secretion in humans is the syndrome of deficient secretion called diabetes insipidus. Because of some deficiency in the hypothalamo–neurophypophyseal system, AVP secretion is low or nonexistent. This syndrome is characterized by an inability to concentrate urine, frequent urination, and excessive thirst. Patients are treated with an AVP analogue that has no V-1 activity, thus the renal effects only predominate.

2. Oxytocin

The primary action of OT is on the myoepithelial cells surrounding the alveoli and ducts of the mammary gland, stimulating contraction of these cells and milk ejection. OT also stimulates rhythmic contraction of myometrial cells in the uterus, aiding in delivery of the fetus. Its uterine effects are not mandatory for the initiation of labor; however, it proceeds more slowly in the absence of OT. A commercial preparation of OT, called *pitocin*, is often employed postpartum to sustain contractions and decrease bleeding. Secondary actions of OT include a central effect to initiate maternal, instinctive behaviors and a pituitary action to stimulate PRL and ACTH secretion. Additionally, there is now evidence that OT may play an important role in fluid and electrolyte homeostasis by actions in the central nervous system and kidney. Within the brain, OT has been demonstrated to be important in the neural mechanisms controlling salt appetite. In the kidney, OT acts to increase water permeability of the renal tubule and therefore influencing the excretion of water and sodium.

Activation of touch receptors in the nipples or vaginal stimulation during intercourse or delivery results in increases in afferent input to the OT-producing cells of the supraoptic and paraventricular nuclei. These spinothalamic inputs are thought to use acetylcholine or dopamine as final neurotransmitters. Estrogens secreted by the developing follicle in the periovulatory period also stimulate OT release. Mild stress (restraint, novel environment, apprehension, and fear) and hemorrhage also stimulate OT secretion. Inhibition of OT release is seen during extreme pain or high-temperature exposure. Disorders in OT secretion are often undetected. Deficient OT secretion results in difficulty nursing because of inadequate milk ejection.

BIBLIOGRAPHY

Daughaday, W. (1985). The anterior pituitary. *In* "Williams Textbook of Endocrinology" (J. W. Wilson and D. W. Foster, eds.). Saunders, Philadelphia.

Ganten, D., and Pfaff, D. (eds.) (1985). "Neurobiology of Vasopressin." Springer-Verlag, Heidelberg.

Ganten, D., and Pfaff, D. (eds.) (1986). "Neurobiology of Oxytocin." Springer-Verlag, Heidelberg.

Imura, H. (ed.) (1994). "The Pituitary Gland," 2nd Ed. Raven, New York.

McMurray, R. W., Allen, S. H., Braun, A. L., Rodriguez, F., and Walker, S. E. (1994). Long-standing hyperprolactinemia associated with systemic lupus erythematosus: Possible hormonal stimulation of an autoimmune disease. *J. Rheumatol.* **21,** 843–850.

Muller, E. E., and Nistico, G. (eds.) (1989). "Brain Messengers and the Pituitary." Academic Press, San Diego.

Reichlin, S. (1985). Neuroendocrinology. *In* "Williams Textbook of Endocrinology" (J. D. Wilson and D. W. Foster, eds.). Saunders, Philadelphia.

Samson, W. K., and Mogg, R. J. (1990). Oxytocin as part of stress response. *In* "Behavioral Aspects of Neuroendocrinology" (D. Ganten and D. Pfaff, eds.). Springer-Verlag, Heidelberg.

Placenta

KURT BENIRSCHKE
University of California, San Diego

GLOSSARY

Amnion Innermost membrane of the placenta, containing the amnionic fluid and the fetus

Blastocyst Early stage of embryonic development when the future embryo becomes enclosed in the trophoblastic shell

Chorion Connective tissue layer beneath the amnion that carries the fetal blood vessels and from which the villi originate

Cotyledon Structural unit of the placenta, made up of numerous villi

Decidua Altered endometrium or inner lining of the uterus, which is changed through pregnancy hormones

Placenta Latin for "cake," the body of the placenta; it is so named because of its oval, flat shape

Trophoblast Epithelial (surface) cells that attach the placenta to the uterus

Villus (villi) Finer ramifications of the chorion; structures covered by trophoblast and containing fetal blood vessels

THE PLACENTA IS A FETAL ORGAN. IT IS INTERPOSED between fetus and mother and serves as the most vital organ to provide the nutritional support to the fetus. At the same time, the placenta transfers gases to and from the fetus, eliminates waste products from fetal metabolism, and also produces proteins and endocrine substances necessary for fetal growth and maintenance of pregnancy. This temporary organ is attached to the inside of the uterus (the womb) and is connected to the fetus by the umbilical cord. After birth of the child, the placenta is delivered. It is for this reason that the placenta is often referred to as the "afterbirth." The placenta is generally discarded unless special circumstances require its examination. Because the placenta provides a diary of events during pregnancy, placental study has recently proved useful for legal adjudication in cases of disputed pregnancy supervision.

The placenta is the most important organ for fetal development. It functions only during pregnancy and it occurs only in mammalian species. The placenta serves as the organ that interfaces between mother and fetus and attaches to the inside of the mother's uterus, her "womb." It is a richly vascular organ with highly specialized function, unusual cell types, and secretory products. It is also the only organ that has no nerve supply. The placenta is attached to the fetus by its umbilical cord. In its membranous sac it contains the fetus with the amnionic fluid (Fig. 1). In most mammals, the mother consumes the placenta after delivery of the infant.

The structure of placentas varies greatly among mammalian species, largely depending on the depth of its uterine invasion. This important feature has led to classification based on a complex anatomic scheme, usually referred to as the Grosser classification of placentation. When invasion is minimal (e.g., in whales), we speak of an epithelio-chorial placenta. Human placentas have achieved the greatest depth and are hemochorial placentas, that is, the chorionic trophoblast is directly in touch with maternal blood.

I. ANATOMY AND DEVELOPMENT OF THE PLACENTA

Shortly after the embryo begins its development in the fallopian tube, two lines of cells become established:

ENCYCLOPEDIA OF HUMAN BIOLOGY, Second Edition, VOLUME 6.

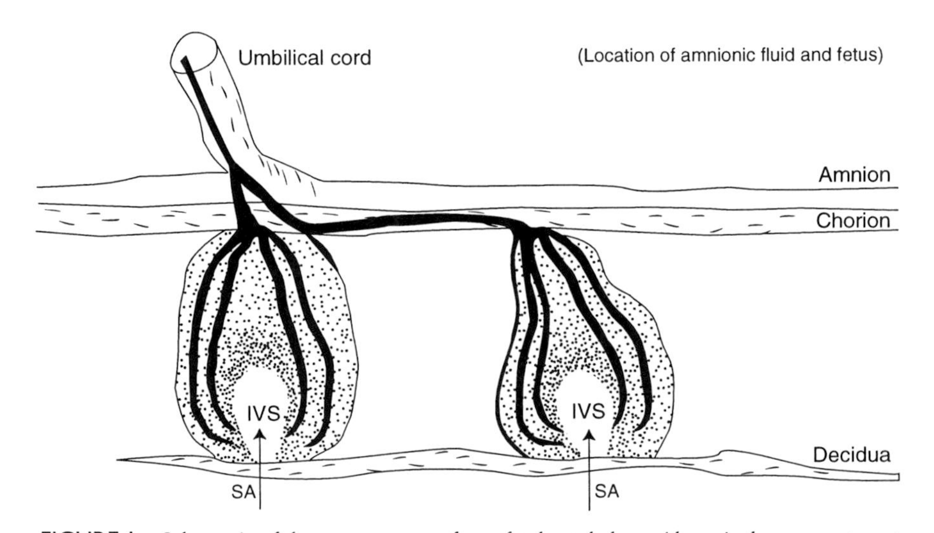

FIGURE 1 Schematic of the arrangement of two fetal cotyledons. Above is the compartment for the amnionic fluid and attachment of the umbilical cord that leads to the fetus. Below is the mother's tissues, the placental "floor." Only a small amount of decidua basalis (gestationally altered endometrium) remains attached to the placenta at birth. The dark lines represent the fetal vasculature in the villous tree, and the light gray shading represents the placental villous tissue. The intervillous space (IVS) is drawn as a vacant space. It is injected by maternal blood through centrally located maternal spiral arterioles (SA). The maternal blood is returned through veins between the cotyledons (not drawn).

those that become the embryo and those that form the placenta, the "shell" of the blastocyst. As an interesting aside, it has recently been found that the *paternal* genes are more responsible for placental development, whereas the *maternal* genes control more of the embryo. This phenomenon is referred to as "imprinting" and has a profound effect on gene expression. Approximately 7 days after fertilization, after the blastocyst had entered the endometrial cavity, the trophoblast attaches itself to the decidual surface and begins its invasion. In the human pregnancy, the blastocyst fully "sinks into" the endometrium, becoming interstitially situated. The cavity then gradually enlarges, pushing the future membranes (the sac that encloses the fluid and embryo) into the uterine cavity (Fig. 2).

The trophoblastic shell deeply invades into the decidua and forms the actual placental disc, after opening the maternal uterine blood vessels and thus creating the intervillous circulation. The continuous growth of villi and fetal capillaries causes an eventually huge surface area (10 m^2) to be formed across which the transfer of gases and nutrients takes place. At term, the placenta weighs approximately one pound, measuring 20 cm across and 2 cm in thickness. The maternal surface of the delivered placenta has a tiny remnant of shed decidua attached to its cotyledons, and the fetal surface has a thin layer of amnion, loosely covering the chorion. The amnion is a continuation of the embryonic ectoderm and is avascular. The chorion carries the fetal blood vessels from the umbilical cord to their finest capillary ramifications in the tertiary villi. Many animals also possess an additional cavity, called the allantois. This structure is very rudimentary in human development and only for technical reasons do we refer to the human placenta as a "chorioallantoic organ." The yolk sac, another important structure in some mammals, is transitory in human development as well.

The villi consist of fetal connective tissue with capillaries and a few macrophages called Hofbauer cells. They are covered by trophoblast, the business end of the placenta. Trophoblast is an unusual group of cells. Innermost (toward the fetal vessels) is the cytotrophoblast, the cell layer in which mitotic activity occurs. The cytotrophoblast is used solely for producing additional elements to become incorporated into the outer layer, the syncytio-trophoblast (Fig. 3). This extraordinary sheath of cytoplasm has no cell boundaries and allows its millions of nuclei to move freely. The syncytium is in contact with the maternal blood and produces all the hormones and serves as the gate check

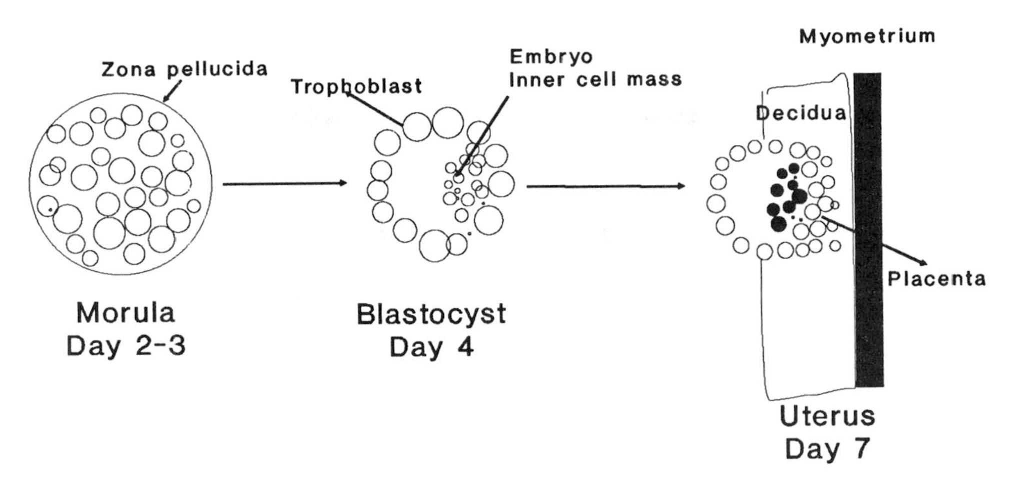

FIGURE 2 Diagrammatic representation of early embryonic and placental development. From a morula that is enclosed in a rigid membrane (the zona pellucida), a blastocyst cavity develops with an embryonic "inner cell mass." The shell of this blastocyst represents the earliest trophoblast differentiation. The zona pellucida is enzymatically removed and, 1 week after fertilization, the blastocyst invades the endometrium. It ultimately becomes "interstitially" implanted, totally surrounded by decidua.

for the transport to and from the fetus of nutrients and other substances. Mitoses never occur here; the syncytium also has no human leukocyte antigen (HLA) surface markers and is thus probably not recognized immunologically by the mother as being "foreign." Small numbers of syncytial cells break off regularly to get swept away by the intervillous maternal circulation; they end up and die in the mother's lung.

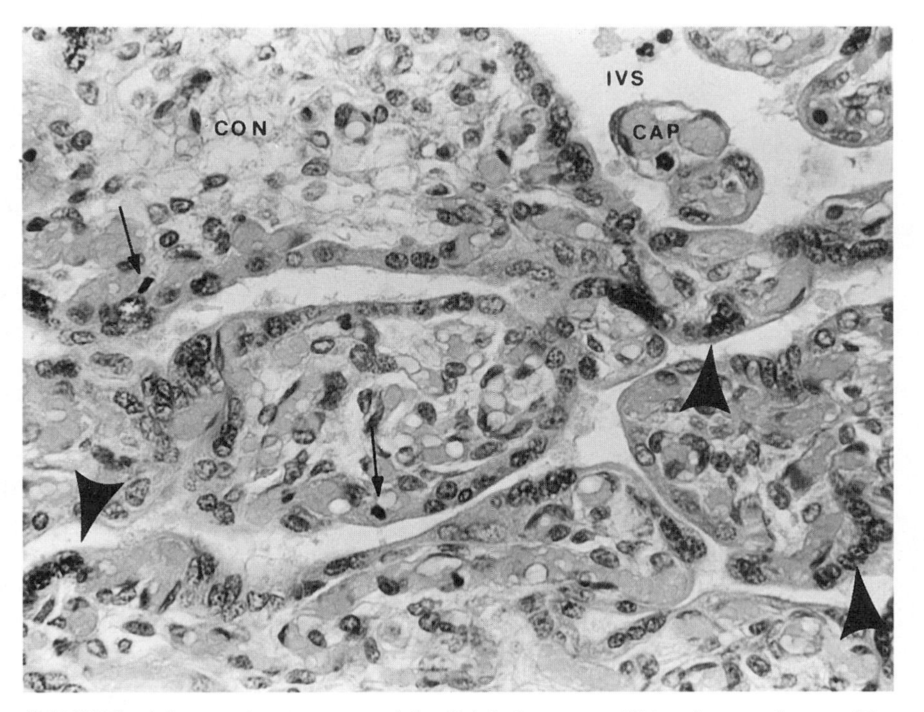

FIGURE 3 Microscopic appearance of the slightly immature villi in a human placenta. The smallest villi are nearly completely filled with fetal capillaries (CAP). The connective tissue of the villi is labeled CON, and the maternal intervillous space is labeled IVS. The arrows indicate the dark-staining mitoses of the cytotrophoblast. The arrowheads denote the syncytio-trophoblast.

II. THE UMBILICAL CORD AND PLACENTAL MEMBRANES

The umbilical cord carries two arteries from the fetus to the placenta and the blood is returned to the fetus through one vein (Fig. 4). The blood is propelled by the fetal heart. The umbilical cord carries these blood vessels in a compressible "Wharton's jelly," which is normally attached near the center of the placenta. The cord normally measures about 50 cm in length, but wide variations exist. The length is determined primarily by fetal movements in early gestation. Infants with reduced ability to move have shorter cords. Excessively long umbilical cords tend to become heavily spiraled and may form lethal knots and entanglements. Normally, the cord has a left spiral course; only some 10% of infants have a right spiral, for an as yet unknown reason.

The membranes are composed of the inner amnion and an outer chorion. A small amount of decidua (capsularis) is peripherally attached. The membranes contain the amnionic fluid in which the fetus is suspended, but they do not produce this fluid. They merely aid in the turnover of solutes and the water. When labor commences at term, the membranes either rupture spontaneously or are artificially broken. This allows the amnionic fluid to escape, the uterus to contract, and labor to ensue. Infectious causes are a frequent reason for premature rupture of membranes. In the case of twins or other multiples, membranous partitions are usually found between the fetuses.

III. FUNCTIONAL ASPECTS

The transfer of water, solutes, nutrients, and gases from the intervillous blood to the fetus is regulated by the trophoblast. The outer layer of trophoblast, the syncytium, "decides" what to transport and in which direction. Much of the transfer is accomplished through minute vesicles; other exchange occurs through osmosis and perhaps even more complex schemes. There are some substances, however, that are never transported, such as large protein molecules and cellular elements. The manner of exchange between mother and fetus is extremely complex and still

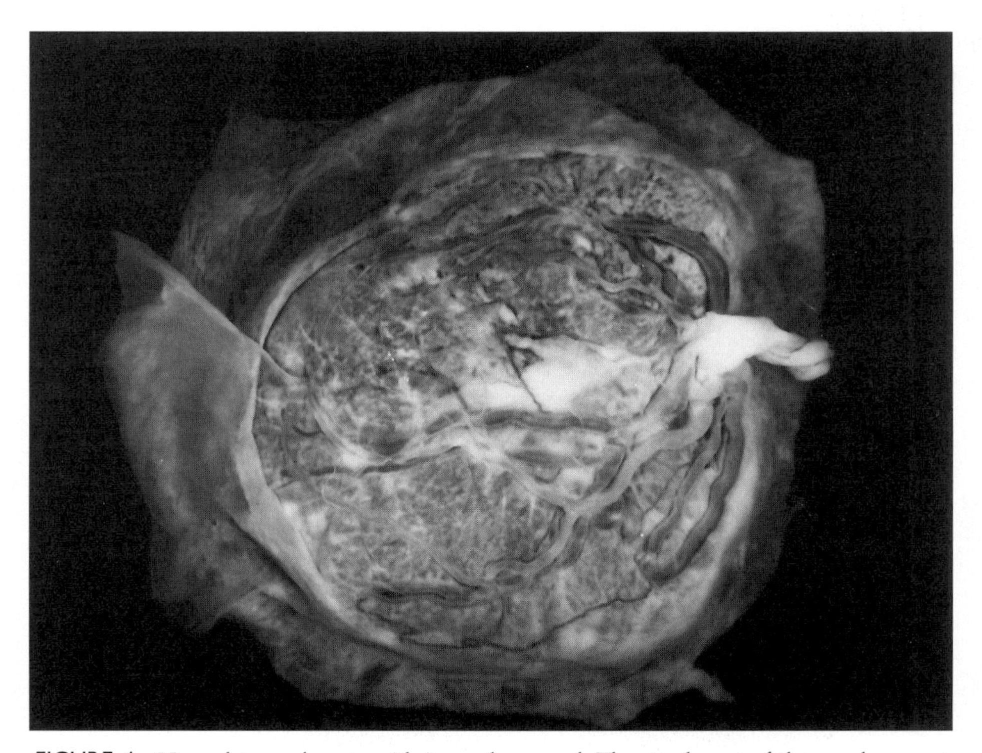

FIGURE 4 Normal term placenta with its cord removed. The attachment of the membranes at the margin is clearly seen. The surface blood vessels that emanate from the cord are the fetal vessels that send the blood into the placenta. Arteries lie on top of the veins; the white patches are maternal clotting products that occur normally in the intervillous space during pregnancy.

under intense study. Presumably, it is in the syncytium where the decision for adequate or subnormal fetal growth is made, and fetal as well as maternal regulatory mechanisms are at work.

A source of continuous investigation is the lack of immunologic rejection of the placenta. Normally, "foreign" tissues, such as skin, would be rejected soon after transplantation of tissues into a competent host. Since the fetus has a different genotype than the mother (it has the father's genetic messages), one would expect the placenta to be rejected. Many mechanisms have been proposed to explain this complexity.

IV. ENDOCRINE PERFORMANCE

The placenta produces two hormones of overriding importance, progesterone and human chorionic gonadotropin (hCG). hCG, a protein hormone, is sent from the placenta into the mother and stimulates the ovarian corpus luteum to produce progesterone. This hormone is essential for the transformation of the endometrium to decidua, thus allowing pregnancy to continue. In the second trimester of pregnancy, there is a drastic diminution of hCG production and placental progesterone supplants its maternal ovarian production. Estrogen also comes from the placenta, but results from the transformation of fetal precursor hormones, or weak androgens. This transformation, as well as many other complex regulatory functions, is accomplished through the action of specific placental enzymes. Aside from these hormones, the placenta produces human placental lactogen (hPL), a hormone whose function is currently unknown, and a variety of specific proteins, especially growth factors. The largest quantity of protein produced is "major basic protein," a substance that is otherwise specific for the granules of eosinophilic leukocytes. In the placenta, it comes from a distinct lineage of trophoblast, called the "X-cell." Its function is currently unknown. [*See* Steroid Hormone Synthesis.]

V. SPECIAL PROBLEMS AND PATHOLOGY

Many abnormalities affect placental development and frequently these interfere with normal fetal development. One of the commonest developmental errors is the absence of one umbilical artery; it occurs in 1% of conceptions and is associated with about 50% anomalies or fetal deaths. A relatively frequent finding is an excessive length of the umbilical cord, presum-

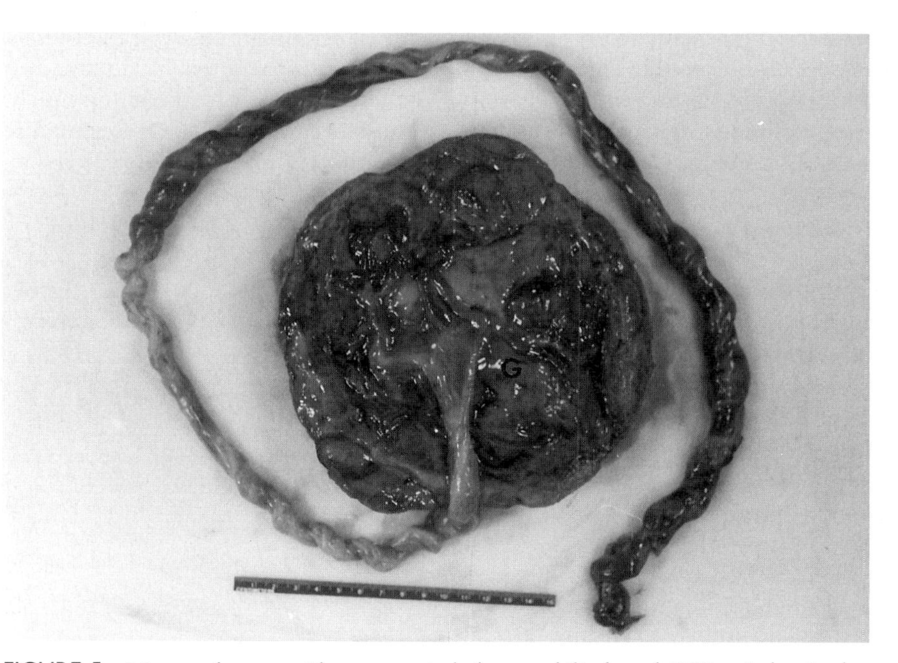

FIGURE 5 Mature placenta with an excessively long umbilical cord (100 cm) that is also heavily spiraled to the right. This spiraling has led to numerous clots in the large surface veins (gray bands, G) and is related to the poor fetal outcome in this pregnancy.

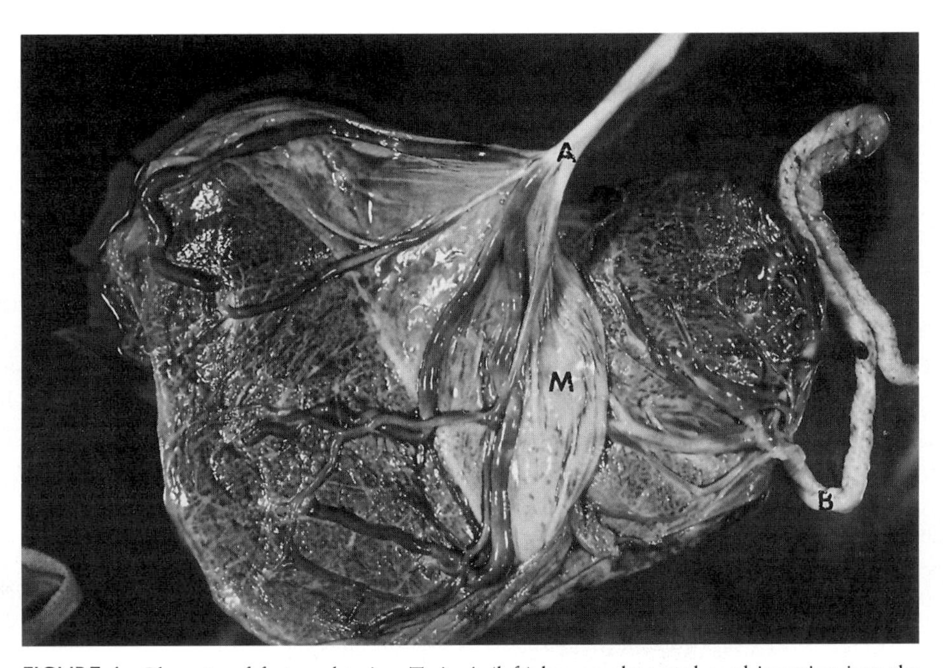

FIGURE 6 Placenta of fraternal twins. Twin A (left) has an abnormal cord insertion into the "dividing" membrane (M); it is a "velamentous cord insertion." This anomaly is often associated with a single umbilical artery, as in this case. The dividing membrane between the two cavities is thick and consists of four separate layers. Twin B has a smaller placenta with marginal cord insertion. This umbilical cord has excessive right spiraling, the presumed reason for the smaller size of placenta and fetus.

ably due to increased fetal movements in early pregnancy. This is generally associated with marked spiraling, and it often leads to fetal death from excessive twisting with consequent cessation of the fetal circulation (Fig. 5). The amnion may rupture in early gestation, leading to amnionic bands that can cause fetal amputations. Premature rupture of membranes and ascending vaginal infections can lead to infection of the amnionic sac, resulting in early delivery. Vascular diseases of the mother lead to infarcts (focal necrosis) of placental tissue, which reduces the amount of villous tissue for transfer of nutrients and thus correlates with growth retardation. Occasional tumors develop in the placenta, and other pathological features take place that make it useful to examine the placenta to better understand the fetal development. [*See* Embryofetopathology.]

The placentas of multiple gestations, such as twins or triplets, present special problems. One can often make the diagnosis as to whether twins are identical or fraternal from a placental study alone (Fig. 6). In identical twins with a monochorionic placenta (about two-thirds of identical twins), blood vessels communicate between the two placental halves. Blood can flow back and forth between the twins through these connections and often cause significant disease. All of these features are becoming much better understood now because of more frequent antenatal sonographic examination. [*See* Embryology and Placentation of Twins.]

BIBLIOGRAPHY

Baldwin, V. J. (1994). "Pathology of Multiple Pregnancy." Springer-Verlag, New York.

Benirschke, K., and Kaufmann, P. (1994). "The Pathology of the Human Placenta," 3rd Ed. Springer-Verlag, New York.

Boyd, J. D., and Hamilton, W. J. (1970). "The Human Placenta." W. Heffer & Sons, Cambridge, England.

Grosser, O. (1927). "Frühentwicklung, Eihautbildung und Placentation des Menschen und der Säugetiere." J. F. Bergmann, Munich.

Page, K. (1993). "The Physiology of the Human Placenta." University College London Press, London.

Plasma Lipoproteins

WILLIAM A. BRADLEY
SANDRA H. GIANTURCO
University of Alabama at Birmingham

GLOSSARY

Cholesterol ester transfer protein (CETP) Plasma protein that facilitates the transfer and exchange of cholesteryl ester and triglyceride among lipoprotein classes

Chylomicron Class of dietary triglyceride-rich lipoproteins produced by intestinal synthesis after eating and introduced into the bloodstream from the thoracic duct

High density lipoproteins (HDL) Class of cholesterol-enriched lipoproteins produced by the liver and highly modified in the plasma

Intermediate density lipoproteins (IDL) Class of lipoproteins produced during lipolysis of VLDL in plasma

Lecithin cholesterol acyl transferase (LCAT) Plasma enzyme that uses the fatty acid of a polar lipid, phosphatidylcholine, and transfers it to the alcohol group of cholesterol to produce a water-insoluble cholesteryl ester

Lipoprotein lipase (LpL) Plasma enzyme that hydrolyzes the neutral fat triglyceride to its water-soluble components, fatty acids and glycerol

Low density lipoproteins (LDL) Class of cholesterol-enriched lipoproteins produced by the action of lipid-modifying enzymes in the plasma on VLDL

Very low density lipoproteins (VLDL) Class of heterogeneous triglyceride-rich lipoproteins synthesized by the liver

PLASMA LIPOPROTEINS ARE DYNAMIC MACROMOLecular complexes of lipids (fat) and protein, primarily synthesized and assembled in the liver and intestine. Lipoproteins transport water-insoluble lipids throughout the body via the bloodstream to provide these essential nutrients to the cells of the body for energy and in specific tissues for steroid production. The structure of a plasma lipoprotein is illustrated in Fig. 1. This model structure indicates that the water-insoluble lipids—triglyceride and cholesterol and cholesteryl esters—are located in the spherical core and are surrounded by a monolayer of polar lipids known as phospholipids and specialized proteins known as apolipoproteins whose "amphipathic" nature allows them to interact with both the water-insoluble lipid core and the surrounding aqueous plasma environment.

I. INTRODUCTION

The lipoproteins were initially classified by their buoyant densities, that is, according to their ability to float instead of sediment in salt solution, under ultracentrifugal force. This unique property has allowed their purification from plasma by simple ultracentrifugal procedures. Since lipoprotein cores contain neutral lipid, which is less dense than water, upon centrifugation the lipoprotein moves (floats) toward the top of the centrifuge tube at a rate that depends on the amount of lipid in the core of the lipoprotein, thus separating the lipoproteins from the remainder of the many other plasma components. The nomenclature of most lipoproteins is based on these flotation characteristics. Chylomicrons are the least dense (these can float to the top of the tube without centrifugation because they are so lipid rich, like the cream in milk), followed by very low density lipoprotein (VLDL), intermediate density lipoprotein (IDL), low density lipo-

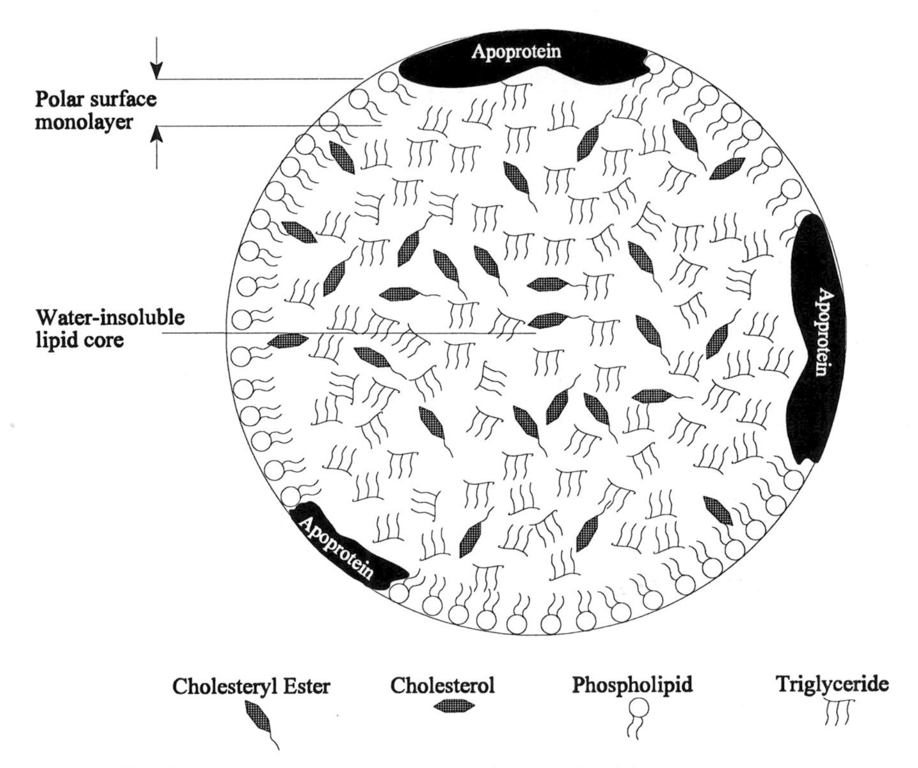

FIGURE 1 Two-dimensional diagram of a generalized lipoprotein structure.

protein (LDL), and the most dense lipoprotein, high density lipoprotein (HDL). Recently, many subspecies within each of these general classes of lipoproteins have been defined by specific physical and metabolic properties and are increasingly important to understanding the normal and pathological metabolism of lipoproteins. Table I provides some general physical properties of the major lipoprotein classes. The largest lipoproteins with the greatest diameters have the largest lipid cores and therefore are the least dense. The smaller the lipoprotein, as determined by the core diameter, the greater its density. All lipoprotein classes are heterogeneous, but the greatest heterogeneity exists within the triglyceride-rich lipoproteins (chylomicrons and VLDL) and is reflected in the large range of particle diameters and compositions indicated in Tables I, II, and III. Table II shows the average percentage component composition of the lipoprotein classes, which in Table III is converted into the number of each kind of lipid molecule found in an individual lipoprotein particle. These tables indicate that the largest lipoproteins are triglyceride-rich on a percentage (and number) basis and that during lipolysis (see Section III) they are converted in blood to cholesterol-enriched lipoproteins. [*See* Lipids; Proteins.]

TABLE I
Physical Properties of Lipoprotein

Lipoprotein	Molecular weight ($\times 10^{-6}$)	Density (g/cm^3)	Diameter (nm)
Chylomicron	31–1000	0.93	75–1000
VLDL	10–31	0.95–0.99	24–74
IDL	3.3	1.01	20–24
LDL	2.2	1.035	20.0
HDL_2	0.36	1.09	10.2
HDL_3	0.18	1.15	7.8

II. APOLIPOPROTEINS

The apolipoproteins, also known as apoproteins, are located on the surface of the lipoproteins (see Fig. 1). Some of the apoproteins (A-I, A-II, A-IV, C-I, C-II, C-III, and E) share the common structural feature consisting of a 22-amino-acid α helix, which may have arisen from a common ancestral gene. This structure

TABLE II
Component Compositions of Human Lipoprotein

Lipoprotein	Protein	Phospholipid	Triglyceride (% by weight)	Cholesteryl ester	Cholesterol
Chylomicron	2	7	84	5	2
VLDL	5.9–11.4	18.5–23.8	41.3–63.6	6.1–15.9	5.9–7.5
IDL	17.7	21	15.5	34.3	11.2
LDL	21	22	11	37	8
HDL_2	41	30	4.5	16	5.4
HDL_3	55	23	4.1	12	2.9

accords these apoproteins with the dual nature of simultaneously interacting with water-insoluble lipid and with the aqueous plasma. The manner in which these apoproteins are able to accomplish this is through their ability to fold into an amphipathic helix. The basic elements of this unique structure consist of an arrangement of amino acid residues forming an α helix with amino acid residues that are lipid soluble on one face of the helix and water soluble on the opposing face.

More than 25 apoproteins are currently known. The apoproteins range in size from small (apoC-I; MW ~6,530) to very large (apoB-100; MW ~512,000). Table IV lists the major apoproteins, the lipoprotein with which each is most commonly found, and alternate functions. In addition to their ability to solubilize/stabilize lipid, some apoproteins have other important physiological functions, usually involved in the metabolism of the lipoprotein class, either as a cofactor for lipid-modifying enzymes or as ligands recognized by distinct, cell-surface receptors that facilitate the lipoprotein interaction with specific cell types of organs like liver and adrenal glands and cells of the artery wall.

Apoproteins can exist in several isoforms, as seen in Table IV. For example, by posttranslational addition of varying amounts of the carbohydrate sialic acid to apoC-III, three major known plasma species of this single gene product are known. ApoE, on the other hand, has three distinct major polymorphic genotypes, designated apoE-2, apoE-3, and apoE-4. (Recently these apoproteins have undergone increased reassessment since an association of the apoE-4 genotype with increased risk for late-onset Alzheimer's disease was identified in several independent studies.) In these polymorphs each is characterized by a single specific change of an amino acid residue (cysteine to arginine) from apoE-2 (no changes) apoE-3 (one change) to apoE-4 (two changes). These subtle changes in structure can change the properties of the apoprotein and the manner in which they are metabolized. ApoB-100 and apoB-48, which is approximately one-half the size (molecular weight) of apoB-100, are derived from the same gene. The unique production of apoB-48 involves RNA editing; an enzyme complex converts a specific nucleotide by deamidation from a cytidine to a uridine of the messenger RNA (mRNA) of the complete apoB-100 mRNA.

TABLE III
Molecules or Lipid per Lipoprotein Particle

Lipoprotein	Phospholipid	Triglyceride	Cholesteryl ester	Cholesterol
Chylomicron	45,160	508,000	27,700	28,840
VLDL	3,071–5,730	4,859–23,530	2,446–5,630	1,938–4,160
IDL	894	602	1,741	955
LDL	653	298	1,310	475
HDL_2	137	19	90	50
HDL_3	51	9.5	32	13

TABLE IV
Apoprotein Properties

Protein	Lipoprotein	Molecular weight	Plasma concentration (mg%)	Function(s)
A-I	HDL, CHYLOS	28,331	120	LCAT activator
A-II	HDL	17,000	38	Hepatic lipase activator
A-III	HDL, LDL, VLDL	~20,000		
A-IV	VLDL, CHYLOS	46,000	15	
B-100	LDL	512,000	83–92	Ligand for LDL receptor
B-48	CHYLOS	246,000		
C-I	VLDL, HDL	6,530		LCAT activator
C-II_1	VLDL, HDL	8,837	4.97	LPL activator
C-II_2	VLDL, HDL(?)	8,837		LPL activator
C-III_0	VLDL, HDL	8,751		LPL inhibitor
C-III_1	VLDL, HDL	8,751		
C-III_2	VLDL, HDL	8,751		
D	HDL, VLDL	22,000		
E-4	VLDL	34,000		Ligand for LDL receptor and receptor family members
E-3	VLDL	34,000	3–10	Ligand for LDL receptor and receptor family members
E-2	VLDL	34,000		Ligand for LDL receptor and receptor family members
F	HDL	26,000–32,000		
G	VLDL, HDL	72,000		
H (β_2-glycoprotein)	VLDL	54,000		LPL activator

This changes the codon from a CAA (coding for a glutamine in apoB-100) to UAA, which is a stop signal for translation. Thus during translation of the protein from the altered mRNA, apoB-48, only half the length of the normal product, is generated. This extensive change in apoB dramatically alters the metabolism of the particles upon which they reside (see Section III).

Modern molecular biology tools have identified many polymorphic forms of the apoproteins. The changes associated with a polymorph can have little or dramatic consequences on the metabolism of the lipoproteins. The study of such natural mutations has improved our understanding of the function of the apoproteins.

III. LIPOPROTEIN METABOLISM

A. Chylomicrons

Chylomicrons are synthesized in the intestine by enterocytes upon absorption of dietary fat from the gut in the form of fatty acid and monoglyceride in bile salt micelles. Little fatty acid is produced by the intestinal enterocyte and so the reesterification of these components into triglyceride, the major component of chylomicrons, reflects the dietary fatty acid profile of the food ingested. In humans, only apoB-48 (and not apoB-100) is synthesized by the enterocyte and it associates with the phospholipid monolayer, stabilizing the nascent lipoprotein particle. In addition, apoA-I, apoA-IV, and apoCs also appear to be *de novo* apoproteins of chylomicrons. The chylomicrons are secreted into the lymph, where they accrue additional apoCs and small amounts of apoE before entering the bloodstream through the thoracic duct. Once in plasma, the chylomicrons are rapidly modified. Lipoprotein lipase (LpL) with apoC-II as its cofactor immediately begins the hydrolysis of the triglyceride to produce free fatty acids that are used for energy in heart and skeletal muscle, where the fatty acids are catabolized, and for storage as triglyceride in adipocytes (fat cells). The lipase is anchored to capillary endothelial cells by cell-surface-bound proteoglycans, which may also facilitate binding of the chylomicrons through apoB and apoE. In addition, the chylomicrons acquire additional apoE, which is essential for their eventual catabolism in the liver, and apoCs through exchange with plasma HDL. These plasma-modified chylomicrons are referred to as chylomicron remnants. The half-life of the chylomicron in plasma is 5–10 min. They are released into the circulation and rapidly entrapped by the liver, and then are internalized by hepatocytes through receptor-specific processes that are predominantly mediated by interaction of the cell-surface receptors with apoE on the remnants. The major hepatic receptors involved are LDL receptors and the genetically related multifunctional receptor known as the LDL receptor-related protein (LRP). The liver is able to reuse the triglyceride and is the one organ that can catabolize the cholesterol in lipoproteins to bile acids for excretion and thus is essential in the body's cholesterol balance. [*See* Cholesterol.]

B. Very Low Density Lipoproteins

VLDL are made by the liver. Their biosynthesis has been studied in greater detail than the intestinal biosynthesis of chylomicrons. Like chylomicrons, VLDL are triglyceride-rich lipoproteins. However, the triglyceride found in VLDL, unlike that of chylomicrons, can be synthesized *de novo* in the liver from acetyl-CoA units produced by carbohydrate utilization, or from free fatty acids brought to the liver by plasma albumin and/or from the hydrolysis and reesterification of the triglyceride derived from the circulating chylomicron remnant uptake by the liver just described. In the liver, the major apoprotein component of human VLDL synthesis is apoB-100. It is required for the assembly and secretion of VLDL by the liver, but the secretion of apoB-containing VLDL is regulated posttranslationally mainly by lipid availability, in particular, by triglyceride availability. Recent reports have identified an endoplasmic reticulum (ER) microsomal triglyceride transfer protein (MTP) that transports triglyceride to the nascent lipoprotein particle and is essential for the secretion of apoB-containing VLDL. This activity is found in the lumen of the ER in both hepatocytes and enterocytes, the known sites of apoB-containing triglyceride-rich lipoprotein synthesis. Since only one molecule of apoB-100 is present on one VLDL particle, with increasing amounts of triglyceride synthesized under different metabolic states, the liver responds by producing larger VLDL particles. In some metabolic states in which there is an overproduction of apoB (certain hyperlipoproteinemias) and apparently a greater availability of lipid, there is a corresponding overproduction of VLDL, in both size and number of particles secreted.

The exact mechanism of VLDL assembly is currently under intense investigation and remains controversial. It is a complicated process involving tight transcriptional regulation of apoB mRNA production, regulation of the translocation of apoB controlled in part by its degradation, which in turn depends on the availability of lipid (or lipid components), and coordinate synthesis of surface and core lipids. ApoCs and apoE are additional apoprotein components synthesized by the liver and associated with these nascent hepatic VLDL.

Upon secretion into the circulation, the newly synthesized VLDL begin a complicated remodeling process. Triglyceride hydrolysis occurs in a manner similar to that for chylomicrons. Again the lipolysis requires apoC-II found on the VLDL surface. Plasma exchange processes cause transfer of apoproteins from HDL to VLDL, the VLDL acquiring apoE while giving up apoCs to the HDL. The transport protein, cholesterol ester transfer protein (CETP), exchanges triglyceride for cholesteryl esters from HDL, enriching VLDL in cholesteryl ester.

As the lipolysis continues, the triglyceride-rich core of the lipoprotein decreases in size and becomes relatively enriched in cholesteryl ester, while losing surface phospholipid and apoproteins to HDL. This cascading lipolytic process of producing smaller VLDL in plasma with concomitant surface component and core lipid changes is one reason for the great heterogeneity seen in the VLDL class of lipoproteins. In normal individuals, a major fraction of VLDL continues to be lipolyzed and remodeled by plasma factors through an intermediate state called IDL until it reaches its end plasma metabolic fate: it becomes LDL.

The VLDL that are not converted to LDL return to the liver for removal after delivering their triglyceride to peripheral sites for energy and storage. The ability of the liver to remove the cholesterol-enriched VLDL is similar to the chylomicron removal. However, VLDL contain two potential ligands for receptor interaction, unlike the chylomicron remnants. Large VLDL (diameters > 34 nm) require the presence of apoE to interact with the LDL receptor, whereas in smaller VLDL (diameters <34 nm) apoB-100 is in a competent conformation and is a primary ligand for the LDL receptor. Many cell-surface-facilitated processes have been described by which the VLDL can initially interact with the liver. These initial localization steps are followed by receptor-mediated uptake of the VLDL in which the components are hydrolyzed and the VLDL cholesterol is used for membrane or lipoprotein synthesis or is converted into bile acid for excretion.

C. Low Density Lipoprotein

LDL is the major apoB-containing lipoprotein and the main carrier of plasma cholesterol in circulation. Most of the LDL is produced by the *in vivo* lipolysis of VLDL as described earlier. Each LDL has on its surface one apoB-100, which is the main apoprotein found in this class of lipoproteins. In the periphery, circulating LDL can be taken up by cells that need cholesterol for membrane synthesis or by cells, like adrenal cells and ovarian cells, that need large amounts of cholesterol for steroid hormone production. Most cells of the body express a receptor for LDL, known as the LDL receptor. This well-characterized receptor specifically binds with high affinity to a region of apoB-100 that is absent in apoB-48. The LDL receptor number on most cells is regulated by the amount of endogenous cellular cholesterol, except in steroidogenic cells. So if the cell has adequate cellular cholesterol, the receptor number is lowered so that additional exogenous LDL cholesterol will not enter the cell. This regulation of the receptor number is tightly governed by intricate control mechanisms of the transcription of the LDL receptor gene, which detect cholesterol (or cholesterol derivative) levels.

The LDL that is not removed by peripheral cells returns to the liver for receptor-mediated removal by the liver LDL receptor pathway.

Lipoprotein(a), Lp(a), is a modified LDL in which a large glycoprotein that has domains (regions) that are highly homologous to the plasma proenzyme, plasminogen, is covalently linked through a disulfide to the apoB-100 of LDL. This unique lipoprotein has some unusual properties and has been associated with the increased risk for several vascular diseases.

D. High Density Lipoprotein

Two major sources of plasma HDL exist. The liver can produce and secrete both a small spherical HDL_3-like particle, containing apoA-I, phospholipid, and cholesteryl ester, and a discoidal structure with no lipid core, consisting mainly of phospholipid, apoA-I, and/or apoE. Upon secretion into the circulation, these HDL precursors can accept cholesterol and phospholipid from a variety of sources, but a major source is the surface components from the triglyceride-rich lipoproteins (chylomicrons and VLDL) undergoing lipolysis. The cholesterol in these phospholipid-rich HDL precursors in an excellent substrate for the enzyme lecithin cholesterol acyl transferase (LCAT), which uses one molecule of fatty acid from HDL phospholipid and one molecule of HDL cholesterol to produce one molecule of cholesteryl ester. This esterification of cholesterol makes it an even less water-soluble molecule that can be transferred to other lipoproteins only through the action of the plasma transport protein, CETP. Thus the action of LCAT immobilizes the cholesteryl esters in the evolving core of HDL, increasing the core size.

As this particle increases in size it must be stabilized by rearrangement of the existing apoA-I on the surface of the particle or by the further addition of transferred apoproteins and phospholipid from the chylomicron remnants and VLDL. This larger HDL represents the second major subfraction, HDL_2. As mentioned, the cholesteryl esters can be transferred from HDL only through the action of CETP. Indeed, this occurs, and the cholesteryl esters of HDL are transferred to VLDL (and LDL) in exchange for triglyceride, thus producing a triglyceride-enriched HDL. Hepatic lipase hydrolyzes the triglyceride of this plasma-modified HDL, shrinking its core size and producing a smaller particle similar in size to HDL_3. This is an elaborate and dynamic process in which many subspecies of HDL are generated. A probable important event in this scheme is the ability of specific HDL particle types to act as acceptor sites for cholesterol from the surface membranes of peripheral cells. The cellular cholesterol is then immobilized in the core of these particles by the action of LCAT. The cholesteryl ester is then transferred to VLDL and LDL by CETP and these particles (or their cholesterol) are taken up by the liver, so that excess peripheral cellular cholesterol can be excreted by the liver as bile salts. The peripheral cells themselves have no mechanism to dispose of cellular cholesterol, except to esterify it. Thus, this is an important system for maintaining cholesterol balance in cells and body, since cellular cholesterol accumulation can be toxic to the cells. This proposed system of transport of cholesterol from cells through the lipoproteins back to the liver for disposal is called the reverse cholesterol transport pathway. HDL can also directly deliver its cholesteryl esters to specific cell types, like liver and adrenal cells, without particle uptake, probably through a receptor-facilitated localization of the HDL on the cell surface with subsequent transfer of the cholesteryl esters by an unknown mechanism.

IV. LIPOPROTEINS AND DISEASE

Lipoproteins have been an important public health issue for more than two decades. Their role in cardiovascular disease has led to some of the most extensive epidemiological and clinical studies undertaken throughout the world.

The original hypothesis driving these clinical and epidemiological studies was that elevated plasma cholesterol levels reflect an increased risk for the occurrence of future heart disease. The concept was driven, in large part, by the pathological findings that in diseased arteries, lesions were filled with lipid, predominantly cholesterol and cholesteryl esters. It is now widely accepted that elevated levels of LDL cholesterol are a strong indicator of risk. Programs to lower plasma cholesterol, particularly LDL cholesterol, and raise HDL cholesterol include dietary recommendations that lower the intake of saturated fat and limit cholesterol intake and drugs that lower apoB-containing lipoprotein production or that accelerate their uptake from plasma. Simply stated, the idea is to reduce the number and/or kind of potentially atherogenic lipoproteins in the plasma to limit their occurrence and hence their interaction with arterial cells involved in heart disease.

BIBLIOGRAPHY

Alaupovic, P. (1996). Significance of apolipoproteins for structure, function, and classification of plasma lipoproteins. *In* "Methods in Enzymology, Plasma Lipoproteins. Part C: Quantitation" (W. A. Bradley, S. H. Gianturco, and J. P. Segrest, eds.), Vol. 263, pp. 32–49. Academic Press, San Diego.

Ginsberg, H. N. (1995). Synthesis and secretion of apolipoprotein B from cultured liver cells. *Curr. Opin. Lipidol.* **6,** 275.

Gotto, A. M., Jr., Pownall, H. J., and Havel, R. J. (1986). Introduction to the plasma lipoproteins. *In* "Methods in Enzymology, Plasma Lipoproteins. Part A: Preparation, Structure, and Molecular Biology" (J. P. Segrest and J. J. Albers, eds.), Vol. 128, pp. 3–40. Academic Press, Orlando, Florida.

Patsch, W., and Gotto, A. M., Jr. (1996). Apolipoproteins: Pathophysiology and clinical implications. *In* "Methods in Enzymology, Plasma Lipoproteins. Part C: Quantitation" (W. A. Bradley, S. H. Gianturco, and J. P. Segrest, eds.), Vol. 263, pp. 32–49. Academic Press, San Diego.

Pease, R. J., and Leiper, J. M. (1996). Regulation of hepatic apolipoprotein-B-containing lipoprotein secretion. *Curr. Opin. Lipidol.* **7,** 132.

Small, D. M. (1992). Structure and metabolism of the plasma lipoproteins. *In* "Current Issues in Endocrinology and Metabolism, Plasma Lipoproteins and Coronary Artery Disease" (R. A. Kreisberg and J. P. Segrest, eds.), pp. 57–91. Blackwell Scientific, Cambridge, Massachusetts.

ISBN 0-12-226976-4